EARTH

TENTH EDITION

EARTH

AN INTRODUCTION TO PHYSICAL GEOLOGY

EDWARD J. TARBUCK

FREDERICK K. LUTGENS

ILLUSTRATED BY

DENNIS TASA

Prentice Hall
Boston   Columbus   Indianapolis   New York
San Francisco   Upper Saddle River   Amsterdam
Cape Town   Dubai   London   Madrid   Milan
Munich   Paris   Montréal   Toronto   Delhi
Mexico City   São Paulo   Sydney   Hong Kong
Seoul   Singapore   Taipei   Tokyo

Library of Congress Cataloging-in-Publication Data

Tarbuck, Edward J.
    Earth : an introduction to physical geology / Edward J. Tarbuck, Frederick
K. Lutgens ; illustrated by Dennis Tasa.—10th ed.
        p. cm.
    ISBN-13: 978-0-321-66304-7
    ISBN-10: 0-321-66304-7
    1.   Physical geology—Textbooks.   I. Lutgens, Frederick K. II. Title.
    QE28.2.T37 2011
    550—dc22

                                                    2009046659

Editor in Chief, Geosciences and Chemistry: *Nicole Folchetti*
Aquisitions Editor: *Andrew Dunaway*
Marketing Manager: *Maureen McLaughlin*
Project Manager: *Crissy Dudonis*
Editorial Assistant: *Kristen Sanchez*
Marketing Assistant: *Nicola Houston*
Managing Editor, Geosciences and Chemistry: *Gina M. Cheselka*
Project Manager, Production: *Edward Thomas*
Full Service/Composition: *Laserwords, Maine*
Production Editor, Full Service: *Patty Donovan*
Art Director: *Maureen Eide*
Interior and Cover Design: *Lisbeth Axell*
Art Editor: *Connie Long*
Illustrations: *Tasa Graphic Arts*
Photo Research Manager: *Elaine Soares*
Photo Researcher: *Kristin Piljay*
Senior Manufacturing and Operations Manager: *Nick Sklitsis*
Operations Supervisor: *Amanda A. Smith*
Senior Media Producer: *Angela Bernhardt*
Media Producer: *Michelle Schreyer*
Senior Media Projection Supervisor: *Liz Winer*
Associate Media Project Manager: *David Chavez*
Cover Photo: *Colin Monteath/Minden Pictures*
Title Page Photo: *Deformed sedimentary strata on Alberta's Mt. Kidd (Photo by Tom Neversely/Photolibrary).*

To Our Grandchildren
Shannon, Amy, Andy, Ali, and Michael
Allison and Lauren
Each is a bright promise for the future.

Pearson Prentice Hall
Pearson Education, Inc.
Upper Saddle River, New Jersey 07458

Printed in the United States of America
10 9 8 7 6 5

ISBN: 0-321-66304-7 / 978-0-321-66304-7 (Student)
ISBN: 0-321-70746-X / 978-0-321-70746-8 (A lá Carte)

Prentice Hall
is an imprint of

PEARSON

Brief Contents

CHAPTER 1
An Introduction to Geology 1

CHAPTER 2
Plate Tectonics: A Scientific Revolution Unfolds 39

CHAPTER 3
Matter and Minerals 73

CHAPTER 4
Magma, Igneous Rocks, and Intrusive Activity 107

CHAPTER 5
Volcanoes and Volcanic Hazards 137

CHAPTER 6
Weathering and Soil 173

CHAPTER 7
Sedimentary Rocks 199

CHAPTER 8
Metamorphism and Metamorphic Rocks 229

CHAPTER 9
Geologic Time 255

CHAPTER 10
Crustal Deformation 279

CHAPTER 11
Earthquakes and Earthquake Hazards 303

CHAPTER 12
Earth's Interior 333

CHAPTER 13
Divergent Boundaries: Origin and Evolution of the Ocean Floor 355

CHAPTER 14
Convergent Boundaries: Origin of Mountains 381

CHAPTER 15
Mass Wasting: The Work of Gravity 405

CHAPTER 16
Running Water 429

CHAPTER 17
Groundwater 461

CHAPTER 18
Glaciers and Glaciation 489

CHAPTER 19
Deserts and Winds 519

CHAPTER 20
Shorelines 541

CHAPTER 21
Global Climate Change 575

CHAPTER 22
Earth's Evolution through Geologic Time 609

CHAPTER 23
Energy and Mineral Resources 643

CHAPTER 24
Planetary Geology 671

APPENDIX A
Metric and English Units Compared 703

GLOSSARY 704

INDEX 716

GEODe: Earth Contents

CH 1 An Introduction to Geology
Earth's Layered Structure
Rock Cycle
Chapter Quiz

CH 2 Plate Tectonics
Introduction
Divergent Boundaries
Convergent Boundaries
Transform Fault Boundaries
Formation and Breakup of Pangaea
Chapter Quiz

CH 3 Matter and Minerals
Introduction
Physical Properties of Minerals
Mineral Groups
Chapter Quiz

CH 4 Igneous Rocks
Introduction to Igneous Rocks
Igneous Compositions
Igneous Textures
Naming Igneous Rocks
Intrusive Igneous Activity
Chapter Quiz

CH 5 Volcanoes
The Nature of Volcanic Eruptions
Materials Extruded during
   an Eruption
Volcanic Structures and Eruptive
   Styles
Chapter Quiz

CH 6 Weathering and Soil
Earth's External Processes
Types of Weathering
Mechanical Weathering
Chemical Weathering
Rates of Weathering
Chapter Quiz

CH 7 Sedimentary Rocks
Introduction to Sedimentary Rocks
Types of Sedimentary Rocks
Sedimentary Environments
Chapter Quiz

CH 8 Metamorphic Rocks
Introduction to Metamorphic Rocks
Agents of Metamorphism
Textural and Mineralogical Changes
Common Metamorphic Rocks
Chapter Quiz

CH 9 Geologic Time
Relative Dating—Key Principles
Dating with Radioactivity
Geologic Time Scale
Chapter Quiz

CH 10 Crustal Deformation
Deformation
Folds
Faults and Fractures
Chapter Quiz

CH 11 Earthquakes
What is an Earthquake?
Seismology
Locating the Source of an Earthquake
Earthquakes at Plate Boundaries
Chapter Quiz

CH 12 Earth's Interior
Earth's Layered Structure
Chapter Quiz

CH 13 Divergent Boundaries
Mapping the Ocean Floor
Features of the Ocean Floor
Oceanic Ridges and Seafloor
   Spreading
Chapter Quiz

CH 14 Convergent Boundaries
Continental Collisions
Crustal Fragments and Mountain
   Building
Chapter Quiz

CH 15 Mass Wasting
Controls and Triggers of Mass
   Wasting
Mass Wasting Processes
Chapter Quiz

CH 16 Running Water
Hydrologic Cycle
Stream Characteristics
Reviewing Valleys and Stream-
   related Features
Chapter Quiz

CH 17 Groundwater
Importance and Distribution
Springs and Wells
Chapter Quiz

CH 18 Glaciers and Glaciation
Introduction
Budget of a Glacier
Reviewing Glacial Features
Chapter Quiz

CH 19 Deserts and Winds
Distribution and Causes of Dry Lands
Common Misconceptions About
   Deserts
Reviewing Landforms and Landscapes
Chapter Quiz

CH 20 Shorelines
Waves and Beaches
Wave Erosion
Chapter Quiz

Table of Contents

Preface xvi

1

An Introduction to Geology 1

The Science of Geology 2
Geology, People, and the Environment 2
Some Historical Notes About Geology 5
   Catastrophism 5
Professional Profile Career Opportunities in the Geosciences 6
   The Birth of Modern Geology 6
   Geology Today 7
Geologic Time 7
   The Magnitude of Geologic Time 7
   Relative Dating and the Geologic Time Scale 8
The Nature of Scientific Inquiry 9
   Hypothesis 12
   Theory 12
   Scientific Methods 12
   Plate Tectonics and Scientific Inquiry 14
Earth's Spheres 14
   Hydrosphere 15
   Atmosphere 15
   Biosphere 16
   Geosphere 16
Earth as a System 17
   Earth System Science 17
   The Earth System 19
Early Evolution of Earth 20
   Origin of Planet Earth 20
   Formation of Earth's Layered Structure 22
 Earth's Internal Structure 23
   Earth's Crust 23
   Earth's Mantle 24
   Earth's Core 24
   How Do We Know What We Know? 24
The Face of Earth 25
   Major Features of the Continents 25
   Major Features of the Ocean Floor 28
Rocks and the Rock Cycle 29
   Basic Rock Types 29
   The Rock Cycle: One of Earth's Subsystems 32
Box 1.1 Understanding Earth Studying Earth from Space 11
Box 1.2 Understanding Earth Do Glaciers Move?
   An Application of the Scientific Method 13

2

Plate Tectonics: A Scientific Revolution Unfolds 39

Continental Drift: An idea before its Time 40
   Evidence: The Continental Jigsaw Puzzle 40
   Evidence: Fossils Match across the Seas 41

   Evidence: Rock Types and Geologic Features 43
   Evidence: Ancient Climates 44
The Great Debate 45
   Rejection of the Continental Drift Hypothesis 45
   Continental Drift and the Scientific Method 45
Continental Drift and Paleomagnetism 45
   Earth's Magnetic Field and Fossil Magnetism 45
   Apparent Polar Wandering 46
A Scientific Revolution Begins 48
   The Seafloor-Spreading Hypothesis 48
   Magnetic Reversals: Evidence for Seafloor Spreading 49
Plate Tectonics 51
   Earth's Major Plates 51
   Plate Boundaries 54
Divergent Boundaries 55
   Oceanic Ridges and Seafloor Spreading 55
   Continental Rifting 56
Convergent Boundaries 56
   Oceanic–Continental Convergence 57
   Oceanic–Oceanic Convergence 58
   Continental–Continental Convergence 59
Transform Fault Boundaries 60
How do Plates and Plate Boundaries Change? 62
Testing the Plate Tectonics Model 62
   Evidence from Ocean Drilling 62
   Hot Spots and Mantle Plumes 64
How is Plate Motion Measured? 65
   Mantle Plumes and Plate Motions 65
   Paleomagnetism and Plate Motions 66
   Measuring Plate Motion from Space 67
What Drives Plate Motions? 67
   Plate–Mantle Convection 67
   Forces that Drive Plate Motion 68
   Models of Plate–Mantle Convection 68

Box 3.2 **People and the Environment** Asbestos: What Are the Risks? 85

Box 3.3 **Understanding Earth** Gemstones 102

4

Magma, Igneous Rocks, and Intrusive Activity 107

Magma: the Parent Material of Igneous Rock 108
  The Nature of Magma 108
  From Magma to Crystalline Rock 108

Igneous Processes 110

Igneous Compositions 110
  Granitic (Felsic) Versus Basaltic (Mafic) Compositions 110
  Other Compositional Groups 111
  Silica Content as an Indicator of Composition 111

Igneous Textures: What can They Tell Us? 112
  Types of Igneous Textures 113

Naming Igneous Rocks 115
  Felsic (Granitic) Igneous Rocks 116
  Intermediate (Andesitic) Igneous Rocks 119
  Mafic (Basaltic) Igneous Rocks 119
  Pyroclastic Rocks 119

Origin of Magma 120
  Generating Magma from Solid Rock 120

How Magmas Evolve 122
  Bowen's Reaction Series and the Composition of Igneous Rocks 122
  Assimilation and Magma Mixing 123

Partial Melting and Magma Composition 124
  Formation of Basaltic Magma 124
  Formation of Andesitic and Granitic Magmas 126

Intrusive Igneous Activity 128
  Nature of Intrusive Bodies 128
  Tablular Intrusive Bodies: Dikes and Sills 128
  Massive Intrusive Bodies: Batholiths and Stocks 131

Box 4.1 **Understanding Earth** Thin Sections and Rock Identification 118

Box 4.2 **Understanding Earth** A Closer Look at Bowen's Reaction Series 127

5

Volcanoes and Volcanic Hazards 137

The Nature of Volcanic Eruptions 138
  Factors Affecting Viscosity 138
  Why Do Volcanoes Erupt? 140

Materials Extruded During an Eruption 142
  Lava Flows 142
  Gases 143
  Pyroclastic Materials 144

Volcanic Structures and Eruptive Styles 146
  Anatomy of a Volcano 146
  Shield Volcanoes 147
  Cinder Cones 149
  Composite Cones 151

The Importance of the Plate Tectonics Theory 69

Box 2.1 **Understanding Earth** Alfred Wegener (1880–1930): Polar Explorer and Visionary 47

Box 2.2 **Understanding Earth** Priority in the Sciences 54

Box 2.3 **Understanding Earth** The Breakup of Pangaea 63

3

Matter and Minerals 73

Minerals: Building Blocks of Rocks 74

Matter and Minerals 00

Atoms: Building Blocks of Minerals 76
  Properties of Protons, Neutrons,and Electrons 76
  Elements: Defined by Their Number of Protons 77

Why Atoms Bond 77
  Ionic Bonds: Electrons Transferred 79
  Covalent Bonds: Electrons Shared 80
  Metallic Bonds: Electrons Free to Move 80
  Other Bonds: Hybrids 80

Isotopes and Radioactive Decay 81

Crystals and Crystallization 81
  How Do Minerals Form? 81
  Crystal Structures 82
  Compositional Variations in Minerals 83
  Structural Variations in Minerals 86

Physical Properties of Minerals 87
  Optical Properties 87
  Crystal Shape or Habit 88
  Mineral Strength 88
  Density and Specific Gravity 89
  Other Properties of Minerals 91

How Are Minerals Named and Classified? 91
  Classifying Minerals 92
  Major Mineral Classes 92

The Silicates 93
  The Silicon–Oxygen Tetrahedron 93
  Joining Silicate Structures 94

Common Silicate Minerals 95
  The Light Silicates 95
  The Dark Silicates 98

Important Nonsilicate Minerals 99

Box 3.1 **People and the Environment** Making Glass from Minerals 76

Living in the Shadow of A Composite Cone 152
  Eruption of Vesuvius AD 79 152
  Nuée Ardente: A Deadly Pyroclastic Flow 153
  Lahars: Mudflows on Active and Inactive Cones 154
Other Volcanic Landforms 156
  Calderas 156
  Fissure Eruptions and Basalt Plateaus 157
  Lava Domes 159
  Volcanic Pipes and Necks 160
Plate Tectonics and Volcanic Activity 160
  Volcanism at Convergent Plate Boundaries 162
  Volcanism at Divergent Plate Boundaries 162
  Intraplate Volcanism 163
Professional Profile Chris Eisinger Studying Active Volcanoes 163
Living with Volcanoes 167
  Volcanic Hazards 167
  Monitoring Volcanic Activity 168
Box 5.1 Understanding Earth Anatomy of an Eruption 141
Box 5.2 Earth as a System Volcanic Air Pollution—
A Hazard in Hawaii 147
Box 5.3 People and the Environment The Lost Continent of
Atlantis 155

6

Weathering and Soil 173

Earth's External Processes 174
Weathering 174
Mechanical Weathering 176
  Frost Wedging 176
  Salt Crystal Growth 176
  Sheeting 176
  Thermal Expansion 178
  Biological Activity 178
Chemical Weathering 179
  Dissolution 179
  Oxidation 181
  Hydrolysis 181
  Spheroidal Weathering 184
Rates of Weathering 185
  Rock Characteristics 185
  Climate 185
  Differential Weathering 186
SOIL 186
  An Interface in the Earth System 186
  What is Soil? 187
Controls of Soil Formation 187
  Parent Material 187
  Time 188
  Climate 188
  Plants and Animals 188
  Topography 189
The Soil Profile 189
Classifying Soils 191

Soil Erosion 191
  How Soil is Eroded 191
  Rates of Erosion 194
  Sedimenation and Chemical Pollution 195
Box 6.1 Understanding Earth The Old Man
of the Mountain 178
Box 6.2 Earth as a System Acid Precipitation—
A Human Impact on the Earth System 182
Box 6.3 People and the Environment Clearing
the Tropical Rain Forest—The Impact on it's
Soils 193

7

Sedimentary Rocks 199

The Importance of Sedimentary Rocks 200
Origins of Sedimentary Rock 200
Detrital Sedimentary Rocks 202
  Shale 203
  Sandstone 204
  Conglomerate and Breccia 206
Chemical Sedimentary Rocks 207
  Limestone 208
  Dolostone 210
  Chert 210
  Evaporites 210
Coal—An Organic Sedimentary Rock 212
Turning Sediment Into Sedimentary Rock: Diagenesis
and Lithification 213
Classification of Sedimentary Rocks 214
Sedimentary Environments 216
  Types of Sedimentary Environments 217
  Sedimentary Facies 220
Sedimentary Structures 221
Box 7.1 Earth as a System The Carbon Cycle
and Sedimentary Rocks 214
Box 7.2 People and the Environment Our Threatened Coral
Reefs 217

8

Metamorphism and Metamorphic Rocks 229

What is Metamorphism? 230
What Drives Metamorphism? 230
  Heat as a Metamorphic Agent 231
  Confining Pressure and Differential Stress 233
  Chemically Active Fluids 234
  The Importance of Parent Rock 234
Metamorphic Textures 235
  Foliation 235
  Foliated Textures 236
  Other Metamorphic Textures 237
Common Metamorphic Rocks 238
  Foliated Rocks 238
  Nonfoliated Rocks 240
Metamorphic Environments 241
  Contact or Thermal Metamorphism 241
  Hydrothermal Metamorphism 242
  Burial and Subduction Zone Metamorphism 244
  Regional Metamorphism 244
  Other Metamorphic Environments 244
Metamorphic Zones 245
  Textural Variations 245
  Index Minerals and Metamorphic Grade 246
Interpreting Metamorphic Environments 248
Box 8.1 **Understanding Earth** Impact Metamorphism and Tektites 247
Box 8.2 **Understanding Earth** Mineral Stability 250

9

Geologic Time 255

Geology Needs A Time Scale 256
Relative Dating—Key Principles 256
  Law of Superposition 257
  Principle of Original Horizontality 257
  Principle of Cross-Cutting Relationships 258
  Inclusions 258
  Unconformities 258
  Using Relative Dating Principles 261
Correlation of Rock Layers 261

Fossils: Evidence of Past Life 262
  Types of Fossils 265
  Conditions Favoring Preservation 266
  Fossils and Correlation 267
Dating With Radioactivity 267
  Reviewing Basic Atomic Structure 268
  Radioactivity 268
  Half-Life 269
  Radiometric Dating 269
  Dating with Carbon-14 270
  Importance of Radiometric Dating 270
The Geologic Time Scale 271
  Structure of the Time Scale 271
  Precambrian Time 271
Difficulties In Dating the Geologic Time Scale 273
Box 9.1 **Understanding Earth** Applying Relative Dating Principles to the Lunar Surface 264
Box 9.2 **Understanding Earth** Terminology and the Geologic Time Scale 273

10

Crustal Deformation 279

Structural Geology: A Study of Earth's Architecture 280
Deformation, Stress, and Strain 280
  Stress: The Force that Deforms Rocks 280
  Strain: A Change in Shape Caused by Stress 281
How Rocks Deform 282
  Elastic, Brittle, and Ductile Deformation 283
  Factors that Affect Rock Strength 284
Structures Formed by Ductile Deformation 285
  Folds 285
Professional Profile Michael Collier Feet in the Fire: Communicating Geologist 286
Structures Formed by Brittle Deformation 290
  Faults 292
  Dip-Slip Faults 292
  Strike-Slip Faults 294
  Joints 296
Mapping Geologic Structures 297
  Strike and Dip 298
Box 10.1 **Understanding Earth** Naming Local Rock Units 298

11

Earthquakes and Earthquake Hazards 303

What is an Earthquake? 304
  Discovering the Causes of Earthquakes 305
  Aftershocks and Foreshocks 306
Faults, Faulting, and Earthquakes 307
  The Nature of Faults 307
  Fault Rupture and Propagation 308
Seismology: The Study of Earthquake Waves 309
Professional Profile Andrea Donnellan Earthquake Forecaster 309
Locating the Source of an Earthquake 312

Measuring the Size of Earthquakes 313
 Intensity Scales 313
 Magnitude Scales 314
Earthquake Belts and Plate Boundaries 316
Earthquake Destruction 317
 Destruction from Seismic Vibrations 317
 Landslides and Ground Subsidence 320
 Fire 320
 What is A Tsunami? 320
Can Earthquakes Be Predicted? 322
 Short-Range Predictions 322
 Long-Range Forecasts 323
Seismic Risk On the San Andreas Fault 326
Earthquakes: Evidence For Plate Tectonics at Plate Boundaries 327
Box 11.1 **Understanding Earth** Wave Amplification and Seismic Risks 318
Box 11.2 **People and the Environment** Earthquakes East of the Rockies 329

12
Earth's Interior   333

Gravity and Layered Planets 334
Probing Earth's Interior: "Seeing" Seismic Waves 334
Earth's Layers 337
 Earth's Crust 337
 Earth's Mantle 339
 Earth's Core 341
Earth's Temperature 342
 How Did Earth Get So Hot? 343
 Heat Flow 343
 Earth's Temperature Profile 344
Earth's Three-Dimensional Structure 345
 Earth's Gravity 346
 Seismic Tomography 347
 Earth's Magnetic Field 347
Box 12.1 **Understanding Earth** Re-creating the Deep Earth 338
Box 12.2 **EARTH AS A SYSTEM** Global Dynamic Connections 350

13
Divergent Boundaries: Origin and Evolution of the Ocean Floor   355

An Emerging Picture of the Ocean Floor 356
 Mapping the Seafloor 356
 Viewing the Ocean Floor From Space 358
 Provinces of the Ocean Floor 358
Continental Margins 358
 Passive Continental Margins 358
Professional Profile Susan DeBari Studying the Deep Roots of Volcanic Arcs 359
 Active Continental Margins 361
Features of Deep-Ocean Basins 362
 Deep-Ocean Trenches 362
 Abyssal Plains 362
 Seamounts, Guyots, and Oceanic Plateaus 363

Anatomy of the Oceanic Ridge 365
Oceanic Ridges and Seafloor Spreading 366
 Seafloor Spreading 366
 Why are Oceanic Ridges Elevated? 367
 Spreading Rates and Ridge Topography 367
The Nature of Oceanic Crust 367
 How does Oceanic Crust Form? 368
 Interactions between Seawater and Oceanic Crust 369
Continental Rifting: The Birth of a New Ocean Basin 371
 Evolution of an Ocean Basin 371
 Mechanisms for Continental Rifting 373
Destruction of Oceanic Lithosphere 375
 Why Oceanic Lithosphere Subducts 375
 Subducting Plates: The Demise of an Ocean Basin 376
Box 13.1 **Understanding Earth:** Explaining Coral Atolls—Darwin's Hypothesis 364
Box 13.2 **Earth as a System:** Deep-Sea Hydrothermal Vents 370

14
Convergent Boundaries: Origin of Mountains   381

Mountain Building 382
Convergence and Subducting Plates 384
 Major Features of Subduction Zones 384
 Dynamics at Subduction Zones 385
Subduction and Mountain Building 386
 Volcanic Island Arcs 386
 Mountain Building along Andean-type Margins 386
 Sierra Nevada, Coast Ranges, and Great Valley 389
Collisional Mountain Belts 390
 Terranes and Mountain Building 390
 Continental Collisions 391
 The Himalayas 391
 The Appalachians 392
Fault-Block Mountains 394
 Basin and Range Province 396
Vertical Movements of the Crust 399
 Isostasy 399

Mantle Convection: A Cause of Vertical Crustal Movement 399

Box 14.1 Understanding Earth: Earthquakes in the Pacific Northwest 387

Box 14.2 Understanding Earth: The Southern Rockies 398

Box 14.3 Understanding Earth: Do Mountains Have Roots? 401

15

Mass Wasting: The Work of Gravity   405

Landslides As Natural Disasters 406

Mass Wasting and Landform Development 406
    The Role of Mass Wasting 406
    Slopes Change Through Time 407

Controls and Triggers of Mass Wasting 407
    The Role of Water 408
    Oversteepened Slopes 409
    Removal of Vegetation 409
    Earthquakes as Triggers 411
    Landslides without Triggers? 412
    The Potential for Landslides 413

Classification of Mass-Wasting Processes 413
    Type of Material 413
    Type of Motion 413
    Rate of Movement 414

Slump 416

Rockslide 417

Debris Flow 419
    Debris Flows in Semiarid Regions 419
    Lahars 419

Professional Profile Bob Rasley Mass Wasting Specialist 421

Earthflow 421

Slow Movements 422
    Creep 422
    Solifluction 422

The Sensitive Permafrost Landscape 423

Submarine Landslides 424

Box 15.1 People and the Environment Landslide Hazards La Conchita, California 410

Box 15.2 People and the Environment The Vaiont Dam Disaster 415

16

Running Water   429

Earth as a System: The Hydrologic Cycle 430

Running Water 432
    Drainage Basins 432
    River Systems 433

Streamflow 434
    Flow Velocity 434
    Gradient and Channel Chracteristics 435
    Discharge 435
    Changes Downstream 435

The Work of Running Water 436
    Stream Erosion 436
    Transport of Sediment by Streams 437
    Deposition of Sediment by Streams 439

Stream Channels 439
    Bedrock Channels 440
    Alluvial Channels 440

Base Level and Graded Streams 441

Shaping Stream Valleys 443
    Valley Deepening 443
    Valley Widening 444
    Incised Meanders and Stream Terraces 445

Depositional Landforms 445
    Deltas 446
    The Mississippi Delta 447
    Natural Levees 449
    Alluvial Fans 449

Drainage Patterns 449
    Formation of a Water Gap 450
    Headward Erosion and Stream Piracy 450

Floods and Flood Control 452
    Types of Floods 452
    Flood Control 454

Box 16.1 People and the Environment: Coastal Wetlands Vanishing on the Mississippi Delta 451

Box 16.2 People and the Environment: Flash Floods 455

17

Groundwater   461

Importance of Groundwater 462

Groundwater—A Basic Resource 464

Distribution of Groundwater 465

The Water Table 466
    Variations in the Water Table 466
    Interaction Between Groundwater and Streams 467

Factors Influencing the Storage and Movement of Groundwater 467
    Porosity 467
    Permeability, Aquitards, and Aquifers 468

Movement of Groundwater 468
    Darcy'S Law 469
    Different Scales of Movement 470

SPRINGS 472

 Hot Springs and Geysers 472

 Wells 474

Artesian Wells 476

Problems Associated with Groundwater Withdrawal 477

Treating Groundwater as a Nonrenewable Resource 477

Subsidence 478

Saltwater Contamination 478

Groundwater Contamination 480

The Geologic Work of Groundwater 482

Caverns 482

Karst Topography 483

Box 17.1 **Earth as a System** Drought Impacts the Hydrologic System 470

Box 17.2 **People and the Environment** Land Subsidence in the San Joaquin Valley 479

18

Glaciers and Glaciation 489

Glaciers: A Part of Two Basic Cycles 490

Valley (Alpine) Glaciers 490

Ice Sheets 490

Other Types of Glaciers 493

What if the Ice Melted? 493

Formation and Movement of Glacial Ice 494

Glacial Ice Formation 494

Movement of A Glacier 494

Rates of Glacial Movement 495

Budget of a Glacier 496

Glacial Erosion 498

Landforms Created By Glacial Erosion 499

Glaciated Valleys 499

Arêtes and Horns 502

Roches Moutonées 503

Glacial Deposits 503

Landforms Made of Till 503

Lateral and Medial Moraines 503

End and Ground Moraines 504

Drumlins 505

Landforms Made of Stratified Drift 506

Outwash Plains and Valley Trains 506

Ice-Contact Deposits 507

Other Effects of Ice-Age Glaciers 507

Crustal Subsidence and Rebound 508

Sea-Level Changes 508

Changes to Rivers and Valleys 508

Ice Dams Create Proglacial Lakes 509

Pluvial Lakes 510

The Glacial Theory and the Ice Age 512

Causes of Glaciation 512

Plate Tectonics 513

Variations in Earth's Orbit 513

Other Factors 514

Box 18.1 **Earth as a System** Glaciers In Retreat 501

Box 18.2 **Understanding Earth** Glacial Lake Missoula, Megafloods, and the Channeled Scablands 511

19

Deserts and Winds 519

Distribution and Causes of Dry Lands 520

Low-Latitude Deserts 520

Middle-Latitude Deserts 521

Geologic Processes In Arid Climates 523

Weathering 524

The Role of Water 524

Basin and Range: The Evolution of a Desert Landscape 526

Transportation of Sediment by Wind 529

Bed Load 529

Suspended Load 530

Wind Erosion 530

Deflation and Blowouts 530

Desert Pavement 531

Ventifacts and Yardangs 532

Wind Deposits 533

Sand Deposits 533

Types of Sand Dunes 534

Loess (Silt) Deposits 536

Box 19.1 **Understanding Earth:** What Is Meant by "Dry"? 523

Box 19.2 **People and the Environment:** The Disappearing Aral Sea—A Large Lake Becomes a Barren Wasteland 524

Box 19.3 **Understanding Earth:** Australia's Mount Uluru 529

20

Shorelines 541

The Shoreline: A Dynamic Interface 542

The Coastal Zone 542

Basic Features 543

Beaches 543

Waves 544

Wave Characteristics 544

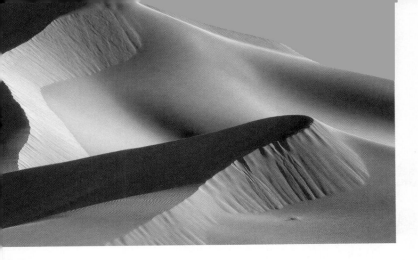

Circular Orbital Motion 545
Waves in the Surf Zone 546

Wave Erosion 547
Sand Movement on the Beach 548
Movement Perpendicular to the Shoreline 548
Wave Refraction 548
Longshore Transport 549
Rip Currents 550
Shoreline Features 550
Professional Profile Rob Thieler Marine Geologist 551
Erosional Features 552
Depositional Features 552
The Evolving Shore 553
Stabilizing the Shore 553
Hard Stabilization 554
Alternatives to Hard Stabilization 556
Erosion Problems Along U.S. Coasts 557
Atlantic and Gulf Coasts 557
Pacific Coast 558
Hurricanes—The Ultimate Coastal Hazard 560
Profile of a Hurricane 562
Hurricane Destruction 562
Coastal Classification 564
Emergent Coasts 565
Submergent Coasts 565
Tides 565
Causes of Tides 566
Monthly Tidal Cycle 566
Tidal Patterns 567
Tidal Currents 568
Tides and Earth's Rotation 570
Box 20.1 **People and the Environment:** The Move
of the Century—Relocating the Cape Hatteras Lighthouse 559
Box 20.2 **People and the Environment:** Hurricane
Forecasting 568

21

Global Climate Change   575

Climate and Geology are Linked 576
The Climate System 577
How is Climate Change Detected? 577
Seafloor Sediment—A Storehouse of Climate Data 578

Oxygen Isotope Analysis 579
Climate Change Recorded In Glacial Ice 580
Tree Rings—Archives of Environmental History 580
Other Types of Proxy Data 582
Some Atmospheric Basics 582
Composition of the Atmosphere 583
Extent and Structure of the Atmosphere 584
Energy from the Sun 585
The Fate of Incoming Solar Energy 586
Heating the Atmosphere: The Greenhouse Effect 586
Natural Causes of Climate Change 587
Volcanic Activity and Climate Change 588
Solar Variability and Climate 589
Professional Profile Michael Mann Climate Change Scientist 590
Human Impact on Global Climate 592
Carbon Dioxide, Trace Gases, and Climate Change 593
CO_2 Levels Are Rising 593
The Atmosphere's Response 593
The Role of Trace Gases 594
Climate-Feedback Mechanisms 596
How Aerosols Influence Climate 598
Some Possible Consequences of Global Warming 599
Sea Level Rise 600
The Changing Artic 602
Increasing Ocean Acidity 603
The Potential for "Surprises" 604
Box 21.1 **Earth as a System:** A Possible Link between
Volcanism and Climate Change in the Geologic Past 592
Box 21.2 **Understanding Earth:** Computer Models
of Climate: Important Yet Perfect Tools 599
Box 21.3 **Earth as a System:** Global Climate Change Impacts
on the United States 602

22

Earth's Evolution Through Geologic Time   609

Is Earth Unique? 610
The Right Planet 610
The Right Location 612
The Right Time 612
Birth of a Planet 614
From Planetesimals to Protoplanets 614
Earth's Early Evolution 614
Origin of the Atmosphere and Oceans 614
Earth's Primitive Atmosphere 614
Oxygen in the Atmosphere 616
Evolution of the Oceans 617
Precambrian History: The Formation of Earth's Continents 618
Earth's First Continents 618
The Making of North America 619
Supercontinents of the Precambrian 620
Geologic History of the Phanerozoic: The Formation of Earth's Modern
Continents 621
Paleozoic History 622

Mesozoic History 623
Cenozoic History 624

Earth's First Life 625

Paleozoic Era: Life Explodes 628
Early Paleozoic Life-Forms 629
Vertebrates Move to Land 630
The Great Permian Extinction 630

Professional Profile Neil Shubin Paleontologist 632

The Mesozoic Era: Age of the Dinosaurs 633

Cenozoic Era: Age of Mammals 634
From Reptiles to Mammals 635
Large Mammals and Extinction 637

Box 22.1 **Understanding Earth:** The Burgess Shale 626
Box 22.2 **Understanding the Earth:** Demise of the Dinosaurs 636

23

Energy and Mineral Resources   643

Renewable and Nonrenewable Resources 644

Energy Resources 645
Coal 645

Oil and Natural Gas 646
Petroleum Formation 646
Oil Traps 647

Oil Sands and Oil Shale—Petroleum for the Future? 649
Oil Sands 649
Oil Shale 650

Alternate Energy Sources 652
Nuclear Energy 652
Solar Energy 653
Wind Energy 654
Hydroelectric Power 655
Geothermal Energy 656

Professional Profile Sally Benson Climate and Energy Scientist 658
Tidal Power 658

Mineral Resources 659

Mineral Resources and Igneous Processes 661
Magmatic Segregation 661
Diamonds 661
Hydrothermal Solutions 662

Mineral Resources and Metamorphic Processes 663

Weathering and Ore Deposits 663
Bauxite 664
Other Deposits 664

Placer Deposits 664

Nonmetallic Mineral Resources 665
Building Materials 665
Industrial Minerals 666

Box 23.1 **Understanding Earth:** Gas Hydrates—A Fuel from Ocean-Floor Sediments 651
Box 23.2 **People and the Environment:** Utah's Bingham Canyon Mine 660

24

Planetary Geology   671

Our Solar System: An Overview 672
Nebular Theory: Formation of the Solar System 672
The Planets: Internal Structures, Atmospheres, and Weather 674

Planetary Impacts 678

Earth's Moon: A Chip Off the Old Block 678
How Did the Moon Form? 679

Planets and Moons 681
Mercury: The Innermost Planet 681
Venus: The Veiled Planet 682
Mars: The Red Planet 683
Jupiter: Lord of the Heavens 686
Saturn: The Elegant Planet 690
Uranus and Neptune: Twins 692

Minor Members of the Solar System 694
Asteroids: Leftover Planetesimals 695
Comets: Dirty Snowballs 696
Meteoroids: Visitors to Earth 698
Dwarf Planets 700

Box 24.1 **Understanding Earth:** *Pathfinder*—The First Geologist on Mars 690
Box 24.2 **Earth as a System:** Is Earth on a Collision Course? 696

APPENDIX A
Metric and English Units Compared 703

Glossary 704

Index 716

Preface

Earth is a very small part of a vast universe, but it is our home. It provides the resources that support our modern society and the ingredients necessary to maintain life. Knowledge of our physical environment is critical to our well-being and vital to our survival. A basic geology course can help a person gain such an understanding. It can also take advantage of the interest and curiosity many of us have about our planet—its landscapes and the processes that create and alter them.

The tenth edition of *Earth: An Introduction to Physical Geology*, like its predecessors, is a college-level text that is intended to be a meaningful, non-technical survey for students taking their first course in geology. In addition to being informative and up-to-date, a major goal of *Earth* is to meet the need of students for a readable and user-friendly text, a book that is a highly usable "tool" for learning the basic principles and concepts of geology.

DISTINGUISHING FEATURES

Readability

The language of this book is straightforward and *written to be understood*. Clear, readable discussions with a minimum of technical language are the rule. The frequent headings and subheadings help students follow discussions and identify the important ideas presented in each chapter. In the tenth edition, improved readability was achieved by examining chapter organization and flow, and writing in a more personal style. Large portions of the text were substantially rewritten in an effort to make the material more accessible and understandable. In particular, those familiar with the previous edition will notice the author's efforts in this regard in Chapters 3, 4, 5, 8, 10, 11, 12, 13, 14, 16, 22, and 24.

The Art Program

Geology is highly visual, so the art program plays a critical role in an introductory book. As in the nine previous editions, Dennis Tasa, a gifted artist and respected geological illustrator, has again worked closely with the authors to plan and produce the diagrams, maps, graphs, and sketches that are so basic to student understanding. Literally hundreds of figures are new or redrawn. The result is art that is clearer and easier to understand. Numerous diagrams and maps are paired with photographs for greater effectiveness. Many new and revised art pieces also have additional labels that "narrate" the process being illustrated and/or "guide" readers as they examine the figure.

Photographs

The tenth edition of *Earth* has more than 250 *new*, high-quality photos and satellite images. In addition to the geologist's sketches that accompany many important images, numerous labels were added to many other photos. Our aim is to get *maximum effectiveness* from the visual component of the book.

We were aided greatly in this quest by Michael Collier. Dozens of his extraordinary aerial photographs were used in the new edition. Michael is an award-winning geologist–photographer. Among his many awards is the American Geological Institute Award for Outstanding Contribution to the Public Understanding of the Geosciences. We are fortunate to have had Michael's assistance in *Earth 10e*. There is more about Michael in the Chapter 10 "Professional Profile" (p. 286).

Focus on Learning

Earth 10e, consists of two products—a traditional college textbook and an interactive Companion Website. Both contain helpful student learning aids.

End-of-Chapter Review   When a chapter has been completed, four useful devices help students review. First, the *Chapter in Review* recaps all the major points. Next, is the checklist of *Key Terms* with page references. Learning the language of geology helps students learn the material. This is followed by the *Questions for Review* section, which helps students examine their knowledge of significant facts and ideas.

Next is a reminder to visit the website for *Earth 10e* (www.mygeoscienceplace.com). It contains many excellent opportunities for review and exploration. In addition, many chapters close with a note about *GEODe: Earth* to remind students about this unique and effective learning aid.

Companion Website with GEODe: Earth   Included with every copy of *Earth* is access to the interactive Companion Website containing quizzes for review, the Pearson eText, and *GEODe: Earth*. Prepared by the authors in conjunction with Dennis Tasa, *GEODe: Earth* is a dynamic program that provides broad coverage of basic principles and reinforces key concepts by using tutorials, interactive exercises, animations, and "In the Lab" activities. In addition to being technically upgraded, the content of the new version has been substantially revised and updated. A table of contents appears in the front matter for easy reference, and a special icon appears throughout the book wherever a text discussion has a corresponding *GEODe: Earth* activity.

Focus on Basic Principles and Instructor Flexibility

Although many topical issues are treated in *Earth*, the main focus of this new edition remains the same as its predecessors—to foster student understanding of basic principles. Whereas student use of the text is a primary concern, the book's adaptability to the needs and desires of the instructor is also important. Realizing the broad diversity of introductory courses in both content and approach, we have designed each chapter to be as self-contained as possible so that material may be taught in a different sequence according to the preference of the instructor or the dictates of the laboratory. Thus, the instructor who wishes to discuss erosional processes prior to earthquakes, plate tectonics, and mountain building may do so without difficulty.

IMPORTANT THEMES

Earth as a System

An important occurrence in modern science has been the realization that Earth is a giant, multidimensional system. Our planet consists of many separate but interacting parts. A change in any one part can produce changes in any or all of the other parts—often in ways that are neither obvious nor immediately apparent. Although it is not possible to study the entire system at once, it is possible to develop an awareness and appreciation for the concept and for many of the system's important interrelationships. Therefore, beginning with the discussion in Chapter 1, the theme of "Earth as a System" recurs at appropriate places throughout the book. It is a thread that "weaves" through the chapters and helps tie them together. Some of the new and revised special-interest boxes relate to "Earth as a System."

People and the Environment

Because knowledge about our planet and how it works is necessary to our survival and well-being, the treatment of environmental issues has always been an important part of *Earth*. Such discussions serve to illustrate the relevance and application of geological knowledge. With each new edition, this focus has been maintained and strengthened. This is certainly the case with the tenth edition. The text integrates a great deal of information about the relationship between people and the natural environment and explores the application of geology to understanding and solving problems that arise from these interactions. In addition to many basic text discussions, more than 20 of the text's special-interest boxes involve the "People and the Environment" theme.

Understanding Earth

As members of a modern society, we are constantly reminded of the benefits derived from science. But what exactly is the nature of scientific inquiry? Developing an understanding of how science is done and how scientists work is another important theme that appears throughout this book, beginning with the revised section on "The Nature of Scientific Inquiry" in Chapter 1. Students will examine some of the difficulties encountered by scientists as they attempt to acquire reliable data about our planet and some of the ingenious methods that have been developed to overcome these difficulties. Students will also explore many examples of how hypotheses are formulated and tested as well as learn about the evolution and development of some major scientific theories. Many basic text discussions, as well as a number of the special-interest boxes on "Understanding Earth," provide the reader with a sense of the observational techniques and reasoning processes involved in developing scientific knowledge. The emphasis is not just on what scientists know, but how they figured it out.

NEW TO THIS EDITION

- Chapter 21 "Global Climate Change" includes a new introduction, a new section entitled "Climate and Geology are Linked," and a revised and expanded section on "Atmospheric Basics."

"Human Impact on Global Climate," the major emphasis of the last half of the chapter, has been *completely revised and updated* to include the latest material and data from the Intergovernmental Panel on Climate Change (IPCC) and from the June 2009 report from the U.S. Global Change Research Program.

- NEW "Professional Profile" boxes. Ten chapters include essays that present profiles of working geologists. These special boxes are intended to give students a sense of what geologists do and a perspective on a variety of geoscience careers.
- NEW "Geologist's Sketch" illustrations. An all new feature in *Earth 10e* is *Geologist's Sketches* which accompany many important photographs and satellite images. Each sketch, which resembles what a geologist might put in a field notebook, helps the student identify important and sometimes subtle aspects of an image. This author-artist collaboration helps make an already strong art program an even stronger, more effective, learning aid.
- More than 250 *new,* high-quality photos and satellite images.
- Thirty percent of the figures have been significantly redrawn.
- NEW animations with assessments on the Companion Website
- *GEODe* now on-line with a new user interface
- NEW Current Earth Events on the Companion Website
- Students can connect to the dynamic nature of geologic events around the world using RSS feeds from USGS Real-Time Resources. Working with data from recent earthquakes, volcanoes, landslides, and extreme weather events, students engage in critical thinking exercises to apply classroom concepts to real world events.

MORE ABOUT THE TENTH EDITION

The Tenth Edition of *Earth* represents a thorough revision. Every part of the book was examined carefully with the dual goals of keeping topics current and improving the clarity of text discussions. People familiar with the preceding edition will see much that is new in the tenth edition. The list of specifics is long. Examples include the following:

- In Chapter 2, "Plate Tectonics: A Scientific Revolution Unfolds," several discussions were substantially revised and rewritten. These include: "Measuring Plate Motion from Space," "What Drives Plate Motions," and "Hot Spots and Mantle Plumes."
- Much of Chapter 3, "Matter and Minerals," was rewritten in an effort to make a sometimes difficult (chemistry-oriented) topic easier for the beginning student to understand. Discussions of basic atomic structure, bonding, crystals and crystallization, compositional and structural variations, and the silicates received the strongest attention.
- Chapter 4, "Magma, Igneous Rocks, and Intrusive Activity," was formerly titled "Igneous Rocks". The new chapter title reflects the reorganization that has occurred. The discussions of intrusive activity that were formerly in Chapter 5 are now the concluding portion of Chapter 4. They logically follow the sections on the "Origin of Magma," and "How Magma Evolves." The entire discussion of "Intrusive Igneous Activity" has been revised and rewritten as has the section "Igneous Textures: What Can They Tell Us?".

- In Chapter 5, "Volcanoes and Volcanic Hazards," the discussion of pyroclastic flows is completely revised and rewritten as is the section on calderas. A box on Hawaii's problem with volcanic air pollution is new.

- Chapter 8, "Metamorphism and Metamorphic Rocks," includes revised and rewritten sections on "What Drives Metamorphism?" and "Hydrothermal Metamorphism."

- In Chapter 9, "Geologic Time," the updated section on the geologic time scale reflects the most recent changes to this basic tool, while a new box clarifies some of the terminology associated with the time scale.

- Chapter 10, "Crustal Deformation," includes an *all new* section on "Deformation, Stress, and Strain," that includes an excellent summary-type art piece.

- Chapter 11, "Earthquakes and Earthquake Hazards," includes an *all new* discussion on "Faults, Faulting, and Earthquakes," that explores the important relationship among these phenomena and includes a new summary art piece. Sections devoted to magnitude scales and earthquake prediction were substantially revised and a new section, "Seismic Risks on the San Andreas Fault," has been added.

- Chapter 12, "Earth's Interior," was *substantially rewritten* in order to make a complex topic more accessible to the beginning student. Significant emphasis was placed on improved readability. In addition, material on heat flow and the discussion of the relationship between the geothermal gradient and melting point curves were revised.

- There is much that is new in Chapter 15, "Mass Wasting: The Work of Gravity," including a new section on "The Potential for Landslides" and new material on the earthquake-triggered landslides that devastated parts of China in May 2008.

- Among the chapters that focus on external processes, Chapter 16, "Running Water," received the greatest attention. Discussions progress in a clearer more logical manner for the beginning student. Topics that received special attention included the nature of drainage basins, streamflow characteristics, shaping stream valleys, and bedrock vs. alluvial channels.

- Chapter 17, "Groundwater," begins with a new introduction and has an *all new* section on "Groundwater—A Basic Resource."

- Chapter 18, "Glaciers and Glaciation," includes a new box on "Glaciers in Retreat" and a new discussion devoted to "Greenland's Glacial Budget."

- Chapter 20, "Shorelines," has new material on rip currents and a completely revised and rewritten section on "Hurricanes—The Ultimate Coastal Hazard" that includes an *all new* discussion on "Profile of a Hurricane." A *new* box on "Hurricane Forecasting" has also been added.

- Chapter 23, "Energy and Mineral Resources," includes updated statistics as well as completely revised and updated discussions on wind power and solar energy.

- All of Chapter 24, "Planetary Geology," has been revised and updated to reflect the most recent research. This is the *most complete* revision of this chapter ever. The subject matter is better organized, more readable, and more up-to-date. Discussions progress in a manner that is easier to follow for the beginning student. The chapter is filled with many new images.

Additional Highlights

- *"Students Sometimes Ask . . ."* This popular feature has been retained and improved in the tenth edition. Instructors and students continue to react favorably and have indicated that the questions and answers that are sprinkled throughout each chapter add interest and relevance to discussions.

- Although there is not a significant change in the number of special interest boxes, several are totally new or substantially revised. As in the previous edition, most are intended to illustrate and reinforce the three themes of "Earth as a System," "People and the Environment," and "Understanding Earth."

THE TEACHING AND LEARNING PACKAGE

The authors and publisher have been pleased to work with a number of talented people who produced an excellent supplements package.

For the Instructor

Instructor Resource Center (IRC) on DVD   The IRC puts all of your lecture resources in one easy-to-reach place:

- *All* of the line art, tables, and photos from the text in .jpg files
- Over 100 animations of key geological processes
- *Images of Earth* photo gallery
- PowerPoint™ presentations
- *Instructor Manual* in Microsoft Word
- *Test Bank File* in Microsoft Word
- TestGen test generation and management software
- Electronic transparency acetates

Animations   The Prentice Hall Geoscience Animation Library includes animations illuminating many difficult-to-visualize topics of physical geology. Created through a unique collaboration among five of Prentice Hall's leading geoscience authors, these animations represent a significant leap forward in lecture presentation aids. They are provided both as Flash files and, for your convenience, pre-loaded into PowerPoint slides.

PowerPoint™ Presentations   Found on the IRC are three PowerPoint files for each chapter. Cut down on your preparation time, no matter what your lecture needs:

- *Exclusively Art*—All of the photos, art, and tables from the text, in order, loaded into PowerPoint slides.
- *Lecture Outline*—This set averages 35 slides per chapter and includes customizable lecture outlines with supporting art.
- *Classroom Response System (CRS) Questions*—Authored for use in conjunction with classroom response systems, these PowerPoints allow you to electronically poll your class for responses to questions, pop quizzes, attendance, and more.

Transparency Acetates   Provided electronically, every table and most of Dennis Tasa's illustrations in *Earth 10e* are available to be printed out as full-color, projection-enhanced transparencies.

Instructor Manual with Printed Test Bank   This contains learning objectives; chapter outlines; answers to end-of-chapter questions; and suggested, short demonstrations to spice up your lecture. The *Test Bank* incorporates art and averages 75 multiple-choice, true/false, short answer, and critical-thinking questions per chapter.

TestGen   Available on the IRC, use this electronic version of the *Printed Test Bank File* to build and customize your tests. Create multiple versions, add or edit questions, add illustrations—your customizable needs are easily addressed by this powerful software.

Blackboard   Already have your own website set up? We will provide a Test Bank in Blackboard or formats for importation. Additional course resources are available on the IRC and are available for use, with permission.

Companion Website (*www.mygeoscienceplace.com*)
Each new copy of *Earth 10e* comes with access to the book's Companion Website, which has been designed to provide students with useful tools needed for online study and review. It contains numerous review quizzes from which students get immediate feedback to guide their study. Other activities and resources, including *GEODe: Earth* help expand student understanding of basic processes and ideas. Somewhere between a text and a tutor, *GEODe: Earth* is a dynamic program that reinforces key concepts by using animations, tutorials, interactive exercises, and review activities. This well-received student-tested product provides broad coverage of basic principles. A special *GEODe: Earth* icon appears throughout the book wherever a text discussion has a corresponding *GEODe: Earth* activity. The Companion Website also includes Pearson eText, which reproduces the complete textbook in a new electronic format. An access code is bound into the front of every new copy of this textbook.

ACKNOWLEDGMENTS

Writing a college textbook requires the talents and cooperation of many individuals. We value the excellent work of Mark Watry and Teresa Tarbuck of Spring Hill College whose talents helped us improve Chapter 3, "Matter and Minerals" and Chapter 24, "Planetary Geology." They helped make both chapters more readable, engaging, and up-to-date.

Working with Dennis Tasa, who is responsible for all of the text's outstanding illustrations and much of the developmental work on *GEODe: Earth,* is always special for us. We not only value his outstanding artistic talents and imagination but his friendship as well.

Sincere thanks to Michael Collier whose contributions as an aerial photographer and geologist added greatly to this project. Collaborating with Michael was a special pleasure.

Great thanks also go to those colleagues who prepared in-depth reviews. Their critical comments and thoughtful input helped guide our work and clearly strengthened the text. Special thanks to: Randal Babcock, Western Washington University; Bill Dupree, University of Houston; Nels Forsman, University of North Dakota; Melida Guiterrez, Missouri State University; Duane Hampton, Western Michigan University; Kevin Hefferan, University of Wisconsin-Stevens Point; Sarah Johnson, Northern Kentucky University; Richard Josephs, University of Iowa; Steve Kadel, Glendale Community College; Michael Kimberley, North Carolina State University; David King, Auburn University; Kyle Mayborn, Western Illinois University; Michael Mann, Pennsylvania State University; Leslie Melim, Western Illinois University; Lauren Neitzke, Rutgers University; Phil Novack-Gottshall, University of West Georgia; Ray Rector, MiraCosta College; and Dean Whitman, Florida International University.

As always we want to acknowledge the team of professionals at Pearson Prentice Hall. We sincerely appreciate the company's continuing strong support for excellence and innovation. All are committed to producing the best textbooks possible. Special thanks to Editor in Chief Nicole Folchetti for stepping in at a critical point to guide the project and to our conscientious Project Manager, Crissy Dudonis, for a job well done. The production team, led by Patty Donovan at Laserwords Maine, once again did an outstanding job. We also appreciate the production assistance of Connie Long and Ed Thomas at Pearson, who helped make the process run smoothly. Kristin Piljay's photo research assistance was also a great help. All are true professionals with whom we are very fortunate to be associated.

Ed Tarbuck
Fred Lutgens

About our Sustainability Initiatives

This book is carefully crafted to minimize environmental impact. The materials used to manufacture this book originated from sources committed to responsible forestry practices. Paper from Forest Stewardship Council™ forests is responsibly harvested to benifit communities, wildlife and the environment.

Pearson closes the loop by recycling every out-of-date text returned to our warehouse. We pulp the books, and the pulp is used to produce items such as paper coffee cups and shopping bags.

The future holds great promise for reducing our impact on Earth's environment, and Pearson is proud to be leading the way. We strive to publish the best books with the most up-to-date and accurate content, and to do so in ways that minimize our impact on Earth.

FSC
www.fsc.org
MIX
Paper from
responsible sources
FSC® C005928

Pearson Prentice Hall
is an imprint of

AN INTRODUCTION TO GEOLOGY

Hawaii's Kilauea Volcano is one of the most active and most studied volcanoes of its kind in the world.

(PHOTO BY CARL SHANEFF/PACIFIC STOCK)

The spectacular eruption of a volcano, the terror brought by an earthquake, the magnificent scenery of a mountain range, and the destruction created by a landslide or flood are all subjects for the geologist (Figure 1.1). The study of geology deals with many fascinating and practical questions about our physical environment. What forces produce mountains? Will there soon be another great earthquake in California? What was the Ice Age like? Will there be another? How were these ore deposits formed? Should we look for water here? Is mining practical in this area? Will plentiful oil be found if a well is drilled at that location?

THE SCIENCE OF GEOLOGY

The subject of this text is **geology,** from the Greek *geo,* "Earth" and *logos,* "discourse." It is the science that pursues an understanding of planet Earth. Geology is traditionally divided into two broad areas—physical and historical. **Physical geology,** which is the primary focus of this book, examines the materials composing Earth and seeks to understand the many processes that operate beneath and upon its surface. The aim of **historical geology,** on the other hand, is to understand the origin of Earth and its development through time. Thus, it strives to establish an orderly chronological arrangement of the multitude of physical and biological changes that have occurred in the geologic past. The study of physical geology logically precedes the study of Earth history because we must first understand how Earth works before we attempt to unravel its past. It should also be pointed out that physical and historical geology are divided into many areas of specialization. Table 1.1 provides a partial list. Every chapter of this book represents one or more areas of specialization in geology.

To understand Earth is challenging because our planet is a dynamic body with many interacting parts and a complex history. Throughout its long existence, Earth has been changing. In fact, it is changing as you read this page and will continue to do so into the foreseeable future. Sometimes the changes are rapid and violent, as when landslides or volcanic eruptions occur. Just as often, change takes place so slowly that it goes unnoticed during a lifetime. Scales of size and space also vary greatly among the phenomena that geologists study. Sometimes they must focus on phenomena that are submicroscopic, and at other times they must deal with features that are continental or global in scale.

Geology is perceived as a science that is done in the out of doors, and rightly so. A great deal of geology is based on observations, measurements, and experiments conducted in the field. But geology is also done in the laboratory where, for example, the study of various Earth materials provides insights into many basic processes. Frequently, geology requires an understanding and application of knowledge and principles from physics, chemistry, and biology. Geology is a science that seeks to expand our knowledge of the natural world and our place in it.

GEOLOGY, PEOPLE, AND THE ENVIRONMENT

The primary focus of this book is to develop an understanding of basic geological principles, but along the way we will explore numerous important relationships between people and the natural environment. Many of the problems and issues addressed by geology are of practical value to people.

FIGURE 1.1 Sheep Mountain, Wyoming, has layers of rock plunging away from its central ridge-line. The rocks on its flanks were once continuous flat sheets of sandstone and shale. Then, about 65 million years ago, mountain-building processes folded the buried strata. Gradually, weathering and erosion peeled back the layers. In his book, *Over the Mountains,* Michael Collier, the geologist-photographer who took this photograph described the scene this way, "I find it impossible to look down and deny that something stupendous happened here—layers of rock thousands of feet thick, flexed like thin pieces of sheet metal!" (Photo by Michael Collier, *Over the Mountains, An Aerial View of Geology.* New York: Mikaya Press, 2007))

Natural hazards are a part of living on Earth. Every day they adversely affect literally millions of people worldwide and are responsible for staggering damages (Figure 1.2). Among the hazardous Earth processes studied by geologists are volcanoes, floods, tsunami, earthquakes, and landslides. Of course, geologic hazards are simply *natural* processes. They become hazards only when people try to live where these processes occur (Figure 1.3).

According to the United Nations, in 2008, for the first time, more people lived in cities than in rural areas. This global trend toward urbanization concentrates millions of people into megacities, many of which are vulnerable to natural hazards. Coastal sites are becoming more vulnerable because development often destroys natural defenses such as wetlands and sand dunes. In addition, there is a growing threat associated with human influences on the Earth system such as sea level rise that is linked to global climate change.* Other megacities are

*The idea of the Earth system is explored later in the chapter. Global climate change and its effects are the focus of Chapter 21.

TABLE 1.1 Different Areas of Geologic Study*	
Archaeological Geology	Ocean Sciences
Biogeosciences	Paleoclimatology
Engineering Geology	Paleontology
Forensic Geology	Petrology
Geochemistry	Planetary Geology
Geomorphology	Sedimentary Geology
Geophysics	Seismology
History of Geology	Structural Geology
Hydrogeology	Tectonics
Medical Geology	Volcanology
Mineralogy	

*Many of these areas of study represent interest sections and specialties of associated societies affiliated with the Geological Society of America (www.geosociety.org) and the American Geophysical Union (www.agu.org), two professional societies to which many geologists belong.

A.

B.

FIGURE 1.2   Natural hazards are part of living on Earth. **A.** Aerial view of earthquake destruction at Muzaffarabad, Pakistan, January 31, 2006. (Photo by Danny Kemp/AFP/Getty Images). **B.** Record-setting floods along the Red River of the North in late March 2009 in the Fargo, North Dakota/Moorhead, Minnesota area. (Photo by Scott Olson/Getty Images)

exposed to seismic (earthquake) and volcanic hazards where inappropriate land use and poor construction practices, coupled with rapid population growth are increasing vulnerability.

Resources represent another important focus of geology that is of great practical value to people. They include water and soil, a great variety of metallic and nonmetallic minerals, and energy (Figure 1.4). Together they form the very foundation of modern civilization. Geology deals not only with the formation and occurrence of these vital resources but also with maintaining supplies and with the environmental impact of their extraction and use.

Complicating all environmental issues is rapid world population growth *and* everyone's aspiration to a better standard of living. This means a ballooning demand for resources and a growing pressure for people to dwell in environments having significant geologic hazards.

FIGURE 1.3   An eruption of Italy's Mt. Etna in late 2006. This volcano towers above Catania, Sicily's second largest city. Geologic hazards are *natural* processes. They only become hazards when people try to live where these processes occur. (Photo by Marco Fulle)

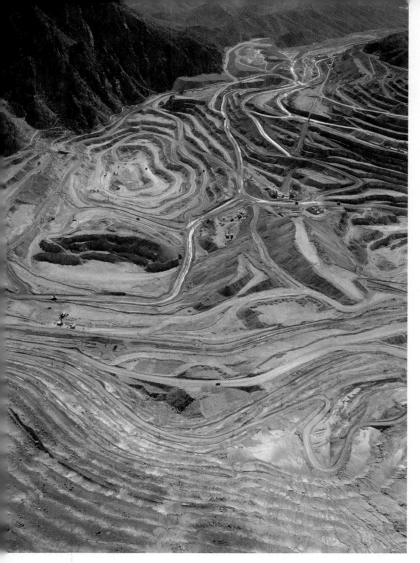

FIGURE 1.4 The open pit copper mine at Morenci, Arizona, is one of the largest producers of copper in the United States. When demand for copper is strong, the mine operates non-stop, processing 700,000 tons of rock each day, and producing about 840 million pounds of copper per year. (Photo by Michael Collier)

Not only do geologic processes have an impact on people but we humans can dramatically influence geologic processes as well. For example, river flooding is natural, but the magnitude and frequency of flooding can be changed significantly by human activities such as clearing forests, building cities, and constructing dams. Unfortunately, natural systems do not always adjust to artificial changes in ways that we can anticipate. Thus, an alteration to the environment that was intended to benefit society often has the opposite effect.

At appropriate places throughout this book, you will have the opportunity to examine different aspects of our relationship with the physical environment. It will be rare to find a chapter that does not address some aspect of natural hazards, environmental issues, or resources. Significant parts of some chapters provide the basic geologic knowledge and principles needed to understand environmental problems. Moreover, a number of the book's special-interest boxes focus on geology, people, and the environment by providing case studies or highlighting a topical issue.

SOME HISTORICAL NOTES ABOUT GEOLOGY

The nature of our Earth—its materials and processes—has been a focus of study for centuries. Writings about such topics as fossils, gems, earthquakes, and volcanoes date back to the early Greeks, more than 2300 years ago.

Certainly the most influential Greek philosopher was Aristotle. Unfortunately, Aristotle's explanations about the natural world were not based on keen observations and experiments. Instead, they were arbitrary pronouncements. He believed that rocks were created under the "influence" of the stars and that earthquakes occurred when air crowded into the ground, was heated by central fires, and escaped explosively. When confronted with a fossil fish, he explained that "a great many fishes live in the earth motionless and are found when excavations are made."

Although Aristotle's explanations may have been adequate for his day, they unfortunately continued to be expounded for many centuries, thus thwarting the acceptance of more up-to-date ideas. Frank D. Adams states in *The Birth and Development of the Geological Sciences* (New York: Dover, 1938) that "throughout the Middle Ages Aristotle was regarded as the head and chief of all philosophers; one whose opinion on any subject was authoritative and final."

Catastrophism

In the mid-1600s, James Ussher, Anglican Archbishop of Armagh, Primate of all Ireland, published a major work that had immediate and profound influences. A respected scholar of the Bible, Ussher constructed a chronology of human and Earth history in which he determined that Earth was only a few thousand years old, having been created in 4004 BC. Ussher's treatise earned widespread acceptance among Europe's scientific and religious leaders, and his chronology was soon printed in the margins of the Bible itself.

During the 17th and 18th centuries the doctrine of **catastrophism** strongly influenced Western thinking about Earth. Briefly stated, catastrophists believed that Earth's landscapes had been shaped primarily by great catastrophes. Features such as mountains and canyons, which today we know take great periods of time to form, were explained as having been produced by sudden and often worldwide disasters produced by unknowable causes that no longer operate. This philosophy was an attempt to fit the rates of Earth processes to the then-current ideas on the age of Earth.

Students Sometimes Ask . . .

What is the current world population and how fast is it growing?

It took until about the year 1800 for world population to reach 1 billion people. In 1970, that number was about 4 billion. By the beginning of 2010, world population is estimated to be about 6.8 billion. The planet is currently adding almost 80 million people per year.

Career Opportunities in the Geosciences

There are many different areas of geologic study. Two places to learn more about what geologists do are the websites noted at the bottom of Table 1.1. At the Geological Society of America site, click on "Sections and Divisions." Go to the section on "Our Science" if you access the site for the American Geophysical Union.

The United States Department of Labor, Bureau of Labor Statistics (http://www.bls.gov) is a source of abundant information and data about careers and employment opportunities. If you were to consult the *Occupational Outlook Handbook, 2008–09 Edition* (http://www.bls.gov/oco), you would gain some useful insights into a career as a geoscientist.* In addition to describing the nature of the work and working conditions, the handbook provides information about the training and education needed, earnings, and expected employment opportunities. Here are a few facts about geoscience careers from the handbook.

- Geoscientists study the composition, structure, and other physical aspects of Earth. They study Earth's geologic past and present by using sophisticated instruments to analyze the composition of various materials. Many geoscientists help search for natural resources such as groundwater, metals, and petroleum. Others work closely with environmental and other scientists to preserve and clean up the environment.
- Some geoscientists spend the majority of their time in an office, but others divide their time between fieldwork, and office and laboratory work.
- A bachelor's degree is adequate for a few entry-level positions, but most geoscientists need a master's degree in geology or Earth science. A master's degree is the preferred educational background for most entry-level research positions in private industry, Federal agencies, and State geological surveys. A Ph.D. is necessary for most high-level research and college teaching positions, but it may not be preferred for other jobs.
- Geoscientists held about 31,000 jobs in 2006. Many more individuals held geoscience faculty positions in colleges and universities, but they are classified as college and university faculty.

*This online handbook is a nationally recognized source of career information designed to provide valuable assistance to individuals making decisions about their future work lives. It is revised every two years. In addition to the section on geoscientists, a related section focuses on environmental scientists and hydrologists.

Collecting rock samples for analysis. (Photo by Peggy/Yoram Kahana/Peter Arnold, Inc.)

- Employment growth of 22 percent for geoscientists is expected between 2006 and 2016, much faster than the average for all occupations. The need for energy, environmental protection, and responsible land and water management will spur employment demand.
- Graduates with a master's degree should have excellent opportunities, especially in scientific and technical consulting and in the engineering services industries. There are few openings for new graduates with only a bachelor's degree in geoscience, but these graduates may have favorable opportunities in related occupations such as high school science teacher or science technician.

"Employment growth of 22 percent for geoscientists is expected between 2006 and 2016, much faster than the average for all occupations."

Scattered through the remaining chapters of this book are several essays that present professional profiles that are intended to give you a perspective on a variety of geoscience careers.

The Birth of Modern Geology

Against this backdrop of Aristotle's views and an Earth created in 4004 BC, a Scottish physician and gentleman farmer named James Hutton published *Theory of the Earth* in 1795 (Figure 1.5).

In this work, Hutton put forth a fundamental principle that is a pillar of geology today: **uniformitarianism**. It states that the *physical, chemical, and biological laws that operate today have also operated in the geologic past.* This means that the forces and processes that we observe presently shaping our planet have been at work for a very long time. Thus, to understand ancient

FIGURE 1.5 James Hutton (1726–1797), a founder of modern geology. (Photo courtesy of The Natural History Museum, London)

rocks, we must first understand present-day processes and their results. This idea is commonly stated as *the present is the key to the past.*

Prior to Hutton's *Theory of the Earth,* no one had effectively demonstrated that geological processes occur over extremely long periods of time. However, Hutton persuasively argued that forces that appear small could, over long spans of time, produce effects that were just as great as those resulting from sudden catastrophic events. Unlike his predecessors, Hutton carefully cited verifiable observations to support his ideas.

For example, when he argued that mountains are sculpted and ultimately destroyed by weathering and the work of running water, and that their wastes are carried to the oceans by processes that can be observed, Hutton said, "We have a chain of facts which clearly demonstrate . . . that the materials of the wasted mountains have traveled through the rivers"; and further, "There is not one step in all this progress . . . that is not to be actually perceived." He then went on to summarize this thought by asking a question and immediately providing the answer: "What more can we require? Nothing but time."

Geology Today

Today the basic tenets of uniformitarianism are just as viable as in Hutton's day. Indeed, we realize more strongly than ever that the present gives us insight into the past and that the physical, chemical, and biological laws that govern geological processes remain unchanging through time. However, we also understand that the doctrine should not be taken too literally. To say that

geological processes in the past were the same as those occurring today is not to suggest that they always had the same relative importance or that they operated at precisely the same rate. Moreover, some important geologic processes are not currently observable, but evidence that they occur is well established. For example, we know that Earth has experienced impacts from large meteorites even though we have no human witnesses. Such events altered Earth's crust, modified its climate, and strongly influenced life on the planet.

The acceptance of uniformitarianism meant the acceptance of a very long history for Earth. Although Earth processes vary in intensity, they still take a very long time to create or destroy major landscape features (Figure 1.6).

For example, geologists have established that mountains once existed in portions of present-day Minnesota, Wisconsin, Michigan, and Manitoba. Today the region consists of low hills and plains. Erosion (processes that wear land away) gradually destroyed these peaks. Estimates indicate that the North American continent is being lowered at a rate of about 3 centimeters per 1000 years. At this rate it would take 100 million years for water, wind, and ice to lower mountains that were 3000 meters (10,000 feet) high.

But even this time span is relatively short on the time scale of Earth history, for the rock record contains evidence that shows Earth has experienced many cycles of mountain building and erosion. Concerning the ever-changing nature of Earth through great expanses of geologic time, Hutton made a statement that was to become his most famous. In concluding his classic 1788 paper published in the *Transactions of the Royal Society of Edinburgh,* he stated, "The results, therefore, of our present enquiry is, that we find no vestige of a beginning—no prospect of an end."

In the chapters that follow, we will be examining the materials that compose our planet and the processes that modify it. It is important to remember that, although many features of our physical landscape may seem to be unchanging over the decades we observe them, they are nevertheless changing, but on time scales of hundreds, thousands, or even many millions of years.

GEOLOGIC TIME

Although Hutton and others recognized that geologic time is exceedingly long, they had no methods to accurately determine the age of Earth. However, in 1896 radioactivity was discovered. Using radioactivity for determining numerical dates was first attempted in 1905 and has been refined ever since. Geologists are now able to assign fairly accurate dates to events in Earth history.* For example, we know the dinosaurs became extinct about 65 million years ago. Today the age of Earth is put at about 4.6 billion years.

The Magnitude of Geologic Time

The concept of geologic time is new to many nongeologists. People are accustomed to dealing with increments of time that are measured in hours, days, weeks, and years. Our history

---

*Chapter 9 is devoted to a much more complete discussion of geologic time.

FIGURE 1.6 Over millions of years, weathering, gravity, and the erosional work of the Colorado River and the streams that flow into it have carved Arizona's Grand Canyon. Geologic processes often act so slowly that changes may not be visible during an entire human lifetime. The relative ages of the rock layers in the canyon can be determined by applying the law of superposition. The youngest rocks are on top, and the oldest are at the bottom. The numerical ages of the rocks in the canyon span hundreds of millions of years. Geologists must routinely deal with ancient materials and events that occurred in the distant geologic past. (Photo by Marc Muench/Muench Photography, Inc.)

books often examine events over spans of centuries, but even a century is difficult to appreciate fully. For most of us, someone or something that is 90 years old is *very old*, and a 1000-year-old artifact is *ancient*.

By contrast, those who study geology must routinely deal with vast time periods—millions or billions (thousands of millions) of years. When viewed in the context of Earth's 4.6-billion-year history, a geologic event that occurred 100 million years ago may be characterized as "recent" by a geologist, and a rock sample that has been dated at 10 million years may be called "young." An appreciation for the magnitude of geologic time is important in the study of geology because many processes are so gradual that vast spans of time are needed before significant changes occur.

How long is 4.6 billion years? If you were to begin counting at the rate of one number per second and continued 24 hours a day, 7 days a week and never stopped, it would take about two lifetimes (150 years) to reach 4.6 billion! Another interesting basis for comparison is as follows:

> Compress, for example, the entire 4.5 billion years of geologic time into a single year. On that scale, the oldest rocks we know date from about mid-March. Living things first appeared in the sea in May. Land plants and animals emerged in late November and the widespread swamps that formed the Pennsylvanian coal deposits flourished for about four days in early December. Dinosaurs became dominant in mid-December, but disappeared on the 26th, at about the time the Rocky Mountains were first uplifted. Manlike creatures appeared sometime dur-

ing the evening of December 31st, and the most recent continental ice sheets began to recede from the Great Lakes area and from northern Europe about 1 minute and 15 seconds before midnight on the 31st. Rome ruled the Western world for 5 seconds from 11:59:45 to 11:59:50. Columbus discovered America 3 seconds before midnight, and the science of geology was born with the writings of James Hutton just slightly more than one second before the end of our eventful year of years.*

The foregoing is just one of many analogies that have been conceived in an attempt to convey the magnitude of geologic time. Although helpful, all of them, no matter how clever, only begin to help us comprehend the vast expanse of Earth history.

Relative Dating and the Geologic Time Scale

During the 19th century, long before the advent of radiometric dating, a geologic time scale was developed using principles of relative dating. **Relative dating** means that events are placed in their proper sequence or order without knowing their age in years. This is done by applying principles such as the **law of superposition** (*super* = over, *positum* = to place). This basic rule applies to materials that were originally deposited at Earth's surface such as layers of sedimentary rock

---

*Don L. Eicher, *Geologic Time,* 2nd ed. (Englewood Cliffs, New Jersey: Prentice Hall, 1978), pp. 18–19. Reprinted by permission. This example uses 4.5 billion years as Earth's age. Current estimates place the figure closer to 4.6 billion years.

and volcanic lava flows. The law simply states that the youngest layer is on top and the oldest layer is on the bottom (assuming that nothing has turned the layers upside down, which sometimes happens). Stated another way, a layer is older than the ones above it and younger than the ones below. Arizona's Grand Canyon provides a fine example in which the oldest rocks are located in the inner gorge and the youngest rocks are found on the rim. So the law of superposition establishes the *sequence* of rock layers—but not, of course, their numerical ages (see Figure 1.6). Today such a proposal appears to be elementary, but 300 years ago it amounted to a major breakthrough in scientific reasoning by establishing a rational basis for relative time measurements.

Fossils, the remains or traces of prehistoric life, were also essential to the development of the geologic time scale (Figure 1.7). Fossils are the basis for the **principle of fossil succession**, which states that *fossil organisms succeed one another in a definite and determinable order, and therefore any time period can be recognized by its fossil content.* This principle was laboriously worked out over decades by collecting fossils from countless rock layers around the world. Once established, it allowed geologists to identify rocks of the same age in widely separated places and to build the geologic time scale shown in Figure 1.8.

Notice that units having the same designations do not necessarily extend for the same number of years. For example, the Cambrian period lasted about 54 million years, whereas the Silurian period spanned only about 28 million years. As we will emphasize again in Chapter 9, this situation exists because the basis for establishing the time scale was not the regular rhythm of a clock but the changing character of life forms through time. Specific numerical dates were added long after the time scale was established. A glance at Figure 1.8 also reveals that the Phanerozoic eon is divided into many more units than earlier eons, even though it encompasses only about 12 percent of Earth history. The meager fossil record for these earlier eons is the primary reason for the lack of detail on this portion of the time scale. Without abundant fossils, geologists lose their primary tool for subdividing geologic time.

In addition to the *law of superposition* and the *principle of fossil succession,* there are a number of other useful methods of relative dating. All are examined and illustrated in Chapter 9.

THE NATURE OF SCIENTIFIC INQUIRY

As members of a modern society, we are constantly reminded of the benefits derived from science. But what exactly is the nature of scientific inquiry? Developing an understanding of how science is done and how scientists work is an important theme that appears throughout this book. You will explore the difficulties in gathering data and some of the ingenious methods that have been developed to overcome these difficulties. You will also see many examples of how hypotheses are formulated and tested, as well as learn about the evolution and development of some major scientific theories.

All science is based on the assumption that the natural world behaves in a consistent and predictable manner that is compre-

FIGURE 1.7 Fossils are important tools for the geologist. In addition to being very important in relative dating, fossils can be useful environmental indicators. **A.** *Archaeopteryx,* a primitive bird that lived during the Jurassic Period (see Geologic Time Scale in Figure 1.8). (Photo by Michael Collier) **B.** A fossil fish of Eocene age from the Green River Formation in Wyoming. (Photo by Francois Gohier/Photo Researchers, Inc.)

hensible through careful, systematic study. The overall goal of science is to discover the underlying patterns in nature and then to use this knowledge to make predictions about what should or should not be expected, given certain facts or circumstances. For example, by knowing how oil deposits form, geologists are able to predict the most favorable sites for exploration and, perhaps as important, how to avoid regions having little or no potential.

The development of new scientific knowledge involves some basic logical processes that are universally accepted. To determine what is occurring in the natural world, scientists collect scientific "*facts*" through observation and measurement. Because some error is inevitable, the accuracy of a particular measurement or observation is always open to question. Nevertheless, these data are essential to science and serve as the springboard for the development of scientific theories (Box 1.1).

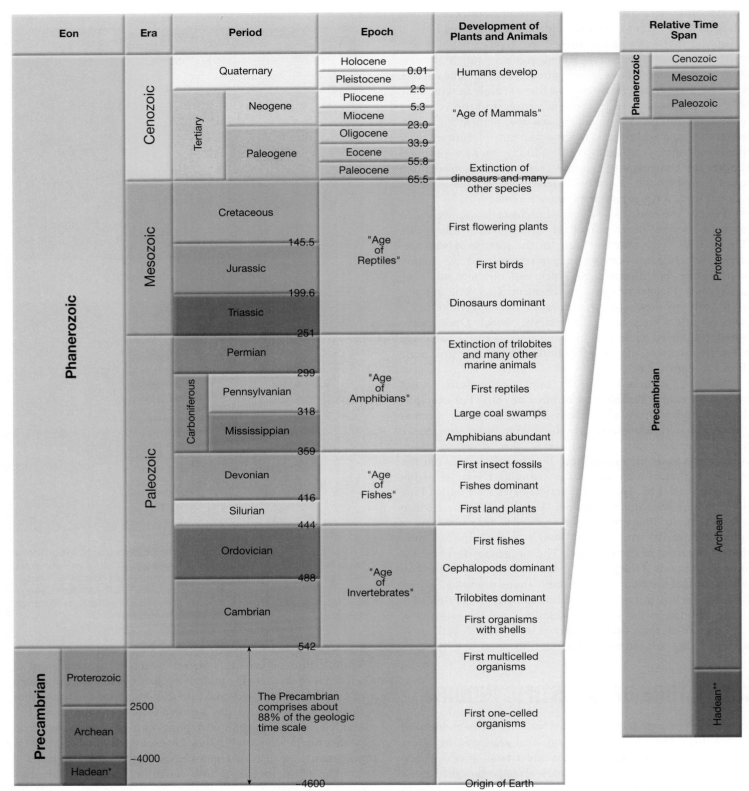

* Hadean is the informal name for the span that begins at Earth s formation and ends with Earth's earliest-known rocks.

FIGURE 1.8   The geologic time scale. Numbers on the time scale represent time in millions of years before the present. These dates were added long after the time scale had been established using relative dating techniques and continue to be revised and updated. The Precambrian accounts for more than 88 percent of geologic time. (Data from International Commission on Stratigraphy and the U.S. Geological Survey)

UNDERSTANDING EARTH

BOX 1.1

Studying Earth from Space

Scientific facts are gathered in many ways, including laboratory studies and field observations and measurements. Satellite images like the ones described here are another useful source of data. Such images provide perspectives that are difficult to gain from more traditional sources. Moreover, the high-tech instruments aboard many satellites enable scientists to gather information from remote regions where data are otherwise scarce.

The image in Figure 1.A was created using satellite radar data from the Antarctic Mapping Mission. It shows the movement of Antarctica's Lambert Glacier. The smaller glaciers that join Lambert Glacier exhibit low velocities, shown in green, of 100–300 meters (330–980 feet) per year. Most of Lambert Glacier itself moves at rates between 400–800 meters (1310–2620 feet) per year. Near its terminus, where the ice spreads out and thins, velocities increase to 1000–1200 meters (3280–3940 feet) per year. Due to the remoteness and extreme weather conditions associated with this region, only a handful of traditional in-situ velocity measurements had previously been reported. Now that

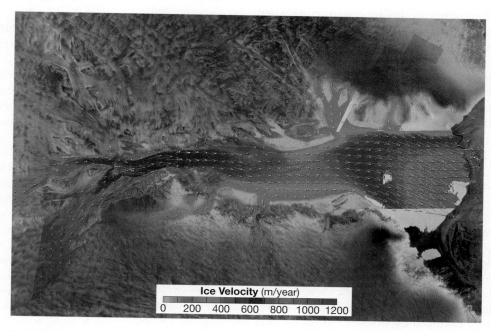

Ice Velocity (m/year)

0   200   400   600   800   1000   1200

FIGURE 1.A   This satellite image provides detailed information about the movement of Antarctica's Lambert Glacier. Such information is basic to understanding changes in the behavior of the glacier over time. The ice velocities are determined from pairs of images obtained 24 days apart, using a technique called radar interferometry. (NASA)

accurate satellite measurements are available, scientists have a quantitative baseline for future comparisons.

The image in Figure 1.B is from NASA's *Tropical Rainfall Measuring Mission (TRMM)*. Rainfall patterns over land have been studied for many years using ground-based radar and other instruments. Now the instruments aboard the *TRMM* satellite have greatly expanded our ability to collect precipitation data. In addition to data for land areas, this satellite provides extremely precise measurements of rainfall over the oceans where conventional land-based instruments cannot see. This is especially important because much of Earth's rain falls in ocean-covered tropical areas, and a great deal of the globe's weather-producing energy comes from heat exchanges involved in the rainfall process. Until the *TRMM*, information on the intensity and amount of rainfall over the tropics was scanty. Such data are crucial to understanding and predicting global climate change.

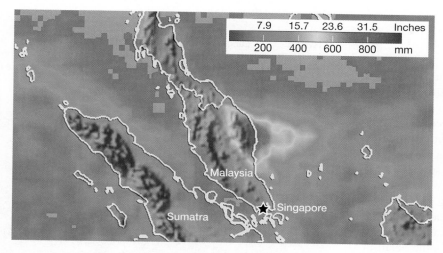

7.9   15.7   23.6   31.5   Inches

200   400   600   800   mm

FIGURE 1.B   This map of rainfall for December 7–13, 2004, in Malaysia was constructed using TRMM data. Over 800 millimeters (32 inches) of rain fell along the east coast of the peninsula (darkest red area). The extraordinary rains caused extensive flooding and triggered many mudflows. (NASA/TRMM image)

Hypothesis

Once facts have been gathered and principles have been formulated to describe a natural phenomenon, investigators try to explain how or why things happen in the manner observed. They often do this by constructing a tentative (or untested) explanation, which is called a scientific **hypothesis**. It is best if an investigator can formulate more than one hypothesis to explain a given set of observations. If an individual scientist is unable to devise multiple hypotheses, others in the scientific community will almost always develop alternative explanations. A spirited debate frequently ensues. As a result, extensive research is conducted by proponents of opposing hypotheses, and the results are made available to the wider scientific community in scientific journals.

Before a hypothesis can become an accepted part of scientific knowledge, it must pass objective testing and analysis. (If a hypothesis cannot be tested, it is not scientifically useful, no matter how interesting it might seem.) The verification process requires that *predictions* be made based on the hypothesis being considered and that the predictions be tested by comparing them against objective observations of nature. Put another way, hypotheses must fit observations other than those used to formulate them in the first place. Those hypotheses that fail rigorous testing are ultimately discarded. The history of science is littered with discarded hypotheses. One of the best known is the Earth-centered model of the universe—a proposal that was supported by the apparent daily motion of the Sun, Moon, and stars around Earth. As the mathematician Jacob Bronowski so ably stated, "Science is a great many things, but in the end they all return to this: Science is the acceptance of what works and the rejection of what does not."

Theory

When a hypothesis has survived extensive scrutiny and when competing ones have been eliminated, a hypothesis may be elevated to the status of a scientific **theory**. In everyday language we may say, "That's only a theory." But a scientific theory is a well-tested and widely accepted view that the scientific community agrees best explains certain observable facts. Some theories that are extensively documented and extremely well supported are comprehensive in scope. For example, the theory of plate tectonics provides the framework for understanding the origin of mountains, earthquakes, and volcanic activity. In addition, plate tectonics explains the evolution of the continents and the ocean basins through time—ideas that are explored in some detail in Chapters 2, 13, and 14.

Scientific Methods

The process just described, in which researchers gather facts through observations and formulate scientific hypotheses and theories, is called the *scientific method*. Contrary to popular belief, the scientific method is not a standard recipe that scientists apply in a routine manner to unravel the secrets of our natural world. Rather, it is an endeavor that involves creativity and insight.

Rutherford and Ahlgren put it this way: "Inventing hypotheses or theories to imagine how the world works and then figuring out how they can be put to the test of reality is as creative as writing poetry, composing music, or designing skyscrapers."*

There is not a fixed path that scientists always follow that leads unerringly to scientific knowledge. Nevertheless, many scientific investigations involve the following steps: (1) the collection of scientific facts through observation and measurement (Figure 1.9); (2) the formulation of questions that relate to the facts and the development of one or more working hypotheses that may answer these questions; (3) development of observations and experiments to test the hypotheses; and (4) the acceptance, modification, or rejection of the hypothesis based on extensive testing (Box 1.2).

Other scientific discoveries may result from purely theoretical ideas, which stand up to extensive examination. Some researchers use high-speed computers to create models that simulate what is happening in the "real" world. These models are useful when dealing with natural processes that occur on very long time scales or take place in extreme or inaccessible locations. Still other scientific advancements are made when a totally unexpected happening occurs during an experiment. These serendipitous discoveries are more than pure luck, for as Louis Pasteur said, "In the field of observation, chance favors only the prepared mind."

Scientific knowledge is acquired through several avenues, so it might be best to describe the nature of scientific inquiry as the methods of science rather than the scientific method. In addition, it should always be remembered that even the most

---

*F. James Rutherford and Andrew Ahlgren, *Science for All Americans* (New York: Oxford University Press, 1990), p. 7.

FIGURE 1.9 Laboratory work is an important part of what many geologists do. These scientists are working with a sediment core. Such cores often contain useful data about the geologic past and Earth's climate history. (Photo by Science Source/Photo Researchers, Inc.)

Do Glaciers Move? An Application of the Scientific Method

In Box 1.1 you learned that specialized instruments aboard satellites allow us to monitor glaciers from space. This, of course, is a recent breakthrough. In the 19th century, learning about the behavior of glaciers was far more difficult and laborious.

The study of glaciers provides an early application of the scientific method. High in the Alps of Switzerland and France, small glaciers exist in the upper portions of some valleys (Figure 1.C). In the late 18th and early 19th centuries, people who farmed and herded animals in these valleys suggested that glaciers in the upper reaches of the valleys had previously been much larger and had occupied downvalley areas. They based their explanation on the fact that the valley floors were littered with angular boulders and other rock debris that seemed identical to the materials that they could see in and near the glaciers at the heads of the valleys.

Although the explanation of these observations seemed logical, others did not accept the notion that masses of ice hundreds of meters thick were capable of movement. The disagreement was settled after a simple experiment was designed and carried out to test the hypothesis that glacial ice can move.

Markers were placed in a straight line completely across an alpine glacier. The position of the line was marked on the valley walls so that if the ice moved, the change in position could be detected. After a year or two the results were clear. The markers on the glacier had advanced down the valley, proving that glacial ice indeed

FIGURE 1.C Glacial ice moves so slowly that careful measurement is necessary to detect it. Some of the earliest attempts were carried out in the Alps. (Photo by Laurent Gillieron/Keystone/Corbis)

moves. In addition, the experiment demonstrated that ice within a glacier does not move at a uniform rate, because the markers in the center advanced farther than did those along the margins. Although most glaciers move too slowly for direct visual detection, the experiment succeeded in demonstrating that movement nevertheless occurs. In the years that followed, this experiment was repeated many times with greater accuracy using more modern surveying techniques. Each time, the basic relationships established by earlier attempts were verified.

The experiment illustrated in Figure 1.D was carried out at Switzerland's Rhône Glacier later in the 19th century. It not only traced the movement of markers within the ice but also mapped the position of the glacier's terminus. Notice that even though the ice within the glacier was advancing, the ice front was retreating. As often occurs in science, experiments and observations designed to test one hypothesis yield new information that requires further analysis and explanation.

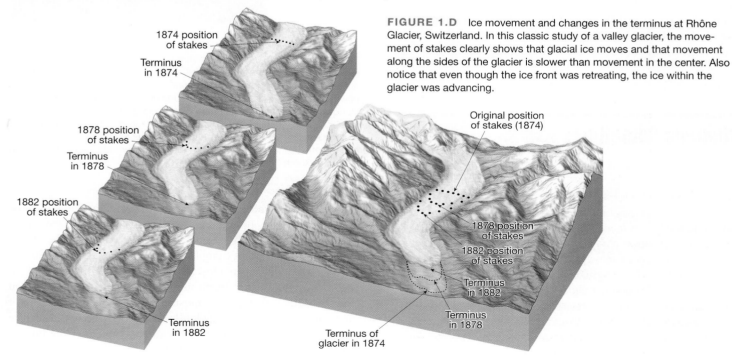

FIGURE 1.D Ice movement and changes in the terminus at Rhône Glacier, Switzerland. In this classic study of a valley glacier, the movement of stakes clearly shows that glacial ice moves and that movement along the sides of the glacier is slower than movement in the center. Also notice that even though the ice front was retreating, the ice within the glacier was advancing.

1874 position of stakes
Terminus in 1874
1878 position of stakes
Terminus in 1878
1882 position of stakes
Terminus in 1882

Original position of stakes (1874)
1878 position of stakes
1882 position of stakes
Terminus in 1882
Terminus in 1878
Terminus of glacier in 1874

compelling scientific theories are still simplified explanations of the natural world.

Plate Tectonics and Scientific Inquiry

There are many opportunities in the pages of this book to develop and reinforce your understanding of how science works and, in particular, how the science of geology works. You will learn about the methods involved in gathering data and develop a sense of the observational techniques and reasoning processes used by geologists. Chapter 2, "Plate Tectonics: A Scientific Revolution Unfolds," is an excellent example.

During the past several decades, a great deal has been learned about the workings of our dynamic planet. This period has seen an unequaled revolution in our understanding of Earth. The revolution began in the early part of the 20th century with the radical proposal of *continental drift*—the idea that the continents moved about the face of the planet. This hypothesis contradicted the established view that the continents and ocean basins are permanent and stationary features on the face of Earth. For that reason, the notion of drifting continents was received with great skepticism and even ridicule. More than 50 years passed before enough data were gathered to transform this controversial hypothesis into a sound theory that wove together the basic processes known to operate on Earth. The theory that finally emerged, called the *theory of plate tectonics*, provided geologists with the first comprehensive model of Earth's internal workings.

As you read Chapter 2, you will not only gain insights into the workings of our planet, you will also see an excellent example of the way geological "truths" are uncovered and reworked.

EARTH'S SPHERES

The classic view of Earth, shown in Figure 1.10A, provided the *Apollo 8* astronauts and the rest of humanity with a unique perspective of our home. Seen from space, Earth is breathtaking in its beauty and startling in its solitude. Such an image reminds us that

FIGURE 1.10   **A.** View that greeted the *Apollo 8* astronauts as their spacecraft emerged from behind the Moon. (NASA) **B.** Africa and Arabia are prominent in this image of Earth taken from *Apollo 17*. The tan cloud-free zones over the land coincide with major desert regions. The band of clouds across central Africa is associated with a much wetter climate that, in places, sustains tropical rain forests. The dark blue of the oceans and the swirling cloud patterns remind us of the importance of the oceans and the atmosphere. Antarctica, a continent covered by glacial ice, is visible at the South Pole. (NASA)

Students Sometimes Ask . . .

In class you compared a hypothesis to a theory. How is each one different from a scientific law?

A scientific *law* is a basic principle that describes a particular behavior of nature that is generally narrow in scope and can be stated briefly—often as a simple mathematical equation. Because scientific laws have been shown time and time again to be consistent with observations and measurements, they are rarely discarded. Laws may, however, require modifications to fit new findings. For example, Newton's laws of motion are still useful for everyday applications (NASA uses them to calculate satellite trajectories), but they do not work at velocities approaching the speed of light. For these circumstances, they have been supplanted by Einstein's theory of relativity.

---

our home is, after all, a planet—small, self-contained, and in some ways even fragile. Bill Anders, one of the *Apollo 8* astronauts, expressed it this way, "We came all this way to explore the Moon, and the most important thing is that we discovered the Earth."

As we look more closely at our planet from space, it becomes apparent that Earth is much more than rock and soil (Figure 1.10B). In fact, the most conspicuous features are not the continents but swirling clouds suspended above the surface and the vast global ocean. These features emphasize the importance of air and water to our planet.

The closer view of Earth from space, shown in Figure 1.10B, helps us appreciate why the physical environment is traditionally divided into three major parts: the water portion of our planet, the *hydrosphere;* Earth's gaseous envelope, the *atmosphere;* and, of course, the solid Earth, or *geosphere.* It needs to be emphasized that our environment is highly integrated and not dominated by rock, water, or air alone. Rather, it is characterized by continuous interactions as air comes in contact with rock, rock with water, and water with air. Moreover, the *biosphere,* which is the totality of all plant and animal life on our planet, interacts with each of the three physical realms and is an equally integral part of the planet. Thus, Earth can be thought of as consisting of four major spheres: the hydrosphere, atmosphere, geosphere, and biosphere.

The interactions among Earth's four spheres are incalculable. Figure 1.11 provides us with one easy-to-visualize example. The shoreline is an obvious meeting place for rock, water, and air. In this scene, ocean waves that were created by the drag of air moving across the water are breaking against the rocky shore. The force of the water can be powerful, and the erosional work that is accomplished can be great.

Hydrosphere

Earth is sometimes called the *blue* planet. Water, more than anything else, makes Earth unique. The **hydrosphere** is a dynamic mass of water that is continually on the move, evaporating from the oceans to the atmosphere, precipitating to the land, and running back to the ocean again. The global ocean is certainly the most prominent feature of the hydrosphere, blanketing nearly 71 percent of Earth's surface to an average depth of about 3800 meters (12,500 feet). It accounts for about 97 percent of Earth's water. However, the hydrosphere also includes the fresh water found underground and in streams, lakes, and glaciers. Moreover, water is an important component of all living things.

Although these latter sources constitute just a tiny fraction of the total, they are much more important than their meager percentage indicates. In addition to providing the fresh water that is so vital to life on land, streams, glaciers, and groundwater are responsible for sculpting and creating many of our planet's varied landforms.

Atmosphere

Earth is surrounded by a life-giving gaseous envelope called the **atmosphere** (Figure 1.12). When we watch a high-flying jet plane cross the sky, it seems that the atmosphere extends upward for a great distance. However, when compared to the thickness (radius) of the solid Earth (about 6400 kilometers or 4000 miles), the atmosphere is a very shallow layer. One half lies below an altitude of 5.6 kilometers (3.5 miles), and 90 percent occurs within just 16 kilometers (10 miles) of

FIGURE 1.11   The shoreline is one obvious example of an *interface*—a common boundary where different parts of a system interact. In this scene, ocean waves (*hydrosphere*) that were created by the force of moving air (*atmosphere*) break against California's rocky Big Sur shore (*geosphere*). The force of the water can be powerful, and the erosional work that is accomplished can be great. (Photo by Kevin Schafer/www.DanitaDelimont)

A.

FIGURE 1.12   **A.** This jet is flying high in the atmosphere at an altitude of more than 9000 meters (30,000 feet). More than two-thirds of the atmosphere is below this height. To someone on the ground, the atmosphere seems to extend for a great distance. however, when compared to the thickness (radius) of the solid Earth, the atmosphere is a very shallow layer. (Photo by Warren Faidley/Weatherstock) **B.** This unique image of Earth's atmosphere merging with the emptiness of space resembles an abstract painting. It was taken over western China in June 2007 by a Space Shuttle crew member. The thin silvery streaks (called noctilucent clouds) high in the blue area are at a height of about 80 kilometers (50 miles). The atmosphere at this altitude is *very* thin. Air pressure here is less than a thousandth that at sea level. The thin reddish zone in the lower portion of the image is the densest part of the atmosphere. It is here, in a layer called the *troposphere,* that practically all weather and cloud formation occur. Ninety percent of Earth's atmosphere occurs within just 16 kilometers (10 miles) of the surface. (NASA)

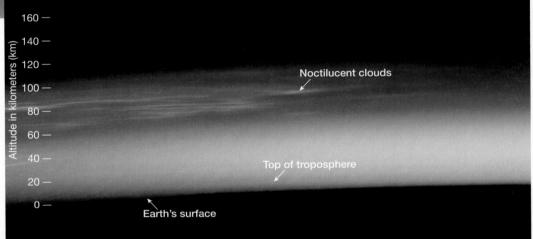

B.

Earth's surface. Despite its modest dimensions, this thin blanket of air is an integral part of the planet. It not only provides the air that we breathe but also protects us from the Sun's intense heat and dangerous ultraviolet radiation. The energy exchanges that continually occur between the atmosphere and Earth's surface and between the atmosphere and space produce the effects we call weather and climate.

If, like the Moon, Earth had no atmosphere, our planet would not only be lifeless but many of the processes and interactions that make the surface such a dynamic place could not operate. Without weathering and erosion, the face of our planet might more closely resemble the lunar surface, which has not changed appreciably in nearly 3 billion years.

Biosphere

The **biosphere** includes all life on Earth. Ocean life is concentrated in the sunlit surface waters of the sea (Figure 1.13). Most life on land is also concentrated near the surface, with tree roots and burrowing animals reaching a few meters underground and flying insects and birds reaching a kilometer or so into the atmosphere. A surprising variety of life

forms are also adapted to extreme environments. For example, on the ocean floor where pressures are extreme and no light penetrates, there are places where vents spew hot, mineral- rich fluids that support communities of exotic life forms. On land, some bacteria thrive in rocks as deep as 4 kilometers (2.5 miles) and in boiling hot springs. Moreover, air currents can carry microorganisms many kilometers into the atmosphere. But even when we consider these extremes, life still must be thought of as being confined to a narrow band very near Earth's surface.

Plants and animals depend on the physical environment for the basics of life. However, organisms do not just respond to their physical environment. Indeed, the biosphere powerfully influences the other three spheres. Without life, the makeup and nature of the geosphere, hydrosphere, and atmosphere would be very different.

Geosphere

Lying beneath the atmosphere and the oceans is the solid Earth, or **geosphere**. The geosphere extends from the surface to the center of the planet, a depth of nearly 6400 kilometers, making it by far

Earth System Science

A simple example of the interactions among different parts of the Earth system occurs in the winter as moisture evaporates from the Pacific Ocean and subsequently falls as rain in the hills and mountains of southern California, triggering destructive landslides. A case study in Chapter 15 explores such an event. The processes that move water from the hydrosphere to the atmosphere and then to the solid Earth have a profound impact on the plants and animals (including humans) that inhabit the affected regions. Figure 1.14 provides another example.

Scientists have recognized that in order to more fully understand our planet, they must learn how its individual components (land, water, air, and life forms) are interconnected. This endeavor, called **Earth system science,** aims to study Earth as a *system* composed of numerous interacting parts, or *subsystems*. Rather than looking through the limited lens of only one of the traditional sciences—geology, atmospheric science, chemistry, biology, and so on—Earth system science attempts to integrate the knowledge of several academic fields. Using an interdisciplinary approach, those engaged in Earth system science attempt to achieve the level of understanding necessary to comprehend and solve many of our global environmental problems.

What Is a System? Most of us hear and use the term *system* frequently. We may service our car's cooling *system,* make use of the city's transportation *system,* and be a participant in the political *system.* A news report might inform us of an approaching weather *system.* Further, we know that Earth is just a small part of a larger system known as the *solar system,* which, in turn, is a subsystem of an even larger system called the Milky Way Galaxy.

Loosely defined, a **system** can be any size group of interacting parts that form a complex whole. Most natural systems are driven by sources of energy that move matter and/or energy from one place to another. A simple analogy is a car's cooling system, which contains a liquid (usually called coolant or antifreeze) that circulates from the engine to the radiator and back again. The role of this system is to transfer heat generated by combustion in the engine to the radiator, where moving air removes it from the system. Hence, the term cooling system.

Systems like a car's cooling system are self-contained with regard to matter and are called **closed systems**. Although energy moves freely in and out of a closed system, no matter (liquid in the case of our auto's cooling system) enters or leaves the system. (This assumes you don't get a leak in your radiator). By

FIGURE 1.13 The hydrosphere contains a significant portion of Earth's biosphere. Modern coral reefs are unique and complex examples and are home to about 25 percent of all marine species. Because of this diversity, they are sometimes referred to as the ocean equivalent of rain forests. For more about coral reefs, see Box 7.2, p. 217. (Photo by Darryl Leniuk/age fotostock)

the largest of Earth's four spheres. Much of our study of the solid Earth focuses on the more accessible surface features. Fortunately, many of these features represent the outward expressions of the dynamic behavior of Earth's interior. By examining the most prominent surface features and their global extent, we can obtain clues to the dynamic processes that have shaped our planet. A first look at the structure of Earth's interior and at the major surface features of the geosphere will come later in the chapter.

Soil, the thin veneer of material at Earth's surface that supports the growth of plants, may be thought of as part of all four spheres. The solid portion is a mixture of weathered rock debris (geosphere) and organic matter from decayed plant and animal life (biosphere). The decomposed and disintegrated rock debris is the product of weathering processes that require air (atmosphere) and water (hydrosphere). Air and water also occupy the open spaces between the solid particles.

EARTH AS A SYSTEM

Anyone who studies Earth soon learns that our planet is a dynamic body with many separate but interacting parts or *spheres.* The hydrosphere, atmosphere, biosphere, and geosphere and all of their components can be studied separately. However, the parts are not isolated. Each is related in some way to the others to produce a complex and continuously interacting whole that we call the *Earth system.*

FIGURE 1.14   This image provides an example of interactions among different parts of the Earth system. Aerial view of Caraballeda, Venezuela, covered by material from a massive debris flow (popularly called a mudslide in the press). In December 1999, extraordinary rains triggered this debris flow and thousands of others along this mountainous coastal zone. Caraballeda was located at the mouth of a steep canyon. An estimated 19,000 lives were lost. (Photo by Kimberly White/Reuters/Corbis/Bettmann)

contrast, most natural systems are **open systems** and are far more complicated than the foregoing example. In an open system both energy and matter flow into and out of the system. In a weather system such as a hurricane, factors such as the quantity of water vapor available for cloud formation, the amount of heat released by condensing water vapor, and the flow of air into and out of the storm can fluctuate a great deal. At times the storm may strengthen; at other times it may remain stable or weaken.

Feedback Mechanisms   Most natural systems have mechanisms that tend to enhance change, as well as other mechanisms that tend to resist change and thus stabilize the system. For example, when we get too hot, we perspire to cool down. This cooling phenomenon works to stabilize our body temperature and is referred to as a **negative feedback mechanism**. Negative feedback mechanisms work to maintain the system as it is or, in other words, to maintain the status quo. By contrast, mechanisms that enhance or drive change are called **positive feedback mechanisms**.

Most of Earth's systems, particularly the climate system, contain a wide variety of negative and positive feedback mechanisms. For example, substantial scientific evidence indicates that Earth has entered a period of global warming. One consequence of global warming is that some of the world's glaciers and ice caps have begun to melt. Highly reflective snow-and ice-covered surfaces are gradually being replaced by brown soils, green trees, or blue oceans, all of which are darker, so they absorb more sunlight. The result is a positive feedback that contributes to the warming.

On the other hand, an increase in global temperature also causes greater evaporation of water from Earth's land–sea surface. One result of having more water vapor in the air is an increase in cloud cover. Because cloud tops are white and highly reflective, more sunlight is reflected back to space, which diminishes the amount of sunshine reaching Earth's surface and thus reduces global temperatures. Further, warmer temperatures tend to promote the growth of vegetation. Plants in turn remove carbon dioxide (CO_2) from the air. Since carbon dioxide is one of the atmosphere's *greenhouse gases*, its removal has a negative impact on global warming.*

In addition to natural processes, we must also consider the human element. Extensive cutting and clearing of the tropical rain forests and the burning of fossil fuels (oil, natural gas, and coal) result in an increase in atmospheric CO_2. Such activity is contributing to the increase in global temperature that our planet is experiencing. One of the daunting tasks of Earth system scientists is to predict what the climate will be like in the future by taking into account many variables, including technological changes, population trends, and the overall impact of the numerous competing positive and negative feedback mechanisms. Chapter 21 on "Global Climate Change" explores this topic in some detail.

---

*Greenhouse gases absorb heat energy emitted by Earth and thus help keep the atmosphere warm.

The Earth System

The Earth system has a nearly endless array of subsystems in which matter is recycled over and over again. One example that you will learn about in Chapter 7 traces the movements of carbon among Earth's four spheres. It shows us, for example, that the carbon dioxide in the air and the carbon in living things and in certain sedimentary rocks is all part of a subsystem described by the *carbon cycle.*

Cycles in the Earth System   A more familiar loop or subsystem is the *hydrologic cycle.* It represents the unending circulation of Earth's water among the hydrosphere, atmosphere, biosphere, and geosphere. Water enters the atmosphere by evaporation from Earth's surface and by transpiration from plants. Water vapor condenses in the atmosphere to form clouds, which in turn produce precipitation that falls back to Earth's surface. Some of the rain that falls onto the land sinks in to be taken up by plants or become groundwater, and some flows across the surface toward the ocean.

Viewed over long time spans, the rocks of the geosphere are constantly forming, changing, and reforming. The loop that involves the processes by which one rock changes to another is called the *rock cycle* and will be discussed at some length later in the chapter. The cycles of the Earth system, such as the hydrologic and rock cycles, are not independent of one another. To the contrary, there are many places where they interface. An **interface** is a common boundary where different parts of a system come in contact and interact. For example, in Figure 1.15, weathering at the surface gradually disintegrates and decomposes solid rock. The work of gravity and running water may eventually move this material to another place and deposit it. Later, groundwater percolating through the debris may leave behind mineral matter that cements the grains together into solid rock (a rock that is often very different from the rock with which we started). This changing of one rock into another could not have occurred without the movement of water through the hydrologic cycle. There are many places where one cycle or loop in the Earth system interfaces with, and is a basic part of, another.

Energy for the Earth System   The Earth system is powered by energy from two sources. The Sun drives external processes that occur in the atmosphere and hydrosphere and at Earth's surface. Weather and climate, ocean circulation, and erosional processes are driven by energy from the Sun. Earth's interior is the second source of energy. Heat remaining from when our planet formed, and heat that is continuously generated by decay of radioactive elements powers the internal processes that produce volcanoes, earthquakes, and mountains.

FIGURE 1.15   This diagram depicts the interface (common boundary) between two important cycles in the Earth system—the hydrologic cycle and the rock cycle.

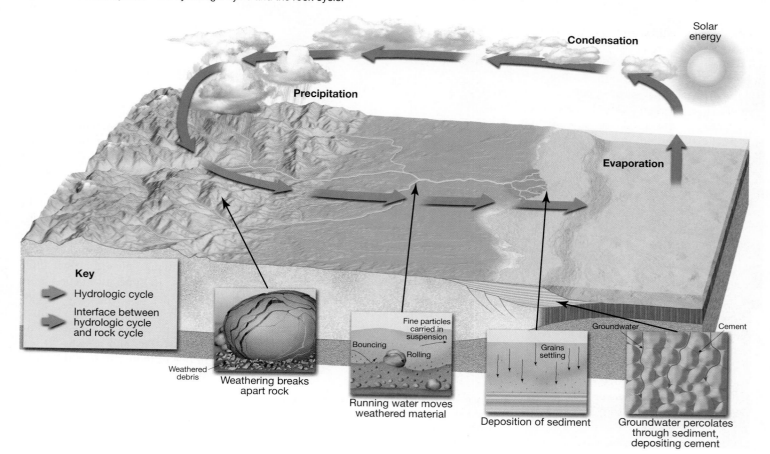

The Parts Are Linked   The parts of the Earth system are linked so that a change in one part can produce changes in any or all of the other parts. For example, when a volcano erupts, lava from Earth's interior may flow out at the surface and block a nearby valley. This new obstruction influences the region's drainage system by creating a lake or causing streams to change course. The large quantities of volcanic ash and gases that can be emitted during an eruption might be blown high into the atmosphere and influence the amount of solar energy that can reach Earth's surface. The result could be a drop in air temperatures over the entire hemisphere.

Where the surface is covered by lava flows or a thick layer of volcanic ash, existing soils are buried. This causes the soil-forming processes to begin anew to transform the new surface material into soil (Figure 1.16). The soil that eventually forms will reflect the interactions among many parts of the Earth system—the volcanic parent material, the climate, and the impact of biological activity. Of course, there would also be significant changes in the biosphere. Some organisms and their habitats would be eliminated by the lava and ash, whereas new settings for life, such as the lake, would be created. The potential climate change could also impact sensitive life forms.

The Earth system is characterized by processes that vary on spatial scales from fractions of millimeters to thousands of kilo-meters. Time scales for Earth's processes range from millisec-onds to billions of years. As we learn about Earth, it becomes increasingly clear that despite significant separations in distance or time, many processes are connected, and a change in one component can influence the entire system.

Humans are *part of* the Earth system, a system in which the living and nonliving components are entwined and intercon-nected. Therefore, our actions produce changes in all of the other parts. When we burn gasoline and coal, dispose of our wastes, and clear the land, we cause other parts of the system to respond, often in unforeseen ways. Throughout this book you will learn about many of Earth's subsystems, including the hydrologic system, the tectonic (mountain-building) system, and the rock cycle, to name a few. Remember that these compo-nents *and we humans* are all part of the complex interacting whole we call the Earth system.

EARLY EVOLUTION OF EARTH

Recent earthquakes caused by displacements of Earth's crust, along with lavas spewed from active volcanoes, represent only the latest in a long line of events by which our planet has attained its present form and structure. The geologic processes operating in Earth's interior can be best understood when viewed in the context of much earlier events in Earth history.

Origin of Planet Earth

This section describes the most widely accepted views on the ori-gin of our solar system. The the-ory described here represents the most consistent set of ideas we have that explains what we know about our solar system today.

Our scenario begins about 14 billion years ago with the *Big Bang,* an incomprehensibly large explo-sion that sent all matter of the uni-verse flying outward at incredible speeds. In time, the debris from this explosion, which was almost entirely hydrogen and helium, began to cool and condense into the first stars and galaxies. It was in one of these galaxies, the Milky Way, that our solar system and planet Earth took form.

Earth is one of eight planets that, along with several dozen moons and numerous smaller

FIGURE 1.16   When Mount St. Helens erupted in May 1980 (inset), the area shown here was buried by a volcanic mudflow. Now plants are reestablished and new soil is forming. (Jack W. Dykinga/inset photo by CORBIS)

bodies, revolve around the Sun. The orderly nature of our solar system leads most researchers to conclude that Earth and the other planets formed at essentially the same time and from the same primordial material as the Sun. The **nebular theory** proposes that the bodies of our solar system evolved from an enormous rotating cloud called the **solar nebula** (Figure 1.17). Besides the hydrogen and helium atoms generated during the Big Bang, the solar nebula consisted of microscopic dust grains and the ejected matter of long-dead stars. (Nuclear fusion in stars converts hydrogen and helium into the other elements found in the universe.)

Nearly 5 billion years ago this huge cloud of gases and minute grains of heavier elements began to slowly contract due to the gravitational interactions among its particles (Figure 1.18). Some external influence, such as a shock wave traveling from a

catastrophic explosion (*supernova*), may have triggered the collapse. As this slowly spiraling nebula contracted, it rotated faster and faster for the same reason ice skaters do when they draw their arms toward their bodies. Eventually the inward pull of gravity came into balance with the outward force caused by the rotational motion of the nebula (see Figure 1.17). By this time the once vast cloud had assumed a flat disk shape with a large concentration of material at its center called the *protosun* (pre-Sun). (Astronomers are fairly confident that the nebular cloud formed a disk because similar structures have been detected around other stars).

During the collapse, gravitational energy was converted to thermal energy (heat), causing the temperature of the inner portion of the nebula to dramatically rise. At these high temperatures, the dust grains broke up into molecules and

FIGURE 1.17 Formation of the solar system according to the nebular theory. **A.** The birth of our solar system began as dust and gases (nebula) started to gravitationally collapse. **B.** The nebula contracted into a rotating disk that was heated by the conversion of gravitational energy into thermal energy. **C.** Cooling of the nebular cloud caused rocky and metallic material to condense into tiny solid particles. **D.** Repeated collisions caused the dust-size particles to gradually coalesce into asteroid-size bodies. Within a few million years these bodies accreted into the planets.

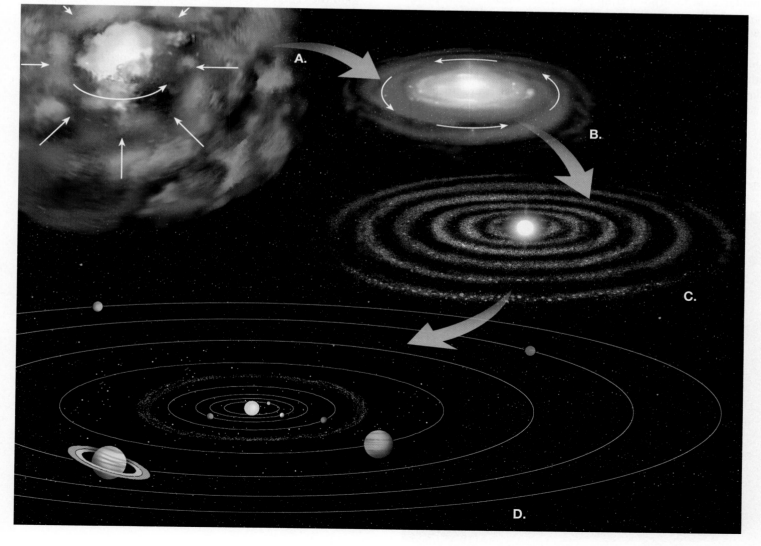

extremely energetic atomic particles. However, at distances beyond the orbit of Mars, the temperatures probably remained quite low. At −200 °C, the tiny particles in the outer portion of the nebula were likely covered with a thick layer of ices made of frozen water, carbon dioxide, ammonia, and methane. (Some of this material still resides in the outermost reaches of the solar system in a region called the *Oort cloud.*) The disk-shaped cloud also contained appreciable amounts of the lighter gases hydrogen and helium.

The formation of the Sun marked the end of the period of contraction and thus the end of gravitational heating. Temperatures in the region where the inner planets now reside began to decline. The decrease in temperature caused those substances with high melting points to condense into tiny particles that began to coalesce (join together). Materials such as iron and nickel and the elements of which the rock-forming minerals are composed—silicon, calcium, sodium, and so forth—formed metallic and rocky clumps that orbited the Sun (see Figure 1.17). Repeated collisions caused these masses to coalesce into larger asteroid-size bodies, called *planetesimals,* which in a few tens of millions of years accreted into the four inner planets we call Mercury, Venus, Earth, and Mars. Not all of these clumps of matter were incorporated into the planetesimals. Those rocky and metallic pieces that remained in orbit are called *meteorites* when they survive an impact with Earth.

As more and more material was swept up by the planets, the high-velocity impact of nebular debris caused the temperature of these bodies to rise. Because of their relatively high temperatures and weak gravitational fields, the inner planets were unable to accumulate much of the lighter components of the nebular

FIGURE 1.18   Lagoon Nebula. It is in glowing clouds like these that gases and dust particles become concentrated into stars. (Courtesy of National Optical Astronomy Observatories)

cloud. The lightest of these, hydrogen and helium, were eventually whisked from the inner solar system by the solar wind.

At the same time that the inner planets were forming, the larger, outer planets (Jupiter, Saturn, Uranus, and Neptune), along with their extensive satellite systems, were also developing. Because of low temperatures far from the Sun, the material from which these planets formed contained a high percentage of ices—water, carbon dioxide, ammonia, and methane—as well as rocky and metallic debris. The accumulation of ices accounts, in part, for the large size and low density of the outer planets. The two most massive planets, Jupiter and Saturn, had a surface gravity sufficient to attract and hold large quantities of even the lightest elements—hydrogen and helium.

Formation of Earth's Layered Structure

As material accumulated to form Earth (and for a short period afterward), the high-velocity impact of nebular debris and the decay of radioactive elements caused the temperature of our planet to steadily increase. During this time of intense heating, Earth became hot enough that iron and nickel began to melt. Melting produced liquid blobs of heavy metal that sank toward the center of the planet. This process occurred rapidly on the scale of geologic time and produced Earth's dense iron-rich core.

The early period of heating resulted in another process of chemical differentiation, whereby melting formed buoyant masses of molten rock that rose toward the surface, where they solidified to produce a primitive crust. These rocky materials were enriched in oxygen and "oxygen-seeking" elements, particularly silicon and aluminum, along with lesser amounts of calcium, sodium, potassium, iron, and magnesium. In addition, some heavy metals such as gold, lead, and uranium, which have low melting points or were highly soluble in the ascending molten masses, were scavenged from Earth's interior and concentrated in the developing crust. This early period of chemical segregation established the three basic divisions of Earth's interior—the iron-rich *core;* the thin *primitive crust;* and Earth's largest layer, called the *mantle,* which is located between the core and crust.

An important consequence of this early period of chemical differentiation is that large quantities of gaseous materials were allowed to escape from Earth's interior, as happens today during volcanic eruptions. By this process a primitive atmosphere gradually evolved. It is on this planet, with this atmosphere, that life as we know it came into existence.

Following the events that established Earth's basic structure, the primitive crust was lost to erosion and other geologic processes, so we have no direct record of its makeup. When and exactly how the continental crust—and thus Earth's first landmasses—came into existence is a matter of ongoing research. Nevertheless, there is general agreement that the continental crust formed gradually over the last 4 billion years. (The oldest rocks yet discovered are isolated fragments found in the Northwest Territories of Canada that have radiometric dates of about 4 billion years.) In addition, as you will see in Chapter 2, Earth is an evolving planet whose continents and ocean basins have continually changed shape and even location during much of this period.

EARTH'S INTERNAL STRUCTURE

AN INTRODUCTION TO GEOLOGY
▶ Earth's Layered Structure

In the preceding section, you learned that the segregation of material that began early in Earth's history resulted in the formation of three major layers defined by their chemical composition—the crust, mantle, and core. In addition to these compositionally distinct layers, Earth can be divided into layers based on physical properties, most notably the lithosphere and the asthenosphere. The physical properties used to define such zones include whether the layer is solid or liquid and how weak or strong it is. Knowledge of both types of layers is essential to our understanding of basic geologic processes, such as volcanism, earthquakes, and mountain building (Figure 1.19).

Earth's Crust

The **crust**, Earth's relatively thin, rocky outer skin, is of two different types—continental crust and oceanic crust. Both share the word "crust," but the similarity ends there. The oceanic crust is roughly 7 kilometers (5 miles) thick and composed of the dark igneous rock *basalt*. By contrast, the continental crust averages 35 to 40 kilometers (25 miles) thick but may exceed 70 kilometers (40 miles) in some mountainous regions such as the Rockies and Himalayas. Unlike the oceanic crust, which has a relatively homogeneous chemical composition, the continental crust consists of many rock types. Although the upper crust has an average composition of a *granitic rock* called *granodiorite,* it varies considerably from place to place.

Continental rocks have an average density of about 2.7 g/cm³, and some have been discovered that are more than 4

FIGURE 1.19   Views of Earth's layered structure. The right side of the large cross section shows that Earth's interior is divided into three different layers based on compositional differences—the crust, mantle, and core. The left side of the large cross section depicts layers of Earth's interior based on physical properties—the lithosphere, asthenosphere, transition zone, lower mantle, D″ layer, outer core, and inner core. The block diagram to the left of the large cross section shows an enlarged view of the upper portion of Earth's interior.

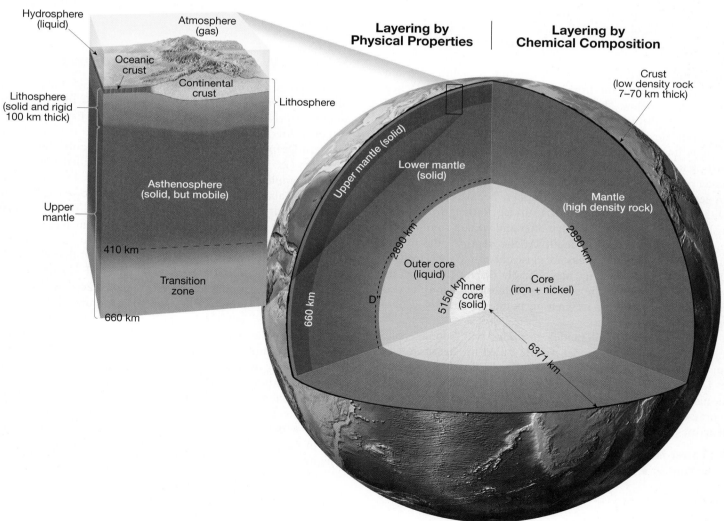

billion years old. The rocks of the oceanic crust are younger (180 million years or less) and denser (about 3.0 g/cm^3) than continental rocks.*

Earth's Mantle

More than 82 percent of Earth's volume is contained in the **mantle**, a solid, rocky shell that extends to a depth of about 2900 kilometers (1800 miles). The boundary between the crust and mantle represents a marked change in chemical composition. The dominant rock type in the uppermost mantle is *peridotite*, which is richer in the metals magnesium and iron than the minerals found in either the continental or oceanic crust.

The Upper Mantle   The upper mantle extends from the crust–mantle boundary down to a depth of about 660 kilometers (410 miles). The upper mantle can be divided into three different parts. The top portion of the upper mantle is part of the stiff *lithosphere,* and beneath that is the weaker *asthenosphere.* The bottom part of the upper mantle is called the *transition zone.*

The **lithosphere** (sphere of rock) consists of the entire crust and uppermost mantle and forms Earth's relatively cool, rigid outer shell (Figure 1.19). Averaging about 100 kilometers in thickness, the lithosphere is more than 250 kilometers thick below the oldest portions of the continents. Beneath this stiff layer to a depth of about 410 kilometers lies a soft, comparatively weak layer known as the **asthenosphere** ("weak sphere"). The top portion of the asthenosphere has a temperature/pressure regime that results in a small amount of melting. Within this very weak zone the lithosphere is mechanically detached from the layer below. The result is that the lithosphere is able to move independently of the asthenosphere, a fact we will consider in the next chapter.

It is important to emphasize that the strength of various Earth materials is a function of both their composition and of the temperature and pressure of their environment. You should not get the idea that the entire lithosphere behaves like a brittle solid similar to rocks found on the surface. Rather, the rocks of the lithosphere get progressively hotter and weaker (more easily deformed) with increasing depth. At the depth of the uppermost asthenosphere, the rocks are close enough to their melting temperature (some melting may actually occur) that they are very easily deformed. Thus, the uppermost asthenosphere is weak because it is near its melting point, just as hot wax is weaker than cold wax.

From about 410 kilometers to about 660 kilometers in depth is the part of the upper mantle called the **transition zone**. The top of the transition zone is identified by a sudden increase in density from about 3.5 to 3.7 gm/cm^3. This change occurs because minerals in the rock peridotite respond to the increase in pressure by forming new minerals with closely packed atomic structures.

The Lower Mantle   From a depth of 660 kilometers to the top of the core, at a depth of 2900 kilometers (1800 miles), is the lower mantle. Because of an increase in pressure (caused by the weight of the rock above) the mantle gradually strengthens with depth. Despite their strength however, the rocks within the lower mantle are very hot and capable of very gradual flow.

In the bottom few hundred kilometers of the mantle is a highly variable and unusual layer called the **D″ layer** (pronounced "dee double-prime"). The nature of this boundary layer between the rocky mantle and the hot liquid iron outer core will be examined in Chapter 12.

Earth's Core

The composition of the **core** is thought to be an iron-nickel alloy with minor amounts of oxygen, silicon, and sulfur—elements that readily form compounds with iron. At the extreme pressure found in the core, this iron-rich material has an average density of nearly 11 g/cm^3 and approaches 14 times the density of water at Earth's center.

The core is divided into two regions that exhibit very different mechanical strengths. The **outer core** is a *liquid layer* 2270 kilometers (1410 miles) thick. It is the movement of metallic iron within this zone that generates Earth's magnetic field. The **inner core** is a sphere having a radius of 1216 kilometers (754 miles). Despite its higher temperature, the iron in the inner core is *solid* due to the immense pressures that exist in the center of the planet.

How Do We Know What We Know?

At this point you may be asking yourself, "How did we learn about the composition and structure of Earth's interior?" You might suspect that the internal structure of Earth has been sampled directly. However, the deepest mine in the world (the Western Deep Levels mine in South Africa) is only about 4 kilometers (2.5 miles) deep, and the deepest drilled hole in the world (completed in the Kola Peninsula of Russia in 1992) goes down only about 12 kilometers (7.5 miles). In essence, humans have never drilled a hole into the mantle (and will never drill into the core) in order to directly sample these materials.

Despite these limitations, theories that describe the nature of Earth's interior have been developed that closely fit most observational data. Thus, our model of Earth's interior represents the best inferences we can make based on the available data. For example, the layered structure of Earth has been established using indirect observations. Every time there is an earthquake, waves of energy (called *seismic waves*) penetrate Earth's interior, much like X-rays penetrate the human body. Seismic waves change their speed and are bent and reflected as they move through zones having different properties. An extensive series of monitoring stations around the world detect and record this energy. With the aid of computers, these data are analyzed and used to determine the structure of Earth's interior. For more about how this is done, see Chapter 12, "Earth's Interior."

What evidence do we have to support the alleged composition of our planet's interior? It may surprise you to learn that rocks that originated in the mantle have been collected at Earth's

---

*Liquid water has a density of 1 g/cm^3; therefore, the density of basalt is three times that of water.

surface. These include diamond-bearing samples, which laboratory studies indicate can form only in very high-pressure environments. Since these rocks must have crystallized at depths exceeding 200 kilometers (120 miles), they are inferred to be samples of the mantle that underwent very little alteration during their ascent to the surface. In addition, we have been able to examine slivers of the uppermost mantle and overlying oceanic crust that have been thrust high above sea level in locations such as Cyprus, Newfoundland, and Oman.

Establishing the composition of the core is another matter altogether. Because of its great depth and high density, not a single sample of the core has reached the surface. Nevertheless, we have significant evidence to suggest that this layer consists primarily of iron.

Surprisingly, meteorites provide important clues to the composition of the core and mantle. (Meteorites are solid, extraterrestrial objects that strike Earth's surface.) Most meteorites are fragments derived from the collisions of larger bodies, principally from the asteroid belt between the orbits of Mars and Jupiter. They are important because they represent samples of the material (*planetesimals*) from which the inner planets, including Earth, formed (Figure 1.20). Meteorites are composed mainly of an iron-nickel alloy (*irons*), silicate minerals (*stones*), or a combination of both (*stony-irons*) of these materials. The stones have an average composition that is close to that calculated for the mantle. Irons, on the other hand, contain a much higher percentage of this metallic material than is presently found in either Earth's crust or mantle. If Earth did indeed form from the same material in the solar nebula that generated the meteorites and the other terrestrial planets, it must contain a much higher percentage of iron than is found in crustal rock. Consequently, we can conclude that the core is greatly enriched in this heavy metal.

This view is also supported by studies of the Sun's composition, which indicate that iron is the most abundant substance found in the solar system that possesses the density that has been calculated for the core. Furthermore, Earth's magnetic field requires that the core be made of a material that conducts electricity, such as iron. Because all of the available evidence points to the core as being composed largely of iron, we take this to be a fact, at least until new evidence tells us differently.

THE FACE OF EARTH

The two principal divisions of Earth's surface are the continents and the ocean basins (Figure 1.21). A significant difference between these two areas is their relative levels. The continents are remarkably flat features that have the appearance of plateaus protruding above sea level. With an average elevation of about 0.8 kilometer (0.5 mile), continental blocks lie close to sea level, except for limited areas of mountainous terrain. By contrast, the average depth of the ocean floor is about 3.8 kilometers (2.4 miles) below sea level, or about 4.5 kilometers (2.8 miles) lower than the average elevation of the continents.

The elevation difference between the continents and ocean basins is primarily the result of differences in their respective densities and thicknesses. Recall that the continents average 35–40 kilometers in thickness and are composed of granitic rocks having a density of about 2.7 g/cm^3. The basaltic rocks that comprise the oceanic crust average only 7 kilometers thick and have an average density of about 3.0 g/cm^3. Thus, the thicker and less dense continental crust is more buoyant than the oceanic crust. As a result, continental crust floats on top of the deformable rocks of the mantle at a higher level than oceanic crust for the same reason that a large, empty (less dense) cargo ship rides higher than a small, loaded (denser) one.

Major Features of the Continents

The largest features of the continents can be grouped into two distinct categories: extensive, flat stable areas that have been eroded nearly to sea level, and uplifted regions of deformed rocks that make up present-day mountain belts. Notice in Figure 1.22 that the young mountain belts tend to be long, narrow topographic features at the margins of continents, and the flat, stable areas are typically located in the interior of the continents.

FIGURE 1.20 Iron meteorite found in Namibia, a country in southwestern Africa. Meteorites are important because they represent samples of the material from which the inner planets formed. (Photo by Neil Setchfield/Alamy)

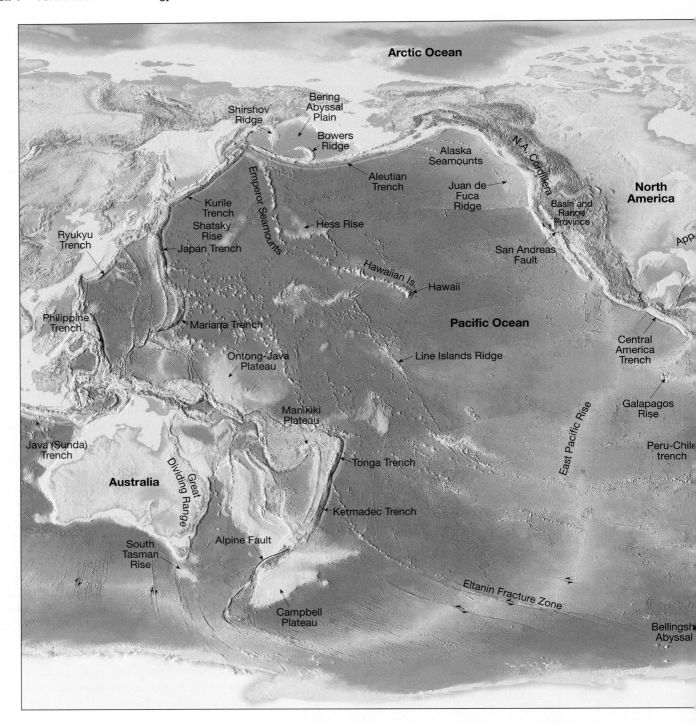

FIGURE 1.21   The topography of Earth's solid surface is shown on these two pages.

Mountain Belts   The most prominent topographic features of the continents are linear mountain belts. Although the distribution of mountains appears to be random, this is not the case. When the youngest mountains are considered (those less than 100 million years old), we find that they are located principally in two major zones. The circum-Pacific belt (the region surrounding the Pacific Ocean) includes the mountains of the western Americas and continues into the western Pacific in the form of volcanic island arcs (see Figure 1.21). Island arcs are active mountainous regions composed largely of volcanic rocks and deformed sedimentary rocks. Examples include the Aleutian Islands, Japan, the Philippines, and New Guinea.

The other major mountainous belt extends eastward from the Alps through Iran and the Himalayas and then dips southward into Indonesia. Careful examination of mountainous terrains reveals that most are places where thick sequences of rocks have been squeezed and highly deformed, as if placed in a gigantic vise. Older mountains are also found on the continents. Examples

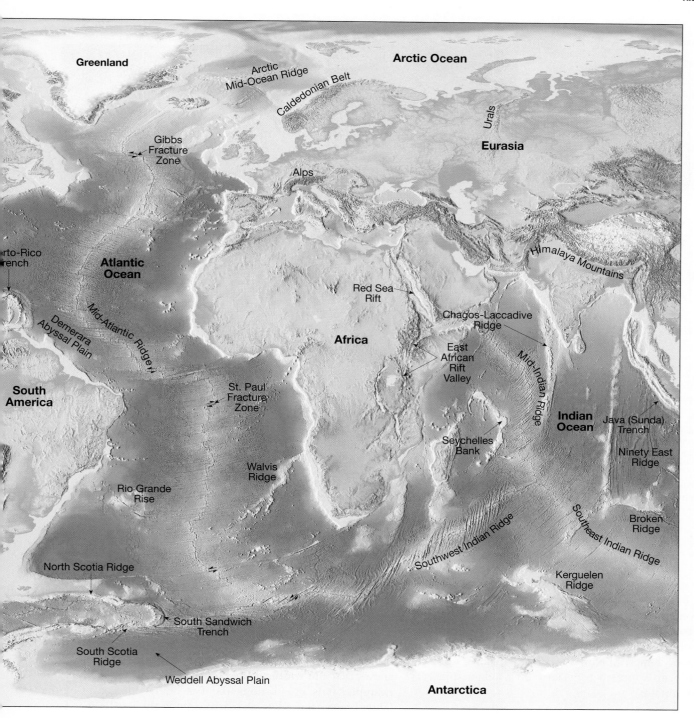

include the Appalachians in the eastern United States and the Urals in Russia. Their once lofty peaks are now worn low, the result of millions of years of weathering and erosion.

The Stable Interior Unlike the young mountain belts, which have formed within the last 100 million years, the interiors of the continents, called **cratons,** have been relatively stable (undisturbed) for the last 600 million years, or even longer. Typically these regions were involved in a mountain-building episode much earlier in Earth's history.

Within the stable interiors are areas known as **shields,** which are expansive, flat regions composed of deformed crystalline rock. Notice in Figure 1.22 that the Canadian shield is exposed in much of the northeastern part of North America. Radiometric dating of various shields has revealed that they are truly ancient regions. All contain Precambrian-age rocks that are more than 1 billion years old, with some samples approaching 4 billion years in age. Even these oldest-known rocks exhibit evidence of enormous forces that have folded, faulted, and metamorphosed them. Thus, we conclude that

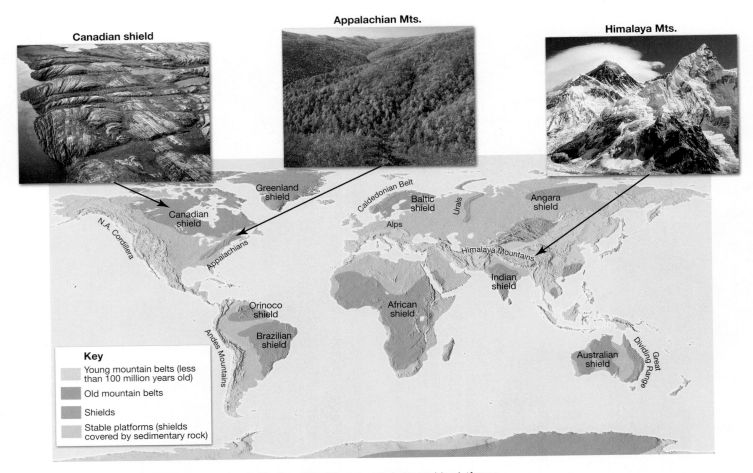

FIGURE 1.22   This map shows the general distribution of Earth's mountain belts, stable platforms, and shields. (Photo by Robert Hildebrand Photography (left); CORBIS (middle); Image Source Pink/Alamy (Right)).

these rocks were once part of an ancient mountain system that has since been eroded away to produce these expansive, flat regions.

Other flat areas of the craton exist in which highly deformed rocks, like those found in the shields, are covered by a relatively thin veneer of sedimentary rocks. These areas are called **stable platforms.** The sedimentary rocks in stable platforms are nearly horizontal except where they have been warped to form large basins or domes. In North America a major portion of the stable platform is located between the Canadian shield and the Rocky Mountains.

Major Features of the Ocean Floor

If all water were drained from the ocean basins, a great variety of features would be seen, including chains of volcanoes, deep canyons, plateaus, and large expanses of monotonously flat plains. In fact, the scenery would be nearly as diverse as that on the continents (see Figure 1.21).

During the past 60 years, oceanographers using modern depth-sounding equipment have slowly mapped significant portions of the ocean floor. From these studies they have defined three major regions: *continental margins, deep-ocean basins,* and *oceanic (mid-ocean) ridges.*

Continental Margins   The **continental margin** is that portion of the seafloor adjacent to major landmasses. It may include the *continental shelf,* the *continental slope,* and the *continental rise.*

Although land and sea meet at the shoreline, this is not the boundary between the continents and the ocean basins. Rather, along most coasts a gently sloping platform of material, called the **continental shelf,** extends seaward from the shore. Because it is underlain by continental crust, it is clearly a flooded extension of the continents. A glance at Figure 1.21 shows that the width of the continental shelf is variable. For example, it is broad along the East and Gulf coasts of the United States but relatively narrow along the Pacific margin of the continent.

The boundary between the continents and the deep-ocean basins lies along the **continental slope,** which is a relatively steep dropoff that extends from the outer edge of the continental shelf to the floor of the deep ocean (see Figure 1.21). Using this as the dividing line, we find that about 60 percent of Earth's surface is represented by ocean basins and the remaining 40 percent by continents.

In regions where trenches do not exist, the steep continental slope merges into a more gradual incline known as the **continental rise.** The continental rise consists of a thick accumulation of sediments that moved downslope from the continental shelf to the deep-ocean floor.

Deep-Ocean Basins   Between the continental margins and oceanic ridges lie the **deep-ocean basins.** Parts of these regions consist of incredibly flat features called **abyssal plains.** The ocean floor also contains extremely deep depressions that are occasionally more than 11,000 meters (36,000 feet) deep. Although these **deep-ocean trenches** are relatively narrow and represent only a small fraction of the ocean floor, they are nevertheless very significant features. Some trenches are located adjacent to young mountains that flank the continents. For example, in Figure 1.21 the Peru–Chile trench off the west coast of South America parallels the Andes Mountains. Other trenches parallel island chains called *volcanic island arcs.*

Dotting the ocean floor are submerged volcanic structures called **seamounts,** which sometimes form long narrow chains. Volcanic activity has also produced several large *lava plateaus,* such as the Ontong Java Plateau located northeast of New Guinea. In addition, some submerged plateaus are composed of continental-type crust. Examples include the Campbell Plateau southeast of New Zealand and the Seychelles Bank northeast of Madagascar.

Oceanic Ridges   The most prominent feature on the ocean floor is the **oceanic** or **mid-ocean ridge.** As shown in Figure 1.21, the Mid-Atlantic Ridge and the East Pacific Rise are parts of this system. This broad elevated feature forms a continuous belt that winds for more than 70,000 kilometers (43,000 miles) around the globe in a manner similar to the seam of a baseball. Rather than consisting of highly deformed rock, such as most of the mountains on the continents, the oceanic ridge system consists of layer upon layer of igneous rock that has been fractured and uplifted.

Being familiar with the topographic features that comprise the face of Earth is essential to understanding the mechanisms that have shaped our planet. What is the significance of the enormous ridge system that extends through all the world's oceans? What is the connection, if any, between young, active mountain belts and oceanic trenches? What forces crumple rocks to produce majestic mountain ranges? These are a few of the questions that will be addressed in the next chapter as we begin to investigate the dynamic processes that shaped our planet in the geologic past and will continue to shape it in the future.

ROCKS AND THE ROCK CYCLE

AN INTRODUCTION TO GEOLOGY
▶ Rock Cycle

Rock is the most common and abundant material on Earth. To a curious traveler, the variety seems nearly endless. When a rock is examined closely, we find that it consists of smaller crystals or grains called minerals. *Minerals* are chemical compounds (or sometimes single elements), each with its own composition and physical properties. The grains or crystals may be microscopically small or easily seen with the unaided eye.

The nature and appearance of a rock is strongly influenced by the minerals that compose it. In addition, a rock's *texture*—the size, shape, and/or arrangement of its constituent minerals—also has a significant effect on its appearance. A rock's mineral composition and texture, in turn, are a reflection of the geologic processes that created it.

Rock characteristics provide geologists with the clues they need to determine the processes that formed them. Such analyses are critical to an understanding of our planet and its history. This understanding also has many practical applications, as in the search for basic mineral and energy resources and developing solutions to environmental problems.

Basic Rock Types

Geologists divide rocks into three major groups: igneous, sedimentary, and metamorphic. What follows is a brief look at these three basic rock groups. As you will learn, each group is linked to the others by the processes that act upon and within the planet.

Igneous Rocks   Igneous rocks (ignis = fire) form when molten rock, called *magma,* cools and solidifies. Magma is melted rock that can form at various levels deep within Earth's crust and upper mantle. As magma cools, crystals of various minerals form and grow. When magma remains deep within the crust, it cools slowly over thousands of years. This gradual loss of heat allows relatively large crystals to develop before the entire mass is completely solidified. Coarse-grained igneous rocks that form far below the surface are called *intrusive.* The cores of many mountains consist of igneous rock that formed in this way. Only subsequent uplift and erosion expose this rock at the surface. A common and important example is *granite* (Figure 1.23). This coarse-grained intrusive rock is rich in the light-colored silicate minerals quartz and feldspar. Granite and related rocks are major constituents of the continental crust.

Sometimes magma breaks through at Earth's surface, as during a volcanic eruption. Because it cools quickly in a surface environment, the molten rock solidifies rapidly, and there is not sufficient time for large crystals to grow. Rather, there is the simultaneous formation of many tiny crystals. Igneous rocks that form at Earth's surface are described as *extrusive* and are usually fine-grained. An abundant and important example is *basalt* (Figure 1.24). This black to dark-green rock is rich in silicate minerals that contain significant iron and magnesium. Because of its higher iron content, basalt is denser than granite. Basalt and related rocks make up the oceanic crust as well as many volcanoes both in the ocean and on the continents.

Sedimentary Rocks   *Sediments,* the raw materials for **sedimentary rocks,** accumulate in layers at Earth's surface. They are materials derived from preexisting rocks by the processes of *weathering.* Some of these processes physically break rock into smaller pieces with no change in composition. Other weathering processes decompose rock—that is, chemically change minerals into new minerals and into substances that readily dissolve in water.

FIGURE 1.23 Granite is an intrusive igneous rock that is especially abundant in Earth's continental crust. Erosion has uncovered this mass of granite in California's Yosemite National Park. The hand sample of granite shows its coarse-grained texture. (Photos by E. J. Tarbuck)

The products of weathering are usually transported by water, wind, or glacial ice to sites of deposition where they form relatively flat layers called *beds*. Sediments are commonly turned into rock or *lithified* by one of two processes. *Compaction* takes place as the weight of overlying materials squeezes sediments into denser masses. *Cementation* occurs as water containing dissolved substances percolates through the open spaces between sediment

FIGURE 1.24 This lava flow at Sunset Crater National Monument near Flagstaff, Arizona, is composed of the black, fine-grained igneous rock called basalt. (Photo by Dennis Tasa) The hand sample of basalt shows its fine-grained texture. The holes, called vesicles, are common in the upper zone of a lava flow. (Photo by E. J. Tarbuck)

grains. Over time the material dissolved in water is chemically precipitated onto the grains and cements them into a solid mass.

Sediments that originate and are transported as solid particles are called *detrital sediments,* and the rocks they form are called *detrital sedimentary rocks.* Particle size is the primary basis for naming the members in this category. Two common examples are *shale* and *sandstone.* Shale is a fine-grained rock consisting of clay-size (less than 1/256 mm) and siltsize (1/256 to 1/16 mm) particles. The deposition of these tiny grains is associated with "quiet" environments such as swamps, river floodplains, and portions of deep-ocean basins. *Sandstone* is the name given sedimentary rocks in which sand-size (1/16 to 2 mm) grains predominate. Sandstones are associated with a variety of environments, including beaches and dunes (Figure 1.25).

Chemical sedimentary rocks form when material dissolved in water precipitates. Unlike detrital sedimentary rocks that are subdivided on the basis of particle size, the primary basis for distinguishing among chemical sedimentary rocks is their mineral composition. Limestone, the most common chemical sedimentary rock, is composed chiefly of the mineral calcite (calcium carbonate, $CaCO_3$). There are many varieties of limestone. The most abundant types have a biochemical origin, meaning that water-dwelling organisms extract dissolved mineral matter and create hard parts such as shells. Later these hard parts accumulate as sediment.

Geologists estimate that sedimentary rocks account for only about 5 percent (by volume) of Earth's outer 16 kilometers (10 miles). However, the importance of this group of rocks is far greater than this percentage implies. If you sampled the rocks exposed at Earth's surface, you would find that the great majority are sedimentary. In fact, about 75 percent of all rock exposures on the continents are sedimentary. Therefore, we can think of sedimentary rocks as comprising a relatively thin and somewhat discontinuous layer in the uppermost portion of the crust. This makes sense because sediment accumulates at the surface.

It is from sedimentary rocks that geologists reconstruct many details of Earth's history. Because sediments are

FIGURE 1.25 This thick layer of sedimentary rock is exposed in the walls of southern Utah's Zion National Park. Known as the Navajo Sandstone, it consists of durable grains of the glassy mineral quartz that once covered this region with mile after mile of drifting sand dunes. (Photo by Jamie & Judy Wild/www.DanitaDelimont) The hand sample provides a closer view of sandstone. (Photo by E. J. Tarbuck)

shale becomes the more compact metamorphic rock *slate*. By contrast, high-grade metamorphism causes a transformation that is so complete that the identity of the parent rock cannot be determined. Furthermore, when rocks at depth (where temperatures are high) are subjected to directed pressure, they gradually deform to generate intricate folds. In the most extreme metamorphic environments, the temperatures approach those at which rocks melt. However, *during metamorphism the rock must remain essentially solid,* for if complete melting occurs, we have entered the realm of igneous activity.

Most metamorphism occurs in one of three settings:

1. When rock is intruded by a magma body, *contact or thermal metamorphism* may take place. Here, change is driven by a rise in temperature within the host rock surrounding an igneous intrusion.

2. *Hydrothermal metamorphism* involves chemical alterations that occur as hot, ion-rich water circulates through fractures in rock. This type of metamorphism is usually associated with igneous activity that provides the heat required to drive chemical reactions and circulate these fluids through rock.

3. During mountain building, great quantities of deeply buried rock are subjected to the directed pressures and high temperatures associated with large-scale deformation called *regional metamorphism.*

deposited in a variety of different settings at the surface, the rock layers that they eventually form hold many clues to past surface environments. They may also exhibit characteristics that allow geologists to decipher information about the method and distance of sediment transport. Furthermore, it is sedimentary rocks that contain fossils, which are vital sources of data in the study of the geologic past.

Metamorphic Rocks **Metamorphic rocks** are produced from pre-existing igneous, sedimentary, or even other metamorphic rocks. Thus, every metamorphic rock has a parent rock—the rock from which it formed. *Metamorphic* is an appropriate name because it literally means to "change form." Most changes occur at the elevated temperatures and pressures that exist deep in Earth's crust and upper mantle (Figure 1.26).

The processes that create metamorphic rocks often progress incrementally, from slight changes (low-grade metamorphism) to substantial changes (high-grade metamorphism). For example, during low-grade metamorphism, the common sedimentary rock

The degree of metamorphism is reflected in the rock's texture and mineral makeup. During regional metamorphism the crystals of some minerals will recrystallize with an orientation that is perpendicular to the direction of the compressional force. The resulting mineral alignment often gives the rock a layered or banded appearance termed a *foliated texture. Schist* and *gneiss* are two examples (Figure 1.27A).

Not all metamorphic rocks have a foliated texture. Such rocks are said to exhibit a *nonfoliated texture.* Metamorphic rocks composed of only one mineral that forms equidimensional crystals are, as a rule, not visibly foliated. For example, limestone, if pure, is composed of only a single mineral, calcite. When a fine-grained limestone is metamorphosed, the small calcite crystals combine to form larger, interlocking crystals. The resulting rock resembles a coarse-grained igneous rock. This nonfoliated metamorphic equivalent of limestone is called *marble* (Figure 1.27B)

FIGURE 1.26 The dark rock shown here is metamorphic. Known as the Vishnu Schist, it is exposed in the inner gorge of the Grand Canyon. Its formation is associated with environments that exist far below Earth's surface where temperatures and pressures are high and with ancient mountain-building processes that occurred in Precambrian time. (Photo by Dennis Tasa)

FIGURE 1.27 Common metamorphic rocks. **A.** The foliated rock gneiss often has a banded appearance and frequently has a mineral composition similar to the igneous rock granite. **B.** Marble is a coarse, crystalline, non-foliated rock whose parent rock is limestone. (Photos by E. J. Tarbuck)

A.

B.

Extensive areas of metamorphic rocks are exposed on every continent. Metamorphic rocks are an important component of many mountain belts, where they make up a large portion of a mountain's crystalline core. Even the stable continental interiors, which are generally covered by sedimentary rocks, are underlain by metamorphic basement rocks. In all of these settings, the metamorphic rocks are usually highly deformed and intruded by igneous masses. Indeed, significant parts of Earth's continental crust are composed of metamorphic and associated igneous rocks.

The Rock Cycle: One of Earth's Subsystems

Earth is a system. This means that our planet consists of many interacting parts that form a complex whole. Nowhere is this idea better illustrated than when we examine the rock cycle (Figure 1.28). The **rock cycle** allows us to view many of the interrelationships among different parts of the Earth system. It helps us understand the origin of igneous, sedimentary, and metamorphic rocks and to see that each type is linked to the others by processes that act upon and within the planet. Learn the rock cycle well; you will be examining its interrelationships in greater detail throughout this book.

The Basic Cycle We begin at the top of Figure 1.28 with magma, molten rock that forms deep beneath Earth's surface. Over time, magma cools and solidifies. This process, called *crystallization,* may occur either beneath the surface or, following a volcanic eruption, at the surface. In either situation, the resulting rocks are called *igneous rocks.*

If igneous rocks are exposed at the surface, they will undergo *weathering,* in which the day-in and day-out influences of the atmosphere slowly disintegrate and decompose rocks. The materials that result are often moved downslope by gravity before being picked up and transported by any of a number of erosional agents, such as running water, glaciers, wind, or waves. Eventually these particles and dissolved substances, called *sediment,* are deposited. Although most sediment ultimately comes to rest in the ocean, other sites of deposition include river floodplains, desert basins, swamps, and sand dunes.

Next the sediments undergo *lithification,* a term meaning "conversion into rock." Sediment is usually lithified into *sedimentary rock* when compacted by the weight of overlying layers or when cemented as percolating groundwater fills the pores with mineral matter.

If the resulting sedimentary rock is buried deep within Earth and involved in the dynamics of mountain building or intruded by a mass of magma, it will be subjected to great pressures and/or intense heat. The sedimentary rock will react to the changing environment and turn into the third rock type, *metamorphic rock.* When metamorphic rock is subjected to addi-

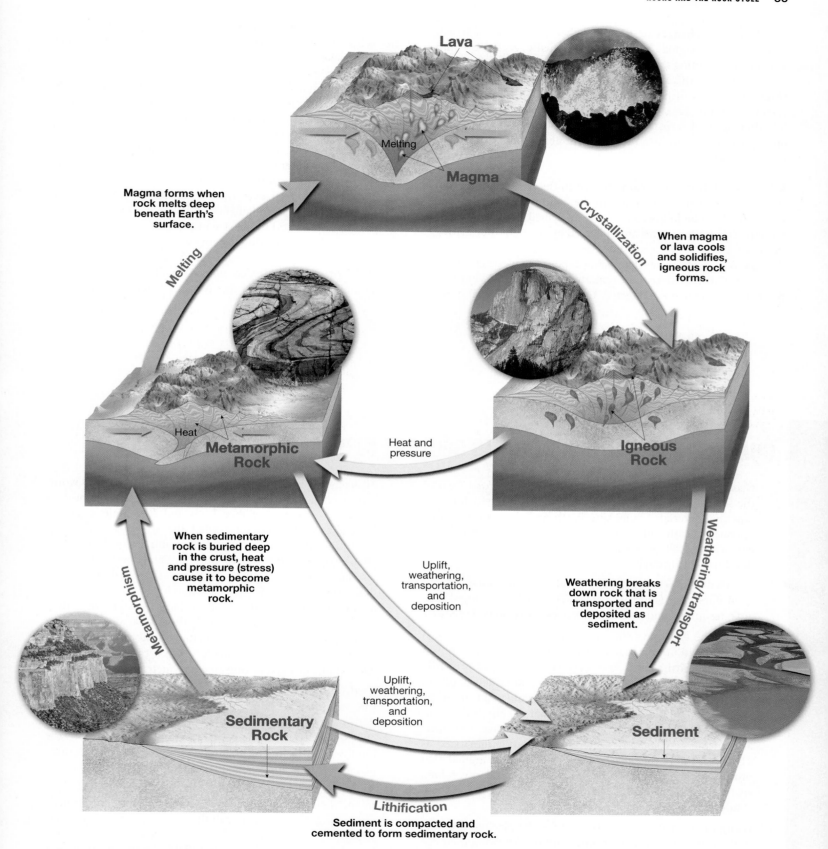

Lava

Melting

Magma

Magma forms when rock melts deep beneath Earth's surface.

Melting

Crystallization

When magma or lava cools and solidifies, igneous rock forms.

Heat and pressure

Metamorphic Rock

Heat

Igneous Rock

When sedimentary rock is buried deep in the crust, heat and pressure (stress) cause it to become metamorphic rock.

Metamorphism

Uplift, weathering, transportation, and deposition

Weathering breaks down rock that is transported and deposited as sediment.

Weathering/transport

Uplift, weathering, transportation, and deposition

Sedimentary Rock

Sediment

Lithification

Sediment is compacted and cemented to form sedimentary rock.

FIGURE 1.28 Viewed over long spans, rocks are constantly forming, changing, and reforming. The rock cycle helps us understand the origin of the three basic rock groups. Arrows represent processes that link each group to the others.

tional pressure changes or to still higher temperatures, it will melt, creating magma, which will eventually crystallize into igneous rock starting the cycle all over again.

Where does the energy that drives Earth's rock cycle come from? Processes driven by heat from Earth's interior are responsible for creating igneous and metamorphic rocks. Weathering and erosion, external processes powered by energy from the Sun, produce the sediment from which sedimentary rocks form.

Alternative Paths   The paths shown in the basic cycle are not the only ones that are possible. To the contrary, other paths are just as likely to be followed as those described in the preceding section. These alternatives are indicated by the light blue arrows in Figure 1.28.

Igneous rocks, rather than being exposed to weathering and erosion at Earth's surface, may remain deeply buried. Eventually these masses may be subjected to the strong compressional forces and high temperatures associated with mountain building. When this occurs, they are transformed directly into metamorphic rocks.

Metamorphic and sedimentary rocks, as well as sediment, do not always remain buried. Rather, overlying layers may be stripped away, exposing the once buried rock. When this happens, the material is attacked by weathering processes and turned into new raw materials for sedimentary rocks.

Although rocks may seem to be unchanging masses, the rock cycle shows that they are not. The changes, however, take time—great amounts of time. We can observe different parts of the cycle operating all over the world. Today new magma is forming beneath the island of Hawaii. When it erupts at the surface, the lava flows add to the size of the island. Meanwhile, the Colorado Rockies are gradually being worn down by weathering and erosion. Some of this weathered debris will eventually be carried to the Gulf of Mexico, where it will add to the already substantial mass of sediment that has accumulated there.

CHAPTER 1 AN INTRODUCTION TO GEOLOGY IN REVIEW

- *Geology* means "the study of Earth." The two broad areas of the science of geology are (1) *physical geology,* which examines the materials composing Earth and the processes that operate beneath and upon its surface; and (2) *historical geology,* which seeks to understand the origin of Earth and its development through time.

- The relationship between people and the natural environment is an important focus of geology. This includes natural hazards, resources, and human influences on geologic processes.

- During the 17th and 18th centuries, *catastrophism* influenced the formulation of explanations about Earth. Catastrophism states that Earth's landscapes have been developed primarily by great catastrophes. By contrast, *uniformitarianism,* one of the fundamental principles of modern geology advanced by *James Hutton* in the late 1700s, states that the physical, chemical, and biological laws that operate today have also operated in the geologic past. The idea is often summarized as "The present is the key to the past." Hutton argued that processes that appear to be slow-acting could, over long spans of time, produce effects that were just as great as those resulting from sudden catastrophic events.

- Using the principles of *relative dating,* the placing of events in their proper sequence or order without knowing their age in years, scientists developed a geologic time scale during the 19th century. Relative dates can be established by applying such principles as the *law of superposition* and the *principle of fossil succession.*

- All science is based on the assumption that the natural world behaves in a consistent and predictable manner. The process by which scientists gather facts and formulate scientific *hypotheses* and *theories* is called the *scientific method.* To determine what is occurring in the natural world, scientists often (1) collect facts, (2) ask questions and develop hypotheses that may answer these questions, (3) develop observations and experiments to test the hypotheses, and (4) accept, modify, or reject hypotheses on the basis of extensive testing. Other discoveries represent purely theoretical ideas that have stood up to extensive examination. Still other scientific advancements have been made when a totally unexpected happening occurred during an experiment.

- Earth's physical environment is traditionally divided into three major parts: the solid Earth or *geosphere;* the water portion of our planet, the *hydrosphere;* and Earth's gaseous envelope, the *atmosphere.* In addition, the *biosphere,* the totality of life on Earth, interacts with each of the three physical realms and is an equally integral part of Earth.

- Although each of Earth's four spheres can be studied separately, they are all related in a complex and continuously interacting whole that we call the *Earth system. Earth system science* uses an interdisciplinary approach to integrate the knowledge of several academic fields in the study of our planet and its global environmental problems.

- A *system* is a group of interacting parts that form a complex whole. *Closed systems* are those in which energy moves freely in and out, but matter does not enter or leave the system. In an *open system*, both energy and matter flow into and out of the system.

- Most natural systems have mechanisms that tend to enhance change, called *positive feedback mechanisms*, and other mechanisms, called *negative feedback mechanisms*, that tend to resist change and thus stabilize the system.

- The two sources of energy that power the Earth system are (1) the Sun, which drives the external processes that occur in the atmosphere, hydrosphere, and at Earth's surface; and (2) heat from Earth's interior, which powers the internal processes that produce volcanoes, earthquakes, and mountains.

- The *nebular hypothesis* describes the formation of the solar system. The planets and Sun began forming about 5 billion years ago from a large cloud of dust and gases. As the cloud contracted, it began to rotate and assume a disk shape. Material that was gravitationally pulled toward the center became the *protosun*. Within the rotating disk, small centers, called *planetesimals*, swept up more and more of the cloud's debris. Because of the high temperatures near the Sun, the inner planets were unable to accumulate many of the elements that vaporize at low temperatures. Because of the very cold temperatures existing far from the Sun, the large outer planets consist of huge amounts of ices and lighter materials. These substances account for the comparatively large sizes and low densities of the outer planets.

- Earth's internal structure is divided into layers based on differences in chemical composition and on the basis of changes in physical properties. Compositionally, Earth is divided into a thin outer *crust*, a solid rocky *mantle*, and a dense *core*. Other layers, based on physical properties, include the *lithosphere, asthenosphere, transition zone, lower mantle, D″ layer, outer core,* and *inner core*.

- Two principal divisions of Earth's surface are the *continents* and *ocean basins*. A significant difference is their relative levels. The elevation differences between continents and ocean basins is primarily the result of differences in their respective densities and thicknesses.

- The largest features of the continents can be divided into two categories: *mountain belts* and the *stable interior*. The ocean floor is divided into three major topographic units: *continental margins, deep-ocean basins,* and *oceanic ridges*.

- Rocks consist of smaller grains or crystals called *minerals*. A rock's mineral composition and *texture* (size, shape, and/or arrangement of the minerals that make it up) reflect the geologic processes that created it. The three basic rock types are *igneous, sedimentary,* and *metamorphic*.

- The *rock cycle* is one of the many cycles or loops of the Earth system in which matter is recycled. The rock cycle is an important way of viewing many of the interrelationships of geology. It illustrates the origin of the three basic rock types and the role of various geologic processes in transforming one rock type into another.

KEY TERMS

abyssal plain (p. 29)
asthenosphere (p. 24)
atmosphere (p. 15)
biosphere (p. 16)
catastrophism (p. 5)
closed system (p. 17)
continental margin (p. 28)
continental rise (p. 28)
continental shelf (p. 28)
continental slope (p. 28)
core (p. 24)
craton (p. 27)
crust (p. 23)
D″ layer (p. 24)

deep-ocean basin (p. 29)
deep-ocean trench (p. 29)
Earth system science (p. 17)
fossil succession, principle of (p. 9)
geology (p. 2)
geosphere (p. 17)
historical geology (p. 2)
hydrosphere (p. 15)
hypothesis (p. 11)
igneous rock (p. 29)
inner core (p. 24)
interface (p. 19)

lithosphere (p. 24)
lower mantle (p. 24)
mantle (p. 24)
metamorphic rock (p. 31)
nebular theory (p. 21)
negative feedback mechanism (p. 18)
oceanic (mid-ocean) ridge (p. 29)
open system (p. 18)
outer core (p. 24)
physical geology (p. 2)
positive feedback mechanism (p. 18)

relative dating (p. 8)
rock cycle (p. 32)
seamount (p. 29)
sedimentary rock (p. 29)
shield (p. 27)
solar nebula (p. 21)
stable platform (p. 28)
superposition, law of (p. 8)
system (p. 17)
theory (p. 11)
transition zone (p. 24)
uniformitarianism (p. 6)

QUESTIONS FOR REVIEW

1. Geology is traditionally divided into two broad areas. Name and describe these two subdivisions.

2. Briefly describe Aristotle's influence on the science of geology.

3. How did the proponents of catastrophism perceive the age of Earth?

4. Describe the doctrine of uniformitarianism. How did the advocates of this idea view the age of Earth?

5. About how old is Earth?

6. The geologic time scale was established without the aid of radiometric dating. What principles were used to develop the time scale?

7. How is a scientific hypothesis different from a scientific theory?

8. List and briefly describe the four "spheres" that constitute our environment.

9. How is an open system different from a closed system?

10. Contrast positive feedback mechanisms and negative feedback mechanisms.

11. Soil is an example of an *interface* in the Earth system. What is an interface? With which of Earth's four spheres is soil associated?

12. What are the two sources of energy for the Earth system?

13. Briefly describe the events that led to the formation of the solar system.

14. List and briefly describe Earth's compositional divisions.

15. Contrast the asthenosphere and the lithosphere.

16. Describe the general distribution of Earth's youngest mountains.

17. Distinguish between shields and stable platforms.

18. List the three major topographic units of the ocean floor.

19. Name each of the rocks described below:
    - light-colored, coarse-grained intrusive rock
    - detrital rock rich in clay-size particles
    - a fine-grained black rock that makes up the oceanic crust
    - nonfoliated rock, for which limestone is its parent rock

20. For each characteristic below, indicate whether it is associated with igneous, sedimentary, or metamorphic rocks:
    - may be intrusive or extrusive
    - lithified by compaction and cementation
    - sandstone is an example
    - some members of this group are foliated
    - this group is divided into detrital and chemical categories
    - gneiss is a member of this group

21. Using the rock cycle, explain the statement "One rock is the raw material for another."

COMPANION WEBSITE

The *Earth 10e* website uses the resources and flexibility of the Internet to aid in your study of the topics in this chapter. Written and developed by the authors and other geology instructors, this site will help improve your understanding of geology. Visit www.mygeoscienceplace.com in order to:

- **Review** key chapter concepts.

- **Read** with links to the eBook and to chapter-specific web resources.
- **Visualize** and comprehend challenging topics using learning activities in *GEODe Earth*.
- **Test** yourself with online quizzes.

GEODe EARTH

GEODe Earth is a valuable and easy-to-use learning aid that can be accessed from your book's Companion Website (www .mygeoscienceplace.com). It is a dynamic instructional tool that promotes understanding and reinforces important concepts by using tutorials, animations, and exercises that actively engage the student.

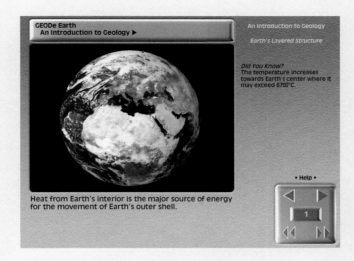

PLATE TECTONICS

A SCIENTIFIC REVOLUTION UNFOLDS

Plate tectonics is the dominant theory to explain the origin of Earth's major features, including mountains, continents, and ocean basins. Grand Teton Range, Wyoming.

(PHOTO BY MICHAEL COLLIER)

Prior to the 1960s most geologists held the view that the ocean basins and continents had fixed geographic positions and were of great antiquity. Less than a decade later researchers came to realize that Earth's continents are not static, instead they gradually migrate across the globe. Because of these movements, blocks of continental material collide, deforming the intervening crust, thereby creating Earth's great mountain chains (Figure 2.1). Furthermore, landmasses occasionally split apart. As the continental blocks separate, a new ocean basin emerges between them. Meanwhile, other portions of the seafloor plunge into the mantle. In short, a dramatically different model of Earth's tectonic processes emerged.*

This profound reversal in scientific thought has been appropriately described as a *scientific revolution*. The revolution began early in the 20th century as a relatively straightforward proposal called *continental drift*. For more than fifty years the idea that continents were capable of movement was categorically rejected by the scientific establishment.** Continental drift was particularly distasteful to North American geologists, perhaps because much of the supporting evidence had been gathered from the continents of Africa, South America, and Australia, with which most North American geologists were unfamiliar.

Following World War II, modern instruments replaced rock hammers as the tools of choice for many researchers. Armed with these more advanced tools, geologists and a new breed of researchers, including *geophysicists* and *geochemists,* made several surprising discoveries that began to rekindle interest in the drift hypothesis. By 1968 these developments led to the unfolding of a far more encompassing explanation known as the *theory of plate tectonics.*

In this chapter, we will examine the events that led to this dramatic reversal of scientific opinion in an attempt to provide insight into how science works. We will also briefly trace the development of the continental drift hypothesis, examine why it was first rejected, and consider the evidence that finally led to the acceptance of its direct descendant—the theory of plate tectonics.

CONTINENTAL DRIFT: AN IDEA BEFORE ITS TIME

The idea that continents, particularly South America and Africa, fit together like pieces of a jigsaw puzzle came about during the 1600s as better world maps became available. However, little significance was given to this notion until 1915, when Alfred Wegener (1880–1930), a German meteorologist and geophysicist, wrote *The Origin of Continents and Oceans.* This book, published in several editions, set forth the basic outline of Wegener's hypothesis called **continental drift**—which dared to challenge the long-held assumption that the continents and ocean basins had fixed geographic positions.

Wegener suggested that a single **supercontinent** consisting of all Earth's landmasses once existed.* He named this giant landmass **Pangaea** (pronounced Pan-jee-ah; meaning "all lands") (Figure 2.2). Wegener further hypothesized that about 200 million years ago, during the early part of the Mesozoic era, this supercontinent began to fragment into smaller landmasses. These continental blocks then "drifted" to their present positions over a span of millions of years. The inspiration for continental drift is believed to have come to Wegener when he observed the breakup of sea ice during a Danish led expedition to Greenland.

Wegener and others who advocated the continental drift hypothesis collected substantial evidence to support their point of view. The fit of South America and Africa and the geographic distribution of fossils and ancient climates all seemed to buttress the idea that these now separate landmasses were once joined. Let us examine some of this evidence.

Evidence: The Continental Jigsaw Puzzle

Like a few others before him, Wegener suspected that the continents might once have been joined when he noticed the remarkable similarity between the coastlines on opposite sides of the

---

*Tectonic processes are those that deform Earth's crust to create major structural features such as mountains, continents, and ocean basins.

**A few geologists, including Alexander du Toit of South Africa and Arthur Homers in England, supported the continental drift hypothesis. But they were clearly in the minority.

---

*Wegener was not the first person to conceive of a long-vanished supercontinent. Edward Suess (1831–1914), a distinguished nineteenth century geologist, pieced together evidence for a giant landmass consisting of the continents of South America, Africa, India, and Australia.

FIGURE 2.1 Climbers camping on a sheer rock face of a mountain known as K7 in Pakistan's Karakoram, a part of the Himalayas. These mountains formed as India collided with Eurasia. (Photo by Jimmy Chin/National Geographic/Getty)

Atlantic Ocean. However, Wegener's use of present-day shorelines to fit these continents together was challenged immediately by other Earth scientists. These opponents correctly argued that shorelines are continually modified by wave erosion and depositional processes. Even if continental displacement had taken place, a good fit today would be unlikely. Because Wegener's original jigsaw fit of the continents was crude, it is assumed that he was aware of this problem (see Figure 2.2B).

Scientists later determined that a much better approximation of the outer boundary of a continent is the seaward edge of its continental shelf, which lies submerged a few hundred meters below sea level. In the early 1960s, Sir Edward Bullard and two associates constructed a map that pieced together the edges of the continental shelves of South America and Africa at a depth of about 900 meters (Figure 2.3). The remarkable fit that was obtained was more precise than even these researchers had expected. As shown in Figure 2.3 there are a few places where the continents overlap. Some of these overlaps are related to the process of stretching and thinning of the continental margins as they drifted apart. Others can be explained by the work of major river systems. For example, since the break-up of Pangaea the Niger River has built an extensive delta that enlarged the continental shelf of Africa.

FIGURE 2.2 Reconstructions of Pangaea as it is thought to have appeared 200 million years ago. **A.** Modern reconstruction. **B.** Wegener's reconstruction redrawn from his book published in 1915.

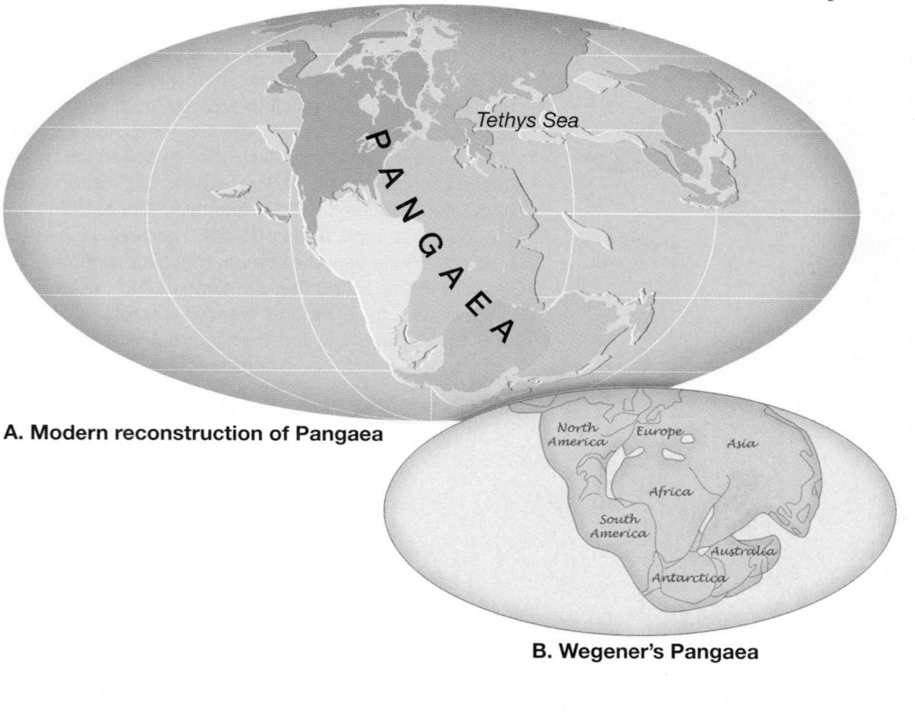

Tethys Sea

PANGAEA

A. Modern reconstruction of Pangaea

North America
Europe
Asia
Africa
South America
Australia
Antarctica

B. Wegener's Pangaea

Evidence: Fossils Match across the Seas

Although the seed for Wegener's hypothesis came from the remarkable similarities of the continental margins on opposite sides of the Atlantic, it was when he learned that identical fossil organisms had been discovered in rocks from both South America and Africa, that his

41

FIGURE 2.3   Drawing which shows the best fit of South America and Africa along the continental slope at a depth of 500 fathoms (about 900 meters). The areas where continental blocks overlap appear in orange. (After A. G. Smith, "Continental Drift," in *Understanding the Earth,* edited by I. G. Gass.)

pursuit of continental drift became more focused. Through a review of the literature, Wegener learned that most paleontologists (scientists who study the fossilized remains of ancient organisms) were in agreement that some type of land connection was needed to explain the existence of similar Mesozoic age life forms on widely separated landmasses. Just as modern life forms native to North America are quite different from

those of Africa and Australia, one would expect that during the Mesozoic era, organisms on widely separated continents would be distinct.

Mesosaurus   To add credibility to his argument, Wegener documented cases of several fossil organisms that were found on different landmasses despite the unlikely possibility that their living forms could have crossed the vast ocean presently separating them (Figure 2.4). A classic example is *Mesosaurus,* an aquatic fish-catching reptile whose fossil remains are limited to black shales of the Permian period (about 260 million years ago) in eastern South America and southwestern Africa (Figure 2.4). If *Mesosaurus* had been able to make the long journey across the South Atlantic, its remains would likely be more widely distributed. As this is not the case, Wegener asserted that South America and Africa must have been joined during that period of Earth history.

How did opponents of continental drift explain the existence of identical fossil organisms in places separated by thousands of kilometers of open ocean? Rafting, transoceanic land bridges (isthmian links), and island stepping stones were the most widely invoked explanations for these migrations (Figure 2.5). We know, for example, that during the Ice Age that ended about 8,000 years ago, the lowering of sea level allowed mammals (including humans) to cross the narrow Bering Strait that separates Russia and Alaska. Was it possible that land bridges once connected Africa and South America but later subsided below sea level? Modern maps of the seafloor substantiate Wegener's contention that if land bridges of this magnitude once existed, their remnants would still lie below sea level.

Glossopteris   Wegener also cited the distribution of the fossil "seed fern" *Glossopteris* as evidence for the existence of Pangaea (see Figure 2.4). This plant, identified by its tongue-shaped leaves and seeds that were too large to be carried by the wind, was known to be widely dispersed among Africa, Aus-

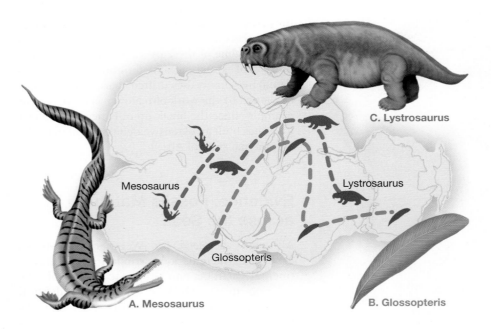

A. Mesosaurus
B. Glossopteris
C. Lystrosaurus
Mesosaurus
Lystrosaurus
Glossopteris

FIGURE 2.4   Fossil evidence supporting continental drift. **A.** Fossils of *Mesosaurus* are found only in nonmarine deposits in eastern South America and western Africa. *Mesosaurus* was a freshwater reptile incapable of swimming the 5000 kilometers of open ocean that now separate these continents. **B.** Remains of *Glossopteris* and related flora are found in Australia, Africa, South America, Antarctica, and India, landmasses which currently have quite varied climates. However, when *Glossopteris* inhabited these regions during the late Paleozoic era, their climates were all subpolar. **C.** Fossils of *Lystrosaurus,* a land-dwelling reptile, are also found on three of these landmasses.

FIGURE 2.5 These sketches by John Holden illustrate various explanations for the occurrence of similar species on landmasses that are presently separated by vast oceans. (Reprinted with permission of John Holden)

a subpolar climate. Therefore, he concluded that when these landmasses were joined, they were located much closer to the South Pole.

Evidence: Rock Types and Geologic Features

Anyone who has worked a jigsaw puzzle knows that its successful completion requires that you fit the pieces together while maintaining the continuity of the picture. The "picture" that must match in the "continental drift puzzle" is one of rock types and geologic features such as mountain belts. If the continents were once together, the rocks found in a particular region on one continent should closely match in age and type those found in adjacent positions on the once adjoining continent. Wegener found evidence of 2.2-billion-year-old igneous rocks in Brazil that closely resembled similarly aged rocks in Africa.

Similar evidence can be found in mountain belts that terminate at one coastline, only to reappear on landmasses across the ocean. For instance, the mountain belt that includes the Appalachians trends northeastward through the eastern United States and disappears off the coast of Newfoundland (Figure 2.6A). Mountains of comparable age and structure are found in the British Isles, and Scandinavia. When these landmasses are reassembled, as in Figure 2.6B, the mountain chains form a nearly continuous belt.

Wegener described how the similarities in geologic features on both sides of the Atlantic linked these landmasses when he said, "It is just as if we were to refit the torn pieces of a newspaper

tralia, India, and South America. Later, fossil remains of *Glossopteris* were also discovered in Antarctica.* Wegener also learned that these seed ferns and associated flora grew only in

*In 1912 Captain Robert Scott and two companions froze to death lying beside thirty-five pounds of rock on their return from a failed attempt to be the first to reach the South Pole. These samples, collected on the moraines of Beardmore Glacier, contained fossil remains of *Glossopteris*.

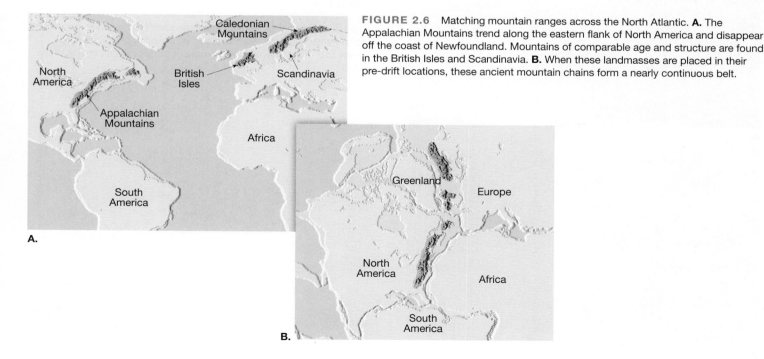

FIGURE 2.6 Matching mountain ranges across the North Atlantic. **A.** The Appalachian Mountains trend along the eastern flank of North America and disappear off the coast of Newfoundland. Mountains of comparable age and structure are found in the British Isles and Scandinavia. **B.** When these landmasses are placed in their pre-drift locations, these ancient mountain chains form a nearly continuous belt.

by matching their edges and then check whether the lines of print run smoothly across. If they do, there is nothing left but to conclude that the pieces were in fact joined in this way."*

Evidence: Ancient Climates

Because Alfred Wegener was a student of world climates, he suspected that paleoclimatic (*paleo* = ancient, *climatic* = climate) data might also support the idea of mobile continents. His assertion was bolstered when he learned that evidence for a glacial period that dated to the late Paleozoic had been discovered in southern Africa, South America, Australia, and India (Figure 2.7A). This meant that about 300 million years ago, vast ice sheets covered extensive portions of the Southern Hemisphere as well as India (Figure 2.7B). Much of the land area that contains evidence of this period of Paleozoic glaciation presently lies within 30 degrees of the equator in subtropical or tropical climates.

How could extensive ice sheets form near the Equator? One proposal suggested that our planet experienced a period of extreme global cooling. Wegener rejected this explanation because during the same span of geologic time, large tropical swamps existed in several locations in the Northern Hemisphere. The lush vegetation in these swamps was eventually

_____
*Alfred Wegener, *The Origin of Continents and Oceans*, translated from the 4th revised German ed. of 1929 by J. Birman (London: Methuen, 1966).

buried and converted to coal. Today these deposits comprise major coal fields in the eastern United States, Northern Europe, and Asia. Many of the fossils found in these coal-bearing rocks were produced by tree ferns that possessed large fronds; a fact consistent with a warm, moist climate. Furthermore, these fern trees lacked growth rings, a characteristic of tropical plants that grow in regions having minimal yearly fluctuations in temperature. By contrast, trees that inhabit the middle-latitudes, like those found in most of the United States, develop multiple tree rings—one for each growing season.

Wegener suggested that a more plausible explanation for the late Paleozoic glaciation was provided by the supercontinent of Pangaea. In this configuration the southern continents are joined together and located near the South Pole (Figure 2.7C). This would account for the conditions necessary to generate extensive expanses of glacial ice over much of these landmasses. At the same time, this geography would place today's northern continents nearer the equator and account for the tropical swamps that generated the vast coal deposits. Wegener was so convinced that his explanation was correct that he wrote, "This evidence is so compelling that by comparison all other criteria must take a back seat."

How does a glacier develop in hot, arid central Australia? How do land animals migrate across wide expanses of the ocean? As compelling as this evidence may have been, 50 years passed before most of the scientific community accepted the concept of continental drift and the logical conclusions to which it led.

A.

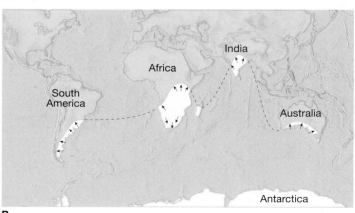
B.

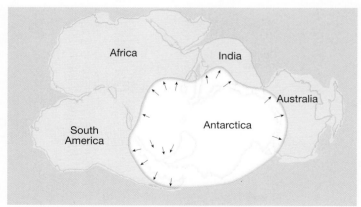
C.

FIGURE 2.7 Paleoclimatic evidence for continental drift. **A.** Glacial striations (scratches) and grooves like these are produced as glaciers drag rock debris across the underlying bedrock. The direction of glacial movement can be deduced from the distinctive patterns of aligned scratches and grooves. (Photo by Gregory S. Springer) **B.** Near the end of the Paleozoic era (about 300 million years ago) ice sheets covered extensive areas of the Southern Hemisphere and India. Arrows show the direction of ice movement that can be inferred from the pattern of glacial striations and grooves found in the bedrock. **C.** The continents restored to their pre-drift positions when they were part of Pangaea. This configuration accounts for the conditions necessary to generate a vast ice sheet and also explains the directions of ice movement that radiated away from an area near the present position of the South Pole.

THE GREAT DEBATE

Wegener's proposal did not attract much open criticism until 1924, when his book was translated into English, French, Spanish, and Russian. From that point until his death in 1930, the drift hypothesis encountered a great deal of hostile criticism. The respected American geologist R. T. Chamberlain stated, "Wegener's hypothesis in general is of the foot-loose type, in that it takes considerable liberty with our globe, and is less bound by restrictions or tied down by awkward, ugly facts than most of its rival theories. Its appeal seems to lie in the fact that it plays a game in which there are few restrictive rules and no sharply drawn code of conduct."

W. B. Scott, former president of the American Philosophical Society, expressed the prevalent American view of continental drift in fewer words when he described the hypothesis as "utter damned rot!"

Rejection of the Continental Drift Hypothesis

One of the main objections to Wegener's hypothesis stemmed from his inability to identify a credible mechanism for continental drift. Wegener proposed that gravitational forces of the Moon and Sun that produce Earth's tides were also capable of gradually moving the continents across the globe. However, the prominent physicist Harold Jeffreys correctly countered that tidal forces of the magnitude needed to displace the continents would bring Earth's rotation to a halt in a matter of a few years.

Wegener also incorrectly suggested that the larger and sturdier continents broke through thinner oceanic crust, much like ice breakers cut through ice. However, no evidence existed to suggest that the ocean floor was weak enough to permit passage of the continents without the continents being appreciably deformed in the process.

In 1930 Wegener made his fourth and final trip to the Greenland Ice Sheet (Box 2.1). Although the primary focus of this expedition was to study the harsh winter polar climate on the ice-covered island, Wegener continued to test his continental drift hypothesis. Like earlier expeditions, he used astronomical methods in an attempt to verify that Greenland had drifted westward with respect to Europe.

While returning from Eismitte (an experimental station located in the center of Greenland), Wegener perished along with a companion. His intriguing idea, however, did not die. Following Wegener's death, Danish workers continued to take measurements of Greenland's position (1936, 1938, and 1948), but found no evidence of drift. Thus, Wegener's final test proved to be a failure. Today the Global Positioning System (GPS) allows scientists to measure the extremely gradual displacement of the continents that Wegener had worked so diligently, but unsuccessfully, to detect.

Continental Drift and the Scientific Method

What went wrong? Why was Wegener unable to overturn the established scientific views of his day? Foremost was the fact that, although the central theme of Wegener's drift hypothesis was correct, it contained some incorrect details. For example, continents do not break through the ocean floor, and tidal energy is much too weak to cause continents to be displaced. Moreover, in order for any comprehensive scientific theory to gain wide acceptance, it must stand up to critical testing from all areas of science. Wegener's great contribution to our understanding of Earth notwithstanding, not *all* of the evidence supported the continental drift hypothesis as he had formulated it.

Although many of Wegener's contemporaries opposed his views, even to the point of open ridicule, some considered his ideas plausible. Among the most notable of this latter group were the eminent South African geologist Alexander du Toit and the well-known Scottish geologist Arthur Holmes. In 1928 Holmes proposed the first plausible driving mechanism for continental drift. In his book *Physical Geology,* Holmes elaborated on this idea by suggesting that the flow of hot material within the mantle was responsible for propelling the continents across the globe. In 1937 du Toit published *Our Wandering Continents,* in which he eliminated some of Wegener's weaker arguments and added a wealth of new evidence in support of this revolutionary idea.

For those geologists who continued the search, the exciting concept of continents adrift held their interest. Others viewed continental drift as a solution to previously unexplainable observations (Figure 2.8). Nevertheless, most of the scientific community, particularly in North America, either categorically rejected continental drift or at least treated it with considerable skepticism.

CONTINENTAL DRIFT AND PALEOMAGNETISM

Little light was shed on the continental drift hypothesis during the two decades following Wegener's death. However, by the mid-1950s two new lines of evidence began to emerge, which seriously questioned science's basic understanding of how Earth works. One line of evidence came from explorations of the seafloor and will be considered later. The other line of evidence came from a comparatively new field—*paleomagnetism.*

Earth's Magnetic Field and Fossil Magnetism

Anyone who has used a compass to find direction knows that Earth's magnetic field has a north and south magnetic pole. Today these magnetic poles align closely, but not exactly, with the geographic poles. (The geographic poles are located where Earth's rotational axis intersects the surface.) Earth's magnetic field is similar to that produced by a simple bar magnet. Invisible lines of force pass through the planet and extend from one magnetic pole to the other (Figure 2.9). A compass needle, itself a small magnet free to rotate on an axis, becomes aligned with the magnetic lines of force and points to the magnetic poles.

Unlike the pull of gravity, we cannot feel Earth's magnetic field, yet its presence is revealed because it deflects a compass needle. In addition, some naturally occurring minerals are magnetic and hence are influenced by Earth's magnetic field. One of the most common is the iron-rich mineral *magnetite* which is

FIGURE 2.8 Plate tectonics, the direct descendant of continental drift, explains the global distribution of Earth's most destructive earthquakes. These ruins in Pisco, Peru, resulted from a powerful quake on August 16, 2007 that was generated along the Peru-Chile subduction zone. (Photo by Sergio Erday/epa/Corbis)

abundant in lava flows of basaltic composition.* Basaltic lavas erupt at the surface at temperatures greater than 1000 °C, exceeding a threshold temperature for magnetism known as the **Curie point** (about 585 °C). Consequently, the magnetite grains in molten lava are non-magnetic. However, as the lava cools these iron-rich grains become magnetized and align themselves in the direction of the existing magnetic lines of force. Once the minerals solidify, the magnetism they possess will usually remain "frozen" in this position. Thus, they act like a compass needle because they "point" toward the position of the magnetic poles at the time of their formation. Rocks that formed thousands or millions of years ago and contain a "record" of the direction of the magnetic poles at the time of their formation are said to possess **fossil magnetism,** or **paleomagnetism.** During the 1950s paleomagnetic data was collected from lava flows around the globe.

Apparent Polar Wandering

A study of rock magnetism conducted in Europe by S. K. Runcorn and his associates led to an interesting discovery. The magnetic alignment of iron-rich minerals in lava flows of different ages

FIGURE 2.9 Earth's magnetic field consists of lines of force much like those a giant bar magnet would produce if placed at the center of Earth.

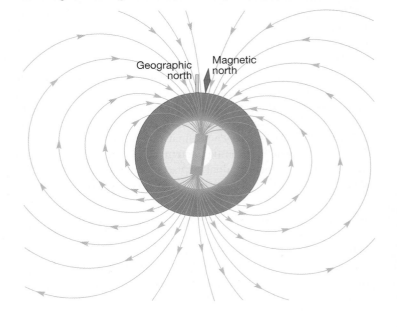

*Some sediments and sedimentary rocks contain enough iron-bearing mineral grains to acquire a measurable amount of magnetization.

BOX 2.1

Alfred Wegener (1880–1930): Polar Explorer and Visionary

Alfred Wegener, polar explorer and visionary, was born in Berlin in 1880. Although he received his doctorate in astronomy (1905), he also developed a strong interest in meteorology. In 1906 he and his brother Kurt set an endurance record for balloon flight by staying aloft for 52 hours, breaking the previous record by 17 hours. Later that year, he joined a Danish expedition to northeastern Greenland where he is thought to have first entertained the possibility of continental drift. That journey marked the beginning of his lifelong dedication to exploring this ice-covered island where he would perish nearly 25 years later.

Following his first expedition to Greenland, Wegener returned to Germany and took an academic position as a lecturer in meteorology and astronomy. During this time, he authored a short paper on continental drift, published in 1912, as well as a book on meteorology. Wegener returned to Greenland—an expedition that distinguished Wegener as the first person to make a scientific crossing of the 1200-kilometer (700-mile) glacial core of the island (Figure 2.A).

Shortly after returning from his second expedition to Greenland, Wegener married Else Köppen, daughter of Wladimir Köppen, an eminent climatologist who developed a classification of world climates that is still used today. Following his marriage, Wegener became a combatant in the First World War, during which he was wounded twice, but remained in the army until the war was over. During his convalescence, Wegener wrote his controversial book on continental drift, entitled *The Origin of Continents and Oceans.* Wegener authored revised editions in 1920, 1922, and 1929.

In addition to his passion for finding supporting evidence for continental drift, Wegener also wrote numerous scientific papers on meteorology and geophysics. In 1924 he coauthored a book on ancient climatics (paleoclimates) with his father-in-law, Köppen.

In the spring of 1930, Wegener departed on his final expedition to his beloved Greenland. One of the goals of the journey was to establish a mid-ice camp (Eismitte Station) 400 kilometers (250 miles) from Greenland's west coast at an elevation of nearly 3000 meters (10,000 feet). Because unusually bad weather hampered attempts to establish this outpost, only a fraction of the supplies that were needed for the two scientists stationed there made it to the camp.

As the expedition leader, Wegener led a relief party consisting of fellow meteorologist Fritz Lowe and 13 Greenlanders to resupply Eismitte Station. Heavy snow and temperatures that fell below −50 °C (−58 °F) caused most of the party to return to base camp. Wegener, Lowe, and one of the Greenlanders, Rasmus Villumsen, trudged on.

Forty days later, on October 30, 1930, Wegener and his two companions reached Eismitte Station. Unable to communicate with base camp, the researchers who were believed to be in dire need of supplies had managed to dig an ice cave for shelter and intended to stretch their supplies through the winter. The heroic supply run had been unnecessary.

Lowe decided to winter at Eismitte due to exhaustion and frostbitten limbs. Wegener, however, was reported to have "looked as fresh, happy and fit as if he had just been on a walk." Two days later, on November 1, 1930, they celebrated Wegener's 50th birthday, and he and Villumsen set off on the downhill trek back to the coast. They never arrived.

Because of the inability to maintain contact between stations during the winter months, the two were thought to have wintered over at Eismitte. While the exact date and cause of Wegener's death remain unknown, a search team located his body beneath the snow, approximately halfway between Eismitte and the coast. Because Wegener was known to be physically fit and his body showed no signs of trauma, starvation, or exposure, it is thought that he may have suffered a fatal heart attack.* Villumsen, Wegener's companion, is assumed to have died during the journey as well, although his remains were never located.

The search team buried Wegener in the position in which he was found and respectfully constructed a snow-block monument. Later, a 20-foot iron cross was erected at the site. All have long since vanished beneath the snow to eventually become part of this great ice sheet.

FIGURE 2.A Alfred Wegener shown waiting out the 1912–1913 Arctic winter during an expedition to Greenland, where he made a 1200-kilometer traverse across the widest part of the island's ice sheet. (Photo courtesy of Bildarchiv Preussischer Kulturbesitz, Berlin)

---

*Wegener was a heavy smoker, which some have suggested contributed to his death.

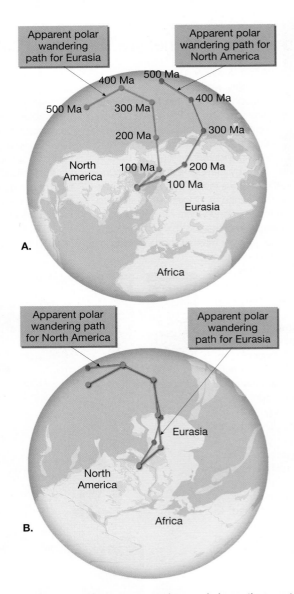

indicated that the position of the paleomagnetic poles had changed through time. A plot of the location of the magnetic north pole with respect to Europe revealed that during the past 500 million years, the pole had gradually wandered from a location near Hawaii northward to its present location near the North Pole (Figure 2.10A). This was strong evidence that either the magnetic poles had migrated, an idea known as *polar wandering,* or that the lava flows moved—in other words, Europe had drifted in relation to the poles.

Although the magnetic poles are known to move in an erratic path around the geographic poles, studies of paleomagnetism from numerous locations show that the positions of the magnetic poles, averaged over thousands of years, correspond closely to the positions of the geographic poles. Therefore, a more acceptable explanation for the apparent polar wandering paths was provided by Wegener's hypothesis. If the magnetic poles remain stationary, their *apparent movement* is produced by continental drift.

Further evidence for continental drift came a few years later when a polar-wandering path was constructed for North America (see Figure 2.10A). For the first 300 million years or so, the paths for North America and Europe were found to be similar in direction, but separated by about 5000 kilometers (3000 miles). Then, during the middle of the Mesozoic era (180 million years ago) they began to converge on the present North Pole. Because the North Atlantic Ocean is about 5000 kilometers wide, the explanation for these curves was that North America and Europe were joined until the Mesozoic, when the Atlantic began to open. From this time forward, these continents continuously moved apart. When North America and Europe are moved back to their pre-drift positions, as shown in Figure 2.10B, these apparent wandering paths coincide.

Although paleomagnetic data—measurements of the magnetic field locked in ancient rocks—strongly supported the continental drift hypothesis, the techniques used in extracting it were relatively new and not universally accepted. Furthermore, most geologists were unfamiliar with the results provided by these studies. Despite these issues, paleomagnetic evidence reinstated continental drift to a respectable subject of scientific inquiry. A new era had begun!

A SCIENTIFIC REVOLUTION BEGINS

Following World War II, marine geologists, equipped with sonar and funding from the U.S. Office of Naval Research, embarked on an unprecedented period of oceanographic exploration. Over the next two decades a much better picture of the seafloor began to emerge. From this work came the discovery of an elevated **oceanic ridge system** that winds through all of the major oceans in a manner similar to the seam on a baseball. One of the segments of this interconnected feature extends down the middle of the Atlantic Ocean and hence was named the *Mid-Atlantic Ridge.*

In other parts of the ocean, new discoveries were also being made. Studies conducted in the western Pacific found that earthquake activity was occurring at much greater depths beneath deep-ocean trenches than in other tectonic settings. Of

FIGURE 2.10 Simplified apparent polar-wandering paths as established from North American and Eurasian paleomagnetic data. **A.** The more westerly path determined from North American data was caused by the westward movement of North America from Eurasia during the breakup of Pangaea. **B.** The positions of the wandering paths when the landmasses are reassembled.

equal importance was the fact that dredging of the seafloor did not bring up any oceanic crust that was older than 180 million years. Further, sediment accumulations in the deep-ocean basins were found to be surprisingly thin, not the thousands of meters that had been predicted. Was the rate of sedimentation in the geologic past much less than the current rate, or was the ocean floor much younger than previously thought?

The Seafloor-Spreading Hypothesis

In the early 1960s, Harry Hess of Princeton University incorporated these newly discovered facts into a hypothesis that was named **seafloor spreading.** In Hess's classic paper, he proposed that oceanic ridges are located above zones of upwelling in the mantle (Figure 2.11).* As rising material reaches the base of the lithosphere it spreads laterally, carrying the seafloor in a conveyor-belt fashion away from the ridge crest. Here, tensional forces fracture the crust and provide pathways for magma to intrude and generate new slivers of oceanic crust. Thus, as the seafloor moves away from the ridge axis, newly formed crust replaces it. Hess further proposed that the descending limbs of these convection currents occur in the vicinity of deep-ocean trenches. These, Hess suggested, are sites where ocean crust is drawn back into Earth's interior. As a consequence, the older portions of the seafloor are gradually consumed as they descend into the mantle. As one researcher summarized, "No wonder the ocean floor was young—it was constantly being renewed!"

Hess referred to his paper as an "essay in geopoetry," which reflected his regard for the speculative nature of this idea. Or, as others have suggested, he may have wanted to deflect criticism from those who were still hostile to continental drift. In either case, his hypothesis provided specific testable ideas, which is one of the hallmarks of good science.

With the seafloor-spreading hypothesis in place, Harry Hess had initiated another phase of this scientific revolution. The conclusive evidence to support his ideas came a few years later from the work of a young Cambridge University graduate student, Fred Vine, and his supervisor, D. H. Matthews (Box 2.2). The significance of the Vine-Matthews hypothesis was that it connected two ideas previously thought to be unrelated: Hess's seafloor-spreading hypothesis and the newly discovered geomagnetic reversals.

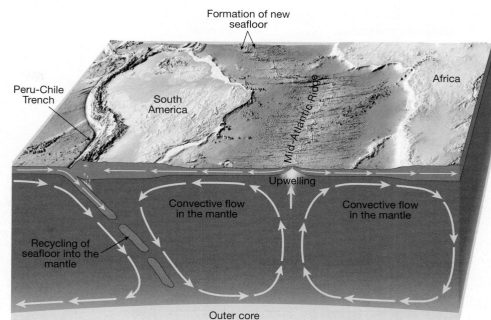

FIGURE 2.11   Seafloor spreading. Harry Hess proposed that upwelling of mantle material along the mid-ocean ridge system created new seafloor. Hess also suggested that the convective motion of mantle material carries the seafloor in a conveyor-belt fashion to deep-ocean trenches, where the seafloor descends into the mantle.

Once the concept of magnetic reversals was confirmed, researchers set out to establish a time scale for these occurrences. The task was to measure the magnetic polarity of hundreds of lava flows and use radiometric dating techniques to establish the age of each flow (Figure 2.12). Figure 2.13 shows the **magnetic time scale** established for the past few million years. The major divisions of the magnetic time scale are called *chrons* and last for roughly 1 million years. As more measurements became available, researchers realized that several, short-lived

Magnetic Reversals: Evidence for Seafloor Spreading

About the time Hess formulated the concept of seafloor spreading, geophysicists discovered that over periods of hundreds of thousands of years, Earth's magnetic field periodically reverses polarity. During a **magnetic reversal** the north magnetic pole becomes the south magnetic pole, and vice versa. Lava solidifying during a period of reverse polarity will be magnetized with the polarity opposite that of volcanic rocks being formed today. When rocks exhibit the same magnetism as the present magnetic field, they are said to possess **normal polarity,** whereas rocks exhibiting the opposite magnetism are said to have **reverse polarity.**

FIGURE 2.12   Schematic illustration of paleomagnetism preserved in lava flows of various ages. Paleomagnetic data from locales around the globe were used to establish the time scale of polarity reversals shown in Figure 2.13.

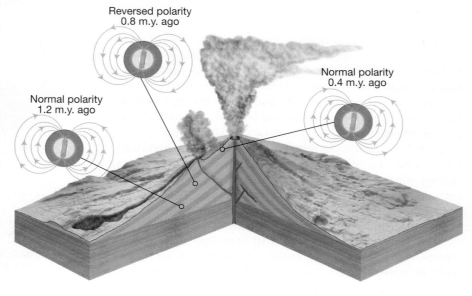

---

*Although Hess proposed that convection in Earth's interior consists of ascending currents of hot material rising from deep within the mantle, it is now clear that these upward flowing currents are shallow features not related to deep convection in the mantle. We will consider this topic further in Chapter 13.

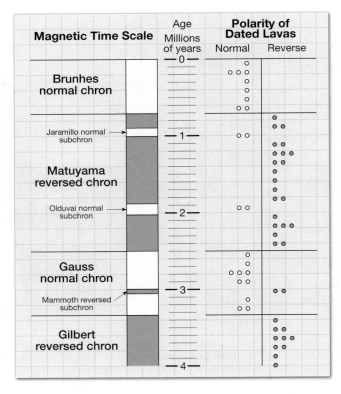

FIGURE 2.13 Time scale of Earth's magnetic field in the recent past. This time scale was developed by establishing the magnetic polarity for lava flows of known age. (Data from Allen Cox and G. B. Dalrymple)

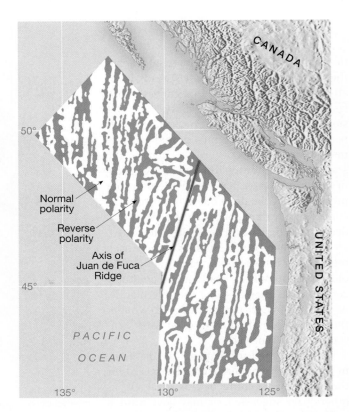

FIGURE 2.14 Pattern of alternating stripes of high- and low-intensity magnetism discovered off the Pacific Coast of North America.

FIGURE 2.15 The ocean floor as a magnetic tape recorder. Magnetic intensities are recorded as a magnetometer is towed across a segment of the oceanic ridge. Notice the symmetrical stripes of low- and high-intensity magnetism that parallel the ridge crest. Vine and Matthews suggested that the stripes of high-intensity magnetism occur where normally magnetized oceanic rocks enhanced the existing magnetic field. Conversely, the low-intensity stripes are regions where the crust is polarized in the reverse direction, which weakens the existing magnetic field.

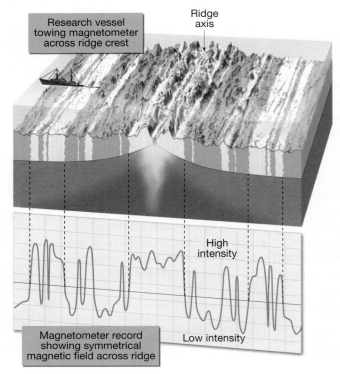

reversals (less than 200,000 years long) often occurred during a single chron.

Meanwhile, oceanographers had begun to do magnetic surveys of the ocean floor in conjunction with their efforts to construct detailed maps of seafloor topography. These magnetic surveys were accomplished by towing very sensitive instruments, called **magnetometers,** behind research vessels. The goal of these geophysical surveys was to map variations in the strength of Earth's magnetic field that arise from differences in the magnetic properties of the underlying crustal rocks.

The first comprehensive study of this type was carried out off the Pacific Coast of North America and had an unexpected outcome. Researchers discovered alternating stripes of high- and low-intensity magnetism as shown in Figure 2.14. This relatively simple pattern of magnetic variation defied explanation until 1963, when Fred Vine and D. H. Matthews demonstrated that the high- and low-intensity stripes supported Hess's concept of seafloor spreading. Vine and Matthews suggested that the stripes of high-intensity magnetism are regions where the paleomagnetism of the ocean crust exhibits normal polarity (Figure 2.15). Consequently, these rocks *enhance* (reinforce) Earth's magnetic field. Conversely, the low-intensity stripes are regions where the ocean crust is polarized in the reverse direction and therefore *weaken* the existing magnetic field. But how do parallel stripes of normally and reversely magnetized rock become distributed across the ocean floor?

Vine and Matthews reasoned that as magma solidifies along narrow rifts at the crest of an oceanic ridge, it is magnetized with the polarity of the existing magnetic field (Figure 2.16).

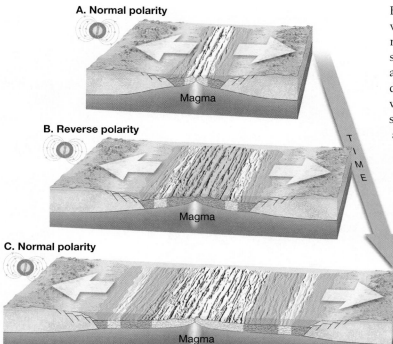

A. Normal polarity

Magma

B. Reverse polarity

Magma

C. Normal polarity

Magma

TIME

FIGURE 2.16 As new basalt is added to the ocean floor at mid-ocean ridges, it is magnetized according to Earth's existing magnetic field. Hence, it behaves much like a tape recorder as it records each reversal of our planet's magnetic field.

FIGURE 2.17 Illustration of Earth's major lithospheric plates.

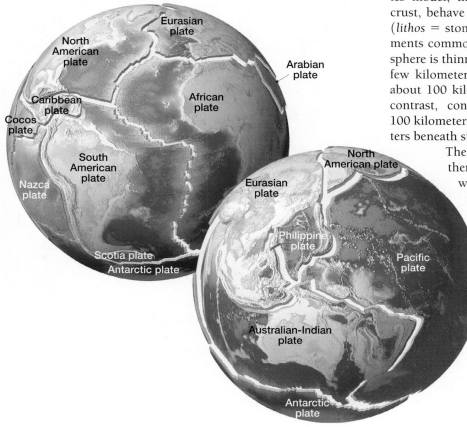

Eurasian plate

North American plate

Arabian plate

Caribbean plate

African plate

Cocos plate

South American plate

Nazca plate

North American plate

Eurasian plate

Philippine plate

Pacific plate

Scotia plate

Antarctic plate

Australian-Indian plate

Antarctic plate

Because of seafloor spreading, this strip of magnetized crust would gradually increase in width. When Earth's magnetic field reverses polarity, any newly formed seafloor (having the opposite polarity) would form in the middle of the old strip. Gradually, the two halves of the old strip are carried in opposite directions away from the ridge crest. Subsequent reversals would build a pattern of normal and reverse magnetic stripes as shown in Figure 2.16. Because new rock is added in equal amounts to both trailing edges of the spreading ocean floor, we should expect that the pattern of stripes (size and polarity) found on one side of an oceanic ridge to be a mirror image of the other side. A few years later a survey across the Mid-Atlantic Ridge just south of Iceland revealed a pattern of magnetic stripes exhibiting a remarkable degree of symmetry to the ridge axis.

By the end of the 1960s the tide of scientific opinion had turned! However, some opposition to plate tectonics continued for at least a decade. Nevertheless, Wegener had been vindicated and the revolution in geology was nearing an end.

PLATE TECTONICS

PLATE TECTONICS

▶ Introduction

By 1968 the concepts of continental drift and seafloor spreading were united into a much more encompassing theory known as **plate tectonics** (*tekton* = to build). According to the plate tectonics model, the uppermost mantle, along with the overlying crust, behave as a strong, rigid layer, known as the **lithosphere** (*lithos* = stone, *sphere* = a ball), which is broken into segments commonly referred to as *plates* (Figure 2.17). The lithosphere is thinnest in the oceans where it varies from as little as a few kilometers along the axis of the oceanic ridge system to about 100 kilometers (60 miles) in the deep-ocean basins. By contrast, continental lithosphere is generally thicker than 100 kilometers and may extend to a depth of 200 to 300 kilometers beneath stable continental cratons.

The lithosphere, in turn, overlies a weak region in the mantle known as the **asthenosphere**(*asthenos* = weak, *sphere* = ball). The temperatures and pressures in the upper asthenosphere (100 to 200 kilometers in depth) are such that the rocks there are very near their melting temperatures and, hence, respond to stress by flowing. As a result, Earth's rigid outer shell is effectively detached from the layers below which permit it to move independently.

Earth's Major Plates

The lithosphere is composed of about 20 segments having irregular sizes and shapes called **lithospheric** or **tectonic plates** that are in constant motion with respect to one another. As shown in Figure 2.18, seven major

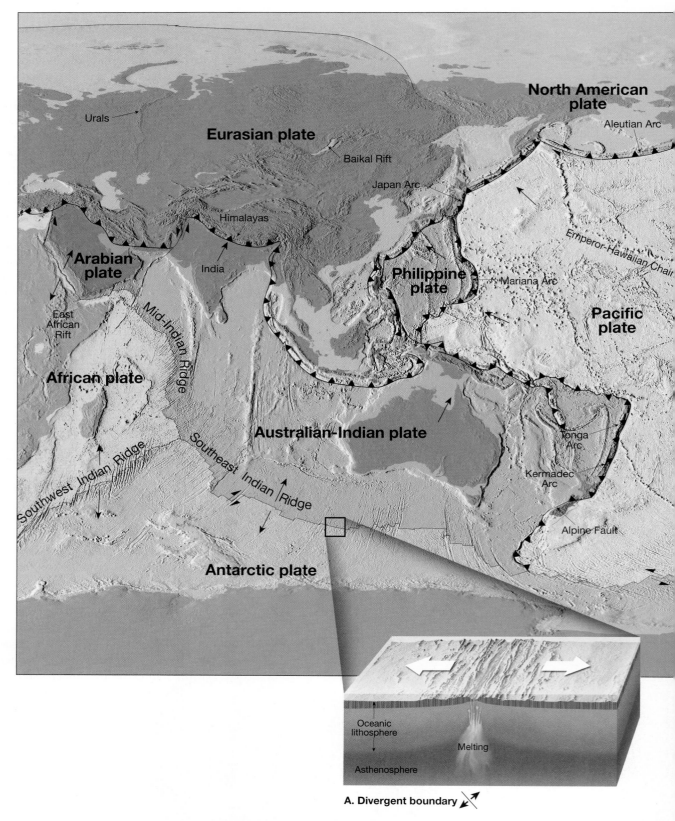

North American
plate

Aleutian Arc

Urals

Eurasian plate

Baikal Rift

Japan Arc

Emperor-Hawaiian Chain

Himalayas

**Arabian
plate**

India

Mariana Arc

**Philippine
plate**

**Pacific
plate**

East
African
Rift

Mid-Indian Ridge

African plate

Southwest Indian Ridge

Southeast Indian Ridge

Australian-Indian plate

Tonga
Arc

Kermadec
Arc

Alpine Fault

Antarctic plate

Oceanic
lithosphere

Melting

Asthenosphere

A. Divergent boundary

FIGURE 2.18   A mosaic of rigid plates constitutes Earth's outer shell. (After W. B. Hamilton, U.S.
Geological Survey)

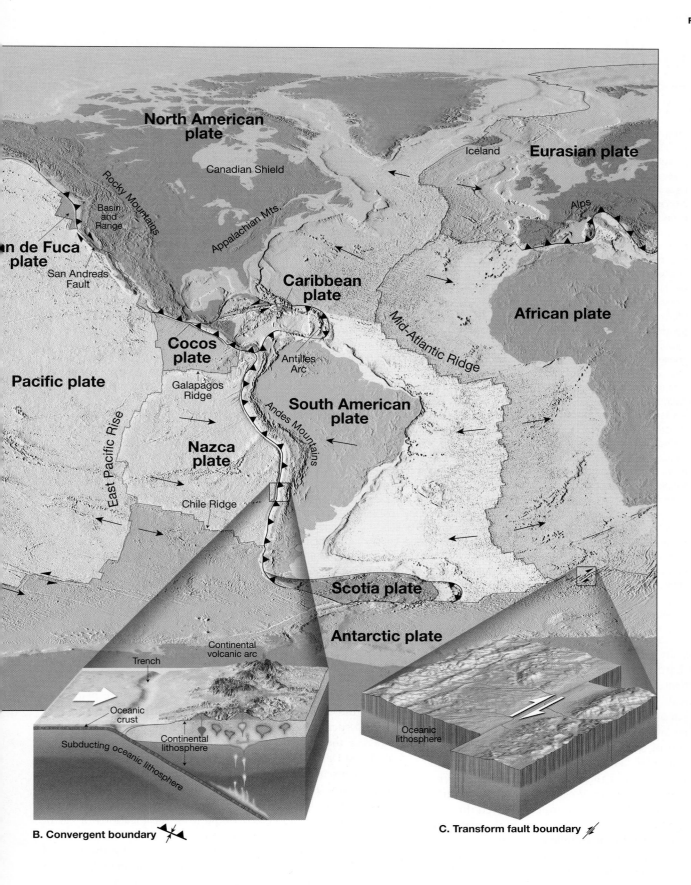

North American plate

Canadian Shield

Rocky Mountains

Basin and Range

n de Fuca plate

San Andreas Fault

Appalachian Mts.

Iceland

Eurasian plate

Alps

African plate

Mid-Atlantic Ridge

Caribbean plate

Cocos plate

Antilles Arc

Pacific plate

Galapagos Ridge

East Pacific Rise

South American plate

Andes Mountains

Nazca plate

Chile Ridge

Scotia plate

Antarctic plate

Continental volcanic arc

Trench

Oceanic crust

Continental lithosphere

Subducting oceanic lithosphere

B. Convergent boundary

Oceanic lithosphere

C. Transform fault boundary

UNDERSTANDING EARTH

Priority in the Sciences

The priority, or credit, for a scientific idea or discovery is *usually* given to the researcher, or group of researchers, who first publish their findings in a scientific journal. However, it is not uncommon for two different researchers to reach similar conclusions almost simultaneously. Well-known examples include the independent discoveries of organic evolution by Charles Darwin and Alfred Wallace and the development of calculus by Isaac Newton and Gottfried W. Leibniz. Similarly, some of the main ideas that led to the plate tectonics revolution in the Earth sciences were also discovered independently by more than one group of investigators.

Although the continental drift hypothesis is rightfully associated with the name Alfred Wegener, he was not the first to suggest continental mobility. In fact, Francis Bacon, in 1620, pointed out the similarities of the outlines of Africa and South America; however, he did not develop this idea further. Nearly three centuries later, in 1910, two years before Wegener made a formal presentation of his ideas, American geologist F. B. Taylor published the first paper to outline the concept we now call continental drift.

Why, then, is Wegener credited with this idea? Because the papers authored by Taylor had relatively little impact among the scientific community, Wegener was not aware of Taylor's work. Hence, it is believed that Wegener independently and simultaneously reached the same conclusion. More important, however, Wegener made great efforts throughout his professional life to provide a wide range of evidence to support his hypothesis. By contrast, Taylor appeared content to simply state, "There are many bonds of union which show that Africa and South America were once joined." Further, whereas Taylor viewed continental drift as a somewhat speculative idea, Wegener was *certain* that the continents had drifted. According to H. W. Menard in his book, *The Ocean of Truth,* Taylor was uncomfortable having his ideas coupled with Wegener's hypothesis. Menard quotes Taylor as writing, "Wegener was a young professor of meteorology. Some of his ideas are very different from mine and he went much further in his speculation."

Another controversy concerning priority came with the development of the seafloor spreading hypothesis. In 1960, Harry Hess of Princeton University wrote a paper that outlined his ideas on seafloor spreading. Rather than rushing it to publication, Hess mailed copies of the manuscript to numerous colleagues, a common practice among researchers. In the meantime, and apparently independently, Robert Dietz of Scripps Institution of Oceanography published a similar paper in *Nature* (1961), titled "Continents and Ocean Basin Evolution by Spreading of the Sea Floor." When Dietz became aware of Hess's earlier, although unpublished paper, he acknowledged priority for the idea of seafloor spreading to Hess. It is interesting to note that the basic ideas in Hess's paper actually appeared in a textbook authored by Arthur Holmes in 1944. Therefore, priority for seafloor spreading may rightfully belong to Holmes. Nevertheless, Dietz and Hess both presented new ideas that were influential to the development of the theory of plate tectonics. Thus, historians associate the names Hess and Dietz with the discovery of seafloor spreading with occasional mention of contributions by Holmes.

Perhaps the most controversial issue of scientific priority related to plate tectonics came in 1963 when Fred Vine and D. H. Matthews published their paper linking the seafloor spreading hypothesis with the newly discovered data on magnetic reversals. However, nine months earlier a similar paper by Canadian geophysicist, L. W. Morley was not accepted for publication. One reviewer of Morley's paper commented, "Such speculation makes interesting talk at cocktail parties, but is not the sort of thing that ought to be published under serious, scientific aegis." Morley's paper was eventually published in 1964, but priority had already been established, and the idea became known as the Vine-Matthews hypothesis. In 1971, N. D. Watkins wrote of Morley's paper, "The manuscript certainly had substantial historical interest, ranking as probably the most significant paper in the earth sciences to ever be denied publication."

---

lithospheric plates are recognized. These plates, which account for 94 percent of Earth's surface area, include the *North American, South American, Pacific, African, Eurasian, Australian-Indian,* and *Antarctic plates.* The largest is the Pacific plate, which encompasses a significant portion of the Pacific Ocean basin. The six other large plates include an entire continent plus a significant amount of ocean floor. Notice in Figure 2.18 that the South American plate encompasses almost all of South America and about one-half of the floor of the South Atlantic. This is a major departure from Wegener's continental drift hypothesis, which proposed that the continents moved through the ocean floor, not with it. Note also that none of the plates are defined entirely by the margins of a single continent.

Intermediate-sized plates include the *Caribbean, Nazca, Philippine, Arabian, Cocos, Scotia,* and *Juan de Fuca plates.* These plates, with the exception of the Arabian plate, are composed mostly of oceanic lithosphere. In addition, there are several smaller plates (*microplates*) that have been identified but are not shown in Figure 2.18.

Plate Boundaries

One of the main tenets of the plate tectonic theory is that plates move as semi-coherent units relative to all other plates. As plates move, the distance between two locations on different plates, such as New York and London, gradually changes whereas the distance between sites on the same plate—New York and Denver, for example—remains relatively constant.

Because plates are in constant motion relative to each other, most major interactions among them (and, therefore, most deformation) occur along their *boundaries.* In fact, plate boundaries were first established by plotting the locations of earthquakes and volcanoes. Plates are bounded by three distinct types of boundaries, which are differentiated by the type of movement they exhibit. These boundaries are depicted at the bottom of Figure 2.18 and are briefly described here:

1. **Divergent boundaries** (*constructive margins*)—where two plates move apart, resulting in upwelling of hot material from the mantle to create new seafloor (Figure 2.18A).

2. **Convergent boundaries** (*destructive margins*)—where two plates move together, resulting in oceanic lithosphere descending beneath an overriding plate, eventually to be reabsorbed into the mantle, or possibly in the collision of two continental blocks to create a mountain system (Figure 2.18B).

3. **Transform fault boundaries** (*conservative margins*)—where two plates grind past each other without the production or destruction of lithosphere (Figure 2.18C).

Divergent and convergent plate boundaries each account for about 40 percent of all plate boundaries. Transform faults account for the remaining 20 percent. In the following sections we will summarize the nature of the three types of plate boundaries.

DIVERGENT BOUNDARIES

PLATE TECTONICS
▶ Divergent Boundaries

Most **divergent** (*di* = apart, *vergere* = to move) **boundaries** are located along the crests of oceanic ridges and can be thought of as *constructive plate margins* since this is where new ocean floor is generated (Figure 2.19). Divergent boundaries are also called **spreading centers**, because seafloor spreading occurs at these boundaries. Here, two adjacent plates are moving away from each other, producing long, narrow fractures in the ocean crust. As a result, hot rock from the mantle below migrates upward to fill the voids left as the crust is being ripped apart. This molten material gradually cools to produce new slivers of seafloor. In a slow, yet unending manner, adjacent plates spread apart and new oceanic lithosphere forms between them.

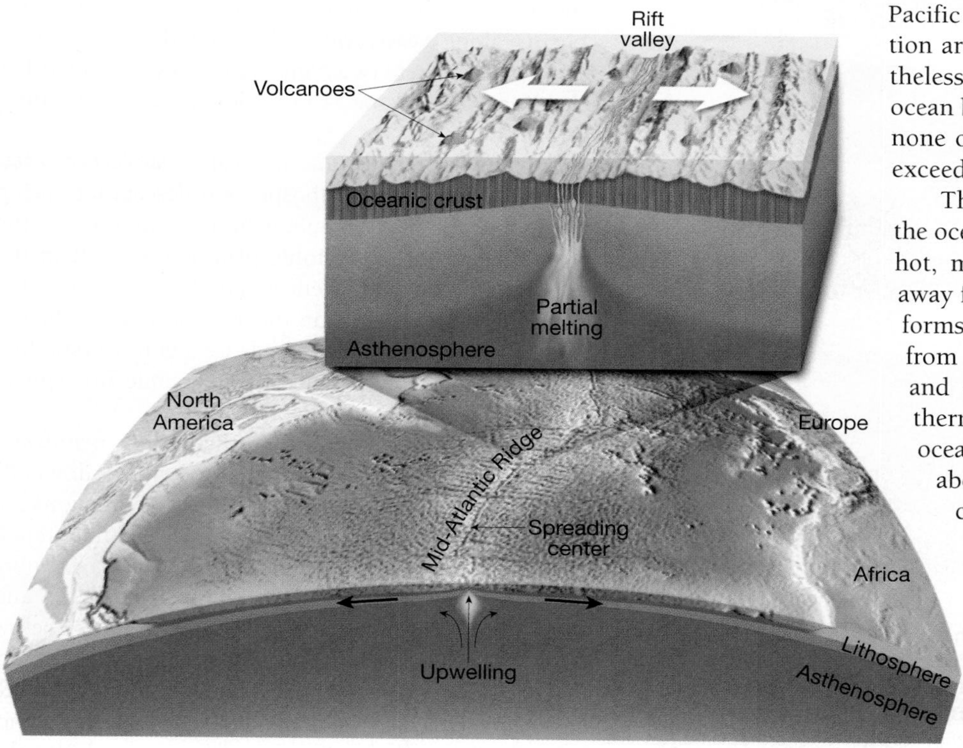

Oceanic Ridges and Seafloor Spreading

Most divergent plate boundaries are associated with *oceanic ridges*: elevated areas of the seafloor that are characterized by high heat flow and volcanism. The global ridge system is the longest topographic feature on Earth's surface, exceeding 70,000 kilometers (43,000 miles) in length. As shown in Figure 2.18 various segments of the global ridge system have been named, including the Mid-Atlantic Ridge, East Pacific Rise, and Mid-Indian Ridge.

Representing 20 percent of Earth's surface, the oceanic ridge system winds through all major ocean basins like the seam on a baseball. Although the crest of the oceanic ridge is commonly 2 to 3 kilometers higher than the adjacent ocean basins, the term "ridge" may be misleading because this feature is not narrow but has widths that vary from 1000 to more than 4000 kilometers. Further, along the axis of some ridge segments is a deep down-faulted structure called a **rift valley**. This structure is evidence that tensional forces are actively pulling the ocean crust apart at the ridge crest.

The mechanism that operates along the oceanic ridge system to create new seafloor is appropriately called *seafloor spreading*. Typical rates of spreading average around 5 centimeters (2 inches) per year. This is roughly the same rate at which human fingernails grow. Comparatively slow spreading rates of 2 centimeters per year are found along the Mid-Atlantic Ridge, whereas spreading rates exceeding 15 centimeters (6 inches) per year have been measured along sections of the East Pacific Rise. Although these rates of seafloor production are slow on a human time scale, they are nevertheless rapid enough to have generated all of Earth's ocean basins within the last 200 million years. In fact, none of the ocean floor that has been dated thus far exceeds 180 million years in age.

The primary reason for the elevated position of the oceanic ridge is that newly created oceanic crust is hot, making it less dense than cooler rocks found away from the ridge axis. As soon as new lithosphere forms, it is slowly yet continually displaced away from the zone of upwelling. Thus, it begins to cool and contract, thereby increasing in density. This thermal contraction accounts for the increase in ocean depths away from the ridge crest. It takes about 80 million years for the temperature of the crust to stabilize and contraction to cease. By this time, rock that was once part of the elevated oceanic ridge system is located in the deep-ocean basin, where it may be buried by substantial accumulations of sediment.

In addition, cooling strengthens the hot material directly below the oceanic crust, thereby adding to the plate's thickness. Stated another way, the thickness of oceanic lithosphere is age-dependent. The older

FIGURE 2.19 Most divergent plate boundaries are situated along the crests of oceanic ridges.

(cooler) it is, the greater its thickness. Oceanic lithosphere that exceeds 80 million years in age is about 100 kilometers thick: about its maximum thickness.

Continental Rifting

Divergent boundaries can also develop within a continent, in which case the landmass may split into two or more smaller segments separated by an ocean basin. Continental rifting occurs where opposing tectonic forces act to pull the lithosphere apart. The initial stage of rifting tends to include mantle upwelling that is associated with broad upwarping of the overlying litho-

sphere (Figure 2.20A). As a result, the lithosphere is stretched, causing the brittle crustal rocks to break into large slabs. As the tectonic forces continue to pull the crust apart, these crustal fragments sink, generating an elongated depression called a **continental rift** (Figure 2.20B).

A modern example of an active continental rift is the East African Rift (see Figure 2.18, left). Whether this rift will eventually result in the break-up of Africa is a topic of continued research. Nevertheless, the East African Rift is an excellent model of the initial stage in the breakup of a continent. Here, tensional forces have stretched and thinned the crust, allowing molten rock to ascend from the mantle. Evidence for recent volcanic activity includes several large volcanic mountains including Mount Kilimanjaro and Mount Kenya, the tallest peaks in Africa. Research suggests that if rifting continues, the rift valley will lengthen and deepen, eventually extending out to the margin of the landmass (Figure 2.20C). At this point, the rift will become a narrow sea with an outlet to the ocean. The Red Sea, which formed when the Arabian Peninsula split from Africa, is a modern example of such a feature. Consequently, the Red Sea provides us with a view of how the Atlantic Ocean may have looked in its infancy (Figure 2.20D).

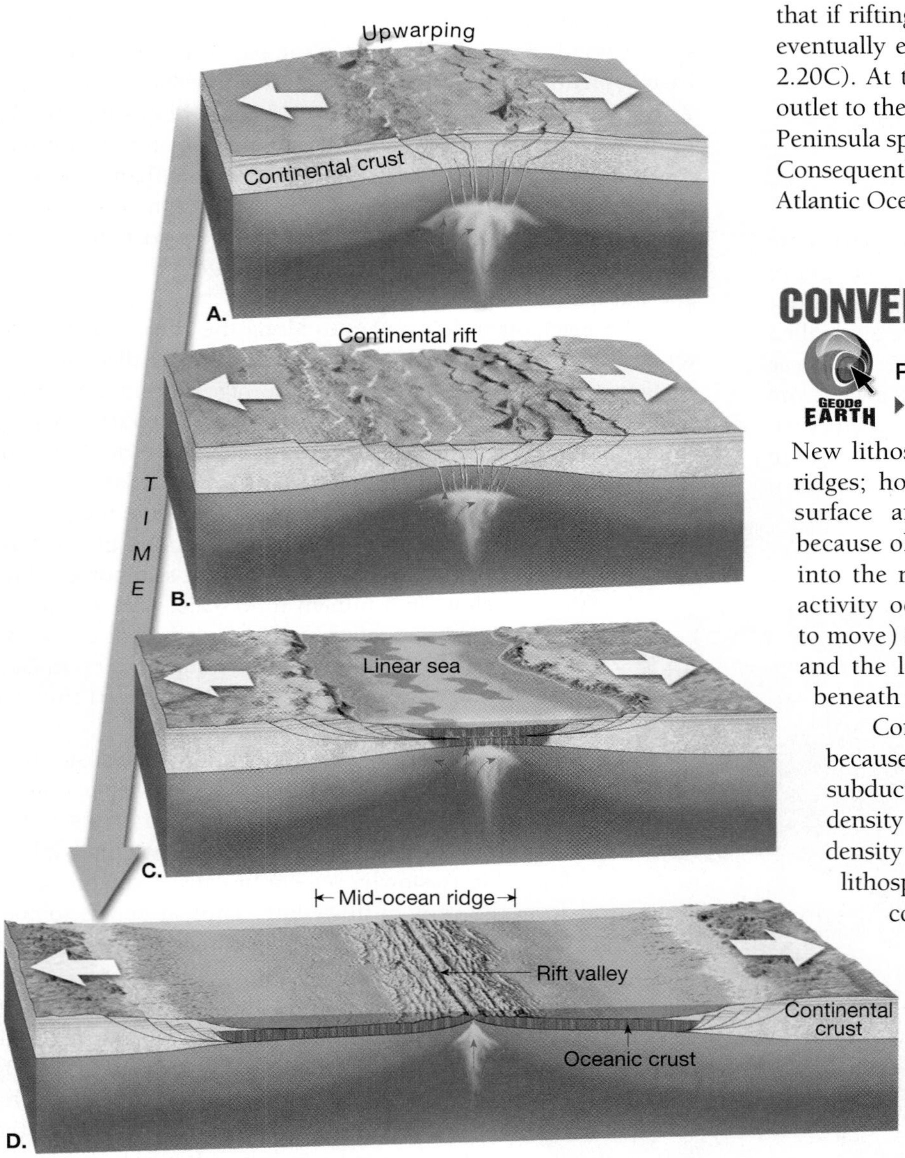

CONVERGENT BOUNDARIES

PLATE TECTONICS
▶ Convergent Boundaries

New lithosphere is constantly being produced at the oceanic ridges; however, our planet is not growing larger—its total surface area remains constant. A balance is maintained because older, denser portions of oceanic lithosphere descend into the mantle at a rate equal to seafloor production. This activity occurs along **convergent** (*con* = together, *vergere* = to move) **boundaries,** where two plates move toward each other and the leading edge of one is bent downward, as it slides beneath the other.

Convergent boundaries are also called **subduction zones,** because they are sites where lithosphere is descending (being subducted) into the mantle. Subduction occurs because the density of the descending tectonic plate is greater than the density of the underlying asthenosphere. In general, oceanic lithosphere is more dense than the asthenosphere, whereas continental lithosphere is less dense and resists subduction. As a consequence, only oceanic lithosphere will subduct to great depths.

Deep-ocean trenches are the surface manifestations produced as oceanic lithosphere descends into the mantle. These large linear depressions are remarkably long and deep. The Peru–Chile trench along the west coast of South America is more than 4500 kilometers (3000 miles) in length and its base is as much as 8 kilometers (5 miles) below sea level. The trenches in the western Pacific, including the Mariana and Tonga trenches, tend to be even deeper than those of the eastern Pacific.

Slabs of oceanic lithosphere descend into the mantle at angles that vary from a few degrees to

FIGURE 2.20 Continental rifting and the formation of a new ocean basin. **A.** The initial stage of continental rifting tends to include upwelling in the mantle that is associated with broad doming of the lithosphere. Tensional forces and buoyant uplifting of the heated lithosphere cause the crust to be broken into large slabs. **B.** As the crust is pulled apart, these large blocks sink, generating a continental rift valley. **C.** Further spreading generates a narrow sea similar to the present-day Red Sea. **D.** Eventually, an expansive deep-ocean basin and oceanic ridge are created.

nearly vertical (90 degrees). The angle at which oceanic lithosphere descends depends largely on its density. For example, when a spreading center is located near a subduction zone, as is the case along the Peru–Chile trench, the subducting lithosphere is young and, therefore, warm and buoyant. Because of this, the angle of descent is small which results in considerable interaction between the descending slab and the overriding plate. Consequently, the region around the Peru–Chile trench experiences great earthquakes, including the 1960 Chilean earthquake—the largest on record.

As oceanic lithosphere ages (gets farther from the spreading center), it gradually cools, which causes it to thicken and increase in density. In parts of the western Pacific, some oceanic lithosphere is 180 million years old. This is the thickest and densest in today's oceans. The very dense slabs in this region typically plunge into the mantle at angles approaching 90 degrees. This largely explains the fact that most trenches in the western Pacific are deeper than trenches in the eastern Pacific.

Although all convergent zones have the same basic characteristics, they are highly variable features. Each is controlled by the type of crustal material involved and the tectonic setting. Convergent boundaries can form between *two oceanic plates, one oceanic and one continental plate,* or *two continental plates.*

Oceanic–Continental Convergence

Whenever the leading edge of a plate capped with continental crust converges with a slab of oceanic lithosphere, the buoyant continental block remains "floating," while the denser oceanic slab sinks into the mantle (Figure 2.21A). When a descending

oceanic slab reaches a depth of about 100 kilometers (60 miles), melting is triggered within the wedge of hot asthenosphere that lies above it. But how does the subduction of a cool slab of oceanic lithosphere cause mantle rock to melt? The answer lies in the fact that water contained in the descending plates acts like salt does to melt ice. That is, "wet" rock in a high-pressure environment, melts at substantially lower temperatures than does "dry" rock of the same composition.

Sediments and oceanic crust contain a large amount of water which is carried to great depths by a subducting plate. As the plate plunges downward, heat and pressure drive water from the voids in the rock. At a depth of roughly 100 kilometers, the wedge of mantle rock is sufficiently hot that the introduction of water from the slab below leads to some melting. This process, called **partial melting,** is thought to generate about 10 percent molten material, which is intermixed with unmelted mantle rock. Being less dense than the surrounding mantle, this hot mobile material gradually rises toward the surface. Depending on the environment, these mantle-derived masses of molten rock may ascend through the crust and give rise to a volcanic eruption. However, much of this material never reaches the surface; rather, it solidifies at depth—a process that thickens the crust.

The volcanoes of the towering Andes are the product of molten rock generated by the subduction of the Nazca plate beneath the South American continent (Figure 2.21B). Mountain systems, such as the Andes, which are produced, in part, by volcanic activity associated with the subduction of oceanic lithosphere, are called **continental volcanic arcs.** The Cascade Range in Washington, Oregon, and California is another

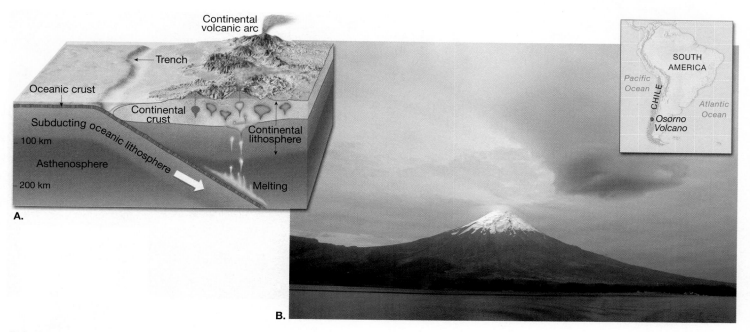

FIGURE 2.21 Oceanic–continental convergent plate boundary. **A.** Illustration of dense, oceanic lithosphere subducting beneath a buoyant continental block. Melting in the asthenosphere generates molten rock that rises toward the surface. This activity produces a chain of volcanic structures built on the overriding landmass, called a continental volcanic arc. **B.** Osorno Volcano is one of the most active volcanoes of the southern Chilean Andes, having erupted 11 times between 1575 and 1869. Located on the shore of Lake Llanquihue, Osorno is similar in appearance to Mount Fuji, Japan. (Photo by Michael Collier)

example that consists of several well-known volcanic mountains, including Mount Rainier, Mount Shasta, and Mount St. Helens. This active volcanic arc also extends into Canada, where it includes Mount Garibaldi, Mount Silverthrone, and others.

Oceanic–Oceanic Convergence

An *oceanic–oceanic convergent boundary* has many features in common with oceanic–continental plate margins. Where two oceanic slabs converge, one descends beneath the other, initiating volcanic activity by the same mechanism that operates at all subduction zones (Figure 2.22A). Water squeezed from the subducting slab of oceanic lithosphere triggers melting in the hot wedge of mantle rock above. In this setting, volcanoes grow up from the ocean floor, rather than upon a continental platform. When subduction is sustained, it will eventually build a chain of volcanic structures large enough to emerge as islands. The volcanic islands tend to be spaced at intervals of about 80 kilometers (50 miles). This newly formed land consisting of an arc-shaped chain of volcanic islands is called a **volcanic island arc,** or simply an **island arc** (Figure 2.22B).

The Aleutian, Mariana, and Tonga islands are examples of relatively young volcanic island arcs. Island arcs are generally located 100 to 300 kilometers (60 to 200 miles) from a deep-ocean trench. Located adjacent to the island arcs just mentioned are the Aleutian trench, the Mariana trench, and the Tonga trench.

Most volcanic island arcs are located in the western Pacific. Only two are located in the Atlantic—the Lesser Antilles arc, on the eastern margin of the Caribbean Sea and the Sandwich Islands located off the tip of South America. The Lesser Antilles are a product of the subduction of the Atlantic seafloor beneath the Caribbean plate. Located within this volcanic arc are the United States and British Virgin Islands as well as the island of Martinique, where Mount Pelée erupted in 1902, destroying the town of St. Pierre and killing an estimated 28,000 people. This chain of islands also includes Montserrat, where there has been recent volcanic activity.*

Relatively young island arcs are fairly simple structures made of numerous volcanic cones that are underlain by oceanic crust that is generally less than 20 kilometers (12 miles) thick. By contrast, older island arcs are more complex and are underlain by highly deformed crust that may reach 35 kilometers in thickness. Examples include the islands that make up the countries of Japan, Indonesia, and the Philippines. These island arcs are built upon material generated by earlier episodes of subduction or on small slivers of continental crust.

---

*More on these volcanic events is found in Chapter 5.

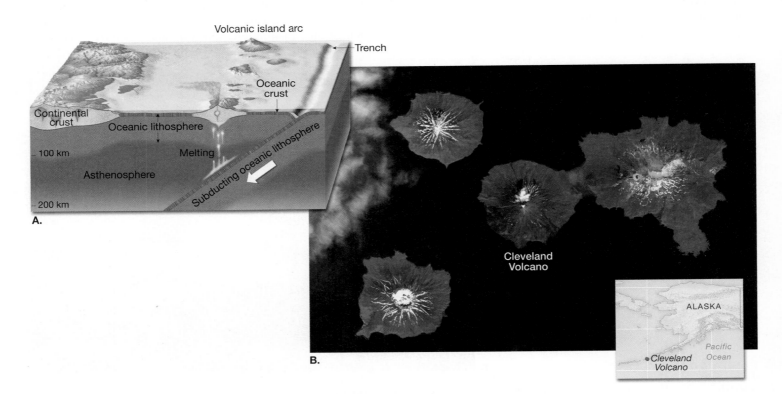

FIGURE 2.22 Oceanic–oceanic convergent plate boundaries. **A.** When oceanic plates converge, one descends beneath the other, initiating volcanic activity in the overriding plate. In this setting a volcanic island arc is built. **B.** These four volcanic structures are part of the Aleutian Islands, a string of both active and dormant volcanoes fed by magma created by the subduction of the Pacific Plate. Steam plumes have recently been observed emanating from Cleveland Volcano (center) evidence of recent tectonicty. (Photo courtesy of NASA)

Continental–Continental Convergence

The third type of convergent boundary results when one land-mass moves toward the margin of another because of subduction of the intervening seafloor (Figure 2.23A). Whereas oceanic lithosphere tends to be dense and sink into the mantle, the buoyancy of continental material inhibits it from being subducted. Consequently, a collision between two converging continental fragments ensues (Figure 2.23C). This event folds and deforms the accumulation of sediments and sedimentary rocks along the continental margins as if they had been placed in a gigantic vise. The result is the formation of a new mountain range composed of deformed sedimentary and metamorphic rocks that often contain slivers of oceanic crust.

Students Sometimes Ask . . .

Someday, will the continents come back together and form a single landmass?

Yes, it is very likely that the continents will eventually come back together, but not anytime soon. Since all of the continents are on the same planetary body, there is only so far a continent can travel before it collides with another landmass. Recent research suggests that a supercontinent may form about once every 500 million years or so. Since it has been about 200 million years since Pangaea broke up, we have only about 300 million years to go before the next supercontinent is assembled.

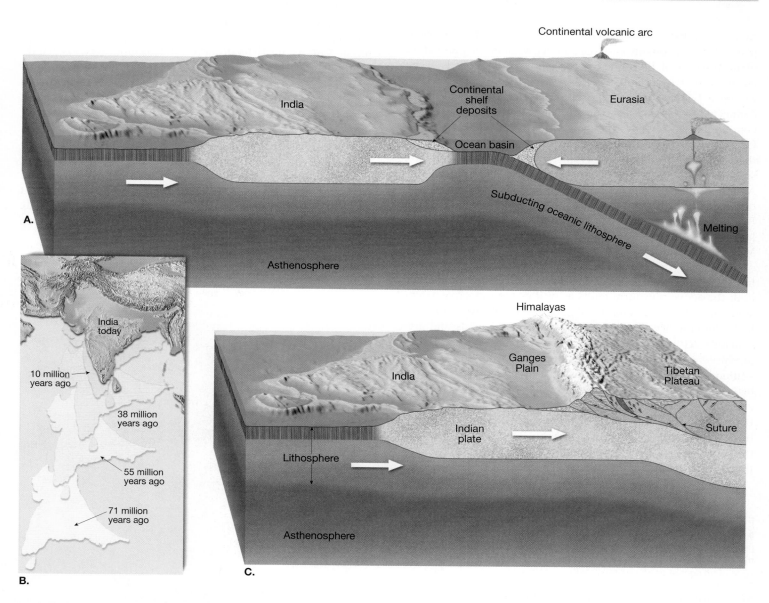

FIGURE 2.23 The ongoing collision of India and Asia began about 50 million years ago—producing the majestic Himalayas. **A.** As India migrated northward the intervening ocean closed as the seafloor was subducted beneath Eurasia. **B.** Position of India in relation to Eurasia at various times. (Modified after Peter Molnar) **C.** Eventually the two landmasses collided, deforming and elevating the sediments that had been deposited along their continental margins. In addition, slices of the crustal rocks were thrust onto the colliding plates.

Such a collision began about 50 million years ago when the subcontinent of India "rammed" into Asia, producing the Himalayas—the most spectacular mountain range on Earth (Figure 2.23C). During this collision, the continental crust buckled and fractured, and was generally shortened and thickened. In addition to the Himalayas, several other major mountain systems, including the Alps, Appalachians, and Urals, formed as continental fragments collided. (This topic will be considered further in Chapter 14.)

TRANSFORM FAULT BOUNDARIES

PLATE TECTONICS
▶ Transform Fault Boundaries

Along a **transform** (*trans* = across, *forma* = form) **fault boundary,** plates slide horizontally past one another without the production or destruction of lithosphere (*conservative plate margins*). The nature of transform faults was discovered in 1965 by Canadian geologist J. Tuzo Wilson, who proposed that these large faults connected two spreading centers (divergent boundaries), or less commonly two trenches (convergent boundaries). Most transform faults are found on the ocean floor (Figure 2.24A). Here they offset segments of the oceanic ridge system, producing a step-like plate margin. Notice that the zigzag shape of the Mid-Atlantic Ridge in Figure 2.18 roughly reflects the shape of the original rifting that caused the beak-up of the supercontinent of Pangaea. (Compare the shapes of the continental margins of the landmasses on both sides of the Atlantic with the shape of the Mid-Atlantic Ridge.)

Typically, transform faults are part of prominent linear breaks in the seafloor known as **fracture zones**, which include both the active transform faults as well as their inactive extensions into the plate interior (Figure 2.24B). Active transform faults lie *only between* the two offset ridge segments and are generally defined by weak, shallow earthquakes. Here seafloor produced at one ridge axis moves in the opposite direction of seafloor produced at an opposing ridge segment. Thus, between the ridge segments these adjacent slabs of oceanic crust are grinding past each other along a transform fault. Beyond the ridge crests are inactive zones, where the fractures are preserved as linear topographic depressions. The trend of these fracture zones roughly parallels the direction of plate motion at the time of their formation. Thus,

these structures are useful in mapping the direction of plate motion in the geologic past.

In another role, transform faults provide the means by which the oceanic crust created at ridge crests can be transported to a site of destruction—the deep-ocean trenches. Figure 2.25 illustrates this situation. Notice that the Juan de Fuca plate moves in a southeasterly direction, eventually being subducted under the west coast of the United States. The southern end of this plate is bounded by a transform fault called the Mendocino Fault. This transform boundary connects the Juan de Fuca ridge to the Cascadia subduction zone. Therefore, it facilitates the movement of the crustal material created at the Juan de Fuca ridge to its destination beneath the North American continent.

Like the Mendocino Fault, most transform fault boundaries are located within the ocean basins, however a few cut through continental crust. Two examples are the earthquake-prone San Andreas Fault of California and New Zealand's Alpine Fault. Notice in Figure 2.25 that the San Andreas Fault connects a spreading center located in the Gulf of California to the Cascadia subduction zone and the Mendocino Fault located along the northwest coast of the United States. Along the San Andreas Fault, the Pacific plate is moving toward the northwest, past the North American plate (Figure 2.26). If this movement continues, that part of California west of the fault zone, including the Baja Peninsula of Mexico, will become an island off the West Coast of the United States and Canada. It could eventually reach Alaska. However, a more immediate concern is the earthquake activity triggered by movements along this fault system.

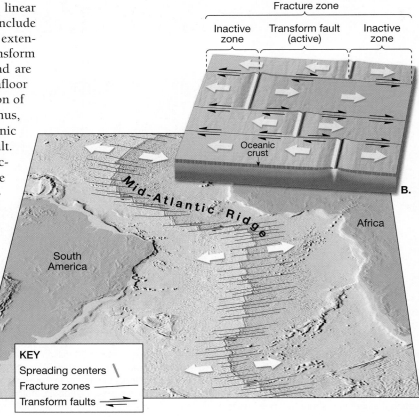

FIGURE 2.24 Transform fault boundaries. **A.** Most transform faults offset segments of a spreading center, producing a step-like plate margin. The zigzag shape of the Mid-Atlantic Ridge roughly reflects the shape of the zone of rifting that produced the break-up of Pangaea. **B.** Fracture zones are long, narrow fractures in the seafloor that are nearly perpendicular to the offset ridge segments. They include both the active transform fault and its "fossilized" trace, where oceanic crust of different ages is juxtaposed. The offsets between ridge segments (transform faults) do not change in length over time.

KEY
Spreading centers
Fracture zones
Transform faults

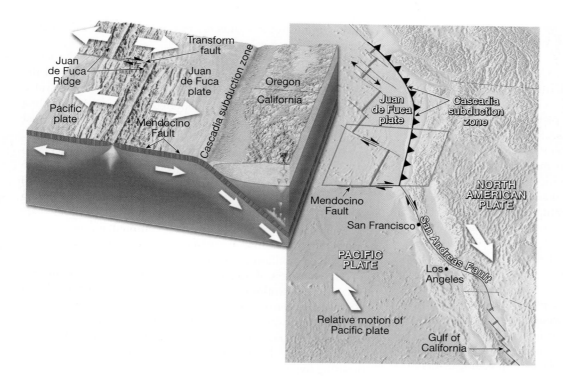

FIGURE 2.25 The Mendocino transform fault permits seafloor generated at the Juan de Fuca ridge to move southeastward past the Pacific plate and beneath the North American plate. Thus, this transform fault connects a spreading center (divergent boundary) to a subduction zone (convergent boundary). Furthermore, the San Andreas Fault, also a transform fault, connects two spreading centers: the Juan de Fuca ridge and a divergent zone located in the Gulf of California.

FIGURE 2.26 Along the San Andreas Fault, the Pacific plate is moving toward the northwest, relative to the North American plate. This aerial view shows the offset in the dry channel of Wallace Creek near Taft, California. (Photo by Michael Collier)

HOW DO PLATES AND PLATE BOUNDARIES CHANGE?

Although the total surface area of Earth does not change, the size and shape of individual plates are constantly changing. For example, the African and Antarctic plates are mainly bounded by divergent boundaries—sites of seafloor production. Thus, the African and Antarctic plates are continually growing in size as new lithosphere is added to their margins. By contrast, the Pacific plate is being consumed into the mantle along its northern and western flanks much faster that it is growing along the East Pacific Rise and thus is diminishing in size.

Another consequence of plate motion is that boundaries also migrate. For example, the position of the Peru–Chile trench, which is the result of the Nazca plate being bent downward as it descends beneath the South American plate, has changed over time (see Figure 2.18). Because of the westward drift of the South American plate relative to the Nazca plate, the position of the Peru–Chile trench has migrated in a westerly direction as well.

Plate boundaries can also be created or destroyed in response to changes in the forces acting on the lithosphere. Recall that the Red Sea is the site of a relatively new spreading center. It came into existence less than 20 million years ago when the Arabian Peninsula began to split from Africa. At other locations, plates carrying continental crust are presently moving toward one another. Eventually these continental fragments may collide and be sutured together. This could occur, for example, in the South Pacific where Australia is moving northward toward southern Asia. Should Australia continue its northward migration, the boundary separating it from Asia will disappear as these plates become one (Box 2.3).

TESTING THE PLATE TECTONICS MODEL

Some of the evidence supporting continental drift and seafloor spreading has already been presented. With the development of the theory of plate tectonics, researchers began testing this new model of how Earth works. Although new supporting data were obtained, it was often new interpretations of already existing data that swayed the tide of opinion.

Evidence from Ocean Drilling

Some of the most convincing evidence for seafloor spreading came from the Deep Sea Drilling Project which operated from 1968 until 1983. One of the early goals was to gather samples of the ocean floor in order to establish its age. To accomplish this, the *Glomar Challenger,* a drilling ship capable of working in water thousands of meters deep, was built. Hundreds of holes were drilled through the layers of sediments that blanket the ocean crust, as well as into the basaltic rocks below. Rather than radiometrically dating the crustal rocks, researchers used the fossil remains of microor-

ganisms found in the sediments resting directly on the crust to date the seafloor at each site.*

When the oldest sediment from each drill site was plotted against its distance from the ridge crest, the plot showed that the sediments increased in age with increasing distance from the ridge (Figure 2.27). This finding supported the seafloor-spreading hypothesis, which predicted that the youngest oceanic crust would be found at the ridge crest, the site of seafloor production, and the oldest oceanic crust would be located adjacent to the continents.

The data collected by the Deep Sea Drilling Project also reinforced the idea that the ocean basins are geologically young because no seafloor with an age in excess of 180 million years was ever found. By comparison, most continental crust exceeds several hundred million years in age and some has been located that exceeds 4 billion years in age.

The thickness of ocean-floor sediments provided additional verification of seafloor spreading. Drill cores from the *Glomar Challenger* revealed that sediments are almost entirely absent on the ridge crest and that sediment thickness increases with increasing distance from the ridge (see Figure 2.27). This pattern of sediment distribution should be expected if the seafloor-spreading hypothesis is correct.

The Ocean Drilling Program, the successor to the Deep Sea Drilling Project, employed a more technologically advanced drilling ship, the *JOIDES Resolution,* to continue the work of the *Glomar Challenger* (Figure 2.28). While the Deep Sea Drilling Project validated many of the major tenets of the theory of plate tectonics, the *JOIDES Resolution* was able to probe deeper into the oceanic crust. This allowed for the study of earthquake-generating zones at convergent plate margins and for the direct examination of oceanic plateaus and seamounts. Sediment cores from the Ocean Drilling Program have also extended our knowledge of long- and short-term climatic changes.

---

*Radiometric dates of the ocean crust itself are unreliable because of the alteration of basalt by seawater.

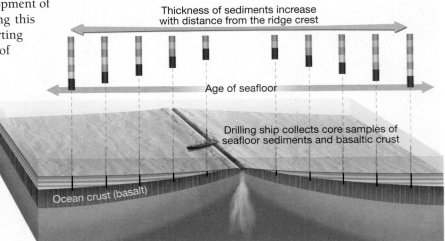

Thickness of sediments increase with distance from the ridge crest

Age of seafloor

Drilling ship collects core samples of seafloor sediments and basaltic crust

Ocean crust (basalt)

FIGURE 2.27 Since 1968 drilling ships have gathered core samples of seafloor sediment and crustal rocks at hundreds of sites. Results from these efforts showed that the ocean floor is indeed youngest at the ridge axis. This was the first direct evidence supporting the seafloor spreading hypothesis and the broader theory of plate tectonics.

UNDERSTANDING EARTH

BOX 2.3

The Breakup of Pangaea

Wegener used evidence from fossils, rock types, and ancient climates to create a jigsaw-puzzle fit of the continents—thereby creating his supercontinent of Pangaea. In a similar manner, but employing modern tools not available to Wegener, geologists have recreated the steps in the breakup of this supercontinent, an event that began nearly 180 million years ago. From this work, the dates when individual crustal fragments separated from one another and their relative motions have become well established (Figure 2.B).

An important consequence of the breakup of Pangaea was the creation of a "new" ocean basin: the Atlantic. As you can see in part B of Figure 2.B, splitting of the supercontinent did not occur simultaneously along the margins of the Atlantic. The first split developed between North America and Africa.

Here, the continental crust was highly fractured, providing pathways for huge quantities of fluid lavas to reach the surface. Remnants of these lavas are found along the Eastern Seaboard of the United States—primarily buried beneath younger sedimentary rocks that form the continental shelf. Radiometric dating of these solidified lavas indicates that rifting began in various stages between 180 million and 165 million years ago. This time span can be used as the "birth date" for this section of the North Atlantic.

By 130 million years ago, the South Atlantic began to open near the tip of what is now South Africa. As this zone of rifting migrated northward, it gradually opened the South Atlantic (compare Figure 2.B, parts B and C). Continued breakup in the Southern Hemisphere led to the separation of Africa and Antarctica and sent India on a northward journey. By the early Cenozoic, about 50 million years ago, Australia had separated from Antarctica, and the South Atlantic had

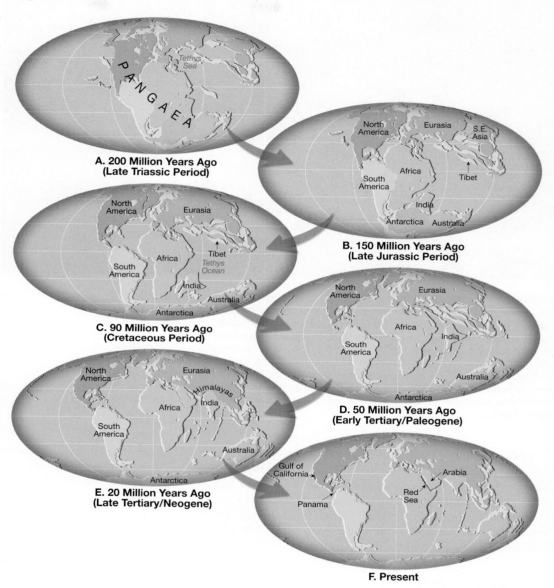

A. 200 Million Years Ago (Late Triassic Period)

B. 150 Million Years Ago (Late Jurassic Period)

C. 90 Million Years Ago (Cretaceous Period)

D. 50 Million Years Ago (Early Tertiary/Paleogene)

E. 20 Million Years Ago (Late Tertiary/Neogene)

F. Present

FIGURE 2.B Several views of the breakup of Pangaea over a period of 200 million years.

emerged as a full-fledged ocean (Figure 2.B, part D).

A modern map (Figure 2.B, part F) shows that India eventually collided with Asia, an event that began about 50 million years ago and created the Himalayas as well as the Tibetan Highlands. About the same time, the separation of Greenland from Eurasia completed the breakup of the northern landmass.

During the last 20 million years or so of Earth history, Arabia rifted from Africa to form the Red Sea, and Baja California separated from Mexico to form the Gulf of California (Figure 2.B, part E). Meanwhile, a sliver of land (now known as Central America) was trapped between North America and South America to produce our globe's familiar, modern appearance.

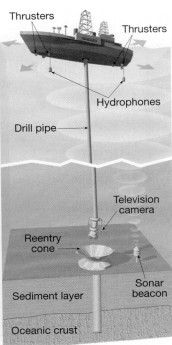

A.

B.

FIGURE 2.28 The *JOIDES Resolution.* **A.** *JOIDES Resolution,* one of the drilling ships of the Integrated Ocean Drilling Program. (Photo courtesy of Ocean Drilling Program) **B.** The *JOIDES Resolution* has a tall metal derrick that is used to conduct *rotary drilling,* while the ship's thrusters hold it in a fixed position at sea. Individual sections of drill pipe are fitted together to make a single string of pipe up to 8200 meters (27,000 feet) long. The drill bit, located at the end of the pipe string, rotates as it is pressed against the ocean bottom and can drill up to 2100 meters (6900 feet) into the seafloor. Like twirling a soda straw into a layer cake, the drilling operation cuts through sediments and rock and retains a cylinder of material (a core sample) on the inside of the hollow pipe, which can then be raised on board the ship and analyzed in state-of-the-art laboratory facilities.

In October 2003, the *JOIDES Resolution* became part of a new program, the Integrated Ocean Drilling Program (IODP). This new international effort uses multiple vessels for exploration, including the massive 210-meter-long (nearly 770-foot-long) *Chikyu,* (meaning "planet Earth" in Japanese) which began operations in 2007. One of the goals of the IODP is to drill a complete section of the ocean crust.

Hot Spots and Mantle Plumes

Mapping volcanic islands and seamounts (submarine volcanoes) in the Pacific Ocean revealed several linear chains of volcanic structures. One of the most studied chains consists of at least 129 volcanoes that extend from the Hawaiian Islands to Midway Island and continue northward toward the Aleutian trench (Figure 2.29). Radiometric dating of this structure, called the *Hawaiian Island–Emperor Seamount chain,* showed that the volcanoes increase in age with increasing distance from the "big island" of Hawaii. The youngest volcanic island in the chain (Hawaii), rose from the ocean floor less than a million years ago, whereas Midway Island is 27 million years old, and Suiko Seamount, near the Aleutian trench, is about 65 million years old (see Figure 2.29).

Most researchers are in agreement that a cylindrically-shaped upwelling of hot rock, called a **mantle plume**, is located beneath the island of Hawaii. As the hot, rocky plume ascends through the mantle, the confining pressure drops, which triggers partial melting. (This process, called *decompression melting,* is discussed in Chapter 5.) The surface manifestation of this activity is a **hot spot,** an area of volcanism, high heat flow, and crustal uplifting that is a few hundred kilometers across. As the Pacific plate moved over the hot spot, a chain of volcanic structures known as a **hot spot track** was built. As shown in Figure 2.29, the age of each volcano indicates how much time has elapsed since it was situated over the mantle plume.

Taking a closer look at the five largest Hawaiian Islands, we see a similar pattern of ages from the volcanically active island of Hawaii, to the inactive volcanoes that make up the oldest island, Kauai (see Figure 2.29). Five million years ago, when Kauai was positioned over the hot spot, it was the *only* Hawaiian Island in existence. Visible evidence of the age of Kauai can be seen by examining its extinct volcanoes, which have been eroded into jagged peaks and vast canyons. By contrast, the relatively young island of Hawaii exhibits many fresh lava flows, and one of its five major volcanoes, Kilauea, remains active today (Figure 2.30).

Research suggests that at least some mantle plumes originate at great depth, perhaps at the core-mantle boundary. Others, however, may have a much shallower origin. Of the 40 or so hot spots that have been identified worldwide, more than a dozen are located near spreading centers. For example, the

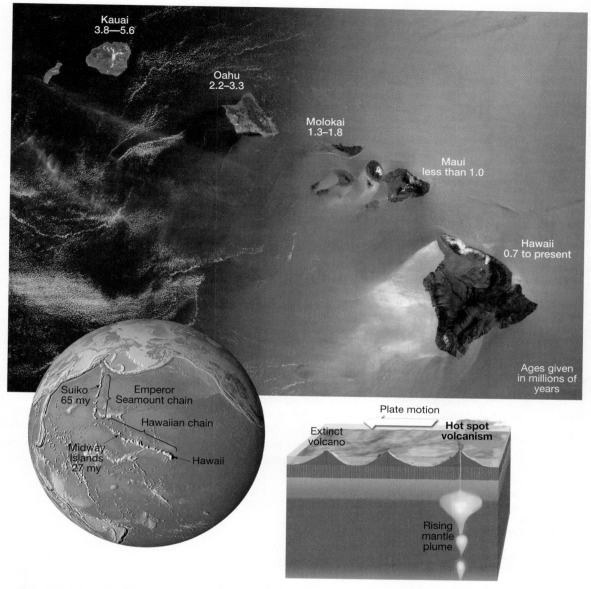

FIGURE 2.29 The chain of islands and seamounts that extends from Hawaii to the Aleutian trench was generated as the Pacific plate moved over a mantle plume (hot spot). Radiometric dating of the Hawaiian Islands shows that the volcanic activity increases in age moving away from the "big island" of Hawaii.

mantle plume located beneath Iceland is responsible for the vast accumulation of volcanic rocks found along this exposed section of the Mid-Atlantic Ridge.

HOW IS PLATE MOTION MEASURED?

PLATE TECTONICS
▶ Formation and Breakup of Pangaea

A number of methods have been employed to establish the direction and rate of plate motion. Paleomagnetism stored in the rocks of the ocean floor is one method used to determine the speeds at which plates move away from the ridge axes where they were generated. In addition, hot spot tracks, such as the Hawaiian Island–Emperor Seamount chain, trace the speed and direction of plate movement relative to the hot plume embedded in the mantle below.

Mantle Plumes and Plate Motions

By measuring the length of a hot spot track and the time interval between the formation of its oldest and youngest volcanic structures, an average rate of plate motion can be calculated. For example, that portion of the Hawaiian Island–Emperor Seamount chain that extends from Hawaii to Suiko Seamount is roughly 6000 kilometers in length and formed over the past 65 million years. Thus, the average rate of movement of the Pacific plate, relative to the mantle plume, was about 9 centimeters (4 inches) per year.

Hot spot tracks can also be useful when establishing the direction a plate is moving. Notice in Figure 2.29 that there is a bend in the Hawaiian Island–Emperor Seamount chain. This

FIGURE 2.30 Hot-spot volcanism, Kilauea Volcano, Hawaii. (Photo courtesy of the U.S. Geological Survey)

bend occurred about 50 million years ago when the motion of the Pacific plate changed from one that was nearly due north to a more northwesterly path. Similarly, hot spots found on the floor of the Atlantic have increased our understanding of the migration of landmasses following the breakup of Pangaea.

The existence of mantle plumes and their association with hot spots is well documented. Most mantle plumes are long-lived features that appear to maintain relatively fixed positions within the mantle. However, recent evidence has shown that some hot spots may slowly migrate. Preliminary results suggest that the Hawaiian hotspot may have migrated southward by as much as 20 degrees latitude. If this is the case, models of past plate motion that were based on a "fixed hot spot" frame of reference will need to be reevaluated.

Paleomagnetism and Plate Motions

Researchers have built a time scale for magnetic reversals going back nearly 200 million years, consequently the rate at which seafloor spreading occurs along various ridge segments can be accurately determined. In the Pacific Ocean, for example, the magnetic stripes are much wider for corresponding time intervals than those of the Atlantic Ocean. Hence, we conclude that the spreading centers found in the Pacific had faster rates of spreading than those that generated the Atlantic basin.

When we apply radiometric dates to these magnetic events, we find that the spreading rate in the North Atlantic is about 2 centimeters (1 inch) per year (Figure 2.31). The rate is only slightly faster in the South Atlantic. By contrast, spreading rates along the East Pacific Rise are mostly between 6 and 12 centime-

FIGURE 2.31 This map illustrates directions and rates of plate motion in centimeters per year. Seafloor-spreading velocities (shown with black arrows and labels) are based on the spacing of dated magnetic strips (anomalies). The red arrows show plate motion at selected locations based on GPS data. (Seafloor data from DeMets and others, GPS data from Jet Propulsion Laboratory)

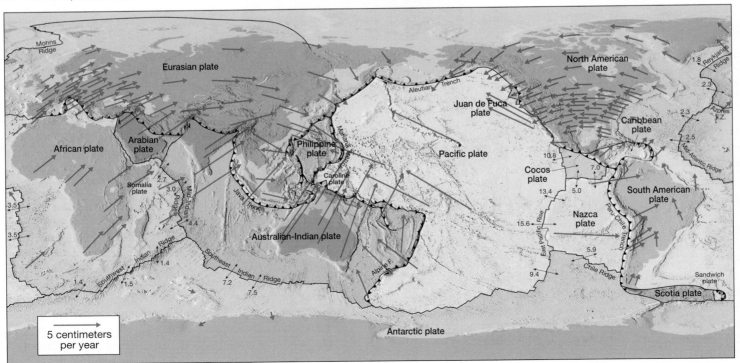

ters (2.5 to 5 inches) a year with a maximum rate of 20 centimeters (8 inches) a year at one location. Thus, the ocean floor acts like a magnetic tape recorder, recording changes in Earth's magnetic field that can be used to calculate rates of seafloor spreading.

In summary, data from the study of hot spot tracks, paleomagnetism, and other indirect techniques, have been useful in measuring the relative speeds at which two plates move toward, away from, or past one another—at least as averaged over millions of years.

Measuring Plate Motion from Space

Plates are not flat surfaces, instead they are curved sections of a sphere which greatly complicates how plate motion is described. In addition, plates usually exhibit some degree of rotational motion, which can cause two locations on the same plate to move at different speeds and in different directions. The latter fact can be illustrated by rotating your dinner plate in a clockwise matter. When doing so you will notice that the items on the left side of the plate move away from you (divergence) as the items on the right side move toward you (convergence). Items in the center will rotate, but their position relative to yours will not change. The complex nature of plate motion makes the task of describing plate motions more difficult than simply establishing the relative motion between two plates along the boundary that separates them. Fortunately, using space-age technology, researchers have recently been able to accurately calculate the absolute motion of hundreds of locations across the globe.

You may be familiar with the Global Positioning System (GPS), which is part of the navigation system used in automobiles to locate one's position and to provide directions to some other location. The Global Positioning System employs two dozen satellites that send radio signals that are intercepted by GPS receivers located at Earth's surface. The exact position of the receiver is determined by simultaneously establishing the distance from the receiver to four or more satellites. Researchers use specifically designed equipment that is able to locate the position of a point on Earth to within a few millimeters (about the diameter of a small pea). To establish plate motion, a particular site is surveyed repeatedly over a number of years.

Data obtained from these and other similar techniques are shown in Figure 2.31. Calculations show that Hawaii is moving

in a northwesterly direction and approaching Japan at 8.3 centimeters per year. A site located in Maryland is retreating from one in England at a speed of 1.7 centimeters per year—a value that is close to the 2.0-centimeters-per-year spreading rate that was established from paleomagnetic evidence. Techniques using GPS devices have also been useful in establishing small-scale crustal movements such as those that occur along faults in regions known to be tectonically active.

WHAT DRIVES PLATE MOTIONS?

The plate tectonics theory *describes* plate motion and the role that this motion plays in generating and modifying the major features of Earth's crust. Therefore, acceptance of plate tectonics does not rely on knowing precisely what drives plate motion. This is fortunate, because none of the models yet proposed can account for all major facets of plate tectonics.

Plate–Mantle Convection

From geophysical evidence, we have learned that although the mantle consists almost entirely of solid rock, it is hot and weak enough to exhibit fluid-like convective flow. The simplest type of convection is analogous to heating a pot of water on a stove (Figure 2.32). Heating the base causes the material to rise in relatively thin sheets or blobs that spread out at the surface and cool. Eventually, the surface layer thickens (increases in density) and sinks back to the bottom where it is reheated until it achieves enough buoyancy to rise again.

Mantle convection is considerably more complex than the model just described. The shape of the mantle does not resemble that of a cooking pot. Rather it is a spherically-shaped zone with a much larger upper boundary (Earth's surface) than lower boundary (core-mantle boundary). Furthermore, mantle convection is driven by a combination of three thermal processes: heating at the bottom by heat loss from Earth's core; heating from within by the decay of radioactive isotopes; and cooling from the top that creates thick, cold lithospheric slabs that sink into the mantle.

When seafloor spreading was first introduced, geologists proposed that the main driving force for plate motion was upwelling

FIGURE 2.32 Convection is a type of heat transfer that involves the actual movement of a substance. Here the stove warms the water in the bottom of a cooking pot. The heated water expands, becomes less dense (more buoyant), and rises. Simultaneously, the cooler, denser water near the top sinks.

Cooler water sinks

Warm water rises

that came from deep in the mantle. Upon reaching the base of the lithosphere, this flow was thought to spread laterally and drag the plates along. Thus, plates were viewed as being carried passively by convective flow in the mantle. Based on physical evidence, however, it became clear that upwelling beneath oceanic ridges is quite shallow and not related to deep circulation in the lower mantle. It is the horizontal movement of lithospheric plates away from the ridge that causes mantle upwelling, not the other way around. Thus, modern models have plates being an integral part of mantle convection and perhaps even its most active component.

Although convection in the mantle is still poorly understood, researchers generally agree on the following:

1. Convective flow in the rocky 2900-kilometer-thick mantle—in which warm, buoyant rock rises and cooler, denser material sinks—is the underlying driving force for plate movement.
2. Mantle convection and plate tectonics are part of the same system. Subducting oceanic plates drive the cold downward-moving portion of convective flow while shallow upwelling of hot rock along the oceanic ridge and buoyant mantle plumes are the upward-flowing arms of the convective mechanism.
3. Convective flow in the mantle is the primary mechanism for transporting heat away from Earth's interior to the surface where it is eventually radiated into space.

What is not known with any high degree of certainty is the exact structure of this convective flow. First, we will look at some of the forces that contribute to plate motion and then we will examine two models that have been proposed to describe plate–mantle convection.

Forces that Drive Plate Motion

There is general agreement that the subduction of cold, dense slabs of oceanic lithosphere is a major driving force of plate motion (Figure 2.33). As these slabs sink into the asthenosphere, they "pull" the trailing plate along. This phenomenon, called **slab pull,** occurs because cold slabs of oceanic lithosphere are more dense than the underlying asthenosphere and hence "sink like a rock."

Another important driving force is **ridge push** (see Figure 2.33). This gravity-driven mechanism results from the elevated position of the oceanic ridge, which causes slabs of lithosphere to "slide" down the flanks of the ridge. Ridge push appears to contribute far less to plate motions than slab pull. The primary evidence for this comes from comparing rates of seafloor spreading along ridge segments having different elevations. For example, despite its greater average height above the seafloor, spreading rates along the Mid-Atlantic

Ridge are considerably less than spreading rates along the less steep East Pacific Rise (Figure 2.31). In addition, fast-moving plates are being subducted along a larger percentage of their margins than slow-moving plates. This fact supports the notion that slab pull is a more significant driving force than ridge push. Examples of fast-moving plates which have extensive subduction zones along their margins include the Pacific, Nazca, and Cocos plates.

Although slab pull and ridge push appear to be the dominant forces acting on plates, they are not the only forces that influence plate motion. Beneath plates, convective flow in the mantle exerts a force, perhaps best described as "mantle drag" (see Figure 2.33). When flow in the asthenosphere is moving at a velocity that exceeds that of the plate, mantle drag enhances plate motion. However, if the asthenosphere is moving more slowly than the plate, or in the opposite direction, this force tends to resist plate motion. Another type of resistance to plate motion occurs along subduction zones. Here friction between the overriding plate and the descending slab generates significant earthquake activity.

Models of Plate–Mantle Convection

Any acceptable model for plate–mantle convection must explain compositional variations known to exist in the mantle. For example, the basaltic lavas that erupt along oceanic ridges, as well as those that are generated by hot-spot volcanism, such as those found in Hawaii, have mantle sources. Yet ocean ridge basalts are very uniform in composition and depleted in certain elements. Hot-spot eruptions, on the other hand, have high concentrations of these elements and tend to have varied compositions. Because basaltic lavas that arise from different tectonic settings have different compositions, they are assumed to be derived from chemically distinct mantle reservoirs.

FIGURE 2.33 Illustration of some of the forces that act on tectonic plates.

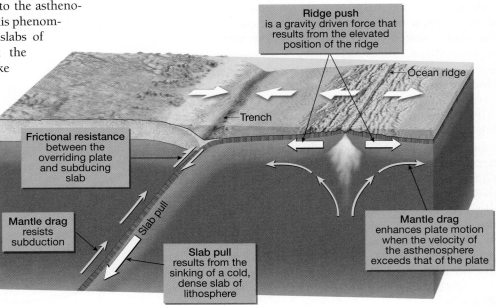

Ridge push is a gravity driven force that results from the elevated position of the ridge

Ocean ridge

Trench

Frictional resistance between the overriding plate and subducing slab

Mantle drag resists subduction

Slab pull

Slab pull results from the sinking of a cold, dense slab of lithosphere

Mantle drag enhances plate motion when the velocity of the asthenosphere exceeds that of the plate

Layering at 660 Kilometers Some researchers argue that the mantle resembles a "giant layer cake" divided at a depth of 660 kilometers. As shown in Figure 2.34A, this layered model has two zones of convection—a thin, dynamic layer in the upper mantle and a thick, sluggish one located below. This model successfully explains why basaltic lavas that erupt along the oceanic ridges have a different chemical make-up than those that erupt in Hawaii as a result of hot-spot activity. The mid-ocean ridge basalts come from the upper convective layer, which is well mixed, whereas the mantle plume that feeds the Hawaiian volcanoes taps a deeper, more primitive magma source that resides in the lower convective layer.

However, data gathered from the study of earthquake waves have shown that at least some subducting oceanic slabs penetrate the 660-kilometer boundary and descend deep into the mantle. The subducting lithosphere should serve to mix the upper and lower layers together, thereby destroying the layered structure proposed in this model.

Whole-Mantle Convection Other researchers favor some type of *whole-mantle convection* in which cold oceanic lithosphere sinks to great depths and stirs the entire mantle (Figure 2.34B). One whole-mantle model suggests that the burial ground for subducting slabs is the core-mantle boundary. Over time, this material melts and buoyantly rises toward the surface as a mantle plume, thereby transporting hot material toward the surface (see Figure 2.34B).

Recent work has predicted that whole-mantle convection would cause the entire mantle to completely mix in a matter of a few hundred million years. This, in turn, would eliminate chemically distinct magma sources—those that are observed in hot-spot volcanism and those associated with volcanic activity along oceanic ridges. Thus, the whole-mantle model also has shortcomings.

Although there is still much to be learned about the mechanisms that cause Earth's tectonic plates to migrate across the globe, one thing is clear. The unequal distribution of heat in Earth's interior generates some type of thermal convection that ultimately drives plate–mantle motion.

THE IMPORTANCE OF THE PLATE TECTONICS THEORY

Plate tectonics is the first theory to provide a comprehensive view of the processes that produced Earth's major surface features, including the continents and ocean basins. As such, it has provided a better understanding of the workings of our dynamic planet by linking many aspects of geology that previously had been considered unrelated. Within the framework of plate tectonics, geologists have found explanations for the geologic distribution of earthquakes, volcanoes, and mountain belts. Further, we are now better able to explain the distributions of plants and animals in the geologic past, as well as to understand why economically significant mineral deposits form in certain tectonic environments.

Despite its usefulness in explaining many of the large-scale geologic processes operating on Earth, plate tectonics is not completely understood. The model that was set forth in 1968 was only a basic framework that left much of the detail to future research. Through critical testing, the initial model has been modified and expanded to become the theory we know today. In a like manner, the current theory will be refined as additional data and observations are obtained. The theory of plate tectonics, although a powerful tool, is nonetheless an evolving model of Earth's dynamic processes.

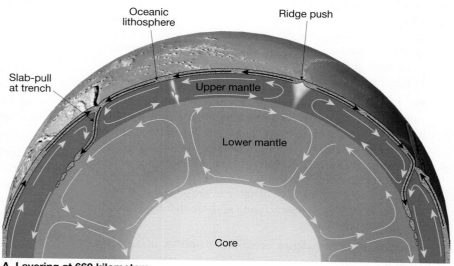

A. Layering at 660 kilometers

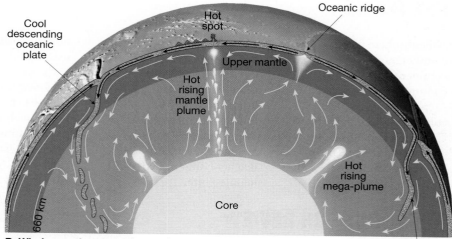

B. Whole mantle convection

FIGURE 2.34 Proposed models for mantle convection. **A.** The "layer cake" model consists of two convection layers—a thin, convective layer above 660 kilometers and a thick one below. **B.** In this whole-mantle convection model, cold oceanic lithosphere descends into the lowermost mantle, while hot mantle plumes transport heat toward the surface.

CHAPTER 2 PLATE TECTONICS IN REVIEW

- In the early 1900s *Alfred Wegener* set forth his *continental drift* hypothesis. One of its major tenets was that a supercontinent called *Pangaea* began breaking apart about 200 million years ago. The rifted continental fragments then "drifted" to their present positions. To support his hypothesis, Wegener used the *fit of South America and Africa, fossil evidence, rock types and structures,* and *ancient climates.* One of the main objections to the continental drift hypothesis was its inability to provide an acceptable mechanism for the movement of continents.

- From the study of *paleomagnetism,* researchers learned that the continents had indeed wandered as Wegener proposed.

- In 1962, Harry Hess formulated the idea of *seafloor spreading,* which states that new seafloor is continually being generated at mid-ocean ridges and old, dense seafloor is being consumed at the deep-ocean trenches. Support for seafloor spreading came with the discovery of alternating stripes of high- and low-intensity magnetism that parallel the ridge crests.

- By 1968, continental drift and seafloor spreading were united into a far more encompassing theory known as *plate tectonics.* According to plate tectonics, Earth's rigid outer layer (*lithosphere*) overlies a weaker region called the *asthenosphere.* Further, the lithosphere is broken into several large and numerous smaller segments, called *plates,* that are in motion and continually changing in shape and size. Plates move as relatively coherent units and are deformed mainly along their boundaries.

- *Divergent plate boundaries* occur where plates move apart, resulting in upwelling of material from the mantle to create new seafloor. Most divergent boundaries occur along the axis of the oceanic ridge system and are associated with seafloor spreading. New divergent boundaries may form within a continent (for example, the East African Rift Valleys), where they may fragment a landmass and develop a new ocean basin.

- *Convergent plate boundaries* occur where plates move together, resulting in the subduction of oceanic lithosphere into the mantle along a deep-ocean trench. Convergence of an oceanic and continental block results in subduction of the oceanic slab and the formation of a *continental volcanic arc* such as the Andes of South America. Oceanic–oceanic convergence results in an arc-shaped chain of volcanic islands called a *volcanic island arc.* When two plates carrying continental crust converge, the buoyant continental blocks collide, resulting in the formation of a mountain belt as exemplified by the Himalayas.

- *Transform fault boundaries* occur where plates grind past each other without the production or destruction of lithosphere. Most transform faults join two segments of a mid-ocean ridge where they provide the means by which oceanic crust created at a ridge crest can be transported to its site of destruction—a deep-ocean trench. Still others, like the San Andreas Fault, cut through continental crust.

- Convective flow in the rocky mantle—in which hot, buoyant rock rises and cooler, denser materials sink—is the underlying driving force for plate tectonics. Mechanisms that contribute to plate motion are slab pull, ridge push, and slab suction. *Slab pull* occurs where cold, dense oceanic lithosphere is subducted and pulls the trailing lithosphere along. *Ridge push* is driven by gravitational forces that set the elevated slabs astride oceanic ridges in motion. *Slab suction* is a force that arises from the drag of a subducting slab on the adjacent mantle.

KEY TERMS

asthenosphere (p. 51)
continental drift (p. 40)
continental rift (p. 56)
continental volcanic arc
  (p. 57)
convergent boundary (p. 56)
Curie point (p. 46)
deep-ocean trench (p. 56)
divergent boundary (p. 55)
fossil magnetism (p. 46)

fracture zone (p. 60)
hot spot (p. 64)
hot spot track (p. 64)
island arc (p. 58)
lithosphere (p. 51)
lithospheric plate (p. 51)
magnetometer (p. 50)
magnetic reversal (p. 49)
magnetic time scale (p. 49)
mantle plume (p. 64)

normal polarity (p. 49)
oceanic ridge system
  (p. 48)
paleomagnetism (p. 46)
Pangaea (p. 40)
partial melting (p. 57)
plate tectonics (p. 51)
reverse polarity (p. 49)
ridge push (p. 68)
rift, or rift valley (p. 55)

seafloor spreading (p. 48)
slab pull (p. 68)
spreading center (p. 55)
subduction zone (p. 56)
supercontinent (p. 40)
tectonic plate (p. 51)
transform fault boundary
  (p. 60)
volcanic island arc (p. 58)

QUESTIONS FOR REVIEW

1. Who is credited with developing the continental drift hypothesis?

2. What was the first line of evidence that led early investigators to suspect that the continents were once connected?

3. What was Pangaea?

4. Describe the four kinds of evidence that Wegener and his supporters gathered to substantiate the continental drift hypothesis.

5. Explain why the discovery of the fossil remains of *Mesosaurus* in both South America and Africa, but nowhere else, supports the continental drift hypothesis.

6. Early in the 20th century, what was the prevailing view of how land animals migrated across vast expanses of open ocean?

7. How did Wegener account for the existence of glaciers in the southern landmasses at a time when areas in North America, Europe, and Asia supported lush tropical swamps?

8. To which two aspects of Wegener's continental drift hypothesis did most Earth scientists object?

9. What is meant by seafloor spreading? Who is credited with formulating this important concept? Where is active seafloor spreading occurring today?

10. Describe how Fred Vine and D. H. Matthews related the seafloor-spreading hypothesis to magnetic reversals.

11. Compare and contrast the lithosphere and the asthenosphere.

12. List the three types of plate boundaries and describe the relative motion at each of them.

13. Where does new lithosphere form? Where is it consumed? Why does the rate at which new lithosphere forms roughly balance the rate at which lithosphere is destroyed?

14. Briefly describe the process of continental rifting. Where is it occurring today?

15. Compare a continental volcanic arc and a volcanic island arc.

16. Why does oceanic lithosphere subduct while continental lithosphere does not?

17. Briefly describe how mountain systems like the Himalayas are formed.

18. Differentiate between transform faults and the two other types of plate boundaries.

19. Some people predict that some day California will sink into the ocean. Is this idea consistent with the theory of plate tectonics?

20. With which of the three types of plate boundaries are the following places or features associated: Himalayas, Aleutian Islands, Red Sea, Andes Mountains, San Andreas Fault, Iceland, Japan, Mount St. Helens?

21. What is the age of the oldest sediments recovered by deep-ocean drilling? How do the ages of these sediments compare to the ages of the oldest continental rocks?

22. Assuming hot spots remain fixed, in what direction was the Pacific plate moving while the Hawaiian Islands were forming? When Suiko Seamount was forming? (See Figure 2.29, p. 65)

23. Describe slab pull and ridge push. Which of these forces appears to contribute the most to plate motion?

24. What role are mantle plumes thought to play in the convective flow of the mantle?

25. Briefly describe the two models proposed for mantle–plate convection. What is lacking in each of these models?

COMPANION WEBSITE

The *Earth 10e* website uses the resources and flexibility of the Internet to aid in your study of the topics in this chapter. Written and developed by the authors and other geology instructors, this site will help improve your understanding of geology. Visit www.mygeoscienceplace.com in order to:

• **Review** key chapter concepts.

• **Read** with links to the eBook and to chapter-specific web resources.

• **Visualize** and comprehend challenging topics using learning activities in *GEODe Earth*.

• **Test** yourself with online quizzes.

GEODe EARTH

GEODe Earth is a valuable and easy to use learning aid that can be accessed from your book's Companion Website (www.mygeoscienceplace.com). It is a dynamic instructional tool that promotes understanding and reinforces important concepts by using tutorials, animations, and exercises that actively engage the student.

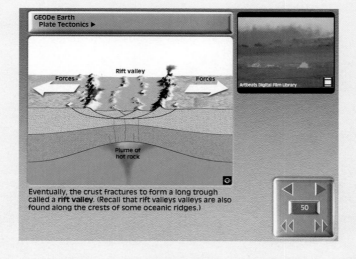

GEODe Earth
Plate Tectonics ▶

Rift valley

Forces   Forces

Plume of
hot rock

Artbeats Digital Film Library

Eventually, the crust fractures to form a long trough called a **rift valley**. (Recall that rift valleys valleys are also found along the crests of some oceanic ridges.)

50

MATTER AND MINERALS*

Crystal shape or habit is a basic mineral property.

(PHOTO BY JEFFREY SCOVIL)

*Professor Mark Watry, Spring Hill College, assisted in the revision of this chapter.

MINERALS: BUILDING BLOCKS OF ROCKS

GEODe EARTH

MATTER AND MINERALS
▶ Introduction

We begin our discussion of Earth materials with an overview of **mineralogy** (*mineral* = mineral, *ology* = the study of), because minerals are the building blocks of rocks. In addition, minerals have been employed by humans for both useful and decorative purposes for thousands of years (Figure 3.1). The first minerals mined were flint and chert, which people fashioned into weapons and cutting tools. As early as 3700 BC, Egyptians began mining gold, silver, and copper; and by 2200 BC humans discovered how to combine copper with tin to make bronze, a tough, hard alloy. The Bronze Age began its decline when the ability to extract iron from minerals such as hematite was discovered. By about 800 BC, iron-working technology had advanced to the point that weapons and many everyday objects were made of iron rather than copper, bronze, or wood. During the Middle Ages, mining of a variety of minerals was common throughout Europe and the impetus for the formal study of minerals was in place.

The term *mineral* is used in several different ways. For example, those concerned with health and fitness extol the benefits of vitamins and minerals. The mining industry typically uses the word when referring to anything taken out of the ground, such as coal, iron ore, or sand and gravel. The guessing game known as *Twenty Questions* usually begins with the question, *Is it animal, vegetable, or mineral?* What criteria do geologists use to determine whether something is a mineral?

Geologists define a **mineral** as *any naturally occurring, inorganic solid that possesses an orderly crystalline structure and a well-defined chemical composition.* Thus, those Earth materials that are classified as minerals exhibit the following characteristics:

1. **Naturally occurring.** Minerals form by natural, geologic processes. Consequently, synthetic diamonds and rubies, as well as a variety of other useful materials produced in a laboratory, are not considered minerals.
2. **Solid substance.** Minerals are solids within the temperature ranges normally experienced at Earth's surface. Thus, ice (frozen water) is considered a mineral, whereas liquid water and water vapor are not.
3. **Orderly crystalline structure.** Minerals are crystalline substances, which means their atoms are arranged in an orderly, repetitive manner. This orderly packing of atoms is reflected in the regularly shaped objects we call crystals (see Figure 3.1). Some naturally

---

Earth's crust and oceans are the source of a wide variety of useful and essential minerals. Most people are familiar with the common uses of many basic metals, including aluminum in beverage cans, copper in electrical wiring, and gold and silver in jewelry. But some people are not aware that pencil lead contains the greasy-feeling mineral graphite and that bath powders and many cosmetics contain the mineral talc. Moreover, many do not know that drill bits impregnated with diamonds are employed by dentists to drill through tooth enamel, or that the common mineral quartz is the source of silicon for computer chips. In fact, practically every manufactured product contains materials obtained from minerals. As the mineral requirements of modern society grow, the need to locate additional supplies of useful minerals also grows, becoming more challenging as well.

In addition to the economic uses of rocks and minerals, all of the processes studied by geologists are in some way dependent on the properties of these basic Earth materials. Events such as volcanic eruptions, mountain building, weathering and erosion, and even earthquakes involve rocks and minerals. Consequently, a basic knowledge of Earth materials is essential to an understanding of all geologic phenomena.

occurring solids, such as volcanic glass (obsidian), lack a repetitive atomic structure and are not considered minerals.

4. **Well-defined chemical composition.** Most minerals are chemical compounds having compositions that are given by their chemical formulas. The mineral pyrite (FeS_2), for example, consists of one iron (Fe) atom, for every two sulfur (S) atoms. In nature, however, it is common for some atoms within a crystal structure to be replaced by others of similar size without changing the internal structure or properties of that mineral. Therefore, the chemical compositions of minerals may vary, but they *vary within specific, well-defined limits.*

5. **Generally inorganic.** Inorganic crystalline solids, as exemplified by ordinary table salt (halite), that are found naturally in the ground are considered minerals. Organic compounds, on the other hand, are generally not. Sugar, a crystalline solid like salt but which comes from sugarcane or sugar beets, is a common example of such an organic compound. However, many marine animals secrete inorganic compounds, such as calcium carbonate (calcite), in the form of shells and coral reefs. If these materials are buried and become part of the rock record, they are considered minerals by geologists.

In contrast to minerals, rocks are more loosely defined. Simply, a **rock** is any solid mass of mineral, or mineral-like matter that occurs naturally as part of our planet. A few rocks are composed almost entirely of one mineral. A common example is the sedimentary rock *limestone,* which consists of impure masses of the mineral calcite. However, most rocks, like the common rock granite shown in Figure 3.2, occur as aggregates of several kinds of minerals. Here, the term *aggregate* implies that the minerals are joined in such a way that the properties of each mineral are retained. Note that you can easily identify the mineral constituents of the granite in Figure 3.2.

A few rocks are composed of nonmineral matter. These include the volcanic rocks *obsidian* and *pumice,* which are noncrystalline glassy substances, and *coal,* which consists of solid organic debris (Box 3.1).

Although this chapter deals primarily with the nature of minerals, keep in mind that most rocks are simply aggregates of minerals. Because the properties of rocks are determined largely by the chemical composition

FIGURE 3.1 Collection of well-developed quartz crystals found near Hot Springs, Arkansas. (Photo by Jeff Scovil)

and crystalline structure of the minerals contained within them, we will first consider these Earth materials. Then, in Chapters 4, 7, and 8 we will take a closer look at Earth's major rock groups.

FIGURE 3.2 Most rocks are aggregates of two or more minerals. Shown here is a hand sample of the igneous rock granite and three of its major constituent minerals. (Photos by E. J. Tarbuck)

PEOPLE AND THE ENVIRONMENT

BOX 3.1

Making Glass from Minerals

Many everyday objects are made of glass, including windowpanes, jars and bottles, and the lenses of some eyeglasses. People have been making glass for at least 2000 years. Glass is manufactured by melting naturally occurring materials and cooling the liquid quickly before the atoms have time to arrange themselves into an orderly crystalline form. (This is the same way that natural glass, called *obsidian,* is generated from lava.)

It is possible to produce glass from a variety of materials, but the primary ingredient (75 percent) of most commercially produced glass is the mineral quartz (SiO_2). Lesser amounts of the minerals calcite (calcium carbonate) and trona (sodium carbonate) are added to the mix. These materials lower the melting temperature and improve the workability of the molten glass.

In the United States, high-quality quartz (usually quartz sandstone) and calcite (limestone) are readily available in many areas. Trona, on the other hand, is mined almost exclusively in the Green River area of southwestern Wyoming. In addition to its use in making glass, trona is used in making detergents, paper, and even baking soda.

Manufacturers can change the properties of glass by adding minor amounts of several other ingredients (Figure 3.A). Coloring agents include iron sulfide (amber), selenium (pink), cobalt oxide (blue), and iron oxides (green, yel-

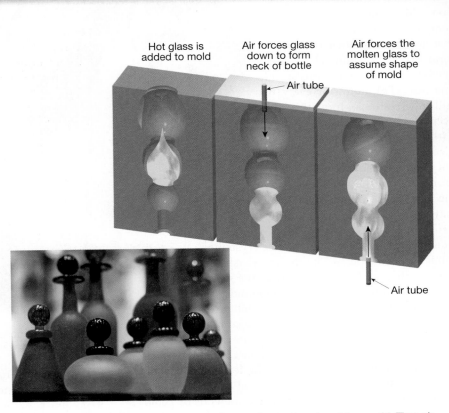

FIGURE 3.A Glass bottles are made by first pouring molten glass into a mold. Then air forces the molten glass to assume the shape of the mold. Metallic compounds are mixed with the raw ingredients to color the glass. (Photo by Cosmo Condina/Stock Connection)

low, brown). The addition of lead imparts clarity and brilliance to glass and is therefore used in the manufacture of fine crystal tableware.

Ovenware, such as Pyrex, owes its heat resistance to boron, whereas aluminum makes glass resistant to weathering.

Students Sometimes Ask . . .

Are the minerals you talked about in class the same as those found in dietary supplements?

Not ordinarily. From a geologic perspective, a mineral must be a *naturally occurring* crystalline solid. Minerals found in dietary supplements are man-made inorganic compounds that contain *elements* needed to sustain life. These dietary minerals typically contain elements that are metals—calcium, potassium, phosphorus, magnesium, and iron—as well as trace amounts of a dozen other elements, such as copper, nickel, and vanadium. Although these two types of "minerals" are different, they are related. The source of the elements used to make dietary supplements is, in fact, the naturally occurring minerals of Earth's crust. It should also be noted that vitamins are *organic compounds* produced by living organisms, not *inorganic compounds,* such as minerals.

ATOMS: BUILDING BLOCKS OF MINERALS

When minerals are carefully examined, even under optical microscopes, the innumerable tiny particles of their internal structures are not discernable. Nevertheless, minerals, as with all matter, are composed of minute building blocks called **atoms**—the smallest particles that cannot be chemically split. Atoms, in turn, contain even smaller particles—*protons* and *neutrons* located in a central **nucleus** that is surrounded by *electrons* (Figure 3.3).

Properties of Protons, Neutrons, and Electrons

Protons and **neutrons** are very dense particles with almost identical masses. By contrast, **electrons** have a negligible mass, about 1/2000th that of a proton. For comparison, if a proton had the

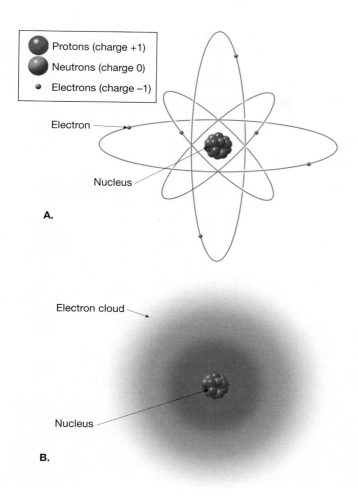

A.

B.

FIGURE 3.3   Two models of the atom. **A.** A very simplified view of the atom. The central nucleus, consists of protons and neutrons, encircled by high-speed electrons. **B.** This model of the atom shows spherically shaped electron clouds (principal shells) surrounding a central nucleus. Note that these models are not drawn to scale. Electrons are minuscule in size compared to protons and neutrons, and the relative space between the nucleus and electron shells is much greater than illustrated.

mass of a baseball an electron would have the mass of a single grain of rice.

Both protons and electrons share a fundamental property called *electrical charge*. Protons have an electrical charge of +1 and electrons have a charge of −1. Neutrons, as the name suggests, have no charge. The charge of protons and electrons are equal in magnitude, but opposite in polarity, so that when these two particles are paired, the charges cancel each other. Since matter typically contains equal numbers of positively charged protons and negatively charged electrons, most substances are electrically neutral.

In illustrations, electrons are sometimes shown orbiting the nucleus in a manner that resembles the planets of our solar system orbiting the Sun. However, electrons do not actually behave this way. A more realistic depiction shows electrons as a cloud of negative charges surrounding the nucleus (Figure 3.3B). Studies of the arrangements of electrons show that they move about the

nucleus in regions called **principal shells**, each with an associated energy level. In addition, each shell can hold a specific number of electrons, with the outermost shell containing the **valence electrons**. These are the electrons that interact with other atoms to form chemical bonds.

Most of the atoms in the universe (except hydrogen and helium) were created inside massive stars by nuclear fusion and released into interstellar space during hot, fiery supernova explosions that mark the deaths of many stars. As this material cooled, the newly formed nuclei attracted electrons to complete their atomic structures. At temperatures below 600 °C, all free atoms have a full complement of electrons—one for each proton in the nucleus.

Elements: Defined by Their Number of Protons

The simplest atoms have only one proton in their nucleus whereas others have more than 100. The number of protons in the nucleus of an atom, called the **atomic number,** determines the chemical nature of an atom. All atoms with the same number of protons have the same chemical and physical properties. Together, a group of the same kind of atoms is called an **element**. There are about 90 naturally occurring elements and 23 that have been synthesized.

You are probably familiar with the names of many elements including copper, iron, oxygen, and carbon. Since the number of protons identifies an element, all carbon atoms have 6 protons and 6 electrons. Likewise, any atom that contains 8 protons is an oxygen atom.

Elements can be organized so that those with similar properties line up in columns. This arrangement, called the **periodic table**, is shown in Figure 3.4. Each element is given a one or two letter symbol. The atomic numbers and masses are also included for each element.

Atoms of the naturally occurring elements are the basic building blocks of Earth's minerals. A few minerals, such as native copper, diamonds, and gold, are made entirely of atoms of only one element (Figure 3.5). However, most elements tend to join with atoms of other elements to form **chemical compounds**. As a result, most minerals are chemical compounds consisting of atoms with two or more elements.

WHY ATOMS BOND

Except for a group of elements known as the noble gases, atoms bond to one another under the conditions (temperatures and pressures) that occur on Earth.* Some atoms bond to form *ionic compounds*, some form *molecules*, and still others form *metallic substances*. Why does this happen? Experiments show that the forces holding atoms together and the forces that bond atoms to each other are electrical. Furthermore, these electrical attractions

---

*Examples of noble gases are helium, neon, and argon. In each case, their outer shell of valence electrons is considered to be full and therefore they have little tendency to participate in chemical reactions.

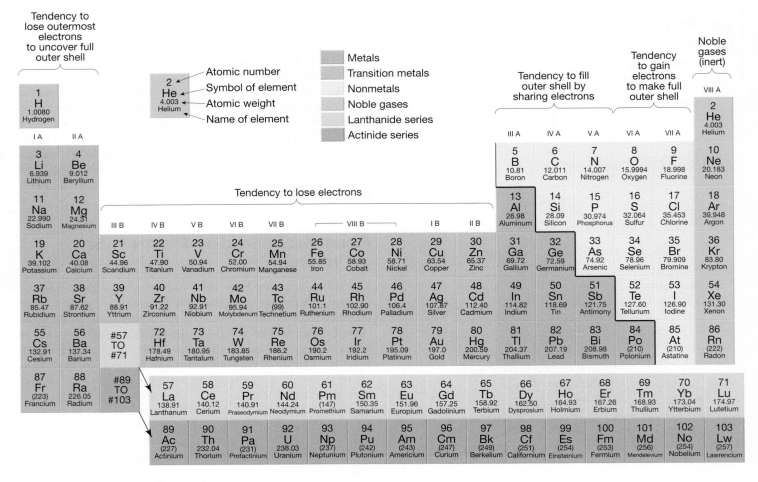

FIGURE 3.4   Periodic table of the elements.

FIGURE 3.5   Gold mixed with quartz. Gold, silver, copper, and diamonds are naturally occurring minerals composed entirely of atoms of a single element. (Photo by Dennis Tasa)

lower the total energy of the bonded atoms. Lower energy states are generally more stable. Consequently, atoms that are bonded in a compound tend to be in a more stable state than atoms that are free (not bonded).

As was noted earlier, valence electrons are the ones generally involved in chemical bonding. Figure 3.6 shows a shorthand way of representing the number of valence (outer shell) electrons. Notice that the elements in Group I have one valence electron, those in Group II have two valence electrons and so on up to eight valence electrons in Group VIII.

The first shell can only hold two electrons. In the second and higher shells, a stable configuration occurs when the valence shell contains eight electrons. The noble gases such as neon and argon have eight electrons in their outer shell and tend not to react. They remain atomic gases well below room temperature.

When an atom's outer shell does not contain eight electrons, it is likely to chemically bond to other atoms to fill its shell. A **chemical bond** is the transfer or sharing of electrons that allows each atom to attain a full valence shell of electrons. Some atoms do this by transferring all of their valence electrons to other atoms so that an inner shell becomes the full valence shell.

Electron Dot Diagrams for Some Representative Elements							
I	II	III	IV	V	VI	VII	VIII
H·							He:
Li·	·Be·	·B·	·C·	·N·	:O·	:F·	:Ne:
Na·	·Mg·	·Al·	·Si·	·P·	:S·	:Cl·	:Ar:
K·	·Ca·	·Ga·	·Ge·	·As·	:Se·	:Br·	:Kr:

FIGURE 3.6   Dot diagrams for some representative elements. Each dot represents a valence electron found in the outermost principal shell.

When valence electrons are transferred between elements to form ions, the bond is an *ionic bond*. When the electrons are shared between atoms, the bond is a *covalent bond*. When valence electrons are shared among all the atoms in a substance, the bonding is *metallic*. In any case, the bonding atoms get stable electron configurations which usually consist of eight electrons in their outermost shells.

Ionic Bonds: Electrons Transferred

Perhaps the easiest type of bond to visualize is the *ionic bond* in which one atom gives up one or more of its valence electrons to another atom to form **ions**—*positively and negatively charged atoms*. The atom that loses electrons becomes a positive ion, and the atom that gains electrons becomes a negative ion. Oppositely charged ions are strongly attracted to one another and join to form ionic compounds.

Consider the ionic bonding that occurs between sodium (Na) and chlorine (Cl) to produce sodium chloride, the mineral halite—common table salt. Notice in Figure 3.7A that sodium gives up its single valence electron to chlorine. As a result, sodium now has a stable configuration with eight electrons in its outermost shell. By acquiring the electron that sodium loses, chlorine—which has seven valence electrons—gains the eighth electron needed to complete its outermost shell. Thus, through the transfer of a single electron, both the sodium and chlorine atoms have acquired a stable electron configuration.

After electron transfer takes place, the atoms are no longer electrically neutral. By giving up one electron, a neutral sodium atom becomes positively charged (with 11

protons and 10 electrons). Similarly, by acquiring one electron, a neutral chlorine atom becomes negatively charged (with 17 protons and 18 electrons). We know that ions with like charges repel, and those with unlike charges attract. Thus, an **ionic bond** is the attraction of oppositely charged ions to one another, producing an electrically neutral compound.

Figure 3.7B illustrates the arrangement of sodium and chlorine ions in ordinary table salt. Notice that salt consists of alternating sodium and chlorine ions, positioned in such a manner that each positive ion is attracted to and surrounded on all sides by negative ions, and vice versa. This arrangement maximizes the attraction between ions with opposite charges while minimizing the repulsion between ions with identical charges. Thus, ionic compounds consist of an orderly arrangement of oppositely charged ions assembled in a definite ratio that provides overall electrical neutrality.

The properties of a chemical compound are dramatically different from the properties of the elements comprising it. For example, sodium is a soft silvery metal that is extremely reactive and poisonous. If you were to consume even a small amount of elemental sodium, you would need immediate medical attention. Chlorine, a green poisonous gas, is so toxic it was used as a chemical weapon during World War I. Together, however,

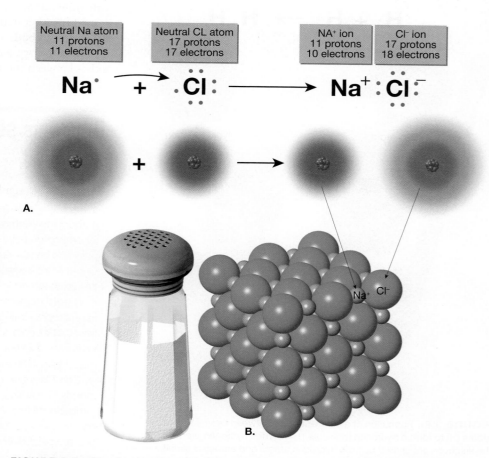

FIGURE 3.7   Chemical bonding of sodium chloride (table salt). **A.** Through the transfer of one electron in the outer shell of a sodium atom to a chlorine atom, sodium becomes a positive ion (cation) and chlorine a negative ion (anion). **B.** Diagram illustrating the arrangement (packing) of sodium and chlorine ions in table salt.

these elements produce sodium chloride, a harmless flavor enhancer that we call table salt. When elements combine to form compounds their properties change dramatically.

Covalent Bonds: Electrons Shared

Sometimes the forces that hold atoms together cannot be understood on the basis of the attraction of oppositely charged ions. One example is the hydrogen molecule (H_2), in which two hydrogen atoms are held together tightly and no ions are present. The strong attractive force that holds two hydrogen atoms together results from a **covalent bond**, *a chemical bond formed by the sharing of a pair of electrons between atoms.*

Imagine two hydrogen atoms (each with one proton and one electron) approaching one another so that their orbitals overlap (Figure 3.8). Once they meet, the electron configuration will change so that both electrons will primarily occupy the space between the atoms. In other words, the two electrons are shared by both hydrogen atoms and attracted simultaneously by the positive charge of the proton in the nucleus of each atom (Figure 3.8). It is the attraction between the electrons and both nuclei that is the force that holds these atoms together. Although ions do not exist in hydrogen molecules, the force that holds these atoms together

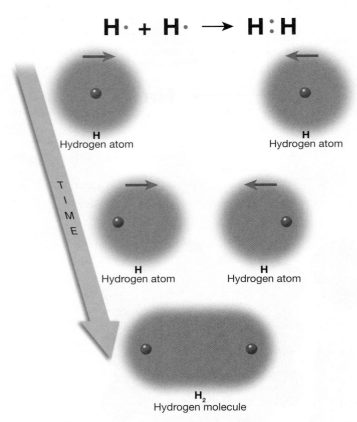

FIGURE 3.8 Formation of a covalent bond between two hydrogen atoms (H) to form a hydrogen molecule (H_2). When hydrogen atoms bond, the electrons are shared by both hydrogen atoms and attracted simultaneously by the positive charge of the proton in the nucleus of each atom. It is the attraction between electrons and both nuclei that is the force that holds (bonds) these atoms together.

arises from the attraction of oppositely charged particles—protons in the nuclei and electrons shared by the atoms.

Other gases that are composed of diatomic molecules include oxygen (O_2), nitrogen (N_2), and chlorine (Cl_2). In addition, the most common mineral group, the silicates, contains silicon atoms covalently bonded to oxygen atoms.

Metallic Bonds: Electrons Free to Move

In **metallic bonds**, the valence electrons are free to move from one atom to another so that all atoms share the available valence electrons. This type of bonding is found in metals such as copper, gold, aluminum, and silver, and in alloys such as brass and bronze. Metallic bonding accounts for the high electrical conductivity of metals, the ease with which metals are shaped, and numerous other special properties of metals.

Other Bonds: Hybrids

We have described the extremes in chemical bonding—complete transfer of electrons and perfect sharing of electrons. As you may suspect, many chemical bonds are actually hybrids that exhibit some degree of electron sharing and some degree of electron transfer. Bonds can be found with almost every possible combination of covalent and ionic character. For example, silicate minerals are composed of silicon and oxygen atoms that are joined together with other elements by bonds that display characteristics of both ionic and covalent bonds.

In summary, a *chemical bond* is a strong attractive force that exists between the atoms in a substance.

Students Sometimes Ask . . .

What's the difference between a carrot, a karat, and a carat?

Like many words in English, *carrot, karat,* and *carat* have the same sound but have different meanings. Such words are called *homonyms* (*homo* = *same, nym* = *name*). A *carrot* is the familiar orange crunchy vegetable, and *karat* and *carat* have to do with gold and gems, respectively. (Both *karat* and *carat* derive from the Greek word *keration* = carob bean, which early Greeks used as a weight standard.)

Karat is a term used to indicate the purity of gold, with 24 karats representing pure gold. Gold less than 24 karats is an alloy (mixture) of gold and another metal, usually copper or silver. For example, 14-karat gold contains 14 parts of gold (by weight) mixed with 10 parts of other metals.

Carat is a unit of weight used for precious gems such as diamonds, emeralds, and rubies. The size of a carat has varied throughout history, but early in the twentieth century it was standardized at 200 milligrams (0.2 gram, or 0.007 ounce). For example, a typical diamond on an engagement ring might range from one half to one carat, and the famous Hope Diamond at the Smithsonian Institution is 45.52 carats.

ISOTOPES AND RADIOACTIVE DECAY

Atoms of the same element can have different masses even though they have the same number of protons and electrons—these atoms are called *isotopes*. In addition, the nuclei of some atoms are unstable and change to become other isotopes of the same element or atoms of another element entirely, a process called *radioactive decay*.

The **mass number** of an atom is simply the total number of its protons and neutrons. Atoms of the same element always have the same number of protons, but the number of neutrons for atoms of the same element can vary. Atoms with the same number of protons, but different numbers of neutrons are **isotopes** of that element. Isotopes of the same element are labeled by placing the mass number after the element's name or symbol. For example, carbon has three well-known isotopes. One has a mass number of 12 (carbon-12), another has a mass number of 13 (carbon-13), and the third, carbon-14, has a mass number of 14. Since all atoms of the same element have the same number of protons, and carbon has six, carbon-12 also has six neutrons to give it a mass number of 12. Carbon-14, on the other hand, has six protons plus eight neutrons to give it a mass number of 14. In chemical behavior, all isotopes of the same element are nearly identical. To distinguish among them is like trying to differentiate identical twins, with one being slightly heavier. Because isotopes of an element behave the same chemically, different isotopes can become parts of the same mineral. For example, when the mineral calcite forms from calcium, carbon, and oxygen, some of its carbon atoms are carbon-12 and some are carbon-14.

The nuclei of most atoms are stable. However, many elements do have isotopes in which the nuclei are unstable (carbon-14 is one example of an unstable isotope). In this context, *unstable* means that the nuclei change through a random process called **radioactive decay**. During radioactive decay, unstable isotopes radiate energy and emit particles. Some of the energy from radioactive decay in Earth's interior contributes to the movements of lithospheric plates. The rates at which unstable isotopes decay are measurable. Therefore, certain radioactive atoms are used to determine the ages of fossils, rocks, and minerals. A discussion of radioactive decay and its applications in dating past geologic events appears in Chapter 9.

CRYSTALS AND CRYSTALLIZATION

Many people associate the word crystal with delicate wine goblets or glassy objects with smooth sides and gem-like shapes. Mineralogists, on the other hand, use the term **crystal** or **crystalline** in reference to *any natural solid with an ordered, repetitive, atomic structure.*

By this definition, crystals may or may not have smooth-sided faces. The specimen shown in Figure 3.1, for example, exhibits the characteristic crystal form associated with the mineral quartz—a six-sided prismatic shape with pyramidal ends. However, the quartz crystals in the sample of granite shown if Figure 3.2 do not display well-defined faces. All minerals are crystalline, but samples with well-developed faces are relatively uncommon. If you picked up one thousand rocks, odds are that none would contain a crystal with well defined faces.

How Do Minerals Form?

Minerals form through the process of **crystallization**, in which molecules and/or ions chemically bond to form an orderly internal structure. Crystallization can occur in many ways and under many different conditions. We will examine three broad classes of crystallization: crystallization of salts from solutions in water; crystallization involving temperature change; and crystallization caused by biological processes.

Perhaps the most familiar type of crystallization results when an aqueous (water) solution containing dissolved ions evaporates. As you might expect, the concentration of dissolved material increases as water is removed through evaporation until there is just enough water to dissolve the material—the solution is saturated. Once saturation occurs, atoms bond together to form solid crystalline substances that precipitate from (settle out of) the solution. The Dead Sea and the Great Salt Lake provide good examples of this process (Figure 3.9). Because they are located in arid regions with high evaporation rates, these water bodies regularly precipitate the minerals halite, sylvite, and gypsum, as well as other soluble salts. Worldwide, extensive salt deposits, some exceeding 300 meters (1000 feet) in thickness, provide evidence of ancient seas that have long since evaporated (see Figure 3.39, p. 101).

Minerals can also precipitate from slowly moving saturated groundwater where crystals form in fractures and voids in rocks

FIGURE 3.9 Aerial view of salt evaporation ponds, Great Salt Lake, Utah. (Photo by Jim Wark/Peter Arnold)

Students Sometimes Ask . . .

According to the textbook, thick beds of halite and gypsum formed when ancient seas evaporated. Has this happened in the recent past?

Yes. During the past 6 million years, the Mediterranean Sea may have dried up and then refilled several times. When 65 percent of seawater evaporates, the mineral gypsum begins to precipitate, meaning it comes out of solution and settles to the bottom. When 90 percent of the water is gone, halite crystals form, followed by salts of potassium and magnesium. Deep-sea drilling in the Mediterranean has encountered the presence of thick beds of gypsum and salt deposits (mostly halite) sitting one atop the other to a maximum depth of 2 kilometers (1.2 miles). These deposits are inferred to have resulted from tectonic events that periodically closed and reopened the connection between the Atlantic Ocean and the Mediterranean Sea (the modern-day Straits of Gibraltar) over the past several million years. During periods when the Mediterranean was cut off from the Atlantic, the warm and dry climate in this region caused the Mediterranean to nearly evaporate. Then, when the connection to the Atlantic was opened, the Mediterranean basin would refill with seawater of normal salinity. This cycle was repeated over and over again, producing the layers of gypsum and salt found on the Mediterranean seafloor.

and sediments. One interesting example, called a *geode*, is a somewhat spherically shaped object with inward-projecting crystals (Figure 3.10). Geodes sometimes contain spectacular crystals of quartz, calcite, or other minerals.

Changes in temperature can also trigger crystallization. A decrease in temperature can cause liquids to crystallize into solids, a process identical to the formation of ice from liquid water. The crystallization of minerals from molten magma, although more complicated, is similar to water freezing. When magma is hot, the atoms are very mobile; but as the molten material cools, the atoms slow and begin to combine. Crystallization of a magma body generates igneous rocks that consist of a mosaic of intergrown crystals that tend to lack well-developed faces (see Figure 3.2, p. 75).

Organisms, such as microscopic bacteria, are also responsible for initiating the crystallization of substantial quantities of mineral matter. For example, the mineral pyrite when found in shale and coal beds, is typically generated by the action of certain bacteria. In addition, mollusks (clams) and other marine invertebrates secrete shells composed of the carbonate minerals calcite and aragonite. The remains of these shells are major components of many limestone beds.

Crystal Structures

The smooth faces and symmetry possessed by well-developed crystals are manifestations of the orderly packing of atoms or molecules that constitute a mineral's internal structure. Although this relationship between crystallization and atomic structure had been proposed much earlier, experimental verification did not occur until 1912 when it was discovered that X-rays are diffracted (scattered) by crystals. X-ray diffraction techniques pass X-rays through a crystal to expose a sheet of photographic film on the far side (Figure 3.11A). The beam interacts with the planes of atoms and is diffracted in a manner that produces a pattern of dark spots on the film (Figure 3.11B). (This is different from the way that medical and dental X-ray images are produced—in those images, the X-rays that strike our bones or other dense material do not make it to the film.) The resulting diffraction pattern is a function of the spacing and distribution of atoms in the crystal, and is unique to each mineral. Hence, X-ray diffraction is an important tool for mineral identification.

With the development of X-ray diffraction techniques, mineralogists were able to directly investigate the internal structures of minerals. What followed was the discovery that minerals have highly ordered atomic arrangements that can be described as spherically shaped atoms, or ions, held together by ionic, covalent, or metallic bonds. The simplest crystal structures are those of native metals such as gold and silver that are composed of only one element. They consist of atoms that are packed together in a rather simple three-dimensional network that minimizes voids. Imagine a group of cannon

Geologist's Sketch

FIGURE 3.10 Geode partially filled with amethyst, a purple variety of quartz, Brazil. Geodes form in cavities in rocks, such as limestone and volcanic rocks. Slowly moving groundwater containing dissolved silicates, or carbonates deposit mineral matter in these structures. (Photo by Jeff Scovil)

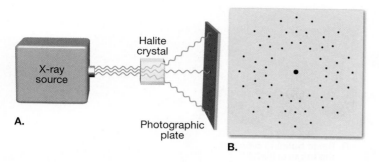

A.

B.

FIGURE 3.11 The principle of X-ray diffraction. **A.** When a parallel beam of X-rays pass through a crystal, they are diffracted (scattered) in such a way as to produce a pattern of dark spots on a photographic plate. **B.** The diffraction pattern shown is for the mineral halite (NaCl).

balls in which the layers are stacked so that the spheres in one layer nestle in the hollows between spheres in the adjacent layers.

The atomic arrangement in most minerals is more complicated than those of native metals, because they consist of at least two different ions (often of very different sizes). Figure 3.12 illustrates the relative sizes of some of the most common ions found in minerals. Notice that *anions*, which are atoms that gain electrons, tend to be larger than *cations*, which lose electrons.

Other things being equal, minerals containing larger cations are typically less dense than those in which the cations are smaller. For example, the density of sylvite (KCl), which contains the relatively large potassium cation, is less than half that of pyrrhotite (FeS), even though their molecular weights are similar. This relationship does not hold true for large cations that have very high atomic weights.

Most crystal structures can be considered three-dimensional arrays of larger spheres (anions) with smaller spheres (cations) located in the spaces between them, so that the positive and negative charges cancel each other out. Consider the mineral halite (NaCl), which has a relatively simple framework composed of an equal number of positively charged sodium ions and negatively charged chlorine ions. Because anions repel anions, and cations repel cations, ions of similar charge are spaced as far apart from each other as possible. Consequently, in the mineral halite, each sodium ion (Na^+) is surrounded on all sides by chlorine ions and vice versa (Figure 3.13). This particular packing arrangement results in basic building blocks, called *unit cells* that have cubic shapes. As shown in Figure 3.13C, these cubic unit cells combine to form cubic-shaped halite crystals. Sodium chloride crystals, including the ones that come out of a salt shaker, are often perfect cubes.

The shape and symmetry of these building blocks relate to the shape and symmetry of the entire crystal. It is important to note, however, that two minerals can be constructed of geometrically similar building blocks yet exhibit different external forms. For example, the minerals fluorite, magnetite, and garnet are all constructed of cubic unit cells. However, cubic cells can join to produce crystals of many shapes. Typically, fluorite crystals are cubes, whereas magnetite crystals are octahedrons, and garnets form dodecahedrons built up of many small cubes as shown in Figure 3.14. Because the building blocks are so small, the resulting crystal faces are smooth and flat.

Despite the fact that natural crystals are rarely perfect, the angles between equivalent crystal faces of the same mineral are remarkably consistent. This observation was first made by Nicolas Steno in 1669. Steno found that the angles between adjacent prism faces of quartz crystals are 120 degrees, regardless of sample size, the size of the crystal faces, or where the crystals were collected (Figure 3.15). This observation is commonly called **Steno's Law**, or the **Law of Constancy of Interfacial Angles**, because it applies to all minerals. Steno's Law states that *angles between equivalent faces of crystals of the same mineral are always the same*. For this reason, crystal shape is frequently a valuable tool in mineral identification.

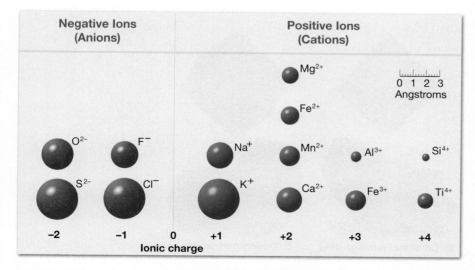

FIGURE 3.12 The relative sizes and ionic charges of various cations and anions commonly found in minerals. Ionic radii are usually expressed in angstroms (1 angstrom equals 10^{-8}cm).

Compositional Variations in Minerals

Using sophisticated analytical techniques, researchers determined that the chemical composition of some minerals varies substantially from sample to sample. These compositional variations are possible because ions of similar size can readily substitute for one another without disrupting a mineral's internal framework. This is analogous to a wall made of bricks of different colors and materials. As long as the bricks are all the same size, the shape of the wall is unaffected; only its composition changes.

Consider the mineral olivine as an example of chemical variability. The chemical formula for olivine is $(Mg,Fe)SiO_4$ where the variable components magnesium and iron are in parentheses.

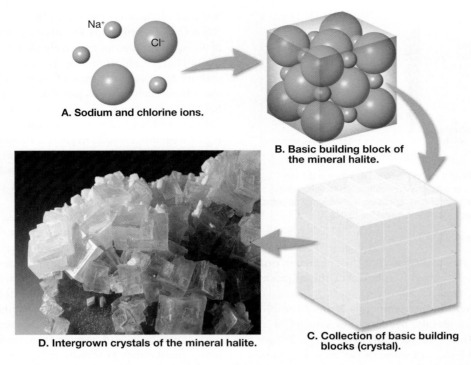

A. Sodium and chlorine ions.

B. Basic building block of the mineral halite.

C. Collection of basic building blocks (crystal).

D. Intergrown crystals of the mineral halite.

FIGURE 3.13   This diagram illustrates the orderly arrangement of sodium and chloride ions in the mineral halite. The arrangement of atoms into basic building blocks having a cubic shape results in regularly shaped cubic crystals. (Photo by Dennis Tasa)

Notice in Figure 3.12 that the cations—magnesium (Mg^{2+}) and iron (Fe^{2+})—are nearly the same size and have the same electrical charge. At one extreme, olivine may contain magnesium without any iron, or vice-versa—but most samples of olivine have some of both of these ions in their structure. Consequently, olivine can be thought of as having a range of combinations from a variety called *fosterite* ($MgSiO_4$) at one end to *fayalite* (Fe_2SiO_4) at the other. Nevertheless, all specimens of olivine have the same internal structure and exhibit very similar, but not identical, properties. For example, iron-rich olivines have a higher density than magnesium-rich specimens, a reflection of the greater atomic weight of iron as compared to magnesium.

In contrast to olivine, minerals such as quartz (SiO_2) and fluorite (CaF_2) tend to have chemical compositions that differ very little from their chemical formulas. However, even these minerals contain extremely small amounts of other less

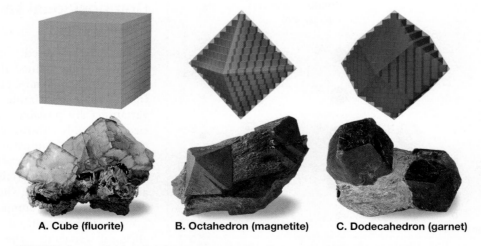

A. Cube (fluorite)   B. Octahedron (magnetite)   C. Dodecahedron (garnet)

FIGURE 3.14   Cubic unit cells stack together in different ways to produce crystals that exhibit different shapes. Fluorite (**A**) tends to display cubic crystals, whereas magnetite crystals (**B**) are typically octahedrons, and garnets (**C**) usually occur as dodecahedrons. (Photos by Dennis Tasa)

PEOPLE AND THE ENVIRONMENT

Asbestos: What Are the Risks?

BOX 3.2

Once considered safe enough to use in toothpaste, asbestos became one of the most feared contaminants on Earth. The asbestos panic in the United States began in 1986 when the Environmental Protection Agency (EPA) instituted the Asbestos Hazard Emergency Response Act. It required inspection of all public and private schools for asbestos. This brought asbestos to public attention and raised parental fears that children could contract asbestos-related cancers because of high levels of airborne fibers in schools.

What is Asbestos?

Asbestos is a commercial term applied to a variety of silicate minerals that readily separate into thin, strong fibers that are highly flexible, heat resistant, and relatively inert (Figure 3.B). These properties make asbestos a desirable material for the manufacture of a wide variety of products, including insulation, fireproof fabrics, cement, floor tiles, and car-brake linings. In addition, wall coatings rich in asbestos fibers were used extensively during the U.S. building boom of the 1950s and early 1960s.

The mineral *chrysotile*, marketed as "white asbestos," belongs to the serpentine mineral group and accounts for the vast majority of asbestos sold commercially. All other forms of asbestos are amphiboles and constitute less than 10 percent of asbestos used commercially. The two most common amphibole asbestos minerals are informally called "brown" and "blue" asbestos, respectively.

Exposure and Risk

Health concerns about asbestos stem largely from claims of high death rates among asbestos miners attributed to asbestosis (lung scarring from asbestos fiber inhalation), mesothelioma (cancer of chest and abdominal cavity), and lung cancer. The degree of concern created by these claims is demonstrated by the growth of an entire industry built around asbestos removal from buildings.

FIGURE 3.B   Chrysotile asbestos. This sample is a fibrous form of the mineral serpentine. (Photo by E. J. Tarbuck)

The stiff, straight fibers of brown and blue (amphibole) asbestos are known to readily pierce, and remain lodged in, the linings of human lungs. The fibers are physically and chemically stable and are not broken down in the human body. These forms of asbestos are,

therefore, a genuine cause for concern. White asbestos, however, being a different mineral, has different properties. The curly fibers of white asbestos are readily expelled from the lungs and, if they are not expelled, can dissolve within a year.

The U.S. Geological Survey has taken the position that the risks from the most widely used form of asbestos (chrysotile or "white asbestos") are minimal to non-existent. They cite studies of miners of white asbestos in northern Italy that show mortality rates from mesothelioma and lung cancer differ very little from the general public. Another study was conducted on people living in the area of Thetford Mines, Quebec, once the largest chrysotile mine in the world. For many years there were no dust controls, so these people were exposed to extremely high levels of airborne asbestos. Nevertheless, they exhibited normal levels of the diseases thought to be associated with asbestos exposure.

Despite the fact that more than 90 percent of all asbestos used commercially is white asbestos, a number of countries have moved to ban the use of asbestos in many applications, because the different mineral forms of asbestos are not distinguished. Very little of this once exalted mineral is presently used in the United States. Perhaps future studies will determine whether the asbestos panic, in which billions of dollars have been spent on testing and removal, was warranted or not.

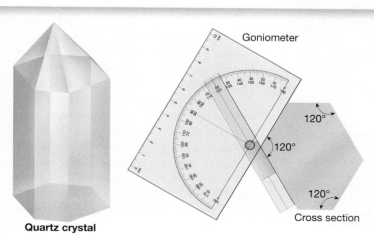

Quartz crystal

Goniometer

120°

120°

120°

Cross section

FIGURE 3.15   Illustration of Steno's Law. Because some faces of a crystal may grow larger than others, two crystals of the same mineral may *not* have identical shapes. Nevertheless, the angles between equivalent faces are remarkably consistent.

common elements, referred to as *trace elements*. Although trace elements have little effect on most mineral properties, they can significantly influence a mineral's color.

Structural Variations in Minerals

It is also common for two minerals with exactly the same chemical composition to have different internal structures and, hence, different external forms. Minerals of this type are called **polymorphs** (*poly* = many, *morph* = form). Graphite and diamond are particularly good examples of polymorphism because, when pure, they are both made up exclusively of carbon atoms—yet they display drastically different properties. Graphite is the soft gray mineral from which pencil "lead" is made, whereas diamond is the hardest known mineral. The differences between these minerals can be attributed to the conditions under which

they were formed. Diamonds form at depths approaching 200 kilometers (nearly 125 miles), where extreme pressures and temperatures produce the compact structure shown in Figure 3.16A. Graphite, on the other hand, forms under comparatively low pressures and consists of sheets of carbon atoms that are widely spaced and weakly bonded (Figure 3.16B). Because the carbon sheets in graphite will easily slide past one another, it has a greasy feel and makes an excellent lubricant.

Scientists have learned that by heating graphite under high confining pressures they can generate synthetic diamonds. Because "man-made" diamonds often contain flaws, they are generally not gem quality. However, due to their hardness, they have many industrial uses. Further, because all diamonds form in environments of extreme pressure and temperature, they are somewhat unstable at Earth's surface where conditions are very different. Fortunately for jewelers, "diamonds are forever" because the rate at which they change to their more stable form, graphite, is infinitesimally slow.

Calcite and aragonite are two other minerals with identical chemical compositions, calcium carbonate ($CaCO_3$), but different internal structures. Calcite forms mainly through biochemical processes and is the major constituent of the sedimentary rock limestone. Aragonite, a less common polymorph of $CaCO_3$ is often deposited by hot springs and is also an important constituent of pearls and the shells of some marine organisms. Because aragonite gradually changes to the more stable crystalline structure of calcite, it is rare in rocks older than 50 million years.

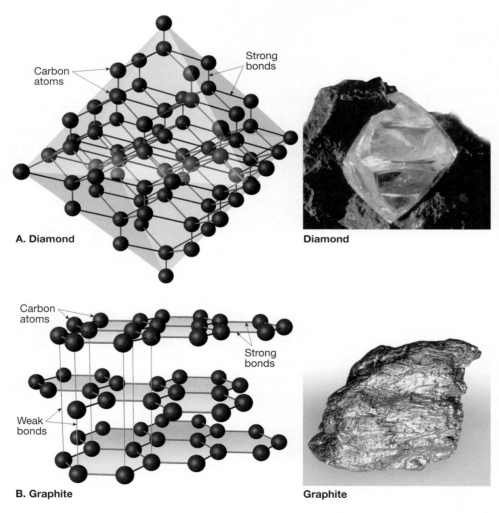

FIGURE 3.16 Comparing the structures of diamond and graphite. Both are natural substances with the same chemical composition—carbon atoms. Nevertheless, their internal structure and physical properties reflect the fact that each formed in a very different environment. **A.** All carbon atoms in diamond are covalently bonded into a compact, three-dimensional framework, which accounts for the extreme hardness of the mineral. **B.** In graphite the carbon atoms are bonded into sheets that are joined in a layered fashion by very weak electrical forces. These weak bonds allow the sheets of carbon to readily slide past each other, making graphite soft and slippery, and thus useful as a dry lubricant. (**A.** Photographer Dane Pendland, courtesy of Smithsonian Institution; **B.** E. J. Tarbuck)

The transformation of one polymorph to another is called a *phase change*. In nature, certain minerals go through phase changes as they move from one environment to another. For example, when a slab of ocean crust composed of olivine-rich basalt is carried to great depths by a subducting plate, olivine changes to a more compact polymorph called spinel.

PHYSICAL PROPERTIES OF MINERALS

MATTER AND MINERALS
▸ Physical Properties of Minerals

Each mineral has a definite crystalline structure and chemical composition that give it a unique set of physical and chemical properties shared by all samples of that mineral. For example, all specimens of halite have the same hardness, the same density, and break in a similar manner. Because the internal structure and chemical composition of a mineral are difficult to determine without the aid of sophisticated tests and equipment, the more easily recognized physical properties are frequently used in identification.

The diagnostic physical properties of minerals are those that can be determined by observation or by performing a simple test. The primary physical properties that are commonly used to identify hand samples are luster, color, streak, crystal shape (habit), tenacity, hardness, cleavage, fracture, and density or specific gravity. Secondary ("special") properties that are exhibited by a limited number of minerals include magnetism, taste, feel, smell, double refraction, and chemical reaction to hydrochloric acid.

Optical Properties

Of the many optical properties of minerals, four—luster, the ability to transmit light, color, and streak—are most frequently used for mineral identification.

Luster   The appearance or quality of light reflected from the surface of a mineral is known as **luster.** Minerals that have the appearance of metals, regardless of color, are said to have a *metallic luster* (Figure 3.17). Some metallic minerals, such as native copper and galena, develop a dull coating or tarnish when exposed to the atmosphere. Because they are not as shiny as samples with freshly broken surfaces, these samples are often said to exhibit a *submetallic luster.*

Most minerals have a *nonmetallic luster* and are described using various adjectives such as *vitreous* or *glassy*. Other nonmetallic minerals are described as having a *dull* or *earthy luster* (a dull appearance like soil), or a *pearly luster* (such as a pearl, or the inside of a clamshell). Still others exhibit a *silky luster* (like satin cloth), or a *greasy luster* (as though coated in oil).

The Ability to Transmit Light   Another optical property used in the identification of minerals is the ability to transmit light. When no light is transmitted, the mineral is described as *opaque;*

FIGURE 3.17   The freshly broken sample of galena (right) displays a metallic luster, while the sample on the left is tarnished and has a submetallic luster. (Photo courtesy of E. J. Tarbuck)

when light but not an image is transmitted through a mineral it is said to be *translucent.* When light and an image are visible through the sample, the mineral is described as *transparent.*

Color   Although **color** is generally the most conspicuous characteristic of any mineral, it is considered a diagnostic property of only a few minerals. Slight impurities in the common mineral quartz, for example, give it a variety of tints including pink, purple, yellow, white, gray, and even black (see Figure 3.34, p. 98). Other minerals, such as tourmaline, also exhibit a variety of hues, with multiple colors sometimes occurring in the same sample. Thus, the use of color as a means of identification is often ambiguous or even misleading.

Streak   Although the color of a sample is not always helpful in identification, **streak**—the color of the powdered mineral—is often diagnostic. The streak is obtained by rubbing the mineral across a piece of unglazed porcelain, termed a *streak plate* and observing the color of the mark it leaves (Figure 3.18). Although

FIGURE 3.18   Although the color of a mineral is not always helpful in identification, the streak, which is the color of the powdered mineral, can be very useful. (Photo by Dennis Tasa)

the color of a mineral may vary from sample to sample, the streak usually does not.

Streak can also help distinguish between minerals with metallic luster and those having nonmetallic luster. Metallic minerals generally have a dense, dark streak, whereas minerals with nonmetallic luster have a streak that is typically light colored.

It should be noted that not all minerals produce a streak when using a streak plate. For example, if the mineral is harder than the streak plate, no streak is observed.

Crystal Shape or Habit

Mineralogists use the term **crystal shape** or **habit** to refer to the common or characteristic shape of a crystal or aggregate of crystals. A few minerals exhibit somewhat regular polygons that are helpful in their identification. Recall that magnetite crystals sometimes occur as octahedrons, garnets often form dodecahedrons, and halite and fluorite crystals tend to grow as cubes or near cubes (see Figure 3.14). While most minerals have only one common habit, a few such as pyrite have two or more characteristic crystal shapes (Figure 3.19).

By contrast, some minerals rarely develop perfect geometric forms. Many of these do, however, develop a characteristic shape that is useful for identification. Some minerals tend to grow equally in all three dimensions, whereas others tend to be elongated in one direction, or flattened if growth in one dimension is suppressed. Commonly used terms to describe these and other crystal habits include *equant* (equidimensional), *bladed, fibrous, tabular, prismatic, platy, blocky,* and *botryoidal* (Figure 3.20).

Mineral Strength

How easily minerals break or deform under stress relates to the type and strength of the chemical bonds that hold the crystals together. Mineralogists use terms including tenacity, hardness,

A. Bladed

B. Prismatic

C. Banded

D. Botryoidal

FIGURE 3.20   Some common crystal habits. **A.** *Bladed.* Elongated crystals that are flattened in one direction. **B.** *Prismatic.* Elongated crystals with faces that are parallel to a common direction. **C.** *Banded.* Minerals that have stripes or bands of different color or texture. **D.** *Botryoidal.* Groups of intergrown crystals resembling a bunch of grapes. (Photos by Dennis Tasa)

cleavage, and fracture to describe mineral strength and how minerals break when stress is applied.

Tenacity   The term **tenacity** describes a mineral's toughness, or its resistance to breaking or deforming. Minerals which are ionically bonded, such as fluorite and halite, tend to be *brittle* and shatter into small pieces when struck. By contrast, minerals with metallic bonds, such as native copper, are *malleable,* or easily hammered into different shapes. Minerals, including gypsum and talc, that can be cut into thin shavings are described as *sectile.* Still others, notably the micas, are *elastic* and will bend and snap back to their original shape after the stress is released.

Hardness   One of the most useful diagnostic properties is **hardness,** a measure of the resistance of a mineral to abrasion or scratching. This property is determined by rubbing a mineral of unknown hardness against one of known hardness, or vice versa. A numerical value of hardness can be obtained by using the **Mohs scale** of hardness, which consists of 10 minerals arranged in order from 1 (softest) to 10 (hardest), as shown in Figure 3.21A. It should be noted that the Mohs scale is a relative ranking, and it does not imply that mineral number 2, gypsum, is twice as hard as mineral 1, talc. In fact, gypsum is only slightly harder than talc as shown in Figure 3.21B.

In the laboratory, other common objects can be used to determine the hardness of a mineral. These include a human fingernail, which has a hardness of about 2.5, a copper penny 3.5, and a piece of glass 5.5. The mineral gypsum, which has a hardness of 2, can be easily scratched with a fingernail. On the other hand, the mineral calcite, which has a hardness of 3, will

FIGURE 3.19   Although most minerals exhibit only one common crystal shape, some, such as pyrite, have two or more characteristic habits. (Photos by Dennis Tasa)

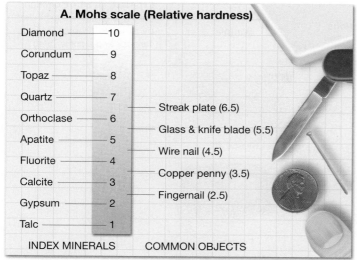

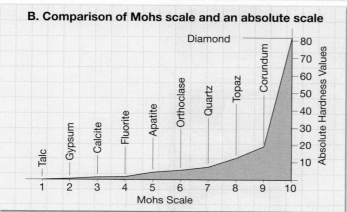

FIGURE 3.21   Hardness scales. **A.** Mohs scale of hardness, with the hardness of some common objects. **B.** Relationship between Mohs relative hardness scale and an absolute hardness scale.

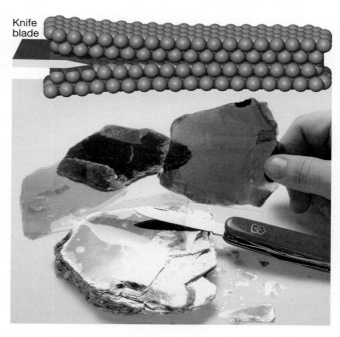

FIGURE 3.22   The thin sheets shown here were produced by splitting a mica (muscovite) crystal parallel to its perfect cleavage. (Photo by Chip Clark)

scratch a fingernail but will not scratch glass. Quartz, one of the hardest common minerals, will easily scratch glass. Diamonds, hardest of all, scratch anything.

Cleavage   In the crystal structure of many minerals, some atomic bonds are weaker than others. It is along these weak bonds that minerals tend to break when they are stressed. **Cleavage** (*Kleiben* = carve) is the tendency of a mineral to break (cleave) along planes of weak bonding. Not all minerals have cleavage, but those that do can be identified by the relatively smooth, flat surfaces that are produced when the mineral is broken.

The simplest type of cleavage is exhibited by the micas (Figure 3.22). Because these minerals have very weak bonds in one direction, they cleave to form thin, flat sheets. Some minerals have excellent cleavage in one, two, three, or more directions, whereas others exhibit fair or poor cleavage, and still others have no cleavage at all. When minerals break evenly in more than one direction, cleavage is described by *the number of cleavage directions and the angle(s) at which they meet* (Figure 3.23).

Each cleavage surface that has a different orientation is counted as a different direction of cleavage. For example, some minerals cleave to form six-sided cubes. Because cubes are defined by three different sets of parallel planes that intersect at 90-degree angles, cleavage is described as *three directions of cleavage that meet at 90 degrees*.

Do not confuse cleavage with crystal shape. When a mineral exhibits cleavage, it will break into pieces that all have the same geometry. By contrast, the smooth-sided quartz crystals shown in Figure 3.1 (p. 75) do not have cleavage. If broken, they fracture into shapes that do not resemble one another or the original crystals.

Fracture   Minerals having chemical bonds that are equally, or nearly equally, strong in all directions exhibit a property called **fracture**. When minerals fracture, most produce uneven surfaces and are described as exhibiting *irregular fracture*. However, some minerals, such as quartz, break into smooth, curved surfaces resembling broken glass. Such breaks are called *conchoidal fractures* (Figure 3.24). Still other minerals exhibit fractures that produce splinters or fibers that are referred to as *splintery* and *fibrous fracture*, respectively.

Density and Specific Gravity

Density is an important property of matter defined as mass per unit volume usually expressed as grams per cubic centimeter. Mineralogists often use a related measure called *specific gravity* to describe the density of minerals. **Specific gravity** is a unitless

FIGURE 3.23 Common cleavage directions exhibited by minerals. (Photos by E. J. Tarbuck and Dennis Tasa)

Number of Cleavage Directions	Shape	Sketch	Directions of Cleavage	Sample
1	Flat sheets			 Muscovite
2 at 90°	Elongated form with rectangle cross section (prism)	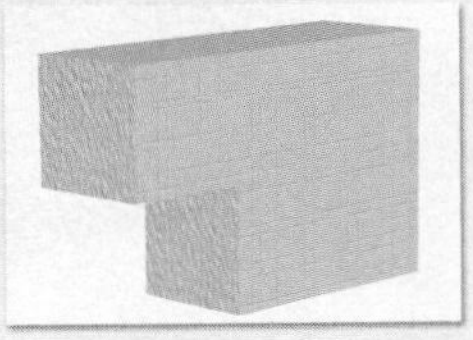	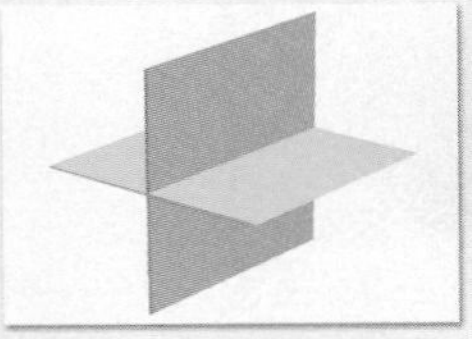	 Feldspar
2 not at 90°	Elongated form with parallelogram cross section (prism)	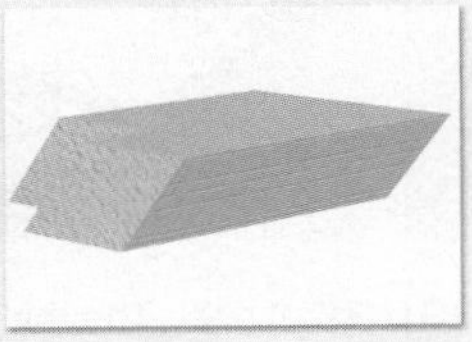		 Hornblende
3 at 90°	Cube	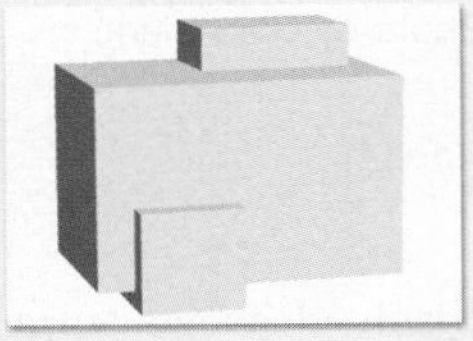	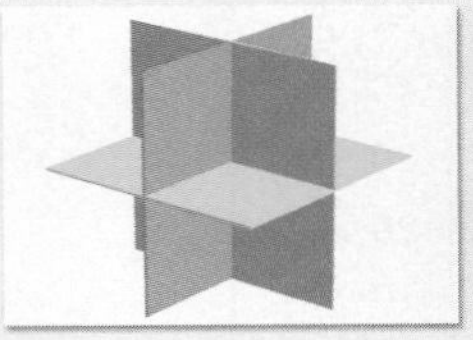	 Halite
3 not at 90°	Rhombohedron	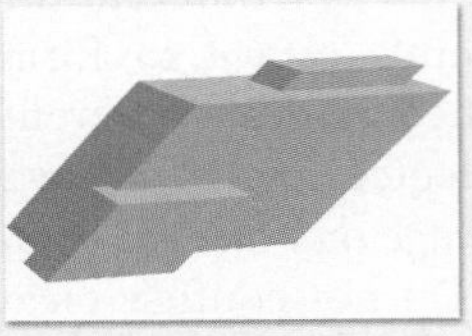		 Calcite
4	Octahedron			 Fluorite

FIGURE 3.24 Conchoidal fracture. The smooth curved surfaces result when minerals break in a glasslike manner. (Photo by E. J. Tarbuck)

number representing the ratio of a mineral's weight to the weight of an equal volume of water.

Most common rock-forming minerals have a specific gravity of between 2 and 3. For example, quartz has a specific gravity of 2.65. By contrast, some metallic minerals such as pyrite, native copper, and magnetite are more than twice as dense as quartz. Galena, which is an ore of lead, has a specific gravity of roughly 7.5, whereas the specific gravity of 24-karat gold is approximately 20.

With a little practice, you can estimate the specific gravity of a mineral by hefting it in your hand. Ask yourself, does this mineral feel about as "heavy" as similar sized rocks you have handled? If the answer is "yes," the specific gravity of the sample will likely be between 2.5 and 3.

Historically, gold would certainly be considered the most prized metal, followed by silver. While both have important industrial uses, gold and silver are better known for their uses in art, jewelry, and coinage—the Canadian Gold Maple Leaf, for example. At the time of this writing, however, the cost of platinum is $1200 USD per troy ounce (31.1 grams), which greatly exceeds that of an equal amount of gold ($900 USD). In addition to its use in jewelry, platinum is used in the making of catalytic converters, which convert harmful emissions from automobile engines into less harmful gases. In 2008, rhodium, a hard silvery metal with very high reflectivity was selling for as much as $10,000 USD per ounce, making it the most expensive precious metal on record. The price of rhodium subsequently dropped to less than one thousand dollars per troy ounce.

FIGURE 3.26 Calcite reacting with a weak acid. (Photo by Chip Clark)

Other Properties of Minerals

In addition to the properties already discussed, some minerals can be recognized by other distinctive properties. For example, halite is ordinary salt, so it can be quickly identified through taste. Talc and graphite both have distinctive feels; talc feels soapy, and graphite feels greasy. Further, the streak of many sulfur-bearing minerals smells like rotten eggs. A few minerals, such as magnetite, have a high iron content and can be picked up with a magnet, while some varieties (lodestone) are natural magnets and will pick up small iron-based objects such as pins and paper clips (see Figure 3.40, p. 103).

Moreover, some minerals exhibit special optical properties. For example, when a transparent piece of calcite is placed over printed material, the letters appear twice. This optical property is known as *double refraction* (Figure 3.25).

One very simple chemical test involves placing a drop of dilute hydrochloric acid from a dropper bottle onto a freshly bro-

ken mineral surface. Certain minerals, called carbonates, will effervesce (fizz) as carbon dioxide gas is released (Figure 3.26). This test is especially useful in identifying common carbonate mineral calcite.

HOW ARE MINERALS NAMED AND CLASSIFIED?

Nearly 4000 minerals have been named, and 30 to 50 new ones are identified each year. Fortunately, for students beginning their study of minerals, no more than a few dozen are abundant! Collectively, these few make up most of the rocks of Earth's crust and, as such, are often referred to as the *rock-forming minerals*.

Although less abundant, many other minerals are used extensively in the manufacture of products that drive our modern society and are called *economic* minerals (see Box 3.2). It should be noted that rock-forming minerals and economic minerals are not mutually exclusive groups. When found in large deposits, some rock-forming minerals are economically significant. One example is the mineral calcite, which is the primary component of the sedimentary rock limestone and has many uses including the production of cement.

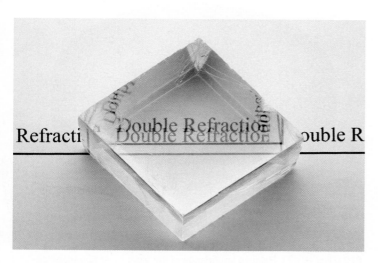

FIGURE 3.25 Double refraction illustrated by the mineral calcite. (Photo by Chip Clark)

TABLE 3.1	Major Mineral Classes		
Class	Anion, Anion Complex, or Elements	Example (Mineral species)	Chemical Formula
silicates	$(SiO_4)^{4-}$	quartz	SiO_2
halides	$Cl^{1-}, F^{1-}, Br^{1-}, I^{1-}$	halite	$NaCl$
oxides	O^{2-}	corundum	Al_2O_3
hydroxides	$(OH)^{1-}$	gibbsite	$Al(OH)_3$
carbonates	$(CO_3)^{2-}$	calcite	$CaCO_3$
nitrates	$(NO_3)^{1-}$	nitratite	$NaNO_3$
sulfates	$(SO_4)^{2-}$	gypsum	$CaSO_4 \cdot 2H_2O$
phosphates	$(PO_4)^{3-}$	apatite	$Ca_5(PO_4)_3(OH, F, Cl)$
native elements	Cu, Ag, S	copper	Cu
sulfides	S^{2-}	pyrite	FeS_2

Determining what is or is not a mineral is in the hands of the International Mineralogical Association. Individuals who believe they have discovered a new mineral must apply to this organization with a detailed description of their findings. If confirmed, the discoverer is allowed to choose a name for the new mineral. Minerals have been named for people, locations where the mineral was discovered, appearance, and chemical composition. Although nearly 50 percent are named in honor of a person, rules prohibit naming a mineral after oneself.

Classifying Minerals

Minerals are placed into categories in much the same way as plants and animals are classified. Mineralogists use the term *mineral species* for a collection of specimens that exhibit similar internal structures and chemical compositions. Some common mineral species include quartz, calcite, galena, and pyrite. However, just as individual plants and animals within a species differ somewhat from one another, so do most specimens within a mineral species.

Mineral species are usually assigned to a *mineral class* based on their anions, or anion complexes as shown in Table 3.1. Some of the important mineral classes include the silicates, (SiO_4^{4-}), carbonates (CO_3^{2-}), halides $(Cl^{1-}, F^{1-}, Br^{1-})$, and sulfates (SO_4^{2-}). Minerals within each class tend to have similar internal structures and, hence, similar properties. For example, minerals that belong to the carbonate class react chemically with acid—albeit to varying degrees—and many exhibit rhombic cleavage. Furthermore, minerals within the same class are often found together in the same rock. The halides, for example, commonly occur together in evaporite deposits.

Mineral classes are further divided based on similarities in atomic structures or compositions. For example, mineralogists group the feldspar minerals (orthoclase, albite, and anorthite) together because they all have similar internal structures. The minerals kyanite, andalusite, and sillimanite are grouped together because they have the same chemical composition (Al_2SiO_5). Recall that minerals having the same composition but different crystal structures are called *polymorphs*. Some

common mineral groups include the feldspars, pyroxenes, amphiboles, olivines, and micas. Two members (species) of the mica group are biotite and muscovite. Although they have similar atomic structures, biotite contains iron and magnesium, which give it a greater density and darker appearance than muscovite.

Some mineral species are further subdivided into *mineral varieties*. For example, pure quartz (SiO_2) is colorless and transparent. However, when small amounts of aluminum are incorporated into its atomic structure, quartz appears quite dark, a variety called *smoky quartz. Amethyst,* another variety of quartz, owes its violet color to the presence of trace amounts of iron.

Major Mineral Classes

Only eight elements make up the bulk of rock-forming minerals and represent more than 98 percent (by weight) of the continental crust (Figure 3.27). These elements, in order of abundance are oxygen (O), silicon (Si), aluminum (Al), iron (Fe), calcium

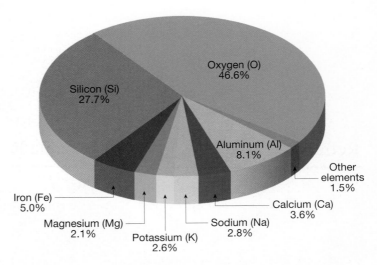

FIGURE 3.27   Relative abundance of the eight most abundant elements in the continental crust.

(Ca), sodium (Na), potassium (K), and magnesium (Mg). As shown in Figure 3.27, silicon and oxygen are by far the most common elements in Earth's crust. These two elements readily combine to form the framework for the most dominant mineral class, the **silicates,** which account for more than 90 percent of Earth's crust. More than 800 species of silicate minerals are known.

Because other mineral classes are far less abundant than the silicates, they are often grouped under the heading **nonsilicates.** Although not as common as the silicates, some nonsilicate minerals are very important economically. They provide us with the iron and aluminum to build our automobiles, gypsum for plaster and drywall to construct our homes, and copper for wire to carry electricity. In addition to their economic importance, these mineral groups include members that are major constituents in sediments and sedimentary rocks.

We will first discuss the most common mineral class, the silicates, and then consider some of the prominent nonsilicate mineral groups.

THE SILICATES

Every silicate mineral contains the two most abundant elements of Earth's crust, oxygen and silicon. Further, most contain one or more of the other common elements. Together, these elements give rise to hundreds of silicate minerals with a wide variety of properties, including hard quartz, soft talc, sheet-like mica, fibrous asbestos, green olivine, and blood-red garnet.

The Silicon–Oxygen Tetrahedron

All silicate minerals have the same fundamental building block, the **silicon–oxygen tetrahedron** (SiO_4^{4-}). This structure consists of four oxygen anions (each $2-$) that are covalently bonded to one comparatively small silicon cation ($4+$), forming a *tetrahedron*— a pyramid shape with four identical faces (Figure 3.28). These tetrahedra are not chemical compounds, but rather complex anions (SiO_4^{4-}) having a net charge of $4-$. To become electrically balanced, these complex anions bond to other positively charged metal ions. Specifically, each O^{2-} has one of its valence electrons bonding with the Si^{4+} located at the center of the tetrahedron. The remaining -1 charge on each oxygen is available to bond with another cation, or with the silicon cation in an adjacent tetrahedron.

Minerals with Independent Tetrahedra One of the simplest silicate structures consists of independent tetrahedra that have their four oxygen ions bonded to cations, such as Mg^{2+}, F^{2+}, and Ca^{2+}. The mineral olivine, with the formula $MgFe_2SiO_4$, is a good example. In olivine, magnesium (Mg^{2+}) and/or iron (Fe^{2+}) cations pack between comparatively large independent SiO_4 tetrahedra, forming a dense three-dimensional structure. Garnet, another common silicate, is also composed of independent tetrahedra ionically bonded by cations. Both olivine and garnet form dense, hard, equidimensional crystals that lack cleavage.

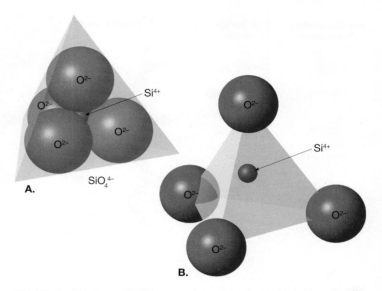

FIGURE 3.28 Two representations of the silicon–oxygen tetrahedron. **A.** The four large spheres represent oxygen ions, and the blue sphere represents a silicon ion. The spheres are drawn in proportion to the radii of the ions. **B.** An expanded view of the tetrahedron that has an oxygen ion at each of the four corners.

Minerals with Chain or Sheet Structures One reason for the great variety of silicate minerals is the ability of SiO_4 tetrahedra to link to one another in a variety of configurations. This important phenomenon, called **polymerization**, is achieved by the sharing of one, two, three, or all four of the oxygen atoms with adjacent tetrahedra. Vast numbers of tetrahedra join together to form single chains, double chains, sheet structures, or three-dimensional frameworks as shown in Figure 3.29.

Students Sometimes Ask . . .

Are these silicates the same materials used in silicon computer chips and silicone breast implants?

Not really, but all three contain the element silicon (Si). Further, the source of silicon for numerous products, including computer chips and breast implants, comes from silicate minerals. Pure silicon (without the oxygen that silicates have) is used to make computer chips, giving rise to the term "Silicon Valley" for the high-tech region of San Francisco, California's South Bay area, where many of these devices are designed. Manufacturers of computer chips engrave silicon wafers with incredibly narrow conductive lines, squeezing millions of circuits into every fingernail-size chip.

Silicone—the material used in breast implants—is a silicon–oxygen polymer gel that feels rubbery and is water repellent, chemically inert, and stable at extreme temperatures. Although concern about the long-term safety of these implants limited their use after 1992, several studies have found no evidence linking them to various diseases.

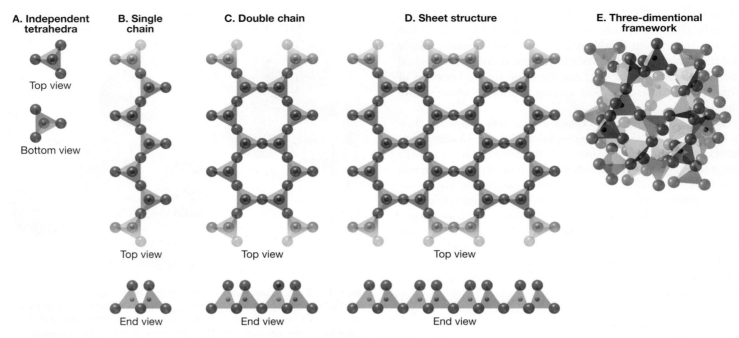

A. Independent tetrahedra

Top view

Bottom view

B. Single chain

Top view

End view

C. Double chain

Top view

End view

D. Sheet structure

Top view

End view

E. Three-dimentional framework

FIGURE 3.29   Five types of silicate structures. **A.** Independent tetrahedra. **B.** Single chains. **C.** Double chains. **D.** Sheet structures. **E.** Three-dimensional framework.

To see how oxygen atoms are shared between adjacent tetrahedra, select one of the silicon ions (small blue spheres) near the middle of the single-chain shown in Figure 3.29B. Notice that this silicon ion is completely surrounded by four larger oxygen ions. Also notice that, of the four oxygen atoms, half are bonded to two silicon atoms, whereas the other two are not shared in this manner. It is the linkage across the shared oxygen ions that join the tetrahedra into a chain structure. Now examine a silicon ion near the middle of the sheet structure (Figure 3.29D) and count the number of shared and unshared oxygen ions surrounding it. As you should have discovered, the sheet structure is the result of three of the four oxygen atoms being shared by adjacent tetrahedra.

Minerals with Three-Dimensional Frameworks   In the most common silicate structure, all four oxygen ions are shared, producing a complex three-dimensional framework (Figure 3.29E). Quartz, a hard, durable mineral, has the simplest structure in which all of the oxygens are shared. Because its structure is electrically neutral, quartz (SiO_2) contains no positive ions—other than silicon.

The ratio of oxygen ions to silicon ions differs in each type of silicate structure. In independent tetrahedra (SiO_4), there are four oxygen ions for every silicon ion. In single chains, the oxygen-to-silicon ratio is 3:1 (SiO_3), and in three-dimensional frameworks as found in quartz the ratio is 2:1 (SiO_2). As more oxygen ions are shared, the percentage of silicon in the structure increases. Silicate minerals are, therefore, described as having a low or high silicon content based on their ratio of oxygen to silicon. Minerals with three-dimensional structures in which all four oxygen ions are shared have the highest silicon content. Minerals composed of independent tetrahedra have the lowest.

This difference in silicon content is important, as you will see in Chapter 4.

Joining Silicate Structures

Except for quartz (SiO_2) the basic structure (chains, sheets, or three-dimensional frameworks) of most silicate minerals has a net negative charge. Therefore, metal cations are required to bring the overall charge into balance and to serve as the "mortar" that holds these structures together. The cations that most often link silicate structures are iron (Fe^{2+}), magnesium (Mg^{2+}), potassium (K^{1+}), sodium (Na^{1+}), aluminum (Al^{3+}), and calcium (Ca^{2+}). These positively charged ions bond with the unshared oxygen ions that occupy the corners of the silicate tetrahedra (Figure 3.30).

As a general rule, the hybrid covalent bonds between silicon and oxygen are stronger than the ionic bonds that hold one silicate structure to the next. Consequently, properties such as cleavage, and to some extent hardness, are controlled by the nature of the silicate framework. Quartz (SiO_2), which has only silicon–oxygen bonds, has great hardness and lacks cleavage, mainly because of equally strong bonds in all directions. By contrast, the mineral talc (the source of talcum powder) with the formula $Mg_3(Si_2O_5)_2(OH)_2$, has a sheet structure. The Mg^{2+} ions lie between the sheets and weakly join them together. The slippery feel of talcum powder is due to the silicate sheets sliding relative to one another, much like the sheets of carbon atoms slide in graphite, giving it its lubricating properties.

Recall that atoms of similar size can substitute freely for one another without altering the mineral's structure. For example,

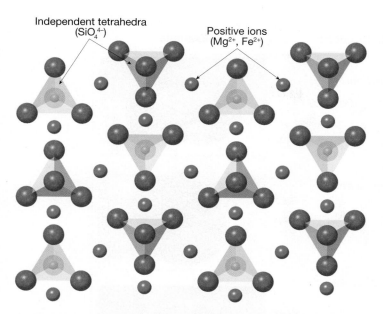

Independent tetrahedra
(SiO_4^{4-})

Positive ions
(Mg^{2+}, Fe^{2+})

FIGURE 3.30 Structure of the mineral olivine $[(Mg, Fe)_2SiO_4]$ in which magnesium (Mg^{2+}) and/or iron (Fe^{2+}) cations pack between relatively large independent SiO_4^{4-} tetrahedra to form a three-dimensional structure. It is the attraction between the metal cations (Mg^{2+}, Fe^{2+}) and the oxygen anions that bond the tetrahedra together. In this simplified drawing, only one layer of tetrahedra is shown—the layers above and below have been omitted for clarity. Further, the structure has been greatly expanded. In olivine, each silicon (Si^{4+}) cation is completely surrounded by four oxygen (O^{2-}) anions and each magnesium (Mg^{2+}) and/or iron (Fe^{2+}) cation is surrounded by six oxygen (O^{2-}) cations. Since each SiO_4^{4-} tetrahedron has a charge of -4 and there are two metal cations (each with a charge of $+2$) for every tetrahedron, the structure is electrically neutral.

in the mineral olivine, iron (Fe^{2+}) and Magnesium (Mg^{2+}) substitute for each other. This also holds true for the third most common element in Earth's crust, aluminum (Al) which often substitutes for silicon (Si) located in the center of the silcon–oxygen tetrahedra. As shown in Figure 3.12 (p. 83), Al^{3+} and Si^{4+} are roughly the same size, but their charges are different. So how can Al^{3+} substitute for Si^{4+}? Consider what happens when one in every four silicon ions are replaced by an aluminum ion in the framework of a silicate like quartz (SiO_2). The result is a complex ion with the formula $AlSi_3O_8^{1-}$ (think of four SiO_2 molecules where one of the four Si atoms has been replaced by Al). Because the resulting structure has a charge of -1, something else needs to be added to make it an electrically neutral compound—a prerequisite for a mineral. One possibility is the addition of a positive ion that has a charge of $+1$. As we can see in Figure 3.12, two common ions, sodium (Na^{1+}) and potassium (K^{1+}), meet this requirement. When potassium is added to the structure it becomes a neutral chemical compound (mineral) with the formula $KAlSi_3O_8$. This mineral is referred to as potassium feldspar, a member of the most common mineral group—the feldspars.

Because most silicate structures will readily accommodate two or more different cations at a given bonding site, individual specimens of a particular mineral may contain varying amounts of certain elements. As a result, many silicate minerals form a *mineral group* that exhibits a range of compositions between two end members. Examples include the olivines, pyroxenes, amphiboles, micas, and feldspars.

COMMON SILICATE MINERALS

MATTER AND MINERALS
▶Mineral Groups

The major groups of silicate minerals and common examples are given in Figure 3.31. The feldspars are, by far, the most plentiful silicate group, comprising more than 50 percent of Earth's crust. Quartz, the second most abundant mineral in the continental crust, is the only common mineral made completely of silicon and oxygen.

Most silicate minerals form when molten rock cools and crystallizes. Cooling can occur at or near Earth's surface (low temperature and pressure) or at great depths (high temperature and pressure). The environment during crystallization and the chemical composition of the molten rock determine, to a large degree, which minerals are produced. For example, the silicate mineral olivine crystallizes at high temperatures, whereas quartz crystallizes at much lower temperatures.

In addition, some silicate minerals form at Earth's surface from the weathered products of other silicate minerals. Still others are formed under the extreme pressures associated with mountain building. Each silicate mineral, therefore, has a structure and a chemical composition that *indicate the conditions under which it formed*. By carefully examining the mineral constituents of rocks, geologists can often determine the circumstances under which the rocks formed.

We will now examine some of the most common silicate minerals, which we divide into two major groups on the basis of their chemical makeup.

The Light Silicates

The **light** (or **nonferromagnesian) silicates** are generally light in color and have a specific gravity of about 2.7, which is considerably less than the dark (ferromagnesian) silicates. These differences are mainly attributable to the presence or absence of iron and magnesium. The light silicates contain varying amounts of aluminum, potassium, calcium, and sodium rather than iron and magnesium.

Feldspar Group *Feldspar*, the most common mineral group, can form under a wide range of temperatures and pressures, a fact that partially accounts for its abundance (Figure 3.32). All feldspars have similar physical properties. They have two planes of cleavage meeting at or near 90-degree angles, are relatively hard (6 on the Mohs scale), and have a luster that ranges from glassy to pearly. As one component in a rock, feldspar crystals

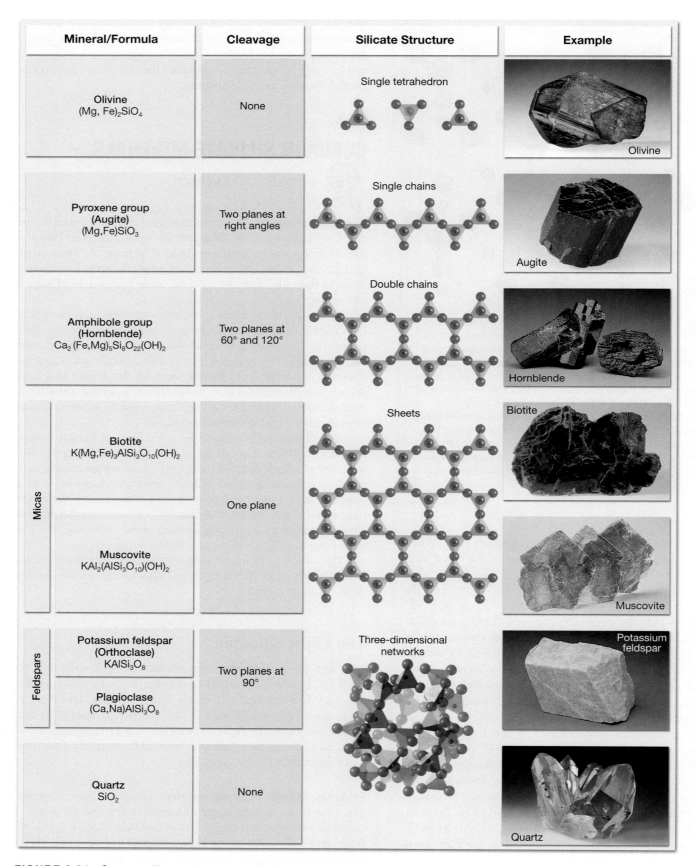

Mineral/Formula	Cleavage	Silicate Structure	Example
Olivine $(Mg, Fe)_2SiO_4$	None	Single tetrahedron	Olivine
Pyroxene group (Augite) $(Mg,Fe)SiO_3$	Two planes at right angles	Single chains	Augite
Amphibole group (Hornblende) $Ca_2(Fe,Mg)_5Si_8O_{22}(OH)_2$	Two planes at 60° and 120°	Double chains	Hornblende
Micas — Biotite $K(Mg,Fe)_3AlSi_3O_{10}(OH)_2$	One plane	Sheets	Biotite
Micas — Muscovite $KAl_2(AlSi_3O_{10})(OH)_2$	One plane	Sheets	Muscovite
Feldspars — Potassium feldspar (Orthoclase) $KAlSi_3O_8$	Two planes at 90°	Three-dimensional networks	Potassium feldspar
Feldspars — Plagioclase $(Ca,Na)AlSi_3O_8$	Two planes at 90°	Three-dimensional networks	
Quartz SiO_2	None		Quartz

FIGURE 3.31 Common silicate minerals. Note that the complexity of the silicate structure increases from top to bottom. (Photos by Dennis Tasa and E. J. Tarbuck)

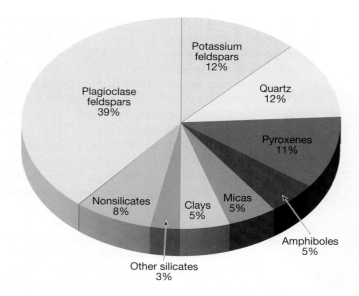

FIGURE 3.32   Estimated percentages (by volume) of the most common minerals in Earth's crust.

can be identified by their rectangular shape and rather smooth shiny faces (see Figure 3.31).

Two different feldspar structures exist. One group of feldspar minerals contains potassium ions in its structure and is therefore referred to as *potassium feldspar*. (*Orthoclase* and *microcline* are common members of the potassium feldspar group.) The other group, called *plagioclase feldspar*, contains both sodium and calcium ions that freely substitute for one another depending on the environment during crystalization.

Potassium feldspar is usually light cream, salmon pink, or occasionally bluish-green in color. The plagioclase feldspars, on the other hand, range in color from white to medium gray. However, color should not be used to distinguish these groups. The only sure way to distinguish the feldspars physically is to look for a multitude of fine parallel lines, called *striations*. Striations are found on some cleavage planes of plagioclase feldspar but are not present on potassium feldspar (Figure 3.33).

Quartz   *Quartz* is the only common silicate mineral consisting entirely of silicon and oxygen. As such, the term *silica* is applied to quartz, which has the chemical formula SiO_2. Because the structure of quartz contains a ratio of two oxygen ions (O^{2-}) for every silicon ion (Si^{4+}), no other positive ions are needed to attain neutrality.

In quartz, a three-dimensional framework is developed through the complete sharing of oxygen by adjacent silicon atoms. Thus, all of the bonds in quartz are of the strong silicon–oxygen type. Consequently, quartz is hard, resistant to weathering, and does not have cleavage. When broken, quartz generally exhibits conchoidal fracture (see Figure 3.24). When pure, quartz is clear and, if allowed to grow without interference, will develop hexagonal crystals that develop pyramid-shaped ends. However, like most other clear minerals, quartz is often colored by inclu-

sions of various ions (impurities) and forms without developing good crystal faces. The most common varieties of quartz are milky (white), smoky (gray), rose (pink), amethyst (purple), and rock crystal (clear) (Figure 3.34).

Muscovite   *Muscovite* is a common member of the mica family. It is light in color and has a pearly luster (see Figure 3.22). Like other micas, muscovite has excellent cleavage in one direction. In thin sheets, muscovite is clear, a property that accounts for its use as window "glass" during the Middle Ages. Because muscovite is very shiny, it can often be identified by the sparkle it gives a rock. If you have ever looked closely at beach sand, you may have seen the glimmering brilliance of the mica flakes scattered among the other sand grains.

Clay Minerals   *Clay* is a term used to describe a category of complex minerals that, like the micas, have a sheet structure. Unlike other common silicates, such as quartz and feldspar, clays do not form in igneous environments. Rather, most clay minerals originate as products of the chemical weathering of other silicate minerals. Thus, clay minerals make up a large percentage of the surface material we call soil. Because of the importance of soil in agriculture, and because of its role as a supporting material for buildings, clay minerals are extremely important to humans. In addition, clays account for nearly half the volume of sedimentary rocks.

Clay minerals are generally very fine grained, which makes identification difficult, unless studied microscopically. Their layered structure and weak bonding between layers give them a characteristic feel when wet. Clays are common in shales, mudstones, and other sedimentary rocks. Although clays are fine grained, they can form very thick beds or layers.

One of the most common clay minerals is *kaolinite*, which is used in the manufacture of fine chinaware and as a coating for *high-gloss paper*, such as that used in this textbook. Further, some clay minerals absorb large amounts of water, which allows

FIGURE 3.33   These parallel lines, called striations, are a distinguishing characteristic of the plagioclase feldspars. (Photo by E. J. Tarbuck)

FIGURE 3.34 Quartz. Some minerals, such as quartz, occur in a variety of colors. These samples include crystal quartz (colorless), amethyst (purple quartz), citrine (yellow quartz), and smoky quartz (gray to black). (Photo courtesy of E. J. Tarbuck)

them to swell to several times their normal size. These clays have been used commercially in a variety of ingenious ways, including as an additive to thicken milkshakes in fast-food restaurants.

The Dark Silicates

The **dark** (or **ferromagnesian**) **silicates** are those minerals containing ions of iron (iron = *ferro*) and/or magnesium in their structure. Because of their iron content, ferromagnesian silicates are dark in color and have a greater specific gravity, between 3.2 and 3.6, than nonferromagnesian silicates. The most common dark silicate minerals are olivine, the pyroxenes, the amphiboles, dark mica (biotite), and garnet.

Olivine Group *Olivine* is a family of high-temperature silicate minerals that are black to olive green in color and have a glassy luster and a conchoidal fracture (see Figure 3.31). Transparent olivine is occasionally used as a gemstone called peridot. Rather than developing large crystals, olivine commonly forms small, rounded crystals that give olivine-rich rocks a granular appearance. Olivine and related forms are thought to constitute up to 50 percent of Earth's upper mantle.

Pyroxene Group The *pyroxenes* are a group of complex minerals that are important components in dark colored igneous rocks. The most common member, *augite,* is a black, opaque

mineral with two directions of cleavage that meet at nearly a 90-degree angle (see Figure 3.31). Its crystalline structure consists of single chains of tetrahedra bonded together by ions of iron and magnesium. Because the silicon–oxygen bonds are stronger than the bonds joining the silicate structures, augite cleaves parallel to the silicate chains, as shown in Figure 3.35. Augite is one of the dominant minerals in basalt, a common igneous rock of the oceanic crust and volcanic areas on the continents (Figure 3.36).

Amphibole Group *Hornblende* is the most common member of a chemically complex group of minerals called *amphiboles* (see Figure 3.31). Hornblende is usually dark green to black in color, and except for its cleavage angles, which are about 60 degrees and 120 degrees, it is very similar in appearance to augite. The double chains of tetrahedra in the hornblende structure account for its particular cleavage (see Figure 3.35). In a rock, hornblende often forms elongated crystals. This helps distinguish it from pyroxene, which forms rather blocky crystals. Hornblende is found in igneous rocks, where it often makes up the dark portion of an otherwise light-colored rock.

Biotite *Biotite* is the dark, iron-rich member of the mica family (see Figure 3.31). Like other micas, biotite possesses a sheet structure that gives it excellent cleavage in one direction. Biotite also has a shiny black appearance that helps distinguish it from the other dark ferromagnesian minerals. Like hornblende, biotite is a common constituent of igneous rocks, including the rock granite.

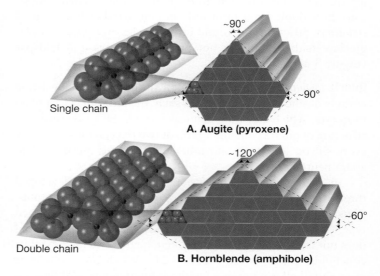

Single chain

A. Augite (pyroxene)

~90°

~90°

Double chain

~120°

~60°

B. Hornblende (amphibole)

FIGURE 3.35 Cleavage angles for augite and hornblende. Because hornblende's double chains are more weakly bonded than augite's single chains, cleavage is better developed in hornblende.

FIGURE 3.36 Researchers walking across hardened lava composed of basalt, Kilauea Volcano, Hawaii. Basalt is a common igneous rock composed of a large percentage of dark silicate minerals. (Photo by Roger Ressmeyer/CORBIS)

IMPORTANT NONSILICATE MINERALS

MATTER AND MINERALS
▶ Mineral Groups

Nonsilicate minerals are typically subdivided into *classes,* based on the anion (negatively charged ion) or complex anion that the members have in common (Table 3.2). For example, the *oxides* contain the negative oxygen ion (O^{2-}), which is bonded to one or more kinds of positive ions. Thus, within each mineral class, the basic structure and type of bonding is similar. As a result, the minerals in each group have similar physical properties that are useful in mineral identification.

Although the nonsilicates make up only about 8 percent of Earth's crust, some minerals, such as gypsum, calcite, and halite, occur as constituents in sedimentary rocks in significant amounts. Furthermore, many others are important economically. Table 3.2 lists some of the nonsilicate mineral classes and a few examples of each. A brief discussion of a few of the more common nonsilicate minerals follows.

Some of the most common nonsilicate minerals belong to one of three classes of minerals—the carbonates (CO_3^{2-}), the sulfates (SO_4^{2-}), and the halides (Cl^{1-}, F^{1-}, Br^{1-}). The carbonate minerals are much simpler structurally than the silicates (Figure 3.38). This mineral group is composed of the carbonate ion (CO_3^{2-}) and one or more kinds of anions. The two most common carbonate minerals are *calcite,* $CaCO_3$ (calcium carbonate), and *dolomite,* $CaMg(CO_3)_2$ (calcium/magnesium carbonate). Because these minerals are similar both physically and chemically, they are difficult to distinguish from each other. Both have a vitreous luster, a hardness between 3 and 4, and nearly perfect rhombic cleavage. They can, however, be distinguished by using dilute hydrochloric acid. Calcite reacts vigorously with this acid, whereas dolomite reacts much more slowly. Calcite and dolomite are usually found together as the primary constituents in the sedimentary rocks limestone and dolostone. When calcite is the dominant mineral, the rock is called *limestone,* whereas *dolostone* results from a predominance of dolomite. Limestone has many uses, including as road

Garnet *Garnet* is similar to olivine in that its structure is composed of individual tetrahedra linked by metallic ions. Also like olivine, garnet has a glassy luster, lacks cleavage, and exhibits conchoidal fracture. Although the colors of garnet are varied, this mineral is most often brown to deep red. Garnet readily forms equidimensional crystals that are most commonly found in metamorphic rocks (Figure 3.37). When garnets are transparent, they may be used as gemstones.

FIGURE 3.37   A deep-red garnet crystal embedded in a light-colored, mica-rich metamorphic rock. (Photo by E. J. Tarbuck)

Students Sometimes Ask . . .

I've seen garnet sandpaper at the hardware store. Is it really made of garnets?

Yes, and it's one of many things at the hardware store that's made of minerals! Hard minerals such as garnet (Mohs hardness = 6.5 to 7.5) and corundum (hardness = 9) make good abrasives. The abundance and hardness of garnets make them suitable for producing abrasive wheels, polishing materials, non-skid surfaces, and in sandblasting applications. Alternatively, those minerals that have low numbers on the Mohs hardness scale are commonly used as lubricants. For example, another mineral found in hardware stores is graphite (hardness = 1), which is used as an industrial lubricant.

aggregate, as building stone, and as the main ingredient in Portland cement.

Two other nonsilicate minerals frequently found in sedimentary rocks are *halite* and *gypsum*. Both minerals are commonly found in thick layers that are the last vestiges of ancient seas that have long since evaporated (Figure 3.39). Like limestone, both are important nonmetallic resources. Halite is the mineral name for common table salt (NaCl). Gypsum ($CaSO_4 \cdot 2 H_2O$), which is calcium sulfate with water bound

TABLE 3.2   Common Nonsilicate Mineral Classes

Mineral Groups (Key ions or elements)	Mineral Name	Chemical Formula	Economic Use
Carbonates (CO_3^{2-})	Calcite	$CaCO_3$	Portland cement, lime
	Dolomite	$CaMg(CO_3)_2$	Portland cement, lime
Halides (Cl^{1-}, F^{1-}, Br^{1-})	Halite	NaCl	Common salt
	Fluorite	CaF_2	Used in steelmaking
	Sylvite	KCl	Fertilizer
Oxides (O^{2-})	Hematite	Fe_2O_3	Ore of iron, pigment
	Magnetite	Fe_3O_4	Ore of iron
	Corundum	Al_2O_3	Gemstone, abrasive
	Ice	H_2O	Solid form of water
Sulfides (S^{2-})	Galena	PbS	Ore of lead
	Sphalerite	ZnS	Ore of zinc
	Pyrite	FeS_2	Sulfuric acid production
	Chalcopyrite	$CuFeS_2$	Ore of copper
	Cinnabar	HgS	Ore of mercury
Sulfates (SO_4^{2-})	Gypsum	$CaSO_4 \cdot 2 H_2O$	Plaster
	Anhydrite	$CaSO_4$	Plaster
	Barite	$BaSO_4$	Drilling mud
Native elements (single elements)	Gold	Au	Trade, jewelry
	Copper	Cu	Electrical conductor
	Diamond	C	Gemstone, abrasive
	Sulfur	S	Sulfa drugs, chemicals
	Graphite	C	Pencil lead, dry lubricant
	Silver	Ag	Jewelry, photography
	Platinum	Pt	Catalyst

FIGURE 3.38 Aragonite crystals. Aragonite is a carbonate mineral and one of the two common polymorphs of calcium carbonate ($CaCO_3$)—the other being calcite. Aragonite can be found in Carlsbad Caverns, and other caves, where it forms decorative stalactites. However, aragonite is most often produced by biological processes and is the main constituent of most mollusk shells. (Photo by Ross Frid)

FIGURE 3.39 Thick bed of halite (salt) at an underground mine in Grand Saline, Texas. Note person for scale. (Photo by Tom Bochsler)

into the structure, is the mineral of which plaster and other similar building materials are composed.

Most nonsilicate mineral classes contain members that are prized for their economic value. This includes the oxides, whose members hematite and magnetite are important ores of iron (Figure 3.40). Also significant are the sulfides, which are

UNDERSTANDING EARTH

BOX 3.3

Gemstones

Precious stones have been prized since antiquity. But misinformation abounds regarding gems and their mineral makeup. This stems partly from the ancient practice of grouping precious stones by color rather than mineral makeup. For example, *rubies* and red *spinels* are very similar in color, but they are completely different minerals. Classifying by color led to the more common spinels being passed off to royalty as rubies. Even today, with modern identification techniques, *yellow quartz* is sometimes sold as *topaz* which is rarer and more valuable.

FIGURE 3.C   Australian sapphires showing variation in cuts and colors. (Photo by Fred Ward, Black Star)

Naming Gemstones

Most precious stones are given names that differ from their parent mineral. For example, *sapphire* is one of two gems that are varieties of the same mineral, *corundum.* Trace elements can produce vivid sapphires of nearly every color (Figure 3.C). Tiny amounts of titanium and iron in corundum produce the most prized blue sapphires. When the mineral corundum contains a sufficient quantity of chromium, it exhibits a brilliant red color, and the gem is called *ruby.* Further, if a specimen is not suitable as a gem, it simply goes by the mineral name *corundum.* Because of its hardness, corundum that is not of gem quality is often crushed and sold as an abrasive.

To summarize, when corundum exhibits a red hue, it is called *ruby,* but if it exhibits any other color, the gem is called *sapphire.* Whereas corundum is the base mineral for two gems, quartz is the parent of more than a dozen gems. Table 3.A lists some well-known gemstones and their parent minerals.

What Constitutes a Gemstone?

When found in their natural state, most gemstones are dull and would be passed over by most people as "just another rock." Gems must be cut and polished by experienced professionals before their true beauty is displayed (Figure 3.C). (One of the methods used to shape a gemstone is *cleaving,* the act of splitting the mineral along one of its planes of weakness, or cleavage.) Only those mineral specimens that are of such quality that they can command a price in excess of the cost of processing are considered gemstones.

Gemstones can be divided into two categories: precious and semiprecious. A *precious* gem has beauty, durability, and rarity, whereas a *semiprecious* gem generally has only one or two of these qualities. The gems traditionally in highest esteem are diamonds, rubies, sapphires, emeralds, and some varieties of opal (Table 3.A). All other gemstones are classified as semiprecious. However, large, high-quality specimens of semiprecious stones often command a very high price.

Today translucent stones with evenly tinted colors are preferred. The most favored hues are red, blue, green, purple, rose, and yellow. The most prized stones are pigeon-blood rubies, blue sapphires, grass-green emeralds, and canary-yellow diamonds. Colorless gems are generally less than desirable except for diamonds that display "flashes of color" known as *brilliance.*

The durability of a gem depends on its hardness; that is, its resistance to abrasion by objects normally encountered in everyday living. For good durability, gems should be as hard or harder than quartz as defined by the Mohs scale of hardness. One notable exception is opal, which is comparatively soft (hardness 5 to 6.5) and brittle. Opal's esteem comes from its "fire," which is a display of a variety of brilliant colors, including greens, blues, and reds.

It seems to be human nature to treasure that which is rare. In the case of gemstones, large, high-quality specimens are much rarer than smaller stones. Thus, large rubies, diamonds, and emeralds, which are rare in addition to being beautiful and durable, command the very highest prices.

TABLE 3.A	Important Gemstones	
Gem	**Mineral Name**	**Prized Hues**
Precious		
Diamond	Diamond	Colorless, yellows
Emerald	Beryl	Greens
Opal	Opal	Brilliant hues
Ruby	Corundum	Reds
Sapphire	Corundum	Blues
Semiprecious		
Alexandrite	Chrysoberyl	Variable
Amethyst	Quartz	Purples
Cat's-eye	Chrysoberyl	Yellows
Chalcedony	Quartz (agate)	Banded
Citrine	Quartz	Yellows
Garnet	Garnet	Reds, greens
Jade	Jadeite or nephrite	Greens
Moonstone	Feldspar	Transparent blues
Peridot	Olivine	Olive greens
Smoky quartz	Quartz	Browns
Spinel	Spinel	Reds
Topaz	Topaz	Purples, reds
Tourmaline	Tourmaline	Reds, blue-greens
Turquoise	Turquoise	Blues
Zircon	Zircon	Reds

A. Magnetite   **B. Hematite**

FIGURE 3.40   Magnetite (**A**) and hematite (**B**) are both oxides and are both important ores of iron. (Photos by E. J. Tarbuck)

basically compounds of sulfur (S) and one or more metals. Examples of important sulfide minerals include galena (lead), sphalerite (zinc), and chalcopyrite (copper). In addition, native elements, including gold, silver, and carbon (diamonds), plus a host of other nonsilicate minerals—fluorite (flux in making steel), corundum (gemstone, abrasive), and uraninite (a uranium source)—are important economically.

CHAPTER 3 MINERALS: BUILDING BLOCKS OF ROCKS IN REVIEW

- A *mineral* is any naturally occurring inorganic solid that possesses an orderly crystalline structure and a well-defined chemical composition. Most *rocks* are aggregates composed of two or more minerals.

- All matter, including minerals, is composed of minute, indivisible particles called *atoms*—the building blocks of minerals. Each atom has a *nucleus*, which contains *protons* (particles with positive electrical charges) and *neutrons* (particles with neutral electrical charges). Surrounding the nucleus of an atom in regions called *principal shells,* are *electrons,* which have negative electrical charges. The number of protons in an atom's nucleus determines its *atomic number* and the name of the element. An *element* is a large collection of electrically neutral atoms, all having the same atomic number.

- Atoms combine with each other to form more complex substances called *compounds.* Atoms bond together by either gaining, losing, or sharing electrons with other atoms. In *ionic bonding,* one or more electrons are transferred from one atom to another, giving the atoms a net positive or negative charge. The resulting electrically charged atoms are called *ions.* Ionic compounds consist of oppositely charged ions assembled in a regular, crystalline structure that allows for the maximum attraction of ions, given their sizes. Another type of bond, the *covalent bond,* is produced when atoms share electrons.

- *Isotopes* are variants of the same element that have a different *mass number* (the total number of neutrons plus protons found in an atom's nucleus). Some isotopes are unstable and disintegrate naturally through a process called *radioactivity.*

- Mineralogists use the term *crystal* or *crystalline* in reference to *any natural solid with an ordered, repetitive, atomic structure.* Minerals form through the process of *crystallization,* which occurs when material precipitates out of a solution, as magma cools and crystallizes, or in high-temperature and high-pressure metamorphic environments.

- The basic building blocks of minerals, called *unit cells,* consist of an array of cations and anions arranged so that the positive and negative charges cancel each other out. These unit cells stack together in a regular manner that relates to the shape and symmetry of a crystal. Thus, the *angles between equivalent faces of crystals of the same mineral are always the same,* an observation known as *Steno's Law.*

- The chemical composition of some minerals varies from sample to sample, because ions of similar size can substitute for one another. It is also common for two minerals with exactly the same chemical composition to have different internal structures, and hence, different external forms. Minerals of this type are called *polymorphs.*

- The properties of minerals include *crystal shape (habit), luster, color, streak, tenacity, hardness, cleavage, fracture,* and *density* or *specific gravity*. In addition, a number of special physical and chemical properties (*taste, smell, elasticity, feel, magnetism, double refraction,* and *chemical reaction to hydrochloric acid*) are useful in identifying certain minerals. Each mineral has a unique set of properties that can be used for identification.

- Of the nearly 4000 minerals, no more than a few dozen make up most of the rocks of Earth's crust and, as such, are classified as rock-forming minerals. Eight elements (oxygen, silicon, aluminum, iron, calcium, sodium, potassium, and magnesium) make up the bulk of these minerals and represent over 98 percent (by weight) of Earth's continental crust.

- The most common mineral group is the *silicates*. All silicate minerals have the negatively charged *silicon–oxygen tetrahedron* as their fundamental building block. In some silicate minerals the tetrahedra are joined in chains (the pyroxene and amphibole groups); in others, the tetrahedra are arranged into sheets (the micas—biotite and muscovite), or three-dimensional networks (the feldspars and quartz). The tetrahedra and various silicate structures are often bonded together by the positive ions of iron, magnesium, potassium, sodium, aluminum, and calcium. Each silicate mineral has a structure and a chemical composition that indicates the conditions under which it formed.

- The *nonsilicate* mineral groups, which contain several economically important minerals, include the *oxides* (e.g., the mineral hematite, mined for iron), *sulfides* (e.g., the mineral sphalerite, mined for zinc, and the mineral galena, mined for lead), *sulfates, halides,* and *native elements* (e.g., gold and silver). The more common nonsilicate rock-forming minerals include the *carbonate minerals,* calcite and dolomite. Two other nonsilicate minerals frequently found in sedimentary rocks are halite and gypsum.

KEY TERMS

atom (p. 76)
atomic number (p. 77)
chemical bond (p. 78)
chemical compound (p. 77)
cleavage (p. 89)
color (p. 87)
Constancy of Interfacial
   Angles, Law of (p. 83)
covalent bond (p. 80)
crystal (p. 81)
crystalline (p. 81)
crystallization (p. 81)
crystal shape (p. 88)

dark silicates (p. 98)
density (p. 89)
electron (p. 76)
element (p. 77)
ferromagnesian silicates (p. 98)
fracture (p. 89)
habit (p. 88)
hardness (p. 88)
ion (p. 79)
ionic bond (p. 79)
isotope (p. 81)
light silicates (p. 95)
luster (p. 87)

mass number (p. 81)
metallic bond (p. 80)
mineral (p. 74)
mineralogy (p. 74)
Mohs scale (p. 88)
neutron (p. 76)
nonferromagnesian silicates
   (p. 95)
nonsilicates (p. 93)
nucleus (p. 76)
periodic table (p. 77)
polymerization (p.93)
polymorph (p. 86)

principal shells (p. 77)
proton (p. 76)
radioactive decay
   (p. 81)
rock (p. 75)
silicate (p. 93)
silicon–oxygen tetrahedron
   (p. 93)
specific gravity (p. 89)
Steno's law (p. 83)
streak (p. 87)
tenacity (p. 88)
valence electron (p. 77)

QUESTIONS FOR REVIEW

1. List five characteristics an Earth material should have in order to be considered a mineral.

2. Define the term *rock*. How are rocks different than minerals?

3. List the three main particles of an atom and explain how they differ from one another.

4. If the number of electrons in a neutral atom is 35 and its mass number is 80, calculate the following:
   a. the number of protons
   b. the atomic number
   c. the number of neutrons

5. What is the significance of valence electrons?

6. Briefly distinguish between ionic and covalent bonding.

7. What occurs in an atom to produce a cation? An anion?

8. What is an isotope?

9. What do minerologists mean when they use the word *crystal*?

10. Describe three ways a mineral might form (crystallize).

11. What is Steno's Law?

12. What are polymorphs? How are they similar?

13. Why might it be difficult to identify a mineral by its color?

14. If you found a glassy-appearing mineral while rock hunting and had hopes that it was a diamond, what simple test might help you make a determination?

15. Explain the use of corundum as given in Table 3.2 (p. 100) in terms of the Mohs hardness scale.

16. Gold has a specific gravity of almost 20. A five-gallon pail of water weighs 40 pounds. How much would a five-gallon pail of gold weigh?

17. What is meant when we refer to a mineral's tenacity? List three terms that describe tenacity.

18. On what basis are minerals placed in mineral classes?

19. List the eight most common elements in Earth's crust in order of abundance (most to least).

20. Explain the difference between the terms *silicon* and *silicate*.

21. Describe the silicon–oxygen tetrahedron.

22. What do ferromagnesian minerals have in common? List examples of ferromagnesian minerals.

23. What do muscovite and biotite have in common? How do they differ?

24. Should color be used to distinguish between orthoclase and plagioclase feldspar? What is the best means of distinguishing between these two types of feldspar?

25. Each of the following statements describes a silicate mineral or mineral group. In each case, provide the appropriate name:

   a. the most common member of the amphibole group

   b. the most common nonferromagnesian member of the mica family

   c. the only common silicate mineral made entirely of silicon and oxygen

   d. a high-temperature silicate with a name that is based on its color

   e. a silicate mineral that is characterized by striations

   f. a silicate mineral that originates as a product of chemical weathering

26. What simple test can be used to distinguish calcite from dolomite?

COMPANION WEBSITE

The *Earth 10e* website uses the resources and flexibility of the Internet to aid in your study of the topics in this chapter. Written and developed by the authors and other geology instructors, this site will help improve your understanding of geology. Visit **www.mygeoscienceplace.com** in order to:

- **Review** key chapter concepts.

- **Read** with links to the eBook and to chapter-specific web resources.

- **Visualize** and comprehend challenging topics using learning activities in *GEODe Earth*.

- **Test** yourself with online quizzes.

GEODe EARTH

GEODe Earth is a valuable and easy to use learning aid that can be accessed from your book's Companion Website (**www.mygeoscienceplace.com**). It is a dynamic instructional tool that promotes understanding and reinforces important concepts by using tutorials, animations, and exercises that actively engage the student.

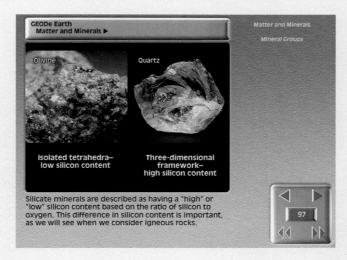

MAGMA, IGNEOUS ROCKS, AND INTRUSIVE ACTIVITY

Half Dome is a large granitic monolith in California's Yosemite National Park.

(PHOTO COREY RICH/ AURORA PHOTOS)

igneous rocks and metamorphic rocks derived from igneous parents" make up about 95 percent of Earth's crust. Furthermore, the mantle, which accounts for more than 82 percent of Earth's volume, is composed entirely of igneous rock. Thus, Earth can be described as a huge mass of igneous rock covered with a thin veneer of sedimentary rock and having a relatively small iron-rich core. Consequently, a basic knowledge of igneous rocks is essential to our understanding of the structure, composition, and internal workings of our planet.

Many prominent landforms are composed of igneous rocks including volcanoes such as Mt. Ranier and Mt. Hood, and the large igneous bodies that make up the Sierra Nevadas. Furthermore, igneous rocks make excellent building stones and are widely used for decorative materials including monuments and household countertops.

MAGMA: THE PARENT MATERIAL OF IGNEOUS ROCK

IGNEOUS ROCKS
▶ Introduction

In our discussion of the rock cycle, it was noted that **igneous rocks** (*ignis* = fire) form as molten rock cools and solidifies. Considerable evidence supports the idea that the parent material for igneous rocks, called **magma,** is formed by melting that occurs at various levels within Earth's crust and upper mantle to depths of perhaps 250 kilometers (about 150 miles).

Once formed, a magma body buoyantly rises toward the surface because it is less dense than the surrounding rocks. (When rock melts it takes up more space and, hence, it becomes less dense than the surrounding solid rock.) Occasionally molten rock reaches Earth's surface where it is called **lava.** Sometimes lava is emitted as fountains that are produced when escaping gasses propel it from a magma chamber. On other occasions, magma is explosively ejected, producing dramatic steam and ash eruptions. However, not all eruptions are violent; many volcanoes emit quiet outpourings of very fluid lava (Figure 4.1).

The Nature of Magma

Magma is completely or partly molten rock, which on cooling solidifies to form an igneous rock composed of silicate minerals. Most magmas consist of three distinct parts—a *liquid component,* a *solid component,* and a *gaseous phase.*

The liquid portion, called **melt,** is composed mainly of mobile ions of the eight most common elements found in Earth's crust—silicon and oxygen, along with lesser amounts of aluminum, potassium, calcium, sodium, iron, and magnesium.

The solid components (if any) in magma are silicate minerals that have already crystallized from the melt. As a magma body cools, the size and number of crystals increase. During the last stage of cooling, a magma body is like a "crystalline mush" with only small amounts of melt.

The gaseous components of magma, called **volatiles,** are materials that will vaporize (form a gas) at surface pressures. The most common volatiles found in magma are water vapor (H_2O), carbon dioxide (CO_2), and sulfur dioxide (SO_2), which are confined by the immense pressure exerted by the overlying rocks. These gases tend to separate from the melt as it moves toward the surface (low-pressure environment). As the gases build up, they may eventually propel magma from the vent. When deeply buried magma bodies crystallize, the remaining volatiles collect as hot, water-rich fluids that migrate through the surrounding rocks. These hot fluids play an important role in metamorphism and will be considered in Chapter 8.

From Magma to Crystalline Rock

To better understand how magma crystallizes, let us consider how a simple crystalline solid melts. Recall that, in any crystalline solid, the ions are arranged in a closely packed regular pattern. However, they are not without some motion—they exhibit a sort of restricted vibration about fixed points. As tem-

FIGURE 4.1 Fluid basaltic lava emitted from Hawaii's Kilauea Volcano. (Photo by G. Brad Lewis/Liaison Agency, Inc.)

perature rises, ions vibrate more rapidly and consequently collide with ever-increasing vigor with their neighbors. Thus, heating causes the ions to occupy more space, which in turn causes the solid to expand. When the ions are vibrating rapidly enough to overcome the force of their chemical bonds, melting occurs. At this stage the ions are able to slide past one another, and the orderly crystalline structure disintegrates. Thus, melting converts a solid consisting of tight, uniformly packed ions into a liquid composed of unordered ions moving randomly about.

In the process called **crystallization**, cooling reverses the events of melting. As the temperature of the liquid drops, ions pack more closely together as their rate of movement slows. When cooled sufficiently, the forces of the chemical bonds will again confine the ions to an orderly crystalline arrangement.

When magma cools, it is generally the silicon and oxygen atoms that link together first to form silicon–oxygen tetrahedra, the basic building blocks of the silicate minerals. As magma continues to lose heat to its surroundings, the tetrahedra join with each other and with other ions to form embryonic crystal nuclei. Slowly each nucleus grows as ions lose their mobility and join the crystalline network.

The earliest formed minerals have space to grow and tend to have better-developed crystal faces than do the later ones that occupy the remaining spaces. Eventually all of the melt is transformed into a solid mass of interlocking silicate minerals that we call an *igneous rock* (Figure 4.2).

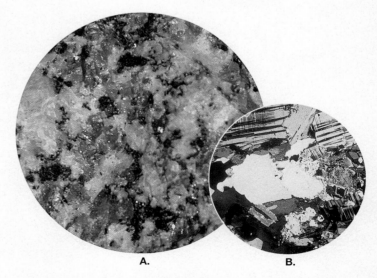

A.   **B.**

FIGURE 4.2 Igneous rock composed of interlocking crystals. **A.** close-up of interlocking crystals in a coarse-grained igneous rock. The largest crystals are about 2 centimeters in length. **B.** Photomicrograph of interlocking crystals in a coarse-grained igneous rock. (Photos by E. J. Tarbuck)

Students Sometimes Ask . . .

Are lava and magma the same thing?

No, but their *composition* might be similar. Both are terms that describe molten or liquid rock: Magma exists beneath Earth's surface, and lava is molten rock that has reached the surface. That's the reason why they can be similar in composition. Lava is produced from magma, but it generally has lost materials that escape as a gas, such as water vapor.

As you will see, the crystallization of magma is much more complex than just described. Whereas a simple compound, such as water, solidifies at a specific temperature, crystallization of magma with its diverse chemistry spans a temperature range of 200 °C, or more. In addition, magmas differ from one another in terms of their chemical composition, the amount of volatiles they contain, and the rate at which they cool. Because all of these factors influence the crystallization process, the appearance and mineral make-up of igneous rocks varies widely.

IGNEOUS PROCESSES

Igneous rocks form in two basic settings. Magma may crystallize at depth or lava may solidify at Earth's surface. When magma loses its mobility before reaching the surface it eventually crystallizes to form **intrusive igneous rocks.** They are also known as **plutonic rocks**—after Pluto, the god of the lower world in classical mythology. Intrusive igneous rocks are coarse-grained and consist of visible mineral crystals. These rocks are observed at the surface in locations where uplifting and erosion have stripped away the overlying rocks. Exposures of intrusive igneous rocks occur in many places, including Mount Washington, New Hampshire; Stone Mountain, Georgia; the Black Hills of South Dakota; and Yosemite National Park, California (Figure 4.3).

Igneous rocks that form when molten rock solidifies *at the surface* are classified as **extrusive igneous rocks.** They are also called **volcanic rocks**—after the Roman fire god, Vulcan. Extrusive igneous rocks form when lava solidifies, in which case they tend to be fine-grained, or when volcanic debris falls to Earth's surface. Extrusive igneous rocks are abundant in western portions of the Americas where they make up the volcanic peaks of the Cascade Range and the Andes Mountains. In addition, many oceanic islands, including the Hawaiian chain and Alaska's Aleutian Islands, are composed almost entirely of extrusive igneous rocks.

IGNEOUS COMPOSITIONS

IGNEOUS ROCKS
▶ Igneous Compositions

Igneous rocks are composed mainly of silicate minerals. Chemical analyses show that silicon and oxygen are by far the most abundant constituents of igneous rocks. These two elements, plus ions of aluminum (Al), calcium (Ca), sodium (Na), potassium (K), magnesium (Mg), and iron (Fe), make up roughly 98 percent, by weight, of most magmas. In addition, magma contains small amounts of many other elements, including titanium and manganese, and trace amounts of much rarer elements such as gold, silver, and uranium.

As magma cools and solidifies, these elements combine to form two major groups of silicate minerals. The *dark* (or *ferromagnesian*) *silicates* are rich in iron and/or magnesium and comparatively low in silica. *Olivine, pyroxene, amphibole,* and *biotite mica* are the common dark silicate minerals of Earth's crust. By contrast, the *light* (or *nonferromagnesian*) *silicates* contain greater amounts of potassium, sodium, and calcium rather than iron and magnesium. As a group, nonferromagnesian minerals are richer in silica than the dark silicates. The light silicates include *quartz, muscovite mica,* and the most abundant mineral group, the *feldspars.* Feldspars make up at least 40 percent of most igneous rocks. Thus, in addition to feldspar, igneous rocks contain some combination of the other light and/or dark silicates listed above.

Granitic (Felsic) Versus Basaltic (Mafic) Compositions

Despite their great compositional diversity, igneous rocks (and the magmas from which they form) can be divided into broad groups according to their proportions of light and dark minerals (Figure 4.4). Near one end of the continuum are rocks composed almost entirely of light-colored silicates—quartz and feldspar. Igneous rocks in which these are the dominant minerals have a **granitic composition.** Geologists also refer to granitic rocks as being **felsic,** a term derived from *feld*spar and *si*lica (quartz). In addition to quartz and feldspar, most granitic rocks contain about 10 percent dark silicate minerals, usually biotite mica and amphibole. Granitic rocks are rich in silica (about 70 percent) and are major constituents of the continental crust.

Rocks that contain substantial dark silicate minerals and calcium-rich plagioclase feldspar (but no quartz) are said to have a **basaltic composition** (see Figure 4.4).

FIGURE 4.3   Mount Rushmore National Memorial, located in the Black Hills of South Dakota, is carved from intrusive igneous rocks. (Photo by Marc Muench)

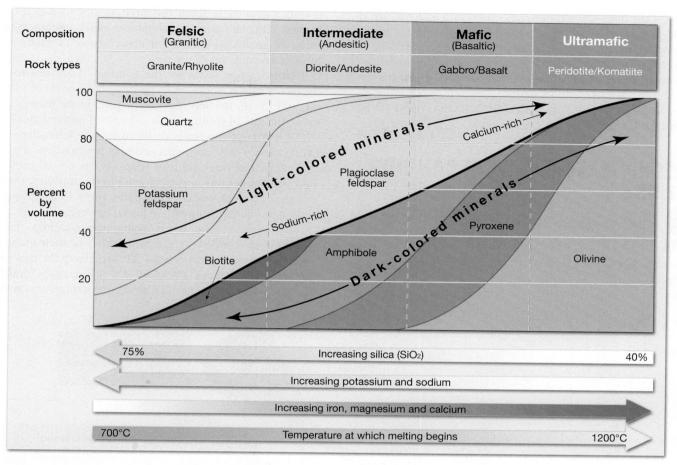

FIGURE 4.4 Mineralogy of common igneous rocks and the magmas from which they form. (After Dietrich, Daily, and Larsen)

Basaltic rocks contain a high percentage of ferromagnesian minerals, so geologists also refer to them as **mafic** (from *magnesium* and *ferrum*, the Latin name for iron). Because of their iron content, mafic rocks are typically darker and denser than granitic rocks. Basaltic rocks make up the ocean floor as well as many of the volcanic islands located within the ocean basins. Basalt also forms extensive lava flows on the continents.

Other Compositional Groups

As you can see in Figure 4.4, rocks with a composition between granitic and basaltic rocks are said to have an **intermediate,** or **andesitic composition** after the common volcanic rock *andesite*. Intermediate rocks contain at least 25 percent dark silicate minerals, mainly amphibole, pyroxene, and biotite mica with the other dominant mineral being plagioclase feldspar. This important category of igneous rocks is associated with volcanic activity that is typically confined to the margins of the continents.

Another important igneous rock, *peridotite,* contains mostly olivine and pyroxene and thus falls on the opposite side of the compositional spectrum from granitic rocks (see Figure 4.4). Because peridotite is composed almost entirely of ferromagnesian minerals, its chemical composition is referred to as **ultramafic.** Although ultramafic rocks are rare at Earth's surface, peridotite is the main constituent of the upper mantle.

Silica Content as an Indicator of Composition

An important aspect of the chemical composition of igneous rocks is silica (SiO_2) content. Typically, the silica content of crustal rocks ranges from a low of about 40 percent in ultramafic rocks to a high of more than 70 percent in granitic rocks (see Figure 4.4). The percentage of silica in igneous rocks actually varies in a systematic manner that parallels the abundance of other elements. For example, rocks that are relatively low in silica contain large amounts of iron, magnesium, and calcium. By contrast, rocks high in silica contain very little iron, magnesium, or calcium but are enriched with sodium and potassium. Consequently, the chemical makeup of an igneous rock can be inferred directly from its silica content.

Further, the amount of silica present in magma strongly influences its behavior. Granitic magma, which has a high silica content, is quite viscous ("thick") and may erupt at temperatures as low as 700 °C. On the other hand, basaltic magmas are low in silica and are generally more fluid. Basaltic magmas also erupt at higher temperatures than granitic magmas—usually at temperatures between 1100 and 1250 °C and are completely solid when cooled to 1000 °C.

In summary, igneous rocks can be divided into broad groups according to the proportions of light and dark minerals they contain. *Granitic (felsic) rocks,* which are composed almost entirely of the light-colored minerals quartz and feldspar, are at one end of the compositional spectrum (see Figure 4.4). *Basaltic (mafic) rocks,* which contain abundant dark silicate minerals in addition to plagioclase feldspar, make up the other major igneous rock group of Earth's crust. Between these groups are rocks with an *intermediate (andesitic) composition. Uultramafic rocks,* which lack light-colored minerals, lie at the far end of the compositional spectrum from granitic rocks.

IGNEOUS TEXTURES: WHAT CAN THEY TELL US?

IGNEOUS ROCKS
▶ Igneous Textures

The term **texture** is used to describe the overall appearance of a rock based on the size, shape, and arrangement of its mineral grains (Figure 4.5). Texture is an important property because it reveals a great deal about the environment in which the rock formed. This fact allows geologists to make inferences about a rock's origin based on careful observations of grain size and other characteristics of the rock.

Three factors influence the textures of igneous rocks: (1) *the rate at which molten rock cools;* (2) *the amount of silica present;* and (3) *the amount of dissolved gases in the magma.* Among these, the rate of cooling tends to be the dominant factor.

A very large magma body located many kilometers beneath Earth's surface will cool over a period of perhaps tens to hundreds of thousands of years. Initially, relatively few crystal nuclei form. Slow cooling permits ions to migrate freely until they eventually join one of the existing crystalline structures. Consequently, slow cooling promotes the growth of fewer, but larger crystals.

On the other hand, when cooling occurs rapidly—for example, in a thin lava flow—the ions quickly lose their mobility and readily combine to form crystals. This results in the development of numerous embryonic nuclei, all of which compete for the available ions. The result is a solid mass of tiny intergrown crystals.

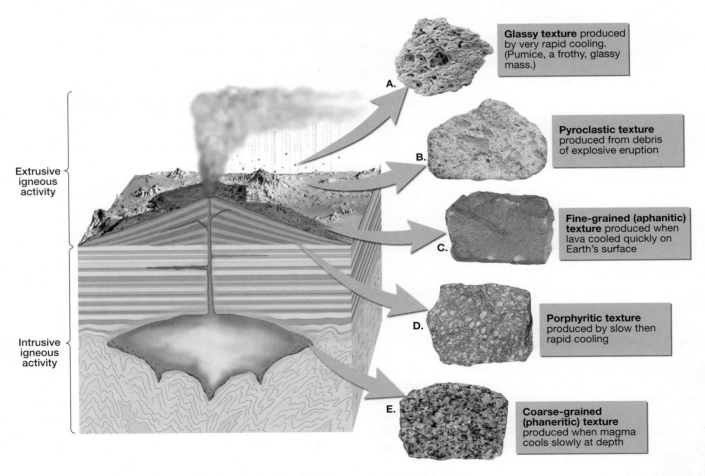

Glassy texture produced by very rapid cooling. (Pumice, a frothy, glassy mass.)

Pyroclastic texture produced from debris of explosive eruption

Fine-grained (aphanitic) texture produced when lava cooled quickly on Earth's surface

Porphyritic texture produced by slow then rapid cooling

Coarse-grained (phaneritic) texture produced when magma cools slowly at depth

Extrusive igneous activity

Intrusive igneous activity

FIGURE 4.5   Igneous rock textures. **A.** During a volcanic eruption in which silica-rich lava is ejected into the atmosphere, a frothy glass called pumice may form. **B.** Rocks that exhibit a *pyroclastic texture* are a result of the consolidation of rock fragments that were ejected during a violent volcanic eruption. **C.** Igneous rocks that crystallize at or near Earths surface cool quickly and often exhibit a *fine-grained (aphanitic) texture.* **D.** A *porphyritic texture* results when magma that already contains some large crystals migrates to a new location where the rate of cooling increases. The resulting rock consists of larger crystals (*phenocrysts*) embedded within a matrix of smaller crystals (*groundmass*). **E.** Coarse-grained (*phaneritic*) igneous rocks form when magma slowly crystallizes at depth. (Photos by E. J. Tarbuck)

FIGURE 4.12 Rocks contain information about the processes that produce them. This massive granitic monolith (El Capitan) located in Yosemite National Park, California, was once a molten mass found deep within Earth. (Photo by Enrique R. Aquirre/Photolibrary)

Granite is a very abundant rock. However, it has become common practice among geologists to apply the term *granite* to any coarse-grained intrusive rock composed predominantly of light silicate minerals. We will follow this practice for the sake of simplicity. You should keep in mind that this use of the term *granite* covers rocks having a wide range of mineral compositions.

Rhyolite Rhyolite is the extrusive equivalent of granite and, like granite, is composed essentially of the light-colored silicates (see Figure 4.11). This fact accounts for its color, which is usually buff to pink or occasionally very light gray. Rhyolite is fine-grained and frequently contains glass fragments and voids, indicating rapid cooling in a surface environment. When rhyolite contains phenocrysts, they are small and composed of either quartz or potassium feldspar. In contrast to granite, which is widely distributed as large plutonic masses, rhyolite deposits are

Students Sometimes Ask . . .

You mentioned that Native Americans used obsidian for making arrowheads and cutting tools. Is this the only material they used?

No. Native Americans used whatever materials were locally available to make tools, including any hard compact rock material that could be shaped. This includes materials such as the metamorphic rocks slate and quartzite, sedimentary deposits made of silica called jasper, chert, opal, flint, and even jade. Some of these deposits have a limited geographic distribution and thus can help anthropologists reconstruct trade routes between different groups.

UNDERSTANDING EARTH

BOX 4.1

Thin Sections and Rock Identification

Igneous rocks are classified on the basis of their mineral composition and texture. When analyzing specimens, geologists examine them closely to identify the minerals present and to determine the size and arrangement of the interlocking crystals. When out in the field, geologists use megascopic techniques to study rocks. The *megascopic* characteristics of rocks are those features that can be determined with the unaided eye or by using a low-magnification (10X) hand lens. When practical to do so, geologists collect hand samples that can be taken back to the laboratory where *microscopic,* or high-magnification, methods can be employed. Microscopic examination is important to identify trace minerals, as well as those textural features that are too small to be visible to the unaided eye.

Because most rocks are not transparent, microscopic work requires the preparation of a very thin slice of rock known as a *thin section* (Figure 4.A, part B). First, a saw containing diamonds embedded in its blade is used to cut a narrow slab from the sample. Next, one side of the slab is polished using grinding powder and then cemented to a microscope slide. Once the mounted sample is firmly in place, the other side of it is ground to a thickness of about 0.03 millimeter. When a slice of rock is that thin, it is usually transparent. But, some minerals, no matter how thin they are ground, remain opaque. These are metallic minerals, usually pyrite and magnetite.

Once produced, thin sections are examined under a specially designed microscope called a *polarizing microscope.* Such an instrument has a light source beneath the stage so that light can be transmitted upward through the thin section. Because minerals have crystalline structures that influence polarized light in a measurable way, this procedure allows for the identification of even the smallest components of a rock. Part C of Figure 4.A is a photomicrograph (photo taken through a microscope) of a thin section of granite shown under polarized light. The mineral constituents are identified by their unique optical properties. In addition to assisting in the study of igneous rocks, microscopic techniques are used with great success in analyzing sedimentary and metamorphic rocks as well.

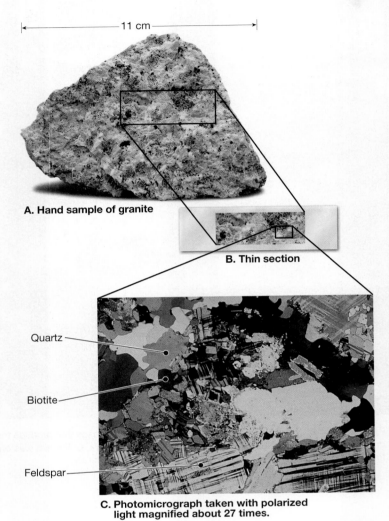

A. Hand sample of granite

B. Thin section

Quartz

Biotite

Feldspar

C. Photomicrograph taken with polarized light magnified about 27 times.

FIGURE 4.A Thin sections are very useful in identifying the mineral constituents in rocks. **A.** A slice of rock is cut from a hand sample using a diamond saw. **B.** This slice is cemented to a microscope slide and ground until it is transparent to light (about 0.03 millimeter thick). This very thin slice of rock is called a *thin section.* **C.** A thin section of granite viewed under polarized light. (Photos by E. J. Tarbuck)

less common and generally less voluminous. Yellowstone Park is one well-known exception. Here, rhyolite lava flows and thick ash deposits of similar composition are extensive.

Obsidian *Obsidian* is a dark-colored glassy rock that usually forms when silica-rich lava is quenched quickly (Figure 4.13). In contrast to the orderly arrangement of ions characteristic of minerals, *the ions in glass are unordered.* Consequently, glassy rocks such as obsidian are not composed of minerals in the same sense as most other rocks.

Although usually black or reddish-brown in color, obsidian has a composition that is more akin to light-colored igneous

rocks such as granite, rather than to dark rocks such as basalt. Obsidian's dark color results from small amounts of metallic ions in an otherwise relatively clear, glassy substance. If you examine a thin edge, obsidian will appear nearly transparent (see Figure 4.7).

Pumice *Pumice* is a volcanic rock with a glassy texture that forms when large amounts of gas escape through silica-rich lava to generate a gray, frothy mass (Figure 4.14). In some samples, the voids are quite noticeable, whereas in others the pumice resembles fine shards of intertwined glass. Because of the large percentage of voids, many samples of pumice will

B. Hand sample of obsidan.

A. Obsidan flow.

FIGURE 4.13 Obsidian is a dark-colored, glassy rock formed from silica-rich lava. The scene in part A shows the leading edge of an obsidian lava flow located south of Mono Lake, California. The large banded block in the lower portion of the image contains white layers of porous pumice-like material. (Photos by E. J. Tarbuck)

float when placed in water. Oftentimes, flow lines are visible in pumice, indicating that some movement occurred before solidification was complete. Moreover, pumice and obsidian can often be found in the same rock mass, where they exist in alternating layers.

Intermediate (Andesitic) Igneous Rocks

Andesite Andesite is a medium-gray, fine-grained rock of volcanic origin. Its name comes from South America's Andes Moun-

← 2 cm →

FIGURE 4.14 Pumice, a glassy rock, is very lightweight because it contains numerous vesicles. (Inset photo by Chip Clark)

tains, where numerous volcanoes are composed of this rock type. In addition to the volcanoes of the Andes and the Cascade Range, many of the volcanic structures occupying the continental margins that surround the Pacific Ocean are of andesitic composition. Andesite commonly exhibits a porphyritic texture (see Figure 4.11). When this is the case, the phenocrysts are often light, rectangular crystals of plagioclase feldspar or black, elongated amphibole crystals. Andesite often resembles rhyolite, so their identification usually requires microscopic examination to verify mineral makeup.

Diorite Diorite is the plutonic equivalent of andesite. It is a phaneritic rock that looks somewhat similar to gray granite. However, it can be distinguished from granite by the absence of visible quartz crystals and because it contains a higher percentage of dark silicate minerals. The mineral makeup of diorite is primarily sodium-rich plagioclase feldspar and amphibole, with lesser amounts of biotite. Because the light-colored feldspar grains and dark amphibole crystals appear to be roughly equal in abundance, diorite has a salt-and-pepper appearance (see Figure 4.11).

Mafic (Basaltic) Igneous Rocks

Basalt Basalt is a very dark green to black, aphaenitic rock composed primarily of pyroxene and calcium-rich plagioclase feldspar, with lesser amounts of olivine and amphibole (see Figure 4.11). When porphyritic, basalt commonly contains small light-colored feldspar phenocrysts or green, glassy-appearing olivine phenocrysts embedded in a dark groundmass.

Basalt is the most common extrusive igneous rock. Many volcanic islands, such as the Hawaiian Islands and Iceland, are composed mainly of basalt. Further, the upper layers of the oceanic crust consist of basalt. In the United States, large portions of central Oregon and Washington were the sites of extensive basaltic outpourings (Figure 4.15). At some locations, these once fluid basaltic flows have accumulated to thicknesses approaching 3 kilometers.

Gabbro Gabbro is the intrusive equivalent of basalt (see Figure 4.11). Like basalt, it tends to be dark green to black in color and composed primarily of pyroxene and calcium-rich plagioclase feldspar. Although gabbro is uncommon in the continental crust, it makes up a significant percentage of oceanic crust.

Pyroclastic Rocks

Pyroclastic rocks are composed of fragments ejected during a volcanic eruption. One of the most common pyroclastic rocks, called *tuff,* is composed mainly of tiny, ash-size fragments that

FIGURE 4.15 The Columbia River basalts. At some locations these once fluid lava flows are more than 3 kilometers thick. Although basalt is usually black, it often weathers to a reddish-brown color. (Photo by Willianborg)

were later cemented together (see Figure 4.5B). In situations where the ash particles remained hot enough to fuse, the rock is called *welded tuff*. Although welded tuff consists mostly of tiny glass shards, it may contain walnut-size pieces of pumice and other rock fragments.

Welded tuffs blanket vast portions of once volcanically active areas of the western United States (Figure 4.16). Some of these tuff deposits are hundreds of feet thick and extend for tens of miles from their source. Most formed millions of years ago as volcanic ash spewed from large volcanic structures (calderas) in an avalanche style, spreading laterally at speeds approaching 100 kilometers per hour. Early investigators of these deposits incorrectly classified them as rhyolite lava flows. Today, we know that silica-rich lava is too viscous (thick) to flow more than a few miles from a vent.

Pyroclastic rocks composed mainly of particles larger than ash are called *volcanic breccia*. The particles in volcanic breccia can consist of streamlined fragments that solidified in air, blocks broken from the walls of the vent, crystals, and glass fragments.

Unlike most igneous rock names, such as granite and basalt, the terms *tuff* and *volcanic breccia* do not imply mineral

Students Sometimes Ask . . .

At a store, I saw a barbecue grill with material the clerk called "lava rock." Is this really a volcanic rock?

Not only is "lava rock" at your hardware store, it's also at home-improvement stores for use as a building and landscaping material and is frequently found in aquarium-supply stores. Geologists call this material *scoria*, which is a red or dark mafic rock characterized by a vesicular texture (full of holes). It's also called *volcanic cinder*. In gas barbecue grills, lava rock is used to absorb and reradiate heat to ensure even cooking.

composition. Thus, they are frequently used with a modifier, as, for example, rhyolite tuff.

ORIGIN OF MAGMA

Most magma originates in the uppermost mantle. The greatest quantities are produced at divergent plate boundaries in association with seafloor spreading. Lesser amounts form at subduction zones, where oceanic lithosphere descends into the mantle. In addition, magma can also originate far from plate boundaries.

Generating Magma from Solid Rock

Based on evidence from the study of earthquake waves, *Earth's crust and mantle are composed primarily of solid, not molten, rock.* Although the outer core is fluid, this iron-rich material is very dense and remains deep within Earth. So where does magma come from?

Increase in Temperature Most magma originates when essentially solid rock, located in the crust and upper mantle, melts. The most obvious way to generate magma from solid rock is to raise the temperature above the rock's melting point.

Workers in underground mines know that temperatures get higher as they go deeper. Although the rate of temperature change varies considerably from place to place, it *averages* about 25 °C per kilometer in the *upper* crust. This increase in

FIGURE 4.16 Outcrop of welded tuff from Valles Caldera near Los Alamos, New Mexico. Tuff is composed mainly of ash-sized particles and may contain larger fragments of pumice or other volcanic rocks. (Photo by Marli Miller)

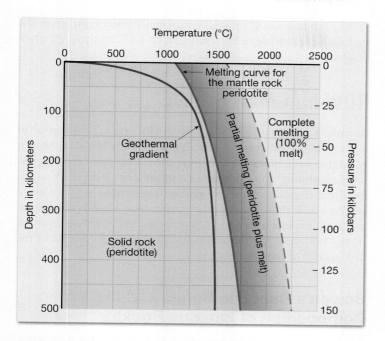

FIGURE 4.17 A schematic diagram illustrating a typical geothermal gradient (increase in temperature with depth) for the crust and upper mantle. Also illustrated is an idealized curve that depicts the melting point temperatures for the mantle rock peridotite. Notice that when the geothermal gradient is compared to the melting point curve for peridotite, the temperature at which peridotite melts is everywhere higher than the geothermal gradient. Thus, under normal conditions the mantle is solid. Special circumstances are required to generate magma.

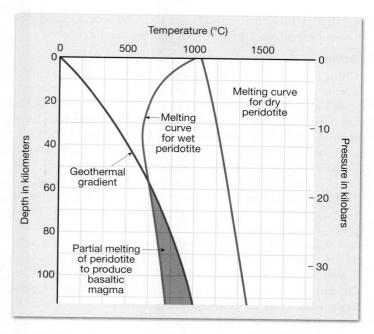

FIGURE 4.18 Idealized melting temperature curves. These curves portray the minimum temperatures required to melt rock within Earth. Notice that dry peridotite melts at higher temperatures with increasing depth. By contrast, the temperature at which wet peridotite begins to melt actually decreases with depth.

temperature with depth, known as the **geothermal gradient**, is somewhat higher beneath the oceans than beneath the continents (Figure 4.17). As shown in Figure 4.17, when a typical geothermal gradient is compared to the melting point curve for the mantle rock peridotite, the temperature at which peridotite melts is everywhere higher than the geothermal gradient. Thus, under normal conditions, the mantle is solid. As you will see, tectonic processes exist that can increase the geothermal gradient sufficiently to trigger melting. In addition, other mechanisms exist that trigger melting by reducing the temperature at which peridotite begins to melt.*

Decrease in Pressure: Decompression Melting   If temperature were the only factor that determined whether or not rock melts, our planet would be a molten ball covered with a thin, solid outer shell. This, of course, is not the case. The reason is that pressure also increases with depth.

Melting, which is accompanied by an increase in volume, *occurs at higher temperatures at depth* because of greater confining pressure (Figure 4.18). Consequently, an increase in confining pressure causes an increase in the rock's melting temperature. Conversely, reducing confining pressure

lowers a rock's melting temperature. When confining pressure drops sufficiently, **decompression melting** is triggered.

Decompression melting occurs where hot, solid mantle rock ascends in zones of convective upwelling, thereby moving into regions of lower pressure. This process is responsible for generating magma along divergent plate boundaries (oceanic ridges) where plates are rifting apart (Figure 4.19). Below the ridge crest, hot mantle rock rises and melts replacing the material

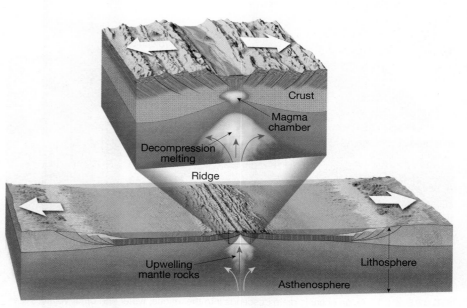

FIGURE 4.19 As hot mantle rock ascends, it continually moves into zones of lower pressure. This drop in confining pressure can initiate *decompression melting*, even without additional heat.

---

*We will consider the heat sources for the geothermal gradient in Chapter 12.

that shifted horizontally away from the ridge axis. Decompression melting also occurs within ascending mantle plumes.

Addition of Volatiles Another important factor affecting the melting temperature of rock is its water content. Water and other volatiles act as salt does to melt ice. That is, volatiles cause rock to melt at lower temperatures. Further, the effect of volatiles is magnified by increased pressure. Deeply buried "wet" rock has a much lower melting temperature than "dry" rock of the same composition (see Figure 4.18). Therefore, in addition to a rock's composition, its temperature, depth (confining pressure), and water content determine whether it exists as a solid or liquid.

Volatiles play an important role in generating magma at convergent plate boundaries where cool slabs of oceanic lithosphere descend into the mantle (Figure 4.20). As an oceanic plate sinks, both heat and pressure drive water from the subducting crustal rocks. These fluids, which are very mobile, migrate into the wedge of hot mantle that lies directly above. The addition of water lowers the melting temperature of peridotite sufficiently to generate some melt. Laboratory studies have shown that the temperature at which peridotite begins to melt can be lowered by as much as 100 °C by the addition of only 0.1 percent water.

Melting of peridotite generates basaltic magma having a temperature of 1200 °C or higher. When enough mantle-derived basaltic magma forms, it will buoyantly rise toward the surface. In a continental setting, basaltic magma may "pond" beneath crustal rocks, which have a lower density and are already near their melting temperature. This may result in some melting of the crust and the formation of a secondary, silica-rich magma.

In summary, magma can be generated three ways: (1) when an *increase in temperature* causes a rock to exceed its melting point; (2) in zones of upwelling *a decrease in pressure* (without the addition of heat) can result in *decompression melting;* and (3) the *introduction of volatiles* (principally water) can lower the melting temperature of hot mantle rock sufficiently to generate magma.

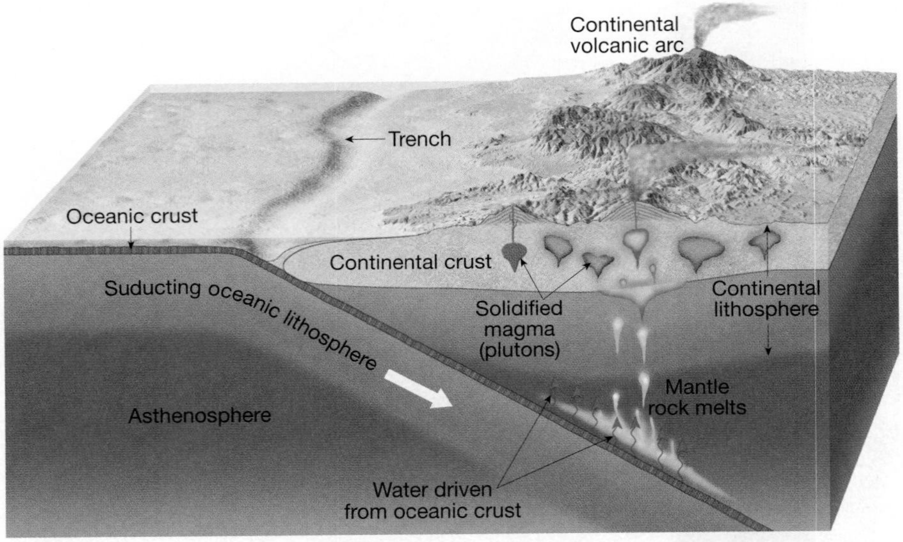

HOW MAGMAS EVOLVE

Because a large variety of igneous rocks exists, it is logical to assume that a wide variety of magmas must also exist. However, geologists have observed that, over time, a volcano may extrude lavas exhibiting quite different compositions. Data of this type led them to examine the possibility that magma might change (evolve) and thus become the parent to a variety of igneous rocks. To explore this idea, a pioneering investigation into the crystallization of magma was carried out by N. L. Bowen in the first quarter of the 20th century.

Bowen's Reaction Series and the Composition of Igneous Rocks

Recall that ice freezes at a single temperature, whereas basaltic magma crystallizes over a range of at least 200 °C of cooling. In a laboratory setting Bowen and his coworkers demonstrated that as a basaltic magma cools, minerals tend to crystallize in a systematic fashion based on their melting points. As shown in Figure 4.21, the first mineral to crystallize is the ferromagnesian mineral olivine. Further cooling generates calcium-rich plagioclase feldspar as well as pyroxene, and so forth down the diagram.

During the crystallization process, the composition of the remaining liquid portion of the magma also continually changes. For example, at the stage when about a third of the magma has solidified, the melt will be nearly depleted of iron, magnesium, and calcium, because these elements are major constituents of the earliest-formed minerals. The removal of these elements causes the melt to become enriched in sodium and potassium. Further, because the original basaltic magma contained about 50 percent silica (SiO_2), the crystallization of the earliest-formed mineral, olivine, which is only about 40 percent silica, leaves the remaining melt richer in SiO_2. Thus, the silica component of the melt becomes enriched as the magma evolves.

Bowen also demonstrated that if the solid components of a magma remain in contact with the remaining melt, they will chemically react and change mineralogy as shown in Figure 4.21. For this reason, this arrangement of minerals became known as **Bowen's reaction series** (Box 4.2). As you will see, in nature the earliest-formed minerals can be separated from the melt, thus halting any further chemical reaction.

The diagram of Bowen's reaction series in Figure 4.21 depicts the sequence that minerals crystallize from a magma of basaltic composition under laboratory conditions. Evidence that this

FIGURE 4.20 As an oceanic plate descends into the mantle, water and other volatiles are driven from the subducting crustal rocks into the mantle above. These volatiles lower the melting temperature of hot mantle rock sufficiently to trigger melting.

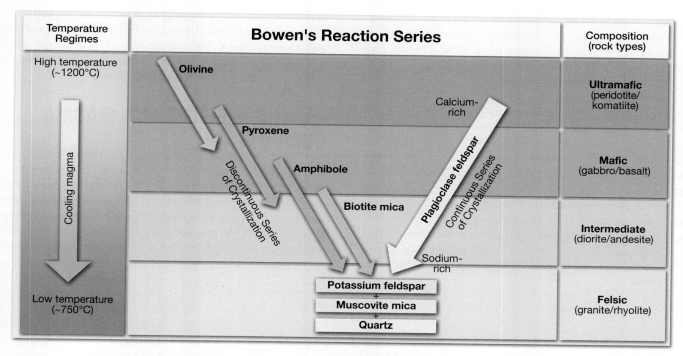

FIGURE 4.21 Bowen's reaction series shows the sequence in which minerals crystallize from a magma. Compare this figure to the mineral composition of the rock groups in Figure 4.10. Note that each rock group consists of minerals that crystallize in the same temperature range.

highly idealized crystallization model approximates what can happen in nature comes from the analysis of igneous rocks. In particular, we find that minerals that form in the same general temperature regime depicted on Bowen's reaction series are found together in the same igneous rocks. For example, notice in Figure 4.21 that the minerals quartz, potassium feldspar, and muscovite, which are located in the same region of Bowen's diagram, are typically found together as major constituents of the plutonic igneous rock *granite.*

Magmatic Differentiation   Bowen demonstrated that minerals crystallize from magma in a systematic fashion. But how do Bowen's findings account for the great diversity of igneous rocks? It has been shown that, at one or more stages during the crystallization of magma, a separation of various components can occur. One mechanism that allows this to happen is called **crystal settling.** This process occurs when the earlier-formed minerals are denser (heavier) than the liquid portion and sink toward the bottom of the magma chamber, as shown in Figure 4.22. (Crystallization can also occur along the cool margins of a magma body.) When the remaining melt solidifies—either in place or in another location if it migrates into fractures in the surrounding rocks—it will form a rock with a minerology much different from the parent magma (Figure 4.22). The formation of one or more secondary magmas from a single parent magma is called **magmatic differentiation.**

A classic example of magmatic differentiation is found in the Palisades Sill, which is a 300-meter-thick tabular mass of dark igneous rock exposed along the west bank of the lower Hudson River. Because of its great thickness and subsequent slow rate of

solidification, crystals of olivine (the first mineral to form) sank and make up about 25 percent of the lower portion of the Palisades Sill. By contrast, near the top of this igneous body, where the last melt crystallized, olivine represents only 1 percent of the rock mass.*

One consequence of magnetic differentiation of basaltic (mafic) magma is that it tends to generate evolved silica-rich felsic melts of low density that can rise toward the top of the magma chamber. The result is a compositionally zoned magma reservoir with dense, basaltic melt stratified below less dense, silica-rich melt. Evidence for compositional zonation comes from thick deposits of pyroclastic material that show a systemic change in composition from highly evolved materials at the base to basaltic compositions at the top (Figure 4.23).

Assimilation and Magma Mixing

Bowen successfully demonstrated that through magmatic differentiation, a parent magma can generate several mineralogically different igneous rocks. However, more recent work indicates that magmatic differentiation cannot, by itself, account for the entire compositional spectrum of igneous rocks.

Once a magma body forms, its composition can also change through the incorporation of foreign material. For example, as magma migrates through the crust, it may incorporate some of the surrounding host rock, a process called

---

*Recent studies indicate that this igneous body was produced by multiple injections of magma and represents more than just a simple case of crystal settling.

assimilation (Figure 4.24A). In a near-surface environment where rocks are brittle, as the magma pushes upward, it causes numerous cracks in the overlying rock. The force of the injected magma is often sufficient to dislodge blocks of "foreign" rock which melt and are incorporated into the magma body.

Another means by which the composition of magma can be altered is **magma mixing**. This process occurs when one magma body intrudes another having a different composition (Figure 4.24C). Once combined, convective flow may stir the two magmas and generate a mass having a composition that is a blend of the two. Magma mixing may occur during the ascent of two chemically distinct magma bodies as the more buoyant mass overtakes the more slowly moving mass.

In summary, Bowen successfully demonstrated that through magmatic differentiation, a single parent magma can generate several mineralogically different igneous rocks. This process, in concert with magma mixing and contamination by crustal rocks, accounts in part for the great diversity of magmas and igneous rocks. We will next look at another important process, partial melting, which is also responsible for generating magmas having varying compositions.

PARTIAL MELTING AND MAGMA COMPOSITION

Recall that the crystallization of basaltic magma occurs over a temperature range of at least 200 °C. As you might expect, melting, the reverse process, spans a similar temperature range. As rock begins to melt, those minerals with the lowest melting temperatures are the first to melt. Should melting continue, minerals with higher melting points begin to melt and the composition of the magma steadily approaches the overall composition of the rock from which it was derived. Most often, however, melting is not complete. The incomplete melting of rocks is known as **partial melting,** a process that produces most magma.

Recall from Bowen's reaction series that rocks with a granitic composition are composed of minerals with the lowest melting (crystallization) temperatures—namely, quartz and potassium feldspar (see Figure 4.21). Also note that as we move up Bowen's reaction series, the minerals have progressively higher melting temperatures and that olivine, which is found at the top, has the highest melting point. When a rock undergoes partial melting, it will form a melt that is enriched in ions from minerals with the lowest melting temperatures. The unmelted crystals are those of minerals with higher melting temperatures. Separation of these two fractions would yield a melt with a chemical composition that is richer in silica and nearer to the granitic (felsic) end of the spectrum than the rock from which it was derived.

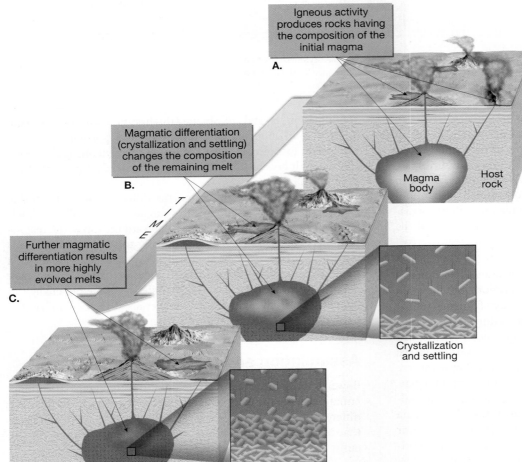

FIGURE 4.22   Illustration of how a magma evolves as the earliest-formed minerals (those richer in iron, magnesium, and calcium) crystallize and settle to the bottom of the magma chamber, leaving the remaining melt richer in sodium, potassium, and silica (SiO₂) **A.** Emplacement of a magma body and associated igneous activity generates rocks having a composition similar to that of the initial magma. **B.** As the magma cools, magmatic differentiation (crystallization and settling) change the composition of the remaining melt, generating rocks having a composition quite different than the original magma. **C.** Further magmatic differentiation results in another, more highly evolved, melt with its associated rock types.

Formation of Basaltic Magma

Most magma that erupts at Earth's surface is basaltic and has a temperature of 1100–1200 °C. Because rocks in the lower crust have temperatures of 700 °C or less, basaltic magma likely forms in the high temperature environ-

Geologist's Sketch

FIGURE 4.23   Ash and pumice ejected during a large eruption of Mount Mazama (Crater Lake) in present day Oregon. Notice the gradation from light-colored, silica-rich ash near the base to dark-colored rocks at the top. It is likely that prior to this eruption the magma began to segregate as the less dense, silica-rich magma migrated toward the top of the magma chamber. The zonation seen in the rocks resulted because a sustained eruption tapped deeper and deeper levels of the magma chamber. Thus, this rock sequence is an inverted representation of the compositional zonation in the magma body; that is, the magma from the top of the chamber erupted first and is found at the base of these ash deposits and vice versa. (Photo by E. J. Tarbuck)

ment found in the mantle. Evidence for a mantle source are inclusions of the rock *peridotite* which basaltic magmas often carry up from the mantle. Experiments show that under the high-pressure conditions calculated for the upper mantle, partial melting of the ultramafic rock peridotite will yield a magma of basaltic composition (Figure 4.25). Although most basaltic magma is thought to originate by partial melting of mantle rock, another source rock may be sediment laden oceanic crust that is recycled back into the mantle at subduction zones.

Basaltic magmas that originate from direct melting of mantle rocks are called *primary* or *primitive* magmas because they have not yet evolved. Partial melting to produce mantle-derived magmas may be triggered by a reduction in confining pressure (decompression melting). This can occur, for example, where hot mantle rock ascends as part of slow-moving convective flow at mid-ocean ridges (see Figure 4.19). Basaltic magmas are also generated at subduction zones, where water driven from the descending slab of oceanic crust promotes partial melting of the mantle rocks above (see Figure 4.20).

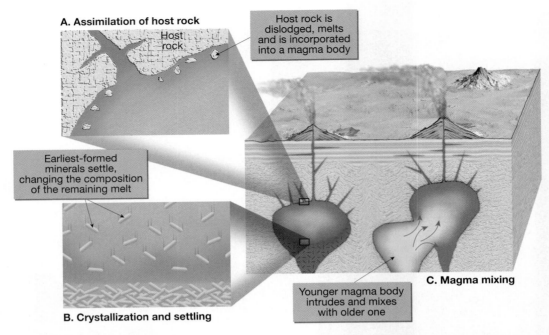

A. Assimilation of host rock

Host rock

Host rock is dislodged, melts and is incorporated into a magma body

Earliest-formed minerals settle, changing the composition of the remaining melt

C. Magma mixing

Younger magma body intrudes and mixes with older one

B. Crystallization and settling

FIGURE 4.24   This illustration shows three ways that the composition of a magma body may be altered: **A.** Assimilation of country rock; **B.** Crystallization and settling (magmatic differentiation); and **C.** Magma mixing.

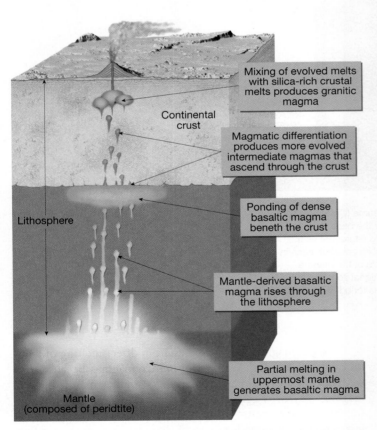

Mixing of evolved melts with silica-rich crustal melts produces granitic magma

Continental crust

Magmatic differentiation produces more evolved intermediate magmas that ascend through the crust

Lithosphere

Ponding of dense basaltic magma beneth the crust

Mantle-derived basaltic magma rises through the lithosphere

Partial melting in uppermost mantle generates basaltic magma

Mantle (composed of peridtite)

FIGURE 4.25   Schematic drawing showing the movement of magma from its source in the upper mantle through the continental crust. During its ascent, mantle-derived basaltic magma evolves through the process of magmatic differentiation and by melting and incorporating continental crust. Magmas that feed volcanoes in a continental setting tend to be silica-rich (viscous) and have a high gas content.

Formation of Andesitic and Granitic Magmas

If partial melting of mantle rocks generates most basaltic magmas, what is the source of the magma that crystallizes to form andesitic (intermediate) and granitic (felsic) rocks? Recall that intermediate and felsic magmas rarely erupt from volcanoes in the deep-ocean basins; rather, they are found mainly within, or adjacent to, the continental margins. This is strong evidence that interactions between mantle-derived basaltic magmas and more silica-rich components of Earth's crust are involved in the creation of these more highly evolved magmas.

Some andesitic magmas are generated when relatively dense, mantle-derived basaltic magmas pond beneath silica-rich crustal rocks (see Figure 4.25). As the basaltic magma slowly crystallizes, the remaining melt evolves by the process of magmatic differentiation. Recall from our discussion of Bowen's reaction series that as basaltic magma solidifies, it is the silica-poor ferromagnesian minerals that crystallize first. If these iron-rich components are separated from the liquid by crystal settling, the remaining melt, now enriched in silica, will have a composition more akin to andesite. These evolved (changed) magmas are termed *secondary magmas*.

Like magmas of andesitic composition, granitic magmas can evolve by magmatic differentiation of mantle-derived basalts. However, only small quantities of granitic rocks are derived by this mechanism. Granitic magma is also generated when andesitic magma migrates up through the crust, melting the silica-rich rocks through which it ascends (see Figure 4.25). This complex process, which involves the mixing of evolved melts with silica-rich melts derived from the crust, is thought to generate granitic rocks that make up a large percentage of the upper continental crust.

Granitic and andesitic melts are higher in silica and thus more viscous (thicker) than basaltic magmas. Therefore, in contrast to basaltic magmas that frequently produce vast outpourings of lava, granitic and andesitic magmas usually lose their mobility before reaching the surface and tend to produce large intrusive igneous structures. On those occasions when silica-rich magmas reach the surface, viscous lava flows and explosive pyroclastic eruptions occur, such as those that occurred at Mount St. Helens (Figure 4.26).

In summary, Bowen's reaction series is a simplified guide to understanding how partial melting produces rocks of various compositions. As rock begins to melt, those minerals with the lowest melting temperatures are the first to melt. As a result, partial melting produces magmas enriched in low-melting temperature components and richer in silica than the parent rock. Partial melting of the ultramafic rock peridotite produces a magma of

UNDERSTANDING EARTH

BOX 4.2

A Closer Look at Bowen's Reaction Series

Based on high-temperature, high-pressure experiments and microscopic studies of igneous rock, Norman Bowen developed the reaction series that bears his name (Figure 4.B). Although it is highly idealized, Bowen's reaction series provides us with a visual representation of the order in which minerals crystallize from a magma of average composition (see Figure 4.21). This model assumes that magma cools slowly at depth in an otherwise unchanging environment. Notice that Bowen's reaction series is divided into two branches—a discontinuous series and a continuous series. In the discontinuous series, each mineral has a different crystalline structure, whereas all of the minerals in the continuous series have the same internal structure.

Discontinuous Reaction Series.

The upper left branch of Bowen's reaction series, called the *discontinuous reaction series,* indicates that as a magma cools, olivine is the first mineral to crystallize. Once formed, olivine will chemically react with the remaining melt to form the mineral pyroxene (see Figure 4.21). In this reaction, olivine, which is composed of individual silicon–oxygen tetrahedra, incorporates more silica into its structure, thereby linking its tetrahedra into single-chain structures of the mineral pyroxene. (Note: pyroxene has a lower crystallization temperature than olivine and is more stable at lower temperatures.) As the magma body cools further, the pyroxene crystals react with the melt to generate the double-chain structure of amphibole, which in turn reacts with the melt to form the sheet-silicate biotite. These reactions do not usually run to completion, so that various amounts of each of the minerals in the series may exist at any given time. In addition, some minerals, such as biotite, may never form.

FIGURE 4.B   Norman Levi Bowen (1887–1956), a Canadian-born geologist, studying igneous rocks with a polarizing microscope. Bowen spent most of his professional life at the Carnegie Institution's Geophysical Laboratory in Washington, DC, where he was considered the leading expert on igneous rocks. (Photo courtesy of the Geophysical Laboratory, Carnegie Institution of Washington)

This branch of Bowen's reaction series is called a *discontinuous reaction series* because at each step a different silicate structure emerges. Olivine, the first mineral in the sequence, is composed of isolated tetrahedra, whereas pyroxene is composed of single chains, amphibole of double chains, and biotite of sheet structures.

Continuous Reaction Series.

The right branch of the reaction series, called the *continuous reaction series,* illustrates that calcium-rich plagioclase feldspar crystals react with the sodium ions in the melt to become progressively more sodium-rich (see Figure 4.21). Here the sodium ions diffuse into the feldspar crystals and displace the calcium ions in the crystal lattice.

During the last stage of crystallization, after much of the magma has solidified, potassium feldspar forms. (Muscovite may form in pegmatites and other plutonic igneous rocks that crystallize at considerable depth.) Finally, if the remaining melt has excess silica, the mineral quartz will form.

Testing Bowen's Reaction Series.

During an eruption of Hawaii's Kilauea Volcano in 1965, basaltic (mafic) lava poured into a pit crater, forming a lava lake that became a natural laboratory for testing Bowen's reaction series. When the surface of the lava lake cooled enough to form a crust, geologists drilled through it into the magma and periodically removed samples that were quenched to preserve the melt as well as minerals that had grown within it. By sampling the lava lake at successive stages of cooling, a history of crystallization was recorded.

As Bowen's reaction series predicts, olivine crystallized first. Samples obtained later confirmed that some of the olivine crystals reacted with the melt to become pyroxene. In addition, the remaining melt changed composition as more and more of the lava crystallized. In contrast to the original basaltic lava, which contained about 50 percent silica (SiO_2), the final melt contained more than 75 percent silica and had a composition more like that of granite.

Although the lava in this setting cooled rapidly compared to rates experienced in deep magma chambers, it was slow enough to verify that minerals do crystallize in a systematic fashion that roughly parallels Bowen's reaction series. Further, had the melt been separated at any stage in the cooling process, it would have formed a rock with a composition much different from the original lava.

basaltic (mafic) composition, while partial melting of the mafic rock basalt generates magma of intermediate (andesitic) composition. Granitic (felsic) magmas evolve from basaltic magmas and/or the melting and assimilation of silica-rich crustal rocks.

INTRUSIVE IGNEOUS ACTIVITY

VOLCANOES
▶ Intrusive Igneous Activity

Although volcanic eruptions can be violent and spectacular events, most magma is emplaced and crystallizes at depth, without fanfare. Therefore, understanding the igneous processes that occur deep underground is as important to geologists as the study of volcanic events.

When magma rises through the crust, it forcefully displaces preexisting crustal rocks referred to as *host* or *country rock*. Invariably, some of the magma will not reach the surface, but instead crystallize or "freeze" at depth where it becomes an intrusive igneous rock. Much of what is known about intrusive igneous activity has come from the study of old, now solid, magma bodies exhumed by erosion.

Nature of Intrusive Bodies

The structures that result from the emplacement of magma into preexisting rocks are called **intrusions** or **plutons**. Because all intrusions form out of view beneath Earth's surface, they are studied primarily after uplifting and erosion have exposed them. The challenge lies in reconstructing the events that generated these structures millions or even hundreds of millions of years ago.

Intrusions are known to occur in a great variety of sizes and shapes. Some of the most common types are illustrated in Figure 4.27. Notice that some plutons have a tabular (tabletop) shape, whereas others are best described as massive. Also, observe that some of these bodies cut across existing structures, such as sedimentary strata; whereas others form when magma is injected between sedimentary layers. Because of these differences, intrusive igneous bodies are generally classified according to their shape as either **tabular** (*tabula* = table) or **massive** and by their orientation with respect to the host rock. Igneous bodies are said to be **discordant** (*discordare* = to disagree) if they cut across existing structures and **concordant** (*concordare* = to agree) if they form parallel to features such as sedimentary strata.

Tablular Intrusive Bodies: Dikes and Sills

Tabular intrusive bodies are produced when magma is forcibly injected into a fracture or zone of weakness, such as a bedding surface (see Figure 4.27). **Dikes** are discordant bodies that cut across bedding surfaces or other structures in the country rock. By contrast, **sills** are nearly horizontal, concordant bodies that form when magma exploits weaknesses between sedimentary beds, or other foliations (Figure 4.28). In general, dikes serve as tabular conduits that transport magma, whereas sills store magma.

Dikes and sills are typically shallow features, occurring where the country rocks are sufficiently brittle to fracture.

FIGURE 4.26 1996 eruption of Mount Ruapehu, Tongariro National Park, New Zealand. Volcanoes that border the Pacific Ocean are fed largely by magmas that have intermediate or felsic compositions. These silica-rich magmas often erupt explosively, generating large plumes of volcanic dust and ash. (Photo by ImageState/Alamy)

Although, they can range in thickness from less than a millimeter, to over a kilometer, most are in the 1 to 20-meter range.

Dikes and sills can occur as solitary bodies, but dikes in particular tend to form in roughly parallel groups called *dike swarms*. These multiple structures reflect the tendency for fractures to form in sets when tensional forces stretch brittle country rock. Dikes can also occur radiating from an eroded volcanic neck, like spokes on a wheel. In these situations the active ascent of magma generated fissures in the volcanic cone out of which lava flowed.

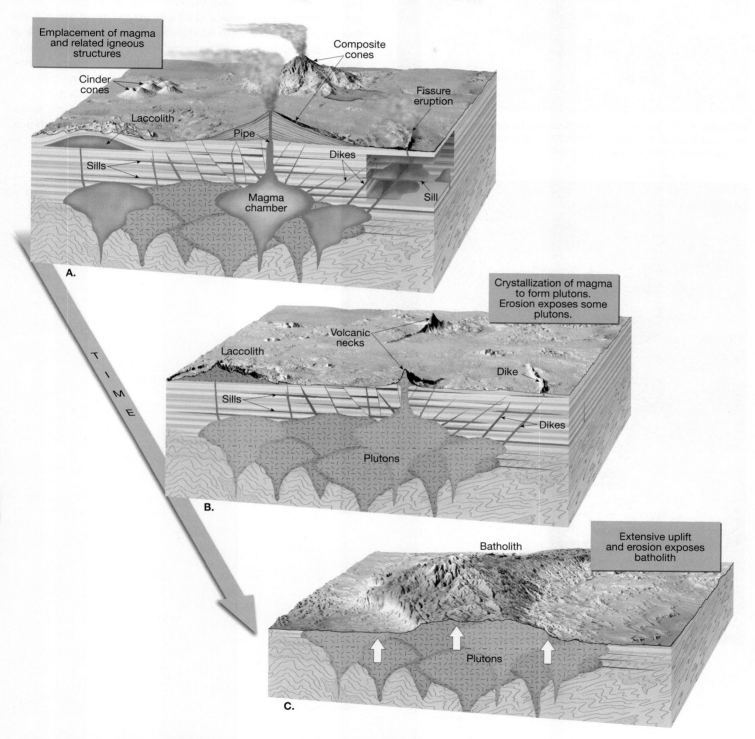

FIGURE 4.27   Illustrations showing basic igneous structures. **A.** This block diagram shows the relationship between volcanism and intrusive igneous activity. **B.** This view illustrates the basic intrusive igneous structures, some of which have been exposed by erosion long after their formation. **C.** After millions of years of uplifting and erosion, a batholith is exposed at the surface.

FIGURE 4.28 Salt River Canyon, Arizona. The dark, essentially horizontal band is a sill of basaltic composition that intruded horizontal layers of sedimenatary rock. (Photo by E. J. Tarbuck)

Dikes frequently weather more slowly than the surrounding rock. Consequently, when exposed by erosion, dikes tend to have a wall-like appearance, as shown in Figure 4.29.

Because dikes and sills are relatively uniform in thickness and can extend for many kilometers they are assumed to be the product of very fluid, and therefore, mobile magmas. One of the largest and most studied of all sills in the United States is the Palisades Sill. Exposed for 80 kilometers along the west bank of the Hudson River in southeastern New York and northeastern New Jersey, this sill is about 300 meters thick. Because it is resistant to erosion, the Palisades Sill forms an imposing cliff that can be easily seen from the opposite side of the Hudson.

In many respects, sills closely resemble buried lava flows. Both are tabular and can have a wide aerial extent and both may exhibit columnar jointing (Figure 4.30). **Columnar joints** form as

FIGURE 4.29 The vertical structure that looks like a stone wall is a dike, which is more resistant to weathering than the surrounding rock. This dike is located west of Granby, Colorado, near Arapaho National Forest. (Photo by R. Jay Fleisher)

Rock surrounding dike is more easily eroded

Dike made of resistant igneous rock

Weathered debris

Geologist's Sketch

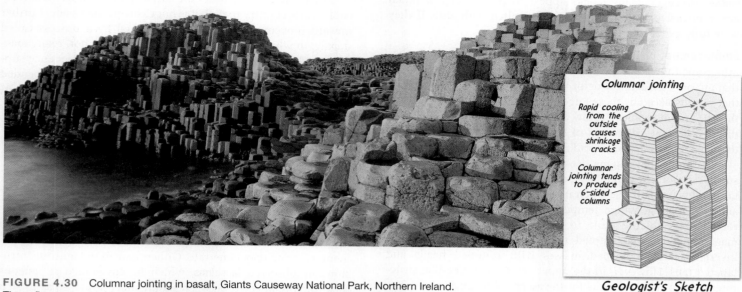

FIGURE 4.30 Columnar jointing in basalt, Giants Causeway National Park, Northern Ireland. These five-to seven-sided columns are produced by contraction and fracturing that results as a lava flow or sill gradually cools. (Photo by Tom Till)

Columnar jointing

Rapid cooling from the outside causes shrinkage cracks

Columnar jointing tends to produce 6-sided columns

Geologist's Sketch

FIGURE 4.31 Granitic batholiths that occur along the western margin of North America. These gigantic, elongated bodies consist of numerous plutons that were emplaced during the last 150 million years of Earth's history.

Coast Range batholith

Pacific Ocean

Idaho batholith

Sierra Nevada batholith

Southern California batholith

igneous rocks cool and develop shrinkage fractures that produce elongated, pillar-like columns. Further, because sills generally form in near-surface environments and may be only a few meters thick, the emplaced magma often cools quickly enough to generate a fine-grained texture. (Recall that most intrusive igneous bodies have a coarse-grained texture.)

Massive Intrusive Bodies: Batholiths and Stocks

By far the largest intrusive igneous bodies are **batholiths** (*bathos* = depth, *lithos* = stone). Batholiths occur as mammoth linear structures several hundreds of kilometers long and up to 100 kilometers wide (Figure 4.31). The Sierra Nevada batholith, for example, is a continuous granitic structure that forms much of the Sierra Nevada, in California. An even larger batholith extends for over 1800 kilometers (1100 miles) along the Coast Mountains of western Canada, and into southern Alaska. Although batholiths can cover a large area, recent gravitational studies indicate that most are less than 10 kilometers (6 miles) thick. Some are even thinner. The Coastal batholith of Peru, for example is essentially a flat slab with an average thickness of only 2–3 kilometers (1–2 miles).

Batholiths are almost always made up of felsic and intermediate rock types and are often referred to as "granite batholiths." Large granite batholiths consist of hundreds of plutons that intimately crowd against or penetrate one another. These bulbous masses were emplaced over spans of millions of years. The intrusive activity that created the Sierra Nevada batholith, for example, occurred nearly continuously over a 130-million-year period that ended about 80 million years ago (see Figure 4.31).

By definition, a plutonic body must have a surface exposure greater than 100 square kilometers (40 square miles) to be considered a batholith. Smaller plutons of this type are termed

stocks. However, many stocks appear to be portions of much larger intrusive bodies that would be called batholiths if they were fully exposed.

Emplacement of Batholiths

How did massive granitic batholiths come to reside within sedimentary and metamorphic rocks that are only moderately deformed? What happened to the rock that was displaced by these huge igneous masses—geologists call this the "room problem". How do magma bodies make their way through several kilometers of solid rock? These are questions that are still being studied and debated by geologists.

We know that magma rises because it is less dense than the surrounding rock, much like a cork held at the bottom of a container of water will rise when it is released. In the upper mantle and lower crust, where temperatures and pressures are high, rock is very ductile (able to flow). In this setting, buoyant magma bodies are assumed to rise in the form of *diapirs,* inverted teardrop shaped masses with rounded heads and tapered tails. However, in the upper crust large dike-like structures may provide conduits for the ascent of magma.

Depending on the tectonic environment, several mechanisms have been proposed to solve the room problem. At great depths, where rock is ductile, a mass of buoyant, rising magma can forcibly make room for itself by pushing aside the overlying rock—a process called *shouldering.* As the magma continues to move upward, some of the host rock that was displaced will fill in the space left by the magma body as it passes.*

---

*An analogous situation occurs when a can of oil-base paint is left in storage. The oily component of the paint is less dense than the pigments used for coloration; thus, oil collects into drops that slowly migrate upward, while the heavier pigments settle to the bottom.

As a magma body nears the surface, it encounters relatively cool, brittle country rock that is not easily pushed aside. Further upward movement may be accomplished by a process called *stoping,* in which blocks of the roof overlying a hot, rising mass become dislodged and sink through the magma. Evidence supporting stoping is found in plutons that contain suspended blocks of country rock called **xenoliths** (*xenos* = a stranger, *lithos* = stone).

Magma may also *melt* and *assimilate* some of the overlying host rock. However, this process is greatly limited by the available thermal energy contained in the magma body. When plutons are emplaced near the surface, the room problem may be solved by "lifting the roof" that overlies the intrusive body.

Laccoliths

A 19th century study by G. K. Gilbert of the U.S. Geological Survey in the Henry Mountains of Utah produced the first clear evidence that igneous intrusions can lift the sedimentary strata they penetrate. Gilbert named the igneous intrusions he observed **laccoliths,** which he envisioned as igneous rock forcibly injected between sedimentary strata, so as to arch the beds above, while leaving those below relatively flat. It is now known that the five major peaks of the Henry Mountains are not laccoliths, but stocks. However, these central magma bodies are the source material for branching offshoots that are true laccoliths, as Gilbert defined them (Figure 4.32).

Numerous other granitic laccoliths have since been identified in Utah. The largest is a part of the Pine Valley Mountains located north of St. George, Utah. Others are found in the La Sal Mountains near Arches National Park and in the Abajo Mountains directly to the south.

FIGURE 4.32 Mount Ellen, the northernmost of five peaks that make up Utah's Henry Mountains. Although the main intrusions in the Henry Mountains are stocks, numerous laccoliths formed as offshoots of these structures. (Photo by Michael DeFreitas North America/ Alamy)

Geologist's Sketch

- *Igneous rocks* form when *magma cools* and solidifies. *Extrusive,* or *volcanic,* igneous rocks result when *lava* cools at the surface. Magma that solidifies at depth produces *intrusive,* or *plutonic,* igneous rocks.

- As magma cools, the ions that compose it arrange themselves into orderly patterns—a process called *crystallization.* Slow cooling results in the formation of relatively large crystals. Conversely, when cooling occurs rapidly, the outcome is a solid mass consisting of tiny intergrown crystals. When molten material is quenched instantly, a mass of unordered atoms, referred to as *glass,* forms.

- The mineral composition of an igneous rock is the consequence of the chemical makeup of the parent magma and the environment of crystallization. Igneous rocks are divided into broad compositional groups based on the percentage of dark and light silicate minerals they contain. *Felsic rocks* (e.g., granite and rhyolite) are composed mostly of the light-colored silicate minerals potassium feldspar and quartz. Rocks of *intermediate* composition, (e.g., andesite and diorite) are rich in plagioclase feldspar and amphibole. *Mafic rocks* (e.g., basalt and gabbro) contain abundant olivine, pyroxene, and calcium-rich plagioclase feldspar. They are high in iron, magnesium, and calcium, low in silica, and are dark gray to black in color.

- The texture of an igneous rock refers to the overall appearance of the rock based on the size, shape, and arrangement of its mineral grains. The most important factor influencing the texture of igneous rocks is the rate at which magma cools. Common igneous rock textures include *aphanitic (fine-grained),* with grains too small to be distinguished without the aid of a microscope; *phaneritic (coarse-grained),* with intergrown crystals that are roughly equal in size and large enough to be identified without the aid of a microscope; *porphyritic,* which has larger crystals (*phenocrysts*) embedded in a matrix of smaller crystals (*groundmass*); and *glassy.*

- The mineral makeup of an igneous rock is ultimately determined by the chemical composition of the magma from which it crystallizes. N. L. Bowen discovered that as magma cools in the laboratory, those minerals with higher melting points crystallize before minerals with lower melting points. *Bowen's reaction series* illustrates the sequence of mineral formation within magma.

- During the crystallization of magma, if the earlier-formed minerals are denser than the liquid portion, they will settle to the bottom of the magma chamber during a process called *crystal settling.* Owing to the fact that crystal settling removes the earlier-formed minerals, the remaining melt will form a rock with a chemical composition that is different from the parent magma. The process of developing more than one magma type from a common magma is called *magmatic differentiation.*

- Once a magma body forms, its composition can change through the incorporation of foreign material, a process termed *assimilation,* or by *magma mixing.*

- Magma originates from essentially solid rock of the crust and mantle. In addition to a rock's composition, its temperature, depth (confining pressure), and water content determine whether it exists as a solid or liquid. Thus, magma can be generated by *raising a rock's temperature,* as occurs when a hot mantle plume "ponds" beneath crustal rocks. A *decrease in pressure* can cause *decompression melting.* Further, the *introduction of volatiles* (water) can lower a rock's melting point sufficiently to generate magma. A process called *partial melting* produces a melt made of the low-melting-temperature minerals, which are higher in silica than the original rock. Thus, magmas generated by partial melting are nearer to the felsic end of the compositional spectrum than are the rocks from which they formed.

- Intrusive igneous bodies are classified according to their *shape* and by their *orientation with respect to the country or host rock,* generally sedimentary or metamorphic rock. The two general shapes are *tabular* (sheet-like) and *massive.* Intrusive igneous bodies that cut across existing sedimentary beds are said to be *discordant;* those that form parallel to existing sedimentary beds are *concordant.*

- *Dikes* are tabular, discordant igneous bodies produced when magma is injected into fractures that cut across rock layers. Nearly horizontal, tabular, concordant bodies, called *sills,* form when magma is injected along the bedding surfaces of sedimentary rocks. In many respects, sills closely resemble buried lava flows. *Batholiths,* the largest intrusive igneous bodies, sometimes make up large linear mountains, as exemplified by the Sierra Nevada. *Laccoliths* are similar to sills but form from less fluid magma that collects as a lens-shaped mass that arches overlying strata upward.

KEY TERMS

andesitic composition
(p. 111)
aphanitic texture (p. 113)
assimilation (p. 124)
basaltic composition (p. 110)
batholiths (p. 131)
Bowen's reaction series
(p. 122)
coarse-grained texture
(p. 113)
columnar joints (p. 130)
concordant (p. 128)
crystallization (p. 109)
crystal settling (p. 123)
decompression melting
(p. 121)

dikes (p. 128)
discordant (p. 128)
extrusive igneous rocks
(p. 110)
felsic (p. 110)
fine-grained texture (p. 113)
fragmental texture (p. 114)
geothermal gradient (p. 121)
glass (p. 113)
glassy texture (p. 113)
groundmass (p. 113)
granitic composition (p. 110)
igneous rocks (p. 108)
intermediate composition
(p. 111)
intrusions (p. 128)

intrusive igneous rocks
(p. 110)
laccoliths (p. 132)
lava (p. 108)
mafic (p. 111)
magma (p. 108)
magma mixing (p. 124)
magmatic differentiation
(p. 123)
massive (p. 128)
melt (p. 108)
partial melting (p. 124)
pegmatite (p. 114)
pegmatitic texture (p. 114)
phaneritic texture (p. 113)
phenocryst (p. 113)

plutonic rocks (p. 110)
plutons (p. 128)
porphyritic texture
(p. 113)
porphyry (p. 113)
pyroclastic texture
(p. 114)
sills (p. 128)
stocks (p. 132)
tabular (p. 128)
texture (p. 112)
ultramafic (p. 111)
vesicular texture (p. 113)
volatiles (p. 108)
volcanic rocks (p. 110)
xenoliths (p. 132)

QUESTIONS FOR REVIEW

1. What is magma?

2. How does lava differ from magma?

3. List the three major components of magma.

4. What is melt? What is a volatile?

5. List the three most common volatiles found in magma.

6. In what basic settings do intrusive and extrusive igneous rocks originate?

7. How does the rate of cooling influence crystal size? What other factors influence the texture of igneous rocks?

8. The statements that follow relate to terms used to describe igneous rock textures. For each statement, identify the appropriate term.

   a. openings produced by escaping gases

   b. obsidian exhibits this texture

   c. a matrix of fine crystals surrounding phenocrysts

   d. crystals are too small to be seen without a microscope

   e. a texture characterized by two distinctly different crystal sizes

   f. coarse-grained, with crystals of roughly equal size

   g. exceptionally large crystals exceeding 1 centimeter in diameter

9. Why are the crystals in pegmatites so large?

10. What does a porphyritic texture indicate about an igneous rock?

11. The classification of igneous rocks is based largely on two criteria. Name these criteria.

12. How are granite and rhyolite different? In what way are they similar?

13. Compare and contrast each of the following pairs of rocks:

   a. granite and diorite

   b. basalt and gabbro

   c. andesite and rhyolite

14. How do tuff and volcanic breccia differ from other igneous rocks such as granite and basalt?

15. What is the geothermal gradient?

16. Describe decompression melting.

17. How does the introduction of volatiles trigger melting?

18. What is magmatic differentiation? How might this process lead to the formation of several different igneous rocks from a single magma?

19. Relate the classification of igneous rocks to Bowen's reaction series.

20. What is partial melting?

21. How does the composition of a melt produced by partial melting compare with the composition of the parent rock?

22. In what tectonic setting are most basaltic magmas generated?

23. Why are rocks of intermediate (andesitic) and felsic (granitic) composition generally *not* found in the ocean basins?

24. Describe each of the four basic intrusive features discussed in the text (dike, sill, laccolith, and batholith)

25. What is the largest of all intrusive igneous bodies? Is it tabular or massive? Concordant or discordant?

26. Describe the mechanisms that have been proposed to explain how large magma bodies come to reside in the space formerly occupied by preexisting rock.

COMPANION WEBSITE

The *Earth 10e* website uses the resources and flexibility of the Internet to aid in your study of the topics in this chapter. Written and developed by the authors and other geology instructors, this site will help improve your understanding of geology. Visit www.mygeoscienceplace.com in order to:

- **Review** key chapter concepts

- **Read** with links to the eBook and to chapter-specific web resources
- **Visualize** and comprehend challenging topics using learning activities in *GEODe Earth*
- **Test** yourself with online quizzes

GEODe EARTH

GEODe Earth is a valuable and easy to use learning aid that can be accessed from your book's Companion Website (www.mygeoscienceplace.com). It is a dynamic instructional tool that promotes understanding and reinforces important concepts by using tutorials, animations, and exercises that actively engage the student.

VOLCANOES AND VOLCANIC HAZARDS

A recent eruption of Italy's Mount Etna

(PHOTO BY ART WOLFE)

On Sunday, May 18, 1980, the most destructive volcanic eruption to occur in North America in historic times transformed a picturesque volcano into a decapitated remnant (Figure 5.1). On this date, in southwestern Washington State, Mount St. Helens erupted with tremendous force. The blast blew out the entire north flank of the volcano, leaving a gaping hole. In one brief moment, a prominent volcano whose summit had been more than 2900 meters (9500 feet) above sea level was lowered by more than 400 meters (1350 feet).

The event devastated a wide swath of timber-rich land on the north side of the mountain (Figure 5.2). Trees within a 400-square-kilometer area lay flattened and intertwined, stripped of their branches, looking like toothpicks strewn about. The accompanying mudflows carried ash, trees, and water-saturated rock debris 29 kilometers (18 miles) down the Toutle River. The eruption claimed 59 lives, some dying from the intense heat and the suffocating cloud of ash and gases, others from being hurled by the blast, and still others from entrapment in the mudflows.

The eruption ejected nearly a cubic kilometer of ash and rock debris. Following the explosion, Mount St. Helens continued to emit great quantities of hot gases and ash. The force of the blast was so strong that some ash was propelled more than 18,000 meters (over 11 miles) into the stratosphere. During the next few days, this very fine-grained material was carried around Earth by strong upper-air winds. Measurable deposits were reported in Oklahoma and Minnesota, with crop damage into central Montana. Meanwhile, ash fallout in the immediate vicinity exceeded 2 meters in depth. The air over Yakima, Washington (130 kilometers to the east), was so filled with ash that residents experienced midnight-like darkness at noon.

Not all volcanic eruptions are as violent as the 1980 Mount St. Helens event. Some volcanoes, such as Hawaii's Kilauea volcano, typically generate relatively quiet outpourings of fluid lavas. These "gentle" eruptions are not without some fiery displays; occasionally fountains of incandescent lava spray hundreds of meters into the air. During Kilauea's most recent active phase, which began in 1983, more than 180 homes and a national park visitor center have been destroyed.

Why do volcanoes like Mount St. Helens erupt explosively, whereas others like Kilauea are relatively quiet? Why do volcanoes occur in chains like the Aleutian Islands or the Cascade Range? Why do some volcanoes form on the ocean floor, while others occur on the continents? This chapter will deal with these and other questions as we explore the nature and movement of magma and lava.

THE NATURE OF VOLCANIC ERUPTIONS

VOLCANOES
▶ The Nature of Volcanic Eruptions

Volcanic activity is commonly perceived as a process that produces a picturesque, cone-shaped structure that periodically erupts in a violent manner, like Mount St. Helens (Box 5.1). Although some eruptions may be very explosive, many are not. What determines whether a volcano extrudes magma violently or "gently"? The primary factors include the magma's *composition*, its *temperature*, and the amount of *dissolved gases* it contains. To varying degrees, these factors affect the magma's mobility, or **viscosity** (*viscos* = sticky). The more viscous the material, the greater its resistance to flow. For example, compare syrup to water—syrup is more viscous and thus, more resistant to flow, than water. Magma associated with an explosive eruption may be five times more viscous than magma that is extruded in a quiescent manner.

Factors Affecting Viscosity

The effect of temperature on viscosity is easily seen. Just as heating syrup makes it more fluid (less viscous), the mobility of lava is strongly influenced by temperature. As lava cools and begins to congeal, its mobility decreases and eventually the flow halts.

A more significant factor influencing volcanic behavior is the chemical composition of the magma. Recall that a major difference among various igneous rocks is their silica (SiO_2) content (Table 5.1). Magmas that produce mafic rocks such as basalt contain about 50 percent silica, whereas magmas that produce felsic rocks (granite and its extrusive equivalent, rhyolite) contain more than 70 percent silica. Intermediate rock types—andesite and diorite—contain about 60 percent silica.

FIGURE 5.1   Before-and-after photographs show the transformation of Mount St. Helens caused by the May 18, 1980, eruption. (Top photo courtesy of U.S. Geological Survey, bottom photo by Michael Collier)

FIGURE 5.2 Douglas fir trees were snapped off or uprooted by the lateral blast of Mount St. Helens on May 18, 1980. (Large photo by Lyn Topinka/AP Photo/U.S. Geological Survey; Inset photo by John M. Burnley/Photo Researchers, Inc.)

Why Do Volcanoes Erupt?

You learned in Chapter 4 that most magma is generated by partial melting of the rock peridotite in the upper mantle to form magma with a basaltic composition. Once formed, the buoyant molten rock will rise toward the surface. Because the density of crustal rocks tends to decrease toward the surface, ascending basaltic magma may reach a level where the rocks above are less dense. Should this occur, the molten material begins to collect or pond, forming a magma chamber. As the magma body cools, minerals having high melting temperatures crystallize first, leaving the remaining melt enriched in silica and other less dense components. Some of this highly evolved material may ascend to the surface to produce a volcanic eruption. In most, but not all, tectonic settings, only a fraction of magma generated at depth ever reaches the surface.

Triggering Hawaiian-Type Eruptions Eruptions that involve very fluid basaltic magmas are often triggered by the arrival of a new batch of melt into a near surface magma reservoir. This can be detected because the summit of the volcano begins to inflate months, or even years, before an eruption begins. The injection of a fresh supply of melt causes the magma chamber to swell and fracture the rock above. This, in turn, mobilizes the magma, which quickly moves upward along the newly formed openings, often generating outpourings of lava for weeks, months, or even years.

The Role of Volatiles in Explosive Eruptions All magmas contain some water and other volatiles that are held in solution by the immense pressure of the overlying rock. Volatiles tend to be most abundant near the tops of magma reservoirs containing highly evolved, silica-rich melts. When magma rises (or the rocks confining the magma fail) a reduction in pressure occurs and the dissolved gases begin to separate from the melt forming tiny bubbles. This is analogous to opening a warm soda and allowing the carbon dioxide bubbles to escape.

When fluid basaltic magmas erupt, the pressurized gases escape with relative ease. At temperatures of 1000° C and low near-surface pressures, these gases can quickly expand to occupy hundreds of times their original volumes. On some occasions,

A magma's viscosity is directly related to its silica content— *the more silica in magma, the greater its viscosity.* Silica impedes the flow of magma because silicate structures start to link together into long chains early in the crystallization process. Consequently, rhyolitic (felsic) lavas are very viscous and tend to form comparatively short, thick flows. By contrast, basaltic lavas which contain less silica are relatively fluid and have been known to travel 150 kilometers (90 miles) or more before congealing.

The amount of **volatiles** (the gaseous components of magma, mainly water) contained in magma also affects its mobility. Other factors being equal, water dissolved in the magma tends to increase fluidity because it reduces polymerization (formation of long silicate chains) by breaking silicon–oxygen bonds. It follows, therefore, that the loss of gases renders magma (lava) more viscous.

TABLE 5.1	**Magma's Different Compositions Cause Properties to Vary**				
Composition	Silica Content	Viscosity	Gas Content	Tendency to Form Pyroclastics	Volcanic Landform
Basaltic (Mafic)	Least (~50%)	Least	Least (1–2%)	Least	Shield Volcanoes Basalt Plateaus Cinder Cones
Andesitic (Intermediate)	Intermediate (~60%)	Intermediate	Intermediate (3–4%)	Intermediate	Composite Cones
Rhyolitic (Felsic)	Most (~70%)	Greatest	Most (4–6%)	Greatest	Pyroclastic Flows Volcanic Domes

UNDERSTANDING EARTH

BOX 5.1

Anatomy of an Eruption

The events leading to the May 18, 1980, eruption of Mount St. Helens began about two months earlier as a series of minor Earth tremors centered beneath the awakening mountain (Figure 5.A, part A). The tremors were caused by the upward movement of magma within the mountain. The first volcanic activity took place a week later, when a small amount of ash and steam rose from the summit. Over the next several weeks, sporadic eruptions of varied intensity occurred. Prior to the main eruption, the primary concern had been the potential hazard of mudflows. These moving lobes of saturated soil and rock are created because heat from magma within the volcano melts the mountain's ice and snow.

The only warning of a potential eruption was a bulge on the volcano's north flank (Figure 5.A, part B). Careful monitoring of this dome-shaped structure indicated a very slow but steady growth rate of a few meters per day. If the growth rate of the bulge changed appreciably, an eruption might quickly follow. Unfortunately, no such variation was detected prior to the explosion. In fact, the seismic activity decreased during the two days preceding the huge blast.

Dozens of scientists were monitoring the mountain when it exploded. "Vancouver, Vancouver, this is it!" was the only warning—and last words from one scientist—that preceded the unleashing of tremendous quantities of pent-up gases. The trigger was a medium-sized earthquake. Its vibrations sent the north slope of the cone plummeting into the Toutle River, removing the overburden that had trapped the magma below (Figure 5.A, part C). With the pressure reduced, the water in the magma vaporized and expanded, causing the mountainside to rupture like an overheated steam boiler. Because the eruption originated around the bulge, several hundred meters below the summit, the initial blast was directed laterally rather than vertically. Had the full force of the

eruption been upward, far less destruction would have occurred.

Mount St. Helens is one of 15 large volcanoes and innumerable smaller ones that comprise the Cascade Range, which extends from British Columbia to northern California. Eight of

the largest volcanoes have been active in the past few hundred years. In addition to Mount St. Helens, the most likely to erupt again are Mount Baker and Mount Rainier in Washington, Mount Shasta and Lassen Peak in California, and Mount Hood in Oregon.

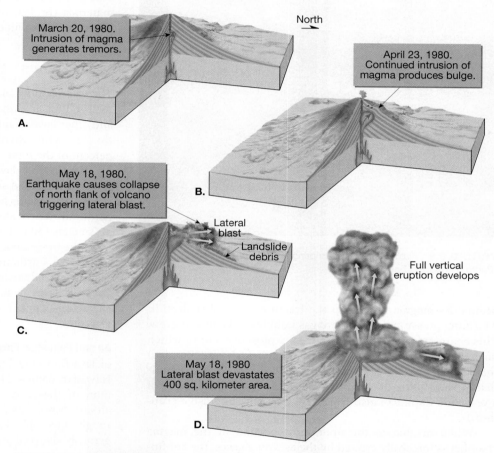

FIGURE 5.A   Idealized diagrams showing the events in the May 18, 1980, eruption of Mount St. Helens. **A.** First, a sizable earthquake indicated that renewed volcanic activity was possible. **B.** Alarming growth of a bulge on the north flank suggested increasing magma pressure below. **C.** Triggered by an earthquake, a giant landslide reduced the confining pressure on the magma body and initiated an explosive lateral blast. **D.** Within seconds a large vertical eruption sent a column of volcanic ash to an altitude of about 18 kilometers (11 miles). This phase of the eruption continued for more than 9 hours.

these expanding gases propel incandescent lava hundreds of meters into the air, producing lava fountains (Figure 5.3). Although spectacular, these fountains are mostly harmless and not generally associated with major explosive events that cause great loss of life and property.

At the other extreme, highly viscous, rhyolitic magmas may produce explosive clouds of hot ash and gases that evolve into buoyant plumes called **eruption columns** that extend thou-

sands of meters into the atmosphere (Figure 5.4). Because of the high viscosity of silica-rich magma, a significant portion of the volatiles remain dissolved until the magma reaches a shallow depth, where tiny bubbles begin to form and grow. Bubbles grow by two processes, continued separation of gases from the melt and expansion of bubbles as the confining pressure drops. Should the pressure of the expanding magma body exceed the strength of the overlying rock, fracturing

FIGURE 5.3 Fluid basaltic lava erupting from Kilauea Volcano, Hawaii. (Photo by Douglas Peebles)

occurs. As magma moves up the fractures, a further drop in confining pressure causes more gas bubbles to form and grow. This chain-reaction may generate an explosive event in which magma is literally blown into fragments (ash and pumice) that are carried to great heights by the hot gases. (As exemplified by the 1980 eruption of Mount St. Helens, the collapse of a volcano's flank can also trigger an energetic explosive eruption.)

When magma in the uppermost portion of the magma chamber is forcefully ejected by the escaping gases, the confining pressure on the molten rock directly below drops suddenly. Thus, rather than a single "bang," volcanic eruptions are really a series of explosions. This process might logically continue until the entire magma chamber is emptied, much like a geyser empties itself of water (see Chapter 17). However, this is generally not the case. It is typically only the magma in the upper part of a magma chamber that has a sufficiently high gas content to trigger a steam-and-ash explosion.

To summarize, the viscosity of magma, plus the quantity of dissolved gases and the ease with which they can escape, largely determine the nature of a volcanic eruption. In general, hot basaltic magmas contain a smaller gaseous component and permit these gases to escape with relative ease as compared to more highly evolved andesitic and rhyolitic magmas. This explains the contrast between "gentle" outflows of fluid basaltic lavas in Hawaii and the explosive and sometimes catastrophic eruptions

of viscous lavas from volcanoes such as Mount St. Helens (1980), Mount Pinatubo in the Philippines (1991), and Soufriere Hills on the island of Montserrat (1995).

MATERIALS EXTRUDED DURING AN ERUPTION

VOLCANOES
▶ Materials Extruded during an Eruption

Volcanoes extrude lava, large volumes of gas, and pyroclastic materials (broken rock, lava "bombs," fine ash, and dust). In this section we will examine each of these materials.

Lava Flows

The vast majority of lava on Earth, more than 90 percent of the total volume, is estimated to be basaltic in composition. Andesites and other lavas of intermediate composition account for most of the rest, while rhyolitic (felsic) flows make up as little as one percent of the total.

Hot basaltic lavas, which are usually very fluid, generally flow in thin, broad sheets or streamlike ribbons. On the island of Hawaii, these lavas have been clocked at 30 kilometers (19 miles) per hour down steep slopes. However, flow rates of 10 to 300 meters (30 to 1000 feet) per hour are more common. By contrast, the movement of silica-rich, rhyolitic lava may be too slow to perceive. Furthermore, most rhyolitic lavas seldom travel more than a few kilometers from their vents. As you might expect, andesitic lavas, which are intermediate in composition, exhibit characteristics that are between the extremes.

Aa and Pahoehoe Flows Two types of lava flows are known by their Hawaiian names. The most common of these, **aa** (pronounced ah-ah) **flows,** have surfaces of rough jagged blocks with dangerously sharp edges and spiny

FIGURE 5.4 Steam and ash eruption column from Mount Augustine, Cook Inlet, Alaska. (Photo by Steve Kaufman/Peter Arnold, Inc.)

A.

B.

FIGURE 5.5   Lava flows **A.** A typical slow-moving, basaltic, aa flow. **B.** A typical fluid pahoehoe
(ropy) lava. Both of these lava flows erupted from a rift on the flank of Hawaii's Kilauea Volcano.
(Photo **A** by J. D. Griggs, U.S. Geological Survey and photo **B** by Frans Lanting/Corbis)

projections (Figure 5.5A). Crossing an aa flow can be a trying and miserable experience. By contrast, **pahoehoe** (pronounced pah-hoy-hoy) **flows** exhibit smooth surfaces that often resemble the twisted braids of ropes (Figure 5.5B). Pahoehoe means "on which one can walk."

Aa and pahoehoe lavas can erupt from the same vent. However, pahoehoe lavas form at higher temperatures and are more fluid than aa flows. In addition, pahoehoe lavas can change into aa lavas flow, although the reverse (aa to pahoehoe) does not occur.

One factor that facilitates the change from pahoehoe to aa is cooling that occurs as the flow moves away from the vent. Cooling increases viscosity and promotes bubble formation. Escaping gas bubbles produce numerous voids and sharp spines in the surface of the congealing lava. As the molten interior advances, the outer crust is broken further, transforming a relatively smooth surface into an advancing mass of rough, clinkery rubble.

Occasionally, basaltic lavas have been known to move at agonizingly slow rates. In 1990, near the Hawaiian village of Kalapana, the residents had to watch for weeks as a pahoehoe flow crept toward their homes at only a few meters per hour. Although most were able to escape injury, they were unable to stop the flow and their houses were eventually incinerated.

Lava Tubes   Hardened basaltic flows commonly contain cavelike tunnels called **lava tubes** that were once conduits carrying lava from the volcanic vent to the flow's leading edge (Figure 5.6). These conduits develop in the interior of a flow where temperatures remain high long after the surface hardens. Lava tubes are important features because they serve as insulated pathways that facilitate the advance of lava great distances from its source.

Lava tubes are associated with volcanoes that emit fluid basaltic lava and are found in most parts of the world. Even the massive volcanoes on Mars have flows that contain numerous lava tubes. Some lava tubes exhibit extraordinary dimensions— one such structure, Kazumura Cave located on the southeast slope of Hawaii's Mauna Loa volcano extends for more than 60 kilometers (40 miles).

Block Lavas   In contrast to fluid basaltic magmas which can travel many kilometers, andesitic and rhyolitic magmas tend to generate relatively short prominent flows, a few hundred meters to a few kilometers long. Their upper surface consists largely of vesicle-free, detached blocks, hence the name **block lava.** Although similar to aa flows, these lavas consist of blocks with slightly curved, smooth surfaces, rather than the rough, clinkery surfaces.

Pillow Lavas   Recall that much of Earth's volcanic output occurs along oceanic ridges (divergent plate boundaries). When outpourings of lava occur on the ocean floor, the flow's outer skin quickly congeals. However, the lava is usually able to move forward by breaking through the hardened surface. This process occurs over and over, as molten basalt is extruded—like toothpaste from a tightly squeezed tube. The result is a lava flow composed of numerous tube-like structures called **pillow lavas,** stacked one atop the other. Pillow lavas are useful in the reconstruction of geologic history because whenever they are observed, they indicate that the lava flow formed in an underwater environment.

Gases

Magmas contain varying amounts of dissolved gases (*volatiles*) held in the molten rock by confining pressure, just as carbon dioxide is held in cans and botles of soft drinks. As with soft drinks, as soon as the pressure is reduced, the gases begin to escape. Obtaining gas samples from an erupting volcano is difficult and dangerous, so geologists usually must estimate the amount of gas originally contained within the magma.

The gaseous portion of most magmas makes up from 1 to 6 percent of the total weight, with most of this in the form of water vapor. Although the percentage may be small, the actual quantity of emitted gas can exceed thousands of tons per day. Occasionally, eruptions emit colossal amounts of volcanic gases that rise high into the atmosphere, where they may reside for several years. Some of these eruptions may have an impact on Earth's climate, a topic we will consider in Chapter 21.

The composition of volcanic gases is important because they contribute significantly to our planet's atmosphere.

Analyses of samples taken during Hawaiian eruptions indicate that the gas component is about 70 percent water vapor, 15 percent carbon dioxide, 5 percent nitrogen, 5 percent sulfur dioxide, with lesser amounts of chlorine, hydrogen, and argon. (The relative proportion of each gas varies significantly from one volcanic region to another.) Sulfur compounds are easily recognized by their pungent odor (Box 5.2). Volcanoes are also natural sources of air pollution—some emit large quantities of sulfur dioxide, which readily combines with atmospheric gases to form sulfuric acid and other sulfate compounds (Figure 5.7).

In addition to propelling magma from a volcano, gases play an important role in creating the narrow conduit that connects the magma chamber to the surface. First, swelling of the magma body fractures the rock above. Then, hot blasts of high-pressure gases expand the cracks and develop a passageway to the surface. Once the passageway is completed, the hot gases, armed with rock fragments, erode its walls, producing a larger conduit. Because these erosive forces are concentrated on any protrusion along the pathway, the volcanic pipes that are produced tend to develop a circular shape. As the conduit enlarges, magma moves upward to produce surface activity. Following an eruptive phase, the volcanic pipe often becomes choked with a mixture of congealed magma and debris that was not thrown clear of the vent. Before the next eruption, a new surge of explosive gases may again clear the conduit.

Pyroclastic Materials

When volcanoes erupt energetically they eject pulverized rock, lava, and glass fragments from the vent. The particles

A.

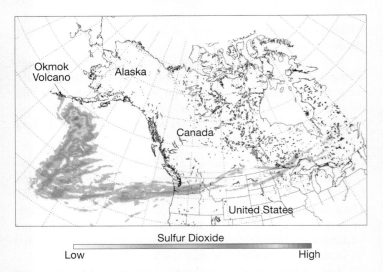

FIGURE 5.6   Lava flows often develop a solid crust while the molten lava below continues to advance in conduits called *lava tubes.* **A.** View of an active lava tube as seen through the collapsed roof. **B.** Thurston Lava Tube, Hawaii Volcanoes National Park. (Photo **A** by G. Brad Lewis/Omjalla Images and Photo **B** by Philip Rosenberg/Pacific Stock)

FIGURE 5.7   Okmok Volcano, in Alaska's Aleutian Islands, erupted in 2009 releasing a plume of ash and gases including sulfur dioxide, which cannot be seen but can be detected by satellites. Sulfur dioxide can affect both human health and climate. Near Earth's surface sulfur dioxide poses hazards, including respiratory ailments, visibility impairments, volcanic smog (vog), and acid rain. At higher altitudes, where this concentration of sulfur dioxide was detected, the gas is transformed into tiny sulfate particles. These aerosol particles create a bright haze that reflects sunlight back to space, preventing it from reaching Earth's surface. (Image courtesy of NASA)

A.

B.

C.

FIGURE 5.8 Pyroclastic materials. **A.** *Volcanic ash* and small pumice fragments (*lapilli*) that erupted from Mount St. Helens in 1980. Inset photo is an image obtained using a scanning electron microscope (SEM). This vesicular ash particle exhibits a glassy texture and is roughly the diameter of a human hair. **B.** Volcanic block. Volcanic blocks are solid fragments that were ejected from a volcano during an explosive eruption. **C.** These basaltic bombs were erupted by Hawaii's Mauna Kea Volcano. Volcanic bombs are blobs of lava that are ejected while still molten and often acquire rounded, aerodynamic shapes as they travel through the air. (Photos courtesy of U.S. Geological Survey)

bombs and blocks usually fall near the vent; however, they are occasionally propelled great distances. For instance, bombs 6 meters (20 feet) long and weighing about 200 tons were blown 600 meters (2000 feet) from the vent during an eruption of the Japanese volcano Asama.

So far we have distinguished various pyroclastic materials based largely on the size of the fragments. Some materials are also identified by their texture and composition. In particular, **scoria** is the name applied to vesicular ejecta that is a product of basaltic magma (Figure 5.10A). These black to reddish-brown fragments are generally found in the size range of lapilli and resemble cinders and clinkers produced by furnaces used to smelt iron. When magmas with intermediate (andesitic) or felsic (rhyolitic) compositions erupt explosively, they emit ash and the vesicular rock **pumice** (Figure 5.10B). Pumice is usually lighter in color and less dense than scoria, and many pumice fragments have so many vesicles that they are light enough to float.

FIGURE 5.9 Volcanic bombs forming during an eruption of Hawaii's Kilauea Volcano. Ejected lava masses take on a streamlined shape as they sail through the air. The bomb in the insert is about 10 centimeters long. (Photo by Arthur Roy/National Audubon Society; inset photo by E. J. Tarbuck)

produced are referred to as **pyroclastic materials** (*pyro* = fire, *clast* = fragment). These fragments range in size from very fine dust and sand-sized volcanic ash (less than 2 millimeters) to pieces that weigh several tons.

Ash and *dust* particles are produced when gas-rich viscous magma erupts explosively (see Figure 5.4). As magma moves up in the vent, the gases rapidly expand, generating a melt that resembles the froth that flows from a bottle of champagne. As the hot gases expand explosively, the froth is blown into very fine glassy fragments (Figure 5.8). When the hot ash falls, the glassy shards often fuse to form a rock called *welded tuff*. Sheets of this material, as well as ash deposits that later consolidate, cover vast portions of the western United States.

Somewhat larger pyroclasts that range in size from small beads to walnuts are known as *lapilli* ("little stones"). These ejecta are commonly called *cinders* (2–64 millimeters). Particles larger than 64 millimeters (2.5 inches) in diameter are called *blocks* when they are made of hardened lava and *bombs* when they are ejected as incandescent lava (see Figure 5.8C). Because bombs are semimolten upon ejection, they often take on a streamlined shape as they hurtle through the air (Figure 5.9). Because of their size,

A. Scoria

B. Pumice

FIGURE 5.10 Scoria and pumice are volcanic rocks that exhibit a vesicular texture. Vesicles are small holes left by escaping gas bubbles. **A.** Scoria is usually a product of mafic (basaltic) magma. **B.** Pumice forms during explosive eruptions of viscous magmas having an intermediate (andesitic) or felsic (rhyolitic) composition. (Photos by E. J. Tarbuck)

VOLCANIC STRUCTURES AND ERUPTIVE STYLES

VOLCANOES
GEODe EARTH ▶ Volcanic Structures and Eruptive Styles

The popular image of a volcano is that of a solitary, graceful, snowcapped cone, such as Mount Hood in Oregon or Japan's Fujiyama. These picturesque, conical mountains are produced by volcanic activity that occurred intermittently over thousands, or even hundreds of thousands, of years. However, many volcanoes do not fit this image. Cinder cones are quite small and form during a single eruptive phase that lasts a few days to a few years. Other volcanic landforms are not volcanoes at all. For example, Alaska's Valley of Ten Thousand Smokes is a flat-topped deposit consisting of 15 cubic kilometers of ash that erupted in less than 60 hours and blanketed a section of river valley to a depth of 200 meters (600 feet).

Volcanic landforms come in a wide variety of shapes and sizes, and each structure has a unique eruptive history. Nevertheless, volcanologists have been able to classify volcanic landforms and determine their eruptive patterns. In this section we will consider the general anatomy of a volcano and look at three major volcanic types: shield volcanoes, cinder cones, and composite cones.

Anatomy of a Volcano

Volcanic activity frequently begins when a fissure (crack) develops in the crust as magma moves forcefully toward the surface. As the gas-rich magma moves up through a fissure, its path is usually localized into a circular **conduit,** or **pipe,** that terminates at a surface opening called a **vent** (Figure 5.11). Successive eruptions of lava, pyroclastic material, or frequently a combination of both, often separated by long periods of inactivity, eventually build the cone-shaped structure we call a **volcano.**

Located at the summit of most volcanoes is a somewhat funnel-shaped depression, called a **crater** (*crater* = a bowl). Volcanoes that are built primarily of pyroclastic materials typically have craters that form by gradual accumulation of volcanic debris on the surrounding rim. Other craters form during explosive eruptions as the rapidly ejected particles erode the crater walls. Craters also form when the summit area of a volcano collapses following an eruption (Figure 5.12). Some volcanoes have very large circular depressions called *calderas* that have diameters greater than one kilometer and in rare cases can exceed 50 kilometers. We will consider the formation of various types of calderas later in this chapter.

During early stages of growth most volcanic discharges come from a central summit vent. As a volcano matures, material also tends to be emitted from fissures that develop along the flanks or at the base of the volcano. Continued activity from a flank eruption may produce a small **parasitic cone** (*parasitus* = one who eats at the table of another). Italy's Mount Etna, for example, has more than 200 secondary vents, some of which

FIGURE 5.11 Anatomy of a "typical" composite cone (see also Figures 5.13 and 5.17 for a comparison with a shield and cinder cone, respectively).

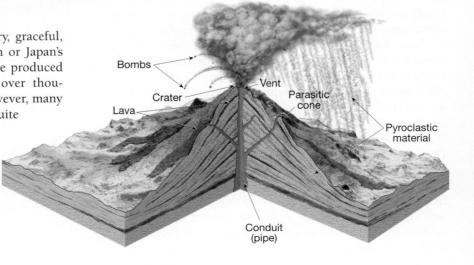

Bombs
Vent
Crater
Parasitic cone
Lava
Pyroclastic material
Conduit (pipe)

EARTH AS A SYSTEM

BOX 5.2

Volcanic Air Pollution—A Hazard in Hawaii

We normally think of air pollution (*smog*) as being a problem in and near large urban and industrial areas. By contrast, our likely perception of the island of Hawaii is that air quality there is usually excellent because it is located in the Pacific Ocean, far from major urban and industrial centers. Surprisingly, this is not always the case. Recently, Hawaii has experienced episodes in which a thick acrid haze covered parts of the island leading to reduced visibility, health complaints, and even damage to sensitive crops.

Such events are not caused by forest fires or the exhausts from cars and factories, but by winds carrying emissions from Kilauea Volcano. The word *vog,* an abbreviation for "volcanic smog," was coined to identify this form of air pollution. Vog is created when sulfur dioxide (SO_2) and other volcanic gases combine and interact chemically in the atmosphere with oxygen, moisture, dust, and sunlight. Vog is a visible haze consisting of gas plus a suspended mixture of tiny liquid and solid particles (aerosols).

Vog was first experienced in 1986. Until that time, the volcano's ongoing eruption, which began three years earlier, consisted of short episodes that often produced spectacular lava fountains. But, beginning in mid-1986, the flow of magma to the surface became more steady, producing a nearly constant but quiet outflow of lava and gas. The vog problem became more serious in March 2008 when a new vent

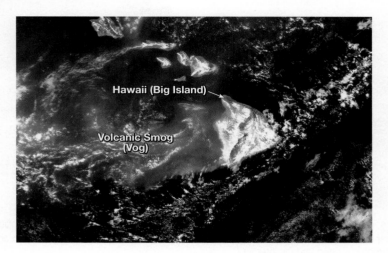

FIGURE 5.B   This gray-white haze known as *vog* (short for volcanic smog) hung low over the Hawaiian Islands in the late fall of 2008, when NASA's *Aqua* satellite captured this image. Though usually not this widespread, vog has become relatively common on the island of Hawaii. It forms when sulfur dioxide from Kilauea Volcano reacts with oxygen and water in the atmosphere. The tiny sulfur oxide particles that result scatter sunlight so that vog shows up easily when viewed from space. The chemicals in vog are harmful to the environment and to the health of plants and animals. (Photo courtesy of NASA)

broke through a wall of Halemaumau crater causing the rate of gas emissions to increase substantially. Figure 5.B is a satellite image that shows an especially thick and widespread episode in December 2008. The vog appears as a haze that is thinner and has more diffuse outlines than nearby white clouds.

In summary, the air quality problems experienced on the island of Hawaii are not human-generated, but rather are natural phenomena created largely by interactions between the geosphere and the atmosphere.

FIGURE 5.12   The crater of Mount Vesuvius, Italy. The city of Naples is located northwest of Vesuvius, while Pompeii, the Roman town that was buried by an eruption in AD 79, is located southeast of the volcano. (NASA; inset image by Krafft/Photo Researchers)

have built parasitic cones. Many of these vents, however, emit only gases and are appropriately called **fumaroles** (*fumus* = smoke).

Shield Volcanoes

Shield volcanoes are produced by the accumulation of fluid basaltic lavas and exhibit the shape of a broad, slightly domed structure that resembles a warrior's shield (Figure 5.13). Most shield volcanoes begin on the ocean floor as seamounts, a few of which grow large enough to form volcanic islands. In fact, with the exception

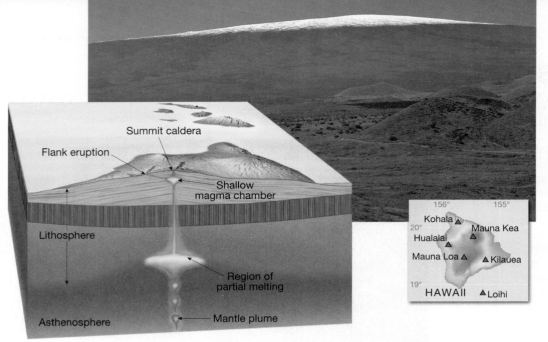

FIGURE 5.13 Mauna Loa is one of five shield volcanoes that collectively make up the island of Hawaii. Shield volcanoes are built primarily of fluid basaltic lava flows and contain only a small percentage of pyroclastic materials. (Photo by Greg Vaughn/Alamy)

of the volcanic islands that form above subduction zones, most other oceanic islands are either a single shield volcano, or more often, the coalescence of two or more shields built upon massive amounts of pillow lavas. Examples include the Canary Islands, the Hawaiian Islands, the Galapagos, and Easter Island. In addition, some shield volcanoes form on continental crust. Included in this group are several volcanic structures located in East Africa.

Mauna Loa: A Classic Shield Volcano Extensive study of the Hawaiian Islands confirms that they are constructed of a myriad of thin basaltic lava flows averaging a few meters thick intermixed with relatively minor amounts of pyroclastic ejecta. Mauna Loa is one of five overlapping shield volcanoes that together comprise the Big Island of Hawaii (see Figure 5.13). From its base on the floor of the Pacific Ocean to its summit, Mauna Loa is over 9 kilometers (6 miles) high, exceeding that of Mount Everest. This massive pile of basaltic rock has a volume of 80,000 cubic kilometers that was extruded over a span of about one million years. The volume of material composing Mauna Loa is roughly 200 times greater than the amount composing a large composite cone such as Mount Rainier (Figure 5.14). Although the shield volcanoes that comprise islands are often quite large, some are more modest in size. In addition, an estimated one million basaltic submarine volcanoes (seamounts) of various sizes dot the ocean floor.

The flanks of Mauna Loa have gentle slopes of only a few degrees. The low-angle results because very hot, fluid lava trav-

els "fast and far" from the vent. In addition, most of the lava (perhaps 80 percent) flows through a well-developed system of lava tubes (see Figure 5.6). This greatly increases the distance lava can travel before it solidifies. Thus, lava emitted near the summit often reaches the sea, thereby adding to the width of the cone at the expense of its height.

Another feature common to many active shield volcanoes is a large, steep-walled caldera that occupies the summit (see Figure 5.29). Calderas on large shield volcanoes form when the roof above the magma chamber collapses. This usually occurs as the magma reservoir empties following a large eruption, or as magma migrates to the flank of a volcano to feed a fissure eruption.

In the final stage of growth, shield volcanoes are more sporadic and pyroclastic ejections are more common. Further, lavas increase in viscosity, resulting in thicker, shorter flows. These eruptions tend to steepen the slope of the summit area, which often becomes capped with clusters of cinder cones. This may explain why Mauna Kea, which is a more mature volcano that has not erupted in historic times, has a steeper summit than Mauna Loa, which erupted as recently as 1984. Astronomers are so certain that Mauna Kea is "over the hill" that they have built an elaborate observatory on its summit, housing some of the world's finest (and most expensive) telescopes.

Kilauea, Hawaii: Eruption of a Shield Volcano Kilauea, the most active and intensely studied shield volcano in the world, is located on the island of Hawaii in the shadow of Mauna Loa. More than 50 eruptions have been witnessed here since record keeping began in 1823. Several months before each eruptive phase, Kilauea inflates as magma gradually migrates upward and accumulates in a central reservoir located a few kilometers below the summit. For up to 24 hours in advance of an eruption, swarms of small earthquakes warn of the impending activity.

Most of the recent activity on Kilauea has occurred along the flanks of the volcano in a region called the East Rift Zone. A rift eruption here in 1960 engulfed the coastal village of Kapoho, located nearly 30 kilometers (20 miles) from the source. The longest and largest rift eruption ever recorded on Kilauea began in 1983 and continues to this day, with no signs of abating.

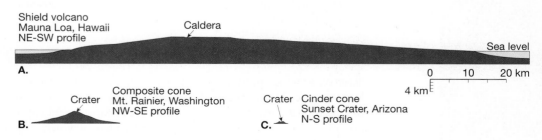

FIGURE 5.14 Profiles comparing scales of different volcanoes. **A.** Profile of Mauna Loa, Hawaii, the largest shield volcano in the Hawaiian chain. Note size comparison with Mount Rainier, Washington, a large composite cone. **B.** Profile of Mount Rainier, Washington. Note how it dwarfs a typical cinder cone. **C.** Profile of Sunset Crater, Arizona, a typical steep-sided cinder cone.

FIGURE 5.15 Lava "curtain" extruded along the East Rift Zone, Kilauea, Hawaii. (Photo by Greg Vaughn/Alamy)

The first discharge began along a 6-kilometer (4-mile) fissure where a 100-meter (300-foot) high "curtain of fire" formed as red-hot lava was ejected skyward (Figure 5.15). When the activity became localized, a cinder and spatter cone, given the Hawaiian name *Puu Oo,* was built. Over the next 3 years the general eruptive pattern consisted of short periods (hours to days) when fountains of gas-rich lava sprayed skyward. Each event was followed by nearly a month of inactivity.

By the summer of 1986 a new vent opened 3 kilometers downrift. Here smooth-surfaced pahoehoe lava formed a lava lake. Occasionally the lake overflowed, but more often lava escaped through tunnels to feed flows that moved down the southeastern flank of the volcano toward the sea. These flows destroyed nearly a hundred rural homes, covered a major roadway, and eventually reached the sea. Lava has been intermittently pouring into the ocean ever since, adding new land to the island of Hawaii.

Situated just 32 kilometers (20 miles) off the southern coast of Kilauea, a submarine volcano, Loihi, is also active. It, however, has another 930 meters (3100 feet) to go before it breaks the surface of the Pacific Ocean.

Evolution of Volcanic Islands Although Mauna Loa and Kilauea are commonly regarded as typical shield volcanoes, other ocean island volcanoes exhibit significant differences. The basaltic shields that comprise the Canary Islands, for example, have been active for as long as 20 million years. Compared to the Hawaiian shields, they have steeper slopes (20–30 degrees) and tend to eject more highly evolved lavas—which accounts for the abundance of pyroclastic material.

Despite these differences, most large oceanic shields have similar origins. They form above a long-lived, rising plume of hot rock anchored in the mantle. Partial melting results in volcanic activity on the ocean floor producing piles of pillow lavas that evolve into a submarine volcano. Eventually, a few of these structures emerge as volcanic islands (Figure 5.16).

A volcanic island will continue to grow as long as it is located over a hot spot. However, as the lithospheric plate moves away from the zone of melting, volcanic activity ceases. Gradually, as the lithosphere cools and contracts, the island begins to subside. Small basaltic islands succumb to both subsidence which causes them to sink below sea level, and erosional forces that flatten their summits. These submarine structures, which are often capped by coral reef deposits, are called **guyots.**

When large seamounts enter subduction zones, they act as hindrances to the descending plate. As a result, they can trigger large earthquakes or underwater debris slides, both of which may generate destructive seismic sea waves called *tsunamis.*

Cinder Cones

As the name suggests, **cinder cones** (also called **scoria cones**) are built from ejected lava fragments that take on the appearance of cinders or clinkers as they begin to harden in flight (Figure 5.17).

FIGURE 5.16 The island of Surtsey was built on an underwater base of pillow lavas by nearly continuous volcanic eruptions between 1963 and 1967. The island has been preserved as a living laboratory, allowing scientists to study how new land areas are colonized by species without the interference of humans. By 2004, the island was host to 60 species of plants, 24 species of fungi, and 89 bird species. (Photo by Arctic Images/Corbis)

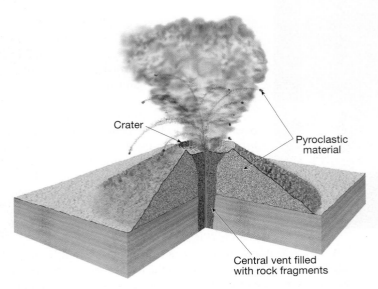

Crater

Pyroclastic material

Central vent filled with rock fragments

FIGURE 5.17 Cinder cones are built from ejected lava fragments (mostly cinders and bombs) and are usually less than 300 meters (1000 feet) in height.

These pyroclastic fragments range in size from fine ash to bombs that may exceed a meter in diameter. However, most of the volume of a cinder cone consists of pea- to walnut-sized lapilli that are markedly vesicular and have a black to reddish-brown color. (Recall that these vesicular rock fragments are called *scoria*.) Although cinder cones are composed mostly of loose pyroclastic material, they sometimes extrude lava. On such occasions the discharges most often come from vents located at or near the base rather than from the summit crater.

Cinder cones have very simple, distinctive shapes determined by the slope that loose pyroclasic material maintains as it comes to rest (Figure 5.18). Because cinders have a high angle of repose (the steepest angle at which material remains stable), cinder cones are steep-sided, having slopes between 30 and 40 degrees. In addition, cinder cones have large, deep craters in relation to the overall size of the structure. Although relatively symmetrical, many cinder cones are elongated, and higher on the side that was downwind during the eruptions.

Most cinder cones are produced by a single, short-lived eruptive event. One study found that half of all cinder cones examined were constructed in less than one month, and that 95 percent formed in less than one year. However, in some cases, they remain active for several years. Parícutin, shown in Figure 5.19, had an eruptive cycle that spanned 9 years. Once the event ceases, the magma in the "plumbing" connecting the vent to the magma source solidifies and the volcano usually does not erupt again. (One exception is Cerro Negro, a cinder cone in Nicaragua, which has erupted more than 20 times since it formed in 1850.) As a consequence of this short life span, cinder cones are small, usually between 30 meters (100 feet) and 300 meters (1000 feet). A few rare examples exceed 700 meters (2100 feet) in height.

Cinder cones number in the thousands around the globe. Some occur in volcanic fields such as the one near Flagstaff, Arizona, which consists of about 600 cones. Others are parasitic cones that are found on the flanks of larger volcanoes.

Parícutin: Life of a Garden-variety Cinder Cone One of the very few volcanoes studied by geologists from its very beginning is the cinder cone called Parícutin, located about 320 kilometers (200 miles) west of Mexico City. In 1943 its eruptive phase began in a cornfield owned by Dionisio Pulido, who witnessed the event as he prepared the field for planting.

For two weeks prior to the first eruption, numerous Earth tremors caused apprehension in the nearby village of Parícutin. Then, on February 20, sulfurous gases began billowing from a small depression that had been in the cornfield for as long as people could remember. During the night, hot, glowing rock fragments were ejected from the vent, producing a spectacular fireworks display. Explosive discharges continued, throwing hot fragments and ash occasionally as high as 6000 meters (20,000 feet) into the air. Larger fragments fell near the crater, some remaining incandescent as they rolled down the slope. These built an aesthetically pleasing cone, while finer ash fell over a much larger area, burning and eventually covering the village of Parícutin. In the first day the cone grew to 40 meters (130 feet), and by the fifth day it was more than 100 meters (330 feet) high.

The first lava flow came from a fissure that opened just north of the cone, but after a few months flows began to emerge from the base of the cone itself. In June 1944, a clinkery aa flow 10 meters (30 feet) thick moved over much of the village of San Juan Parangaricutiro, leaving only the church steeple exposed (see Figure 5.19). After 9 years of intermittent pyroclastic explosions and nearly continuous discharge of lava from vents at its base, the activity ceased almost as quickly as it had

Lava flow

Volcanic bomb

FIGURE 5.18 SP Crater, a cinder cone located in the San Francisco Peaks volcanic field north of Flagstaff, Arizona. The lava flow in the upper part of the image originated from the base of the cinder cone. Inset photo shows a large bomb on the slopes of SP Crater. (Photos by Michael Collier)

FIGURE 5.19 The village of San Juan Parangaricutiro engulfed by aa lava from Parícutin, shown erupting in the background. Only the church towers remain. (Photo by Tad Nichols)

begun. Today, Parícutin is just another one of the scores of cinder cones dotting the landscape in this region of Mexico. Like the others, it will not erupt again.

Composite Cones

Earth's most picturesque yet potentially dangerous volcanoes are **composite cones** or **stratovolcanoes.** Most are located in a relatively narrow zone that rims the Pacific Ocean, appropriately called the *Ring of Fire* (see Figure 5.37). This active zone consists of a chain of continental volcanoes that are distributed along the west coast of the Americas, including the large cones of the Andes in South America and the Cascade Range of the western United States and Canada. The latter group includes Mount St. Helens, Mount Shasta, and Mount Garibaldi (Figure 5.20). The most active

regions in the Ring of Fire are located along curved belts of volcanic cones situated adjacent to the deep-ocean trenches of the northern and western Pacific. This nearly continuous chain of volcanoes stretches from the Aleutian Islands to Japan and the Philippines and to the North Island of New Zealand. These impressive volcanic structures are manifestations of processes that occur in the mantle in association with *subduction zones.*

The classic composite cone is a large, nearly symmetrical structure consisting of alternating layers of explosively erupted cinders and ash interbedded with lava flows. A few composite cones, notably Italy's Etna and Stromboli, display very persistent eruption activity, and molten lava has been observed in their summit craters for decades. Stromboli is so well known for eruptions that eject incandescent blobs of lava that it has been referred to as the "Lighthouse of the Mediterranean." Mount Etna, on the other hand, has erupted, on average, once every two years since 1979 (see Chapter Opening photo).

Just as shield volcanoes owe their shape to fluid basaltic lavas, composite cones reflect the viscous nature of the material from which they are made. In general, composite cones are the product of gas-rich magma having an andesitic composition. However, many composite cones also emit various amounts of fluid basaltic lava and occasionally, pyroclastic material having rhyolitic composition. Relative to shields, the silica-rich magmas typical of composite cones generate thick viscous lavas that travel less than a few kilometers. In addition, composite cones are noted for generating explosive eruptions that eject huge quantities of pyroclastic material.

A conical shape, with a steep summit area and more gradually sloping flanks, is typical of many large composite cones. This classic profile, which adorns calendars and postcards, is partially a consequence of the way viscous lavas and pyroclastic ejecta contribute to the growth of the cone. Coarse fragments ejected from the summit crater tend to accumulate near their source. Because of their high angle of repose, coarse materials contribute to the steep slopes of the summit area. Finer ejecta, on the other hand, are deposited as a thin layer over a large area. This acts to flatten the flank of the cone. In addition, during the early stages of growth,

FIGURE 5.20 Aerial view of Mount Shasta, California, one of the largest composite cones in the Cascade Range. (Photo courtesy of the U.S. Geological Survey)

FIGURE 5.21 Japan's Fujiyama exhibits the classic form of a composite cone—steep summit and gently sloping flanks. (Photo by Koji Nakano/Getty Images/Sebun)

Many composite cones have numerous small, parasitic cones on their flanks, while others, such as Crater Lake, have been truncated by the collapse of their summit (Figure 5.22). Still others have a lake in their crater which may be hot and muddy. Such lakes are often highly acidic because of the influx of sulfur and chlorine gases that react with water to produce sulfuric (H_2SO_4) and hydrochloric acid (HCl).

LIVING IN THE SHADOW OF A COMPOSITE CONE

More than 50 volcanoes have erupted in the United States in the past 200 years (Figure 5.23). Fortunately, the most explosive of these eruptions occurred in sparsely inhabited regions of Alaska. On a global scale many destructive eruptions have occurred during the past few thousand years, a few of which may have influenced the course of human civilization (Box 5.3).

Eruption of Vesuvius AD 79

One well documented event of historic proportions was the AD 79 eruption of the Italian volcano we now call Vesuvius. Prior to this eruption, Vesuvius had been dormant for centuries and had vineyards adorning its sunny slopes. On August 24, however, the tranquility ended, and in less than 24 hours the city of Pompeii (near Naples) and more than 2000 of its 20,000 residents perished. Some were entombed beneath a layer of pumice nearly 3 meters (10 feet) thick, while others were encased within a layer of ash (Figure 5.24B). They remained this way for nearly 17 centuries, until the city was excavated, giving archaeologists a superbly detailed picture of ancient Roman life (Figure 5.24A).

By reconciling historical records with detailed scientific studies of the region, volcanologists have pieced together the chronology of the destruction of Pompeii. The eruption most likely began as steam discharges on the morning of August 24. By early afternoon fine ash and pumice fragments formed a tall eruptive cloud. Shortly thereafter, debris from this cloud began to shower Pompeii, which was located 9 kilometers (6 miles) downwind of the volcano. Many people fled during this early phase of the eruption. For the next several hours pumice fragments as large as 5 centimeters (2 inches) fell on Pompeii. One historical record of the eruption states that some people tied pillows to their heads in order to fend off the flying fragments.

The rain of pumice continued for several hours, accumulating at the rate of 12 to 15 centimeters (5 to 6 inches) per hour. Most of the roofs in Pompeii eventually gave way. Despite the accumulation of more than 2 meters of pumice, many of the people that had not evacuated Pompeii were probably still alive the next morning. Then, suddenly and unexpectedly, a surge of

lavas tend to be more abundant and flow greater distances from the vent than lavas do later in the volcano's history. This contributes to the cone's broad base. As the volcano matures, the shorter flows that come from the central vent serve to armor and strengthen the summit area. Consequently, steep slopes exceeding 40 degrees are sometimes possible. Two of the most perfect cones—Mount Mayon in the Philippines and Fujiyama in Japan—exhibit the classic form we expect of a composite cone, with its steep summit and gently sloping flanks (Figure 5.21).

Despite the symmetrical forms of many composite cones, most have complex histories. Huge mounds of volcanic debris surrounding these structures provide evidence that large sections of these volcanoes slid down slope as massive landslides. Others develop horseshoe-shaped depressions at their summits as a result of explosive lateral eruptions—as occurred during the 1980 eruption of Mount St. Helens. Often, so much rebuilding has occurred since these eruptions that no trace of the amphitheater-shaped scars remain.

FIGURE 5.22 Oregon's Crater Lake formed in a caldera following a prehistoric eruption and collapse of a once prominent composite cone we now refer to as Mount Mazama. (Photo by Michael Collier)

searing hot ash and gas swept rapidly down the flanks of Vesuvius. This blast killed an estimated 2000 people who had somehow managed to survive the pumice fall. Most died instantly as a result of inhaling the hot, ash-laden gases. Their remains were quickly buried by the falling ash. Rain then caused the ash to become rock hard before their bodies had time to decay. The subsequent decomposition of the bodies produced cavities in the hardened ash that replicated their forms and, in some cases, even preserved facial expressions. Nineteenth-century excavators found these cavities and created casts of the corpses by pouring plaster of Paris into the voids (Figure 5.24B).

Today, Vesuvius towers over the Naples skyline. Such an image should prompt us to consider how volcanic crises might be managed in the future.

Nuée Ardente: A Deadly Pyroclastic Flow

One of the most destructive forces of nature are **pyroclastic flows** which consist of hot gases infused with incandescent ash and larger lava fragments. Also referred to as **nuée ardentes** (*glowing*

A.

B.

FIGURE 5.24 The Roman city of Pompeii was destroyed in AD 79 during an eruption of Mount Vesuvius. **A.** Ruins of Pompeii. Excavation began in the 18th century and continues today. **B.** Plaster casts of several victims of the AD 79 eruption of Mount Vesuvius. (Photo **A** by Roger Ressmeyer, Photo **B** by Leonard von Matt/Photo Researchers, Inc.)

avalanches), these fiery flows are capable of racing down steep volcanic slopes at speeds that can exceed 200 kilometers (125 miles) per hour (Figure 5.25). Nuée ardentes are composed of two parts—a low-density cloud of hot expanding gases containing fine ash particles, and a ground-hugging portion that contains most of the material in the flow.

Driven by gravity, pyroclastic flows tend to move in a manner similar to snow avalanches. They are mobilized by volcanic gases released from the lava fragments and by the expansion of heated air that is overtaken and trapped in the moving front. These gases reduce friction between the fragments and the ground. Strong turbulent flow is another important mechanism which aids in the transport of ash and pumice fragments down slope in a nearly frictionless environment (see Figure 5.25). This helps explain why some nuée ardente deposits are found more than 100 kilometers (60 miles) from their source.

Sometimes, powerful hot blasts that carry small amounts of ash separate from the main body of a pyroclastic flow. These low-density clouds, called *surges*, can be deadly, but seldom have sufficient force to destroy buildings in their paths. Nevertheless, on June 3, 1991, a hot ash cloud from Japan's Unzen Volcano engulfed and burned hundreds of homes and moved cars as much as 80 meters (250 feet).

Pyroclastic flows may originate in a variety of volcanic settings. Some occur when a powerful eruption blasts pyroclastic material out of the side of a volcano—the lateral eruption of

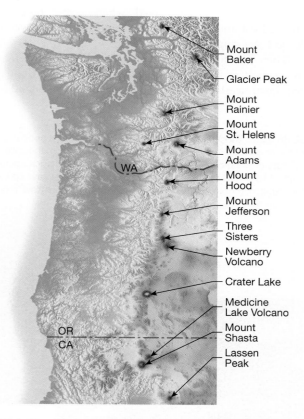

Mount Baker

Glacier Peak

Mount Rainier

Mount St. Helens

Mount Adams

Mount Hood

Mount Jefferson

Three Sisters

Newberry Volcano

Crater Lake

Medicine Lake Volcano

Mount Shasta

Lassen Peak

FIGURE 5.23 Of the 13 potentially active volcanoes in the Cascade Range, 11 have erupted in the past 4000 years and 7 in just the past 200 years. More than 100 eruptions, most of which were explosive, have occurred in the past 4000 years. Mount St. Helens is the most active volcano in the Cascades. Its eruptions have ranged from relatively quiet outflows of lava to explosive events much larger than that of May 18, 1980.

FIGURE 5.25 Pyroclastic flows. **A.** Illustration of a fiery ash and pumice flow racing down the slope of a volcano. **B.** Pyroclastic flow moving rapidly down the forested slopes of Mt. Unzen toward a Japanese village. (Photo by Yomiuri/AP Photo)

Mount St. Helens in 1980, for example. More frequently, however, nuée ardentes are generated by the collapse of tall eruption columns during an explosive event. When gravity eventually overcomes the initial upward thrust provided by the escaping gases, the ejecta begin to fall, sending massive amounts of incandescent blocks, ash, and pumice cascading down slope.

In summary, pyroclastic flows are a mixture of hot gases and pyroclastic materials moving along the ground, driven primarily by gravity. In general, flows that are fast and highly turbulent can transport fine particles for distances of 100 kilometers or more.

The Destruction of St. Pierre In 1902, an infamous nuée ardente and associated surge from Mount Pelée, a small volcano on the Caribbean island of Martinique, destroyed the port town of St. Pierre. Although the main pyroclastic flow was largely confined to the valley of Riviere Blanche, the fiery surge spread south of the river and quickly engulfed the entire city. The destruction happened in moments and was so devastating that almost all of St. Pierre's 28,000 inhabitants were killed. Only one person on the outskirts of town—a prisoner protected in a dungeon—and a few people on ships in the harbor were spared (Figure 5.26).

Within days of this calamitous eruption, scientists arrived on the scene. Although St. Pierre was mantled by only a thin layer of volcanic debris, they discovered that masonry walls nearly a meter thick were knocked over like dominoes; large trees were uprooted and cannons were torn from their mounts. A further reminder of the destructive force of this

nuée ardente is preserved in the ruins of the mental hospital. One of the immense steel chairs that had been used to confine alcoholic patients can be seen today, contorted, as though it were made of plastic.

Lahars: Mudflows on Active and Inactive Cones

In addition to violent eruptions, large composite cones may generate a type of very fluid mudflow referred to by its Indonesian name **lahar.** These destructive flows occur when volcanic debris becomes saturated with water and rapidly moves down steep volcanic slopes, generally following gullies and stream valleys. Some lahars may be triggered when magma is emplaced near the surface, causing large volumes of ice and snow to melt. Others are generated when heavy rains saturate weathered volcanic deposits. Thus, lahars may occur even when a volcano is *not* erupting.

When Mount St. Helens erupted in 1980, several lahars were generated. These flows and accompanying flood waters raced down nearby river valleys at speeds exceeding 30 kilome-

FIGURE 5.26 The photo on the left shows St. Pierre as it appeared shortly after the eruption of Mount Pelée, 1902. (Reproduced from the collection of the Library of Congress) The photo on the right shows St. Pierre before the eruption. Many vessels are anchored offshore, as was the case on the day of the eruption. (Photo courtesy of The Granger Collection, New York)

PEOPLE AND THE ENVIRONMENT

BOX 5.3

The Lost Continent of Atlantis

Anthropologists have proposed that a catastrophic eruption on the island of Santorini (also called Thera) contributed to the collapse of the advanced Minoan civilization that was centered around Crete in the Aegean Sea (Figure 5.C). This event also gave rise to the enduring legend of the lost continent of Atlantis. According to an account by the Greek philosopher Plato, an island empire named Atlantis was swallowed up by the sea in a single day and night. Although the connection between Plato's Atlantis and the Minoan civilization is somewhat tenuous, there is no doubt that a catastrophic eruption took place on Santorini about 1600 BC.

This eruption generated a tall, billowing *eruption column* composed of huge quantities of pyroclastic materials. Ash and pumice rained from this plume for days, eventually blanketing the surrounding landscape to a

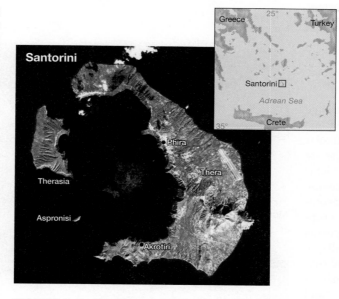

FIGURE 5.C   Satellite view showing the remains of the volcanic island of Santorini after the top of the cone collapsed into the emptied magma chamber following an explosive eruption. The location of the recently excavated Minoan town of Akrotiri is shown. Volcanic eruptions over the last 500 years built the central islands. Despite the fact that another destructive eruption is likely, the city of Phira was built on the flanks of the caldera. (Photo courtesy of NASA)

maximum depth of 60 meters (200 feet). One nearby Minoan city, now called Akrotiri, was buried and its remains sealed until 1967, when archaeologists began investigating the area. The excavation of beautiful ceramic jars and elaborate wall paintings indicate that Akrotiri was home to a wealthy and sophisticated society.

Following the eruption of this large quantity of material, the summit of Santorini collapsed, producing a caldera 8 kilometers (5 miles) across. This once majestic volcano presently consists of five small islands. The eruption and collapse of Santorini generated large sea waves (*tsunami*) that caused widespread destruction to the coastal villages of Crete and to nearby islands to the north.

Was this eruption the main cause of the dispersal of this great civilization as some scholars suggest, or only one of many contributing factors? Was Santorini the island continent of Atlantis described by Plato? Whatever the answers to these questions, clearly volcanism can dramatically change the course of human events.

---

ters per hour. These raging rivers of mud destroyed or severely damaged nearly all the homes and bridges along their paths. Fortunately, the area was not densely populated (Figure 5.27).

In 1985, deadly lahars were produced during a small eruption of Nevado del Ruiz, a 5300-meter (17,400-foot) volcano in the Andes Mountains of Colombia. Hot pyroclastic material melted ice and snow that capped the mountain (*nevado* means *snow* in Spanish) and sent torrents of ash and debris down three major river valleys that flank the volcano. Reaching speeds of 100 kilometers (60 miles) per hour, these mudflows tragically took 25,000 lives.

Mount Rainier, Washington, is considered by many to be America's most dangerous volcano because, like Nevado del Ruiz, it has a thick, year-round mantle of snow and glacial ice. Adding to the risk is the fact that more than 100,000 people live in the valleys around Rainier, and many homes are built on deposits left by lahars that flowed down the volcano hundreds or thousands of years ago. A future eruption, or perhaps just a period of extraordinary rainfall, may produce lahars that could take similar paths.

FIGURE 5.27   Lahars are mudflows that originate on volcanic slopes. This lahar raced down the Muddy River, located southeast of Mount St. Helens, following the May 18, 1980 eruption. The height of this mudflow was recorded on the tree trunks. (Photo by Lyn Topinka/U.S. Geological Survey)

OTHER VOLCANIC LANDFORMS

The most obvious volcanic structure is a cone. But other distinctive and important landforms are also associated with volcanic activity.

Calderas

Calderas (*caldaria* = a cooking pot) are large depressions with diameters that exceed one kilometer and have a somewhat circular form. (Those less than a kilometer across are called *collapse pits* or *craters*.) Most calderas are formed by one of the following processes: (1) the collapse of the summit of a large composite volcano following an explosive eruption of silica-rich pumice and ash fragments (*Crater Lake-type calderas*); (2) the collapse of the top of a shield volcano caused by subterranean drainage from a central magma chamber (*Hawaiian-type calderas*); and (3) the collapse of a large area, caused by the discharge of colossal volumes of silica-rich pumice and ash along ring fractures (*Yellowstone-type calderas*).

Crater Lake-Type Calderas   Crater Lake, Oregon, is situated in a caldera that has a maximum diameter of 10 kilometers (6 miles) and is 1175 meters (more than 3800 feet) deep. This caldera formed about 7000 years ago when a composite cone, later named Mount Mazama, violently extruded 50 to 70 cubic kilometers of pyroclastic material (Figure 5.28). With the loss of support, 1500 meters (nearly a mile) of the summit of this once prominent cone collapsed. After the collapse, rainwater filled the caldera (Figure 5.28). Later volcanic activity built a small cinder cone in the caldera. Today this cone, called Wizard Island, provides a mute reminder of past activity.

Hawaiian-Type Calderas   Although some calderas are produced by *collapse following an explosive eruption,* many are not. For example, Hawaii's active shield volcanoes, Mauna Loa and Kilauea, both have large calderas at their summits. Kilauea's measures 3.3 by 4.4 kilometers (about 2 by 3 miles) and is 150 meters (500 feet) deep (Figure 5.29). The walls of this caldera are almost vertical, and as a result, it looks like a vast, nearly flat-bottomed pit. Kilauea's caldera formed by gradual subsidence as magma slowly drained laterally from the underlying magma chamber to the East Rift Zone, leaving the summit unsupported.

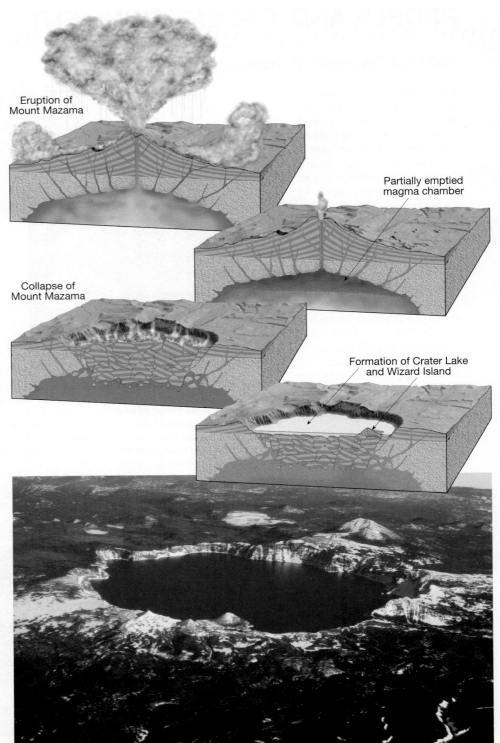

Eruption of Mount Mazama

Partially emptied magma chamber

Collapse of Mount Mazama

Formation of Crater Lake and Wizard Island

FIGURE 5.28   Sequence of events that formed Crater Lake, Oregon. About 7000 years ago a violent eruption partly emptied the magma chamber, causing the summit of former Mount Mazama to collapse. Rainfall and groundwater contributed to form Crater Lake, the deepest lake in the United States. Subsequent eruptions produced the cinder cone called Wizard Island. (After H. Williams, *The Ancient Volcanoes of Oregon.* Photo courtesy of the U.S. Geological Survey)

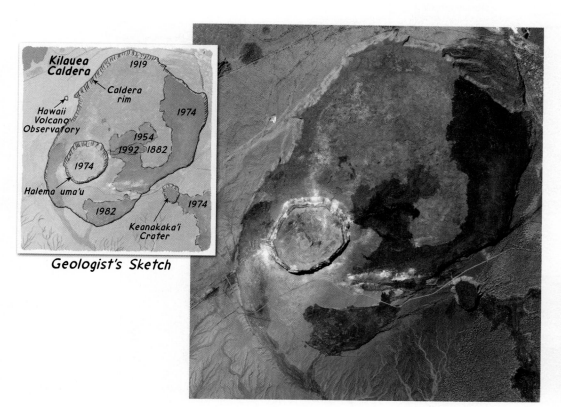

Geologist's Sketch

Yellowstone-Type Calderas Historic and destructive eruptions such as Mount St. Helens and Vesuvius pale in comparison to what happened 630,000 years ago in the region now occupied by Yellowstone National Park, when approximately 1000 cubic kilometers of pyroclastic material erupted. This super-eruption sent showers of ash as far as the Gulf of Mexico and resulted in the eventual development of a caldera 70 kilometers (43 miles) across. It also gave rise to the Lava Creek Tuff, a hardened ash deposit that is 400 meters (more than 1200 feet) thick in some places. Vestiges of this event are the many hot springs and geysers in the Yellowstone region.

Based on the extraordinary volume of erupted material, researchers have determined that the magma chambers associated with Yellowstone-type calderas must also be similarly monstrous. As more and more magma accumulates, the pressure within the magma chamber begins to exceed the pressure exerted by the weight of the overlying rocks (Figure 5.30A). An eruption occurs when the gas-rich magma raises the overlying strata enough to create vertical fractures that extend to the surface. Magma surges upward along these cracks forming a ring-shaped eruption (Figure 5.30B). With a loss of support, the roof of the magma chamber collapses forcing even more gas-rich magma toward the surface.

Caldera-forming eruptions are of colossal proportions, ejecting huge volumes of pyroclastic materials, mainly in the form of ash and pumice fragments. Typically, these materials form pyroclastic flows that sweep across the landscape, destroying most living things in their paths. Upon coming to rest, the hot fragments of ash and pumice fuse together, forming a welded tuff that closely resembles a solidified lava flow. Despite the immense size of these calderas, their eruptions are brief, lasting hours to perhaps a few days.

A distinctive characteristic of large caldera eruptions is a slow upheaval, or *resurgence,* of the floor of the caldera that follows the collapse (Figure 5.30C). As a consequence, these structures consist of large, somewhat circular depressions containing a central elevated region. Large calderas tend to exhibit a complex eruptive history. In the Yellowstone region, for example, three caldera-forming episodes are known to have occurred over the past 2.1 million years. The most recent event (630,000 years ago) was followed by episodic outpourings of rhyolitic and basaltic lavas. Geological evidence suggests that a magma reservoir still exists beneath Yellowstone; thus, another caldera-forming eruption is likely, but not imminent.

Unlike calderas associated with shield volcanoes or composite cones, these depressions are so large and poorly defined that many remained undetected until high-quality aerial and satellite images became available (Figure 5.31). Other examples of large calderas located in the United States are California's Long Valley Caldera, LaGarita Caldera, located in the San Juan Mountains of southern Colorado, and the Valles Caldera west of Los Alamos, New Mexico. These and similar calderas found around the globe are the largest volcanic structures on Earth. Volcanologists have compared their destructive force with that of the impact of a small asteroid. Fortunately, no eruption of this type has occurred in historic times.

Fissure Eruptions and Basalt Plateaus

The greatest volume of volcanic material is extruded from fractures in the crust called **fissures** (*fissura* = to split). Rather than building a cone, these long, narrow cracks tend to emit

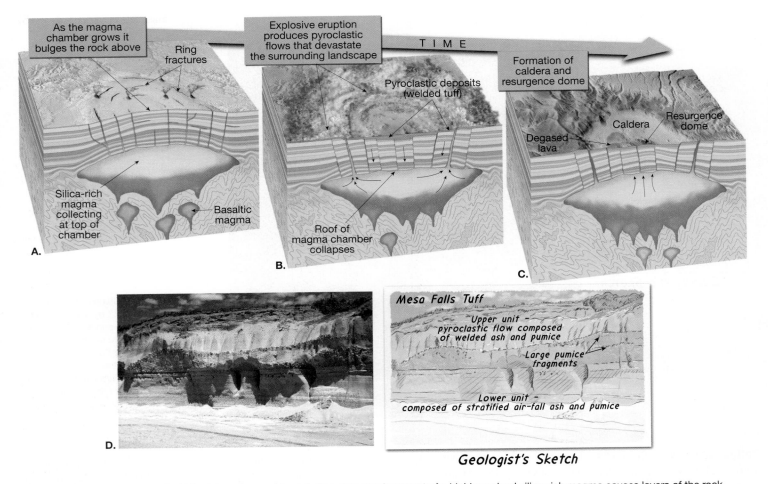

FIGURE 5.30 Formation of a Yellowstone-type caldera. **A.** The slow emplacement of a highly evolved silica-rich magma causes layers of the rock above to bulge and crack, producing a set of *ring fractures.* **B.** Suddenly hundreds to thousands of cubic kilometers of magma erupts, mostly along ring fractures. As the magma surges upward it produces a fiery cloud of ash and gases—called a *pyroclastic flow.* This flow can travel hundreds of kilometers, devastating the surrounding landscape, often burying it under tens of meters of ash. With a loss of support, the roof of the magma chamber collapses. **C.** After the main eruption, degassed lava often erupts in the caldera. Most large calderas also experience a long period of slow upheaval that produces an elevated central region, called a *resurgence dome.* **D.** Mesa Falls Tuff is the middle of three, mostly welded, rhyolitic pyroclastic flow deposits that were generated by eruptions in the Yellowstone area. Based on radiometric dating, this extensive ash and pumice sheet was deposited by a massive eruption that occurred about 1.2 million years ago. In places, the Mesa Falls Tuff is more than 100 meters (300 feet) thick. (Photo by Robert L. Christianson/U.S. Geological Survey)

FIGURE 5.31 Long Valley Caldera. This view is toward the east looking across the northern portion of the caldera. Long Valley Caldera was created about 760,000 years ago by the enormous explosive eruption of 600 cubic kilometers of magma—mostly in the form of hot pyroclastic ash flows. Lookout Mountain, located behind the tree, is a volcanic structure of rhyolitic (felsic) composition that formed after the collapse of the caldera. Glass Creek flow (in the left foreground) is about 600 years old. Long Valley Caldera has been seismically active since 1983. Geologist's sketch shows the caldera boundary and location of the resurgence dome. (Photo by S. R. Brantly/U.S. Geological Survey)

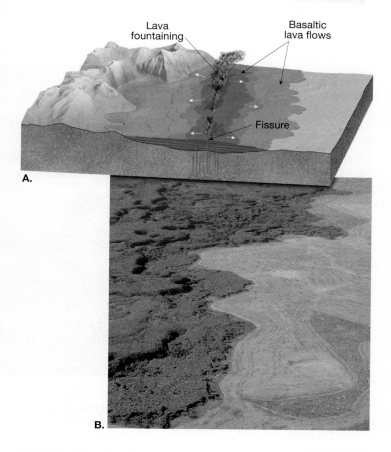

FIGURE 5.32 Basaltic fissure eruption. **A.** Lava fountaining from a fissure and formation of fluid lava flows called *flood basalts*. **B.** Photo of basalt flows near Idaho Falls. (Photo by John S. Shelton)

low-viscosity basaltic lavas that blanket a wide area (Figure 5.32).

The Columbia Plateau in the northwestern United States is the product of this type of activity (Figure 5.33). Numerous **fissure eruptions** have buried the landscape creating a lava plateau nearly a mile thick. Some of the lava remained molten long enough to flow 150 kilometers (90 miles) from its source. The term **flood basalts** appropriately describes these deposits.

Massive accumulations of basaltic lava, similar to those of the Columbia Plateau, occur elsewhere in the world. One of the largest examples is the Deccan Traps, a thick sequence of flat-lying basalt flows covering nearly 500,000 square kilometers (195,000 square miles) of west central India. When the Deccan Traps formed about 66 million years ago, nearly 2 million cubic kilometers of lava were extruded in less than 1 million years. Several other huge deposits of flood basalts, including the Ontong Java Plateau, are found on the floor of the ocean (see Figure 5.42).

Lava Domes

In contrast to mafic lavas, silica-rich felsic lavas are so viscous they hardly flow at all. As the thick lava is "squeezed" out of the vent, it often produces a dome-shaped mass called a **lava dome** (Figure 5.34). Most lava domes are only a few tens of meters high, but some are more than one kilometer high.

Lava domes come in a variety of shapes that range from pancake-like flows to steep-sided plugs that were pushed upward like pistons. Most develop over a period of several years following an explosive eruption of gas-rich magma. A recent example is the dome that continues to grow in the crater of Mount St. Helens (Figure 5.34). A second dome building event began in October, 2004. Although these eruptive phases produced some ash plumes, they were benign compared to the May 18, 1980 eruption.

Although lava domes often form on the summit of a composite cone they can also develop on the flanks of volcanoes. In addition, some domes occur as

FIGURE 5.33 Volcanic areas that comprise the Columbia Plateau in the Pacific Northwest. **A.** The Columbia River basalts cover an area of nearly 200,000 square kilometers (80,000 square miles). Activity here began about 17 million years ago as lava began to pour out of large fissures, eventually producing a basalt plateau with an average thickness of more than 1 kilometer. **B.** Basalt flows exposed along Dry Falls in eastern Washington State. (Photo by E. J. Tarbuck)

FIGURE 5.34   This lava dome began to develop following the May, 1980, eruption of Mount St. Helens. (Photo by Lyn Topinka/U.S. Geological Survey)

FIGURE 5.35   These lava domes are part of a 17 kilometer-long (10 mile-long) chain of rhyolitic domes that form Mono Craters, California. The blocky nature of the domes is hidden beneath a cover of pyroclastic debris. The last eruption here occurred about 600 years ago. (Photo by E. J. Tarbuck)

isolated features, whereas others form linear chains. One example is the line of rhyolitic and obsidian domes at Mono Craters, California (Figure 5.35).

Volcanic Pipes and Necks

Most volcanoes are fed magma through short conduits, called *pipes,* that connect a magma chamber to the surface. One rare type of pipe, called a *diatreme* extends to depths that exceed 200 kilometers (125 miles). Magmas that migrate upward through diatremes travel rapidly enough that they undergo very little alteration during their ascent. Geologists consider these unusually deep pipes to be "windows" into Earth that allow us to view rock normally found only at great depths.

The best-known volcanic pipes are the diamond-bearing structures of South Africa. The rocks filling these pipes originated at depths of at least 150 kilometers (90 miles), where pressure is high enough to generate diamonds and other high-pressure minerals. The process of transporting essentially unaltered magma (along with diamond inclusions) through 150 kilometers of solid rock is exceptional. This fact accounts for the scarcity of natural diamonds.

Volcanoes on land are continually being lowered by weathering and erosion. Cinder cones are easily eroded because they are composed of unconsolidated materials. However, all volcanoes will eventually succumb to erosion. As erosion progresses, the rock occupying a volcanic pipe is often more resistant and may remain standing above the surrounding terrain long after most of the cone has vanished. Shiprock, New Mexico, is a classic example of this structure which geologists call a **volcanic neck** (Figure 5.36). Higher than many skyscrapers, Ship Rock is but one of many such landforms that protrude conspicuously from the red desert landscapes of the American Southwest.

PLATE TECTONICS AND VOLCANIC ACTIVITY

Geologists have known for decades that the global distribution of volcanism is not random. Most active volcanoes are located along the margins of the ocean basins—notably within the circum-Pacific belt known as the *Ring of Fire* (Figure 5.37). These

FIGURE 5.36 Shiprock, New Mexico, is a volcanic neck. This structure, which stands over 420 meters (1380 feet) high, consists of igneous rock which crystallized in the vent of a volcano that has long since been eroded away. (Photo by Dennis Tasa)

volcanoes consist mainly of composite cones that emit volatile-rich magma having an intermediate (andesitic) composition and that occasionally produce awe-inspiring eruptions.

A second group includes the basaltic shields that emit very fluid lavas. These volcanic structures comprise most of the islands of the deep ocean basins, including the Hawaiian Islands, the Galapagos Islands, and Easter Island. In addition, this group includes many active submarine volcanoes that dot the ocean floor; particularly notable are the innumerable small seamounts that occur along the axis of the mid-ocean ridge. At these depths,

the pressures are so great that the gases that are emitted quickly dissolve in the seawater and never reach the surface. Thus, firsthand knowledge of these eruptions is limited, coming mainly from deep-diving submersibles.

A third group includes volcanic structures that appear to be somewhat randomly distributed in the interiors of the continents. None are found in Australia nor in the eastern two-thirds of North and South America. Africa is notable because it has many potentially active volcanoes including Mount Kilimanjaro, the highest point on the continent (5895 meters, 19,454 feet).

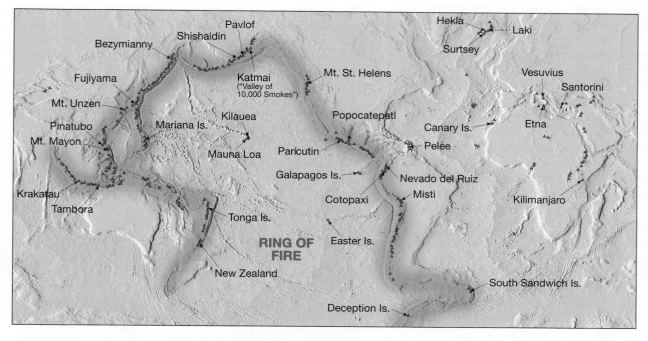

FIGURE 5.37 Locations of some of Earth's major volcanoes.

Volcanism on continents is the most diverse, ranging from eruptions of very fluid basaltic lavas, like those that generated the Columbia Plateau to explosive eruptions of silica-rich rhyolitic magma as occurred in Yellowstone.

Until the late 1960s, geologists had no explanation for the apparently haphazard distribution of continental volcanoes, nor were they able to account for the almost continuous chain of volcanoes that circles the margin of the Pacific basin. With the development of the theory of plate tectonics, the picture was greatly clarified. Recall that most primary (unaltered) magma originates in the upper mantle and that the mantle is essentially solid, *not molten* rock. The basic connection between plate tectonics and volcanism is that *plate motions provide the mechanisms by which mantle rocks melt to generate magma.*

We will examine three zones of igneous activity and their relationship to plate boundaries. These active areas are located (1) along convergent plate boundaries where plates move toward each other and one sinks beneath the other; (2) along divergent plate boundaries, where plates move away from each other and new seafloor is created; and (3) areas within the plates proper that are not associated with any plate boundary. These three volcanic settings are depicted in Figure 5.38. (If you are unclear as to how magma is generated, study the section entitled "Origin of Magma" in Chapter 4 before proceeding.)

Volcanism at Convergent Plate Boundaries

Recall that at convergent plate boundaries slabs of oceanic crust are bent as they descend into the mantle, generating a deep-ocean trench. As a slab sinks deeper into the mantle, the increase in temperature and pressure drives volatiles (mostly water) from the oceanic crust. These mobile fluids migrate upward into the wedge-shaped piece of mantle located between the subducting slab and overriding plate (Figure 5.39). Once the sinking slab reaches a depth of about 100 kilometers, these water-rich fluids reduce the melting point of hot mantle rock sufficiently to trigger some melting. The partial melting of mantle rock (peridotite) generates magma with a basaltic composition. After a sufficient quantity of magma has accumulated, it slowly migrates upward.

Volcanism at a convergent plate margin results in the development of a slightly curved chain of volcanoes called a *volcanic arc*. These volcanic chains develop roughly parallel to the associated trench—at distances of 200 to 300 kilometers (100 to 200 miles). Volcanic arcs can be constructed on oceanic, or continental, lithosphere. Those that develop within the ocean and grow large enough for their tops to rise above the surface are labeled *island archipelagos* in most atlases. Geologists prefer the more descriptive term **volcanic island arcs,** or simply **island arcs** (Figure 5.38A). Several young volcanic island arcs border the western Pacific basin, including the Aleutians, the Tongas, and the Marianas.

Volcanism associated with convergent plate boundaries may also develop where slabs of oceanic lithosphere are subducted under continental lithosphere to produce a **continental volcanic arc** (Figure 5.39). The mechanisms that generate these mantle-derived magmas are essentially the same as those operating at island arcs. The major difference is that continental crust is much thicker and is composed of rocks having a higher silica content than oceanic crust. Hence, through the assimilation of silica-rich crustal rocks, plus extensive magmatic differentiation, a mantle-derived magma may become highly evolved as it rises through continental crust. Stated another way, the primary magmas generated in the mantle may change from a comparatively dry, fluid basaltic magma to a viscous andesitic or rhyolitic magma having a high concentration of volatiles as it moves up through the continental crust. The volcanic chain of the Andes Mountains along the western margin of South America is perhaps the best example of a mature continental volcanic arc.

Since the Pacific basin is essentially bordered by convergent plate boundaries and associated subduction zones, it is easy to see why the irregular belt of explosive volcanoes we call the Ring of Fire formed in this region (Figure 5.40). The volcanoes of the Cascade Range in the northwestern United States, including Mount Hood, Mount Rainier, and Mount Shasta, are included in this group (Figure 5.39).

Volcanism at Divergent Plate Boundaries

The greatest volume of magma (perhaps 60 percent of Earth's total yearly output) is produced along the oceanic ridge system in association with seafloor spreading (see Figure 5.38B). Below the ridge axis where lithospheric plates are continually being pulled apart, the solid yet mobile mantle responds to the decrease in overburden and rises to fill the rift. Recall that as rock rises, it experiences a decrease in confining pressure and undergoes melting without the addition of heat. This process, called *decompression melting,* is the most common process by which mantle rocks melt.

Partial melting of mantle rock at spreading centers produces basaltic magma. Because this newly formed magma is less dense than the mantle rock from which it was derived, it rises and collects in reservoirs located just beneath the ridge crest. About 10 percent of this melt eventually migrates upward along fissures to erupt on the ocean floor. This activity continuously adds new basaltic rock to plate margins, temporarily welding them together, only to break again as spreading continues. Along some ridges, outpourings of bulbous pillow lavas build numerous small seamounts.

Students Sometimes Ask . . .

I've heard that explosive volcanic eruptions might cause global air temperatures to drop. Is that possible?

Yes, it is possible. When you view an image such as the one of Mount Augustine in Figure 5.4, it might seem that the cloud of volcanic ash should block the sunlight and cause temperatures to drop. Although this occurs, the effect is very short-lived because volcanic ash quickly settles from the atmosphere. However, the gases emitted by volcanoes, especially sulfur dioxide and carbon dioxide may indeed trigger more enduring episodes of climate change. These possibilities are described at some length in Chapter 21 "Global Climate Change."

Studying Active Volcanoes

Chris Eisinger describes volcanoes the same way he might discuss old, temperamental friends. "For the most part, volcanoes are very approachable," he says, "as long as you know their cycles." As a graduate student in the Department of Geological Sciences at Arizona State University, Eisinger had 10 years of first-hand experience with volcanoes under

> "For the most part, volcanoes are very approachable as long as you know their cycles."

his belt. He began as an undergraduate and spent time at the Hawaiian Volcano Observatory as well as in Indonesia, a region rife with volcanic activity.

But the summit of an active volcano is anything but a breezy locale. On the contrary, volcanoes emit malodorous gases, as Eisinger can attest.

"The most distinct things you notice are the fumes," he says. "You get a strong sulfur smell."

If the volcano is near enough to the ocean for lava to flow into the sea, Eisinger also says the runoff creates a steam that is highly acidic, thanks to the high chlorine content of seawater. "Your eyes will water," he said. "It's typically difficult to breathe and you end up coughing if you're not wearing a gas mask."

Rainstorms can further cloud the air, creating steam when the falling water strikes the hot surface of the lava. At times, Eisinger says, visibility for volcanologists studying at the summit is reduced to a few feet.

The heavy influx of gases doesn't affect only the eyes and nose, either. The taste, if one isn't wearing a mask, can be "pretty nasty," Eisinger says, adding that the smell adheres to one's clothing as well.

For volcanoes that are in a constant state of eruption, Eisinger said there are distinct patterns in the emission cycles. Volcanoes he has visited in Indonesia, for example, would emit a stream of gaseous material every 20 to 60 minutes.

But while most volcanologists are able to remain safe by paying close attention to eruption cycles, fatalities have happened. Recently, in August 2000, two volcanologists died at the summit of Semeru on the island of Java in Indonesia when the volcano erupted with no warning.

Active volcanoes also emit low, rumbling sounds, Eisinger said, due largely to subterranean explosions that tend to be very muffled. Hawaiian volcanoes are unique in that they also emit an intense hissing sound, "like a jet engine," from the expulsion of gases. Eruptions also are often foreshadowed by seismic activity that contributes to the rumble of the explosions. Though Eisinger himself has never witnessed a volcanic eruption of historic magnitude, he noted that accounts of such colossal eruptions as that of Krakatau in 1883 produced reports of a boom that could be heard thousands of miles away.

—Chris Wilson

Volcanologist Chris Eisinger uses a crowbar to hit some molten sulfur on Kawah Ijen volcano in Indonesia. The 200 °C (390 °F) molten sulfur is still red hot and will cool to a yellow or green color. (Courtesy of Chris Eisinger)

Although most spreading centers are located along the axis of an oceanic ridge, some are not. In particular, the East African Rift is a site where continental lithosphere is being pulled apart (see Figure 5.38F). In this setting, magma is generated by decompression melting in the same manner as along the oceanic ridge system. Vast outpourings of fluid lavas as well as basaltic shield volcanoes are common in this region.

Intraplate Volcanism

We know why igneous activity is initiated along plate boundaries, but why do eruptions occur in the interiors of plates? Hawaii's Kilauea is considered the world's most active volcano, yet it is situated thousands of kilometers from the nearest plate boundary in the middle of the vast Pacific plate (Figure 5.38C).

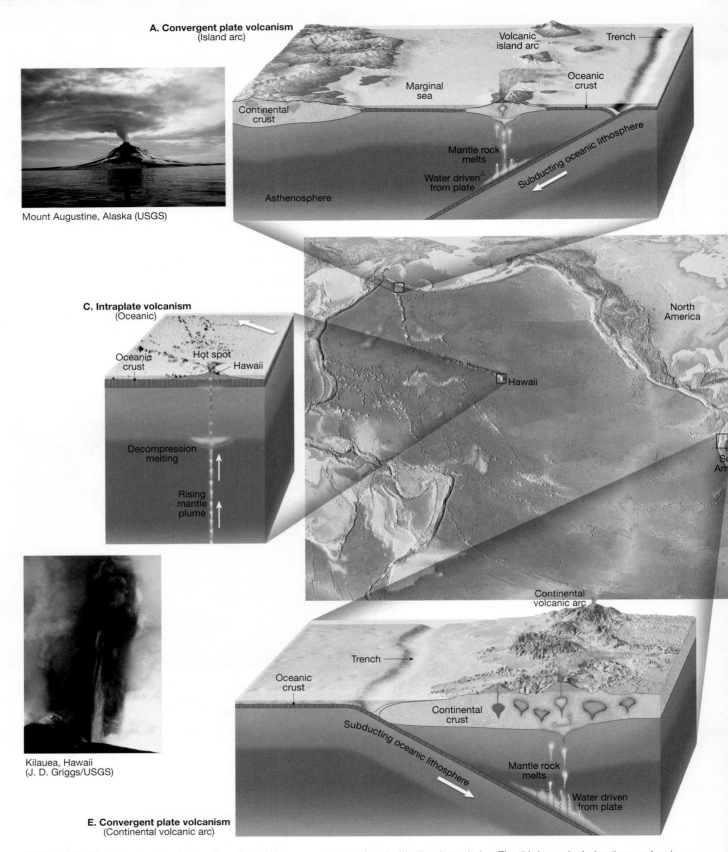

A. Convergent plate volcanism
(Island arc)

Mount Augustine, Alaska (USGS)

C. Intraplate volcanism
(Oceanic)

Kilauea, Hawaii
(J. D. Griggs/USGS)

E. Convergent plate volcanism
(Continental volcanic arc)

FIGURE 5.38   Three zones of volcanism. Two of the zones are associated with plate boundaries. The third zone includes those volcanic structures that are irregularly distributed in the interiors of plates. (Photo F by ABPL/Daryl Balfour/Animals Animals–Earth Scenes)

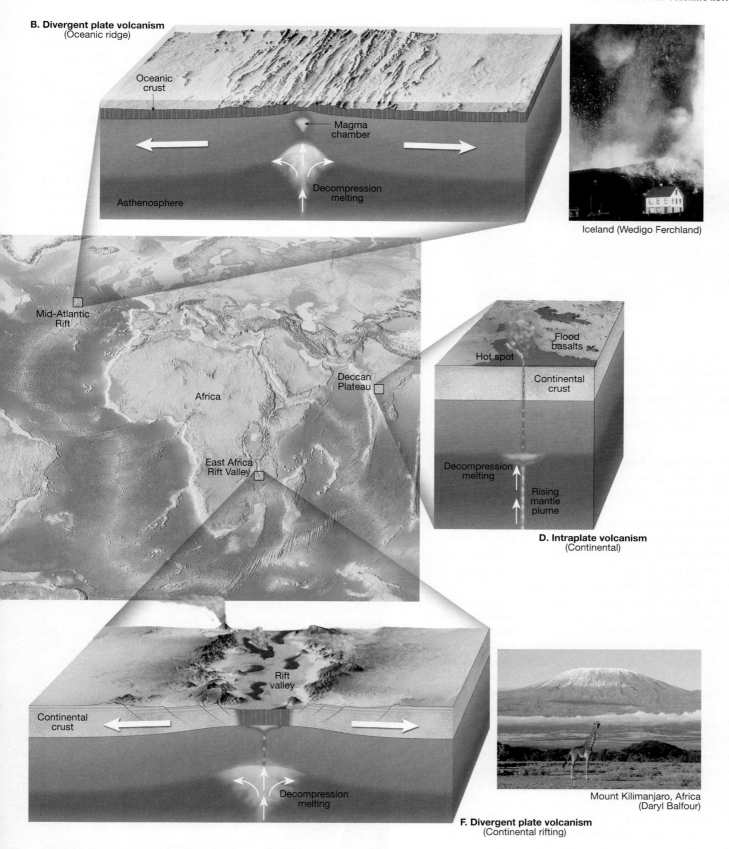

B. Divergent plate volcanism
(Oceanic ridge)

Oceanic crust

Magma chamber

Asthenosphere

Decompression melting

Iceland (Wedigo Ferchland)

Mid-Atlantic Rift

Africa

Deccan Plateau

East Africa Rift Valley

Hot spot

Flood basalts

Continental crust

Decompression melting

Rising mantle plume

D. Intraplate volcanism
(Continental)

Rift valley

Continental crust

Decompression melting

Mount Kilimanjaro, Africa
(Daryl Balfour)

F. Divergent plate volcanism
(Continental rifting)

FIGURE 5.38 (continued)

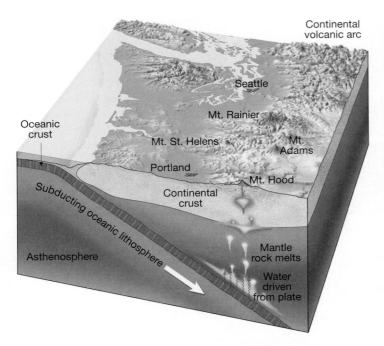

FIGURE 5.39 Volcanic activity along a convergent plate boundary off the Pacific coast of North America. As an oceanic plate descends into the mantle, water and other volatiles are driven from the subducting crustal rocks. These volatiles lower the melting temperature of mantle rock sufficiently to generate magma. The result was the formation of the continental volcanic arc we call the Cascade Range.

Other sites of **intraplate volcanism** (meaning "within the plate") include the Canary Islands, Yellowstone, and several volcanic centers that you may be surprised to learn are located in the Sahara Desert of Africa.

Geologists now recognize that most intraplate volcanism occurs where a mass of hotter than normal mantle material called a **mantle plume** ascends toward the surface (Figure 5.41). Although the depth at which (at least some) mantle plumes originate is still hotly debated, some appear to form deep within Earth at the core–mantle boundary. These plumes of solid yet mobile mantle rock rise toward the surface in a manner similar to the blobs that form within a lava lamp. (These are the trendy lamps that contain two immiscible liquids in a glass container. As the base of the lamp is heated, the denser liquid at the bottom becomes buoyant and forms blobs that rise to the top.) Like the blobs in a lava lamp, a mantle plume has a bulbous head that draws out a narrow stalk beneath it as it rises. Once the plume head nears the top of the mantle, decompression melting generates basaltic magma that may eventually trigger volcanism at the surface.

The result is a localized volcanic region a few hundred kilometers across called a **hot spot** (see Figure 5.41). More than 40 hot spots have been identified, and most have persisted for millions of years. The land surface surrounding a hot spot is often elevated because it is buoyed up by the rising plume of warm low-density material. Furthermore, by measuring the heat flow in these regions, geologists have determined that the mantle beneath hot spots must be 100–150 °C hotter than normal mantle material.

Mantle plumes are responsible for the vast outpourings of basaltic lava that created the large basalt plateaus including the Siberian Traps in Russia, India's Deccan Plateau, and the Ontong Java Plateau in the western Pacific (Figure 5.42). The most widely accepted explanation for these eruptions, which emit extremely large volumes of basaltic lava over relatively short time intervals, involves a plume with a monsterous head and a long, narrow tail (Figure 5.41A). Upon reaching the base of the lithosphere, these unusually hot, massive heads begin to melt. Melting progresses rapidly, causing the burst of volcanism that emits voluminous outpourings of lava to form a huge basalt plateau in a matter of a million or so years (Figure 5.41B). The comparatively short initial eruptive phase is followed by tens of millions of years of less voluminous activity, as the plume tail slowly rises to the surface.

FIGURE 5.40 Mount Unzen, Japan. This composite cone produced several fiery pyroclastic flows between 1900 and 1995. The most destructive eruption occurred in 1991 when a searing surge killed 43 people, burned over 500 homes and one school. Pumice- and ash-flow deposits (light color) can be seen in the valley leading down the volcano. Note the protective channels (lower part of the image) constructed to divert flows away from the surrounding villages. (Photo by Michael S. Yamashita/CORBIS)

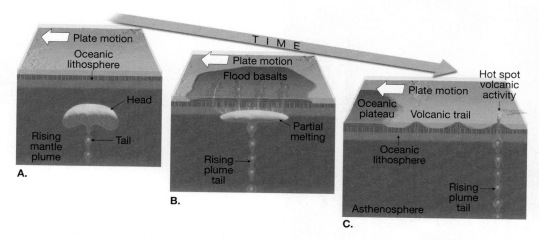

FIGURE 5.41 Model of hot-spot volcanism thought to explain the formation of oceanic plateaus and the volcanic islands associated with these features. **A.** A rising mantle plume with large bulbous head and narrow tail. **B.** Rapid decompression melting of the head of a mantle plume produces vast outpourings of basalt to generate the oceanic plateau. Large basaltic plateaus can also form on continental crust—examples include the Columbia Plateau in the northwestern United States and India's Deccan Plateau. **C.** Later, less voluminous activity caused by the rising plume tail produces a linear volcanic chain on the seafloor.

Extending away from most large flood basalt provinces is a chain of volcanic structures, similar to the Hawaiian chain, that terminates over an active hot spot marking the current position of the remaining tail of the plume (Figure 5.41C).

LIVING WITH VOLCANOES

About 10 percent of Earth's population lives in the vicinity of an active volcano. In fact, several major cities including Seattle, Washington; Mexico City, Mexico; Tokyo, Japan; Naples, Italy; and Quito, Ecuador, are located on or near a volcano (Figure 5.43).

Until recently, the dominant view of Western societies was that humans possess the wherewithal to subdue volcanoes and other types of catastrophic natural hazards. Today it is apparent that volcanoes are not only very destructive but unpredictable as well. With this awareness, a new attitude is developing—"How do we live with volcanoes?"

Volcanic Hazards

Volcanoes produce a wide variety of potential hazards that can kill people and wildlife, as well as destroy property (Figure 5.44). Perhaps the greatest threats to life are pyroclastic flows. These

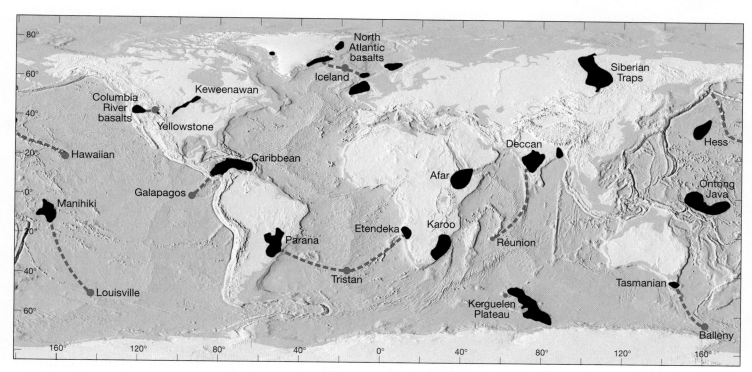

FIGURE 5.42 Global distribution of flood basalt provinces (shown in black) and associated hotspots (shown as red dots). The red dashed lines are hot-spot tracks, which appear as lines of volcanic structures mostly found on the ocean floor. The Keweenawan and Siberian Traps formed along large continental rifts where the crust had been greatly thinned. Whether there is a connection between the Columbia River basalts and the Yellowstone hot spot is still a matter of ongoing research.

FIGURE 5.43 Seattle, Washington, with Mount Rainier in the background. (Photo by Ken Straiton/Corbis)

hot mixtures of gas, ash, and pumice that sometimes exceed 800° C, race down the flanks of volcanoes, giving people little chance to escape.

Lahars, which can occur even when a volcano is quiet, are perhaps the next most dangerous volcanic hazard. These mixtures of volcanic debris and water can flow for tens of kilometers down steep volcanic slopes at speeds that may exceed 100 kilo-meters (60 miles) per hour. Lahars pose a potential threat to many communities downstream from glacier-clad volcanoes such as Mount Rainier. Other potentially destructive mass-wasting events include the rapid collapse of the volcano's summit or flank.

Other obvious hazards include explosive eruptions that can endanger people and property hundreds of miles from a volcano. During the past 15 years, at least 80 commercial jets have been damaged by inadvertently flying into clouds of volcanic ash (Figure 5.45). One of these was a near crash that occurred in 1989 when a Boeing 747, with more than 300 passengers aboard, encountered an ash cloud from Alaska's Redoubt Volcano. All four engines stalled after they became clogged with ash. Fortunately, the engines were restarted at the last minute and the aircraft managed to land safely in Anchorage.

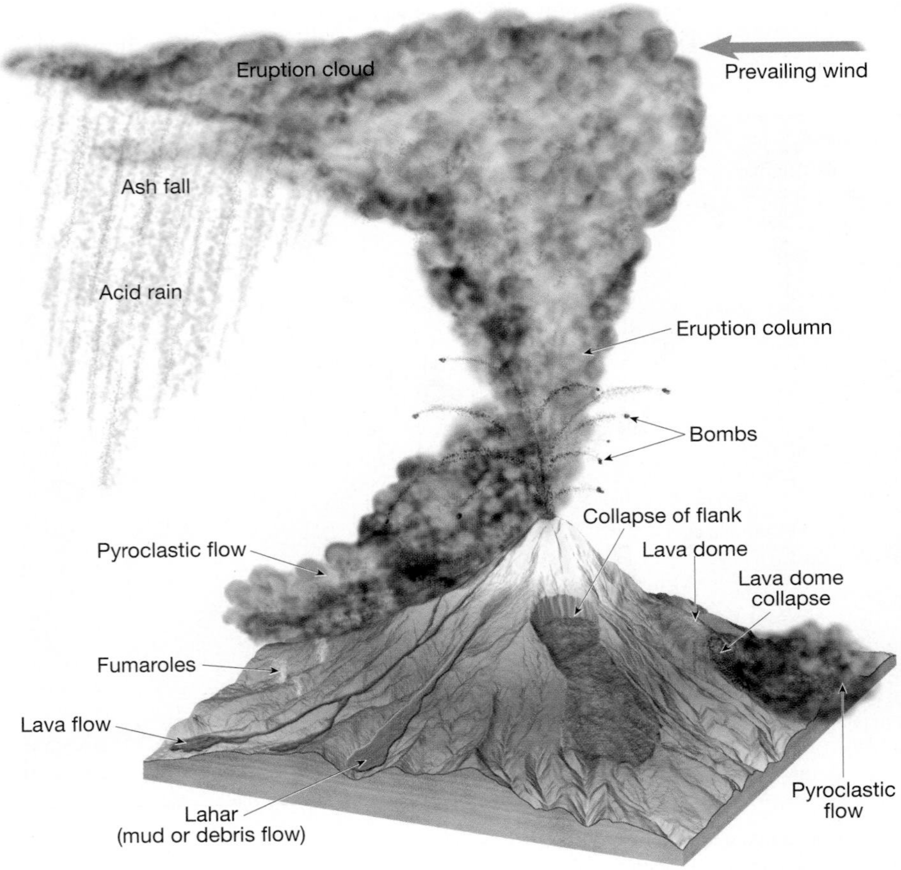

Monitoring Volcanic Activity

Today a number of volcano monitoring techniques are employed, with most of them aimed at detecting the movement of magma from a subterranean reservoir (typically several kilometers deep) toward the surface. The four most noticeable changes in a volcanic landscape caused by the migration of magma are: (1) changes in the pattern of volcanic earthquakes; (2) expansion of a near-surface magma chamber which leads to inflation of the volcano; (3) changes in the amount and/or composition of the gases that are released from a volcano; and (4) an increase in ground temperature caused by the implacement of new magma.

Almost a third of all volcanoes that have erupted in historic times are now monitored using seismographs, instruments that detect earthquake tremors. In general, a sharp increase in seismic unrest followed by a period of relative quiet has been shown to be a precursor for many volcanic eruptions. However, some large volcanic struc-

FIGURE 5.44 Simplified drawing showing a wide variety of natural hazards associated with volcanoes. (After U.S. Geological Survey)

Eruption cloud

Prevailing wind

Ash fall

Acid rain

Eruption column

Bombs

Collapse of flank

Lava dome

Lava dome collapse

Pyroclastic flow

Fumaroles

Lava flow

Lahar (mud or debris flow)

Pyroclastic flow

168

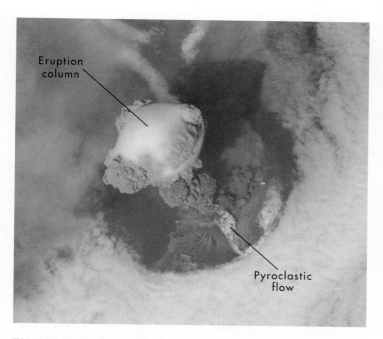

FIGURE 5.46 Monitoring South Sister Volcano, Cascade Range, Oregon. This geologist is measuring the degree of inflation of the volcano's surface for potential eruptive activity.

FIGURE 5.45 Photo taken by an astronaut aboard the International Space Station of a steam-and-ash eruption emitted from Sarychev Peak located northwest of Japan. Commercial airline flights were quickly diverted from the region to minimize the danger of engine failures caused by ash intake. (Photo courtesy of NASA)

tures have exhibited lengthy periods of seismic unrest. For example, Rabaul Caldera in New Guinea recorded a strong increase in seismicity in 1981. This activity lasted 13 years and finally culminated with an eruption in 1994. Occasionally, a large earthquake triggers a volcanic eruption, or at least disturbs the volcano's plumbing. Kilauea, for example, began to erupt after the Kalapana earthquake of 1975.

The roof of a volcano may rise as new magma accumulates in its interior—a phenomenon which precedes many volcanic eruptions. Because the accessibility of many volcanoes is limited, remote sensing devices, including lasers, Doppler radar, and Earth orbiting satellites, are often used to determine whether or not a volcano is swelling. The recent discovery of ground doming at Three Sisters Volcanoes in Oregon was first detected using radar images obtained from satellites (Figure 5.46).

Volcanologists also frequently monitor the gases that are released from volcanoes in an effort to detect even minor changes in their amount and/or composition. Some volcanoes show an increase in sulfur dioxide (SO_2) emissions months or years prior to an eruption. On the other hand, a few days prior to the 1991 eruption of Mount Pinatubo, emissions of carbon dioxide (CO_2) dropped dramatically.

The development of remote sensing devices has greatly increased our ability to monitor volcanoes. These instruments and techniques are particularly useful for monitoring eruptions in progress. Photographic images and infrared (heat) sensors can detect lava flows and volcanic columns rising from a volcano (Figure 5.45). Furthermore, satellites can detect ground deformation as well as monitor SO_2 emissions (see Figure 5.7).

The overriding goal of all monitoring is to discover precursors that may warn of an imminent eruption. This is accomplished by first diagnosing the current condition of a volcano and then using this baseline data to predict its future behavior. Stated another way, a volcano must be observed over an extended period to recognize significant changes from its "resting state."

CHAPTER 5 VOLCANOES AND VOLCANIC HAZARDS IN REVIEW

- The primary factors that determine the nature of volcanic eruptions include the magma's *composition,* its *temperature,* and the *amount of dissolved gases* it contains. As lava cools, it begins to congeal and, as *viscosity* increases, its mobility decreases. The *viscosity of magma is also directly related to its silica content.* Rhyolitic (felsic) lava, with its high silica content (over 70 percent), is very viscous and forms short, thick flows. Basaltic (mafic) lava, with a lower silica content

(about 50 percent), is more fluid and may travel a long distance before congealing. Dissolved gases tend to make magma more fluid and, as they expand, provide the force that propels molten rock from the volcano.

- The materials associated with a volcanic eruption include (1) *lava flows* (*pahoehoe* flows, which resemble twisted braids; and *aa* flows, consisting of rough, jagged blocks; both form from basaltic lavas); (2) *gases* (primarily *water* vapor); and (3) *pyroclastic material* (pulverized rock and lava fragments blown from the volcano's vent, which include *ash, pumice, lapilli, cinders, blocks,* and *bombs*).

- Successive eruptions of lava from a central vent result in a mountainous accumulation of material known as a *volcano.* Located at the summit of many volcanoes is a steep-walled depression called a *crater. Shield cones* are broad, slightly domed volcanoes built primarily of fluid, basaltic lava. *Cinder cones* have steep slopes composed of pyroclastic material. *Composite cones,* or *stratovolcanoes,* are large, nearly symmetrical structures built of interbedded lavas and pyroclastic deposits. Composite cones produce some of the most violent volcanic activity. Often associated with a violent eruption is a *nuée ardente,* a fiery cloud of hot gases infused with incandescent ash that races down steep volcanic slopes. Large composite cones may also generate a type of mudflow known as a *lahar.*

- Most volcanoes are fed by *conduits* or *pipes.* As erosion progresses, the rock occupying the pipe, which is often more resistant, may remain standing above the surrounding terrain as a *volcanic neck.* The summits of some volcanoes have large, nearly circular depressions called *calderas* that result from collapse. Calderas also form on shield volcanoes by subterranean drainage from a central magma chamber, and the largest calderas form by the discharge of colossal volumes of silica-rich pumice along ring fractures. Although volcanic eruptions from a central vent are the most familiar, by far the largest amounts of volcanic material are extruded from cracks in the crust called *fissures.* The term *flood basalts* describes the fluid basaltic lava flows that cover an extensive region in the northwestern United States known as the Columbia Plateau. When silica-rich magma is extruded, *pyroclastic flows,* consisting largely of ash and pumice fragments, usually result.

- *Most active volcanoes are associated with plate boundaries.* Active areas of volcanism are found along mid-ocean ridges where seafloor spreading is occurring (*divergent plate boundaries*), in the vicinity of ocean trenches where one plate is being subducted beneath another (*convergent plate boundaries*), and in the interiors of plates themselves (intraplate volcanism). Rising plumes of hot mantle rock are the source of most intraplate volcanism.

KEY TERMS

aa flow (p. 142)
block lavas (p. 143)
caldera (p. 156)
cinder cone (p. 149)
composite cone (p. 151)
conduit (p. 146)
continental volcanic arc (p. 162)
crater (p. 146)
eruption column (p. 141)

fissure (p. 157)
fissure eruption (p. 159)
flood basalt (p. 159)
fumarole (p. 147)
guyots (p. 149)
hot spot (p. 166)
intraplate volcanism (p. 166)
island arc (p. 162)
lahar (p. 154)
lava dome (p. 159)

lava tube (p. 143)
mantle plume (p. 166)
nuée ardente (p. 153)
pahoehoe flow (p. 143)
parasitic cone (p. 146)
pillow lava (p. 143)
pipe (p. 146)
pumice (p. 145)
pyroclastic flow (p. 153)
pyroclastic material (p. 145)

scoria (p. 145)
scoria cone (p. 149)
shield volcano (p. 147)
stratovolcano (p. 151)
vent (p. 146)
viscosity (p. 138)
volcanic island arc (p. 162)
volcanic neck (p. 160)
volcano (p. 146)
volatiles (p. 140)

QUESTIONS FOR REVIEW

1. What event triggered the May 18, 1980, eruption of Mount St. Helens? (See Box 5.1.)

2. List three factors that determine the nature of a volcanic eruption. What role does each play?

3. Why is a volcano fed by highly viscous magma likely to be a greater threat to life and property than a volcano supplied with very fluid magma?

4. Describe pahoehoe and aa lava.

5. List the main gases released during a volcanic eruption. What role do gases have in eruptions?

6. How do volcanic bombs differ from blocks of pyroclastic debris?

7. What is scoria? How is scoria different from pumice?

8. Compare a volcanic crater to a caldera.

9. Compare and contrast the three main types of volcanoes (size, composition, shape, and eruptive style).

10. Name a prominent volcano for each of the three types of volcanoes.

11. Briefly compare the eruptions of Kilauea and Parícutin.

12. Contrast the destruction of Pompeii with the destruction of St. Pierre (time frame, volcanic material, and nature of destruction).

13. Describe the formation of Crater Lake. Compare it to the caldera found on shield volcanoes, such as Kilauea.

14. What are the largest volcanic structures on Earth?

15. What is Shiprock, New Mexico, and how did it form?

16. How do the eruptions that created the Columbia Plateau differ from eruptions that create large composite cones?

17. Where are fissure eruptions most common?

18. Extensive pyroclastic flow deposits are associated with which volcanic structures?

19. Volcanism at divergent plate boundaries is associated with which rock type? What causes rocks to melt in these regions?

20. What is the Ring of Fire?

21. What type of plate boundary is associated with the Ring of Fire?

22. Are volcanoes in the Ring of Fire generally described as relatively quiet or violent? Name a volcano that would support your answer.

23. How is magma generated along convergent plate boundaries?

24. What is the source of magma for intraplate volcanism?

25. What is meant by hot-spot volcanism?

26. At which type of plate boundary is the greatest quantity of magma generated?

27. The Hawaiian Islands and Yellowstone are associated with which of the three zones of volcanism? Cascade Range? Flood basalt provinces?

28. What are the four changes in a volcanic area that are monitored in order to detect the movement of magma?

29. List at least two reasons why an eruption of Mount Rainier similar to the one that occurred at Mount St. Helens in 1980 would be considerably more destructive.

30. Describe four natural hazards associated with volcanoes.

COMPANION WEBSITE

The *Earth 10e* website uses the resources and flexibility of the Internet to aid in your study of the topics in this chapter. Written and developed by the authors and other geology instructors, this site will help improve your understanding of geology. Visit www.mygeoscienceplace.com in order to:

- **Review** key chapter concepts

- **Read** with links to the eBook and to chapter-specific web resources
- **Visualize** and comprehend challenging topics using learning activities in *GEODe Earth*
- **Test** yourself with online quizzes

GEODe EARTH

GEODe Earth is a valuable and easy to use learning aid that can be accessed from your book's Companion Website (www.mygeoscienceplace.com). It is a dynamic instructional tool that promotes understanding and reinforces important concepts by using tutorials, animations, and exercises that actively engage the student.

GEODe Earth
Volcanoes ▶

More about this image...
Aa flow.

Another common type of basaltic lava, called **aa**, has a rough, jagged surface. Active aa flows are relatively cool and thick, and advance more slowly than pahoehoe flows.

19

WEATHERING AND SOIL

Rock weathering contributed to the sculpting of these balanced rocks in Arizona's Marble Canyon.

(PHOTO BY MICHAEL COLLIER)

Earth's surface is constantly changing. Rock is disintegrated and decomposed, moved to lower elevations by gravity, and carried away by water, wind, or ice. In this manner Earth's physical landscape is sculptured (Figure 6.1). This chapter focuses on the first step of this never-ending process—weathering. What causes solid rock to crumble, and why does the type and rate of weathering vary from place to place? Soil, an important product of the weathering process and a vital resource, is also examined.

EARTH'S EXTERNAL PROCESSES

WEATHERING AND SOIL
▶ Earth's External Processes

Weathering, mass wasting, and erosion are called **external processes** because they occur at or near Earth's surface and are powered by energy from the Sun. External processes are a basic part of the rock cycle because they are responsible for transforming solid rock into sediment.

To the casual observer, the face of Earth may appear to be without change, unaffected by time. In fact, 200 years ago most people believed that mountains, lakes, and deserts were permanent features of an Earth that was thought to be no more than a few thousand years old. Today we know that Earth is about 4.6 billion years old and that mountains eventually succumb to weathering and erosion, lakes fill with sediment or are drained by streams, and deserts come and go with changes in climate.

Earth is a dynamic body. Some parts of Earth's surface are gradually elevated by mountain building and volcanic activity. These **internal processes** derive their energy from Earth's interior. Meanwhile, opposing external processes are continually breaking rock apart and moving the debris to lower elevations. The latter processes include:

1. **Weathering**—the physical breakdown (disintegration) and chemical alteration (decomposition) of rock at or near Earth's surface.
2. **Mass wasting**—the transfer of rock and soil downslope under the influence of gravity.
3. **Erosion**—the physical removal of material by mobile agents such as water, wind, or ice.

In this chapter we will focus on rock weathering and the products generated by this activity. However, weathering cannot be easily separated from mass wasting and erosion because as weathering breaks rocks apart, erosion and mass wasting remove the rock debris. This transport of material by erosion and mass wasting further disintegrates and decomposes the rock.

WEATHERING

WEATHERING AND SOIL
▶ Types of Weathering

Weathering goes on all around us, but it seems like such a slow and subtle process that it is easy to underestimate its importance. Yet it is worth remembering that weathering is a basic part of the rock cycle and thus a key process in the Earth system. Weathering is also important to humans—even to those of us who are not studying geology. For example, many of the life-sustaining minerals and elements found in soil, and ultimately in

FIGURE 6.1 Arizona's Monument Valley. When weathering accentuates differences in rocks, spectacular landforms are sometimes created. As the rock gradually disintegrates and decomposes, mass wasting and erosion remove the products of weathering. (Photo by Michael Collier)

the food we eat, were freed from solid rock by weathering processes. As the chapter-opening photo, Figure 6.1, and many other images in this book illustrate, weathering also contributes to the formation of some of Earth's most spectacular scenery. Of course these same processes are also responsible for causing the deterioration of many of the structures we build (Figure 6.2).

All materials are susceptible to weathering. Consider, for example, the fabricated product concrete, which closely resembles a sedimentary rock called conglomerate. A newly poured concrete sidewalk has a smooth, fresh, unweathered look. However, not many years later the same sidewalk will appear chipped, cracked, and rough, with pebbles exposed at the surface. If a tree is nearby, its roots may heave and buckle the concrete as well. The same natural processes that eventually break apart a concrete sidewalk also act to disintegrate rock.

Weathering occurs when rock is mechanically fragmented (disintegrated) and/or chemically altered (decomposed). **Mechanical weathering** is accomplished by physical forces that break rock into smaller and smaller pieces without changing the rock's mineral composition. **Chemical weathering** involves a chemical transformation of rock into one or more new compounds. These two concepts can be illustrated with a piece of paper. The paper can be disintegrated by tearing it into smaller and

smaller pieces, whereas decomposition occurs when the paper is set afire and burned.

Why does rock weather? Simply, weathering is the response of Earth materials to a changing environment. For instance,

FIGURE 6.2 Even the most "solid" monuments that people erect eventually yield to the day in and day out attack of weathering processes. Temple of Olympian Zeus, Athens, Greece. (Photo by CORBIS)

after millions of years of uplift and erosion, the rocks overlying a large, intrusive igneous body may be removed, exposing it at the surface. This mass of crystalline rock—formed deep below ground where temperatures and pressures are high—is now subjected to a very different and comparatively hostile surface environment. In response, this rock mass will gradually change. This transformation of rock is what we call weathering.

In the following sections we will discuss the various modes of mechanical and chemical weathering. Although we will consider these two categories separately, keep in mind that mechanical and chemical weathering processes usually work simultaneously in nature and reinforce each other.

MECHANICAL WEATHERING

WEATHERING AND SOIL
▸ Mechanical Weathering

When a rock undergoes *mechanical weathering,* it is broken into smaller and smaller pieces, each retaining the characteristics of the original material. The end result is many small pieces from a single large one. Figure 6.3 shows that breaking a rock into smaller pieces increases the surface area available for chemical attack. An analogous situation occurs when sugar is added to a liquid. In this situation, a cube of sugar will dissolve much slower than an equal volume of sugar granules because the cube has much less surface area available for dissolution. Hence, by breaking rocks into smaller pieces, mechanical weathering increases the amount of surface area available for chemical weathering.

In nature, four important physical processes lead to the fragmentation of rock: frost wedging, expansion resulting from unloading, thermal expansion, and biological activity. In addition, although the work of erosional agents such as wind, glacial ice, and running water is usually considered separately from mechanical weathering, it is nevertheless important. As these mobile agents move rock debris, they relentlessly disintegrate these materials.

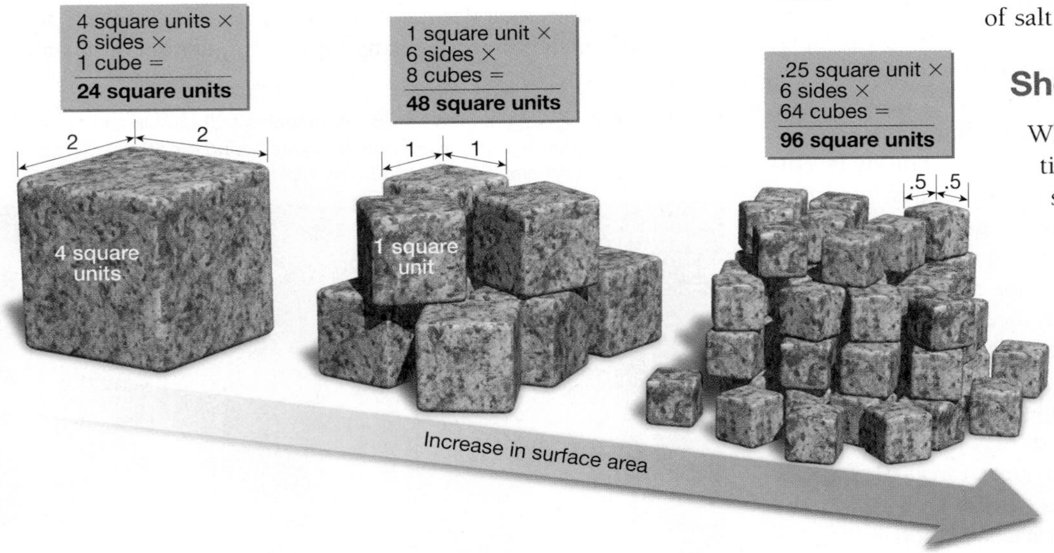

Frost Wedging

If you leave a glass bottle of water in the freezer a bit too long, you will find the bottle fractured. The bottle breaks because liquid water has the unique property of expanding about 9 percent upon freezing. This is also the reason that poorly insulated or exposed water pipes rupture during frigid weather. You might also expect this same process to fracture rocks in nature. This is, in fact, the basis for the traditional explanation of **frost wedging.** After water works its way into the cracks in rock, the freezing water enlarges the cracks and angular fragments are eventually produced. (Figure 6.4).

For many years, the conventional wisdom was that most frost wedging occurred in this way. Recently, however, research has shown that frost wedging can also occur in a different way.* It has long been known that when moist soils freeze, they expand or *frost heave* due to the growth of ice lenses. These masses of ice grow larger because they are supplied with water migrating from unfrozen areas as thin liquid films. As more water accumulates and freezes, the soil is heaved upward. A similar process occurs within the cracks and pore spaces of rocks. Lenses of ice grow larger as they attract liquid water from surrounding pores. The growth of these ice masses gradually weakens the rock, causing it to fracture (Box 6.1).

Salt Crystal Growth

Another expansive force that can split rocks is created by the growth of salt crystals. Rocky shorelines and arid regions are common settings for this process. It begins when sea spray from breaking waves or salty groundwater penetrates crevices and pore spaces in rock. As this water evaporates, salt crystals form. As these crystals gradually grow larger, they weaken the rock by pushing apart the surrounding grains or enlarging tiny cracks.

This same process can also contribute to crumbling roadways where salt is spread to melt snow and ice in winter. The salt dissolves in water and seeps into cracks that quite likely originated from frost action. When the water evaporates, the growth of salt crystals further breaks the pavement.

Sheeting

When large masses of igneous rock, particularly granite, are exposed by erosion, concentric slabs begin to break loose. The process generating these onionlike layers is called **sheeting.** It is thought that this occurs, at least in part, because of the great reduction in pressure when the overlying rock is eroded away, a process called *unloading.* Accompanying this unloading, the outer layers expand more than the rock below and thus separate from the rock body (Figure 6.5). Continued

4 square units ×
6 sides ×
1 cube =
24 square units

1 square unit ×
6 sides ×
8 cubes =
48 square units

.25 square unit ×
6 sides ×
64 cubes =
96 square units

2   2

4 square units

1   1

1 square unit

.5 .5

Increase in surface area

FIGURE 6.3 Chemical weathering can occur only to those portions of a rock that are exposed to the elements. Mechanical weathering breaks rock into smaller and smaller pieces, thereby increasing the surface area available for chemical attack

*Bernard Hallet, "Why Do Freezing Rocks Break?," *Science,* Vol. 314, 17 November 2006, pp. 1092–93.

FIGURE 6.4 Traditional depiction of frost wedging. As water freezes, it expands, exerting a force great enough to break rock. When frost wedging occurs in a setting such as this, the broken rock fragments fall to the base of the cliff and create a cone-shaped accumulation of angular rock fragments known as a talus slope. (Photo by Marli Miller)

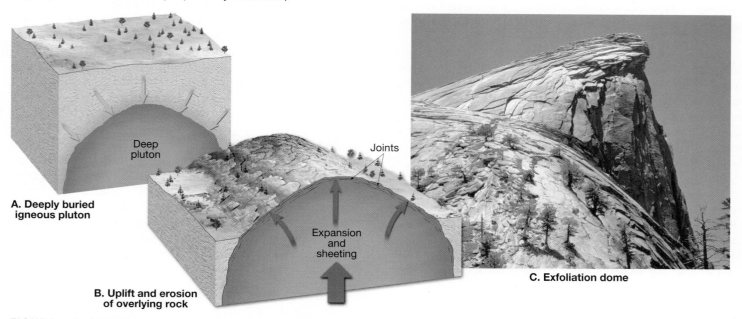

FIGURE 6.5 Sheeting is caused by the expansion of crystalline rock as erosion removes the overlying material. When the deeply buried pluton in **A** is exposed at the surface following uplift and erosion in **B,** the igneous mass fractures into thin slabs. The photo in **C** is of the summit of Half Dome in Yosemite National Park, California. It is an exfoliation dome and illustrates the onionlike layers created by sheeting. (Photo by Breck P. Kent)

UNDERSTANDING EARTH

BOX 6.1

The Old Man of the Mountain

The Old Man of the Mountain, also known as The Great Stone Face or simply The Profile, was one of New Hampshire's (*The Granite State*) best-known and most enduring symbols. Beginning in 1945, it appeared at the center of the official state emblem. It was a natural rock formation sculpted from Conway Red Granite that, when viewed from the proper location, gave the appearance of an old man (Figure 6.A, left). Each year hundreds of thousands of people traveled to view the Old Man, which protruded from high on Cannon Mountain, 360 meters (1200 feet) above Profile Lake in northern New Hampshire's Franconia Notch State Park.

On Saturday morning, May 3, 2003, the people of New Hampshire learned that the famous landmark had succumbed to nature and collapsed (Figure 6.A, right). The collapse ended decades of efforts to protect the state symbol from the same natural processes that created it in the first place. Ultimately, frost wedging and other weathering processes prevailed.

FIGURE 6.A   **(left)** The Old Man of the Mountain, high above Franconia Notch in New Hampshire's White Mountains, as it appeared prior to May 3, 2003. (Jim Cole/AP Photo) The inset shows the state emblem of New Hampshire. **(right)** The famous granite outcrop after it collapsed on May 3, 2003. The natural processes that sculpted the Old Man ultimately destroyed it. (Jim Cole/AP Photo)

weathering eventually causes the slabs to separate and spall off, creating **exfoliation domes** (*ex* = off, *folium* = leaf). Excellent examples of exfoliation domes are Stone Mountain, Georgia, and Half Dome and Liberty Cap in Yosemite National Park (see Figure 6.5).

Deep underground mining provides us with another example of how rocks behave once the confining pressure is removed. Large rock slabs have been known to explode off the walls of newly cut mine tunnels because of the abruptly reduced pressure. Evidence of this type (plus the fact that fracturing occurs parallel to the floor of a quarry when large blocks of rock are removed) strongly supports the process of unloading as the cause of sheeting.

Although many fractures are created by expansion, others are produced by contraction during the crystallization of magma (see Figure 4.30, p. 131), and still others by tectonic forces during mountain building. Fractures produced by these activities generally form a definite pattern and are called *joints* (Figure 6.6). Joints are important rock structures that allow water to penetrate to depth and start the process of weathering long before the rock is exposed.

Thermal Expansion

The daily cycle of temperature may weaken rocks, particularly in hot deserts where daily variations may exceed 30° C. Heating a

rock causes expansion, and cooling causes contraction. Repeated swelling and shrinking of minerals with different expansion rates should logically exert some stress on the rock's outer shell.

Although this process was once thought to be of major importance in the disintegration of rock, laboratory experiments have not substantiated this. In one test, unweathered rocks were heated to temperatures much higher than those normally experienced on Earth's surface and then cooled. This procedure was repeated many times to simulate hundreds of years of weathering, but the rocks showed little apparent change.

Nevertheless, pebbles in desert areas do show evidence of shattering that may have been caused by temperature changes (Figure 6.7). A proposed solution to this dilemma suggests that rocks must first be weakened by chemical weathering before they can be broken down by thermal activity. Further, this process may be aided by the rapid cooling of a desert rainstorm. Additional data are needed before a definite conclusion can be reached as to the impact of temperature variation on rock disintegration.

Biological Activity

Weathering is also accomplished by the activities of organisms, including plants, burrowing animals, and humans. Plant roots in search of nutrients and water grow into fractures, and as the roots

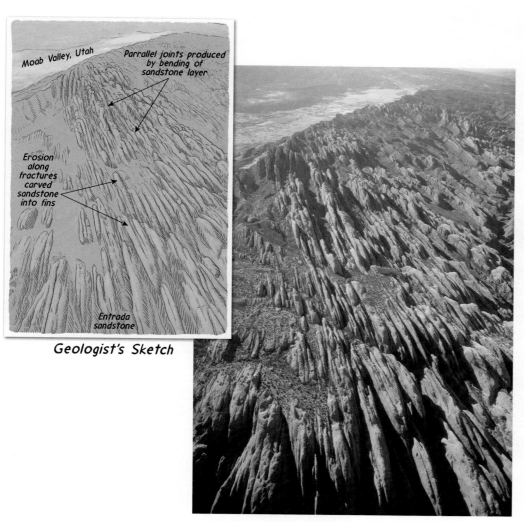

Geologist's Sketch

Moab Valley, Utah — Parrallel joints produced by bending of sandstone layer. Erosion along fractures carved sandstone into fins. Entrada sandstone.

FIGURE 6.6 Aerial view of nearly parallel joints near Moab, Utah. (Photo by Michael Collier)

FIGURE 6.7 These stones were once rounded stream gravels; however, long exposure in a hot desert climate disintegrated them. (Photo by C. B. Hunt, U.S. Geological Survey)

grow, they wedge the rock apart (Figure 6.8). Burrowing animals further break down rock by moving fresh material to the surface, where physical and chemical processes can more effectively attack it. Decaying organisms also produce acids that contribute to chemical weathering. Where rock has been blasted in search of minerals or for road construction, the impact of humans is particularly noticeable.

CHEMICAL WEATHERING

WEATHERING AND SOIL
▶ Chemical Weathering

In the preceding discussion of mechanical weathering, you learned that breaking rock into smaller pieces aids chemical weathering by increasing the surface area available for chemical attack. It should also be pointed out that chemical weathering contributes to mechanical weathering. It does so by weakening the outer portions of some rocks, which, in turn, makes them more susceptible to being broken by mechanical weathering processes.

Chemical weathering involves the complex processes that break down rock components and internal structures of minerals. Such processes convert the constituents to new minerals or release them to the surrounding environment. During this transformation, the original rock decomposes into substances that are stable in the surface environment. Consequently, the products of chemical weathering will remain essentially unchanged as long as they remain in an environment similar to the one in which they formed.

Water is by far the most important agent of chemical weathering. Pure water alone is a good solvent, and small amounts of dissolved materials result in increased chemical activity for weathering solutions. The major processes of chemical weathering are dissolution, oxidation, and hydrolysis. Water plays a leading role in each.

Dissolution

Perhaps the easiest type of decomposition to envision is the process of **dissolution.** Just as sugar dissolves in water, so too do certain minerals. One of the most water-soluble minerals is halite (common salt), which as you may recall, is composed of sodium and chloride ions. Halite readily dissolves in water because, although this compound maintains overall electrical neutrality, the individual ions retain their respective charges.

FIGURE 6.8 Root wedging widens fractures in rock and aids the process of mechanical weathering. Mt. Sanitas, Boulder, Colorado. (Photo by Kristin Piljay)

Moreover, the surrounding water molecules are polar—that is, the oxygen end of the molecule has a small residual negative charge; the end with hydrogen has a small positive charge. As the water molecules come in contact with halite, their negative ends approach sodium ions and their positive ends cluster about chloride ions. This disrupts the attractive forces in the halite crystal and releases the ions to the water solution (Figure 6.9).

Although most minerals are, for all practical purposes, insoluble in pure water, the presence of even a small amount of acid dramatically increases the corrosive force of water. (An acidic solution contains the reactive hydrogen ion, H^+.) In nature, acids are produced by a number of processes. For example, carbonic acid is created when carbon dioxide in the atmosphere dissolves in raindrops. As acidic rainwater soaks into the ground, carbon dioxide in the soil may increase the acidity of the weathering solution. Various organic acids are also released into the soil as organisms decay, and sulfuric acid is produced by the weathering of pyrite and other sulfide minerals.

Regardless of the source of the acid, this highly reactive substance readily decomposes most rocks and produces certain products that are water soluble. For example, the mineral calcite, $CaCO_3$, which composes the common building stones marble and limestone, is easily attacked by even a weakly acidic solution.

The overall reaction by which calcite dissolves in water containing carbon dioxide is:

$$CaCO_3 + (H^+ + HCO_3^-) \longrightarrow Ca^{2+} + 2\,HCO_3^-$$

calcite   carbonic acid   calcium ion   bicarbonate ion

During this process, the insoluble calcium carbonate is transformed into soluble products. In nature, over periods of thousands of years, large quantities of limestone are dissolved and carried away by underground water. This activity is clearly evidenced by the large number of subsurface caverns found in every one of the contiguous 48 states (Figure 6.10). Monuments and buildings made of limestone or marble are also subjected to the corrosive work of acids, particularly in industrial areas that have smoggy, polluted air (Box 6.2).

The soluble ions from reactions of this type are retained in our underground water supply. It is these dissolved ions that are responsible for the so-called hard water found in many locales. Simply, hard water is undesirable because the active ions react with soap to produce an insoluble material that renders soap nearly useless in removing dirt. To solve this problem, a water softener can be used to remove these ions, generally by replacing them with others that do not chemically react with soap.

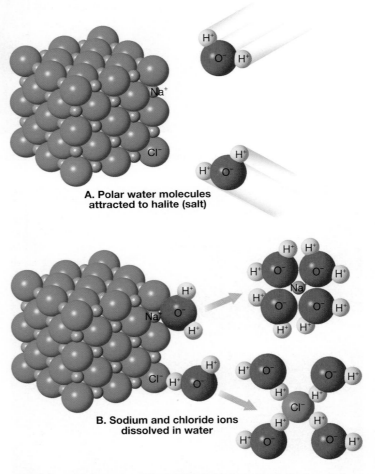

A. Polar water molecules attracted to halite (salt)

B. Sodium and chloride ions dissolved in water

FIGURE 6.9 Illustration of halite dissolving in water. A. Sodium and chloride ions are attacked by the polar water molecules. B. Once removed, these ions are surrounded and held by a number of water molecules as shown.

FIGURE 6.10 The dissolving power of carbonic acid plays an important part in forming limestone caverns. Carlsbad Caverns National Park, New Mexico. (Photo by INTERFOTO/Alamy)

Oxidation

Everyone has seen iron and steel objects that rusted when exposed to water (Figure 6.11). The same thing can happen to iron-rich minerals. The process of rusting occurs when oxygen combines with iron to form iron oxide as follows:

$$\underset{\text{iron}}{4\,\text{Fe}} \;+\; \underset{\text{oxygen}}{3\,\text{O}_2} \;\longrightarrow\; \underset{\text{iron oxide (hematite)}}{2\,\text{Fe}_2\text{O}_3}$$

This type of chemical reaction, called **oxidation,*** occurs when electrons are lost from one element during the reaction. In this case, we say that iron was oxidized because it lost electrons to oxygen. Although the oxidation of iron progresses very slowly in a dry environment, the addition of water greatly speeds the reaction.

Oxidation is important in decomposing such ferromagnesian minerals as olivine, pyroxene, and hornblende. Oxygen readily combines with the iron in these minerals to form the reddish-brown iron oxide called *hematite* (Fe_2O_3) or in other cases a yellowish-colored rust called *limonite* [FeO(OH)]. These products are responsible for the rusty color on the surfaces of dark igneous rocks, such as basalt, as they begin to weather. However, oxidation can occur only after iron is freed from the silicate structure by another process, called hydrolysis.

Another important oxidation reaction occurs when sulfide minerals such as pyrite decompose. Sulfide minerals are major constituents in many metallic ores, and pyrite is frequently asso-

ciated with coal deposits as well. In a moist environment, chemical weathering of pyrite (FeS_2) yields sulfuric acid (H_2SO_4) and iron oxide [FeO(OH)]. In many mining locales this weathering process creates a serious environmental hazard, particularly in humid areas where abundant rainfall infiltrates spoil banks (waste material left after coal or other minerals are removed). This so-called *acid mine drainage* eventually makes its way to streams, killing aquatic organisms and degrading aquatic habitats (Figure 6.12).

Hydrolysis

The most common mineral group, the silicates, is decomposed primarily by the process of **hydrolysis** (*hydro* = water, *lysis* = a loosening), which basically is the reaction of any substance with water. Ideally, the hydrolysis of a mineral could take place in pure water as some of the water molecules dissociate to form the very reactive hydrogen (H^+) and hydroxyl (OH^-) ions. It is the hydrogen ion that attacks and replaces other positive ions found in the crystal lattice. With the introduction of hydrogen ions into the crystalline structure, the original orderly arrangement of atoms is destroyed and the mineral decomposes.

In nature, water usually contains other substances that contribute additional hydrogen ions, thereby greatly accelerating hydrolysis. The most common of these substances is carbon dioxide, CO_2, which dissolves in water to form carbonic acid, H_2CO_3. Rain dissolves some carbon dioxide in the atmosphere, and additional amounts, released by decaying organic matter, are acquired as the water percolates through the soil.

FIGURE 6.11 Iron reacts with oxygen to form iron oxide as seen on these rusted barrels. (Photo by Steven Robertson/iStockphoto)

---

*The reader should note that *oxidation* is a term referring to any chemical reaction in which a compound or radical loses electrons. The element oxygen is not necessarily present.

EARTH AS A SYSTEM

Acid Precipitation—A Human Impact on the Earth System

Humans are part of the complex interacting whole we call the Earth system. As such, our actions cause changes to all the other parts of the system. For example, by going about our normal routine, we humans modify the composition of the atmosphere. These atmospheric modifications in turn cause unintended and unwanted changes to occur in the hydrosphere, biosphere, and solid Earth. Acid precipitation is one small but significant example.

Decomposed stone monuments and structures are common sights in many cities (Figure 6.B). Although we expect rock to gradually decompose, many of these monuments have succumbed prematurely. An important cause for this accelerated chemical weathering is acid precipitation.

Rain is naturally somewhat acidic (Figure 6.C). When carbon dioxide from the atmosphere dissolves in water, the product is weak carbonic acid. However, the term *acid precipitation* refers to precipitation that is much more acidic than natural, unpolluted rain and snow.

As a consequence of burning large quantities of fossil fuels, primarily coal and petroleum products, millions of tons of sulfur and nitrogen oxides are released into the atmosphere each year in the United States. In 2007, the total was 37 million tons.* The major sources of these emissions include power-generating plants, industrial processes such as ore smelting and petroleum refining, and vehicles of all kinds. Through a series of complex chemical reactions, some of these pollutants are converted into acids that then fall to Earth's surface as rain or snow. Another portion is deposited in dry form and subsequently converted into acid after coming in contact with precipitation, dew, or fog.

Northern Europe and eastern North America have experienced widespread acid rain for some time. Studies have also shown that acid rain occurs in many other regions, including western North America, Japan, China, Russia, and South America. In addition to local pollution sources, a portion of the acidity found in the northeastern United States and eastern Canada originates hundreds of kilometers away in industrialized regions to the south and southwest. This situation occurs because many pollutants remain in the atmosphere as long as five days, dur-

FIGURE 6.B   Acid rain accelerates the chemical weathering of stone monuments and structures. (Photo by Adam Hart-Davis/Science Photo Library/Photo Researchers, Inc.)

---

*This figure has been steadily falling. Data from the EPA show that in 1990 the combined total of sulfur dioxide and nitrogen oxides emitted in the United States was 48 million tons. In 2000, it was 38 million tons.

ing which time they may be transported great distances.

The damaging environmental effects of acid rain are thought to be considerable in some

FIGURE 6.12   This water seeping from an abandoned mine in Colorado is an example of *acid mine drainage*. Acid mine drainage is water with a high concentration of sulfuric acid (H_2SO_4) produced by the oxidation of sulfide minerals such as pyrite. When such acid-rich water migrates from its source, it may pollute surface waters and groundwater and cause significant ecological damage. (Photo by Tim Haske/Profiles West/Photolibrary)

In water, carbonic acid ionizes to form hydrogen ions (H^+) and bicarbonate ions (HCO_3^-). To illustrate how a rock undergoes hydrolysis in the presence of carbonic acid, let's examine the chemical weathering of granite, a common continental rock. Recall that granite consists mainly of quartz and potassium feldspar. The weathering of the potassium feldspar component of granite is as follows:

$$2\ KAlSi_3O_8 + 2(H^+ + HCO_3^-) + H_2O \longrightarrow$$

potassium feldspar   carbonic acid   water

$$Al_2Si_2O_5(OH)_4 + 2\ K^+ + 2\ HCO_3^- + 4\ SiO_2$$

kaolinite (residual clay)   potassium ion   bicarbonate ion   silica

in solution

In this reaction, the hydrogen ions (H^+) attack and replace potassium ions (K^+) in the feldspar structure, thereby disrupting the crystalline network. Once removed, the potassium is

areas and imminent in others (Figure 6.D). The best-known effect is an increased acidity in thousands of lakes in Scandinavia and eastern North America. Accompanying this have been substantial increases in dissolved aluminum leached from the soil by the acidic water, which is toxic to fish. Consequently, some lakes are virtually devoid of fish, and others are approaching this condition. Ecosystems are characterized by many interactions at many levels of organization, which means that evaluating the effects of acid precipitation on these complex systems is difficult and expensive and far from complete.

Even within small areas, effects of acid precipitation can vary significantly from one lake to another. Much of this variation is related to the composition of the of the soil and rock materials in the area surrounding the lake. Because minerals such as calcite can neutralize acid solutions, lakes surrounded by calcite-rich materials are less likely to become acidic. In contrast, lakes that lack this buffering material can be severely affected. Even so, over a period of time the pH of lakes that have not yet been acidified may drop as the buffering material in the surrounding soil becomes depleted.

In addition to the many lakes that can no longer support fish, research indicates that acid precipitation may also reduce agricultural crop yields and impair the productivity of forests. Acid rain not only harms the foliage but also damages roots and leaches nutrient minerals from the soil. Finally, acid precipitation promotes the corrosion of metals and contributes to the destruction of stone structures.

FIGURE 6.D Damage to forests by acid precipitation is well documented in Europe and eastern North America. These trees in the Great Smoky Mountains have been injured by acid-laden clouds. (Photo by Doug Locke/Dembinsky Photo Associates)

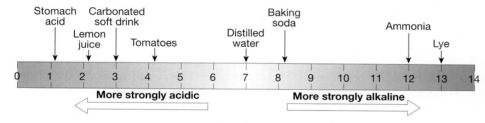

FIGURE 6.C The *pH scale* is a common measure of the degree of acidity or alkalinity of a solution. The scale ranges from 0 to 14, with a value of 7 denoting a solution that is neutral. Values below 7 indicate greater acidity, whereas numbers above 7 indicate greater alkalinity. The pH values of some familiar substances are shown on the diagram. Although distilled water is neutral (pH 7), rainwater is naturally acidic. It is important to note that the pH scale is logarithmic; that is, each whole number increment indicates a tenfold difference. Thus, pH 4 is 10 times more acidic than pH 5 and 100 times (10 × 10) more acidic than pH 6.

available as a nutrient for plants or becomes the soluble salt potassium bicarbonate ($KHCO_3$), which may be incorporated into other minerals or carried to the ocean.

The most abundant product of the chemical breakdown of potassium feldspar is the clay mineral kaolinite. Clay minerals are the end products of weathering and are very stable under surface conditions. Consequently, clay minerals make up a high percentage of the inorganic material in soils. Moreover, the most abundant sedimentary rock, shale, contains a high proportion of clay minerals.

In addition to the formation of clay minerals during the weathering of potassium feldspar, some silica is removed from the feldspar structure and carried away by groundwater. This dissolved silica will eventually precipitate, producing nodules of chert or flint, or it will fill in the pore spaces between grains of sediment, or it will be carried to the ocean, where microscopic animals will remove it from the water to build hard silica shells.

To summarize, the weathering of potassium feldspar generates a residual clay mineral, a soluble salt (potassium bicarbonate), and some silica, which enters into solution.

Quartz, the other main component of granite, is very resistant to chemical weathering; it remains substantially unaltered when attacked by weak acidic solutions. As a result, when granite weathers, the feldspar crystals dull and slowly turn to clay, releasing the once interlocked quartz grains, which still retain their fresh, glassy appearance. Although some of the quartz remains in the soil, much is eventually transported to the sea or to other sites of deposition, where it becomes the main constituent of such features as sandy beaches and sand dunes. In time these quartz grains may be lithified to form the sedimentary rock sandstone.

Table 6.1 lists the weathered products of some of the most common silicate minerals. Remember that silicate minerals make up most of Earth's crust and that these minerals are essentially composed of only eight elements. When chemically weathered, the silicate minerals yield sodium, calcium, potassium, and magnesium ions that form soluble products, which may be removed from groundwater. The element iron combines with oxygen, producing relatively insoluble iron oxides, most notably hematite and limonite, which give soil a reddish-brown

TABLE 6.1	Products of Weathering	
Mineral	Residual Products	Material in Solution
Quartz	Quartz grains	Silica
Feldspars	Clay minerals	Silica, K^+, Na^+, Ca^{2+}
Amphibole (hornblende)	Clay minerals Limonite Hematite	Silica Ca^{2+}, Mg^{2+}
Olivine	Limonite Hematite	Silica Mg^{2+}

Students Sometimes Ask . . .

Is the clay created by chemical weathering the same clay that's used in making ceramics?

Yes. Kaolinite, the clay described in the section on hydrolysis, is called *china clay* and is used for high-quality porcelain. However, far greater quantities of this clay are used as a coating in the manufacture of high-quality paper, such as that used in this book.

Weathering actually creates many different clay minerals that have many different uses. Clay minerals are used in making bricks, tiles, sewer pipes, and cement. Clays are used as lubricants in the bore holes of oil-drilling rigs and are a common ingredient in paint. Products as varied as your car's catalytic converter and filters used in beer- and wine-making rely on clay minerals.

or yellowish color. Under most conditions the three remaining elements—aluminum, silicon, and oxygen—join with water to produce residual clay minerals. However, even the highly insoluble clay minerals are very slowly removed by subsurface water.

Spheroidal Weathering

In addition to altering the internal structure of minerals, chemical weathering causes physical changes as well. For instance, when angular rock masses are chemically weathered as water enters along joints, the boulders take on a spherical shape. The gradual rounding of the corners and edges of angular blocks is illustrated in Figure 6.13. The corners are attacked most readily

because of the greater surface area for their volume as compared to the edges and faces. This process, called **spheroidal weathering,** gives the weathered rock a more rounded or spherical shape (Figure 6.13D).

Sometimes during the formation of spheroidal boulders, successive shells separate from the rock's main body (Figure 6.14). Eventually the outer shells break off, allowing the chemical weathering activity to penetrate deeper into the boulder. This spherical scaling results because, as the minerals in the rock weather to clay, they increase in size through the addition of water

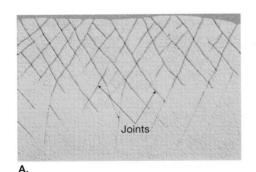

A.

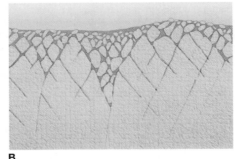

B.

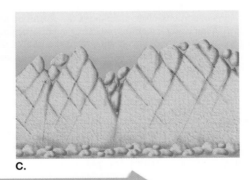

C.

T I M E

D.

FIGURE 6.13 Spheroidal weathering of extensively jointed rock. Chemical weathering associated with water moving through the joints enlarges them. Because the rocks are attacked more on the corners and edges, they take on a spherical shape. The photo shows spheroidal weathering in Joshua Tree National Park, California. (Photo by E. J. Tarbuck)

FIGURE 6.14 Successive shells are loosened as the weathering process continues to penetrate ever deeper into the rock. (Photo by Martin Schmidt, Jr.)

to their structure. This increased bulk exerts an outward force that causes concentric layers of rock to break loose and fall off.

Hence, chemical weathering does produce forces great enough to cause mechanical weathering. This type of spheroidal weathering in which shells spall off should not be confused with the phenomenon of sheeting discussed earlier. In sheeting, the fracturing occurs as a result of unloading, and the rock layers that separate from the main body are largely unaltered at the time of separation.

RATES OF WEATHERING

WEATHERING AND SOIL
GEODe EARTH ▸ Rates of Weathering

Several factors influence the type and rate of rock weathering. We have already seen how mechanical weathering affects the rate of weathering. By breaking rock into smaller pieces, the amount of surface area exposed to chemical weathering increases. Other important factors examined here include the roles of rock characteristics and climate.

Rock Characteristics

Rock characteristics encompass all of the chemical traits of rocks, including mineral composition and solubility. In addition, any physical features, such as joints (cracks), can be important because they influence the ability of water to penetrate rock.

The variations in weathering rates due to the mineral constituents can be demonstrated by comparing old headstones made from different rock types. Headstones of gran-

ite, which is composed of silicate minerals, are relatively resistant to chemical weathering. We can see this by examining the inscriptions on the headstones shown in Figure 6.15. In contrast, the marble headstone shows signs of extensive chemical alteration over a relatively short period. Marble is composed of calcite (calcium carbonate), which readily dissolves even in a weakly acidic solution.

The silicates, the most abundant mineral group, chemically weather in essentially the same order as their order of crystallization. By examining Bowen's Reaction Series (see Figure 4.21, p. 123), you can see that olivine crystallizes first and is therefore least resistant to chemical weathering, whereas quartz, which crystallizes last, is the most resistant.

Climate

Climatic factors, particularly temperature and moisture, are crucial to the rate of rock weathering. One important example from mechanical weathering is that the frequency of freeze-thaw cycles greatly affects the amount of frost wedging. Temperature and moisture also exert a strong influence on rates of chemical weathering and on the kind and amount of vegetation present. Regions with lush vegetation often have a thick mantle of soil rich in decayed organic matter from which chemically active fluids such as carbonic acid and humic acids are derived.

The optimum environment for chemical weathering is a combination of warm temperatures and abundant moisture. In polar regions chemical weathering is ineffective because frigid temperatures keep the available moisture locked up as ice, whereas in arid regions there is insufficient moisture to foster rapid chemical weathering.

Human activities can influence the composition of the atmosphere, which in turn can impact the rate of chemical weathering. Box 6.2 examines one well-known example—acid rain.

FIGURE 6.15 An examination of headstones in the same cemetary reveals the rate of chemical weathering on diverse rock types. The granite headstone (left) was erected four years before the marble headstone (right). The inscription date of 1872 on the marble monument is nearly illegible. (Photos by E. J. Tarbuck)

Differential Weathering

Masses of rock do not weather uniformly. Take a moment to look back at the photo of the dike in Figure 4.29 (p. 130) and of Shiprock; New Mexico, in Figure 5.36 (p. 161). These durable igneous masses stand above the surrounding terrain like stone walls. A glance at Figure 6.1 shows an additional example of this phenomenon, called **differential weathering.** The results vary in scale from the rough, uneven surface of the marble headstone in Figure 6.15 to the boldly sculpted pinnacle in Figure 6.16.

Many factors influence the rate of rock weathering. Among the most important are variations in the composition of the rock. More resistant rock protrudes as ridges (see Figure 14.18, p. 396) or pinnacles, or as steeper cliffs on an irregular hillside (see Figure 7.4, p. 204). The number and spacing of joints can also be a significant factor (see Figures 6.6 and 6.13). Differential weathering and subsequent erosion are responsible for creating many unusual and sometimes spectacular rock formations and landforms.

SOIL

Soil covers most land surfaces. Along with air and water, it is one of our most indispensable resources (Figure 6.17). Also like air and water, soil is taken for granted by many of us. The following quote helps put this vital layer in perspective.

> Science, in recent years, has focused more and more on the Earth as a planet, one that for all we know is unique—where

FIGURE 6.16 Differential weathering is illustrated by these sculpted rock pinnacles in Utah's Bryce Canyon National Park. (Photo by Joe Cornish/Photolibrary)

FIGURE 6.17 Soil is an essential resource that we often take for granted. Soil is not a living entity, but it contains a great deal of life. Moreover, this complex medium supports nearly all plant life, which in turn supports animal life. The pie chart shows the composition (by volume) of a soil in good condition for plant growth. Although the percentages vary, each soil is composed of mineral and organic matter, water, and air. (Photo by Colin Molyneux/Getty Images)

a thin blanket of air, a thinner film of water, and the thinnest veneer of soil combine to support a web of life of wondrous diversity in continuous change.*

Soil has accurately been called "the bridge between life and the inanimate world." All life—the entire biosphere—owes its existence to a dozen or so elements that must ultimately come from Earth's crust. Once weathering and other processes create soil, plants carry out the intermediary role of assimilating the necessary elements and making them available to animals, including humans.

An Interface in the Earth System

When Earth is viewed as a system, soil is referred to as an *interface*—a common boundary where different parts of a system interact. This is an appropriate designation because soil forms where the geosphere, the atmosphere, the hydrosphere, and the biosphere meet. Soil is a material that develops in response to complex environmental interactions among different parts of the Earth system. Over time, soil gradually evolves to a state of equilibrium or balance with the environment. Soil is dynamic and sensitive to almost every aspect of its surroundings. Thus, when environmental changes occur, such as climate, vegetative cover, and animal (including human) activity, the soil responds. Any such change produces a gradual alteration of soil characteristics until a new balance is reached. Although thinly distributed over the land surface, soil functions as a fundamental interface, providing an

*Jack Eddy, "A Fragile Seam of Dark Blue Light," in *Proceedings of the Global Change Research Forum*. U.S. Geological Survey Circular 1086, 1993, p. 15.

excellent example of the integration among many parts of the Earth system.

What is Soil?

With few exceptions, Earth's land surface is covered by **regolith** (*rhegos* = blanket, *lithos* = stone), the layer of rock and mineral fragments produced by weathering. Some would call this material soil, but soil is more than an accumulation of weathered debris. **Soil** is a combination of mineral and organic matter, water, and air—that portion of the regolith that supports the growth of plants. Although the proportions of the major components in soil vary, the same four components always are present to some extent (see Figure 6.17). About one-half of the total volume of a good-quality surface soil is a mixture of disintegrated and decomposed rock (mineral matter) and **humus,** the decayed remains of animal and plant life (organic matter). The remaining half consists of pore spaces among the solid particles where air and water circulate.

Although the mineral portion of the soil is usually much greater than the organic portion, humus is an essential component. In addition to being an important source of plant nutrients, humus enhances the soil's ability to retain water. Because plants require air and water to live and grow, the portion of the soil consisting of pore spaces that allow for the circulation of these fluids is as vital as the solid soil constituents.

Soil water is far from "pure" water; instead, it is a complex solution containing many soluble nutrients. Soil water not only provides the necessary moisture for the chemical reactions that sustain life, it also supplies plants with nutrients in a form they can use. The pore spaces not filled with water contain air. This air is the source of necessary oxygen and carbon dioxide for most microorganisms and plants that live in the soil.

the parent material is bedrock, the soils are termed *residual soils*. By contrast, those developed on unconsolidated sediment are called *transported soils* (Figure 6.18). It should be pointed out that transported soils form *in place* on parent materials that have been carried from elsewhere and deposited by gravity, water, wind, or ice.

The nature of the parent material influences soils in two ways. First, the type of parent material will affect the rate of weathering and thus the rate of soil formation. Also, because unconsolidated deposits are already partly weathered, soil development on such material will likely progress more rapidly than when bedrock is the parent material. Second, the chemical makeup of the parent material will affect the soil's fertility. This influences the character of the natural vegetation the soil can support.

At one time the parent material was thought to be the primary factor causing differences among soils. However, soil scientists

CONTROLS OF SOIL FORMATION

Soil is the product of the complex interplay of several factors, including parent material, time, climate, plants and animals, and topography. Although all these factors are interdependent, their roles will be examined separately.

Parent Material

The source of the weathered mineral matter from which soils develop is called the **parent material** and is a major factor influencing a newly forming soil. Gradually it undergoes physical and chemical changes as the processes of soil formation progress. Parent material can either be the underlying bedrock or a layer of unconsolidated deposits. When

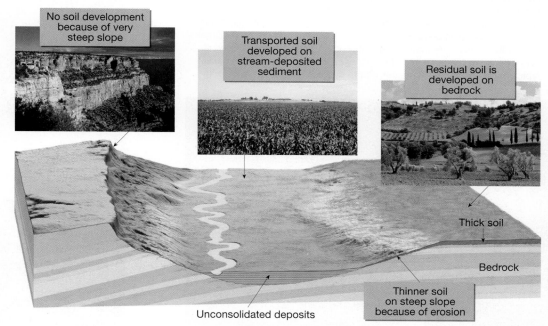

No soil development because of very steep slope

Transported soil developed on stream-deposited sediment

Residual soil is developed on bedrock

Thick soil

Bedrock

Thinner soil on steep slope because of erosion

Unconsolidated deposits

FIGURE 6.18 The parent material for residual soils is the underlying bedrock, whereas transported soils form on unconsolidated deposits. Also note that as slopes become steeper, soil becomes thinner. (Left and center photos by E. J. Tarbuck; right photo by Grilly Bernard/Getty Images, Inc./Stone Allstock)

came to understand that other factors, especially climate, are more important. In fact, it was found that similar soils often develop from different parent materials and that dissimilar soils have developed from the same parent material. Such discoveries reinforce the importance of other soil-forming factors.

Time

Time is an important component of *every* geological process, and soil formation is no exception. The nature of soil is strongly influenced by the length of time that processes have been operating. If weathering has been going on for a comparatively short time, the character of the parent material strongly influences the characteristics of the soil. As weathering processes continue, the influence of parent material on soil is overshadowed by the other soil-forming factors, especially climate. The amount of time required for various soils to evolve cannot be listed because the soil-forming processes act at varying rates under different circumstances. However, as a rule, the longer a soil has been forming, the thicker it becomes and the less it resembles the parent material.

Climate

Climate is considered to be the most influential control of soil formation. Temperature and precipitation are the elements that exert the strongest impact. Variations in temperature and precipitation determine whether chemical or mechanical weathering will predominate and also greatly influence the rate and depth of weathering. For instance, a hot, wet climate may produce a thick layer of chemically weathered soil in the same

amount of time that a cold, dry climate produces a thin mantle of mechanically weathered debris. Also, the amount of precipitation influences the degree to which various materials are removed from the soil by percolating waters (a process called *leaching*), thereby affecting soil fertility. Finally, climatic conditions are an important control on the type of plant and animal life present.

Plants and Animals

Plants and animals play a vital role in soil formation. The types and abundance of organisms present have a strong influence on the physical and chemical properties of a soil (Figure 6.19). In fact, for well-developed soils in many regions, the significance of natural vegetation in influencing soil type is frequently implied in the description used by soil scientists. Such phrases as *prairie soil, forest soil,* and *tundra soil* are common.

Plants and animals furnish organic matter to the soil. Certain bog soils are composed almost entirely of organic matter, whereas desert soils might contain as little as a small fraction of 1 percent. Although the quantity of organic matter varies substantially among soils, it is the rare soil that completely lacks it.

B.

FIGURE 6.19   The nature of the vegetation in an area can have a significant influence on soil formation. **A.** The northern coniferous forest. The organic litter received by the soil from the conifers is high in acid resins, which contributes to an accumulation of acid in the soil. As a result, intensive acid leaching is an important soil-forming process here. (Photo by Bill Brooks/Alamy) **B.** The relatively meager vegetation in Arizona's Sonoran Desert is very different in character from the northern coniferous forest. Desert soils typically lack much organic matter. (Photo by Russ Bishop/age fotostock)

A.

The primary source of organic matter in soil is plants, although animals and an infinite number of microorganisms also contribute. When organic matter is decomposed, important nutrients are supplied to plants, as well as to animals and microorganisms living in the soil. Consequently, soil fertility is in part related to the amount of organic matter present. Furthermore, the decay of plant and animal remains causes the formation of various organic acids. These complex acids hasten the weathering process. Organic matter also has a high water-holding ability and thus aids water retention in a soil.

Microorganisms, including fungi, bacteria, and single-called protozoa, play an active role in the decay of plant and animal remains. The end product is humus, a material that no longer resembles the plants and animals from which it is formed. In addition, certain microorganisms aid soil fertility because they have the ability to convert atmospheric nitrogen into soil nitrogen.

Earthworms and other burrowing animals act to mix the mineral and organic portions of a soil. Earthworms, for example, feed on organic matter and thoroughly mix soils in which they live, often moving and enriching many tons per acre each year. Burrows and holes also aid the passage of water and air through the soil.

Topography

The lay of the land can vary greatly over short distances. Such variations in topography can lead to the development of a variety of localized soil types. Many of the differences exist because the length and steepness of slopes have a significant impact on the amount of erosion and the water content of soil.

On steep slopes, soils are often poorly developed. In such situations the quantity of water soaking in is slight; as a result, the moisture content of the soil may not be sufficient for vigorous plant growth. Further, because of accelerated erosion on steep slopes, the soils are thin or in some cases nonexistent (see Figure 6.18).

In contrast, poorly drained and waterlogged soils found in bottomlands have a much different character. Such soils are usually thick and dark. The dark color results from the large quantity of organic matter that accumulates because saturated conditions retard the decay of vegetation. The optimum terrain for soil development is a flat-to-undulating upland surface. Here we find good drainage, minimum erosion, and sufficient infiltration of water into the soil.

Slope orientation, or the direction the slope is facing, is another consideration. In the midlatitudes of the Northern Hemisphere, a south-facing slope will receive a great deal more sunlight than a north-facing slope. In fact, a steep north-facing slope may receive no direct sunlight at all. The difference in the amount of solar radiation received will cause differences in soil temperature and moisture, which in turn influence the nature of the vegetation and the character of the soil.

Although this section dealt separately with each of the soil-forming factors, remember that all of them work together to form soil. No single factor is responsible for a soil's character; rather, it is the combined influence of parent material, time, climate, plants and animals, and topography that determines this character.

THE SOIL PROFILE

Because soil-forming processes operate from the surface downward, variations in composition, texture, structure, and color gradually evolve at varying depths. These vertical differences, which usually become more pronounced as time passes, divide the soil into zones or layers known as **horizons.** If you were to dig a trench in soil, you would see that its walls are layered. Such a vertical section through all of the soil horizons constitutes the **soil profile** (Figure 6.20).

Figure 6.21 presents an idealized view of a well-developed soil profile in which five horizons are identified. From the surface downward, they are designated as *O, A, E, B,* and *C.* These five horizons are common to soils in temperate regions. The characteristics and extent of development of horizons vary in different environments. Thus, different localities exhibit soil profiles that can contrast greatly with one another.

The *O* soil horizon consists largely of organic material. This is in contrast to the layers beneath it, which consist mainly of mineral matter. The upper portion of the *O* horizon is primarily plant litter, such as loose leaves and other organic debris that are still recognizable. By contrast, the lower portion of the *O* horizon is made up of partly decomposed organic matter (humus) in which plant structures can no longer be identified. In addition to plants, the *O* horizon is teeming with microscopic life, including bacteria, fungi, algae, and insects. All of these organisms contribute oxygen, carbon dioxide, and organic acids to the developing soil.

Underlying the organic-rich *O* horizon is the *A* horizon. This zone is largely mineral matter, yet biological activity is high and humus is generally present—up to 30 percent in some instances. Together the *O* and *A* horizons make up what is commonly called the *topsoil.* Below the *A* horizon, the *E* horizon is a light-colored layer that contains little organic material. As water percolates downward through this zone, finer particles are carried away. This washing out of fine soil components is termed **eluviation** (*elu* = get away from, *via* = a way). Water percolating downward also dissolves soluble inorganic soil components

Students Sometimes Ask . . .

I was digging in my yard the other day and came across a deep "hardpan" layer that was really difficult to penetrate. How does a hardpan form?

Hardpans are created by the process of eluviation. As water percolates through the soil, fine clay-size particles from the upper soil layers are moved by eluviation and concentrated in the subsoil (*B* horizon). Over time the accumulation of these clay-size particles creates a nearly impenetrable layer, which is what you found. Sometimes hardpans are so impermeable that they serve as effective barriers to water movement, preventing further infiltration of water. Hardpans are also called *adobe layers,* because their high clay content makes them suitable for use as construction bricks.

A.

B.

FIGURE 6.20 A soil profile is a vertical cross section from the surface through all of the soil's horizons and into the parent material. **A.** This profile shows a well-developed soil in southeastern South Dakota. (Photo by E. J. Tarbuck) **B.** The boundaries between horizons in this soil in Puerto Rico are indistinct, giving it a relatively uniform appearance. (Photo courtesy of Soil Science Society of America)

and carries them to deeper zones. This depletion of soluble materials from the upper soil is termed **leaching.**

Immediately below the *E* horizon is the *B* horizon, or *subsoil*. Much of the material removed from the *E* horizon by eluviation is deposited in the *B* horizon, which is often referred to as the *zone of accumulation*. The accumulation of the fine clay

particles enhances water retention in the subsoil. The *O, A, E,* and *B* horizons together constitute the **solum,** or "true soil." It is in the solum that the soil-forming processes are active and that living roots and other plant and animal life are largely confined.

Below the solum and above the unaltered parent material is the *C* horizon, a layer characterized by partially altered parent material. Whereas the *O, A, E,* and *B* horizons bear little resemblance to the parent material, it is easily identifiable in the *C* horizon. Although this material is undergoing changes that will eventually transform it into soil, it has not yet crossed the threshold that separates regolith from soil.

The characteristics and extent of development can vary greatly among soils in different environments. The boundaries between soil horizons may be sharp, or the horizons may blend gradually from one to another. Consequently, a well-developed soil profile indicates that environmental conditions have been relatively stable over an extended time span and that the soil is *mature*. By contrast, some soils lack horizons altogether.

Such soils are called *immature* because soil building has been going on for only a short time. Immature soils are also characteristic of steep slopes, where erosion continually strips away the soil, preventing full development.

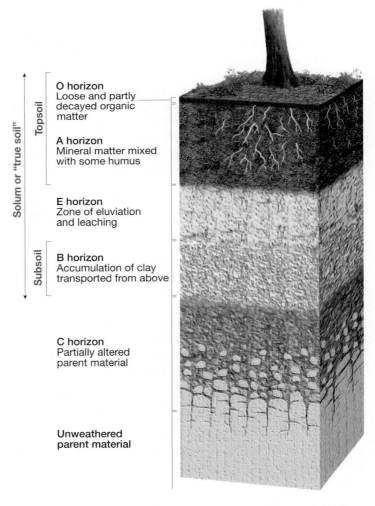

O horizon
Loose and partly decayed organic matter

A horizon
Mineral matter mixed with some humus

E horizon
Zone of eluviation and leaching

B horizon
Accumulation of clay transported from above

C horizon
Partially altered parent material

Unweathered parent material

Topsoil / Subsoil / Solum or "true soil"

FIGURE 6.21 Idealized soil profile from a humid climate in the middle latitudes. The topsoil and subsoil together comprise the solum or "true soil."

CLASSIFYING SOILS

There are many variations from place to place and from time to time among the factors that control soil formation. These differences lead to a bewildering variety of soil types. To cope with such variety, it is essential to devise some means of classifying the vast array of data to be studied. By establishing groups consisting of items that have certain important characteristics in common, order and simplicity are introduced. Bringing order to large quantities of information not only aids comprehension and understanding but also facilitates analysis and explanation.

In the United States, soil scientists have devised a system for classifying soils known as the **Soil Taxonomy.** It emphasizes the physical and chemical properties of the soil profile and is organized on the basis of observable soil characteristics. There are six hierarchical categories of classification, ranging from *order,* the broadest category, to *series,* the most specific category. The system recognizes 12 soil orders and more than 19,000 soil series.

The names of the classification units are combinations of syllables, most of which are derived from Latin or Greek. The names are descriptive. For example, soils of the order Aridosol (from the Latin *aridus,* dry, and *solum,* soil) are characteristically dry soils in arid regions. Soils in the order Inceptisols (from Latin *inceptum,* beginning, and *solum,* soil) are soils with only the beginning or inception of profile development.

Brief descriptions of the 12 basic soil orders are provided in Table 6.2. Figure 6.22 shows the complex worldwide distribution pattern of the Soil Taxonomy's 12 soil orders (see Box 6.3). Like many classification systems, the Soil Taxonomy is not suitable for every purpose. It is especially useful for agricultural and related land-use purposes, but it is not a useful system for engineers who are preparing evaluations of potential construction sites.

SOIL EROSION

Soils are just a tiny fraction of all Earth materials, yet they are vital. Soils are necessary for the growth of rooted plants and thus are the very foundation of the human life-support system. Because soil forms slowly, it must be thought of as a finite resource. Just as human ingenuity can increase the agricultural productivity of soils through fertilization and irrigation, soils can be damaged or destroyed by careless activities. Despite their basic role in providing food, fiber, and other basic materials, soils are among our most abused resources.

Perhaps this neglect and indifference has occurred because a substantial amount of soil seems to remain even where soil erosion is serious. Nevertheless, although the loss of fertile topsoil may not be obvious to the untrained eye, it is a growing problem as human activities expand and disturb more and more of Earth's surface.

How Soil is Eroded

Soil erosion is a natural process; it is part of the constant recycling of Earth materials that we call the *rock cycle.* Once soil forms, erosional forces, especially water and wind, move soil components from one place to another. Every time it rains, raindrops strike the land with surprising force (Figure 6.23). Each drop acts like a tiny

TABLE 6.2	Basic Soil Orders
Alfisols	Moderately weathered soils that form under boreal forests or broadleaf deciduous forests, rich in iron and aluminum. Clay particles accumulate in a subsurface layer in response to leaching in moist environments. Fertile, productive soils, because they are neither too wet nor too dry.
Andisols	Young soils in which the parent material is volcanic ash and cinders, deposited by recent volcanic activity.
Aridosols	Soils that develop in dry places; insufficient water to remove soluble minerals, may have an accumulation of calcium carbonate, gypsum, or salt in subsoil; low organic content (see Figure 6.19B).
Entisols	Young soils having limited development and exhibiting properties of the parent material. Productivity ranges from very high for some formed on recent river deposits to very low for those forming on shifting sand or rocky slopes.
Gelisols	Young soils with little profile development that occur in regions with permafrost. Low temperatures and frozen conditions for much of the year; slow soil-forming processes.
Histosols	Organic soils with little or no climatic implications. Can be found in any climate where organic debris can accumulate to form a bog soil. Dark, partially decomposed organic material commonly referred to as *peat.*
Inceptisols	Weakly developed young soils in which the beginning (inception) of profile development is evident. Most common in humid climates, they exist from the Arctic to the tropics. Native vegetation is most often forest.
Mollisols	Dark, soft soils that have developed under grass vegetation, generally found in prairie areas. Humus-rich surface horizon that is rich in calcium and magnesium. Soil fertility is excellent. Also found in hardwood forests with significant earthworm activity. Climatic range is boreal or alpine to tropical. Dry seasons are normal (see Figure 6.20A).
Oxisols	Soils that occur on old land surfaces unless parent materials were strongly weathered before they were deposited. Generally found in the tropics and subtropical regions. Rich in iron and aluminum oxides, oxisols are heavily leached; hence are poor soils for agricultural activity (see Figure 6.20B).
Spodosols	Soils found only in humid regions on sandy material. Common in northern coniferous forests (see Figure 6.19A) and cool humid forests. Beneath the dark upper horizon of weathered organic material lies a light-colored horizon of leached material, the distinctive property of this soil.
Ultisols	Soils that represent the products of long periods of weathering. Water percolating through the soil concentrates clay particles in the lower horizons (argillic horizons). Restricted to humid climates in the temperate regions and the tropics, where the growing season is long. Abundant water and a long frost-free period contribute to extensive leaching, hence poorer soil quality.
Vertisols	Soils containing large amounts of clay, which shrink upon drying and swell with the addition of water. Found in subhumid to arid climates, provided that adequate supplies of water are available to saturate the soil after periods of drought. Soil expansion and contraction exert stresses on human structures.

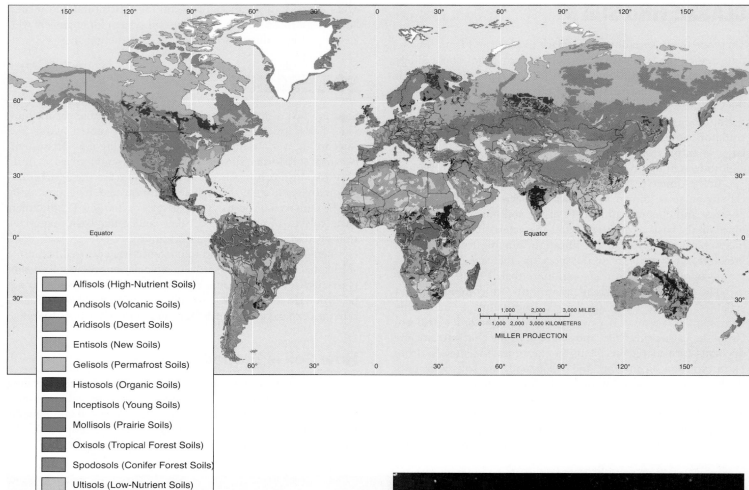

FIGURE 6.22 Global soil regions. Worldwide distribution of the Soil Taxonomy's 12 soil orders. (After U.S. Department of Agriculture, Natural Resources Conservation Service, World Soil Resources Staff)

Alfisols (High-Nutrient Soils)
Andisols (Volcanic Soils)
Aridisols (Desert Soils)
Entisols (New Soils)
Gelisols (Permafrost Soils)
Histosols (Organic Soils)
Inceptisols (Young Soils)
Mollisols (Prairie Soils)
Oxisols (Tropical Forest Soils)
Spodosols (Conifer Forest Soils)
Ultisols (Low-Nutrient Soils)
Vertisols (Swelling Clay Soils)
Rock Land
Shifting Sands
Ice/Glacier

bomb, blasting movable soil particles out of their positions in the soil mass. Then, water flowing across the surface carries away the dislodged soil particles. Because the soil is moved by thin sheets of water, this process is termed *sheet erosion*.

After flowing as a thin, unconfined sheet for a relatively short distance, threads of current typically develop, and tiny channels called *rills* begin to form. Still deeper cuts in the soil, known as *gullies,* are created as rills enlarge (Figure 6.24). When normal farm cultivation cannot eliminate the channels, we know the rills have grown large enough to be called gullies. Although most dislodged soil particles move only a short distance during each rainfall, substantial quantities eventually leave the fields and make their way downslope to a stream. Once in the stream channel, these soil particles, which can now be called *sediment,* are transported downstream and eventually deposited.

FIGURE 6.23 When it is raining, millions of water drops are falling at velocities approaching 10 meters per second (35 kilometers per hour). When water drops strike an exposed surface, soil particles may splash as high as 1 meter into the air and land more than a meter away from the point of raindrop impact. Soil dislodged by splash erosion is more easily moved by sheet erosion. (Photo courtesy of U.S.D.A./Natural Resources Conservation Service)

PEOPLE AND THE ENVIRONMENT

Clearing the Tropical Rain Forest—The Impact on its Soils

BOX 6.3

FIGURE 6.E Clearing the tropical rain forest in West Kalimantan (Borneo), Indonesia. The thick soil is highly leached. (Photo by Wayne Lawler/Photo Researchers, Inc.)

Thick red soils are common in the wet tropics and subtropics. They are the end product of extreme chemical weathering. Because lush tropical rain forests are associated with these soils, we might assume they are fertile and have great potential for agriculture. However, just the opposite is true—they are among the poorest soils for farming. How can this be?

Because rain forest soils develop under conditions of high temperature and heavy rainfall, they are severely leached. Not only does leaching remove the soluble materials such as calcium carbonate but the great quantities of percolating water also remove much of the silica, with the result that insoluble oxides of iron and aluminum become concentrated in the soil. Iron oxides give the soil its distinctive orange color. Because bacterial activity is very high in the tropics, rain forest soils contain practically no humus. Moreover, leaching destroys fertility because most plant nutrients are removed by the large volume of downward-percolating water. Therefore, even though the vegetation may be dense and luxuriant, the soil itself contains few available nutrients.

Most nutrients that support the rain forest are locked up in the trees themselves. As vegetation dies and decomposes, the roots of the rain forest trees quickly absorb the nutrients before they are leached from the soil. The

nutrients are continuously recycled as trees die and decompose.

Therefore, when forests are cleared to provide land for farming or to harvest the timber, most of the nutrients are removed as well (Figure 6.E). What remains is a soil that contains little to nourish planted crops.

The clearing of rain forests not only removes plant nutrients but also accelerates

erosion. When vegetation is present, its roots anchor the soil, and its leaves and branches provide a canopy that protects the ground by deflecting the full force of the frequent heavy rains.

The removal of vegetation also exposes the ground to strong direct sunlight. When baked by the Sun, these tropical soils can harden to a bricklike consistency and become practically impenetrable to water and crop roots. In only a few years, soils in a freshly cleared area may no longer be cultivable.

The term *laterite*, which is often applied to these soils, is derived from the Latin word *latere*, meaning "brick," and was first applied to the use of this material for brickmaking in India and Cambodia. Laborers simply excavated the soil, shaped it, and allowed it to harden in the Sun. Ancient but still well-preserved structures built of laterite remain standing today in the wet tropics (Figure 6.F). Such structures have withstood centuries of weathering because all of the original soluble materials were already removed from the soil by chemical weathering. Laterites are therefore virtually insoluble and very stable.

In summary, we have seen that some rain forest soils are highly leached products of extreme chemical weathering in the warm, wet tropics. Although they may be associated with lush tropical rain forests, these soils are unproductive when vegetation is removed. Moreover, when cleared of plants, these soils are subject to accelerated erosion and can be baked to bricklike hardness by the Sun.

FIGURE 6.F This ancient temple at Angkor Wat, Cambodia, was built of bricks made of laterite. (Photo by R. Ian Lloyd/The Stock Market)

A.

B.

FIGURE 6.24 **A.** Soil erosion from this field in northeastern Wisconsin is obvious. Just 1 millimeter of soil lost from a single acre of land amounts to about five tons. (Photo by D. P. Burnside/Photo Researchers, Inc.) **B.** Gully erosion is severe in this poorly protected soil in southern Colombia. (Photo by Carl Purcell/Photo Researchers, Inc.)

Rates of Erosion

We know that soil erosion is the ultimate fate of practically all soils. In the past, erosion occurred at slower rates than it does today because more of the land surface was covered and protected by trees, shrubs, grasses, and other plants. However, human activities such as farming, logging, and construction, which remove or disrupt the natural vegetation, have greatly accelerated the rate of soil erosion. Without the stabilizing effect of plants, the soil is more easily swept away by the wind or carried downslope by sheet wash.

Natural rates of soil erosion vary greatly from one place to another and depend on soil characteristics as well as such factors as climate, slope, and type of vegetation. Over a broad area, erosion caused by surface runoff may be estimated by determining the sediment loads of the streams that drain the region. When studies of this kind were made on a global scale, they indicated that prior to the appearance of humans, sediment transport by rivers to the ocean amounted to just over 9 billion metric tons per year. By contrast, the amount of material currently transported to the sea by rivers is about 24 billion metric tons per year, or more than two and a half times the earlier rate.

It is estimated that flowing water is responsible for about two-thirds of the soil erosion in the United States. Much of the remainder is caused by wind. When dry conditions prevail, strong winds can remove large quantities of soil from unprotected fields (Figure 6.25). Such was the case in the 1930s in portions of the Great Plains.

During a span of dry years in the 1930s, large dust storms plagued the Great Plains. Because of the size and severity of these storms, the region came to be called the Dust Bowl, and the time period the Dirty Thirties. The heart of the Dust Bowl was nearly 100 million acres in the panhandles of Texas and Oklahoma and adjacent parts of Colorado, New Mexico, and Kansas (Figure 6.26). At times, dust storms were so severe that they were called "black blizzards" and "black rollers" because visibility was sometimes reduced to only a few feet.

What caused the Dust Bowl? Clearly the fact that portions of the Great Plains experience some of North America's strongest winds is important. However, it was the expansion of agriculture during an unusually wet period that set the stage for the disastrous period of soil erosion. Mechanization allowed the rapid transformation of the grass-covered prairies of this semiarid region into farms. As long as precipitation was adequate, the soil remained in place. However, when a prolonged drought struck in the 1930s, the unprotected soils were vulnerable to the wind. The result was severe soil loss, crop failure, and economic hardship.*

In many regions the rate of soil erosion is significantly greater than the rate of soil formation. This means that a renewable

---

*For a readable and engaging account of this time you can read *The Worst Hard Times, The Untold Story of those Who Survived the Great American Dust Bowl*, by Timothy Egan, Houghton Mifflin Co., 2006.

FIGURE 6.25 When the land is dry and largely unprotected by anchoring vegetation, soil erosion by wind can be significant. The man is pointing to where the ground surface was when the grasses began to grow. Later wind erosion of the soil lowered the land surface to the level of his feet. (Photo courtesy of U.S.D.A./Natural Resources Conservation Service)

FIGURE 6.26 A dust storm blackens the Colorado sky in this historic image from the Dust Bowl of the 1930s. (Photo courtesy of U.S.D.A./Natural Resources Conservation Service)

resource has become nonrenewable in these places. At present, it is estimated that topsoil is eroding faster than it forms on more than one-third of the world's croplands. The result is lower productivity, poorer crop quality, reduced agricultural income, and an ominous future.

Sedimentation and Chemical Pollution

Another problem related to excessive soil erosion involves the deposition of sediment. Each year in the United States hundreds of millions of tons of eroded soil are deposited in lakes, reservoirs, and streams. The detrimental impact of this process can be significant. For example, as more and more sediment is deposited in a reservoir, the capacity of the reservoir is diminished, limiting its usefulness for flood control, water supply, and/or hydroelectric power generation. In addition, sedimentation in streams and other waterways can restrict navigation and lead to costly dredging operations.

In some cases soil particles are contaminated with pesticides used in farming. When these chemicals are introduced into a lake or reservoir, the quality of the water supply is threatened and aquatic organisms may be endangered. In addition to pesticides, nutrients found naturally in soils as well as those added by agricultural fertilizers make their way into streams and lakes, where they stimulate the growth of plants. Over a period of time, excessive nutrients accelerate the process by which plant growth leads to the depletion of oxygen and an early death of the lake.

The availability of good soils is critical if the world's rapidly growing population is to be fed. On every continent, unnecessary soil loss is occurring because appropriate conservation measures are not being used. Although it is a recognized fact that soil erosion can never be completely eliminated, soil conservation programs can substantially reduce the loss of this basic resource (Figure 6.27). Windbreaks (rows of trees), terracing, and plowing along the contours of hills are some of the effective measures, as are special tillage practices and crop rotation.

FIGURE 6.27 These images illustrate some of the ways that soil erosion can be reduced. **A.** In this scene from a farm in northeastern Iowa, corn (tan) and hay (green) have been planted in strips that follow the contours of the hillsides. This pattern of crop planting can reduce soil loss due to water erosion by significantly slowing the rate at which water runs off. **B.** Windbreaks protecting wheat fields in North Dakota. These flat expanses are susceptible to wind erosion especially when the fields are bare. The rows of trees slow the wind and deflect it upward, which decreases the loss of fine soil particles. (Photos courtesy of U.S.D.A./Natural Resources Conservation Service)

A.

B.

CHAPTER 6 WEATHERING AND SOIL IN REVIEW

- External processes include (1) *weathering*—the disintegration and decomposition of rock at or near Earth's surface; (2) *mass wasting*—the transfer of rock material downslope under the influence of gravity; and (3) *erosion*—the removal of material by a mobile agent, usually water, wind, or ice. They are called *external processes* because they occur at or near Earth's surface and are powered by energy from the Sun. By contrast, *internal processes,* such as volcanism and mountain building, derive their energy from Earth's interior.

- *Mechanical weathering* is the physical breaking up of rock into smaller pieces. Rocks can be broken into smaller fragments by *frost wedging* (where water works its way into cracks or voids in rock and, upon freezing, expands and enlarges the openings), *salt crystal growth, sheeting* (expansion and breaking due to a great reduction in pressure when the overlying rock is eroded away), *thermal expansion* (weakening of rock as the result of expansion and contraction as it heats and cools), and *biological activity* (by humans, burrowing animals, plant roots, etc.).

- *Chemical weathering* alters a rock's chemistry, changing it into different substances. Water is by far the most important agent of chemical weathering. *Dissolution* occurs when water-soluble minerals such as halite become dissolved in water. Oxygen dissolved in water will *oxidize* iron-rich minerals. When carbon dioxide (CO_2) is dissolved in water, it forms *carbonic acid,* which accelerates the decomposition of silicate minerals by *hydrolysis.* The chemical weathering of silicate minerals frequently produces (1) soluble products containing sodium, calcium, potassium, and magnesium ions, and silica in solution; (2) insoluble iron oxides; and (3) clay minerals.

- The rate at which rock weathers depends on such factors as (1) *particle size*—small pieces generally weather faster than large pieces; (2) *mineral makeup*—calcite readily dissolves in mildly acidic solutions, and silicate minerals that form first from magma are least resistant to chemical weathering; and (3) *climatic factors,* particularly temperature and moisture.

Frequently, rocks exposed at Earth's surface do not weather at the same rate. This *differential weathering* of rocks is influenced by such factors as mineral makeup and degree of jointing.

- *Soil* is a combination of mineral and organic matter, water, and air—the portion of the *regolith* (the layer of rock and mineral fragments produced by weathering) that supports the growth of plants. About half of the total volume of a good-quality soil is a mixture of disintegrated and decomposed rock (mineral matter) and *humus* (the decayed remains of animal and plant life); the remaining half consists of pore spaces, where air and water circulate. The most important factors that control soil formation are *parent material, time, climate, plants and animals,* and *slope.*

- Soil-forming processes operate from the surface downward and produce zones or layers in the soil that are called *horizons.* From the surface downward, the soil horizons are respectively designated as *O* (largely organic matter), *A* (largely mineral matter), *E* (where the fine soil components and soluble materials have been removed by *eluviation* and *leaching*), *B* (*subsoil,* often referred to as the *zone of accumulation*), and *C* (partially altered parent material). Together the *O* and *A* horizons make up what is commonly called the *topsoil.*

- In the United States, soils are classified using a system known as the *Soil Taxonomy.* It is based on physical and chemical properties of the soil profile and includes six hierarchical categories. The system is especially useful for agricultural and related land-use purposes.

- Soil erosion is a natural process; it is part of the constant recycling of Earth materials that we call the rock cycle. Once in a stream channel, soil particles are transported downstream and eventually deposited. *Rates of soil erosion* vary from one place to another and depend on the soil's characteristics as well as such factors as climate, slope, and type of vegetation. Human activities which remove or disrupt natural vegetation, greatly accelerate the rate of soil erosion.

KEY TERMS

chemical weathering (p. 175)
differential weathering (p. 186)
dissolution (p. 179)
eluviation (p. 189)
erosion (p. 174)
exfoliation dome (p. 178)

external process (p. 174)
frost wedging (p. 176)
horizon (p. 189)
humus (p. 187)
hydrolysis (p. 181)
internal process (p. 174)
leaching (p. 190)

mass wasting (p. 174)
mechanical weathering (p. 175)
oxidation (p. 181)
parent material (p. 187)
regolith (p. 187)
sheeting (p. 176)

soil (p. 187)
soil profile (p. 189)
Soil Taxonomy (p. 191)
solum (p. 190)
spheroidal weathering (p. 184)
weathering (p. 174)

QUESTIONS FOR REVIEW

1. Describe the role of external processes in the rock cycle.

2. If two identical rocks were weathered, one mechanically and the other chemically, how would the products of weathering for the two rocks differ?

3. Describe the formation of an exfoliation dome. Give an example of such a feature.

4. How does mechanical weathering add to the effectiveness of chemical weathering?

5. Granite and basalt are exposed at the surface in a hot, wet region.

   a. Which type of weathering will predominate?

   b. Which of these rocks will weather most rapidly? Why?

6. Heat speeds up a chemical reaction. Why then does chemical weathering proceed slowly in a hot desert?

7. How is carbonic acid (H_2CO_3) formed in nature? What results when this acid reacts with potassium feldspar?

8. List some possible environmental effects of acid precipitation (see Box 6.2).

9. What is the difference between soil and regolith?

10. What factors might cause different soils to develop from the same parent material, or similar soils to form from different parent materials?

11. Which of the controls of soil formation is most important? Explain.

12. How can topography influence the development of soil? What is meant by the term *slope orientation*?

13. List the characteristics associated with each of the horizons in a well-developed soil profile. Which of the horizons constitute the solum? Under what circumstances do soils lack horizons?

14. The tropical soils described in Box 6.3 support luxuriant rain forests yet are considered to have low fertility. Explain.

15. List three detrimental effects of soil erosion other than the loss of topsoil from croplands.

16. Briefly describe the conditions that led to the Dust Bowl of the 1930s.

COMPANION WEBSITE

The *Earth 10e* website uses the resources and flexibility of the Internet to aid in your study of the topics in this chapter. Written and developed by the authors and other geology instructors, this site will help improve your understanding of geology. Visit **www.mygeoscienceplace.com** in order to:

- **Review** key chapter concepts

- **Read** with links to the eBook and to chapter-specific web resources

- **Visualize** and comprehend challenging topics using learning activities in *GEODe Earth*

- **Test** yourself with online quizzes

GEODe EARTH

GEODe Earth is a valuable and easy-to-use learning aid that can be accessed from your book's Companion Website (**www.mygeoscienceplace.com**). It is a dynamic instructional tool that promotes understanding and reinforces important concepts by using tutorials, animations, and exercises that actively engage the student.

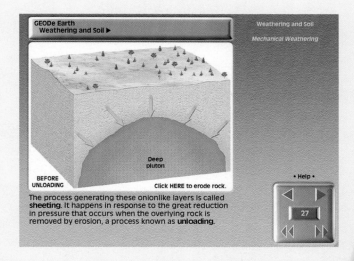

SEDIMENTARY ROCKS

This view of Antelope Canyon near Page, Arizona, has the appearance of an abstract sculpture. The canyon is carved into a rock formation called the Navajo Sandstone. The grains of sand composing this rock were deposited by the wind as dunes more than 150 million years ago.

(PHOTO BY MICHAEL COLLIER)

The preceding

chapter provided

the background you

[ne]eded to understand the

[ori]gin of sedimentary rocks.

[Re]call that weathering of exist-

[ing] rocks begins the process.

[Ne]xt, gravity and agents of ero-

[sio]n such as running water, wind,

[an]d glacial ice, remove the products

[of] weathering and carry them to a

[ne]w location where they are deposited.

[Us]ually the particles are broken down

[fur]ther during this transport phase.

[Fol]lowing deposition, this material,

[wh]ich is now called sediment, becomes

[lith]ified (turned to rock).

THE IMPORTANCE OF SEDIMENTARY ROCKS

Most of the solid Earth consists of igneous and metamorphic rocks. Geologists estimate these two categories represent 90 to 95 percent of the outer 16 kilometers (10 miles) of the crust. Nevertheless, most of Earth's solid surface consists of either sediment or sedimentary rock! About 75 percent of land areas are covered by sediments and sedimentary rocks. Across the ocean floor, which represents about 70 percent of Earth's solid surface, virtually everything is covered by sediment. Igneous rocks are exposed only at the crest of mid-ocean ridges and at some volcanic areas. Thus, while sediment and sedimentary rocks make up only a small percentage of Earth's crust, they are concentrated at or near the surface—the interface among the geosphere, hydrosphere, atmosphere, and biosphere. Because of this unique position, sediments and the rock layers that they eventually form contain evidence of past conditions and events at the surface. Furthermore, it is sedimentary rocks that contain fossils, which are vital tools in the study of the geologic past. This group of rocks provides geologists with much of the basic information they need to reconstruct the details of Earth history (Figure 7.1).

Such study is not only of interest for its own sake but has practical value as well. Coal, which provides a significant portion of our electrical energy, is classified as a sedimentary rock. Moreover, other major energy sources—oil, natural gas, and uranium—are derived from sedimentary rocks. So are major sources of iron, aluminum, manganese, and phosphate fertilizer, plus numerous materials essential to the construction industry such as cement and aggregate. Sediments and sedimentary rocks are also the primary reservoir of groundwater. Thus, an understanding of this group of rocks and the processes that form and modify them is basic to locating additional supplies of many important resources.

ORIGINS OF SEDIMENTARY ROCK

SEDIMENTARY ROCKS
▶ Introduction

Figure 7.2 illustrates the portion of the rock cycle that occurs near Earth's surface—the part that pertains to sediments and sedimentary rocks. A brief overview of these processes provides a useful perspective:

- Weathering begins the process. It involves the physical disintegration and chemical decomposition of pre-existing igneous, metamorphic, and sedimentary rocks. Weathering generates a variety of products, including various solid particles and ions in solution. These are the raw materials for sedimentary rocks.
  - Soluble constituents are carried away by runoff and

groundwater. Solid particles are frequently moved downslope by gravity, a process termed mass wasting, before running water, groundwater, wind, and glacial ice remove them. Transportation moves these materials from the sites where they originated to locations where they accumulate. The transport of sediment is usually intermittent. For example, during a flood, a rapidly moving river moves large quantities of sand and gravel. As the flood waters recede, particles are temporarily deposited, only to be moved again by a subsequent flood.

- Deposition of solid particles occurs when wind and water currents slow down and as glacial ice melts. The word *sedimentary* actually refers to this process. It is derived from the Latin *sedimentum*, which means "to settle," a ref-

erence to solid material settling out of a fluid (water or air). The mud on the floor of a lake, a delta at the mouth of a river, a gravel bar in a stream bed, the particles in a desert sand dune, and even household dust are examples.

- The deposition of material dissolved in water is not related to the strength of wind or water currents. Rather, ions in solution are removed when chemical or temperature changes cause material to crystallize and precipitate or when organisms remove dissolved material to build shells.
- As deposition continues, older sediments are buried beneath younger layers and gradually converted to sedimentary rock (lithified) by compaction and cementation. This and other changes are referred to as *diagenesis* (*dia* = change; *genesis* = origin), a collective term for

FIGURE 7.1 Sedimentary rocks are exposed at the surface more than igneous and metamorphic rocks. Because they contain fossils and other clues about the geologic past, sedimentary rocks are important in the study of Earth history. Vertical changes in rock types represent environmental changes through time. Capital Reef National Park, Utah. (Photo by Scott T. Smith/CORBIS)

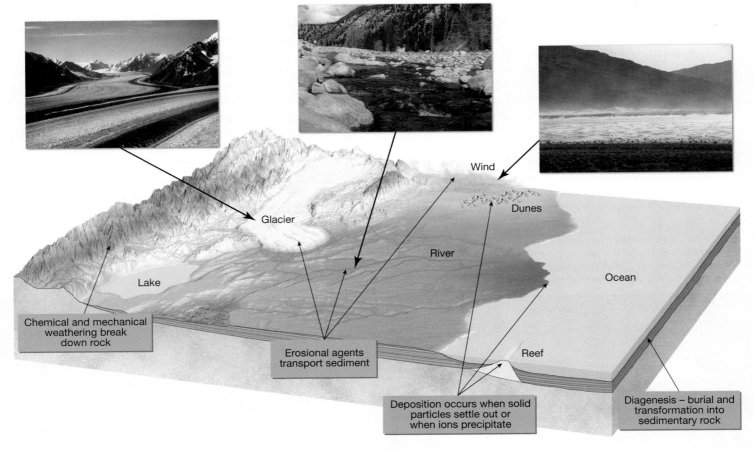

FIGURE 7.2 This diagram outlines the portion of the rock cycle that pertains to the formation of sedimentary rocks. Weathering, transportation, deposition, and diagenesis represent the basic process involved. (Photos by E. J. Tarbuck)

all of the changes (short of metamorphism) that take place in texture, composition, and other physical properties after sediments are deposited.

Because there are a variety of ways that the products of weathering are transported, deposited, and transformed into solid rock, three categories of sedimentary rocks are recognized. As the overview reminded us, sediment has two principal sources. First, it may be an accumulation of material that originates and is transported as solid particles derived from both mechanical and chemical weathering. Deposits of this type are termed *detrital,* and the sedimentary rocks that they form are called **detrital sedimentary rocks.**

The second major source of sediment is soluble material produced largely by chemical weathering. When these ions in solution are precipitated by either inorganic or biologic processes, the material is known as chemical sediment, and the rocks formed from it are called **chemical sedimentary rocks.**

The third category is **organic sedimentary rocks.** The primary example is coal. This black combustible rock consists of organic carbon from the remains of plants that died and accumulated on the floor of a swamp. The bits and pieces of undecayed plant material that constitute the "sediments" in coal are quite unlike the weathering products that make up detrital and chemical sedimentary rocks.

DETRITAL SEDIMENTARY ROCKS

SEDIMENTARY ROCKS
▶ Types of Sedimentary Rocks

Though a wide variety of minerals and rock fragments (*clasts*) may be found in detrital rocks, clay minerals and quartz are the chief constituents of most sedimentary rocks in this category. Recall from Chapter 6 that clay minerals are the most abundant product of the chemical weathering of silicate minerals, especially the feldspars. Clays are fine-grained minerals with sheetlike crystalline structures similar to the micas. The other common mineral, quartz, is abundant because it is extremely durable and very resistant to chemical weathering. Thus, when igneous rocks such as granite are attacked by weathering processes, individual quartz grains are freed.

Other common minerals in detrital rocks are feldspars and micas. Because chemical weathering rapidly transforms these minerals into new substances, their presence in sedimentary rocks indicates that erosion and deposition were fast enough to preserve some of the primary minerals from the source rock before they could be decomposed.

Particle size is the primary basis for distinguishing among various detrital sedimentary rocks. Table 7.1 presents the size categories for particles making up detrital rocks. Particle size is not

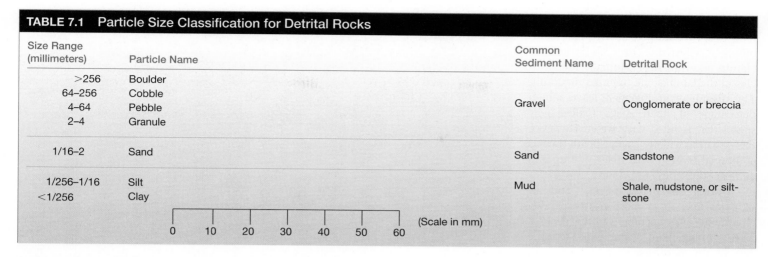

TABLE 7.1   Particle Size Classification for Detrital Rocks

Size Range (millimeters)	Particle Name	Common Sediment Name	Detrital Rock
>256	Boulder		
64–256	Cobble	Gravel	Conglomerate or breccia
4–64	Pebble		
2–4	Granule		
1/16–2	Sand	Sand	Sandstone
1/256–1/16	Silt	Mud	Shale, mudstone, or silt-stone
<1/256	Clay		

(Scale in mm) 0 10 20 30 40 50 60

only a convenient method of dividing detrital rocks, the sizes of the component grains also provide useful information about environments of deposition. Currents of water or air sort the particles by size; the stronger the current, the larger the particle size carried. Gravels, for example, are moved by swiftly flowing rivers as well as by landslides and glaciers. Less energy is required to transport sand; thus, it is common to such features as windblown dunes and some river deposits and beaches. Very little energy is needed to transport clay, so it settles very slowly. Accumulation of these tiny particles is generally associated with the quiet water of a lake, lagoon, swamp, or certain marine environments.

Common detrital sedimentary rocks, in order of increasing particle size, are shale, sandstone, and conglomerate or breccia. We will now look at each type and how it forms.

Shale

Shale is a sedimentary rock consisting of silt- and clay-size particles (Figure 7.3). These fine-grained detrital rocks account for well over half of all sedimentary rocks. The particles in these rocks are so small that they cannot be readily identified without great magnification and for this reason make shale more difficult to study and analyze than most other sedimentary rocks.

Much of what can be learned is based on particle size. The tiny grains in shale indicate that deposition occurs as the result

of gradual settling from relatively quiet, nonturbulent currents. Such environments include lakes, river floodplains, lagoons, and portions of the deep-ocean basins. Even in these "quiet" environments, there is usually enough turbulence to keep clay-size particles suspended almost indefinitely. Consequently, much of the clay is deposited only after the individual particles coalesce to form larger aggregates.

Sometimes the chemical composition of the rock provides additional information. One example is black shale, which is black because it contains abundant organic matter (carbon). When such a rock is found, it strongly implies that deposition occurred in an oxygen-poor environment such as a swamp, where organic materials do not readily oxidize and decay.

Students Sometimes Ask . . .

According to Table 7.1, clay is a term used to indicate a microscopic particle size. I thought clays were a group of platy silicate minerals. Which one is right?

Both are. In the context of detrital particle size, the term *clay* refers only to those grains less than 1/256 millimeter, thus being microscopic in size. It does not indicate that these particles are of a particular composition. However, the term *clay* is also used to denote a specific composition: namely, a group of silicate minerals related to the micas. Although most of these clay minerals are of clay size, not all clay-size sediment consists of clay minerals!

FIGURE 7.3   Shale is a fine-grained detrital rock that is by far the most abundant of all sedimentary rocks. Dark shales containing plant remains are relatively common. (Photo courtesy of E. J. Tarbuck)

As silt and clay accumulate, they tend to form thin layers, which are commonly referred to as *laminae* (lamin = a thinsheet). Initially the particles in the laminae are oriented randomly. This disordered arrangement leaves a high percentage of open space (called *pore space*) that is filled with water. However, this situation usually changes with time as additional layers of sediment pile up and compact the sediment below.

During this phase the clay and silt particles take on a more nearly parallel alignment and become tightly packed. This rearrangement of grains reduces the size of the pore spaces and forces out much of the water. Once the grains are pressed closely together, the tiny spaces between particles do not readily permit solutions containing cementing material to circulate. Therefore, shales are often described as being weak because they are poorly cemented and therefore not well lithified.

The inability of water to penetrate its microscopic pore spaces explains why shale often forms barriers to the subsurface movement of water and petroleum. Indeed, rock layers that contain groundwater are commonly underlain by shale beds that block further downward movement. The opposite is true for underground reservoirs of petroleum. They are often capped by shale beds that effectively prevent oil and gas from escaping to the surface.*

It is common to apply the term *shale* to all fine-grained sedimentary rocks, especially in a nontechnical context. However, be aware that there is a more restricted use of the term. In this narrower usage, shale must exhibit the ability to split into thin layers along well-developed, closely spaced planes. This property is termed **fissility** (*fissilis* = that which can be cleft or split). If the rock breaks into chunks or blocks, the name *mudstone* is applied. Another fine-grained sedimentary rock that, like mudstone, is often grouped with shale but lacks fissility is *siltstone*. As its name implies, siltstone is composed largely of silt-size particles and contains less clay-size material than shale and mudstone.

Although shale is far more common than other sedimentary rocks, it does not usually attract as much notice as other less abundant members of this group. The reason is that shale does not form prominent outcrops as sandstone and limestone often do. Rather, shale crumbles easily and usually forms a cover of soil that hides the unweathered rock below. This is illustrated nicely in the Grand Canyon, where the gentler slopes of weathered shale are quite inconspicuous and overgrown with vegetation, in sharp contrast with the bold cliffs produced by more durable rocks (Figure 7.4).

Although shale beds may not form striking cliffs and prominent outcrops, some deposits have economic value. Certain shales are quarried to obtain raw material for pottery, brick, tile, and china. Moreover, when mixed with limestone, shale is used

FIGURE 7.4   Sedimentary rock layers exposed in the walls of the Grand Canyon, Arizona. Beds of resistant sandstone and limestone produce bold cliffs. By contrast, weaker, poorly cemented shale crumbles and produces a gentler slope of weathered debris in which some vegetation is growing. (Photo by Pixtal/age fotostock)

to make portland cement. In the future, one type of shale, called oil shale, may become a valuable energy resource. This possibility will be explored in Chapter 23.

Sandstone

Sandstone is the name given rocks in which sand-size grains predominate (Figure 7.5). After shale, sandstone is the most abundant sedimentary rock, accounting for approximately 20 percent of the entire group. Sandstones form in a variety of environments and often contain significant clues about their origin, including sorting, particle shape, and composition.

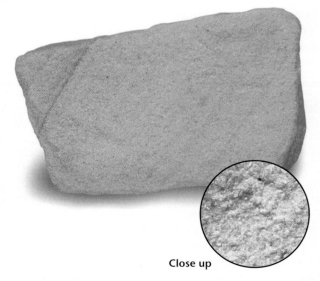

Close up

FIGURE 7.5   Quartz sandstone. After shale, sandstone is the most abundant sedimentary rock. (Photos by E. J. Tarbuck)

---

*The relationship between impermeable beds and the occurrence and movement of groundwater is examined in Chapter 17. Shale beds as cap rocks in oil traps are discussed in Chapter 23.

A. Sorting

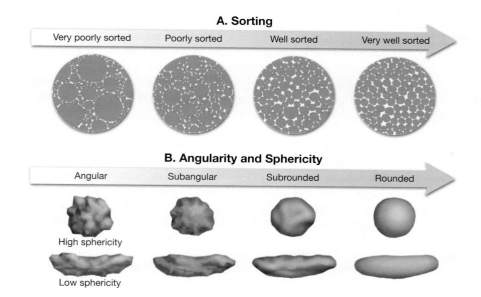

Very poorly sorted   Poorly sorted   Well sorted   Very well sorted

B. Angularity and Sphericity

Angular   Subangular   Subrounded   Rounded

High sphericity

Low sphericity

FIGURE 7.6 **A.** Detrital rocks commonly have a variety of different size clasts. *Sorting* refers to the range of sizes present. Rocks with clasts that are nearly all the same size are considered "well-sorted." When sediments are "very poorly sorted," there is a wide range of different sizes. When a rock contains larger clasts surrounded by much smaller ones, the mass of smaller clasts is often referred to as the *matrix.* **B.** Geologists describe a particle's shape in terms of its *angularity* (degree to which the clast's edges and corners are rounded) and *sphericity* (how close the shape of the clast is to a sphere). Transportation reduces the size and angularity of clasts but does not change their general shape.

Sorting and Particle Shape   **Sorting** is the degree of similarity in particle size in a sedimentary rock. For example, if all the grains in a sample of sandstone are about the same size, the sand is considered *well-sorted.* Conversely, if the rock contains mixed large and small particles, the sand is said to be *poorly sorted* (Figure 7.6A).

By studying the degree of sorting, we can learn much about the depositing current. Deposits of wind-blown sand are usually better sorted than deposits sorted by wave activity (Figure 7.7). Particles washed by waves are commonly better sorted than materials deposited by streams. Sediment accumulations that exhibit poor

A.

B.

FIGURE 7.7   **A.** The orange and yellow cliffs of Utah's Zion National Park expose thousands of feet of Navajo Sandstone. The vast sand sea from which the sandstone hardened is estimated to have once covered up to 400,000 square kilometers (156,000 square miles), an area the size of California. (Photo by Dennis Tasa) **B.** The quartz grains composing the Navajo Sandstone were deposited by wind as dunes similar to these in Colorado's Great Sand Dunes National Park. These sand grains are considered to be well-sorted because they are all practically the same size. (Photo by David Muench/David Muench Photography, Inc.)

Students Sometimes Ask . . .

Why are many of the sedimentary rocks pictured in this chapter so colorful?

In the western and southwestern United States, steep cliffs and canyon walls made of sedimentary rocks often exhibit a brilliant display of different colors. (For example, see the chapter-opening photo and Figures 7.1, 7.4, 7.7A, 7.13B, and 7.16.) In the walls of Arizona's Grand Canyon we can see layers that are red, orange, purple, gray, brown, and buff. Some of the sedimentary rocks in Utah's Bryce Canyon are a delicate pink color (see Figure 6.16). Sedimentary rocks in more humid places are also colorful, but they are usually covered by soil and vegetation.

The most important "pigments" are iron oxides, and only very small amounts are needed to color a rock. Hematite tints rocks red or pink, whereas limonite produces shades of yellow and brown. When sedimentary rocks contain organic matter, it often colors them black or gray (see Figure 7.3).

sorting usually result when particles are transported for only a relatively short time and then rapidly deposited. For example, when a turbulent stream reaches the gentler slopes at the base of a steep mountain, its velocity is quickly reduced, and poorly sorted sands and gravels are deposited.

The shapes of sand grains can also help decipher the history of a sandstone (Figure 7.6B). When streams, winds, or waves move sand and other larger sedimentary particles, the grains lose their sharp edges and corners and become more rounded as they collide with other particles during transport. Thus, rounded grains likely have been airborne or waterborne. Further, the degree of rounding indicates the distance or time involved in the transportation of sediment by currents of air or water. Highly rounded grains indicate that a great deal of abrasion and hence a great deal of transport has occurred.

Very angular grains, on the other hand, imply two things: that the materials were transported only a short distance before they were deposited, and that some other medium may have transported them. For example, when glaciers move sediment, the particles are usually made more irregular by the crushing and grinding action of the ice.

In addition to affecting the degree of rounding and the amount of sorting that particles undergo, the length of transport by turbulent air and water currents also influences the mineral composition of a sedimentary deposit. Substantial weathering and long transport lead to the gradual destruction of weaker and less stable minerals, including the feldspars and ferromagnesians. Because quartz is very durable, it is usually the mineral that survives the long trip in a turbulent environment.

The preceding discussion has shown that the origin and history of sandstone can often be deduced by examining the sorting, roundness, and mineral composition of its constituent grains. Knowing this information allows us to infer that a well-sorted, quartz-rich sandstone consisting of highly rounded grains must be the result of a great deal of transport. Such a

rock, in fact, may represent several cycles of weathering, transport, and deposition. We may also conclude that a sandstone containing significant amounts of feldspar and angular grains of ferromagnesian minerals underwent little chemical weathering and transport and was probably deposited close to the source area of the particles.

Composition   Owing to its durability, quartz is the predominant mineral in most sandstones. When this is the case, the rock may simply be called *quartz sandstone*. When a sandstone contains appreciable quantities of feldspar (25 percent or more), the rock is called *arkose*. In addition to feldspar, arkose usually contains quartz and sparkling bits of mica. The mineral composition of arkose indicates that the grains were derived from granitic source rocks. The particles are generally poorly sorted and angular, which suggests short-distance transport, minimal chemical weathering in a relatively dry climate, and rapid deposition and burial.

A third variety of sandstone is known as *graywacke*. Along with quartz and feldspar, this dark-colored rock contains abundant rock fragments and matrix. More than 15 percent of graywacke's volume is matrix. The poor sorting and angular grains characteristic of graywacke suggest that the particles were transported only a relatively short distance from their source area and then rapidly deposited. Before the sediment could be reworked and sorted further, it was buried by additional layers of material. Graywacke is frequently associated with submarine deposits made by dense sediment-choked torrents called turbidity currents.

Conglomerate and Breccia

Conglomerate consists largely of gravels (Figure 7.8). As Table 7.1 indicates, these particles can range in size from large boulders to particles as small as garden peas (Figure 7.9). The particles are commonly large enough to be identified as distinctive rock types; thus, they can be valuable in identifying the source areas of sediments. More often than not, conglomerates are poorly sorted

FIGURE 7.8   Conglomerate is composed primarily of rounded gravel-size particles. (Photo by E. J. Tarbuck)

FIGURE 7.9   Gravel deposits along Carbon Creek in Grand Canyon National Park, Arizona. If these poorly-sorted sediments were lithified, the rock would be conglomerate. (Photo by Michael Collier)

because the opening between the large gravel particles contains sand or mud.

Gravels accumulate in a variety of environments and usually indicate the existence of steep slopes or very turbulent currents. The coarse particles in a conglomerate may reflect the action of energetic mountain streams or result from strong wave activity along a rapidly eroding coast. Some glacial and landslide deposits also contain plentiful gravel.

If the large particles are angular rather than rounded, the rock is called *breccia* (Figure 7.10). Because large particles abrade and become rounded very rapidly during transport, the pebbles and cobbles in a breccia indicate that they did not travel far from their source area before they were deposited. Thus, as with many sedimentary rocks, conglomerates and breccias contain clues to their history. Their particle sizes reveal the strength of the currents that transported them, whereas the degree of rounding indicates how far the particles

FIGURE 7.10   When the gravel-size particles in a detrital rock are angular, the rock is called breccia. (Photo by E. J. Tarbuck)

traveled. The fragments within a sample identify the source rocks that supplied them.

CHEMICAL SEDIMENTARY ROCKS

SEDIMENTARY ROCKS
▶ Types of Sedimentary Rocks

In contrast to detrital rocks, which form from the solid products of weathering, chemical sediments derive from ions that are carried *in solution* to lakes and seas. This material does not remain dissolved in the water indefinitely, however. Some of it precipitates to form chemical sediments. These become rocks such as limestone, chert, and rock salt.

This precipitation of material occurs in two ways. *Inorganic* (*in* = not, *organicus* = life) processes such as evaporation and chemical activity can produce chemical sediments. *Organic* (life) processes of water-dwelling organisms also form chemical sediments, said to be of **biochemical** origin.

One example of a deposit resulting from inorganic chemical processes is the dripstone that decorates many caves (Figure 7.11).

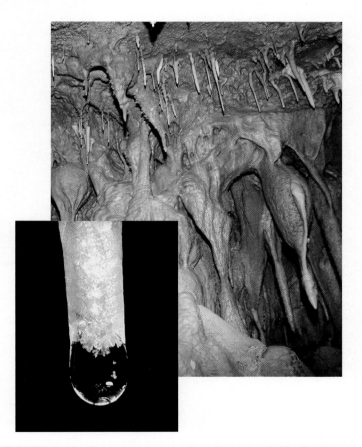

FIGURE 7.11   Because many cave deposits are created by the seemingly endless dripping of water over long time spans, they are commonly called *dripstone*. The material being deposited is calcium carbonate ($CaCO_3$) and the rock is a form of limestone called *travertine*. The calcium carbonate is precipitated when some dissolved carbon dioxide escapes from a water drop. (Photo by Guillen Photography; Inset photo by Clifford Stroud/Wind Cave National Park)

Close up

FIGURE 7.12   This rock, a variety of limestone called coquina, consists of shell fragments; therefore, it has a biochemical origin. (Photo by E. J. Tarbuck)

Carbonate Reefs   Corals are one important example of organisms that are capable of creating large quantities of marine limestone (see Figure 1.13 p. 17). These relatively simple invertebrate animals secrete a calcareous (calcium carbonate) external skeleton. Although they are small, corals are capable of creating massive structures called *reefs* (Figure 7.13A). Reefs consist of coral colonies made up of great numbers of individuals that live side by side on a calcite structure secreted by the animals. In addition, calcium carbonate-secreting algae live with the corals and help cement the entire structure into a solid mass. A wide variety of other organisms also live in and near the reefs (Box 7.2).

Certainly the best-known modern reef is Australia's Great Barrier Reef, 2000 kilometers (1240 miles) long, but many lesser reefs also exist. They develop in the shallow, warm waters of the tropics and subtropics equatorward of about 30° latitude. Striking examples exist in the Bahamas, Hawaii, and the Florida Keys.

Another is the salt left behind as a body of seawater evaporates. In contrast, many water-dwelling animals and plants extract dissolved mineral matter to form shells and other hard parts. After the organisms die, their skeletons collect by the millions on the floor of a lake or ocean as biochemical sediment (Figure 7.12).

Limestone

Representing about 10 percent of the total volume of all sedimentary rocks, *limestone* is the most abundant chemical sedimentary rock. It is composed chiefly of the mineral calcite ($CaCO_3$) and forms either by inorganic means or as the result of biochemical processes (see Box 7.1). Regardless of its origin, the mineral composition of all limestone is similar, yet many different types exist. This is true because limestones are produced under a variety of conditions. Those forms having a marine bio-chemical origin are by far the most common.

A.

B.

FIGURE 7.13   **A.** Australia's Great Barrier Reef is the largest reef system in the world. Located off the coast of Queensland, it extends for 2600 kilometers (1600 miles) and consists of more than 2900 individual reefs. (Photo by Ingo Schulz/Photolibrary) **B.** This image in Guadalupe Mountains National Park, Texas, shows just a small portion of a huge Permian-age reef complex that once formed a 600-kilometer (nearly 375-mile) loop around the margin of the Delaware Basin. The reef consisted of a rich community of sponges, bryozoans, crinoids, gastropods, calcareous algae, and rare corals. (Photo by Michael Collier)

Modern corals were not the first reef builders. Earth's first reef-building organisms were photosynthesizing bacteria living during Precambrian time more than 2 billion years ago. From fossil remains, it is known that a variety of organisms have constructed reefs, including bivalves (clams and oysters), bryozoans (coral-like animals), and sponges. Corals have been found in fossil reefs as old as 500 million years, but corals similar to the modern colonial varieties have constructed reefs only during the last 60 million years.

In the United States, reefs of Silurian age (416 to 444 million years ago) are prominent features in Wisconsin, Illinois, and Indiana. In west Texas and adjacent southeastern New Mexico, a massive reef complex that formed during the Permian period (251 to 299 million years ago) is strikingly exposed in Guadalupe Mountains National Park (see Figure 7.13B)

Coquina and Chalk   Although much limestone is the product of biological processes, this origin is not always evident, because shells and skeletons may undergo considerable change before becoming lithified into rock. However, one easily identified biochemical limestone is *coquina,* a coarse rock composed of poorly cemented shells and shell fragments (see Figure 7.12). Another less obvious but nevertheless familiar example

Students Sometimes Ask . . .

Are diatoms an ingredient in diatomaceous earth, which is used in swimming-pool filters?

Not only are diatoms used in swimming-pool filters, they are also used in a variety of everyday products, including toothpaste (yes, you're brushing your teeth with the remains of dead microscopic organisms!). Diatoms secrete walls of silica in a great variety of forms that accumulate as sediments in enormous quantities. Because it is lightweight, chemically stable, has high surface area, and is highly absorbent, diatomaceous earth has many practical uses. The main uses of diatoms include filters (for refining sugar, straining yeast from beer, and filtering swimming-pool water); mild abrasives (in household cleaning and polishing compounds and facial scrubs); and absorbents (for chemical spills).

is *chalk,* a soft, porous rock made up almost entirely of the hard parts of microscopic marine organisms. Among the most famous chalk deposits are those exposed along the southeast coast of England (Figure 7.14).

FIGURE 7.14   The White Chalk Cliffs, Sussex, England. This prominent deposit underlies large portions of southern England as well as parts of northern France. (Photo by Art Wolfe)

Inorganic Limestones   Limestones having an inorganic origin form when chemical changes or high water temperatures increase the concentration of calcium carbonate to the point that it precipitates. *Travertine,* the type of limestone commonly seen in caves, is an example (see Figure 7.11). When travertine is deposited in caves, groundwater is the source of the calcium carbonate. As water droplets become exposed to the air in a cavern, some of the carbon dioxide dissolved in the water escapes, causing calcium carbonate to precipitate.

Another variety of inorganic limestone is *oolitic limestone.* It is a rock composed of small spherical grains called *ooids.* Ooids form in shallow marine waters as tiny "seed" particles (commonly small shell fragments) are moved back and forth by currents. As the grains are rolled about in the warm water, which is supersaturated with calcium carbonate, they become coated with layer upon layer of the chemical precipitate (Figure 7.15).

Dolostone

Closely related to limestone is *dolostone,* a rock composed of the calcium-magnesium carbonate mineral dolomite [$CaMg(CO_3)_2$]. Although dolostone and limestone sometimes closely resemble one another, they can be easily distinguished

FIGURE 7.15   Oolitic limestone consists of *ooids,* which are small spherical grains formed by the chemical precipitation of calcium carbonate around a tiny nucleus. The carbonate is added in concentric layers as the spheres are rolled back and forth by currents in a warm, shallow marine setting. (Photos by Marli Miller)

by observing their reaction to dilute hydrochloric acid. When a drop of acid is placed on limestone, the reaction (fizzing) is obvious. However, unless dolostone is powdered, it will not visibly react to the acid.

The origin of dolostone is not altogether clear and remains a subject of discussion among geologists. No marine organisms produce hard parts of dolomite and the chemical precipitation of dolomite from seawater occurs only under conditions of unusual water chemistry in certain nearshore sites. Yet dolostone is abundant in many ancient sedimentary rock successions.

It appears that significant quantities of dolostone are produced when magnesium-rich waters circulate through limestone and convert calcite to dolomite by the replacement of some calcium ions with magnesium ions (a process called *dolomitization*). However, other dolostones lack evidence that they formed by such a process and their origin remains uncertain.

Chert

Chert is a name used for a number of very compact and hard rocks made of microcrystalline quartz (SiO_2). One well-known form is *flint,* whose dark color results from the organic matter it contains. *Jasper,* a red variety, gets its bright color from the iron oxide it contains. The banded form is usually referred to as *agate* (Figure 7.16). Like glass, most chert has a conchoidal fracture. Its hardness, ease of chipping, and ability to hold a sharp edge made chert a favorite of Native Americans for fashioning "points" for spears and arrows. Because of chert's durability and extensive use, "arrowheads" are found in many parts of North America.

Chert deposits are commonly found in one of two situations: as layered deposits referred to as *bedded cherts* and as *nodules,* somewhat spherical masses varying in diameter from a few millimeters (pea size) to a few centimeters. Most water-dwelling organisms that produce hard parts make them of calcium carbonate. But some, such as diatoms and radiolarians, produce glasslike silica skeletons. These tiny organisms are able to extract silica even though seawater contains only tiny quantities. It is from their remains that most bedded cherts are believed to originate. Some bedded cherts occur in association with lava flows and layers of volcanic ash. For these occurrences it is probable that the silica was derived from the decomposition of the volcanic ash and not from biochemical sources. Chert nodules are sometimes referred to as *secondary* or *replacement cherts* and most often occur within beds of limestone. They form when silica, originally deposited in one place, dissolves, migrates, and then chemically precipitates elsewhere, replacing older material.

Evaporites

Very often evaporation is the mechanism triggering deposition of chemical precipitates. Minerals commonly precipitated in this fashion include halite (sodium chloride, NaCl), the chief

A. Agate

B. Flint   **C.** Jasper

D. Chert arrowhead

FIGURE 7.16   *Chert* is a name used for a number of dense, hard rocks made of microcrystalline quartz. **A.** *Agate* is the banded variety. (Photo by Jeffrey A. Scoville) **B.** The dark color of *flint* results from organic matter. (Photo by E. J. Tarbuck) **C.** The red variety, called *jasper,* gets its color from iron oxide. (Photo by E. J. Tarbuck) **D.** Native Americans frequently made arrowheads and sharp tools from chert. (Photo by LA VENTA/ CORBIS SYGMA)

component of *rock salt,* and gypsum (hydrous calcium sulfate, $CaSO_4 \cdot 2\ H_2O$), the main ingredient of *rock gypsum.* Both have significant importance. Halite is familiar to everyone as the common salt used in cooking and seasoning foods. Of course, it has many other uses, from melting ice on roads to making hydrochloric acid, and has been considered important enough that people have sought, traded, and fought over it for much of human history. Gypsum is the basic ingredient in plaster of Paris. This material is used most extensively in the construction industry for wallboard and interior plaster.

In the geologic past, many areas that are now dry land were basins, submerged under shallow arms of a sea that had only narrow connections to the open ocean. Under these conditions, seawater continually moved into the bay to replace water lost by evaporation. Eventually the waters of the bay became saturated and salt deposition began. Such deposits are called **evaporites.**

When a body of seawater evaporates, the minerals that precipitate do so in a sequence that is determined by their solubility. Less soluble minerals precipitate first, and more soluble minerals precipitate later as salinity increases (Figure 7.17). For example, gypsum precipitates when about 80 percent of the seawater has evaporated, and halite settles out when 90 percent of the water has been removed. During the last stages of this process, potassium and magnesium salts precipitate. One of these last-formed salts, the mineral *sylvite,* is mined as a significant source of potassium ("potash") for fertilizer.

On a smaller scale, evaporite deposits can be seen in such places as Death Valley, California. Here, following rains or periods

FIGURE 7.17   Each year about 30 percent of the world's supply of salt is extracted from seawater. In this process, salt water is held in shallow ponds, while solar energy evaporates the water. The nearly pure salt deposits that eventually form are essentially artificial evaporite deposits. At the southern end of San Francisco Bay, it takes nearly 38,000 liters (10,000 gallons) of water to produce 900 kilograms (1 ton) of salt. (Photo by William E. Townsend Jr./Photo Researchers, Inc.)

FIGURE 7.18 The Bonneville salt flats in Utah are a well-known example of evaporite deposits. (Photo by Creatas/Photolibrary)

of snowmelt in the mountains, streams flow from the surrounding mountains into an enclosed basin (see Figure 19.9, p. 000). As the water evaporates, **salt flats** form when dissolved materials are precipitated as a white crust on the ground (Figure 7.18).

COAL—AN ORGANIC SEDIMENTARY ROCK

Coal is quite different from other rocks. Unlike limestone and chert, which are calcite- and silica-rich, coal is made of organic matter. Close examination of coal under a magnifying glass often reveals plant structures such as leaves, bark, and wood that have been chemically altered but are still identifiable. This supports the conclusion that coal is the end product of large amounts of plant material, buried for millions of years (Figure 7.19).

The initial stage in coal formation is the accumulation of large quantities of plant remains. Special conditions are required for such accumulations, because dead plants readily decompose when exposed to the atmosphere or other oxygen-rich environments. One important environment that allows for the buildup of plant material is a swamp.

Stagnant swamp water is oxygen-deficient, so complete decay (oxidation) of the plant material is not possible. Instead, the plants are attacked by certain bacteria that partly decompose the organic material and liberate oxygen and hydrogen. As these elements escape, the percentage of carbon gradually increases. The bacteria are not able to finish the job of decomposition because they are destroyed by acids liberated from the plants.

The partial decomposition of plant remains in an oxygen-poor swamp creates a layer of *peat,* a soft brown material in which plant structures are still easily recognized. With shallow burial, peat slowly changes to *lignite,* a soft brown coal. Burial increases the temperature of sediments as well as the pressure on them.

The higher temperatures bring about chemical reactions within the plant materials and yield water and organic gases (volatiles). As the load increases from more sediment on top of the developing coal, the water and volatiles are pressed out and the proportion of *fixed carbon* (the remaining solid combustible material) increases. The greater the carbon content, the greater the coal's energy ranking as a fuel. During burial, the coal also becomes increasingly compact. For example, deeper burial transforms lignite into a harder, more compacted black rock called *bituminous* coal. Compared to the peat from which it formed, a bed of bituminous coal may be only 1/10 as thick.

Lignite and bituminous coals are sedimentary rocks. However, when sedimentary layers are subjected to the folding and deformation associated with mountain building, the heat and pressure cause a further loss of volatiles and water, thus increasing the concentration of fixed carbon. This metamorphoses bituminous coal into *anthracite,* a very hard, shiny, black *metamorphic* rock. Although anthracite is a clean-burning fuel, only a relatively small amount is mined. Anthracite is not widespread and is more difficult and expensive to extract than the relatively flat-lying layers of bituminous coal.

Coal is a major energy resource. Its role as a fuel and some of the problems associated with burning coal are discussed in Chapter 23.

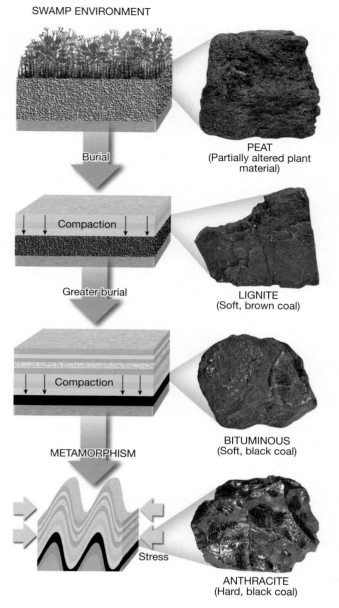

SWAMP ENVIRONMENT

Burial

PEAT
(Partially altered plant material)

Compaction

Greater burial

LIGNITE
(Soft, brown coal)

Compaction

METAMORPHISM

BITUMINOUS
(Soft, black coal)

Stress

ANTHRACITE
(Hard, black coal)

FIGURE 7.19 Successive stages in the formation of coal. (Photos by E. J. Tarbuck)

TURNING SEDIMENT INTO SEDIMENTARY ROCK: DIAGENESIS AND LITHIFICATION

A great deal of change can occur to sediment from the time it is deposited until it becomes a sedimentary rock and is subsequently subjected to the temperatures and pressures that convert it to metamorphic rock. The term **diagenesis** (*dia* = change, *genesis* = origin) is a collective term for all of the chemical, physical, and biological changes that take place after sediments are deposited and during and after lithification.

Burial promotes diagenesis because as sediments are buried, they are subjected to increasingly higher temperatures and pres-

sures. Diagenesis occurs within the upper few kilometers of Earth's crust at temperatures that are generally less than 150° to 200 °C. Beyond this somewhat arbitrary threshold, metamorphism is said to occur.

One example of diagenetic change is *recrystallization,* the development of more stable minerals from less stable ones. It is illustrated by the mineral aragonite, the less stable form of calcium carbonate ($CaCO_3$). Aragonite is secreted by many marine organisms to form shells and other hard parts, such as the skeletal structures produced by corals. In some environments, large quantities of these solid materials accumulate as sediment. As burial takes place, aragonite recrystallizes to the more stable form of calcium carbonate, calcite, the main constituent in the sedimentary rock limestone.

Another example of diagenesis was provided in the preceding discussion of coal. It involved the chemical alteration of organic matter in an oxygen-poor environment. Instead of completely decaying, as would occur in the presence of oxygen, the organic matter is slowly transformed to solid carbon.

Diagenesis includes **lithification,** the processes by which unconsolidated sediments are transformed into solid sedimentary rocks (*lithos* = stone, *fic* = making). Basic lithification processes include compaction and cementation.

The most common physical diagenetic change is **compaction.** As sediment accumulates, the weight of overlying material compresses the deeper sediments. The deeper a sediment is buried, the more it is compacted and the firmer it becomes. As the grains are pressed closer and closer, there is considerable reduction in pore space (the open space between particles). For example, when clays are buried beneath several thousand meters of material, the volume of clay may be reduced by as much as 40 percent. As pore space decreases, much of the water that was trapped in the sediments is driven out. Because sands and other coarse sediments are less compressible, compaction is most significant as a lithification process in fine-grained sedimentary rocks.

Cementation is the most important process by which sediments are converted to sedimentary rock. It is a diagenetic change that involves the crystallization of minerals among the individual sediment grains. Groundwater carries ions in solution. Gradually, the crystallization of new minerals from these ions takes place in the pore spaces, cementing the clasts together. Just as the amount of pore space is reduced during compaction, the addition of cement into a sedimentary deposit reduces its porosity as well.

Calcite, silica, and iron oxide are the most common cements. It is often a relatively simple matter to identify the cementing material. Calcite cement will effervesce with dilute hydrochloric acid. Silica is the hardest cement and thus produces the hardest sedimentary rocks. An orange or dark red color in a sedimentary rock means that iron oxide is present.

Most sedimentary rocks are lithified by means of compaction and cementation. However, some initially form as solid masses of intergrown crystals rather than beginning as accumulations of separate particles that later become solid. Other crystalline sedimentary rocks do not begin that way but are transformed into masses of interlocking crystals sometime after the sediment is deposited.

EARTH AS A SYSEM

BOX 7.1

The Carbon Cycle and Sedimentary Rocks

To illustrate the movement of material and energy in the Earth system, let us take a brief look at the *carbon cycle* (Figure 7.A). Pure carbon is relatively rare in nature. It is found predominantly in two minerals: diamond and graphite. Most carbon is bonded chemically to other elements to form compounds such as carbon dioxide, calcium carbonate, and the hydrocarbons found in coal and petroleum. Carbon is also the basic building block of life as it readily combines with hydrogen and oxygen to form the fundamental organic compounds that compose living things.

In the atmosphere, carbon is found mainly as carbon dioxide (CO_2). Atmospheric carbon dioxide is significant because it is a greenhouse gas, which means it is an efficient absorber of energy emitted by Earth and thus influences the heating of the atmosphere.* Because many of the processes that operate on Earth involve carbon dioxide, this gas is constantly moving into and out of the atmosphere (Figure 7.B). For example, through the process of photosynthesis, plants absorb carbon dioxide from the atmosphere to produce the essential organic compounds needed for growth. Animals that consume these plants (or consume other animals that eat plants)

*For more on this idea, see the discussion "Heating the Atmosphere: The Greenhouse Effect" in Chapter 21.

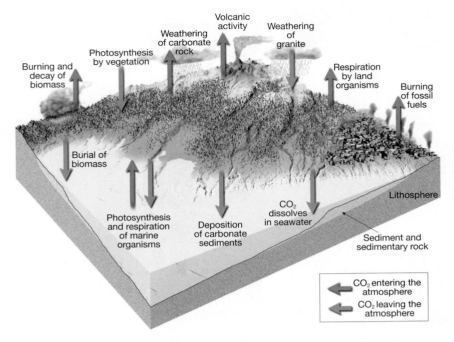

FIGURE 7.A   Simplified diagram of the carbon cycle, with emphasis on the flow of carbon between the atmosphere and the hydrosphere, lithosphere, and biosphere. The colored arrows show whether the flow of carbon is into or out of the atmosphere.

use these organic compounds as a source of energy and, through the process of respiration, return carbon dioxide to the atmosphere. (Plants also return some CO_2 to the atmosphere via respiration.) Further, when plants

die and decay or are burned, this biomass is oxidized, and carbon dioxide is returned to the atmosphere.

Not all dead plant material decays immediately back to carbon dioxide. A small percentage

---

For example, with time and burial, loose sediment consisting of delicate calcareous skeletal debris may be recrystallized into a relatively dense crystalline limestone. Because crystals grow until they fill all the available space, pore spaces are frequently lacking in crystalline sedimentary rocks. Unless the rocks later develop joints and fractures, they will be relatively impermeable to fluids like water and oil.

CLASSIFICATION OF SEDIMENTARY ROCKS

The classification scheme in Figure 7.20 divides sedimentary rocks into two major groups: detrital and chemical/organic. Further, we can see that the main criterion for subdividing the detrital rocks is particle size, whereas the primary basis for dis-

tinguishing among different rocks in the chemical group is their mineral composition.

As is the case with many (perhaps most) classification of natural phenomena, the categories presented in Figure 7.20 are more rigid than the actual state of nature. In reality, many of the sedimentary rocks classified into the chemical group also contain at least small quantities of detrital sediment. Many limestones, for example, contain varying amounts of mud or sand, giving them a "sandy" or "shaly" quality. Conversely, because practically all detrital rocks are cemented with material that was originally dissolved in water, they too are far from being "pure."

As was the case with the igneous rocks examined in Chapter 4, *texture* is a part of sedimentary rock classification. There are two major textures used in the classification of sedimentary rocks: clastic and nonclastic. The term **clastic** is taken

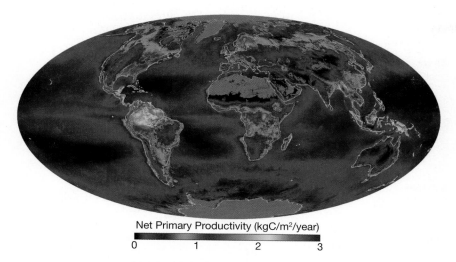

Net Primary Productivity (kgC/m²/year)

0   1   2   3

FIGURE 7.B   This map was created using space-based measurements of a range of plant properties and shows the net productivity of vegetation on land and in the oceans in 2002. It is calculated by determining how much CO_2 is taken up by vegetation during photosynthesis minus how much is given off during respiration. Scientists expect this global measure of biological activity to yield new insights into Earth's complex carbon cycle. (NASA image)

is deposited as sediment. Over long spans of geologic time, considerable biomass is buried with sediment. Under the right conditions, some of these carbon-rich deposits are converted to fossil fuels—coal, petroleum, or natural gas. Eventually some of the fuels are recovered (mined or pumped from a well) and burned to run factories and fuel our transportation system. One result of fossil-fuel combustion is the release of huge quantities of CO_2 into the atmosphere. Certainly one of the most active parts of the carbon cycle is the movement of CO_2 from the atmosphere to the biosphere and back again.

Carbon also moves from the geosphere and hydrosphere to the atmosphere and back again. For example, volcanic activity early in Earth's history is thought to be the source of much of the carbon dioxide found in the atmosphere. One way that carbon dioxide makes its way back to the hydrosphere and then to the solid Earth is by first combining with water to form carbonic acid (H_2CO_3), which then attacks the rocks that compose Earth's crust. One product of this chemical weathering of solid rock is the soluble bicarbonate ion ($2HCO_3$) which is carried by groundwater and streams to the ocean. Here water-dwelling organisms extract this dissolved material to produce hard parts of calcium carbonate ($CaCO_3$). When the organisms die, these skeletal remains settle to the ocean floor as biochemical sediment and become sedimentary rock. In fact, the crust is by far Earth's largest depository of carbon, where it is a constituent of a variety of rocks, the most abundant being limestone. Eventually the limestone may be exposed at Earth's surface, where chemical weathering will cause the carbon stored in the rock to be released to the atmosphere as CO_2.

In summary, carbon moves among all four of Earth's major spheres. It is essential to every living thing in the biosphere. In the atmosphere carbon dioxide is an important greenhouse gas. In the hydrosphere, carbon dioxide is dissolved in lakes, rivers, and the ocean. In the geosphere carbon is contained in carbonate sediments and sedimentary rocks and is stored as organic matter dispersed through sedimentary rocks and as deposits of coal and petroleum.

from a Greek word meaning "broken." Rocks that display a clastic texture consist of discrete fragments and particles that are cemented and compacted together. Although cement is present in the spaces between particles, these openings are rarely filled completely. All detrital rocks have a clastic texture. In addition, some chemical sedimentary rocks exhibit this texture. For example, coquina, the limestone composed of shells and shell fragments, is obviously as clastic as a conglomerate or sandstone. The same applies for some varieties of oolitic limestone.

Some chemical sedimentary rocks have a **nonclastic** or **crystalline** texture in which the minerals form a pattern of interlocking crystals. The crystals may be microscopically small or large enough to be visible without magnification. Common examples of rocks with nonclastic textures are those deposited when seawater evaporates (Figure 7.21). The materials that make up many other nonclastic rocks may actually have originated as detrital deposits. In these instances, the particles probably consisted of shell fragments and other hard parts rich in calcium carbonate or silica. The clastic nature of the grains was subsequently obliterated or obscured because the particles recrystallized when they were consolidated into limestone or chert.

Nonclastic rocks consist of intergrown crystals, and some may resemble igneous rocks, which are also crystalline. The two rock types are usually easy to distinguish because the minerals contained in nonclastic sedimentary rocks are quite unlike those found in most igneous rocks. For example, rock salt, rock gypsum, and some forms of limestone consist of intergrown crystals, but the minerals within these rocks (halite, gypsum, and calcite) are seldom associated with igneous rocks.

Detrital Sedimentary Rocks

ClasticTexture (particle size)		Sediment Name	Rock Name
Coarse (over 2 mm)		Gravel (Rounded particles)	Conglomerate
		Gravel (Angular particles)	Breccia
Medium (1/16 to 2 mm)		Sand (If abundant feldspar is present the rock is called **Arkose**)	Sandstone
Fine (1/16 to 1/256 mm)		Mud	Siltstone
Very fine (less than 1/256 mm)		Mud	Shale or Mudstone

Chemical and Organic Sedimentary Rocks

Composition	Texture	Rock Name	
Calcite, CaCO₃	Nonclastic: Fine to coarse crystalline	Crystalline Limestone	
		Travertine	
	Clastic: Visible shells and shell fragments loosely cemented	Coquina	Biochemical Limestone
	Clastic: Various size shells and shell fragments cemented with calcite cement	Fossiliferous Limestone	
	Clastic: Microscopic shells and clay	Chalk	
Quartz, SiO₂	Nonclastic: Very fine crystalline	Chert (light colored) Flint (dark colored)	
Gypsum CaSO₄•2H₂O	Nonclastic: Fine to coarse crystalline	Rock Gypsum	
Halite, NaCl	Nonclastic: Fine to coarse crystalline	Rock Salt	
Altered plant fragments	Nonclastic: Fine-grained organic matter	Bituminous Coal	

FIGURE 7.20 Identification of sedimentary rocks. Sedimentary rocks are divided into three groups, detrital, chemical, and organic. The main criterion for naming detrital sedimentary rocks is particle size, whereas the primary basis for distinguishing among chemical sedimentary rocks is their mineral composition.

Close up

FIGURE 7.21 Like other evaporites, this sample of rock salt is said to have a nonclastic texture because it is composed of intergrown crystals. (Photos by E. J. Tarbuck)

SEDIMENTARY ENVIRONMENTS

SEDIMENTARY ROCKS
▶ Sedimentary Environments

Sedimentary rocks are important in the interpretation of Earth history. By understanding the conditions under which sedimentary rocks form, geologists can often deduce the history of a rock, including information about the origin of its component particles, the method of sediment transport, and the nature of the place where the grains eventually came to rest; that is, the environment of deposition (Figure 7.22).

An **environment of deposition** or **sedimentary environment** is simply a geographic setting where sediment is accumulating. Each site is characterized by a particular combination of geologic processes and environmental conditions. Some sediments, such as the chemical sediments that precipitate in water bodies, are solely the product of their sedimentary environment. That is, their component minerals originated and were deposited in the same place. Other sediments originate far from the site where they accumulate. These materials are

PEOPLE AND THE ENVIRONMENT

Our Threatened Coral Reefs

BOX 7.2

Modern coral reefs are unique ecosystems of plants, animals, and their associated geological framework. They are home to about 25 percent of all marine species. Because of this great diversity, they are sometimes referred to as the ocean equivalent of rain forests. The tiny colonial animals that build these intricate limestone masses are dying at an alarming rate. If the trend continues, many of the world's reefs will be significantly damaged in the next 20 years.

To better understand some of the reasons that coral reefs are threatened, it is useful to have a little background about these tiny invertebrates that are related to jellyfish and anemones. Corals derive nourishment by catching plankton (microscopic plants and animals) and other suspended food particles with arm-like tentacles which feed a centrally located mouth (Figure 7.C). Most reef-building corals also host symbiotic algae.* These photosynthetic algae live within the coral's tissues and provide the host with important nutrients. Because sunlight is essential for photosynthesis, reef-building corals require clear, sunlit water in which to grow.

Why are coral reefs in peril? Coral reefs can be damaged by natural processes, such as storms, but they are increasingly at risk from human activities. Coral reef health can be seriously affected and the growth of new colonies hampered when runoff transports fine-grained

FIGURE 7.C Coral polyps are tiny animals that are typically found in colonies. They secrete calcium carbonate to form hard skeletons. Although reef-building corals obtain some food by catching tiny organisms using stinging cells on their tentacles, these animals obtain a large part of their nourishment from photosynthetic algae that live in their tissues. (Photo by Jez Tryner/Image Quest Marine)

sediment from land areas into coastal waters. The amount of sediment can be greatly increased by nearby land developments for agriculture, industry, and housing. Such developments often remove anchoring vegetation, leaving the surface more vulnerable to erosion. The sunlight necessary for photosynthesis by the algae living in corals is diminished in waters made cloudy by suspended sediment, reducing coral productivity.

When corals are stressed, they often expel the symbiotic algae that are critical to their well being in a process called *coral bleaching*. One known cause of coral bleaching is an increase in ocean water temperatures. Regional increases in sea-surface temperatures can occur naturally. In addition, ocean temperatures worldwide may be changing as a result of global warming, for which humans are largely responsible.**

Other human activities are known to directly and indirectly harm coral reefs. Oil spills and pollutants can threaten entire reefs. Excessive nutrients discharged into the ocean, such as sewage outfall and agricultural fertilizers, promote the growth of forms of algae that can smother corals. Such algae also thrive when the fish that normally graze on them are overharvested. Other organisms harmful to corals also multiply when the species that prey on them are removed by overfishing.

---

*A *symbiotic relationship* is one between two organisms which is mutually beneficial for both participants.

**For more on global warming, see Chapter 21 "Global Climate Change."

---

transported great distances from their source by some combination of gravity, water, wind, and ice.

At any given time, the geographic setting and environmental conditions of a sedimentary environment determine the nature of the sediments that accumulate. Therefore, geologists carefully study the sediments in present-day depositional environments because the features they find can also be observed in ancient sedimentary rocks.

By applying a thorough knowledge of present-day conditions, geologists attempt to reconstruct the ancient environments and geographical relationships of an area at the time a particular set of sedimentary layers were deposited. Such analyses often lead to the creation of maps, which depict the geographic distribution of land and sea, mountains and river valleys, deserts and glaciers, and other environments of deposition. The foregoing description is an excellent example of the

application of a fundamental principle of modern geology, namely that "the present is the key to the past."*

Types of Sedimentary Environments

Sedimentary environments are commonly placed into one of three categories: continental, marine, or transitional (shoreline). Each category includes many specific subenvironments. Figure 7.22 is an idealized diagram illustrating a number of important sedimentary environments associated with each category. Realize that this is just a sampling of the great diversity of depositional environments. The remainder of this section provides a brief overview of each category. Later, Chapters 16

---

*For more on this idea, see "The Birth of Modern Geology" in Chapter 1.

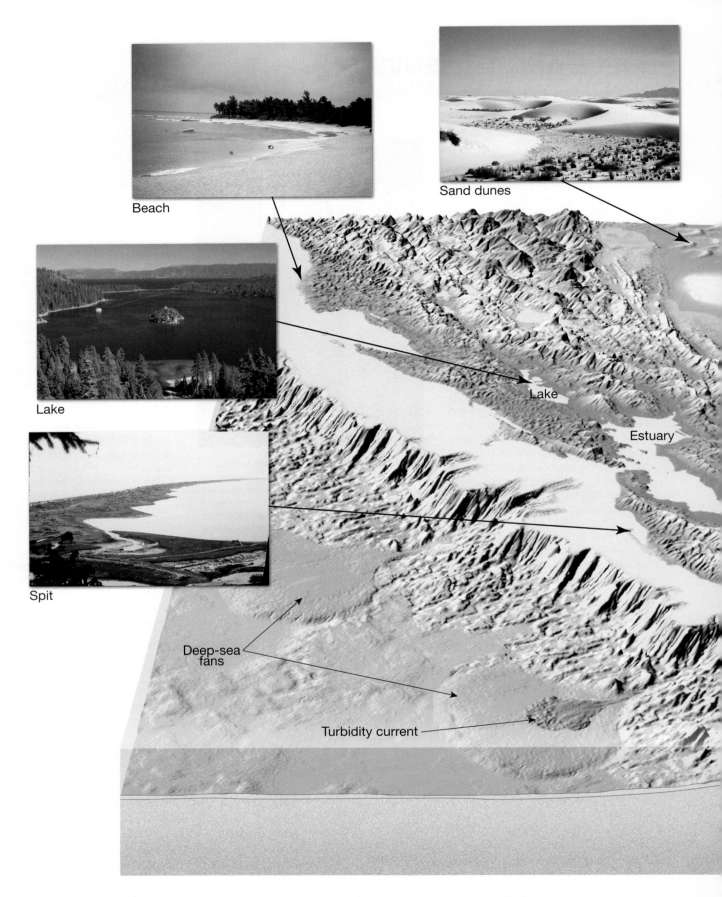

Beach

Sand dunes

Lake

Lake

Estuary

Spit

Deep-sea
fans

Turbidity current

FIGURE 7.22 Sedimentary environments are those places where sediment accumulates. Each is characterized by certain physical, chemical, and biological conditions. Because each sediment contains clues about the environment in which it was deposited, sedimentary rocks are important in the interpretation of Earth history. A number of important continental, transitional, and marine sedimentary environments are represented in this idealized diagram. (Lake photo by Jon Arnold Images/www.DanitaDelimont; Alluvial Fan photo by Marli Miller; Swamp photo by Raymond Gehman/National Geographic; all others by E. J. Tarbuck)

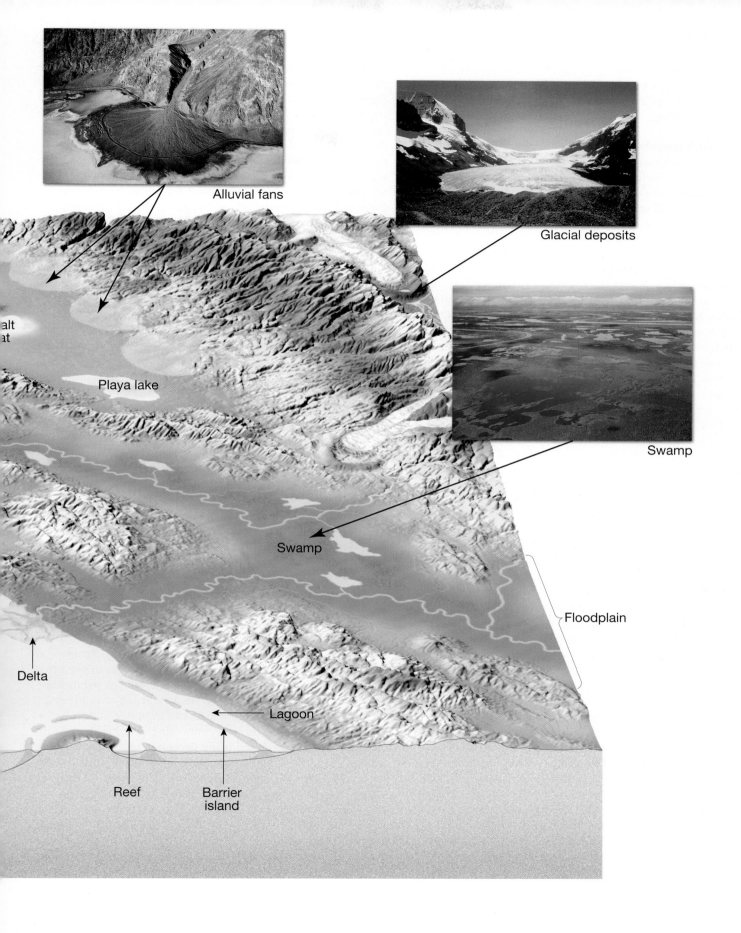

Alluvial fans

Glacial deposits

Salt
flat

Playa lake

Swamp

Swamp

Floodplain

Delta

Lagoon

Reef

Barrier
island

through 20 will examine many of these environments in greater detail. Each is an area where sediment accumulates and where organisms live and die. Each produces a characteristic sedimentary rock or assemblage that reflects prevailing conditions.

Continental Environments Continental environments are dominated by the erosion and deposition associated with streams. In some cold regions, moving masses of glacial ice replace running water as the dominant process. In arid regions (as well as some coastal settings) wind takes on greater importance. Clearly, the nature of the sediments deposited in continental environments is strongly influenced by climate.

Streams are the dominant agent of landscape alteration, eroding more land and transporting and depositing more sediment than any other process. In addition to channel deposits, large quantities of sediment are dropped when flood waters periodically inundate broad, flat valley floors (called *floodplains*). Where rapid streams emerge from a mountainous area onto a flatter surface, a distinctive cone-shaped accumulation of sediment known as an *alluvial fan* forms.

In frigid, high-latitude or high-altitude settings, glaciers pick up and transport huge volumes of sediment. Materials deposited directly from ice are typically unsorted mixtures of particles that range in size from clay to boulders. Water from melting glaciers transports and redeposits some of the glacial sediment, creating stratified, sorted accumulations.

The work of wind and its resulting deposits are referred to as *eolian,* after Aeolus, the Greek god of wind. Unlike glacial deposits, eolian sediments are well sorted. Wind can lift fine dust high into the atmosphere and transport it great distances. Where winds are strong and the surface is not anchored by vegetation, sand is transported closer to the ground, where it accumulates in *dunes*. Deserts and coasts are common sites for this type of deposition.

In addition to being areas where dunes sometimes develop, desert basins are the sites where shallow *playa lakes* occasionally form following heavy rains or periods of snowmelt in adjacent mountains. They rapidly dry up, sometimes leaving behind evaporites and other characteristic deposits. Figure 19.9 on page 528 illustrates such an environment. In humid regions lakes are more enduring features, and their quiet waters are excellent sediment traps. Small deltas, beaches, and bars form along the lakeshore, with finer sediments coming to rest on the lake floor.

Marine Environments Marine depositional environments are divided according to depth. The *shallow marine* environment reaches to depths of about 200 meters (nearly 700 feet) and extends from the shore to the outer edge of the continental shelf. The *deep marine* environment lies seaward of the continental shelf in waters deeper than 200 meters.

The shallow marine environment borders all of the world's continents. Its width varies greatly, from practically nonexistent in some places to broad expanses extending as far as 1500 kilometers (more than 900 miles) in other locations. On the average this zone is about 80 kilometers (50 miles) wide. The kind of sediment deposited here depends on several factors, including distance from shore, the elevation of the adjacent land area, water depth, water temperature, and climate.

Due to the ongoing erosion of the adjacent continent, the shallow marine environment receives huge quantities of land-derived sediment. Where the influx of such sediment is small and the seas are relatively warm, carbonate-rich muds may be the predominant sediment. Most of this material consists of the skeletal debris of carbonate-secreting organisms mixed with inorganic precipitates. Coral reefs are also associated with warm, shallow marine environments. In hot regions where the sea occupies a basin with restricted circulation, evaporation triggers the precipitation of soluble materials and the formation of marine evaporite deposits.

Deep marine environments include all the floors of the deep ocean. Far from landmasses, tiny particles from many sources remain adrift for long spans. Gradually these small grains "rain" down on the ocean floor, where they accumulate very slowly. Significant exceptions are thick deposits of relatively coarse sediment that occur at the base of the continental slope. These materials move down from the continental shelf as turbidity currents—dense gravity-driven masses of sediment and water (see Figure 7.26, p. 222).

Transitional Environments The shoreline is the transition zone between marine and continental environments. Here we find the familiar deposits of sand or gravel called *beaches*. Mud-covered *tidal flats* are alternately covered with shallow sheets of water and then exposed to the air as tides rise and fall. Along and near the shore, the work of waves and currents distributes sand, creating *spits, bars,* and *barrier islands*. Offshore bars and reefs create *lagoons*. The quieter waters in these sheltered areas are another site of deposition in the transition zone.

Deltas are among the most significant deposits associated with transitional environments. The complex accumulations of sediment build outward into the sea when rivers experience an abrupt loss of velocity and drop their load of detrital debris.

Sedimentary Facies

When a series of sedimentary layers is studied, we can see the successive changes in environmental conditions that occurred at a particular place with the passage of time. Changes in past environments may also be seen when a single unit of sedimentary rock is traced laterally. This is true because at any one time many different depositional environments can exist over a broad area. For example, when sand is accumulating in a beach environment, finer muds are often being deposited in quieter, offshore waters. Still farther out, perhaps in a zone where biological activity is high and land-derived sediments are scarce, the deposits consist largely of the calcareous remains of small organisms. In this example, different sediments are accumulating adjacent to one another at the same time. Each unit possesses a distinctive set of characteristics reflecting the conditions in a particular environment. To describe such sets of sediments, the term **facies** is used. When a sedimentary unit is examined in cross section from one end to the other, each facies grades laterally into another that formed at the same time but which exhibits different characteristics (Figure 7.23). Commonly, the merging of adjacent facies tends to be a gradual transition rather than a sharp boundary, but abrupt changes do sometimes occur.

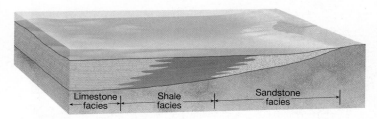

FIGURE 7.23 When a single sedimentary layer is traced laterally, we may find that it is made up of several different rock types. This can occur because many sedimentary environments can exist at the same time over a broad area. The term *facies* is used to describe such sets of sedimentary rocks. Each facies grades laterally into another that formed at the same time but in a different environment.

SEDIMENTARY STRUCTURES

In addition to variations in grain size, mineral composition, and texture, sediments exhibit a variety of structures. Some, such as graded beds, are created when sediments are accumulating and are a reflection of the transporting medium. Others, such as *mud cracks,* form after the materials have been deposited and result from processes occurring in the environment. When present, sedimentary structures provide additional information that can be useful in the interpretation of Earth history.

Sedimentary rocks form as layer upon layer of sediment accumulates in various depositional environments. These layers, called **strata,** or **beds,** are probably *the single most common and characteristic feature of sedimentary rocks.* Each stratum is unique. It may be a coarse sandstone, a fossil-rich limestone, a black shale, and so on. When you look at Figure 7.24, or look back through this chapter at Figures 7.1 (p. 201) and 7.4 (p. 204), you will see many such layers, each different from the others. The variations in texture, composition, and thickness reflect the different conditions under which each layer was deposited.

The thickness of beds ranges from microscopically thin to tens of meters thick. Separating the strata are **bedding planes,** relatively flat surfaces along which rocks tend to separate or break. Changes in the grain size or in the composition of the sediment being deposited can create bedding planes. Pauses in deposition can also lead to layering because chances are slight that newly deposited material will be exactly the same as previously deposited sediment. Generally, each bedding plane marks the end of one episode of sedimentation and the beginning of another.

Because sediments usually accumulate as particles that settle from a fluid, most strata are originally deposited as horizontal layers. There are circumstances, however, when sediments do not accumulate in horizontal beds. Sometimes

FIGURE 7.24 This outcrop of sedimentary strata illustrates the characteristic layering of this group of rocks. Minnewaska State Park, New York. (Photo by Garrett Drapala/Alamy)

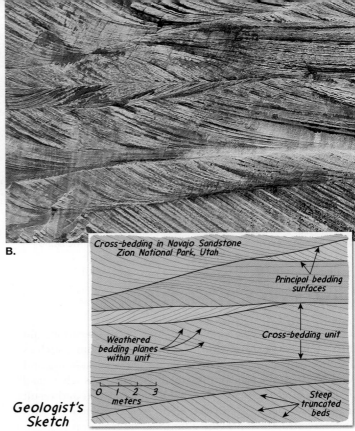

A.

FIGURE 7.25 **A.** The cutaway section of this sand dune shows cross-bedding. (Photo by John S. Shelton) **B.** The cross-bedding of this sandstone indicates it was once a sand dune. (Photo by Dennis Tasa)

B.

Cross-bedding in Navajo Sandstone Zion National Park, Utah

Principal bedding surfaces

Cross-bedding unit

Weathered bedding planes within unit

0 1 2 3 meters

Steep truncated beds

Geologist's Sketch

when a bed of sedimentary rock is examined, we see layers within it that are inclined to the horizontal. When this occurs, it is called **cross-bedding** and is most characteristic of sand dunes, river deltas, and certain stream channel deposits (Figure 7.25).

Graded beds represent another special type of bedding. In this case the particles within a single sedimentary layer gradually change from coarse at the bottom to fine at the top. Graded beds are most characteristic of rapid deposition from water containing sediment of varying sizes. When a current experiences a rapid energy loss, the largest particles settle first, followed by successively smaller grains. The deposition of a

graded bed is most often associated with a turbidity current, a mass of sediment-choked water that is denser than clear water and that moves downslope along the bottom of a lake or ocean (Figure 7.26).

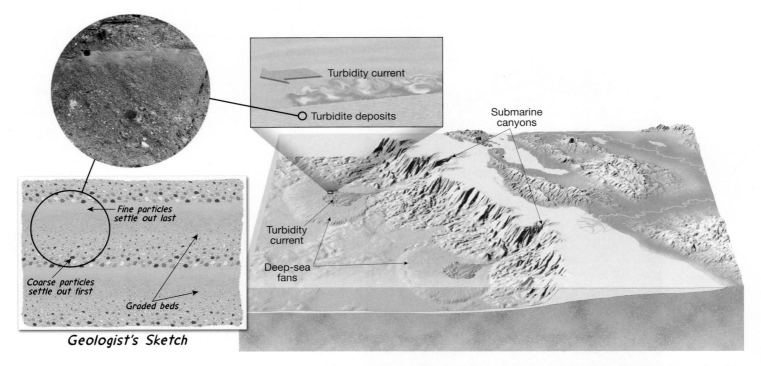

Turbidity current

Turbidite deposits

Submarine canyons

Turbidity current

Deep-sea fans

Fine particles settle out last

Coarse particles settle out first

Graded beds

Geologist's Sketch

FIGURE 7.26 Turbidity currents are downslope movements of dense, sediment-laden water. They are created when sand and mud on the continental shelf and slope are dislodged and thrown into suspension. Because such mud-choked water is denser than normal seawater, it flows downslope, eroding and accumulating more sediment. Beds deposited by these currents are called *turbidites.* Each event produces a single bed characterized by a decrease in sediment size from bottom to top, a feature known as a *graded bed.* (Photo by Marli Miller)

As geologists examine sedimentary rocks, much can be deduced. A conglomerate, for example, may indicate a high-energy environment, such as a surf zone or rushing stream, where only coarse materials settle out and finer particles are kept suspended (Figure 7.27). If the rock is arkose, it may signify a dry climate, where little chemical alteration of feldspar is possible. Carbonaceous shale is a sign of a low-energy, organic-rich environment, such as a swamp or lagoon.

Other features found in some sedimentary rocks also give clues to past environments. Ripple marks are such a feature. **Ripple marks** are small waves of sand that develop on the surface of a sediment layer by the action of moving water or air (Figure 7.28). The ridges form at right angles to the direction of motion. If the ripple marks were formed by air or water moving in essentially one direction, their form will be asymmetrical. These *current ripple marks* will have steeper sides in

A.

FIGURE 7.27   In a turbulent stream channel, only large particles settle out. Finer sediments remain suspended and continue their down-stream journey. (Photo by Kharlamov Igor Viktorovich/Shutterstock)

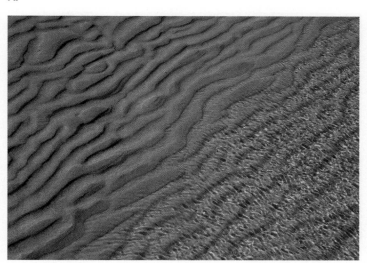

B.

FIGURE 7.28   **A.** Ripple marks preserved in sedimentary rock (Photo by Stephen Trimble). **B.** Curret ripple marks in a stream bed. (Photo by Suzanne Long/Alamy)

the downcurrent direction and more gradual slopes on the upcurrent side. Ripple marks produced by a stream flowing across a sandy channel or by wind blowing over a sand dune are two common examples of current ripples. When present in solid rock, they may be used to determine the direction of movement of ancient wind or water currents. Other ripple marks have a symmetrical form. These features, called *oscillation ripple marks,* result from the back-and-forth movement of surface waves in a shallow nearshore environment.

Mud cracks (Figure 7.29) indicate that the sediment in which they were formed was alternately wet and dry. When exposed to air, wet mud dries out and shrinks, producing cracks. Mud cracks are associated with such environments as tidal flats, shallow lakes, and desert basins.

Fossils, the remains or traces of prehistoric life, are important inclusions in sediment and sedimentary rocks (Figure 7.30). They are important tools for interpreting the geologic past. Knowing the nature of the life forms that existed at a particular time helps researchers understand past environmental conditions. Further, fossils are important time indicators and play a key role in correlating rocks of similar ages that are from different places.*

FIGURE 7.29   Mud cracks form when wet mud or clay dries out and shrinks. (Photo by Gary Yeowell/Getty Images)

*The section entitled "Fossils: Evidence of Past Life" in Chapter 9 contains a more detailed discussion of the role of fossils in the interpretation of Earth history.

A.

B.

FIGURE 7.30   Only a tiny fraction of the organisms that lived in the geologic past have been preserved as fossils. The fossil record is slanted toward organisms with hard parts that lived in environments where rapid burial was possible. **A.** Trilobites were sea-dwelling animals that were very abundant during the Paleozoic era. (Photo by Kevin Schafer/www.DanitaDelimo) **B.** These dinosaur footprints near Cameron, Arizona were originally made in mud that eventually became a sedimentary rock. Footprints such as these are considered to be *trace fossils* (see Chapter 9). Dinosaurs flourished during the Mesozoic era. (Photo by Tom Bean/Corbis)

CHAPTER 7: SEDIMENTARY ROCKS IN REVIEW

- Sedimentary rocks account for about 5 to 10 percent of Earth's outer 16 kilometers (10 miles). Because they are concentrated at Earth's surface, the importance of this group is much greater than this percentage implies. Sedimentary rocks contain much of the basic information needed to reconstruct Earth history. In addition, this group is associated with many important energy and mineral resources.

- *Sedimentary rock* consists of *sediment* that in most cases has been *lithified* into solid rock by the processes of *compaction* and *cementation*. Sediment has two principal sources: (1) as *detrital material,* which originates and is transported as solid particles from both mechanical and chemical weathering, which, when lithified, forms detrital sedimentary rocks; and (2) from soluble material produced largely by chemical weathering, which, when precipitated, forms *chemical sedimentary rocks.* Coal is the primary example of a third group called *organic sedimentary rocks,* which consist of organic carbon from the remains of partially altered plant material.

- *Particle size* is the primary basis for distinguishing among various detrital sedimentary rocks. The size of the particles in a detrital rock indicates the energy of the medium that transported them. For example, gravels are moved by swiftly flowing rivers, whereas less energy is required to transport sand. Common detrital sedimentary rocks include *shale* (silt- and clay-size particles), *sandstone,* and *conglomerate* (rounded, gravel-size particles) or *breccia* (angular, gravel-size particles).

- Precipitation of chemical sediments occurs in two ways: (1) by *inorganic processes,* such as evaporation and chemical activity; or by (2) *organic processes* of water-dwelling organisms that produce sediments of *biochemical origin. Limestone,* the most abundant chemical sedimentary rock, consists of the mineral calcite ($CaCO_3$) and forms either by inorganic means or as the result of biochemical processes. Inorganic limestones include *travertine,* which is commonly seen in caves, and *oolitic limestone,* consisting of small spherical grains of calcium carbonate. Other common chemical sedimentary rocks include *dolostone* (composed of the calcium-magnesium carbonate mineral dolomite), *chert* (made of microcrystalline quartz), and *evaporites* (such as rock salt and rock gypsum).

- *Diagenesis* refers to all of the physical, chemical, and biological changes that occur after sediments are deposited and during and after the time they are turned into sedimentary rock. Burial promotes diagenesis. Diagenesis includes lithification.

- *Lithification* refers to the processes by which unconsolidated sediments are transformed into solid sedimentary rock. Most sedimentary rocks are lithified by means of *compaction* and/or *cementation.* Compaction occurs when the weight of overlying materials compresses the deeper sediments. Cementation, the most important process by which sediments are converted to sedimentary rocks, occurs when soluble cementing materials, such as *calcite, silica,* and *iron oxide,* are precipitated onto sediment grains, fill open spaces, and join the particles. Although most sedimentary rocks are lithified by compaction and cementation, certain chemical rocks, such as the evaporites, initially form as solid masses of intergrown crystals.

- Sedimentary rocks are divided into three groups: *detrital, chemical,* and *organic.* All detrital rocks have a *clastic texture,* which consists of discrete fragments and particles that are cemented and compacted together. The main criterion for subdividing the detrital rocks is particle size. Common detrital rocks include *conglomerate, sandstone,* and *shale.* The primary basis for distinguishing among different rocks in the chemical group is their mineral composition. Some chemical rocks, such as those deposited when seawater evaporates, have a *nonclastic (crystalline) texture* in which the minerals form a pattern of interlocking crystals. However, in reality, many of the sedimentary rocks classified into the chemical group also contain at least small quantities of detrital sediment. Common chemical rocks include *limestone, chert,* and *rock gypsum.* Coal is the primary example of an organic sedimentary rock.

- Sedimentary environments are those places where sediment accumulates. They are grouped into continental, marine, and transitional (shoreline) environments. Each is characterized by certain physical, chemical, and biological conditions. Because sediment contains clues about the environment in which it was deposited, sedimentary rocks are important in the interpretation of Earth history.

- Sedimentary rocks are particularly important in interpreting Earth history because, as layer upon layer of sediment accumulates, each records the nature of the environment at the time the sediment was deposited. These layers, called *strata,* or *beds,* are probably the single most characteristic feature of sedimentary rocks. Other features found in some sedimentary rocks, such as *ripple marks, mud cracks, cross-bedding,* and *fossils,* also provide clues to past environments.

KEY TERMS

bedding plane (p. 221)
beds (strata) (p. 221)
biochemical (p. 207)
cementation (p. 213)
chemical sedimentary rock
  (p. 202)
clastic texture (p. 214)
compaction (p. 213)
cross-bedding (p. 222)

crystalline texture (p. 215)
detrital sedimentary rock
  (p. 202)
diagenesis (p. 213)
environment of deposition
  (p. 216)
evaporite deposit (p. 211)
facies (p. 220)

fissility (p. 204)
fossil (p. 224)
graded bed (p. 222)
lithification (p. 213)
mud crack (p. 224)
nonclastic texture (p. 215)
organic sedimentary rock
  (p. 202)

ripple mark (p. 223)
salt flat (p. 212)
sedimentary environment
  (p. 216)
sorting (p. 205)
strata (beds) (p. 221)

QUESTIONS FOR REVIEW

1. How does the volume of sedimentary rocks in Earth's crust compare with the volume of igneous rocks in the crust? Are sedimentary rocks evenly distributed throughout the crust?

2. List and briefly distinguish among the three basic sedimentary rock categories.

3. What minerals are most common in detrital sedimentary rocks? Why are these minerals so abundant?

4. What is the primary basis for distinguishing among various detrital sedimentary rocks?

5. Why does shale often crumble easily?

6. How are the degree of sorting and the amount of rounding related to the transportation of sand grains?

7. Distinguish between conglomerate and breccia.

8. Distinguish between the two categories of chemical sedimentary rocks.

9. What are evaporite deposits? Name a rock that is an evaporite.

10. When a body of seawater evaporates, minerals precipitate in a certain order. What determines this order?

11. Each of the following statements describes one or more characteristics of a particular sedimentary rock. For each statement, name the sedimentary rock that is being described.
    a. An evaporite used to make plaster.
    b. A fine-grained detrital rock that exhibits *fissility*.
    c. The primary example of an organic sedimentary rock.

    d. The most abundant chemical sedimentary rock.
    e. A dark-colored, hard rock made of microcrystalline quartz.
    f. A variety of limestone composed of small spherical grains.

12. What is the primary basis for distinguishing among different chemical sedimentary rocks?

13. What is diagenesis? Give an example.

14. Compaction is most important as a lithification process with which sediment size?

15. List three common cements for sedimentary rocks. How might each be identified?

16. Distinguish between clastic and nonclastic textures. What type of texture is common to all detrital sedimentary rocks?

17. Some nonclastic sedimentary rocks closely resemble igneous rocks. How might the two be distinguished easily?

18. List three categories of sedimentary environments. Provide one or more examples for each category.

19. What is probably the single most characteristic feature of sedimentary rocks?

20. Distinguish between cross-bedding and graded bedding.

21. How do current ripple marks differ from oscillation ripple marks?

COMPANION WEBSITE

The *Earth 10e* website uses the resources and flexibility of the Internet to aid in your study of the topics in this chapter. Written and developed by the authors and other geology instructors, this site will help improve your understanding of geology. Visit www.mygeoscienceplace.com in order to:

- **Review** key chapter concepts

- **Read** with links to the eBook and to chapter-specific web resources
- **Visualize** and comprehend challenging topics using learning activities in *GEODe Earth*
- **Test** yourself with online quizzes

GEODe EARTH

GEODe Earth is a valuable and easy-to-use learning aid that can be accessed from your book's Companion Website (www.mygeoscienceplace.com). It is a dynamic instructional tool that promotes understanding and reinforces important concepts by using tutorials, animations, and exercises that actively engage the student.

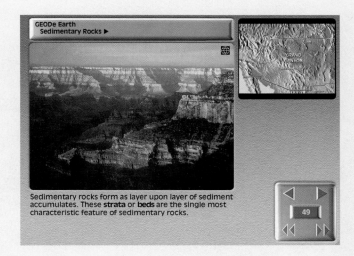

GEODe Earth
Sedimentary Rocks ▶

Sedimentary rocks form as layer upon layer of sediment accumulates. These **strata** or **beds** are the single most characteristic feature of sedimentary rocks.

49

8

METAMORPHISM AND METAMORPHIC ROCKS

Aerial view of metamorphic rocks exposed in the Canadian Shield.

(PHOTO BY ROBERT HILDEBRAND)

229

The folded and metamorphosed rocks shown in Figure 8.1 were once flat-lying sedimentary strata. Compressional forces of unimaginable magnitude and temperatures hundreds of degrees above surface conditions prevailed for perhaps thousands or millions of years to produce the deformation displayed by these rocks. Under such extreme conditions, solid rock responds by folding, fracturing, and often by flowing. This chapter looks at the tectonic forces that forge metamorphic rocks and how these rocks change in appearance, mineralogy, and sometimes even in overall chemical composition.

Extensive areas of metamorphic rocks are exposed on every continent in the relatively flat regions known as shields (see chapter opening photo). These metamorphic regions are found in eastern Canada, Brazil, Africa, India, Australia, and Greenland. Moreover, metamorphic rocks are an important component of many mountain belts, including the Alps and the Appalachians. Even those portions of the stable interiors of continents that are covered by sedimentary rocks are underlain by metamorphic basement rocks. In these settings the metamorphic rocks are highly deformed and intruded by large igneous masses. Indeed, significant parts of Earth's continental crust are composed of metamorphic and associated igneous rocks.

Unlike some igneous and sedimentary processes that take place in surface or near-surface environments, metamorphism most often occurs deep within Earth, beyond our direct observation. Notwithstanding this significant obstacle, geologists have developed techniques that have allowed them to learn a great deal about the conditions under which metamorphic rocks form. In turn, the study of metamorphic rocks provides important insights into the tectonic processes that operate within Earth's crust and upper mantle.

WHAT IS METAMORPHISM?

METAMORPHIC ROCKS
▶ Introduction

Recall from the discussion of the rock cycle that metamorphism is the transformation of one rock type into another. Metamorphic rocks are produced from preexisting igneous, sedimentary, or even from other metamorphic rocks. Thus, every metamorphic rock has a **parent rock**—the rock from which it was formed.

Metamorphism, which means to "change form," is a process that leads to changes in the mineralogy, texture, and sometimes the chemical composition of rocks. Metamorphism takes place where preexisting rock is subjected to new conditions, usually elevated temperatures and pressures, that are significantly different from those in which it initially formed. In response to these new conditions, the rock gradually changes until a state of equilibrium with the new environment is achieved.

The intensity of metamorphism can vary substantially from one environment to another. For example, in low-grade metamorphic environments, the common sedimentary rock *shale* becomes the more compact metamorphic rock *slate*. Hand samples of these rocks are sometimes difficult to distinguish, illustrating that the transition from sedimentary to metamorphic is often gradual and the changes can be subtle.

In more extreme environments, metamorphism causes a transformation so complete that the identity of the parent rock cannot be determined. In high-grade metamorphism, such features as bedding planes, fossils, and vesicles that existed in the parent rock are obliterated. Further, when rocks deep in the crust (where temperatures are high) are subjected to directed pressure, the entire mass may deform, producing large-scale structures, mainly folds (Figure 8.2).

In the most extreme metamorphic environments, the temperatures approach those at which rocks melt. However, *during metamorphism the rock remains essentially solid*—when complete melting occurs, we have entered the realm of igneous activity.

WHAT DRIVES METAMORPHISM?

METAMORPHIC ROCKS
▶ Agents of Metamorphism

The agents of metamorphism include *heat, pressure (stress), and chemically active fluids*. During metamorphism, rocks are usually subjected to all three metamorphic agents simultaneously. However, the degree of metamorphism and the contribution of each agent vary greatly from one environment to another.

FIGURE 8.1 Deformed metamorphic rocks exposed in a road cut in the Eastern Highland of Connecticut. (Photo by Phil Dombrowski)

Heat as a Metamorphic Agent

The most important factor driving metamorphism is *heat* because it provides the energy needed to drive the chemical reactions that result in the recrystallization of existing minerals and/or the formation of new minerals. Recall from the discussion of

igneous rocks that an increase in temperature causes the ions within a mineral to vibrate more rapidly. Even in a crystalline solid, where ions are strongly bonded, this elevated level of activity allows individual atoms to migrate more freely between sites in the crystalline structure.

Changes Caused by Heat When Earth materials are heated, especially those that form in low-temperature environments, they are affected in two ways. First, heating promotes recrystallization of mineral grains. This is particularly true of sedimentary and volcanic rocks that are composed of fine-grained clay and silt sized particles. Higher temperatures promote crystal growth where fine particles join together to form larger grains of the same mineralogy.

Second, when rocks are heated, they eventually reach a temperature at which one or more minerals become chemically unstable. When this occurs, the constituent atoms begin to arrange themselves into crystalline structures that are more stable in the new high-temperature environment. These chemical reactions create new minerals with stable configurations that have an overall composition roughly equivalent to that of the original rock. (In some environments ions may actually migrate into or out of a rock, thereby changing its overall chemical composition.)

To summarize, imagine being a rock collector who is crossing a region where metamorphic rocks have been uplifted and then exposed by erosional processes. If you are traveling from an area where metamorphism was less intense to one where it had been more intense, you would expect to observe two

FIGURE 8.2 Deformed and folded gneiss, Anza Borrego Desert State Park, California. (Photo by A. P. Trujillo/APT Photos)

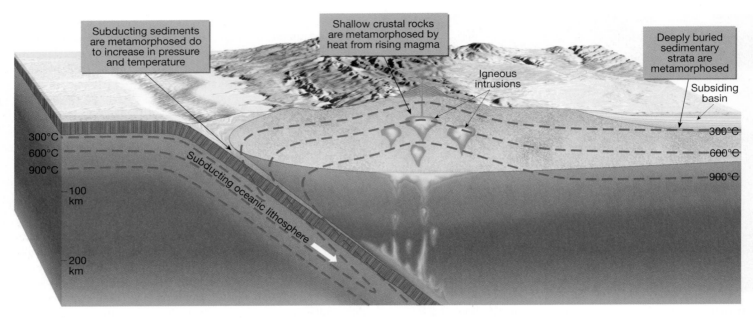

FIGURE 8.3   The geothermal gradient and its role in metamorphism. Notice how the geothermal gradient is lowered by the subduction of relatively cool oceanic lithosphere. By contrast, thermal heating is evident where magma intrudes the upper crust.

changes that were largely attributable to temperature. The average size of the crystals in the rock samples you collected would increase and your analysis of the rocks would show a change in mineralogy.

What is the Source of Heat?   Earth's internal heat comes mainly from energy that is continually being released by radioactive decay and thermal energy that remains from the time when our planet was forming. Recall that temperatures increase with depth at a rate known as the *geothermal gradient.* In the upper crust, this increase in temperature averages about 25 °C per kilometer (Figure 8.3). Thus, rocks that formed at Earth's surface will experience a gradual increase in temperature if they are transported to greater depths. When buried to a depth of about 8 kilometers (5 miles), where temperatures are about 200 °C, clay minerals tend to become unstable and begin to recrystallize into new minerals, such as chlorite and muscovite, that are stable in this environment. Chlorite is a micalike mineral formed by the metamorphism of dark (iron and magnesium rich) silicate minerals. However, many silicate minerals, particularly those found in crystalline igneous rocks—quartz and feldspar for example—remain stable at these temperatures. Thus, metamorphic changes in these minerals generally occur at much greater depths.

Environments where rocks may be carried to great depths and heated include convergent plate boundaries where slabs of sediment-laden oceanic crust are being subducted. Rocks may also become deeply buried in large basins where gradual subsidence results in very thick accumulations of sediment (see Figure 8.3). These basins, exemplified by the Gulf of Mexico, are known to develop low-grade metamorphic conditions near the base of the pile. In addition, continental collisions, which result in crustal thickening by folding and faulting, cause some rocks

to be uplifted while others are thrust downward where elevated temperatures may cause metamorphism.

Heat may also be transported from the mantle into even the shallowest layers of the crust by igneous intrusions (Figure 8.4). Rising mantle plumes, upwelling at mid-ocean ridges, and magma generated by partial melting of mantle rock at subduction zones are three examples. Anytime magma forms and buoyantly rises toward the surface, metamorphism occurs. When magma intrudes relatively cool rocks at shallow depths, the host rock is "baked." This process, called *contact metamorphism,* will be considered later in the chapter.

FIGURE 8.4   Earth's interior is the source of heat that drives metamorphism. Lava Lake in Pu'O'o Crater, Hawaii. (Photo by Frans Lanting/CORBIS)

Students Sometimes Ask . . .

How hot is it deep in the crust?

The increase in temperature with depth, based on the geothermal gradient, can be expressed as *the deeper one goes, the hotter it gets.* This relationship has been observed by miners in deep mines and in deeply drilled wells. In the deepest mine in the world (the Western Deep Levels mine in South Africa, which is 4 kilometers or 2.5 miles deep), the temperature of the surrounding rock is so hot that it can scorch human skin! In fact, miners often work in groups of two: one to mine the rock, and the other to operate a large fan that keeps the other worker cool.

The temperature is even higher at the bottom of the deepest well in the world, which was completed in the Kola Peninsula of Russia in 1992 and goes down a record 12.3 kilometers (7.7 miles). At this depth it is 245 °C (473 °F), much greater than the boiling point of water. What keeps water from boiling is the high confining pressure at depth.

direction perpendicular to that stress. As a result, the rocks involved are often *folded* or *flattened* (similar to stepping on a rubber ball). Along convergent plate boundaries the greatest differential stress is directed roughly horizontal in the direction of plate motion, and the least pressure is in the vertical direction. Consequently, in these settings the crust is greatly shortened (horizontally) and thickened (vertically). Although, differential stresses are generally small when compared to confining pressure, they are important in creating the various large scale structures and textures exhibited by metamorphic rocks.

In surface environments where temperatures are comparatively low, rocks are *brittle* and tend to fracture when subjected to differential stress. Continued deformation grinds and pulverizes the mineral grains into small fragments. By contrast, in high-temperature environments rocks are *ductile*. When rocks exhibit ductile behavior, their mineral grains tend to flatten and

Confining Pressure and Differential Stress

Pressure, like temperature, also increases with depth as the thickness of the overlying rock increases. Buried rocks are subjected to **confining pressure,** which is analogous to water pressure, in which the forces are applied equally in all directions (Figure 8.5A). The deeper you go in the ocean, the greater the confining pressure. The same is true for buried rock. Confining pressure causes the spaces between mineral grains to close, producing a more compact rock having a greater density (Figure 8.5A). Furthermore, as confining pressure increases some minerals recrystallize into new minerals that have the same chemical composition but a more compact crystalline form. Confining pressure does *not,* however, fold and deform rocks like those shown in Figure 8.1.

In addition to confining pressure, rocks may be subjected to directed pressure. This occurs, for example, at convergent plate boundaries where slabs of lithosphere collide. Here the forces that deform rock are unequal in different directions and are referred to as **differential stress.** (A more in-depth discussion of *differential stress* is provided in Chapter 10).

Unlike confining pressure, which "squeezes" rock equally in all directions, differential stresses are greater in one direction than in others. As shown in Figure 8.5B, rocks subjected to differential stress are shortened in the direction of greatest stress and elongated, or lengthened, in the

FIGURE 8.5 Confining pressure and differential stress as metamorphic agents. **A.** In a depositional environment, as confining pressure increases, rocks deform by decreasing in volume. **B.** During mountain building, rocks subjected to differential stress are shortened in the direction that pressure is applied, and lengthened in the direction perpendicular to that force.

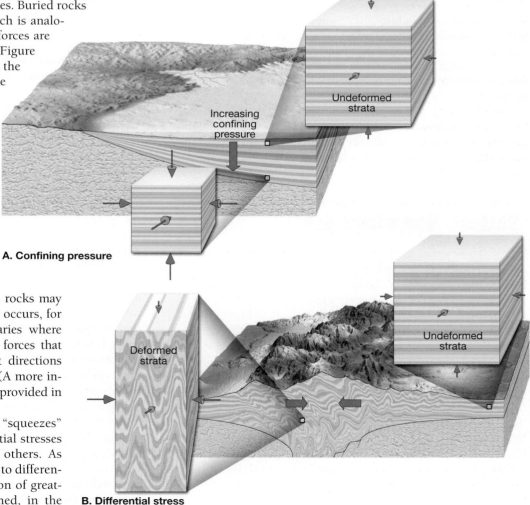

A. Confining pressure

B. Differential stress

FIGURE 8.6 Metaconglomerate, also called stretched pebble conglomerate. These once nearly spherical pebbles have been heated and flattened into elongated structures. (Photo by E. J. Tarbuck)

elongate when subjected to differential stress (Figure 8.6). This accounts for their ability to deform by flowing (rather than fracturing) to generate intricate folds.

Chemically Active Fluids

Many minerals, including clays, micas, and amphiboles, are hydrated—meaning they contain water in their crystalline structures. Elevated temperatures and pressures cause the dehydration of these minerals. Once expelled, these hot fluids promote recrystallization by enhancing the migration of mineral matter.

As discussed earlier, the metamorphism of shale to slate involves clay minerals that recrystallize to form mica and chlorite minerals. Hot fluids enhance this process by dissolving and transporting ions from one site in the crystal structure to another. In increasingly hotter environments these fluids become correspondingly more reactive.

In some metamorphic environments, hot fluids transport mineral matter over considerable distances. This occurs, for example, when hot, mineral-rich fluids are expelled from a magma body as it cools and solidifies. If the rocks that surround the pluton differ markedly in composition from the invading fluids, there may be an exchange of ions between the fluids and host rocks. When this occurs, the overall chemical composition of the surrounding rock changes. When substantial chemical change accompanies metamorphism the process is called **metasomatism.**

The Importance of Parent Rock

Most metamorphic rocks have the same overall chemical composition as the parent rock from which they formed, except for the possible loss or acquisition of volatiles such as water (H_2O) and carbon dioxide (CO_2). Therefore, when trying to establish the parent material from which metamorphic rocks were derived, the most important clue comes from their chemical composition.

Consider the large exposures of the metamorphic rock marble found high in the Alps of southern Europe. Because marble and the common sedimentary rock limestone have the same mineralogy (calcite, $CaCO_3$), it seems reasonable to conclude that limestone is the parent rock of marble. Furthermore, because limestone usually forms in warm, shallow marine environments we can surmise that considerable deformation must have occurred to convert limy deposits in a shallow sea into marble crags in the lofty Alps.

The mineral makeup of the parent rock also largely determines the degree to which each metamorphic agent will cause change. For example, when magma forces its way into surrounding rock, high temperatures and hot fluids may alter the host rock. If the host rock is composed of minerals that are comparatively unreactive, such as quartz grains in sandstone, any alterations that may occur will be confined to a narrow zone next to the pluton. However, when the host rock is limestone, which is highly reactive, the zone of metamorphism may extend far from the intrusion.

METAMORPHIC TEXTURES

METAMORPHIC ROCKS
▶ Textural and Mineralogical Changes

Recall that the term **texture** is used to describe the size, shape, and arrangement of grains within a rock. Most igneous and many sedimentary rocks consist of mineral grains that have a random orientation and thus appear the same when viewed from any direction. By contrast, deformed metamorphic rocks that contain platy minerals (micas) and/or elongated minerals (amphiboles), typically display some kind of *preferred orientation* in which the mineral grains exhibit a parallel to sub-parallel alignment. Like a fistful of pencils, rocks containing elongated minerals that are oriented parallel to each other will appear different when viewed from the side than when viewed head-on. A rock that exhibits a preferred orientation of its minerals is said to possess *foliation*.

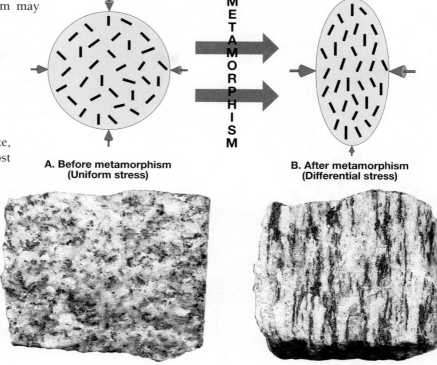

A. Before metamorphism (Uniform stress)

B. After metamorphism (Differential stress)

FIGURE 8.7 Mechanical rotation of platy or elongated mineral grains. **A.** Existing mineral grains keep their random orientation if force is uniformly applied. **B.** As differential stress causes rocks to flatten, mineral grains rotate toward the plane of flattening. (Photos by E. J. Tarbuck)

Foliation

The term **foliation** refers to any planar (nearly flat) arrangement of mineral grains or structural features within a rock. Although foliation may occur in some sedimentary and even a few types of igneous rocks, it is a fundamental characteristic of regionally metamorphosed rocks—that is, rock units that have been strongly deformed, mainly by folding. In metamorphic environments, foliation is ultimately driven by compressional stresses that shorten rock units, causing mineral grains in preexisting rocks to develop parallel, or nearly parallel, alignments. Examples of foliation include the parallel alignment of platy minerals; the parallel alignment of flattened pebbles; compositional banding in which the separation of dark and light minerals generates a layered appearance; and rock cleavage where rocks can be easily split into tabular slabs. These diverse types of foliation can form in many different ways, including:

1. Rotation of platy and/or elongated mineral grains into a parallel or nearly parallel orientation.
2. Recrystallization that produces new minerals with grains that exhibit a preferred orientation.
3. Mechanisms that change spherically shaped grains into elongated shapes that are aligned in a preferred orientation.

The rotation of existing mineral grains is the easiest of these mechanisms to envision. Figure 8.7 illustrates the mechanics by which platy or elongated minerals are rotated. Note that the new alignment is roughly perpendicular to the direction of maxi-

mum shortening. Although physical rotation of platy minerals contributes to the development of foliation in low-grade metamorphism, other mechanisms dominate in more extreme environments.

Recall that recrystallization is the creation of new mineral grains out of old ones. When recrystallization occurs as rock is being subjected to differential stresses, any elongated and platy minerals that form tend to recrystallize perpendicular to the direction of maximum stress. Thus, the newly formed mineral grains will possess a parallel alignment and the metamorphic rock containing them will exhibit foliation.

Mechanisms that change the shapes of existing grains are especially important for the development of preferred orientations in rocks that contain minerals such as quartz, calcite, and olivine—minerals that normally develop roughly spherical crystals.

These processes operate in metamorphic environments where differential stresses exist. A change in grain shape can occur as units of a mineral's crystalline structure slide relative to one another along discrete planes, thereby distorting the grain as shown in Figure 8.8. This type of gradual solid-state flow involves slippage that disrupts the crystal lattice as atoms shift positions. This process involves the breaking of existing chemical bonds and the formation of new ones.

The shape of a mineral may also change as ions move from one location along the margin of the grain that is highly stressed

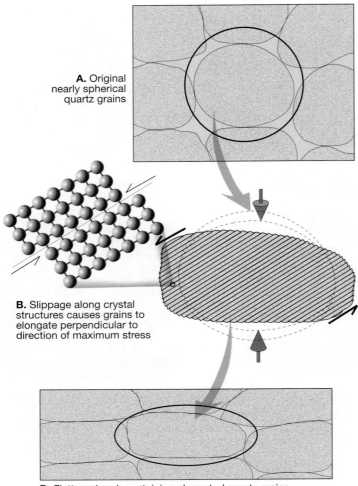

A. Original nearly spherical quartz grains

B. Slippage along crystal structures causes grains to elongate perpendicular to direction of maximum stress

D. Flattened rock containing elongated quartz grains

FIGURE 8.8   Development of preferred orientations of minerals that have roughly spherical crystals, such as quartz, calcite, and olivine can occur in one of two ways. The first mechanism, shown in this illustration, occurs as units of a mineral's crystalline structure slide relative to one another. The other mechanism is shown in Figure 8.9.

to a less-stressed position on the same grain (Figure 8.9). Mineral matter dissolves where grains are in contact with each other (areas of high stress) and precipitates in pore spaces (areas of low stress). As a result, the mineral grains tend to become elongated in the direction of maximum stress. This mechanism is aided by hot, chemically active fluids.

Foliated Textures

Various types of foliation exist, depending largely upon the grade of metamorphism and the mineralogy of the parent rock. We will look at three: *rock* or *slaty cleavage, schistosity,* and *gneissic texture.*

Rock or Slaty Cleavage   **Rock cleavage** refers to closely spaced, flat surfaces along which rocks split into thin slabs when hit with a hammer. Rock cleavage develops in various metamorphic rocks but is best displayed in slates that exhibit an excellent splitting property called **slaty cleavage.**

Depending on the metamorphic environment and the composition of the parent rock, rock cleavage develops in a number of different ways. In a low-grade metamorphic environment, rock cleavage is known to develop where beds of shale (and related sedimentary rocks) are strongly folded and metamorphosed to form slate. The process begins as platy grains are kinked and bent—generating microscopic folds having limbs (sides) that are roughly aligned (Figure 8.10). With further deformation, this new alignment is enhanced as old grains break down and recrystallize preferentially in the direction of the newly developed orientation. In this manner the rock develops narrow parallel zones where mica flakes are concentrated. These features alternate with zones containing quartz and other mineral grains that do not exhibit a pronounced linear orientation. It is along these very thin zones of platy mineral that slate splits (Figure 8.11).

Because slate typically forms during the low-grade metamorphism of shale, evidence of the original sedimentary bedding planes is often preserved. However, as Figure 8.10D illustrates, the orientation of slate's cleavage usually develops at an oblique

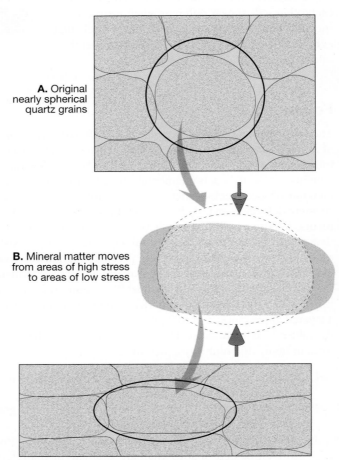

A. Original nearly spherical quartz grains

B. Mineral matter moves from areas of high stress to areas of low stress

D. Flattened rock containing elongated quartz grains

FIGURE 8.9   This mechanism for changing the shape of mineral grains involves dissolving material from areas of high stress and depositing that material in locations of low stress. This mechanism, as well as the one shown in Figure 8.8, changes the shape of mineral grains, but not their volume and composition.

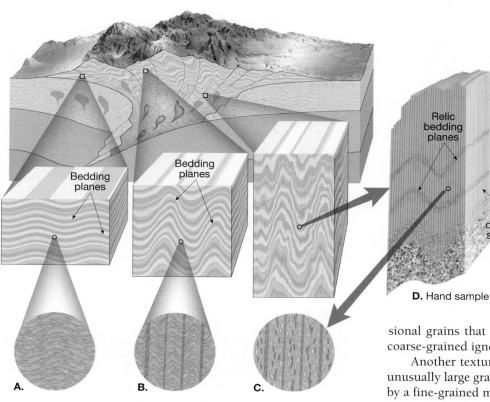

D. Hand sample

FIGURE 8.10 Development of rock cleavage. As shale is strongly folded (**A, B**) and metamorphosed to form slate, the developing mica flakes are bent into microfolds. **C.** Further metamorphism results in the recrystallization of mica grains along the limbs of these folds to enhance the foliation. **D.** This hand sample of slate illustrates rock cleavage and its orientation to relic bedding surfaces.

Other Metamorphic Textures

Not all metamorphic rocks exhibit a foliated texture. Those that *do not* are referred to as **nonfoliated.** Nonfoliated metamorphic rocks typically develop in environments where deformation is minimal and the parent rocks are composed of minerals that exhibit equidimensional crystals, such as quartz or calcite. For example, when a fine-grained limestone (made of calcite) is metamorphosed by the intrusion of a hot magma body, the small calcite grains recrystallize to form larger interlocking crystals. The resulting rock, *marble,* exhibits large, equidimensional grains that are randomly oriented, similar to those in a coarse-grained igneous rock.

Another texture common to metamorphic rocks consists of unusually large grains, called *porphyroblasts,* that are surrounded by a fine-grained matrix of other minerals. **Porphyroblastic textures** develop in a wide range of rock types and metamorphic environments when minerals in the parent rock recrystallize to form new minerals. During recrystallization certain metamorphic

angle to the original sedimentary layers. Thus, unlike shale, which splits along bedding planes, slate often splits across them. Other metamorphic rocks, such as schists and gneisses, may also split along planar surfaces and exhibit rock cleavage.

Schistosity   Under higher temperature-pressure regimes, the minute mica and chlorite crystals in slate begin to grow. When these platy minerals are large enough to be discernible with the unaided eye and exhibit a planar or layered structure, the rock is said to exhibit a type of foliation called **schistosity.** Rocks having this texture are referred to as *schist.* In addition to platy minerals, schist often contains deformed quartz and feldspar crystals that appear flat or lens-shaped and hidden among the mica grains.

Gneissic Texture   During high-grade metamorphism, ion migration can result in the segregation of minerals as shown in Figure 8.12. Notice that the dark biotite crystals and light silicate minerals (quartz and feldspar) have separated, giving the rock a banded appearance called **gneissic texture.** A metamorphic rock with this texture is called *gneiss* (pronounced "nice"). Although foliated, gneisses will not usually split as easily as slates and some schists.

FIGURE 8.11   Excellent slaty cleavage is exhibited by the rock in this slate quarry in Tanzania. Because slate breaks into flat slabs, it has many uses. (Photo by Randy Olson/NGS Image Collection) The inset photo shows the use of slate for the roof of this house in Switzerland. (Photo by E. J. Tarbuck)

FIGURE 8.12   This rock displays a gneissic texture. Notice that the dark biotite flakes and light silicate minerals are segregated, giving the rock a banded or layered appearance. (Photo by E. J. Tarbuck)

minerals, including garnet, staurolite, and andalusite, often develop *a small number of very large crystals*. By contrast, minerals such as muscovite, biotite, and quartz typically form *a large number of very small grains*. As a result, when metamorphism generates the minerals garnet, biotite, and muscovite in the same setting, the rock will contain large crystals (porphyroblasts) of garnet embedded in a finer-grained matrix of biotite and muscovite (Figure 8.13).

Close up
of porphyroblast

FIGURE 8.13   Garnet-mica schist. The dark red garnet crystals (porphyroblasts) are embedded in a matrix of fine-grained micas. (Photo by E. J. Tarbuck)

COMMON METAMORPHIC ROCKS

METAMORPHIC ROCKS
▶ Common Metamorphic Rocks

Recall that metamorphism causes many changes in rocks, including increased density, change in grain size, reorientation of mineral grains into a planar arrangement known as foliation, and the transformation of low-temperature minerals into high-temperature minerals. Moreover, the introduction of ions may generate new minerals, some of which are economically important.

The major characteristics of some common metamorphic rocks are summarized in Figure 8.14. Notice that metamorphic rocks can be broadly classified by the type of foliation exhibited and to a lesser extent on the chemical composition of the parent rock.

Foliated Rocks

Slate   Slate is a very fine-grained (less than 0.5 millimeter) foliated rock composed of minute mica flakes that are too small to be visible. Thus, slate generally appears dull and closely resembles shale. A noteworthy characteristic of slate is its excellent rock cleavage, or tendency to break into flat slabs (see Figure 8.11).

Slate is most often generated by the low-grade metamorphism of shale, mudstone, or siltstone. Less frequently it is produced when volcanic ash is metamorphosed. Slate's color depends on its mineral constituents. Black (carbonaceous) slate contains organic material, red slate gets its color from iron oxide, and green slate usually contains chlorite.

Phyllite   Phyllite represents a gradation in the degree of metamorphism between slate and schist. Its constituent platy minerals are larger than those in slate but not yet large enough to be readily identifiable with the unaided eye. Although phyllite appears similar to slate, it can be easily distinguished from slate by its glossy sheen and its sometimes wavy surface (Figure 8.15). Phyllite usually exhibits rock cleavage and is composed mainly of very fine crystals of either muscovite, chlorite, or both.

Schist   Schists are medium- to coarse-grained metamorphic rocks in which platy minerals predominate. These flat components commonly include the micas (muscovite and biotite), which display a planar alignment that gives the rock its foliated texture. In addition, schists contain smaller amounts of other minerals, often quartz and feldspar. Schists composed mostly of dark minerals (amphiboles) are known. Like slate, the parent rock of many schists is shale, which has undergone medium- to high-grade metamorphism during major mountain-building episodes.

The term *schist* describes the texture of a rock and as such it is used to describe rocks having a wide variety of chemical compositions. To indicate the composition, mineral names are used. For example, schists composed primarily of muscovite and

Rock Name	Texture		Grain Size	Comments	Original Parent Rock
Slate	Foliated (Increasing Metamorphism)		Very fine	Excellent rock cleavage, smooth dull surfaces	Shale, mudstone, or siltstone
Phyllite			Fine	Breaks along wavy surfaces, glossy sheen	Shale, mudstone, or siltstone
Schist			Medium to Coarse	Micaceous minerals dominate, scaly foliation	Shale, mudstone, or siltstone
Gneiss			Medium to Coarse	Compositional banding due to segregation of minerals	Shale, granite, or volcanic rocks
Migmatite			Medium to Coarse	Banded rock with zones of light-colored crystalline minerals	Shale, granite, or volcanic rocks
Mylonite	Weakly Foliated		Fine	When very fine-grained, resembles chert, often breaks into slabs	Any rock type
Metaconglomerate			Coarse-grained	Stretched pebbles with preferred orientation	Quartz-rich conglomerate
Marble	Nonfoliated		Medium to coarse	Interlocking calcite or dolomite grains	Limestone, dolostone
Quartzite			Medium to coarse	Fused quartz grains, massive, very hard	Quartz sandstone
Hornfels			Fine	Usually, dark massive rock with dull luster	Any rock type
Anthracite			Fine	Shiny black rock that may exhibit conchoidal fracture	Bituminous coal
Fault breccia			Medium to very coarse	Broken fragments in a haphazard arrangement	Any rock type

FIGURE 8.14 Classification of common metamorphic rocks.

A.

B.

FIGURE 8.15 Phyllite (**A.**) can be distinguished from slate (**B.**) by its glossy sheen and wavy surface. (Photos by E. J. Tarbuck)

FIGURE 8.16 Mica schist. This sample of schist is composed mostly of muscovite and biotite. (Photo by E. J. Tarbuck)

biotite are called *mica schist* (Figure 8.16). Depending upon the degree of metamorphism and composition of the parent rock, mica schists often contain *accessory minerals,* some of which are unique to metamorphic rocks. Some common accessory minerals that occur as porphyroblasts include *garnet, staurolite,* and *sillimanite,* in which case the rock is called *garnet-mica schist, staurolite-mica schist,* and so forth (see Figure 8.13).

In addition, schists may be composed largely of the minerals chlorite or talc, in which case they are called *chlorite schist* and *talc schist,* respectively. Both chlorite and talc schists can form when rocks with a basaltic composition undergo metamorphism. Others contain the mineral *graphite,* which is used as pencil "lead," graphite fibers (used in fishing rods), and lubricant (commonly for locks).

Gneiss *Gneiss* is the term applied to medium- to coarse-grained banded metamorphic rocks in which granular and elongated (as opposed to platy) minerals predominate. The most common minerals in gneiss are quartz, potassium feldspar, and sodium-rich plagioclase feldspar. Most gneisses also contain lesser amounts of biotite, muscovite, and amphibole that develop a preferred orientation. Some gneisses will split along the layers of platy minerals, but most break in an irregular fashion.

Recall that during high-grade metamorphism the light and dark components separate, giving gneisses their characteristic banded or layered appearance. Thus, most gneisses consist of alternating bands of white or reddish feldspar-rich zones and layers of dark ferromagnesian minerals (see Figure 8.12). These banded gneisses often exhibit evidence of deformation, including folds and sometimes faults (see Figure 8.1).

Most gneisses have a felsic composition and are often derived from granite or its fine-grained equivalent, rhyolite. However, many form from the high-grade metamorphism of shale. In this instance, gneiss represents the last rock in the sequence of shale, slate, phyllite, schist, and gneiss. Like schists, gneisses may also include large crystals of accessory minerals such as garnet and staurolite. Gneisses made up primarily of dark minerals such as those that compose basalt also occur. For example, an amphibole-rich rock that exhibits a gneissic texture is called *amphibolite.*

Nonfoliated Rocks

Marble Marble is a coarse, crystalline metamorphic rock whose parent was limestone or dolostone (Figure 8.17). Pure marble is white and composed essentially of the mineral calcite. Because of its relative softness (hardness of 3), marble is easy to cut and shape. White marble is particularly prized as a stone from which to create monuments and statues, such as the Lincoln Memorial in Washington, D.C. (Figure 8.18). Unfortunately, marble's composition of calcium carbonate causes it to weather when exposed to acid rain.

The parent rocks from which most marbles form contain impurities that color the stone. Thus, marble can be pink, gray, green, or even black and may contain a variety of accessory minerals (chlorite, mica, garnet, and wollastonite). When marble forms from limestone interbedded with shales, it will appear banded and exhibit visible foliation. When deformed, these banded marbles may develop highly contorted mica-rich folds that give the rock a rather artistic design. Hence, these decorative marbles have been used as a building stone since prehistoric times (Figure 8.18).

Quartzite Quartzite is a very hard metamorphic rock formed from quartz sandstone (Figure 8.19). Under moderate- to high-

Photomicrograph (6.5x)

FIGURE 8.17 Marble, a crystalline rock formed by the metamorphism of limestone. Photomicrograph shows interlocking calcite crystals using polarized light. (Photos by E. J. Tarbuck)

A.

B.

FIGURE 8.18 Marble, because of its workability, is a widely used building stone. **A.** The white exterior of the Lincoln Memorial in Washington, D.C. is constructed mainly of marble that was quarried in Marble, Colorado. Inside, pink Tennessee marble was used for the floors, Alabama marble for the ceilings, and Georgia marble for Lincoln's statue. (Photo by Ryan McGinnis/Alamy). **B.** The exterior of the Taj Mahal is constructed primarily of the metamorphic rock marble. (Photo by Steve Vider/Superstock)

grade metamorphism, the quartz grains in sandstone fuse together (inset in Figure 8.19). The recrystallization is often so complete that when broken, quartzite will split through the quartz grains rather than along their boundaries. In some instances, sedimentary features such as cross-bedding are preserved and give the rock a banded appearance. Pure quartzite is white, but iron oxide may produce reddish or pinkish stains, while dark mineral grains may impart a gray color.

METAMORPHIC ENVIRONMENTS

There are many environments in which metamorphism occurs. Most are in the vicinity of plate margins, and several are associated with igneous activity. We will consider the following types of metamorphism: (1) *contact* or *thermal metamorphism;* (2) *hydrothermal metamorphism;* (3) *burial and subduction zone metamorphism;* (4) *regional metamorphism;* (5) *impact metamorphism;* and (6) *metamorphism along faults.* With the exception of impact metamorphism, there is considerable overlap among the other types.

Contact or Thermal Metamorphism

Contact or **thermal metamorphism** occurs when rocks immediately surrounding a molten igneous body are "baked" and therefore altered from their original state. The altered rocks occur in a zone called a metamorphic **aureole** (Figure 8.20). The emplacement of small intrusions such as dikes and sills typically form aureoles only a few centimeters thick. While large igneous plutons that generate batholiths can produce aureoles that extend outward for several kilometers.

In addition to the size of the magma body, the mineral composition of the host rock and the availability of water greatly affect the size of the aureole produced. In chemically active rock such as limestone, the zone of alteration can be 10 kilometers

Photomicrograph (26.6x)

FIGURE 8.19 Quartzite is a nonfoliated metamorphic rock formed from quartz sandstone. The photomicrograph shows the interlocking quartz grains typical of quartzite. (Photos by E. J. Tarbuck)

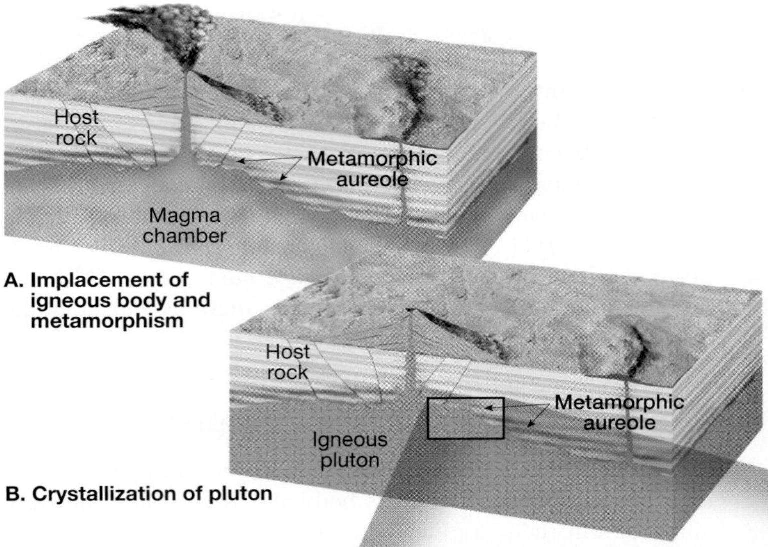

sure, rocks found within a metamorphic aureole are usually not foliated.

During contact metamorphism of mudstones and shales, the clay minerals are baked as if placed in a kiln. The result is a very hard, fine-grained metamorphic rock called *hornfels* (Figure 8.21). Hornfels can form from a variety of materials including volcanic ash and basalt. In some cases, large grains of metamorphic minerals, such as garnet and staurolite, may form, giving the hornfels a porphyroblastic texture (see Figure 8.13).

Hydrothermal Metamorphism

When hot, ion-rich fluids circulate through fissures and cracks in rock, a chemical alteration called **hydrothermal metamorphism** occurs (Figure 8.22). This type of metamorphism is often closely associated with the emplacement of magma. As large magma bodies cool and solidify, silica-rich fluids (mainly water) are driven into the host rocks. When the host rock is highly fractured, mineral matter contained in these **hydrothermal solutions** may precipitate to form a variety of minerals, some of which are economically important. If the host rocks are permeable and highly reactive, such as the carbonate rock limestone, silicate-rich hydrothermal solutions react to produce a variety of calcium-rich silicate minerals. Recall that a metamorphic process that alters the overall chemical composition of a rock unit is called *metasomatism*.

As our understanding of plate tectonics grew, it became clear that the most widespread occurrence of hydrothermal

FIGURE 8.20 Contact metamorphism produces a zone of alteration called an *aureole* around an intrusive igneous body. In the photo, the dark layer, called a *roof pendant,* consists of metamorphosed host rock adjacent to the upper part of the light-colored igneous pluton. The term *roof pendant* implies that the rock was once the roof of a magma chamber. Sierra Nevada, near Bishop, California. (Photo by John S. Shelton)

A. Implacement of igneous body and metamorphism

B. Crystallization of pluton

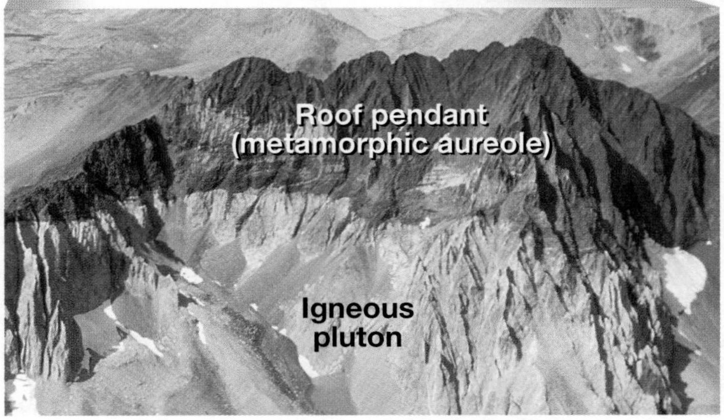

C. Uplift and erosion expose pluton and metamorphic cap rock

(6 miles) thick. These large aureoles often consist of distinct *zones of metamorphism*. Near the magma body, high-temperature minerals such as garnet may form, whereas farther away low-grade minerals such as chlorite are produced.

Although contact metamorphism is not entirely restricted to shallow crustal depths, it is most easily recognized when it occurs in this setting. Here, the temperature contrast between the molten body and the surrounding host rock is large. Because contact metamorphism does not involve directed pres-

Students Sometimes Ask . . .

Recently I helped a friend move, who had a billiard table that was really heavy. He said its surface was made of slate. Is this true?

Yes, and it must have cost your friend quite a bit of money. Only the best-quality billiard tables have surfaces made of slate. Slate—a very fine-grained foliated rock composed of microscopic mica flakes—has the ability to split easily along its rock cleavage planes, producing flat slabs of smooth rock. It is highly prized for use as billiard-table surfaces as well as for building materials such as floor and roof tiles.

FIGURE 8.21 Contact metamorphism of shale yields hornfels, while contact metamorphism of quartz sandstone and limestone produces quartzite and marble, respectively.

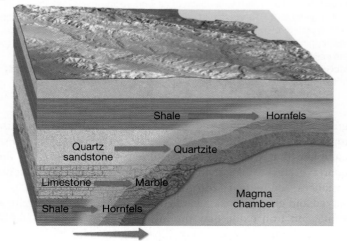

Increasing metamorphic grade

FIGURE 8.22 Hydrothermal metamorphism can occur at shallow crustal depths in regions where hot springs and geysers are active. (Photo by Philippe Clement/Nature Picture Library)

metamorphism is along the axis of the mid-ocean ridge system. As plates move apart, upwelling magma from the mantle generates new seafloor. As seawater percolates through the young, hot oceanic crust, it is heated and chemically reacts with the newly formed basaltic rocks (Figure 8.23). The result is the conversion of ferromagnesian minerals, such as olivine and pyroxene, into hydrated silicates, such as serpentine, chlorite, and talc. In addition, calcium-rich plagioclase feldspars in basalt become increasingly sodium enriched as the salt (NaCl) in seawater exchanges Na ions for Ca ions.

Hydrothermal solutions circulating through the seafloor also remove large amounts of metals, such as iron, cobalt, nickel, silver, gold, and copper, from the newly formed crust. These hot, metal-rich fluids eventually rise along fractures and gush from the seafloor at temperatures of about 350 °C, generating particle-filled clouds called *black smokers*. Upon mixing with the cold seawater, sulfides and carbonate minerals containing these heavy metals precipitate to form metallic deposits, some of which are economically valuable. This process is believed to be the origin of the copper ores mined today on the Mediterranean island of Cyprus.

FIGURE 8.23 Hydrothermal metamorphism along a mid-ocean ridge. (Photo by R. Ballard/Woods Hole)

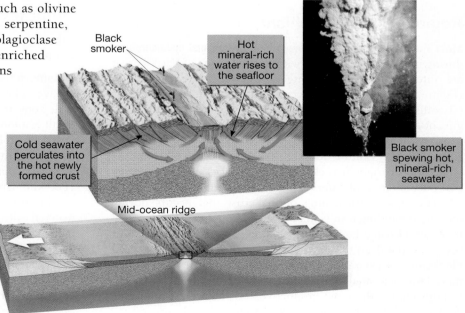

Black smoker

Hot mineral-rich water rises to the seafloor

Cold seawater perculates into the hot newly formed crust

Black smoker spewing hot, mineral-rich seawater

Mid-ocean ridge

Burial and Subduction Zone Metamorphism

Burial metamorphism tends to occur where massive amounts of sedimentary or volcanic material accumulates in a subsiding basin (see Figure 8.3). Here, low-grade metamorphic conditions may be attained within the deepest layers. Confining pressure and geothermal heat drive the recrystallization of the constituent minerals—changing the texture and/or mineralogy of the rock without appreciable deformation.

The depth required for burial metamorphism varies from one location to another, depending mainly on the prevailing geothermal gradient. Metamorphism typically begins at depths of about 8 kilometers (5 miles), where temperatures are about 200 °C. However, in areas that exhibit large geothermal gradients and where molten rock has been emplaced near the surface, such as near the Salton Sea in California and in northern New Zealand, drilling operations have collected metamorphic minerals from depths of only a few kilometers.

Rocks and sediments can also be carried to great depths along convergent boundaries where oceanic lithosphere is being subducted. This phenomenon, called **subduction zone metamorphism,** differs from burial metamorphism in that differential stresses play a major role in deforming rock as it is metamorphosed. Furthermore, metamorphic rocks that form along subduction zones are often further metamorphosed by the collision of two continental blocks. A discussion of this process is found later in the chapter in the section entitled "Interpreting Metamorphic Environments."

Regional Metamorphism

Most metamorphic rock is produced by **regional metamorphism** during mountain building when large segments of Earth's crust are intensely deformed along convergent plate boundaries (Figure 8.24). This activity occurs most often during continental collisions. Sediments and crustal rocks that form the margins of the colliding continental blocks are folded and faulted, causing them to shorten and thicken like a rumpled carpet (Figure 8.24). Continental collisions also involve crystalline continental basement rocks, as well as slices of oceanic crust that once floored the intervening ocean basin.

The general thickening of the crust that occurs during mountain building results in buoyant lifting, in which deformed rocks are elevated high above sea level. Crustal thickening also results in the deep burial of large quantities of rock as crustal blocks are thrust one beneath another. Deep in the roots of mountains, elevated temperatures caused by deep burial are responsible for the most productive and intense metamorphic activity within a mountain belt. Often, these deeply buried

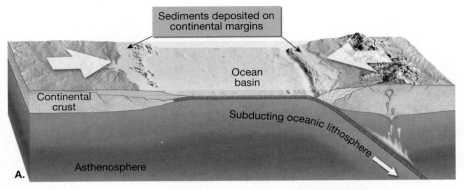

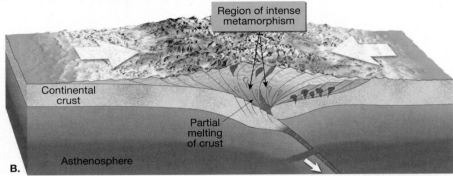

FIGURE 8.24   Regional metamorphism occurs where rocks are squeezed between two converging lithospheric plates during mountain building.

rocks become heated to their melting point. As a result, magma collects until it forms bodies large enough to buoyantly rise and intrude the overlying metamorphic and sedimentary rocks (see Figure 8.24). Consequently, the cores of many mountain ranges consist of folded and faulted metamorphic rocks, often intertwined with igneous bodies. Over time, these deformed rock masses are uplifted, and erosion removes the overlying material to expose the igneous and metamorphic rocks that comprise the central core of the mountain range.

Other Metamorphic Environments

Other types of metamorphism that generate relatively small amounts of metamorphic rock tend to be localized.

Metamorphism along Fault Zones   Near the surface, rock behaves like a brittle solid. Consequently, movement along a fault zone fractures and pulverizes rock (Figure 8.25A). The result is a loosely coherent rock called *fault breccia* that is composed of broken and crushed rock fragments (Figure 8.25A). Displacements along California's San Andreas Fault have created a zone of fault breccia and related rock types more than 1000 kilometers long and up to 3 kilometers wide.

In some shallow fault zones a soft, uncemented claylike material called *fault gouge* is also produced. Fault gouge is formed by the crushing and grinding of rock material during fault movement. The resulting crushed material is further altered by groundwater that infiltrates the porous fault zone.

Much of the deformation associated with fault zones occurs at great depth and thus at high temperatures. In this environment preexisting minerals deform by ductile flow (Figure 8.25B).

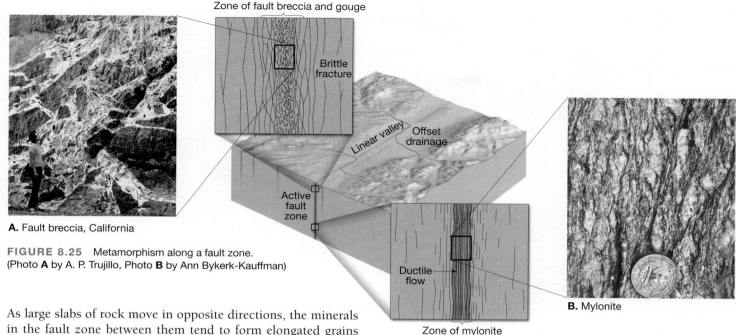

A. Fault breccia, California

FIGURE 8.25 Metamorphism along a fault zone. (Photo **A** by A. P. Trujillo, Photo **B** by Ann Bykerk-Kauffman)

B. Mylonite

As large slabs of rock move in opposite directions, the minerals in the fault zone between them tend to form elongated grains that give the rock a foliated or lineated appearance. Rocks formed in these zones of intense ductile deformation are termed *mylonites* (*mylo* = a mill, *ite* = a stone).

Impact Metamorphism **Impact** (or **shock**) **metamorphism** occurs when high-speed projectiles called *meteorites* (fragments of comets or asteroids) strike Earth's surface. Upon impact the energy of the once rapidly moving meteorite is transformed into heat energy and shock waves that pass through the surrounding rocks. The result is pulverized, shattered, and sometimes melted rock.

The products of these impacts, called *impactiles,* include mixtures of fused fragmented rock plus glass-rich ejecta that resemble volcanic bombs (see Box 8.1). In some cases, a very dense form of quartz (*coesite*) and minute *diamonds* are found. These high-pressure minerals provide convincing evidence that pressures and temperatures as great as those existing in the upper mantle must have been attained for at least a brief moment.

METAMORPHIC ZONES

In areas affected by metamorphism, there are usually systematic variations in the mineralogy and texture of the rocks that can be observed as we traverse the region. These differences are clearly related to variations in the degree of metamorphism experienced in each metamorphic zone.

Textural Variations

When we begin with a clay-rich sedimentary rock such as shale or mudstone, a gradual increase in metamorphic intensity is accompanied by a general coarsening of the grain size. Thus, we observe shale changing to a fine-grained slate, which then forms phyllite and, through continued recrystallization, generates a coarse-grained schist (Figure 8.26).

FIGURE 8.26 Idealized illustration of progressive regional metamorphism. From left to right, we progress from low-grade metamorphism (slate) to high-grade metamorphism (gneiss). (Photos by E. J. Tarbuck)

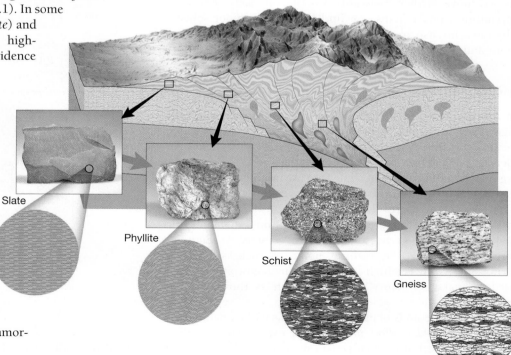

Slate

Phyllite

Schist

Gneiss

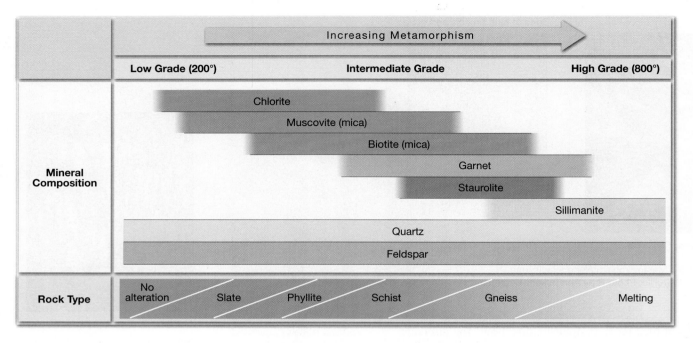

FIGURE 8.27   The typical transition in mineralogy that results from the progressive metamorphism of shale.

Under more intense conditions a gneissic texture that exhibits layers of dark and light minerals may develop. This systematic transition in metamorphic textures can be observed as we approach the Appalachian Mountains from the west. Beds of shale, which once extended over large areas of the eastern United States, still occur as nearly flat-lying strata in Ohio. However, in the broadly folded Appalachians of central Pennsylvania, the rocks that once formed flat-lying beds are folded and display a preferred orientation of platy mineral grains as exhibited by well-developed slaty cleavage. As we move farther eastward into the intensely deformed crystalline Appalachians, we find large outcrops of schists. The most intense zones of metamorphism are found in Vermont and New Hampshire, where gneissic rocks outcrop.

Index Minerals and Metamorphic Grade

In addition to textural changes, we encounter corresponding changes in mineralogy as we shift from regions of low-grade metamorphism to regions of high-grade metamorphism. An idealized transition in mineralogy that results from the regional metamorphism of shale is shown in Figure 8.27. The first new mineral to form as shale changes to slate is chlorite. At higher temperatures flakes of muscovite and biotite begin to dominate. Under more extreme conditions, metamorphic rocks may contain garnet and staurolite crystals. At temperatures approaching the melting point of rock, sillimanite forms. Sillimanite is a high-temperature metamorphic mineral used to make refractory porcelains such as those used in spark plugs.

Through the study of metamorphic rocks in their natural settings (called *field studies*) and through experimental studies, researchers have learned that certain minerals, such as those in Figure 8.27, are good indicators of the metamorphic environment in which they formed. Using these **index minerals,** geologists distinguish among different zones of regional metamorphism. For example, the mineral chlorite begins to form when temperatures are relatively low, less than 200 °C (Figure 8.28). Thus, rocks that contain chlorite (usually slates) are referred to as *low-grade*. By

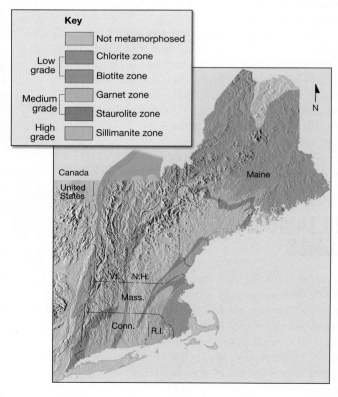

FIGURE 8.28   Zones of metamorphic intensities in New England.

UNDERSTANDING EARTH

BOX 8.1

Impact Metamorphism and Tektites

It is now clear that comets and asteroids have collided with Earth far more frequently than was once assumed. The evidence: More than 100 giant impact structures have been identified to date. Previously many of these structures were thought to have resulted from some poorly understood volcanic process. Most impact structures, like Manicouagan in Quebec, are so old and weathered that they no longer resemble an impact crater. A notable exception is the very fresh-looking Meteor Crater in Arizona (Figure 8.A).

One earmark of impact craters is *shock metamorphism.* When high-velocity projectiles (comets, asteroids) impact Earth's surface, pressures reach millions of atmospheres and temperatures exceed 2000 °C momentarily. The result is pulverized, shattered, and melted rock. Where impact craters are relatively fresh, shock-melted ejecta and rock fragments ring the impact site. Although most material is deposited close to its source, some ejecta can travel great distances. One example is *tektites,* beads of silica-rich glass, some of which have been aerodynamically shaped like teardrops during flight (Figure 8.B). Most tektites are no more than a few centimeters across and are jet-black to dark green or yellowish. In Australia, millions of tektites are strewn over an area seven times the size of Texas. Several such tektite groupings have been identified worldwide, one stretching nearly halfway around the globe.

No tektite falls have been observed, so their origin is not known with certainty. Because tektites are much higher in silica than volcanic glass (obsidian), a volcanic origin is unlikely. Most researchers agree that tektites are the result of impacts of large projectiles.

One hypothesis suggests an extraterrestrial origin for tektites. Asteroids may have struck the Moon with such force that ejecta "splashed" outward fast enough to escape the Moon's gravity. Others argue that tektites are terrestrial, but an objection is that some groupings, such as Australia's, lack an identifiable impact crater. However, the object that

FIGURE 8.A Meteor Crater, located west of Winslow, Arizona. (Photo by Michael Collier)

produced the Australian tektites might have struck the continental shelf, leaving the remnant crater out of sight below sea level. Evidence supporting a terrestrial origin includes tektite falls in western Africa that appear to be the same age as a crater in the same region.

FIGURE 8.B Tektites recovered from Nullarbor Plain, Australia. (Photo by Brian Mason/Smithsonian Institution)

FIGURE 8.29 Migmatite. The lightest-colored layers are igneous rock composed of quartz and feldspar, whereas the darker layers have a metamorphic origin. (Photo by Stephen Trimble)

contrast, the mineral sillimanite only forms in extreme environments where temperatures exceed 600 °C, and rocks containing it are considered *high-grade*. By mapping the occurrences of index minerals, geologists are in effect mapping zones of varying metamorphic grade. *Grade* is a term used in a relative sense to refer to the conditions of temperature (or sometimes pressure) to which a rock has been subjected.

Migmatites   In the most extreme environments, even the highest-grade metamorphic rocks undergo change. For example, gneissic rocks may be heated sufficiently to cause melting to begin. However, recall from our discussion of igneous rocks that different minerals melt at different temperatures. The light-colored silicates, usually quartz and potassium feldspar, have the lowest melting temperatures and begin to melt first, whereas the mafic silicates, such as amphibole and biotite, remain solid. When this partially melted rock cools, the light bands will be composed of igneous, or igneous-appearing components, while the dark bands will consist of unmelted metamorphic material. Rocks of this type are called **migmatites** (*migma* = mixture, *ite* = a stone) (Figure 8.29). The light-colored bands in migmatites often form tortuous folds and may contain tabular inclusions of the dark components.

Migmatites serve to illustrate the fact that some rocks are transitional and do not clearly belong to any one of the three basic rock groups.

INTERPRETING METAMORPHIC ENVIRONMENTS

About a century ago geologists came to realize that groups of associated minerals could be used to determine the pressures and temperatures at which rocks undergo metamorphism (see Box 8.2). This discovery led Finnish geologist Pennti Eskola to propose the concept of metamorphic facies. Simply, metamorphic rocks containing the same assemblage of minerals belong to the same **metamorphic facies**—implying that they formed in very similar metamorphic environments. The concept of metamorphic facies is analogous to using a group of plants to define climatic zones—regions that experience similar conditions of precipitation and temperature exhibit similar plants. For instance, forests consisting of scrawny spruce, fir, larch, and birch trees identify the subarctic or taiga climate zones on Earth.

There are several common metamorphic facies shown in Figure 8.30. These include the *hornfels, zeolite, greenschist, amphibolite, granulite, blueschist,* and *eclogite facies.* The name for each facies is based on the minerals that define them. For example, rocks of the amphibolite facies are characterized by the mineral hornblende (a common amphibole); and the greenschist facies consists of schists in which the green minerals chlorite, epidote, and serpentine are prominent. Similar suites of minerals are found in rocks of all ages and in all parts of the world. Thus, the concept of metamorphic facies is useful in the interpretation of Earth history. Rocks that belong to the same metamorphic facies all formed under the same conditions of

FIGURE 8.30   Metamorphic facies and corresponding temperature and pressure conditions. Note the equivalent metamorphic rocks produced from regional metamorphism of basalt and shale parent rocks.

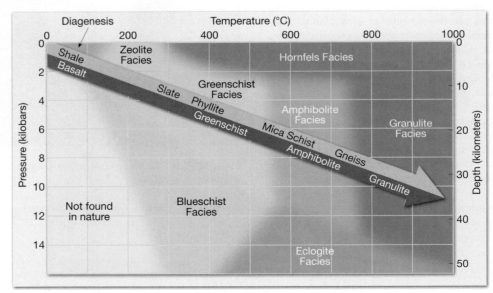

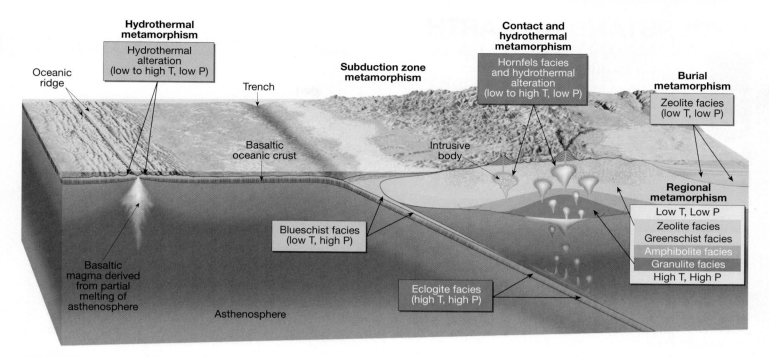

FIGURE 8.31 The association of metamorphic facies with plate tectonic environments.

temperature and pressure, and therefore in similar tectonic settings, regardless of their location or age.

It should be noted that the name for each metamorphic facies refers to a metamorphic rock derived specifically from a basaltic parent. This occurs because Pennti Eskola concentrated on the metamorphism of basalts, and his basic terminology, although now slightly modified, remains. The names of Eskola's facies serve as convenient labels for a particular combination of temperatures and pressures no matter what the mineral composition. In other words, even if a nonbasaltic parent rock produces different indicator minerals under a given set of metamorphic conditions, the facies names shown in Figure 8.30 are used for the purpose of denoting the temperature and pressure ranges embodied by that metamorphic rock.

How the concept of facies fits into the context of plate tectonics is shown in Figure 8.31. Near deep-ocean trenches, slabs of relatively cool oceanic lithosphere are subducted. As the lithosphere descends, sediments and crustal rocks are subjected to steadily increasing temperatures and pressures (Figure 8.31). However, temperatures in the slab remain cooler than the surrounding mantle because rock is a poor conductor of heat and therefore warms slowly. The metamorphic facies associated with this type of high-pressure, low-temperature environment is called the *blueschist facies,* because of the presence of the blue-colored variety of amphibole called *glaucophane* (Figure 8.32A). The rocks of the Coast Range of California belong to the blueschist facies. Highly deformed rocks that were once deeply buried have been uplifted because of a change in the plate boundary. In some areas subduction carries rocks to even greater depths, producing the *eclogite*

facies that is diagnostic of very high temperatures and pressures (Figure 8.32B).

Along some convergent zones, continental blocks collide to form extensive mountain belts (see Figure 8.24). This activity results in large areas of regional metamorphism that include zones of contact, hydrothermal, and subduction zone metamorphism. The increasing temperatures and pressures associated with regional metamorphism are recorded by the *greenschist-amphibolite-granulite facies* sequence shown in Figure 8.30.

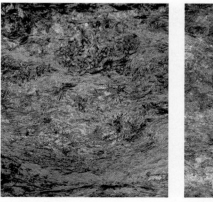
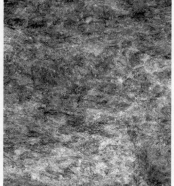

A. Blueschist   **B.** Eclogite

FIGURE 8.32 Rocks produced by subduction zone metamorphism. **A.** Blueschist, representing low temperature, high pressure conditions (note blue-colored amphibole called glaucophane). **B.** Eclogite, representing high temperature, high-pressure conditions of the mantle. Note pink grains of garnet and green grains of pyroxene. (Photos by C. Tsujita)

UNDERSTANDING EARTH

BOX 8.2

Mineral Stability

In most tectonic environments, such as along subduction zones, rocks experience an increase in *both* pressure and temperature simultaneously. An increase in pressure causes minerals to contract which favors the formation of high density minerals. However, an increased temperature results in expansion, so that mineral phases which occupy greater volume (are less dense) tend to be more stable at high temperatures. Thus, determining the conditions of temperature and pressure at which a mineral is stable (does not change) is not an easy task. To help in this endeavor, researchers have turned to the laboratory. Here materials of various composition are heated and placed under pressures that approximate conditions at various depths within Earth. From these experiments we can determine what minerals are likely to form in various metamorphic environments.

It turns out that some minerals, quartz for example, are stable over a wide range of metamorphic settings. Fortunately, other suites of minerals do provide useful estimates of conditions during metamorphism. One of the most important of these groups includes the minerals *kyanite, andalucite,* and *sillimanite.* All three minerals have identical chemical compositions (Al_2SiO_5) but different crystalline structures, which makes them *polymorphs* (see Chapter 3). Figure 8.C is a *phase diagram* that shows the specific range of pressures and temperatures at which each of these aluminum-rich silicates is stable.

Because shales and mudstones, which are very common, contain the elements found in these minerals, metamorphic products of shale (slate, schist, and gneiss) often contain varying amounts of either kyanite, andalucite, or sillimanite. For example, if shale were buried to a depth of about 35 kilometers (10 kbars), where the temperature was 550 °C, the mineral kyanite would form (see the "X" in Figure 8.C).

In general, andalusite is produced by contact metamorphism in near surface environments where temperatures are high but pressures are relatively low. Kyanite is considered the high-pressure polymorph that forms during subduction and deep burial associated with mountain building. Sillimanite, on the other hand, forms only at high temperatures, the result of contact with a very hot magma body and/or very deep burial. Knowing the ranges of temperatures and pressures a rock experienced during metamorphism provides geologists with valuable data needed to interpret past tectonic environments.

A. Andalusite   **B.** Kyanite   **C.** Sillimanite

FIGURE 8.C   Phase diagram illustration of the conditions of pressure and temperature at which the three Al_2SiO_5 minerals are stable. (Photos by **A.** Harry Taylor/Dorling Kindersley Media Library, **B.** Dennis Tasa, and **C.** Biophoto Associates/Photo Researchers, Inc.)

CHAPTER 8 METAMORPHISM AND METAMORPHIC ROCKS IN REVIEW

- *Metamorphism* is the transformation of one rock type into another. *Metamorphic rocks* form from preexisting rocks (either igneous, sedimentary, or other metamorphic rocks) that have been altered by the agents of metamorphism, which include *heat, pressure (stress),* and *chemically active fluids.* During metamorphism the material essentially remains solid. The changes that occur in metamorphosed rocks are textural as well as mineralogical.

- The mineral makeup of the parent rock determines, to a large extent, the degree to which each metamorphic agent will cause change. Heat is the most important agent because it provides the energy to drive chemical reactions that result in the recrystallization of minerals. Pressure, like temperature, also increases with depth. When subjected to *confining pressure,* minerals may recrystallize into more compact forms. During mountain building, rocks are subjected to *differential stress,* which tends to shorten them in the direction pressure is applied and lengthen them in the direction perpendicular to that force. At depth, rocks are warm and *ductile,* which accounts for their ability to deform by flowing when subjected to differential stresses. Chemically active fluids, most commonly water containing ions in solution, also enhance the metamorphic process by dissolving minerals and aiding the migration and precipitation of this material at other sites.

- *The grade of metamorphism is reflected in the texture and mineralogy of metamorphic rocks.* During regional metamorphism, rocks typically display a *preferred orientation* called *foliation* in which their platy and elongated minerals are aligned. Foliation develops as platy or elongated minerals are rotated into parallel alignment, recrystallize to form new grains that exhibit a preferred orientation, or are plastically deformed into flattened grains that exhibit a planar alignment. *Rock cleavage* is a type of foliation in which rocks split cleanly into thin slabs along surfaces where platy minerals are aligned. *Schistosity* is a type of foliation defined by the parallel alignment of medium- to coarse-grained platy minerals. During high-grade metamorphism, ion migrations can cause minerals to segregate into distinct layers or bands. Metamorphic rocks with a banded texture are called *gneiss.*

- Metamorphic rocks composed of only one mineral forming equidimensional crystals often appear *nonfoliated. Marble* (metamorphosed limestone) is often nonfoliated. Further, metamorphism can cause the transformation of low-temperature minerals into high-temperature minerals and, through the introduction of ions from *hydrothermal solutions,* generate new minerals, some of which form economically important metallic ore deposits.

- Common foliated metamorphic rocks include *slate, phyllite,* various types of *schists* (e.g., garnet-mica schist), and *gneiss.* Nonfoliated rocks include *marble* (parent rock—limestone) and *quartzite* (most often formed from quartz sandstone).

- The four geologic environments in which metamorphism commonly occurs are (1) *contact* or *thermal metamorphism,* (2) *hydrothermal metamorphism,* (3) *burial and subduction zone metamorphism,* and (4) *regional metamorphism.* Contact metamorphism occurs when rocks are in contact with an igneous body, resulting in the formation of zones of alteration around the magma called *aureoles.* Most contact metamorphic rocks are fine-grained, dense, tough rocks of various chemical compositions. Because directional pressure is not a major factor, these rocks are not generally foliated. Hydrothermal metamorphism occurs where hot, ion-rich fluids circulate through rock and cause chemical alteration of the constituent minerals. Most hydrothermal alteration occurs along the mid-ocean ridge system, where seawater migrates through hot oceanic crust and chemically alters newly formed basaltic rocks. Metallic ions that are removed from the crust are eventually carried to the floor of the ocean, where they precipitate from black smokers to form metallic deposits, some of which may be economically important. Regional metamorphism takes place at considerable depths over an extensive area and is associated with the process of mountain building. A gradation in the degree of change usually exists in association with regional metamorphism, in which the intensity of metamorphism (low- to high-grade) is reflected in the texture and mineralogy of the rocks. In the most extreme metamorphic environments, rocks called *migmatites* fall into a transition zone *somewhere between* "true" igneous rocks and "true" metamorphic rocks.

KEY TERMS

aureole (p. 241)
burial metamorphism (p. 244)
confining pressure (p. 233)
contact metamorphism (p. 241)
differential stress (p. 233)
foliation (p. 235)
gneissic texture (p. 237)

hydrothermal metamorphism (p. 242)
hydrothermal solution (p. 242)
impact metamorphism (p. 245)
index mineral (p. 246)
metamorphic facies (p. 248)
metamorphism (p. 230)

metasomatism (p. 234)
migmatite (p. 248)
nonfoliated texture (p. 237)
parent rock (p. 230)
porphyroblastic texture (p. 237)
regional metamorphism (p. 244)
rock cleavage (p. 236)

schistosity (p. 237)
shock metamorphism (p. 245)
slaty cleavage (p. 236)
subduction zone metamorphism (p. 244)
texture (p. 235)
thermal metamorphism (p. 241)

QUESTIONS FOR REVIEW

1. What is metamorphism? What are the agents that change rocks?

2. Why is heat considered the most important agent of metamorphism?

3. How is confining pressure different than differential stress?

4. What role do chemically active fluids play in metamorphism?

5. In what two ways can the parent rock affect the metamorphic process?

6. What is foliation? Distinguish between *slaty cleavage, schistosity,* and *gneissic* textures.

7. Briefly describe the three mechanisms by which minerals develop a preferred orientation.

8. List some changes that might occur to a rock in response to metamorphic processes.

9. Slate and phyllite resemble each other. How might you distinguish one from the other?

10. Each of the following statements describes one or more characteristics of a particular metamorphic rock. For each statement, name the metamorphic rock that is being described.

   a. calcite-rich and often nonfoliated

   b. loosely coherent rock composed of broken fragments that formed along a fault zone

   c. represents a grade of metamorphism between slate and schist

   d. very fine-grained and foliated; excellent rock cleavage

   e. foliated and composed predominately of platy minerals

   f. composed of alternating bands of light and dark silicate minerals

   g. hard, nonfoliated rock resulting from contact metamorphism

11. Distinguish between contact metamorphism and regional metamorphism. Which creates the greatest quantity of metamorphic rock?

12. Where does most hydrothermal metamorphism occur?

13. Describe burial metamorphism.

14. How do geologists use index minerals?

15. Briefly describe the textural changes that occur in the transformation of slate to phyllite to schist and then to gneiss.

16. How are gneisses and migmatites related?

17. With which type of plate boundary is regional metamorphism associated?

18. Why do the cores of Earth's major mountain chains contain metamorphic rocks?

19. Briefly describe the tectonic environment that produces each of these metamorphic facies—hornfels, blueschist, and granulite facies.

COMPANION WEBSITE

The *Earth 10e* website uses the resources and flexibility of the Internet to aid in your study of the topics in this chapter. Written and developed by the authors and other geology instructors, this site will help improve your understanding of geology. Visit **www.mygeoscienceplace.com** in order to:

- **Review** key chapter concepts

- **Read** with links to the eBook and to chapter-specific web resources

- **Visualize** and comprehend challenging topics using learning activities in *GEODe Earth*

- **Test** yourself with online quizzes

GEODe EARTH

GEODe Earth is a valuable and easy to use learning aid that can be accessed from your book's Companion Website (www .mygeoscienceplace.com). It is a dynamic instructional tool that promotes understanding and reinforces important concepts by using tutorials, animations, and exercises that actively engage the student.

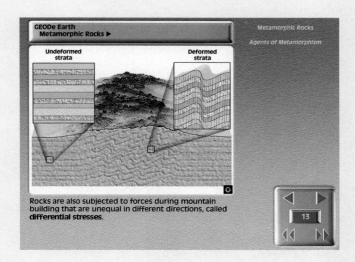

GEOLOGIC TIME

Paleontologists excavating the remains of an Ice Age mammoth, a prehistoric relative of modern elephants, near Hot Springs in the Black Hills of South Dakota. Fossils are important tools in deciphering Earth history.

(PHOTO BY PHIL DEGGINGER/ALAMY)

In the late 18th century, James Hutton recognized the immensity of Earth history and the importance of time as a component in all geological processes. In the 19th century, Sir Charles Lyell and others effectively demonstrated that Earth had experienced many episodes of mountain building and erosion, which must have required great spans of geologic time. Although these pioneering scientists understood that Earth was very old, they had no way of knowing its true age. Was it tens of millions, hundreds of millions, or even billions of years old? Rather, a geologic time scale was developed that showed the sequence of events based on relative dating principles. What are these principles? What part do fossils play? With the discovery of radioactivity and radiometric dating techniques, geologists now can assign fairly accurate dates to many of the events in Earth history. What is radioactivity? Why is it a good "clock" for dating the geologic past?

GEOLOGY NEEDS A TIME SCALE

In 1869, John Wesley Powell, who was later to head the U.S. Geological Survey, led a pioneering expedition down the Colorado River and through the Grand Canyon (Figure 9.1). Writing about the rock layers that were exposed by the downcutting of the river, Powell noted that "the canyons of this region would be a Book of Revelations in the rock-leaved Bible of geology." He was undoubtedly impressed with the millions of years of Earth history exposed along the walls of the Grand Canyon (Figure 9.2).

Powell realized that the evidence for an ancient Earth is concealed in its rocks. Like the pages in a long and complicated history book, rocks record the geological events and changing life forms of the past. The book, however, is not complete. Many pages, especially in the early chapters, are missing. Others are tattered, torn, or smudged. Yet enough of the book remains to allow much of the story to be deciphered.

Interpreting Earth history is a prime goal of the science of geology. Like a modern-day sleuth, the geologist must interpret the clues found preserved in the rocks. By studying rocks, especially sedimentary rocks, and the features they contain, geologists can unravel the complexities of the past.

Geological events by themselves, however, have little meaning until they are put into a time perspective. Studying history, whether it be the Civil War or the age of dinosaurs, requires a calendar. Among geology's major contributions to human knowledge are the *geologic time scale* and the discovery that Earth history is exceedingly long.

RELATIVE DATING—KEY PRINCIPLES

GEOLOGIC TIME
▶ Relative Dating—Key Principles

The geologists who developed the geologic time scale revolutionized the way people think about time and perceive our planet. They learned that Earth is much older than anyone had previously imagined and that its surface and interior have been changed over and over again by the same geological processes that operate today.

During the late 1800s and early 1900s, attempts were made to determine Earth's age. Although some of the methods appeared promising at the time, none of these early efforts proved to be reliable. What these scientists were seeking was a **numerical date.** Such dates specify the actual number of years that have passed since an event occurred. Today, our understanding of radioactivity allows us to accurately determine numerical dates for rocks that represent important events in Earth's distant past. We will study radioactivity

FIGURE 9.1   **FIGURE 9.1**   The start of the Powell expedition from Green River Station, Wyoming, is depicted in this drawing from Powell's 1875 book. The inset photo is of Major John Wesley Powell, pioneering geologist and second director of the U.S. Geological Survey. (Courtesy of the U.S. Geological Survey, Denver)

later in this chapter. Prior to the discovery of radioactivity, geologists had no reliable method of numerical dating and had to rely solely on relative dating.

Relative dating means that rocks are placed in their proper *sequence of formation*—which formed first, second, third, and so on. Relative dating cannot tell us how long ago something took place, only that it followed one event and preceded another. The relative dating techniques that were developed are valuable and still widely used. Numerical dating methods did not replace these techniques; they simply supplemented them. To establish a relative time scale, a few basic principles or rules had to be discovered and applied. Although they may seem obvious to us today, they were major breakthroughs in thinking at the time, and their discovery was an important scientific achievement.

Law of Superposition

Nicolaus Steno, a Danish anatomist, geologist, and priest (1638–1686), is credited with being the first to recognize a sequence of historical events in an outcrop of sedimentary rock layers. Working in the mountains of western Italy, Steno applied a very simple rule that has come to be the most basic principle of relative dating—the **law of superposition** (*super* = above; *positum* = to place). The law simply states that in an undeformed sequence of sedimentary rocks, each bed is older than the one above and younger than the one

below. Although it may seem obvious that a rock layer could not be deposited with nothing beneath it for support, it was not until 1669 that Steno clearly stated this principle.

This rule also applies to other surface-deposited materials, such as lava flows and beds of ash from volcanic eruptions. Applying the law of superposition to the beds exposed in the upper portion of the Grand Canyon (Figure 9.3), we can easily place the layers in their proper order. Among those that are pictured, the sedimentary rocks in the Supai Group are the oldest, followed in order by the Hermit Shale, Coconino Sandstone, Toroweap Formation, and Kaibab Limestone.

Principle of Original Horizontality

Steno is also credited with recognizing the importance of another basic principle, called the **principle of original horizontality.** Simply stated, it

FIGURE 9.2   This hiker is resting atop the Kaibab Formation, the uppermost layer in the Grand Canyon. Hundreds of millions of years of Earth history is contained in the strata that lay beneath him. This is a view of Cape Royal on the Grand Canyon's North Rim. (Photo by Michael Collier)

Kaibab Limestone—shallow marine limestone that rims much of the canyon

Toroweap Formation—shallow marine, thin-to-medium bedded sandy limestone

Coconino Sandstone—cliff-forming cross-bedded sandstone

Hermit Shale—red, slope-forming thinly-bedded shales and siltstones

Supai Group— alternating layers of sandstone, siltstone and shale

Geologist's Sketch

FIGURE 9.3 Applying the law of superposition to these layers exposed in the upper portion of the Grand Canyon, the Supai Group is oldest and the Kaibab Limestone is youngest. (Photo by E. J. Tarbuck)

means that layers of sediment are generally deposited in a horizontal position. Thus, if we observe rock layers that are flat, it means they have not been disturbed and still have their *original* horizontality. The layers in the Grand Canyon illustrate this in Figures 9.2 and 9.3. But if they are folded or inclined at a steep angle, they must have been moved into that position by crustal disturbances sometime *after* their deposition (Figure 9.4).

Principle of Cross-Cutting Relationships

When a fault cuts through other rocks, or when magma intrudes and crystallizes, we can assume that the fault or intrusion is younger than the rocks affected.* For example, in

---

*Faults are fractures in the crust along which appreciable displacement has taken place. Faults are discussed in some detail in Chapter 10.

FIGURE 9.4 Most layers of sediment are deposited in a nearly horizontal position. Thus, when we see rock layers that are folded or tilted, we can assume that they must have been moved into that position by crustal disturbances *after* their deposition. These folds are at Agio Pavlos on the Mediterranean island of Crete. (Photo by Marco Simoni/Robert Harding)

Figure 9.5, the faults and dikes clearly must have occurred after the sedimentary layers were deposited.

This is the **principle of cross-cutting relationships.** By applying the cross-cutting principle, you can see that fault A occurred *after* the sandstone layer was deposited because it "broke" the layer. Likewise, fault A occurred *before* the conglomerate was laid down because that layer is unbroken.

We can also state that dike B and its associated sill are older than dike A because dike A cuts the sill. In the same manner, we know that the batholith was emplaced after movement occurred along fault B but before dike B was formed. This is true because the batholith cuts across fault B, while dike B cuts across the batholith.

Inclusions

Sometimes inclusions can aid the relative dating process. **Inclusions** (*includere* = to enclose) are fragments of one rock unit that have been enclosed within another. The basic principle is logical and straightforward. The rock mass adjacent to the one containing the inclusions must have been there first in order to provide the rock fragments. Therefore, the rock mass containing inclusions is the younger of the two. Figure 9.6 provides an example. Here, the inclusions of intrusive igneous rock in the adjacent sedimentary layer indicate that the sedimentary layer was deposited on top of a weathered igneous mass rather than being intruded from below by magma that later crystallized.

Unconformities

When we observe layers of rock that have been deposited essentially without interruption, we call them **conformable.** Particular sites exhibit conformable beds representing certain spans of geologic time. However, no place on Earth has a complete set of conformable strata.

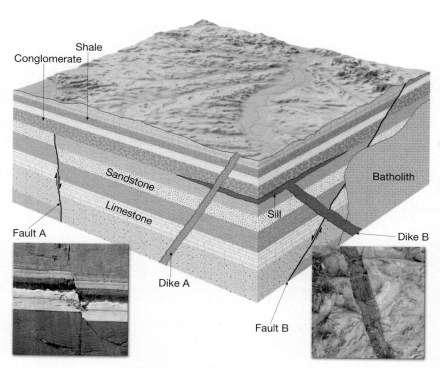

Shale
Conglomerate
Sandstone
Limestone
Fault A
Dike A
Fault B
Sill
Batholith
Dike B

FIGURE 9.5 Cross-cutting relationships represent one principle used in relative dating. An intrusive rock body is younger than the rocks it intrudes. A fault is younger than the rock layers it cuts.

history of sedimentation in a variety of environments—advancing seas, rivers and deltas, tidal flats and sand dunes. But the record is not continuous. Unconformities represent vast amounts of time that have not been recorded in the canyon's layers. Figure 9.7 is a geologic cross section of the Grand Canyon. Refer to it as you read about the three basic types of unconformities: angular unconformities, disconformities, and nonconformities.

Angular Unconformity   Perhaps the most easily recognized unconformity is an **angular unconformity.** It consists of tilted or folded sedimentary rocks that are overlain by younger, more flat-lying strata. An angular unconformity indicates that during the pause in deposition, a period of deformation (folding or tilting) and erosion occurred (Figure 9.8).

When James Hutton studied an angular unconformity in Scotland more than 200 years ago, it was clear to him that it represented a major episode of geologic

Throughout Earth history, the deposition of sediment has been interrupted over and over again. All such breaks in the rock record are termed unconformities. An **unconformity** represents a long period during which deposition ceased, erosion removed previously formed rocks, and then deposition resumed. In each case, uplift and erosion are followed by subsidence and renewed sedimentation. Unconformities are important features because they represent significant geologic events in Earth history. Moreover, their recognition helps us identify what intervals of time are not represented by strata and thus are missing from the geologic record.

The rocks exposed in the Grand Canyon of the Colorado River represent a tremendous span of geologic history. It is a wonderful place in which to take a trip through time. The canyon's colorful strata record a long

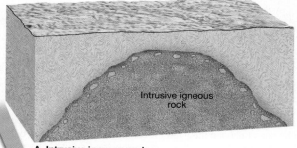

A. Intrusive igneous rock

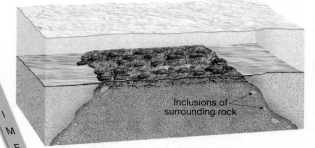

B. Exposure and weathering of intrusive igneous rock

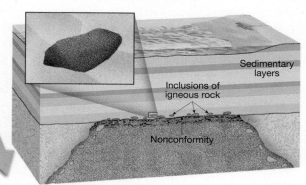

C. Deposition of sedimentary layers

FIGURE 9.6 These diagrams illustrate two ways that inclusions can form, as well as a type of unconformity termed a nonconformity. In diagram **A,** the inclusions in the igneous mass represent unmelted remnants of the surrounding host rock that were broken off and incorporated at the time the magma was intruded. In diagram **C,** the igneous rock must be older than the overlying sedimentary beds because the sedimentary beds contain inclusions of the igneous rock. When older intrusive igneous rocks are overlain by younger sedimentary layers, a *noncomformity* is said to exist. The inset shows a close-up view of an inclusion.

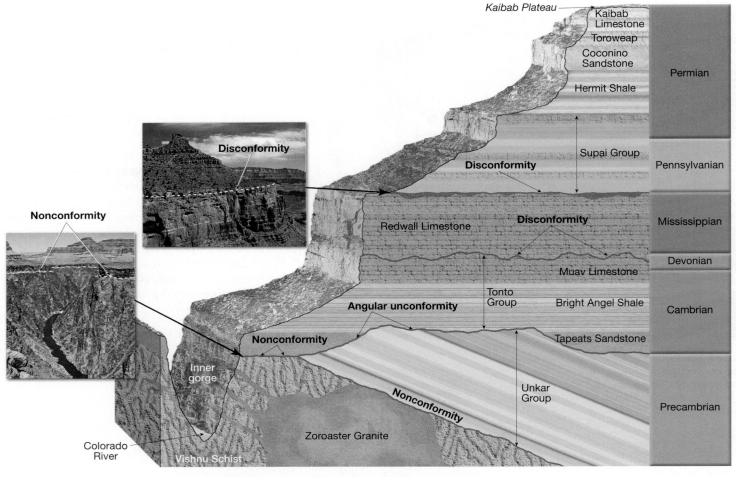

FIGURE 9.7   This cross section of the Grand Canyon illustrates the three basic types of unconformities. An angular unconformity can be seen between the tilted Precambrian Unkar Group and the Cambrian strata above. Two disconformities are marked, above and below the Redwall Limestone. A nonconformity occurs between the igneous and metamorphic rocks exposed in the inner gorge and the sedimentary strata of the Unkar Group. A nonconformity, highlighted by a photo, also occurs between the rocks of the inner gorge and Tapeats Sandstone.

Students Sometimes Ask . . .

You mentioned early attempts at determining Earth's age that proved unreliable. How did 19th-century scientists go about making such calculations?

One method that was attempted several times involved the rate at which sediment is deposited. Some reasoned that if they could determine the rate that sediment accumulates and could further ascertain the total thickness of sedimentary rock that had been deposited during Earth history, they could estimate the length of geologic time. All that was necessary was to divide the rate of sediment accumulation into the total thickness of sedimentary rock.

Estimates of Earth's age varied each time this method was attempted. The age of Earth as calculated by this method ranged from 3 million to 1.5 billion years! Obviously this method was riddled with difficulties. Can you suggest what some might have been?

activity (Figure 9.9).* He and his colleagues also appreciated the immense time span implied by such relationships. When a companion later wrote of their visit to this site, he stated that "the mind seemed to grow giddy by looking so far into the abyss of time."

Disconformity   When contrasted with angular unconformities, **disconformities** are more common but usually far less conspicuous because the strata on either side are essentially parallel. Many disconformities are difficult to identify because the rocks above and below are similar and there is little evidence of erosion. Such a break often resembles an ordinary bedding plane. Other disconformities are easier to identify because the ancient erosion surface is cut deeply into the older rocks below.

Nonconformity   The third basic type of unconformity is a **nonconformity.** Here the break separates older metamorphic or intrusive igneous rocks from younger sedimentary strata (see

---

*This pioneering geologist is discussed in the section on the birth of modern geology in Chapter 1.

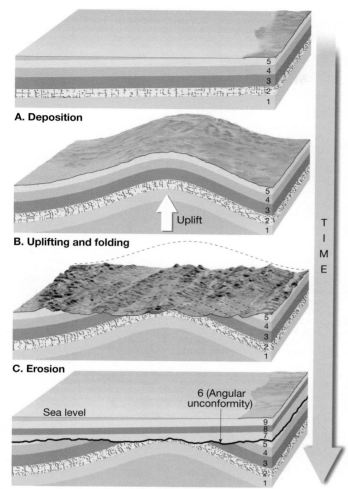

A. Deposition

B. Uplifting and folding

Uplift

TIME

C. Erosion

6 (Angular unconformity)

Sea level

D. Subsidence and renewed deposition

FIGURE 9.8 Formation of an angular unconformity. An angular unconformity represents an extended period during which deformation and erosion occurred.

Figures 9.6 and 9.7). Just as angular unconformities and disconformities imply crustal movements, so too do nonconformities. Intrusive igneous masses and metamorphic rocks originate far below the surface. Thus, for a nonconformity to develop, there must be a period of uplift and the erosion of overlying rocks. Once exposed at the surface, the igneous or metamorphic rocks are subjected to weathering and erosion prior to subsidence and the renewal of sedimentation.

Using Relative Dating Principles

If you apply the principles of relative dating to the hypothetical geologic cross section in Figure 9.10, you can place in proper sequence the rocks and the events they represent. The statements within the figure summarize the logic used to interpret the cross section.

In this example, we establish a relative time scale for the rocks and events in the area of the cross section. Remember that this method gives us no idea how many years of Earth history are represented, for we have no numerical dates. Nor do we know how this area compares to any other (see Box 9.1).

CORRELATION OF ROCK LAYERS

To develop a geologic time scale that is applicable to the entire Earth, rocks of similar age in different regions must be matched up. Such a task is referred to as **correlation.**

Within a limited area, correlating rocks of one locality with those of another may be done simply by walking along the

Above the unconformity lie gently dipping beds of redish sandstone and conglomerate

Angular unconformity

Rock hammer

Below the unconformity lie nearly vertical sandstones and shales

Geologist's Sketch

FIGURE 9.9 This angular unconformity at Siccar Point, Scotland was first described by James Hutton more than 200 years ago. (Photo by Marli Miller)

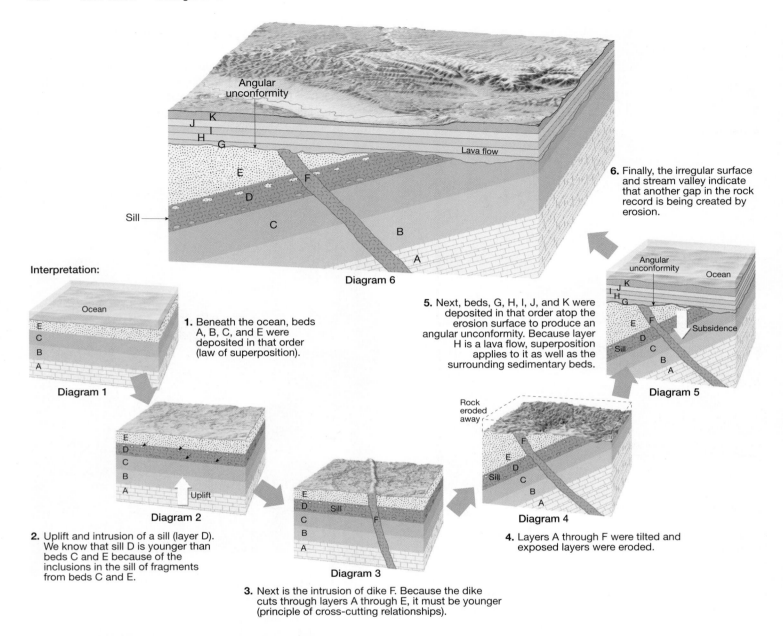

FIGURE 9.10 Geologic cross section of a hypothetical region.

Labels and captions within the figure:

Angular unconformity

J K I H G

Lava flow

E F

Sill

D

C

B

A

Diagram 6

Interpretation:

Ocean

E C B A

Diagram 1

1. Beneath the ocean, beds A, B, C, and E were deposited in that order (law of superposition).

2. Uplift and intrusion of a sill (layer D). We know that sill D is younger than beds C and E because of the inclusions in the sill of fragments from beds C and E.

E D C B A Uplift

Diagram 2

3. Next is the intrusion of dike F. Because the dike cuts through layers A through E, it must be younger (principle of cross-cutting relationships).

E D Sill C B A F

Diagram 3

Rock eroded away

F E D Sill C B A

Diagram 4

4. Layers A through F were tilted and exposed layers were eroded.

5. Next, beds, G, H, I, J, and K were deposited in that order atop the erosion surface to produce an angular unconformity. Because layer H is a lava flow, superposition applies to it as well as the surrounding sedimentary beds.

Angular unconformity   Ocean

K J I H G

E F D Sill C B A Subsidence

Diagram 5

6. Finally, the irregular surface and stream valley indicate that another gap in the rock record is being created by erosion.

outcropping edges. However, this may not be possible when the rocks are mostly concealed by soil and vegetation. Correlation over short distances is often achieved by noting the position of a bed in a sequence of strata. Or a layer may be identified in another location if it is composed of distinctive or uncommon minerals.

By correlating the rocks from one place to another, a more comprehensive view of the geologic history of a region is possible. Figure 9.11, for example, shows the correlation of strata at three sites on the Colorado Plateau in southern Utah and northern Arizona. No single locale exhibits the entire sequence, but correlation reveals a more complete picture of the sedimentary rock record.

Many geologic studies involve relatively small areas. Although they are important in their own right, their full value is realized only when they are correlated with other regions. Although the methods just described are sufficient to trace a rock formation over relatively short distances, they are not adequate for matching up rocks that are separated by great distances. When correlation between widely separated areas or between continents is the objective, geologists must rely on fossils.

FOSSILS: EVIDENCE OF PAST LIFE

Fossils, the remains or traces of prehistoric life, are important inclusions in sediment and sedimentary rocks. They are basic and important tools for interpreting the geologic past. The scientific study of fossils is called **paleontology.** It is an interdisciplinary

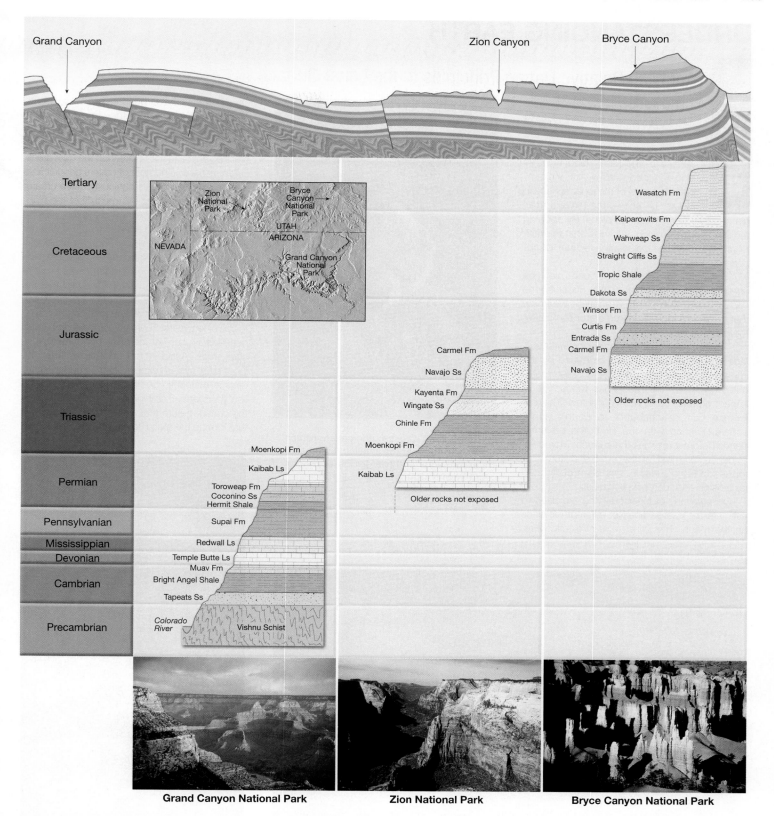

FIGURE 9.11 Correlation of strata at three locations on the Colorado Plateau provides a more complete picture of sedimentary rocks in the region. The diagram at the top is a geologic cross section of the region. (After U.S. Geological Survey; photos by E. J. Tarbuck)

UNDERSTANDING EARTH

BOX 9.1

Applying Relative Dating Principles to the Lunar Surface

Just as we use relative dating principles to determine the sequence of geological events on Earth, so too can we apply such principles to the surface of the Moon (and to other planetary bodies as well). For example, the image of the lunar surface in Figure 9.A shows the forward margin of a lava flow "frozen" in place. By applying the law of superposition, we know that this flow is younger than the adjacent layer that disappears beneath it.

Cross-cutting relationships can also be used. In Figure 9.B, when we observe one impact crater that overlaps another, we know that the continuous unbroken crater came after the one that it cuts across.

The most obvious features on the lunar surface are craters. Most were produced by the impact of rapidly moving objects called meteorites. Whereas the Moon has thousands of impact craters, Earth has only a few. This difference can be attributed to Earth's atmosphere. Friction with the air burns up small debris before it reaches the surface. Moreover, evidence for most of the sizable craters that formed in Earth's history has been obliterated by erosion and tectonic processes.

Observations of lunar cratering are used to estimate the relative ages of different locations on the Moon. The principle is straightforward.

FIGURE 9.B Cross-cutting relationships allow us to say that the smaller, unbroken crater formed after the larger crater. (Photo courtesy of NASA)

Older regions have been exposed to meteorite impact longer and therefore have more craters. Using this technique in conjunction with Figure 9.C, we can infer that the highly cratered highlands are older than the dark areas, called maria. The number of craters per unit area (called *crater density*) is obviously much greater in the highlands. Does this mean that the highlands are *much* older? Although this may seem a logical conclusion, the answer is no. Remember that we are dealing with a principle of *relative* dating. Both the highlands and maria are very old. Radiometric dating of Moon rocks brought back from the *Apollo* missions showed that the age of the highlands is more than 4 billion years, whereas the maria have ages ranging from 3.2 to 3.9 billion years. Thus, the very different crater densities are *not* just the result of different exposure times. Astronomers now realize that the inner solar system experienced a sudden sharp drop in meteoritic bombardment about 3.9 billion years ago. The highlands received most of the craters before that time, and the lava flows that formed the maria solidified afterward.

FIGURE 9.C Crater density. Younger regions have fewer craters than older regions. The densely cratered highlands are older than the dark areas, called maria. (UCO/Lick Observatory Image)

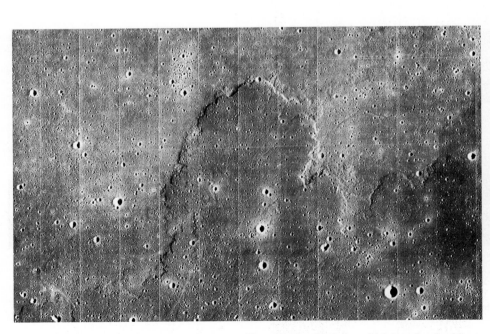

FIGURE 9.A By applying the law of superposition, you can determine which lava flow is older. (Photo courtesy of National Space Data Center)

science that blends geology and biology in an attempt to understand all aspects of the succession of life over the vast expanse of geologic time. Knowing the nature of the life forms that existed at a particular time helps researchers understand past environmental conditions. Further, fossils are important time indicators and play a key role in correlating rocks of similar ages that are from different places.

Types of Fossils

Fossils are of many types. The remains of relatively recent organisms may not have been altered at all. Such objects as teeth, bones, and shells are common examples (Figure 9.12). Far less common are entire animals, flesh included, that have been preserved because of rather unusual circumstances. Remains of prehistoric elephants called mammoths that were frozen in the Arctic tundra of Siberia and Alaska are examples, as are the mummified remains of sloths preserved in a dry cave in Nevada.

Given enough time, the remains of an organism are likely to be modified. Often fossils become *petrified* (literally, "turned into stone"), meaning that the small internal cavities and pores of the original structure are filled with precipitated mineral matter (Figure 9.13A). In other instances *replacement* may occur. Here the cell walls and other solid material are removed and replaced with mineral matter. Sometimes the microscopic details of the replaced structure are faithfully retained.

Molds and casts constitute another common class of fossils. When a shell or other structure is buried in sediment and then dissolved by underground water, a *mold* is created. The mold faithfully reflects only the shape and surface marking of the organism; it does not reveal any information concerning its internal structure. If these hollow spaces are subsequently filled with mineral matter, *casts* are created (Figure 9.13B).

A type of fossilization called *carbonization* is particularly effective in preserving leaves and delicate animal forms. It occurs when fine sediment encases the remains of an organism. As time passes, pressure squeezes out the liquid and gaseous components and leaves behind a thin residue of carbon (Figure 9.13C). Black shales deposited as organic-rich mud in oxygen-poor environments often contain abundant carbonized remains. If the film of carbon is lost from a fossil preserved in fine-grained sediment, a replica of the surface, called an *impression,* may still show considerable detail (Figure 9.13D).

Delicate organisms, such as insects, are difficult to preserve, and consequently they are relatively rare in the fossil record. Not only must they be protected from decay but they must not be subjected to any pressure that would crush them. One way in which some insects have been preserved is in *amber,* the hardened resin of ancient trees. The fly in Figure 9.13E was preserved after being trapped in a drop of sticky resin. Resin sealed off the insect from the atmosphere and protected the remains from damage by water and air. As the resin hardened, a protective pressure-resistant case was formed.

In addition to the fossils already mentioned, there are numerous other types, many of them only traces of prehistoric life. Examples of such indirect evidence include:

1. Tracks—animal footprints made in soft sediment that was later lithified (see Figure 7.30B).
2. Burrows—tubes in sediment, wood, or rock made by an animal. These holes may later become filled with mineral matter and preserved. Some of the oldest-known fossils are believed to be worm burrows.
3. Coprolites—fossil dung and stomach contents that can provide useful information pertaining to food habits of organisms (Figure 9.13F).
4. Gastroliths—highly polished stomach stones that were used in the grinding of food by some extinct reptiles.

A.

FIGURE 9.12 A. Excavating bones from Pit 91 at the La Brea tar pits in Los Angeles. It is a site rich in the unaltered remains of Ice Age organisms. Scientists have been digging here since 1915. (Reed Saxon/AP Photo) **B.** Fossils of many relatively recent organisms are unaltered remains. The skeleton of the mammoth from the La Brea tar pits is a spectacular example. (Martin Shields/Alamy) **B.**

FIGURE 9.13 There are many types of fossilization. Six examples are shown here. **A.** Petrified wood in Petrified Forest National Park, Arizona. **B.** This trilobite photo illustrates mold and cast. **C.** A fossil bee preserved as a thin carbon film. **D.** Impressions are common fossils and often show considerable detail. **E.** Insect in amber. **F.** A coprolite is fossil dung. (Photo A by David Muench; Photos B, D, and F by E. J. Tarbuck; Photo C by Florissant Fossil Beds National Monument; Photo E by Breck P. Kent)

Conditions Favoring Preservation

Only a tiny fraction of the organisms that have lived during the geologic past have been preserved as fossils. Normally, the remains of an animal or plant are destroyed. Under what circumstances are they preserved? Two special conditions appear to be necessary: rapid burial and the possession of hard parts.

When an organism perishes, its soft parts usually are quickly eaten by scavengers or decomposed by bacteria. Occasionally, however, the remains are buried by sediment. When this occurs, the remains are protected from the environment, where destructive processes operate. Rapid burial, therefore, is an important condition favoring preservation.

In addition, animals and plants have a much better chance of being preserved as part of the fossil record if they have hard parts. Although traces and imprints of soft-bodied animals such as jellyfish, worms, and insects exist, they are not common. Flesh usually decays so rapidly that preservation is exceedingly unlikely. Hard parts such as shells, bones, and teeth predominate in the record of past life.

Because preservation is contingent on special conditions, the record of life in the geologic past is biased. The fossil record of those organisms with hard parts that lived in areas of sedimentation is quite abundant. However, we get only an occasional glimpse of the vast array of other life forms that did not meet the special conditions favoring preservation.

Students Sometimes Ask . . .

How is paleontology different from archaeology?

People frequently confuse these two areas of study because a common perception of both paleontologists and archaeologists is of scientists carefully extracting important clues about the past from layers of rock or sediment. While it is true that scientists in both disciplines "dig" a lot, the focus of each is different. Paleontologists study fossils and are concerned with *all* life forms in the geologic past. By contrast, archaeologists focus on the material remains of past human life. These remains include both the objects used by people long ago, called *artifacts,* and the buildings and other structures associated with where people lived, called *sites.* Archaeologists help us learn about how our human ancestors met the challenges of life in the past.

Fossils and Correlation

The existence of fossils had been known for centuries, yet it was not until the late 1700s and early 1800s that their significance as geologic tools was made evident. During this period, an English engineer and canal builder, William Smith, discovered that each rock formation in the canals he worked on contained fossils unlike those in the beds either above or below. Further, he noted that sedimentary strata in widely separated areas could be identified—and correlated—by their distinctive fossil content.

Based on Smith's classic observations and the findings of many geologists who followed, one of the most important and basic principles in historical geology was formulated: *Fossil organisms succeed one another in a definite and determinable order, and therefore any time period can be recognized by its fossil content.* This has come to be known as the **principle of fossil succession.** In other words, when fossils are arranged according to their age, they do not present a random or haphazard picture. To the contrary, fossils document the evolution of life through time.

For example, an Age of Trilobites is recognized quite early in the fossil record. Then, in succession, paleontologists recognize an Age of Fishes, an Age of Coal Swamps, an Age of Reptiles, and an Age of Mammals. These "ages" pertain to groups that were especially plentiful and characteristic during particular time periods. Within each of the "ages" there are many subdivisions based, for example, on certain species of trilobites and certain types of fish, reptiles, and so on. This same succession of dominant organisms, never out of order, is found on every continent.

When fossils were found to be time indicators, they became the most useful means of correlating rocks of similar age in different regions. Geologists pay particular attention to certain fossils called **index fossils.** These fossils are widespread geographically and are limited to a short span of geologic time, so their presence provides an important method of matching rocks of the same age. Rock formations, however, do not always contain a specific index fossil. In such situations, groups of fossils are used to establish the age of the bed. Figure 9.14 illustrates how an assemblage of fossils may be used to date rocks more precisely than could be accomplished by the use of any one of the fossils.

In addition to being important, and often essential, tools for correlation, fossils are important environmental indicators. Although much can be deduced about past environments by studying the nature and characteristics of sedimentary rocks, a close examination of the fossils present can usually provide a great deal more information. For example, when the remains of certain clam shells are found in limestone, the geologist quite reasonably assumes that the region was once covered by a shal-

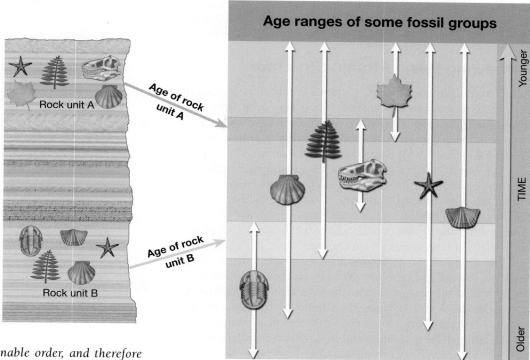

FIGURE 9.14   Overlapping ranges of fossils help date rocks more exactly than using a single fossil.

low sea. Also, by using what we know of living organisms, we can conclude that fossil animals with thick shells, capable of withstanding pounding and surging waves, inhabited shorelines.

On the other hand, animals with thin, delicate shells probably indicate deep, calm offshore waters. Hence, by looking closely at the types of fossils, the approximate position of an ancient shoreline may be identified. Further, fossils can be used to indicate the former temperature of the water. Certain kinds of present-day corals must live in warm and shallow tropical seas like those around Florida and the Bahamas. When similar types of coral are found in ancient limestones, they indicate the marine environment that must have existed when they were alive. These examples illustrate how fossils can help unravel the complex story of Earth history.

DATING WITH RADIOACTIVITY

GEOLOGIC TIME
▶ Dating with Radioactivity

In addition to establishing relative dates by using the principles described in the preceding sections, it is also possible to obtain reliable numerical dates for events in the geologic past. For example, we know that Earth is about 4.6 billion years old and that the dinosaurs became extinct about 65 million years ago. Dates that are expressed in millions and billions of years truly stretch our imagination because our personal calendars involve time measured in hours, weeks, and years. Nevertheless, the vast expanse of geologic time is a reality, and it is radiometric dating that allows us to measure it. In this section you will learn about radioactivity and its application in radiometric dating.

Reviewing Basic Atomic Structure

Recall from Chapter 3 that each atom has a *nucleus* containing protons and neutrons and that the nucleus is orbited by electrons. *Electrons* have a negative electrical charge, and *protons* have a positive charge. A *neutron* is actually a proton and an electron combined, so it has no charge (it is neutral).

The *atomic number* (each element's identifying number) is the number of protons in the nucleus. Every element has a different number of protons and thus a different atomic number (hydrogen = 1, carbon = 6, oxygen = 8, uranium = 92, etc.). Atoms of the same element always have the same number of protons, so the atomic number stays constant.

Practically all of an atom's mass (99.9 percent) is in the nucleus, indicating that electrons have virtually no mass at all. So, by adding the protons and neutrons in an atom's nucleus, we derive the atom's *mass number.* The number of neutrons can vary, and these variants, or *isotopes,* have different mass numbers.

To summarize with an example, uranium's nucleus always has 92 protons, so its atomic number always is 92. But its neutron population varies, so uranium has three isotopes: uranium-234 (protons + neutrons = 234), uranium-235, and uranium-238. All three isotopes are mixed in nature. They look the same and behave the same in chemical reactions.

Radioactivity

The forces that bind protons and neutrons together in the nucleus usually are strong. However, in some isotopes, the nuclei are unstable because the forces binding protons and neutrons together are not strong enough. As a result, the nuclei spontaneously break apart, or decay, a process called **radioactivity.**

What happens when unstable nuclei break apart? Three common types of radioactive decay are illustrated in Figure 9.15 and can be summarized as follows:

1. *Alpha particles* (α particles) may be emitted from the nucleus. An alpha particle is composed of 2 protons and 2 neutrons. Thus, the emission of an alpha particle means that the mass number of the isotope is reduced by 4 and the atomic number is lowered by 2.

2. When a *beta particle* (β particle), or electron, is given off from a nucleus, the mass number remains unchanged, because electrons have practically no mass. However, because the electron has come from a neutron (remember, a neutron is a combination of a proton and an electron), the nucleus contains one more proton than before. Therefore, the atomic number increases by 1.

3. Sometimes an electron is captured by the nucleus. The electron combines with a proton and forms a neutron. As in the last example, the mass number remains unchanged. However, since the nucleus now contains one less proton, the atomic number decreases by 1.

An unstable radioactive isotope is referred to as the *parent,* and the isotopes resulting from the decay of the parent are termed

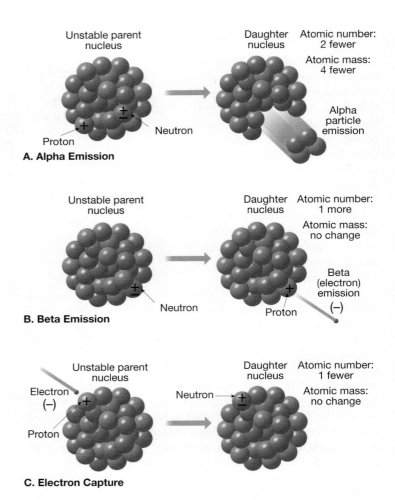

FIGURE 9.15 Common types of radioactive decay. Notice that in each case the number of protons (atomic number) in the nucleus changes, thus producing a different element.

the *daughter products.* Figure 9.16 provides an example of radioactive decay. Here it can be seen that when the radioactive parent, uranium-238 (atomic number 92, mass number 238) decays, it follows a number of steps, emitting 8 alpha particles and 6 beta particles before finally becoming the stable daughter product lead-206 (atomic number 82, mass number 206). One of the unstable daughter products produced during this decay series is radon.

Certainly among the most important results of the discovery of radioactivity is that it provided a reliable means of calculating the ages of rocks and minerals that contain particular radioactive isotopes. The procedure is called **radiometric dating.** Why is radiometric dating reliable? Because the rates of decay for many isotopes have been precisely measured and do not vary under the physical conditions that exist in Earth's outer layers. Therefore, each radioactive isotope used for dating has been decaying at a fixed rate since the formation of the minerals in which it occurs, and the products of decay have been accumulating at a corresponding rate. For example, when uranium is incorporated into a mineral that crystallizes from magma, there is no lead (the stable daughter product) from previous decay. The radiometric "clock" starts at this point. As the uranium in this newly formed mineral disintegrates, atoms of the daughter product are trapped, and measurable amounts of lead eventually accumulate.

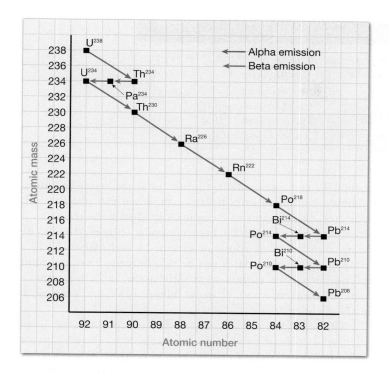

FIGURE 9.16 The most common isotope of uranium (U-238) is an example of a radioactive decay series. Before the stable end product (Pb-206) is reached, many different isotopes are produced as intermediate steps.

Half-Life

The time required for half of the nuclei in a sample to decay is called the **half-life** of the isotope. Half-life is a common way of expressing the rate of radioactive disintegration. Figure 9.17 illustrates what occurs when a radioactive parent decays directly into its stable daughter product. When the quantities of parent and daughter are equal (ratio 1:1), we know that one half-life has transpired. When one-quarter of the original parent atoms remain and three-quarters have decayed to the daughter product, the parent/daughter ratio is 1:3 and we know that two half-lives have passed. After three half-lives, the ratio of parent atoms to daughter atoms is 1:7 (one parent atom for every seven daughter atoms).

If the half-life of a radioactive isotope is known and the parent/daughter ratio can be determined, the age of the sample can be calculated. For example, assume that the half-life of a hypothetical unstable isotope is 1 million years and the parent/daughter ratio in a sample is 1:15. Such a ratio indicates that four half-lives have passed and that the sample must be 4 million years old.

Radiometric Dating

Notice that the *percentage* of radioactive atoms that decay during one half-life is always the same: 50 percent. However, the *actual number* of atoms that decay with the passing of each half-life continually decreases. Thus, as the percentage of radioactive parent atoms declines, the proportion of stable daughter atoms rises, with the increase in daughter atoms just matching the drop in parent atoms. This fact is the key to radiometric dating.

Of the many radioactive isotopes that exist in nature, five have proved particularly useful in providing radiometric ages for ancient rocks (Table 9.1). Rubidium-87, thorium-232, and the two isotopes of uranium are used only for dating rocks that are millions of years old, but potassium-40 is more versatile.

Potassium-Argon Although the half-life of potassium-40 is 1.3 billion years, analytical techniques make possible the detection of tiny amounts of its stable daughter product, argon-40, in some rocks that are younger than 100,000 years. Another important reason for its frequent use is that potassium is an abundant constituent of many common minerals, particularly micas and feldspars.

Although potassium (K) has three natural isotopes ^{39}K, ^{40}K, and ^{41}K, only ^{40}K is radioactive. When ^{40}K decays, it does so in two ways. About 11 percent changes to argon-40 (^{40}Ar) by means of electron capture (see Figure 9.15C). The remaining 89 percent of ^{40}K decays to calcium-40 (^{40}Ca) by beta emission (see Figure 9.15B). The decay of ^{40}K to ^{40}Ca, however, is not useful for radiometric dating because the ^{40}Ca produced by radioactive disintegration cannot be distinguished from calcium that may have been present when the rock formed.

The potassium-argon clock begins when potassium-bearing minerals crystallize from a magma or form within a metamorphic rock. At this point the new minerals will contain ^{40}K but will be free of ^{40}Ar, because this element is an inert gas that does not chemically combine with other elements. As time passes, the ^{40}K steadily decays by electron capture. The ^{40}Ar produced by this process remains trapped within the mineral's crystal lattice.

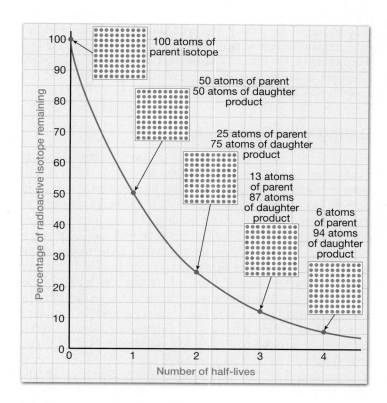

FIGURE 9.17 The radioactive-decay curve shows change that is exponential. Half of the radioactive parent remains after one half-life. After a second half-life one-quarter of the parent remains, and so forth.

TABLE 9.1	Isotopes Frequently Used in Radiometric Dating	
Radioactive Parent	Stable Daughter Product	Currently Accepted Half-life Values
Uranium-238	Lead-206	4.5 billion years
Uranium-235	Lead-207	713 million years
Thorium-232	Lead-208	14.1 billion years
Rubidium-87	Strontium-87	47.0 billion years
Potassium-40	Argon-40	1.3 billion years

Because no ^{40}Ar was present when the mineral formed, all of the daughter atoms trapped in the mineral must have come from the decay of ^{40}K. To determine a sample's age, the ^{40}K/^{40}Ar ratio is measured precisely, and the known half-life for ^{40}K applied.

Sources of Error   It is important to realize that an accurate radiometric date can be obtained only if the mineral remained a closed system during the entire period since its formation. A correct date is not possible unless there was neither the addition nor loss of parent or daughter isotopes. This is not always the case. In fact, an important limitation of the potassium-argon method arises from the fact that argon is a gas and it may leak from minerals, throwing off measurements. Indeed, losses can be significant if the rock is subjected to relatively high temperatures.

Of course, a reduction in the amount of ^{40}Ar leads to an underestimation of the rock's age. Sometimes temperatures are high enough for a sufficiently long period that all argon escapes. When this happens, the potassium-argon clock is reset, and dating the sample will give only the time of thermal resetting, not the true age of the rock. For other radiometric clocks, a loss of daughter atoms can occur if the rock has been subjected to weathering or leaching. To avoid such a problem, one simple safeguard is to use only fresh, unweathered material and not samples that may have been chemically altered.

Dating with Carbon-14

To date very recent events, carbon-14 is used. Carbon-14 is the radioactive isotope of carbon. The process is often called **radiocarbon dating.** Because the half-life of carbon-14 is only 5730 years, it can be used for dating events from the historic past as well as those from very recent geologic history. In some cases carbon-14 can be used to date events as far back as 70,000 years.

Carbon-14 is continuously produced in the upper atmosphere as a consequence of cosmic-ray bombardment. Cosmic rays (high-energy nuclear particles) shatter the nuclei of gas atoms, releasing neutrons. Some of the neutrons are absorbed by nitrogen atoms (atomic number 7, mass number 14), causing each nucleus to emit a proton. As a result, the atomic number decreases by 1 (to 6), and a different element, carbon-14, is created (Figure 9.18A). This isotope of carbon quickly becomes incorporated into carbon dioxide, which circulates in the atmosphere and is absorbed by living matter. As a result, all organisms contain a small amount of carbon-14, including you.

As long as an organism is alive, the decaying radiocarbon is continually replaced, and the proportions of carbon-14 and carbon-12 remain constant. Carbon-12 is the stable and most common isotope of carbon. However, when any plant or animal dies, the amount of carbon-14 gradually decreases as it decays to nitrogen-14 by beta emission (Figure 9.18B). By comparing the proportions of carbon-14 and carbon-12 in a sample, radiocarbon dates can be determined. It is important to emphasize that carbon-14 is only useful in dating organic materials such as wood, charcoal, bones, flesh, and even cloth made of cotton fibers.

Although carbon-14 is only useful in dating the last small fraction of geologic time, it has become a very valuable tool for anthropologists, archaeologists, and historians, as well as for geologists who study very recent Earth history. In fact, the development of radiocarbon dating was considered so important that the chemist who discovered this application, Willard F. Libby, received a Nobel Prize in 1960.

Importance of Radiometric Dating

Bear in mind that although the basic principle of radiometric dating is simple, the actual procedure is quite complex. The analysis that determines the quantities of parent and daughter must be painstakingly precise. In addition, some radioactive materials do not decay directly into the stable daughter product, as was the case with our hypothetical example, a fact that may further complicate the analysis. In the case of uranium-238,

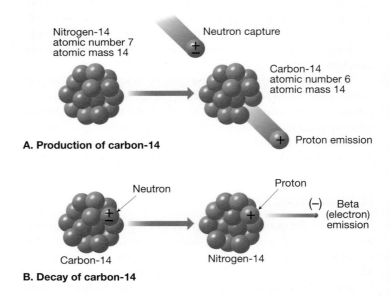

Nitrogen-14
atomic number 7
atomic mass 14

Neutron capture

Carbon-14
atomic number 6
atomic mass 14

Proton emission

A. Production of carbon-14

Neutron

Proton

(−) Beta (electron) emission

Carbon-14

Nitrogen-14

B. Decay of carbon-14

FIGURE 9.18   A. Production and **B.** decay of carbon-14. These sketches represent the nuclei of the respective atoms.

there are 13 intermediate unstable daughter products formed before the 14th and last daughter product, the stable isotope lead-206, is produced (see Figure 9.16).

Radiometric dating methods have produced literally thousands of dates for events in Earth history. Rocks exceeding 3.5 billion years in age are found on all of the continents. Earth's oldest rocks (so far) are gneisses from northern Canada near Great Slave Lake that have been dated at 4.03 billion years (b.y.). Rocks from western Greenland have been dated at 3.7 to 3.8 b.y. and rocks nearly as old are found in the Minnesota River Valley and northern Michigan (3.5 to 3.7 b.y.), in southern Africa (3.4 to 3.5 b.y.) and in western Australia (3.4 to 3.6 b.y.). It is important to point out that these ancient rocks are not from any sort of "primordial crust" but originated as lava flows, igneous intrusions, and sediments deposited in shallow water—an indication that Earth history began *before* these rocks formed. Even older mineral grains have been dated. Tiny crystals of the mineral zircon having radiometric ages as old as 4.3 b.y. have been found in younger sedimentary rocks in western Australia. The source rocks for these tiny durable grains either no longer exist or have not yet been found.

Radiometric dating has vindicated the ideas of Hutton, Darwin, and others, who more than 150 years ago inferred that geologic time must be immense. Indeed, modern dating methods have proved that there has been enough time for the processes we observe to have accomplished tremendous tasks.

THE GEOLOGIC TIME SCALE

GEOLOGIC TIME
▶ The Geologic Time Scale

Geologists have divided the whole of geologic history into units of varying magnitude. Together, they comprise the **geologic time scale** of Earth history (Figure 9.19). The major units of the time scale were delineated during the 19th century, principally by workers in Western Europe and Great Britain. Because radiometric dating was unavailable at that time, the entire time scale was created using methods of relative dating. It was only in the 20th century that radiometric methods permitted numerical dates to be added.

Structure of the Time Scale

The geologic time scale subdivides the 4.6-billion-year history of Earth into many different units and provides a meaningful time frame within which the events of the geologic past are arranged. As shown in Figure 9.19, **eons** represent the greatest expanses of time. The eon that began about 542 million years ago is the **Phanerozoic,** a term derived from Greek words meaning *visible life.* It is an appropriate description because the rocks and deposits of the Phanerozoic eon contain abundant fossils that document major evolutionary trends.

Another glance at the time scale reveals that eons are divided into **eras.** The three eras within the Phanerozoic are the **Paleozoic** (*paleo* = ancient, *zoe* = life), the **Mesozoic** (*meso* = middle, *zoe* = life), and the **Cenozoic** (*ceno* = recent, *zoe* = life). As the names imply, these eras are bounded by profound worldwide changes in life forms.*

Each era of the Phaneroic eon is subdivided into time units known as **periods.** The Paleozoic has seven, and the Mesozoic and Cenozoic each have three. Each of these periods is characterized by a somewhat less profound change in life forms as compared with the eras (Box 9.2).

Each of the periods is divided into still smaller units called **epochs.** As you can see in Figure 9.19, seven epochs have been named for the periods of the Cenozoic. The epochs of other periods usually are simply termed *early, middle,* and *late.*

Precambrian Time

Notice that the detail of the geologic time scale does not begin until about 542 million years ago, the date for the beginning of the Cambrian period. The nearly 4 billion years prior to the Cambrian are divided into two eons, the **Archean**

*Major changes in life forms are discussed in Chapter 22 "Earth's Evolution through Geologic Time."

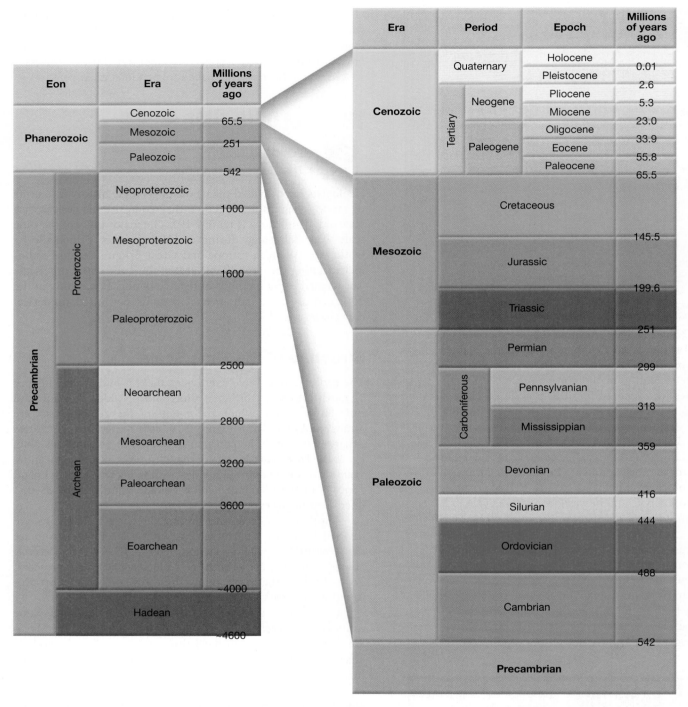

FIGURE 9.19   The geologic time scale. The numerical dates were added long after the time scale had been established using relative dating techniques. The time scale is a dynamic tool. Advances in the geosciences require that updates be made from time to time. For more about the geologic time scale, read Box 9.2.

(*archaios* = ancient), and the **Proterozoic** (*proteros* = before, *zoe* = life) which are divided into four eras. It is also common for this vast expanse of time to simply be referred to as the **Precambrian.** Although it represents about 88 percent of Earth history, the Precambrian is not divided into nearly as many smaller time units as the Phanerozoic eon.

Why is the huge expanse of Precambrian time not divided into numerous eras, periods, and epochs? The reason is that Precambrian history is not known in great enough detail. The quan-

tity of information that geologists have deciphered about Earth's past is somewhat analogous to the detail of human history. The further back we go, the less that is known. Certainly more data and information exist about the past 10 years than for the first decade of the 20th century; the events of the 19th century have been documented much better than the events of the 1st century A.D.; and so on. So it is with Earth history. The more recent past has the freshest, least disturbed, and more observable record. The further back in time the geologist goes, the more fragmented

UNDERSTANDING EARTH

BOX 9.2

Terminology and the Geologic Time Scale

There are some terms that are associated with the geologic time scale, but are not "officially" recognized as being a part of it. The best known, and most common, example is *Precambrian*—the informal name for the eons that came before the current Phanerozoic Eon. Although the term *Precambrian* has no formal status on the geologic time scale, it has been traditionally used as though it did.

Hadean is another informal term that is found on some versions of the geologic time scale and is used by some geologists. It refers to the earliest interval (eon) of Earth history—before the oldest known rocks. When the term was coined in 1972, the age of Earth's oldest rocks was about 3.8 billion years. Today that number stands at slightly greater than 4 billion, and, of course, is subject to revision. The name *Hadean* derives from *Hades,* Greek for *underworld*—a reference to the "hellish" conditions that prevailed on Earth early in its history.

Effective communication in the geosciences requires that the geologic time scale consist of standardized divisions and dates. So, who determines which names and dates on the geologic time scale are "official"? The organization that is largely responsible for maintaining and updating this important document is the International Committee on Stratigraphy (ICS), a committee of the International Union of Geological Sciences.* Advances in the geosciences require that the scale be periodically updated to include changes in unit names and boundary age estimates.

For example, the geologic time scale shown in Figure 9.19 was updated as recently as July 2009. After considerable dialogue among geologists who focus on very recent Earth history, the ICS changed the date for the start of the Quaternary Period and the Pleistocene Epoch from

*To view the current version of the ICS time scale, go to http://www.stratigraphy.org. *Stratigraphy* is the branch of geology that studies rock layers (strata) and layering (stratification), thus its primary focus is sedimentary and layered volcanic rocks.

1.8 million to 2.6 million years ago. Who knows, perhaps by the time you read this, other changes will have been made.

If you were to examine a geologic time scale from just a few years ago, it is quite possible that you would see the Cenozoic Era divided into the Tertiary and Quaternary periods. However, on more recent versions the space formerly designated as Tertiary is divided into the Paleogene and Neogene periods. As our understanding of this time span changed, so too did its designation on the geologic time scale. Today, the Tertiary Period is considered as an "historic" name and is given no official status on the ICS version of the time scale. Many time scales still contain references to the Tertiary period, including Figure 9.19. One reason for this is that a great deal of past (and some current) geological literature uses this name.

For those who study historical geology, it is important to realize that the geologic time scale is a dynamic tool that continues to be refined as our knowledge and understanding of Earth history evolves.

---

the record and clues become. There are other reasons to explain our lack of a detailed time scale for this vast segment of Earth history.

1. The first abundant fossil evidence does not appear in the geologic record until the beginning of the Cambrian period. Prior to the Cambrian, simple life forms such as algae, bacteria, fungi, and worms predominated. All of these organisms lack hard parts, an important condition favoring preservation. For this reason, there is only a meager Precambrian fossil record. Many exposures of Precambrian rocks have been studied in some detail, but correlation is often difficult when fossils are lacking.

2. Because Precambrian rocks are very old, most have been subjected to a great many changes. Much of the Precambrian rock record is composed of highly distorted metamorphic rocks. This makes the interpretation of past environments difficult, because many of the clues present in the original sedimentary rocks have been destroyed.

Radiometric dating has provided a partial solution to the troublesome task of dating and correlating Precambrian rocks. But untangling the complex Precambrian record still remains a daunting task.

Students Sometimes Ask . . .

Was there ever a time in Earth history when dinosaurs and humans coexisted?

Although some old movies and cartoons have depicted people and dinosaurs living side by side, this was never the case. Dinosaurs flourished during the Mesozoic era and became extinct about 65 million years ago. By contrast, humans and their close ancestors did not appear on the scene until very late in the Cenozoic era, more than 60 million years *after* the demise of the dinosaurs.

DIFFICULTIES IN DATING THE GEOLOGIC TIME SCALE

Although reasonably accurate numerical dates have been worked out for the periods of the geologic time scale (see Figure 9.19), the task is not without difficulty. The primary difficulty in assigning numerical dates to units of time is the fact that not all rocks can be dated by radiometric methods. Recall that for a radiometric date to be useful, all the minerals in the rock must have formed at about the same time. For this reason, radioactive isotopes can be used to determine when minerals in an igneous rock crystallized and when pressure and heat created new minerals in a metamorphic rock.

FIGURE 9.20   A useful numerical date for this conglomerate is not possible because the gravel that composes it was derived from rocks of diverse ages. (Photo by E. J. Tarbuck)

However, samples of sedimentary rock can only rarely be dated directly by radiometric means. Although a detrital sedimentary rock may include particles that contain radioactive isotopes, the rock's age cannot be accurately determined because the grains composing the rock are not the same age as the rock in which they occur. Rather, the sediments have been weathered from rocks of diverse ages (Figure 9.20).

Radiometric dates obtained from metamorphic rocks may also be difficult to interpret because the age of a particular mineral in a metamorphic rock does not necessarily represent the time when the rock initially formed. Instead, the date might indicate any one of a number of subsequent metamorphic phases.

If samples of sedimentary rocks rarely yield reliable radiometric ages, how can numerical dates be assigned to sedimentary layers? Usually the geologist must relate the strata to datable igneous masses, as in Figure 9.21. In this example, radiometric dating has determined the ages of the volcanic ash bed within the Morrison Formation and the dike cutting the Mancos Shale and Mesaverde Formation. The sedimentary beds below the ash are obviously older than the ash, and all the layers above the ash are younger. The dike is younger than the Mancos Shale and the Mesaverde Formation but older than the Wasatch Formation because the dike does not intrude the Tertiary rocks.

From this kind of evidence, geologists estimate that a part of the Morrison Formation was deposited about 160 million years ago, as indicated by the ash bed. Further, they conclude that the Tertiary period began after the intrusion of the dike, 66 million years ago. This is one example of literally thousands that illustrate how datable materials are used to bracket the various episodes in Earth history within specific time periods. It shows the necessity of combining laboratory dating methods with field observations of rocks.

FIGURE 9.21   Numerical dates for sedimentary layers are usually determined by examining their relationship to igneous rocks. (After U.S. Geological Survey)

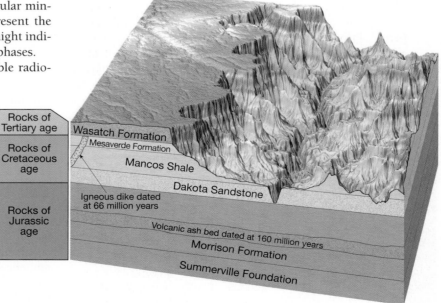

CHAPTER 9 GEOLOGIC TIME IN REVIEW

- The two types of dates used by geologists to interpret Earth history are (1) *relative dates,* which put events in their *proper sequence of formation,* and (2) *numerical dates,* which pinpoint the *time in years* when an event occurred.

- Relative dates can be established using the *law of superposition* (in an undeformed sequence of sedimentary rocks or surface-deposited igneous rocks, each bed is older than the one above, and younger than the one below), *principle of original horizontality* (most layers are deposited in a horizontal position), *principle of cross-cutting relationships* (when a fault or intrusion cuts through another rock, the fault or intrusion is younger than the rocks cut through), and *inclusions* (the rock mass containing the inclusion is younger than the rock that provided the inclusion).

- *Unconformities* are gaps in the rock record. Each represents a long period during which deposition ceased, erosion removed previously formed rocks, and then deposition resumed. The three basic types of unconformities are *angular unconformities* (tilted or folded sedimentary rocks that are overlain by younger, more flat-lying strata), *disconformities* (the strata on either side of the unconformity are essentially parallel), and *nonconformities* (where a break separates older metamorphic or intrusive igneous rocks from younger sedimentary strata).

- *Correlation,* the matching up of two or more geologic phenomena of similar age in different areas, is used to develop a geologic time scale that applies to the whole Earth.

- Fossils are the remains or traces of prehistoric life. The special conditions that favor preservation are *rapid burial* and the possession of *hard parts,* such as shells, bones, or teeth.

- Fossils are used to *correlate* sedimentary rocks that are from different regions by using the rocks' distinctive fossil content and applying the *principle of fossil succession.* It is based on the work of *William Smith* in the late 1700s and states that fossil organisms succeed one another in a definite and determinable order, and therefore any time period can be recognized by its fossil content. The use of *index fossils,* those that are widespread geographically and are limited to a short span of geologic time, provides an important method for matching rocks of the same age.

- Each atom has a nucleus containing *protons* (positively charged particles) and *neutrons* (neutral particles). Orbiting the nucleus are negatively charged *electrons.* The *atomic number* of an atom is the number of protons in the nucleus. The *mass number* is the number of protons plus the number of neutrons in an atom's nucleus. *Isotopes* are variants of the same atom, but with a different number of neutrons and hence a different mass number.

- *Radioactivity* is the spontaneous breaking apart (decay) of certain unstable atomic nuclei. Three common types of radioactive decay are (1) emission of *alpha particles* from the nucleus, (2) emission of *beta particles* from the nucleus, and (3) *capture of electrons* by the nucleus.

- An unstable *radioactive isotope,* called the *parent,* will decay and form stable *daughter products.* The length of time for half of the nuclei of a radioactive isotope to decay is called the *half-life* of the isotope. If the half-life of the isotope is known, and the parent/daughter ratio can be measured, the age of a sample can be calculated. An accurate radiometric date can be obtained only if the mineral containing the radioactive isotope has remained in a closed system during the entire period since its formation.

- The *geologic time scale* divides Earth's history into units of varying magnitude. It is commonly presented in chart form, with the oldest time and event at the bottom and the youngest at the top. The principle subdivisions of the geologic time scale, called *eons,* include the *Archean, Proterozoic* (together, these two eons are commonly referred to as the *Precambrian*), and, beginning about 542 million years ago, the *Phanerozoic.* The Phanerozoic (meaning "visible life") eon is divided into the following *eras: Paleozoic* ("ancient life"), *Mesozoic* ("middle life"), and *Cenozoic* ("recent life").

- A significant problem in assigning numerical dates is that *not all rocks can be radiometrically dated.* A sedimentary rock may contain particles of many ages that have been weathered from different rocks that formed at various times. One way geologists assign numerical dates to sedimentary rocks is to relate them to datable igneous masses, such as volcanic ash beds.

KEY TERMS

angular unconformity (p. 259)
Archean eon (p. 271)
Cenozoic era (p. 271)
conformable (p. 258)
correlation (p. 261)
cross-cutting relationships, principle of (p. 258)
disconformity (p. 260)

eon (p. 271)
epoch (p. 271)
era (p. 271)
fossil (p. 262)
fossil succession, principle of (p. 267)
geologic time scale (p. 271)
half-life (p. 269)
inclusions (p. 258)

index fossil (p. 267)
Mesozoic era (p. 271)
nonconformity (p. 260)
numerical date (p. 256)
original horizontality, principle of (p. 257)
paleontology (p. 262)
Paleozoic era (p. 271)
period (p. 271)

Phanerozoic eon (p. 271)
Precambrian (p. 272)
Proterozoic eon (p. 272)
radioactivity (p. 268)
radiocarbon dating (p. 270)
radiometric dating (p. 260)
relative dating (p. 257)
superposition, law of (p. 257)
unconformity (p. 259)

QUESTIONS FOR REVIEW

1. Distinguish between numerical and relative dating.

2. What is the law of superposition? How are cross-cutting relationships used in relative dating?

3. Refer to Figure 9.5 (p. 259) and answer the following questions:

   a. Is fault A older or younger than the sandstone layer?

   b. Is dike A older or younger than the sandstone layer?

   c. Was the conglomerate deposited before or after fault A?

   d. Was the conglomerate deposited before or after fault B?

   e. Which fault is older, A or B?

   f. Is dike A older or younger than the batholith?

4. When you observe an outcrop of steeply inclined sedimentary layers, what principle allows you to assume that the beds were tilted after they were deposited?

5. A mass of granite is in contact with a layer of sandstone. Using a principle described in this chapter, explain how you might determine whether the sandstone was deposited on top of the granite or whether the granite was intruded from below after the sandstone was deposited.

6. Distinguish among angular unconformity, disconformity, and nonconformity.

7. What is meant by the term *correlation?*

8. Describe William Smith's important contribution to the science of geology.

9. List and briefly describe at least five different types of fossils.

10. List two conditions that improve an organism's chances of being preserved as a fossil.

11. Why are fossils such useful tools in correlation?

12. Figure 9.22 is a block diagram of a hypothetical area in the American Southwest. Place the lettered features in the proper sequence, from oldest to youngest. Identify an angular unconformity and a nonconformity.

13. If a radioactive isotope of thorium (atomic number 90, mass number 232) emits 6 alpha particles and 4 beta particles during the course of radioactive decay, what are the atomic number and mass number of the stable daughter product?

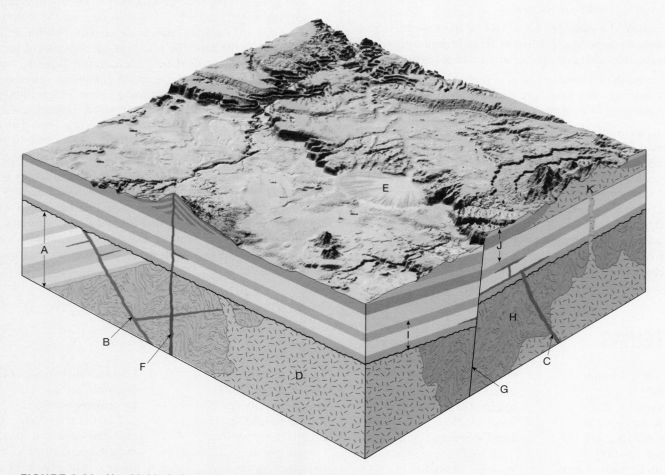

FIGURE 9.22 Use this block diagram in conjunction with Review Question 12.

14. Why is radiometric dating the most reliable method of dating the geologic past?

15. A hypothetical radioactive isotope has a half-life of 10,000 years. If the ratio of radioactive parent to stable daughter product is 1:3, how old is the rock containing the radioactive material?

16. To provide a reliable radiometric date, a mineral must remain in a closed system from the time of its formation until the present. Why is this true?

17. What precautions are taken to ensure reliable radiometric dates?

18. To make calculations easier, let us round the age of Earth to 5 billion years.

    a. What fraction of geologic time is represented by recorded history (assume 5000 years for the length of recorded history)?

    b. The first abundant fossil evidence does not appear until the beginning of the Cambrian period (540 million years ago). What percent of geologic time is represented by abundant fossil evidence?

19. What subdivisions make up the geologic time scale?

20. Explain the lack of a detailed time scale for the vast span known as the Precambrian.

21. Briefly describe the difficulties in assigning numerical dates to layers of sedimentary rock.

COMPANION WEBSITE

The *Earth 10e* website uses the resources and flexibility of the Internet to aid in your study of the topics in this chapter. Written and developed by the authors and other geology instructors, this site will help improve your understanding of geology. Visit **www.mygeoscienceplace.com** in order to:

- **Review** key chapter concepts

- **Read** with links to the eBook and to chapter-specific web resources

- **Visualize** and comprehend challenging topics using learning activities in *GEODe Earth*

- **Test** yourself with online quizzes

GEODe EARTH

GEODe Earth is a valuable and easy to use learning aid that can be accessed from your book's Companion Website (**www.mygeoscienceplace.com**). It is a dynamic instructional tool that promotes understanding and reinforces important concepts by using tutorials, animations, and exercises that actively engage the student.

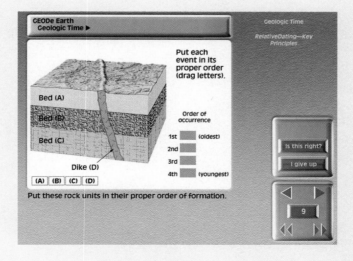

10
CRUSTAL DEFORMATION

Tilted strata forming hogback ridges east of Capitol
Reef National Park, Utah.

(PHOTO BY MICHAEL COLLIER)

Earth is a dynamic planet. Shifting lithospheric plates continually change the face of our planet by moving continents across the globe. The results of this tectonic activity are perhaps most strikingly apparent in Earth's major mountain belts. Rocks containing fossils of marine organisms are found thousands of meters above sea level and massive rock units are bent, contorted, overturned, and sometimes rife with fractures.

In the Canadian Rockies, for example, rock strata have been thrust hundreds of kilometers over other layers. On a smaller scale, crustal movements of a few meters occur along faults during major earthquakes. Even in the stable interiors of the continents, rocks reveal a history of deformation that shows they have been uplifted from much deeper levels in the crust.

STRUCTURAL GEOLOGY: A STUDY OF EARTH'S ARCHITECTURE

Structural geology is the study of the architecture of Earth's crust and "how it got this way." The basic geologic features that form as a result of the forces generated by the interactions of tectonic plates, called **rock** or **tectonic structures**, include folds, faults, joints, and small-scale structures associated with metamorphic rocks such as foliation and rock cleavage (Figure 10.1). By studying the orientations of faults, folds, and tilted sedimentary formations, structural geologists can often reconstruct the original geologic setting and the nature of the forces that generated these structures. In this way, the complex events of Earth's geologic history are unraveled.

An understanding of rock structures is also basic to our economic well-being. Most occurrences of oil and natural gas, for example, are associated with geologic structures that trap these valuable fluids in reservoirs (see Chapter 23). Furthermore, rock fractures are sites of hydrothermal mineralization, the source for many important metallic ore deposits. Moreover, the orientation of fractures and bedding surfaces (zones of weakness in rocks) must be considered when selecting sites for major construction projects such as bridges, hydroelectric dams, and nuclear power plants. In short, a working knowledge of the crust's architecture is essential to our modern way of life.

In this chapter we will examine the forces that deform rock and the rock structures that result. Foliation and rock cleavage were examined in Chapter 8, this chapter will be devoted to the other major structural features of Earth's crust and the tectonic forces that produce them.

DEFORMATION, STRESS, AND STRAIN

CRUSTAL DEFORMATION
▶ Deformation

Every body of rock, no matter how strong, has a point at which it will fracture or flow. **Deformation** (*de* = out, *forma* = form) is a general term that refers to all changes in the shape, position, or orientation of a rock mass. Significant amounts of crustal deformation occur along plate margins. Plate motions and the interactions along plate boundaries generate the tectonic forces that cause rock to deform.

Stress: The Force that Deforms Rocks

From everyday experience you know that if a door is stuck you must expend energy, called *force,*

FIGURE 10.1 Uplifted and folded sedimentary strata at Stair Hole, near Lulworth, Dorset, England. These layers of Jurassic-age rock, originally deposited in horizontal beds, have been folded as a result of the collision between the African and European crustal plates. (Photo by Tom Bean/Corbis)

to open it. Structural geologists use the term **stress** to describe the forces that deform rocks. Whenever the stresses acting on a rock body exceed its strength, the rock will deform—usually by flowing, folding, fracturing, or faulting.

The magnitude of stress is not simply a function of the amount of force applied, but also relates to the area on which the force acts. For example, if you are walking barefoot on a hard surface, the force (weight) of your body is distributed across your entire foot, so the stress acting on any one point of your foot is low. However, if you step on a small pointed rock (ouch!), the stress concentration at that point on your foot will be high. Thus, you can think of stress as a measure of how much force is applied over a particular area.

Compressional Stress   As you saw in Chapter 8, stress may be applied uniformly in all directions, this is called **confining pressure**, or it may be applied non-uniformly. When stress is applied unequally in different directions, it is termed **differential stress.** Differential stress that squeezes and shortens a rock mass is known as **compressional stress** (*com* = together, *premere* = to press) (Figure 10.2A). Compressional stresses are most often associated with convergent plate boundaries. When plates collide, Earth's crust is generally shortened and thickened, producing mountainous terrain.

Tensional Stress   Stress that pulls apart or elongates a rock unit is known as **tensional stress** (*tendere* = to stretch) (Figure

10.2B). At divergent plate boundaries, where plates are rifted apart, tensional stresses stretch and lengthen rock bodies in the upper crust by displacement along faults. By contrast, displacement is accomplished at depth by ductile flow. In the Basin and Range Province in the western United States, tensional forces have stretched the crust by as much as twice its original width.

Shear Stress   Differential stress can also cause rock to **shear**, which involves the movement of one part of a rock body past another (Figure 10.2C). Shear is similar to the slippage that occurs between individual playing cards when the top of the deck is moved relative to the bottom (Figure 10.3). When rocks are deformed, slippage often occurs on closely spaced parallel surfaces of weakness, such as foliation surfaces and microscopic fractures. By contrast, at transform fault boundaries, such as the San Andreas Fault, shear stresses cause large segments of Earth's crust to slip horizontally past one another.

Strain: A Change in Shape Caused by Stress

We are now ready to expand our definition of deformation to state that it refers to changes in the shape, position, or orientation of a rock body *resulting from the application of differential stress.* When flat-lying sedimentary layers are uplifted and

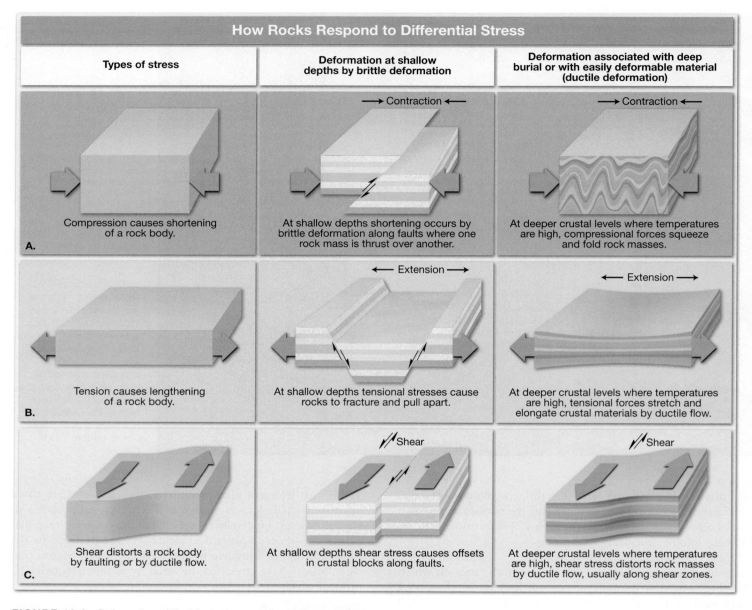

How Rocks Respond to Differential Stress

Types of stress	Deformation at shallow depths by brittle deformation	Deformation associated with deep burial or with easily deformable material (ductile deformation)
A. Compression causes shortening of a rock body.	← Contraction ← At shallow depths shortening occurs by brittle deformation along faults where one rock mass is thrust over another.	→ Contraction ← At deeper crustal levels where temperatures are high, compressional forces squeeze and fold rock masses.
B. Tension causes lengthening of a rock body.	← Extension → At shallow depths tensional stresses cause rocks to fracture and pull apart.	← Extension → At deeper crustal levels where temperatures are high, tensional forces stretch and elongate crustal materials by ductile flow.
C. Shear distorts a rock body by faulting or by ductile flow.	Shear At shallow depths shear stress causes offsets in crustal blocks along faults.	Shear At deeper crustal levels where temperatures are high, shear stress distorts rock masses by ductile flow, usually along shear zones.

FIGURE 10.2   Deformation of Earth's crust caused by the three types of stresses that result from the movement of lithospheric plates—compressional, tensional, and shear. Brittle deformation (fracturing and faulting) dominates in the upper crust where the temperatures are comparatively cool. By contrast, at depths greater than about 10 kilometers, where temperatures are high, rock deforms by ductile flow and folding. **A.** Compressional stresses associated with convergent plate boundaries tend to shorten and thicken Earth's crust by folding, flowing, and faulting. **B.** Tensional stresses at divergent plate boundaries tend to lengthen rock bodies by displacement along faults in the upper crust and by ductile flow at depth. **C.** Shear stresses at transform plate boundaries tend to produce offsets along fault zones and ductile flow at depth.

tilted, their orientations change, but their shapes are often retained. Differential stresses can also cause a change in the shape of a rock body, referred to as **strain.** Like the circle shown in Figure 10.3B, *strained bodies lose their original configuration during deformation.* A small-scale example of strain is illustrated by the deformed fossil shown in Figure 10.4. When studying rock units that are deformed, geologists attempt to answer the question, "What do these deformed structures indicate about the original arrangement of these rocks, and what was the nature of the forces that caused this deformation?"

In summary, deformation is a general term that refers to changes in the shape, position, or orientation of a rock body.

We can think of *stress* as the force that acts to deform a body of rock, while *strain* is the distortion or change in shape that results.

HOW ROCKS DEFORM

When rocks are subjected to stresses greater than their strength, they begin to deform, usually by flowing or fracturing. It is easy to visualize how rocks break, because we normally think of them as being brittle. But how can masses of rock be *bent* into intricate folds without fracturing in the process? To determine

A. Deck of playing cards

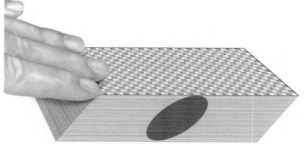

B. Shearing occurs when hand pushes top of deck

FIGURE 10.3 Illustration of shearing and the resulting deformation (strain). **A.** An ordinary deck of playing cards with a circle embossed on its side. **B.** By sliding the top of the deck relative to the bottom, we can illustrate the type of shearing that commonly occurs along closely spaced planes of weakness in rocks. Notice that the circle becomes an ellipse, which can be used to measure the amount and type of strain. Additional displacement (shearing) of the cards would result in further strain and would be indicated by a change in the shape of the ellipse.

FIGURE 10.4 Deformed trilobite, a very common Paleozoic life-form. Compare this image to the one in Figure 22.18 (p. 622) to see the extent of change. (Photo courtesy of Smithsonian Institution)

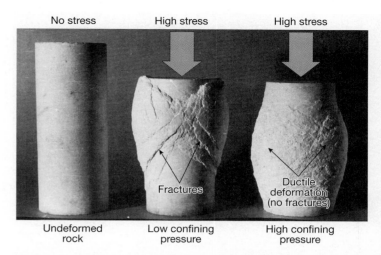

FIGURE 10.5 A marble cylinder deformed in the laboratory by applying thousands of pounds of load from above. Each sample was deformed in an environment that duplicated the confining pressure found at various depths. Notice that when the confining pressure was low, the sample deformed by brittle fracture. When the confining pressure was high, the sample deformed plastically. (Photo courtesy of M. S. Patterson, Australian National University)

this, structural geologists performed laboratory experiments in which rocks were subjected to differential stresses under conditions that simulate those existing at various depths within the crust (Figure 10.5).

Elastic, Brittle, and Ductile Deformation

Although each rock type deforms somewhat differently, the general characteristics of rock deformation were determined from such experiments. Geologists learned that when stress is gradually applied, rocks first respond by deforming elastically. Changes that result from **elastic deformation** are recoverable; that is, like a rubber band, the rock will return to nearly its original size and shape when the stress is removed. During elastic deformation the chemical bonds of the minerals within a rock are stretched, but do not break. You will see in the next chapter that the energy for most earthquakes comes from stored elastic energy that is released as rock snaps back to its original shape.

Once the elastic limit (strength) of a rock is surpassed, it either flows or fractures. Rocks that break into smaller pieces exhibit **brittle deformation** (*bryttian* = to shatter). From our everyday experience, we know that glass objects, wooden pencils, china plates, and even our bones exhibit brittle failure once their strength is surpassed. Brittle deformation occurs when stress causes the chemical bonds that hold a material together to break.

Ductile deformation, on the other hand, is a type of solid-state flow that produces a change in the shape of an object without fracturing (Figure 10.6). Ordinary objects that display ductile behavior include modeling clay, beeswax, taffy, and some metals. For example, a copper penny placed on a railroad track will be flattened and deformed (without breaking) by the force applied by a passing train. In rocks, ductile deformation is the result of some chemical bonds breaking, while others are forming, allowing minerals to change shape.

FIGURE 10.6 Rocks exhibiting ductile deformation. These rocks were deformed at great depth and were subsequently exposed at the surface. Vishnu Schist, Grand Canyon National Park, Arizona. (Photo by Michael Collier)

Factors that Affect Rock Strength

The major factors that influence the strength of a rock and how it will deform include temperature, confining pressure, rock type, and time.

The Role of Temperature   The effect of temperature on the strength of a material can be easily demonstrated with a piece of glass tubing commonly found in a chemistry lab. If the tubing is dropped on a hard surface, it will shatter. However, if the tubing is heated over a Bunsen burner, it can be easily bent into a variety of shapes. Rocks respond similarly to heat. Where temperatures are high (deep in Earth's crust), rocks tend to deform ductilely and flow. Likewise, where temperatures are low (at or near the surface), rocks tend to behave like brittle solids and fracture.

The Role of Confining Pressure   Recall from Chapter 8 that pressure, like temperature, increases with depth as the thickness of the overlying rock increases. Buried rocks are subjected to confining pressure, which is much like water pressure, where the forces are applied equally in all directions. The deeper you go in the ocean, the greater the confining pressure. The same is true for rock that is buried. Confining pressure "squeezes" the materials in Earth's crust. Therefore, rocks that are deeply buried are "held together" by the immense pressure and tend to flow, rather than fracture.

The Influence of Rock Type   In addition to the physical environment, the mineral composition and texture of rock greatly influence how it will deform. For example, crystalline rocks composed of minerals that have strong internal molecular bonds tend to fail by brittle fracture. By contrast, sedimentary rocks that are weakly cemented, or metamorphic rocks that contain zones of weakness, such as foliation, are more susceptible to ductile deformation.

Weak rocks that are most likely to behave in a ductile manner (flow or fold) when subjected to differential stress, include

rock salt, shale, limestone, and schist. In fact, rock salt is so weak that large masses of it often rise through overlying beds of sedimentary rocks much like hot magma rises toward the surface. Perhaps the weakest naturally occurring solid to exhibit ductile flow is glacial ice.

Igneous and some metamorphic rocks tend to be strong and brittle. In a near-surface environment, strong, brittle rocks will fail by fracturing when subjected to stresses that exceed their strength. At increasing depths, however, the strength of all rock types decreases significantly.

Some outcrops consist of a sequence of interbedded weak and strong rock layers that have been moderately deformed—for example, interlayered shale and well-cemented sandstone beds. In these settings, the strong sandstone layers are often highly fractured, while the weak shale beds form broad undulating folds. The formation of these diverse structures, one brittle and one ductile, can be illustrated by placing a Milky Way® or similar chocolate-over-caramel bar into a refrigerator. When the cool candy bar is slowly bent, it will exhibit brittle deformation in the chocolate and ductile deformation in the caramel.

Time as a Factor   One key factor that researchers are unable to duplicate in the laboratory is how rocks respond to small stresses applied gradually over long spans of *geologic time*. However, insights into the effects of time on deformation are provided in everyday settings. For example, marble benches have been known to sag under their own weight over a span of 100 years or so, and wooden bookshelves may bend after being loaded with books for a relatively short period.

In general, when tectonic forces are applied slowly over long time spans, rocks tend to display ductile behavior and deform by flowing and folding. An analogous situation occurs when you take a taffy bar and slowly move the two ends together. The taffy will deform by folding. However, if you swiftly hit the taffy against the edge of a table, it will break into two or more pieces, exhibiting brittle failure.

Likewise, rocks tend to deform in a ductile manner by folding when stress builds gradually. These same rocks may fracture if the stress increases suddenly. As a consequence, folding and faulting may occur simultaneously in the same rock body (Figure 10.7).

To review, the processes by which rocks deform occur along a continuum that ranges from brittle fracture at one end to ductile flow at the other. The processes of deformation generate geologic

FIGURE 10.7   Deformed sedimentary strata exposed in a road cut near Palmdale, California. In addition to the obvious folding, light-colored beds are offset along a fault located on the right side of the photograph. (Photo by E. J. Tarbuck)

Geologist's Sketch

would form if you were to hold the ends of a sheet of paper and then push them together. In nature, folds come in a wide variety of sizes and configurations. Some folds are broad flexures in which strata hundreds of meters thick have been slightly warped. Others are very tight microscopic structures found in metamorphic rocks. Size differences notwithstanding, most folds are the result of *compressional stresses that result in a shortening and thickening of the crust.*

To aid our understanding of folds and folding, we need to become familiar with the terminology used to name the parts of a fold. As shown in Figure 10.8, the two sides of a fold are called *limbs*. A line drawn along the points of maximum curvature of

changes on many scales. At one extreme are Earth's major mountain systems. At the other extreme are minor fractures in bedrock created by highly localized stresses. All of these phenomena, from the largest folds in the Alps to the smallest fractures in a slab of rock, are considered to be *rock* or *tectonic structures.*

STRUCTURES FORMED BY DUCTILE DEFORMATION

CRUSTAL DEFORMATION

▸ Folds

Because folds are common features of deformed sedimentary rocks, we know rocks can bend without breaking. Ductile deformation is often accomplished by gradual slippage along planes of weakness within the atomic structure of mineral grains (see Figure 8.8, p. 236). This microscopic form of gradual solid-state flow involves slippage facilitated by chemical bonds breaking in one location as new ones form at another site. Rocks that display evidence of ductile flow usually were deformed at great depth and may exhibit contorted folds that give the impression that the strength of the rock was akin to soft putty (see Figure 10.6).

Folds

Along convergent plate boundaries, flat-lying sedimentary and volcanic rocks are often bent into a series of wavelike undulations called **folds.** Folds in sedimentary strata are much like those that

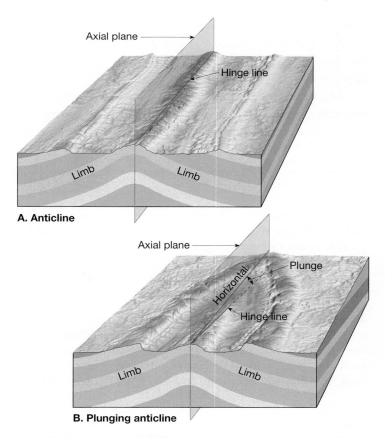

A. Anticline

B. Plunging anticline

FIGURE 10.8   Idealized sketches illustrating the features associated with symmetrical folds. The hinge line of the fold in **A** is horizontal, whereas the hinge line of the fold in **B** is plunging.

Feet in the Fire: Communicating Geologist

Once, after an eight-hour flight to Chile and another two hours to the Falkland Islands, I found myself aboard a boat headed toward Antarctica. It was early evening and even though the December sky was dark, familiar Northern Hemisphere landmarks were nowhere to be seen. I was on the top deck, leaning backward over the railing, when a fierce gust of loneliness broadsided me: home was seven thousand miles away. Leaning precariously over that rail, I took comfort in the sight of Orion, 'upright' because I was upside down. I've always needed to know where I fit in, either upon the Earth or in the sky.

A few years earlier, on a winter river trip through Utah's Cataract Canyon, we were boating alongside ice on the Colorado when I realized for the first time that I was exactly in the world where I wanted to be. I was a budding photographer, capturing landscape images throughout the American West. I was learning to love the color and cadence of stories that inevitably shine through the photographs—rivers rolling, waves crashing, mountains gleaming. These stories had grand themes of creation, metamorphosis, and destruction. I had unexpectedly warmed to a geologic way of looking at the world.

I first studied "gee-whiz geology" at a college near Grand Canyon, where all we undergraduates wanted to do was hike our brains out. Then, as a graduate student in California, I was thrown into an unexpectedly mathematical appreciation of structural folds and faults. Feet in the fire, head in the freezer: it was a balanced education that nicely prepared me for life as a communicating geologist.

After finishing school, I rowed boats in Grand Canyon for a living. We would spend almost two weeks drifting from Lees Ferry to the take-out at Diamond Creek. My masters thesis had examined a curious fold

> **"Feet in the fire, head in the freezer: it was a balanced education that nicely prepared me for life as a communicating geologist."**

in the Muav Limestone created when lava had temporarily flooded the Canyon a few thousand years earlier. My passengers, leaning over the raft's rubber side and trailing their fingers absentmindedly in the water, were a captive audience to my geologic stories about time and rock and those Muav folds. They didn't seem to mind.

Along the way, I learned to fly and kayak. The two motions resonated: three-dimensional activities in fluid media. Like a paddling friend says, sometimes you're supposed to be upside down in a rapid. And now, after 5500 hours aloft, I've found the same to be true in the air. It doesn't seem unnatural when my plane is sideways howling around a curve at the bottom of some narrow canyon. You're supposed to be sideways. Otherwise you'll smack into the wall that's filling the windshield.

> **"It doesn't seem unnatural when my plane is sideways howling around a curve at the bottom of some narrow canyon."**

My plane is a fifty-five-year-old Cessna 180 that's happiest when operating off dirt strips in the middle of nowhere.

---

each layer is termed the *hinge line,* or simply the *hinge.* In some folds, as Figure 10.8A illustrates, the hinge is horizontal, or parallel to the surface. However, in more complex folding, the hinge is often inclined at an angle known as the *plunge* (Figure 10.8B). Further, the *axial plane* is an imaginary surface that divides a fold as symmetrically as possible.

Anticlines and Synclines The two most common types of folds are anticlines and synclines (Figure 10.9). **Anticlines** usually arise by upfolding, or arching, of sedimentary layers and are sometimes spectacularly displayed along highways that have been cut through deformed strata (Figure 10.10).* Almost

always found in association with anticlines are downfolds, or troughs, called **synclines** (Figure 10.11). Notice in Figure 10.9 that the limb of an anticline is also a limb of the adjacent syncline.

Depending on their orientation, these basic folds are described as *symmetrical* when the limbs are mirror images of each other and *asymmetrical* when they are not. An asymmetrical fold is said to be *overturned* if one or both limbs are tilted beyond the vertical (Figure 10.12). An overturned fold can also "lie on its side" so a plane extending through the axis of the fold is horizontal. These *recumbent* folds are common in highly deformed mountainous regions such as the Alps (Figure 10.13).

Folds do not continue forever; rather their ends die out much like the wrinkles in cloth. Some folds *plunge* because the axis of the fold penetrates the ground (see Figure 10.8B).

---

*By strict definition, an anticline is a structure in which the oldest strata are found in the center. A syncline is a structure in which the youngest strata are found in the center.

Geologist-pilot Michael Collier and his Cessna 180. He is also an award-winning photographer and writer. Michael classifies himself as a "communicating geologist." You will see dozens of his outstanding images in the pages of this book. (Photo by Patty DiRienzo)

My wife swears that the plane's tail wags when I walk up to it. Together my plane and I have seen the continent from Tegucigalpa to Tanana, from Portland to Portland, from Baja to Boston. I use the plane to study and photograph landscapes. From the sky, I've grown intimate with the ground. From above, I see stories that the Earth has to tell; I've learned where I fit in. Once the pilot of a large business plane was shocked when I told him that upon take-off, I might know where I'm going to land only two times out of three. I go where the stories lead, where the light draws my camera.

I started out publishing small books about the geology of national parks—Grand Canyon, of course, Capitol Reef, Denali, and Death Valley. Regardless of how much scientific research I had found in libraries, I always insisted on developing a personal sense of each new landscape—hiking, scrambling, driving, and flying before beginning to write. Without a feel for the land, my words and photographs were doomed to be vacuous.

I now write books about geologic processes for lay audiences—how mountains are built, how glaciers move, how faults whipsaw the Earth, how climate behaves, how rivers flow. Each requires the assimilation of new information about unfamiliar topics. Each book introduces me to accomplished scientists, once again humbling me in the presence of experts in their chosen work. Each new project is a great opportunity to learn and grow. Each makes *my* tail wag.

But here's an odd twist . . . after a while I became dissatisfied with the mercurial life of an itinerant geologist, boatman, pilot, writer, and freelance photographer. I wanted to be able to walk up and shake a person's hand, look them in the eye and ask how could I help. So, in the 1980s I detoured into medical school and, after residency, became a family practice physician. Thirty years later, I still fly and write and photograph, but now I also spend half my life as a doctor. We only go around once, so we might as well pack it in while we can. Doctoring has been profoundly satisfying because it balances a life spent outdoors, studying beautiful landscapes and listening to stories that are as large as the Earth itself.

—**Michael Collier**

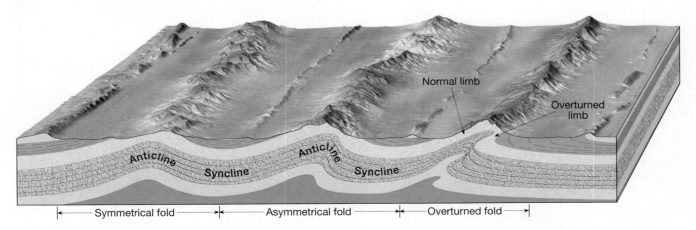

FIGURE 10.9 Block diagram of principal types of folded strata. The upfolded, or arched, structures are *anticlines.* The downfolds, or troughs, are *synclines.* Notice that the limb of an anticline is also the limb of the adjacent syncline.

FIGURE 10.10   An asymmetrical anticline in which one limb dips more steeply than the other. (Photo by E. J. Tarbuck)

Figure 10.14 shows an example of a plunging anticline and the pattern produced when erosion removes the upper layers of the structure and exposes its interior. Note that the outcrop pattern of an anticline points in the direction it is plunging. The opposite is true for a syncline. A good example of the kind of topography that results when erosional forces attack folded sedimentary strata is found in the Valley and Ridge Province of the Appalachians (see Figure 14.18, p. 396).

It is important to realize that ridges are not necessarily associated with anticlines, nor are valleys related to synclines. Rather, ridges and valleys result because of differential weathering and erosion. For example, in the Valley and Ridge Province, resistant sandstone beds remain as imposing ridges separated by valleys cut into more easily eroded shale or limestone beds.

Domes and Basins   Broad upwarps in basement rock may deform the overlying cover of sedimentary strata and generate

FIGURE 10.12   Overturned fold, East Fork of Toklat River, Alaska. Overturned folds have one or both limbs tilted beyond vertical. (Photo by Michael Collier)

FIGURE 10.13   Recumbent folds in the Swiss Alps. (Photo by Mike Andrews/Animals Animals–Earth Scenes)

FIGURE 10.11   A nearly symmetrical syncline formed in limestone and siltstone strata. (Photo by E. J. Tarbuck)

FIGURE 10.14   Plunging anticline, Sheep Mountain, Wyoming. In a plunging anticline the outcrop pattern "points" in the direction of plunge, the opposite is true of plunging synclines. (Photo by Michael Collier)

large folds. When this upwarping produces a circular or slightly elongated structure, the feature is called a **dome** (Figure 10.15A). Downwarped structures having a similar shape are termed **basins** (Figure 10.15B).

The Black Hills of western South Dakota is a large domed structure generated by upwarping. Here erosion has stripped away the highest portions of the overlying sedimentary beds, exposing older igneous and metamorphic rocks in the center (Figure 10.16).

Domes can also be formed by the intrusion of magma (laccoliths) as shown in Figure 4.32, p. 132. In addition, the upward migration of salt formations can produce salt domes like those common to the Gulf of Mexico.

Several large basins exist in the United States (Figure 10.17). The basins of Michigan and Illinois have gently sloping beds similar to saucers. These basins are thought to be the result of large accumulations of sediment, whose weight caused the crust to subside (see section on isostasy in Chapter 14). A few structural basins may have been the result of giant asteroid impacts.

Because large basins contain sedimentary beds sloping at low angles, they are usually identified by the age of the rocks composing them. The youngest rocks are found near the center, and the oldest rocks are at the flanks. This is just the opposite order of a domed structure, such as the Black Hills, where the oldest rocks form the core.

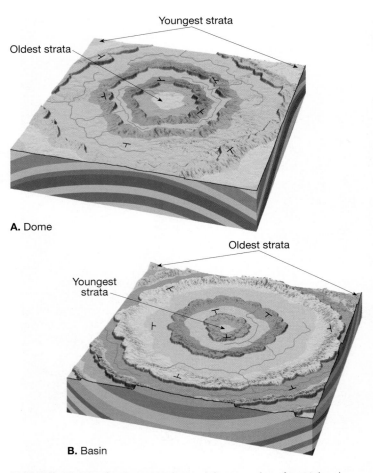

A. Dome

B. Basin

FIGURE 10.15 Gentle upwarping and downwarping of crustal rocks produce domes **A.** and basins **B.** Erosion of these structures results in an outcrop pattern that is roughly circular or elongate.

Monoclines Although we have separated our discussion of folds and faults, in the real world folds can be intimately coupled with faults. Examples of this close association are broad, regional features called *monoclines*. Particularly prominent features of the Colorado Plateau, **monoclines** (*mono* = one, *kleinen* = incline) are large, steplike folds in otherwise horizontal sedimentary strata (Figure 10.18). These folds appear to be the result of the reactivation of ancient, steep dipping faults located in basement rocks beneath the plateau. As large blocks of basement rock were displaced upward, the comparatively ductile sedimentary strata above responded by folding. Displacement along these reactivated faults often exceeds 1 kilometer (Figure 10.18).

FIGURE 10.16 The Black Hills of South Dakota, a large domal structure with resistant igneous and metamorphic rocks exposed in the core.

Examples of monoclines found on the Colorado Plateau include the East Kaibab Monocline, Raplee Anticline, Waterpocket Fold, and the San Rafael Swell. The inclined strata shown in Figure 10.18 once extended over the sedimentary layers that are now exposed at the surface—evidence that a tremendous volume of rock has been eroded from this area. Differential erosion has left the more resistant beds outcropping as prominent angular ridges called **hogbacks**. Because hogbacks can form whenever resistant strata are steeply inclined, they are associated with most types of folds.

STRUCTURES FORMED BY BRITTLE DEFORMATION

CRUSTAL DEFORMATION
▶ Faults and Fractures

You have probably seen a drinking glass drop on a hard surface, shattering into pieces. What you witnessed is analogous to brittle deformation that occurs in Earth's crust. In nature, brittle deformation occurs when stresses exceed the strength of a rock, causing it to break or fracture.

In the upper crust, to depths of about 10 kilometers, rocks tend to exhibit brittle behavior by fracturing, or by faulting. When tectonic forces cause upwarping of the crust, rocks near the surface are stretched and pulled apart to form fractures called *joints*. *Faults,* on the other hand, are fracture surfaces in which shear stresses cause the rocks on one side of the fault to move relative to the rocks on the other side. Faults involve slippage, joints do not.

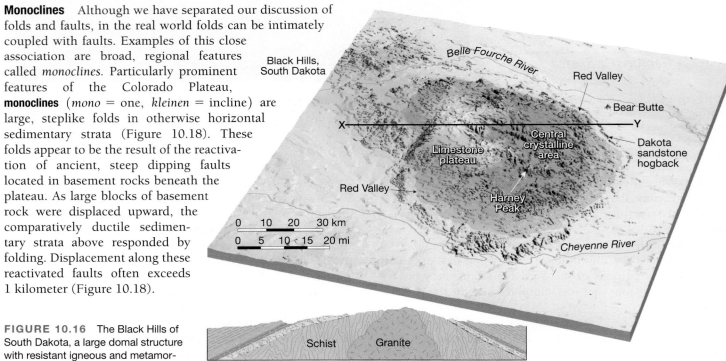

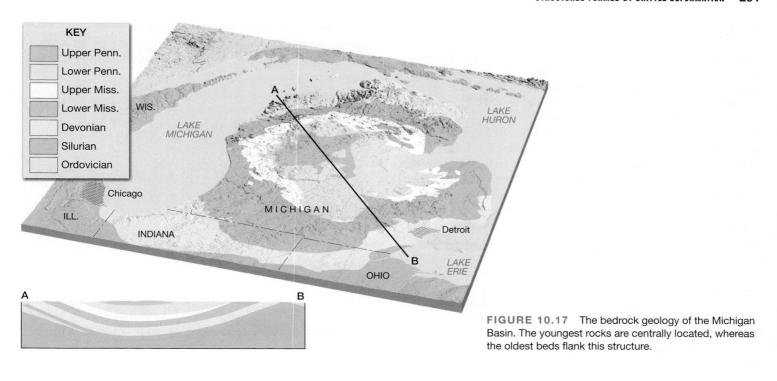

KEY
- Upper Penn.
- Lower Penn.
- Upper Miss.
- Lower Miss.
- Devonian
- Silurian
- Ordovician

FIGURE 10.17 The bedrock geology of the Michigan Basin. The youngest rocks are centrally located, whereas the oldest beds flank this structure.

FIGURE 10.18 The San Rafael Swell, Utah. This monocline consists of bent sedimentary beds that were deformed by faulting in the bedrock below. The thrust fault in this sketch is called a *blind thrust* because it does not reach the surface. (Photo by Stephen Trimble)

Faults

Faults are fractures in the crust along which appreciable displacement has taken place. Occasionally, small faults can be recognized in road cuts where sedimentary beds have been offset a few meters, as shown in Figure 10.19. Faults of this scale usually occur as single discrete breaks. By contrast, large faults, like the San Andreas Fault in California, have displacements of hundreds of kilometers and consist of many interconnecting fault surfaces. These structures, best described as *fault zones*, can be several kilometers wide and are often easier to identify from high-altitude photographs than at ground level.

Sudden movements along faults are the cause of most earthquakes. However, the vast majority of faults are inactive and thus are remnants of past deformation. The rock along fault zones often becomes broken and pulverized as crustal blocks on opposite sides of a fault grind past one another. The loosely coherent, clayish material that results from this activity is called *fault gouge*.

On some fault surfaces the rocks become highly polished and striated, or grooved, as the crustal blocks slide past one another. These polished and striated surfaces, called *slickensides* (*slicken* = smooth), provide geologists with evidence for the direction of the most recent displacement along the fault.

Dip-Slip Faults

Faults in which movement is primarily parallel to the *dip* (or inclination) of the fault surface are called **dip-slip faults**. It has become common practice to call the rock surface that is immediately above the fault the **hanging wall block** and to call the rock surface below, the **footwall block** (Figure 10.20). This nomenclature arose from prospectors and miners who excavated shafts and tunnels along fault zones that were sites of ore deposits. In these tunnels, the miners would walk on the rocks below the mineralized fault zone (the footwall block)

Students Sometimes Ask . . .

How do geologists determine which side of a fault has moved?

Surprisingly, for many faults this question cannot be answered definitively. For example, in the picture of a fault shown in Figure 10.19, did the left side move down, or did the right side move up? Because the surface (at the top of the photo) has been eroded flat, either side could have moved; or both could have moved, with one side moving more than the other (for instance, they both may have moved up, but the right side just moved up more than the left side). That's why geologists talk about *relative* motion across faults. In this case, the left side moved down *relative to* the right side, and the right side moved up *relative to* the left side (note arrows on photo).

and hang their lanterns on the rocks above (the hanging wall block).

Vertical displacements along dip-slip faults may produce long, low cliffs called **fault scarps** (*scarpe* = a slope). Fault scarps, such as the one shown in Figure 10.21, are produced by displacements that generate earthquakes.

Normal Faults   Dip-slip faults are classified as **normal faults** when the hanging wall block moves down relative to the footwall

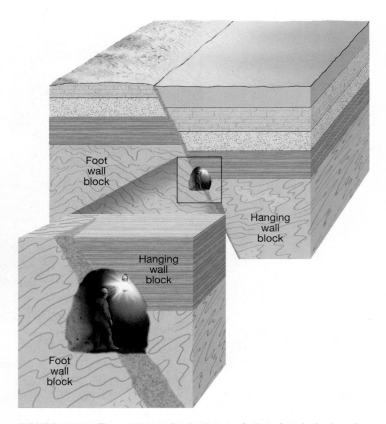

FIGURE 10.20   The rock immediately above a fault surface is the *hanging wall block*, and that below is called the *footwall block*. These names came from miners who excavated ore deposits that formed along fault zones. The miners hung their lanterns on the rocks above the fault trace (hanging wall block) and walked on the rocks below the fault trace (footwall block).

FIGURE 10.19   Faulting caused the vertical displacement of these strata. (Photo by Fletcher & Baylis/Photo Researchers, Inc.)

FIGURE 10.21 This fault scarp formed north of Landers, California, during an earthquake in 1992. It is the largest of several scarps that formed during this event. Note the geologist for scale. The scarp was created when land on one side of the fault moved down relative to the land on the opposite side. (Photo by Roger Ressmeyer/CORBIS)

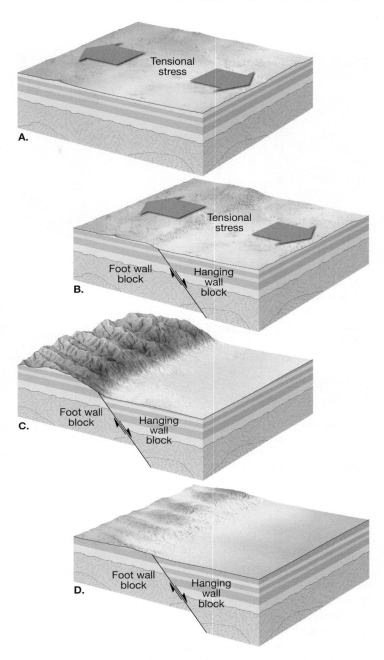

FIGURE 10.22 Block diagrams illustrating a normal fault. **A.** Rock strata prior to faulting. **B.** The relative movement of displaced blocks. Displacement may continue in a fault-block mountain range over millions of years and consist of many widely spaced episodes. **C.** How erosion might alter the upfaulted block. **D.** Eventually the period of deformation ends and erosion becomes the dominant geologic process.

block (Figure 10.22). Because of the downward motion of the hanging wall block, normal faults accommodate lengthening, or extension, of the crust.

Normal faults are found in a variety of sizes, some are small, having displacements of only a meter or so, like the one shown in the road cut in Figure 10.19. Others extend for tens of kilometers where they may sinuously trace the boundary of a mountain front. Most large, normal faults have relatively steep dips which tend to flatten out with depth.

In the western United States, large normal faults are associated with structures called **fault-block mountains**. Excellent examples of fault-block mountains are found in the Basin and Range Province, a region that encompasses Nevada and portions of the surrounding states (Figure 10.23). Here the crust has been elongated and broken to create more than 200 relatively small

mountain ranges. Averaging about 80 kilometers in length, the ranges rise 900 to 1500 meters above the adjacent down-faulted basins.

The topography of the Basin and Range Province evolved in association with a system of roughly north-south trending normal faults. Movements along these faults produced alternating uplifted fault blocks called **horsts** and down-dropped blocks called **grabens** (*graben* = ditch). Horsts generate elevated topography, whereas grabens form basins. As Figure 10.23 illustrates, structures called **half-grabens**, which are titled

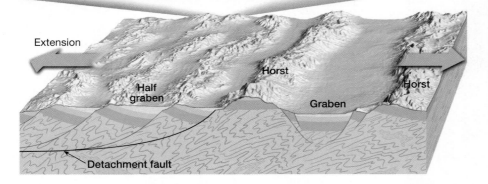

FIGURE 10.23   Normal faulting in the Basin and Range Province. Here, tensional stresses have elongated and fractured the crust into numerous blocks. Movement along these fractures has tilted the blocks, producing parallel mountain ranges called fault-block mountains. The down-faulted blocks (grabens) form basins, whereas the upfaulted blocks (horsts) erode to form rugged mountainous topography. In addition, numerous tilted blocks (half-grabens) form both basins and mountains. (Photo by Michael Collier)

fault blocks, also contribute to the alternating topographic highs and lows in the Basin and Range Province. The horsts and higher ends of the tilted fault blocks are the source of sediments that have accumulated in the basins created by the grabens and lower ends of the tilted blocks.

Notice in Figure 10.23 that the slopes of the normal faults decrease with depth and eventually join to form a nearly horizontal fault called a **detachment fault.** These faults form a major boundary between the rocks below, which exhibit ductile deformation, and the rocks above, which exhibit brittle deformation.

Fault motion provides geologists with a method of determining the nature of the tectonic forces at work within Earth. Normal faults are associated with tensional forces that pull the crust apart. This "pulling apart" can be accomplished either by uplifting that causes the surface to stretch and break or by opposing horizontal forces.

Reverse and Thrust Faults   **Reverse faults** are dip-slip faults in which the hanging wall block moves up relative to the footwall block (Figure 10.24). **Thrust faults** are reverse faults having dips less than 45 °, so the overlying block moves nearly horizontally over the underlying block.

Whereas normal faults occur in tensional environments, reverse faults result from strong compressional stresses. Because

the hanging wall block moves up and over the footwall block, reverse and thrust faults accommodate horizontal shortening of the crust.

Most high-angle reverse faults are small and accommodate local displacements in regions dominated by other types of faulting. Thrust faults, on the other hand, exist at all scales with some large thrust faults having displacements on the order of tens to hundreds of kilometers.

Montana's Glacier National Park is a classic site of thrust faulting (Figure 10.25). Mountain peaks that provide the park's majestic scenery have been carved from Precambrian rocks that were displaced over much younger Cretaceous strata. At the eastern edge of Glacier National Park, there is a peak called Chief Mountain. This well-known landmark is an isolated remnant of a thrust sheet that was severed by the erosional forces of glacial ice and running water. An isolated block, such as Chief Mountain, is called a **klippe** (*klippe* = cliff) (Figure 10.25). Examples of mountainous belts produced by this type of compressional tectonics include the Alps, Northern Rockies, Himalayas, and Appalachians.

Thrust faulting is most pronounced along convergent plate boundaries. Compressional forces associated with colliding plates generally create folds as well as thrust faults that thicken and shorten the crust to produce mountainous topography (Figure 10.26).

Strike-Slip Faults

A fault in which the dominant displacement is horizontal and parallel to the *trend* or *strike* of the fault surface is called a **strike-slip fault** (Figure 10.27). The earliest scientific records of strike-slip faulting were made following surface ruptures that produced large earthquakes. One of the most noteworthy of these was the great San Francisco earthquake of 1906. During this strong earthquake, structures such as fences that were built across the San Andreas Fault were displaced as much as 4.7 meters (15 feet). Because movement along the San Andreas causes the crustal block on the

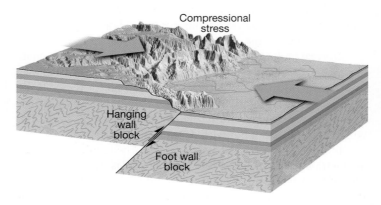

FIGURE 10.24   Block diagram showing the relative displacement along a reverse fault.

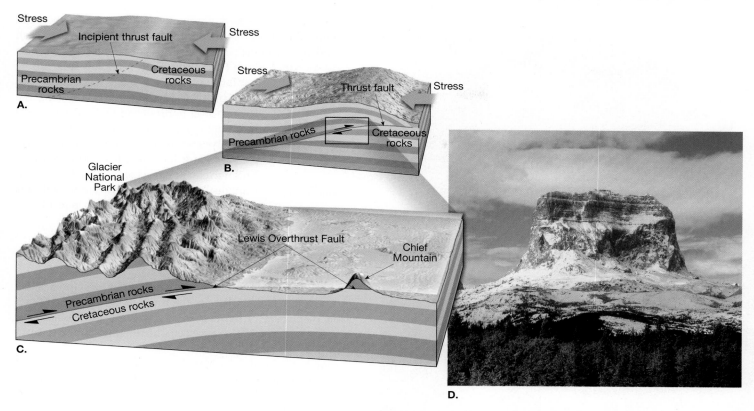

FIGURE 10.25 Development of Lewis Overthrust Fault in Montana. **A.** Geologic setting prior to deformation. **B.** Large-scale movement along a thrust fault displaced Precambrian rock over Cretaceous strata in the region of Glacier National Park, Montana. **C.** Erosion by glacial ice and running water sculptured the thrust sheet into a majestic landscape and isolated a remnant of the thrust sheet called Chief Mountain. **D.** Chief Mountain, Glacier National Park, Montana, is a klippe. (Photo by David Muench)

opposite side of the fault to move to the right as you face the fault, it is called a *right-lateral* strike-slip fault.

The Great Glen Fault in Scotland is a well-known example of a *left-lateral* strike-slip fault, which exhibits the opposite sense of displacement. The total displacement along the Great Glen Fault is estimated to exceed 100 kilometers (60 miles). Also associated with this fault trace are numerous lakes, including Loch Ness, home of the legendary monster.

Rather than a single fracture along which movement takes place, large strike-slip faults consist of a zone of roughly parallel fractures. The zone may be up to several kilometers wide. The most recent movement, however, is often along a strand only a few meters wide, which may offset features such as stream chan-

nels (Figure 10.28). Crushed and broken rocks produced during faulting are more easily eroded, so linear valleys or sag ponds often mark the locations of strike-slip faults. Because of their large size and linear nature, many strike-slip faults produce a trace that is visible over a great distance.

FIGURE 10.27 Aerial view of strike-slip (right-lateral) fault in southern Nevada. Notice the ridge of white rock in the top right portion of the photo was offset to the right relative to the portion of the same ridge that appears on the bottom left of the image. (Photo by Marli Miller)

FIGURE 10.26 Compressional tectonics produces folds and thrust faults that thicken and shorten the crust resulting in mountainous topography.

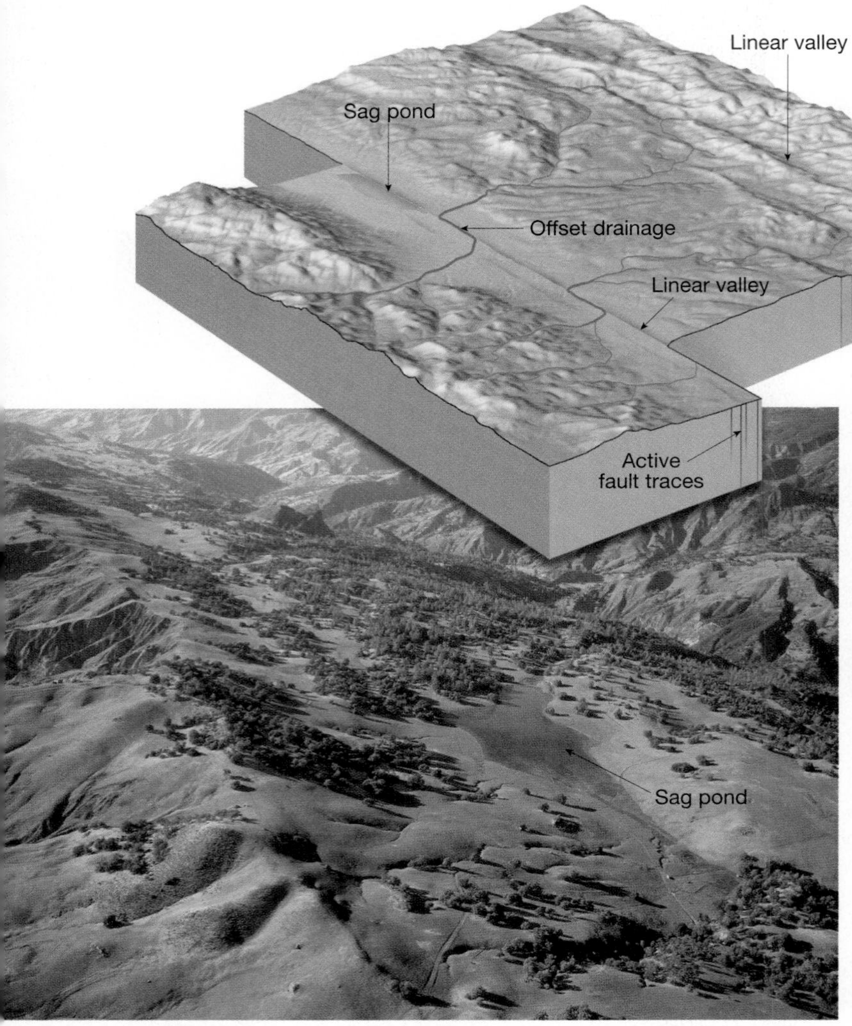

oceanic ridges. Others accommodate displacement between continental plates that move horizontally with respect to each other. One of the best-known transform faults is California's San Andreas Fault. This plate-bounding fault can be traced for about 950 kilometers (600 miles) from the Gulf of California to a point along the Pacific Coast north of San Francisco, where it heads out to sea. Since its formation about 30 million years ago, displacement along the San Andreas Fault has exceeded 560 kilometers. This movement has accommodated the northward displacement of southwestern California and the Baja Peninsula of Mexico in relation to the remainder of North America.

Joints

Among the most common rock structures are fractures called joints. Unlike faults, **joints** are fractures along which no appreciable displacement has occurred. Although some joints have a random orientation, most occur in roughly parallel groups (Figure 10.30).

We have already considered two types of joints. In Chapter 4 we learned that *columnar joints* form when igneous rocks cool and develop shrinkage fractures that produce elongated, pillar-

FIGURE 10.28 Strike-slip faulting. The block diagram illustrates the features associated with large strike-slip faults. Notice how the stream channels have been offset by fault movement. Photo of sag ponds along the trace of the San Andreas Fault. (Photo by Michael Collier)

Some strike-slip faults cut through the lithosphere and accommodate motion between two large tectonic plates. Recall this special kind of strike-slip fault is called a **transform fault** (*trans* = across, *forma* = form) (Figure 10.29). Numerous transform faults cut the oceanic lithosphere and link spreading

Students Sometimes Ask . . .

Has anyone ever seen a fault scarp forming?

Amazingly, yes. There have been several instances in which people have been, fortuitously, at the appropriate place and time to observe the creation of a fault scarp—and have lived to tell about it. In Idaho a large earthquake in 1983 created a 3-meter (10-foot) fault scarp that was witnessed by several people, many of whom were knocked off their feet. More often, though, fault scarps are noticed *after* they form. For example, a 1999 earthquake in Taiwan created a fault scarp that formed a new waterfall and destroyed a nearby bridge.

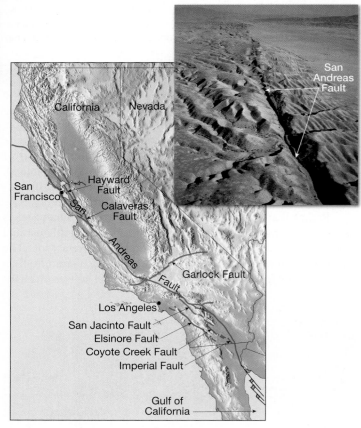

FIGURE 10.29 Map showing the extent of the San Andreas Fault system. Inset is an aerial view of the San Andreas Fault. (Photo by D. Parker/Photo Researchers)

FIGURE 10.30 Nearly parallel joints in the Navajo Sandstone, Arches National Park, Utah. (Photo by Michael Collier)

FIGURE 10.31 Devil's Tower National Monument is in eastern Wyoming. Established in September 1906, it was our nation's first national monument. This nearly vertical monolith rises more than 380 meters (1265 feet) above the surrounding grasslands and pine forests. This imposing structure exhibits columnar joints. Columnar joints form as igneous rocks cool and develop shrinkage fractures that produce five- to seven-sided elongated, pillarlike structures. (Photo by Geoff Renner/Robert Harding Picture Library/agefotostock)

like columns (Figure 10.31). Also recall from Chapter 6 that sheeting produces a pattern of gently curved joints that develop more or less parallel to the surface of large exposed igneous bodies such as batholiths. Here the jointing results from the gradual expansion that occurs when erosion removes the overlying load (see Figure 6.5, p. 177).

Most joints are produced when rocks in the outermost crust are deformed as tensional stresses cause the rock to fail by brittle fracture. Extensive joint patterns often develop in response to relatively subtle and often barely perceptible regional upwarping and downwarping of the crust. In many cases, the cause for jointing at a particular locale is not readily apparent.

Many rocks are broken by two or even three sets of intersecting joints that slice the rock into numerous regularly shaped blocks (see Figure 10.30). These joint sets often exert a strong influence on other geologic processes. For example, chemical weathering tends to be concentrated along joints, and in many areas groundwater movement and the resulting dissolution in soluble rocks is controlled by the joint pattern. Moreover, a system of joints can influence the direction stream courses follow. The rectangular drainage pattern described in Chapter 16 is such a case.

Students Sometimes Ask . . .

Do faults exhibit only strike-slip or dip-slip motion?

No. Strike-slip faults and dip-slip faults are on the opposite ends of a spectrum of fault structures. Faults that exhibit a combination of dip-slip and strike-slip movements are called *oblique-slip faults.* Although most faults could technically be classified as oblique-slip, they predominantly exhibit either strike-slip or dip-slip motion.

Joints may also be significant from an economic standpoint. Some of the world's largest and most important mineral deposits are located along joint systems. Hydrothermal solutions, which are basically mineralized fluids, can migrate into fractured host rocks and precipitate economically important amounts of copper, silver, gold, zinc, lead, and uranium.

Highly jointed rocks present a risk to the construction of engineering projects, including highways and dams. On June 5, 1976, 14 lives were lost and nearly $1 billion in property damage occurred when the Teton Dam in Idaho failed. This earthen dam was constructed of very erodible clays and silts and was situated on highly fractured volcanic rocks. Although attempts were made to fill the voids in the jointed rock, water gradually penetrated the subsurface fractures and undermined the dam's foundation. Eventually the moving water cut a tunnel into the easily erodible clays and silts. Within minutes the dam failed, sending a 20-meter-high wall of water down the Teton and Snake rivers.

MAPPING GEOLOGIC STRUCTURES

When conducting a study of a region, geologists identify and describe the dominant tectonic structures. A structure often is so large that only a small portion is visible from any particular

UNDERSTANDING EARTH

BOX 10.1

Naming Local Rock Units

One of the primary goals of geology is to reconstruct Earth's long and complex history through the systematic study of rocks. In most areas, exposures of rocks are not continuous over great distances. Consequently, the study of rock layers must be done locally and then correlated with data from adjoining areas to produce a larger and more complete picture. The first step in assessing past geologic events is describing and mapping rock units exposed in local outcrops.

Describing anything as complex as a thick sequence of rocks requires subdividing the layers into units of manageable size. The most basic rock division is called a *formation,* which is simply a distinctive series of strata that originated through the same geologic processes. More precisely, a formation is a mappable rock unit that has definite boundaries (or contacts with other units) and certain obvious characteristics (rock type) by which it may be traced from place to place and distinguished from other rock units.

Figure 10.A shows several named formations that are exposed in the walls of the Grand Canyon. Just as these rock strata in the Grand Canyon were subdivided, geologists subdivide rock sequences throughout the world into formations.

Those who have had the opportunity to travel to some of the national parks in the West may already be familiar with the names of certain formations. Well-known formations include the Navajo Sandstone in Zion National Park, the Redwall Limestone in the Grand Canyon, the Entrada Sandstone in Arches National Park, and the Wasatch Formation in Bryce Canyon National Park.

Although formations can consist of igneous or metamorphic rocks, the vast majority are sedimentary rocks. A formation may be relatively thin and composed of a single rock type, for example, a 1-meter-thick layer of limestone. At the other extreme, formations can be thousands of meters thick and consist of an interbedded sequence of rock types such as sandstones and shales. The most important condition to be met when establishing a formation is that *it constitutes a unit of rock produced by uniform or uniformly alternating conditions.*

In most regions of the world, the name of each formation consists of two parts—for example, the Oswego Sandstone and the Carmel Formation. The first part of the name is generally taken from a geologic structure or a locality where the formation is clearly and completely exposed. For instance, the expansive Morrison Formation is well exposed at Morrison, Colorado. As a result, this particular exposure is known as the *type locality.* Ideally, the second part of the name indicates the dominant rock type as exemplified by such names as the Dakota Sandstone, the Kaibab Limestone, and the Burgess Shale. When no single rock type dominates, the term *formation* is used. One such formation is Chinle Formation, which is exposed in Arizona's Petrified Forest National Park.

In summary, describing and naming formations is an important first step in the process of organizing and simplifying the study and analysis of Earth history.

FIGURE 10.A   Grand Canyon with a few of its rock formations named. (Photo by E. J. Tarbuck)

vantage point. In many situations, most of the bedrock is concealed by vegetation or buried by recent sedimentation. Consequently, the reconstruction must be done using data gathered from a limited number of *outcrops,* which are sites where bedrock is exposed at the surface (Box 10.1). Despite these challenges, a number of mapping techniques enable geologists to reconstruct the orientation and shape of existing structures. In recent years this work has been aided by advances in aerial photography, satellite imagery, and the development of the global positioning system (GPS). In addition, seismic reflection profiling (see Chapter 12) and drill holes provide data on the composition and structure of rocks that lie at depth.

Because sediments are usually deposited in horizontal layers, geologic mapping is most easily accomplished where sedimentary strata are exposed. If the sedimentary rock layers are still horizontal, geologists assume the area is undisturbed structurally. Inclined, bent, or broken strata indicate a period of deformation occurred following deposition.

Strike and Dip

Geologists use measurements called *strike* (trend) and *dip* (inclination) to help determine the orientation or attitude of a rock layer or fault surface (Figure 10.32). By knowing the strike

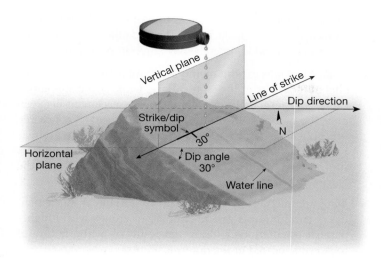

FIGURE 10.32 Strike and dip of a rock layer.

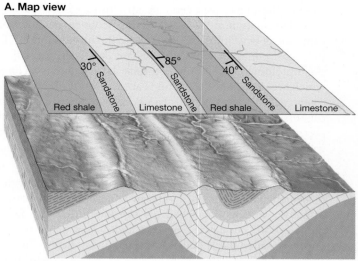

B. Block diagram

FIGURE 10.33 By establishing the strike and dip of outcropping sedimentary beds on a map **A.** geologists can infer the orientation of the structure below ground **B.**

and dip of rocks at the surface, geologists can predict the nature and structure of rock units and faults that are hidden beneath the surface.

Strike is the compass direction of the line produced by the intersection of an inclined rock layer or fault, with a horizontal plane (Figure 10.32). The strike, or compass bearing, is generally expressed as an angle relative to north. For example, N10 °E means the line of strike is 10 ° to the east of north. The strike of the rock units illustrated in Figure 10.32 is approximately north 75 ° east (N75 °E).

Dip is the angle of inclination of the surface of a rock unit or fault measured from a horizontal plane. Dip includes both an angle of inclination and a direction toward which the rock is

inclined. In Figure 10.32 the dip angle of the rock layer is 30 °. A good way to visualize dip is to imagine that water will always run down the rock surface parallel to the dip. The direction of dip will always be at a 90 ° angle to the strike.

In the field, geologists measure the strike (trend) and dip (inclination) of sedimentary strata at as many outcrops as practical. These data are then plotted on a topographic map or an aerial photograph along with a color-coded description of the rock. From the orientation of the strata, an inferred orientation and shape of the structure can be established, as shown in Figure 10.33. Using this information, the geologist can reconstruct the pre-erosional structures and begin to interpret the region's geologic history.

CHAPTER 10 CRUSTAL DEFORMATION IN REVIEW

- *Deformation* refers to changes in the shape and/or volume of a rock body and is most pronounced along plate margins. To describe the forces that deform rocks, geologists use the term *stress*, which is the amount of force applied to a given area. Stress that is uniform in all directions is called *confining pressure*, whereas *differential stresses* are applied unequally in different directions. Differential stresses that shorten a rock body are *compressional stresses;* those that elongate a rock unit are *tensional stresses*. *Strain* is the change in size and shape of a rock unit caused by stress.

- Rocks deform differently depending on the environment (temperature and confining pressure), the composition and

texture of the rock, and the length of time stress is maintained. Rocks first respond by deforming *elastically* and will return to their original shape when the stress is removed. Once their elastic limit (strength) is surpassed, rocks either deform by ductile flow or they fracture. *Ductile deformation* is a solid-state flow that changes the shape of an object without fracturing. Ductile flow may be accomplished by gradual slippage and recrystallization along planes of weakness within the crystal lattice of mineral grains. Ductile deformation tends to occur in a high-temperature/high-pressure environment. In a near-surface environment, most rocks deform by *brittle failure*.

- Among the most basic geologic structures associated with rock deformation are *folds* (flat-lying sedimentary and volcanic rocks bent into a series of wavelike undulations). The two most common types of folds are *anticlines,* formed by the upfolding, or arching, of rock layers, and *synclines,* which are downfolds. Most folds are the result of horizontal *compressional stresses.* Folds can be *symmetrical, asymmetrical,* or, if one limb has been tilted beyond the vertical, *overturned.* *Domes* (upwarped structures) and *basins* (downwarped structures) are circular or somewhat elongated folds formed by vertical displacements of strata.

- *Faults* are fractures in the crust along which appreciable displacement has occurred. Faults in which the movement is primarily vertical are called *dip-slip faults.* Dip-slip faults include both *normal* and *reverse faults.* Low-angle reverse faults are called *thrust faults.* Normal faults indicate *tensional stresses* that pull the crust apart. Along divergent plate boundaries, a central block bounded by normal faults called a *graben,* may form as the plates separate.

- Reverse and thrust faulting indicate that *compressional forces* are at work. Large *thrust faults* are associated with subduction zones and other convergent boundaries where plates collide.

In mountainous regions such as the Alps, Northern Rockies, Himalayas, and Appalachians, thrust faults have displaced strata as far as 50 kilometers over adjacent rock units.

- *Strike-slip faults* exhibit mainly horizontal displacement parallel to the strike of the fault surface. Large strike-slip faults, called *transform faults,* accommodate displacement between plate boundaries. Most transform faults cut the oceanic lithosphere and link spreading centers. The San Andreas Fault cuts the continental lithosphere and accommodates the northward displacement of the Pacific plate past the North American plate.

- *Joints* are fractures along which no appreciable displacement has occurred. Joints generally occur in groups with roughly parallel orientations and are the result of brittle failure of rock units located in the outermost crust.

- The orientation of rock units or fault surfaces is established with measurements called strike and dip. *Strike* is the compass direction of a line produced by the intersection of an inclined rock layer or fault with a horizontal plane. *Dip* is the angle of inclination of the surface of a rock unit or fault measured from a horizontal plane.

KEY TERMS

anticline (p. 286)
basin (p. 289)
brittle deformation (p. 283)
compressional stress (p. 281)
confining pressure (p. 281)
deformation (p. 280)
detachment fault (p. 294)
differential stress (p. 281)
dip (p. 299)
dip-slip fault (p. 292)

dome (p. 289)
ductile deformation (p. 283)
elastic deformation (p. 283)
fault (p. 292)
fault-block mountain (p. 293)
fault scarp (p. 292)
fold (p. 285)
footwall block (p. 292)
graben (p. 293)

half-graben (p. 293)
hanging wall block (p. 292)
hogback (p. 290)
horst (p. 293)
joint (p. 296)
klippe (p. 294)
monocline (p. 290)
normal fault (p. 292)
reverse fault (p. 294)
rock structure (p. 280)

shear (p. 281)
strain (p. 282)
stress (p. 281)
strike (p. 299)
strike-slip fault (p. 294)
syncline (p. 286)
tectonic structures (p. 280)
tensional stress (p. 281)
thrust fault (p. 294)
transform fault (p. 296)

QUESTIONS FOR REVIEW

1. List three (3) rock structures that are associated with deformation.

2. What is rock deformation? How might a rock body change during deformation?

3. Contrast compressional and tensional stresses.

4. Compare stress and strain.

5. Describe elastic deformation.

6. How is brittle deformation different from ductile deformation?

7. List and describe the four factors that affect rock strength.

8. Distinguish between anticlines and synclines, domes and basins, anticlines and domes.

9. The Black Hills of South Dakota is a good example of what type of structural feature?

10. Describe the formation of a monocline.

11. What is the hanging wall block? Footwall block?

12. Contrast the movements that occur along normal and reverse faults. What type of stress is indicated by each fault?

13. Is the fault shown in Figure 10.19 a normal or a reverse fault?

14. Describe a horst and a graben.

15. What type of faults are associated with fault-block mountains?

16. How are reverse faults different from thrust faults? In what way are they the same?

17. Describe the relative movement along a strike-slip fault.

18. The San Andreas Fault is an excellent example of a special strike-slip fault called a _____.

19. With which of the three types of plate boundaries does normal faulting predominate? Thrust faulting? Strike-slip faulting?

20. How are joints different from faults?

21. What is an outcrop?

22. What two measurements are used to establish the orientation of deformed strata? Distinguish between them.

COMPANION WEBSITE

The *Earth 10e* website uses the resources and flexibility of the Internet to aid in your study of the topics in this chapter. Written and developed by the authors and other geology instructors, this site will help improve your understanding of geology. Visit **www.mygeoscienceplace.com** in order to:

- **Review** key chapter concepts

- **Read** with links to the eBook and to chapter-specific web resources
- **Visualize** and comprehend challenging topics using learning activities in *GEODe Earth*
- **Test** yourself with online quizzes

GEODe EARTH

GEODe Earth is a valuable and easy to use learning aid that can be accessed from your book's Companion Website (**www .mygeoscienceplace.com**). It is a dynamic instructional tool that promotes understanding and reinforces important concepts by using tutorials, animations, and exercises that actively engage the student.

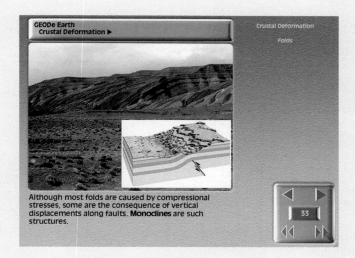

EARTHQUAKES AND EARTHQUAKE HAZARDS

11

Aftermath of a devastating tsunami at the coastal town of Banda Aceh on the Indonesian island of Sumatra, December 26, 2004.

(PHOTO BY MARK PEARSON/ALAMY)

On October 17, 1989, at 5:04 PM Pacific daylight time, millions of television viewers around the world were settling in to watch the third game of the World Series. Instead, they saw their television sets go black as tremors hit San Francisco's Candlestick Park, where power and communication lines were severed. Although the earthquake was centered in a remote section of the Santa Cruz Mountains, 100 kilometers to the south, major damage occurred in the Marina District of San Francisco.

The most tragic result of the violent shaking was the collapse of some double-decked sections of Interstate 880. The ground motions caused the upper deck to sway, shattering the concrete support columns along a one-mile section of the freeway. As a result, the upper deck collapsed onto the lower roadway, flattening cars as if they were aluminum beverage cans. This earthquake, named the Loma Prieta quake for its point of origin, claimed 67 lives.

In mid-January 1994, less than five years after the Loma Prieta event, a major earthquake struck the Northridge area of Los Angeles. Although not the fabled "Big One," this moderate, 6.7-magnitude earthquake left 57 dead, more than 5000 injured, and tens of thousands of households without water and electricity. The quake, attributed to a previously unknown fault that ruptured 18 kilometers (11 miles) beneath the city of Northridge, caused damage that exceeded $40 billion.

The Northridge earthquake began at 4:31 AM (PST) and lasted roughly 40 seconds. During this brief period the quake terrorized the entire Los Angeles area. In the three-story Northridge Meadows apartment complex, 16 people died when sections of the upper floors suddenly dropped onto the first-floor units. An investigation revealed that improperly attached floor joists contributed to the collapse. Nearly 300 schools were seriously damaged, and a dozen major roadways buckled. Among these were two of California's major arteries—the Golden State Freeway (Interstate 5), where an overpass collapsed completely and blocked the roadway, and sections of the Santa Monica Freeway. Fortunately, these roadways had practically no traffic at this early morning hour.

In nearby Granada Hills, broken gas lines caused multiple fires while streets flooded from broken water mains. Seventy homes burned in the Sylmar district of Los Angeles. A 64-car freight train derailed, including some cars carrying hazardous cargo. It is remarkable that the destruction was not greater. Unquestionably, the upgrading of structures to meet the requirements of building codes developed for this earthquake-prone area helped minimize what could have been a more significant human tragedy.

WHAT IS AN EARTHQUAKE?

EARTHQUAKES
▶ What Is an Earthquake?

Earthquakes are natural geologic phenomena caused by the sudden and rapid movement of a large volume of rock (Figure 11.1). The violent shaking and destruction caused by earthquakes are the result of rupture and slippage along fractures in Earth's crust called **faults.** Larger quakes result from the rupture of larger fault segments. The origin of an earthquake occurs at depths between 5 and 700 kilometers, at the **focus** (foci = a point) or **hypocenter.** The point at the surface directly above the focus is called the **epicenter** (Figure 11.2).

During large earthquakes, a massive amount of energy is released as heat and **seismic waves**—a form of elastic energy that causes vibrations in the material that transmits them. Seismic waves are analogous to waves produced when a stone is dropped into a calm pond. Just as the impact of the stone creates a pattern of waves in motion, an earthquake generates waves that radiate outward in all directions from the focus. Even though seismic energy dissipates rapidly with increasing distance, sensitive instruments located around the world detect and record these events.

More than 30,000 earthquakes that are strong enough to be felt by humans occur worldwide annually. Fortunately, most are minor tremors that have little impact on our lives. Nevertheless, over geologic time earthquakes help shape our landscape.

FIGURE 11.1 Destruction caused by a major earthquake that struck northwestern Turkey on August 17, 1999. More than 17,000 people perished. (Photo by Yann Arthus-Bertrand/Peter Arnold, Inc.)

Only about 75 strong earthquakes are recorded each year, and many of these occur in remote regions. Occasionally, a large earthquake is triggered near a major population center. Such events are among the most destructive natural forces on Earth. The shaking of the ground, coupled with the liquefaction of soils, wreaks havoc on buildings, roadways, and other structures. In addition, when a quake occurs in a populated area, power and gas lines are often ruptured, causing numerous fires.

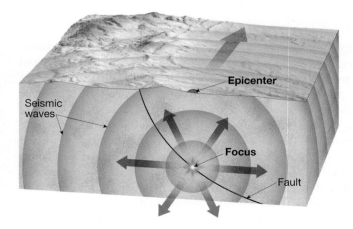

FIGURE 11.2 Earthquake focus and epicenter. The *focus* is the zone within Earth where the initial displacement occurs. The *epicenter* is the surface location directly above the focus.

In the famous 1906 San Francisco earthquake, much of the damage was caused by fires that became uncontrollable when broken water mains left firefighters with only trickles of water (Figure 11.3).

Discovering the Causes of Earthquakes

The energy released by atomic explosions or by the movement of magma in Earth's crust can generate earthquake-like waves, but these events are generally quite weak. What mechanism produces a destructive earthquake? As you have learned, Earth is not a static planet. We know that large sections of Earth's crust have been thrust upward, because fossils of marine organisms have been discovered thousands of meters above sea level. Other regions exhibit evidence of extensive subsidence. In addition to these vertical displacements, offsets in fence lines, roads, and other structures indicate that horizontal movements are also common (Figure 11.4).

The actual mechanism of earthquake generation eluded geologists until H. F. Reid of Johns Hopkins University conducted a study following the great 1906 San Francisco earthquake. The earthquake was accompanied by horizontal surface displacements of several meters along the northern portion of the San Andreas Fault. Field studies determined that during this single earthquake, the Pacific plate lurched as much as 4.7 meters (15 feet) northward past the adjacent North American plate.

305

FIGURE 11.3   San Francisco in flames after the 1906 earthquake. (Reproduced from the collection of the Library of Congress) Inset photo shows fire triggered when a gas line ruptured during the Northridge earthquake in Southern California in 1994. (AFP/Getty Images)

FIGURE 11.4   Slippage along a fault produced an offset in this orange grove east of Calexico, California. (Photo by John S. Shelton) Inset photo shows a fence offset 2.5 meters (8.5 feet) during the 1906 San Francisco earthquake. (Photo by G. K. Gilbert, U.S. Geological Survey)

What Reid concluded from his investigations is illustrated in Figure 11.5. Tectonic stresses acting over tens to hundreds of years slowly deform the crustal rocks on both sides of a fault. When deformed by differential stress, rocks bend and store elastic energy, much like a wooden stick does if bent (Figure 11.5B). Eventually, the frictional resistance holding the rocks in place is overcome. Slippage allows the deformed (strained) rock to "snap back" to its original, stress-free, shape (Figure 11.5C, D). The "springing back" was termed **elastic rebound** by Reid because the rock behaves elastically, much like a stretched rubber band does when it is released. The vibrations we know as an earthquake are generated by the rock elastically returning to its original shape.

In summary, *earthquakes are produced by the rapid release of elastic energy stored in rock that has been deformed by differential stresses. Once the strength of the rock is exceeded, it suddenly ruptures, causing the vibrations of an earthquake.*

Aftershocks and Foreshocks

Strong earthquakes are followed by numerous smaller tremors, called **aftershocks,** that gradually diminish in frequency and intensity over a period of several months. Within 24 hours of the massive 1964 Alaskan earthquake, 28 aftershocks were recorded, 10 of which had magnitudes that exceeded 6. More than 10,000 aftershocks with magnitudes of 3.5 or above occurred in the following 69 days and thousands of minor tremors were recorded over a span of 18 months. Because aftershocks happen mainly on the section of the fault that has slipped, they provide geologists with data that is useful in establishing the dimensions of the rupture surface.

Although aftershocks are weaker than the main earthquake, they can trigger the destruction of already weakened structures. This occurred in northwestern Armenia (1988) where many people lived in large apartment buildings constructed of brick and

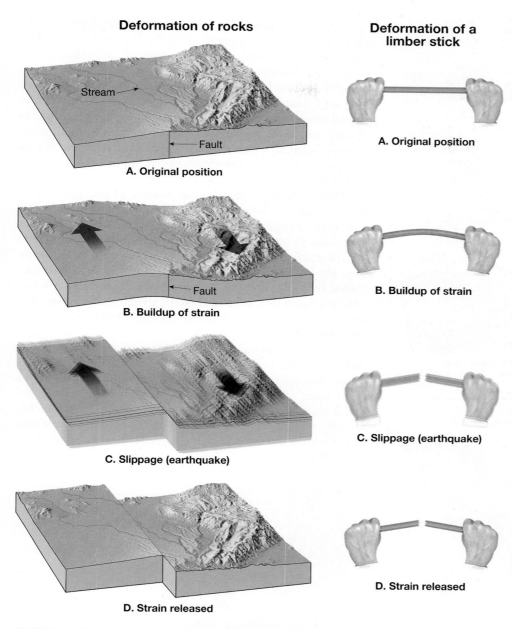

Deformation of rocks

Stream →

← Fault

A. Original position

← Fault

B. Buildup of strain

C. Slippage (earthquake)

D. Strain released

Deformation of a limber stick

A. Original position

B. Buildup of strain

C. Slippage (earthquake)

D. Strain released

FIGURE 11.5 Elastic rebound. As rock is deformed, it bends, storing elastic energy. Once strained beyond its breaking point, the rock cracks, releasing the stored-up energy in the form of earthquake waves.

example is the San Andreas Fault, which is the transform fault boundary that separates two great sections of Earth's lithosphere: the North American plate and the Pacific plate. Other faults are small and capable of producing only minor earthquakes.

Most of the displacement that occurs along faults can be satisfactorily explained by the plate tectonics theory which states that large slabs of Earth's lithosphere are in continual slow motion. These mobile plates interact with neighboring plates, straining and deforming the rocks at their margins. Faults associated with plate boundaries are the source of most large earthquakes.

The Nature of Faults

Recall from Chapter 10 that three basic types of faults are recognized: *normal faults, reverse (including thrust) faults,* and *strike-slip faults* (Figure 11.6). Normal faults occur where the crust is stretched and elongated, therefore most earthquakes along normal faults are associated with divergent plate boundaries, mainly seafloor spreading centers and continental rifts. Reverse faults, including low angle thrust faults are most often found along convergent plate boundaries. These faults are generated by compressional forces associated with subduction zones and continental collisions. Strike-slip faults are the result of shear stresses that cause large segments of Earth's crust to slip horizontally past each other. Recall that large strike-slip faults are called *transform faults* when they form a plate boundary.

Most large faults are not perfectly straight or continuous; instead, they consist of numerous branches and smaller fractures that display kinks and offsets. Such a pattern is displayed in Figure 11.7, which shows the San Andreas Fault as a system of several faults of various sizes.

concrete slabs. After a moderate earthquake of magnitude 6.9 weakened the buildings, a strong aftershock of magnitude 5.8 completed the demolition.

In contrast to aftershocks, small earthquakes called **fore-shocks** often precede a major earthquake by days, or in some cases, by several years. Monitoring of foreshocks to predict forthcoming earthquakes has been attempted with limited success.

FAULTS, FAULTING, AND EARTHQUAKES

Earthquakes take place along faults both new and old that occur in places where differential stresses have ruptured Earth's crust. Some faults are large and capable of generating major earthquakes. One

The San Andreas is undoubtedly the most studied fault system in the world. Over the years, research has shown that displacement occurs along discrete segments that behave somewhat differently from one another. A few sections of the San Andreas exhibit a slow, gradual displacement known as **fault creep,** which occurs without the build up of significant strain. These sections produce only minor seismic shaking. Other segments slip at regular intervals, producing small-to-moderate earthquakes. Still other segments remain locked and store energy for a few hundred years before rupturing in great earthquakes. Earthquakes that occur along locked segments of the San Andreas Fault tend to be repetitive. As soon as one is over, the continuous motion of the plates begins building strain anew. Decades or centuries later, the fault fails again.

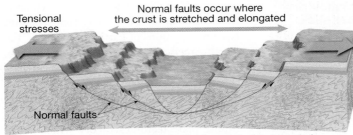

Tensional stresses

Normal faults occur where the crust is stretched and elongated

Normal faults

Normal faults are associated with divergent plate boundaries, mainly seafloor spreading centers and continental rifting

A. Normal faults

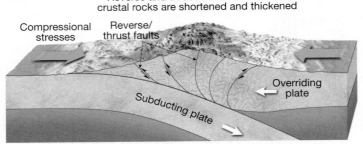

Reverse and thrust faults occur where crustal rocks are shortened and thickened

Compressional stresses

Reverse/ thrust faults

Overriding plate

Subducting plate

Reverse and thrust faults are associated with subduction zones and continental collisions

B. Reverse and thrust faults

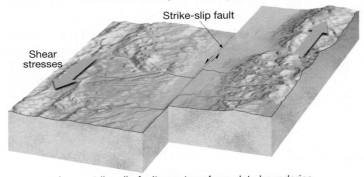

Strike-slip faults occur where large segments of Earth's crust slip horizontally past each other

Strike-slip fault

Shear stresses

Large strike-slip faults are transform plate boundaries

C. Strike-slip faults

FIGURE 11.6   Major fault types. **A.** Normal faults are associated with tensional forces that stretch and elongate the crust. **B.** Reverse (thrust) faults are produced by compressional forces that shorten and thicken crustal rocks. **C.** Strike-slip faults are the result of shear stresses that cause large segments of Earths crust to slide horizontally past each other.

Fault Rupture and Propagation

We know that the forces that cause sudden slippage along faults are ultimately the result of the interactions of Earth's tectonic plates. It is also clear that most faults are locked, except for brief, abrupt movements that accompany an earthquake rupture. Faults are locked because the confining pressure exerted by the overlying crust is enormous, causing these fractures to be "squeezed shut." Whenever the elastic energy stored in the rock exceeds the frictional resistance, slippage occurs along a segment of the fault, producing an earthquake.

During an earthquake, the initial rupture begins at the focus and propagates (travels) along the fault surface, at 2 to 3 kilometers per second. The propagation of the rupture zone along a 300-kilometer-long fault, for example, takes about 1.5 minutes as compared to about 30 seconds for a 100-kilometer-long fault. Although a rupture sometimes advances in both horizontal directions along the fault plane, more often it propagates in only one direction. Slippage in one area adds strain to the adjacent segment which may also rupture. As this zone of rupture advances, it can slow down, speed up, or even jump to a nearby fault segment.

During small earthquakes the amount of displacement on the fault surface, called **fault slip,** is typically less than one meter and occurs along a relatively small fault, or a small segment of a large fault. By contrast, following the largest earthquakes, fault slips of nearly 20 meters (60 feet) have been observed. In addition, the fault surface that ruptures during a large quake can exceed 800 kilometers in length. Therefore, the strong vibrations produced by a large earthquake are not only stronger, but also last longer than the vibrations produced by slippage along a small fault segment.

Why do earthquake ruptures stop rather than continuing along the entire fault? Evidence suggests that slippage usually stops when the rupture reaches a section of the fault where the rocks have not been sufficiently strained to overcome frictional resistance, such as in a section of the fault that has recently experienced an earthquake. The rupture may also stop if it encounters a sufficiently large kink, or an offset along the fault plane.

CALIFORNIA

Hayward Fault

San Francisco

Calaveras Fault

San Andreas Fault

Eastern California Shear Zone

Garlock Fault

Los Angeles

San Jacinto Fault

Imperial Fault

FIGURE 11.7   Map showing the extent of the San Andreas Fault system and the Eastern California Shear Zone.

Earthquake Forecaster

Dr. Andrea Donnellan, Deputy Manager of NASA's Jet Propulsion Laboratory Earth and Space Sciences Division, leads a team of scientists working on earthquake forecasting. She was co-winner of the Women in Aerospace 2003 Outstanding Achievement Award and twice a finalist in the NASA astronaut selection Process. (NASA Headquarters)

> **"I look at the quiet part of the earthquake cycle,"** Donnellan said. **"Each earthquake changes where the next earthquake will be."**

Andrea Donnellan talks about earthquakes the same way meteorologists talk about the weather: as dynamic, interconnected systems with their own peculiar set of rules. So it is only appropriate that she refers to her work as earthquake "forecasting."

Specifically, Donnellan and her colleagues have installed hundreds of high-precision global positioning system (GPS) receivers across southern California, a network known as the Southern California Integrated GPS Network (SCIGN), in order to study the movements of tectonic plates. The receivers, which Donnellan says are more highly sophisticated than commercial GPS devices such as those used in cell phones, can measure millimeter-scale slips for faults and give scientists valuable data for understanding how and why earthquakes take place. Donnellan's work goes far beyond studying just the earthquakes themselves. In fact, most of the work occurs in between all the moving and shaking.

"I look at the quiet part of the earthquake cycle," Donnellan said. "Each earthquake changes where the next earthquake will be."

Like meteorologists modeling a weather system, Donnellan and her colleagues have reproduced earthquake systems using supercomputers in order to understand how they change over time.

> **"My focus is on modeling earthquake systems," she said. "We want to treat it like weather, where the system is always changing."**

"My focus is on modeling earthquake systems," she said. "We want to treat it like weather, where the system is always changing."

But Donnellan's research consists of far more than just sitting in front of a computer all day. In fact, her studies have taken her to places around the globe, from Antarctica to Mongolia to Bolivia, to name a few. Her current research has taken her across much of southern California, where she and her team are studying the San Andreas Fault.

Back in the lab, Donnellan is also involved in an ambitious project called QuakeSim, in which scientists are developing sophisticated, state-of-the-art computer models of earthquake systems.

Donnellan said the results of the project will eventually be accessible by anyone, and that schools are already using some of the software for educational purposes.

In terms of the practical implications for Donnellan's work, her research is already giving scientists a much clearer idea of where earthquakes may strike next, allowing localities to prepare much further in advance. Although, like the weather, earthquakes will probably never be totally predictable, a close study of the ever-shifting activity beneath the ground by Donnellan and her colleagues is making the picture a lot clearer.

--- Chris Wilson

SEISMOLOGY: THE STUDY OF EARTHQUAKE WAVES

EARTHQUAKES
▶ Seismology

The study of earthquake waves, **seismology,** dates back to attempts made by the Chinese almost 2000 years ago to determine the direction from which these waves originated. One of the instruments developed by the Chinese was a large hollow jar that contained a weight suspended from the top (Figure 11.8).

The suspended weight (similar to a clock pendulum) was connected to the jaws of several large dragon figurines that encircled the container. The jaws of each dragon held a metal ball. When earthquake waves reached the instrument, the relative motion between the suspended mass and the jar would dislodge some of the metal balls into the waiting mouths of frogs directly below.

The Chinese were aware that the first strong ground motion from an earthquake is directional, and when it is strong enough, all poorly supported items will topple over in the same direction. The Chinese used this fact, in addition to the position of the dislodged balls, to detect the direction of an earthquake's source.

FIGURE 11.8 Ancient Chinese seismograph. During an Earth tremor, the dragons located in the direction of the main vibrations would drop a ball into the mouths of the frogs below. (Photo by James E. Patterson)

In principle, modern **seismographs** are similar to the instruments used by the Chinese. Seismographs have a weight freely suspended from a support that is securely attached to bedrock (Figure 11.9). When vibrations from an earthquake reach the instrument, the **inertia** of the weight keeps it relatively stationary while Earth and the support move.*

Earthquakes cause both vertical and horizontal ground motion; therefore, more than one type of seismograph is employed. The instrument shown in Figure 11.9 detects horizontal ground motion. Vertical ground motion is detected if the weight is suspended from a spring, as shown in Figure 11.10.

To detect very weak earthquakes, or a great earthquake that has occurred in another part of the world, most seismographs are designed to amplify ground motion. Other instruments are designed to withstand the violent shaking that occurs very near the focus.

The records obtained from seismographs, called **seismograms,** provide useful information about the nature of seismic waves. Seismograms reveal that two main groups of seismic waves are generated by the slippage of a rock mass. One of these wave types, called **surface waves,** travels along the outer part of Earth. Others travel through Earth's interior and are called **body waves.** Body waves are divided into two types—called **primary,** or **P, waves** and **secondary** or **S, waves.**

Body waves are identified by their mode of travel through intervening materials. P waves are "push-pull" waves—they momentarily push (squeeze) and pull (stretch) rocks in the direction the wave is traveling (Figure 11.11A). This wave motion is similar to that generated by human vocal cords as they move air to create sound. Solids, liquids, and gases resist a change in volume when compressed and will elastically spring back once the force is removed. Therefore, P waves, can travel through all these materials.

On the other hand, S waves "shake" the particles at right angles to their direction of travel. This can be illustrated by fastening one end of a rope and shaking the other end, as shown in Figure 11.11C. Unlike P waves, which temporarily change the *volume* of intervening material by alternately squeezing and stretching it, S waves change the *shape* of the material that transmits them. Because fluids (gases and liquids) do not resist

FIGURE 11.9 Principle of the seismograph. **A.** The inertia of the suspended weight tends to keep it motionless while the recording drum, which is anchored to bedrock, vibrates in response to seismic waves. The stationary weight provides a reference point from which to measure the amount of displacement occurring as a seismic wave passes through the ground. **B.** Seismograph recording earthquake tremors. (Photo courtesy of Zephyr/Photo Researchers, Inc.)

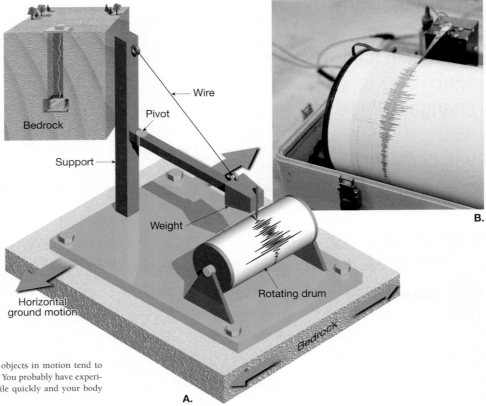

*Inertia: Simply stated, objects at rest tend to stay at rest, and objects in motion tend to remain in motion unless either is acted upon by an outside force. You probably have experienced this phenomenon when you tried to stop your automobile quickly and your body continued to move forward.

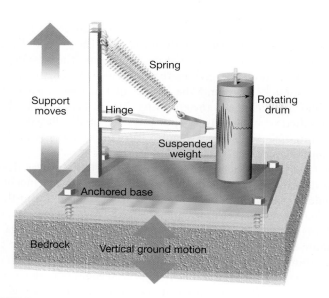

FIGURE 11.10 Seismograph designed to record vertical ground motion.

stresses that cause changes in shape—meaning fluids will not return to their original shape once the stress is removed—they will not transmit S waves.

The motion of surface waves is somewhat more complex. As surface waves travel along the ground, they cause the ground and anything resting upon it to move, much like ocean swells toss a ship. In addition to their up-and-down motion, surface waves have a side-to-side motion similar to an S wave oriented in a horizontal plane. This latter motion is particularly damaging to the foundations of structures.

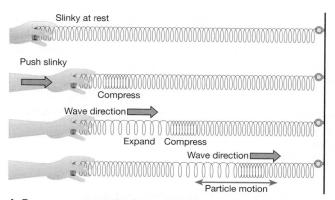

A. P waves generated using a slinky

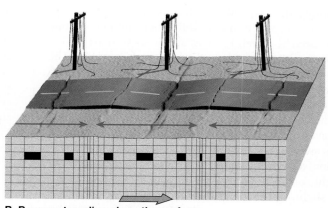

B. P waves traveling along the surface

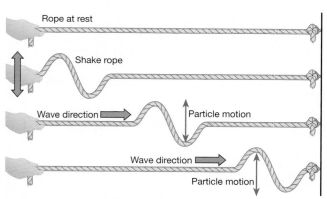

C. S waves generated using a rope

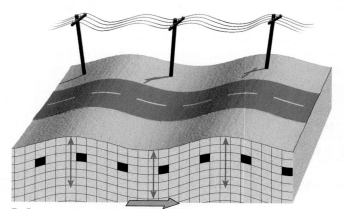

D. S waves traveling along the surface

FIGURE 11.11 Types of seismic waves and their characteristic motion. (Note that during a strong earthquake, ground shaking consists of a combination of various kinds of seismic waves.) **A.** As illustrated by a slinky, P waves are compressional waves that alternately compress and expand the material through which they pass. **B.** The back-and-forth motion produced as compressional waves travel along the surface can cause the ground to buckle and fracture, and may cause power lines to break. **C.** S waves cause material to oscillate at right angles to the direction of wave motion. **D.** Because S waves can travel in any plane, they produce up-and-down and sideways shaking of the ground.

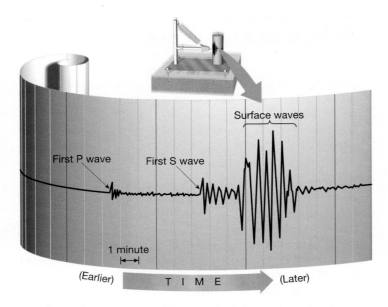

FIGURE 11.12 Typical seismogram. Note the time interval (about 5 minutes) between the arrival of the first P wave and the arrival of the first S wave.

By examining the "typical" seismic record shown in Figure 11.12, you can see a major difference among seismic waves—their speed of travel. P waves are the first to arrive at a recording station, then S waves, and finally surface waves. The velocity of P waves through the crustal rock granite is about 6 kilometers per second and increases to nearly 13 kilometers per second at the base of the mantle. P waves can pass through Earth's mantle in about 20 minutes. Generally, in any solid material, P waves travel about 1.7 times faster than S waves, and surface waves are roughly 10 percent slower than S waves.

In addition to velocity differences, notice in Figure 11.12 that the height, or *amplitude,* of these wave types also varies. S waves have slightly greater amplitudes than P waves, while surface waves exhibit even greater amplitudes. Surface waves also retain their maximum amplitude longer than P and S waves. As a result, surface waves tend to cause greater ground shaking, and hence, greater destruction, than either P or S waves.

Seismic waves are useful in determining the location and magnitude of earthquakes. In addition, seismic waves provide an important tool for probing Earth's interior (see Chapter 12).

LOCATING THE SOURCE OF AN EARTHQUAKE

EARTHQUAKES
▶ Locating the Source of an Earthquake

When analyzing an earthquake, the first task seismologists undertake is determining its *epicenter,* the point on Earth's surface directly above the focus (see Figure 11.2). The method used for locating an earthquake's epicenter relies on the fact that P waves travel faster than S waves.

The method is analogous to the results of a race between two autos, one faster than the other. The first P wave, like the

faster auto, always wins the race, arriving ahead of the first S wave. The greater the length of the race, the greater the difference in their arrival times at the finish line (the seismic station). Therefore, the greater the interval between the arrival of the first P wave and the arrival of the first S wave, the greater the distance to the epicenter. Figure 11.13 shows three simplified seismograms for the same earthquake. Based on the P-S interval, which city, Nagpur, Darwin, or Paris is farthest from the epicenter?

The system for locating earthquake epicenters was developed by using seismograms from earthquakes whose epicenters could be easily pinpointed from physical evidence. From these seismograms, travel-time graphs were constructed (Figure 11.14). Using the sample seismogram for Nagpur, India in Figure 11.13A and the travel-time curve in Figure 11.14, we can determine the distance separating the recording station from the earthquake in two steps: (1) Using the seismogram, determine the time interval between the arrival of the first P wave and the arrival of the first S wave, and (2) using the travel-time graph, find the P–S interval on the vertical axis and use that information to determine the distance to the epicenter on the horizontal axis. Following this procedure, we can determine that the earthquake occurred 3400 kilometers (2100 miles) from the recording instrument in Nagpur, India.

Now we know the *distance,* but what about *direction?* The epicenter could be in any direction from the seismic station. Using a method called *triangulation,* the precise location can be determined when the distance is known from three or more seismic stations (Figure 11.15). On a globe, a circle is drawn around each seismic station. The radius of these circles is equal to the distance from the seismic station to the epicenter. The point where the three circles intersect is the epicenter of the quake.

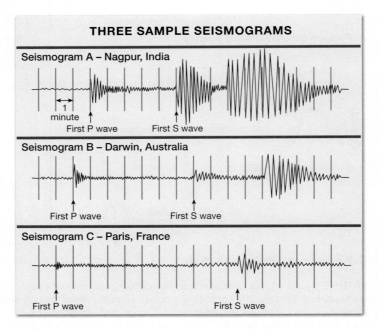

FIGURE 11.13 Simplified seismograms of the same earthquake recorded in three different cities. **A.** Nagpur, India. **B.** Darwin, Australia. **C.** Paris, France.

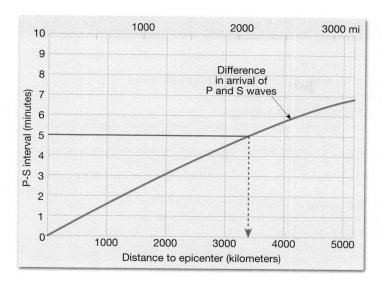

FIGURE 11.14 A travel-time graph is used to determine the distance to an earthquake epicenter. The difference in arrival time between the first P wave and the first S wave in the example is 5 minutes. Thus, the epicenter is roughly 3400 kilometers (2100 miles) away.

MEASURING THE SIZE OF EARTHQUAKES

Historically, seismologists have employed a variety of methods to determine two fundamentally different measures that describe the size of an earthquake—intensity and magnitude. The first of these to be used was **intensity**—a measure of the degree of earthquake shaking at a given locale based on observed damage. Later, with the development of seismographs, it became possible to measure ground motion using instruments. This quantitative measurement, called **magnitude,** relies on data gleaned from seismic records to estimate the amount of energy released at an earthquake's source.

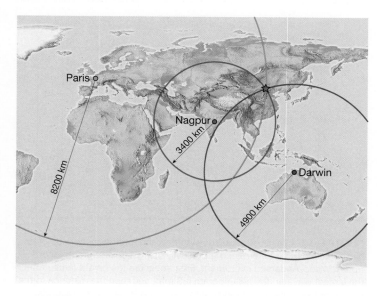

FIGURE 11.15 Determining an earthquake epicenter using the distances obtained from three or more seismic stations—a method called *triangulation.*

Intensity and magnitude provide useful, though different, information about earthquake strength. Consequently, both measures are used to describe earthquake severity.

Intensity Scales

Until a little more than a century ago, historical records provided the only accounts of the severity of earthquake shaking and destruction. Perhaps the first attempt to "scientifically" describe the aftermath of an earthquake came following the great Italian earthquake of 1857. By systematically mapping effects of the earthquake, a measure of the intensity of ground shaking was established. The map generated by this study employed lines to connect places of equal damage and hence equal ground shaking. Using this technique, zones of intensity were identified, with the zone of highest intensity located near the center of maximum ground shaking and often (but not always) the earthquake epicenter (Figure 11.16).

In 1902, Giuseppe Mercalli developed a more reliable intensity scale, which in a modified form is still used today. The **Modified Mercalli Intensity Scale,** shown in Table 11.1, was developed using California buildings as its standard. For example, if some well-built wood structures and most masonry buildings were destroyed by an earthquake, the affected area would be assigned an intensity of X (ten) on the Mercalli scale (Table 11.1).

Despite their usefulness in providing seismologists with a tool to compare earthquake severity, intensity scales have significant drawbacks. Such intensity scales are based on effects

FIGURE 11.16 Zones of destruction associated with an earthquake that struck Japan in 1925. Intensity levels are based on the Modified Mercalli Intensity Scale. Roman numerals represent the intensity categories. The zone of maximum intensity during this event roughly corresponded to the epicenter, but this is not always the case.

TABLE 11.1   Modified Mercalli Intensity Scale

I	Not felt except by a very few under especially favorable circumstances.
II	Felt only by a few persons at rest, especially on upper floors of buildings.
III	Felt quite noticeably indoors, especially on upper floors of buildings, but many people do not recognize it as an earthquake.
IV	During the day felt indoors by many, outdoors by few. Sensation like heavy truck striking building.
V	Felt by nearly everyone, many awakened. Disturbances of trees, poles, and other tall objects sometimes noticed.
VI	Felt by all; many frightened and run outdoors. Some heavy furniture moved; few instances of fallen plaster or damaged chimneys. Damage slight.
VII	Everybody runs outdoors. Damage negligible in buildings of good design and construction; slight-to-moderate in well-built ordinary structures; considerable in poorly built or badly designed structures.
VIII	Damage slight in specially designed structures; considerable in ordinary substantial buildings with partial collapse; great in poorly built structures. (Fall of chimneys, factory stacks, columns, monuments, walls.)
IX	Damage considerable in specially designed structures. Buildings shifted off foundations. Ground cracked conspicuously.
X	Some well-built wooden structures destroyed. Most masonry and frame structures destroyed. Ground badly cracked.
XI	Few, if any, (masonry) structures remain standing. Bridges destroyed. Broad fissures in ground.
XII	Damage total. Waves seen on ground surfaces. Objects thrown upward into air.

(largely destruction) that depend not only on the severity of ground shaking but also on factors such as building design and the nature of surface materials. For example, the modest 6.9-magnitude 1988 Armenian earthquake mentioned earlier was extremely destructive, mainly because of inferior building practices. A quake that struck Mexico City in 1985 was deadly because of the soft sediment upon which part of the city rests. Thus, the destruction wrought by an earthquake is frequently not a good measure of the amount of energy that was unleashed.

Magnitude Scales

In order to more accurately compare earthquakes across the globe, a measure was needed that does not rely on parameters that vary considerably from one part of the world to another. As a consequence, a number of magnitude scales were developed.

Richter Magnitude   In 1935 Charles Richter of the California Institute of Technology developed the first magnitude scale using seismic records. As shown in Figure 11.17 (top), the **Richter scale** is based on the amplitude of the largest seismic wave (P, S, or surface wave) recorded on a seismogram. Because seismic waves weaken as the distance between the focus and the seismograph increases, Richter developed a method that accounts for the decrease in wave amplitude with increasing distance. Theoretically, as long as equivalent instruments are used, monitoring stations at various locations will obtain the same Richter magnitude for each recorded earthquake. In practice, however, different recording stations often obtain slightly different Richter magnitudes for the same earthquake—a consequence of the variations in rock types through which the waves travel.

Earthquakes vary enormously in strength, and great earthquakes produce wave amplitudes that are thousands of times larger than those generated by weak tremors (Figure 11.18). To accommodate this wide variation, Richter used a *logarithmic scale* to express magnitude, in which a *tenfold* increase in wave amplitude corresponds to an increase of 1 on the magnitude scale. Thus, the degree of ground shaking for a 5-magnitude earthquake is 10 times greater than that produced by an earthquake having a Richter magnitude of 4.

In addition, each unit of Richter magnitude equates to roughly a *32-fold energy increase*. Thus, an earthquake with a magnitude of 6.5 releases 32 times more energy than one with a magnitude of 5.5, and roughly 1000 times (32×32) more energy than a 4.5-magnitude quake. A major earthquake with a magnitude of 8.5 releases millions of times more energy than the smallest earthquakes felt by humans (see Figure 11.18).

Although the Richter scale has no upper limit, the largest magnitude recorded was 8.9. Great shocks such as these release

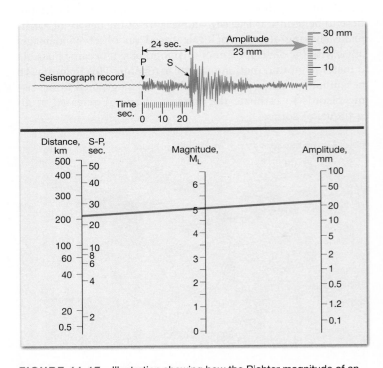

FIGURE 11.17   Illustration showing how the Richter magnitude of an earthquake can be determined graphically using a seismograph record from a Wood-Anderson instrument. First, measure the height (amplitude) of the largest wave on the seismogram (23 mm) and then the distance to the epicenter using the time interval between S and P waves (24 seconds). Next, draw a line between the distance scale (left) and the wave amplitude scale (right). By doing this, you should obtain the Richter magnitude (M_L) of 5. (Data from California Institute of Technology)

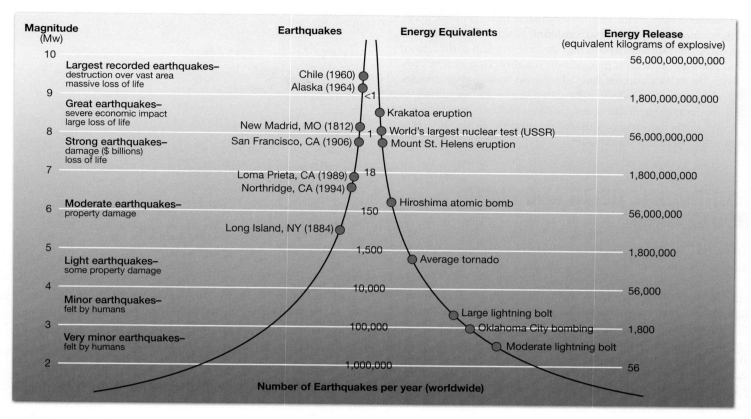

FIGURE 11.18   The size or magnitude of an earthquake (left side) compared to the number of earthquakes of various magnitudes that occur worldwide each year. The largest earthquakes occur less than once a year, whereas strong earthquakes happen more than once a month and weak quakes, those less than magnitude 2, occur hundreds of times per day. (Data from IRIS Consortium, www.iris.edu)

an amount of energy that is roughly equivalent to the detonation of 1 billion tons of explosives. Conversely, earthquakes with a Richter magnitude of less than 2.0 are generally not felt by humans.

Richter's original goal was modest in that he only attempted to rank shallow earthquakes in southern California into groups of large, medium, and small magnitude. Hence, Richter magnitude was designed to classify relatively local earthquakes and is designated by the symbol (M_L)—where M is for *magnitude* and L is for *local*.

The convenience of describing the size of an earthquake by a single number that can be calculated quickly from seismograms makes the Richter scale a powerful tool. Further, unlike intensity scales that can only be applied to populated areas of the globe, Richter magnitudes could be assigned to earthquakes in more remote regions and even to events that occur in the ocean basins. In time, seismologists modified Richter's work and developed new Richter-like magnitude scales.

Despite its usefulness, the Richter scale is not adequate for describing very large earthquakes. For example, the 1906 San Francisco earthquake and the 1964 Alaskan earthquake had roughly the same Richter magnitudes. However, based on the relative size of the affected areas and the associated tectonic changes, the Alaskan earthquake released considerably more energy than the San Francisco quake. Thus, the Richter scale is said to be *saturated* for major earthquakes because it cannot distinguish among them.

Moment Magnitude   In recent years, seismologists have come to favor a newer measure called **moment magnitude** (M_W), which determines the strain energy released from the entire fault surface. Because moment magnitude estimates the total energy released, it is better for measuring or describing very large earthquakes. In light of this, seismologists have recalculated the magnitudes of older, strong earthquakes using the moment magnitude scale. For example, the 1964 Alaskan earthquake was originally given a Richter magnitude of 8.3, but a recent recalculation using the moment magnitude scale resulted in an upgrade to 9.2. Similarly, the 1906 San Francisco earthquake which had a Richter magnitude of 8.3 was downgraded to a M_W 7.9. The strongest earthquake on record is the 1960 Chilean subduction zone earthquake, with a moment magnitude of 9.5.

Students Sometimes Ask . . .

Do moderate earthquakes decrease the chances of a major quake in the same region?

No. This is due to the vast increase in release of energy associated with higher-magnitude earthquakes (see Figure 11.18). For instance, an earthquake with a magnitude of 8.5 releases millions of times more energy than the smallest earthquakes felt by humans. Similarly, thousands of moderate tremors would be needed to release the huge amount of energy equal to one "great" earthquake.

Moment magnitude can be calculated from geologic field-work by measuring the average amount of slip on the fault, the area of the fault surface that slipped, and the strength of the faulted rock. The area of the fault plane can be roughly calculated by multiplying the surface-rupture length by the depth of the aftershocks. This method is most effective for determining the magnitude of large earthquakes generated along strike-slip faults in which the ruptures reach the surface. Moment magnitude can also be calculated using data from seismograms.

EARTHQUAKE BELTS AND PLATE BOUNDARIES

About 95 percent of the energy released by earthquakes originates in the few relatively narrow zones shown in Figure 11.19. The zone of greatest seismic activity, called *circum-Pacific belt,* encompasses the coastal regions of Chile, Central America, Indonesia, Japan, and Alaska, including the Aleutian Islands (Figure 11.19). Most earthquakes in the circum-Pacific belt occur along convergent plate boundaries where one plate slides at a low angle beneath another. The zone of contact between the subducting and overlying plates forms a huge thrust fault called a *megathrust,* along which Earth's largest earthquakes are generated (Figure 11.20).

There are more than 40,000 kilometers of subduction boundaries in the circum-Pacific belt where displacement is dominated by thrust faulting. Ruptures occasionally occur along segments that are 1000 kilometers or more in length, generating catastrophic *megathrust earthquakes* having magnitudes of (M_W) 8, or greater. Because subduction zone earthquakes usually happen beneath the ocean they may also generate destructive waves called *tsunami.* For example, the 2004 quake off the coast of Sumatra produced a tsunami that claimed an estimated 230,000 lives.

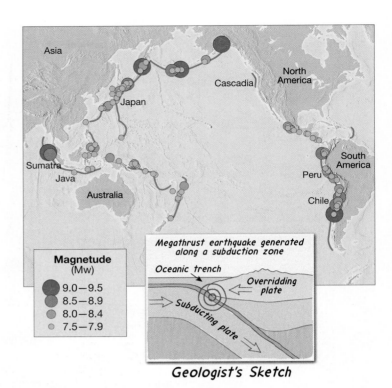

FIGURE 11.20 Distribution of known earthquakes of magnitude 7.5 or greater since 1900.

Another major concentration of strong seismic activity, referred to as the *Alpine-Himalayan belt,* runs through the mountainous regions that flank the Mediterranean Sea and extends past the Himalayan Mountains (see Figure 11.19). Tectonic activity in this region is mainly attributed to the collision of the African plate with Eurasia and the collision of the Indian plate with southeast Asia. These plate interactions created many thrust and strike-slip faults which remain active. In addition, numerous faults located away from these plate boundaries have been reactivated as India continues its northward advance into Asia. For example, slippage on a complex fault system in 2008 in the Sichuan Province of China killed at least 70,000 people and left 1.5 million others homeless. The "culprit" is the Indian subcontinent which shoves the Tibetan Plateau eastward against the rocks of the Sichuan Basin.

Figure 11.19 shows another continuous earthquake belt that extends for thousands of kilometers through the world's oceans. This zone coincides with the oceanic ridge system, which is an area of frequent but low-intensity seismic activity. As tensional forces pull the plates apart during seafloor spreading, displacement along normal faults generates most of the earth-

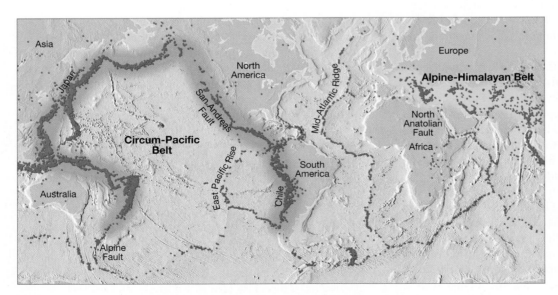

FIGURE 11.19 Distribution of nearly 15,000 earthquakes with magnitudes equal to or greater than 5 for a 10-year period. (Data from U. S. Geological Survey)

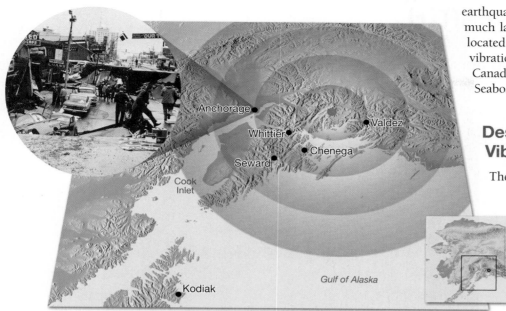

FIGURE 11.21   Region most affected by the Good Friday earthquake of 1964, the strongest earthquake ever recorded in North America. Note the location of the epicenter (red dot). Inset photo shows the collapse of a street in Anchorage, Alaska. (Photo courtesy of AP Photo)

earthquake of 1811, the area of influence can be much larger. The epicenter of this earthquake was located directly south of Cairo, Illinois, and the vibrations were felt from the Gulf of Mexico to Canada, and from the Rockies to the Atlantic Seaboard.

Destruction from Seismic Vibrations

The 1964 Alaskan earthquake provided geologists with insights into the role of ground shaking as a destructive force. As the energy released by an earthquake travels along Earth's surface, it causes the ground to vibrate in a complex manner by moving up and down as well as from side to side. The amount of damage to manmade structures attributable to the vibrations depends on several factors, including (1) the intensity and (2) the duration of the vibrations, (3) the nature of the material upon which the structure rests, and (4) the nature of building materials and the construction practices of the region.

All of the multistory structures in Anchorage were damaged by the vibrations. The more flexible wood-frame residential buildings fared best. However, many homes were destroyed when the ground failed. A striking example of how construction variations affect earthquake damage is shown in Figure 11.22. You can see that the steel-frame building on the left withstood the vibrations, whereas the poorly designed J.C. Penney building was badly damaged. Engineers have learned that buildings built of blocks and bricks that are not reinforced with steel rods are the most serious safety threat in earthquakes.

Most large structures in Anchorage were damaged, even though they were built according to the earthquake provisions of the Uniform Building Code. Perhaps some of that destruction

quakes. The remaining seismic activity in this zone is associated with slippage along transform faults located between ridge segments.

Transform faults and smaller strike-slip faults also run through continental crust where they may generate large earthquakes that tend to occur on a cyclical basis. Examples include California's San Andreas Fault, New Zealand's Alpine Fault, and Turkey's North Anatolian Fault that produced a deadly earthquake in 1999.

EARTHQUAKE DESTRUCTION

The most violent earthquake ever recorded in North America—the Good Friday Alaskan earthquake—occurred at 5:36 PM on March 27, 1964. Felt throughout that state, the earthquake had a moment magnitude (M_W) of 9.2 and lasted 3 to 4 minutes. This event left 131 people dead, thousands homeless, and the economy of the state badly disrupted. Had schools and business districts been open, the toll surely would have been higher. Within 24 hours of the initial shock, 28 aftershocks were recorded, 10 of which exceeded a magnitude of 6. The location of the epicenter and the towns that were hardest hit by the quake are shown in Figure 11.21.

Many factors determine the degree of destruction that will accompany an earthquake. The most obvious is the magnitude of the earthquake and its proximity to a populated area. During an earthquake, the region within 20 to 50 kilometers (12 to 30 miles) of the epicenter will ordinarily experience roughly the same degree of ground shaking, but beyond this limit vibrations deteriorate rapidly. Occasionally, during earthquakes that occur in the stable continental interior, such as the New Madrid, Missouri

FIGURE 11.22   Damage caused to the five-story J.C. Penney Co. building, Anchorage, Alaska. Very little structural damage was incurred by the adjacent building. (Courtesy of NOAA/Seattle)

UNDERSTANDING EARTH

Wave Amplification and Seismic Risks

The 1985 Mexican earthquake provided dramatic evidence of how surface materials can amplify ground shaking. A part of Mexico City was built on sand and clay that was deposited in a valley once occupied by Lake Texcoco. The lake was drained after the Spanish conquest of Montezuma and his Aztec warriors in the early sixteenth century to allow development of the city. Later, the modern Mexico City was built mainly on higher ground surrounding the area where buildings were situated on poorly consolidated lake sediments.

During the 1985 quake, seismic waves traveled toward Mexico City from the source, spread out, and substantially decreased in amplitude. The tremors that shook the firm surface materials in the higher parts of the city were mild, causing only minor damage. However, despite being more than 350 kilometers (225 miles) from the epicenter, nearly 500 multistory buildings collapsed or were severely damaged in the area built on sediments from Lake Texcoco (Figure 11.A). The death toll approached 9000 and 30,000 people were injured. Nearly 500,000 residents were left homeless when their adobe dwellings were destroyed.

To understand why this happened, recall that as seismic waves pass through crustal rocks they cause the intervening material to vibrate much like a tuning fork that has been struck. Although most objects can be "forced" to vibrate over a wide range of frequencies, each has a natural period of vibration that is preferred. Different Earth materials, like different-length tuning forks, also have different natural periods of vibration.*

Ground-motion amplification is enhanced when the supporting material has a natural period of vibration (frequency) that matches that of the seismic waves. A common example of this phenomenon occurs when someone pushes a child sitting on a swing. When periodically pushed in rhythm with the frequency of the swing, the child moves back and forth in a greater and greater arc (amplitude). By chance, the column of sediment beneath Mexico City had a natural period of vibration of about 2 seconds, matching that of the strongest seismic waves. Thus, when the seismic waves began shaking the soft sediments,

FIGURE 11.A   During the 1985 Mexican earthquake, multistory buildings swayed back and forth as much as 1 meter. Many, including the hotel shown here, collapsed or were seriously damaged. (Photo by James L. Beck)

a *resonance* developed, which greatly increased the amplitude of the vibrations. The level of ground shaking was too intense for many poorly designed buildings. In addition, intermediate-height structures (5 to 15 stories) swayed back and forth with a period between 1 and 2 seconds. Thus, resonance also developed between these buildings and the ground, resulting in the failure of most of the intermediate height structures (see Figure 11.A). In contrast, shorter buildings and higher buildings, such as the 37 story high Latin American Tower, were not structurally damaged.

Sediment-induced wave amplification also contributed significantly to the failure of the two-tiered section of Interstate 880 during the 1989 Loma Prieta earthquake (Figure 11.B). Studies conducted on the 1.4-kilometer section that collapsed showed that it was built on mud from San Francisco Bay. An adjacent section of this highway was damaged, but did not collapse because it was constructed on firmer alluvial materials.

*To demonstrate the natural period of vibration of an object, hold a ruler over the edge of a table so that most of it is not supported. Start it vibrating, and notice the noise it makes. By changing the length of the unsupported portion of the ruler, the natural period of vibration will change accordingly.

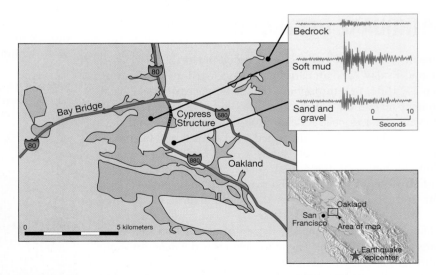

FIGURE 11.B   The portion of the Cypress Freeway structure in Oakland, California, that stood on soft mud (dashed red line) collapsed during the 1989 Loma Prieta earthquake. Adjacent parts of the structure (solid red) that were built on firmer ground remained standing. Seismograms from an aftershock (upper right) show that shaking is greatly amplified in the soft mud as compared to the firmer materials.

FIGURE 11.23 Effects of liquefaction. This tilted building rests on unconsolidated sediment that behaved like quicksand during the 1985 Mexican earthquake. (Photo by James L. Beck)

bedrock. Thus, the buildings in Anchorage that were situated on unconsolidated sediments experienced heavy structural damage. By contrast, most of the town of Whittier, although much nearer the epicenter, rested on a firm foundation of solid bedrock and suffered much less damage from seismic vibrations. Following the quake, however, Whittier was damaged by a tsunami—a phenomenon that will be described later in the chapter.

Liquefaction In areas where unconsolidated materials are saturated with water, earthquake vibrations can turn stable soil into a mobile fluid, a phenomenon known as **liquefaction.** As a result, the ground is not capable of supporting buildings and underground storage tanks and sewer lines may literally float toward the surface (Figure 11.23). During the 1989 Loma Prieta earthquake, in San Francisco's Marina District, foundations failed and geysers of sand and water shot from the ground, indicating that liquefaction had occurred (Figure 11.24).

Seiches The effects of great earthquakes may be felt thousands of kilometers from their source. Ground motion may generate *seiches,* the rhythmic sloshing of water in lakes, reservoirs, and enclosed basins such as the Gulf of Mexico. The 1964 Alaskan earthquake, for example, generated 2-meter waves off the coast of Texas, which damaged small craft, while much smaller waves were noticed in swimming pools in both Texas and Louisiana.

Seiches can be particularly dangerous when they occur in reservoirs retained by earthen dams. These waves have been known to slosh over reservoir walls and weaken the structures thereby endangering the lives of those downstream.

can be attributed to the unusually long duration of this earthquake. Most quakes involve tremors that last less than a minute. For example, the 1994 Northridge earthquake was felt for about 40 seconds, and the strong vibrations of the 1989 Loma Prieta earthquake lasted less than 15 seconds. But the Alaska quake reverberated for 3 to 4 minutes.

Amplification of Seismic Waves Although the region near the epicenter will experience about the same intensity of ground shaking, destruction may vary considerably within this area (see Box 11.1). Such differences are usually attributable to the nature of the ground on which the structures are built. Soft sediments, for example, generally amplify the vibrations more than solid

FIGURE 11.24 Liquefaction. **A.** These "mud volcanoes" were produced by the Loma Prieta earthquake of 1989. They formed when geysers of sand and water shot from the ground, an indication that liquefaction occurred. (Photo by Richard Hilton, courtesy of Dennis Fox) **B.** Students experiencing the nature of liquefaction. (Photo by Marli Miller)

A.

B.

Landslides and Ground Subsidence

The greatest damage to structures is often caused by landslides and ground subsidence triggered by earthquake vibrations. This was the case during the 1964 Alaskan earthquake in Valdez and Seward, where the violent shaking caused deltaic sediments to slump, carrying both waterfronts away. In Valdez, 31 people on a dock died when it slid into the sea. Because of the threat of recurrence, the entire town of Valdez was relocated to more stable ground about 7 kilometers away.

Much of the damage in the city of Anchorage was attributed to landslides. Homes were destroyed in Turnagain Heights when a layer of clay lost its strength and over 200 acres of land slid toward the ocean (Figure 11.25). A portion of this spectacular landslide was left in its natural condition as a reminder of this destructive event. The site was appropriately named "Earthquake Park." Downtown Anchorage was also disrupted as sections of the main business district dropped by as much as 3 meters (10 feet).

A recent example of major earthquake-triggered landslides occurred in China's Sichuan Province in May 2008. An image

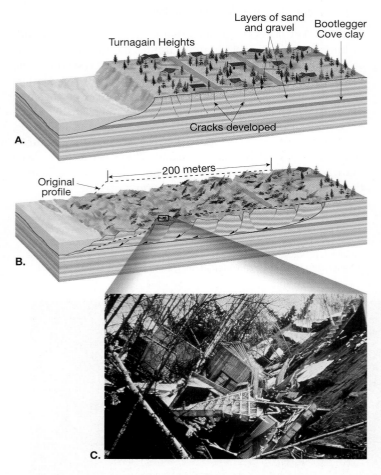

FIGURE 11.25 Turnagain Heights slide caused by the 1964 Alaskan earthquake. **A.** Vibrations from the earthquake caused cracks to appear near the edge of the bluff. **B.** Within seconds blocks of land began to slide toward the sea on a weak layer of clay. In less than 5 minutes, as much as 200 meters of the Turnagain Heights bluff area had been destroyed. **C.** Photo of a small portion of the Turnagain Heights slide. (Photo courtesy of U.S. Geological Survey)

(Figure 15.8, p. 412) and description can be found in "Earthquakes as Triggers" in Chapter 15.

Fire

More than 100 years ago, San Francisco was the economic center of the western United States, largely because of gold and silver mining. Then, at dawn on April 18, 1906, a violent earthquake struck unexpectedly, triggering an enormous firestorm. Much of the city was reduced to ashes and ruins. It is estimated that 3000 people died and 225,000 of the city's 400,000 residents were left homeless.

That historic earthquake reminds us of the formidable threat of fire. The central city contained mostly large, older wooden structures and brick buildings. Although many of the unreinforced brick buildings were extensively damaged by vibrations, the greatest destruction was caused by fires, which started when gas and electrical lines were severed. The fires raged out of control for three days and devastated over 500 blocks of the city. The initial ground shaking, which broke the city's water lines into hundreds of disconnected pieces, made controlling the fires virtually impossible.

The fires were finally contained when buildings were dynamited along a wide boulevard to provide a fire break, similar to the strategy used in fighting forest fires. Only a few deaths were attributed to the San Francisco fires, but other earthquake-initiated fires have been more destructive and claimed many more lives. For example, a 1923 earthquake in Japan triggered an estimated 250 fires, which devastated the city of Yokohama and destroyed more than half the homes in Tokyo. More than 100,000 deaths were attributed to the fires, which were driven by unusually high winds.

What is a Tsunami?

Large undersea earthquakes occasionally set in motion massive waves that scientists call **seismic sea waves.** You may be more familiar with the Japanese term **tsunami** which is frequently used to describe these destructive phenomena. Because of Japan's location along the circum-Pacific belt and its expansive coastline, it is especially vulnerable to tsunami destruction.

Most tsunami are caused by the vertical displacement of a slab of seafloor along a thrust fault on the ocean floor, or a large submarine landslide triggered by an earthquake (Figure 11.26). Once generated, a tsunami resembles the ripples formed when a pebble is dropped into a pond. In contrast to ripples, tsunami advance across the ocean at amazing speeds, between 500 and 950 kilometers per hour. Despite this striking characteristic, a tsunami in the open ocean can pass undetected because its height (amplitude) is usually less than 1 meter and the distance between wave crests is great, ranging from 100 to 700 kilometers. However, upon entering shallow coastal waters, these destructive waves "feel bottom" and slow, causing the water to pile up (see Figure 11.26). A few exceptional tsunami have reached 30 meters (100 feet) in height. As the crest of a tsunami approaches the shore, it appears as a rapid rise in sea level with a turbulent and chaotic surface (Figure 11.27A).

The first warning of an approaching tsunami is a rapid withdrawal of water from beaches. Some inhabitants of the Pacific

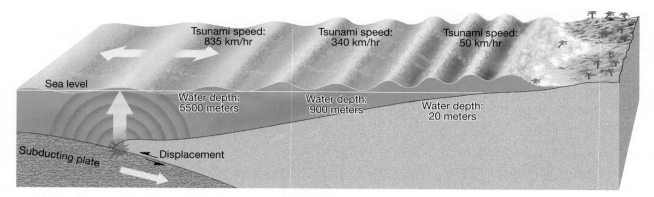

FIGURE 11.26 Schematic drawing of a tsunami generated by displacement of the ocean floor. The speed of a wave correlates with ocean depth. Waves moving in deep water advance at speeds in excess of 800 kilometers per hour. Speed gradually slows to 50 kilometers per hour at depths of 20 meters. Decreasing depth slows the movement of the wave. As waves slow in shallow water, they grow in height until they topple and rush onto shore with tremendous force. The size and spacing of these swells are not to scale.

basin have learned to heed this warning and move to higher ground. Approximately 5 to 30 minutes after the retreat of water, a surge capable of extending hundreds of meters inland occurs. In a successive fashion, each surge is followed by a rapid oceanward retreat of the sea.

Alaska 1964 The tsunami generated during the 1964 Alaskan earthquake inflicted heavy damage to the communities in the vicinity of the Gulf of Alaska, completely destroying the town of Chenega. Whittier and Kodiak were also heavily damaged when seismic sea waves carried most of their fishing fleets into the towns. The deaths of 107 persons were attributed to this event. By contrast, only nine people died in Anchorage as a result of the collapse of buildings.

Tsunami damage following the Alaskan earthquake extended along much of the west coast of North America, and despite a one-hour warning, 12 people perished over 4000 kilometers (2500 miles) away in Crescent City, California. The first wave crested about 4 meters (13 feet) above low tide and was followed by three progressively smaller waves. Believing that the tsunami had ceased, people returned to the shore, only to be met by the fifth and largest wave.

FIGURE 11.27 A massive earthquake (M_W 9.1) off the Indonesian island of Sumatra sent a tsunami racing across the Indian Ocean and Bay of Bengal on December 26, 2004. **A.** Unsuspecting foreign tourists, who at first walked on the sand after the water receded, now rush toward shore as the first of six tsunami roll toward Hat Rai Lay Beach near Krabi in southern Thailand. (AFP/Getty Images Inc.)
B. Tsunami survivors walk among the debris from this earthquake-triggered event. (Photo by Kimmasa Mayama/Reuters/CORBIS)

A.

B.

Tsunami Damage from the 2004 Indonesian Earthquake A massive undersea earthquake of moment magnitude 9.1 occurred near the island of Sumatra on December 26, 2004, and sent waves of water racing across the Indian Ocean and Bay of Bengal. It was one of the deadliest natural disasters of any kind in modern times, claiming more than 230,000 lives. As water surged several kilometers inland, cars and trucks were flung around like toys in a bathtub, and fishing boats were rammed into homes. In some locations, the backwash of water dragged bodies and huge amounts of debris out to sea.

The destruction was indiscriminate, destroying luxury resorts and poor fishing hamlets on the Indian Ocean coast (see Figure 11.27B). Damages were reported as far away as the Somalia coast of Africa, 4100 kilometers (2500 miles) west of the earthquake epicenter.

The killer waves generated by this massive quake achieved heights as great as 10 meters (33 feet) and struck many unprepared areas during a three-hour span following the earthquake. Although the Pacific basin had a tsunami warning system in place, the Indian Ocean unfortunately did not. The rarity of tsunami in the Indian Ocean also contributed to a lack of preparedness. It should come as no surprise that a tsunami warning system for the Indian Ocean was subsequently established.

Tsunami Warning System In 1946, a large tsunami struck the Hawaiian Islands without warning. A wave more than 15 meters (50 feet) high left several coastal villages in shambles. This destruction motivated the U.S. Coast and Geodetic Survey to establish a tsunami warning system for coastal areas of the Pacific. Seismic observatories throughout the region report large earthquakes to the Tsunami Warning Center in Honolulu. Scientists at the Center use deep-sea buoys equipped with pressure sensors to detect energy released by an earthquake. In addition, tidal gauges measure the rise and fall in sea level that accompany tsunami, resulting in warnings issued within the hour. Although tsunami travel very rapidly, there is sufficient time to evacuate all but the areas nearest the epicenter. For example, a tsunami generated near the Aleutian Islands would take five hours to reach Hawaii, and one generated near the coast of Chile would travel 15 hours before reaching Hawaii (Figure 11.28).

CAN EARTHQUAKES BE PREDICTED?

The vibrations that shook Northridge, California, in 1994 caused 57 deaths and more than $40 billion in damage (Figure 11.29). This level of destruction was the result of an earthquake of moderate intensity (M_W 6.7). Seismologists warn that other earthquakes of comparable or greater strength can be expected along the San Andreas system, which cuts a 1300-kilometer (800-mile) path through the state. The obvious question is: Can these earthquakes be predicted?

Short-Range Predictions

The goal of short-range earthquake prediction is to provide a warning of the location and magnitude of a large earthquake within a narrow time frame. Substantial efforts to achieve this objective have been attempted in Japan, the United States, China,

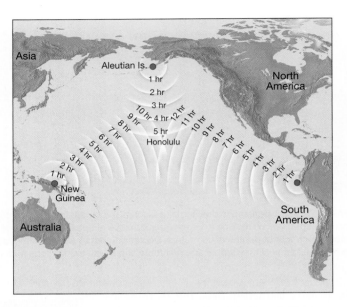

FIGURE 11.28 Tsunami travel times to Honolulu, Hawaii, from selected locations throughout the Pacific. (Data from NOAA)

and Russia—countries where earthquake risks are high (Table 11.2). This research has concentrated on monitoring possible *precursors*—events or changes that precede a forthcoming earthquake and thus may provide a warning. In California, for example, seismologists are monitoring changes in ground elevation and variations in strain levels near active faults. Other researchers are measuring changes in groundwater levels, while still others are trying to predict earthquakes based on an increase in the frequency of foreshocks that precede some, but not all earthquakes.

Japanese and Chinese scientists have been known to monitor anomalous animal behavior. A few days before the May 12, 2008, earthquake in China's Sichuan Province, the streets of a village near the fault were filled with toads migrating from the mountains. Was this a warning? Perhaps Walter Mooney, a USGS seismologist put it best—"Everyone hopes that animals can tell us something we don't know . . . but animal behavior is way too unreliable." Although precursors may exist, we have yet to determine a successful way to interpret and utilize the information.

One claim of a successful short-range prediction was made by the Chinese government after the February 4, 1975, earthquake in Liaoning Province. According to reports, very few people were killed—even though more than 1 million lived near the epicenter—because the earthquake was predicted and the residents were evacuated. Some Western seismologists have questioned this claim and suggest instead that an intense swarm of foreshocks, which began 24 hours before the main earthquake, may have caused many people to evacuate on their own accord.

One year after the Liaoning earthquake, an estimated 240,000 people died in the Tangshan, China, earthquake, which was *not* predicted. There were no foreshocks. Predictions can also lead to false alarms. In a province near Hong Kong, people reportedly evacuated their dwellings for over a month, but no earthquake followed.

In order for a short-range prediction scheme to warrant general acceptance, it must be both accurate and reliable. Thus, *it must have a small range of uncertainty in regard to location and timing, and it must produce few failures or false alarms.* Can you imagine the debate that would precede an order to evacuate a large city in the United States, such as Los Angeles or San Francisco? The cost of evacuating millions of people, arranging for living accommodations, and providing for their lost work time and wages would be staggering.

Currently, no reliable method exists for making short-range earthquake predictions. In fact, except for a brief period of optimism during the 1970s, the leading seismologists of the past 100 years have generally concluded that short-range earthquake prediction is not feasible.

Long-Range Forecasts

In contrast to short-range predictions, which aim to predict earthquakes within a time frame of hours, or at most, days, long-range forecasts give the probability of a certain magnitude earthquake occurring on a time scale of 30 to 100 years or more. These forecasts give statistical estimates of the expected intensity of ground motion for a given area, over a specified time frame. Although

long-range forecasts are not as informative as we might like, these data are useful for providing important guides for building codes so that buildings, dams, and roadways are constructed to withstand expected levels of ground shaking (Box 11.2).

For example, in the 1970s, before the 800-mile-long Trans-Alaskan oil pipeline was built, geologists did a hazards study of the Denali Fault system—a major tectonic structure across Alaska. It was determined that during a magnitude 8 earthquake on the Denali Fault, it would experience a 6-meter (20-foot) horizontal displacement. As a result of this investigation, the pipeline was designed to allow it to slide horizontally without breaking (Figure 11.30). In 2002, the Denali Fault ruptured producing a 7.9 magnitude earthquake. Although the total displacement along the fault was about 5 meters (18 feet), there was no oil spill. The Trans-Alaskan pipeline carries nearly 20 percent of the domestic oil supply of the United States—roughly 600,000 barrels per day—with a degree of scientific reassurance that it will withstand future displacement.

Long-range forecasts are based on evidence that many large faults break repeatedly, producing similar quakes at roughly

FIGURE 11.29 Damage to Interstate 5 caused by the January 17, 1994, Northridge earthquake. (Photo by Tom McHugh/Photo Researchers, Inc.)

TABLE 11.2   Some Notable Earthquakes

Year	Location	Deaths(est.)	Magnitude[†]	Comments
1556	Shensi, China	830,000		Possibly the greatest natural disaster.
1755	Lisbon, Portugal	70,000		Tsunami damage extensive.
*1811–1812	New Madrid, Missouri	Few	7.9	Three major earthquakes.
*1886	Charleston, South Carolina	60		Greatest historical earthquake in the eastern United States.
*1906	San Francisco, California	3,000	7.8	Fires caused extensive damage.
1908	Messina, Italy	120,000		
1923	Tokyo, Japan	143,000	7.9	Fire caused extensive destruction.
1960	Southern Chile	5,700	9.5	The largest-magnitude earthquake ever recorded.
*1964	Alaska	131	9.2	Greatest North American earthquake.
1970	Peru	70,000	7.9	Great rockslide.
*1971	San Fernando, California	65	6.5	Damage exceeded $1 billion.
1975	Liaoning Province, China	1,328	7.5	First major earthquake to be predicted.
1976	Tangshan, China	255,000	7.5	Not predicted.
1985	Mexico City	9,500	8.1	Major damage occurred 400 km from epicenter.
1988	Armenia	25,000	6.9	Poor construction practices.
*1989	San Francisco Bay area	62	7.1	Damages exceeded $6 billion.
1990	Iran	50,000	7.4	Landslides and poor construction practices caused great damage.
1993	Latur, India	10,000	6.4	Located in stable continental interior.
*1994	Northridge, California	51	6.7	Damages in excess of $15 billion.
1995	Kobe, Japan	5,472	6.9	Damages estimated to exceed $100 billion.
1999	Izmit, Turkey	17,127	7.4	Nearly 44,000 injured and more than 250,000 displaced.
1999	Chi-Chi, Taiwan	2,300	7.6	Severe destruction; 8,700 injuries.
2001	Bhuj, India	25,000+	7.9	Millions homeless.
2003	Bam, Iran	41,000+	6.6	Ancient city with poor construction.
2004	Indian Ocean (Sumatra)	230,000	9.1	Devastating tsunami damage.
2005	Pakistan/Kashmir	86,000	7.6	Many landslides; 4 million homeless.
2008	Sichuan, China	70,000	7.9	Millions homeless, some towns will not be rebuilt.

SOURCE: U.S. Geological Survey

*U.S. earthquakes.

[†]Widely differing magnitudes have been estimated for some of these earthquakes. When available, moment magnitudes are used.

similar intervals. In other words, as soon as a section of a fault ruptures, the continuing motions of Earth's plates begin to build strain in the rocks again until they fail once more. This led seismologists to study historical records of earthquakes to see if there are any discernible patterns so that the probability of recurrence might be established.

Seismic Gaps   With this concept in mind, seismologists began to plot the distribution of rupture zones associated with great earthquakes around the globe. The maps revealed that individual rupture zones tended to occur adjacent to one another without appreciable overlap, thereby tracing out a plate boundary. Because plates are moving at known velocities, the rate at which strain builds can also be estimated.

When these researchers studied historical records, they discovered that some seismic zones had not produced a large earthquake in more than a century, and in some locations, for several centuries. These quiet zones, called **seismic gaps,** are believed to be inactive zones that are storing strain for future major quakes.

An area of recent interest to seismologists is the northern edge of the Indian plate, which is colliding with Asia (Figure 11.31).

Although this area had historically been seismically quiet, three major earthquakes have struck the plate boundary since 2004. The most destructive was the previously described December 2004 Sumatra earthquake (M_W 9.1). Three months later, in March 2005, a second strong earthquake (M_W 8.6) struck Indonesia on the same fault system directly south of the deadly 2004 event. Fortunately, the second quake was much less destructive, because no substantial tsunami was generated. This difference can be explained by comparing the displacement that occurred during the two events. The 2004 quake ruptured nearly 1600 kilometers of the Sunda megathrust, nearly four times that of the 2005 quake. More importantly, the area of greatest displacement occurred under deep water at the site of the December 2004 rupture, but on land or shallow water where the March 2005 event occurred.

In October 2005, the Pakistan/Kashmir earthquake struck, claiming 86,000 lives. The severity of the destruction caused by this quake was attributed to severe thrusting, coupled with poor construction practices (Figure 11.32).

Regrettably, as the map in Figure 11.31 illustrates, several mature seismic gaps (shown in white) are located along this plate margin. One of these lies on the Sunda megathrust, just south of the March 2005 rupture. Did the displacement in 2005 transfer

FIGURE 11.30 The Trans-Alaskan oil pipeline was designed and built to withstand several meters of horizontal displacement where it crosses the Denali Fault. During a magnitude 7.9 earthquake in 2002, the pipeline moved as predicted and no oil spill occurred. This illustrates the importance of estimating potential ground motion and designing structures to mitigate the risks. (Photo courtesy of USGS)

quake, occurred on this segment of the San Andreas Fault in 1857, roughly 150 years ago. If earthquakes are cyclic, a major event in southern California appears to be imminent.

Using other paleoseismology techniques, researchers determined that several powerful earthquakes (magnitude of 8 or larger) have repeatedly struck the coastal Pacific Northwest over the past several thousand years. The most recent event occurred about 300 years ago. As a result of these findings, public officials have taken steps to strengthen some of the region's existing buildings, dams, bridges, and water systems. Even the private sector responded. The U.S. Bancorp building in Portland, Oregon, was strengthened at a cost of $8 million.

In summary, *the best prospects for making useful earthquake predictions involve forecasting magnitudes and locations on time scales of years or perhaps even decades. These forecasts are important because they provide information that can be used in the design of structures and to assist in land-use planning in order to reduce injuries and loss of life and property.*

FIGURE 11.31 Seismic gaps. Map of the northern boundary of the Indian plate, where it is moving toward Asia. Shown in red are the rupture zones for the three large earthquakes that occurred along this plate boundary since 2000. Large earthquakes that occurred between 1905 and 1999 are shown in black and seismic gaps are shown in white. Seismic gaps are "quiet zones" thought to be inactive zones that are storing elastic strain that will eventually produce major earthquakes.

sufficient stress to nudge the neighboring region toward failure?

Other seismic gaps are located within the continent along the margins of the Himalayan Mountains. One of these sites is located adjacent to the area of slippage that produced the October 2005 Pakistan/Kashmir quake. Another is a 600-kilometer-long region on the central Himalaya that has apparently not ruptured since 1505.

Paleoseismology Studies Another method of long-term forecasting involves *paleoseismology* (*palaois* = ancient, *seismos* = shake, *ology* = the study of). One investigation that employed this method focused on a segment of the southern San Andreas Fault east of Los Angeles. At this site, the drainage of Pallet Creek has been repeatedly disturbed by successive ruptures along the fault zone. Ditches excavated across the creek bed have exposed sediments that have been displaced by several large earthquakes over a span of 1500 years. From these data it was determined that strong earthquakes occur an average of once every 135 years. The last major event, called the Fort Tejon earth-

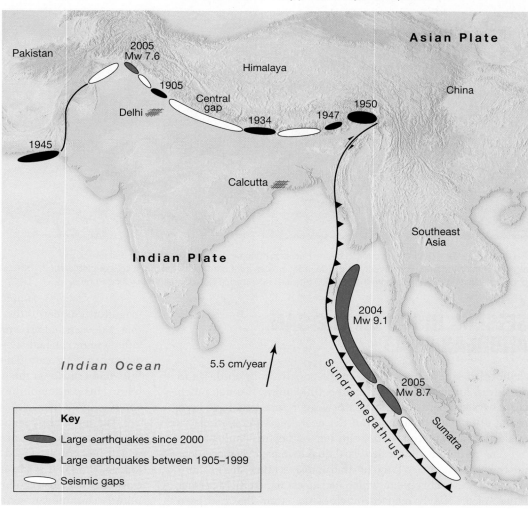

building during the previous 200 years or so. Consequently, the U.S. Geological Survey predicted that there is less than a 21 percent probability that a major earthquake will occur along the northern section of this fault system in the next 30 years.

Despite this encouraging news, another study recently released by the U.S. Geological Survey concluded that between 2003 and 2032 there is a 62 percent probability of at least one magnitude 6.7 or greater earthquake striking somewhere in the San Francisco Bay area (Figure 11.34). If it occurs, this earthquake would be roughly comparable to the 1989 Loma Prieta quake (M_W 6.9) and capable of causing significant damage somewhere in the area. The map in Figure 11.34 shows some, but not all, of the fault segments that might rupture during a future earthquake.

Located just south of the 1906 rupture is a section of the San Andreas Fault that exhibits *fault creep* (see Figure 11.33). When plates gradually slide past each other, as they do in this zone, less strain accumulates than when the fault is locked, diminishing the prospects of an eventual large earthquake.

As shown in Figure 11.33, a large break in the San Andreas happened east of Los Angeles in 1857. This rupture produced that great Fort Tejon earthquake that had an estimated magnitude similar to the 1906 San Francisco earthquake. Recall that paleoseismology studies indicate that this section of the fault system has had substantial displacement on the average of once every 135 years. This means that a reenactment of the 1857 Fort Tejon quake in the near future is within the realm of possibility. This portion of the fault has been given a 60 percent probability of producing a major earthquake in the next 30 years.

The most likely source of the next major quake on the San Andreas may well be its southernmost 200 kilometers (120 miles) which has not produced a large event in about 300 years (see Figure 11.33). A study using radar imaging from space found evidence that this section of the fault is mostly locked and continues to accumulate significant amounts of strain. The potential

FIGURE 11.32 Destruction caused by the Pakistan/Kashmir earthquake (M_W 7.6) that struck the region in October, 2005. The severity of the destruction was attributed to severe thrusting, coupled with poor construction practices. More than 86,000 fatalities occurred. (Photo by AP Wide World Photos)

SEISMIC RISKS ON THE SAN ANDREAS FAULT

California's San Andreas Fault runs diagonally from southeast to northwest for nearly 1300 kilometers (800 miles) through much of the western part of the state. For years researchers have been trying to predict the location of the next "Big One"—an earthquake with a magnitude of 8, or greater—along this fault system.

The 1906 San Francisco earthquake caused displacement on the 450-kilometer-long northernmost section of the fault (Figure 11.33). This event, which had an estimated Richter magnitude of 8.3, likely relieved much of the strain that had been

FIGURE 11.33 Map showing the portions of the San Andreas Fault that ruptured during the great 1906 San Francisco earthquake and the 1857 Fort Tejon earthquake. Located directly south of the 1906 rupture is the creeping segment where the plates are gradually sliding past each other. The southernmost 200 kilometers of the fault has not produced a large quake in about 300 years. This segment is given the greatest probability of producing the next "Big One."

slip rate, the pace plates are moving along the fault, has been estimated to be about 2 to 3 centimeters a year. This means that during the last 300 "dormant years" the fault has accumulated approximately 6 to 8 meters of slip. If this strain is released in a single quake, it would result in a magnitude 8 event.

The southernmost section of the San Andreas system cuts through the populated areas of Palm Springs, San Bernardino, Riverside, and Borrego Spring. Should a major quake occur, it would be felt throughout much of southern California including the densely populated metropolitan areas of Los Angeles and San Diego.

EARTHQUAKES: EVIDENCE FOR PLATE TECTONICS

EARTHQUAKES

▶ Earthquakes at Plate Boundaries

Once the basic outline of the plate tectonics theory was formulated, researchers from various branches of the geosciences began to test its validity. One of the first efforts was undertaken by a group of seismologists who were able to demonstrate a good fit between the plate tectonics model and the global distribution of earthquakes shown in Figure 11.35. In particular, these scien-

tists were able to account for the close association between deep-focus earthquakes and oceanic trenches where slabs of oceanic lithosphere are being subducted into the mantle.

When earthquake data were plotted according to geographic location and depth, several interesting observations were noted. Rather than a random mixture of shallow and deep earthquakes, some very definite patterns emerged (Figure 11.35). Earthquakes generated along the oceanic ridge system always have a shallow focus, and none are very strong. Further, it was noted that almost all deep-focus earthquakes occurred in the circum-Pacific belt, particularly in regions situated landward of oceanic trenches.

Based on our understanding of the mechanism that generates most earthquakes, it is logical to conclude that earthquakes should occur only in Earth's cool, rigid, outermost layer. Recall that as these rocks are deformed, they bend and store elastic energy—like a stretched rubber band. Once the rock is strained sufficiently, it ruptures, releasing the stored energy as the vibrations of an earthquake. By contrast, the hot mobile rocks of Earths' mantle are not capable of storing elastic energy and, therefore, should not generate earthquakes. Yet earthquakes have occurred to depths of 700 kilometers (435 miles).

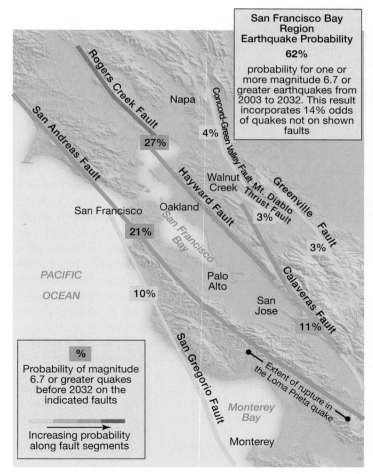

FIGURE 11.34 The San Francisco Bay area is given a 62 percent probability of experiencing a magnitude of 6.7 or greater earthquake before 2032. Some of the faults that could potentially rupture to produce this event are shown. (After the U.S. Geological Survey Fact Sheet 039-03)

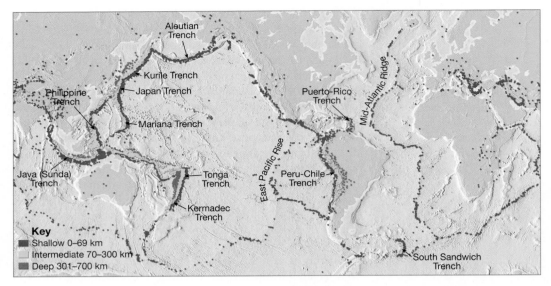

FIGURE 11.35   Distribution of shallow-, intermediate-, and deep-focus earthquakes. Note that deep-focus earthquakes occur only in association with convergent plate boundaries and subduction zones. (Data from NOAA)

The unique connection between deep-focus earthquakes and oceanic trenches was established through studies conducted in the Tonga Islands. When the depth of earthquake foci and their location within the Tonga Arc were plotted, the pattern shown in Figure 11.35 emerged. Most shallow-focus earthquakes occur within or adjacent to the trench, whereas intermediate- and deep-focus tremors occur toward the Tonga Islands. This zone of increasing foci depths away from the trench is called a **Wadati–Benioff Zone** after the two scientists who first identified them.

In the plate tectonics model, oceanic trenches form where dense slabs of cool oceanic lithosphere plunge into the mantle (Figure 11.36). Shallow-focus earthquakes are produced in response to the bending and fracturing of the lithosphere as it begins its descent, or as the subducting slab interacts with the overriding plate. Farther into the mantle, deep-focus earthquakes are generated. These earthquakes occur in the cool subducting slab rather than the mantle because rocks are poor conductors of heat. Consequently, the lithosphere remains colder and stronger than the mantle, which is too hot and weak to accumulate enough elastic strain to cause a rupture.

Additional evidence supporting the plate tectonics model came from observations that *only* shallow-focus earthquakes occur along divergent and transform fault boundaries. For example, along the San Andreas Fault most earthquakes occur in the upper 20 kilometers (12 miles) of the crust. Below that depth, rocks behave in a ductile manner (much like putty) and are incapable of storing elastic strain. Because oceanic trenches are the only places where cold slabs of oceanic crust plunge to

great depths, these should be the only sites of deep-focus earthquakes. Indeed, the absence of deep-focus earthquakes along oceanic ridges and transform faults supports the theory of plate tectonics.

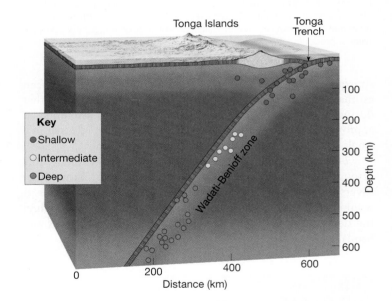

FIGURE 11.36   Idealized distribution of earthquake foci where the Pacific plate is being subducted beneath the Tonga Islands. Note that the intermediate- and deep-focus earthquakes occur only within the sinking oceanic lithosphere. (Modified after B. Isacks, J. Oliver, and L. R. Sykes)

PEOPLE AND THE ENVIRONMENT

BOX 11.2

Earthquakes East of the Rockies

Most earthquakes occur near plate boundaries in places such as California and Japan. However, areas distant from plate boundaries are not immune. A team of seismologists estimated that the probability of a damaging earthquake east of the Rocky Mountains during the next 30 years is roughly two-thirds as likely as an earthquake of comparable strength in California. This prediction is based in part on the geographic distribution and average rate of earthquake occurrences in these regions.

At least six major earthquakes have occurred in the central and eastern United States since colonial times. Three were centered near the Mississippi River Valley in southeastern Missouri and had estimated Richter magnitudes of 7.5, 7.3, and 7.8. Occurring over a three-month span between December 1811 and February 1812, these earthquakes, and numerous smaller tremors, destroyed the town of New Madrid, Missouri. They also triggered massive landslides, inflicted damage in six states, altered the course of the Mississippi River, and enlarged Tennessee's Reelfoot Lake.

The distance over which these earthquakes were felt is truly remarkable. Chimneys were downed as far away as Cincinnati. Hundreds of miles to the northeast, Boston residents reported feeling the tremor. Although destruction from the New Madrid earthquakes was slight compared to the Loma Prieta earthquake of 1989, it is only because the Midwest was sparsely populated in the early 1800s. Memphis, Tennessee, located near the epicenter, had not yet been established, and St. Louis was a small frontier town. Today, the Memphis metropolitan area has a population of more than one million people. In addition, it rests on unconsolidated floodplain deposits, so its buildings are particularly susceptible to earthquake damage. A federal study concluded that a 7.6-magnitude earthquake in this region could cause an estimated 2500 deaths, collapse 3000 structures, cause $25 billion in damages, and displace a quarter of a million people.

The greatest historical earthquake in the eastern states occurred in Charleston, South Carolina, in 1886. This one-minute event caused 60 deaths, numerous injuries, and great economic loss within 200 kilometers (120 miles) of Charleston. Minutes after the quake, strong vibrations shook the upper floors of buildings in Chicago and St. Louis, causing people to rush outdoors. In Charleston more than a hundred buildings were destroyed, and 90 percent of the remaining structures were damaged (Figure 11.C).

New England and adjacent areas have experienced sizable shocks, including the Massachusetts quakes in Plymouth (1683) and Cambridge (1755). Since records have been kept, New York State has experienced more than 300 earthquakes large enough to be felt by humans. Other damaging earthquakes include one that was centered near Aurora, Illinois (1909), and another that shook Valentine, Texas (1931).

Earthquakes in the central and eastern states occur far less frequently than in California. Yet shocks east of the Rockies generally produce structural damage over larger areas than tremors of similar magnitude in California. The reason for this difference is related to the underlying bedrock. In the central and eastern United States the bedrock is older and more rigid than in the west. As a result, seismic waves travel greater distances with less attenuation than in the western United States. When similar earthquakes are compared, the region of maximum ground motion in the East may be up to 10 times larger than in the West. Consequently, the greater earthquake frequency in the West is partly balanced by the potential for more widespread damage in the East.

FIGURE 11.C   Damage to Charleston, South Carolina, caused by the August 31, 1886, earthquake. Damage ranged from toppled chimneys and broken plaster to total collapse. (Photo courtesy of U.S. Geological Survey)

CHAPTER 11 EARTHQUAKES AND EARTHQUAKE HAZARDS IN REVIEW

- *Earthquakes* are vibrations of Earth produced by the rapid release of energy from rocks that rupture because they have been subjected to stresses that exceed their strength. This energy, which takes the form of *seismic waves*, radiates in all directions from the earthquake's source, called the *focus*. The movements that produce most large earthquakes occur along large fractures, called *faults*, that are usually associated with plate boundaries.

- Along a fault, rocks store energy as they are bent. As slippage occurs at the weakest point (the focus), displacement will exert stress farther along a fault, where additional slippage will occur until most of the built-up strain is released. An earthquake occurs as the rock elastically returns to its original shape. The "springing back" of the rock is termed *elastic rebound*. Small earthquakes, called *foreshocks*, often precede a major earthquake. The adjustments that follow a major earthquake often generate smaller earthquakes called *aftershocks*.

- Two main types of *seismic waves* are generated during an earthquake: (1) *surface waves*, which travel along the outer layer of Earth, and (2) *body waves*, which travel through Earth's interior. Body waves are further divided into *primary*, or *P, waves*, which push (squeeze) and pull (stretch) rocks in the direction the wave is traveling, and *secondary*, or *S, waves*, which "shake" the particles in rock at right angles to their direction of travel. P waves can travel through solids, liquids, and gases. Fluids (gases and liquids) will not transmit S waves. In any solid material, P waves travel about 1.7 times faster than S waves.

- The location on Earth's surface directly above the focus of an earthquake is the *epicenter*. Using the difference in arrival times between P and S waves, the distance separating a recording station from the earthquake epicenter can be determined. When the distances are known from three or more seismic stations, the epicenter can be located using a method called *triangulation*.

- Seismologists use two fundamentally different measures to describe the size of an earthquake—intensity and magnitude. *Intensity* is a measure of the degree of ground shaking at a given locale based on the amount of damage. The *Modified Mercalli Intensity Scale* uses damages to buildings in California to estimate the intensity of ground shaking for a local earthquake. *Magnitude* is calculated from seismic records and estimates the amount of energy released at the source of an earthquake. Using the *Richter scale*, the magnitude of an earthquake is estimated by measuring the *amplitude* (maximum displacement) of the largest seismic wave recorded. A logarithmic scale is used to express magnitude, in which a tenfold increase in ground shaking corresponds to an increase of 1 on the magnitude scale. *Moment magnitude* is currently used to estimate the size of moderate and large earthquakes. It can be calculated using the amount of slip on the fault surface, the area of the fault surface, and the strength of the faulted rock.

- *A close correlation exists between earthquake epicenters and plate boundaries.* The greatest energy is released by earthquakes along the margin of the Pacific Ocean, known as the *circum-Pacific belt,* and the mountainous regions that flank the Mediterranean Sea and continue past the Himalayan complex. Another zone of comparatively weak seismisity runs through the world's oceans along the *oceanic ridge system.*

- The primary factors that determine the amount of destruction accompanying an earthquake are the magnitude of the earthquake and the proximity of the quake to a populated area. Structural damage attributable to ground shaking depends on several factors, including (1) the intensity and (2) the duration of the vibrations, (3) the nature of the material upon which the structure rests, and (4) the design of the structure. Secondary effects of earthquakes include *tsunami*, landslides, ground subsidence, and fire.

- Substantial research to predict earthquakes is under way in Japan, the United States, China, and Russia—countries where earthquake risk is high. No reliable method of short-range prediction has yet been devised. Long-range forecasts are based on the premise that earthquakes are repetitive or cyclical. Seismologists study the history of earthquakes for patterns so their occurrences might be predicted. Long-range forecasts are important because they provide information used to develop the Uniform Building Code and to assist in land-use planning.

- The distribution of earthquakes provides strong evidence for the theory of plate tectonics, especially the close association between deep-focus earthquakes and subduction zones.

KEY TERMS

aftershock (p. 306)
body wave (p. 310)
earthquake (p. 304)
elastic rebound (p. 306)
epicenter (p. 304)
fault (p. 304)
fault creep (p. 307)
fault slip (p. 308)

focus (p. 304)
foreshock (p. 307)
hypocenter (p. 304)
inertia (p. 310)
intensity (p. 313)
liquefaction (p. 319)
magnitude (p. 313)
Modified Mercalli Intensity Scale (p. 313)

moment magnitude (p. 315)
primary (P) waves (p. 310)
Richter scale (p. 314)
secondary (S) waves (p. 310)
seismic gaps (p. 324)
seismic sea wave (p. 320)

seismic waves (p. 304)
seismogram (p. 310)
seismograph (p. 310)
seismology (p. 309)
surface wave (p. 310)
tsunami (p. 320)
Wadati–Benioff zone (p. 328)

QUESTIONS FOR REVIEW

1. Define *earthquakes* in geologic terms.
2. How are faults, foci, and epicenters related?
3. Who first explained the actual mechanism by which earthquakes are generated?
4. Explain what is meant by *elastic rebound*.
5. Faults that are "locked" may be considered "safe." Rebut or defend this statement.
6. List some reasons why some earthquakes are much stronger than others?
7. Describe the principle of a seismograph.
8. List the major differences between P and S waves.
9. Explain why P waves move through solids, liquids, and gases, whereas S waves move only through solids.
10. Which type of seismic wave causes the greatest destruction in shallow focus earthquakes?
11. Using Figure 11.14, determine the distance between an earthquake and a seismic station if the first S wave arrives 3 minutes after the first P wave.
12. Most of the very largest earthquakes occur in a zone on the globe known as the _____.
13. Distinguish between the Mercalli scale and the Richter scale.
14. For each increase of 1 on the Richter scale, seismic wave amplitude increases _____ times.
15. An earthquake measuring 7 on the Richter scale releases about _____ times more energy than an earthquake with a magnitude of 6.
16. Explain why the moment magnitude scale has gained popularity over the Richter scale among seismologists.
17. List four factors that affect the amount of destruction caused by seismic vibrations.
18. What factor contributed most to the extensive damage that occurred in the central portion of Mexico City during the 1985 earthquake? (see Box 11.1)
19. In addition to the destruction created directly by seismic vibrations, list three other types of destruction associated with earthquakes.
20. Describe the process of liquefaction.
21. What is a tsunami? How is one generated?
22. Cite some reasons why an earthquake with a moderate magnitude might cause more extensive damage than a quake with a high magnitude.
23. What prominent features on the ocean floor are associated with deep-focus earthquakes?
24. Are accurate short-range earthquake predictions possible using modern seismic instruments?
25. What is the value of long-range earthquake forecasts?
26. Briefly describe how earthquakes can be used as evidence for the theory of plate tectonics.

COMPANION WEBSITE

The *Earth 10e* website uses the resources and flexibility of the Internet to aid in your study of the topics in this chapter. Written and developed by the authors and other geology instructors, this site will help improve your understanding of geology. Visit **www.mygeoscienceplace.com** in order to:

- **Review** key chapter concepts.
- **Read** with links to the eBook and to chapter-specific web resources.
- **Visualize** and comprehend challenging topics using learning activities in *GEODe Earth*.
- **Test** yourself with online quizzes.

GEODe EARTH

GEODe Earth is a valuable and easy-to-use learning aid that can be accessed from your book's Companion Website (**www.mygeoscienceplace.com**). It is a dynamic instructional tool that promotes understanding and reinforces important concepts by using tutorials, animations, and exercises that actively engage the student.

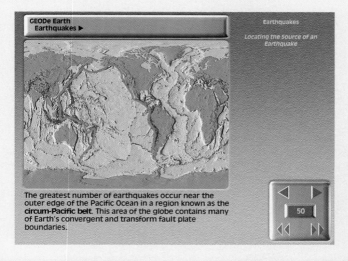

The greatest number of earthquakes occur near the outer edge of the Pacific Ocean in a region known as the **circum-Pacific belt**. This area of the globe contains many of Earth's convergent and transform fault plate boundaries.

EARTH'S INTERIOR*

Volcanic eruptions such as this one at Italy's Mount Etna, provide data about the nature of Earth's interior.

(PHOTO BY MARCO FULLE)

*This chapter was originally prepared by Professor Michael Wysession, Washington University.

If you could look inside a planet, you would quickly notice its distinct layers (Figure 12.1). The heaviest materials (metals) appear in the center, lighter solids (rocks) occupy the middle, and less dense liquids and gases are found at the top. For planet Earth, we know these layers as the iron core, the rocky mantle and crust, the liquid ocean, and the gaseous atmosphere. More than 95 percent of the variations in composition and temperature within Earth are due to this seemingly simple layered structure. However, Earth's interior is much more complex, sustaining what might otherwise be a lifeless cinder floating in space.

In addition to Earth's layers, small horizontal variations in composition and temperature at various depths verify that the interior of our planet is a dynamic place. Thanks to plate tectonics, the rocks of the mantle and crust are constantly moving. In addition, material is continuously recycling between the surface and the deep interior. It is also from Earth's deep interior that the water and air of our oceans and atmosphere are replenished, allowing life to exist at the surface.

Discovering and identifying the patterns of Earth's deep motions requires a variety of scientific research techniques. Light does not travel through rock so we must find other ways to "see" into our planet. The seismic waves produced by earthquakes are one way to investigate Earth's interior. Other techniques include mineral physics experiments that recreate the conditions of extreme temperature and pressure deep inside planets, and gravity measurements that establish variations in the distribution of mass. Examining Earth's magnetic field provides clues to the flow patterns of liquid iron in the core. Taken together, these methods give us a picture of Earth as a churning, varied, complex body that continues to change and evolve over time.

GRAVITY AND LAYERED PLANETS

If a bottle filled with clay, iron filings, water, and air was shaken, it would appear to have a single, muddy composition. If that bottle was then allowed to sit undisturbed, the different materials would separate and settle into layers. The iron filings, which are the densest, would be the first to sink to the bottom. Above the iron would be a layer of clay, then water, then air. This is similar to what happens inside planets. At their birth, planets accumulate huge quantities of nebular debris that melts and quickly segregates into layers. The iron sinks to form the core, rocky substances form the mantle and crust, and gases rise to create an atmosphere. All large bodies in the solar system have iron cores and rocky mantles, including the predominantly gaseous bodies Jupiter, Saturn, and the Sun. For both the bottle of muddy water and the planets, the force of gravity is responsible for the layering.

Another effect of gravity is that density changes occur not only between layers, but also within layers; because materials compress when they are squeezed. Rocks in the upper mantle have a density of about 3.3 g/cm^3 (grams per cubic centimeter). However, the density of these same rocks, when taken to the base of the mantle increases to 5.6 g/cm^3, nearly twice the original value.

This increase in density occurs partly because atoms shrink and occupy less space when subjected to immense pressure. In addition, atoms compress at various rates, and it is easier to compress negative ions than positive ions. Negative ions have more electrons than protons, and tend to be "fluffier" than positive ions. For example, when rocks are squeezed, the negative ions (such as O^{-2}) compress more easily than the positive ions (such as Si^{+4} and Mg^{+2}), so the ratios of ionic sizes change. As these ratios change, the structure may eventually become unstable, and the atoms rearrange into a more stable and denser structure. This is called a *mineral phase change*.

At depths between 300 and 400 kilometers, the intense pressure of the overlying rocks causes the mineral *olivine* to become unstable. As a result, the atoms in olivine rearrange into a denser and stable crystalline structure to form the mineral *spinel*. The increase in density of mantle rocks is due both to the compression of existing minerals and to the formation of new "high-pressure" minerals.

PROBING EARTH'S INTERIOR: "SEEING" SEISMIC WAVES

The ideal way to learn about Earth's interior would be to dig or drill a hole and examine it. Unfortu-

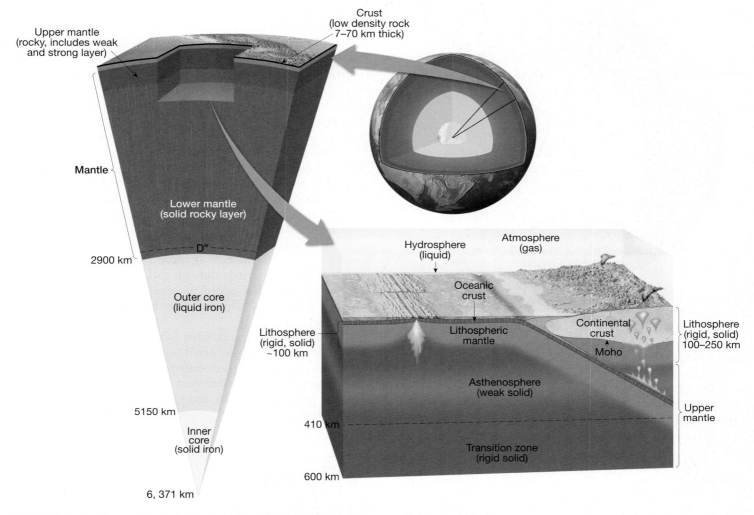

FIGURE 12.1 Views of Earth's layered structure. The study of seismic waves and other geophysical techniques have shown Earth to be a dynamic planet with many interacting parts. The properties of Earth's layers include the physical state of the material (solid, liquid, or gas) as well as how stiff the material is (for example, the distinction between the lithosphere and asthenosphere). These studies have shown that Earth's layers are mainly determined by density, with the heaviest materials (iron) at the center and the lightest ones on the outside (gases and liquids).

nately, this is only possible at shallow depths. The deepest a drill has ever penetrated is only 12.3 kilometers (7.5 miles), about 1/500 of the way to Earth's center! Even this was an extraordinary accomplishment due to the rapid increases in temperature and pressure with depth.

Fortunately for seismologists, many earthquakes are large enough that their seismic waves travel all the way through Earth and can be detected on the other side (Figure 12.2). This property of seismic waves is similar to how medical X-rays image our bones and organs. There are about 100 to 200 earthquakes each year that are large enough (about Mw > 6) to be recorded by seismographs around the globe. These large quakes provide the means to "see" into our planet and have been the source of most of the data that allow us to more fully understand Earth's interior.

Interpreting the waves recorded on seismograms is challenging because seismic waves usually do not travel along

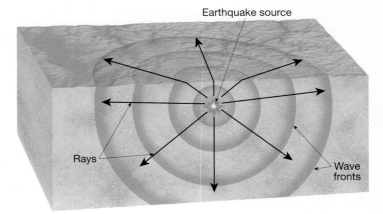

FIGURE 12.2 When traveling through a medium with uniform properties, seismic waves spread out from an earthquake source (focus) as spherically shaped features called *wave fronts*. It is common practice, however, to consider the paths taken by these waves as *seismic rays*, lines drawn perpendicular to the wave front as shown in this diagram.

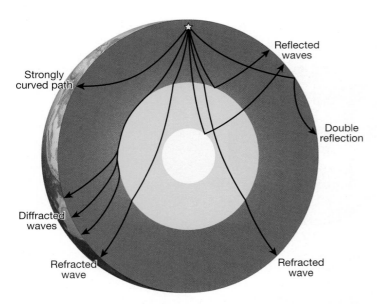

FIGURE 12.3 A few of the many possible paths that seismic rays follow through Earth.

straight paths. Instead, seismic waves are reflected, refracted, and diffracted as they pass through our planet (Figure 12.3). They reflect from boundaries between layers, and refract when passing from one layer to another—similar to how light is refracted (bent) as it passes from air to water. In addition, they diffract around obstacles they encounter (see Figure 12.2).

As Figure 12.4 illustrates, changes in the composition or structure of rock cause seismic waves to reflect off boundaries between layers. This characteristic of waves is especially important in the exploration for oil and natural gas where arti-

ficially generated seismic waves are used to probe the crust. Petroleum tends to get trapped in certain kinds of geological structures, and these structures are identified by mapping the layering of the upper crust. Gasoline would be considerably more expensive without seismic imaging because a huge number of wells would have to be randomly drilled to find oil. Likewise, seismic waves are useful in locating the boundaries between the crust, mantle, outer core, inner core, and other layers that have been discovered within the mantle.

The speed at which seismic waves travel through Earth's layers depends largely on the properties of the materials encountered. In general, seismic waves travel faster when rock is *stiff* (*rigid*) or *less compressible*. These properties of stiffness and compressibility are then used to interpret the composition and temperature of the rock. For instance, when rock is heated it becomes less stiff (imagine warming a frozen chocolate bar), and waves travel through it more slowly. Likewise, waves travel at different speeds through rocks having different compositions. Thus, the speed at which seismic waves travel through a layer can help determine both the type of rock and its temperature.

One of the most noticeable behaviors of seismic waves is that they follow strongly curved (refracted) paths because their velocities generally increase with depth (Figure 12.5). Within a particular layer, the speed of seismic waves increases with depth because pressure increases and squeezes the rock into a more compact, rigid material.

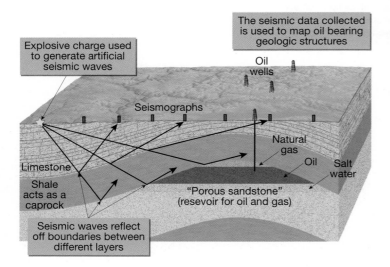

FIGURE 12.4 Reflected seismic waves are used to search for oil and natural gas underground. The seismic waves from explosions reflect off the boundaries between layers of different composition. Using computer programs, the data shows the geometry of the strata, including folds and faults. Using this information, geologists map potential petroleum reservoirs in Earth's crust.

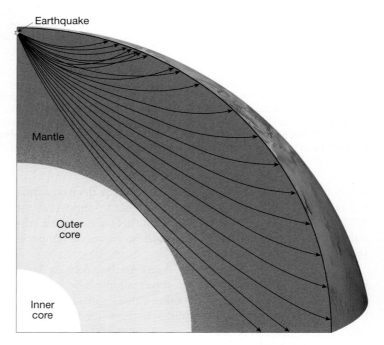

FIGURE 12.5 Slice through Earth's mantle showing some of the ray paths that seismic waves from an earthquake might follow. The rays follow curved (refracting) rather than straight paths because the seismic velocity of rocks increases with depth in the mantle, a result of increasing pressure with depth. Notice the complicated ray paths in the upper mantle, with some even crossing one another. This is due to the sudden seismic velocity increases that result from mineral phase changes at increasing pressures.

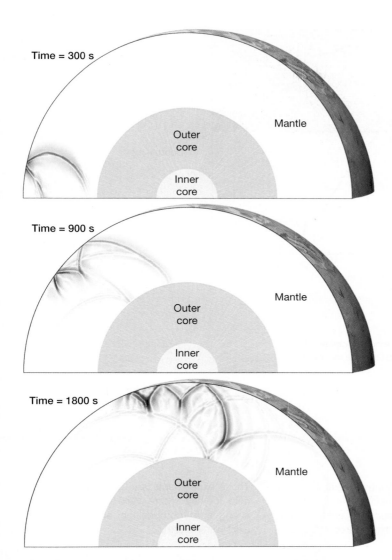

FIGURE 12.6   Three snapshots in time showing the locations of S waves within Earth's mantle following an earthquake. In addition to refracting and diffracting, S waves reflect from boundaries such as the core-mantle boundary. Note that S waves do not penetrate the outer core because they do not travel through liquids.

Within Earth's mantle, where there are both sharp boundaries and gradual seismic velocity changes, the pattern of seismic waves is complex. Figure 12.6 shows what S waves look like when they travel from a deep earthquake through the mantle. Note how the single wave from the shock is soon broken into many different waves that appear on seismograms as separate signals.

EARTH'S LAYERS

EARTH'S INTERIOR
▶ Earth's Layered Structure

Combining the data obtained from seismological studies and mineral physics experiments has given us a layer-by-layer understanding of the composition of Earth (Box 12.1). Seismic

velocities, as a function of depth, are shown in Figure 12.7. By examining the behavior of a variety of rocks at the pressures corresponding to these depths, geologists have made important discoveries about the compositions of Earth's crust, mantle, and core.

Earth's Crust

Earth's **crust** consists of two distinct types—continental crust and oceanic crust. Continental crust and oceanic crust have very different compositions, histories, and ages. In fact, oceanic crust is compositionally more similar to the mantle than to the continental crust.

Oceanic Crust   The ocean crust averages about 7 kilometers (4.5 miles) thick and forms at mid-ocean ridges, which separate two diverging tectonic plates. Ocean crust has P wave velocities of about 5–7 km/s (kilometers per second) and a density of about 3.0 g/cm^3, which compares to experimental values for the rocks basalt and gabbro. The composition and formation of ocean crust is discussed in greater detail in Chapter 13.

Continental Crust   While oceanic crust is fairly uniform, no two continental regions have the same structure or composition. Continental crust averages about 40 kilometers (25 miles) in thickness, but can be more than 70 kilometers (45 miles) thick in mountainous regions such as the Himalayas and the Andes. The thinnest crust in North America is beneath the Basin and Range region in the western United States, where the crust is as thin as 20 kilometers (12 miles). The thickest North American crust, beneath the Rockies, is more than 50 kilometers (30 miles) thick.

Seismic velocities within continents are quite variable, suggesting that the composition of continental crust must also vary greatly. Continents have an average density of about 2.7g/cm^3, which is much lower than both oceanic crust and mantle rock. This low density explains why continents are buoyant—acting like giant rafts, floating atop tectonic plates, and why they cannot be subducted into the mantle.

Discovering Boundaries: The Moho   The boundary between the crust and mantle, called the **Moho,** was one of the first features of Earth's interior discovered using seismic waves. In 1909, Croatian seismologist Andrija Mohorovičić discovered this boundary which now bears his name. At the base of the continents, P waves travel about 6 km/s but abruptly increase to 8 km/s at a slightly greater depth.

Mohorovičić cleverly used this large jump in seismic velocity to make his discovery. He noticed two different sets of seismic waves were recorded at seismographs located within a few hundred kilometers of an earthquake. One set of waves moved across the ground at about 6 km/s, and the other set of waves traveled about 8 km/s. From these two waves, Mohorovičić correctly determined that the different waves were traveling through two different layers.

During a shallow earthquake, *direct waves* travel along a straight path through the crust as shown in Figure 12.8. Other seismic waves follow a path through the crust and

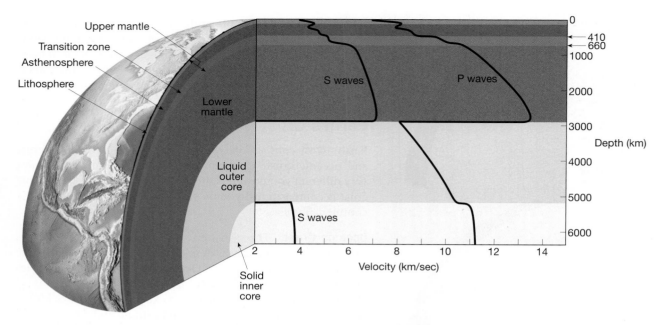

FIGURE 12.7   Cross-section of Earth showing its different layers and the average velocities of P and S waves at each depth. S waves are an indication of how rigid the material is—the inner core is less rigid than the mantle, and the liquid outer core has no rigidity.

UNDERSTANDING EARTH

BOX 12.1

Re-creating the Deep Earth

Seismology alone cannot determine the nature of materials deep in Earth's interior. Additional information must be obtained by other techniques. *Mineral physics* experiments can measure physical properties of rocks and minerals such as stiffness, compressibility, and density while simulating the extreme conditions of the mantle and core.

Most mineral physics experiments are conducted using presses with hard carbonized steel tips that simulate the enormous pressures at depth by squeezing mineral samples. The highest pressures are obtained using diamond-anvil presses like the one shown in Figure 12.A. These take advantage of two important properties of diamonds—hardness and transparency. The tips of two diamonds are cut off, and a small sample of mineral or rock is placed between them. By squeezing two diamonds together, pressures as high as those in Jupiter's interior have been simulated. High temperatures are achieved by firing a laser beam through the diamond and into the mineral sample.

One experiment determines the temperature at which minerals begin to melt under various pressures. Another examines the temperatures and pressures at which one mineral phase will become unstable and convert into a new "high-pressure" phase.

These experiments are useful because they help identify where vertical and horizontal changes in composition and temperature exist within Earth.

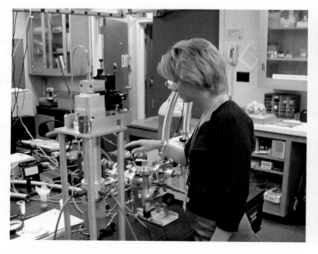

FIGURE 12.A   High-pressure experiments inside a diamond anvil cell (left photo) simulate the conditions at the center of a planet. The apparatus is small, and fits on a tabletop. High pressures are generated by cutting the tips off of high-quality diamonds (right photo), putting a small sample of rock between, squeezing the diamonds together, and heating the sample with a laser. (Left photo courtesy of Lawrence Livermore National Laboratory; right photo by Douglass L. Peck Photography)

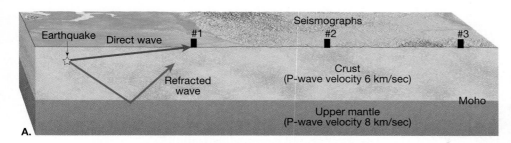

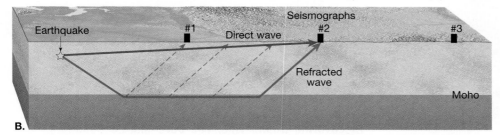

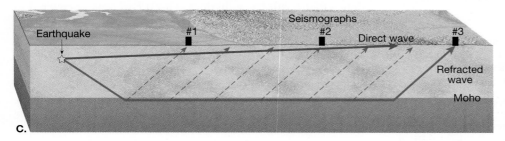

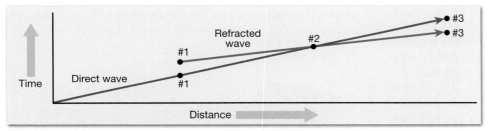

FIGURE 12.8 Seismic waves from an earthquake arriving at three different seismographs. Over a short distance, such as at seismograph #1, the direct wave arrives first. For greater distances, such as at seismograph #3, the refracted wave arrives first. At the cross-over point, which in this diagram occurs at seismograph #2, both waves arrive at the same time. The distance to the cross-over point increases with the depth of the Moho, and therefore can be used to determine the thickness of the crust.

along the top of the mantle. These are called *refracted waves* because they are bent, or refracted, as they enter the mantle. Seismographs near the epicenter record the direct waves first. However, seismographs further from the epicenter record the refracted waves first. The point at which both waves arrive at the same time, called the *cross-over,* can be used to determine the depth of the Moho. Thus, using data from these two sets of waves and seismographs at various distances from the epicenter, the thickness of the crust for any location can be calculated.

The difference between travel times for direct and refracted waves is comparable to driving to a destination on local roads versus interstate highways. For short distances you will typically arrive sooner if you drive the most direct route using local roads. For long distances, the trip may take less time if you take a less direct route that involves mostly interstate highways. The cross-over point, where both routes take an equal amount of time, is directly related to how far you must drive before reaching the interstate highway. Applied to determining the depth of the Moho, the cross-over is related to how far seismic waves travel through the crust (slow layer) before they reach the mantle (fast layer)—the greater the cross-over distance, the deeper the Moho.

Earth's Mantle

More than 82 percent of Earth's volume is contained within the **mantle,** a nearly 2900-kilometer-thick shell extending from the base of the crust (Moho) to the liquid outer core (see Figure 12.1). Because S waves readily travel through the mantle, we know that it is a solid rocky layer composed of silicate minerals that are rich in iron and magnesium. However, despite its solid nature, rock in the mantle is quite hot and capable of flow, albeit at very slow velocities.

The Upper Mantle   The **upper mantle** extends from the Moho to a depth of about 660 kilometers and can be divided into three shells. The uppermost mantle and the crust define the rigid **lithosphere.** Beneath the **lithospheric mantle** is a weak layer called the **asthenosphere.** Differences between these two layers are the result of Earth's temperature structure, and will be discussed later in this chapter. The lower portion of the upper mantle is called the *transition zone.*

Rocks brought to the surface by volcanism and other geological processes have provided geologists with valuable information about the upper mantle (Figure 12.9). The seismic velocities observed for the mantle are consistent with the rock peridotite. Mantle *peridotite,* an ultramafic rock composed mostly of the minerals *olivine* and *pyroxene,* is richer in iron and magnesium than rocks found in either the continental or oceanic crust.

Transition Zone   The **transition zone** lies within the upper mantle at depths between 410 kilometers and 660 kilometers. The top of the transition zone is identified by a sudden increase in density from about 3.5 to 3.7 g/cm^3. Like the Moho, this boundary reflects seismic waves. Unlike the Moho, this boundary is due to a change in mineral phase rather than a change in chemical composition. When the mineral olivine, which is stable in the uppermost mantle, is subjected to greater pressure in the

FIGURE 12.9   This sample of peridotite (green mineral) was carried up from the mantle and provides clues to the composition of Earth's interior. The mantle fragment (xenolith) was contained within a volcanic bomb from Vulken-Eifel, Germany. One Euro coin for scale. (Photo by Woudloper)

transition zone it collapses into denser structures. In the top half of the transition zone, olivine converts to a phase called β-spinel, and in the bottom half, β-spinel converts to a more compact spinel structure called ringwoodite.

Water cycles slowly through the planet, brought into the mantle with subducting oceanic lithosphere and carried upward by rising plumes of mantle rock. Mineral physics experiments have revealed that the transition zone is capable of holding a great deal of water, up to 2 percent by weight. This is considerably more than for the rocks of the upper mantle, which can hold only about 0.1 percent of its weight as water. Because the transition zone represents 10 percent of Earth's volume, it could potentially hold up to five times the volume of Earth's oceans. How much water is actually contained within the transition zone has not been determined.

The Lower Mantle   The **lower mantle** lies between the transition zone (660 kilometers) and the liquid core (2900 kilometers). Beneath the 660-kilometer discontinuity, both olivine and pyroxene take the form of the mineral *perovskite* (Fe, Mg) SiO_3. Because the lower mantle is undoubtedly Earth's largest layer, occupying 56 percent of the volume of the planet, perovskite is the single most abundant material within Earth.

The D″ Layer   In the bottom few hundred kilometers of the mantle, is a highly variable and unusual layer called the **D″ layer** (pronounced "dee double-prime") which is a boundary layer between the rocky mantle and the liquid iron outer core (Figure 12.10). The D″ layer is similar in many ways to the lithosphere, which is the boundary layer between the mantle and the ocean/atmosphere layer. Both the lithosphere and D″ layer have large variations in composition and temperature.

The difference in temperature between rocks of the hot mid-ocean ridges and those of the cold abyssal seafloor is more than 1000 °C. The horizontal changes in temperature within the D″ layer are similar. The composition of the lithosphere varies greatly, with either continental or oceanic crust embedded in it. The D″ layer, in turn, is thought to be the graveyard of some subducted oceanic lithosphere and the birthplace of some mantle plumes. Therefore, large slabs of differing rock types appear to be contained within D″.

The very base of D″, the part of the mantle directly in contact with the hot liquid iron core, is like Earth's surface in that there are "upside-down mountains" of rock that protrude into the core. Furthermore, in some regions of the core–mantle boundary, the base of D″ may be hot enough to be partially molten. Evidence for partial melting comes from zones at the very base of the mantle where S-wave velocities decrease by 30 percent.

Discovering Boundaries: The Core–Mantle Boundary   Evidence that Earth has a distinct central core was uncovered in 1906 by British geologist Richard Dixon Oldham. (In 1914, Beno Gutenberg calculated 2900 kilometers to the core boundary depth, which remains the accepted value.) At distances beyond approximately 100° from the epicenter of a large earthquake, Oldham observed that P and S waves were absent or very weak. In other words, Oldham found evidence for a central core that produced a "*shadow zone*" for seismic waves (Figure 12.11).

As Oldham predicted, Earth's core exhibits markedly different elastic properties from the mantle above, which causes considerable refraction of P waves—similar to how light is refracted

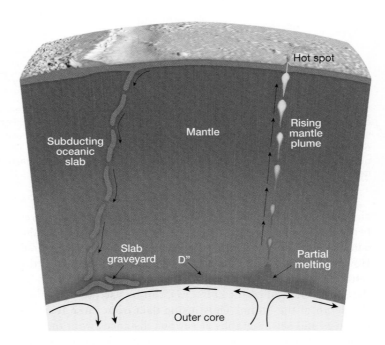

FIGURE 12.10   Schematic of the variable and unusual D″ layer at the base of the mantle. Like the lithosphere at the top of the mantle, the D″ layer contains large, horizontal variations in both temperature and composition. Many scientists believe that D″ is the graveyard of some subducted oceanic lithosphere and the birthplace of some mantle plumes.

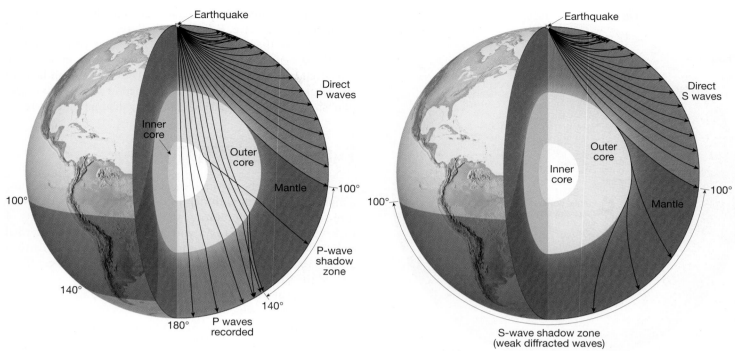

FIGURE 12.11 Two views of Earth's interior showing the effects of the outer and inner cores on the ray paths of P and S waves. **A.** When P waves interact with the slow-velocity liquid iron of the outer core, their rays are refracted downward. This creates a shadow zone where no direct P waves are recorded (although diffracted P waves travel there). The P waves that travel through the core are called PKP waves. The "K" represents the path through the core, and comes from the German word for core, which is *kern*. The increase in seismic velocity at the top of the inner core can refract waves sharply so some arrive within the shadow zone, shown here as a single ray. **B.** The core is an obstacle to S waves, because they cannot pass through liquids. Therefore, a large shadow zone exists for S waves. However, some S waves diffract around the core and are recorded on the other side of the planet.

as it passes from air to water. In addition, because the outer core is liquid iron, it blocks the transmission of S waves (recall S waves do not travel through liquids).

The locations of the P and S wave shadow zones and how their paths are affected by the core are shown in Figure 12.11. Whereas some P and S waves still arrive in the shadow zone, they differ greatly from those expected within a planet without a core.

Earth's Core

The Outer Core The boundary between Earth's mantle and outer core, called the *core–mantle boundary,* is significant because of its varied material properties. P wave velocities drop from 13.7 to 8.1 km/s at the core–mantle boundary, and S waves drop dramatically from 7.3 km/s to zero. Because S waves do not pass through liquids, their absence in the **outer core** indicates its liquid state. The change in density from 5.6 to 9.9 g/cm^3 is even larger than the rock–air difference observed at Earth's surface.

Based on our knowledge of the composition of meteorites and the Sun, geologists expect Earth to contain a significantly higher percentage of iron than is observed in rocks found in the crust and mantle. This fact, coupled with the great density of the core, indicates that the outer core consists mostly of

iron and lesser amounts of nickel, which has a density similar to iron. Density and seismic studies suggest that about 15 percent of the outer core consists of lighter elements. Based on mineral physics experiments, these are likely to include sulfur, oxygen, silicon, and hydrogen. For instance, an iron-sulfur mixture melts at a much lower temperature than pure iron. As Earth was forming, high-velocity impacts of nebular debris and the decay of radioactive elements caused the temperature of our planet to increase. When heated sufficiently, iron in the presence of sulfur began to melt. Melting produced liquid blobs of an iron-sulfur alloy that sank to form the core. In a similar manner, other light elements were dragged down into the core.

The **core** accounts for about 1/6 of Earth's volume, but 1/3 of its mass because it is composed mostly of iron, which is the most dense of the common elements. In fact, iron is Earth's most abundant element when measured by mass.

The Inner Core At the center of the core is a solid sphere of iron with trace amounts of other elements called the **inner core.** Because the inner core is a sphere whereas Earth's other layers are shells, drawings make the inner core appear much larger than it really is (see Figure 12.1). The inner core is actually relatively small, only 1/142 of the volume of Earth (less than one percent). The inner core did not exist early in Earth's history, when our planet was hotter. However, as Earth cooled,

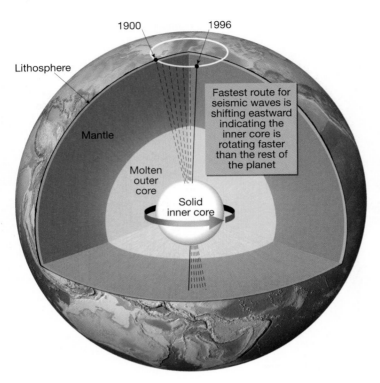

Fastest route for seismic waves is shifting eastward indicating the inner core is rotating faster than the rest of the planet

FIGURE 12.12 The solid inner core is separated from the mantle by the liquid outer core, and moves independently. Slight variations in the travel times of seismic waves through the core, measured over many decades, suggest that the inner core actually rotates faster than the mantle. The reason for this is not yet understood.

iron began to crystallize at the center to form the solid inner core. Even today, the inner core continues to grow as the planet cools.

Separated from the mantle by the liquid outer core, the solid inner core is free to move independently. Recent studies suggest that the inner core is actually rotating faster than the crust and mantle, lapping them every few hundred years (Figure 12.12). The inner core's small size and great distance from the surface make it Earth's most difficult region to examine.

Discovering Boundaries: The Inner Core–Outer Core Boundary

The boundary between the solid inner core and liquid outer core was discovered in 1936 by Danish seismologist Inge Lehman. Unable to determine whether or not the inner core was actually solid, she used basic trigonometry to assert that some P waves were being strongly refracted (bent) by a sudden increase in seismic velocities at the inner core–outer core boundary. This is the opposite of what occurs to produce the P-wave shadow zone. When seismic velocities suddenly decrease, such as at the mantle–outer core boundary, waves bend toward

Earth's interior producing a shadow zone at the surface where no direct waves arrive. When seismic waves suddenly increase in velocity, as they do at the inner core–outer core boundary, waves are bent toward the surface which results in several P waves emerging at the same location. These waves can be bent enough to arrive within the P-wave shadow zone. This situation, shown in Figure 12.11A, confirms a distinct inner core.

EARTH'S TEMPERATURE

One way to describe the interior of a planet is to examine how temperature changes with depth. Deciphering Earth's temperature structure is important for determining the movements of rock within our planet. As you are probably aware, thermal energy flows from hotter regions toward colder regions. Earth is about 5500 °C at its center and 0 °C at its surface, so heat flows toward the surface.

The rate at which Earth is cooling can be estimated by determining the rate at which heat escapes Earth's surface—a mere 87 milliwatts per square meter. At this rate, it would require all the energy emitted from about 690 square meters, roughly the size of a baseball diamond, to power one 60-watt light bulb. However, because Earth's surface is so large, heat leaves at a rate about three times greater than the world rate of energy consumption.

Figure 12.13 illustrates that heat does not leave Earth's surface at the same rate in all locations. Heat flow is highest near mid-ocean ridges where hot magma is consistently rising toward the surface. The rate of heat flow is also high in continental regions where the rocks are enriched in radioactive isotopes. Heat flow is lowest in the deep abyssal plains, which are areas of old, cold oceanic seafloor.

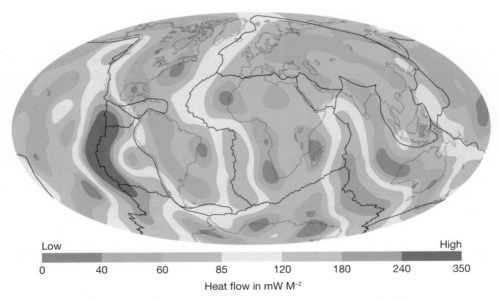

Low   High

| 0 | 40 | 60 | 85 | 120 | 180 | 240 | 350 |

Heat flow in mW M^{-2}

FIGURE 12.13 A map of the rate of heat flow out of Earth as it gradually cools over time, measured in milliwatts per square meter. Earth loses most of its heat near mid-ocean ridges, where magma rises toward the surface to fill the cracks formed when tectonic plates pull apart. Continents lose heat faster than old oceanic seafloor because they contain higher amounts of heat-producing radioactive isotopes.

How did Earth Get So Hot?

Like all planets in our solar system, Earth has experienced two thermal stages. The first stage occurred during Earth's formation and lasted about 50 million years—a relatively short time span in geologic terms—and involved a rapid increase in internal temperature. The second stage has been the very slow process of cooling down that has taken the remaining 4.5 billion years of Earth history.

Several factors contributed to the early increase in temperature. As discussed in Chapter 1, Earth formed through a very violent process involving the collisions of countless planetesimals ("baby planets") during the birth of our solar system. With each collision, the kinetic energy of motion was converted into thermal energy. As the early Earth grew in size, its temperature rapidly increased. Our young planet also contained many short-lived radioactive isotopes, such as aluminum-26 and calcium-41. As these isotopes decayed to stable forms, they released a great deal of energy, called *radiogenic heat.*

Another significant event that heated our planet was the collision of a Mars-sized object with Earth that led to the formation of the Moon. At this time the entire core, and most, if not all, of the mantle was molten. From that point, about 4.5 billion years ago, to the present, Earth has gradually cooled.

If Earth's only source of heat was from its early formation and the decay of short-lived radioactive isotopes, our planet would have cooled to a frozen cinder long ago. However, the

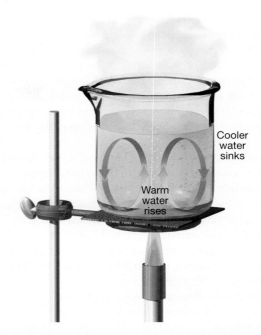

FIGURE 12.15 A simple example of convection, which is heat transfer that involves the actual movement of a substance. Here the flame warms the water in the bottom of the beaker. This heated water expands, becomes less dense (more buoyant), and rises. Simultaneously, the cooler, denser water near the top sinks.

mantle and crust also contain long-lived radioactive isotopes that keep our planet cooking as if on a slow burner. As shown in Table 9.1 (p. 270), the half-lives of the four main isotopes—uranium-235, uranium-238, thorium-232, and potassium-40—are billions of years long. Therefore, radioactivity plays two vital roles in geology. It provides the means for determining the ages of rocks, as discussed in Chapter 9. More importantly, however, it is the source of radiogenic heat that has kept mantle convection and plate tectonics active for billions of years.

Heat Flow

Heat travels from Earth's interior to space via three different mechanisms: *radiation, convection,* and *conduction.* The layers where these mechanisms contribute significantly to the outward flow of thermal energy are shown in Figure 12.14. Only two of these processes, convection and conduction, operate within Earth's interior.

Convection The transfer of heat by moving material in a fluid-like manner in which hot materials displace those that are cooler (or vice-versa) is called **convection.** It is the primary means of heat transfer within Earth. You are familiar with convection if you have ever watched boiling water in a pot. The water appears to be rolling—rising up in the middle of the pot, then down the sides (Figure 12.15). This pattern, called a *convection cycle,* occurs within Earth's mantle and outer core, and possibly within the inner core as well.

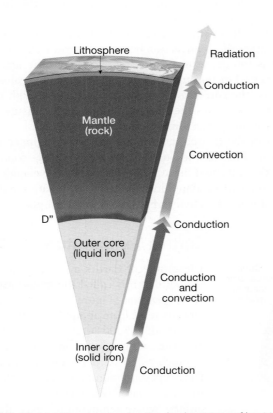

FIGURE 12.14 Diagram showing the dominant type of heat transfer at various depths within Earth as the planet cools. Earth ultimately loses its heat to space through radiation. However, heat travels from Earth's interior to the surface, through the processes of convection and conduction.

Convection occurs because of several factors—thermal expansion, gravity induced buoyancy, and fluidity. When water at the bottom of a pot is heated, it expands and rises, replacing the cooler, denser water at the top. Gravity is the driving force for convection. Hot water is less dense (more buoyant) than cold water, so it rises while the cold water sinks to take its place. If you tried to boil water in outer space, with no strong gravity present, you would find that the pot of water would not convect.

Materials must also be weak enough to flow. Scientists usually measure a material's fluidity in terms of its *resistance* to flow, called its **viscosity.** Water flows easily and has a low viscosity. The liquid iron of Earth's outer core likely has a viscosity similar to water, and convects easily as well. Materials that are viscous do not flow as easily as water, but can still flow. For example, catsup flows although it is thousands of times more viscous than water. Rock in the lower mantle is 10 trillion trillion (10^{25}) times more viscous than water, but yet it flows.

Temperature differences between the top and bottom of a convection cycle determine how vigorous the convection will be. Earth's surface is extremely cold compared to the interior, so newly formed oceanic lithosphere cools rapidly (Figure 12.16). This causes oceanic lithosphere to contract, become denser and heavier, and, in time, it sinks back into the mantle at subduction zones. As these cold sinking slabs descend, they absorb heat in the process. Some of this crustal material eventually becomes warm and buoyant enough to return to the surface. Oceanic lithosphere can therefore be thought of as the top of the mantle convection cycle, whereas the warm buoyant rocks called mantle plumes are the upward flowing arms of this convection cycle (Figure 12.16).

Convection sometimes occurs when differences in density arise through chemical rather than thermal means. *Chemical convection* is an important mechanism operating in the outer core. As iron crystallizes and sinks to form the solid inner core, it leaves behind a molten material that contains a higher percentage of lighter elements. Because this liquid is more buoyant than the surrounding iron-rich material, it rises and contributes to convective flow in the outer core.

Conduction   The flow of heat *through a material* is called **conduction.** Heat conducts in two ways: (1) through the collisions of atoms and (2) through the flow of electrons. In rocks, atoms are locked in place but are constantly oscillating. If one side of a rock is heated, the atoms here will oscillate more energetically. This increases the intensity of the collisions with neighboring atoms, and like a domino effect, the energy will slowly propagate through the rock. Conduction occurs much more quickly in metals than rocky substances. Although the atoms in metals are also locked in place, some of their electrons are free to move through the material, and these electrons can carry heat quickly from one side of a metallic object to another.

Materials conduct at vastly different rates. For example, heat conducts about 40,000 times more easily through a diamond than through air. Most rocks are poor conductors of heat, so conduction is not an efficient way to move heat through most of Earth. However, it is an important mechanism in places such as the lithosphere, the D″ layer, and the core.

Heat Flow in Earth's Interior   The dominant types of heat transfer in Earth's interior are illustrated in Figure 12.14. Conduction is thought to be the most important process in the solid inner core. When heat conducts from the inner core to the outer core, convection begins to play a more significant role in carrying heat to the top of the core.

The transfer of energy from the core to the mantle is by conduction rather than convection because the iron-rich material of the core is too dense to intrude (convect) into the less dense mantle rocks. For thermal energy to leave the core, it must conduct across the core–mantle boundary and up through the D″ layer. Once it reaches the lower mantle, thermal energy is carried toward the surface through mantle convection.

Most of the thermal energy that reaches the upper mantle makes its final journey to the surface by slowly conducting across the stiff, rigid lithosphere. The remaining energy is carried to the surface along divergent plate boundaries and other sites of volcanic activity where molten rock erupts.

Earth's Temperature Profile

The profile of Earth's average temperature at each depth is called the **geothermal gradient** or **geotherm** (Figure 12.17A). Temperatures increase from about 0 °C at the surface to more than 5000 °C at the center. Within the crust, temperatures increase rapidly—as much as 30 °C per kilometer of depth. The deepest diamond mines in South Africa are more than 3 kilometers below Earth's surface, where temperatures exceed 50 °C (120 °F). However, temperature increases do not continue at this rapid rate; otherwise, our planet would be molten below a depth of 100 kilometers.

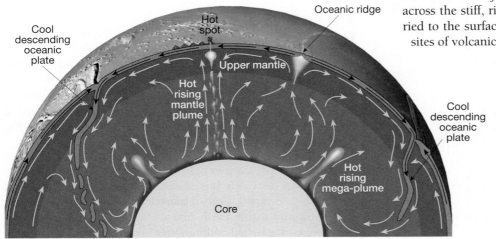

FIGURE 12.16   Convection within Earth's mantle. The entire mantle is in motion, driven by the sinking of cold oceanic lithosphere back into the deep mantle. This is like stirring a pot of stew with downward strokes of a spoon. The upward flow of rock likely occurs through a combination of mantle plumes and a broad return flow of rock to replace the ocean lithosphere that leaves the surface at subduction zones.

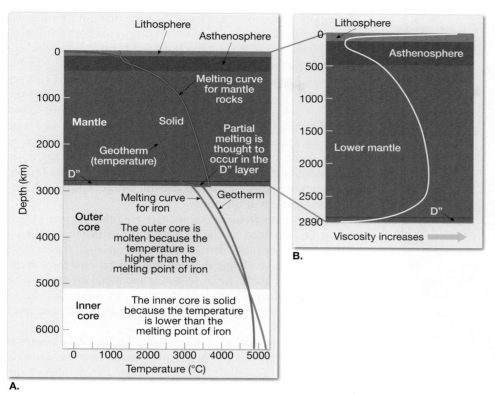

A.

FIGURE 12.17 These graphs illustrate how the viscosity of Earth materials at various depths is related to Earth's geotherm and their melting points. **A.** Earth's temperature profile with depth, or *geotherm*. Note that Earth's temperature increases gradually in most places. Within Earth's two major thermal boundary layers, the lithosphere and the D″ layer at the core–mantle boundary, temperature increases rapidly over short distances. Also shown is the melting point curve for the materials (rock or metal) found at various depths. Where the geotherm crosses above (to the right of) the melting point curve, as in the outer core, the material is molten. **B.** This graph shows how viscosity (resistance to flow) changes with depth from Earth's surface to the bottom of the mantle. High viscosities, as in the crust and lithosphere, show rock that is stiffer and flows less easily. If you compare these two figures, you can see that rocks are weakest and flow more easily at depths where the temperatures of rocks are close to melting (the asthenosphere and D″ layer).

At the base of the lithosphere, about 100 kilometers down, the temperature is roughly 1400 °C. You would need to descend to the bottom of the mantle before the temperature doubled to 2800 °C. For most of the mantle, the temperature increases very slowly—about 0.3 °C per kilometer. The exception to this pattern is the D″ layer, which acts as a thermal boundary, where temperatures increase by more than 1000 °C from top to bottom. Through the outer and inner cores, temperatures increase gradually. Determining temperatures deep within Earth is difficult, and uncertainties remain. The temperature at Earth's center has been estimated to be as high as 8000 °C. Mineral physics experiments have been useful in establishing the geotherm shown in Figure 12.17A. For example, high pressure experiments are used to determine the temperature at which liquid and solid iron would coexist at the boundary between the inner and outer core.

Also plotted in Figure 12.17A is the curve for the melting point of material at each depth. The melting point curves generally increase gradually with depth as a result of the continual increase in pressure. Squeezing a material makes it more difficult to melt because liquids usually take up more volume than solids. As a result, high pressures result in higher melting temperatures.

Considered together, the geotherm and melting point curve are valuable tools for investigating the behavior of Earth's materials. In layers where the geotherm (temperature at depth) is greater than the melting temperature, material is molten, a situation that occurs in the outer core (see Figure 12.17A). The relationship between the geotherm and the melting temperature determines not only whether or not a material is molten, but also indicates its stiffness, or viscosity. Notice how viscosity is directly related to the proximity of the geotherm curves to the melting point curves in Figure 12.17A. When rock approaches its melting point, it begins to soften and weaken. Figure 12.17B shows the viscosity of material in the crust and mantle. High-viscosity regions, like the lithosphere, are very stiff. Low-viscosity regions, such as the asthenosphere and D″, are much weaker.

Most of the lower mantle is very stiff, so rock moves sluggishly (Figure 12.17B). Researchers have determined that convective flow is several times slower in the lower mantle than in the upper mantle. However, at the very base of the mantle, temperature increases rapidly with depth. Rock in the D″ layer is relatively weak, flows more easily, and may experience some melting.

In the core, temperature increases much more slowly than the pressure. From the core–mantle boundary to Earth's center, the temperature increases by only about 40 percent, or from 4000 to 5500 °C. Pressure, however, nearly triples, increasing from 1.36 to 3.64 megabars. Although iron is cooler in the outer core than it is in the inner core, it remains a liquid because it is under less pressure. Conversely, iron in the inner core remains solid because it is under extreme pressure.

In summary, *Earth's different layers behave according to their particular conditions. The lithosphere is stiff because its temperature is much colder than its melting temperature. The asthenosphere is weaker and softer because it is very close to its melting temperature, and partial melting likely occurs in some places.*

EARTH'S THREE-DIMENSIONAL STRUCTURE

As you have seen, Earth is not perfectly layered. At the surface a variety of compositional and structural differences exist: oceans, continents, mountains, valleys, trenches, and mid-ocean ridges. Geophysical observations show that horizontal variations are not limited to the surface—they also occur within Earth, and are directly related to the process of mantle

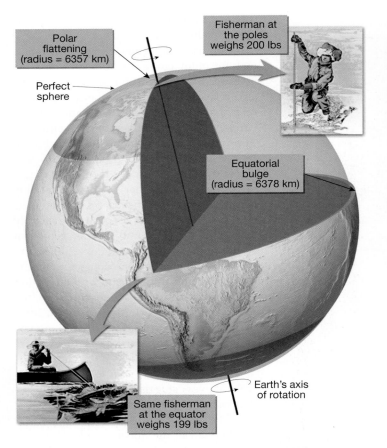

FIGURE 12.18 Because Earth rotates, it bulges at the equator and flattens at the poles. The combination of Earth's elliptical shape and its daily rotation actually cause the force of gravity to be slightly weaker at the equator than at the poles. This difference is large enough to be measured on a bathroom scale. Imagine two fishermen of equal mass both standing at sea level. If the one at the North Pole weighs 200 pounds, the one at the equator would weigh only 199 pounds.

of rotation. In a manner similar to the force that throws you sideways in a vehicle going too quickly around a curve or corner, centrifugal force acts to throw objects outward at the equator, where the force is greatest.

Earth's rotation also affects its shape, with the equator slightly further from Earth's center (6378 kilometers) than the poles (6357 kilometers) (Figure 12.18). Earth, therefore, is not a perfect sphere, but instead bulges at the equator—a shape called an *oblate ellipsoid*. This difference causes the force of gravity to be slightly weaker at the equator than at the poles because gravitational attraction is less when objects are further apart. In fact, your weight at the equator will actually be 0.5 percent less than it is at the poles.

Gravity measurements show there are other variations that cannot be explained by Earth's rotation. For instance, when a large body of unusually dense rock is underground, the increase in mass will cause a larger than average gravitational force at the surface directly above. Because metals and metal ores tend to be much denser than silicate rocks, local *gravity anomalies* (differences from the expected) have long been used to help prospect for ore deposits.

A map of regional gravity anomalies for the United States is shown in Figure 12.19. A narrow *positive gravity anomaly* (stronger than expected) that runs down the middle of the

FIGURE 12.19 Gravity anomalies beneath the continental United States. Changing elevation changes the strength of Earth's gravity, so values are calculated for what would be measured if you were at sea level at each location. This allows the gravity anomalies to be compared across the map. The negative anomalies (blue) beneath the Rockies and Appalachians show us that the crust has deep roots beneath the mountains there. The negative anomaly (blue) in the Basin and Range Province is the result of hotter, tectonically active crust (rifting and volcanoes). The narrow positive anomaly (red) that runs in a line down the middle of the country is the mid-continent rift, where denser volcanic rocks entered the crust more than a billion years ago.

convection and plate tectonics. Three-dimensional structures within Earth have been identified by studying variations in Earth's gravitational and magnetic fields and imaging called *seismic tomography*.

Earth's Gravity

Earth's rotation is the most significant cause for the differences in the force of gravity observed at the surface. Because Earth rotates around its axis once every day, the acceleration due to gravity* is less at the equator (9.78 m/s^2) than at the poles (9.83 m/s^2). Two reasons account for this phenomenon. Earth's rotation causes a centrifugal force that is in proportion to the distance from the axis

---

*The force of gravity causes objects, such as an apple, to accelerate as it falls to the ground, hence the expression "acceleration due to gravity."

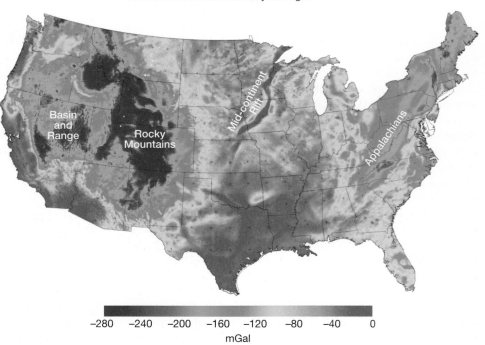

−280   −240   −200   −160   −120   −80   −40   0

mGal

country is the mid-continent rift (red), where thick, dense volcanic rocks filled a rupture in the crust more than a billion years ago. The *negative gravity anomaly* (blue) in the Basin and Range Province is the result of hot, low-density crust that has been stretched and thinned as it was intruded by buoyant magma bodies.

Some large-scale differences in density deep beneath the surface have also been detected using satellites. These gravity anomalies are a result of the large upwellings and downwellings of mantle convection. Areas of upwelling are associated with hot mantle plumes, whereas downwelling occurs where cold, oceanic slabs descend into the mantle.

Seismic Tomography

The three-dimensional changes in composition and density that are detected with gravity measurements can also be viewed using seismology. The technique, called **seismic tomography,** involves collecting signals from many different earthquakes recorded at many seismograph stations, in order to "see" all parts of Earth's interior. Seismic tomography is similar to medical tomography, in which doctors use techniques such as CT scans to make three-dimensional images of humans' internal structures.

Seismic tomography identifies regions where P or S waves travel faster or slower than average for a particular depth. These *seismic velocity anomalies* are then interpreted as variations in material properties such as temperature, composition, mineral phase, or water content. For instance, increasing the temperature of rock about 100 °C can decrease S-wave velocities about 1 percent, so images from seismic tomography are often interpreted in terms of temperature variations.

A seismic tomography cross-section for the mantle centered beneath North America is shown in Figure 12.20. Regions where waves travel slower than average (negative anomalies) are red, and regions where waves travel faster than average (positive anomalies) are blue. Significant patterns can be observed in this diagram. For example, the lithosphere located beneath the interiors of North America and Africa exhibits faster seismic velocities than oceanic lithosphere, because it is older and has been cooling for billions of years. Seismic imaging also shows that continental lithosphere (deep blue areas) can be quite thick, extending more than 300 kilometers into the mantle. Conversely, oceanic ridges such as the Mid-Atlantic Ridge, exhibit slow seismic velocities because they are extremely hot (see Figure 12.20).

In the mantle beneath North America, you can see a sloping zone of fast seismic velocities (blue/green) representing a sheet of descending oceanic lithosphere known as the Farallon

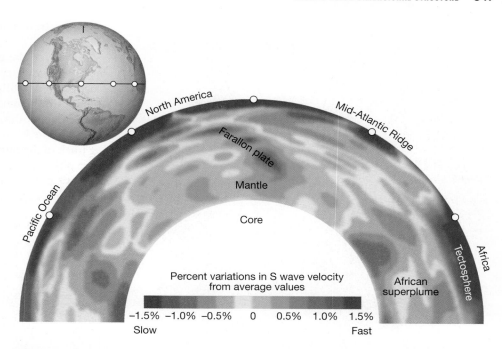

FIGURE 12.20   A seismic tomographic slice through Earth showing mantle structure. Colors show variations in the speed of S waves from their average values. Older portions of continents such as eastern North America and Africa are cold and stiff, so their blue colors show fast S wave speeds. The western United States is hotter and tectonically active, making that portion of the continent warmer and weaker, which slows S waves. The large blue structure extending far below North America is a sheet of cold, dense, ancient Pacific seafloor that is sinking toward the base of the mantle. The large orange structures beneath western Africa and the Pacific Ocean are thought to be megaplumes of warm material that are rising toward the surface.

plate. The segment of this former slab of seafloor seen in Figure 12.20 is currently sinking and warming as it moves toward the core–mantle boundary. Given enough time, this slab will become hot and buoyant enough to begin to rise back to the surface.

The large region of slow seismic velocities beneath Africa (the large reddish orange region at the lower right of Figure 12.20) is called the *African superplume*—a region of upward flow in the mantle. These slow velocities are likely due to both unusually high temperatures and rock that is highly enriched in iron. The rising rock cannot easily break through the thick African crust, so it seems to be deflected to both sides of the continent, perhaps supplying magma to both the Mid-Atlantic and Indian Ocean spreading centers.

Images from seismic tomography, such as the one in Figure 12.20, reveal many features associated with mantle convection. Sheets of cold, ancient, ocean sea floor sink to the base of the mantle, where they warm, expand, and rise toward the surface again.

Earth's Magnetic Field

Convection of liquid iron in the outer core is vigorous, which makes the outer core *appear* uniform when viewed with seismic waves. In *reality,* however, patterns of flow in the outer core create variations in Earth's magnetic field which are measurable at the surface, and tell a different story.

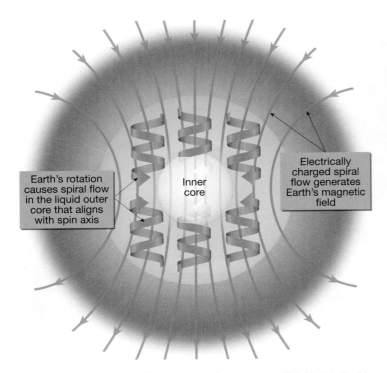

FIGURE 12.21 Illustration of convection patterns within Earth's liquid iron outer core that could give rise to the magnetic field we measure at the surface. It is thought that convection takes the form of cylindrical gyres of rotating molten iron that are aligned in the direction of Earth's axis of rotation.

Convective flow in the outer core is thought to be driven by three main mechanisms:

1. As heat conducts out of the core into the overlying mantle, the material in the outermost core cools, becomes denser, and sinks. This is a form of top-down, thermally driven convection.
2. Crystallization of solid iron near the bottom of the outer core, sinks to form the inner core, and leaves behind fluids that are depleted in iron. As this buoyant fluid rises, away from the inner core boundary, it drives a type of chemical convection.
3. There are also thought to be radioactive isotopes, such as potassium-40, within the core that could provide additional heat to drive thermal convection.

Geoscientists have not yet determined the relative importance of each of these mechanisms.

The Geodynamo As fluid in the core rises, its path becomes twisted because of Earth's rotation. As a result, the fluid moves in spiraling columns that align with Earth's axis of rotation (Figure 12.21). Because the iron-rich fluid is electrically charged, it generates a magnetic field similar to an electromagnet—a phenomenon called a **geodynamo.** If a wire is wrapped around an iron nail and an electric current passes through it, the nail generates a magnetic field that resembles the one that surrounds a bar magnet (Figure 12.22A, B). This type of magnetic field is called a *dipolar field* because it has two poles (a north and south magnetic pole). The magnetic field that

emanates from Earth's outer core has the same dipolar form (Figure 12.22).

However, the convection in the outer core is considerably more complex (see Box 12.2). More than 90 percent of Earth's magnetic field is dipolar, but the remainder is the result of more complicated patterns of convection. In addition, some of the features of Earth's magnetic field change over time. For centuries, sailors have used compasses to navigate, primarily keeping track of the direction in which compass needles point. From these observations, it was determined that the positions of the magnetic poles gradually change. Understanding this phenomenon requires an examination of how the magnetic field is measured.

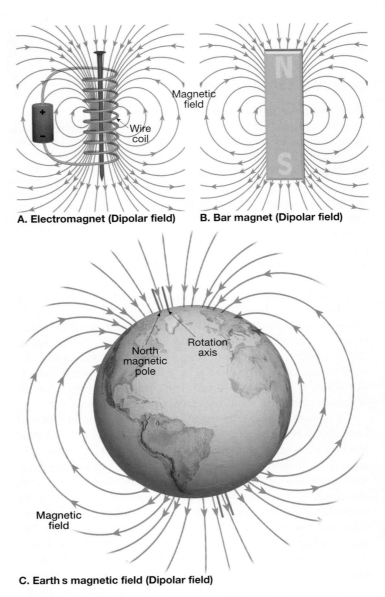

A. Electromagnet (Dipolar field)   **B. Bar magnet (Dipolar field)**

C. Earth s magnetic field (Dipolar field)

FIGURE 12.22 Demonstration of the similarity of Earth's magnetic field to that of an electromagnet **(A),** which consists of an electrical current passed through a coil of wire, or bar magnet **(B).** While it was once thought that Earth's core acts like a large bar magnet, scientists now think that Earth's magnetic field **(C)** is more like an electromagnet, and that the cylinders of spiraling liquid iron shown in Figure 12.21 behave like the coil of current passing through the wires of an electromagnet.

Anywhere on Earth's surface, the direction of the magnetic field is measured with two angles, called *declination* and *inclination*. The declination measures the direction to the magnetic north pole with respect to the direction to the geographic North Pole (Earth's axis of rotation). The inclination measures the downward tilt of the magnetic lines of force at any location—what a compass would show if tilted on its side. At the magnetic north pole the field points directly downward, while at the equator it is horizontal (Figure 12.23). In the central United States it tilts downward at an intermediate angle.

Recent studies have shown that the locations of the magnetic poles change significantly over time. Earth's magnetic north pole was previously located in Canada, but moved northward into the Arctic Ocean during the past decade. Currently, it is moving rapidly toward the north geographic pole at a rate of about 20 kilometers per year (Figure 12.24). The process is not symmetrical. Though the magnetic north pole has been moving towards the geographic North Pole, the magnetic south pole has been moving *away* from the geographic South Pole, passing from Antarctica to the Pacific Ocean.

Magnetic Reversals   Although the pattern of convection in the core changes over time, causing the magnetic poles to move, the locations of the magnetic poles averaged over thousands of years align with Earth's axis of rotation (geographic poles). One major exception occurs during periods of magnetic field reversals. At apparently random times, Earth's magnetic field reverses polarity so the *north* needle on a compass would point *south*. (The importance of these reversals in the study of paleomagnetism was

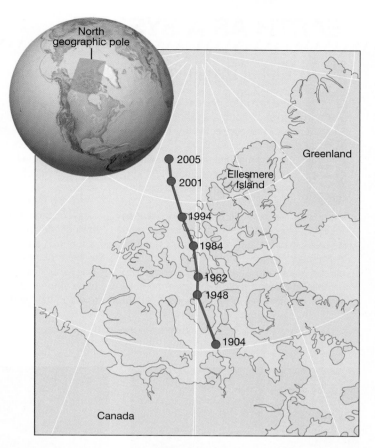

FIGURE 12.24   Map showing the change in measured locations of the north magnetic pole over time. The patterns of convection within the outer core change fast enough that we can see the magnetic field change significantly over our lifetimes.

described in Chapter 2). During a reversal, the strength of the magnetic field decreases to about 10 percent of normal and the locations of the poles begin to wander, going so far as to cross the equator (Figure 12.25). When the strength of the magnetic field returns to normal levels, the field is regenerated with reverse polarity. The entire process takes only a few thousand years.

The rate at which the magnetic field reverses is evidence that convection patterns in the outer core change over relatively short time spans. This complex process is now being modeled using high-speed computers (Figure 12.25). In addition, Figure 12.25 illustrates how the magnetic field lines twist in a complex manner before returning to a more uniform, simpler dipolar pattern.

The existence of magnetic reversals has been extremely important to geoscientists in providing the foundation for the theory of plate tectonics. However, magnetic reversals have potentially harmful consequences for Earth's land dwellers. An atmospheric magnetic layer, known as the *magnetosphere*, surrounds our planet and protects Earth's surface from bombardment by ionized particles, called *solar wind*, emitted by the Sun. If the strength of the magnetic field decreases significantly during a reversal, the increased amounts of ionized particles reaching Earth's surface could cause health hazards for humans and other life forms.

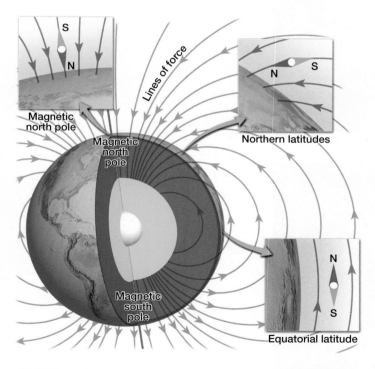

FIGURE 12.23   This drawing shows the direction of the magnetic field at different locations along Earth's surface. Though a compass measures only the horizontal direction of the magnetic field (the declination), at most locations the field also dips in or out of the surface at a variable angle (inclination).

EARTH AS A SYSTEM

Global Dynamic Connections

The layers of planet Earth are not isolated from one another, rather they are connected by their thermally driven motions. These connected motions do not necessarily interact in a predictable, steady manner—rather, they tend to occur episodically or in pulses. One example is the possible connection between magnetic reversals, hot-spot volcanism, and the breakup of the supercontinent of Pangaea.

Pangaea began to break up about 200 million years ago, and the accompanying increase in plate motion resulted in the subduction of large amounts of seafloor. About 80 million years later, the core reversal process shut down, preventing Earth's magnetic field from reversing for 35 million years. During this period, several enormous outpourings of lava have been linked to the arrival of new hot-spot mantle plumes at the surface.

In one hypothesis, these three events are closely connected. In the 80 million years following the breakup of Pangaea, a large amount of subducted lithosphere is thought to have plunged to the base of the mantle. This would have displaced hot rock at the base of the mantle, causing much of it to rise toward the surface and erupt as flood basalts. India's Deccan Traps are an example of such a process. At the same time the sudden infusion of comparatively cold ocean lithosphere next to the hot core at the core–mantle boundary would have chilled the uppermost core. This increased the temperature gradient—hotter at the bottom, cooler at the top—in the outer core. The result was more vigorous convection that prevented the magnetic field from weakening and reversing. This hypothesis, if deemed accurate, is an important reminder that Earth is a complex, churning, pulsing planet that is active in a variety of geological functions.

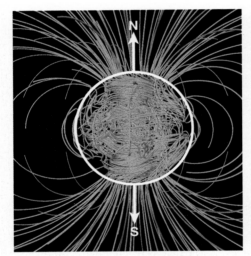

A. Normal orientation of magnetic field

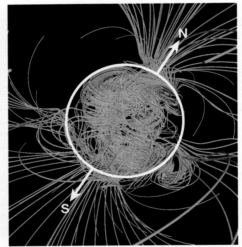

B. Magnetic field weakens and poles begin to wander

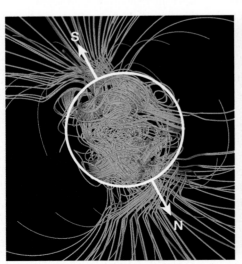

C. Poles wander across the equator

D. Reversal complete with north pole pointing south

FIGURE 12.25 Computer simulations showing how Earth's magnetic field could reverse direction. The white circle represents the core-mantle boundary and the arrows point to the north (N) and south (S) magnetic poles respectively. During a reversal the strength of the magnetic field weakens and the poles begin to wander greatly, going so far as to cross the equator. When the strength of the field returns to normal levels, the field is regenerated with reverse polarity.

CHAPTER 12 EARTH'S INTERIOR IN REVIEW

- Earth is layered with the densest materials at the center and lightest materials forming the outer layer. This layering is a result of gravity, and is similar for all planets. Earth's layers consist of the *inner core* (solid iron), *outer core* (liquid iron), *mantle* (dense rock), *crust* (low-density rock), *ocean* (water), and *atmosphere* (gas). Within layers, the density of materials increases with depth due to compression resulting from increasing pressure. Within the mantle, increases in density also occur because of mineral phase changes.

- Because it is impossible to drill deep into Earth, *seismic waves* are used to probe Earth's interior. The patterns of seismic waves are complicated because their behavior is influenced by different structures inside the planet before returning to the surface. Seismic waves travel faster through cold rock and slower through hot rock. Seismic waves reflect off layers composed of different materials. The results of seismic imaging of Earth's interior can be interpreted through comparison with mineral physics experiments. These experiments recreate temperatures and pressures within Earth, and allow scientists to see what rocks and metals are like at various depths.

- *Oceanic crust* and *continental crust* are very different. Oceanic crust is created at mid-ocean ridges, and is similar in composition and thickness throughout. Continental crust is highly variable, has many different compositions, and is formed in a variety of ways. Oceanic crust averages about 7 kilometers thick, but continents can be thicker than 70 kilometers. The boundary between the crust and mantle is called the *Moho*.

- The *mantle* comprises about 82 percent of Earth's volume. The *upper mantle* extends from the Moho to a depth of 660 kilometers. The upper mantle includes the lower part of the stiff lithosphere, the weak asthenosphere, and the transition zone. The lower mantle extends from 660 kilometers down to the core–mantle boundary, 2891 kilometers beneath the surface. At the base of the lower mantle is the complex D″ layer.

- The core is largely an alloy of iron and nickel that contains about 15 percent lighter elements. Because iron is dense, the core makes up one-third of Earth's mass, and iron is Earth's most abundant element by mass. The solid inner core grows over time as Earth cools.

- Earth's temperature increases from about 0 °C at the surface to roughly 5500 °C at the center (though the exact temperature is unknown). Heat flows unevenly from Earth's interior, with most of the heat loss occurring along the oceanic ridge system. Earth became very hot early in its history, largely due to the impacts of planetesimals and heat released by radioactive decay. Since then, Earth has slowly cooled. Earth is still geologically active because of the radiogenic heat supplied by long-lived radioactive isotopes.

- Heat flows from Earth's hot interior to its surface, and does so primarily by convection and conduction. *Convection* transfers heat through the movement of material. *Conduction* transfers heat by collisions between atoms or the motions of electrons. Convection is very important within Earth's mantle and outer core, whereas conduction is most important in the inner core, lithosphere, and D″. Within the asthenosphere and the base of D″, the temperature is close enough to the melting point that some partial melting may occur and the rock is weak enough to flow more easily than elsewhere in the mantle.

- Rotation causes Earth's shape to take the form of an *oblate ellipsoid,* meaning its equator bulges slightly. The combination of Earth's rotation and its ellipsoidal shape cause gravity to vary slightly—from the equator to the poles. Gravity also varies around Earth's surface due to the presence of rocks of different densities.

- Three-dimensional images of structure variations within the mantle are made from large numbers of seismic records using *seismic tomography.* These images show cold subducted oceanic lithosphere sinking to the base of the mantle and large superplumes of hot rock rising from the core-mantle boundary. The motions are evidence that convection occurs throughout the mantle.

- Convection of the liquid iron in the outer core causes a magnetic *geodynamo* that is responsible for Earth's magnetic field. Convection takes the form of spiraling cylinders that align with Earth's rotational axis. The magnetic field is primarily dipolar; that is, it resembles the field from a bar magnet or electromagnet. The patterns of convection in the outer core change rapidly enough that the magnetic field varies noticeably over our lifetimes.

- The magnetic field randomly reverses, with the north and south poles swapping positions. A reversal takes only a few thousand years and involves a significant decrease in the strength of the dipolar field. This is important because the magnetic field creates a magnetosphere around Earth that protects our planet from much of the Sun's solar wind that would otherwise bombard it. If the magnetosphere weakens, life on land would be adversely affected.

KEY TERMS

asthenosphere (p. 339)
conduction (p. 344)
convection (p. 343)
core (p. 341)
crust (p. 337)
D″ layer (p. 340)

geodynamo (p. 348)
geothermal gradient,
   or geotherm (p. 344)
inner core (p. 341)
lithosphere (p. 339)
lithospheric mantle (p. 339)

lower mantle (p. 340)
mantle (p. 339)
Moho (p. 337)
outer core (p. 341)
seismic tomography
  (p. 347)

transition zone (p. 339)
upper mantle (p. 339)
viscosity (p. 344)

QUESTIONS FOR REVIEW

1. What role does gravity play in the layering of planets?

2. What are the two major reasons for the increase in density with depth within Earth's mantle?

3. Why is seismology responsible for gasoline prices being more affordable than they might be without seismology?

4. List three ways that oceanic crust and continental crust differ. Where is the thickest crust found? The thinnest crust?

5. What is the importance of determining the *cross-over* distance? (See Figure 12.8.)

6. How do S waves "tell us" the mantle is solid?

7. If there were a lot of water in Earth's mantle, in what layer would it most likely reside?

8. What mineral phase changes occur at the top and bottom of the transition zone?

9. What layer of Earth has the greatest volume?

10. How is the D″ layer similar to the lithosphere?

11. True or False: No seismic waves arrive in the *shadow zone*? Explain.

12. Why is Earth's core one-sixth of Earth's volume but one-third of its mass?

13. Describe how Earth's inner core grows in size.

14. Why is heat flow from Earth's surface not evenly distributed?

15. Describe the heat sources that caused Earth to get very hot early in its history?

16. What prevents Earth from being a cold, motionless sphere of totally solid rock and metal?

17. Distinguish between conduction and convection.

18. Why is convection an inefficient means of heat transfer in materials with high viscosity?

19. Why is conduction more important than convection within Earth's crust?

20. What happens to rock as the geotherm approaches its melting temperature?

21. What happens to rock in regions where the geotherm crosses above the melting point curve?

22. Why would tectonic plates have a hard time moving if it were not for the existence of the asthenosphere?

23. Why is the lithosphere stiffer than the asthenosphere?

24. Earth once rotated much faster than it currently does. How would Earth's shape have been different in the past?

25. Would you expect to find a large layer of iron ore underground in a region with a positive or negative gravity anomaly? Explain.

26. Why does the mid-Atlantic ridge appear as a slow seismic velocity anomaly in Figure 12.20?

27. What are the three driving forces of convection in the outer core?

28. What occurs during a magnetic reversal?

29. Why might a magnetic reversal be dangerous to humans?

COMPANION WEBSITE

The *Earth 10e* Web site uses the resources and flexibility of the Internet to aid in your study of the topics in this chapter. Written and developed by the authors and other geology instructors, this site will help improve your understanding of geology. Visit www.mygeoscienceplace.com in order to:

- **Review** key chapter concepts.

- **Read** with links to the eBook and to chapter-specific web resources.

- **Visualize** and comprehend challenging topics using learning activities in *GEODe Earth*.

- **Test** yourself with online quizzes.

GEODe EARTH

GEODe Earth is a valuable and easy to use learning aid that can be accessed from your book's Companion Web site (www .mygeoscienceplace.com). It is a dynamic instructional tool that promotes understanding and reinforces important concepts by using tutorials, animations, and exercises that actively engage the student.

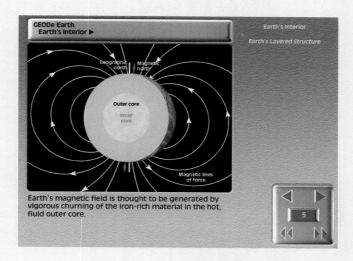

DIVERGENT BOUNDARIES

ORIGIN AND EVOLUTION OF THE OCEAN FLOOR

The Chikyu is a state of the art scientific drilling vessel.

(AP PHOTO/ITSUO INOUYE)

The ocean is Earth's most prominent feature, covering more than 70 percent of its surface. Yet, prior to 1940s, information about the ocean floor s extremely limited. Recall that Wegener's ntinental drift hypothesis was initially rejected the scientific community, in part because little s known about the ocean floor. Until the 20th ntury, weighted lines were used to measure ter depth, a task that took hours to perform and ld be wildly inaccurate, especially in deep water.

With the development of modern instruments our derstanding of the diverse topography of the ocean or improved dramatically. Particularly significant was e discovery of the global oceanic ridge system. This ad elevated landform, which stands 2 to 3 kilometers her than the adjacent deep-ocean basins, is the longest opographic feature on Earth.

Today we know that oceanic ridges mark divergent plate argins where new oceanic lithosphere is born. Oceano-aphic studies also discovered deep-ocean trenches, where eanic lithosphere descends into the mantle. Because the ocesses of plate tectonics continuously create oceanic crust at d-ocean ridges and consume it at subduction zones, oceanic ust is perpetually renewed and recycled.

In this chapter we will examine the topography of the ocean floor d look at the processes that generate its varied features. You will rn about the composition, structure, and origin of oceanic crust. In dition, you will examine those processes that recycle oceanic litho-here and consider how this activity causes Earth's landmasses to grate about the face of the globe.

AN EMERGING PICTURE OF THE OCEAN FLOOR

DIVERGENT BOUNDARIES
GEODe EARTH ▶ Mapping the Ocean Floor

If all water could be drained from the ocean basins, a great variety of features would be observed, including volcanic peaks, deep trenches, extensive plains, linear ridges, and large plateaus. In fact, the topography would be nearly as diverse as that on the continents.

Mapping the Seafloor

The complex nature of ocean-floor topography did not unfold until the historic three-and-a-half-year voyage of the HMS *Challenger* (Figure 13.1). From December 1872 to May 1876, the *Challenger* expedition made the first comprehensive study of the global ocean ever attempted. During the 127,500-kilometer (79,200-mile) voyage, the ship and its crew of scientists traveled to every ocean except the Arctic. Throughout the voyage, they sampled a multitude of ocean properties, including water depth, which was accomplished by laboriously lowering long weighted lines overboard. The knowledge gained by the *Challenger* of the ocean's great depth and varied topography expanded with the laying of transatlantic telegraph cables. A far better understanding of the seafloor emerged with the development of modern instruments that measure ocean depths. **Bathymetry** (*bathos* = depth, *metry* = measurement) is the measurement of ocean depths and the charting of the shape or topography of the ocean floor.

Modern Bathymetric Techniques Today, sound energy is used to measure water depths. The basic approach employs **sonar,** an acronym for *so*und *na*vigation and *r*anging. The first devices that used sound to measure water depth, called **echo sounders,** were developed early in the twentieth century. Echo sounders work by transmitting a sound wave (called a *ping*) into the water in order to produce an echo when it bounces off any object, such as a large marine organism or the ocean floor (Figure 13.2A). A sensitive receiver intercepts the reflected echo and a clock precisely measures the travel time to fractions of a second. By knowing the speed of sound waves in water—about 1500 meters (4900 feet) per second—and the time required for the energy pulse to reach the ocean floor and return, depth can be calculated. Depths determined from continuous monitoring of these echoes are plotted to obtain a profile of the ocean floor. By laboriously combining profiles, a chart of the seafloor was produced.

Following World War II, the U.S. Navy developed *sidescan sonar* to look for explosive devices that had been deployed in

FIGURE 13.1   The first systematic bathymetric measurements of the ocean were made aboard the HMS *Challenger* during its historic three-and-a-half-year voyage. Inset shows route of the HMS *Challenger,* which departed England in December of 1872 and returned in May 1876. (From C. W. Thompson and Sir John Murray, *Report on the Scientific Results of the Voyage of the HMS Challenger,* Vol. 1, Great Britain: Challenger Office, 1895, Plate 1. Library of Congress)

shipping lanes (Figure 13.2B). These torpedo-shaped instruments can be towed behind ships where they send out a fan of sound extending to either side of the ship's path. By combining swaths of sidescan sonar data, oceanographers produced the first photograph-like images of the seafloor. Although sidescan sonar provides valuable views of the seafloor, it does not provide bathymetric (water depth) data.

This drawback was resolved in the 1990s with the development of *high-resolution multibeam* instruments. These systems use hull-mounted sound sources that send out a fan of sound, then record reflections from the seafloor through a set of narrowly focused receivers aimed at different angles. Rather than obtaining the depth of a single point every few seconds, this technique allows a survey ship to map a swath of ocean floor tens of kilometers wide (Figure 13.3). In addition, these systems collect bathymetric data of such high resolution that they can distinguish depths that differ by less than a meter. When multibeam sonar is used to map sections of seafloor, the ship travels in a regularly spaced back-and-forth pattern known as "mowing the lawn."

Despite their greater efficiency and enhanced detail, research vessels equipped with multibeam sonar travel at a mere 10 to 20 kilometers (6 to 12 miles) per hour. It would take at least 100 vessels outfitted with this equipment hundreds of years to map the entire seafloor. This explains why only about 5 percent of the seafloor has been mapped in detail—and why large areas of the seafloor have not yet been mapped with sonar at all.

Seismic Reflection Profiles   Marine geologists are also interested in viewing the rock structure beneath the sediments that blanket much of the seafloor. This is accomplished by making a **seismic reflection profile.** To construct such a profile, strong, low-frequency sounds are produced by explosions (depth charges) or air guns. The sound waves penetrate the seafloor and reflect off the boundaries between rock layers and fault surfaces. Figure 13.4 shows a seismic profile of a portion of the Madeira abyssal plain in the eastern Atlantic. Although the seafloor is flat, the image allows us to see the irregular ocean crust buried by a thick accumulation of sediments.

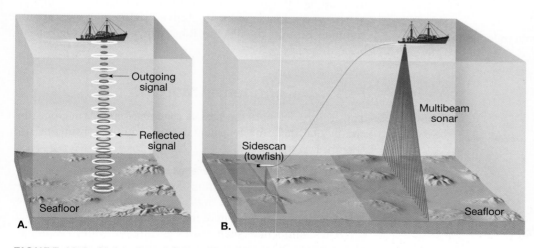

FIGURE 13.2   Various types of sonar. **A.** An echo sounder determines the water depth by measuring the time interval required for an acoustic wave to travel from a ship to the seafloor and back. The speed of sound in water is 1500 m/sec. Therefore, depth = $\frac{1}{2}$ (1500 m/sec × echo travel time). **B.** Modern multibeam sonar and sidescan sonar obtain an "image" of a narrow swath of seafloor every few seconds.

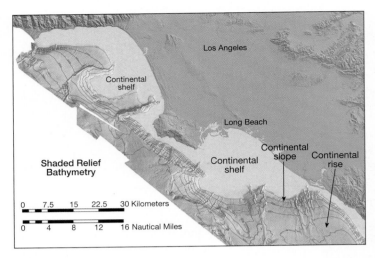

FIGURE 13.3   Color-enhanced perspective map of the seafloor and coastal landforms in the Los Angeles area of California. The ocean floor portion of this map was constructed from data collected using a high-resolution mapping system. (U.S. Geological Survey)

Viewing the Ocean Floor from Space

Another technological breakthrough that led to an enhanced understanding of the seafloor involves measuring the shape of the ocean surface from space. After compensating for waves, tides, currents, and atmospheric effects, it was discovered that the water's surface is not perfectly "flat." Because massive structures such as seamounts and ridges exert stronger than average gravitational attraction, they produce elevated areas on the ocean surface. Conversely, canyons and trenches create slight depressions.

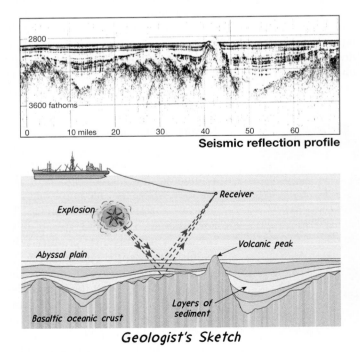

FIGURE 13.4   Seismic cross section and matching sketch across a portion of the Madeira abyssal plain in the eastern Atlantic Ocean, showing the irregular oceanic crust buried by sediments. (Image courtesy of Charles Hollister, Woods Hole Oceanographic Institution)

Satellites equipped with *radar altimeters* are able to measure subtle differences in sea level by bouncing microwaves off the sea surface (Figure 13.5). These devices can measure variations as small as a few centimeters. Such data have added greatly to our knowledge of ocean-floor topography. Combined with traditional sonar depth measurements, the data are used to produce detailed ocean-floor maps, such as the one in Figure 1.21 (p. 26).

Provinces of the Ocean Floor

Oceanographers studying the topography of the ocean floor identify three major areas: *continental margins, deep-ocean basins,* and *oceanic (mid-ocean) ridges.* The map in Figure 13.6 outlines these provinces for the North Atlantic Ocean, and the profile at the bottom shows the varied topography. Profiles of this type usually have their vertical dimension exaggerated many times (40 times in this case) to make topographic features more conspicuous. Vertical exaggeration, however, makes slopes appear *much* steeper than they actually are.

CONTINENTAL MARGINS

 DIVERGENT BOUNDARIES
**GEODe
EARTH**   ▸ Features of the Ocean Floor

Two types of **continental margins** have been identified—*passive* and *active.* Passive margins are found along most of the coastal areas that surround the Atlantic and Indian oceans, including the east coasts of North and South America, as well as the coastal areas of Europe and Africa. Passive margins consist of continental crust capped with weathered materials eroded from adjacent landmasses.

By contrast, active continental margins occur where oceanic lithosphere subducts into the mantle beneath the edge of a continent. As a result, active margins are narrow and consist of highly deformed sediments that were scraped from the descending lithospheric slab and plastered against the margin of the overriding continent. Active continental margins are common around the Pacific Rim, where they parallel deep-ocean trenches.

Passive Continental Margins

The features comprising **passive continental margins** include the continental shelf, the continental slope, and the continental rise (Figure 13.7).

Continental Shelf   The **continental shelf** is a gently sloping, submerged surface extending from the shoreline toward the deep-ocean basin. Because it is underlain by continental crust, it is clearly a flooded extension of the continents.

The continental shelf varies greatly in width. Although almost nonexistent along some continents, the shelf extends seaward more than 1500 kilometers (930 miles) along others. On average, the continental shelf is about 80 kilometers (50 miles) wide and its seaward edge is about 130 meters (425 feet) deep.

Studying the Deep Roots of Volcanic Arcs

I discovered geology the summer I worked doing trail maintenance in the North Cascade mountains of Washington State. I had just finished my freshman year in college and had never before studied Earth science. But a coworker (now my best friend) began to describe the geological features of the mountains that we were hiking in—the classic cone shape of Mount Baker volcano, the U-shaped glacial valleys, the advance of active glaciers, and other wonders. I was hooked and went back to college that fall with a pas-

> **"I was hooked and went back to college that fall with a passion for geology that hasn't abated."**

sion for geology that hasn't abated. As an undergraduate, I worked as a field assistant to a graduate student and did a senior thesis project on rocks from the Aleutian island arc.

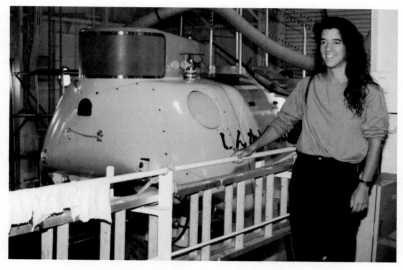

Susan DeBari photographed with the Japanese submersible, *Shinkai 6500,* which she used to collect rock samples from the Izu Bonin trench. (Photo courtesy of Susan DeBari)

From that initial spark, island arcs have remained my top research interest, on through Ph.D. research at Stanford University, postdoctoral work at the University of Hawaii, and as a faculty member at San Jose State University and Western Washington University. Of most interest was the deep crust of arcs, the material that lies close to the Mohorovičić discontinuity (fondly known as the Moho).

What kinds of processes are occurring down there at the base of the crust in island arcs? What is the source of magmas that make their way to the surface—the mantle, or the deep crust itself? How do these magmas interact with the crust as they make their way upward? What do these magmas look like chemically? Are they very different from what erupts at the surface?

Obviously, geologists cannot go down to the base of the crust (typically 20 to 40 kilometers beneath Earth's surface). So what they do is play a bit of a detective game. They must use rocks that are *now exposed at the surface* that were originally formed in the deep crust of an island arc. The rocks must have been brought to the surface rapidly along fault zones to preserve their original features. Thus, I can walk on rocks of the deep crust without really leaving Earth's surface! There are a few places around the world where these rare rocks are exposed. Some of the places that I have worked are the Chugach Mountains of Alaska, the Sierras Pampeanas of Argentina, the Karakorum Range in Pakistan, Vancouver Island's west coast, and the North Cascades of Washington. Fieldwork has most commonly involved hiking on foot, along with extensive use of mules and trucks.

I also went looking for exposed pieces of the deep crust of island arcs in a less obvious place—the Izu Bonin trench, one of the deepest oceanic trenches of the world. Here I dove into the ocean in a sub-

mersible called the *Shinkai 6500* (pictured to my right in the background). The *Shinkai 6500* is a Japanese submersible that has the capability to dive to 6500 meters below the surface of the ocean (approximately 4 miles). My plan was to take rock samples from the wall of the trench at its deepest levels using the submersible's mechanical arm. Because preliminary data suggested that vast amounts of rock were exposed for several kilometers in a vertical sense, this could be a great way to sample the deep arc basement. I dove in the submersible three times, reaching a maximum depth of 6497 meters. Each dive lasted nine hours, and was spent in a space no bigger than

> **"Each dive lasted nine hours, and was spent in a space no bigger than the front seat of a Honda, shared with two Japanese pilots . . ."**

the front seat of a Honda, shared with two Japanese pilots who controlled the submersible's movements. It was an exhilarating experience!

I am now on the faculty at Western Washington University, where I continue to do research on the deep roots of volcanic arcs, and get students involved as well. I am also involved in science education training for K–12 teachers, hoping to get young people motivated to ask questions about the fascinating world that surrounds them!

--- Susan DeBari

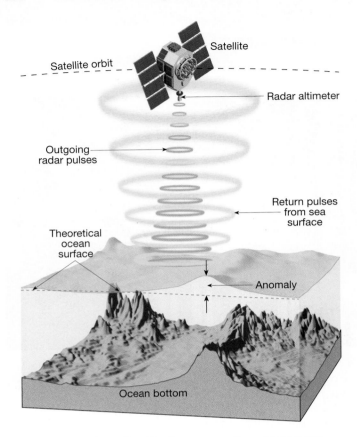

FIGURE 13.5 A satellite altimeter measures the variation in sea surface elevation, which is caused by gravitational attraction and mimics the shape of the seafloor. The sea surface anomaly is the difference between the measured and theoretical ocean surface.

As a result, the average inclination of the continental shelf is only about one-tenth of 1 degree, a slope so slight that it would appear to an observer to be a horizontal surface.

The continental shelf tends to be relatively featureless; however, some areas are mantled by extensive glacial deposits and thus are quite rugged. In addition, some continental shelves are dissected by large valleys running from the coastline into deeper waters. Many of these *shelf valleys* are the seaward extensions of river valleys on the adjacent landmass. They were eroded during the last Ice Age (Pleistocene epoch) when enormous quantities of water were stored in vast ice sheets on the continents causing sea levels to drop at least 100 meters (330 feet). Because of this drop, rivers extended their courses, and land-dwelling plants and animals migrated to the newly exposed portions of the continents. Dredging off the coast of North America has retrieved the ancient remains of numerous land dwellers, including mammoths, mastodons, and horses; further evidence that portions of the continental shelves were once above sea level.

Although continental shelves represent only 7.5 percent of the total ocean area, they have economic and political significance because they contain important mineral deposits and support important fishing grounds (Figure 13.8). According to the United Nations Convention on the Law of the Sea, all countries that ratified the treaty had until 2009 to claim any extension of their continental shelf beyond the normal 200 nautical miles. The extension could be no more than 350 nautical miles from land and required scientific evidence to show that the seafloor in question was, indeed, continental shelf. In 2001, Russia was the first country to submit a continental-shelf claim, based

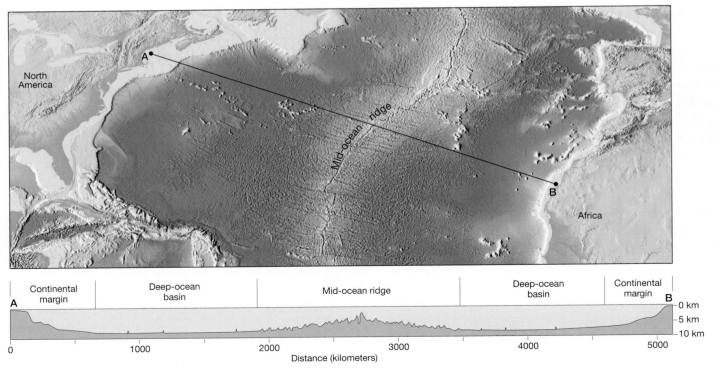

FIGURE 13.6 Major topographic divisions of the North Atlantic and a profile from New England to the coast of North Africa.

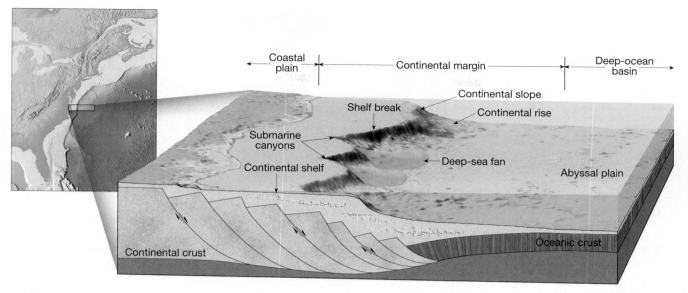

FIGURE 13.7 Schematic view showing the major features of a passive continental margin. Note that the slopes shown for the continental shelf and continental slope are greatly exaggerated. The continental shelf has an average slope of one-tenth of 1 degree, while the continental slope has an average slope of about 5 degrees.

mainly on known reservoirs of oil and natural gas. Many other nations have followed suit.

Continental Slope Marking the seaward edge of the continental shelf is the **continental slope,** a relatively steep structure that

FIGURE 13.8 Offshore drilling rigs are used to tap the oil and natural gas reserves of the continental shelf. This platform is in the North Sea. (Photo by Peter Bowater/Photo Researchers, Inc.)

marks the boundary between continental crust and oceanic crust. Although the inclination of the continental slope varies greatly from place to place, it averages about 5 degrees and in places exceeds 25 degrees.

Continental Rise The continental slope merges into a more gradual incline known as the **continental rise** that may extend seaward for hundreds of kilometers. The continental rise consists of a thick accumulation of sediment that has moved down the continental slope and onto deep-ocean floor. Most of the sediments are delivered to the seafloor by *turbidity currents* that periodically flow down *submarine canyons* (see Figure 7.26, p. 222). When these muddy slurries emerge from the mouth of a canyon onto the relatively flat ocean floor, they deposit sediment that forms a **deep-sea fan** (see Figure 13.7). As fans from adjacent submarine canyons grow, they merge laterally to produce a continuous wedge of sediment at the base of the continental slope forming the continental rise.

Active Continental Margins

Along active continental margins the continental shelf is very narrow, if it exists at all, and the continental slope descends abruptly into a deep-ocean trench. In these settings, the landward wall of a trench and the continental slope are essentially the same feature.

Active continental margins are located primarily around the Pacific Ocean in areas where oceanic lithosphere is being subducted beneath the leading edge of a continent (Figure 13.9). In such settings, sediments from the ocean floor and pieces of oceanic crust are scraped from the descending oceanic plate and plastered against the edge of the overriding continent. This chaotic accumulation of deformed sediment and scraps of oceanic crust is called an **accretionary wedge** (*ad* = toward, *crescere* = to grow). Prolonged plate subduction can produce massive accumulations of sediment along active continental margins.

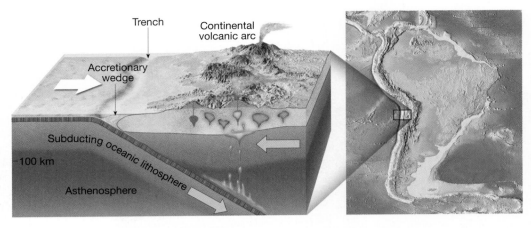

FIGURE 13.9   Active continental margin. Sediments from the ocean floor are scraped from the descending plate and added to the continental crust as an accretionary wedge.

Some active margins have little or no sediment accumulation, indicating that material is being carried into the mantle with the subducting plate. This tends to occur where old oceanic lithosphere is subducting nearly vertically into the mantle. In these locations the continental margin is very narrow, as the trench may lie a mere 50 kilometers (31 miles) offshore.

FEATURES OF DEEP-OCEAN BASINS

DIVERGENT BOUNDARIES
▸ Features of the Ocean Floor

Between the continental margin and the oceanic ridge lies the **deep-ocean basin** (see Figure 13.6). The size of this region—almost 30 percent of Earth's surface—is roughly comparable to the percentage of land above sea level. This region includes *deep-ocean trenches,* which are extremely deep linear depressions in the ocean floor; remarkably flat areas known as *abyssal plains;*

tall volcanic peaks called *seamounts* and *guyots;* and large flood basalt provinces called *oceanic plateaus.*

Deep-Ocean Trenches

Deep-ocean trenches are long, relatively narrow creases in the seafloor that represent the deepest parts of the ocean floor (Table 13.1). Most trenches are located along the margins of the Pacific Ocean (Figure 13.10), where many exceed 10 kilometers (6 miles) in depth. A portion of one trench—the Challenger Deep in the Mariana Trench—has been measured at 11,022 meters (36,163 feet) below sea level, making it the deepest known part of the world ocean. Two trenches are located in the Atlantic—the Puerto Rico Trench adjacent to the Lesser Antilles arc and the South Sandwich Trench.

Trenches are sites of plate convergence where slabs of oceanic lithosphere subduct and plunge back into the mantle. In addition to earthquakes being created as one plate "scrapes" against another, volcanic activity is also associated with these regions. Thus, trenches are often paralleled by an arc-shaped row of active volcanoes called a *volcanic island arc.* Furthermore, *continental volcanic arcs,* such as those making up portions of the Andes and Cascades, are located parallel to trenches that lie adjacent to continental margins. The volcanic activity associated with the trenches that surround the Pacific Ocean explains why this region is called the *Ring of Fire.*

Abyssal Plains

Abyssal plains (*a* = without, *byssus* = bottom) are deep, flat features—in fact, they are likely the most level places on Earth. The abyssal plain found off the coast of Argentina, for example,

TABLE 13.1 Dimensions of Some Deep-Ocean Trenches			
Trench	Depth (kilometers)	Average Width (kilometers)	Length (kilometers)
Aleutian	7.7	50	3700
Central America	6.7	40	2800
Japan	8.4	100	800
Java	7.5	80	4500
Kurile–Kamchatka	10.5	120	2200
Mariana	11.0	70	2550
Peru–Chile	8.1	100	5900
Philippine	10.5	60	1400
Puerto Rico	8.4	120	1550
South Sandwich	8.4	90	1450
Tonga	10.8	55	1400

Students Sometimes Ask ...

Have humans ever explored the deepest ocean trenches? Could anything live there?

Humans have indeed visited the deepest part of the oceans—where there is crushing high pressure, complete darkness, and near-freezing water temperatures—more than 50 years ago! In January 1960, U.S. Navy Lt. Don Walsh and explorer Jacques Piccard descended to the bottom of the Challenger Deep region of the Mariana Trench in the deep-diving bathyscaphe *Trieste.* At 9906 meters (32,500 feet), the men heard a loud cracking sound that shook the cabin. Although they were unable to see it, a 7.6-centimeter (3-inch) Plexiglas viewing port had cracked (miraculously, it held for the rest of the dive). More than five hours after leaving the surface, they reached the bottom at 10,912 meters (35,800 feet)—a record human descent that has not been broken since. They observed life forms adapted to the deep: a small flatfish, a shrimp, and some jellyfish.

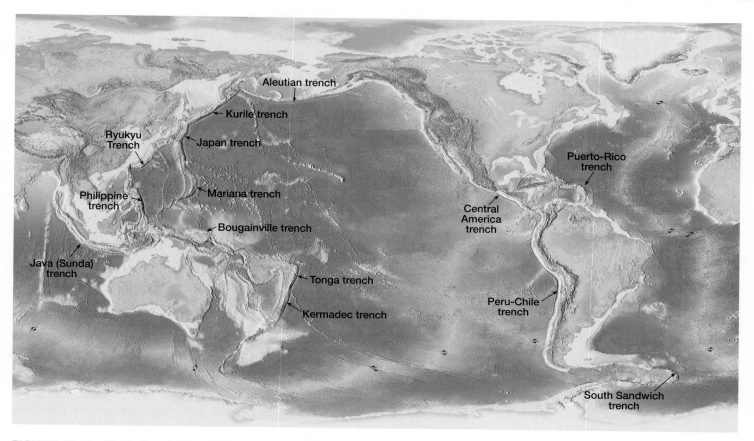

FIGURE 13.10   Distribution of the world's deep-ocean trenches.

has less than 3 meters (10 feet) of relief over a distance exceeding 1300 kilometers (800 miles). The monotonous topography of abyssal plains is occasionally interrupted by the protruding summit of a partially buried volcanic peak.

Using *seismic profilers,* researchers have determined that the relatively featureless topography of abyssal plains is due to thick accumulations of sediment that have buried an otherwise rugged ocean floor (see Figure 13.4). The nature of the sediment indicates that these plains consist primarily of fine sediments transported far out to sea by turbidity currents, deposits that have precipitated out of seawater, and shells and skeletons of microscopic marine organisms.

Abyssal plains are found in all oceans. However, the Atlantic Ocean has the most extensive abyssal plains because it has few trenches to act as traps for sediment carried down the continental slope.

Seamounts, Guyots, and Oceanic Plateaus

Dotting the ocean floor are submarine volcanoes called **seamounts,** which may rise hundreds of meters above the surrounding topography. It is estimated that more than a million exist. Some grow large enough to become oceanic islands, but most do not have a sufficiently long eruptive history to build a structure above sea level. Although seamounts are found on the floors of all the oceans, they are most common in the Pacific.

Some, like the Hawaiian Island–Emperor Seamount chain which stretches from the Hawaiian Islands to the Aleutian trench, form over volcanic hot spots in association with mantle plumes (see Figure 2.29 p. 65). Others are born near oceanic ridges. If the volcano is large enough before it is carried from the magma source by plate movement, the structure may emerge as an island. Examples in the Atlantic include Azores, Ascension, Tristan da Cunha, and St. Helena.

During the time they exist as islands, some of these volcanic structures are lowered to near sea level by the forces of weathering and erosion. In addition, islands gradually sink and disappear below the water surface as the moving plate slowly carries them away from the elevated oceanic ridge or hot spot where they originated (Box 13.1). Submerged, flat topped seamounts which formed in this manner are called **guyots** or **tablemounts.***

The ocean floor also contains several massive **oceanic plateaus,** which resemble flood basalt provinces on the continents. Oceanic plateaus, which in some cases are more than 30 kilometers thick, were generated from vast outpourings of fluid basaltic lavas. Some oceanic plateaus appear to have formed quickly in geologic terms. Examples include the Ontong Java which formed in less than 3 million years and the Kerguelen Plateau in 4.5 million years (Figure 13.11).

---

*The term *guyot* is named after Princeton University's first geology professor. It is pronounced "GEE-oh" with a hard *g* as in "give."

UNDERSTANDING EARTH

BOX 13.1

Explaining Coral Atolls—Darwin's Hypothesis

Coral *atolls* are ring-shaped structures that often extend from slightly above sea level to depths of several thousand meters (Figure 13.A). What causes atolls to form, and how do they attain such thicknesses?

Corals are tiny animals that generally appear in large numbers, that when linked form colonies. Most corals create a hard external skeleton made of calcium carbonate. Some build large calcium carbonate structures, called *reefs*, where new colonies grow atop the strong skeletons of previous colonies. Sponges and algae may attach to the reef, enlarging it further.*

Reef-building corals grow best in waters with an average annual temperature of about 24 °C (75 °F). They cannot survive prolonged exposure to temperatures below 18 °C (64 °F) or above 30 °C (86 °F). In addition, reef-builders require clear, sunlit water. Consequently, the depth of most active reef growth is limited to no more than about 45 meters (150 feet).

The strict environmental conditions required for coral growth create an interesting paradox: How can corals—which require warm, shallow, sunlit water no deeper than a few dozen meters—create thick structures such as coral atolls that extend to great depths?

---

*For more about coral reefs see Box 7.2, "Our Threatened Coral Reefs," p. 217.

FIGURE 13.A　An aerial view of Tetiaroa Atoll in the Pacific. The light blue waters of the relatively shallow lagoon contrast with the dark blue color of the deep ocean surrounding the atoll. (Photo by Douglas Peebles Photography)

The naturalist Charles Darwin was one of the first to formulate a hypothesis on the origin of ringed-shaped atolls. From 1831 to 1836 he sailed aboard the British ship HMS *Beagle* during its famous global circumnavigation. In various places that Darwin visited, he noticed a progression of stages in coral reef development from (1) a *fringing reef* along the margins of a volcano to (2) a *barrier reef* with a volcano in the middle to (3) an *atoll,* consisting of a continuous or broken ring of coral reef surrounding a central lagoon (Figure 13.B). The essence of Darwin's hypothesis, illus-

trated in Figure 13.B, was that as a volcanic island slowly sinks, corals continue to build the reef complex upward. During Darwin's time, however, there was no plausible mechanism to account for how an island might sink.

Currently, plate tectonics helps explain how volcanic islands become extinct and sink to great depths over long periods of time. Some volcanic islands form over a relatively stationary mantle plume, which causes the lithosphere to be buoyantly uplifted. Over a span of millions of years, these volcanic islands become inactive and gradually sink as the moving plate carries them away from the region of hot-spot volcanism (Figure 13.B).

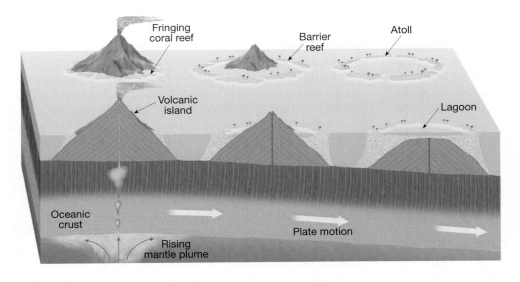

FIGURE 13.B　Formation of a coral atoll due to the gradual sinking of oceanic crust and upward growth of the coral reef. A fringing coral reef forms around an active volcanic island generated by a mantle plume. As the volcanic island moves away from the region of hotspot activity it sinks, and the fringing reef gradually becomes a barrier reef. Eventually, the volcano is completely submerged and a coral atoll remains.

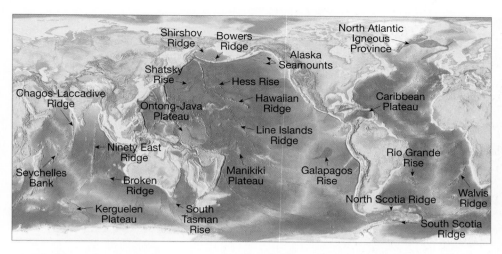

FIGURE 13.11 Distribution of oceanic plateaus, hot spot tracks, and other submerged crustal fragments.

ANATOMY OF THE OCEANIC RIDGE

DIVERGENT BOUNDARIES

▶ Oceanic Ridges and Seafloor Spreading

Along well-developed divergent plate boundaries, the seafloor is elevated, forming a broad linear swell called the **oceanic ridge,** or **mid-ocean ridge.** Our knowledge of the oceanic ridge system comes from soundings of the ocean floor, core samples from deep-sea drilling, visual inspection using deep-diving submersibles (Figure 13.12), and even firsthand inspection of slices of ocean floor that have been thrust onto dry land during continental collisions. At oceanic ridges we find extensive normal and strike-slip faulting, earthquakes, high heat flow, and volcanism.

The oceanic ridge system winds through all major oceans in a manner similar to the seam on a baseball, and is the longest topographic feature on Earth, exceeding 70,000 kilometers (43,000 miles) in length (Figure 13.13). The crest of the ridge typically stands 2 to 3 kilometers above the adjacent deep-ocean basins and marks the plate boundary where new oceanic crust is created.

Notice in Figure 13.13 that large sections of the oceanic ridge system have been named based on their locations within the various ocean basins. Some ridges run through the middle of ocean basins, where they are appropriately called *mid-ocean* ridges. The Mid-Atlantic Ridge and the Mid-Indian Ridge are examples. By contrast, the East Pacific Rise is *not* a "mid-ocean" feature. Rather, as its name implies, it is located in the eastern Pacific, far from the center of the ocean.

The term *ridge* is somewhat misleading, because these features are not narrow and steep as the term implies, but have widths of from 1000 to 4000 kilometers and the appearance of broad, elongated swells that exhibit varying degrees of ruggedness. Furthermore, the ridge system is broken into segments that range from a few tens to hundreds of kilometers in length. Each segment is offset from the adjacent segment by a transform fault.

Oceanic ridges are as high as some mountains on the continents; but the similarities end there. Whereas most mountain ranges on land form when the compressional forces associated with continental collisions fold and metamorphose thick sequences of sedimentary rocks, oceanic ridges form where upwelling from the mantle generates new oceanic crust. Oceanic ridges consist of layers and piles of newly formed basaltic rocks that are buoyantly uplifted by the hot mantle rocks from which they formed.

Along the axis of some segments of the oceanic ridge system are deep, down-faulted structures called **rift valleys** because of their striking similarity to the continental rift valleys found in East Africa (Figure 13.14). Some rift valleys, including those along the rugged Mid-Atlantic Ridge, are typically 30 to 50 kilometers wide and have walls that tower 500 to 2500 meters above the valley floor. This makes them comparable to the deepest and widest part of Arizona's Grand Canyon.

FIGURE 13.12 The deep-diving submersible *Alvin* is 7.6 meters long, weighs 16 tons, has a cruising speed of 1 knot, and can reach depths as great as 4000 meters. A pilot and two scientific observers are along during a normal 6- to 10-hour dive. (Courtesy of Rod Catanach/Woods Hole Oceanographic Institution)

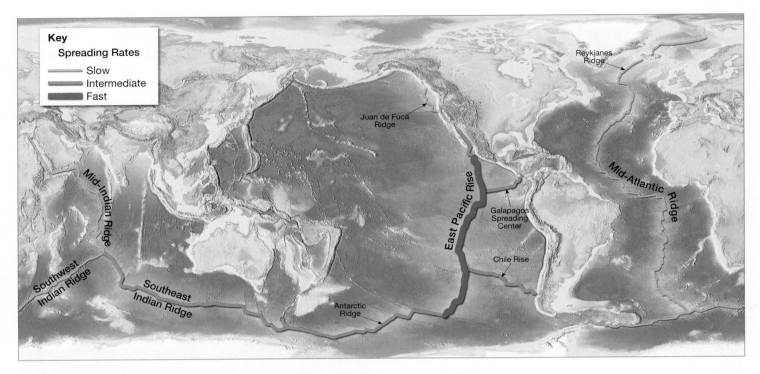

FIGURE 13.13   Distribution of the oceanic ridge system. The map shows ridge segments that exhibit slow, intermediate, and fast spreading rates.

OCEANIC RIDGES AND SEAFLOOR SPREADING

DIVERGENT BOUNDARIES

▶ Oceanic Ridges and Seafloor Spreading

The greatest volume of magma (more than 60 percent of Earth's total yearly output) is produced along the oceanic ridge system in association with seafloor spreading. As plates diverge, fractures created in the oceanic crust fill with molten rock that gradually wells up from the hot mantle below. This molten material slowly cools and crystallizes, producing new slivers of seafloor. This process repeats itself in episodic bursts, generating new lithosphere that moves away from the ridge crest in a conveyor belt fashion.

Seafloor Spreading

Harry Hess of Princeton University formulated the concept of seafloor spreading in the early 1960s. Later, geologists were able to verify Hess's view that seafloor spreading occurs along the crests of oceanic ridges where hot mantle rock rises to replace the material that has shifted horizontally. Recall from Chapter 4 that as rock rises it experiences a decrease in confining pressure that may lead to *decompression melting*. Partial melting of mantle rock produces basaltic magma that has a surprisingly consistent chemical composition. The newly

FIGURE 13.14   The axis of some segments of the oceanic ridge system contains deep downfaulted structures called *rift valleys* that may exceed 30–50 kilometers in width and from 500 to 2500 meters in depth.

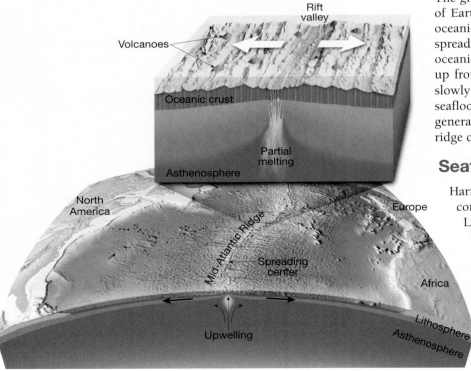

formed melt separates from the mantle rock and rises toward the surface. Along some ridge segments, the melt collects in small elongated reservoirs located just beneath the ridge crest. Eventually, about 10 percent migrates upward along fissures to erupt as lava flows on the ocean floor (see Figure 13.14). This activity continuously adds new basaltic rock to diverging plate margins, temporarily welding them together, only to be broken as spreading continues. Along some ridges, outpourings of bulbous lavas build submerged shield volcanoes (seamounts) as well as elongated lava ridges. At other locations, more voluminous lava flows create a relatively subdued topography.

Why are Oceanic Ridges Elevated?

The primary reason for the elevated position of the ridge system is that newly created oceanic lithosphere is hot and therefore less dense than cooler rocks of the deep-ocean basin. As the newly formed basaltic crust travels away from the ridge crest, it is cooled from above as seawater circulates through the pore spaces and fractures in the rock. In addition, it cools because it gets farther and farther from the zone of hot mantle upwelling. As a result, the lithosphere gradually cools, contracts, and becomes more dense. This thermal contraction accounts for the greater ocean depths that occur away from the ridge. It takes almost 80 million years of cooling and contraction for rock that was once part of an elevated ocean-ridge system to relocate to the deep-ocean basin.

As lithosphere is displaced away from the ridge crest, cooling also causes a gradual increase in lithospheric thickness. This happens because the boundary between the lithosphere and asthenosphere is a thermal (temperature) boundary. Recall that the lithosphere is Earth's cool, stiff, outer layer, whereas the asthenosphere is a comparatively hot and weak layer. As material in the uppermost asthenosphere ages (cools), it becomes stiff and rigid. Thus, the upper portion of the asthenosphere is gradually converted to lithosphere simply by cooling. Oceanic lithosphere continues to thicken until it is about 80–100 kilometers thick. Thereafter, its thickness remains relatively unchanged until it is subducted.

Spreading Rates and Ridge Topography

When researchers studied various segments of the oceanic ridge system, it was clear that there were topographic differences. Many appear to be controlled by spreading rates—which largely determine the amount of melt generated at a rift zone. At fast spreading centers more magma wells up from the mantle than at slow spreading centers. This difference in output causes differences in the structure and topography of various ridge segments.

Oceanic ridges that exhibit slow spreading rates from 1 to 5 centimeters per year have prominent rift valleys and rugged topography (Figure 13.15A). The Mid-Atlantic and Mid-Indian ridges are examples. The vertical displacement of large slabs of oceanic crust along normal faults is responsible for the steep walls of the rift valleys. Furthermore, volcanism produces numerous cones in the rift valley which enhance the rugged topography of the ridge crest.

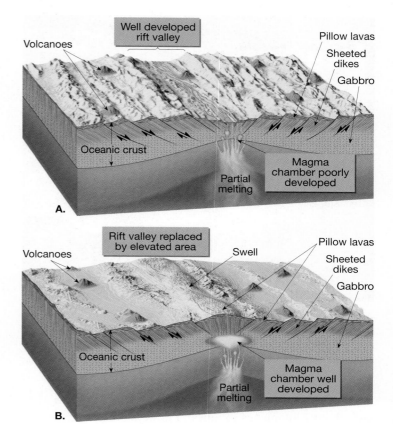

FIGURE 13.15 Topography of the crest of an oceanic ridge. **A.** At slow spreading rates a prominent rift valley develops along the ridge crest, and the topography is typically rugged. **B.** Along fast spreading centers no median rift valleys develop, and the topography is comparatively smooth.

By contrast, along the Galápagos ridge, an intermediate spreading rate of 5 to 9 centimeters per year is the norm. As a result, the rift valleys that develop are relatively shallow—often less than 200 meters deep. In addition, their topography is more subdued when compared to ridges that have slower spreading rates.

At fast spreading centers (greater than 9 centimeters per year), such as along much of the East Pacific Rise, rift valleys are generally absent (Figure 13.15B). Instead, the ridge axis is elevated. These elevated structures, called *swells*, are built from lava flows up to 10 meters (30 feet) thick that have incrementally paved the ridge crest with volcanic rocks (Figure 13.16). In addition, because the depth of the ocean depends largely on the age of the seafloor, ridge segments that exhibit faster spreading rates tend to have more gradual profiles than ridges that have slower spreading rates (Figure 13.17). Because of these differences in topography, the gently sloping, less rugged portions of fast spreading ridges are called *rises*.

THE NATURE OF OCEANIC CRUST

An interesting aspect of oceanic crust is that its thickness and structure are remarkably consistent throughout the entire ocean basin. Seismic soundings indicate that its thickness averages

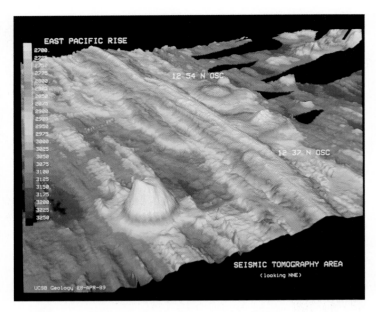

FIGURE 13.16   False-color sonar image of a segment of the East Pacific Rise. The linear pink area is the swell that formed above the ridge axis. Note also the large volcanic cone in the lower left portion of the image. (Courtesy of Dr. Ken C. Macdonald)

only about 7 kilometers (5 miles). Furthermore, it is composed almost entirely of mafic (basaltic) rocks that are underlain by a layer of the ultramafic rock peridotite, which forms the lithospheric mantle.

Although most oceanic crust forms out of view, far below sea level, geologists have been able to examine the structure of the ocean floor firsthand. In such locations as Newfoundland, Cyprus, Oman, and California, slivers of oceanic crust have

been thrust high above sea level. From these exposures, and core samples collected by deep-sea drilling ships, researchers conclude that the ocean crust consists of four distinct layers (Figure 13.18):

- Layer 1: The upper layer is a sequence of unconsolidated sediments. Sediments are very thin near the axes of oceanic ridges, but may be several kilometers thick next to continents.
- Layer 2: Below the layer of sediments is a rock unit composed mainly of basaltic lavas that contain abundant pillowlike structures called *pillow basalts*.
- Layer 3: The middle, rocky layer is made up of numerous interconnected dikes having a nearly vertical orientation, called the *sheeted dike complex*. These dikes are former pathways where magma rose to feed lava flows on the ocean floor.
- Layer 4: The lowest unit is mainly gabbro, the coarse-grained equivalent of basalt, which crystallized deeper in the crust without erupting.

This sequence of layers composing the oceanic crust is called an **ophiolite complex** (see Figure 13.18). From studies of various ophiolite complexes around the globe and related data, geologists have pieced together a scenario for the formation of the ocean floor.

How does Oceanic Crust Form?

The molten rock that goes into the making of new oceanic crust originates from partial melting of the mantle rock peridotite at depths greater than 40 kilometers. This process generates a melt having the composition of basalt, which is less dense than the

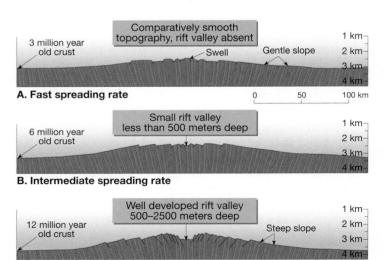

FIGURE 13.17   Schematic of ridge segments that exhibit fast, intermediate, and slow spreading rates. Fast spreading centers have gentle slopes and lack a rift valley. By contrast, ridges that have slow spreading rates have well-developed rift valleys and steep flanks. The slopes of all these profiles are greatly exaggerated.

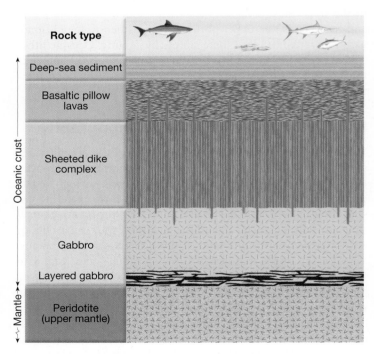

FIGURE 13.18   The four layers that make up a typical section of oceanic crust—based on data obtained from ophiolite complexes, seismic profiling, and core samples obtained from deep-sea drilling expeditions.

surrounding solid rock. The newly formed melt rises through the upper mantle along thousands of tiny conduits that feed into a few dozen larger, elongated channels, perhaps 100 meters (300 feet), or more wide. These structures, in turn, feed lens-shaped magma chambers located directly beneath the ridge crest. With the addition of melt from below, the pressure inside the chambers steadily increases. As a result, the rocks above these reservoirs periodically fracture, allowing the melt to ascend into the young oceanic crust above.

The molten rock surges upward along numerous vertical fractures that develop in the ocean crust. Some cools and solidifies to form dikes. New dikes intrude older dikes, which are still warm and weak, to form a **sheeted dike complex.** This portion of the oceanic crust is usually 1 to 2 kilometers thick.

Roughly 10 percent of the melt eventually erupts on the ocean floor. Because the surface of a submarine lava flow is chilled quickly by seawater, it generally travels no more than a few kilometers before completely solidifying. The forward motion occurs as lava accumulates behind the congealed margin and then breaks through. This process occurs repeatedly, as molten basalt is extruded—like toothpaste from a tightly squeezed tube (Figure 13.19). The result is protuberances resembling large bed pillows stacked one atop the other, hence the name **pillow basalts** (Figure 13.20).

In some settings, pillow lavas may build volcano-size mounds that resemble shield volcanoes, whereas in other situations they form elongated ridges tens of kilometers long. These structures are eventually separated from their supply of magma as they are carried away from the ridge crest by seafloor spreading.

The lowest unit of the ocean crust develops from crystallization within the central magma chamber itself. The first minerals to crystallize are olivine, pyroxene, and occasionally chromite

FIGURE 13.20 Pillow lava exposed along a sea cliff, Cape Wanbrow, New Zealand. Notice that each "pillow" shows an outer, rapidly cooled, dark glassy layer enclosing a dark gray basalt interior. (Photo by G.R. Roberts/Photo Researchers, Inc.)

(chromium oxide), which settle through the magma to form a layered zone near the floor of the reservoir. The remaining melt tends to cool along the walls of the chamber to form massive amounts of coarse-grained gabbro. This portion accounts for up to 5 of the 7 kilometers of ocean crust thickness.

Although molten rock rises continuously from the asthenosphere toward the surface, seafloor spreading occurs in pulse-like bursts. As the melt begins to accumulate in the lens-shaped reservoirs, it is blocked from continuing upward by the stiff overlying rocks. As the amount of melt entering the magma reservoirs increases, pressure rises. Periodically, the pressure exceeds the strength of the overlying rocks, which fracture and initiate a short episode of seafloor spreading.

Interactions Between Seawater and Oceanic Crust

In addition to serving as a mechanism for the dissipation of Earth's internal heat, the interaction between seawater and the newly formed basaltic crust alters both the seawater and the crust. The permeable and highly fractured lava of the upper oceanic crust allows seawater to penetrate to depths of 2 to 3 kilometers. As seawater circulates through the hot crust, it is heated and chemically reacts with the basaltic rock by a process called *hydrothermal* (hot water) *metamorphism* (see Chapter 8, p. 242). This alteration causes the dark silicates (olivine and pyroxene) in basalt to form new metamorphic minerals such as chlorite and serpentine. Simultaneously, the hot seawater dissolves ions of silica, iron, copper, and occasionally silver and gold from the hot basalts. When the water temperature reaches a few hundred degrees Celsius, these mineral-rich fluids buoyantly rise along fractures and eventually spew out on the ocean floor (Box 13.2).

Studies conducted by submersibles along the Juan de Fuca Ridge have photographed these metallic-rich solutions as they gushed from the seafloor to form particle-filled clouds called **black smokers.** As the hot liquid (up to 400 °C) mixes with the

FIGURE 13.19 A photograph taken from the *Alvin* during Project FAMOUS shows lava extrusions in the rift valley of the Mid-Atlantic Ridge. Large toothpastelike extrusions such as this were common features. A mechanical arm is sampling an adjacent blisterlike extrusion. (Photo courtesy of Woods Hole Oceanographic Institution)

EARTH AS A SYSTEM

BOX 13.2

Deep-Sea Hydrothermal Vents*

Sitting in a computer-filled, darkened room, a group of geologists, biologists, and chemists peer intently at video monitors showing astonishing imagery of giant, smoke-billowing, chimney-like rock formations and an abundance of bizarre animals. Is this a scene from Hollywood's latest sci-fi blockbuster? No, it's a typical scene from an actual research vessel located 250 kilometers southwest of Vancouver Island. The images are being relayed to the ship by the Canadian, remotely controlled vehicle *ROPOS (Remotely Operated Platform for Ocean Science),* which is busy working more than 2 kilometers below the surface along the Juan de Fuca Ridge. This seafloor mountain chain is actively being created by the rifting and pulling apart of the Pacific and Juan de Fuca tectonic plates and the production of new oceanic crust by upwelling magma.

Along ridges such as the Juan de Fuca, cold seawater circulates several hundreds of meters down into the highly fractured basaltic crust, where it is heated by magmatic sources. Along the way, the hot water strips metals and elements such as sulfur from surrounding rock. This heated fluid eventually becomes buoyant and rises along conduits and fractures along the ridges. When it reaches the surface of the crust,

the fluid can be more than 400 °C, but it does not boil because of the extremely high pressures exerted by the water column above the vents. When this hydrothermal fluid comes into contact with the much colder chemical-rich seawater, minerals rapidly precipitate to form shimmering smoke-like clouds called "*black smokers*" (Figure 13.C). Some minerals immediately solidify and contribute to the formation of spectacular chimney-like structures, which can be as tall as a 15-story building, and are appropriately given names like *Godzilla* and *Inferno.* In some cases, these chimneys and related deposits contain concentrated amounts of iron, copper, zinc, lead, silver, and occasionally gold.

The Juan de Fuca vents are also remarkable for the biology that they support. In these environments, completely devoid of sunlight, microorganisms utilize the mineral-rich hydrothermal fluid to perform chemosynthesis. The microbial communities, in turn, support larger, more complex animals such as fish, crabs, worms, mussels, and clams. Some species are found exclusively at these vents. The most famous of these, and perhaps the most unique, is the tubeworm (Figure 13.D). With their white chitinous tubes and bright red plumes, these conspicuous creatures rely entirely on bacteria growing in their *trophosome,* an internal organ designed for harvesting bacteria.

FIGURE 13.D Tube worms up to 3 meters (10 feet) in length are among the organisms found in the extreme environment of hydrothermal vents along the crest of the oceanic ridge, where sunlight is nonexistent. These organisms obtain their food from internal microscopic bacteria-like organisms, which acquire their nourishment and energy through the processes of chemosynthesis. (Photo by Al Giddings Images, Inc.)

These *symbiotic* bacteria rely on the tubeworm to provide them with a suitable habitat and, in return, they provide carbon-based building blocks to the tubeworms.

There is considerable concern over the possibility of damaging these unique ecosystems by sampling activities of scientists, increasing ecotourism, or the potential exploitation of biological and mineral resources. These issues have recently led the Canadian government to designate a portion of the Juan de Fuca hydrothermal vents as Canada's first Marine Protected Area (MPA).

FIGURE 13.C A black smoker spewing hot, mineral-rich water along the East Pacific Rise. As heated solutions meet cold seawater, sulfides of copper, iron, and zinc precipitate immediately, forming mounds of minerals around these vents. (Photo by Dudley Foster, Woods Hole Oceanographic Institution)

*This box is based on material prepared by Richard Leveille, a research scientist at the University of Quebec, Montreal, Canada.

cold, mineral-laden seawater, the dissolved minerals precipitate to form massive, metallic sulfide deposits, some of which are economically important. Occasionally these deposits grow upward to form underwater chimney-like structures equivalent in height to skyscrapers.

CONTINENTAL RIFTING: THE BIRTH OF A NEW OCEAN BASIN

The reason why Pangaea began to split apart nearly 200 million years ago is a subject still debated among geoscientists. Nevertheless, the event illustrates that ocean basins originate when continents break apart. This was, undoubtedly, the case for the Atlantic Ocean, which formed as the Americas drifted from Europe and Africa.

Evolution of an Ocean Basin

The opening of a new ocean basin begins with the formation of a **continental rift,** an elongated depression along which the entire lithosphere is stretched and thinned. Where the lithosphere is thick, cool, and strong, rifts tend to be narrow—often less than a few hundred kilometers wide. Modern examples of narrow continental rifts include the East African Rift, the Baikal Rift (south central Siberia), and the Rhine Valley (northwestern Europe). By contrast, where the crust is thin, hot, and weak, rifts can be more than 1000 kilometers wide, as exemplified by the Basin and Range Province in the western United States.

In those settings where rifting continues, the rift system evolves into a young, narrow ocean basin, such as the present-day Red Sea. Continued seafloor spreading eventually results in the formation of a mature ocean basin bordered by rifted continental margins. The Atlantic Ocean is such a feature. What follows is an overview of ocean basin evolution using modern examples to represent the various stages of rifting.

East African Rift The East African Rift is a continental rift that extends through eastern Africa for approximately 3000 kilometers (2000 miles). It consists of several interconnected rift valleys that split into eastern and western sections around Lake Victoria (Figure 13.21). Whether this rift will eventually develop into a spreading center, with the Somali subplate separating from the continent of Africa, is uncertain.

The most recent period of rifting began about 20 million years ago as upwelling in the mantle intruded the base of the lithosphere (Figure 13.22A). Buoyant uplifting of the heated lithosphere led to doming and stretching of the crust. Consequently, the upper crust was broken along high-angle normal faults, producing downfaulted

blocks, or *grabens,* while the lower crust deformed by ductile stretching (Figure 13.22B).

In the early stages of rifting, magma generated by decompression melting of the rising mantle rocks intruded the crust. Occasionally, some of the magma migrated upward along fractures and erupted at the surface. This activity produced extensive basaltic flows within the rift as well as volcanic cones—some forming more than 100 kilometers from the rift axis. Examples include Mount Kenya and Mount Kilimanjaro, the highest point in Africa, rising almost 6000 meters (20,000 feet) above the Serengeti Plain.

Red Sea Research suggests that if spreading continues, a rift valley will lengthen and deepen, eventually extending to the margin of the continent (Figure 13.22C). At this point, the continental rift becomes a narrow linear sea with an outlet to the ocean, similar to the Red Sea.

The Red Sea formed when the Arabian Peninsula rifted from Africa beginning about 30 million years ago. Steep fault scarps that rise as much as 3 kilometers above sea level flank the margins of this water body. Thus, the escarpments surrounding the Red Sea are similar to the steep cliffs that border the East African Rift. Although the Red Sea reaches oceanic depths (up to 5 kilometers) in only a few locations, symmetrical magnetic stripes indicate that typical seafloor spreading has been occurring for at least the past 5 million years.

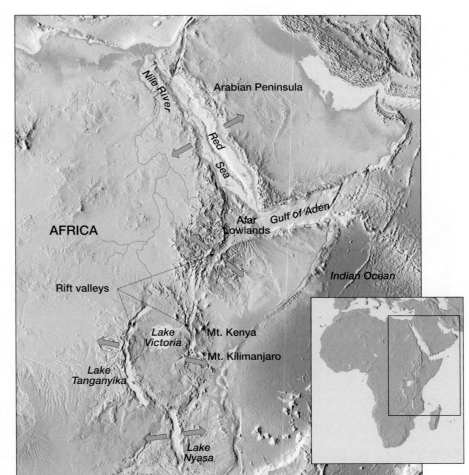

FIGURE 13.21 East African rift valleys and associated features.

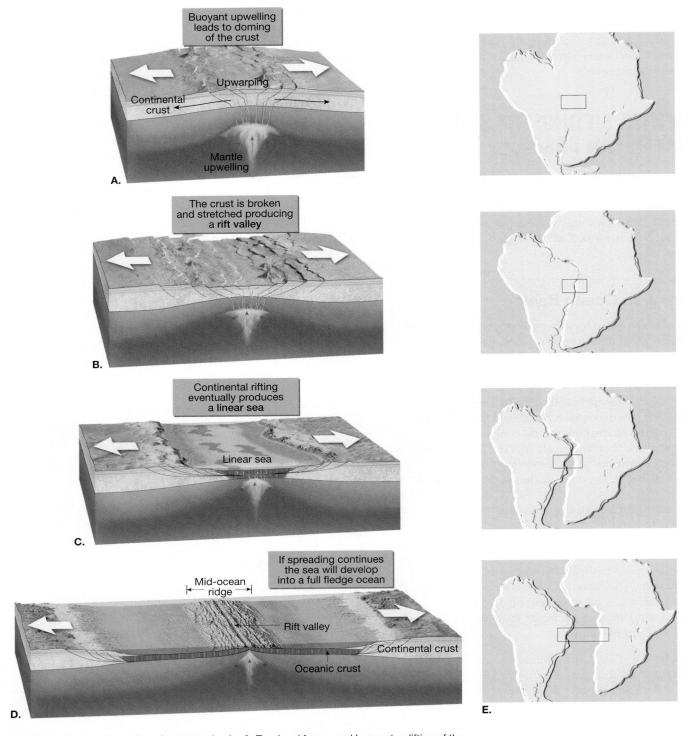

FIGURE 13.22 Formation of an ocean basin. **A.** Tensional forces and buoyant uplifting of the heated lithosphere cause the upper crust to be broken along normal faults, while the lower crust deforms by ductile stretching. **B.** As the crust is pulled apart, large slabs of rock sink, generating a rift zone. **C.** Further spreading generates a narrow sea. **D.** Eventually, an expansive ocean basin and ridge system are created. **E.** Illustration of the separation of South America and Africa to form the South Atlantic.

Atlantic Ocean   If spreading continues, the Red Sea will grow wider and develop an elevated oceanic ridge similar to the Mid-Atlantic Ridge (Figure 13.22D). As new oceanic crust is added to the diverging plates, the rifted continental margins gradually recede from the region of upwelling. As a result, they cool, contract, and sink.

Over time, continental margins subside below sea level and material eroded from the adjacent highlands blanket this once

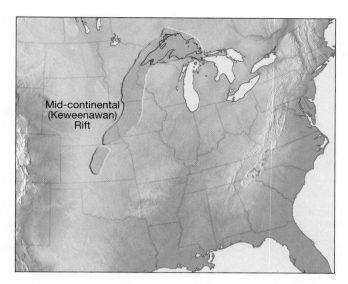

FIGURE 13.23 Map showing the location of a failed rift extending from Lake Superior to Kansas.

rugged topography. The result is a *passive continental margin* consisting of a rifted continental crust that has been covered by a thick wedge of relatively undisturbed sediment and sedimentary rock.

Not all continental rift valleys develop into full-fledged spreading centers. In the central United States a failed rift extends from Lake Superior into central Kansas (Figure 13.23). This once active rift valley is filled with sediments and volcanic rock that was extruded onto the crust more than a billion years ago. Why one rift valley develops into a full-fledged active spreading center while others are abandoned is not fully understood.

Mechanisms for Continental Rifting

Geologists know that supercontinents existed in the geologic past. Pangaea, the most recent of these, was assembled into a supercontinent between 450 and 230 million years ago, only to break up shortly after it formed. Based on numerous studies, geologists have concluded that the formation of supercontinents followed by continental splitting is an integral part of plate tectonics. It must involve major changes in the direction and nature of the forces that drive plate motion. In other words, over long periods of geologic time, the forces that drive plate motions tend to organize crustal fragments into a single supercontinent, only to change directions and disperse them again. Mechanisms that are thought to contribute to continental rifting include: plumes of hot mobile rock rising from deep in the mantle; upwelling from shallow levels in the asthenosphere; and forces that arise from plate motions.

Mantle Plumes and Hot-Spot Volcanism   Recall that a *mantle plume* consists of hotter than normal mantle rock that has a large mushroom-shaped head hundreds of kilometers in diameter attached to a long, narrow, trailing tail. As the plume head nears the base of the rigid lithosphere, it spreads laterally. Decompression melting within the plume generates huge

volumes of basaltic magma that rises and triggers *hot-spot volcanism* at the surface.

Research suggests that mantle plumes tend to concentrate beneath a supercontinent, because once assembled, a large landmass forms an insulating "blanket" that traps heat in the mantle. The resulting temperature increase leads to the formation of mantle plumes that serve to dissipate heat.

Evidence that mantle plumes play a role in the breakup of at least some landmasses can be observed in modern passive continental margins. In several regions on both sides of the Atlantic, continental rifting was preceded by crustal uplift and massive outpourings of basaltic lava. Examples include the Etendeka flood basalts of southwest Africa and the Paraná basalt province of South America (Figure 13.24A).

About 130 million years ago, when South America and Africa were a single landmass, vast outpourings of lava produced a large continental basalt plateau (Figure 13.24B). Next, the South Atlantic began to open, splitting the basalt province into two parts—the Etendeka and Paraná basalt plateaus. As the ocean basin grew, the tail of the plume produced a string of seamounts

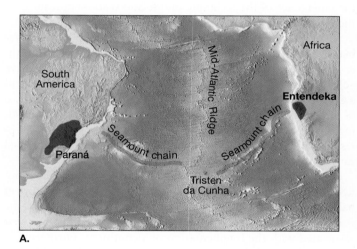

A.

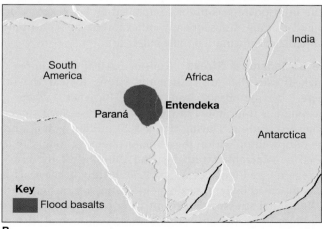

B.

FIGURE 13.24 Evidence for the role that mantle plumes might play in continental rifting. **A.** Relationship of the Paraná and Etendeka basalt plateaus to the Tristan da Cunha hotspot. **B.** Location of these flood-basalt plateaus 130 million years ago, just before the South Atlantic began to open.

on each side of the newly formed ridge (Figure 13.24A). The modern area of hot-spot activity is centered around the volcanic island of Tristan da Cunha, on the Mid-Atlantic Ridge.

When hot, buoyant mantle plumes reach the base of the lithosphere, they cause the overlying crust to dome and weaken. Doming, of perhaps as much as 1000 meters, tends to produce *three rifts* or *arms,* that join in the area above the rising plume—called a *triple junction.* Frequently, continental breakup and the formation of an ocean basin occur along two of the rift arms, whereas the third arm may be less developed, and constitute a failed rift that becomes filled with sediments. Figure 13.25 illustrates the location of a few mantle plumes that have generated large flood basalt plateaus and were presumably involved in the breakup of Pangaea.

One of the mantle plumes is currently located beneath Iceland near the crest of the Mid-Atlantic Ridge (Figure 13.26). Evidence of vast outpourings of basaltic lava, that began about 55 million years ago, occurs in eastern Greenland, as well as across the Atlantic in the Hebridean Islands of northern Scotland. The oldest magnetic stripes between Greenland and Europe are the same age, supporting the connection between the emergence of the Icelandic plume and seafloor spreading in the North Atlantic.

The Afar plume, which is associated with the split of the Arabian Peninsula from Africa, is located beneath a region of northeastern Ethiopia called the Afar Lowlands, an area of extensive volcanism (see Figure 13.21). This plume generated a typical rift system consisting of three arms that meet at a triple junction. Two of these rifts, the Red Sea and the Gulf of Aden, are active spreading centers. The third arm is the East African Rift, which may represent the initial stage in the breakup of a continent as described earlier, or it may be destined to become a failed rift.

It is important to note that hot-spot volcanism does not necessarily lead to rifting. For example, vast outpourings of basaltic lava that constitute the Columbia River basalts in the Pacific Northwest, as well as Russia's Siberian Traps, are not associated with the fragmentation of a continent. Furthermore, along some rifted continental margins stretching and thinning of the lithosphere was not accompanied by large-scale volcanism and melting. Consequently, other forces that contribute to continental fragmentation must exist.

Role of Tensional Stress Continental rifting requires tensional stresses that are sufficiently strong to tear the lithosphere. In the Basin and Range Province, where the lithosphere is thin, hot, and weak, small stresses are sufficient to cause spreading. During the last 20 million years, a broad zone of upwelling within the asthenosphere is thought to have caused considerable stretching and thinning of the crust in this region (see Figure 14.20, p. 397). Rifting in these settings is accompanied by large-scale melting and volcanism.

Tensional stresses that result from plate motions are also thought to be particularly significant in continental rifting. In settings where a continent is attached to a subducting slab of oceanic lithosphere, it will be pulled along by the descending slab. However, continents overlie thick sections of lithospheric mantle. As a result, they tend to resist being towed, which creates tensional stresses that may be sufficient to tear

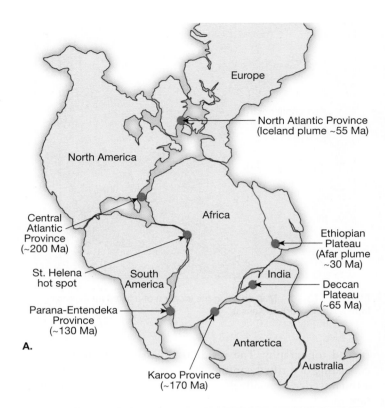

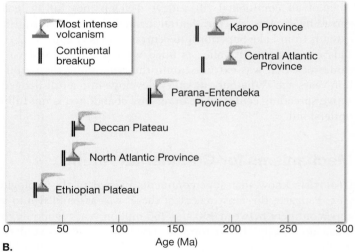

FIGURE 13.25 Reconstruction of Pangaea showing the locations of a few mantle plumes that have generated large basalt plateaus (provinces) and are thought to be involved in the fragmentation of that supercontinent. **A.** Approximate surface location of mantle plumes prior to the breakup of Pangaea. The location of the plume that produced the Central Atlantic Province is unknown and may have involved a superplume that was deflected by the unusually thick lithosphere beneath western Africa. The Central Atlantic Province includes lava flows, sills, and dikes in northeastern South America, northwestern Africa, southwestern Europe, and eastern North America. **B.** Timing of the breakup of Pangaea along various rift zones and the plume volcanism that was associated with each period of continental fragmentation. In most cases, volcanism appears to precede breakup by a few million years, or more. (Data after Courtillot, et al.)

the landmass. The location of zones of rifting in the fragmentation of a supercontinent may be influenced by a preexisting weakness, such as a suture zone that formed when the supercontinent was assembled.

In summary, continental rifting occurs when a landmass experiences tensional stresses that are sufficient to overcome the strength of the lithosphere. Thin, hot, and weak lithosphere requires less stress in order to be torn, whereas thick, cold, and strong lithosphere requires more. Plate motions that generate tensional stresses and hot mantle plumes that weaken and stretch the lithosphere are important mechanisms for rifting.

DESTRUCTION OF OCEANIC LITHOSPHERE

Although new lithosphere is continually being produced at divergent plate boundaries, Earth's surface area is not growing larger. In order to balance the amount of newly created lithosphere, there must be a process whereby plates are destroyed.

Why Oceanic Lithosphere Subducts

The process of plate subduction is complex, and the ultimate fate of oceanic lithosphere is still being debated. What is known with some certainty is that oceanic lithosphere will resist subduction unless its overall density is greater than that of the underlying mantle. It takes about 15 million years for a young slab of oceanic lithosphere to become cooler and denser than the supporting asthenosphere. In parts of the western Pacific, some oceanic lithosphere is nearly 180 million years old, the thickest and densest in today's oceans. The subducting slabs in this region typically descend into the mantle at angles approaching 90 degrees (Figure 13.27A). By contrast, when a spreading center is located near a subduction zone, the oceanic lithosphere is still young, and, therefore, warm and buoyant. In these settings, the slab's angle of descent is small (Figure 13.27B).

It is important to note that the *lithospheric mantle,* which makes up about 80 percent of the descending oceanic slab, drives subduction. Even when the *oceanic crust* is quite old, its density is still less than the underlying asthenosphere. Subduction, therefore, depends on lithospheric mantle which is colder and denser than the asthenosphere that supports it.

When an oceanic slab descends to about 400 kilometers, mineral phase changes (the transition from low-density mineral to high-density), enhance subduction (see Chapter 12). At this depth, the transition from olivine (low-density) to spinel (its high-density form) increases the density of the slab which helps pull the plate into the subduction zone.

At some locations, the oceanic crust is unusually thick and buoyant because it is capped by large outpourings of basaltic lava, or other thick crustal fragments. In these settings subduction may be modified, or even prevented. This appears

FIGURE 13.26 Volcanism associated with the Icelandic plume. Iceland's largest fishing port was damaged extensively in 1973 by the eruption of Heimaey Volcano. (Photo by Bettman/CORBIS)

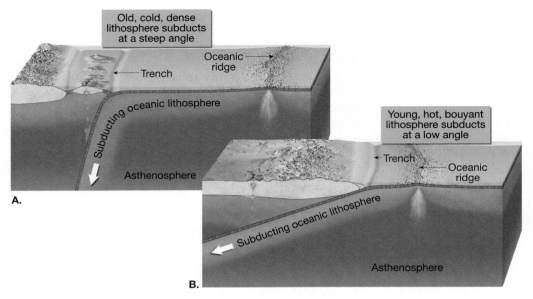

FIGURE 13.27 The angle at which oceanic lithosphere descends into the asthenosphere depends on its density. **A.** In parts of the Pacific, some oceanic lithosphere is nearly 180 million years old and typically descends into the mantle at angles approaching 90 degrees. **B.** Young oceanic lithosphere is warm and buoyant, hence it tends to subduct at a low angle.

It has also been determined that unusually thick units of oceanic crust, those that are greater than 30 kilometers in thickness, are not likely to subduct. The Ontong Java Plateau, for example, is a thick oceanic plateau, about the size of Alaska, located in the western Pacific. About 20 million years ago this plateau reached the trench that forms the boundary between the subducting Pacific plate and the overriding Australian–Indian plate. Apparently too buoyant to subduct, the Ontong Java Plateau clogged the trench. We will consider the fate of crustal fragments that are too buoyant to subduct in the next chapter.

Subducting Plates: The Demise of an Ocean Basin

to be the situation in two areas along the Peru–Chile Trench, where the angle of descent is shallow—about 10 to 15 degrees. Low dip angles often result in a strong interaction between the descending slab and the overriding plate. Consequently, the regions near the Peru–Chile Trench experience frequent, great earthquakes.

In the 1970s, geologists began using magnetic stripes and fracture zones on the ocean floor to reconstruct the last 200 million years of plate movement. This research showed that parts of, or even entire, ocean basins have been destroyed along subduction zones. For example, during the breakup of Pangaea shown in Figure 2.B (p. 63), notice that the African plate moves northward eventually colliding with Eurasia. During this event, the floor of the inter-

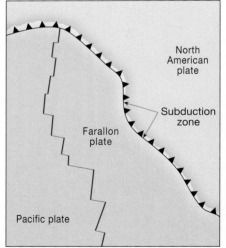

A. 56 million years ago

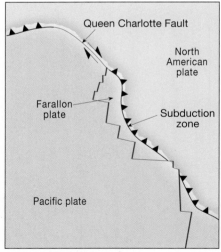

B. 37 million years ago

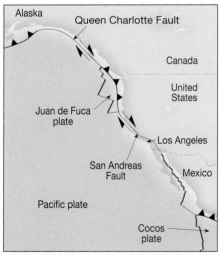

C. Today

FIGURE 13.28 Simplified illustration of the demise of the Farallon plate, which once ran along the western margin of the Americas. Because the Farallon plate was subducting faster than it was being generated, it got smaller and smaller. The remaining fragments of the once mighty Farallon plate are the Juan de Fuca, Cocos, and Nazca plates.

vening Tethys Ocean was almost entirely consumed into the mantle, leaving behind a small remnant—the Mediterranean Sea.

Reconstructions of the breakup of Pangaea also helped investigators understand the demise of the Farallon plate—a large oceanic plate that once occupied much of the eastern Pacific basin. The Farallon plate was once situated on the eastern side of a spreading center opposite the Pacific plate as shown in Figure 13.28A. The modern remnant of this spreading center, which generated both the Farallon and Pacific plates, is the East Pacific Rise.

Beginning about 180 million years ago, the Americas were propelled westward by seafloor spreading in the Atlantic. Therefore, the Farallon plate, which was subducting beneath the Americas faster than it was being generated, decreased in size (Figure 13.28B). As its surface area shrank, it broke into smaller pieces, some of which subducted entirely. The remaining fragments of the once extensive Farallon plate are the Juan de Fuca, Cocos, and Nazca plates.

The westward migration of North America also caused a section of the East Pacific Rise to enter the subduction zone that once lay off the coast of California (Figure 13.28B). As this spreading center subducted, it was destroyed and replaced by a newly generated transform fault system that currently accommodates the differential motion between the North American and Pacific plates. As more of the ridge subducted, the transform fault system, which we now call the San Andreas Fault, increased in length (Figure 13.28C). A similar event generated the Queen Charlotte transform fault located off the west coast of Canada and southern Alaska.

Students Sometimes Ask ...

If oceanic lithosphere cools sufficiently to be denser than the underlying asthenosphere in about 15 million years, why doesn't it begin to subduct at that time?

Plate–mantle convection is much more complicated than classic convective flow that develops when a liquid is heated from below. In a convective liquid, as soon as the material at the top cools and becomes denser than the material below, it begins to sink. In plate–mantle convection, the upper boundary layer—the lithosphere—is a rigid solid. For a new subduction zone to develop, an area of weakness needs to exist somewhere in the lithospheric slab. In addition, the density of the lithosphere must be sufficient enough to overcome the strength of the cool rigid plate. In other words, in order for part of a plate to subduct, the forces acting on the plate must be large enough to bend the plate.

Today, the southern end of the San Andreas Fault connects to a young spreading center that is generating the Gulf of California (Figure 13.29). Because of this change in plate geometry, the Pacific plate has captured a sliver of North America (the Baja Peninsula and a portion of southern California) and carries it northwestward toward Alaska at a rate of about 6 centimeters per year.

Geologist's Sketch

FIGURE 13.29 Satellite image showing the separation between the Baja Penninsula and North America. (Image courtesy of NASA)

CHAPTER 13 DIVERGENT BOUNDARIES: ORIGIN AND EVOLUTION OF THE OCEAN FLOOR IN REVIEW

- *Ocean bathymetry is determined using echo sounders and multibeam sonars,* which bounce sonic signals off the ocean floor. Ship-based receivers record the reflected echoes and accurately measure the time interval of the signals. With this information, ocean depths are calculated and plotted to produce maps of oceanfloor topography. Recently, *satellite measurements* of the ocean surface have provided a new type of data that can be used to map the ocean floor.

- Oceanographers studying the topography of ocean basins have delineated three major units: *continental margins, deep-ocean basins,* and *oceanic (mid-ocean) ridges.*

- The zones that collectively make up a *passive continental margin* include the *continental shelf* (a gently sloping, submerged surface extending from the shoreline toward the deep-ocean basin; the *continental slope* (the true edge of the continent, which has a steep slope that leads from the continental shelf into deep water); and the *continental rise* (a gradual incline composed of sediments that have moved downslope from the continental shelf to the deep-ocean floor).

- Most *active continental margins* are located around the Pacific Ocean in areas where the leading edge of a continent is overrunning oceanic lithosphere. At these sites, sediment scraped from the descending oceanic plate is plastered against the continent to form a collection of sediments called an *accretionary wedge.* An active continental margin generally has a narrow continental shelf, which grades into a deep-ocean trench.

- The deep-ocean basin lies between the continental margin and the oceanic ridge system. Its features include *deep-ocean trenches* (long, narrow depressions that are the deepest parts of the ocean and are located where moving crustal plates descend back into the mantle); *abyssal plains* (among the most level places on Earth, consisting of thick accumulations of sediments that were deposited atop the low, rough portions of the ocean floor by turbidity currents); *seamounts* (volcanic peaks on the ocean floor that originate near oceanic ridges or in association with volcanic hot spots);

and *oceanic plateaus* (large, thick, flood basalt provinces similar to those found on the continents).

- *Oceanic (mid-ocean) ridges,* the sites of seafloor spreading, are found in all major oceans and represent more than 20 percent of Earth's surface. They are the most prominent features in the oceans and form an almost continuous swell that rises 2 to 3 kilometers above the adjacent ocean basin floor. Ridges are characterized by an *elevated position, extensive faulting,* and *volcanic structures* that have developed on newly formed oceanic crust. Most of the geologic activity associated with ridges occurs along a narrow region on the ridge crest, called the *rift zone,* where magma from the asthenosphere moves upward to create new slivers of oceanic crust. The topography of the oceanic ridge is controlled by the rate of seafloor spreading.

- New oceanic crust is formed in a continuous manner by the process of seafloor spreading. The upper crust is composed of *pillow lavas* of basaltic composition. Below this layer are numerous interconnected dikes (*sheeted dike complex*) that are underlain by a thick layer of gabbro. This entire sequence is called an *ophiolite complex.*

- The development of a new ocean basin begins with the formation of a *continental rift* similar to the East African Rift. In those settings where rifting continues, a narrow ocean basin develops, exemplified by the Red Sea. Eventually, seafloor spreading creates an ocean basin bordered by rifted continental margins similar to the present-day Atlantic Ocean. Mechanisms that drive continental rifting include: hot mantle plumes, upwelling from shallow levels in the mantle, and forces that arise from plate motions.

- Oceanic lithosphere subducts because its overall density is greater than the underlying asthenosphere. The subduction of oceanic lithosphere may result in the destruction of sections of or even entire, ocean basins. A classic example is the Farallon plate, most of which subducted beneath the Americas as these continents were displaced westward by seafloor spreading in the Atlantic.

KEY TERMS

abyssal plain (p. 362)
accretionary wedge (p. 361)
active continental margin (p. 361)
bathymetry (p. 356)
black smokers (p. 369)
continental margin (p. 358)
continental rift (p. 371)

continental rise (p. 361)
continental shelf (p. 358)
continental slope (p. 361)
deep-ocean basin (p. 362)
deep-ocean trench (p. 362)
deep-sea fan (p. 361)
echo sounder (p. 356)
guyot (p. 363)

mid-ocean ridge (p. 365)
oceanic plateau (p. 363)
oceanic ridge (p. 365)
ophiolite complex (p. 368)
passive continental margin (p. 358)
pillow basalts (p. 369)
rift valley (p. 365)

seamount (p. 363)
seismic reflection profile (p. 357)
sheeted dike complex (p. 369)
sonar (p. 356)
tablemount (p. 363)

QUESTIONS FOR REVIEW

1. Assuming that the average speed of sound waves in water is 1500 meters per second, determine the water depth if the signal sent out by an echo sounder requires 6 seconds to strike bottom and return to the recorder (see Figure 13.2).

2. Describe how satellites orbiting Earth can determine features on the seafloor without being able to directly observe them beneath several kilometers of seawater.

3. What are the three major topographic provinces of the ocean floor?

4. List the three major features that comprise a passive continental margin. Which of these features is considered a flooded extension of the continent? Which one has the steepest slope?

5. Describe the differences between active and passive continental margins. Include how various features relate to plate tectonics and give a geographic example of each type of margin.

6. Why are abyssal plains more extensive on the floor of the Atlantic than on the floor of the Pacific?

7. How does a flat-topped *seamount*, or *guyot*, form?

8. Briefly describe the oceanic ridge system.

9. Although oceanic ridges can be as tall as some mountains found on the continents, how are these features different?

10. What is the source of magma for seafloor spreading?

11. What is the primary reason for the elevated position of the oceanic ridge system?

12. How does hydrothermal metamorphism alter the basaltic rocks that make up the seafloor? How is seawater changed during this process?

13. What is a black smoker?

14. Compare and contrast a slow spreading center such as the Mid-Atlantic Ridge with one that exhibits a faster spreading rate, such as the East Pacific Rise.

15. Briefly describe the four layers of the ocean crust.

16. How does the *sheeted dike complex* form?

17. Name a modern example of a continental rift.

18. What role are mantle plumes thought to play in the rifting of a continent?

19. What evidence suggests that hot spot volcanism does not always lead to the breakup of a continent?

20. Explain why oceanic lithosphere subducts even though the oceanic crust is less dense than the underlying asthenosphere.

21. Why does the lithosphere thicken as it moves away from the ridge as a result of seafloor spreading?

22. What happened to the Farallon plate? Name the remaining parts.

COMPANION WEBSITE

The *Earth 10e* Web site uses the resources and flexibility of the Internet to aid in your study of the topics in this chapter. Written and developed by the authors and other geology instructors, this site will help improve your understanding of geology. Visit **www.mygeoscienceplace.com** in order to:

- **Review** key chapter concepts.

- **Read** with links to the eBook and to chapter-specific web resources.

- **Visualize** and comprehend challenging topics using learning activities in *GEODe Earth*.

- **Test** yourself with online quizzes.

GEODe EARTH

GEODe Earth is a valuable and easy to use learning aid that can be accessed from your book's Companion Website (**www.mygeoscienceplace.com**). It is a dynamic instructional tool that promotes understanding and reinforces important concepts by using tutorials, animations, and exercises that actively engage the student.

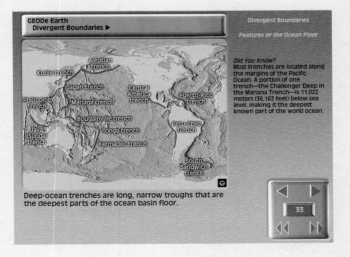

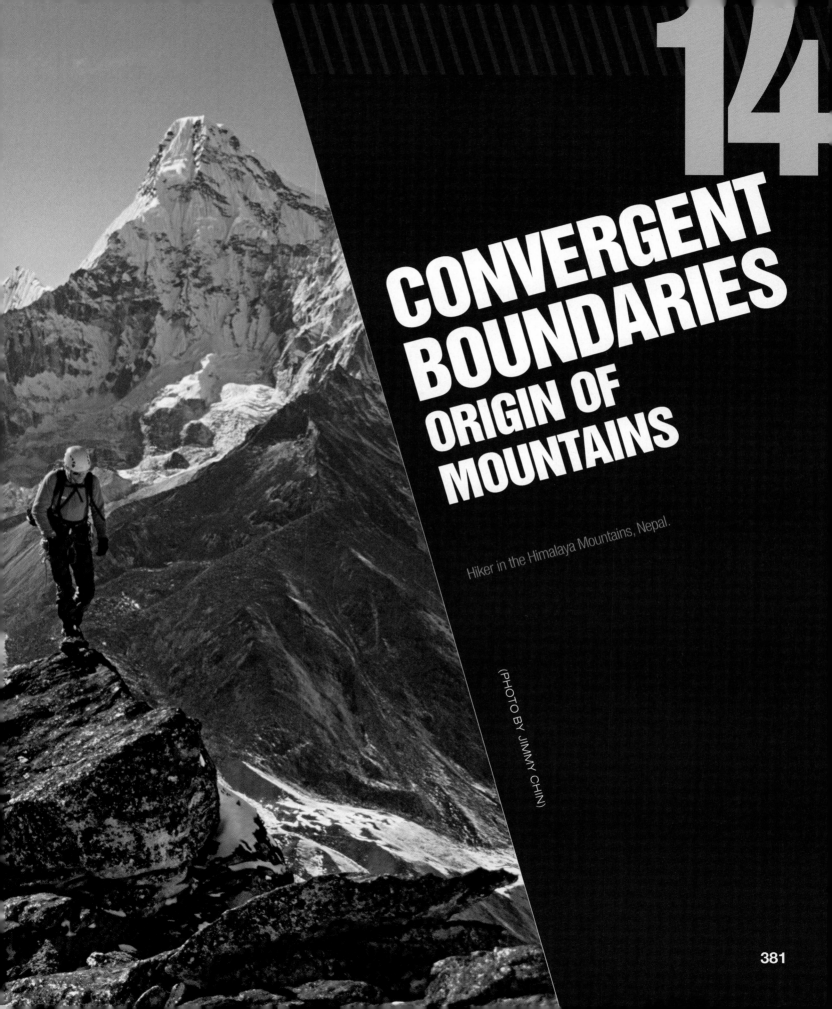

14

CONVERGENT BOUNDARIES
ORIGIN OF MOUNTAINS

Hiker in the Himalaya Mountains, Nepal.

(PHOTO BY JIMMY CHIN)

1

381

Mountains are spectacular features that rise abruptly above the surrounding terrain. These lofty structures can exert enormous influences on climate, affecting not only a region's temperature, but also wind patterns and the distribution of rainfall. It follows, then, that the rise of mountain ranges in the past, or their subsequent demise, triggered profound changes in climate. This, in turn, altered the surrounding landscape and greatly influenced the types of habitats for living organisms.

Some mountains occur as isolated masses; the volcanic cone Kilimanjaro stands almost 6000 meters (20,000 feet) above sea level, overlooking the expansive grasslands of East Africa. Other peaks are parts of extensive mountain belts, such as the American Cordillera, which runs almost continuously from the southern tip of South America through Alaska. Chains such as the Himalayas consist of geologically young, towering peaks that are still rising, whereas others, including the Appalachian Mountains in the eastern United States, are much older and have eroded far below their once lofty heights.

Most major mountain belts show evidence of enormous horizontal forces that have folded, faulted, and generally deformed large sections of Earth's crust. Although tectonic processes contribute to the majestic appearance of mountains, much of the credit for their beauty must be given to weathering and the work of running water and glacial ice, which sculpt these uplifted masses in an unending effort to lower them to sea level. In this chapter, we will examine the nature of mountains and the processes that determine their location and structure.

MOUNTAIN BUILDING

Mountain building has occurred in the recent geologic past at several locations around the world (Figure 14.1). Young mountain belts include the American Cordillera, which runs along the western margin of the Americas from Cape Horn at the tip of South America to Alaska and includes the Andes and Rocky Mountains; the Alpine–Himalaya chain, that extends along the margin of the Mediterranean, through Iran to northern India, and into Indochina; and the mountainous terrains of the western Pacific, which include volcanic island arcs that comprise Japan, the Philippines, and Sumatra. Most of these young mountain belts have come into existence within the last 100 million years (Figure 14.2). Some, including the Himalayas, began their growth as recently as 50 million years ago.

In addition to these young mountain belts, there are several chains of Paleozoic-age mountains found on Earth. Although these older structures are deeply eroded and topographically less prominent, they exhibit the same structural features found in younger mountains. The Appalachians in the eastern United States and the Urals in Russia are classic examples of this group of older and well worn mountain belts.

The term for the processes that collectively produce a mountain belt is **orogenesis,** (*oros* = mountain, *genesis* = to come into being). Most major mountain belts display striking visual evidence of great horizontal forces that have shortened and thickened the crust. These **compressional mountains** contain large quantities of preexisting sedimentary and crystalline rocks that have been faulted and contorted into a series of folds (Figure 14.3). Although folding and thrust faulting are often the most conspicuous signs of orogenesis, varying degrees of metamorphism and igneous activity are always present.

How do mountain belts form? As early as the ancient Greeks, this question has intrigued some of the greatest philosophers and scientists. One early proposal suggested that mountains are simply wrinkles in Earth's crust, produced as the planet cooled from its original semimolten state. According to this idea, Earth contracted and shrank as it lost heat, which caused the crust to deform in a manner similar to how an orange peel wrinkles as the fruit dries out. However, neither this nor any other early hypothesis withstood scientific scrutiny.

With the development of the theory of plate tectonics, a model for orogenesis with excellent explanatory power has emerged. Its premise asserts that the tectonic processes that generate Earth's major mountainous terrains begin at convergent plate boundaries, where oceanic lithosphere descends into the mantle.

FIGURE 14.1 Denali, which means the "the great one" in native Athabaskan language, is the highest peak in North America, rising 6000 meters (20,320 feet) above sea level. Located in the Alaskan Range, Denali (the peak on the left) officially goes by the name Mt. McKinley after President William McKinley. (Photo by Michael Collier)

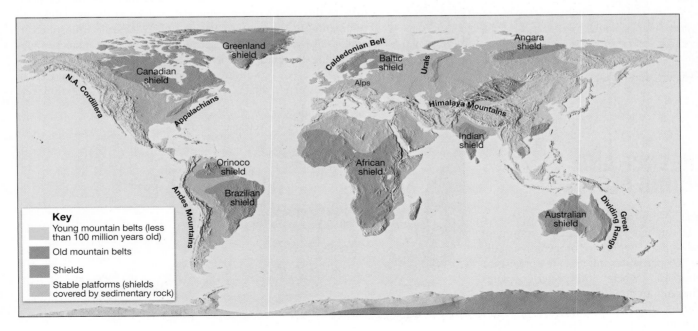

FIGURE 14.2 Earth's major mountain belts.

FIGURE 14.3 Highly deformed sedimentary strata exposed on the face of Alberta's Mount Kidd. These sedimentary rocks are continental shelf deposits that were displaced toward the interior of Canada by low-angle thrust faults. (Photo by Tom Neversely/Photolibrary)

Students Sometimes Ask ...

You mentioned that most mountains are the result of crustal deformation. Are there areas that are mountainous but have been produced without crustal deformation?

Yes. Plateaus—areas of high-standing rocks that are essentially horizontal—are features that can be deeply dissected by erosional forces into rugged, mountainlike landscapes. Although these highlands resemble mountains topographically, they lack the structures associated with orogenesis. The reverse situation also exists. For instance, the Piedmont section of the eastern Appalachians exhibits topography that is nearly as subdued as that seen in the Great Plains. Yet because this region is composed of folded, faulted, and deformed metamorphic rocks, it is clearly part of the Appalachian Mountains.

CONVERGENCE AND SUBDUCTING PLATES

In their quest to unravel the events that produce mountains, researchers examine ancient mountain structures as well as sites where orogenesis is currently active. Of particular interest are convergent plate boundaries, where lithospheric plates subduct. The subduction of oceanic lithosphere generates Earth's strongest earthquakes and most explosive volcanic eruptions, as well as playing a pivotal role in generating many of Earth's mountain belts.

Major Features of Subduction Zones

Subduction zones can be roughly divided into four regions that include: (1) a *volcanic arc,* which is built upon the overlying plate; (2) a *deep-ocean trench,* which forms where a subducting slab of oceanic lithosphere bends and descends into the asthenosphere; (3) a *forearc region,* located between the trench and the volcanic arc and; (4) a *backarc region,* on the side of the volcanic arc opposite the trench. Although all subduction zones exhibit these features, large variations occur along the length of an individual subduction zone as well as among different subduction zones.

Volcanic Arcs Perhaps the most obvious structure generated by subduction is a *volcanic arc* (Figure 14.4). Where two oceanic slabs converge, one is subducted beneath the other, initiating partial melting of the mantle wedge located above the subducting plate. This eventually leads to the growth of a **volcanic island arc,** or simply an **island arc,** on the ocean floor. Examples of active island arcs include the Mariana, Tonga, and Aleutian arcs in the Pacific (Figure 14.5).

When oceanic lithosphere is subducted beneath a continental block, a **continental volcanic arc** results. Continental volcanic arcs build upon the topography of older continental rocks, resulting in volcanic peaks that may reach 6000 meters (nearly 20,000 feet) above sea level.

Deep-Ocean Trenches Deep-ocean trenches are created where oceanic lithosphere bends as it descends into the mantle. Trench depth is strongly related to the age, and therefore the temperature, of the subducting oceanic slab as well as the availability of sediments. In the western Pacific, where oceanic lithosphere is cold, dense oceanic slabs descend into the mantle at steep angles producing trenches with average depths of 7 to 8 kilometers below sea level. A well-known example is the Mariana Trench, where the deepest area is an amazing 11,000 meters (36,000 feet)

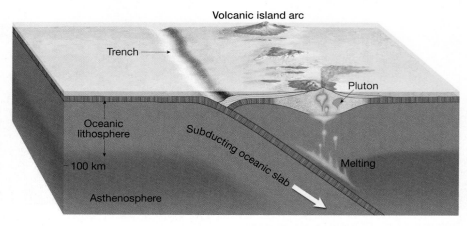

FIGURE 14.4 The development of a volcanic arc by the convergence of two oceanic plates. Continuous subduction results in the development of thick units of continental-type crust.

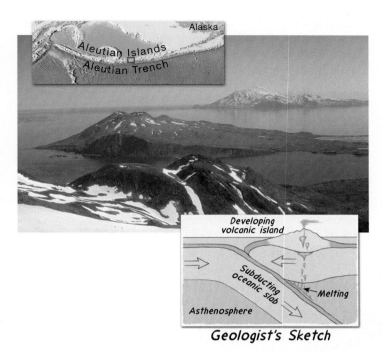

FIGURE 14.5 Three of the many volcanic islands that comprise the Aleutian arc. This narrow band of volcanism results from the subduction of the Pacific plate. In the distance is the Great Sitkin volcano (1772 meters), which the Aleuts call the "Great Emptier of Bowels" because of its frequent activity. (Photo by Bruce D. Marsh)

below sea level. By contrast, the Cascadia subduction zone off the coasts of Washington and Oregon lacks a well-defined trench partly because the warm, buoyant Juan de Fuca plate subducts at a very low angle. In addition, a massive amount of sediment from the Columbia River basin mostly fills what would otherwise be a shallow trench 2 to 3 kilometers deep.

Forearc and Backarc Regions

Located between developing volcanic arcs and deep-ocean trenches are *forearc* regions where pyroclastic material from the volcanic arc, as well as sediments eroded from the adjacent landmass, accumulate. Ocean-floor sediments are also carried to forearc regions by subducting plates.

Another site where sediments and volcanic debris accumulate is the *backarc* region, located on the side of the volcanic arc opposite the trench. In these regions, tensional forces are often prevalent, causing the crust to be stretched and thinned.

Dynamics at Subduction Zones

Because subduction zones form where two plates converge, it is logical to assume that large compressional forces deform the plate margins. However, convergent margins are *not always* regions dominated by compressional forces.

Extension and Backarc Spreading

Along some convergent plate margins, tensional stresses act on the overlying plates, which cause stretching and thinning of the crust. But how do extensional processes operate where two plates are moving together?

The age of the subducting oceanic slab is thought to play a significant role in determining the dominant forces acting on the overriding plate. When a relatively cold, dense slab

subducts, it does *not* follow a fixed path into the asthenosphere. Rather, it sinks vertically as it descends, causing the trench to retreat, or "roll back," as shown in Figure 14.6. As the subducting plate sinks, it creates a flow called *slab suction* in the asthenosphere that "pulls" the upper plate toward the retreating trench. (Visualize what would have happened if you were sitting in a lifeboat near the *Titanic* as it sank!) As a result, the overriding plate is elongated and thinned, and many eventually develop a **backarc basin.**

Recall from Chapter 13 that thinning and rifting of the lithosphere results in upwelling of hot mantle rock and accompanying decompression melting and volcanism. Continued extension may initiate seafloor spreading that increases the size of the newly formed backarc basin. Seafloor spreading is currently enlarging the backarc basins found landward of the Mariana and Tonga islands.

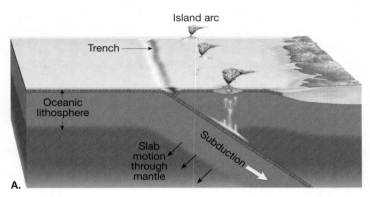

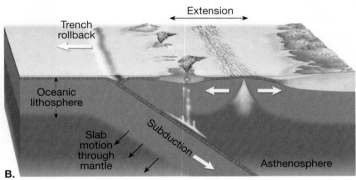

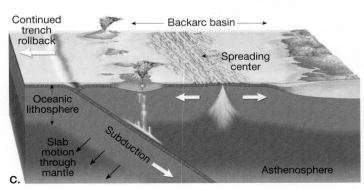

FIGURE 14.6 Model showing the formation of an active backarc basin. Subduction and "roll back" of an oceanic slab creates flow in the mantle that "pulls" the upper (non-subducting) plate toward the retreating trench. Rifting can result in the development of a spreading center.

SUBDUCTION AND MOUNTAIN BUILDING

The subduction of oceanic lithosphere gives rise to two different types of tectonic structures. Where oceanic lithosphere subducts beneath an oceanic plate, a *volcanic island arc* and related tectonic features develop. Subduction beneath a continental block, on the other hand, results in the formation of a volcanic arc along the margin of a continent (Box 14.1). Plate boundaries that generate continental volcanic arcs are referred to as **Andean-type plate margins**.

Volcanic Island Arcs

Island arcs result from the steady subduction of oceanic lithosphere, which may last for 200 million years or more. Periodic volcanic activity, the emplacement of igneous plutons at depth, and the accumulation of sediment that is scraped from the subducting plate gradually increase the volume of crustal material capping the upper plate. Some large volcanic island arcs, such as Japan, owe their size to having been built upon a preexisting fragment of continental crust.

The continued growth of a volcanic island arc can result in the formation of mountainous topography consisting of belts of igneous and metamorphic rocks. This activity, however, is viewed as just one phase in the development of a major mountain belt. As you will see later, some volcanic arcs are carried by a subducting plate to the margin of a large continental block, where they become involved in a large-scale mountain-building episode.

Mountain Building along Andean-type Margins

The first stage in the development of an Andean-type mountain belt occurs along a *passive continental margin* prior to the formation of the subduction zone. The East Coast of the United States provides a modern example of a passive continental margin where sedimentation has produced a thick platform of shallow-water sandstones, limestones, and shales (Figure 14.7A). At some point, the forces that drive plate motions change and a subduction zone develops along the margin of the continent. It is along these *active continental margins* that the three distinct structural elements of a developing mountain belt gradually take form: volcanic arcs, accretionary wedges, and forearc basins (Figure 14.7B).

Building a Volcanic Arc   Recall that as oceanic lithosphere descends into the mantle, increasing temperatures and pressures drive volatiles (mostly water) from the crustal rocks. These mobile fluids migrate upward into the wedge-shaped piece of mantle between the subducting slab and upper plate. Once the sinking slab reaches a depth of about 100 kilometers (60 miles), these water-rich fluids reduce the melting point of hot mantle rock sufficiently to trigger some melting (Figure 14.7B). Partial melting of the ultramafic rock peridotite generates *primary magmas,* with mafic (basaltic) compositions. Because these newly formed basaltic magmas are less dense

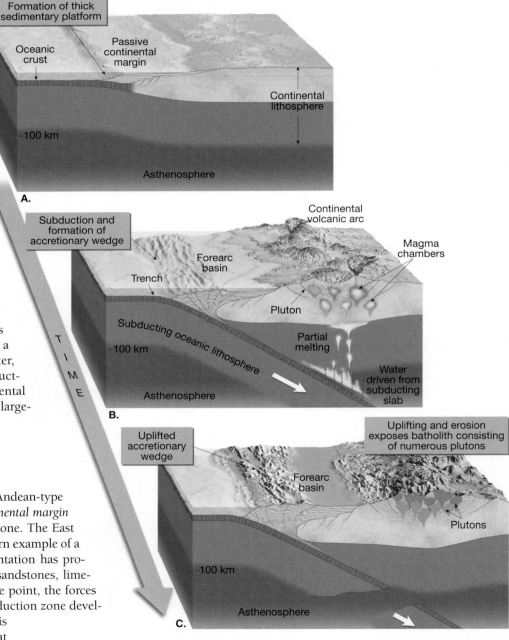

FIGURE 14.7   Mountain building along an Andean-type subduction zone. **A.** Passive continental margin with an extensive platform of sediments and sedimentary rocks. **B.** Plate convergence generates a subduction zone, and partial melting produces a volcanic arc. Continued convergence and igneous activity further deform and thicken the crust, elevating the mountain belt, while an accretionary wedge grows. **C.** Subduction ends and is followed by a period of uplift and erosion.

UNDERSTANDING EARTH

BOX 14.1

Earthquakes in the Pacific Northwest

Seismic studies have shown that the Cascadia subduction zone off the coasts of Washington and Oregon experiences less earthquake activity than any other subduction zone that rims the Pacific basin (Figure 14.A). Does this mean that earthquakes pose no major threat to the population centers of the Pacific Northwest? For some time, that was the conventional wisdom. However, the discovery of marshes and coastal forests that were likely buried by rapid subsidence during multiple large earthquakes has altered that view.

The Cascadia subduction zone is very similar to the convergent margin in central Chile, where the oceanic slab descends at a shallow angle of about 10–15 degrees. In Chile, large compressional stresses strain the rock until it ruptures, triggering earthquakes. Earth's strongest quake occurred there in 1960. Research indicates that subduction at shallow dip angles results in an environment conducive to great earthquakes (Mw 8.0 or greater). This phenomenon is, in part, explained by the large contact area between the overriding plate and the subducting slab. This, and related evidence, has led some researchers to conclude that the Cascadia subduction zone is primed for a great earthquake.

However, contrary evidence indicates that a potentially catastrophic event is not likely, at least not in the near future. Studies conducted

FIGURE 14.A Scene outside a historic Pioneer Square building following the February 28, 2001, Seattle earthquake. (Photo by Tim Crosby/Newsmakers/Liaison Agency, Inc.)

along coastal areas in the Pacific Northwest over the past few decades indicate that elastic strain is not accumulating to the extent required for a great earthquake.

Which view is correct? Is a major earthquake in the Pacific Northwest imminent or unlikely? Continued research seeks to resolve this question. In the meantime, those living in the region bordering the Cascadia subduction zone would be wise to become acquainted with the safety precautions that can be taken to lessen the effects of a great earthquake.

---

than the rocks from which they originated, they will buoyantly rise. Upon reaching the base of the low-density materials of the continental crust, they typically collect, or pond. (When the crust is stretched and thinned, or if the magma is especially hot [mantle plumes], large outpourings of basaltic magma are common.)

Continued ascent through the thick continental crust is generally achieved through magmatic differentiation, in which heavy iron-rich minerals crystallize and settle out of the magma, leaving the remaining melt enriched in silica and other "light" components (see Chapter 4). Hence, through magmatic differentiation, a comparatively dense basaltic magma can generate low-density, buoyant melts.

Volcanism along continental arcs is dominated by the eruption of lavas and pyroclastic materials of andesitic composition, with lesser amounts of basaltic and rhyolitic materials present. Because water driven from the subducting plate is necessary for melting, these mantle-derived magmas are enriched in water and other volatiles (the gaseous component of magma). These gas-laden andesitic magmas produce the explosive eruptions that characterize volcanic arcs (Figure 14.8).

Emplacement of Batholiths Because of its low density and great thickness, continental crust significantly impedes the ascent of molten rock. Consequently, a high percentage of the magma that intrudes the crust never reaches the surface—instead, it crystallizes at depth to form massive igneous plutons called *batholiths*. The result of this activity is to thicken the crust. The emplacement of these hot igneous bodies acts to metamorphose the host rock by a process called contact metamorphism (see Chapter 8).

Eventually, uplifting and erosion exhume the batholiths which consist of numerous interconnected plutons (see Figure 14.7C). The American Cordillera contains several large batholiths, including the Sierra Nevada of California, the Coast Range Batholith of western Canada, and several large igneous bodies in the Andes (Figure 14.9). Most batholiths consist of intrusive igneous rocks that range in composition from granite to diorite.

FIGURE 14.8 Redoubt Volcano is located west of Cook Inlet, Alaska. This subduction zone volcano has erupted five times since 1900, most recently in 2009. An eruption in 1989 caught Royal Dutch Airlines Flight 867 in its plume. Although all four engines temporarily failed, the aircraft landed safely at Anchorage. (Photo by Michael Collier)

Development of an Accretionary Wedge During the development of volcanic arcs, unconsolidated sediments that are carried on the subducting plate, as well as fragments of oceanic crust, may be scraped off and plastered against the edge of the overriding plate. The resulting chaotic accumulation of deformed and thrust-faulted sediments and scraps of ocean crust is called an **accretionary wedge** (see Figure 14.7B). The processes that deform these sediments are comparable to a wedge of soil as it is scraped and pushed in front of an advancing bulldozer.

Some of the sediments that comprise an accretionary wedge are muds that accumulated on the ocean floor and were subsequently carried to the subduction zone by plate motion (Figure 14.10). Additional materials are derived from the adjacent volcanic arc and consist of volcanic ash and other pyroclastic materials.

In regions where sediment is plentiful, prolonged subduction may thicken a developing accretionary wedge sufficiently so it protrudes above sea level. This has occurred along the southern end of the Puerto Rico Trench, where the Orinoco River Basin of Venezuela is a major source region. The resulting wedge emerges to form the island of Barbados. By contrast, some subduction zones have small accretionary wedges, or none at all. The Mariana Trench, for example, has a poorly developed accretionary wedge, in part because of its distance from a significant source of sediment.

Not all the available sediment becomes part of the accretionary wedge; some is carried into the subduction zone atop the descending slab. As these sediments descend, pressure steadily increases. However, temperatures within the sediments remain relatively low because they are in contact with the cool, plunging plate. This activity generates a suite of high-pressure, low-temperature metamorphic minerals. Because of their low density, some of the subducted sediments and associated metamorphic components buoyantly rise toward the surface. This "backflow" tends to mix and churn the sediment within the accretionary wedge. Thus, an accretionary wedge evolves into a complex structure consisting of faulted and folded sedimentary rocks and scraps of oceanic crust that may be intermixed with metamorphic rocks formed during the subduction process. The unique structure of accretionary wedges has allowed geologists to more accurately reconstruct the events that generated our modern continents.

Forearc Basins As an accretionary wedge grows upward, it tends to act as a barrier to the movement of sediment from the volcanic arc to the trench. As a result, sediments begin to collect

Beginning about 30 million years ago, subduction gradually ceased along much of the margin of North America as the spreading center that produced the Farallon plate entered the California trench (see Figure 13.26). The uplifting and erosion that followed removed most of the evidence of past volcanic activity and exposed a core of crystalline igneous and associated metamorphic rocks that make up the Sierra Nevada. The Coast Ranges were uplifted only recently, as evidenced by the young, unconsolidated sediments that currently mantle portions of these highlands.

California's Great Valley is a remnant of the forearc basin that formed between the Sierra Nevada and the Coast Ranges. Throughout much of its history, portions of the Great Valley lay below sea level. This sediment-laden basin contains thick marine deposits and debris eroded from the continental volcanic arc.

In summary, *the growth of mountain belts at subduction zones is a response to crustal thickening caused by the addition of mantle-derived igneous rocks and sediments scraped from the descending oceanic slab.*

Geologist's Sketch

Light rock is a batholith composed of granite

Dark rock is roof pendant—metamorphosed host rock

Torres de Plaine, southern Chile

FIGURE 14.9 Torres del Paine in southern Chile. This mountainous area east of the high central Andes consists mainly of a large granite batholith (light color) covered by a layer of metamorphic rocks (dark color). (Photo by Michael Collier)

between the accretionary wedge and volcanic arc. This region, which is composed of relatively undeformed layers of sediment and sedimentary rocks, is called a **forearc basin** (see Figure 14.7B). Subsidence and continued sedimentation in forearc basins can generate a sequence of nearly horizontal, sedimentary strata that can attain thicknesses of several kilometers.

Sierra Nevada, Coast Ranges, and Great Valley

California's Sierra Nevada, Coast Ranges, and Great Valley are excellent examples of the structures that are typically generated along an Andean-type subduction zone (Figure 14.11). The brief overview that follows examines the processes that drive mountain building in these settings.

The Sierra Nevada and Coast Ranges (Figure 14.11) are mountain belts produced by the subduction of a portion of the Pacific basin (Farallon plate) under the western margin of California (see Figure 13.28, p. 376). The Sierra Nevada batholith is a remnant of the continental volcanic arc that was produced by many surges of magma during a time span that exceeded 100 million years. The Coast Ranges were built from the vast accumulation of sediments (accretionary wedge) that collected along the continental margin.

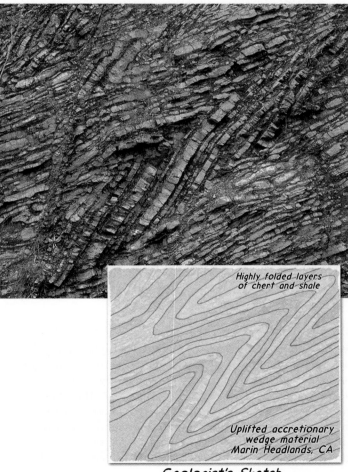

Geologist's Sketch

Highly folded layers of chert and shale

Uplifted accretionary wedge material Marin Headlands, CA

FIGURE 14.10 Interbedded layers of chert and shale that have been strongly folded by compressional stresses during the growth of an accretionary wedge. A recent period of uplift has exposed these deformed strata near Marin Headlands, north of San Francisco. (Photo by Michael Collier)

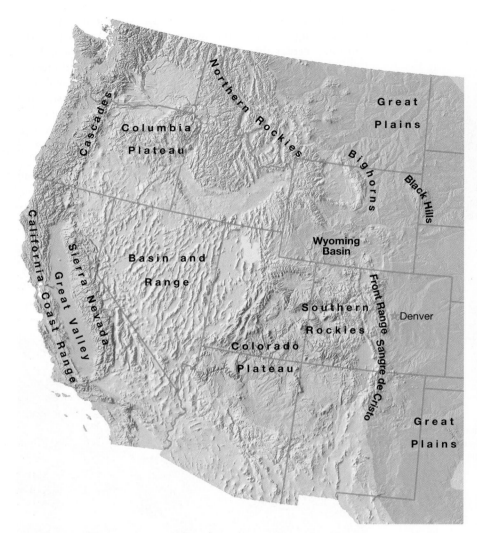

FIGURE 14.11 Map of mountains and landform regions in the western United States. (After Thelin and Pike, U.S. Geological Survey)

COLLISIONAL MOUNTAIN BELTS

CONVERGENT BOUNDARIES
GEODe EARTH
▶ Continental Collisions
▶ Crustal Fragments and Mountain Building

Most major mountain belts are generated when one or more buoyant crustal fragments collide with a continental margin as a result of subduction. Oceanic lithosphere, which is relatively dense, readily subducts, whereas continental lithosphere which contains significant amounts of low-density crustal rocks, is too buoyant to undergo subduction. Consequently, the arrival of a crustal fragment at a trench results in a collision with the margin of the adjacent continental block and an end to subduction.

Terranes and Mountain Building

The process of collision and accretion (joining together) of comparatively small crustal fragments to a continental margin has generated many of the mountainous regions that rim the Pacific. Geologists refer to these accreted crustal blocks as ter-

ranes. **Terrane** refers to any crustal fragment that has a geologic history distinct from that of the adjoining terranes.

The Nature of Terranes What is the nature of these crustal fragments, and where did they originate? Research suggests that prior to their accretion to a continental block, some of these fragments may have been **microcontinents** similar to the modern-day island of Madagascar, located east of Africa in the Indian Ocean. Many others were island arcs similar to Japan, the Philippines, and the Aleutian Islands. Still others may have been submerged oceanic plateaus created by massive outpourings of basaltic lavas associated with mantle plumes (see Figure 13.11, p. 365). More than 100 of these relatively small crustal fragments are presently known to exist.

Accretion and Orogenesis As oceanic plates move, they carry embedded oceanic plateaus, volcanic island arcs, and microcontinents to an Andean-type subduction zone. When an oceanic plate contains small seamounts, these structures are generally subducted along with the descending oceanic slab. However, large thick units of oceanic crust, such as the Ontong Java Plateau, which is the size of Alaska, or an island arc composed of abundant "light" igneous rocks, render the oceanic lithosphere too buoyant to subduct. In these situations, a collision between the crustal fragment and the continental margin occurs.

The sequence of events that happen when an island arc reaches an Andean-type margin is shown in Figure 14.12. Rather than subduct, the upper crustal layers of these thickened zones are peeled from the descending plate and thrust in relatively thin sheets upon the adjacent continental block. Because subduction often continues for 100 million years or longer, several crustal fragments can be transported to the continental margin. Each collision displaces earlier accreted terranes further inland, adding to the zone of deformation as well as to the thickness and lateral extent of the continental margin.

tinents will likewise be accreted to active continental margins, producing new orogenic belts.

Continental Collisions

The Himalayas, Appalachians, Urals, and Alps represent mountain belts that were formed by the closure of major ocean basins. Continental collisions result in the development of mountains characterized by shortened and thickened crust achieved through folding and faulting. Some mountainous regions have crustal thicknesses that exceed 70 kilometers (40 miles).

Noteworthy features of most collisional mountain ranges are **fold-and-thrust belts.** These mountainous zones result from the deformation of thick sequences of shallow marine sedimentary rocks like those currently found along passive continental margins of the Atlantic. During a continental collision, these sedimentary rocks are pushed inland, away from the core of the developing mountain belt and over the stable continental interior. In essence, crustal shortening is achieved by displacement along thrust faults where once relatively flat-lying strata are "sliced" into thick layers that are stacked one upon another. During this displacement, material caught between the thrust faults is often folded, thereby forming the other major structure of a fold-and-thrust belt. Excellent examples of fold-and-thrust belts are found in the Appalachian Valley and Ridge Province, the Canadian Rockies, the Lesser (southern) Himalayas, and the northern Alps.

The zone where two continents collide and are "welded" together is called the **suture.** This portion of the mountain belt often preserves slivers of oceanic lithosphere that were trapped between the colliding plates. As a result of their unique ophiolite structure (see Figure 13.18, p. 368), these pieces of oceanic lithosphere help identify the location of the collision boundary.

Next, we will take a closer look at two examples of collision mountains—the Himalayas and the Appalachians. The Himalayas are the youngest collision mountains on Earth and are still rising. The Appalachians are a much older mountain belt, in which active mountain building ceased about 250 million years ago.

The Himalayas

The mountain-building episode that created the Himalayas began roughly 50 million years ago when India began to collide with Asia. Prior to the breakup of Pangaea, India was located between Africa and Antarctica in the Southern Hemisphere (see Figure 2.B, p. 63). As Pangaea fragmented, India moved rapidly, geologically speaking, a few thousand kilometers in a northward direction.

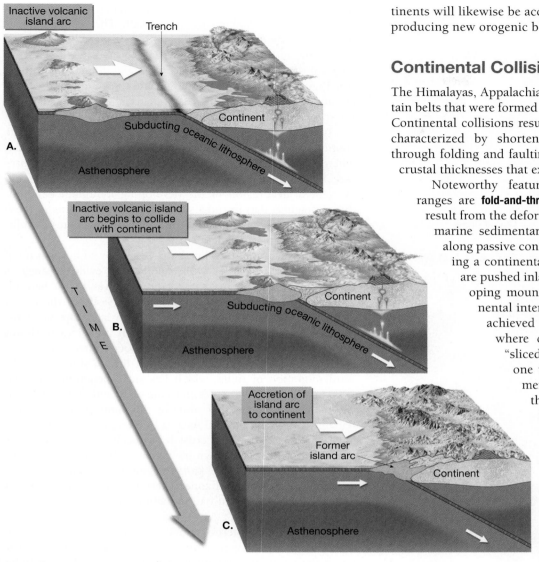

FIGURE 14.12 Sequence of events showing the collision and accretion of an island arc to a continental margin.

The North American Cordillera The relationship between mountain building and the accretion of crustal fragments arose primarily from studies conducted in the North American Cordillera (Figure 14.13). Researchers determined that some of the rocks in the orogenic belts of Alaska and British Columbia contained fossil and paleomagnetic evidence that indicated these strata previously lay much closer to the equator.

It is now known that many of the terranes that make up the North American Cordillera were scattered throughout the eastern Pacific, like the island arcs and oceanic plateaus currently distributed in the western Pacific. During the breakup of Pangaea, the eastern portion of the Pacific basin (Farallon plate) began to subduct under the western margin of North America. This activity resulted in the piecemeal addition of crustal fragments to the entire Pacific margin of the continent—from Mexico's Baja Peninsula to northern Alaska (see Figure 14.13). Geologists expect that many modern microcon-

FIGURE 14.13 Map showing terranes that have been added to western North America during the past 200 million years. Paleomagnetic studies and fossil evidence indicate that some of these terranes originated thousands of kilometers to the south of their present locations. (After D. R. Hutchinson and others)

The subduction zone that facilitated India's northward migration was near the southern margin of Asia (Figure 14.14A). Continued subduction along Asia's margin created an Andean-type plate margin that contained a well-developed continental volcanic arc and an accretionary wedge. India's northern margin, on the other hand, was a passive continental margin consisting of a thick platform of shallow-water sediments and sedimentary rocks.

Geologists have determined that one, or perhaps more, small continental fragments were positioned on the subducting plate somewhere between India and Asia. During the closing of the intervening ocean basin, a small crustal fragment, which

now forms southern Tibet, reached the trench. This event was followed by the docking of India itself. The tectonic forces involved in the collision of India with Asia were immense and caused the more deformable materials located on the seaward edges of these landmasses to become highly folded and faulted (see Figure 14.14B). The shortening and thickening of the crust elevated great quantities of crustal material, thereby generating the spectacular Himalayan mountains (Figure 14.15).

In addition to uplift, crustal shortening caused rocks at the "bottom of the pile" to become deeply buried—an environment where they experienced elevated temperatures and pressures (see Figure 14.14B). Partial melting within the deepest and most deformed region of the developing mountain belt produced magmas that intruded the overlying rocks. It is in these environments that the metamorphic and igneous cores of collisional mountains are generated.

The formation of the Himalayas was followed by a period of uplift that raised the Tibetan Plateau. Seismic evidence suggests that a portion of the Indian subcontinent was thrust beneath Tibet a distance of perhaps 400 kilometers. If so, the added crustal thickness would account for the lofty landscape of southern Tibet, which has an average elevation higher than Mount Whitney, the highest point in the contiguous United States.

The collision with Asia slowed but did not stop the northward migration of India, which has since penetrated at least 2000 kilometers (1200 miles) into the mainland of Asia. Crustal shortening accommodated some of this motion. Much of the remaining penetration into Asia caused lateral displacement of large blocks of the Asian crust by a mechanism described as *continental escape*. As shown in Figure 14.16, when India continued its northward trek, parts of Asia were "squeezed" eastward out of the collision zone. These displaced crustal blocks include much of present-day Indochina and sections of mainland China.

Why was the interior of Asia deformed to such a large extent while India has remained essentially intact? The answer lies in the nature of these diverse crustal blocks. Much of India is a continental shield composed mainly of Precambrian rocks (see Figure 14.2). This thick, cold slab of crustal material has been intact for more than 2 billion years. By contrast, Southeast Asia was assembled more recently from the collision of several smaller crustal fragments. Consequently, it is still relatively "warm and weak" from recent periods of mountain building (Figure 14.16B)

The Appalachians

The Appalachian Mountains provide great scenic beauty near the eastern margin of North America from Alabama to Newfoundland. In addition, mountains of similar origin that formed during the same period are found in the British Isles, Scandinavia, northwestern Africa, and Greenland (see Figure 2.6, p. 43). The orogeny that generated this extensive mountain system that presently lies on both sides of the North Atlantic, lasted a few hundred million years and was one of the stages in assembling the supercontinent of Pangaea. Detailed studies of the Appalachians indicate that the formation of this mountain belt was complex and resulted from three distinct episodes of mountain building.

FIGURE 14.14 Diagrams illustrating the collision of India with the Eurasian plate, producing the spectacular Himalayas.

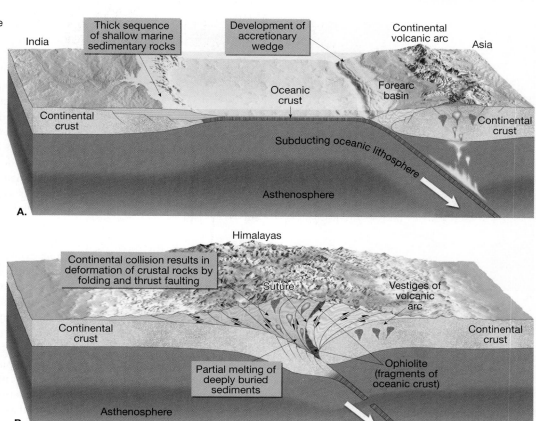

Thick sequence of shallow marine sedimentary rocks

Development of accretionary wedge

Continental volcanic arc

India

Asia

Continental crust

Oceanic crust

Forearc basin

Continental crust

Subducting oceanic lithosphere

Asthenosphere

A.

Himalayas

Continental collision results in deformation of crustal rocks by folding and thrust faulting

Suture

Vestiges of volcanic arc

Continental crust

Continental crust

Partial melting of deeply buried sediments

Ophiolite (fragments of oceanic crust)

Asthenosphere

B.

FIGURE 14.15 Hikers camping in Pakistan's Charakusa Valley with the bold peaks of the Karakoram Range, part of the "Greater Himalayas," in the background. The second highest peak in the world, K2, is located in the Karakoram Range. (Photo by Jimmy Chin/National Geographic/Getty)

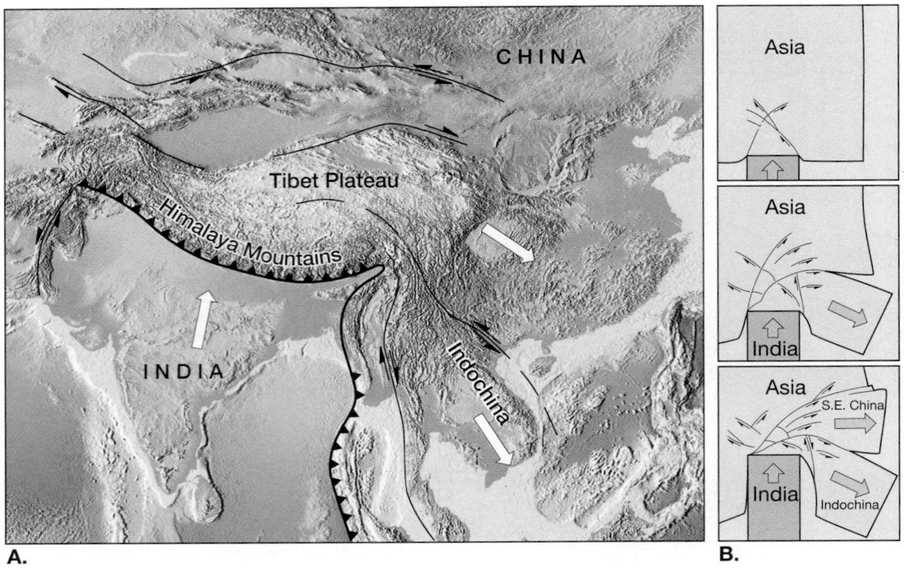

FIGURE 14.16 The collision between India and Asia that generated the Himalayas and Tibetan Plateau also severely deformed much of Southeast Asia. **A.** Map view of some of the major structural features of Southeast Asia thought to be related to this episode of mountain building. **B.** Re-creation of the deformation of Asia, with a rigid block representing India pushed into a mass of deformable modeling clay.

Our simplified overview begins roughly 750 million years ago with the breakup of a pre-Pangaea supercontinent called Rodinia, which rifted North America from Europe and Africa. This episode of continental rifting and seafloor spreading generated the ancestral North Atlantic. Located within this developing ocean basin was a fragment of continental crust that had been rifted from North America (Figure 14.17A).

About 600 million years ago, plate motion dramatically changed and the ancestral North Atlantic began to close. Two subduction zones likely formed—one located seaward of the coast of Africa that gave rise to a volcanic arc and another that developed along the margin of the continental fragment that lay off the coast of North America (Figure 14.17A).

Between 450 and 500 million years ago, the marginal sea between the crustal fragment and North America began to close. The collision that ensued, called the *Taconic Orogeny,* deformed the continental shelf and sutured the crustal fragment to the North American plate. The metamorphosed remnants of the continental fragment are recognized today as the crystalline rocks of the Blue Ridge and western Piedmont regions of the Appalachians (Figure 14.17B). In addition to the pervasive regional metamorphism, numerous magma bodies intruded the crustal rocks along the entire continental margin, particularly in what we know today as New England.

A second episode of mountain building, called the *Acadian Orogeny,* began about 400 million years ago. The continued closing of the ancestral North Atlantic resulted in the collision of the developing island arc with North America (Figure 14.17C). Evidence for this event is visible in the Carolina Slate Belt of the eastern Piedmont, which contains metamorphosed sedimentary and volcanic rocks characteristic of an island arc.

The final orogeny occurred between 250 and 300 million years ago, when Africa collided with North America. The result

was landward displacement of the Blue Ridge and Piedmont provinces by as much as 250 kilometers (155 miles). This event also displaced and further deformed the shelf sediments and sedimentary rocks that had once flanked the eastern margin of North America (Figure 14.17D). Today these folded and thrust-faulted sandstones, limestones, and shales make up the largely unmetamorphosed rocks of the Valley and Ridge Province. Outcrops of the folded and thrust-faulted structures that characterize collision mountains are found as far inland as central Pennsylvania and western Virginia (Figure 14.18).

Following the collision of Africa and North America, the Appalachians lay in the interior of Pangaea. About 180 million years ago, this newly formed supercontinent began to break into smaller fragments, a process that ultimately created the modern Atlantic Ocean. Because this new zone of rifting occurred east of the suture that formed when Africa and North America collided, remnants of Africa remain "welded" to the North American plate (Figure 14.17E).

Other mountain ranges that exhibit evidence of continental collisions include the Alps and the Urals. The Alps formed as Africa and Europe collided during the closing of the Tethys Sea. Similarly, the Urals, were uplifted during the assembly of Pangaea when northern Europe and northern Asia collided forming a major portion of Eurasia.

FAULT-BLOCK MOUNTAINS

Most mountain belts form in compressional environments, as evidenced by the predominance of large thrust faults and folded strata. However, other tectonic processes, such as continental rifting, can also produce topographic mountains. Recall that continental rifting occurs when tensional forces stretch and thin the lithosphere, resulting in upwelling of hot mantle rock. Upwelling heats the thinned lithosphere which becomes less

Students Sometimes Ask ...

What are some modern examples of features that may become terranes in the future?

The southwestern Pacific Ocean is a good place to find island arcs, oceanic plateaus, and microcontinents that will one day be accreted to the edge of a continent. Another potential terrane is the area west of the San Andreas Fault, which includes southwestern California and Mexico's Baja California peninsula. The "California Terrane," as it is sometimes called, is moving to the northwest and will probably detach from North America in about 50 million years. Continued movement toward the northwest will bring it to southern Alaska, where it will become another in a long procession of terranes that have "docked" there during the past 200 million years.

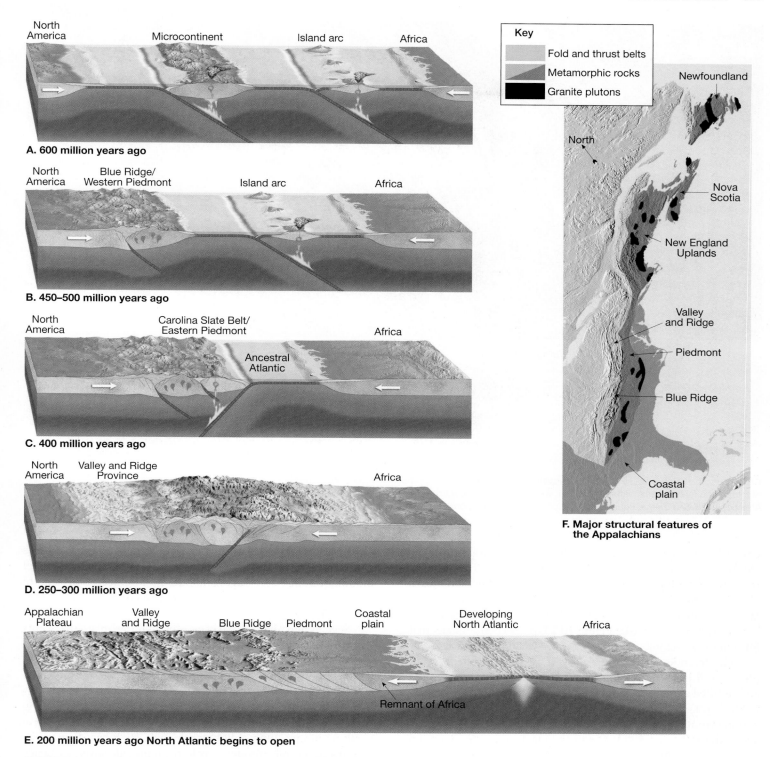

A. 600 million years ago

B. 450–500 million years ago

C. 400 million years ago

D. 250–300 million years ago

E. 200 million years ago North Atlantic begins to open

F. Major structural features of the Appalachians

Key
Fold and thrust belts
Metamorphic rocks
Granite plutons

FIGURE 14.17 These simplified diagrams depict the development of the southern Appalachians as the ancient North Atlantic was closed during the formation of Pangaea. Three separate stages of mountain-building activity spanned more than 300 million years. (After Zve Ben-Avraham, Jack Oliver, Larry Brown, and Frederick Cook)

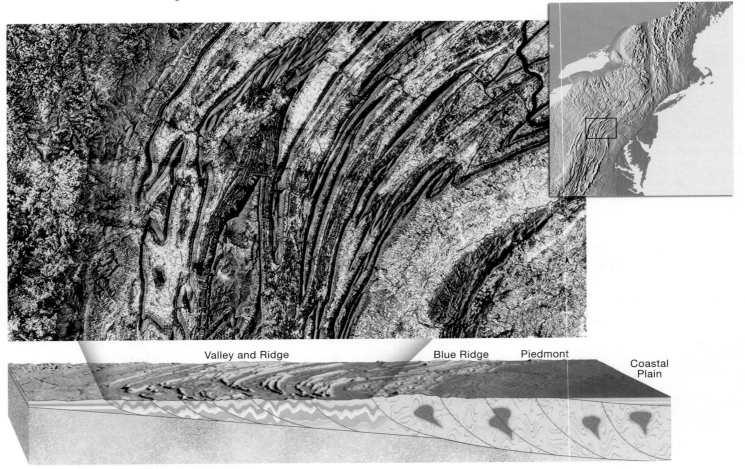

FIGURE 14.18   The Valley and Ridge Province. This portion of the Appalachian Mountains consists of folded and faulted sedimentary strata that were displaced landward with the closing of the proto-Atlantic. (LANDSAT image, courtesy of Phillips Petroleum Company, Exploration Projects Section)

dense (more buoyant) and rises. This accounts, in part, for the elevated topography associated with continental rifts. Simultaneously, stretching elongates the rigid upper crust, which breaks into large crustal blocks that are bounded by high-angle normal-faults. Continued rifting causes the blocks to tilt with one edge rising as the other drops (see Figure 10.23, p. 294). Mountains that form in these tectonic settings are termed **fault-block mountains.**

The Teton Range in western Wyoming is an excellent example of fault-block mountains. This lofty structure was faulted and uplifted along its eastern flank as the block tilted downward to the west. Looking west from Jackson Hole, Wyoming, the eastern front of this mountain rises more than 2 kilometers above the valley making it one of the most imposing mountain fronts in the United States (Figure 14.19).

Basin and Range Province

Located between the Sierra Nevada and Rocky Mountains is one of Earth's largest regions of fault-block mountains—the Basin and Range Province (Figure 14.20). This region extends in a roughly north-south direction for nearly 3000 kilometers (2000 miles) and encompasses all of Nevada and portions of the surrounding states, as well as parts of southern Canada and western Mexico. In this region, the brittle upper crust has literally been broken into hundreds of fault blocks. Tilting of these faulted structures called *half-grabens,* gave rise to nearly parallel mountain ranges, averaging about 80 kilometers in length, which rise above adjacent sediment-filled basins (see Figure 10.23, p. 294).

Several hypotheses have been proposed to explain the events that generated the Basin and Range (see Figure 14.20). One proposal suggests that about 20 million years ago, tensional forces became dominant in the western United States and stretched the crust to twice its original width. Figure 14.21 shows a rough outline of the boundaries of the western states prior to and following this event. The extension thinned and weakened the lithosphere resulting in upwelling of hot mantle rock. High heat flow in the region and several episodes of volcanism provide strong evidence that mantle upwelling accompanied crustal uplift.

Another model suggests that the cold, dense lithospheric mantle located beneath the Basin and Range decoupled (separated) from the overlying crustal layer and slowly sank into the mantle. This process, called *delamination,* resulted in upwelling and lateral spreading of hot mantle rock that produced tensional forces that stretched and thinned the crust. According to this view, these elevated crustal blocks began to gravitationally slide from their lofty perches to generate the fault-block topography of the Basin and Range Province (see Figure 14.20).

FIGURE 14.19   The mountains of Wyoming's Teton Range are fault-block mountains. (Photo by Stafano Amantini/Atlantide Phototravel/CORBIS)

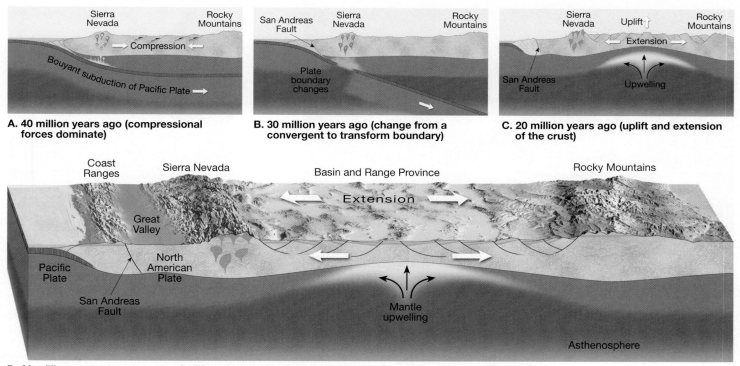

D. 20 million years ago to present (uplift and extension creates the topography of the Basin and Range)

FIGURE 14.20   The Basin and Range Province consists of numerous fault-block mountains that were generated during the past 20 million years. Upwelling of hot mantle rock and perhaps gravitational collapse (crustal sliding) contributed to considerable stretching and thinning of the crust.

UNDERSTANDING EARTH

BOX 14.2

The Southern Rockies

The portion of the Rocky Mountains that extends from southern Montana to New Mexico was produced during a period of deformation known as the *Laramide Orogeny.* This event, which created some of the most picturesque scenery in the United States, peaked about 60 million years ago (Figure 14.B). The mountain ranges generated during the Laramide Orogeny include the Bighorns of Wyoming, the Front Range of Colorado, and the Sangre de Cristo of Colorado and New Mexico.

These mountain ranges formed when deeply buried crystalline rock was lifted nearly vertically along steeply dipping faults, upwarping the overlying layers of younger sedimentary rocks. Uplifting accelerated the processes of weathering and erosion, which removed much of the sedimentary cover from the highest portions of the uplifted blocks. The resulting mountainous topography consists of dissected blocks of Precambrian igneous and metamorphic rocks that are separated by sediment-filled basins. Remnants of the large uplifted blocks include a number of steep summits, such as Pike's Peak and Long's Peak in Colorado's Front Range. In many areas, the sedimentary strata that previously covered this region form prominent angular ridges, called *hogbacks,* that flank the crystalline cores of the mountains (Figure 14.C).

Until recently, geologists assumed the middle and southern Rockies stood tall because the crust had been thickened by past tectonic events. However, seismic studies conducted across this region showed that the crustal thickness was no greater than that of the adjacent Great Plains. These data eliminated crustal buoyancy as the cause for the abrupt 2-kilometer (1.2-mile) jump in elevation that occurs where the Great Plains meet the Rockies.

Although the southern Rockies have been studied extensively for more than a century, there is still considerable debate regarding the mechanisms that led to uplift. One hypothesis that is illustrated in Figure 14.20 parts A-C, p. 397, proposes that this period of uplift was associated with the low-angle subduction of the Farallon plate eastward beneath North America. As the subducted slab decended beneath the continent, compressional forces initiated a period of tectonic activity. Eventually, the comparatively cool Farallon plate sank, and was replaced by hot rock that upwelled from the mantle. Thus, according to this scenario, the hot mantle provided the buoyancy to raise the southern Rockies, as well as the Colorado Plateau and the mountains of the Basin and Range.

Others disagree, maintaining that plate convergence and the collision of one or more microcontinents with the western margin of North America was the driving force behind the Laramide Orogeny (see the section entitled "Terranes and Mountain Building").

It should be pointed out that neither of these proposals has gained widespread support. As one geologist familiar with this region put it, "We just don't know."

FIGURE 14.B   The spectacular Maroon Bells are part of the Colorado Rockies. (Photo by Peter Saloutos/The Stock Market)

FIGURE 14.C   Hogback ridges in the Rocky Mountains of Colorado. This is a view looking along the east flank of the Front Range. These upturned sedimentary rocks are remnants of strata that once covered the Precambrian igneous and metamorphic core of the mountains to the west. (Photo by Fabio Somenzi)

FIGURE 14.21 Extension in the Basin and Range Province has "stretched" the crust in some locations by as much as twice its original width. Shown here is a rough outline of the western states before (left) and after (right) extension.

VERTICAL MOVEMENTS OF THE CRUST

Gradual up-and-down motions of the continental crust occur at many locations around the globe. Unfortunately, the reasons for these changes are not easily determined.

Isostasy

During the 1840s, researchers discovered that Earth's low-density crust floats on top of the high-density, deformable rocks of the mantle. The concept of a floating crust in gravitational balance is called **isostasy** (*iso* = equal, *stasis* = standing). One way to explore this concept is to envision a series of wooden blocks of different heights floating in water, as shown in Figure 14.22. Note that the thicker wooden blocks float higher than the thinner blocks. Similarly, compressional mountains stand high above the surrounding terrain because crustal thickening creates buoyant crustal "roots" that extend deep into the supporting material below (see Box 14.3). Thus, lofty mountains behave much like the thicker wooden blocks shown in Figure 14.22.

Isostatic Adjustment   Visualize what would happen if another small block of wood were placed atop one of the blocks in Figure 14.22. The combined block would sink until it reached a new isostatic (gravitational) balance. At this point, the top of the combined block would be higher than before, and the bottom would be lower. This process of establishing a new level of gravitational equilibrium is called **isostatic adjustment.**

Applying the concept of isostatic adjustment, we should expect that when weight is added to the crust, it will respond by subsiding, and when weight is removed, it will rebound. (Visualize what happens when a ship's cargo is loaded or unloaded.) Evidence for crustal subsidence followed by crustal rebound is provided by Ice Age glaciers. When continental ice sheets occupied portions of North America during the Pleistocene epoch, the added weight of 3-kilometer-thick masses of ice caused downwarping of Earth's crust by hundreds of meters. In the 8000 years since this last ice sheet melted, uplifting of as much as 330 meters (1000 feet) has occurred in Canada's Hud-

son Bay region, where the thickest ice had accumulated (see Figure 18.28, p. 508).

One of the consequences of isostatic adjustment is that, as erosion lowers the summits of mountains, the crust rises in response to the reduced load (Figure 14.23). The processes of uplift and erosion continue until the mountain block reaches "normal" crustal thickness. When this occurs, these once elevated structures will be near sea level, and the once deeply buried interior of the mountain will be exposed at the surface. In addition, as mountains are worn down, the eroded sediment is deposited on adjacent landscapes, causing these areas to subside (Figure 14.23).

How High Is Too High?   Where compressional forces are great, such as those driving India into Asia, lofty mountains such as the Himalayas result. But is there a limit on how high a mountain can rise? As mountaintops are elevated, gravity-driven processes such as erosion and mass wasting are accelerated, carving the deformed strata into rugged landscapes. Equally important, however, is the fact that gravity also acts on the rocks within these massive structures. The higher the mountain, the greater the downward force on the rocks near the base. Eventually, the rocks deep within the developing mountain, which are comparatively warm and weak, begin to flow laterally, as shown in Figure 14.24. This is analogous to a ladle of very thick pancake batter poured on a hot griddle. Similarly, mountains are altered by a process called **gravitational collapse,** which involves ductile spreading at depth and normal faulting and subsidence in the upper, brittle portion of the crust.

Considering these factors, a seemingly logical question follows—"What keeps the Himalayas standing?" Simply, the horizontal compressional forces that are driving India into Asia are greater than the vertical force of gravity. However, once India's northward trek ends, the downward pull of gravity will become the dominant force acting on this mountainous region.

Mantle Convection: A Cause of Vertical Crustal Movement

Based on studies of Earth's gravitational field, it became apparent that up-and-down convective flow in the mantle also affects the elevation of Earth's major landforms. The buoyancy of hot rising

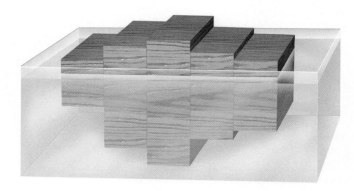

FIGURE 14.22   This drawing illustrates how wooden blocks of different thicknesses float in water. In a similar manner, thick sections of crustal material float higher than thinner crustal slabs.

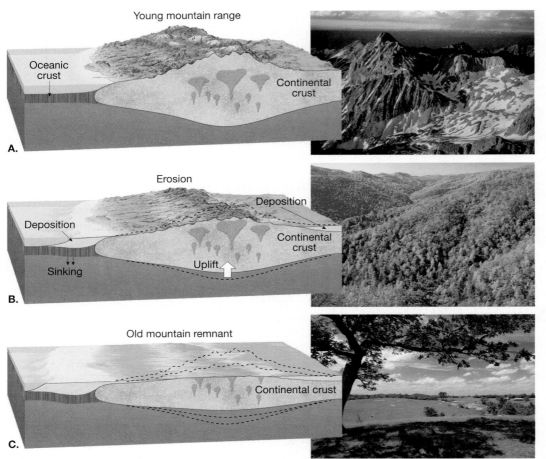

FIGURE 14.23 This sequence illustrates how the combined effects of erosion and isostatic adjustment result in a thinning of the crust in mountainous regions. **A.** When mountains are young, the continental crust is thickest. (Photo by Michael Collier) **B.** As erosion lowers the mountains, the crust rises in response to the reduced load. (Photo by Mark Karrass/CORBIS) **C.** Erosion and uplift continue until the mountains reach "normal" crustal thickness. (Photo by Pat & Chuck Blackley/Alamy)

elevation of nearly 1500 meters (5000 feet)—much higher than what would be predicted for a stable continental platform.

Evidence from seismic tomography (see Figure 12.20, p. 347) indicates that a large mass of hot mantle rock is centered below the southern tip of Africa. This structure, called a *superplume,* extends upward about 2900 kilometers (1800 miles) from the mantle–core boundary and has a lateral expanse of a few thousand kilometers. Researchers have determined that the upward flow associated with this huge mantle plume is sufficient to elevate southern Africa.

Crustal Subsidence Extensive areas of downwarping also occur at Earth's surface. In the United States, large, nearly circular basins are found in Michigan and Illinois. Similar structures are known on other continents as well.

The cause of the downwarping that created these basins may be linked to the subduction of slabs of oceanic lithosphere. One proposal suggests that when subduction ceases, the descending slab detaches from the trailing lithosphere and continues its descent into the mantle. As this detached lithospheric slab sinks, it creates a downward flow in its wake that tugs at the base of the overriding continent. In some settings, the crust is apparently pulled down sufficiently to produce a large basin that eventually fills with sediments. As the oceanic slab sinks deeper into the mantle, the pull of the trailing wake weakens and the continent "floats" back into isostatic balance.

material accounts for broad upwarping in the overlying lithosphere, whereas downward flow causes downwarping.

Uplifting Whole Continents Southern Africa is one region where large-scale vertical motion is evident. The topography of the region consists of an expansive plateau having an average

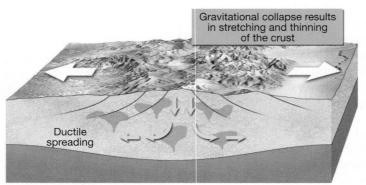

A. Horizontal compressional forces dominate

B. Gravitational forces dominate

FIGURE 14.24 Block diagram of a mountain belt that is collapsing under its own "weight." Gravitational collapse involves normal faulting in the upper, brittle portion of the crust and ductile spreading in the warm, weak rocks at depth.

UNDERSTANDING EARTH

BOX 14.3

Do Mountains Have Roots?

One of the major advances in determining the structure of mountains occurred in the 1840s when Sir George Everest (for whom Mount Everest is named) conducted the first topographic survey in India. This survey measured the distance between the towns of Kalianpur and Kaliana, located south of the Himalayas, using two different methods. One method employed the conventional surveying technique of triangulation, while the other method calculated the distance astronomically. Although the two techniques were expected to produce similar results, the astronomical calculations placed the towns nearly 150 meters closer to each other than the triangulation survey.

The discrepancy was attributed to the gravitational attraction exerted by the massive Himalayas on the plumb bob used for leveling the astronomical instrument. (A plumb bob is a metal weight suspended by a cord, used to determine a vertical orientation.) It was suggested that the deflection of the plumb bob would be greater at Kaliana than at Kalianpur because of its proximity to the mountains (Figure 14.D).

A few years later, researcher J. H. Pratt estimated the mass of the Himalayas and calculated the magnitude of the deflection attributable to the gravitational influence of these

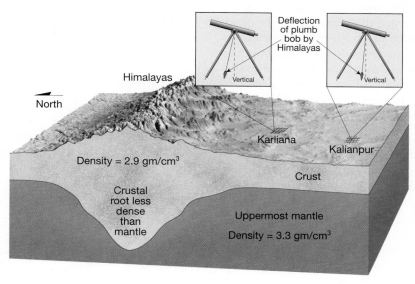

FIGURE 14.D During the first survey of India, an error in measurement occurred because the plumb bob on an instrument was deflected by the massive Himalayas. Later work by George Airy predicted that the mountains have roots of light crustal rocks. Airy's model explained why the plumb bob was deflected much less than expected.

massive mountains. To his surprise, Pratt discovered that the mountains should have deflected the plumb blob three times the amount that actually occurred. Simply stated, the mountains were not "pulling their weight." It was as if they had a hollow central core.

The hypothesis which explained the "missing" mass was developed by another scientist, George Airy. Airy suggested that Earth's lighter crustal rocks float on the denser, more easily deformed mantle. Further, he correctly argued that the crust must be thicker under mountains than beneath the adjacent lowlands. In other words, mountainous terrains are supported by light crustal material that extends as "roots" into the denser mantle (see Figure 14.D). This phenomenon is exhibited by icebergs, which are buoyed by the weight of the displaced water. If the Himalayas have roots of light crustal rocks that extend far beneath them, then the massive structures would exert less gravitational attraction than was predicted. Hence, Airy's model explained why the plumb bob was deflected much less than expected.

More recently, seismological and gravitational studies have confirmed the existence of crustal roots under most mountain ranges. The thickness of continental crust is normally about 35 kilometers, but crustal thicknesses twice that amount have been determined for some mountain belts.

CHAPTER 14 CONVERGENT BOUNDARIES: ORIGIN OF MOUNTAINS IN REVIEW

- The name for the processes that collectively produce a *compressional mountain belt* is *orogenesis.* Most compressional mountains consist of folded and faulted sedimentary and volcanic rocks, portions of which have been strongly metamorphosed and intruded by younger igneous bodies.

- Plate convergence can result in a subduction zone consisting of four regions: (1) a *deep-ocean trench* that forms where a subducting slab of oceanic lithosphere bends and descends into the asthenosphere; (2) a *volcanic arc,* which is built upon the overlying plate; (3) a *forearc region* located between the trench and the volcanic arc; and (4) a *backarc*

region on the side of the volcanic arc opposite the trench. Along some subduction zones backarc spreading results in the formation of a *backarc basin.*

- Subduction of oceanic lithosphere beneath a continental block gives rise to an *Andean-type plate margin* that is characterized by a continental volcanic arc and associated igneous plutons. In addition, sediment derived from the land, as well as material scraped from the subducting plate, becomes plastered against the landward side of the trench, forming an *accretionary wedge.*

- Mountain belts can develop as a result of the collision and merger of one or more crustal fragments, including island arcs and oceanic plateaus to a continental block. Many of the mountain belts of the North American Cordillera were generated in this manner.

- Continued subduction of oceanic lithosphere beneath an Andean-type continental margin will eventually close an ocean basin. The result will be a *continental collision* and the development of compressional mountains that are characterized by shortened and thickened crust. The development of a major mountain belt is often complex, involving two or more distinct episodes of mountain building. Common features of compressional mountains are *fold-and-thrust belts*. Continental collisions have generated most of Earth's major mountain belts, including the Himalayas, Alps, Urals, and Appalachians.

- Although most mountains form along convergent plate boundaries, other tectonic processes, such as continental rifting, can produce topographic mountains. The *fault-block mountains* that form in these settings are bounded by high-angle normal faults. The Basin and Range Province in the western United States consists of hundreds of tilted, faulted blocks that give rise to nearly parallel mountain ranges that stand above sediment-laden basins.

- Earth's crust floats on top of the deformable rocks of the mantle, much like wooden blocks float in water. The concept of a floating crust in gravitational balance is called *isostasy*. Most mountains consist of crustal rocks that have been shortened and thickened producing deep crustal roots that isostatically support them. As erosion lowers the peaks, *isostatic adjustment* gradually raises the mountains in response. The processes of uplift and erosion will continue until the mountain block reaches "normal" crustal thickness. Gravity can also cause these elevated structures to collapse under their own "weight."

- Convective flow in the mantle contributes to the up-and-down bobbing of the crust. The upward flow of a large superplume located beneath southern Africa is thought to have elevated this region by as much as 1500 meters. Crustal subsidence has produced large basins.

KEY TERMS

accretionary wedge (p. 388)	forearc basin (p. 389)
Andean-type plate margin (p. 386)	gravitational collapse (p. 399)
backarc basin (p. 385)	island arc (p. 384)
compressional mountains (p. 382)	isostasy (p. 399)
continental volcanic arc (p. 384)	isostatic adjustment (p. 399)
fault-block mountains (p. 396)	microcontinent (p. 390)
fold-and-thrust belts (p. 391)	orogenesis (p. 382)
	suture (p. 391)
	terrane (p. 390)
	volcanic island arc (p. 384)

QUESTIONS FOR REVIEW

1. In the plate tectonics model, which type of plate boundary is most directly associated with Earth's major mountain belts?

2. List the four main regions of a subduction zone, and describe where each is located relative to the others.

3. Briefly describe how backarc basins form.

4. Describe the process that generates most basaltic magma at subduction zones.

5. How are magmas that exhibit an intermediate to felsic composition thought to be produced from mantle-derived basaltic magmas at Andean-type plate margins?

6. What is a batholith? In what modern tectonic setting are batholiths being generated?

7. In what ways are the Sierra Nevada and the Andes similar?

8. What is an accretionary wedge? Briefly describe its formation.

9. What is a passive margin? Give an example. Provide an example of an active continental margin.

10. The formation of mountainous topography at a volcanic island arc is considered just one phase in the development of a major mountain belt. Explain.

11. In what way are the Coast Ranges of California related to subduction of oceanic lithosphere?

12. Suture zones are often described as the place where continents are "welded" together. Why might that statement be misleading?

13. During the formation of the Himalayas, the continental crust of Asia was deformed more than India proper. Why was this the case?

14. Where might magma be generated in a newly formed collision mountain belt?

15. Suppose a sliver of oceanic crust was discovered in the interior of a continent. Would this support or refute the theory of plate tectonics? Explain.

16. How can the Appalachian Mountains be considered a collision-type mountain range when the nearest continent is 5000 kilometers (3000 miles) away?

17. How does the plate tectonics theory help explain the existence of fossil marine life in rocks atop compressional mountains?

18. In your own words, briefly describe the stages in the formation of a major mountain belt according to the plate tectonics model.

19. How is *terrane* different from *terrain*?

20. In addition to microcontinents, what other structures are carried by the oceanic lithosphere and eventually accreted to a continent?

21. Briefly describe the major differences between the evolution of the Appalachian Mountains and the North American Cordillera.

22. Compare the processes that generate fault-block mountains to those associated with most other major mountain belts.

23. Give one example of evidence that supports the concept of crustal uplift.

24. What happens to a floating object when weight is added? Subtracted? How does this principal apply to changes in the elevations of mountains? What term is applied to crustal uplift of this type?

25. How do some researchers explain the elevated position of southern Africa?

COMPANION WEBSITE

The *Earth 10e* Web site uses the resources and flexibility of the Internet to aid in your study of the topics in this chapter. Written and developed by the authors and other geology instructors, this site will help improve your understanding of geology. Visit www.mygeoscienceplace.com in order to:

- **Review** key chapter concepts.
- **Read** with links to the eBook and to chapter-specific web resources.

- **Visualize** and comprehend challenging topics using learning activities in *GEODe Earth.*
- **Test** yourself with online quizzes.

GEODe EARTH

GEODe Earth is a valuable and easy to use learning aid that can be accessed from your book's Companion Website (www.mygeoscienceplace.com). It is a dynamic instructional tool that promotes understanding and reinforces important concepts by using tutorials, animations, and exercises that actively engage the student.

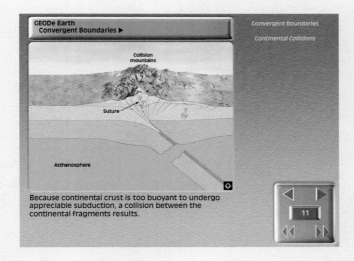

MASS WASTING
THE WORK OF GRAVITY

Landslide debris atop Buckskin Glacier in Alaska's Denali National Park. In the evolution of many landforms, gravity-driven mass wasting processes transfer the products of weathering downslope where an erosional agent, in this case a valley glacier, carries the debris away.

(PHOTO BY MICHAEL COLLIER)

Earth's surface is never perfectly flat, but instead consists of slopes of many different varieties. Some are steep and precipitous; others are moderate or gentle. Some are long and gradual; others are short and abrupt. Slopes can be mantled with soil and covered by vegetation or consist of barren rock and rubble. Taken together, slopes are the most common elements in our physical landscape. Although most slopes may appear to be stable and unchanging, the force of gravity causes material to move downslope. At one extreme, the movement may be gradual and practically imperceptible. At the other extreme, it may consist of a roaring debris flow or a thundering rock avalanche. Landslides are a worldwide natural hazard. When these hazardous processes lead to loss of life and property, they become natural disasters.

LANDSLIDES AS NATURAL DISASTERS

Even in areas with steep slopes, catastrophic landslides are relatively rare occurrences. As a result, people living in susceptible areas often do not appreciate the risk of living where they do. However, media reports remind us that such events occur with some regularity around the world (Figure 15.1). The three examples described here occurred during a span of just four months.

On October 8, 2005, a magnitude 7.6 earthquake struck the Kashmir region between India and Pakistan. Compounding the tragic effects caused directly by the severe ground shaking were hundreds of landslides triggered by the quake and its many aftershocks. Rockfalls and debris slides thundered down steep mountain slopes and were focused into narrow valleys where many people made their homes. The landslides also blocked roads and trails, delaying attempts to reach those in need.

Just three days earlier, on October 5, 2005, torrential rains from Hurricane Stan triggered mudflows in Guatemala. A slurry of mud 1 kilometer wide and up to 12 meters (40 feet) deep buried the village of Panabaj. Death toll estimates for the area approached 1400 people. Flows such as this can travel at speeds of 50 kilometers (30 miles) per hour down rugged mountain slopes.

On February 17, 2006, only a few months after the tragedy in Central America, a lethal mudflow, triggered by extraordinary rains, buried a small town on the Philippine island of Leyte. A mass of mud engulfed this remote coastal area to depths as great as 10 meters (30 feet). Although an accurate count of fatalities was difficult, as many as 1800 people perished. This region is prone to such events, due in part to the fact that deforestation has denuded the nearby mountain slopes. In the pages that follow you will take a closer look at rapid mass-wasting events in an attempt to better understand their causes and effects.

MASS WASTING AND LANDFORM DEVELOPMENT

Landslides are spectacular examples of a basic geologic process called mass wasting. **Mass wasting** refers to the downslope movement of rock, regolith, and soil under the direct influence of gravity. It is distinct from the erosional processes that are examined in subsequent chapters because mass wasting does not require a transporting medium such as water, wind, or glacial ice.

The Role of Mass Wasting

In the evolution of most landforms, mass wasting is the step that follows weathering. By itself, weathering does not produce

FIGURE 15.1 **A.** On October 8, 2005, a major earthquake in Kashmir triggered hundreds of landslides including the one shown here. (AP Photo/Burhan Ozblici) **B.** In February 2006, heavy rains triggered this mudflow that buried a small town on the Philippine island of Leyte. (AP Photo/Pat Roque)

the significance of mass-wasting processes in supplying material to streams. This is illustrated by the Grand Canyon (Figure 15.2). The walls of the canyon extend far from the Colorado River, owing to the transfer of weathered debris downslope to the river and its tributaries by mass-wasting processes. In this manner, streams and mass wasting combine to modify and sculpt the surface. Of course, glaciers, groundwater, waves, and wind are also important agents in shaping landforms and developing landscapes.

Slopes Change Through Time

It is clear that if mass wasting is to occur, there must be slopes that rock, soil, and regolith can move down. It is Earth's mountain building and volcanic processes that produce these slopes through sporadic changes in the elevations of landmasses and the ocean floor. If dynamic internal processes did not continually produce regions having higher elevations, the system that moves debris to lower elevations would gradually slow and eventually cease.

Most rapid and spectacular mass-wasting events occur in areas of rugged, geologically young mountains. Newly formed mountains are rapidly eroded by rivers and glaciers into regions characterized by steep and unstable slopes. It is in such settings that massive destructive landslides, such as those described at the beginning of the chapter, occur. As mountain building subsides, mass wasting and erosional processes lower the land. Through time, steep and rugged mountain slopes give way to gentler, more subdued terrain. Thus, as a landscape ages, massive and rapid mass-wasting processes give way to smaller, less dramatic downslope movements.

significant landforms. Rather, landforms develop as products of weathering are removed from the places where they originate. Once weathering weakens and breaks rock apart, mass wasting transfers the debris downslope, where a stream, acting as a conveyor belt, usually carries it away. Although there may be many intermediate stops along the way, the sediment is eventually transported to its ultimate destination: the sea.

The combined effects of mass wasting and running water produce stream valleys, which are the most common and conspicuous of Earth's landforms. If streams alone were responsible for creating the valleys in which they flow, the valleys would be very narrow features. However, the fact that most river valleys are much wider than they are deep is a strong indication of

CONTROLS AND TRIGGERS OF MASS WASTING

MASS WASTING
▶ Controls and Triggers of Mass Wasting

Gravity is the controlling force of mass wasting, but several factors play an important role in overcoming inertia and creating downslope movements. Long before a landslide occurs, various processes work to weaken slope material, gradually making it more and more susceptible to the pull of gravity. During this span, the slope remains stable but gets closer and closer to being unstable. Eventually, the strength of the slope is weakened to the point that something causes it to cross the threshold from stability to instability. Such an event that initiates downslope movement is called a *trigger*. Remember that the trigger is not the sole cause of the mass-wasting event but just the last of many causes. Among the common factors that trigger mass-wasting processes are saturation of material with water, oversteepening of slopes, removal of anchoring vegetation, and ground vibrations from earthquakes.

Students Sometimes Ask . . .

It seems as though you have used the term "landslide" to refer to several different things—from mudflows to rock avalanches. What exactly is the definition of "landslide"?

Although many people, including geologists, frequently use the word *landslide,* the term has no specific definition in geology. Rather, it is a popular, nontechnical term used to describe any or all relatively rapid forms of mass wasting.

Colorado River →

Sedimentary layers

Material eroded by running water

ls
ss
Shale
ss
Shale

ls

Shale
ss
Shale

Metamorphic rocks

Weathered debris moved downslope by mass wasting

Colorado River

Geologist's Sketch

FIGURE 15.2 The walls of the Grand Canyon extend far from the channel of the Colorado River. This results primarily from the transfer of weathered debris downslope to the river and its tributaries by mass wasting processes. (Photo by Bryan Brazil/Shutterstock)

The Role of Water

Mass wasting is sometimes triggered when heavy rains or periods of snow-melt saturate surface materials. This was the case in October 1998 when torrential downpours associated with Hurricane Mitch triggered devastating mudflows in Central America (Figure 15.3). Box 15.1 presents a case study of another event that occurred at La Conchita, California, in January 2005.

When the pores in sediment become filled with water, the cohesion among particles is destroyed, allowing them to slide past one another with relative ease. For example, when sand is

FIGURE 15.3 Many large debris flows and floods were triggered by the torrential rains that accompanied Hurricane Mitch when it struck Honduras in October 1998. It was that country's worst natural disaster in 200 years. Pictured here is the El Berrinche landslide one of two that destroyed portions of the city of Tegucigalpa, killing more than 1000 people and damming the Rio Choluteca. The mass of debris flow material was estimated to be 6 million cubic meters, enough to fill nearly 300 thousand average dump trucks! (Photos by Michael Collier)

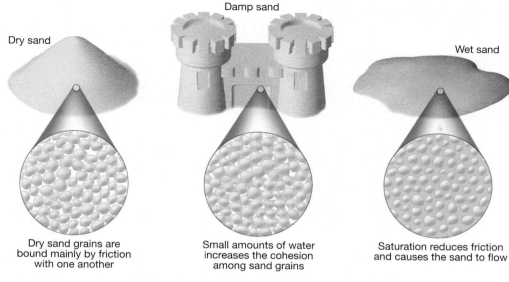

Dry sand

Damp sand

Wet sand

Dry sand grains are bound mainly by friction with one another

Small amounts of water increases the cohesion among sand grains

Saturation reduces friction and causes the sand to flow

FIGURE 15.4   The effect of water on mass wasting can be great. When little or no water is present, friction among the closely packed soil particles on the slope holds them in place. When the soil is saturated, the grains are forced apart and friction is reduced, allowing the soil to move downslope.

slightly moist, it sticks together quite well. However, if enough water is added to fill the openings between the grains, the sand will ooze out in all directions (Figure 15.4). Thus, saturation reduces the internal resistance of materials, which are then easily set in motion by the force of gravity. When clay is wetted, it becomes very slick—another example of the "lubricating" effect of water. Water also adds considerable weight to a mass of material. The added weight in itself may be enough to cause the material to slide or flow downslope.

Oversteepened Slopes

Oversteepening of slopes is another trigger of many mass movements. There are many situations in nature where oversteepening takes place. A stream undercutting a valley wall and waves pounding against the base of a cliff are but two familiar examples. Furthermore, through their activities, people often create oversteepened and unstable slopes that become prime sites for mass wasting (Figure 15.5).

Unconsolidated, granular particles (sand-size or coarser) assume a stable slope called the **angle of repose** (*reposen* = to be at rest). This is the steepest angle at which material remains stable (Figure 15.6). Depending on the size and shape of the particles, the angle varies from 25 to 40 degrees. The larger, more angular particles maintain the steepest slopes. If the angle is increased, the rock debris will adjust by moving downslope.

Oversteepening is not just important because it triggers movements of unconsolidated granular materials. Oversteepening also produces unstable slopes and mass movements in cohesive soils, regolith, and bedrock. The response will not be immediate, as with loose, granular material, but sooner or later, one or more mass-wasting processes will eliminate the oversteepening and restore stability to the slope.

Removal of Vegetation

Plants protect against erosion and contribute to the stability of slopes because their root systems bind soil and regolith together. In addition, plants shield the soil surface from the erosional effects of raindrop impact (see Figure 6.23, p. 192). Where plants are lacking, mass wasting is enhanced, especially if slopes are steep and water is plentiful. When anchoring vegetation is removed by forest fires or by people (for timber, farming, or development), surface materials frequently move downslope.

An unusual example illustrating the anchoring effect of plants occurred several decades ago on steep slopes near Menton, France. Farmers replaced olive trees, which have deep roots, with a more profitable but shallow-rooted crop: carnations. When the less stable slope failed, the landslide took 11 lives.

In July 1994 a severe wildfire swept Storm King Mountain west of Glenwood Springs, Colorado, denuding the slopes of vegetation. Two months later, heavy rains resulted in numerous debris

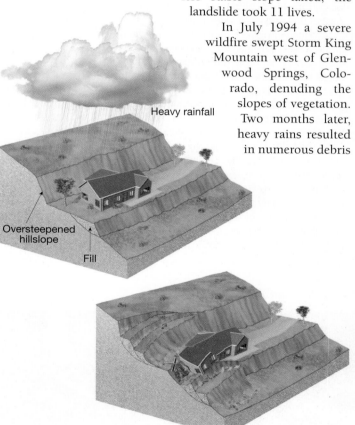

Heavy rainfall

Oversteepened hillslope

Fill

FIGURE 15.5   When slopes are oversteepened and made unstable, they are prime sites for mass wasting. Natural processes such as stream and wave erosion can oversteepen slopes. Changing the slope to accommodate a new house or road can also lead to instability and a destructive mass wasting event.

PEOPLE AND THE ENVIRONMENT

BOX 15.1

Landslide Hazards at La Conchita, California*

Southern California lies astride a major plate boundary defined by the San Andreas Fault and numerous other related faults that are spread across the region. It is a dynamic environment characterized by rugged mountains and steep-walled canyons. Unfortunately, this scenic landscape presents serious geologic hazards. Just as tectonic forces are steadily pushing the landscape upward, gravity is relentlessly pulling it downward. When gravity prevails, landslides occur.

As you might expect, some of the region's landslides are triggered by earthquakes. Many others, however, are related to periods of prolonged and intense rainfall (Figure 15.A). A tragic example of the latter situation occurred on January 10, 2005, when a massive debris flow (popularly called a *mudslide*) swept through La Conchita, California, a small town located about 80 kilometers (50 miles) northwest of Los Angeles (Figure 15.B).

Although the rapid torrent of mud took many of the town's inhabitants by surprise, such an event should not have been unexpected. Let's briefly examine the factors that contributed to the deadly debris flow at La Conchita.

The town is situated on a narrow coastal strip about 250 meters (800 feet) wide between the shoreline and a steep 180-meter (600-foot) bluff. The bluff consists of poorly sorted marine sediments and weakly cemented layers of shale, siltstone, and sandstone.

The deadly 2005 debris flow involved little or no newly failed material, but rather consisted of the remobilization of a portion of a large landslide that destroyed several homes in 1995. In fact, historical accounts dating back to 1865

indicate that landslides in the immediate area have been a regular occurrence. Further, geologic evidence shows that landsliding of a variety of types and scales has probably been occurring at La Conchita for thousands of years.

The most significant contributing factor to the tragic 2005 debris flow was prolonged and intense rain. The event occurred at the end of a span that produced near record amounts of rainfall in Southern California. Wintertime rainfall at nearby Ventura totaled 49.3 centimeters (19.4 inches) as compared to an average value of just 12.2 centimeters (4.8 inches). As Figure 15.A indicates, much of that total fell during the two weeks immediately preceding the debris flow.

This was not the first destructive landslide to strike La Conchita, nor is it likely to be the last. The town's geologic setting and history of rapid mass-wasting events clearly support this notion. When the amount and intensity of rainfall is sufficient, debris flows are to be expected. The concluding paragraph from a U.S. Geological Survey report puts it this way:

The La Conchita area has experienced, and will likely continue to experience, a rather bewildering variety of landslide hazards. Different landslide scenarios are more or less likely to occur as a result of different specific rainfall conditions, and no part of the community can be considered safe from landslides. Unfortunately, we currently lack the understanding to accurately forecast what might happen in each possible rainfall scenario. Prudence would certainly dictate, however, that we anticipate renewed landslide activity during or after future periods of prolonged and/or

intense rainfall. Future earthquakes, of course, also could trigger landsliding in the area.**

*Based in part on material prepared by the U.S. Geological Survey.

**Jibsen, Randall W. "Landslide Hazards at La Conchita, California," *U.S. Geological Survey Open-File Report 2005–1067*, p. 11.

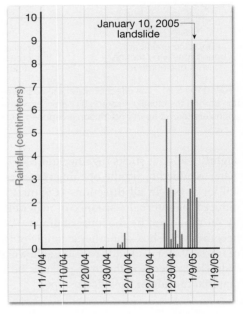

FIGURE 15.A   Daily rainfall at the nearby town of Ventura during the weeks leading up to the January 2005 La Conchita event. Each line on the bar graph shows rain for a particular day. The 2005 debris flow occurred at the culmination of the heaviest rainfall of the season. About 80 percent of the season's exceptional total fell in this short span. (After National Weather Service)

FIGURE 15.6   The angle of repose for this granular material is about 30 °. (Photo by G. Leavens/Photo Researchers)

flows, one of which blocked Interstate 70 and threatened to dam the Colorado River. A 5-kilometer (3-mile) length of the highway was inundated with tons of rock, mud, and burned trees. The closure of Interstate 70 imposed costly delays on this major highway.

Wildfires are inevitable in the western United States and fast-moving, highly destructive debris flows triggered by intense rainfall are one of the most dangerous post-fire hazards (Figure 15.7). Such events are particularly dangerous because they tend to occur with little warning. Their mass and speed make them particularly destructive. Post-fire debris flows are most common in the 2 years after a fire. Some of the largest debris-flow events have been triggered by the very first intense rain event following the wildfire. It

FIGURE 15.B   Views of the La Conchita debris flow taken shortly after it occurred in January 2005. The light-colored exposed rock in the upper part of the photo on the left is the main scarp of a slide that occurred 10 years earlier in 1995. The January 2005 event was a remobilization of the 1995 event. The flow was quite thick (viscous) and moved houses in its path rather than flowing around them. (AP Wide World Photos)

takes much less rain to trigger debris flows in burned areas than in unburned areas. In southern California, as little as 7 millimeters (0.3 inch) of rain in 30 minutes has triggered debris flows.

How large can these flows be? According to the U.S. Geological Survey, documented debris flows from burned areas in southern California and other western states have ranged in volume from as small as 600 cubic meters to as much as 300,000 cubic meters. This larger volume is enough material to cover a football field with mud and rocks to a depth of about 65 meters (almost 215 feet)!

In addition to eliminating plants that anchor the soil, fire can promote mass wasting in other ways. Following a wildfire, the upper part of the soil may become dry and loose. As a result, even in dry weather, the soil tends to move down steep slopes.

Moreover, fire can also "bake" the ground, creating a water-repellant layer at a shallow depth. This nearly impermeable barrier prevents or slows the infiltration of water, resulting in increased surface runoff during rains. The consequence can be dangerous torrents of viscous mud and rock debris.

Earthquakes as Triggers

Conditions that favor mass wasting may exist in an area for a long time without movement occurring. An additional factor is sometimes necessary to trigger the movement. Among the more important and dramatic triggers are earthquakes. An earthquake and its aftershocks can dislodge enormous

FIGURE 15.7 During summer, wildfires are common occurrences in many parts of the West. Millions of acres are burned each year. The loss of anchoring vegetation sets the stage for accelerated mass wasting. (Photo by Raymond Gehman)

volumes of rock and unconsolidated material. The event in the Kashmir region described near the beginning of the chapter is one tragic example.

Examples from California and China   A memorable United States example occurred in January 1994 when a quake struck the Los Angeles region of southern California. Named for its epicenter in the town of Northridge, the 6.7-magnitude event produced estimated losses of $20 billion. Some of the losses were the result of more than 11,000 landslides in an area of about 10,000 square kilometers (3900 square miles) that were set in motion by the quake. Most were shallow rock falls and slides, but some were much larger and filled canyon bottoms with jumbles of soil, rock, and plant debris. The debris in canyon bottoms created a secondary threat because it can mobilize during rainstorms, producing debris flows. Such flows are common and often disastrous in southern California.

On May 12, 2008, a magnitude 7.9 earthquake struck near the city of Chengdu in China's Sichuan Province. Compounding the tragic effects caused directly by the severe ground shaking were hundreds of landslides triggered by the quake and its many aftershocks (Figure 15.8). Rock avalanches and debris slides thundered down steep mountain slopes burying buildings and blocking roads and rail lines. The landslides also dammed rivers creating more than two dozen lakes. Earthquake-created lakes present a dual danger. Apart from the upstream floods that occur as the lake builds behind the natural dam, the piles of rubble that form the dam may be unstable. Another quake, or simply the pressure of water behind it, could burst the dam, sending a wall of water downstream. Such floods may also occur when water begins to cascade over the top of the dam. The largest of the lakes created by the May 12 earthquake, Tangjiashan Lake, threatened roughly 1.3 million people. In this instance, a disaster was avoided when Chinese engineers

successfully breached the landslide dam and safely drained the lake.

Liquefaction   Intense ground shaking during earthquakes can cause water-saturated surface materials to lose their strength and behave as fluidlike masses that flow. This process, called *liquefaction*, was a major cause of property damage in Anchorage, Alaska, during the massive 1964 Good Friday earthquake described in Chapter 11.

Landslides without Triggers?

Do rapid mass-wasting events always require some sort of trigger such as heavy rains or an earthquake? The answer is no; such events sometimes occur without being triggered. For example, on the afternoon of May 9, 1999. a landslide killed 10 hikers and injured many others at Sacred Falls State Park near Hauula on the

FIGURE 15.8 When a magnitude 7.9 earthquake struck west northwest of Chengdu in China's Sichuan Province on May 12, 2008, it triggered hundreds of landslides that destroyed roads and bridges. Buildings and rail lines were buried by landslide debris. Several rivers were blocked by debris dams creating lakes which threatened the safety of thousands of people living downstream. This aerial view shows the huge volume of landslide debris that dammed a river and created a lake. Note the heavy equipment for scale. (Photo by MARK/epa/Corbis)

north shore of Oahu, Hawaii. The tragic event occurred when a mass of rock from a canyon wall plunged 150 meters (500 feet) down a nearly vertical slope to the valley floor. Because of safety concerns, the park was closed so that landslide specialists from the U.S. Geological Survey could investigate the site. Their study concluded that the landslide occurred *without triggering* from any discernible external conditions.

Many rapid mass-wasting events occur without a discernible trigger. Slope materials gradually weaken over time under the influence of longterm weathering, infiltration of water, and other physical processes. Eventually, if the strength falls below what is necessary to maintain slope stability, a landslide will occur. The timing of such events is random, and thus accurate prediction is impossible (see Box 15.2).

The Potential for Landslides

Figure 15.9 shows the landslide potential for the contiguous United States. All states experience some damage from rapid mass-wasting processes, but it is obvious that not all areas of the country have the same landslide hazard potential. As you might expect, there are greater landslide risks in mountain areas. In the East, landslides are most common in the Appalachian Mountains. In the mountainous parts of the Pacific Northwest, water from heavy rains and melting snow often trigger rapid forms of mass wasting. Coastal California's steep slopes have a high landslide potential. Here mass-wasting events may be triggered by winter storms or the ground shaking associated with earthquakes. Landslides are also triggered when strong wave activity undercuts and oversteepens coastal cliffs.

A glance at the map shows that Florida and the adjacent Atlantic and Gulf coastal plains have some of the lowest landslide potentials because steep slopes are largely absent. In the center of the country, the plains states are relatively flat, so landslide potential is mostly low-to-moderate. High-potential areas occur along the steep bluffs associated with river valleys.

CLASSIFICATION OF MASS-WASTING PROCESSES

There is a broad array of different processes that geologists call mass wasting. Generally, the different types are classified based on the type of material involved, the kind of motion displayed, and the velocity of the movement.

Type of Material

The classification of mass-wasting processes on the basis of the material involved in the movement depends upon whether the descending mass began as unconsolidated material or as bedrock. If soil and regolith dominate, terms such as debris, mud, or earth are used in the description. In contrast, when a mass of bedrock breaks loose and moves downslope, the term rock may be part of the description.

Type of Motion

In addition to characterizing the type of material involved in a mass-wasting event, the way in which the material moves may also be important. Generally, the kind of motion is described as either a fall, a slide, or a flow.

Fall When the movement involves the freefall of detached individual pieces of any size, it is termed a **fall.** Fall is a common form of movement on slopes that are so steep that loose material cannot remain on the surface. The rock may fall directly to the base of the slope or move in a series of leaps and bounds over other rocks along the way. Rockfalls are the primary way in which *talus slopes* are built and maintained (Figure 15.10). Many falls result when freeze and thaw cycles and/or the action of plant roots loosen rock to the point that gravity takes over. Although signs along bedrock cuts on highways warn of falling rock, few of us have actually witnessed such an event. However, as Figure 15.11 illustrates, they do indeed occur.

When large masses of rock plunge from great heights, they hit the ground with enormous force and often trigger additional mass-wasting events. One deadly example occurred in Peru. In May 1970, an earthquake caused a huge mass of rock and ice to break free from the precipitous north face of Nevados Huascaran, the loftiest peak in the Peruvian Andes. The material plunged

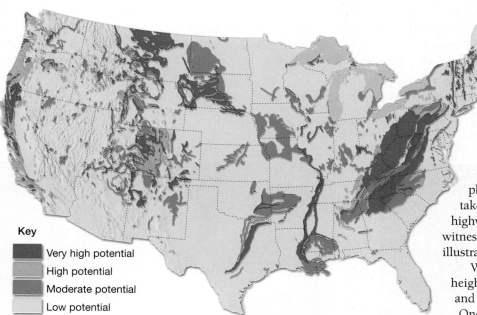

Key

■ Very high potential
■ High potential
■ Moderate potential
□ Low potential

FIGURE 15.9 Landslide potential for the contiguous United States. Risks are clearly greatest in mountain areas where slopes are steepest. Rapid forms of mass wasting occur and can cause damage in all 50 states. Heavy rains, earthquakes, volcanic activity, coastal wave attack, and wildfires can all contribute to widespread slope instability. (After U.S. Geological Survey)

FIGURE 15.10 Talus is a slope built of angular rock fragments. Mechanical weathering, especially frost wedging, loosens the pieces of bedrock, which then fall to the base of the cliff. With time, a series of steep, coneshaped accumulations build up at the base of the vertical slope. These talus cones are in Banff National Park, Alberta, Canada. (Photo by Marli Miller)

nearly a kilometer and pulverized on impact. The rock avalanche that followed rushed down the mountainside, made fluid by trapped air and ice. Along the way it ripped loose millions of tons of additional debris that ultimately and tragically buried more than 20,000 people in the towns of Yungay and Ranrahirca.

A different effect triggered by a rockfall occurred in Yosemite National Park on July 10, 1996. When two large rock masses broke loose from steep cliffs and fell about 500 meters (1640 feet) to the floor of Yosemite Valley the impacts were great enough to be recorded at seismic stations 200 kilometers (125 miles) from the site. As the dislodged rock masses struck the ground, they generated atmospheric pressure waves that

were comparable in velocity to a tornado or hurricane. The force of the airblasts uprooted and snapped more than a thousand trees, including some that were 40 meters tall.

Slide Many mass-wasting processes are described as **slides.** The term refers to mass movements in which there is a distinct zone of weakness separating the slide material from the more stable underlying material. Two basic types of slides are recognized. *Rotational slides* are those in which the surface of rupture is a concave-upward curve that resembles the shape of a spoon and the descending material exhibits a downward and outward rotation. By contrast, a *translational slide* is one in which a mass of material moves along a relatively flat surface such as a joint, fault, or bedding plane. Such slides exhibit little rotation or backward tilting.

Flow The third type of movement common to mass-wasting processes is termed **flow.** Flow occurs when material moves downslope as a viscous fluid. Most flows are saturated with water and typically move as lobes or tongues.

Rate of Movement

Some of the events that have been described so far in this chapter involved very rapid rates of movement. For example, it is estimated that the debris that rushed down the slopes of Peru's Nevados Huascaran moved at speeds in excess of 200 kilometers (125 miles) per hour. This most rapid type of mass move-

FIGURE 15.11 On August 6, 2006, photographer Herb Dunn witnessed this rockfall along the Merced River in California's Yosemite National Park. It appears to have been an event that had no discernible trigger (see "Landslides without Triggers" on p. 412) **A.** The impact of falling rock produces an explosion of dust and debris. **B.** The rockfall triggers a chain reaction. A debris avalanche moves down the slope knocking down trees along the way. (© 2006 DunnRight Photograph)

PEOPLE AND THE ENVIRONMENT

The Vaiont Dam Disaster

BOX 15.2

Most often, landslides are triggered by a natural event such as heavy rains or an earthquake. However, sometimes the rapid downslope movement of surface material is caused by the actions of people. Such is the case with the example discussed in this box. Unfortunately, it had disastrous consequences.

In 1960 a large dam, almost 265 meters (870 feet) tall, was built across Vaiont Canyon in the Italian Alps. It was engineered without good geological input, and the result was a disaster only three years later.

The bedrock in Vaiont Canyon slanted steeply downward toward the lake impounded behind the dam. The bedrock was weak, highly fractured limestone strata with beds of clay and numerous solution cavities. As the reservoir filled behind the completed dam, rocks became saturated and the clays became swollen and more plastic. The rising water reduced the internal friction that had kept the rock in place.

Measurements made shortly after the reservoir was filled hinted at the problem, because they indicated that a portion of the mountain was slowly creeping downhill at the rate of 1 centimeter per week. In September 1963 the rate increased to 1 centimeter per day, then 10–20 centimeters per day, and eventually as much as 80 centimeters on the day of the disaster.

Finally, the mountainside let loose. In just an instant, 240 million cubic meters of rock and rubble slid down the mountainside and filled nearly 2 kilometers of the gorge to heights of 150 meters above the reservoir level (Figure 15.C). This pushed the water completely over the dam in a wave more than 90 meters high. More than 1.5 kilometers downstream, the wall of water was still 70 meters high, destroying everything in its path.

The entire event lasted less than seven minutes, yet it claimed an estimated 2600 lives. Although this is known as the worst dam disaster in history, the Vaiont Dam itself remained intact. And while the catastrophe was triggered by human interference with the Vaiont River, the slide eventually would have occurred on its own; however, the effects would not have been nearly as tragic.

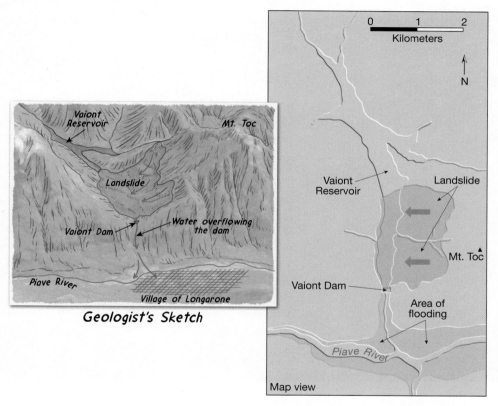

FIGURE 15.C Sketch map of the Vaiont River area showing the limits of the landslide, the portion of the reservoir that was filled with debris, and the extent of flooding downstream.

---

ment is termed a **rock avalanche** (*avaler* = to descend). Researchers now understand that rock avalanches, such as the one that produced the scene in Figure 15.12, must literally "float on air" as they move downslope. That is, high velocities result when air becomes trapped and compressed beneath the falling mass of debris, allowing it to move as a buoyant, flexible sheet across the surface.

Most mass movements, however, do not move with the speed of a rock avalanche. In fact, a great deal of mass wasting is imperceptibly slow. One process that we will examine later, termed *creep,* results in particle movements that are usually measured in millimeters or centimeters per year. Thus, as you

Student Sometimes Ask . . .

How difficult is it to walk up a talus slope?

Very. It might be more accurately described as a scramble because of its steepness. Ascending a talus slope of coarser material involves climbing from boulder to boulder. A talus slope composed of finer material is more difficult to climb because you can cause the material to slide as you ascend. Often, this tiring activity results in sliding about one-half step backward for each step you take.

FIGURE 15.12 Aerial view of the Blackhawk landslide, a prehistoric event that occurred on the north slope of California's San Bernardino Mountains. It is considered to be one of the largest known landslides in North America. This 8-kilometer long tongue of rubble is about 3 kilometers wide and 9 to 30 meters thick. Research has shown that the mass of debris raced downslope to its resting place on a cushion of compressed air. (Photo by Michael Collier)

can see, rates of movement can be spectacularly sudden or exceptionally gradual. Although various types of mass wasting are often classified as either rapid or slow, such a distinction is highly subjective because a wide range of rates exists between the two extremes. Even the velocity of a single process at a particular site can vary considerably.

SLUMP

MASS WASTING
▶ Types of Mass Wasting

Slump refers to the downward sliding of a mass of rock or unconsolidated material moving as a unit along a curved surface (Figure 15.13). Usually the slumped material does not travel

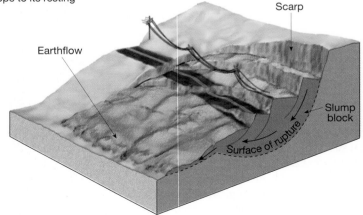

FIGURE 15.13 Slump occurs when material slips downslope en masse along a curved surface of rupture. It is an example of a rotational slide. Earthflows frequently form at the base of the slump.

spectacularly fast nor very far. This is a common form of mass wasting, especially in thick accumulations of cohesive materials such as clay. The ruptured surface is characteristically spoon-shaped and concave upward or outward. As the movement occurs, a crescent-shaped scarp is created at the head, and the block's upper surface is sometimes tilted backward. Although slump may involve a single mass, it often consists of multiple blocks. Sometimes water collects between the base of the scarp and the top of the tilted block. As this water percolates downward along the surface of rupture, it may promote further instability and additional movement.

Slump commonly occurs because a slope has been over-steepened. The material on the upper portion of a slope is held in place by the material at the bottom of the slope. As this anchoring material at the base is removed, the material above is made unstable and reacts to the pull of gravity. One relatively common example is a valley wall that becomes oversteepened by a meandering river. The photo in Figure 15.14 provides another example in which a coastal cliff has been undercut by wave action at its base. Slumping may also occur when a slope is overloaded, causing internal stress on the material below. This type of slump often occurs where weak, clay-rich material underlies layers of stronger, more resistant rock such as sandstone. The seepage of water through the upper layers reduces the strength of the clay below, resulting in slope failure.

ROCKSLIDE

MASS WASTING
▶ Types of Mass Wasting

Rockslides occur when blocks of bedrock break loose and slide down a slope (Figure 15.15). If the material involved is largely unconsolidated, the term **debris slide** is used instead. Such events

Students Sometimes Ask . . .

Are snow avalanches considered a type of mass wasting?

Sure. Sometimes these thundering downslope movements of snow and ice move large quantities of rock, soil, and trees. Of course, snow avalanches are very dangerous, especially to skiers on high mountain slopes and to buildings and roads at the bottom of slopes in avalanche-prone regions.

About 10,000 snow avalanches occur each year in the mountainous western United States. In an average year they claim between 15 and 25 lives in the United States and Canada. They are a growing problem as more people become involved in winter sports and recreation.

Geologist's Sketch

Slump blocks · Scarps · Wave action undercutting cliff and steepening slope

FIGURE 15.14 Slump at Point Fermin, California. Slump is often triggered when slopes become oversteepened by erosional processes such as wave action. (Photo by John S. Shelton)

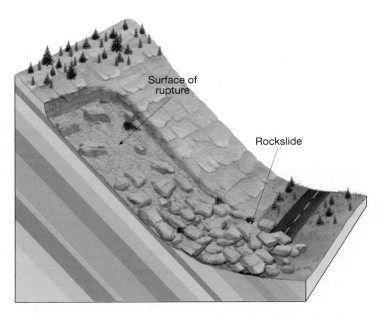

FIGURE 15.15 Rockslides and debris slides are rapid movements that are classified as translational slides in which the material moves along a relatively flat surface with little or no rotation or backward tilting.

are among the fastest and most destructive mass movements. Usually rockslides take place in a geologic setting where the rock strata are inclined, or where joints and fractures exist parallel to the slope. When such a rock unit is undercut at the base of the slope, it loses support and the rock eventually gives way. Sometimes the rockslide is triggered when rain or melting snow lubricates the underlying surface to the point that friction is no longer sufficient to hold the rock unit in place. As a result, rockslides tend to be more common during the spring, when heavy rains and melting snow are most prevalent.

As was mentioned earlier, earthquakes can trigger rockslides and other mass movements. There are many well-known examples. The 1811 earthquake at New Madrid, Missouri, caused slides in an area of more than 13,000 square kilometers (5000 square miles) along the Mississippi River valley. On August 17, 1959, a severe earthquake west of Yellowstone National Park triggered a massive slide in the canyon of the Madison River in southwestern Montana. In a matter of moments an estimated 27 million cubic meters of rock, soil, and trees slid into the canyon. The debris dammed the river and buried a campground and highway. More than 20 unsuspecting campers perished.

Not far from the site of the Madison Canyon slide, the legendary Gros Ventre rockslide occurred 34 years earlier. The Gros Ventre River flows west

from the northernmost part of the Wind River Range in northwestern Wyoming, through Grand Teton National Park, and eventually empties into the Snake River. On June 23, 1925, a massive rockslide took place in its valley, just east of the small town of Kelly. In the span of just a few minutes a great mass of sandstone, shale, and soil crashed down the south side of the valley, carrying with it a dense pine forest. The volume of debris, estimated at 38 million cubic meters (50 million cubic yards), created a 70-meter-high dam on the Gros Ventre River. Because the river was completely blocked, a lake was formed. It filled so quickly that a house that had been 18 meters (60 feet) above the river was floated off its foundation 18 hours after the slide. In 1927 the lake overflowed the dam, partially draining the lake and resulting in a devastating flood downstream.

Why did the Gros Ventre rockslide take place? Figure 15.16 shows a diagrammatic cross-sectional view of the geology of the valley. You will notice that (1) the sedimentary strata in this area dip (tilt) 15–21 degrees; (2) underlying the bed of sandstone is a relatively thin layer of clay; and (3) at the bottom of the valley the river had cut through much of the sandstone layer. During the spring of 1925, water from heavy rains and melting snow seeped through the sandstone, saturating the clay below. Because much of the sandstone layer had been cut through by the Gros

FIGURE 15.16 Part **A** shows a cross-sectional view of the Gros Ventre rockslide. The slide occurred when the tilted and undercut sandstone bed could no longer maintain its position atop the saturated bed of clay. As the photo in part **B** illustrates, even though the Gros Ventre rockslide occurred in 1925, the scar left on the side of Sheep Mountain is still a prominent feature. (Part A after W. C. Alden, "Landslide and Flood at Gros Ventre, Wyoming," Transactions (AIME) 76 (1928): 348; part B photo by Stephen Trimble)

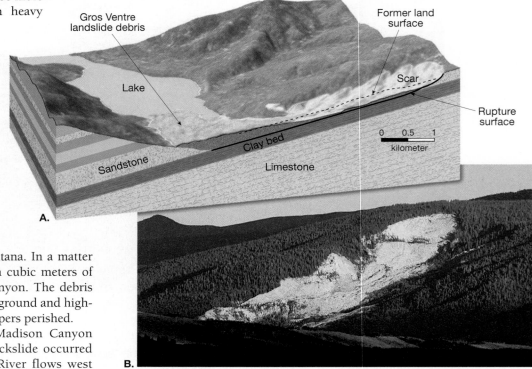

Ventre River, the layer had virtually no support at the bottom of the slope. Eventually the sandstone could no longer hold its position on the wetted clay, and gravity pulled the mass down the side of the valley. The circumstances at this location were such that the event was inevitable.

DEBRIS FLOW

MASS WASTING
▶ Types of Mass Wasting

Debris flow is a relatively rapid type of mass wasting that involves a flow of soil and regolith containing a large amount of water (Figure 15.17). Several are pictured in this chapter—see Figure 15.1B, Figure 15.3, and the photos in Box 15.1. Debris flows are sometimes called **mudflows** when the material is primarily fine-grained. Although they can occur in many different climatic settings, they probably occur more frequently in semi-arid mountainous regions. Debris flows, called *lahars,* are also common on the steep slopes of some volcanoes. Because of their fluid properties, debris flows frequently follow canyons and stream channels. In populated areas, debris flows can pose a significant hazard to life and property.

Debris Flows in Semiarid Regions

When a cloudburst or rapidly melting mountain snows create a sudden flood in a semiarid region, large quantities of soil and regolith are washed into nearby stream channels because there is usually little vegetation to anchor the surface material. The end product is a flowing tongue of well-mixed mud, soil, rock, and water. Its consistency may range from that of wet concrete to a soupy mixture not much thicker than muddy water. The

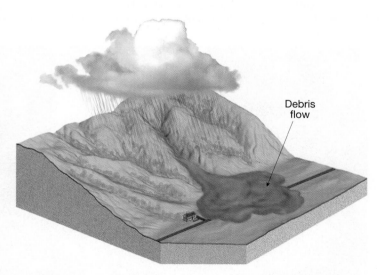

FIGURE 15.17 Debris flow is a moving tongue of well-mixed mud, soil, rock, and water. Its consistency may range from that of wet concrete to a soupy mixture not much thicker than muddy water.

rate of flow, therefore, depends not only on the slope but also on the water content. When dense, debris flows are capable of carrying or pushing large boulders, trees, and even houses with relative ease.

Debris flows pose a serious hazard to development in relatively dry mountainous areas such as southern California. The construction of homes on canyon hillsides and the removal of native vegetation by brush fires and other means have increased the frequency of these destructive events. Moreover, when a debris flow reaches the end of a steep, narrow canyon, it spreads out, covering the area beyond the mouth of the canyon with a mixture of wet debris. This material contributes to the buildup of fanlike deposits, called *alluvial fans,* at canyon mouths. The fans are relatively easy to build on, often have nice views, and are close to the mountains; in fact, like the nearby canyons, many have become preferred sites for development. Because debris flows occur only sporadically, the public is often unaware of the potential hazard of such sites.

Lahars

Debris flows composed mostly of volcanic materials on the flanks of volcanoes are called **lahars.** The word originated in Indonesia, a volcanic region that has experienced many of these often destructive events. Historically, lahars have been one of the deadliest volcano hazards. They can occur either during an eruption or when a volcano is quiet. They take place when highly unstable layers of ash and debris become saturated with water and flow down steep volcanic slopes, generally following existing stream channels (Figure 15.18). Heavy rainfalls often trigger these flows. Others are initiated when large volumes of ice and snow are melted by heat flowing to the surface from within the volcano or by the hot gases and near-molten debris emitted during a violent eruption.

When Mount St. Helens erupted in May 1980, several lahars were created. The flows and accompanying floods raced down the valleys of the north and south forks of the Toutle River at speeds that were often in excess of 30 kilometers (20 miles) per hour. Fortunately, the affected area was not densely settled. Nevertheless, more than 200 homes were destroyed or severely damaged (Figure 15.19). Most bridges met a similar fate. According to the U.S. Geological Survey:

> Even after traveling many tens of miles from the volcano and mixing with cold waters, the mudflows maintained temperatures in the range of about 84° to 91 °C; they undoubtedly had higher temperatures closer to the eruption source . . . Locally the mudflows surged up the valley walls as much as 360 feet and over hills as high as 250 feet. From the evidence left by the "bathtub-ring" mudlines, the larger mudflows at their peak averaged from 33 to 66 feet deep.*

Eventually the lahars in the Toutle River drainage area carried more than 50 million cubic meters of material to the lower Cowlitz and Columbia Rivers. The deposits temporarily

*Robert I. Tilling, *Eruptions of Mount St. Helens: Past, Present, and Future.* Washington, DC: U.S. Government Printing Office, 1987.

FIGURE 15.18 The eruption of Mount Redoubt, a volcano southwest of Anchorage on Alaska's Kenai Peninsula, sent large debris flows (lahars) down the Drift River Valley in April 2009. Channels form a branching pattern just west of Cook Inlet. The dark color of the lahar contrasts sharply with the surrounding snow-covered landscape. (NASA)

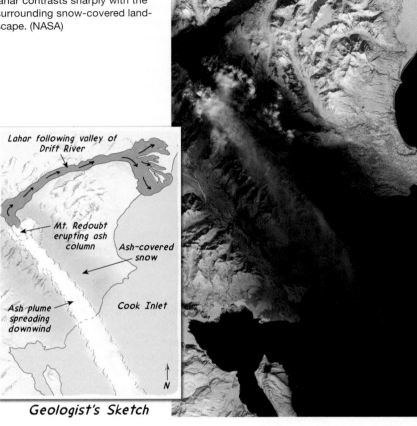

Geologist's Sketch

Lahar following valley of Drift River

Mt. Redoubt erupting ash column

Ash-covered snow

Ash plume spreading downwind

Cook Inlet

N

feet) of the peak, producing torrents of hot viscous mud, ash, and debris. The lahars moved outward from the volcano, following the valleys of three rain-swollen rivers that radiate from the peak. The flow that moved down the valley of the Lagunilla River was the most destructive. It devastated the town of Armero, 48 kilometers (30 miles) from the mountain. Most of the more than 25,000 deaths caused by the event occurred in this once thriving agricultural community.

Death and property damage due to the lahars also occurred in 13 other villages within the 180-square-kilometer (70-square-mile) disaster area. Although a great deal of pyroclastic material was explosively ejected from Nevado del Ruiz, it was the lahars triggered by this eruption that made this such a devastating natural disaster. In fact, it was the worst volcanic disaster since 28,000 people died following the 1902 eruption of Mount Pelée on the Caribbean island of Martinique.**

**A discussion of the Mount Pelée eruption can be found in Chapter 5.

reduced the water-carrying capacity of the Cowlitz River by 85 percent, and the depth of the Columbia River navigational channel was decreased from 12 meters to less than 4 meters.

In November 1985, lahars were produced during the eruption of Nevado del Ruiz, a 5300-meter (17,400-foot) volcano in the Andes Mountains of Colombia. The eruption melted much of the snow and ice that capped the uppermost 600 meters (2000

FIGURE 15.19 A house damaged by a lahar along the Toutle River, west-northwest of Mount St. Helens. The end section of the house was torn free and lodged against trees. (Photo by D. R. Crandell, U.S. Geological Survey)

Students Sometimes Ask . . .

How much do landslides cost in terms of monetary losses?

According to the U.S. Geological Survey, it is estimated that the total dollar losses from landslides is between $2 billion and $4 billion dollars (2009 dollars) per year. This figure is a conservative estimate, as there is no uniform method or overall agency that keeps track or reports landslide losses. Landslides result in extremely high monetary losses in other countries, but there is no overall estimate as to the exact amount.

Mass Wasting Specialist

Bob Rasely's line of work may be the very definition of "down and dirty." As a Utah-based geologist with the U.S. Department of Agriculture's Natural Resources Conservation Service, Rasely specializes in predicting the likelihood of rapid mass wasting events in the wake of a wildfire.

Although most of the situations Rasely encounters are not emergencies, there are times when the possibility of a rapid mass wasting process, especially a debris flow, poses an immediate threat to a populated area.

"If there is a town down below, we have to rush to get ahead of the next storm," Rasely said. "In the worst-case scenario, we place someone halfway up the mountain with a radio."

Geologist and mass-wasting specialist Bob Rasely recently retired from the U.S. Department of Agriculture's Natural Resources Conservation Service. He is standing on the hummocky surface of the Thistle landslide in central Utah. Losses from this 1983 event exceeded $400 million, making it the most costly single landslide in U.S. history. It dammed the Spanish Fork River, inundating the town of Thistle, and obliterated a highway and the mainline of the Denver and Rio Grande Railroad. (Courtesy of Robert C. Rasely)

In such cases, Rasely said, they often set up evacuation plans that may remain in effect for long periods of time.

The probability and scale of a potential disaster depends on many factors. Rasely classifies most wildfires or "burns" as either "light" or "heavy." Light burns only destroy the surface growth and have a recovery span of about 1 year. By contrast, heavy burns destroy the roots and require 3 to 5 years for recovery.

"First we want to determine the condition of the burn, if the roots are intact," he said. Rasely mentioned that light burns increase the erosion rate on a given surface by 3 to 5 times, whereas heavy burns increase the rate 6 to 10 times.

Other factors to consider include the steepness and length of the slopes and whether established paths for water to flow already exist on the surface. Steep slopes and established channels are a bad sign because water will run off at a much faster rate. Pine needles and oaks also leave a waxy residue when they burn which can further expedite the flow, Rasely said.

Once the general risk of mass wasting has been assessed, Rasely and his team have a variety of techniques they can employ that are intended to minimize potential danger to people and damage to property. One involves erecting fencing made out of mesh material called "geofabric." Another approach involves digging what they call a "debris basin" to intercept landslide debris. "If there is any kind of little terrace, you can stop a lot of stuff," he said.

> "If there is a town down below, we have to rush to get ahead of the next storm."

Rasely's job has taken him to all but two of Utah's mountain ranges, and he frequently works in state and national parks, of which there are many in Utah. And while the severity of each situation can vary greatly, time is always of the essence in this line of work.

—Chris Wilson

EARTHFLOW

MASS WASTING
▶ Types of Mass Wasting

We have seen that debris flows are frequently confined to channels in semiarid regions. In contrast, **earthflows** most often form on hillsides in humid areas during times of heavy precipitation or snowmelt. When water saturates the soil and regolith on a hillside, the material may break away, leaving a scar on the slope and forming a tongue- or teardrop-shaped mass that flows downslope (Figure 15.20).

The materials most commonly involved are rich in clay and silt and contain only small proportions of sand and coarser particles. Earthflows range in size from bodies a few meters long, a few meters wide, and less than a meter deep to masses more than 1 kilometer long, several hundred meters wide, and more than 10 meters deep. Because earthflows are quite viscous,

Geologist's Sketch

FIGURE 15.20   This small, tongue-shaped earthflow occurred on a newly formed slope along a recently constructed highway in Central Illinois. It formed in clay-rich material following a period of heavy rain. Notice the small slump at the head of the earthflow. (Photo by E. J. Tarbuck)

they generally move at slower rates than the more fluid debris flows described in the preceding section. They are characterized by a slow and persistent movement and may remain active for periods ranging from days to years. Depending on the steepness of the slope and the material's consistency, measured velocities range from less than a millimeter a day up to several meters a day. Over the time span that earthflows are active, movement is typically faster during wet periods than during drier times. In addition to occurring as isolated hillside phenomena, earthflows commonly take place in association with large slumps. In this situation, they may be seen as tonguelike flows at the base of the slump block.

SLOW MOVEMENTS

MASS WASTING
▶ Types of Mass Wasting

Movements such as rockslides, rock avalanches, and lahars are certainly the most spectacular and catastrophic forms of mass wasting. These dangerous events deserve intensive study to enable more effective prediction, timely warnings, and better controls to save lives. However, because of their large size and spectacular nature, they give us a false impression of their importance as a mass-wasting process. Indeed, sudden movements are responsible for moving less material than the slower and far more subtle action of creep. Whereas rapid types of mass wasting are characteristic of mountains and steep hillsides, creep takes place on both steep and gentle slopes and is thus much more widespread.

Creep

Creep is a type of mass wasting that involves the gradual downhill movement of soil and regolith. One factor that contributes to creep is the alternate expansion and contraction of surface material caused by freezing and thawing or wetting and drying. As shown in Figure 15.21, freezing or wetting lifts particles at right angles to the slope, and thawing or drying allows the particles to fall back to a slightly lower level. Each cycle therefore moves the material a tiny distance downslope. Creep is aided by anything that disturbs the soil, such as raindrop impact and disturbance by plant roots and burrowing animals. Creep is also promoted when the ground becomes saturated with water. Following a heavy rain or snowmelt, a water-logged soil may lose its internal cohesion, allowing gravity to pull the material downslope. Because creep is imperceptibly slow, the process cannot be observed in action. What can be observed, however, are the effects of creep. Creep causes fences and utility poles to tilt and retaining walls to be displaced (Figure 15.22).

Solifluction

When soil is saturated with water, the soggy mass may flow downslope at a rate of a few millimeters or a few centimeters per day or per year. Such a process is called **solifluction** (literally, "soil

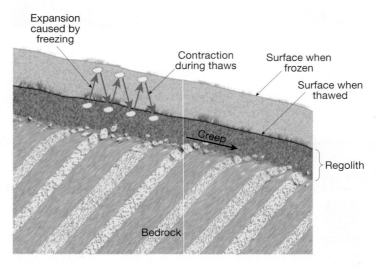

FIGURE 15.21   The repeated expansion and contraction of the surface material causes a net downslope migration of rock particles—a process called *creep*.

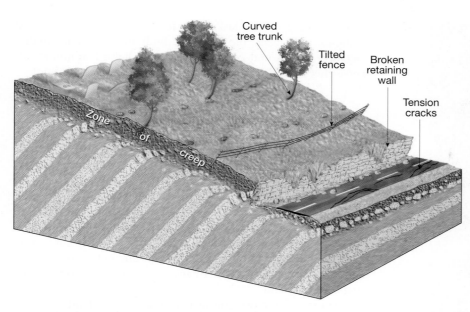

FIGURE 15.22   Although creep is an imperceptibly slow movement, its effects are often visible.

flow"). It is a type of mass wasting that is common wherever water cannot escape from the saturated surface layer by infiltrating to deeper levels. A dense clay hardpan in soil or an impermeable bedrock layer can promote solifluction.

Solifluction is also common in regions underlain by *permafrost*. Permafrost refers to the permanently frozen ground that occurs in association with Earth's harsh tundra and ice-cap climates. (There is more about permafrost in the next section.) Solifluction occurs in a zone above the permafrost called the *active layer*, which thaws to a depth of about a meter during the brief high-latitude summer and then refreezes in winter. During the summer season, water is unable to percolate into the impervious permafrost layer below. As a result, the active layer becomes saturated and slowly flows. The process can occur on slopes as gentle as 2 to 3 degrees. Where there is a well-developed mat of vegetation, a solifluction sheet may move in a series of well-defined lobes or as a series of partially overriding folds (Figure 15.23).

THE SENSITIVE PERMAFROST LANDSCAPE

Many of the mass-wasting disasters described in this chapter had sudden and disastrous impacts on people. When the activities of people cause ice contained in permanently frozen ground to melt, the impact is more gradual and less deadly. Nevertheless, because permafrost regions are sensitive and fragile landscapes, the scars resulting from poorly planned actions can remain for generations.

Permanently frozen ground, known as **permafrost,** occurs where summers are too cool to melt more than a shallow surface layer. Deeper ground remains frozen year-round. Strictly speaking, permafrost is defined only on the basis of temperature; that is, it is ground with temperatures that have remained below 0 °C (32 °F) continuously for two years or more. The degree to which ice is present in the ground strongly affects the behavior of the surface material. Knowing how much ice is present and where it is located is very important when it comes to constructing roads, buildings, and other projects in areas underlain by permafrost.

Permafrost is extensive in the lands surrounding the Arctic Ocean. It covers more than 80 percent of Alaska, about 50 percent of Canada, and a substantial portion of northern Siberia (Figure 15.24). Near the southern margins of the region, the permafrost consists of relatively thin, isolated masses. Farther north, the area and thickness gradually increase to the point where the permafrost is essentially continuous and its thickness may approach or even exceed 500 meters. In the discontinuous zone, land-use planning is frequently more difficult than in the continuous zone farther north because the occurrences of permafrost are patchy and difficult to predict.

When people disturb the surface, such as by removing the insulating vegetation mat or by constructing roads and buildings,

FIGURE 15.23   Solifluction lobes northeast of Fairbanks, Alaska. Solifluction occurs in permafrost regions when the active layer thaws in summer. (Photo by James E. Patterson)

Geologist's Sketch

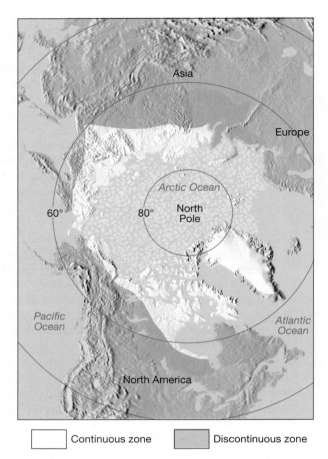

FIGURE 15.24   Distribution of permafrost in the Northern Hemisphere. More than 80 percent of Alaska and about 50 percent of Canada are underlain by permafrost. Two zones are recognized. In the continuous zone, the only ice-free areas are beneath deep lakes or rivers. In the higher-latitude portions of the discontinuous zone, there are only scattered islands of thawed ground. Moving southward, the percentage of unfrozen ground increases until all the ground is unfrozen. (After the U.S. Geological Survey)

the delicate thermal balance is disturbed, and the permafrost can thaw (Figure 15.25A). Thawing produces unstable ground that may slide, slump, subside, and undergo severe frost heaving. When a heated structure is built directly on permafrost that contains a high proportion of ice, thawing creates soggy material into which a building can sink. One solution is to place buildings and other structures on piles, like stilts. Such piles allow subfreezing air to circulate between the floor of the building and the soil and thereby keep the ground frozen.

When oil was discovered on Alaska's North Slope, many people were concerned about the building of a pipeline linking the oil fields at Prudhoe Bay to the ice-free port of Valdez 1300 kilometers to the south. There was serious concern that such a massive project might damage the sensitive permafrost environment. Many also worried about possible oil spills.

Because oil must be heated to about 60 °C to flow properly, special engineering procedures had to be developed to isolate this heat from the permafrost. Methods included insulating the pipe, elevating portions of the pipeline above ground level, and even placing cooling devices in the ground to keep it frozen

(Figure 15.25B). The Alaskan pipeline is clearly one of the most complex and costly projects ever built in the Arctic tundra. Detailed studies and careful engineering helped minimize adverse effects resulting from the disturbance of frozen ground.

SUBMARINE LANDSLIDES

As you might imagine, mass-wasting processes are not confined to land. The development of high-quality instruments that perform ocean-floor imaging has allowed us to determine that submarine mass wasting is a common and widespread phenomenon. For example, studies reveal enormous submarine landslides on the flanks of the Hawaiian chain as well as along the continental shelf and slope of the U.S. mainland. In fact, many submarine landslides, mostly in the form of slumps and debris avalanches, appear to be much larger than any similar mass-wasting events that occur on land.

Among the most spectacular underwater landslides are those that occur on the flanks of submarine volcanoes (called *seamounts*) and volcanic islands such as Hawaii. On the submerged flanks of the Hawaiian Islands dozens of major land-

FIGURE 15.25   **A.** When a rail line was built across this permafrost landscape in Alaska, the ground subsided. (Photo by Lynn A. Yehle, U.S. Geological Survey) **B.** In places in Alaska, a pipeline is suspended above ground to prevent melting of delicate permafrost. (Tom & Pat Lesson/Photo Researchers)

slides more than 20 kilometers (13 miles) long have been identified. Some have truly spectacular dimensions. One of the largest yet mapped, known as the Nuuanu debris avalanche, is on the north-eastern side of Oahu. It extends for nearly 25 kilometers (15 miles) across the ocean floor, then rises up a 300-meter (nearly 1000-foot) slope at its terminus, indicating that it must have had great power and momentum. Giant blocks many kilometers across were transported by this giant landslide. It is probable that when such large and rapid events occur, they produce giant sea waves called tsunami that race across the Pacific.*

The massive submarine slides discovered on the flanks of the Hawaiian Islands are almost certainly related to the movement of magma while a volcano is active. As huge quantities of lava are added to the seaward margin of a volcano, the buildup of material eventually triggers a great landslide. In the Hawaiian

chain, it appears that this process of growth and collapse is repeated at intervals of from 100,000 to 200,000 years while the volcano is active.

Along the continental margins of the U.S. mainland, large slumps and debris-flow scars mark the continental slope. These processes are triggered by the rapid buildup of unstable sediments or by such forces as storm waves and earthquakes. Submarine mass wasting is especially active near deltas, which are massive deposits of sediment at the mouths of rivers. Here, as great loads of water-saturated clay and organic-rich sediments accumulate, they become unstable and readily flow down even gentle slopes. Some of these movements have been forceful enough to damage large, offshore drilling platforms.

Mass wasting appears to be an integral part of the growth of passive continental margins. Sediments supplied to the continental shelf by rivers move across the shelf to the upper continental slope. From here slumps, slides, and debris flows move sediment down to the continental rise and sometimes beyond.

---

*For more on these destructive waves, see the section on tsunamis in Chapter 11.

CHAPTER 15 MASS WASTING IN REVIEW

- *Mass wasting* refers to the downslope movement of rock, regolith, and soil under the direct influence of gravity. In the evolution of most landforms, mass wasting is the step that follows weathering. The combined effects of mass wasting and erosion by running water produce stream valleys.

- *Gravity is the controlling force of mass wasting.* Other factors that influence or trigger downslope movements are saturation of the material with water, oversteepening of slopes beyond the *angle of repose,* removal of vegetation, and ground shaking by earthquakes.

- The various processes included under the name of mass wasting are divided and described on the basis of (1) the type of material involved (debris, mud, earth, or rock); (2) the type of motion (fall, slide, or flow); and (3) the rate of movement (rapid or slow).

- The more rapid forms of mass wasting include *slump,* the downward sliding of a mass of rock or unconsolidated material moving as a unit along a curved surface; *rockslide,* blocks

- of bedrock breaking loose and sliding downslope; *debris flow,* a relatively rapid flow of soil and regolith containing a large amount of water; and *earthflow,* an unconfined flow of saturated clay-rich soil that most often occurs on a hill-side in a humid area following heavy precipitation or snowmelt.

- The slowest forms of mass wasting include *creep,* the gradual downhill movement of soil and regolith; and *solifluction,* the gradual flow of a saturated surface layer that is underlain by an impermeable zone. Common sites for solifluction are regions underlain by *permafrost* (permanently frozen ground associated with tundra and ice-cap climates).

- *Permafrost,* permanently frozen ground, covers large portions of North America and Siberia. Thawing produces unstable ground that may slide, slump, subside, and undergo severe frost heaving.

- Mass wasting is not confined to land; it also occurs underwater. Many *submarine landslides,* mostly slumps and debris avalanches, are much larger than those that occur on land.

KEY TERMS

angle of repose (p. 409)
creep (p. 422)
debris flow (p. 419)
debris slide (p. 417)

earthflow (p. 421)
fall (p. 413)
flow (p. 414)
lahar (p. 419)

mass wasting (p. 406)
mudflow (p. 419)
permafrost (p. 423)
rock avalanche (p. 415)

rockslide (p. 417)
slide (p. 414)
slump (p. 416)
solifluction (p. 422)

QUESTIONS FOR REVIEW

1. Describe how mass-wasting processes contribute to the development of stream valleys.

2. What is the controlling force of mass wasting?

3. How does water affect mass-wasting processes?

4. Describe the significance of the angle of repose.

5. How might the removal of vegetation by fire or logging promote mass wasting?

6. How are earthquakes linked to landslides?

7. How did the building of a dam contribute to the Vaiont Canyon disaster? Was the disaster avoidable? (See Box 15.2, p. 415).

8. Distinguish among fall, slide, and flow.

9. Why can rock avalanches move at such great speeds?

10. Both slump and rockslide move by sliding. In what ways do these processes differ?

11. What factors led to the massive rockslide at Gros Ventre, Wyoming?

12. Explain why building a home on an alluvial fan might not be a good idea.

13. Compare and contrast mudflow and earthflow.

14. Describe the mass wasting that occurred at Mount St. Helens during its active period in 1980 and at Nevado del Ruiz in 1985.

15. Because creep is an imperceptibly slow process, what evidence might indicate that this phenomenon is affecting a slope?

16. What is permafrost? What portion of Earth's land surface is affected?

17. During what season does solifluction occur in permafrost regions?

COMPANION WEBSITE

The *Earth 10e* website uses the resources and flexibility of the Internet to aid in your study of the topics in this chapter. Written and developed by the authors and other geology instructors, this site will help improve your understanding of geology. Visit www.mygeoscienceplace.com in order to:

• **Review** key chapter concepts

• **Read** with links to the eBook and to chapter-specific web resources

• **Visualize** and comprehend challenging topics using learning activities in *GEODe Earth*

• **Test** yourself with online quizzes

GEODe EARTH

GEODe Earth is a valuable and easy to use learning aid that can be accessed from your book's Companion Website (www.mygeoscienceplace.com). It is a dynamic instructional tool that promotes understanding and reinforces important concepts by using tutorials, animations, and exercises that actively engage the student.

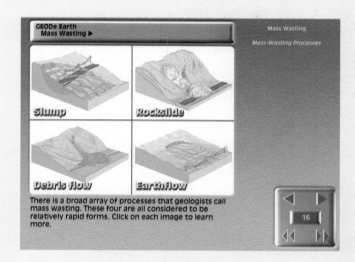

RUNNING WATER

Rainstorm over Toklat River, Denali National Park, Ala

(PHOTO BY MICHAEL COLLIER)

EARTH AS A SYSTEM: THE HYDROLOGIC CYCLE

RUNNING WATER
▶ Hydrologic Cycle

rs have always played an

portant role in human affairs. All

he early great civilizations rose on

of large rivers: Mesopotamia on

of the Tigris and Euphrates; Egypt

le valley of the Nile; the Yellow River,

Mother River," was the cradle of

ese civilization; and the Indus River on

ntinent of India was the birthplace of

y Civilization. Because rivers change nat-

time, as a result of climatic and tectonic

, the "engineers" of ancient history tried to

he effects of floods and channel changes.

rn river engineers are challenged by similar

Although their carefully designed structures

successful in managing flood control, navigation,

nel stabilization, they are costly to build and main-

ately, through these efforts we have learned that

not necessarily be controlled by humans.

der the bittersweet relationship we have with rivers.

vital economic tools—used as highways to move

urces of water for irrigation and energy—as well as

ations for sport and recreation. When considered as

e Earth system, rivers and streams represent a funda-

k in the constant cycling of our planet's water. Running

he dominant agent of landscape alteration, eroding more

d transporting more sediment than any other process

.1). However, because human populations gravitate toward

oding in these otherwise valuable valleys represents enormous

or destruction.

chapter provides an overview of: the hydrologic cycle; the nature

stems; the types of river channels and the factors that produce

influences of running water on our planet's landscape; and the

floods and their impact on humans.

We live on a planet that is unique in the solar system—it is in just the right location and is just the right size (see Chapter 22). If Earth were appreciably closer to the Sun, water would exist only as a vapor. Conversely, water would be forever frozen if our planet was much farther away. Moreover, Earth is large enough to have a hot mantle that supports conductive flow which carries water to the surface through volcanism. Water that rose from Earth's interior through mantle convection generated our planet's oceans and atmosphere. Thus, by coincidence of favorable size and location, Earth is the only planet in the solar system with a global ocean and a hydrologic cycle.

Water is found almost everywhere on Earth—in the oceans, glaciers, rivers, lakes, air, soil, and in living tissue. All of these "reservoirs" constitute Earth's hydrosphere which contains about 1.36 billion cubic kilometers (326 million cubic miles) of water. The vast majority of it, about 97 percent, is stored in the global ocean (Figure 16.2). Ice sheets and glaciers account for slightly more than 2 percent, leaving less than one percent to be divided among lakes, streams, subsurface water, and the atmosphere.

> All the rivers run into the sea; yet the sea is not full; unto the place from whence the rivers come, thither they return again.

> (Ecclesiastes 1:7)

As the perceptive writer of Ecclesiastes implied, water is constantly moving among Earth's different spheres—the *hydrosphere,* the *atmosphere,* the *geosphere,* and the *biosphere.* This unending circulation of water, called the **hydrologic cycle,** describes what happens as water evaporates from the ocean, plants, and soil, moves through the atmosphere, and eventually falls as precipitation (Figure 16.3). Precipitation that falls onto the ocean has completed its cycle and is ready to begin another.

When precipitation falls on land, it either soaks into the ground, a process called **infiltration,** or flows over the surface as **runoff** or it immediately evaporates. Much of the water that infiltrates or runs off eventually finds its way back to the atmosphere via evaporation from soil, lakes, and streams. In addition, some of the water that soaks into the ground is absorbed by plants, which later release it into the atmosphere. This process is called **transpiration** (*trans* = across, *spiro* = to breathe). Because both evaporation and transpiration involve the transfer of water from the surface directly to the atmosphere, they are often considered together as the combined process of **evapotranspiration.**

More water falls on land as precipitation than is lost by

FIGURE 16.1   Stream in New York's Lake George Wild Forest. (Photo by Radius Images/Photolibrary)

evapotranspiration. The excess is carried back to the ocean mainly by streams—less than one percent returns as groundwater. However, much of the water that flows in rivers is not transmitted directly into river channels after falling as precipitation. Instead, a large percentage first soaks into the soil and then gradually flows as groundwater to river channels. In this manner, groundwater provides a form of storage which sustains the flow of streams between storms and during periods of drought.

When precipitation falls in very cold areas—at high elevations or high latitudes—the water may not immediately soak in, run off, or evaporate. Instead, it becomes part of a snowfield or a glacier. In this way, glaciers store large quantities of water. If present-day glaciers were to melt and release their stored water, sea level would rise by several tens of meters worldwide and submerge many heavily populated coastal areas. As you will see in Chapter 18, over the past 2 million years, huge ice sheets have formed and melted on several occasions, each time changing the balance of the hydrologic cycle.

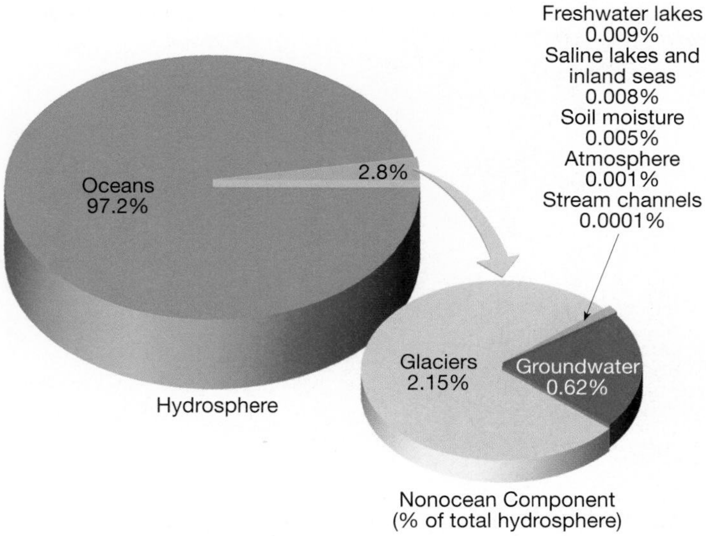

Freshwater lakes
0.009%
Saline lakes and
inland seas
0.008%
Soil moisture
0.005%
Atmosphere
0.001%
Stream channels
0.0001%

Oceans
97.2%

2.8%

Hydrosphere

Glaciers
2.15%

Groundwater
0.62%

Nonocean Component
(% of total hydrosphere)

FIGURE 16.2   Distribution of Earth's water.

Students Sometimes Ask . . .

Is there a significant amount of water vapor that plants emit into the atmosphere through transpiration?

Use this example to judge for yourself. Each year a field of crops may transpire the equivalent of a water layer 60 cm (2 ft) deep over the entire field. The same area of trees may pump twice this amount into the atmosphere.

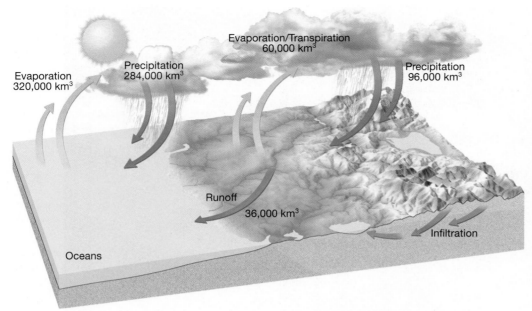

FIGURE 16.3 Earth's water balance. Each year, solar energy evaporates about 320,000 cubic kilometers of water from the oceans, while evaporation from the land (including lakes and streams) contributes 60,000 cubic kilometers of water. Of this total of 380,000 cubic kilometers of water, about 284,000 cubic kilometers fall back to the ocean, and the remaining 96,000 cubic kilometers fall on the land surface. Of that 96,000 cubic kilometers, only 60,000 cubic kilometers of water return to the atmosphere by evaporation and transpiration, leaving 36,000 cubic kilometers of water to erode the land during the journey back to the oceans.

Figure 16.3 also shows that Earth's hydrologic cycle is balanced. Each year, solar energy evaporates about 320,000 cubic kilometers of water from the oceans, but only 284,000 cubic kilometers return to the oceans as precipitation. A balance is achieved by the 36,000 cubic kilometers that are carried to the ocean as runoff. Although runoff makes up a small percentage of the total, running water is, nevertheless, *the single most important erosional agent sculpting Earth's land surface.*

To summarize, the hydrologic cycle represents the continuous movement of water from the oceans to the atmosphere, from the atmosphere to the land, and from the land back to the sea. The wearing down of Earth's land surface is largely attributable to the last of these steps and is the primary focus of the remainder of this chapter.

RUNNING WATER

RUNNING WATER
▶ Stream Characteristics

Recall that most of the precipitation that falls on land either enters the soil (infiltration) or remains at the surface, moving downslope as runoff. The amount of water that runs off rather than soaking into the ground depends on several factors: (1) intensity and duration of rainfall; (2) amount of water already in the soil; (3) nature of the surface material; (4) slope of the land, and; (5) the extent and type of vegetation. When the surface material is highly impermeable, or when it becomes saturated, runoff is the dominant process. Runoff is also high in urban areas because large areas are covered by impermeable buildings, roads, and parking lots.

Runoff initially flows in broad, thin sheets across hillslopes by a process called **sheet flow.** This thin, unconfined flow eventually develops threads of current that form tiny channels called **rills.** Rills meet to form gullies, which join to form brooks, creeks, or streams—then, when they reach an undefined size, they are called rivers. Although the terms *river* and *stream* are often used interchangeably, geologists define **stream** as water that flows in a channel, regardless of size. **River,** on the other hand, is a general term for streams that carry substantial amounts of water and have numerous tributaries.

In humid regions, the water to support stream flow comes from two sources, overland flow that sporadically enters the stream and groundwater that slowly, yet continuously, enters the channel. In areas where the bedrock is composed of soluable rocks such as limestone, large openings may exist that facilitate the transport of groundwater to streams. In arid regions, however, the *water table* may be below the level of the stream channel, in which case the stream loses water to the groundwater system by outflow through the streambed.

Drainage Basins

Every stream drains an area of land called a **drainage basin** (Figure 16.4). Each drainage basin is bounded by an imaginary line called a **divide,** something that is clearly visible as a sharp

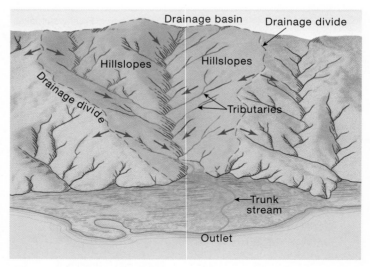

FIGURE 16.4 The drainage basin is the area of land drained by a river and its tributaries.

ridge in some mountainous areas, but can be rather difficult to discern in more subdued topographies. The outlet, where the stream exits the drainage basin, is at a lower elevation than the rest of the basin.

Drainage divides range in scale from a small ridge separating two gullies on a hillside to a *continental divide,* that splits an entire continent into enormous drainage basins. The Mississippi River has the largest drainage basin in North America collecting and carrying 40 percent of the flow in the United States (Figure 16.5).

By looking at the drainage basin in Figure 16.4, it should be obvious that hillslopes cover most of the area. Water erosion on the hillslopes is aided by the impact of falling rain and surface flow, which moves downslope as sheets or in rills toward a stream channel. Hillslope erosion is the main source of fine particles (clays and fine sand) carried in stream channels.

River Systems

Rivers drain much of the land area, with the exception of extremely arid regions or polar areas that are permanently frozen. To a large extent, the variety of rivers that exist is a reflection of the different environments in which they are found. For example, Uruguay's La Plata River drains roughly the same size area as the Nile in Egypt, yet the La Plata carries nearly 10 times more water to the ocean. Because its drainage basin is in a tropical climate, the La Plata has a huge runoff. By contrast, the Nile, which also originates in a humid region, flows through an expansive arid landscape where significant amounts of water evaporate or are withdrawn to sustain agriculture. Thus, climatic differences and human intervention can significantly influence the character of a river. Later, we will examine other factors that contribute to stream variability.

River systems not only involve a network of stream channels, but the entire drainage basin. Based on the dominant processes operating within them, rivers systems can be divided into three zones; *sediment production*—where erosion dominates, *sediment transport,* and *sediment deposition* (Figure 16.6). It is important to recognize that sediment is being eroded, transported, and deposited along the entire length of a stream, regardless of which process is dominant within each zone.

The zone of *sediment production,* where most of the water and sediment is derived, is located in the headwater region of the river system. Much of the sediment carried by streams begins as bedrock that is subsequently broken down by weathering, then transported downslope by mass wasting and overland flow. Bank erosion can also contribute significant amounts of sediment. In addition, scouring of the channel bed deepens the channel and adds to the stream's sediment load.

Sediment acquired by a stream is then transported through the channel network along sections referred to as *trunk streams.* When trunk streams are in balance, the amount of sediment eroded from their banks equals the amount deposited elsewhere in the channel. Although trunk streams rework their channels over time, they are not a source of sediment nor do they accumulate or store it.

When a river reaches the ocean, or another large body of water, it slows and the energy to transport sediment is greatly reduced. Most of the sediments either accumulate at the mouth of the river to form a delta, are reconfigured by wave action to form a variety of coastal features, or are moved far offshore by ocean currents. Because coarse sediments tend to be deposited upstream, it is primarily the fine sediments (clay, silt, and fine sand) that eventually reach the ocean. Taken together,

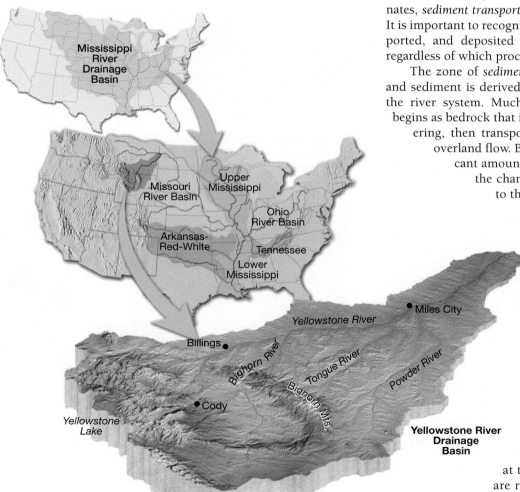

FIGURE 16.5 Drainage basins and divides exist for all streams, regardless of size. The drainage basin of the Yellowstone River is one of many that contribute water to the Missouri River, which in turn, is one of many that make up the drainage basin of the Mississippi River. The drainage basin of the Mississippi River, North America's largest, covers about 3 million square kilometers.

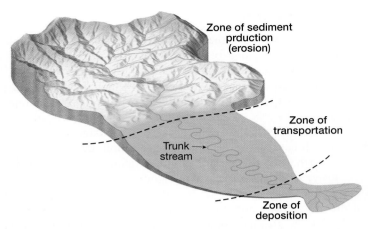

FIGURE 16.6 A river system can be divided into three zones based on the dominant processes operating within each zone. These are the zones of sediment production (erosion), sediment transport, and sediment deposition. (Adapted from Shumm)

erosion, transportation, and deposition are the processes by which rivers move Earth's surface materials and sculpt landscapes (see Figure 16.6).

STREAMFLOW

RUNNING WATER

▶ Stream Characteristics

The water in river channels travels downslope under the influence of gravity. In very slow flowing streams, water moves in roughly straight-line paths parallel to the stream channel, called **laminar flow.** However, streams typically exhibit **turbulent flow,** where the water moves in an erratic fashion characterized by a series of horizontal and vertical swirling motions. Strong, turbulent behavior occurs in whirlpools and eddies, as well as in rolling whitewater rapids. Even streams that appear smooth on the surface often exhibit turbulent flow near the bottom and sides of the channel where flow resistance is greatest. Turbulence contributes to the stream's ability to erode its channel because it acts to lift sediment from the streambed.

Flow Velocity

Flow velocities can vary significantly from place to place along a stream channel, as well as over time, in response to variations in the amount and intensity of precipitation. If you have ever waded into a stream, you know that velocity increases as you move into deeper parts of the channel. This is the result of frictional resistance which is greatest near the banks and beds of stream channels.

Scientists determine flow velocities at gaging stations by averaging measurements taken at various locations across the stream's channel (Figure 16.7C, D). Some sluggish streams have flow velocities of less than 1 kilometer per hour, whereas stretches of some fast-flowing rivers may exceed 30 kilometers per hour.

The ability of a stream to erode and transport material is directly related to its flow velocity. Even slight variations in flow rate can lead to significant changes in the sediment load transported by a stream. Several factors influence flow velocities and, therefore, control a stream's potential to do "work." These factors include: (1) channel slope or gradient; (2) channel size and cross-sectional shape; (3) channel roughness, and; (4) the amount of water flowing in the channel.

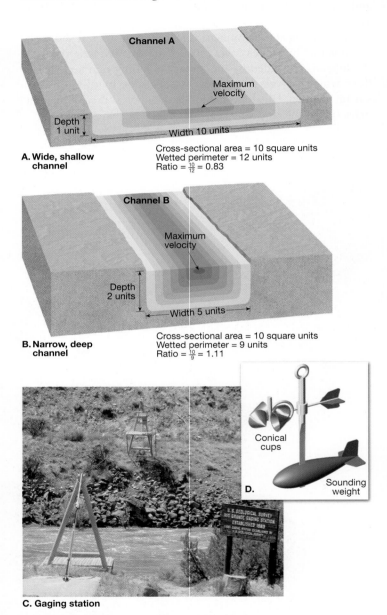

FIGURE 16.7 Influence of channel shape on velocity. **A.** Stream A has a wide, shallow channel and a large, wetted perimeter. **B.** The cross-sectional area of channel B is the same as stream A, but has less water in contact with its channel and therefore less frictional drag. Thus, water will flow more rapidly in channel B, all other factors being equal. **C.** Continuous records of stage and discharge are collected by the U.S. Geological Survey at more than 7000 gaging stations in the United States. Average velocities are determined by using measurements from several spots across the stream. This station is on the Rio Grande south of Taos, New Mexico. (Photo by E. J. Tarbuck) **D.** Current meter used to measure stream velocity at a gaging station.

Gradient and Channel Characteristics

The slope of a stream channel, expressed as the vertical drop of a stream over a specified distance, is called **gradient.** Portions of the lower Mississippi River have very low gradients, about 10 centimeters or less per kilometer. By contrast, some mountain streams have channels that drop at a rate of more than 40 meters per kilometer—a gradient 400 times steeper than the lower Mississippi (Figure 16.8). Gradient also varies along the length of a particular channel. When the gradient is steeper, more gravitational energy is available to drive channel flow.

As water in a stream channel moves downslope it encounters a significant amount of frictional resistance. The *cross-sectional shape* (a slice taken across the channel) determines, to a large extent, the amount of flow in contact with the banks and bed of the channel. This measure is referred to as the **wetted perimeter.** The most efficient channel is one with the least wetted perimeter for its cross-sectional area. Figure 16.7 compares two channels that differ only in shape—channel A is wide and shallow, channel B is narrow and deep. Although the cross-sectional area of both is identical, channel B has less water in contact with the channel, and therefore, less frictional drag. As a result, if all other factors are equal, the water will flow more efficiently and at a higher velocity in channel B than in channel A (see Figure 16.7).

Water depth also affects the frictional resistance the channel exerts on flow. Maximum flow velocity occurs when a stream is *bankfull,* before water starts to inundate the floodplain. At this stage, the channel's ratio of the cross-sectional area to wetted perimeter is highest and stream flow is most efficient. Similarly, an increase in channel size increases the ratio of cross-sectional area to wetted perimeter and therefore, increases channel efficiency. All other factors being equal, flow velocities are higher in large channels than in small channels.

Most streams have channels that can be described as *rough.* Elements such as boulders, irregularities in the channel bed, and woody debris create turbulence that significantly impede flow.

Discharge

Streams vary in size from small headwater creeks less than a meter wide, to large rivers with widths of several kilometers. The size of a stream channel is largely determined by the amount of water supplied from the drainage basin. The measure most often used to compare the size of streams is **discharge**—the volume of water flowing past a certain point in a given unit of time. Discharge, usually measured in cubic meters per second or cubic feet per second, is determined by multiplying a stream's cross-sectional area by its velocity.

Table 16.1 lists the world's largest rivers in terms of discharge. The largest river in North America, the Mississippi, has a discharge that averages 17,300 cubic meters per second. Nevertheless, that amount is dwarfed by South America's mighty Amazon, which discharges 12 times more water than the Mississippi. In fact, it has been estimated that the flow of the Amazon accounts for about 15 percent of all the fresh water transported to the ocean by all of the world's rivers. Just one day's discharge would supply the water needs of New York City for nine years!

The discharge of a river system changes over time because of variations in the amount of precipitation received by the drainage basin. Studies show that when discharge increases, the width, depth, and flow velocity of the channel all increase predictably. As we saw earlier, when the size of the channel increases, proportionally less water is in contact with the bed and banks of the channel. Thus, friction, which acts to retard the flow, is reduced, resulting in an increase in the rate of flow.

Changes Downstream

One useful way of studying a stream is to examine its **longitudinal profile.** Such a profile is simply a cross-sectional view of a stream from its source area (called the **head** or **headwaters**) to its **mouth,** the point downstream where it empties into another water body—a river, lake, or ocean. By examining Figure 16.9, the

FIGURE 16.8 Rapids are common in mountain streams because channels are rough and irregular. (Photo by Fogstock Llc/Photolibrary)

			Drainage Area		Average Discharge	
Rank	River	Country	Square kilometers	Square miles	Cubic meters per sec.	Cubic feet per sec.
1	Amazon	Brazil	5,778,000	2,231,000	212,400	7,500,000
2	Congo	Rep. of Congo	4,014,500	1,550,000	39,650	1,400,000
3	Yangtze	China	1,942,500	750,000	21,800	770,000
4	Brahmaputra	Bangladesh	935,000	361,000	19,800	700,000
5	Ganges	India	1,059,300	409,000	18,700	660,000
6	Yenisei	Russia	2,590,000	1,000,000	17,400	614,000
7	Mississippi	United States	3,222,000	1,244,000	17,300	611,000
8	Orinoco	Venezuela	880,600	340,000	17,000	600,000
9	Lena	Russia	2,424,000	936,000	15,500	547,000
10	Parana	Argentina	2,305,000	890,000	14,900	526,000

TABLE 16.1 World's Largest Rivers Ranked by Discharge

most obvious feature of a typical longitudinal profile is its concave shape—a result of the decrease in slope that occurs from the headwaters to the mouth. In addition, local irregularities exist in the profiles of most streams—the flatter sections may be associated with lakes or reservoirs, and the steeper sections are sites of rapids or waterfalls.

The change in slope observed on most stream profiles is usually accompanied by an increase in discharge, channel size, and a reduction in sediment particle size (Figure 16.10). For example, data from successive gaging stations along most rivers shows that, in humid regions, discharge increases toward the mouth. This should come as no surprise because, as we move downstream, more and more tributaries contribute water to the main channel. In the case of the Amazon, for example, about 1000 tributaries join the main river along its 6500-kilometer course across South America.

In order to accommodate the growing volume of water, channel size typically increases downstream as well. Recall that flow velocities are higher in large channels compared to small channels. Furthermore, observations show a general decline in sediment size downstream, making the channel smoother and more efficient.

Although the channel slope decreases toward a stream's mouth, the flow velocity generally increases. This fact contradicts our intuitive assumptions of swift, narrow headwater streams and wide, placid rivers flowing across more subtle topography. Increases in channel size and discharge, and decreases in channel roughness that occur downstream compensate for the decrease in slope—thereby making the stream more efficient (see Figure 16.10). Thus, the average flow velocity is typically lower in headwater streams than in wide, placid rivers just "rollin' along."

THE WORK OF RUNNING WATER

Streams are Earth's most important erosional agents. Not only do they have the ability to downcut and widen their channels, but streams also have the capacity to transport enormous quantities of sediment delivered to them by overland flow, mass wasting, and groundwater. Eventually, much of this material is deposited to create a variety of landforms.

Stream Erosion

A stream's ability to accumulate and transport soil and weathered rock is aided by the work of raindrops, which knock sediment particles loose (see Figure 6.23 on p. 192). When the ground is saturated, rainwater cannot infiltrate, so it flows downslope, transporting some of the material it dis-

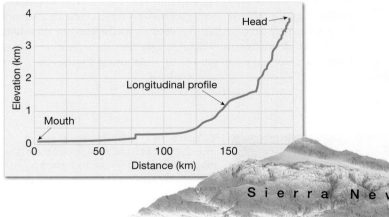

FIGURE 16.9 Longitudinal profile of California's Kings River. It originates in the Sierra Nevada and flows westward into the San Joaquin Valley. A *longitudinal profile* is a cross-section along the length of a stream. Note the concave-upward curve of the profile, with a steeper gradient upstream and a gentler gradient downstream.

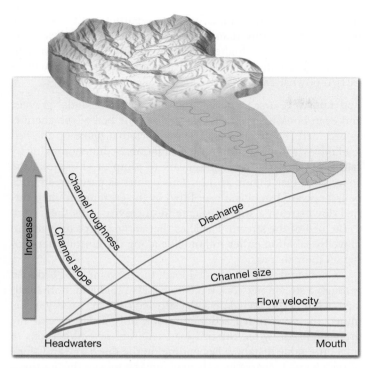

FIGURE 16.10 Graph showing how various properties of a stream channel change from its headwaters to its mouth. Although the gradient decreases toward the mouth, increases in channel size and discharge, and decreases in roughness more than offset the decrease in slope. Consequently, stream flow velocity usually increases toward the mouth.

lodges. On barren slopes the flow of muddy water (sheet flow) often produces small channels (rills), which in time can evolve into larger gullies (see Figure 6.24 on p. 194).

Once flow is confined in a channel, the erosional power of a stream is related to its slope and discharge. The rate of erosion, however, is also dependent on the relative resistance of the bank and bed material. In general, channels composed of unconsolidated materials are more easily eroded than channels cut into bedrock.

When the channel is sandy, the particles are easily dislodged from the bed and banks, and lifted into the moving water. Moreover, banks consisting of sandy material are often undercut, dumping even more loose debris into the water to be carried downstream. Banks that consist of coarse gravels or cohesive clay and silt particles tend to be relatively resistant to erosion. Thus, channels with cohesive silty banks are generally narrower than comparable ones with sandy banks.

Streams cut channels into bedrock through three main processes, *quarrying*, *abrasion*, and *corrosion*. **Quarrying** involves the removal of blocks from the bed of the channel. This process is aided by fracturing and weathering that loosen the blocks sufficiently so they are moveable during times of high flow rates. Quarrying is mainly the result of the impact forces exerted by flowing water.

Abrasion is the process by which the bed and banks of a bedrock channel are ceaselessly bombarded by particles carried into the flow. In addition, the individual sediment grains are also abraded by their many impacts with the channel and with one another. Thus, by scraping, rubbing, and bumping, abrasion erodes a bedrock channel and simultaneously smoothes and rounds the abrading particles. That is why smooth, rounded rocks and pebbles are found in streams. Abrasion also results in a reduction in the size of the sediments transported by streams.

Features common to some bedrock channels are circular depressions known as **potholes** created by the abrasive action of particles swirling in fast-moving eddies (Figure 16.11). The rotational motion of sand and pebbles acts like a drill to bore the holes. As the particles wear down to nothing, they are replaced by new ones that continue to drill the stream bed. Eventually, smooth depressions several meters across and deep may result.

Bedrock channels formed in soluble rock such as limestone are susceptible to **corrosion**—a process in which rock is gradually dissolved by the flowing water. Corrosion is a type of chemical weathering between the solutions in the water and the mineral matter that makes up the bedrock.

Transport of Sediment by Streams

All streams, regardless of size, transport some rock material (Figure 16.12). Streams also sort the solid sediment they transport because finer, lighter material is carried more readily than larger, heavier particles. Streams transport their load of sediment in three ways: (1) in solution (**dissolved load**), (2) in suspension (**suspended load**), and (3) sliding or rolling along the bottom (**bed load**).

Dissolved Load Most of the dissolved load is brought to a stream by groundwater and is dispersed throughout the flow. When water percolates through the ground, it acquires soluble soil compounds. Then, it seeps through cracks and pores in bedrock, dissolving additional mineral matter. Eventually much of this mineral-rich water finds its way into streams.

The velocity of streamflow has essentially no effect on a stream's ability to carry its dissolved load—material in the solution goes wherever the stream goes. Precipitation of the dissolved mineral matter occurs when the chemistry of the water changes, or when the water enters an inland "sea," located in an arid climate where the rate of evaporation is high.

FIGURE 16.11 Potholes in the bed of a stream. The rotational motion of swirling pebbles acts like a drill to create potholes. (Photo by Elmari Joubert/Alamy)

Suspended Load Most streams carry the largest part of their load in *suspension*. Indeed, the muddy appearance created by suspended sediment is the most obvious portion of a stream's load (Figure 16.13). Usually only very fine sand, silt, and clay particles are carried this way, but during flood stage, larger particles can also be transported in suspension. During flood stage, the total quantity of material carried in suspension also increases dramatically, verifiable by people whose homes have become sites for the deposition of this material. For example, during its flood stage, the Yellow River (Hwang Ho) of China is reported to carry an amount of sediment equal in weight to the water that carries it. Rivers such as this are appropriately described as "too thick to drink but too thin to cultivate."

The type and amount of material carried in suspension are controlled by two factors: the flow velocity and the settling velocity of each sediment grain. **Settling velocity** is defined as the speed at which a particle falls through a still fluid. The larger the particle, the more rapidly it settles toward the stream bed. In addition to size, the shape and specific gravity of particles also influence settling velocity. Flat grains sink through water more

438

slowly than spherical grains, and dense particles fall toward the bottom more rapidly than less dense particles. The slower the settling velocity and higher the flow velocity, the longer a sediment particle will stay in suspension, and the farther it will be carried downstream.

Bed Load Coarse material, including coarse sands, gravels, and even boulders typically move along the bed of the channel as bed load (Figure 16.14). The particles that make up the bed load move by rolling, sliding, and saltation. Sediment moving by **saltation** (*saltare* = to leap) appears to jump or skip along the stream bed (see Figure 16.12). This occurs as particles are propelled upward by collisions or lifted by the current and then carried downstream a short distance until gravity pulls them back to the bed of the stream. Particles that are too large or heavy to move by saltation either roll or slide along the bottom, depending on their shapes.

Compared with suspended load, the movement of bed load through a stream network tends to be less rapid and more localized. A study conducted on a glacially fed river in Norway determined that suspended sediments took only a day to exit the drainage basin, while the bed load required several decades to travel the same distance. Depending on the discharge and slope of the channel, coarse gravels may only be moved during times of high flow, while boulders move only during exceptional floods. Once set in motion, large particles are usually carried short distances. Along some stretches of a stream, bed load cannot be carried at all until it is broken into smaller particles.

Capacity and Competence A stream's ability to carry solid particles is described using two criteria—*capacity* and *competence*. **Capacity** is the maximum load of solid particles a stream can transport per unit time. The greater the discharge, the greater the stream's capacity for hauling sediment. Consequently, large rivers with high flow velocities have large capacities.

Competence is a measure of a stream's ability to transport particles based on size rather than quantity. Flow velocity is key—swift streams have greater competencies than slow streams, regardless of channel size. A stream's competence increases proportionately to the square of its velocity. Thus, if the velocity of a stream doubles,

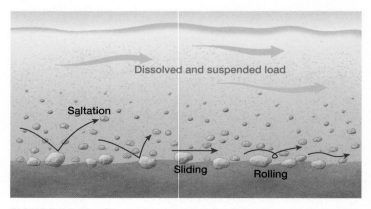

FIGURE 16.12 Streams transport their load of sediment in three ways. The dissolved and suspended loads are carried in the general flow. The bed load includes coarse sand, gravel, and boulders that move by rolling, sliding, and saltation.

mechanism by which solid particles of various sizes are separated. This process, called **sorting,** explains why particles of similar size are deposited together.

The general term for sediment deposited by streams is **alluvium.** Many different depositional features are composed of alluvium. Some occur within stream channels, some occur on the valley floor adjacent to the channel, and some are found at the mouth of the stream. We will consider the nature of these features later in the chapter.

STREAM CHANNELS

A basic characteristic that distinguishes streamflow from overland flow is that it is confined in a channel. A stream channel can be thought of as an open conduit consisting of the streambed and banks that act to confine flow except, of course, during floods.

Although somewhat oversimplified, we can divide stream channels into two basic types. *Bedrock channels* are those in which the streams are actively cutting into solid rock. In contrast, when the bed and banks are composed mainly of unconsolidated sediment or alluvium, the channel is called an *alluvial channel.*

FIGURE 16.14   Although the bed load of many rivers consists of sand, the bed load of this stream is made up of boulders and is easily seen during periods of low water. During floods, the seemingly immovable rocks in this channel are rolled along the bed of the stream. The maximum-size particle a stream can move is determined by the velocity of the water. (Photo by E. J. Tarbuck)

FIGURE 16.13   Colorado River, Grand Canyon National Park. The muddy appearance is a result of suspended sediment. (Photo by Michael Collier)

the impact force of the water increases four times; if the velocity triples, the force increases nine times, and so forth. Hence, large boulders that are often visible during low-water and seem immovable can, in fact, be transported during exceptional floods because of the stream's increased competence (see Figure 16.14).

By now it should be clear why the greatest erosion and transportation of sediment occur during periods of high water associated with floods. The increase in discharge results in greater capacity; the increased velocity produces greater competency. Rising velocity makes the water more turbulent, and larger particles are set in motion. In the course of a few days, or perhaps a few hours, a stream at flood stage can erode and transport more sediment than it does during several months of normal flow.

Deposition of Sediment by Streams

Deposition occurs whenever a stream slows, causing a reduction in competence. As its velocity decreases, sediment begins to settle, largest particles first. Thus, stream transport provides a

Bedrock Channels

As the name suggests, **bedrock channels** are cut into the underlying strata and typically form in the headwaters of river systems where streams have steep slopes. The energetic flow tends to transport coarse particles that actively abrade the bedrock channel. Potholes are often visible evidence of the erosional forces at work.

Steep bedrock channels often develop a sequence of *steps* and *pools,* relatively flat segments (pools) where alluvium tends to accumulate, and steep segments (steps) where bedrock is exposed. The steep areas contain rapids or, occasionally, waterfalls.

The channel pattern exhibited by streams cutting into bedrock is controlled by the underlying geologic structure. Even when flowing over rather uniform bedrock, streams tend to exhibit winding or irregular patterns rather than flowing in straight channels. Anyone who has gone white-water rafting has observed the steep, winding nature of a stream flowing in a bedrock channel.

Alluvial Channels

Alluvial channels form in sediment that was previously deposited in the valley. When the valley floor reaches sufficient width, material deposited by the stream can form a *floodplain* that borders the channel. Because the banks and beds of alluvial channels are composed of unconsolidated sediment (alluvium) they can undergo major changes in shape as material is continually being eroded, transported, and redeposited. The major factors affecting the shapes of these channels are the average size of the sediment being transported, the channel's gradient, and discharge.

Alluvial channel patterns reflect a stream's ability to transport its load at a uniform rate while expending the least amount of energy. Thus, the size and type of sediment being carried help determine the nature of the stream channel. Two common types of alluvial channels are *meandering channels* and *braided channels.*

Meandering Channels   Streams that transport much of their load in suspension generally move in sweeping bends called **meanders** (Figure 16.15). These streams flow in relatively deep, smooth channels and primarily transport mud (silt and clay), sand, and occasionally fine gravel. The lower Mississippi River exhibits this type of channel.

Meandering channels evolve over time as individual bends migrate across the floodplain. Most of the erosion is focused at the outside of the meander, where velocity and turbulence are greatest. In time, the outside bank is undermined, especially during periods of high water. Because the outside of a meander is a zone of active erosion, it is often referred to as the **cut bank** (Figure 16.16). Debris acquired by the stream at the cut bank moves downstream where the coarser material is generally deposited as **point bars** on the insides of bends. In this manner, meanders migrate laterally by eroding the outside of the bends and depositing sediment on the inside without appreciably changing their shape.

In addition to migrating laterally, the bends in a channel also migrate down the valley. This occurs because erosion is more effective on the downstream (downslope) side of the meander. Sometimes the downstream migration of a meander is slowed when it reaches a more resistant bank material. This allows the next meander upstream to gradually erode the material between the two meanders as shown in Figure 16.17. Eventually, the river may erode through the narrow neck of land forming a new, shorter channel segment called a **cutoff.** Because of its shape, the abandoned bend is called an **oxbow lake** (Figure 16.17).

Braided Channels   Some streams consist of a complex network of converging and diverging channels that thread their way among numerous islands or gravel bars (Figure 16.18). Because these channels have an interwoven appearance, they are said to be **braided.** Braided channels form where a large portion of a stream's load consists of coarse material (sand and gravel) and the stream has a highly variable discharge. Because the bank material is readily erodable, braided channels are wide and shallow.

One setting in which braided streams form is at the end of glaciers where there is a large seasonal variation in discharge. During the summer, large amounts of ice-eroded sediment are dumped into the meltwater streams flowing away from the glacier. However, when flow is sluggish,

FIGURE 16.15   Meanders on the White River near Augusta, Arkansas. (Photo by Michael Collier)

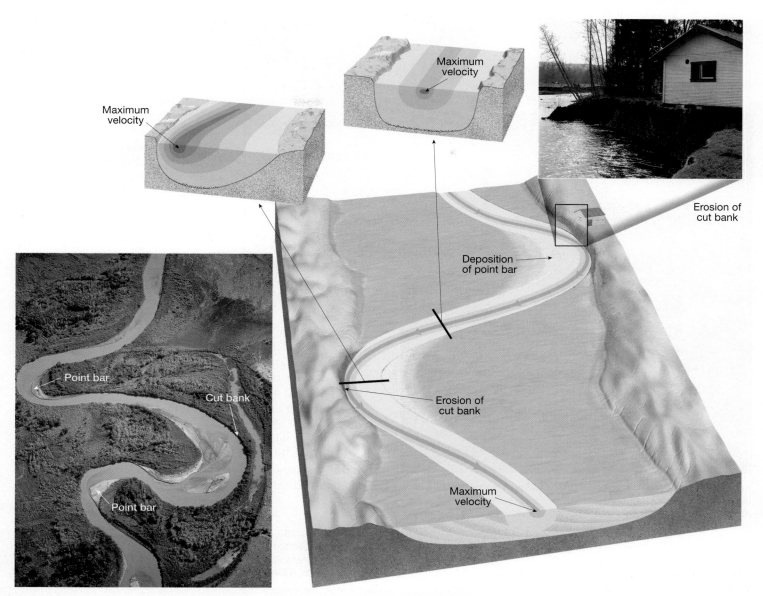

FIGURE 16.16 When a stream meanders, its zone of maximum speed shifts toward the outer bank. The point bars shown here are on the White River near Vernal, Utah. The point bars shown here are on the White River near Vernal, Utah. The black and white photo shows erosion of a cut bank along the Newaukum River in Washington State. By eroding its outer bank and depositing material on the inside of the bend, a stream is able to shift its channel. (Point bar photo by Michael Collier; cut bank photo by P. A. Glancy, U.S. Geological Survey)

the stream deposits the coarsest material as elongated structures called *bars*. This process causes the flow to split into several paths around the bars. During the next period of high flow, the laterally shifting channels erode and redeposit much of this coarse sediment, thereby transforming the entire streambed. In some braided streams, the bars have built semipermanent islands anchored by vegetation.

In summary, meandering channels develop where the load consists largely of fine-grained unconsolidated particles that are transported as suspended load in deep, relatively smooth channels. By contrast, wide, shallow braided channels develop where coarse-grained alluvium is transported mainly as bed load.

BASE LEVEL AND GRADED STREAMS

In 1875 John Wesley Powell, the pioneering geologist who first explored the Grand Canyon and later headed the U.S. Geological Survey, introduced the concept of a downward limit to stream erosion, which he called **base level.** A fundamental concept in the study of stream activity, base level is defined as the lowest elevation to which a stream can erode its channel. Essentially this is the level at which the mouth of a stream enters the ocean, a lake, or a trunk stream. Powell determined that two

Geologist's Sketch

FIGURE 16.17 Oxbow lakes occupy abandoned meanders. As they fill with sediment, oxbow lakes gradually become swampy meander scars. Aerial view of an oxbow lake created by the meandering Green River near Bronx, Wyoming. (Photo by Michael Collier)

its sediment-transporting ability. As a result, the stream deposits material, thereby building up its channel. This process continues until the stream again has a gradient sufficient to carry its load. The profile of the new channel will be similar to the old, but somewhat higher.

If, on the other hand, base level is lowered by a drop in sea level, the stream will have excess energy and downcut its channel to establish a balance with its new base level. Erosion first occurs near the mouth, then progresses upstream creating a new stream profile.

Observing streams that adjust their profiles to changes in base level led to the concept of a graded stream. A **graded stream** has the necessary slope and other channel characteristics to maintain the minimum velocity required to transport the material supplied to it. On average, a graded system is neither eroding nor depositing material but simply transporting it. When a stream reaches equilibrium, it becomes a self-regulating system in which a change in one characteristic causes a change in the others to counteract the effect.

Consider what would happen if displacement along a fault raises a layer of resistant rock along the course of a graded stream. As shown in Figure 16.20, the resistant rock forms a waterfall and serves as a temporary base level for the stream.

FIGURE 16.18 The Knik River, a classic braided stream with multiple channels separated by migrating gravel bars. The Knik is choked with sediment from the meltwater of four glaciers in the Chugach Mountains north of Anchorage, Alaska. (Photo by Michael Collier)

types of base level exist: "We may consider the level of the sea to be a grand base level, below which the dry lands cannot be eroded; but we may also have, for local and temporary purposes, other base levels of erosion."*

Sea level, which Powell called "grand base level," is now referred to as **ultimate base level.** **Local** or **temporary base levels** include lakes, resistant layers of rock, and rivers that act as base levels for their tributaries. All limit a stream's ability to downcut its channel.

Changes in base level cause corresponding adjustments in the "work" that streams perform. When a dam is built along a stream course, the reservoir that forms behind it raises the base level of the stream (Figure 16.19). Upstream from the reservoir the stream gradient is reduced, lowering its velocity and, hence,

Exploration of the Colorado River of the West (Washington, D.C.: Smithsonian Institution, 1875), p. 203.

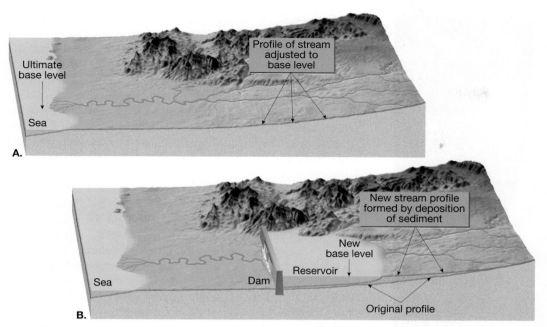

FIGURE 16.19 When a dam is built and a reservoir forms, the stream's base level is raised. This reduces the stream's velocity and leads to deposition and a reduction of the gradient upstream from the reservoir.

In some arid regions, where weathering is slow and rock is particularly resistant, narrow valleys having nearly vertical walls are also found (Figure 16.21). Stream valleys exist on a continuum from narrow, steep-sided valleys to those that are so flat and wide that the valley walls are not discernable.

Streams, with the aid of weathering and mass wasting, shape the landscape through which they flow. As a result, streams continuously modify the valleys they occupy.

Valley Deepening

When a stream's gradient is steep and the channel is well above base level, downcutting is the dominant activity. Abrasion caused by bed load sliding and rolling along the bottom, and the hydraulic power of fast-moving water, slowly

Because of the increased gradient, the stream concentrates its erosive energy on the resistant rock along an area called a *knickpoint*. Eventually, the river erases the knickpoint from its path and reestablishes a smooth profile.

SHAPING STREAM VALLEYS

RUNNING WATER
▶ Reviewing Valleys and Stream-related Features

A stream valley consists of a channel and the surrounding terrain that directs water to the stream. It includes the *valley floor,* which is the lower, flatter area that is partially or totally occupied by the stream channel, and the sloping *valley walls* that rise above the valley floor on both sides. Alluvial channels often flow in valleys that have wide valley floors consisting of sand and gravel deposited in the channel, and clay and silt deposited by floods. Bedrock channels, on the other hand, tend to be located in narrow V-shaped valleys.

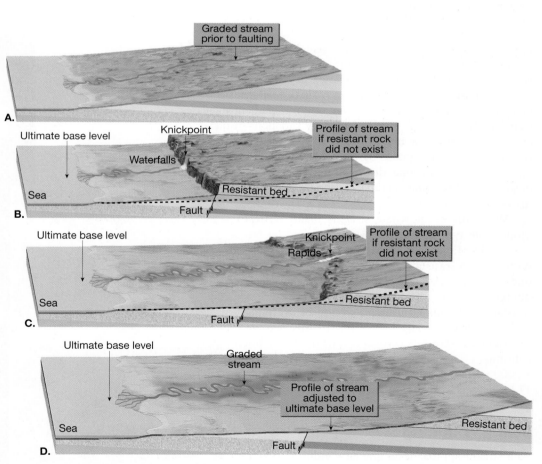

FIGURE 16.20 Changes in base level. A resistant layer of rock acts as a temporary base level. Because of the increased gradient, the stream concentrates its erosive energy on the resistant rock at the knickpoint. Eventually, the river erases the knickpoint and reestablishes a smooth profile.

FIGURE 16.21 This steep-sided gorge in Zion National Park, Utah, has been cut through solid rock by stream erosion. Canyons such as this are narrow due to rapid downcutting by the stream and very slow weathering of the canyon walls. (Photo by Zandria Muench Beraldo/CORBIS)

lowers the streambed. The result is usually a V-shaped valley with steep sides. A classic example of a V-shaped valley is the section of the Yellowstone River shown in Figure 16.22.

The most prominent features of V-shaped valleys are *rapids* and *waterfalls*. Both occur where the stream's gradient increases significantly, a situation usually caused by variations in the erodability of the bedrock into which a stream channel is cutting. Resistant beds create rapids by acting as a temporary base level upstream while allowing downcutting to continue downstream. Recall that, in time, erosion usually eliminates the resistant rock.

Waterfalls occur where streams make vertical drops. One type of waterfall is exemplified by Niagara Falls (Figure 16.23). Here, the falls are supported by a resistant bed of dolostone that is underlain by less resistant shale. As the water plunges over the lip of the falls, it erodes the less resistant shale, undermining a section of overlying rock, which eventually breaks off. In this manner the waterfall retains its vertical cliff while slowly, but continually, retreating upstream. Over the past 12,000 years, Niagara Falls has retreated more than 11 kilometers (7 miles) upstream.

Valley Widening

As a stream approaches a graded condition, downcutting becomes less dominant. At this point the stream's channel takes on a meandering pattern, and more of its energy is directed from side to side. As a

result, the valley widens as the river cuts away at one bank and then the other (Figure 16.24). The continuous lateral erosion caused by shifting meanders gradually produces a broad, flat valley floor covered with alluvium (Figure 16.25). This feature, called a **floodplain,** is appropriately named because when a river overflows its banks during flood stage, it inundates the floodplain. Over time the floodplain will widen to a point where the stream is only actively eroding the valley walls in a few places. In the case of the lower Mississippi River, for example, the distance from one valley wall to another sometimes exceeds 160 kilometers (100 miles).

When a river erodes laterally and creates a floodplain as described, it is called an *erosional floodplain*. Floodplains can be depositional as well. *Depositional floodplains* are produced by major fluctuations in conditions, such as changes in base level or climate. The floodplain in California's Yosemite Valley is one such feature; it was produced when a glacier gouged the valley floor about 300 meters (1000 feet) deeper than its former level. After the glacial ice melted, running water refilled the valley with alluvium. The Merced River currently winds across a relatively flat floodplain that forms much of the floor of Yosemite Valley.

FIGURE 16.22 V-shaped valley of the Yellowstone River. The rapids and waterfalls indicate that the river is vigorously downcutting. (Photo by Art Wolfe, Inc.)

Geologist's Sketch

FIGURE 16.23 Retreat of Niagara Falls. The river plunges over the falls and erodes the shale beneath the more resistant Lockport Dolostone. As a section of dolostone is undercut, it loses support and breaks off. During the past 12,000 years, Niagara Falls has retreated more than 11 kilometers (7 miles) upstream (Photo by Howard Sandler/iStockphoto)

Incised Meanders and Stream Terraces

We usually find streams with highly meandering courses on floodplains in wide valleys. However, some rivers have meandering channels that flow in steep, narrow, bedrock valleys. Such meanders are called **incised meanders** (*incisum* = to cut into) (Figure 16.26).

How do these features form? Originally the meanders probably developed on the floodplain of a stream that was in balance with its base level. Then, a change in base level caused the stream to begin downcutting. One of two events likely occurred—either base level dropped or the land upon which the river was flowing was uplifted. For example, regional uplifting of the Colorado Plateau in the southwestern United States generated incised meanders on several rivers. As the plateau gradually rose, meandering rivers began downcutting because of their steepening gradient (Figure 16.26).

After a river has adjusted in this manner, it may once again produce a floodplain at a level below the old one. The remnants of a former floodplain are sometimes present as relatively flat surfaces called **terraces** (Figure 16.27).

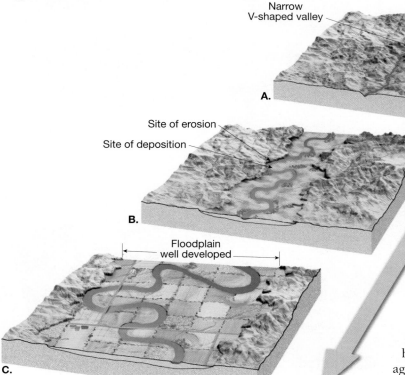

FIGURE 16.24 Development of an erosional floodplain.

DEPOSITIONAL LANDFORMS

Recall that streams continually pick up sediment in one part of their channel and deposit it downstream. These small-scale channel deposits are called **bars.** Such features, however, are only temporary, as the material will be picked up again and eventually carried to the ocean. In addition to sand and gravel bars, streams also create other depositional features that have somewhat longer life spans. These include *deltas, natural levees,* and *alluvial fans.*

FIGURE 16.25   Lateral erosion by meanders increases the size of the floodplain. Red Deer River, Alberta, Canada. (Photo by Georg Gerster)

Deltas

Deltas form where sediment-charged streams enter the relatively still waters of a lake, an inland sea, or the ocean (Figure 16.28A). As the stream's forward motion is slowed, sediments are deposited by the dying current, producing three types of beds. *Foreset beds* are composed of coarse particles that drop almost immediately upon entering the water to form layers that slope downcurrent from the delta front. The foreset beds are usually covered by thin, horizontal *topset beds* deposited during flood stage. The finer silts and clays settle away from the mouth in nearly horizontal layers called *bottomset beds*.

As a delta grows outward from the shoreline, the stream's gradient continually decreases. This circumstance eventually causes the channel to become choked with sediment. As a consequence, the river seeks shorter, higher-gradient routes to base level. Figure 16.28B shows the main channel dividing into several smaller ones, called **distributaries** that carry water away from the main channel in varying paths to base level. After numerous

A.

B. Before

C. After

FIGURE 16.26   Incised meanders. Aerial view of incised meanders of the Colorado River on the Colorado Plateau. As the plateau gradually rose, these meanders downcut because of the steepening gradient. (Photo by Michael Collier)

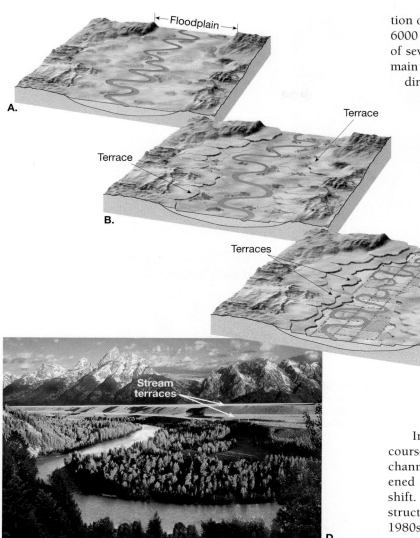

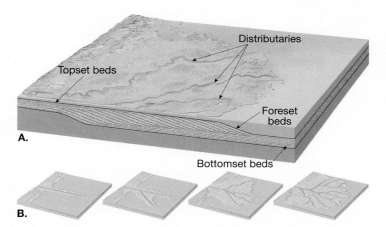

FIGURE 16.27 Stream terraces. Parts **A, B,** and **C** show the development of multiple stream terraces. Terraces can form when a stream downcuts through previously deposited alluvium. This may occur in response to a lowering of base level or as a result of regional uplift. The photo in **D** shows terraces of the Snake River in Wyoming. (Photo by Lester Lefkowitz/CORBIS)

tion of the Mississippi delta that has accumulated over the past 6000 years. As the figure illustrates, the delta is actually a series of seven coalescing subdeltas. Each subdelta formed when the main flow was diverted from one channel to a shorter, more direct path to the Gulf of Mexico. The individual subdeltas intertwine and partially cover each other to produce a complex structure. It is also apparent from Figure 16.29 that after each channel was abandoned, coastal erosion modified the newly formed subdelta. The present subdelta (number 7 in Figure 16.29), called a *bird-foot* delta because of the configuration of its distributaries, has been built by the Mississippi over the past 500 years.

At present, this active bird-foot delta has extended seaward almost as far as natural forces will allow. In fact, for many years the river has been "struggling" to cut through a narrow neck of land and shift its course to the Atchafalaya River (see inset in Figure 16.29). If this were to happen, the Mississippi would abandon the lowermost 500-kilometers of its channel in favor of the Atchafalaya's much shorter 225-kilometer route to the Gulf.

In a concerted effort to keep the Mississippi on its present course, a damlike structure was erected at the site where the channel was trying to break through. A flood in 1973 weakened the control structure, and the river again threatened to shift. This event caused the U.S. Corps of Engineers to construct a massive auxiliary dam that was completed in the mid-1980s. For the time being, at least, the inevitable has been avoided, and the Mississippi River continues to flow past Baton Rouge and New Orleans on its way to the Gulf of Mexico (see Box 16.1).

shifts in the main flow from one distributary to the next, a delta may grow into the triangular shape of the Greek letter delta (Δ), although several other shapes exist.

In the ocean, deltas form where the supply of sediment exceeds the rate of marine erosion. Differences in the configurations of shorelines, and variations in the nature and strength of wave activity are responsible for the shape and structure of each delta. Many of the world's great rivers have created massive deltas, each with its own peculiarities and typically more complex than the one illustrated in Figure 16.28A.

The Mississippi Delta

The Mississippi River delta resulted from the accumulation of huge quantities of sediment derived from the vast region drained by the river and its tributaries. New Orleans currently rests where there was once ocean. Figure 16.29 shows the por-

FIGURE 16.28 Formation of a simple delta. **A.** Structure of a simple delta that forms in relatively quiet waters. **B.** Growth of a simple delta. As a stream extends its channel, the gradient is reduced. Frequently, during flood stage the river is diverted to a higher-gradient route, forming a new distributary. Old, abandoned distributaries are gradually invaded by aquatic vegetation and filled with sediment.

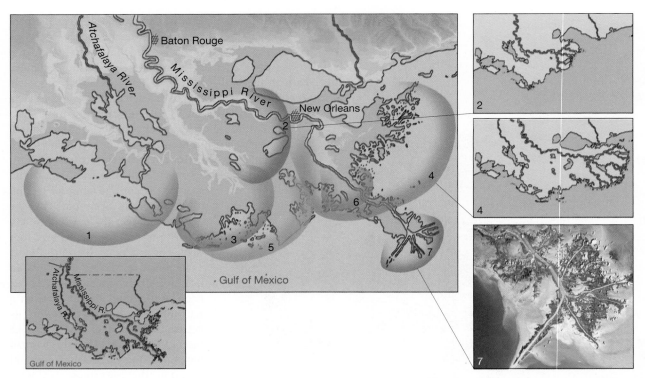

FIGURE 16.29 During the past 6000 years, the Mississippi River has built a series of seven coalescing subdeltas. The numbers indicate the order in which the subdeltas were deposited. The present bird-foot delta (number 7) represents the activity of the past 500 years. (Image courtesy of JPL/Cal Tech/NASA) Without ongoing human efforts, the present course will shift and follow the path of the Atchafalaya River. The left inset shows the point where the Mississippi may someday break through (arrow) and the shorter path it would take to the Gulf of Mexico. (After C. R. Kolb and J. R. Van Lopik)

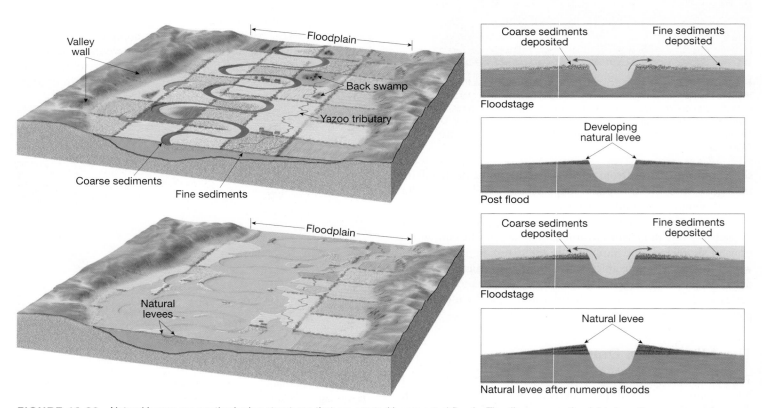

FIGURE 16.30 Natural levees are gently sloping structures that are created by repeated floods. The diagrams on the right show the sequence of development. Because the ground next to the stream channel is higher than the adjacent floodplain, back swamps and yazoo tributaries may develop.

Natural Levees

Meandering rivers that occupy valleys with broad floodplains tend to build **natural levees** (*lever* = to raise) that parallel their channels on both banks (Figure 16.30). Natural levees are built by years of successive floods. When a stream overflows onto the floodplain, the water flows over the surface as a broad sheet. Because the flow velocity drops significantly, the coarser portion of the suspended load is immediately deposited adjacent to the channel. As the water spreads across the floodplain, a thin layer of fine sediment is laid down over the valley floor. This uneven distribution of material produces the gentle, almost imperceptible, slope of the natural levee (Figure 16.30).

The natural levees of the lower Mississippi River rise 6 meters (20 feet) above the adjacent valley floor. The area behind the levee is characteristically poorly drained for the obvious reason that water cannot flow over the levee and into the river. Marshes called **back swamps** often result. When a tributary stream enters a river valley having a substantial natural levee, it often flows for many kilometers through the back swamp before finding an opening where it enters the main river. Such streams are called **yazoo tributaries,** after the Yazoo River, which parallels the lower Mississippi River for more than 300 kilometers.

Alluvial Fans

As Figure 16.31 illustrates, **alluvial fans** are fan-shaped deposits that accumulate along steep mountain fronts. When mountain streams emerge onto a relatively flat lowland their gradient drops and they deposit a large portion of their sediment load. Although alluvial fans are more prevalent in arid climates, they are occasionally found in humid regions.

Mountain streams, because of their steep gradients, carry much of their sediment load as coarse sand and gravel. Because alluvial fans are composed of these same coarse materials, water that flows across them readily soaks in. When a stream emerges from its valley onto an alluvial fan, its flow divides itself into several distributary channels. The fan shape is produced because the main flow swings back and forth between the distributaries from a fixed point where the stream exits the mountain.

Between rainy periods in deserts, little or no water flows across an alluvial fan, evident by the many dry channels that

FIGURE 16.31 Alluvial fans are built from sediments deposited at the mouth of a valley that emerges from a mountainous or upland area onto a relatively flat lowland. The surface of the fan slopes outward in a broad arc from the apex at the mouth of the stream valley. Usually, coarse material is dropped near the apex of the fan, while finer materials are carried toward the base of the deposit. California's Death Valley has many large alluvial fans. As adjacent fans grow larger, they may coalesce to form a steep apron of sediment called a *bajada*. See Figure 19.9, p. 528. (Photo by Michael Collier)

cross its surface. Thus, fans in dry regions grow intermittently, receiving considerable water and sediment only during wet periods. As you learned in Chapter 15, steep canyons in dry regions are prime locations for debris flows. Therefore, as would be expected, many alluvial fans have debris-flow deposits interbedded with the coarse alluvium.

DRAINAGE PATTERNS

Drainage systems are interconnected networks of streams that form a variety of patterns. Variances occur primarily in response to the type of material on which the streams developed and/or the structural pattern of fractures, faults, and folds.

The most common, the **dendritic drainage pattern,** is characterized by irregular branching of tributary streams that resemble the branching pattern of a deciduous tree (Figure 16.32A). In fact, the word *dendritic* means "treelike." This drainage pattern develops whenever underlying bedrock is relatively uniform, such as flat-lying sedimentary strata or massive igneous rocks. Because

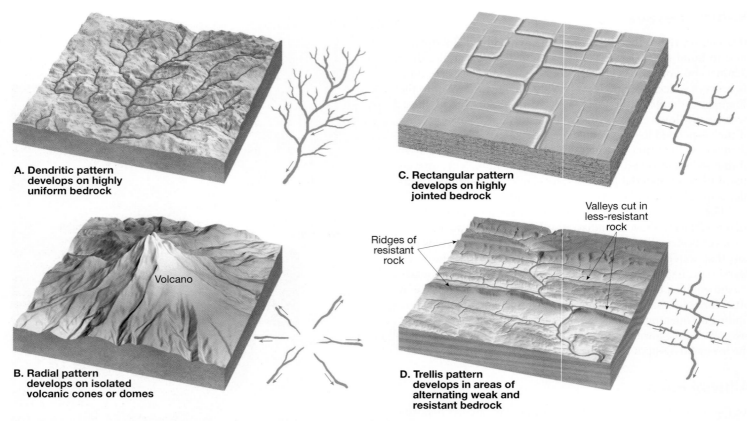

FIGURE 16.32   Drainage patterns. **A.** Dendritic. **B.** Radial. **C.** Rectangular. **D.** Trellis.

the underlying material is essentially uniform in its resistance to erosion, it does not control the pattern of streamflow. Rather, the pattern is primarily determined by the slope of land.

When streams diverge from a central area, like spokes from the hub of a wheel, it is called a **radial drainage pattern** (Figure 16.32B). This pattern typically develops on isolated volcanic cones and domal uplifts.

Figure 16.32C illustrates a **rectangular drainage pattern,** with many right-angle bends. This pattern typically develops where the bedrock is crisscrossed by a series of joints. Because fractured rock tends to weather and erode more easily than unbroken rock, the geometric pattern of joints guides the direction of streams as they carve their valleys.

Illustrated in Figure 16.32D is a **trellis drainage pattern,** a rectangular pattern in which tributary streams are nearly parallel to one another and have the appearance of a garden trellis. This pattern forms in areas underlain by alternating bands of resistant and less resistant rock and is particularly well displayed in the folded Appalachian Mountains, where both weak and strong strata outcrop in nearly parallel belts (see Figure 14.18, page 396).

Formation of a Water Gap

To fully understand the drainage pattern displayed by streams, we must consider their entire histories. For example, river valleys occasionally cut through a ridge or mountainous topography that lies across their path. The steep-walled notch followed by the river through a tectonic structure is called a **water gap** (Figure 16.33).

Why do streams cut across such structures and not flow around them? One possibility is that the stream existed before the ridge or mountain was uplifted. In this situation, the stream, called an **antecedent stream,** downcut its bed at a pace equal to the rate of uplift. That is, the stream would maintain its course as folding or faulting raised the structure across its path.

A second possibility is that a **superposed stream** eroded its channel into an existing structure (see Figure 16.33). This can occur when folded beds or resistant rocks are buried beneath layers of relatively flat-lying sediments or sedimentary strata. Streams originating on the overlying strata establish their courses without regard to the structures below. Then, as the valley deepens the river continues to cut its valley into the structure below. The folded Appalachians feature several superposed rivers including the Potomac and the Susquehanna, which cut their channels through folded strata on their way to the Atlantic.

Headward Erosion and Stream Piracy

Streams lengthen their courses by **headward erosion,** that is, by extending the heads of their valleys upslope. In their headwater sections, streams tend to downcut their channels. This leads to increased rates of erosion on the steeper slopes which lie in the opposite direction of stream flow. Thus, through headward erosion, the valley is extended into previously undissected terrain.

Headward erosion by streams plays a major role in the dissection of upland areas. In addition, sufficient headward erosion

PEOPLE AND THE ENVIRONMENT

Coastal Wetlands Vanishing on the Mississippi Delta

BOX 16.1

Coastal wetlands include swamps, tidal flats, coastal marshes, and bayous. They are rich in wildlife and provide nesting grounds and important stopovers for waterfowl and migratory birds, as well as spawning areas and valuable habitats for fish.

The delta of the Mississippi River in Louisiana contains about 40 percent of all coastal wetlands in the lower 48 states. Louisiana's wetlands are naturally sheltered from the wave action of hurricanes and winter storms by low-lying offshore barrier islands. Both the wetlands and the barrier islands have formed as a result of sediments carried to the Gulf of Mexico by the Mississippi River during the past 6000 years.

The dependence of Louisiana's coastal wetlands and offshore islands on the Mississippi River and its distributaries as a source of sediment, leaves them vulnerable to changes in the river system. Moreover, the reliance on barrier islands for protection from storm waves leaves coastal wetlands vulnerable should these narrow offshore islands erode.

The coastal wetlands of Louisiana are disappearing at an alarming rate, accounting for 80 percent of the wetland loss in the lower 48 states. According to the U.S. Geological Survey, Louisiana lost nearly 5000 square kilometers (1900 square miles) of coastal land between 1932 and 2000. Continuing at this rate, another 3000 square kilometers will vanish under the Gulf of Mexico by the year 2050.* Global climate change could increase the severity of the problem because rising sea level and stronger tropical storms accelerate rates of coastal erosion. Unfortunately, this was especially obvious during the extraordinary 2005 hurricane season when hurricanes Katrina and Rita devastated portions of the Gulf Coast.

The delta, its wetlands, and adjacent barrier islands are naturally dynamic features. Whenever the river shifted, zones of delta growth and destruction also shifted. Over the millen-

nia, as sediment accumulated and built the delta in one area, erosion and subsidence caused losses elsewhere (Figure 16.A). However, the Mississippi has not shifted in more than 500 years. So, why are Louisiana's wetlands shrinking at such an alarming rate?

Before Europeans settled the delta, the Mississippi River regularly overflowed its banks in seasonal floods. Huge quantities of sediment were deposited on top of the delta which acted to keep it elevated above sea level. However, with settlement came flood-control efforts and the desire to maintain and improve navigation on the river. Artificial levees were constructed to contain the rising river during flood stage. Over time the levees were extended all the way to the mouth of the Mississippi to keep the channel open for navigation.

The effects have been straightforward. The levees prevent sediment from being dispersed onto the wetlands allowing the river to carry its load to the Gulf. Meanwhile, the natural processes of compaction, subsidence, and wave erosion continue. Sediment was not added in sufficient quantity to offset these forces, thereby

shrinking the size of the delta and the extent of its wetlands.

The problem has been exacerbated by a decline in the sediment transported by the Mississippi, which has decreased approximately 50 percent over the past 100 years. A substantial portion of the reduction results from sediment being trapped by the construction of dams on many tributaries of the Mississippi.

Another factor contributing to wetland decline is the 13,000 kilometers (8000 miles) of navigation channels and canals that lace the delta. These artificial openings to the sea allow salty waters of the Gulf of Mexico to flow far inland. The invasion of saltwater and tidal action have caused massive marsh die-offs referred to as *brownouts*.

Understanding and modifying the human impact is necessary for any successful plan to reduce the loss of Mississippi delta wetlands. The U.S. Geological Survey estimates it will cost $14 billion over the next 40 years to restore Louisiana's coasts.

*See "Louisiana's Vanishing Wetlands: Going, Going . . ." in *Science,* vol. 289, 15 September 2000, pp. 1860–63. Also see Elizabeth Kolbert, "Watermark—Can Southern Louisiana be Saved?" *The New Yorker,* February 27, 2006, pp. 46–57.

FIGURE 16.A Dead cypress trees, known as ghost forests, killed by encroaching salt water in Terrebonne Parish, Louisiana. (Photo by Robert Caputo/Aurora Photos)

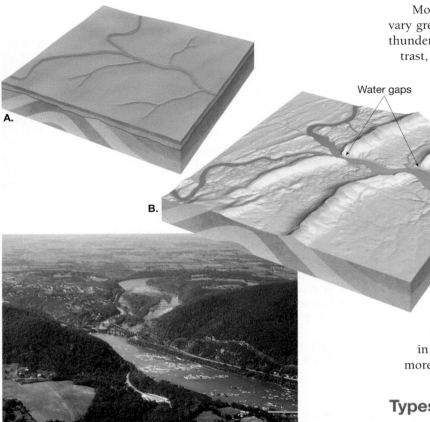

Water gaps

FIGURE 16.33 Development of a superposed stream. **A.** The river establishes its course on relatively uniform strata. **B.** It then encounters and cuts through the underlying structure. **C.** Harpers Ferry gap at the confluence of the Shenandoah and Potomac rivers near the West Virginia-Maryland border. Water gaps such as this one are common in parts of the Appalachians. (Photo by John S. Shelton)

can result in a stream entering the drainage basin of an adjacent stream. This process, called **stream piracy,** is the capture of drainage that previously flowed to another stream.

Numerous situations are possible that can lead to stream piracy. In Figure 16.34, the upstream portion of stream A was captured when headward erosion by the tributary to stream B breached the divide.

Stream piracy also explains the existence of narrow, steep-sided gorges that do not have active streams running through them. These abandoned water gaps (called *wind gaps*) form when the stream that cut the notch had its course changed by a pirate stream. In Figure 16.34, one water gap created by stream *A* became a wind gap as a result of stream piracy.

FLOODS AND FLOOD CONTROL

Floods occur when the flow of a stream becomes so great that it exceeds the capacity of its channel and overflows its banks (Figure 16.35). Among the most deadly and most destructive of all geologic hazards, floods are, nevertheless, simply part of the *natural* behavior of streams.

Most floods are caused by atmospheric processes that can vary greatly in both time and space. An hour or less of intense thunderstorm rainfall can trigger floods in small valleys. By contrast, major floods in large river valleys are often the result of an extraordinary series of precipitation events over a broad region for many days or weeks.

Land use planning in river basins requires an understanding of the frequency and magnitude of floods. For any particular river, a relationship exists between the size of a flood and the frequency with which it occurs. The larger the flood, the less often it is expected to occur. You have probably heard of a *100-year flood*. This describes the *recurrence interval* or *return period* which is an estimate of how often a flood of a given size can be expected to occur. A 25-year event would be much smaller, but would be four times more likely to occur than a 100-year flood. The relationship between the frequency and magnitude of floods differs from one region to another. For example, in arid climates, large, 100-year floods are typically much more extreme than those in humid areas.

Types of Floods

Floods can be the result of several naturally-occurring and human-induced factors. Common flood types include *regional floods, flash floods, ice-jam floods,* and *dam-failure floods.*

Regional Floods Most regional floods are seasonal. Rapid melting of snow in spring and/or heavy spring rains often overwhelm rivers. For example, the extensive 1997 flood along the Red River of the North was preceded by an especially snowy winter and an early spring blizzard. Early April brought rapidly rising temperatures, melting the snow in a matter of days, causing a record-breaking 500-year flood. Roughly 4.5 million acres

Students Sometimes Ask . . .

Does plate tectonics influence rivers?

Yes, in many ways. For example, the existence of a major river is due largely to a continent's position in a climatic zone where precipitation is plentiful. These factors are influenced by the movement of plates. Over millions of years, rivers appeared and disappeared due to continents moving across different climatic zones.

Plate tectonics also affects rivers in other ways. Mountain building at convergent boundaries strongly influences regional slopes and modifies precipitation patterns. Folding and faulting associated with tectonic processes affects drainage patterns, whereas extensive lava flows created by plate-related volcanic activity can radically change river systems.

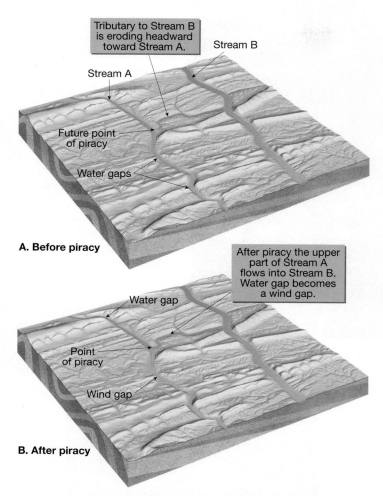

Stream A

Tributary to Stream B is eroding headward toward Stream A.

Stream B

Future point of piracy

Water gaps

A. Before piracy

After piracy the upper part of Stream A flows into Stream B. Water gap becomes a wind gap.

Water gap

Point of piracy

Wind gap

B. After piracy

FIGURE 16.34 Stream piracy and the formation of a wind gap. A tributary of stream **B** erodes headward until it eventually captures and diverts a portion of the drainage of stream **A**. A water gap through which stream **A** flowed is abandoned because of the piracy. As a result, this feature becomes a wind gap. In this valley and ridge-type setting, the softer rocks in the valleys are eroded more easily than the resistant ridges.

were underwater, and the losses in the Grand Forks, North Dakota region exceeded $3.5 billion.*

Regional floods can also be the result of numerous heavy rain events. The extensive and costly June 2008 floods in parts of the Midwest were the result of record-breaking rainfall on already waterlogged soils. Indiana experienced its costliest weather disaster in history, but Iowa suffered even greater losses with 83 of its 99 counties declared disaster areas. Nine of the state's rivers were at or above previous record flood levels and millions of acres of productive farmland were submerged. Thousands of people were evacuated, mostly in Cedar Rapids where more than 400 city blocks were under water (see Figure 16.35). Persistent wet weather patterns led to a similar period of exceptional rains and devastating floods in the upper Mississippi River Valley during the summer of 1993 (Figure 16.36).

Flash Floods Flash floods occur with little warning and are potentially deadly because they produce rapid rises in water levels and can have devastating flow velocities (Box 16.2). Rainfall intensity and duration, surface conditions, and topography are among the factors that influence flash flooding. Mountainous areas are especially susceptible because steep slopes can funnel runoff into narrow canyons with disastrous consequences, such as the Big Thompson River flood of July 31, 1976, in Colorado (see Figure 16.C in Box 16.2). During a four-hour period, more than 30 centimeters (12 inches) of rain fell, overwhelming its small drainage basin. The flash flood in this narrow canyon lasted only a few hours but claimed 139 lives and caused tens of millions of dollars in damages.

Urban areas are also susceptible to flash floods because a high percentage of the surface area is composed of impervious roofs, streets, and parking lots, where infiltration is minimal and runoff is rapid. In fact, a recent study determined the area of impervious surfaces in the United States (excluding Alaska and Hawaii) amounts to more than 112,600 square kilometers (nearly 44,000 square miles), which is slightly less than the area of the state of Ohio.*

The effect of urbanization on streamflow can be observed by examining Figure 16.37. A hypothetical hydrograph shows the time relationship between a rainstorm and the occurrence of flooding (Figure 16.37, top). Notice the water level in the stream does not rise at the onset of precipitation, because time is required for water to move from the place where it fell to the stream. This time difference is called the *lag time*. The bottom hydrograph in Figure 16.37 depicts the same hypothetical area and rainfall event *after* urbanization. The peak discharge during a flood is greater and the lag time between precipitation and flood peak is shorter than before urbanization. The explanation for this effect is simple—streets, parking lots, and buildings cover the ground that once soaked up water. Thus, less water infiltrates and the quantity and rate of runoff increases.

Ice-Jam Floods Frozen rivers are especially susceptible to ice-jam floods. As the level of a stream rises, it breaks up ice and creates ice flows that can accumulate on channel obstructions. Jams of this nature create temporary ice dams across the channel. Water trapped upstream can rise rapidly and overflow the channel banks. When an ice dam fails, water behind the dam is often released with sufficient force to inflict considerable damage downstream.

Dam-Failure Floods Human interference with stream systems can also cause floods. A prime example is the failure of a dam or an artificial levee designed to contain small or moderate floods. When larger floods occur, the dam or levee may fail, resulting in the water behind it being released as a flash flood. The bursting of a dam in 1889 on the Little Conemaugh River caused the devastating Johnstown, Pennsylvania, flood that took over 2200 lives.

---

*Ice jams also contribute to floods on the Red River of the North. See the section on "Ice-Jam Floods" on this page.

*C. D. Elvidge, et al. "U.S. Constructed Area Approaches the Size of Ohio," in *EOS, Transactions, American Geophysical Union*, vol. 85, no. 24, (June 15, 2004) p. 233.

FIGURE 16.35 The flooding Cedar River covers a large portion of Cedar Rapids, Iowa on June 14, 2008. (AP Photo/Jeff Roberson)

volume of its reservoir, reducing the long-term effectiveness of this flood-control measure.

Channelization *Channelization* involves altering a stream channel in order to make the flow more efficient. This may simply involve clearing a channel of obstructions or dredging a channel to make it wider and deeper. Another alteration involves straightening, and thus shortening a channel by creating *artificial cutoffs*. Shortening the stream increases the channel's flow velocity.

Between 1929 and 1942, the Army Corps of Engineers removed 16 meander bends on the lower Mississippi for the purpose of increasing the

Flood Control

Several strategies have been devised to eliminate or lessen the catastrophic impact of floods on our lives and environment. Engineering efforts include the construction of artificial levees, the building of flood-control dams, and river channelization.

Artificial Levees *Artificial levees* are earthen mounds built on river banks to increase the volume of water the channel can hold. Levees, used since ancient times, are the most common stream-containment structures. In some locations, concrete floodwalls are constructed that function as artificial levees.

Many artificial levees were not built to withstand periods of extreme flooding. For example, numerous levees failed during the summer of 1993, when the upper Mississippi and many of its tributaries experienced record flooding (Figure 16.38). During that event, floodwalls at St. Louis, Missouri, created a bottleneck for the river that led to increased flooding upstream of the city.

Flood-Control Dams *Flood-control dams* are built to store floodwater and then release it slowly, in a controlled manner. Since the 1920s, thousands of dams have been built on nearly every major river in the United States. Many dams have significant nonflood-related functions such as providing water for irrigated agriculture and for hydroelectric power generation. Many reservoirs are also major regional recreational facilities.

Although dams are effective in reducing flooding and provide other benefits, their construction and maintenance also have significant costs and consequences. For example, reservoirs created by dams may cover valuable farmland, forests, historic sites, and scenic valleys. Large dams can also cause significant damage to river ecosystems that have developed over thousands of years.

Furthermore, building dams is not a permanent solution to flooding. Sedimentation behind a dam gradually diminishes the

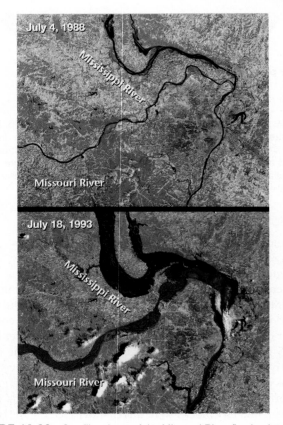

FIGURE 16.36 Satellite views of the Missouri River flowing into the Mississippi River. St. Louis is just south of their confluence. The upper image shows the rivers during a drought that occurred in summer 1988. The lower image depicts the peak of the record-breaking 1993 flood. Exceptional rains produced the wettest spring and early summer of the 20th century in the upper Mississippi River basin. In all, nearly 14 million acres were inundated, displacing at least 50,000 people. (Courtesy of Spaceimaging.com)

Flash Floods

BOX 16.2

Tornadoes and hurricanes are nature's most awesome storms, and thus, periodically become the focus of media coverage and public attention. Surprisingly, however, in most years these dreaded events do not claim the greatest number of storm-related deaths—that distinction is reserved for floods.

Flash floods are local floods of great volume and short duration (Figure 16.B). The rapidly rising surge of water usually occurs with little advance warning and can destroy roads, bridges, homes, and other substantial structures. Discharges quickly reach a maximum and diminish almost as rapidly. Flood flows often contain large quantities of sediment and debris as they surge through channels.

Frequently, flash floods result from torrential rains associated with slow-moving severe thunderstorms, or take place when a series of thunderstorms repeatedly strike the same location. Sometimes they are triggered by heavy rains from hurricanes and tropical storms. Occasionally, floating debris or ice can accumulate at natural or artificial obstructions and restrict the flow of water. When such temporary dams fail, torrents of water can be released as a flash flood.

Flash floods can take place almost anywhere. They are particularly common in mountainous terrains, where steep slopes can quickly channel runoff into narrow valleys. The hazard is most acute when the soil is already nearly saturated from earlier rains or consists of impermeable materials. A disaster in Shadydale, Ohio, demonstrates what can happen when even moderately heavy rains fall on saturated ground with steep slopes.

> On the evening of 14 June 1990, 26 people lost their lives as rains estimated to be in the range of 3 to 5 inches fell on saturated soil, which generated flood waves in streams that reached tens of feet in height, destroying near-bank residences and businesses. Preceding months of above-normal rainfall had generated soil moisture contents of near saturation. As a result, moderate amounts of rainfall caused large amounts of surface and near-surface runoff. Steep valleys with practically vertical walls channeled the floods, creating very fast, high, and steep wave crests.*

Why do so many people perish in flash floods? The element of surprise is a factor. For example, when floods occur at night most people are asleep. More importantly, most

*"Prediction and Mitigation of Flash Floods: A Policy Statement of the American Meteorological Society," *Bulletin of the American Meteorological Society,* vol. 74, no. 8 (1993), p. 1586.

people do not fully comprehend or respect the power of moving water. Figure 16.C illustrates the force of a flood wave. A mere 15 centimeters (6 inches) of fast-moving floodwater can cause an average-sized person to loose footing or fall. Most automobiles will float and be swept away in only 0.6 meter (2 feet) of water. More than half of all U.S. flash-flood fatalities are auto related! Clearly, people should never attempt to drive over a flooded road because water depth and current strength are often impossible to judge. Flash floods are calamities with potential for high death tolls and huge property losses. Although efforts are being made to improve warnings, flash floods remain elusive natural killers.

FIGURE 16.B Flash flooding in Las Vegas, Nevada, in August 2003. Parts of the city received nearly half the average annual rainfall in a matter of hours. Here, firefighters are rescued from a firetruck that was caught in a torrent of water. (Photo by John Locher/*Las Vegas Review-Journal*)

FIGURE 16.C The disastrous nature of flash floods is illustrated by the Big Thompson River flood of July 31, 1976, in Colorado. During a four-hour span more than 30 centimeters (12 inches) of rain fell on portions of the river's small drainage basin. This amounted to nearly three-quarters of the average yearly total. The flash flood in the narrow canyon lasted only a few hours but cost 139 people their lives. Damages were estimated at $39 million. (U.S. Geological Survey, Denver)

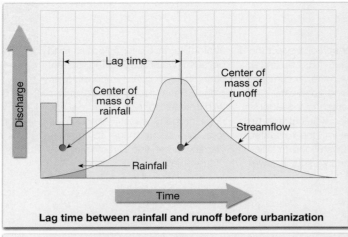

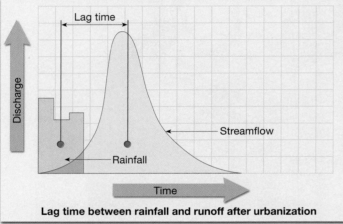

FIGURE 16.37 When an area changes from rural to urban, the lag time between rainfall and flood peak is shortened. The flood peak is also higher following urbanization. (After L. B. Leopold, U.S. Geological Survey)

FIGURE 16.38 Water rushes through a break in an artificial levee in Monroe County, Illinois. During the record-breaking 1993 Midwest floods, many artificial levees could not withstand the force of the floodwaters. Sections of many weakened structures were overtopped or simply collapsed. (Photo by James A. Finley/AP/Wide World Photos)

slope of the channel and, thereby reducing the threat of flooding. This river, a vital transportation corridor, was shortened about 240 kilometers (150 miles). These efforts have been somewhat successful in reducing the maximum height of the river during flood stage. However, channel shortening led to increased gradients and accelerated erosion of bank material, both of which necessitated further intervention. Following the creation of artificial cutoffs, massive bank protection was installed along several stretches of the lower Mississippi.

A similar case in which artificial cutoffs accelerated bank erosion occurred on the Blackwater River in Missouri, whose meandering course was shortened in 1910. Among the many effects of this project was a significant increase in the channel's width due to increased velocity. One particular bridge over the river collapsed in 1930 because of bank erosion. During the following 17 years the bridge was replaced three times, each requiring a longer span.

A Nonstructural Approach   All of the flood-control measures described so far employ structural solutions to "control" rivers. These solutions are typically expensive and often give those who reside on the floodplain a false sense of security.

Currently, many scientists and engineers advocate a nonstructural approach to flood control. They suggest that sound floodplain management is an alternative to artificial levees, dams, and channelization. By identifying high-risk flood areas, appropriate zoning regulations can be implemented to minimize development and promote safer, more appropriate land use.

CHAPTER 16 RUNNING WATER IN REVIEW

- The *hydrologic cycle* describes the continuous interchange of water among the oceans, atmosphere, and continents. Powered by energy from the Sun, it is a global system in which the atmosphere provides the link between the oceans and continents. The processes involved in the hydrologic cycle include *precipitation, evaporation, infiltration* (the movement of water into rocks or soil through cracks and pore spaces), *runoff* (water that flows over the land), and *transpiration* (the release of water vapor to the atmosphere by plants). *Running water is the single most important agent sculpting Earth's land surface.*

- Initially, runoff flows as broad, thin sheets across the ground, called *sheet flow.* After a short distance, threads of current typically develop, and tiny channels called *rills* form.

- The land area that contributes water to a stream is its *drainage basin.* Drainage basins are separated by imaginary lines called *divides.*

- River systems consist of three main parts: the zones of *sediment production, sediment transport,* and *sediment deposition.*

- The factors that determine a stream's *flow velocity* are *gradient* (slope of the stream channel), *shape, size,* and *roughness* of the channel, and the stream's *discharge* (amount of water passing a given point per unit of time). Most often, the gradient and roughness of a stream decrease downstream, while width, depth, discharge, and velocity increase.

- Streams transport their load of sediment in solution (*dissolved load*), in suspension (*suspended load*), and along the bottom of the channel (*bed load*). Much of the dissolved load is contributed by groundwater. Most streams carry the greatest part of their load in suspension.

- A stream's ability to transport solid particles is described using two criteria: *capacity* (the maximum load of solid particles a stream can carry) and *competence* (the maximum particle size a stream can transport). Competence increases as the square of stream velocity, so if velocity doubles, water's force increases fourfold.

- Streams deposit sediment when velocity slows and competence is reduced. Stream deposits are called *alluvium* and may occur as channel deposits called *bars*, as floodplain deposits, such as *natural levees*, and as *deltas* or *alluvial fans* at the mouths of streams.

- Stream channels are of two basic types: *bedrock channels* and *alluvial channels*. Bedrock channels are most commonly found in headwaters regions where gradients are steep. Rapids and waterfalls are common features. Two types of alluvial channels are *meandering channels* and *braided channels*.

- The two general types of *base level* (the lowest point to which a stream may erode its channel) are (1) *ultimate base level* and (2) *temporary*, or *local base level*. Any change in base level will cause a stream to adjust and establish a new balance. Lowering base level will cause a stream to downcut, whereas raising base level results in deposition of material.

- Although many gradations exist, the two general types of stream valleys are (1) *narrow V-shaped valleys* and (2) *wide valleys with flat floors*. Because the dominant activity is downcutting toward base level, narrow valleys often contain *waterfalls* and *rapids*.

- When a stream has cut its channel closer to base level, its energy is directed from side to side, and erosion produces a flat valley floor, or *floodplain*. Streams that flow upon floodplains often move in sweeping bends called *meanders*. Widespread meandering may result in shorter channel segments, called *cutoffs*, and abandoned bends, called *oxbow lakes*.

- Common *drainage patterns* (the form of a network of streams) produced by a main channel and its tributaries include (1) *dendritic*, (2) *radial*, (3) *rectangular*, and (4) *trellis*.

- *Headward erosion* lengthens a stream course by extending the head of its channel upslope. This process can lead to *stream piracy* (the diversion of the drainage of one stream by another).

- *Floods* are triggered by heavy rains and/or rapid snowmelt. Sometimes human interference can worsen or even cause floods. Flood-control measures include the building of *artificial levees* and *dams*, as well as *channelization*, which can involve creating *artificial cutoffs*. Many scientists and engineers advocate a nonstructural approach to flood control that involves more appropriate land use.

KEY TERMS

abrasion (p. 437)
alluvial fan (p. 449)
alluvial channels (p. 440)
alluvium (p. 439)
antecedent stream (p. 450)
back swamp (p. 449)
bar (p. 445)
base level (p. 441)
bed load (p. 437)
bedrock channels (p. 440)
braided channels (p. 440)
capacity (p. 438)
competence (p. 438)
corrosion (p. 437)
cut bank (p. 440)
cutoff (p. 440)
delta (p. 446)
dendritic drainage pattern (p. 449)

discharge (p. 435)
dissolved load (p. 437)
distributary (p. 446)
divide (p. 432)
drainage basin (p. 432)
evapotranspiration (p. 430)
flood (p. 452)
floodplain (p. 444)
graded stream (p. 442)
gradient (p. 435)
head (headwaters) (p. 435)
headward erosion (p. 450)
hydrologic cycle (p. 430)
incised meander (p. 445)
infiltration (p. 430)
laminar flow (p. 434)
local (temporary) base level (p. 442)

longitudinal profile (p. 435)
meanders (p. 440)
mouth (p. 435)
natural levee (p. 449)
oxbow lake (p. 440)
point bar (p. 440)
pothole (p. 437)
quarrying (p. 437)
radial drainage pattern (p. 450)
rectangular drainage pattern (p. 450)
rills (p. 432)
rivers (p. 432)
runoff (p. 430)
saltation (p. 438)
settling velocity (p. 438)
sheet flow (p. 432)

sorting (p. 439)
stream (p. 432)
stream piracy (p. 452)
stream valley (p. 443)
superposed stream (p. 450)
suspended load (p. 437)
temporary (local) base level (p. 442)
terrace (p. 445)
transpiration (p. 430)
trellis drainage pattern (p. 450)
turbulent flow (p. 434)
ultimate base level (p. 442)
water gap (p. 450)
wetted perimeter (p. 435)
yazoo tributary (p. 449)

QUESTIONS FOR REVIEW

1. Describe the movement of water through the hydrologic cycle. Once precipitation has fallen on land, what paths might the water take?

2. Over the oceans, evaporation exceeds precipitation, yet sea level does not drop. Can you explain why?

3. List several factors that influence infiltration.

4. What are the three main parts (zones) of a river system?

5. A stream originates at 2000 meters above sea level and travels 250 kilometers to the ocean. What is its average gradient in meters per kilometer?

6. Suppose the stream mentioned in Question 5 developed extensive meanders that lengthened its course to 500 kilometers. Calculate the new gradient. How does meandering affect gradient?

7. When the discharge of a river increases, what happens to the river's velocity?

8. What typically happens to channel width, channel depth, velocity, and discharge from the point where a stream begins to the point where it ends? Briefly explain why these changes take place.

9. In what three ways does a stream transport its load?

10. If you collect a jar of water from a stream, what part of its load will settle to the bottom of the jar? What portion will remain in the water indefinitely? What part of the stream's load would probably not be present in your sample?

11. Explain the difference between capacity and competency.

12. What is settling velocity? What factors influence settling velocity?

13. Are bedrock channels more likely to be found near the head or near the mouth of a stream?

14. Describe a situation that might cause a stream channel to become braided.

15. Define *base level*. Name the main river in your area. For what streams does it act as base level? What is base level for the Mississippi River? The Missouri River?

16. Describe two situations that would trigger the formation of incised meanders.

17. Briefly describe the formation of a natural levee. How is this feature related to back swamps and yazoo tributaries?

18. List two major depositional features, other than natural levees, that are associated with streams. Under what circumstances does each form?

19. How has the construction of artificial levees and dams on the Mississippi River and its tributaries contributed to a shrinking of the Mississippi's delta and extensive wetlands? (see Box 16.1)

20. Each of the following statements refers to a particular drainage pattern. Identify the pattern.

    a. Streams diverging from a central high area, such as a dome.

    b. Branching treelike pattern.

    c. A pattern that develops when bedrock is crisscrossed by joints and faults.

21. Describe how a water gap might form.

22. Contrast regional floods and flash floods. Which type is deadliest?

23. List and briefly describe three basic flood-control strategies. What are some drawbacks of each?

COMPANION WEBSITE

The *Earth 10e* Web site uses the resources and flexibility of the Internet to aid in your study of the topics in this chapter. Written and developed by the authors and other geology instructors, this site will help improve your understanding of geology. Visit www.mygeoscienceplace.com in order to:

- **Review** key chapter concepts.

- **Read** with links to the eBook and to chapter-specific web resources.

- **Visualize** and comprehend challenging topics using learning activities in *GEODe Earth*.

- **Test** yourself with online quizzes.

GEODe EARTH

GEODe Earth is a valuable and easy-to-use learning aid that can be accessed from your book's Companion Web site (www.mygeoscienceplace.com). It is a dynamic instructional tool that promotes understanding and reinforces important concepts by using tutorials, animations, and exercises that actively engage the student.

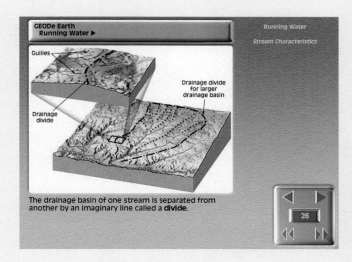

GROUNDWATER

Sunset along the Li River in China's Guilin District. This part of southern China exhibits tower karst development in which groundwater has dissolved large volumes of limestone leaving only these residual towers.

(PHOTO BY MOODBOARD/CORBIS)

Although it is hidden from view, vast quantities of water exist in the cracks, crevices, and pore spaces of rock and soil. It occurs almost everywhere beneath Earth's surface and is a major source of water worldwide. Groundwater is a valuable natural resource that provides about half of our drinking water and is essential to the vitality of agriculture and industry (Figure 17.1). In addition to human uses, groundwater plays a crucial role in sustaining streamflow between precipitation events—especially during protracted dry periods. Many ecosystems depend on groundwater discharge into streams, lakes, and wetlands. In some regions, large-scale development has caused groundwater levels to decline resulting in water shortages, streamflow depletion, land subsidence, contamination by saltwater, and increased pumping costs. Groundwater pollution is also a serious issue in some places.

IMPORTANCE OF GROUNDWATER

GROUNDWATER
GEODe EARTH ▶ Importance and Distribution of Groundwater

Groundwater is one of our most important and widely available resources, yet people's perceptions of the subsurface environment from which it comes are often unclear and incorrect. The reason is that the groundwater environment is largely hidden from view except in caves and mines, and the impressions people gain from these subsurface openings are misleading. Observations on the land surface give an impression that Earth is "solid." This view remains when we enter a cave and see water flowing in a channel that appears to have been cut into solid rock.

Because of such observations, many people believe that groundwater occurs only in underground "rivers." In reality, most of the subsurface environment is not "solid" at all. It includes countless tiny *pore spaces* between grains of soil and sediment, plus narrow joints and fractures in bedrock. Together, these spaces add up to an immense volume. It is in these small openings that groundwater collects and moves.

Considering the entire hydrosphere, or all of Earth's water, only about six-tenths of 1 percent occurs underground. Nevertheless, this small percentage, stored in the rocks and sediments beneath Earth's surface, is a vast quantity. When the oceans are excluded and only sources of fresh water are considered, the significance of groundwater becomes more apparent.

Table 17.1 contains estimates of the distribution of fresh water in the hydrosphere. Clearly the largest volume occurs as glacial ice. Second in rank is groundwater, with slightly more than 14 percent of the total. However, when ice is excluded and just liquid water is considered, more than 94 percent of all fresh water is groundwater. Without question, *groundwater represents the largest reservoir of fresh water that is readily available to humans.* Its value in terms of economics and human well-being is incalculable.

Geologically, groundwater is important as an erosional agent. The dissolving action of groundwater slowly removes soluble rock such as limestone, allowing surface depressions known as *sinkholes* to form as well as creating subterranean caverns (Figure 17.2). Groundwater is also an equalizer of streamflow. Much of the water that flows in rivers is not direct runoff from rain and snowmelt. Rather, a large percentage of precipitation soaks in and then moves slowly underground to stream channels. Groundwater is thus a form of storage that sustains streams during periods when rain does not fall. Therefore, when we see water flowing in a river during a dry period, it is rain that fell at some earlier time and was stored underground.

462

FIGURE 17.1   A. Satellite image of circular crop fields irrigated by center pivot irrigation. Resembling a work of modern art, crop circles cover what was once short grass prairie in southwestern Kansas. Three different crops are being grown—corn, wheat, and sorghum. Each is at a different point of development which accounts for the varying shades of green and yellow. (NASA) **B.** Groundwater provides nearly 216 billion liters (57 billion gallons) per day in support of the agricultural economy of the United States. (Photo by Clark Dunbar/Corbis)

TABLE 17.1   Fresh Water of the Hydrosphere

Parts of the Hydrosphere	Volume of Fresh Water (km³)	Share of Total Volume of Fresh Water (percent)	Share of total Volume of Liquid Fresh Water (percent)
Ice sheets and glaciers	24,000,000	84.945	0
Groundwater	4,000,000	14.158	94.05
Lakes and reservoirs	155,000	0.549	3.64
Soil moisture	83,000	0.294	1.95
Water vapor in the atmosphere	14,000	0.049	0.33
River water	1,200	0.004	0.03
Total	28,253,200	100.00	100

Source: U.S. Geological Survey Water Supply Paper 2220, 1987.

FIGURE 17.2   A view of the interior of a cavern. The dissolving action of acidic groundwater created the caverns. Later, groundwater deposited the limestone decorations. Carlsbad Caverns National Park, New Mexico. (Photo by Clint Farlinger)

GROUNDWATER—A BASIC RESOURCE

Water is basic to life. It has been called the "bloodstream" of both the biosphere and society. Each day in the United States we use about 345 billion gallons of fresh water.* About 76 percent comes from surface sources. Groundwater provides the remaining 24 percent (Figure 17.3). One of the advantages of groundwater is that it exists almost everywhere across the country and, thus, is often available in places that lack reliable surface sources such as lakes and rivers. Water in a groundwater system is stored in subsurface pore spaces and fractures. As water is withdrawn from a well, the connected pore spaces and fractures act as a "pipeline" that allows water to gradually move from one part of the hydrologic system to where it is being withdrawn.

Although people have been digging and drilling wells for thousands of years, extensive use of groundwater is a relatively recent phenomenon—growing rapidly with the development of rural electrification and more effective pumping technologies during the past 80 years. The 83 billion gallons of groundwater withdrawn each day in the United States represents only about 8 percent of the estimated 1 trillion gallons per day of natural replenishment (termed *recharge*). Therefore, our groundwater resource appears to be ample. However, this is misleading because

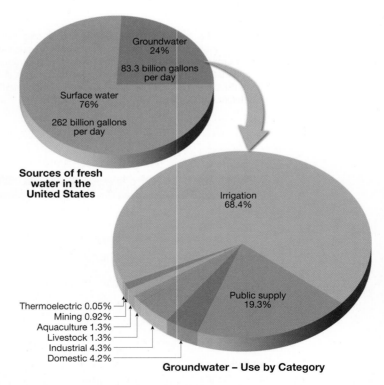

Groundwater – Use by Category

FIGURE 17.3   Each day in the United States we use 345 billion gallons of fresh water. Groundwater is the source of nearly one quarter of the total. More groundwater is used for irrigation than for all other uses combined. (Data from U.S. Geological Survey)

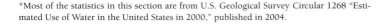

*Most of the statistics in this section are from U.S. Geological Survey Circular 1268 "Estimated Use of Water in the United States in 2000," published in 2004.

groundwater availability varies widely. For example, in the dry western states, groundwater demand is great, but rainfall to replenish the supply is scarce.

What are the primary ways that we use groundwater? The U.S. Geological Survey identifies several categories that are shown in Figure 17.3. More groundwater is used for irrigation than for all other uses combined. There are nearly 62 million acres (nearly 251,000 square kilometers or about 96,900 square miles) of irrigated land in the United States. That is an area nearly the size of the state of Wyoming. The vast majority (75 percent) of the irrigated land is in the seventeen conterminous western states where annual precipitation is typically less than 20 inches. About 42 percent of the water used for irrigation is groundwater.

Public and domestic uses include water used for indoor and outdoor household purposes as well as water used for commercial purposes. Common indoor uses include drinking, cooking, bathing, washing clothes and dishes, and flushing toilets. If you are curious about how much water an average American uses each day for indoor domestic purposes, look at Figure 17.4. Major outdoor uses are watering lawns and gardens. Water for domestic use may come from a public supply or be self-supplied.* For those who are self-supplied, practically all (98 percent) rely on groundwater.

Another category, aquaculture, involves water used for fish hatcheries, fish farms, and shellfish farms. Many mining operations require significant quantities of water, as do industrial processes that include petroleum refining and the manufacture of chemicals, plastics, paper, steel, and concrete.

DISTRIBUTION OF GROUNDWATER

GROUNDWATER
▶ Importance and Distribution of Groundwater

When rain falls, some of the water runs off, some returns to the atmosphere by evaporation and transpiration, and the remainder soaks into the ground. This last path is the primary source of

---

*According to the U.S. Geological Survey, *public supply* refers to water withdrawn by suppliers that furnish water to at least 25 people or have at least 15 connections.

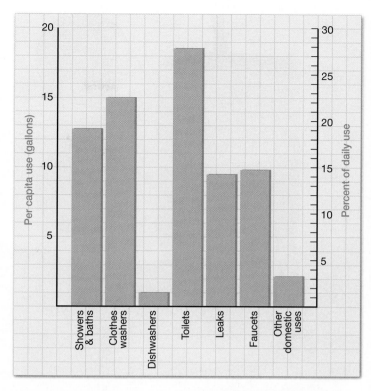

FIGURE 17.4 According to the American Water Works Association, daily indoor per capita use in an average single family home is 69.3 gallons. By installing more efficient water fixtures and regularly checking for leaks, households could reduce this figure by about one-third.

practically all subsurface water. The amount of water that takes each of these paths, however, varies greatly both in time and space. Influential factors include steepness of slope, nature of surface material, intensity of rainfall, and type and amount of vegetation. Heavy rains falling on steep slopes underlain by impervious materials will obviously result in a high percentage of the water running off. Conversely, if rain falls steadily and gently upon more gradual slopes composed of materials that are easily penetrated by the water, a much larger percentage of water soaks into the ground.

Some of the water that soaks in does not travel far, because it is held by molecular attraction as a surface film on soil particles. This near-surface zone is called the **zone of soil moisture.** It is crisscrossed by roots, voids left by decayed roots, and animal and worm burrows that enhance the infiltration of rainwater into the soil. Soil water is used by plants in life functions and transpiration. Some water also evaporates directly back into the atmosphere.

Water that is not held as soil moisture percolates downward until it reaches a zone where all of the open spaces in sediment and rock are completely filled with water (Figure 17.5). This is the **zone of saturation** (also called the *phreatic zone*). Water within it is called **groundwater.** The upper limit of this zone is known as the **water table.** Extending upward from the water table is the **capillary fringe** (*capillus* = hair). Here groundwater is held by surface tension in tiny passages between grains of soil or sediment. The area above the water table that includes the capillary fringe and the zone of soil moisture is called the **unsaturated zone**

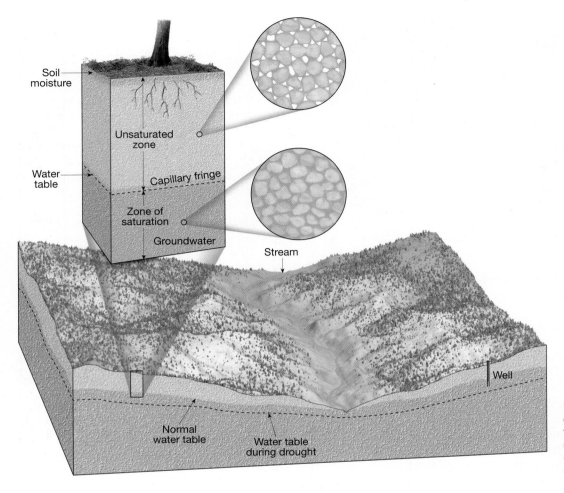

FIGURE 17.5 Distribution of underground water. The shape of the water table is usually a subdued replica of the surface topography. During periods of drought, the water table falls, reducing streamflow and drying up some wells.

(also known as the *vadose zone*). The pore spaces in this zone contain both air and water. Although a considerable amount of water can be present in the unsaturated zone, this water cannot be pumped by wells because it clings too tightly to rock and soil particles. By contrast, below the water table the water pressure is great enough to allow water to enter wells, thus permitting groundwater to be withdrawn for use. We will examine wells more closely later in the chapter.

THE WATER TABLE

GROUNDWATER
▶ Importance and Distribution of Groundwater

The water table, the upper limit of the zone of saturation, is a very significant feature of the groundwater system. The water table level is important in predicting the productivity of wells, explaining the changes in the flow of springs and streams, and accounting for fluctuations in the levels of lakes.

Variations in the Water Table

The depth of the water table is highly variable and can range from zero, when it is at the surface, to hundreds of meters in some places. An important characteristic of the water table is that its configuration varies seasonally and from year to year because the addition of water to the groundwater system is closely related to the quantity, distribution, and timing of precipitation. Except where the water table is at the surface, we cannot observe it directly. Nevertheless, its elevation can be mapped and studied in detail where wells are numerous because the water level in wells coincides with the water table (Figure 17.6). Such maps reveal that the water table is rarely level, as we might expect a table to be. Instead, its shape is usually a subdued replica of the surface topography, reaching its highest elevations beneath hills and then descending toward valleys (see Figure 17.5). Where a wetland (swamp) is encountered, the water table is right at the surface. Lakes and streams generally occupy areas low enough that the water table is above the land surface.

Several factors contribute to the irregular surface of the water table. One important influence is the fact that groundwater moves very slowly and at varying rates under different conditions. Because of this, water tends to "pile up" beneath high areas between stream valleys. If rainfall were to cease completely, these water-table "hills" would slowly subside and gradually approach the level of the valleys. However, new supplies of rainwater are usually added frequently enough to prevent this. Nevertheless, in times of extended drought (Box 17.1), the water table may drop enough to dry up shallow wells (see

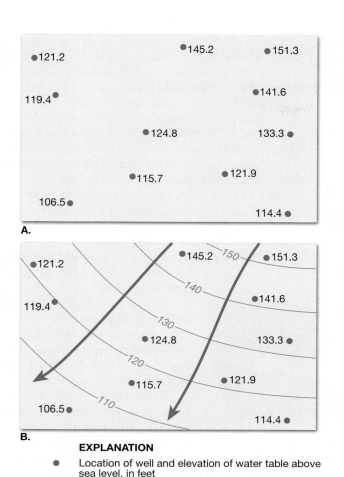

EXPLANATION

● Location of well and elevation of water table above sea level, in feet

⌒120⌒ Water table contour shows elevation of water table, contour interval 10 feet

← Groundwater flow line

FIGURE 17.6 Preparing a map of the water table. The water level in wells coincides with the water table. **A.** First, the locations of wells and the elevation of the water table above sea level are plotted on a map. **B.** These data points are used to guide the drawing of water-table contour lines at regular intervals. On this sample map the interval is 10 feet. Groundwater flow lines can be added to show water movement in the upper portion of the zone of saturation. Groundwater tends to move approximately perpendicular to the contours and down the slope of the water table. (After U.S. Geological Survey)

Figure 17.5). Other causes for the uneven water table are variations in rainfall and permeability from place to place.

Interaction Between Groundwater and Streams

The interaction between the groundwater system and streams is a basic link in the hydrologic cycle. It can take place in one of three ways. Streams may gain water from the inflow of groundwater through the streambed. Such streams are called **gaining streams** (Figure 17.7A). For this to occur, the elevation of the water table must be higher than the level of the surface of the stream. Streams may lose water to the groundwater system by outflow through the streambed. The term **losing stream**

is applied to this situation (Figure 17.7B,C). When this happens, the elevation of the water table must be lower than the surface of the stream. The third possibility is a combination of the first two—a stream gains in some sections and loses in others.

Losing streams can be connected to the groundwater system by a continuous saturated zone or they can be disconnected from the groundwater system by an unsaturated zone. Compare parts B and C in Figure 17.7. When the stream is disconnected, the water table may have a discernible bulge beneath the stream if the rate of water movement through the streambed and unsaturated zone is greater than the rate of groundwater movement away from the bulge.

In some settings, a stream might always be a gaining stream or always be a losing stream. However, in many situations flow direction can vary a great deal along a stream; some sections receive groundwater and other sections lose water to the groundwater system. Moreover, the direction of flow can change over a short time span as the result of storms adding water near the streambank or when temporary flood peaks move down the channel.

Groundwater contributes to streams in most geologic and climatic settings. Even where streams are primarily losing water to the groundwater system, certain sections may receive groundwater inflow during some seasons. In one study of 54 streams in all parts of the United States, the analysis indicated that 52 percent of the streamflow was contributed by groundwater. The groundwater contribution ranged from a low of 14 percent to a maximum of 90 percent. Groundwater is also a major source of water for lakes and wetlands.

FACTORS INFLUENCING THE STORAGE AND MOVEMENT OF GROUNDWATER

The nature of subsurface materials strongly influences the rate of groundwater movement and the amount of groundwater that can be stored. Two factors are especially important—porosity and permeability.

Porosity

Water soaks into the ground because bedrock, sediment, and soil contain countless voids or openings. These openings are similar to those of a sponge and are often called *pore spaces*. The quantity of groundwater that can be stored depends on the **porosity** of the material, which is the percentage of the total volume of rock or sediment that consists of pore spaces. Voids most often are spaces between sedimentary particles, but also common are joints, faults, cavities formed by the dissolving of soluble rock such as limestone, and vesicles (voids left by gases escaping from lava).

Variations in porosity can be great. Sediment is commonly quite porous, and open spaces may occupy 10 to 50 percent of the sediment's total volume. Pore space depends on the size and shape of the grains, how they are packed together, the degree of sorting,

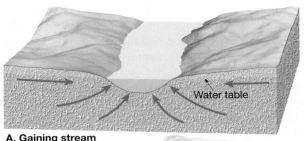

A. Gaining stream

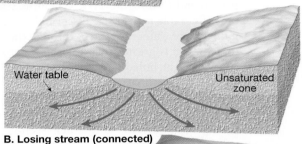

B. Losing stream (connected)

FIGURE 17.7   Interaction between the groundwater system and streams. **A.** Gaining streams receive water from the groundwater system. **B.** Losing streams lose water to the groundwater system. **C.** When losing streams are separated from the groundwater system by the unsaturated zone, a bulge may form in the water table. (After U.S. Geological Survey)

C. Losing stream (disconnected)

and in sedimentary rocks, the amount of cementing material. For example, clay may have a porosity as high as 50 percent, whereas some gravels may have only 20 percent voids.

Where sediments are poorly sorted, the porosity is reduced because the finer particles tend to fill the openings among the larger grains (see Figure 7.6 on p. 205). Most igneous and metamorphic rocks, as well as some sedimentary rocks, are composed of tightly interlocking crystals so the voids between the grains may be negligible. In these rocks, fractures must provide the voids.

Permeability, Aquitards, and Aquifers

Porosity alone cannot measure a material's capacity to yield groundwater. Rock or sediment may be very porous yet still not allow water to move through it. The pores must be *connected* to allow water flow, and they must be *large enough* to allow flow. Thus, the **permeability** (*permeare* = to penetrate) of a material—its ability to *transmit* a fluid—is also very important.

Groundwater moves by twisting and turning through small interconnected openings. The smaller the pore spaces, the slower the water moves. This idea is clearly illustrated by examining the information about the water-

yielding potential of different materials in Table 17.2. Here groundwater is divided into two categories: (1) the portion that will drain under the influence of gravity (called *specific yield*) and (2) the part that is retained as a film on particle and rock surfaces and in tiny openings (called *specific retention*). Specific yield indicates how much water is actually available for use, whereas specific retention indicates how much water remains bound in the material. For example, clay's ability to store water is great, owing to its high porosity, but its pore spaces are so small that water is unable to move through it. Thus, clay's porosity is high but because its permeability is poor, clay has a very low specific yield.

Impermeable layers that hinder or prevent water movement are termed **aquitards** (*aqua* = water, *tard* = low). Clay is a good example. On the other hand, larger particles, such as sand or gravel, have larger pore spaces. Therefore, the water moves with relative ease. Permeable rock strata or sediment that transmit groundwater freely are called **aquifers** (*aqua* = water, *fer* = carry). Sands and gravels are common examples.

In summary, you have seen that porosity is not always a reliable guide to the amount of groundwater that can be produced, and permeability is significant in determining the rate of groundwater movement and the quantity of water that might be pumped from a well.

MOVEMENT OF GROUNDWATER

The movement of water in the atmosphere and on the land surface is relatively easy to visualize, but the movement of groundwater is not. Near the beginning of the chapter we mentioned the common misconception that groundwater occurs in underground rivers that resemble surface streams. Although subsurface streams do exist, they are *not* common. Rather, as you learned in the preceding sections, groundwater exists in the

TABLE 17.2   Selected Values of Porosity, Specific Yield, and Specific Retention*

Material	Porosity	Specific Yield	Specific Retention
Clay	50	2	48
Sand	25	22	3
Gravel	20	19	1
Limestone	20	18	2
Sandstone (semiconsolidated)	11	6	5
Granite	0.1	0.09	0.01
Basalt (fresh)	11	8	3

*Values in percent by volume.
SOURCE: U.S. Geological Survey Water Supply Paper 2220, 1987.

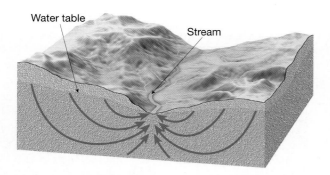

FIGURE 17.8 Arrows indicate groundwater movement through uniformly permeable material. The looping curves may be thought of as a compromise between the downward pull of gravity and the tendency of water to move toward areas of reduced pressure.

pore spaces and fractures in rock and sediment. Thus, contrary to any impressions of rapid flow that an underground river might evoke, the movement of most groundwater is exceedingly slow, from pore to pore. By exceedingly slow, we mean typical rates of a few centimeters each day.

Figure 17.8 depicts a simple example of a *groundwater flow system*—a three-dimensional body of Earth material saturated with moving groundwater. It shows groundwater moving along flow paths from areas of recharge to a zone of discharge along a stream. Discharge also occurs at springs, lakes, or wetlands, and in coastal areas, as groundwater seeps into bays or the ocean. Transpiration by plants whose roots extend to near the water table represents another form of groundwater discharge.

The energy that makes groundwater move is provided by the force of gravity. In response to gravity, water moves from areas where the water table is high to zones where the water table is lower. Although some water takes the most direct path down the slope of the water table, much of the water follows long, curving paths.

Figure 17.8 shows water percolating into a stream from all possible directions. Some paths clearly turn upward, apparently against the force of gravity, and enter through the bottom of the channel. This is easily explained: The deeper you go into the zone of saturation, the greater is the water pressure. Thus, the looping curves followed by water in the saturated zone may be thought of as a compromise between the downward pull of gravity and the tendency of water to move toward areas of reduced pressure. As a result, water at any given height is under greater pressure beneath a hill than beneath a stream channel, and the water tends to migrate toward points of lower pressure.

Darcy's Law

The foundations of our modern understanding of groundwater movement began in the mid-19th century with the work of the French engineer Henri Darcy. During this time, Darcy made measurements and conducted experiments in an attempt to determine whether the water needs of the city of Dijon in east–central France could be met by tapping local supplies of groundwater. Among the experiments carried out by Darcy was one that showed that the velocity of groundwater flow is proportional to the slope of the water table—the steeper the slope, the faster the water moves (because the steeper the slope, the greater the pressure difference between two points). The water-table slope is known as the **hydraulic gradient** and can be expressed as follows:

$$\text{hydraulic gradient} = \frac{h_1 - h_2}{d}$$

where h_1 is the elevation of one point on the water table, h_2 the elevation of a second point, and d is the horizontal distance between the two points (Figure 17.9).

Darcy also experimented with different materials such as coarse sand and fine sand by measuring the rate of flow through sediment-filled tubes that were tilted at varying angles. He found that the flow velocity varied with the permeability of the sediment—groundwater flows more rapidly through sediments having greater permeability than through materials having lower permeability. This factor is known as **hydraulic conductivity** and is a coefficient that takes into account the permeability of the aquifer and the viscosity of the fluid.

To determine discharge (Q), that is, the actual volume of water that flows through an aquifer in a specified time, the following equation is used:

$$Q = \frac{KA(h_1 - h_2)}{d}$$

where $\dfrac{h_1 - h_2}{d}$ is the hydraulic gradient, K is the coefficient that represents hydraulic conductivity, and A is the cross-sectional area of the aquifer. This expression has come to be called **Darcy's law** in honor of the pioneering French scientist-engineer.

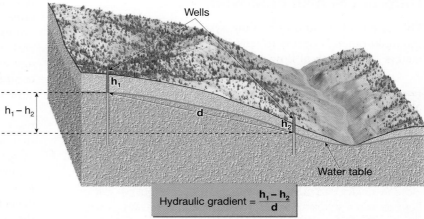

FIGURE 17.9 The hydraulic gradient is determined by measuring the difference in elevation between two points on the water table ($h_1 - h_2$) divided by the distance between them, d. Wells are used to determine the height of the water table.

EARTH AS A SYSTEM

BOX 17.1

Drought Impacts the Hydrologic System*

Drought is a period of abnormally dry weather that persists long enough to produce a significant hydrologic imbalance such as crop damage or water supply shortages. Drought severity depends upon the degree of moisture deficiency, its duration, and the size of the affected area.

During the summer of 2008, the Oklahoma Panhandle and portions of neighboring states were experiencing *exceptional drought* (Figure 17.A). It was the result of many months of meager rainfall coupled with high temperatures and desiccating winds. This level of drought severity is considered a "once in fifty years" occurrence. Conditions in the western Panhandle were as dry or drier than the conditions the area experienced during the Dust Bowl of the 1930s. Crops failed and rangeland grasses were in such poor condition that many ranchers were forced to sell their herds.

Although natural disasters such as floods and hurricanes usually generate more attention, droughts can be just as devastating and carry a bigger price tag. On the average, droughts cost the United States $6 to $8 billion annually compared to $2.4 billion for floods

and $1.2 to $4.8 billion for hurricanes. Direct economic losses from a severe drought in 1988 were estimated at $61 billion (2002 dollars).

Drought is different from other natural hazards in several ways. First, it occurs in a gradual, "creeping" way, making its onset and end difficult to determine. The effects of drought accumulate slowly over an extended time span and sometimes linger for years after the drought has ended. Second, there is not a precise and universally accepted definition of drought. This adds to the confusion about whether or not a drought is actually occurring and if it is, its severity. Third, drought seldom produces structural damages, so its social and economic effects are less obvious than damages from other natural disasters.

Definitions reflect four basic approaches to measuring drought: meteorological, agricultural, hydrological, and socioeconomic. *Meteorological drought* deals with the degree of dryness

*Based in part on material prepared by the National Drought Mitigation Center (http://drought.unl.edu).

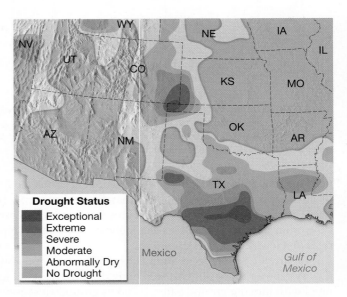

FIGURE 17.A   Drought-status map from the National Drought Mitigation Center for early August 2008. The preceding year had been the driest in the Oklahoma Panhandle since records began in 1921. Central and southern Texas were also suffering extreme to exceptional drought. Maps such as this are based on a blend of several criteria including precipitation, soil moisture, water levels in reservoirs, and satellite-based maps of vegetation health. (NASA)

based on the departure of precipitation from normal values and the duration of the dry period. *Agricultural drought* is usually linked to a deficit of soil moisture. A plant's need for water depends on prevailing weather conditions, bio-

Different Scales of Movement

The extent of groundwater flow systems varies from a few square kilometers or less to tens of thousands of square kilometers. The length of flow paths ranges from a few meters to tens and sometimes hundreds of kilometers. Figure 17.10 is a cross section of a hypothetical region in which a deep groundwater flow system is overlain by and connected to several, more shallow local flow systems. The subsurface geology exhibits a complicated arrangement of high hydraulic-conductivity aquifer units and low hydraulic-conductivity aquitard units.

Starting near the top of Figure 17.10, the blue arrows represent water movement in several local groundwater systems that occur in the upper water table aquifer. They are separated by groundwater divides at the center of the hills and discharge into the nearest surface water body. Beneath these most shallow systems, red arrows show water movement in a somewhat deeper system in which groundwater does not discharge into the nearest surface water body, but into a more distant one. Finally, the black arrows show groundwater movement in a deep regional system that lies beneath the more shallow ones and is connected to them. The horizontal scale of the figure could range from tens to hundreds of kilometers.

Students Sometimes Ask . . .

How does the rate of groundwater movement compare to the velocity of streams?

Velocities of groundwater flow are generally orders of magnitude less than velocities of streamflow. A velocity of 1 foot per day or greater is a high rate of movement for groundwater, and groundwater velocities can be as low as 1 foot per year or 1 foot per decade. In contrast, rates of streamflow are generally measured in feet per second. A velocity of 1 foot per second equals about 16 miles per day.

logical characteristics of the particular plant, its stage of growth, and various soil properties. *Hydrological drought* refers to deficiencies in surface and subsurface water supplies. It is measured as streamflow and as lake, reservoir, and groundwater levels. There is a time lag between the onset of dry conditions and a drop in streamflow, or the lowering of lakes, reservoirs, or groundwater levels. So hydrological measurements are not the earliest indicators of drought. *Socioeconomic drought* is a reflection of what happens when a physical water shortage affects people. Socioeconomic drought occurs when the demand for an economic good exceeds supply as a result of a shortfall in water supply. For example, drought can result in significantly reduced hydroelectric power production, which in turn may require conversion to more expensive fossil fuels and/or significant energy shortfalls.

There is a sequence of impacts associated with meteorological, agricultural, and hydrological drought (Figure 17.B). When meteorological drought begins, the agricultural sector is usually the first to be affected because of its heavy dependence on soil moisture. Soil moisture is rapidly depleted during extended dry periods. If precipitation deficiencies continue, those dependent on rivers, reservoirs, lakes, and groundwater may be affected.

When precipitation returns to normal, meteorological drought comes to an end. Soil moisture is replenished first, followed by streamflow, reservoirs and lakes, and finally groundwater. Thus, drought impacts may diminish rapidly in the agricultural sector because of its reliance on soil moisture but linger for months or years in other sectors that depend on stored surface or subsurface water supplies. Groundwater users, who are often the last to be affected following the onset of meteorological drought, may also be the last to experience a return to normal water levels. The length of the recovery period depends upon the intensity of the meteorological drought, its duration, and the quantity of precipitation received when the drought ends.

The impacts suffered because of drought are the product of both the meteorological event and the vulnerability of society to periods of precipitation deficiency. As demand for water increases as a result of population growth and regional population shifts, future droughts can be expected to produce greater impacts whether or not there is any increase in the frequency or intensity of meteorological drought.

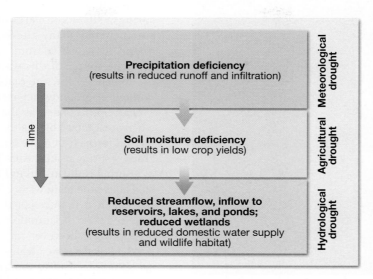

FIGURE 17.B Sequence of drought impacts. After the onset of meteorological drought, agriculture is affected first, followed by reductions in streamflow and water levels in lakes, reservoirs, and underground. When meteorological drought ends, agricultural drought ends as soil moisture is replenished. It takes a considerably longer time span for hydrological drought to end.

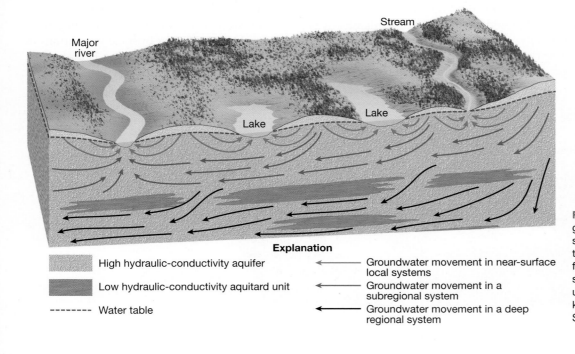

Explanation

High hydraulic-conductivity aquifer

Low hydraulic-conductivity aquitard unit

-------- Water table

⟵ Groundwater movement in near-surface local systems

⟵ Groundwater movement in a subregional system

⟵ Groundwater movement in a deep regional system

FIGURE 17.10 A hypothetical groundwater flow system that includes subsystems at different scales. Variations in surface topography and subsurface geology can produce a complex situation. The horizontal scale of the figure could range from tens to hundreds of kilometers. (After U.S. Geological Survey)

FIGURE 17.11 Thousand Springs along the Snake River in Hagerman Valley, Idaho. (Photo by David R. Frazier/Alamy)

SPRINGS

GROUNDWATER
▶ Springs and Wells

Springs have aroused the curiosity and wonder of people for thousands of years. The fact that springs were, and to some people still are, rather mysterious phenomena is not difficult to understand, for here is water flowing freely from the ground in all kinds of weather in seemingly inexhaustible supply, but with no obvious source.

Not until the middle of the 17th century did the French physicist Pierre Perrault invalidate the age-old assumption that precipitation could not adequately account for the amount of water emanating from springs and flowing in rivers. Over several years Perrault computed the quantity of water that fell on France's Seine River basin. He then calculated the mean annual runoff by measuring the river's discharge. After allowing for the loss of water by evaporation, he showed that there was sufficient water remaining to feed the springs. Thanks to Perrault's pioneering efforts and the measurements by many afterward, we now know that the source of springs is water from the zone of saturation and that the ultimate source of this water is precipitation.

Whenever the water table intersects Earth's surface, a natural outflow of groundwater results, which we call a **spring** (Figure 17.11). Springs such as the one pictured in Figure 17.11 form when an aquitard blocks the downward movement of groundwater and forces it to move laterally. Where the permeable bed outcrops, a spring results. Another situation leading to the formation of a spring is illustrated in Figure 17.12. Here an aquitard is situated above the main water table. As water percolates downward, a portion of it is intercepted by the aquitard, thereby creating a localized zone of saturation and a **perched water table.**

Springs, however, are not confined to places where a perched water table creates a flow at the surface. Many geological situations lead to the formation of springs because subsurface conditions vary greatly from place to place. Even in areas underlain by impermeable crystalline rocks, permeable zones may exist in the form of fractures or solution channels. If these openings fill with water and intersect the ground surface along a slope, a spring will result.

HOT SPRINGS AND GEYSERS

By definition, the water in **hot springs** is 6–9 °C (10–15 °F) warmer than the mean annual air temperature for the localities where they occur. In the United States alone, there are more than 1000 such springs (Figure 17.13).

Temperatures in deep mines and oil wells usually rise with increasing depth, an average of about 2 °C per 100 meters (1 °F per 100 feet). Therefore, when groundwater circulates at great depths, it becomes heated. If it rises to the surface, the water may emerge as a hot spring. The water of some hot springs in the eastern United States is heated in this manner. However, the

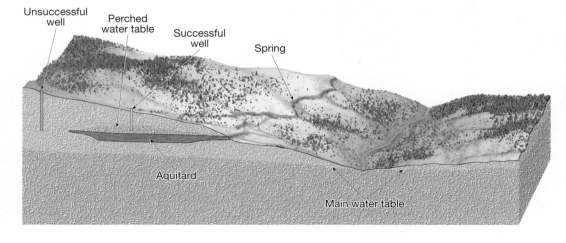

FIGURE 17.12 When an aquitard is situated above the main water table, a localized zone of saturation may result. Where the perched water table intersects the side of the valley, a spring flows. The perched water table also caused the well on the right to be successful, whereas the well on the left will be unsuccessful unless it is drilled to a greater depth.

Unsuccessful well   Perched water table   Successful well   Spring   Aquitard   Main water table

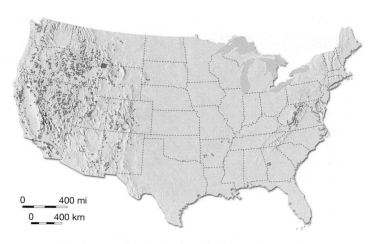

0   400 mi

0   400 km

FIGURE 17.13   Distribution of hot springs and geysers in the United States. Note the concentration in the West, where igneous activity has been most recent. (After G. A. Waring, U.S. Geological Survey Professional Paper 492, 1965)

thunderous roar. Perhaps the most famous geyser in the world is Old Faithful in Yellowstone National Park (Figure 17.14). The great abundance, diversity, and spectacular nature of Yellowstone's geysers and other thermal features undoubtedly was the primary reason for its becoming the first national park in the United States. Geysers are also found in other parts of the world, notably New Zealand and Iceland. In fact, the Icelandic word *geysa,* to gush, gives us the name *geyser.*

Geysers occur where extensive underground chambers exist within hot igneous rocks. How they operate is shown in Figure 17.15. As relatively cool groundwater enters the chambers, it is heated by the surrounding rock. At the bottom of the chambers, the water is under great pressure because of the weight of the overlying water. This great pressure prevents the water from boiling at the normal surface temperature of 100 °C (212 °F). For example, water at the bottom of a 300-meter (1000-foot) water-filled chamber must attain nearly 230 °C before it will boil. The heating causes the water to expand, with the result that some is

great majority (more than 95 percent) of the hot springs (and geysers) in the United States are found in the West (Figure 17.13). The reason for such a distribution is that the source of heat for most hot springs is cooling igneous rock, and it is in the West that igneous activity has occurred more recently.

Geysers are intermittent hot springs or fountains in which columns of water are ejected with great force at various intervals, often rising 30–60 meters (100–200 feet) into the air. After the jet of water ceases, a column of steam rushes out, usually with a

FIGURE 17.14   Yellowstone's Old Faithful is one of the world's most famous geysers. Each eruption lasts roughly 1.5 to 5 minutes. It emits up to 32,000 liters (8,400 gallons) of hot water and steam skyward as high as 55 meters (more than 180 feet). (Photo by Jeff Vanuga/Corbis)

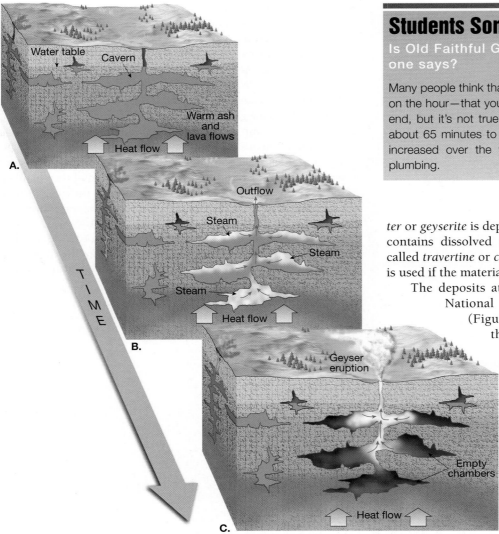

A.

B.

C.

TIME

Water table

Cavern

Warm ash and lava flows

Heat flow

Outflow

Steam

Steam

Steam

Heat flow

Geyser eruption

Empty chambers

Heat flow

FIGURE 17.15 Idealized diagrams of a geyser. A geyser can form if the heat is not distributed by convection. **A.** In this figure, the water near the bottom is heated to near its boiling point. The boiling point is higher there than at the surface because the weight of the water above increases the pressure. **B.** The water higher in the geyser system is also heated; therefore, it expands and flows out at the top, reducing the pressure on the water at the bottom. **C.** At the reduced pressure on the bottom, boiling occurs. Some of the bottom water flashes into steam, and the expanding steam causes an eruption.

forced out at the surface. This loss of water reduces the pressure on the remaining water in the chamber, which lowers the boiling point. A portion of the water deep within the chamber quickly turns to steam, and the geyser erupts (Figure 17.15). Following eruption, cool groundwater again seeps into the chamber and the cycle begins anew.

When groundwater from hot springs and geysers flows out at the surface, material in solution is often precipitated, producing an accumulation of chemical sedimentary rock. The material deposited at any given place commonly reflects the chemical makeup of the rock through which the water circulated. When the water contains dissolved silica, a material called *siliceous sin-*

ter or *geyserite* is deposited around the spring. When the water contains dissolved calcium carbonate, a form of limestone called *travertine* or *calcareous tufa* is deposited. The latter term is used if the material is spongy and porous.

The deposits at Mammoth Hot Springs in Yellowstone National Park are more spectacular than most (Figure 17.16). As the hot water flows upward through a series of channels and then out at the surface, the reduced pressure allows carbon dioxide to separate and escape from the water. The loss of carbon dioxide causes the water to become supersaturated with calcium carbonate, which then precipitates. In addition to containing dissolved silica and calcium carbonate, some hot springs contain sulfur, which gives water a poor taste and unpleasant odor. Undoubtedly, Rotten Egg Spring, Nevada, is such a situation.

WELLS

GROUNDWATER

▶ Springs and Wells

The most common method for removing groundwater is the **well,** a hole bored into the zone of saturation (Figure 17.17). Wells serve as small reservoirs into which groundwater migrates and from which it can be pumped to the surface. The use of wells dates back many centuries and continues to be an important method of obtaining water today. By far the single greatest use of this water in the United States is irrigation for agriculture. More than 65 percent of the groundwater used each year is for this purpose.

The water-table level may fluctuate considerably during the course of a year, dropping during the dry seasons and rising following periods of rain. Therefore, to ensure a continuous supply of water, a well must penetrate below the water table. Whenever water is withdrawn from a well, the water table around the well is lowered. This effect, termed **drawdown,** may substantially lower the water table in an area and cause nearby shallow wells to become dry. Figure 17.18 illustrates this situation.

Digging a successful well is a familiar problem for people in areas where groundwater is the primary source of supply. One well may be successful at a depth of 10 meters (33 feet), whereas a neighbor may have to go twice as deep to find an adequate supply. Still others may be forced to go deeper or try a different site altogether. When subsurface materials are heterogeneous, the amount of water a well is capable of providing may vary a great deal over short distances. For example, when two nearby wells are drilled to the same level and only one is successful, it may be caused by the presence of a perched water table beneath one of them. Such a case is shown in Figure 17.12. Massive igneous and metamorphic rocks provide a second example. These crystalline rocks are usually not very permeable except where they are cut by many intersecting joints and fractures. Therefore, when a well drilled into such rock does not intersect an adequate network of fractures, it is likely to be unproductive.

FIGURE 17.16 Mammoth Hot Springs at Yellowstone National Park. Although most of the deposits associated with geysers and hot springs in Yellowstone Park are silica-rich geyserite, the deposits at Mammoth Hot Springs consist of a form of limestone called travertine. (Photo by Stephen Trimble)

decreases with increasing distance from the well. The result is a depression in the water table, roughly conical in shape, known as a **cone of depression** (Figure 17.18). Because the cone of depression increases the hydraulic gradient near the well, groundwater will flow more rapidly toward the opening. For most smaller domestic wells, the cone of depression is negligible. However, when wells are heavily pumped for irrigation or for industrial purposes, the withdrawal of water can be great enough to create a very wide and steep cone of depression. This

Students Sometimes Ask . . .

I have heard people say that supplies of groundwater can be located using a forked stick. Can this actually be done?

What you describe is a practice called "water dowsing." In the classic method, a person holding a forked stick walks back and forth over an area. When water is detected, the bottom of the "Y" is supposed to be attracted downward.

Geologists and engineers are dubious, to say the least. Case histories and demonstrations may seem convincing, but when dowsing is exposed to scientific scrutiny, it fails. Most "successful" examples of water dowsing occur in places where water would be hard to miss. In a region of adequate rainfall and favorable geology, it is difficult to drill and not find water!

FIGURE 17.17 Wells are the most common means by which people obtain groundwater. (Photo by ASP/YPP/agefotostock)

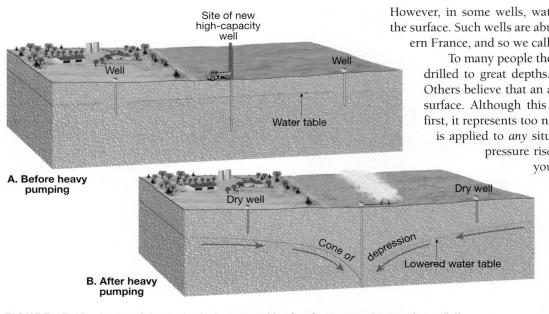

A. Before heavy pumping

B. After heavy pumping

FIGURE 17.18   A cone of depression in the water table often forms around a pumping well. If heavy pumping lowers the water table, the shallow wells may be left dry.

ARTESIAN WELLS

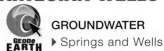

GROUNDWATER
▶ Springs and Wells

In most wells, water cannot rise on its own. If water is first encountered at 30 meters depth, it remains at that level, fluctuating perhaps a meter or two with seasonal wet and dry periods.

However, in some wells, water rises, sometimes overflowing at the surface. Such wells are abundant in the *Artois* region of northern France, and so we call these self-rising wells *artesian*.

To many people the term *artesian* is applied to any well drilled to great depths. This use of the term is incorrect. Others believe that an artesian well must flow freely at the surface. Although this is a more correct notion than the first, it represents too narrow a definition. The term **artesian** is applied to *any* situation in which groundwater under pressure rises above the level of the aquifer. As you will see, this does not always mean a free-flowing surface discharge.

For an artesian system to exist, two conditions usually are met (Figure 17.19): (1) water is confined to an aquifer that is inclined so that one end can receive water; and (2) aquitards, both above and below the aquifer, must be present to prevent the water from escaping. Such an aquifer is called a **confined aquifer.** When such a layer is tapped, the pressure created by the weight of the water above will force the water to rise. If there were no friction, the water in the well would rise to the level of the water at the top of the aquifer. However, friction reduces the height of the pressure surface. The greater the distance from the recharge area (where water enters the inclined aquifer), the greater the friction and the less the rise of water.

In Figure 17.19, well 1 is a **nonflowing artesian well,** because at this location the pressure surface is below ground level. When

FIGURE 17.19   Artesian systems occur when an inclined aquifer is surrounded by impermeable beds. Such an aquifer is called a *confined aquifer*. The photo shows a flowing artesian well. (Photo by James E. Patterson)

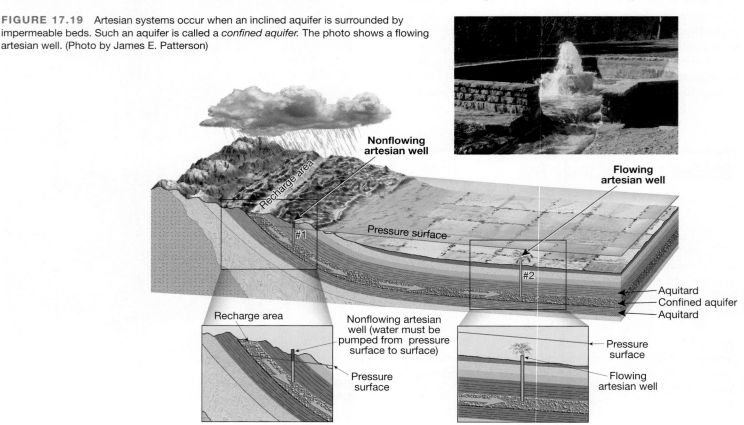

FIGURE 17.20 A "gusherlike" flowing artesian well at Woonsocket, South Dakota, circa 1900. The stream of water from a 3-inch pipe reached a height of about 90 feet. Thousands of additional wells now tap the same confined aquifer; thus, the pressure has dropped to the point that many wells stopped flowing altogether and have to be pumped. (Photo by N. H. Darton, U.S. Geological Survey)

the pressure surface is above the ground and a well is drilled into the aquifer, a **flowing artesian well** is created (well 2, Figure 17.19). Not all artesian systems are wells. *Artesian springs* also exist. Here groundwater may reach the surface by rising along a natural fracture such as a fault rather than through an artificially produced hole. In deserts, artesian springs are sometimes responsible for creating an oasis.

Artesian systems act as conduits, often transmitting water great distances from remote areas of recharge to points of discharge. A well-known artesian system in South Dakota is a good example of this. In the western part of the state, the edges of a series of sedimentary layers have been bent up to the surface along the flanks of the Black Hills. One of these beds, the permeable Dakota Sandstone, is sandwiched between impermeable strata and gradually dips into the ground toward the east. When the aquifer was first tapped, water poured from the ground surface, creating fountains many meters high (Figure 17.20). In some places the force of the water was sufficient to power waterwheels. Scenes such as the one pictured in Figure 17.20, however, can no longer occur, because thousands of additional wells now tap the same aquifer. This depleted the reservoir, and the water table in the recharge area was lowered. As a consequence, the pressure dropped to the

point where many wells stopped flowing altogether and had to be pumped.

On a different scale, city water systems can be considered examples of artificial artesian systems (Figure 17.21). The water tower, into which water is pumped, would represent the area of recharge, the pipes the confined aquifer, and the faucets in homes the flowing artesian wells.

PROBLEMS ASSOCIATED WITH GROUNDWATER WITHDRAWAL

As with many of our valuable natural resources, groundwater is being exploited at an increasing rate. In some areas, overuse threatens the groundwater supply. In other places, groundwater withdrawal has caused the ground and everything resting on it to sink. Still other localities are concerned with the possible contamination of the groundwater supply.

Treating Groundwater as a Nonrenewable Resource

Many natural systems tend to establish a condition of equilibrium. The groundwater system is no exception. The water table's height reflects a balance between the rate of infiltration and the rate of discharge and withdrawal. Any imbalance will either raise or lower the water table. Longterm imbalances can lead to a significant drop in the water table if there is either a decrease in recharge due to prolonged drought, or an increase in groundwater discharge or withdrawal.

For many, groundwater appears to be an endlessly renewable resource because it is continually replenished by rainfall and melting snow. But in some regions, groundwater has been and continues to be treated as a *nonrenewable* resource. Where this occurs, the water available to recharge the aquifer falls significantly short of the amount being withdrawn.

The High Plains Aquifer provides one example (Figure 17.22). Underlying about 111 million acres (450,000 square

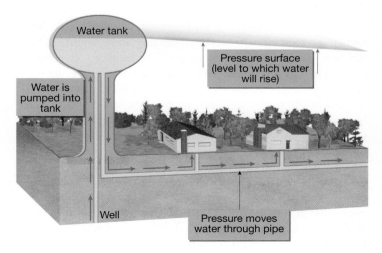

FIGURE 17.21 City water systems can be considered to be artificial artesian systems.

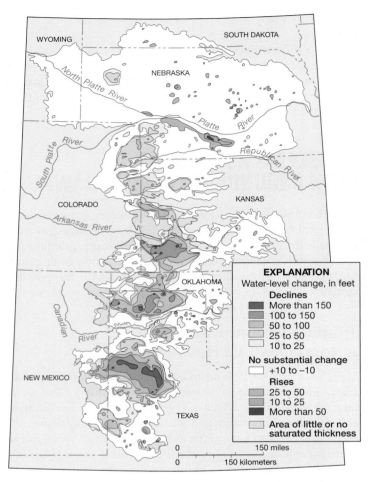

FIGURE 17.22   Changes in groundwater levels in the High Plains aquifer from predevelopment to 2005. Extensive pumping for irrigation has led to water level declines in excess of 100 feet in parts of Kansas, Oklahoma, Texas, and New Mexico. Water level rises have occurred where surface water is used for irrigation, such as along the Platte River in Nebraska. (After U.S. Geological Survey)

kilometers or 174,000 square miles) in parts of eight western states, it is one of the largest and most agriculturally significant aquifers in the United States. It accounts for about 30 percent of all groundwater withdrawn for irrigation in the country. Mean annual precipitation is modest—ranging from about 40 centimeters (16 inches) in western portions to about 71 centimeters (28 inches) in eastern parts. Evaporation rates, on the other hand, are high—ranging from about 150 centimeters (60 inches) in the cooler northern parts of the region to 265 centimeters (105 inches) in the warmer southern parts. Because evaporation rates are high relative to precipitation, there is little rain water to recharge the aquifer. Thus, in some parts of the region, where intense irrigation has been practiced for an extended period, depletion of groundwater has been severe. Figure 17.22 bears this out. The U.S. Geological Survey estimates that during the past 50 to 60 years, water in storage in the High Plains aquifer declined about 200 million acre feet (about 65 trillion gallons) with 62 percent of the total decline occurring in Texas.

Subsidence

As you will see later in this chapter, surface subsidence can result from natural processes related to groundwater. However, the ground may also sink when water is pumped from wells faster than natural recharge processes can replace it. This effect is particularly pronounced in areas underlain by thick layers of unconsolidated sediments. As the water is withdrawn, the water pressure drops and the weight of the overburden is transferred to the sediment. The greater pressure packs the sediment grains tightly together, and the ground subsides. The size of the area affected by such subsidence is significant. In the contiguous United States, it amounts to an estimated 26,000 square kilometers (more than 10,000 square miles)—an area about the size of Massachusetts!

Many areas may be used to illustrate land subsidence caused by the excessive pumping of groundwater from relatively loose sediment. A classic example in the United States occurred in the San Joaquin Valley of California and is discussed in Box 17.2. Many other cases of land subsidence due to groundwater pumping exist in the United States, including Las Vegas, Nevada; New Orleans and Baton Rouge, Louisiana; portions of southern Arizona; and the Houston—Galveston area of Texas. In the low-lying coastal area between Houston and Galveston, land subsidence ranges from 1.5 to 3 meters (5 to 9 feet). The result is that about 78 square kilometers (30 square miles) are permanently flooded.

Outside the United States, one of the most spectacular examples of subsidence occurred in Mexico City, a portion of which is built on a former lake bed. In the first half of the 20th century, thousands of wells were sunk into the water-saturated sediments beneath the city. As water was withdrawn, portions of the city subsided by as much as 6 to 7 meters. In some places buildings have sunk to such a point that access to them from the street is located at what used to be the second-floor level!

Saltwater Contamination

In many coastal areas the groundwater resource is being threatened by the encroachment of saltwater. To understand this problem, let us examine the relationship between fresh groundwater and salty groundwater. Figure 17.23A is a diagrammatic cross section that illustrates this relationship in a coastal area underlain by permeable homogeneous materials. Fresh water is less dense than saltwater, so it floats on the saltwater and forms a large lens-shaped body that may extend to considerable depths below sea level. In such a situation, if the water table is 1 meter above sea level, the base of the freshwater body will extend to a depth of about 40 meters below sea level. Stated another way, the depth of the fresh water below sea level is about 40 times greater than the elevation of the water table above sea level. Thus, when excessive pumping lowers the water table by a certain amount, the bottom of the freshwater zone will rise by 40 times that amount. Therefore, if groundwater withdrawal continues to exceed recharge, there will come a time when the elevation of the saltwater will be sufficiently high to be drawn into wells, thus contaminating the freshwater supply (Figure 17.23B). Deep wells and wells near the shore are usually the first to be affected.

PEOPLE AND THE ENVIRONMENT

Land Subsidence in the San Joaquin Valley

BOX 17.2

The San Joaquin Valley is a broad structural basin that contains a thick fill of sediments. The size of Maryland, it constitutes the southern two-thirds of California's Central Valley, a flatland separating two mountain ranges—the Coast Ranges to the west and the Sierra Nevada to the east (Figure 17.C). The valley's aquifer system is a mixture of alluvial materials derived from the surrounding mountains. Sediment thicknesses average about 870 meters (about half a mile). The valley's climate is arid to semiarid, with average annual precipitation ranging from 12 to 35 centimeters (5 to 14 inches).

The San Joaquin Valley has a strong agricultural economy that requires large quantities of water for irrigation. For many years up to 50 percent of this need was met by groundwater. In addition, nearly every city in the region uses groundwater as its principal source for homes and industry.

Although development of the valley's groundwater for irrigation began in the late 1800s, land subsidence did not begin until the mid-1920s when withdrawals were substantially increased. By the early 1970s, water levels had declined up to 120 meters (400 feet). The resulting ground subsidence exceeded 8.5 meters (29 feet) at one place in the region (Figure 17.D). At that time, areas within the valley were subsiding faster than 0.3 meter (1 foot) per year.

Then, because surface water was being imported and groundwater pumping was being decreased, water levels in the aquifer recovered and subsidence ceased. However, during a drought in 1976–1977, heavy groundwater pumping led to renewed subsidence. This time water levels dropped much faster because of the reduced storage capacity caused by earlier compaction of sediments. In all, half the entire valley was affected by subsidence. According to the U.S. Geological Survey:

> Subsidence in the San Joaquin Valley probably represents one of the greatest single manmade alterations in the configuration of the Earth's surface. . . . It has caused serious and costly problems in construction and maintenance of water-transport structures, highways, and highway structures; also many millions of dollars have been spent on the repair or replacement of deep-water wells. Subsidence, besides changing the gradient and course of valley creeks and streams, has caused unexpected flooding, costing farmers many hundreds of thousands of dollars in recurrent land leveling.*

Similar effects have been documented in the San Jose area of the Santa Clara Valley, California, where between 1916 and 1966 subsidence approached 4 meters. Flooding of lands bordering the southern part of San

Francisco Bay was one of the results. As was the case in the San Joaquin Valley, the subsidence stopped when imports of surface water were increased, allowing groundwater withdrawal rates to be decreased.

FIGURE 17.D The marks on this utility pole indicate the level of the surrounding land in preceding years. Between 1925 and 1975 this part of the San Joaquin Valley subsided almost 9 meters because of the withdrawal of groundwater and the resulting compaction of sediments. (Photo courtesy of U.S. Geological Survey)

FIGURE 17.C The shaded area shows California's San Joaquin Valley.

---

*Ireland R. L., J. F. Poland, and F. S. Riley, *Land Subsidence in the San Joaquin Valley, California, as of 1980*, U.S. Geological Survey Professional Paper 437-I (Washington, D.C.: U.S. Government Printing Office, 1984), p. 11.

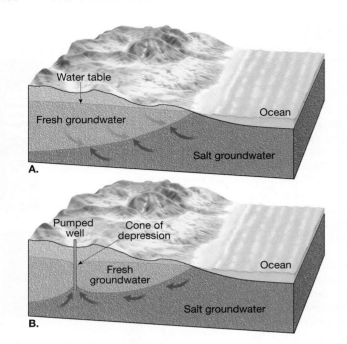

FIGURE 17.23 **A.** Because fresh water is less dense than saltwater, it floats on the saltwater and forms a lens-shaped body that may extend to considerable depths below sea level. **B.** When excessive pumping lowers the water table, the base of the freshwater zone will rise by 40 times that amount. The result may be saltwater contamination of wells.

In urbanized coastal areas, the problems created by excessive pumping are compounded by a decrease in the rate of natural recharge. As more and more of the surface is covered by streets, parking lots, and buildings, infiltration into the soil is diminished.

In an attempt to correct the problem of saltwater contamination of groundwater resources, a network of recharge wells may be used. These wells allow wastewater to be pumped back into the groundwater system. A second method of correction is accomplished by building large basins. These basins collect surface drainage and allow it to seep into the ground. On New York's Long Island, where the problem of saltwater contamination was recognized more than 50 years ago, both of these methods have been employed with considerable success.

Contamination of freshwater aquifers by saltwater is primarily a problem in coastal areas, but it can also threaten noncoastal locations. Many ancient sedimentary rocks of marine origin were deposited when the ocean covered places that are now far inland. In some instances significant quantities of seawater were trapped and still remain in the rock. These strata sometimes contain quantities of fresh water and may be pumped for use by people. However, if fresh

water is removed more rapidly than it is replenished, saline water may encroach and render the wells unusable. Such a situation threatened users of a deep (Cambrian age) sandstone aquifer in the Chicago area. To counteract this, water from Lake Michigan was allocated to the affected communities to offset the rate of withdrawal from the aquifer.

Groundwater Contamination

The pollution of groundwater is a serious matter, particularly in areas where aquifers provide a large part of the water supply. One common source of groundwater pollution is sewage. Its sources include an ever increasing number of septic tanks, as well as inadequate or broken sewer systems and farm wastes.

If sewage water that is contaminated with bacteria enters the groundwater system, it may become purified through natural processes. The harmful bacteria may be mechanically filtered by the sediment through which the water percolates, destroyed by chemical oxidation, and/or assimilated by other organisms. For purification to occur, however, the aquifer must be of the correct composition. For example, extremely permeable aquifers (such as highly fractured crystalline rock, coarse gravel, or cavernous limestone) have such large openings that contaminated groundwater may travel long distances without being cleansed. In this case, the water flows too rapidly and is not in contact with the surrounding material long enough for purification to occur. This is the problem at well 1 in Figure 17.24A.

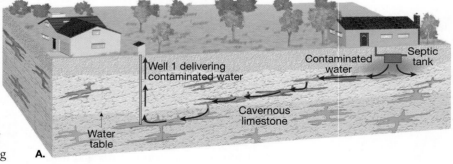

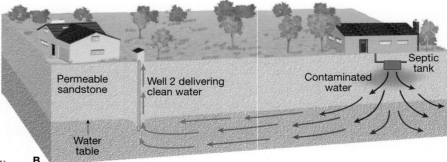

FIGURE 17.24 **A.** Although the contaminated water has traveled more than 100 meters before reaching well 1, the water moves too rapidly through the cavernous limestone to be purified. **B.** As the discharge from the septic tank percolates through the permeable sandstone, it is purified in a relatively short distance.

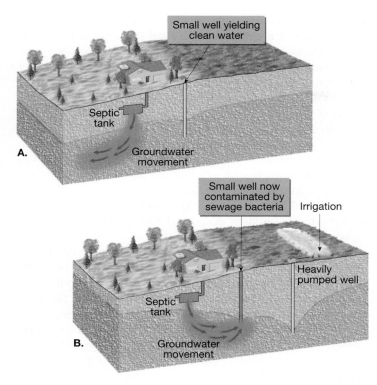

FIGURE 17.25 **A.** Originally the outflow from the septic tank moved away from the small well. **B.** The heavily pumped well changed the slope of the water table, causing contaminated groundwater to flow toward the small well.

On the other hand, when the aquifer is composed of sand or permeable sandstone, it can sometimes be purified after traveling only a few dozen meters through it. The openings between sand grains are large enough to permit water movement, yet the movement of the water is slow enough to allow ample time for its purification (well 2, Figure 17.24B).

Sometimes sinking a well can lead to groundwater pollution problems. If the well pumps a sufficient quantity of water, the cone of depression will locally increase the slope of the water table. In some instances the original slope may even be reversed. This could lead to the contamination of wells that yielded unpolluted water before heavy pumping began (Figure 17.25). Also recall that the rate of groundwater movement increases as the slope of the water table steepens. This could produce problems because a faster rate of movement allows less time for the water to be purified in the aquifer before it is pumped to the surface.

Other sources and types of contamination also threaten groundwater supplies (Figure 17.26). These include widely used substances such as highway salt, fertilizers that are spread across the land surface, and pesticides. In addition, a wide array of chemicals and industrial materials may leak from pipelines, storage tanks, landfills, and holding ponds. Some of these pollutants are classified as *hazardous,* meaning that they are either flammable, corrosive, explosive, or toxic. In land disposal, potential contaminants are heaped onto mounds or spread directly over the ground. As rainwater oozes through the refuse, it may dissolve a variety of organic and inorganic materials. If the leached material reaches the water table, it will mix with the groundwater and contaminate the supply. Similar problems may result from leakage of shallow excavations called holding ponds into which a variety of liquid wastes are disposed.

Because groundwater movement is usually slow, polluted water can go undetected for a long time. In fact, contamination is sometimes discovered only after drinking water has been affected and people become ill. By this time, the volume of polluted water may be very large, and even if the source of contamination is removed immediately, the problem is not solved. Although the sources of groundwater contamination are numerous, there are relatively few solutions.

A.   **B.**

FIGURE 17.26 Sometimes agricultural chemicals **A.** and materials leached from landfills **B.** find their way into the groundwater. These are two of the potential sources of groundwater contamination. (Photo **A** by Michael Collier; Photo **B** by F. Rossotto/Corbis/The Stock Market)

Once the source of the problem has been identified and eliminated, the most common practice is simply to abandon the water supply and allow the pollutants to be flushed away gradually. This is the least costly and easiest solution, but the aquifer must remain unused for many years. To accelerate this process, polluted water is sometimes pumped out and treated. Following removal of the tainted water, the aquifer is allowed to recharge naturally, or in some cases the treated water or other fresh water is pumped back in. This process is costly, time-consuming, and it may be risky because there is no way to be certain that all of the contamination has been removed. Clearly, the most effective solution to groundwater contamination is prevention.

THE GEOLOGIC WORK OF GROUNDWATER

Groundwater dissolves rock. This fact is key to understanding how caverns and sinkholes form. Because soluble rocks, especially limestone, underlie millions of square kilometers of Earth's surface, it is here that the groundwater carries on its important role as an erosional agent. Limestone is nearly insoluble in pure water but is quite easily dissolved by water containing small quantities of carbonic acid, and most groundwater contains this acid. It forms because rainwater readily dissolves carbon dioxide from the air and from decaying plants. Therefore, when groundwater comes in contact with limestone, the carbonic acid reacts with the calcite (calcium carbonate) in the rocks to form calcium bicarbonate, a soluble material that is then carried away in solution.

Caverns

The most spectacular results of groundwater's erosional handiwork are limestone **caverns.** In the United States alone about 17,000 caves have been discovered and new ones are being found every year. Although most are relatively small, some have spectacular dimensions. Mammoth Cave in Kentucky and Carls-

bad Caverns in southeastern New Mexico are famous examples. The Mammoth Cave system is the most extensive in the world, with more than 540 kilometers of interconnected passages. The dimensions at Carlsbad Caverns are impressive in a different way. Here we find the largest and perhaps most spectacular single chamber. The Big Room at Carlsbad Caverns has an area equivalent to 14 football fields and enough height to accommodate the U.S. Capitol building.

Most caverns are created at or just below the water table in the zone of saturation. Here acidic groundwater follows lines of weakness in the rock, such as joints and bedding planes. As time passes, the dissolving process slowly creates cavities and gradually enlarges them into caverns. Material that is dissolved by the groundwater is eventually discharged into streams and carried to the ocean.

In many caves, development has occurred at several levels, with the current cavern-forming activity occurring at the lowest elevation. This situation reflects the close relationship between the formation of major subterranean passages and the river valleys into which they drain. As streams cut their valleys deeper, the water table drops as the elevation of the river drops. Consequently, during periods when surface streams are rapidly downcutting, surrounding groundwater levels drop rapidly and cave passages are abandoned by the water while the passages are still relatively small in cross-sectional area. Conversely, when the entrenchment of streams is slow or negligible, there is time for large cave passages to form.

Certainly the features that arouse the greatest curiosity for most cavern visitors are the stone formations that give some caverns a wonderland appearance. These are not erosional features, like the cavern itself, but depositional features created by the seemingly endless dripping of water over great spans of time. The calcium carbonate that is left behind produces the limestone we call travertine. These cave deposits, however, are also commonly called *dripstone,* an obvious reference to their mode of origin. Although the formation of caverns takes place in the zone of saturation, the deposition of dripstone is not possible until the caverns are above the water table in the unsaturated zone. As soon as the chamber is filled with air, the stage is set for the decoration phase of cavern building to begin.

The various dripstone features found in caverns are collectively called **speleothems** (*spelaion* = cave, *them* = put), no two of which are exactly alike (Figure 17.27). Perhaps the most familiar speleothems are **stalactites** (*stalaktos* = trickling). These icicle-like pendants hang from the ceiling of the cavern and form where water seeps through cracks above. When the water reaches air in the cave, some of the dissolved carbon dioxide escapes from the drop, and calcite precipitates. Deposition occurs as a ring around the edge of the water drop. As drop after drop follows, each leaves an infinitesimal trace of calcite behind, and a hollow limestone tube is created. Water then moves through the tube, remains suspended momentarily at the end, contributes a tiny ring of calcite, and falls to the cavern floor. The stalactite just described is appropriately called a *soda straw* (Figure 17.28). Often the hollow tube of the soda straw becomes plugged or its supply of water increases. In either

FIGURE 17.27 Speleothems are of many types, including stalactites, stalagmites, and columns. Chinese Theater, Carlsbad Caverns National Park. (Photo by David Muench/David Muench Photography)

limestone areas of Florida, Kentucky, and southern Indiana, there are literally tens of thousands of these depressions varying in depth from just a meter or two to a maximum of more than 50 meters.

Sinkholes commonly form in two ways. Some develop gradually over many years without any physical disturbance to the rock. In these situations the limestone immediately below the soil is dissolved by downward-sweeping rainwater that is freshly charged with carbon dioxide. With time, the bedrock surface is lowered and the fractures into which the water seeps are enlarged. As the fractures grow in size, soil subsides into the widening voids, from which it is removed by groundwater flowing in the passages below. These depressions are usually shallow and have gentle slopes.

By contrast, sinkholes can also form abruptly and without warning when the roof of a cavern collapses under its own weight. Typically, the depressions created in this manner are steep-sided and deep. When they form

case, the water is forced to flow and hence deposit along the outside of the tube. As deposition continues, the stalactite takes on the more common conical shape.

Speleothems that form on the floor of a cavern and reach upward toward the ceiling are called **stalagmites** (*stalagmos* = dropping). The water supplying the calcite for stalagmite growth falls from the ceiling and splatters over the surface. As a result, stalagmites do not have a central tube and are usually more massive in appearance and rounded on their upper ends than stalactites. Given enough time, a downward-growing stalactite and an upward-growing stalagmite may join to form a *column*.

Karst Topography

Many areas of the world have landscapes that, to a large extent, have been shaped by the dissolving power of groundwater. Such areas are said to exhibit **karst topography,** named for the Krs Plateau in Slovenia located along the northeastern shore of the Adriatic Sea where such topography is strikingly developed. In the United States, karst landscapes occur in many areas that are underlain by limestone, including portions of Kentucky, Tennessee, Alabama, southern Indiana, and central and northern Florida (Figure 17.29). Generally, arid and semiarid areas are too dry to develop karst topography. When solution features exist in such regions, they are likely to be remnants of a time when rainier conditions prevailed.

Karst areas typically have irregular terrain punctuated with many depressions, called **sinkholes** or **sinks** (Figure 17.30). In the

FIGURE 17.28 "Live" soda straw stalactite in Chinn Springs Cave, Independence County, Arkansas. To view another example, see Figure 7.11, p. 207. (Photo by Dante Fenolio/Photo Researchers, Inc.)

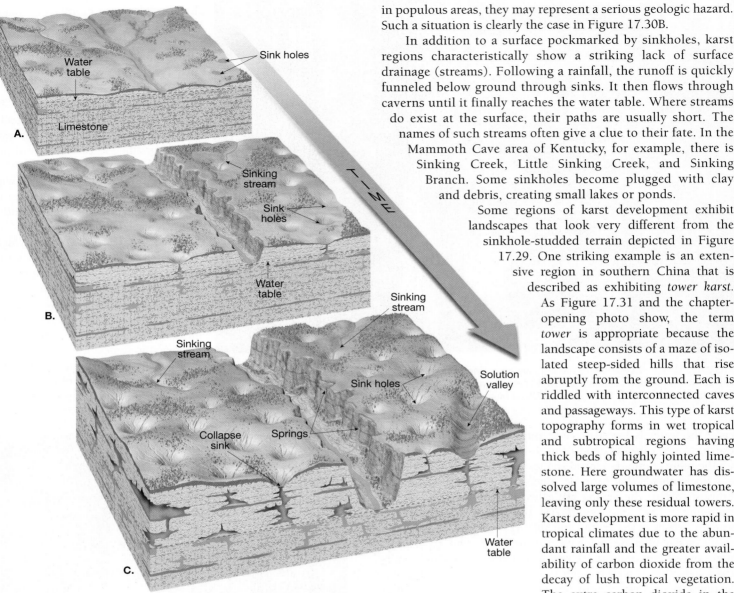

FIGURE 17.29 Development of a karst landscape. **A.** During early stages, groundwater percolates through limestone along joints and bedding planes. Solution activity creates and enlarges caverns at and below the water table. **B.** In this view, sinkholes are well developed and surface streams are funneled below ground. **C.** With the passage of time, caverns grow larger and the number and size of sinkholes increase. Collapse of caverns and coalescence of sinkholes form larger, flat-floored depressions. Eventually, solution activity may remove most of the limestone from the area, leaving only isolated remnants as in Figure 17.31.

in populous areas, they may represent a serious geologic hazard. Such a situation is clearly the case in Figure 17.30B.

In addition to a surface pockmarked by sinkholes, karst regions characteristically show a striking lack of surface drainage (streams). Following a rainfall, the runoff is quickly funneled below ground through sinks. It then flows through caverns until it finally reaches the water table. Where streams do exist at the surface, their paths are usually short. The names of such streams often give a clue to their fate. In the Mammoth Cave area of Kentucky, for example, there is Sinking Creek, Little Sinking Creek, and Sinking Branch. Some sinkholes become plugged with clay and debris, creating small lakes or ponds.

Some regions of karst development exhibit landscapes that look very different from the sinkhole-studded terrain depicted in Figure 17.29. One striking example is an extensive region in southern China that is described as exhibiting *tower karst*. As Figure 17.31 and the chapter-opening photo show, the term *tower* is appropriate because the landscape consists of a maze of isolated steep-sided hills that rise abruptly from the ground. Each is riddled with interconnected caves and passageways. This type of karst topography forms in wet tropical and subtropical regions having thick beds of highly jointed limestone. Here groundwater has dissolved large volumes of limestone, leaving only these residual towers. Karst development is more rapid in tropical climates due to the abundant rainfall and the greater availability of carbon dioxide from the decay of lush tropical vegetation. The extra carbon dioxide in the soil means there is more carbonic acid for dissolving lime-stone. Other tropical areas of advanced karst development include portions of Puerto Rico, western Cuba, and northern Vietnam.

Students Sometimes Ask . . .

Is limestone the only rock type that develops karst features?

No. For example, karst development can occur in other carbonate rocks such as marble and dolostone. In addition, evaporites such as gypsum and salt (halite) are highly soluble and are readily dissolved to form karst features including sinkholes, caves, and disappearing streams. This latter situation is termed *evaporite karst*.

A.

B.

FIGURE 17.30 **A.** Groundwater was responsible for creating these sinkholes in a limestone plateau north of Jajce, Bosnia and Herzegovina. (Photo by Jerome Wyckoff) **B.** This small sinkhole formed suddenly in 1991 when the roof of a cavern collapsed, destroying this home in Frostproof, Florida. (Photo by *St. Petersburg Times*/Getty Images, Inc/Liaison)

A.

B.

FIGURE 17.31 **A.** Painting of Chinese tower karst, "Peach Garden Land of Immortals," by Qiu Ying. (Asian Art and Archaeology, Inc./CORBIS—NY) **B.** One of the best-known and most distinctive regions of tower karst development is the Guilin District of southeastern China. (Photo by Topham/The Image Works)

CHAPTER 17 GROUNDWATER IN REVIEW

- As a resource, *groundwater* represents the largest reservoir of fresh water that is readily available to humans. Geologically, the dissolving action of groundwater produces *caves* and *sinkholes*. Groundwater is also an equalizer of streamflow.

- Each day in the United States we use about 345 billion gallons of fresh water. Groundwater provides about 83 billion gallons or 24 percent of the total. More groundwater is used for irrigation than for all other uses combined.

- Groundwater is water that completely fills the pore spaces in sediment and rock in the subsurface *zone of saturation*. The upper limit of this zone is the *water table*. The *unsaturated zone* is above the water table where the soil, sediment, and rock are not saturated.

- The interaction between streams and groundwater takes place in one of three ways: streams gain water from the inflow of groundwater (*gaining stream*); they lose water through the streambed to the groundwater system (*losing stream*); or they do both, gaining in some sections and losing in others.

- Materials with very small pore spaces (such as clay) hinder or prevent groundwater movement and are called *aquitards*. *Aquifers* consist of materials with larger pore spaces (such as sand) that are permeable and transmit groundwater freely.

- Groundwater moves in looping curves that are a compromise between the downward pull of gravity and the tendency of water to move toward areas of reduced pressure.

- The primary factors influencing the velocity of groundwater flow are the slope of the water table (*hydraulic gradient*) and the permeability of the aquifer (*hydraulic conductivity*).

- *Springs* occur whenever the water table intersects the land surface and a natural flow of groundwater results. *Wells*, openings bored into the zone of saturation, withdraw groundwater and create roughly conical depressions in the water table known as *cones of depression*. *Artesian wells* occur when water rises above the level at which it was initially encountered.

- When groundwater circulates at great depths, it becomes heated. If it rises, the water may emerge as a *hot spring*. *Geysers* occur when groundwater is heated in underground chambers, expands, and some water quickly changes to steam, causing the geyser to erupt. The source of heat for most hot springs and geysers is hot igneous rock.

- Some of the current environmental problems involving groundwater include (1) *overuse* by intense irrigation, (2) *subsidence* caused by groundwater withdrawal, (3) *saltwater contamination,* and (4) contamination by pollutants.

- Most *caverns* form in limestone at or below the water table when acidic groundwater dissolves rock along lines of weakness, such as joints and bedding planes. The various *dripstone* features found in caverns are collectively called *speleothems*. Landscapes that, to a large extent, have been shaped by the dissolving power of groundwater exhibit *karst topography*, an irregular terrain punctuated with many depressions, called *sinkholes* or *sinks*.

KEY TERMS

aquifer (p. 468)
aquitard (p. 468)
artesian (p. 476)
capillary fringe (p. 465)
cavern (p. 482)
cone of depression (p. 475)
confined aquifer (p. 476)
Darcy's law (p. 469)
drawdown (p. 474)

flowing artesian well
  (p. 477)
gaining stream (p. 467)
geyser (p. 473)
groundwater (p. 465)
hot spring (p. 472)
hydraulic conductivity
  (p. 469)
hydraulic gradient (p. 469)

karst topography (p. 483)
losing stream (p. 467)
nonflowing artesian well
  (p. 476)
perched water table (p. 472)
permeability (p. 468)
porosity (p. 467)
sinkhole (sink) (p. 483)
speleothem (p. 482)

spring (p. 472)
stalactite (p. 482)
stalagmite (p. 483)
unsaturated zone (p. 465)
water table (p. 465)
well (p. 474)
zone of saturation (p. 465)
zone of soil moisture
  (p. 465)

QUESTIONS FOR REVIEW

1. What percentage of fresh water is groundwater? If glacial ice is excluded and only liquid fresh water is considered, about what percentage is groundwater?

2. Geologically, groundwater is important as an erosional agent. Name another significant geological role for groundwater.

3. Compare and contrast the unsaturated zone and the zone of saturation. Which zone contains groundwater?

4. Although we usually think of tables as being flat, the water table generally is not. Explain.

5. Although meteorological drought may have ended, hydrological drought may still continue. Explain (see Box 17.1).

6. Contrast a gaining stream and a losing stream.

7. Distinguish between porosity and permeability.

8. What is the difference between an aquitard and an aquifer?

9. Under what circumstances can a material have a high porosity but not be a good aquifer?

10. As illustrated in Figure 17.8, groundwater moves in looping curves. What factors cause the water to follow such paths?

11. Briefly describe the important contribution to our understanding of groundwater movement made by Henri Darcy.

12. When an aquitard is situated above the main water table, a localized saturated zone may be created. What term is applied to such a situation?

13. What is the source of heat for most hot springs and geysers? How is this reflected in the distribution of these features?

14. Two neighbors each dig a well. Although both wells penetrate to the same depth, one neighbor is successful and the other is not. Describe a circumstance that might explain what happened.

15. What is meant by the term *artesian?*

16. In order for artesian wells to exist, two conditions must be present. List these conditions.

17. When the Dakota Sandstone was first tapped, water poured freely from many artesian wells. Today these wells must be pumped. Explain.

18. What problem is associated with the pumping of groundwater for irrigation in the southern part of the High Plains?

19. Briefly explain what happened in the San Joaquin Valley as the result of excessive groundwater withdrawal (see Box 17.2).

20. In a particular coastal area, the water table is 4 meters above sea level. Approximately how far below sea level does the fresh water reach?

21. Why does the rate of natural groundwater recharge decrease as urban areas develop?

22. Which aquifer would be most effective in purifying polluted groundwater: coarse gravel, sand, or cavernous limestone?

23. What is meant when a groundwater pollutant is classified as hazardous?

24. Name two common speleothems and distinguish between them.

25. Areas whose landscapes largely reflect the erosional work of groundwater are said to exhibit what kind of topography?

26. Describe two ways in which sinkholes are created.

COMPANION WEBSITE

The *Earth 10e* website uses the resources and flexibility of the Internet to aid in your study of the topics in this chapter. Written and developed by the authors and other geology instructors, this site will help improve your understanding of geology. Visit www.mygeoscienceplace.com in order to:

• **Review** key chapter concepts

• **Read** with links to the eBook and to chapter-specific web resources

• **Visualize** and comprehend challenging topics using learning activities in *GEODe Earth*

• **Test** yourself with online quizzes

GEODe EARTH

GEODe Earth is a valuable and easy to use learning aid that can be accessed from your book's Companion Website (www.mygeoscienceplace.com). It is a dynamic instructional tool that promotes understanding and reinforces important concepts by using tutorials, animations, and exercises that actively engage the student.

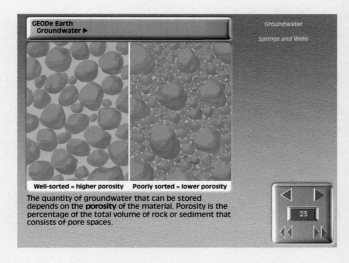

GLACIERS AND GLACIATION

Kennicott Glacier is a 43-kilometer (27-mile) long valley glacier in Alaska's Wrangell-St. Elias National Park. The dark stripes of sediment are called medial moraines.

(PHOTO BY MICHAEL COLLIER)

Climate has a strong influence on the nature and intensity of Earth's external processes. This fact is dramatically illustrated in this chapter because the existence and extent of glaciers is largely controlled by Earth's changing climate.

Like the running water and groundwater that were the focus of the preceding two chapters, glaciers represent a significant erosional process. These moving masses of ice are responsible for creating many unique landforms and are part of an important link in the rock cycle in which the products of weathering are transported and deposited as sediment.

Today glaciers cover nearly 10 percent of Earth's land surface; however, in the recent geologic past, ice sheets were three times more extensive, covering vast areas with ice thousands of meters thick. Many regions still bear the mark of these glaciers. The basic character of such diverse places as the Alps, Cape Cod, and Yosemite Valley was fashioned by now vanished masses of glacial ice (Figure 18.1). Moreover, Long Island, the Great Lakes, and the fiords of Norway and Alaska all owe their existence to glaciers. Glaciers, of course, are not just a phenomenon of the geologic past. As you will see, they are still sculpting and depositing debris in many regions today.

GLACIERS: A PART OF TWO BASIC CYCLES

GLACIERS AND GLACIATION
▶ Introduction

Glaciers are a part of two fundamental cycles in the Earth system—the hydrologic cycle and the rock cycle. Earlier you learned that the water of the hydrosphere is constantly cycled through the atmosphere, biosphere, and geosphere. Time and time again, water evaporates from the oceans into the atmosphere, precipitates upon the land, and flows in rivers and underground back to the sea. However, when precipitation falls at high elevations or high latitudes, the water may not immediately make its way toward the sea. Instead, it may become part of a glacier. Although the ice will eventually melt, allowing the water to continue its path to the sea, water can be stored as glacial ice for many tens, hundreds, or even thousands of years.

A **glacier** is a thick ice mass that forms over hundreds or thousands of years. It originates on land from the accumulation, compaction, and recrystallization of snow. A glacier appears to be motionless, but it is not—glaciers move very slowly. Like running water, groundwater, wind, and waves, glaciers are dynamic erosional agents that accumulate, transport, and deposit sediment. As such, glaciers are among the processes that perform a basic function in the rock cycle. Although glaciers are found in many parts of the world today, most are located in remote areas, either near Earth's poles or in high mountains.

Valley (Alpine) Glaciers

Literally thousands of relatively small glaciers exist in lofty mountain areas, where they usually follow valleys that were originally occupied by streams. Unlike the rivers that previously flowed in these valleys, the glaciers advance slowly, perhaps only a few centimeters per day. Because of their setting, these moving ice masses are termed **valley glaciers** or **alpine glaciers** (Figure 18.2). Each glacier actually is a stream of ice, bounded by precipitous rock walls, that flows downvalley from an accumulation center near its head. Like rivers, valley glaciers can be long or short, wide or narrow, single or with branching tributaries. Generally, the widths of alpine glaciers are small compared to their lengths. Some extend for just a fraction of a kilometer, whereas others go on for many tens of kilometers. The west branch of the Hubbard Glacier, for example, runs through 112 kilometers (nearly 70 miles) of mountainous terrain in Alaska and the Yukon Territory.

Ice Sheets

In contrast to valley glaciers, **ice sheets** exist on a much larger scale. The low total annual

FIGURE 18.1   Glaciers continue to sculpt the Swiss Alps. (Photo by Arno Balzarini/epa/Corbis)

FIGURE 18.2   An aerial view of Taku Glacier near Juneau, Alaska. Valley glaciers continue to modify this mountainous landscape. (Photo by Michael Collier)

solar radiation reaching the poles makes these regions eligible for great ice accumulations. Although many ice sheets have existed in the past, just two achieve this status at present (Figure 18.3). In the area of the North Pole, Greenland is covered by an imposing ice sheet that occupies 1.7 million square kilometers, or about 80 percent of this large island. Averaging nearly 1500 meters (5000 feet) thick, the ice extends 3000 meters (10,000 feet) above the island's bedrock floor in some places.

In the Southern Hemisphere, the huge Antarctic Ice Sheet attains a maximum thickness of about 4300 meters (14,000 feet) and covers nearly the entire continent, an area of more than 13.9 million square kilometers (5.4 million square miles). Because of the proportions of these huge features, they often are called *continental ice sheets*. Indeed, the combined areas of present-day continental ice sheets represent almost 10 percent of Earth's land area.

These enormous masses flow out in all directions from one or more snow-accumulation centers and completely obscure all but the highest areas of underlying terrain. Even sharp variations in the topography beneath the glacier usually appear as relatively subdued undulations

491

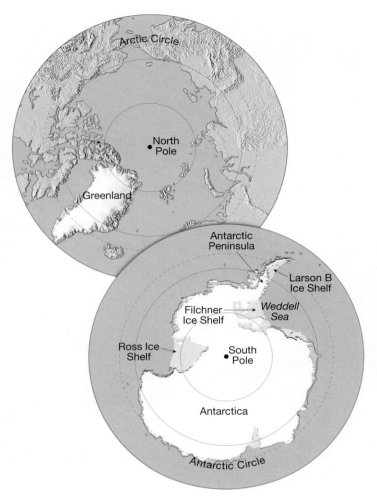

FIGURE 18.3   The only present-day continental ice sheets are those covering Greenland and Antarctica. Their combined areas represent almost 10 percent of Earth's land area. Greenland's ice sheet occupies 1.7 million square kilometers, or about 80 percent of the island. The area of the Antarctic Ice Sheet is almost 14 million square kilometers. Ice shelves occupy an additional 1.4 million square kilometers adjacent to the Antarctic Ice Sheet.

Students Sometimes Ask . . .

It's hard to imagine how large the Greenland Ice Sheet really is. Could you provide a comparison that we might be able to relate to?

Try this. The ice sheet is long enough to extend from Key West, Florida, to 100 miles north of Portland, Maine. Its width could reach from Washington, D.C. to Indianapolis, Indiana. Put another way, the ice sheet is 80 percent as big as the entire United States east of the Mississippi River. The area of Antarctica's ice sheet is more than eight times as great!

to their bases. Antarctica's ice shelves extend over approximately 1.4 million square kilometers (0.6 million square miles). The Ross and Filchner ice shelves are the largest, with the Ross Ice Shelf alone covering an area approximately the size of Texas (see Figure 18.3). In recent years, satellite monitoring has shown that some ice shelves are unstable and breaking apart. For example, during a 35-day span in February and March 2002 an ice shelf on the eastern side of the Antarctic Peninsula, known as the Larsen B Ice Shelf, broke apart and separated from the continent. Thousands of icebergs were set adrift in the adjacent Weddell Sea. The event was captured on satellite imagery (Figure 18.4). This was not an isolated happening but part of a trend. Since 1974 the extent of seven ice shelves surrounding the Antarctic Peninsula declined by about 13,500 square kilometers (nearly 5300 square miles). Scientists attribute the breakup of the ice shelves to a strong regional climate warming.

FIGURE 18.4   These satellite images document the breakup of the Larsen B Ice Shelf in early 2002. Look at the lower portion of Figure 18.3 to see the location of the ice shelf adjacent to the Antarctic Peninsula. The total area of the ice shelf that collapsed was 3250 square kilometers (1267 square miles). The average thickness of the ice was about 220 meters (720 feet). This event liberated more than 700 billion metric tons of icebergs. (NASA)

on the surface of the ice. Such topographic differences, however, do affect the behavior of the ice sheets, especially near their margins, by guiding flow in certain directions and creating zones of faster and slower movement.

Along portions of the Antarctic coast, glacial ice flows into the adjacent ocean, creating features called **ice shelves.** They are large, relatively flat masses of floating ice that extend seaward from the coast but remain attached to the land along one or more sides. The shelves are thickest on their landward sides, and they become thinner seaward. They are sustained by ice from the adjacent ice sheet as well as being nourished by snowfall and the freezing of seawater

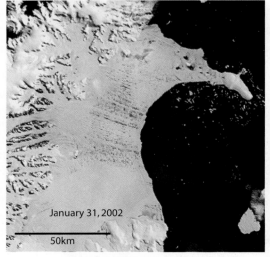
Before

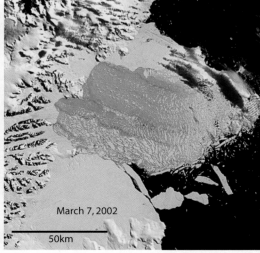
After

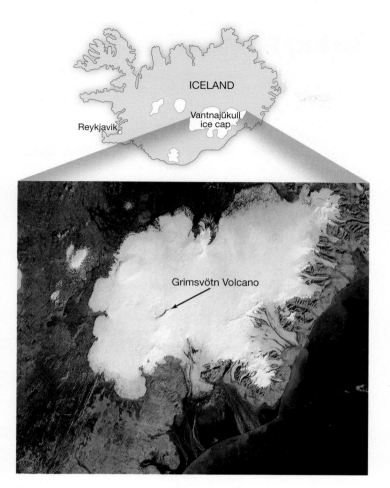

FIGURE 18.5 The ice cap in this satellite image is the Vantnajükull in southeastern Iceland (*jükull* means "ice cap" in Danish). In 1996 the Grimsvötn Volcano erupted beneath the ice cap, producing large quantities of melted glacial water that created floods. (*Landsat* image from NASA)

Other Types of Glaciers

In addition to valley glaciers and ice sheets, other types of glaciers are also identified. Covering some uplands and plateaus are masses of glacial ice called **ice caps.** Like ice sheets, ice caps completely bury the underlying landscape but are much smaller than the continental-scale features. Ice caps occur in many places, including Iceland and several of the large islands in the Arctic Ocean (Figure 18.5).

Often ice caps and ice sheets feed **outlet glaciers.** These tongues of ice flow down valleys extending outward from the

Students Sometimes Ask . . .

Do you have some idea about how many glaciers there are worldwide?

It is estimated that approximately 160,000 glaciers presently occupy Earth's polar and high mountain environments. Present-day glaciers cover only about one-third the area that was covered at the height of the most recent Ice Age.

margins of these larger ice masses. The tongues are essentially valley glaciers that are avenues for ice movement from an ice cap or ice sheet through mountainous terrain to the sea. Where they encounter the ocean, some outlet glaciers spread out as floating ice shelves. Often large numbers of icebergs are produced.

Piedmont glaciers occupy broad lowlands at the bases of steep mountains and form when one or more alpine glaciers emerge from the confining walls of mountain valleys. Here the advancing ice spreads out to form a broad sheet. The size of individual piedmont glaciers varies greatly. Among the largest is the broad Malaspina Glacier along the coast of southern Alaska. It covers thousands of square kilometers of the flat coastal plain at the foot of the lofty St. Elias range (Figure 18.6).

What if the Ice Melted?

How much water is stored as glacial ice? Estimates by the U.S. Geological Survey indicate that only slightly more than 2 percent of the world's water is tied up in glaciers. But even 2 percent of a vast quantity is still large. The total volume of just valley glaciers is about 210,000 cubic kilometers, comparable to the combined volume of the world's largest saline and freshwater lakes.

As for ice sheets, Antarctica's represents 80 percent of the world's ice and an estimated 65 percent of Earth's fresh water. It covers an area almost one and a half times the area of the entire United States. If this ice melted, sea level would rise an estimated 60 to 70 meters and the ocean would inundate many densely populated coastal areas (Figure 18.7).

The hydrologic importance of the Antarctic ice can be illustrated in another way. If the ice sheet were melted at a uniform rate, it could feed (1) the Mississippi River for more than 50,000

FIGURE 18.6 Malaspina Glacier in southern Alaska is considered a classic example of a piedmont glacier. Piedmont glaciers occur where valley glaciers exit a mountain range onto broad lowlands, are no longer laterally confined, and spread to become wide lobes. Malaspina Glacier is actually a compound glacier, formed by the merger of several valley glaciers, the most prominent of which seen here are Agassiz Glacier (left) and Seward Glacier (right). In total, Malaspina Glacier is up to 65 kilometers (40 miles) wide and extends up to 45 kilometers (28 miles) from the mountain front nearly to the sea. (Image from NASA/JPL)

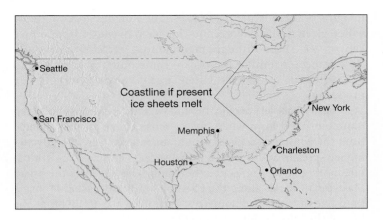

FIGURE 18.7 This map of a portion of North America shows the present-day coastline compared to the coastline that would exist if present ice sheets on Greenland and Antarctica melted.

years, (2) all the rivers in the United States for about 17,000 years, (3) the Amazon River for approximately 5000 years, or (4) all the rivers of the world for about 750 years.

FORMATION AND MOVEMENT OF GLACIAL ICE

GLACIERS AND GLACIATION
▶ Budget of a Glacier

Snow is the raw material from which glacial ice originates; therefore, glaciers form in areas where more snow falls in winter than melts during the summer. Glaciers develop in the high latitude polar realm because, even though annual snowfall totals are modest, temperatures are so low that little of the snow melts. Glaciers can form in mountains because temperatures drop with an increase in altitude. So even near the equator, glaciers may form at elevations above about 5000 meters (16,400 feet). For example, Tanzania's Mount Kilimanjaro, which is located practically astride the equator at an altitude of 5895 meters (19,336 feet), has glaciers at its summit. The elevation above which snow remains throughout the year is called the **snowline.** As you would expect, the elevation of the snowline varies with latitude. Near the equator, it occurs high in the mountains, whereas in the vicinity of the 60th parallel it is at or near sea level. Before a glacier is created, snow must be converted into glacial ice.

Glacial Ice Formation

When temperatures remain below freezing following a snowfall, the fluffy accumulation of delicate hexagonal crystals soon changes. As air infiltrates the spaces between the crystals, the extremities of the crystals evaporate and the water vapor condenses near the centers of the crystals. In this manner, snowflakes become smaller, thicker, and more spherical, and the large pore spaces disappear. By this process air is forced out and what was once light, fluffy snow is recrystallized into a much denser mass of small grains having the consistency of coarse

sand. This granular recrystallized snow is called **firn** and is commonly found making up old snowbanks near the end of winter. As more snow is added, the pressure on the lower layers gradually increases, compacting the ice grains at depth. Once the thickness of ice and snow exceeds 50 meters (160 feet), the weight is sufficient to fuse firn into a solid mass of interlocking ice crystals. Glacial ice has now been formed.

The rate at which this transformation occurs varies. In regions where the annual snow accumulation is great, burial is relatively rapid and snow may turn to glacial ice in a matter of a decade or less. Where the yearly addition of snow is less abundant, burial is slow and the transformation of snow to glacial ice may take hundreds of years.

Movement of a Glacier

The movement of glacial ice is generally referred to as *flow.* The fact that glacial movement is described in this way seems paradoxical—how can a solid flow? The way in which ice flows is complex and is of two basic types. The first of these, **plastic flow,** involves movement *within* the ice. Ice behaves as a brittle solid until the pressure upon it is equivalent to the weight of about 50 meters (165 feet) of ice. Once that load is surpassed, ice behaves as a plastic material and flow begins. Such flow occurs because of the molecular structure of ice. Glacial ice consists of layers of molecules stacked one upon the other. The bonds between layers are weaker than those within each layer. Therefore, when a stress exceeds the strength of the bonds between the layers, the layers remain intact and slide over one another.

A second and often equally important mechanism of glacial movement consists of the entire ice mass slipping along the ground. With the exception of some glaciers located in polar regions where the ice is probably frozen to the solid bedrock floor, most glaciers are thought to move by this sliding process, called **basal slip.** In this process, meltwater probably acts as a hydraulic jack and perhaps as a lubricant helping the ice over the rock. The source of the liquid water is related in part to the fact that the melting point of ice decreases as pressure increases. Therefore, deep within a glacier the ice may be at the melting point even though its temperature is below 0 °C.

In addition, other factors may contribute to the presence of meltwater deep within the glacier. Temperatures may be increased by plastic flow (an effect similar to frictional heating), by heat added from the Earth below, and by the refreezing of meltwater that has seeped down from above. This last process relies on the property that as water changes state from liquid to solid, heat (termed latent heat of fusion) is released.

Figure 18.8 illustrates the effects of these two basic types of glacial motion. This vertical profile through a glacier also shows that not all the ice flows forward at the same rate. Frictional drag with the bedrock floor causes the lower portions of the glacier to move more slowly.

In contrast to the lower portion of the glacier, the upper 50 meters or so are not under sufficient pressure to exhibit plastic flow. Rather, the ice in this uppermost zone is brittle and is appropriately referred to as the **zone of fracture.** The ice in the zone of fracture is carried along "piggyback" style by the ice below. When the glacier moves over irregular terrain, the zone of fracture is subjected to tension, resulting in cracks called **crevasses** (Figure 18.9). These gaping cracks may extend to depths of 50 meters and can make travel across glaciers dangerous. Below this depth, plastic flow seals them off.

Rates of Glacial Movement

Unlike streamflow, glacial movement is not obvious. If we could watch an alpine glacier move, we would see that, like the water in a river, all of the ice in the valley does not move at an equal rate. Just as friction with the valley floor slows the movement of the ice at the bottom of the glacier, the drag created by the valley walls causes the rate of flow to be greatest in the center of the glacier. This was first demonstrated by experiments during the nineteenth century, in which markers were carefully placed in a straight line across the top of a valley glacier. Periodically, the positions of the stakes were recorded, revealing the type of movement just described. More about these experiments may be found in Box 1.2 on page 13.

How rapidly does glacial ice move? Average velocities vary considerably from one glacier to another.* Some move so slowly that trees and other vegetation may become well established in the debris that has accumulated on the glacier's surface, whereas others may move at rates of up to several meters per day. For example, Byrd Glacier, an outlet glacier in Antarctica that was the subject of a 10-year study using satellite images, moved at an average rate of 750 to 800 meters per year (about 2 meters per day). Other glaciers in the study advanced at one-fourth that rate.

The advance of some glaciers is characterized by periods of extremely rapid movements called **surges.** Glaciers that exhibit such movement may flow along in an apparently normal manner, then speed up for a relatively short time before returning to

*Specialized instruments aboard satellites allow us to monitor some glaciers from space. Figure 1.B in Box 1.1 (p. 11) provides one example of monitoring the movement of Antarctica's Lambert Glacier.

FIGURE 18.9 Crevasses form in the brittle ice of the zone of fracture. They can extend to depths of 50 meters and can obviously make travel across glaciers dangerous. (Photo by Bo Tornvig/age footstock)

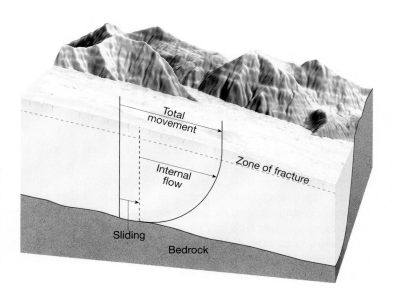

FIGURE 18.8 Vertical cross section through a glacier to illustrate ice movement. Glacial movement is divided into two components. Below about 50 meters (160 feet), ice behaves plastically and flows. In addition, the entire mass of ice may slide along the ground. The ice in the zone of fracture is carried along "piggyback" style. Notice that the rate of movement is slowest at the base of the glacier, where frictional drag is greatest.

the normal rate again. The flow rates during surges are as much as 100 times the normal rate. Evidence indicates that many glaciers may be of the surging type.

It is not yet clear whether the mechanism that triggers these rapid movements is the same for each surging-type glacier. However, researchers studying the Variegated Glacier shown in Figure 18.10 have determined that the surges of this ice mass take the form of a rapid increase in basal sliding that is caused by increases in water pressure beneath the ice. The increased water pressure at the base of the glacier acts to reduce friction between the underlying bedrock and the moving ice. The pressure buildup, in turn, is related to changes in the system of passageways that conduct water along the glacier's bed and deliver it as outflow to the terminus.

Budget of a Glacier

Snow is the raw material from which glacial ice originates; therefore, glaciers form in areas where more snow falls in winter than melts during the summer. Glaciers are constantly gaining and losing ice.

August 1964

August 1965

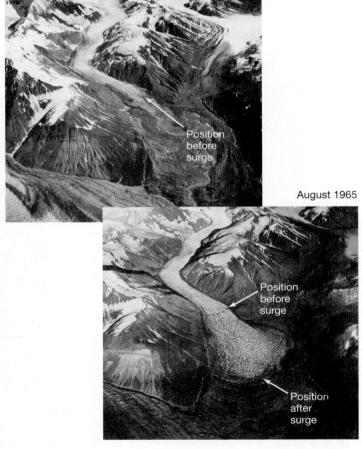

FIGURE 18.10 The surge of Variegated Glacier, a valley glacier near Yakutat, Alaska, northwest of Juneau, is captured in these two aerial photographs taken one year apart. During a surge, ice velocities in Variegated Glacier are 20 to 50 times greater than during a quiescent phase. (Photos by Austin Post, U.S. Geological Survey)

Glacial Zones Snow accumulation and ice formation occur in the **zone of accumulation.** Its outer limits are defined by the snowline. As noted earlier, the elevation of the snowline varies greatly, from sea level in polar regions to altitudes approaching 5000 meters near the equator. Above the snowline, in the zone of accumulation, the addition of snow thickens the glacier and promotes movement. Beyond the snowline is the **zone of wastage.** Here there is a net loss to the glacier as all of the snow from the previous winter melts, as does some of the glacial ice (Figure 18.11).

In addition to melting, glaciers also waste as large pieces of ice break off the front of the glacier in a process called **calving.** Calving creates *icebergs* in places where the glacier has reached the sea or a lake (Figure 18.12). Because icebergs are just slightly less dense than seawater, they float very low in the water, with more than 80 percent of their mass submerged. Along the margins of Antarctica's ice shelves, calving is the primary means by which these masses lose ice. The relatively flat icebergs produced here can be several kilometers across and 600 meters thick. By comparison, thousands of irregularly shaped icebergs are produced by outlet glaciers flowing from the margins of the Greenland Ice Sheet. Many drift southward and find their way into the North Atlantic, where they can pose a hazard to navigation.

Whether the margin of a glacier is advancing, retreating, or remaining stationary depends on the budget of the glacier. The **glacial budget** is the balance, or lack of balance, between accumulation at the upper end of the glacier and loss at the lower end. This loss is termed **ablation.** If ice accumulation exceeds ablation, the glacial front advances until the two factors balance. When this happens, the terminus of the glacier becomes stationary.

If a warming trend increases ablation and/or if a drop in snowfall decreases accumulation, the ice front will retreat (Box 18.1). As the terminus of the glacier retreats, the extent of the zone of wastage diminishes. Therefore, in time a new balance will be reached between accumulation and wastage, and the ice front will again become stationary.

Whether the margin of a glacier is advancing, retreating, or stationary, the ice within the glacier continues to flow forward. In the case of a receding glacier, the ice still flows forward, but not rapidly enough to offset ablation. This point is illustrated well in Figure 1.D on page 13. As the line of stakes within Rhone Glacier continued to move downvalley, the terminus of the glacier slowly retreated upvalley.

Greenland's Glacial Budget Is the Greenland Ice Sheet growing or shrinking? Answering this relatively simple question is not a simple task and is much more difficult for ice sheets than for valley glaciers. Accurate answers require large quantities of data. Conducting field studies on an ice sheet poses significant challenges— sampling a huge area with an extreme climate in a remote part of the world over an extended time span. Data can only be collected for short periods each year at widely scattered sites. The information gathered by such studies, though useful, provided only a glimpse into what was happening on the Greenland Ice Sheet. However, in recent years satellite observations have started to provide scientists with the data they need to begin to determine whether the margins of the ice sheet are melting faster than the interior is gaining ice (net loss) or whether the interior is adding new ice faster than the edges are wasting away (net gain).

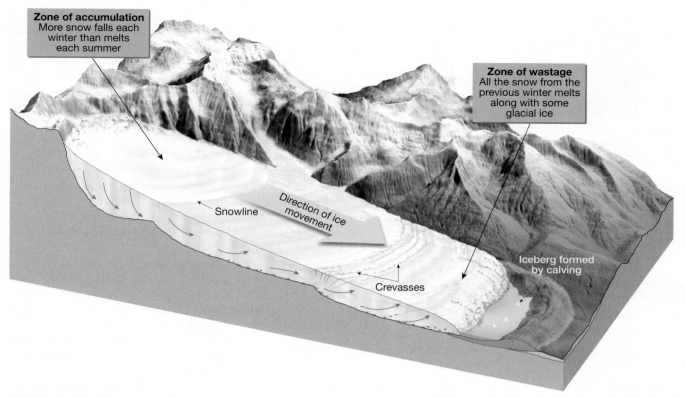

Zone of accumulation
More snow falls each winter than melts each summer

Zone of wastage
All the snow from the previous winter melts along with some glacial ice

Direction of ice movement

Snowline

Crevasses

Iceberg formed by calving

FIGURE 18.11 The snowline separates the zone of accumulation and the zone of wastage. Above the snowline, more snow falls each winter than melts each summer. Below the snowline, the snow from the previous winter completely melts as does some of the underlying ice. Whether the margin of a glacier advances, retreats, or remains stationary depends on the balance between accumulation and wastage (ablation). When a glacier moves across irregular terrain, *crevasses* form in the brittle portion.

Geologist's Sketch

FIGURE 18.12 Icebergs are created when large pieces calve from the front of a glacier after it reaches a water body. In the large image, ice is calving from the terminus of Hubbard Glacier in Wrangell-St. Elias National Park, Alaska. (Photo by Bernhard Edmaier/Photo Researchers, Inc.) As the drawing illustrates, only about 20 percent (or less) of an iceberg protrudes above the waterline. The smaller photo shows an iceberg in Canada's Labrador Sea. Note the red kayak for scale. (Photo by Radius Images/Photolibrary)

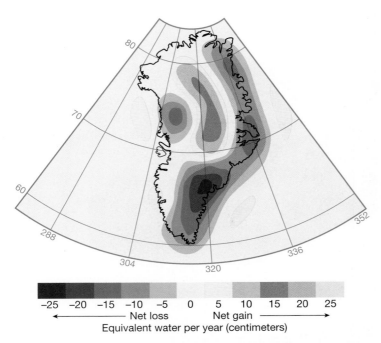

-25 -20 -15 -10 -5   0   5   10 15 20 25

← Net loss   Net gain →

Equivalent water per year (centimeters)

FIGURE 18.13 This map was compiled using observations from NASA's Gravity Recovery and Climate Experiment (GRACE) satellites. It shows where Greenland gained mass during the 2003–2005 study period and where it lost mass. While the equivalent of 10 to 15 centimeters of water per year accumulated over the core of the island (red and orange areas), an even larger area experienced losses (blue) of between 5 and 25 centimeters per year. Overall, Greenland lost 20 percent more mass than it received in snowfall each year. These results are consistent with trends in ice loss that other types of observations of Greenland have documented. (NASA)

The map in Figure 18.13 was prepared using satellite data for the years 2003 to 2005. During that span, the island's coastal area lost 155 gigatons (41 cubic miles) of ice per year, while snow accumulation in the interior of the ice sheet was only 54 gigatons per year. Clearly, during this span, the processes resulting in a net loss of mass were dominant.

GLACIAL EROSION

Glaciers are capable of great erosion. For anyone who has observed the terminus of an alpine glacier, the evidence of its erosive force is clear (Figure 18.14). You can witness firsthand the release of rock material of various sizes from the ice as it melts. All signs lead to the conclusion that the ice has scraped, scoured, and torn rock from the floor and walls of the valley and carried it downvalley. It should be pointed out, however, that in mountainous regions mass-wasting processes also make substantial contributions to the sediment load of a glacier. A glance back at the Chapter 15 opening photo on page 405 provides a striking example.

Once rock debris is acquired by the glacier, the enormous competency

of ice will not allow the debris to settle out like the load carried by a stream or by the wind. Indeed, as a medium of sediment transport, ice has no equal. Consequently, glaciers can carry huge blocks that no other erosional agent could possibly budge (Figure 18.15). Although today's glaciers are of limited importance as erosional agents, many landscapes that were modified by the widespread glaciers of the most recent Ice Age still reflect, to a high degree, the work of ice.

Glaciers erode the land primarily in two ways—plucking and abrasion. First, as a glacier flows over a fractured bedrock surface, it loosens and lifts blocks of rock and incorporates them into the ice. This process, known as **plucking,** occurs when meltwater penetrates the cracks and joints of bedrock beneath a glacier and freezes. As the water expands, it exerts tremendous leverage that pries the rock loose. In this manner, sediment of all sizes becomes part of the glacier's load.

The second major erosional process is **abrasion** (Figure 18.16). As the ice and its load of rock fragments slide over bedrock, they function like sandpaper to smooth and polish the surface below. The pulverized rock produced by the glacial "grist mill" is appropriately called **rock flour.** So much rock flour may be produced that meltwater streams flowing out of a glacier often have the grayish appearance of skim milk and offer visible evidence of the grinding power of ice.

When the ice at the bottom of a glacier contains large rock fragments, long scratches and grooves called **glacial striations** may even be gouged into the bedrock (Figure 18.16A). These linear grooves provide clues to the direction of ice flow. By mapping the striations over large areas, patterns of glacial flow can often be reconstructed. On the other hand, not all abrasive action produces striations. The rock surfaces over which the glacier moves may also become highly polished by the ice and its load of finer particles. The broad expanses of smoothly polished granite in Yosemite National Park provide an excellent example (Figure 18.16B).

As is the case with other agents of erosion, the rate of glacial erosion is highly variable. This differential erosion by ice

FIGURE 18.14 Glaciers are capable of great erosion. As the terminus of this Alaskan glacier wastes away, it deposits large quantities of unsorted sediment called *till.* (Photos by Michael Collier)

FIGURE 18.15   A large, glacially transported boulder in Denali National Park, Alaska. Such boulders are called *glacial erratics*. (Photo by Michael Collier)

is largely controlled by four factors: (1) rate of glacial movement; (2) thickness of the ice; (3) shape, abundance, and hardness of the rock fragments contained in the ice at the base of the glacier; and (4) the erodibility of the surface beneath the glacier. Variations in any or all of these factors from time to time and/or from place to place mean that the features, effects, and degree of landscape modification in glaciated regions can vary greatly.

LANDFORMS CREATED BY GLACIAL EROSION

GLACIERS AND GLACIATION
▶ Reviewing Glacial Features

The erosional effects of valley glaciers and ice sheets are quite different. A visitor to a glaciated mountain region is likely to see a sharp and angular topography. The reason is that as alpine glaciers move downvalley, they tend to accentuate the irregularities of the mountain landscape by creating steeper canyon walls and making bold peaks even more jagged. By contrast, continental ice sheets generally override the terrain and hence subdue rather than accentuate the irregularities they encounter. Although the erosional potential of ice sheets is enormous, landforms carved by these huge ice masses usually do not inspire the same wonderment and awe as do the erosional features created by valley glaciers. Much of the rugged mountain scenery so celebrated for its majestic beauty is the product of erosion by alpine glaciers. Figure 18.17 shows a hypothetical mountain area before, during, and after glaciation. You will refer to this often in the following discussion.

Glaciated Valleys

A hike up a glaciated valley reveals a number of striking ice-created features. The valley itself is often a dramatic sight. Unlike streams, which create their own valleys, glaciers take the path of least resistance by following the course of existing stream valleys. Prior to glaciation, mountain valleys are characteristically narrow and V-shaped because streams are well above base level and are therefore downcutting. However, during glaciation these narrow valleys undergo a transformation as the glacier widens and deepens them, creating a U-shaped **glacial trough** (Figure 18.17C and Figure 18.18). In addition to producing a broader and deeper valley, the glacier also straightens the valley. As ice flows around sharp curves, its great erosional force removes the spurs of land that extend into the valley. The results of this activity are triangular-shaped cliffs called **truncated spurs** (Figure 18.17).

The amount of glacial erosion that takes place in different valleys in a mountainous area varies. Prior to glaciation, the mouths of tributary streams join the main (trunk) valley at the elevation of

FIGURE 18.16   **A.** Glacial abrasion created the scratches and grooves in this bedrock. Cove Glacier, Prince William Sound, northeast of Whittier, Alaska. **B.** Glacially polished granite in California's Yosemite National Park. (Photos by Michael Collier)

A.

B.

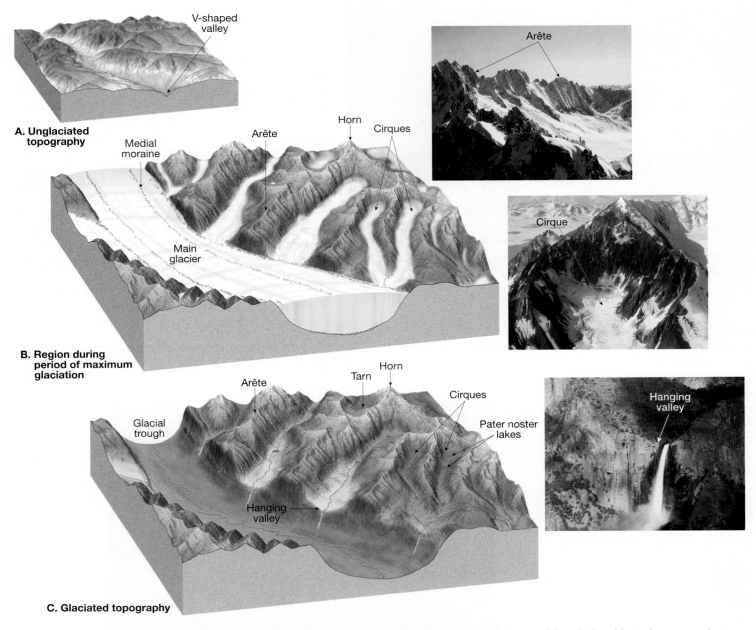

FIGURE 18.17   Prior to glaciation, a mountain valley is typically narrow and V-shaped. During glaciation, an alpine glacier widens, deepens, and straightens the valley, creating a U-shaped glacial trough. These diagrams of a hypothetical area show the development of erosional landforms created by alpine glaciers. The unglaciated landscape in part **A** is modified by valley glaciers in part **B**. After the ice recedes, in part **C**, the terrain looks very different than before glaciation. (Cirque photo by Marli Miller, Hanging Valley photo by Marc Muench, Arête photo by James E. Patterson)

the stream in that valley. During glaciation, the amount of ice flowing through the main valley can be much greater than the amount advancing down each tributary. Consequently, the valley containing the trunk glacier is eroded deeper than the smaller valleys that feed it. Thus, when the glaciers eventually recede, the valleys of tributary glaciers are left standing above the main glacial trough and are termed **hanging valleys** (Figure 18.17). Rivers flowing through hanging valleys may produce spectacular waterfalls, such as those in Yosemite National Park (Figure 18.17).

As hikers walk up a glacial trough, they may pass a series of bedrock depressions on the valley floor that were probably created by plucking and then scoured by the abrasive force of the ice. If these depressions are filled with water, they are called

pater noster lakes (Figure 18.17). The Latin name means "our Father" and is a reference to a string of rosary beads.

At the head of a glacial valley is a very characteristic and often imposing feature called a **cirque.** As the photo in Figure 18.17 illustrates, these bowl-shaped depressions have precipitous walls on three sides but are open on the downvalley side. The cirque is the focal point of the glacier's growth because it is the area of snow accumulation and ice formation. Cirques begin as irregularities in the mountainside that are subsequently enlarged by the frost wedging and plucking that occur along the sides and bottom of the glacier. After the glacier has melted away, the cirque basin is often occupied by a small lake called a **tarn** (Figure 18.17).

EARTH AS A SYSTEM

BOX 18.1

Glaciers in Retreat

Glaciers are sensitive to changes in temperature and precipitation and therefore can provide clues about changes in climate. In 1991, the mummified remains of a Neolithic man were discovered by hikers in the Alps. The remains had been locked and preserved in glacial ice for more than 5000 years. The discovery and subsequent analysis provided a fascinating glimpse of a very different time and fascinated the world. The discovery also meant that the

glacier had reached a 5000 year minimum. The shrinking glacier that held the "ice man" is not unique.

In North America, one of the most visited glaciers is Athabasca Glacier, a 7-kilometer tongue of ice that spills down from the Columbia Icefield in the rugged Canadian Rockies. People who return to the area a few years after their first visit will likely notice a change in the position of the end of the glacier (Figure 18.A). Over the past 125 years, Athabasca Glacier has lost about half of its volume and receded more than 1.5 kilometers (nearly 1 mile), leaving a thick and uneven layer of rocky debris in its place.

With few exceptions, glaciers around the world have been retreating at unprecedented rates over the last century (Figure 18.B). Some have disappeared altogether. One study that used satellite data, reported that between 1961 and 2004, glaciers around the world (outside of the Greenland and Antarctic ice sheets) lost an estimated 8000 cubic kilometers (1900 cubic miles) of ice.* That is approximately enough ice to cover a 2-kilometer wide swath of land between New York and Los Angeles with a glacier 1-kilometer thick.

*Mark B. Dyurgerov and Mark F. Meier. *Glaciers and the Changing Earth System,* Boulder, Colorado: INSTAAR Occasional Paper No. 58, 2005.

FIGURE 18.A Athabasca Glacier in Jasper National Park in the Canadian Rockies. This photo was taken in August 2005. The date marker shows where the snout of the glacier was in 1992. The glacier is currently receding at a rate of 2–3 meters per year. (Photo by Hughrocks)

FIGURE 18.B Two images taken 63 years apart from the same spot in Alaska's Glacier Bay National Park. Muir Glacier, which is prominent in the 1941 photo, has retreated out of the field of view in the 2004 image. Also Riggs Glacier (upper right) has thinned and retreated significantly. (Photos courtesy of National Snow and Ice Data Center)

FIGURE 18.18 Prior to glaciation, a mountain valley is typically narrow and V-shaped. During glaciation, an alpine glacier widens, deepens, and straightens the valley, creating the U-shaped glacial trough seen here. Romsdal Valley, Norway. (Photo by Hal Beral/Corbis)

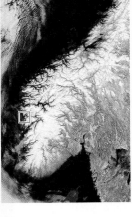

FIGURE 18.19 The coast of Norway is known for its many fiords. Frequently these ice-sculpted inlets of the sea are hundreds of meters deep. (Satellite images courtesy of NASA; photo by Wolfgang Meier/Photolibrary)

Sometimes, when two glaciers exist on opposite sides of a divide, each flowing away from the other, the dividing ridge between their cirques is largely eliminated as plucking and frost action enlarge each one. When this occurs, the two glacial troughs come to intersect, creating a gap or pass from one valley into the other. Such a feature is termed a **col**. Some important and well-known mountain passes that are cols include St. Gotthard Pass in the Swiss Alps, Tioga Pass in California's Sierra Nevada, and Berthoud Pass in the Colorado Rockies.

Before leaving the topic of glacial troughs and their associated features, one more rather well-known feature should be discussed: fiords. **Fiords** are deep, often spectacular steep-sided inlets of the sea that are present at high latitudes where mountains are adjacent to the ocean (Figure 18.19). They are drowned glacial troughs that became submerged as the ice left the valley and sea level rose following the Ice Age. The depths of fiords may exceed 1000 meters (3300 feet). However, the great depths of these flooded troughs are only partly explained by the post–Ice Age rise in sea level. Unlike the situation governing the downward erosional work of rivers, sea level does

502

not act as base level for glaciers. As a consequence, glaciers are capable of eroding their beds far below the surface of the sea. For example, a 300-meter-thick (1000-foot-thick) glacier can carve its valley floor more than 250 meters (820 feet) below sea level before downward erosion ceases and the ice begins to float. Norway, British Columbia, Greenland, New Zealand, Chile, and Alaska all have coastlines characterized by fiords.

Arêtes and Horns

A visit to the Alps and the Northern Rockies, or many other scenic mountain landscapes carved by valley glaciers, reveals not only glacial troughs, cirques, pater noster lakes, and the other related features just discussed. You also are likely to see sinuous sharp-edged ridges called **arêtes** (French for *knife-edge*) and sharp pyramid-like peaks called **horns** projecting above the surroundings. Both features can originate from the same basic process, the enlargement of cirques produced by plucking and frost action (Figure 18.17). In the case of the spires of rock called horns, groups of cirques around a single high mountain are responsible. As the cirques enlarge and converge, an isolated horn is produced. The most famous example is the Matterhorn in the Swiss Alps (Figure 18.20).

Arêtes can be formed in a similar manner, except that the cirques are not clustered around a point but rather exist on opposite sides of a divide. As the cirques grow, the divide separating them is reduced to a narrow knifelike partition. An arête, however, may also be created in another way. When two glaciers occupy parallel valleys, an arête can form when the divide sepa-

FIGURE 18.20 Horns are sharp, pyramidlike peaks that are fashioned by alpine glaciers. This example is the famous Matterhorn in the Swiss Alps. (Photo by Gavriel Jecan/Art Wolfe, Inc.)

FIGURE 18.21 Roche Moutonnée, in Yosemite National Park, California. The gentle slope was abraded and the steep side was plucked. The ice moved from right to left. (Photo by E. J. Tarbuck)

Geologist's Sketch

rating the moving tongues of ice is progressively narrowed as the glaciers scour and widen their adjacent valleys.

Roches Moutonnées

In many glaciated landscapes, but most frequently where continental ice sheets have modified the terrain, the ice carves small streamlined hills from protruding bedrock knobs. Such as asymmetrical knob of bedrock is called a **roche moutonnée** (French for *sheep rock*). They are formed when glacial abrasion smoothes the gentle slope facing the oncoming ice sheet and plucking steepens the opposite side as the ice rides over the knob (Figure 18.21). Roches moutonnées indicate the direction of glacial flow because the gentler slope is generally on the side from which the ice advanced.

GLACIAL DEPOSITS

GLACIERS AND GLACIATION
▶ Reviewing Glacial Features

Glaciers pick up and transport a huge load of debris as they slowly advance across the land. Ultimately these materials are deposited when the ice melts. In regions where glacial sediment is deposited, it can play a truly significant role in forming the physical landscape. For example, in many areas once covered by the continental ice sheets of the recent Ice Age, the bedrock is rarely exposed because glacial deposits that are tens or even hundreds of meters thick completely mantle the terrain. The general effect of these deposits is to reduce the local relief and thus level the topography. Indeed, rural country scenes that are familiar to many of us—rocky pastures in New England, wheat fields in the Dakotas, rolling farmland in the Midwest—result directly from glacial deposition.

Long before the theory of an extensive Ice Age was ever proposed, much of the soil and rock debris covering portions of Europe was recognized as coming from somewhere else. At the time, these "foreign" materials were believed to have been "drifted" into their present positions by floating ice during an ancient flood. As a consequence, the term *drift* was applied to this sediment. Although rooted in an incorrect concept, this term was so well established by the time the true glacial origin of the debris became widely recognized that it remained in the basic glacial vocabulary. Today **glacial drift** is an all-embracing term for sediments of glacial origin, no matter how, where, or in what shape they were deposited.

One of the features distinguishing drift from sediments laid down by other erosional agents is that glacial deposits consist primarily of mechanically weathered rock debris that underwent little or no chemical weathering prior to deposition. Thus, minerals that are notably prone to chemical decomposition, such as hornblende and the plagioclase feldspars, are often conspicuous components in glacial sediments.

Glacial drift is divided by geologists into two distinct types: (1) materials deposited directly by the glacier, which are known as *till*, and (2) sediments laid down by glacial meltwater, called *stratified drift*. We will now look at landforms made of each type.

Landforms Made of Till

Till is deposited as glacial ice melts and drops its load of rock fragments. Unlike moving water and wind, ice cannot sort the sediment it carries; therefore, deposits of till are characteristically unsorted mixtures of many particle sizes (Figure 18.22). A close examination of this sediment shows that many of the pieces are scratched and polished as the result of being dragged along by the glacier. Such pieces help distinguish till from other deposits that are a mixture of different sediment sizes, such as material from a debris flow or a rockslide.

Boulders found in the till or lying free on the surface are called **glacial erratics** if they are different from the bedrock below (see Figure 18.15). Of course, this means that they must have been derived from a source outside the area where they are found. Although the locality of origin for most erratics is unknown, the origin of some can be determined. In many cases, boulders were transported as far as 500 kilometers (300 miles) from their source area and, in a few instances, more than 1000 kilometers (600 miles). Therefore, by studying glacial erratics as well as the mineral composition of the remaining till, geologists are sometimes able to trace the path of a lobe of ice.

In portions of New England and other areas, erratics dot pastures and farm fields. In fact, in some places these large rocks were cleared from fields and piled to make fences and walls. Keeping the fields clear, however, is an ongoing chore because each spring newly exposed erratics appear. Wintertime frost heaving has lifted them to the surface.

Lateral and Medial Moraines

The most common term for landforms made of glacial deposits is *moraine*. Originally this term was used by French peasants when referring to the ridges and embankments of rock debris found

FIGURE 18.22 Glacial till is an unsorted mixture of many different sediment sizes. A close examination often reveals cobbles that have been scratched as they were dragged along by the glacier. (Photos by E. J. Tarbuck)

Close up
of cobble

The second type of moraine that is unique to alpine glaciers is the **medial moraine.** Medial moraines are created when two alpine glaciers coalesce to form a single ice stream. The till that was once carried along the sides of each glacier joins to form a single dark stripe of debris within the newly enlarged glacier. Figure 18.23 illustrates this nicely. The creation of these dark stripes within the ice stream is one obvious proof that glacial ice moves, because the moraine could not form if the ice did not flow downvalley. It is quite common to see several medial moraines within a single large alpine glacier, because a streak will form whenever a tributary glacier joins the main valley.

End and Ground Moraines

An **end moraine** is a ridge of till that forms at the terminus of a glacier, and is characteristic of ice sheets and valley glaciers alike. These relatively common landforms are deposited when a state of equilibrium is attained between ablation and ice accumulation. That is, the end moraine forms when the ice is melting and evaporating near the end of the glacier at a rate equal to the forward advance of the glacier from its region of nourishment. Although the terminus of the glacier is now stationary, the ice continues to flow forward, delivering a continuous supply of sediment in the same manner a conveyor belt delivers goods to the end of a production line. As the ice melts, the till is dropped and the end moraine grows. The longer the ice front remains stable, the larger the ridge of till will become.

Eventually the time comes when ablation exceeds nourishment. At this point, the front of the glacier begins to recede in the direction from which it originally advanced. However, as the ice front retreats, the conveyor-belt action of the glacier continues to provide fresh supplies of sediment to the terminus. In this manner a large quantity of till is deposited as the ice melts away, creating a rock-strewn, undulating plain. This gently rolling layer of till deposited as the ice front recedes is termed **ground moraine.** Ground moraine has a leveling effect, filling in low spots and clogging old stream channels, often leading to a derangement of the existing drainage system. In areas where this layer of till is still relatively fresh, such as the northern Great Lakes region, poorly drained swampy lands are quite common.

Periodically, a glacier will retreat to a point where ablation and nourishment once again balance. When this happens, the ice front stabilizes and a new end moraine forms.

The pattern of end moraine formation and ground moraine deposition may be repeated many times before the glacier has completely vanished. Such a pattern is illustrated by Figure 18.24. It should be pointed out that the outermost end moraine marks the limit of the glacial advance. Because of its special status, this end moraine is also called the **terminal moraine.** On the other hand, the end moraines that were created as the ice front occasionally stabilized during retreat are termed **recessional moraines.** Note that both terminal and recessional moraines are essentially alike; the only difference between them is their relative positions.

End moraines deposited by the most recent major stage of Ice Age glaciation are prominent features in many parts of the

near the margins of glaciers in the French Alps. Today, however, moraine has a broader meaning, because it is applied to a number of landforms, all of which are composed primarily of till.

Alpine glaciers produce two types of moraines that occur exclusively in mountain valleys. The first of these is called a **lateral moraine.** As we learned earlier, when an alpine glacier moves downvalley, the ice erodes the sides of the valley with great efficiency. In addition, large quantities of debris are added to the glacier's surface as rubble falls or slides from higher up on the valley walls and collects on the margins of the moving ice (Figure 18.23). When the ice eventually melts, this accumulation of debris is dropped next to the valley walls. These ridges of till paralleling the sides of the valley constitute the lateral moraines.

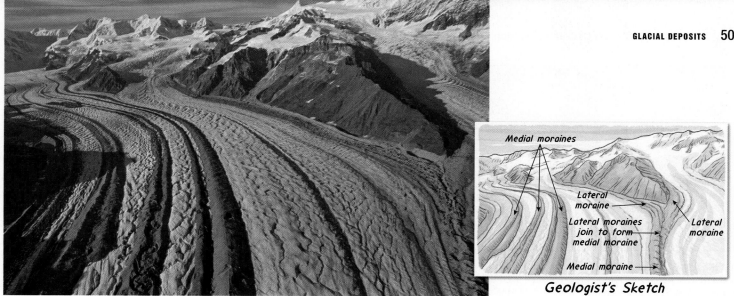

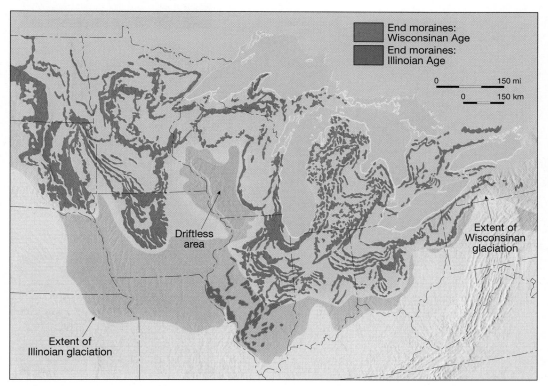

FIGURE 18.23 Medial moraines form when the lateral moraines of merging valley glaciers join. Saint Elias National Park, Alaska. (Photo by Tom Bean/Alamy)

Midwest and Northeast. In Wisconsin the wooded, hilly terrain of the Kettle Moraine near Milwaukee is a particularly picturesque example. A well-known example in the Northeast is Long Island. This linear strip of glacial sediment that extends northeastward from New York City is part of an end moraine complex that stretches from eastern Pennsylvania to Cape Cod, Massachusetts (Figure 18.25). The end moraines that make up Long Island represent materials that were deposited by a continental ice sheet in the relatively shallow waters off the coast and built up many meters above sea level. Long Island Sound, the narrow body of water separating the island and the mainland,

was not built up as much by glacial deposition and was therefore subsequently flooded by the rising sea following the Ice Age.

Figure 18.26 represents a hypothetical area during glaciation and following the retreat of ice sheets. It shows the moraines that were described in this section, as well as the depositional features that are discussed in the sections that follow. This figure depicts landscape features similar to what might be encountered if you were traveling in the upper Midwest or New England. As you read upcoming sections dealing with other glacial deposits, you will be referred to this figure several times.

Drumlins

Moraines are not the only landforms deposited by glaciers. In some areas that were once covered by continental ice sheets, a special variety of glacial landscape exists—one characterized by smooth, elongate, parallel hills called **drumlins** (Figure 18.26). Certainly one of the best-known drumlins is Bunker Hill in Boston, the site of the famous Revolutionary War battle in 1775.

An examination of Bunker Hill or other less famous drumlins would reveal that drumlins are streamlined, asymmetrical hills composed largely of till. They range in height from about 15 to 50 meters and may be up to 1 kilometer long. The steep side of the hill faces the direction from which the ice advanced, whereas the gentler, longer slope points in the direction the ice moved. Drumlins are not found as isolated landforms but rather occur in clusters called *drumlin fields*

FIGURE 18.24 End moraines of the Great Lakes region. Those deposited during the most recent (Wisconsinan) stage are most prominent.

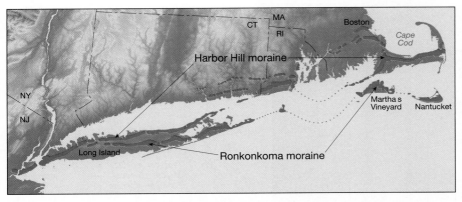

FIGURE 18.25　End moraines make up substantial parts of Long Island, Cape Cod, Martha's Vineyard, and Nantucket. Although portions are submerged, the Ronkonkoma Moraine (a terminal moraine) extends through central Long Island, Martha's Vineyard, and Nantucket. It was deposited about 20,000 years ago. The recessional Harbor Hill Moraine, which formed about 14,000 years ago, extends along the north shore of Long Island, through southern Rhode Island and Cape Cod.

glaciers advance over previously deposited drift and reshape the material.

LANDFORMS MADE OF STRATIFIED DRIFT

As the name implies, **stratified drift** is sorted according to the size and weight of the particles. Because ice is not capable of such sorting activity, these materials are not deposited directly by the glacier as till is, but instead reflect the sorting action of glacial meltwater. Accumulations of stratified drift often consist largely of sand and gravel—that is, bed-load material—because the finer rock flour remains suspended and therefore is commonly carried far from the glacier by the meltwater streams.

(Figure 18.27). One such cluster, east of Rochester, New York, is estimated to contain about 10,000 drumlins. Although drumlin formation is not fully understood, their streamlined shape indicates that they were molded in the zone of plastic flow within an active glacier. It is believed that many drumlins originate when

Outwash Plains and Valley Trains

At the same time that an end moraine is forming, water from the melting glacier cascades over the till, sweeping some of it out in front of the growing ridge of unsorted debris. Meltwater gener-

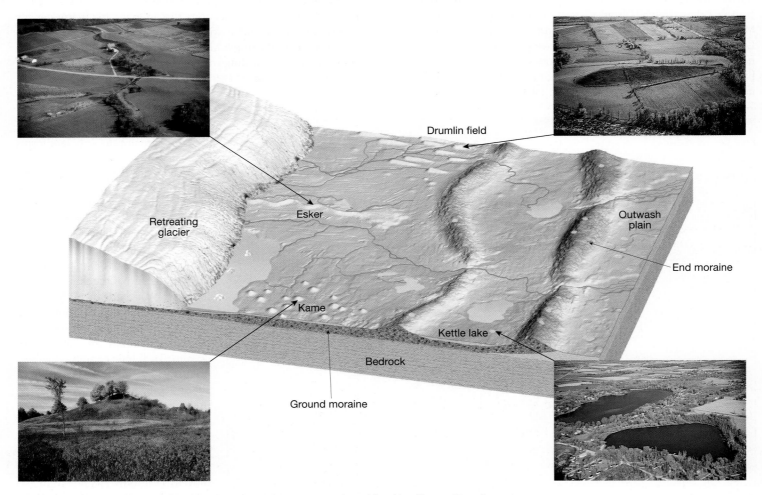

FIGURE 18.26　This hypothetical area illustrates many common depositional landforms. (Drumlin photo courtesy of Ward's Natural Science Establishment; kame, esker, and kettle photos by Richard R. Jacobs/JLM Visuals)

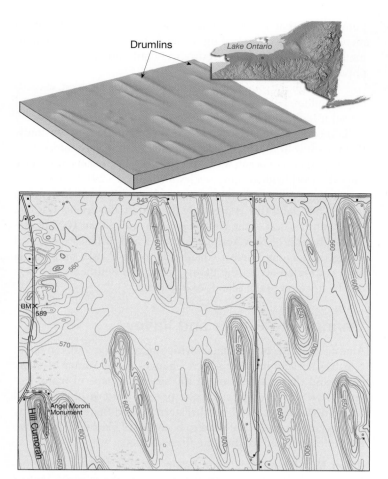

FIGURE 18.27 Portion of a drumlin field shown on the Palmyra, New York, 7.5-minute topographic map. North is at the top. The drumlins are steepest on the north side, indicating that the ice advanced from this direction.

ally emerges from the ice in rapidly moving streams that are often choked with suspended material and carry a substantial bed load as well. As the water leaves the glacier, it moves onto the relatively flat surface beyond and rapidly loses velocity. As a consequence, much of its bed load is dropped and the meltwater begins weaving a complex pattern of braided channels (Figure 18.26). In this way, a broad, ramplike surface composed of stratified drift is built adjacent to the downstream edge of most end moraines. When the feature is formed in association with an ice sheet, it is termed an **outwash plain,** and when largely confined to a mountain valley, it is usually referred to as a **valley train.**

Outwash plains and valley trains often are pockmarked with basins or depressions known as **kettles** (Figure 18.26). Kettles also occur in deposits of till. Kettles are formed when blocks of stagnant ice become wholly or partly buried in drift and eventually melt, leaving pits in the glacial sediment. Although most kettles do not exceed 2 kilometers in diameter, some with diameters exceeding 10 kilometers occur in Minnesota. Likewise, the typical depth of most kettles is less than 10 meters, although the vertical dimensions of some approach 50 meters. In many cases water eventually fills the depression and forms a pond or lake. One well-known example is Walden Pond near Concord, Massachusetts. It is here that Henry David Thoreau lived alone for

two years in the 1840s and about which he wrote his famous book *Walden; or, Life in the Woods.*

Ice-Contact Deposits

When the melting terminus of a glacier shrinks to a critical point, flow virtually stops and the ice becomes stagnant. Meltwater that flows over, within, and at the base of the motionless ice lays down deposits of stratified drift. Then, as the supporting ice melts away, the stratified sediment is left behind in the form of hills, terraces, and ridges. Such accumulations are collectively termed **ice-contact deposits** and are classified according to their shapes.

When the ice-contact stratified drift is in the form of a mound or steep-sided hill, it is called a **kame** (Figure 18.26). Some kames represent bodies of sediment deposited by meltwater in openings within, or depressions on top of, the ice. Others originate as deltas or fans built outward from the ice by meltwater streams. Later, when the stagnant ice melts away, these various accumulations of sediment collapse to form isolated, irregular mounds.

When glacial ice occupies a valley, **kame terraces** may be built along the sides of the valley. These features commonly are narrow masses of stratified drift laid down between the glacier and the side of the valley by streams that drop debris along the margins of the shrinking ice mass.

A third type of ice-contact deposit is a long, narrow, sinuous ridge composed largely of sand and gravel. Some are more than 100 meters high with lengths in excess of 100 kilometers. The dimensions of many others are far less spectacular. Known as **eskers,** these ridges are deposited by meltwater rivers flowing within, on top of, and beneath a mass of motionless, stagnant glacial ice (Figure 18.26). Many sediment sizes are carried by the torrents of meltwater in the ice-banked channels, but only the coarser material can settle out of the turbulent stream.

OTHER EFFECTS OF ICE-AGE GLACIERS

In addition to the massive erosional and depositional work carried on by Pleistocene glaciers, the ice sheets had other effects, sometimes profound, on the landscape. For example, as the ice advanced and retreated, animals and plants were forced to migrate. This led to stresses that some organisms could not tolerate. Hence, a number of plants and animals became extinct.

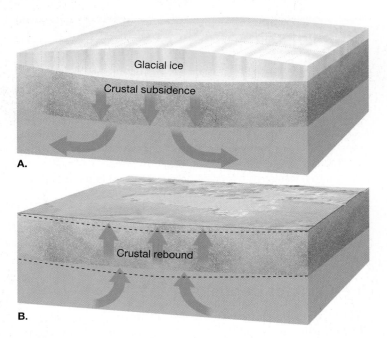

FIGURE 18.28   Simplified illustration showing crustal subsidence and rebound resulting from the addition and removal of continental ice sheets. **A.** In northern Canada and Scandinavia, where the greatest accumulation of glacial ice occurred, the added weight caused downwarping of the crust. **B.** Ever since the ice melted, there has been gradual uplift or rebound of the crust.

Other effects of Ice-Age glaciers that are described in this section involve adjustments in Earth's crust due to the addition and removal of ice and sea-level changes associated with the formation and melting of ice sheets. The advance and retreat of ice sheets also led to significant changes in the routes taken by rivers. In some regions, glaciers acted as dams that created large lakes. When these ice dams failed, the effects on the landscape were profound. In areas that today are deserts, lakes of another type, called pluvial lakes, formed during the Pleistocene.

Crustal Subsidence and Rebound

In areas that were major centers of ice accumulation, such as Scandinavia and the Canadian Shield, the land has been slowly rising over the past several thousand years. Uplifting of almost 300 meters (1000 feet) has occurred in the Hudson Bay region. This, too, is the result of the continental ice sheets. But how can glacial ice cause such vertical crustal movement? We now understand that the land is rising because the added weight of the 3-kilometer-thick (2-mile-thick) mass of ice caused downwarping of Earth's crust. Following the removal of this immense load, the crust has been adjusting by gradually rebounding upward ever since (Figure 18.28).*

Sea-Level Changes

One of the more interesting and perhaps dramatic effects of the Ice Age was the fall and rise of sea level that accompanied the advance and retreat of the glaciers. Earlier in the chapter it was

*For a more complete discussion of this concept, termed *isostatic adjustment,* see the discussion of isostasy in the section on "Vertical Movements of the Crust" in Chapter 14.

pointed out that sea level would rise by an estimated 60 or 70 meters if the water locked up in the Antarctic Ice Sheet were to melt completely. Such an occurrence would flood many densely populated coastal areas.

Although the total volume of glacial ice today is great, exceeding 25 million cubic kilometers, during the Ice Age the volume of glacial ice amounted to about 70 million cubic kilometers, or 45 million cubic kilometers more than at present. Because we know that the snow from which glaciers are made ultimately comes from the evaporation of ocean water, the growth of ice sheets must have caused a worldwide drop in sea level (Figure 18.29). Indeed, estimates suggest that sea level was as much as 100 meters lower than today. Thus, land that is presently flooded by the oceans was dry. The Atlantic Coast of the United States lay more than 100 kilometers to the east of New York City; France and Britain were joined where the famous English Channel is today; Alaska and Siberia were connected across the Bering Strait; and Southeast Asia was tied by dry land to the islands of Indonesia.

Changes to Rivers and Valleys

Among the effects associated with the advance and retreat of North American ice sheets were changes in the routes of many rivers and the modification in the size and shape of many valleys. If we are to understand the present pattern of rivers and lakes in the central and northeastern United States (and many other places as well), we need to be aware of glacial history. Two examples are used to illustrate these effects.

Upper Mississippi Drainage Basin   Figure 18.30A shows the familiar present-day pattern of rivers in the central United

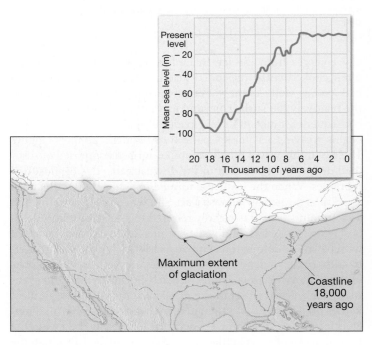

FIGURE 18.29   Changing sea level during the past 20,000 years. About 18,000 years ago, when the most recent ice advance was at a maximum, sea level was nearly 100 meters lower than at present. Thus, land that is presently covered by the ocean was exposed and the shoreline looked very different than today. As the ice sheets melted, sea level rose and the shoreline shifted.

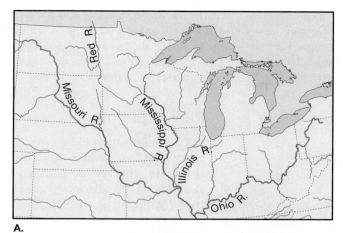

A.

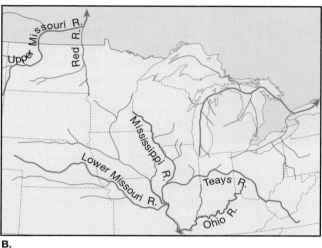

B.

FIGURE 18.30   A. This map shows the Great Lakes and the familiar present-day pattern of rivers in the central United States. Pleistocene ice sheets played a major role in creating this pattern. **B.** Reconstruction of drainage systems in the central United States prior to the Ice Age. The pattern was very different from today, and there were no Great Lakes.

States, with the Missouri, Ohio, and Illinois rivers as major tributaries to the Mississippi. Figure 18.30B depicts drainage systems in this region prior to the Ice Age. The pattern is *very* different from the present. This remarkable transformation of river systems resulted from the advance and retreat of the ice sheets.

Notice that prior to the Ice Age, a significant part of the Missouri River drained north toward Hudson Bay. Moreover, the Mississippi River did not follow the present Iowa-Illinois boundary but rather flowed across west-central Illinois, where the lower Illinois River flows today. The preglacial Ohio River barely reached to the present-day state of Ohio, and the rivers that today feed the Ohio in western Pennsylvania flowed north and drained into the North Atlantic. The Great Lakes were created by glacial erosion during the Ice Age. Prior to the Pleistocene, the basins occupied by these huge lakes were lowlands with rivers that ran eastward to the Gulf of St. Lawrence.

The large Teays River was a significant feature prior to the Ice Age (Figure 18.30B). It flowed from West Virginia across Ohio, Indiana, and Illinois, where it discharged into the Missis-

sippi River not far from present-day Peoria. This river valley, which would have rivaled the Mississippi in size, was completely obliterated during the Pleistocene, buried by glacial deposits hundreds of feet thick. Today the sands and gravels in the buried Teays valley make it an important aquifer.

New York's Finger Lakes   The recent geologic history of west-central New York State south of Lake Ontario was dominated by ice sheets. We have already noted the drumlins in the vicinity of Palmyra (see Figure 18.27). Many other depositional features and erosional effects occur in the region. Perhaps the best known are the Finger Lakes. They consist of 11 long, narrow, roughly parallel water bodies oriented north-south as fingers on a pair of outstretched hands (Figure 18.31). Prior to the Ice Age, the Finger Lakes area consisted of a series of river valleys that were oriented parallel to the direction of ice movement. Multiple episodes of glacial erosion transformed these preglacial valleys into deep, steep-walled lakes. Two of the lakes are very deep—the greatest depth of Seneca Lake is more than 180 meters (600 feet) whereas Cayuga Lake is nearly 135 meters (450 feet) deep and the beds of both lakes are below sea level. The depth to which the glaciers carved these basins is much greater. There are hundreds of feet of glacial sediment in the deep rock troughs below the lake beds.

Ice Dams Create Proglacial Lakes

Ice sheets and alpine glaciers can act as dams to create lakes by trapping glacial meltwater and blocking the flow of rivers. Some of these lakes are relatively small, short-lived impoundments. Others can be large and exist for hundreds or thousands of years.

Figure 18.32 is a map of Lake Agassiz—the largest lake to form during the Ice Age in North America. With the retreat of the ice sheet came enormous volumes of meltwater. The Great Plains generally slope upward to the west. As the terminus of the ice sheet receded northeastward, meltwater was trapped between the ice on one side and the sloping land on the other, causing Lake Agassiz to deepen and spread across the landscape. It came into existence about 12,000 years ago and lasted for about 4500 years. Such water bodies are termed *proglacial lakes,* referring to their position just beyond the outer limits of a glacier or ice sheet. The lake's history is complicated by the dynamics of the ice sheet, which, at various times, readvanced and affected lake levels and drainage systems. Where drainage occurred depended upon the water level of the lake and the position of the ice sheet.

Lake Agassiz left marks over a broad region. Former beaches, many kilometers from any water, mark former

Students Sometimes Ask . . .
How large are the Great Lakes?

The Great Lakes constitute the largest body of fresh water on Earth. Formed between about 10,000 and 12,000 years ago, the lakes currently contain about 19 percent of Earth's surface fresh water.

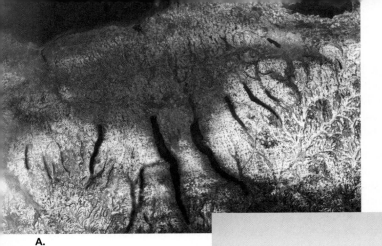

A.

Research shows that the shifting of glaciers and the failure of ice dams can cause the rapid release of huge volumes of water. Such events occurred during the history of Lake Agassiz. One of the most dramatic examples of such glacial outbursts occurred in the Pacific Northwest and is described in Box 18.2.

Pluvial Lakes

While the formation and growth of ice sheets was an obvious response to significant changes in climate, the existence of the glaciers themselves triggered important climatic changes in the regions beyond their margins. In arid and semiarid areas on all of the continents, temperatures were lower and thus evaporation rates were lower, but at the same time moderate precipitation totals were experienced. This cooler, wetter climate formed many **pluvial lakes** (from the Latin term *pluvia*, meaning *rain*). In North America the greatest concentration of pluvial lakes occurred in the vast Basin and Range region of Nevada and Utah (Figure 18.33). By far the largest of the lakes in this region was Lake Bonneville. With maximum depths exceeding 300 meters and an area of 50,000 square kilometers,

FIGURE 18.31 The long, narrow basins occupied by the Finger Lakes in west central New York, were created when river valleys were scoured into deep troughs by ice sheets. **A.** Satellite view of the Finger Lakes region. (NASA) **B.** Canandaigua Lake is one of the Finger Lakes. (Photo by James Schwabel/Alamy)

B.

shorelines. Several modern river valleys, including the Red River and the Minnesota River were originally cut by water entering or leaving the lake. Present-day remnants of Lake Agassiz include Lakes Winnipeg, Manitoba, Winnipegosis, and Lake of the Woods. The sediments of the former lake basin are now fertile agricultural land.

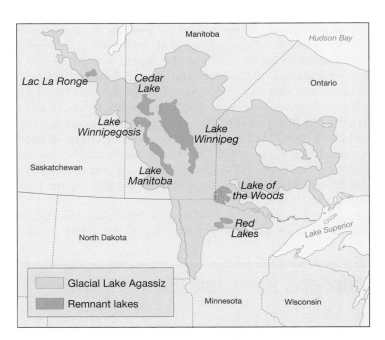

FIGURE 18.32 Map showing the extent of glacial Lake Agassiz. It was an immense feature—bigger than all of the present-day Great Lakes combined. The modern-day remnants of this proglacial water body are still major landscape features.

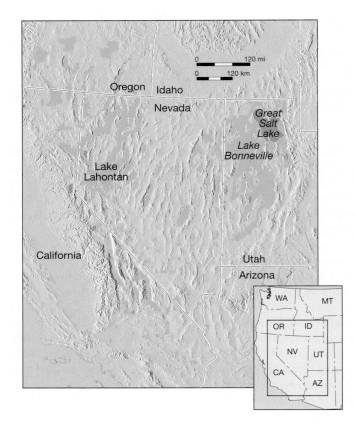

FIGURE 18.33 Pluvial lakes of the western United States. (After R. F. Flint, *Glacial and Quaternary Geology,* New York: John Wiley & Sons)

UNDERSTANDING **EARTH**

BOX 18.2

Glacial Lake Missoula, Megafloods, and the Channeled Scablands

Lake Missoula was a prehistoric proglacial lake in western Montana that existed as the Pleistocene ice age was drawing to a close between about 15,000 and 13,000 years ago. It was a time when the climate was gradually warming and the Cordilleran ice sheet, which covered much of western Canada and portions of the Pacific Northwest, was melting and retreating.

The lake was the result of an ice dam that formed when a southward protruding mass of glacial ice called the Purcell Lobe blocked the Clark Fork River (Figure 18.C). As the water rose behind the 600 meter- (2000 foot-) high dam, it flooded the valleys of western Montana. At its greatest extent, Lake Missoula extended eastward for more than 300 kilometers (200 miles). Its volume exceeded 2500 cubic kilometers (600 cubic miles)—greater than present day Lake Ontario.

Eventually, the lake became so deep that the ice dam began to float. That is, the rising waters behind the dam lifted the buoyant ice so that it no longer functioned as a dam. The result was a catastrophic flood as Lake Missoula's waters suddenly poured out under the failed dam. The outburst rushed across

FIGURE 18.D   The landscape of the Channeled Scablands was sculpted by the megafloods associated with glacial Lake Missoula. (Photo by Michael Collier)

the lava plains of eastern Washington and down the Columbia River to the Pacific Ocean. Based on the sizes of boulders moved during the event, the velocity of the torrent approached or exceeded 70 kilometers (nearly 45 miles) per hour. The entire lake was emptied in a matter of a few days. Due to the temporary impounding of water behind narrow gaps along the flood's path, flooding probably continued through the devastated region for a few weeks.

The erosional and depositional results of such a megaflood were dramatic. The towering mass of rushing water stripped away thick layers of sediment and soil and cut deep canyons (*coulees*) into the underlying basalt. Today the region is called the *Channeled Scablands*—a landscape consisting of a bizarre assemblage of landforms. Perhaps the most striking features are the table-like lava mesas called *scabs* left between braided interlocking channels (Figure 18.D).

It was not a single flood from glacial Lake Missoula that created this extraordinary landscape. What makes Lake Missoula remarkable is that it alternately filled and emptied in a cycle that was repeated more than 40 times over a span of 1500 years. After each flood, the glacial lobe would again block the valley and create a new ice dam, and the cycle of lake growth, dam failure, and megaflood would follow at intervals of 20 to 60 years.

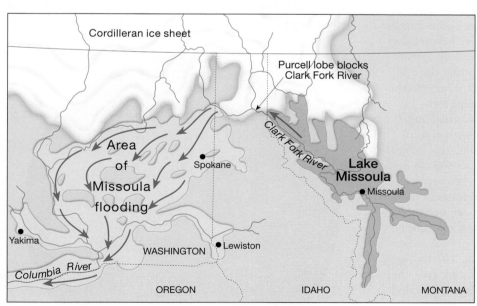

FIGURE 18.C   Lake Missoula was a proglacial lake created when the Purcell Lobe of the Cordilleran ice sheet formed an ice dam on the Clark Fork River. Periodically, the ice dam would fail sending a huge torrent of water flooding across the landscape of eastern Washington.

Lake Bonneville was nearly the same size as present-day Lake Michigan. As the ice sheets waned, the climate again grew more arid, and the lake levels lowered in response. Although most of the lakes completely disappeared, a few small remnants of Lake Bonneville remain, the Great Salt Lake being the largest and best known.

THE GLACIAL THEORY AND THE ICE AGE

In the preceding pages, we mentioned the Ice Age, a time when ice sheets and alpine glaciers were far more extensive than they are today. As noted, there was a time when the most popular explanation for what we now know to be glacial deposits was that the materials had been drifted in by means of icebergs or perhaps simply swept across the landscape by a catastrophic flood. What convinced geologists that an extensive ice age was responsible for these deposits and many other glacial features?

In 1821 a Swiss engineer, Ignaz Venetz, presented a paper suggesting that glacial landscape features occurred at considerable distances from the existing glaciers in the Alps. This implied that the glaciers had once been larger, and occupied positions farther downvalley. Another Swiss scientist, Louis Agassiz, doubted the proposal of widespread glacial activity put forth by Venetz. He set out to prove that the idea was not valid. Ironically, his 1836 fieldwork in the Alps convinced him of the merits of his colleague's hypothesis. In fact, a year later Agassiz hypothesized a great ice age that had extensive and far-reaching effects—an idea that was to give Agassiz widespread fame.

The proof of the glacial theory proposed by Agassiz and others constitutes a classic example of applying the principle of uniformitarianism. Realizing that certain features are produced by no other known process but glacial action, they were able to begin reconstructing the extent of now vanished ice sheets based on the presence of features and deposits found far beyond the margins of present-day glaciers. In this manner, the development and verification of the glacial theory continued during the 19th century, and through the efforts of many scientists, a knowledge of the nature and extent of former ice sheets became clear.

By the beginning of the 20th century, geologists had largely determined the extent of the Ice Age glaciation. Further, during the course of their investigations, they discovered that many glaciated regions had not one layer of drift but several. Moreover, close examination of these older deposits showed well-developed zones of chemical weathering and soil formations as well as the remains of plants that require warm temperatures. The evidence was clear; there had been not just one glacial advance but many, each separated by extended periods when climates were as warm or warmer than the present. The Ice Age had not simply been a time when the ice advanced over the land, lingered for a while, and then receded. Rather, the period was a very complex event, characterized by a number of advances and withdrawals of glacial ice.

By the early 20th century, a fourfold division of the Ice Age had been established for both North America and Europe. The divisions were based largely on studies of glacial deposits. In North America each of the four major stages was named for the midwestern state where deposits of that stage were well exposed

and/or were first studied. These are, in order of occurrence, the Nebraskan, Kansan, Illinoian, and Wisconsinan. These traditional divisions remained in place for many years, until it was learned that sediment cores from the ocean floor contain a much more complete record of climate change during the Ice Age. Unlike the glacial record on land, which is punctuated by many unconformities, seafloor sediments provide an uninterrupted record of climatic cycles for this period. Studies of these seafloor sediments showed that glacial/interglacial cycles had occurred about every 100,000 years. About 20 such cycles of cooling and warming were identified for the span we call the Ice Age.

During the glacial age, ice left its imprint on almost 30 percent of Earth's land area, including about 10 million square kilometers of North America, 5 million square kilometers of Europe, and 4 million square kilometers of Siberia (Figure 18.34). The amount of glacial ice in the Northern Hemisphere was roughly twice that of the Southern Hemisphere. The primary reason is that the southern polar ice could not spread far beyond the margins of Antarctica. By contrast, North America and Eurasia provided great expanses of land for the spread of ice sheets.

Today we know that the Ice Age began between 2 million and 3 million years ago. This means that most of the major glacial stages occurred during a division of the geologic time scale called the **Pleistocene epoch.** Although the Pleistocene is commonly used as a synonym for the Ice Age, note that this epoch does not encompass all of the last glacial period. The Antarctic ice sheet, for example, probably formed at least 30 million years ago.

CAUSES OF GLACIATION

A great deal is known about glaciers and glaciation. Much has been learned about glacier formation and movement, the extent of glaciers past and present, and the features created by glaciers,

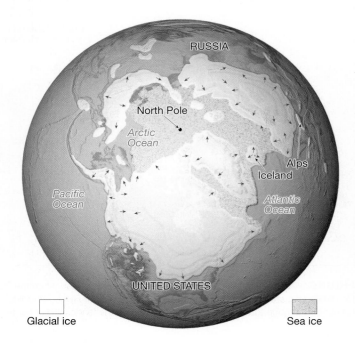

FIGURE 18.34   Maximum extent of ice sheets in the Northern Hemisphere during the Ice Age.

both erosional and depositional. However, the causes of glacial ages are not completely understood.

Although widespread glaciation has been rare in Earth's history, the Ice Age that encompassed the Pleistocene epoch is not the only glacial period for which a record exists. Earlier glaciations are indicated by deposits called **tillite,** a sedimentary rock formed when glacial till becomes lithified. Such deposits, found in strata of several different ages, usually contain striated rock fragments, and some overlie grooved and polished bedrock surfaces or are associated with sandstones and conglomerates that show features of outwash deposits. For example, in the discussion of evidence in support of the continental drift hypothesis in Chapter 2, mention was made of a glacial period that occurred in late Paleozoic time (see Figure 2.7A, p. 44). Two Precambrian glacial episodes have been identified in the geologic record, the first approximately 2 billion years ago and the second about 600 million years ago.

Any theory that attempts to explain the causes of glacial ages must successfully answer two basic questions. (1) *What causes the onset of glacial conditions?* For continental ice sheets to have formed, average temperature must have been somewhat lower than at present and perhaps substantially lower than throughout much of geologic time. Thus, a successful theory would have to account for the cooling that finally leads to glacial conditions. (2) *What caused the alternation of glacial and interglacial stages that have been documented for the Pleistocene epoch?* The first question deals with long-term trends in temperature on a scale of millions of years, but this second question relates to much shorter-term changes.

Although the literature of science contains many hypotheses relating to the possible causes of glacial periods, we will discuss only a few major ideas to summarize current thought.

Plate Tectonics

Probably the most attractive proposal for explaining the fact that extensive glaciations have occurred only a few times in the geologic past comes from the theory of plate tectonics. Because glaciers can form only on land, we know that landmasses must exist somewhere in the higher latitudes before an ice age can commence. Many scientists suggest that ice ages have occurred only when Earth's shifting crustal plates have carried the continents from tropical latitudes to more poleward positions.

Glacial features in present-day Africa, Australia, South America, and India indicate that these regions, which are now tropical or subtropical, experienced an ice age near the end of the Paleozoic era, about 250 million years ago. However, there is no evidence that ice sheets existed during this same period in what are today the higher latitudes of North America and Eurasia. For many years this puzzled scientists. Was the climate in these relatively tropical latitudes once like it is today in Greenland and Antarctica? Why did glaciers not form in North America and Eurasia? Until the plate tectonics theory was formulated, there had been no reasonable explanation.

Today scientists understand that the areas containing these ancient glacial features were joined together as a single supercontinent located at latitudes far to the south of their present positions. Later this landmass broke apart, and its pieces, each moving on a different plate, migrated toward their

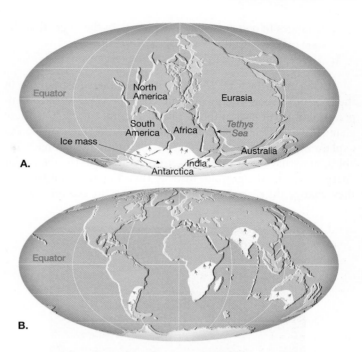

FIGURE 18.35 **A.** The supercontinent Pangaea showing the area covered by glacial ice near the end of the Paleozoic Era. **B.** The continents as they are today. The white areas indicate where evidence of the former ice sheets exists.

present locations (Figure 18.35). Now we know that during the geologic past plate movements accounted for many dramatic climate changes as landmasses shifted in relation to one another and moved to different latitudinal positions. Changes in oceanic circulation also must have occurred, altering the transport of heat and moisture and consequently the climate as well. Because the rate of plate movement is very slow—a few centimeters annually—appreciable changes in the positions of the continents occur only over great spans of geologic time. Thus, climate changes triggered by shifting plates are extremely gradual and happen on a scale of millions of years.

Variations in Earth's Orbit

Because climatic changes brought about by moving plates are extremely gradual, the plate tectonics theory cannot be used to explain the alternation between glacial and interglacial climates that occurred during the Pleistocene epoch. Therefore, we must look to some other triggering mechanism that might cause climate change on a scale of thousands rather than millions of years. Today many scientists strongly suspect that the climate oscillations that characterized the Pleistocene are linked to variations in Earth's orbit. This hypothesis was first developed and strongly advocated by the Serbian astrophysicist Milutin Milankovitch and is based on the premise that variations in incoming solar radiation are a principal factor in controlling Earth's climate.

Milankovitch formulated a comprehensive mathematical model based on the following elements (Figure 18.36):

1. Variations in the shape (*eccentricity*) of Earth's orbit about the Sun;

2. Changes in *obliquity*; that is, changes in the angle that the axis makes with the plane of Earth's orbit;
3. The wobbling of Earth's axis, called *precession*.

Using these factors, Milankovitch calculated variations in the receipt of solar energy and the corresponding surface temperature of Earth back into time in an attempt to correlate these changes with the climate fluctuations of the Pleistocene. In explaining climate changes that result from these three variables, note that they cause little or no variation in the *total* solar energy reaching the ground. Instead, their impact is felt because they change the degree of contrast between the seasons. Some-

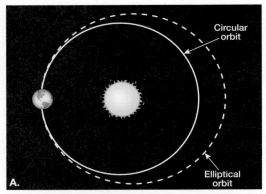

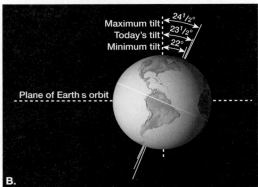

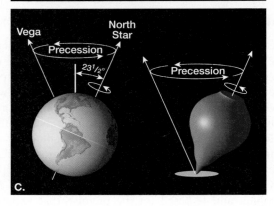

FIGURE 18.36 Orbital variations. **A.** The shape of Earth's orbit changes during a cycle that spans about 100,000 years. It gradually changes from nearly circular to one that is more elliptical and then back again. This diagram greatly exaggerates the amount of change. **B.** Today the axis of rotation is tilted about 23.5° to the plane of Earth's orbit. During a cycle of 41,000 years, this angle varies from 21.5° to 24.5°. **C.** Precession. Earth's axis wobbles like that of a spinning top. Consequently, the axis points to different spots in the sky during a cycle of about 26,000 years.

what milder winters in the middle to high latitudes mean greater snowfall totals, whereas cooler summers would bring a reduction in snowmelt.

Among the studies that have added credibility to the astronomical hypothesis of Milankovitch is one in which deep-sea sediments containing certain climatically sensitive microorganisms were analyzed to establish a chronology of temperature changes going back nearly one-half million years.* This time scale of climate change was then compared to astronomical calculations of eccentricity, obliquity, and precession to determine if a correlation did indeed exist. Although the study was very involved and mathematically complex, the conclusions were straightforward. The researchers found that major variations in climate over the past several hundred thousand years were closely associated with changes in the geometry of Earth's orbit; that is, cycles of climate change were shown to correspond closely with the periods of obliquity, precession, and orbital eccentricity. More specifically, the authors stated: "It is concluded that changes in the earth's orbital geometry are the fundamental cause of the succession of Quaternary ice ages."**

Let us briefly summarize the ideas that were just described. The theory of plate tectonics provides an explanation for the widely spaced and nonperiodic onset of glacial conditions at various times in the geologic past, whereas the theory proposed by Milankovitch and supported by the work of J. D. Hays and his colleagues furnishes an explanation for the alternating glacial and interglacial episodes of the Pleistocene.

Other Factors

Variations in Earth's orbit correlate closely with the timing of glacial–interglacial cycles. However, the variations in solar energy reaching Earth's surface caused by these orbital changes do not adequately explain the magnitude of the temperature changes that occurred during the most recent Ice Age. Other factors must also have contributed. One factor involves variations in the composition of the atmosphere. Other influences involve changes in the reflectivity of Earth's surface and in ocean circulation. Let's take a brief look at these factors.

Chemical analyses of air bubbles that become trapped in glacial ice at the time of ice formation indicate that the ice-age atmosphere contained less of the gases carbon dioxide and methane than the post Ice-Age atmosphere (Figure 18.37). Carbon dioxide and methane are important "greenhouse" gases, which means that they trap radiation emitted by Earth and contribute to the heating of the atmosphere. When the amount of carbon dioxide and methane in the atmosphere increases, global temperatures rise, and when there is a reduction in these gases, as occurred during the Ice Age, temperatures fall. Therefore, reductions in the concentrations of greenhouse gases help explain the magnitude of the temperature drop that occurred during glacial times. Although scien-

---

*J. D. Hays, John Imbrie, and N. J. Shackelton, "Variations in the Earth's Orbit: Pacemaker of the Ice Ages," *Science* 194 (1976): 1121–32.

**J. D. Hays et al., p. 1131. The term *Quaternary* refers to the period on the geologic time scale that encompasses the last 2.6 million years.

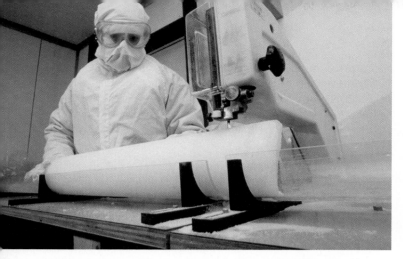

FIGURE 18.37 Scientist slicing an ice core sample from Antarctica for analysis. He is wearing protective clothing and a mask to minimize contamination of the sample. Chemical analysis of ice cores can provide important data about past climates. (Photo by British Antarctic Survey/Photo Researchers, Inc.)

tists know that concentrations of carbon dioxide and methane dropped, they do not know what caused the drop. As often occurs in science, observations gathered during one investigation yield information and raise questions that require further analysis and explanation.

Obviously, whenever Earth enters an ice age, extensive areas of land that were once ice free are covered with ice and snow. In

addition, a colder climate causes the area covered by sea ice (frozen surface sea water) to expand as well. Ice and snow reflect a large portion of incoming solar energy back to space. Thus, energy that would have warmed Earth's surface and the air above is lost and global cooling is reinforced.*

Yet another factor that influences climate during glacial times relates to ocean currents. Research has shown that ocean circulation changes during ice ages. For example, studies suggest that the warm current that transports large amounts of heat from the tropics toward higher latitudes in the North Atlantic was significantly weaker during the Ice Age. This would lead to a colder climate in Europe, amplifying the cooling attributable to orbital variations.

In conclusion, we emphasize that the ideas just discussed do not represent the only possible explanations for glacial ages. Although interesting and attractive, these proposals are certainly not without critics, nor are they the only possibilities currently under study. Other factors may be, and probably are, involved.

---

*Recall from Chapter 1 that something that reinforces (adds to) the initial change is called a *positive feedback mechanism*. To review this idea, see the discussion on feedback mechanisms in the section on "Earth As a System" in Chapter 1.

CHAPTER 18 GLACIERS AND GLACIATION IN REVIEW

- A *glacier* is a thick mass of ice originating on land as a result of the compaction and recrystallization of snow, and it shows evidence of past or present flow. Today, *valley* or *alpine glaciers* are found in mountain areas where they usually follow valleys that were originally occupied by streams. *Ice sheets* exist on a much larger scale, covering most of Greenland and Antarctica.

- Near the surface of a glacier, in the *zone of fracture,* ice is brittle. However, below about 50 meters, pressure is great, causing ice to *flow* like *plastic material.* A second important mechanism of glacial movement consists of the entire ice mass *slipping* along the ground.

- The average velocity of glacial movement is generally quite slow but varies considerably from one glacier to another. The advance of some glaciers is characterized by periods of extremely rapid movements called *surges.*

- Glaciers form in areas where more snow falls in winter than melts during summer. Snow accumulation and ice formation occur in the *zone of accumulation.* Its outer limits are defined by the *snowline.* Beyond the snowline is the *zone of wastage,* where there is a net loss to the glacier. The *glacial budget* is the balance, or lack of balance, between accumulation at the upper end of the glacier, and loss, called *ablation,* at the lower end.

- Glaciers erode land and acquire debris by *plucking* (lifting pieces of bedrock out of place) and *abrasion* (grinding and scraping of a rock surface). Mass-wasting processes also make significant contributions to the load of many alpine glaciers. Erosional features produced by valley glaciers include *glacial troughs, hanging valleys, pater noster lakes, fiords, cirques, arêtes, horns,* and *roches moutonnées.*

- Any sediment of glacial origin is called *drift.* The two distinct types of glacial drift are (1) *till,* which is unsorted sediment deposited directly by the ice; and (2) *stratified drift,* which is relatively well-sorted sediment laid down by glacial meltwater.

- The most widespread features created by glacial deposition are layers or ridges of till, called *moraines.* Associated with valley glaciers are *lateral moraines,* formed along the sides of the valley, and *medial moraines,* formed between two valley glaciers that have joined. *End moraines,* which mark the former position of the front of a glacier, and *ground moraines,* undulating layers of till deposited as the ice front retreats, are common to both valley glaciers and ice sheets. An *outwash plain* is often associated with the end moraine of an ice sheet. A *valley train* may form when the glacier is confined to a valley. Other depositional features include *drumlins* (streamlined, asymmetrical hills composed of till), *eskers* (sinuous ridges composed largely of sand and gravel deposited by streams flowing in tunnels beneath the ice, near the terminus of a glacier), and *kames* (steep-sided hills consisting of sand and gravel).

- In addition to massive erosional and depositional work, other effects of Ice Age glaciers included the *forced migration of organisms, changes in stream courses, formation of large proglacial lakes, adjustment of the crust* by rebounding after the removal of the immense load of ice, and *climate changes* caused by the existence of the glaciers themselves. In the sea, the most far-reaching effect of the Ice Age was the *worldwide change* in *sea level* that accompanied each advance and retreat of the ice sheets.

- The *Ice Age,* which began about 2 million years ago, was a very complex period characterized by a number of advances and withdrawals of glacial ice. Most of the major glacial episodes occurred during a division of the geologic time scale called the *Pleistocene epoch.* Perhaps the most convincing evidence for the occurrence of several glacial advances during the Ice Age is the widespread existence of *multiple layers of drift* and an uninterrupted record of climate cycles preserved in *seafloor sediments.*

- Any theory that attempts to explain the causes of glacial ages must answer two basic questions: (1) What causes the onset of glacial conditions? and (2) What caused the alternating glacial and interglacial stages that have been documented for the Pleistocene epoch? Two of the many hypotheses for the cause of glacial ages involve (1) plate tectonics and (2) variations in Earth's orbit. Other factors that are related to climate change during glacial ages include: changes in atmospheric composition, variations in the amount of sunlight reflected by Earth's surface, and changes in ocean circulation.

KEY TERMS

ablation (p. 496)	glacial budget (p. 496)	kame terrace (p. 507)	rock flour (p. 498)
abrasion (p. 498)	glacial drift (p. 503)	kettle (p. 507)	snowline (p. 494)
alpine glacier (p. 490)	glacial erratic (p. 503)	lateral moraine (p. 504)	stratified drift (p. 506)
arête (p. 502)	glacial striations (p. 498)	medial moraine (p. 504)	surge (p. 495)
basal slip (p. 494)	glacial trough (p. 499)	outlet glacier (p. 493)	tarn (p. 500)
calving (p. 496)	glacier (p. 490)	outwash plain (p. 507)	terminal moraine (p. 504)
cirque (p. 500)	ground moraine (p. 504)	pater noster lakes (p. 500)	till (p. 503)
col (p. 502)	hanging valley (p. 500)	piedmont glacier (p. 493)	tillite (p. 513)
crevasse (p. 495)	horn (p. 502)	plastic flow (p. 494)	truncated spur (p. 499)
drumlin (p. 505)	ice cap (p. 493)	Pleistocene epoch (p. 512)	valley glacier (p. 490)
end moraine (p. 504)	ice-contact deposit (p. 507)	plucking (p. 498)	valley train (p. 507)
esker (p. 507)	ice sheet (p. 490)	pluvial lake (p. 510)	zone of accumulation (p. 496)
fiord (p. 502)	ice shelf (p. 492)	recessional moraine (p. 504)	zone of fracture (p. 495)
firn (p. 494)	kame (p. 507)	roche moutonnée (p. 503)	zone of wastage (p. 496)

QUESTIONS FOR REVIEW

1. Where are glaciers found today? What percentage of Earth's land area do they cover? How does this compare to the area covered by glaciers during the Pleistocene?

2. Describe how glaciers fit into the hydrologic cycle. What role do they play in the rock cycle?

3. Each statement below refers to a particular type of glacier. Name the type of glacier.

   a. The term *continental* is often used to describe this type of glacier.

   b. This type of glacier is also called an *alpine glacier.*

   c. This is a stream of ice leading from the margin of an ice sheet through the mountains to the sea.

   d. This is a glacier formed when one or more valley glaciers spreads out at the base of a steep mountain front.

   e. Greenland is the only example in the Northern Hemisphere.

4. Describe the two components of glacial flow. At what rates do glaciers move? In a valley glacier, does all of the ice move at the same rate? Explain.

5. Why do crevasses form in the upper portion of a glacier but not below 50 meters (160 feet)?

6. Under what circumstances will the front of a glacier advance? Retreat? Remain stationary?

7. Describe the processes of glacial erosion.

8. How does a glaciated mountain valley differ in appearance from a mountain valley that was not glaciated?

9. List and describe the erosional features you might expect to see in an area where valley glaciers exist or have recently existed.

10. What is glacial drift? What is the difference between till and stratified drift? What general effect do glacial deposits have on the landscape?

11. List the four basic moraine types. What do all moraines have in common? What is the significance of terminal and recessional moraines?

12. Why are medial moraines proof that valley glaciers must move?

13. How do kettles form?

14. What direction was the ice sheet moving that affected the area shown in Figure 18.27? Explain how you were able to determine this.

15. What are ice-contact deposits? Distinguish between kames and eskers.

16. Describe at least four effects of Ice Age glaciers aside from the formation of major erosional and depositional features.

17. The development of the glacial theory is a good example of applying the principle of uniformitarianism. Explain briefly.

18. During the Pleistocene epoch the amount of glacial ice in the Northern Hemisphere was about twice as great as in the Southern Hemisphere. Briefly explain why this was the case.

19. How might plate tectonics help explain the cause of ice ages? Can plate tectonics explain the alternation between glacial and interglacial climates during the Pleistocene?

COMPANION WEBSITE

The *Earth 10e* website uses the resources and flexibility of the Internet to aid in your study of the topics in this chapter. Written and developed by the authors and other geology instructors, this site will help improve your understanding of geology. Visit **www.mygeoscienceplace.com** in order to:

- **Review** key chapter concepts

- **Read** with links to the eBook and to chapter-specific web resources
- **Visualize** and comprehend challenging topics using learning activities in *GEODe Earth*
- **Test** yourself with online quizzes

GEODe EARTH

GEODe Earth is a valuable and easy to use learning aid that can be accessed from your book's Companion Website (**www.mygeoscienceplace.com**). It is a dynamic instructional tool that promotes understanding and reinforces important concepts by using tutorials, animations, and exercises that actively engage the student.

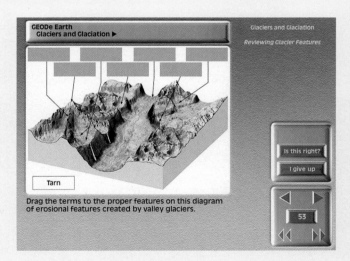

DESERTS AND WINDS

Wind commonly deposits sand in mounds or ridges called dunes. These dunes are in Death Valley National Park, California.

(PHOTO BY MICHAEL COLLIER)

Climate has a strong influence on the nature and intensity of Earth's external processes. This was clearly demonstrated in the preceding chapter on glaciers. Another excellent example of the strong link between climate and geology is seen when we examine the development of arid landscapes. The word *desert* literally means deserted or unoccupied. For many dry regions this is an appropriate description, although where water is available in deserts, plants and animals thrive. Nevertheless, the world's dry regions are probably the least familiar land areas on Earth outside of the polar realm.

Desert landscapes frequently appear stark. Their profiles are not softened by a carpet of soil and abundant plant life. Instead, barren rocky outcrops with steep, angular slopes are common. At some places the rocks are tinted orange and red. At others they are gray and brown and streaked with black. For many visitors, desert scenery exhibits a striking beauty; to others, the terrain seems bleak. No matter which feeling is elicited, it is clear that deserts are very different from the more humid places where most people live.

As you will see, arid regions are not dominated by a single geologic process. Rather, the effects of tectonic forces, running water, and wind are all apparent. Because these processes combine in different ways from place to place, the appearance of desert landscapes varies a great deal as well (Figure 19.1).

DISTRIBUTION AND CAUSES OF DRY LANDS

DESERTS AND WINDS
▶ Distribution and Causes of Dry Lands

The dry regions of the world encompass about 42 million square kilometers, a surprising 30 percent of Earth's land surface. No other climate category covers so large a land area. Within these water-deficient regions, two climatic types are commonly recognized: **desert,** or arid, and **steppe,** or semiarid. The two share many features; their differences are primarily a matter of degree (Box 19.1). The steppe is a marginal and more humid variant of the desert and is a transition zone that surrounds the desert and separates it from bordering humid climates. The world map showing the distribution of desert and steppe regions reveals that dry lands are concentrated in the subtropics and in the middle latitudes (Figure 19.2).

Low-Latitude Deserts

The heart of the low-latitude dry climates lies in the vicinities of the Tropics of Cancer and Capricorn. Figure 19.2 shows a virtually unbroken desert environment stretching for more than 9300 kilometers (5800 miles) from the Atlantic coast of North Africa to the dry lands of northwestern India. In addition to this single great expanse, the Northern Hemisphere contains another, much smaller area of tropical desert and steppe in northern Mexico and the southwestern United States.

In the Southern Hemisphere, dry climates dominate Australia. Almost 40 percent of the continent is desert, and much of the remainder is steppe. In addition, arid and semiarid areas occur in southern Africa and make a limited appearance in coastal Chile and Peru.

What causes these bands of low-latitude desert? The answer is the global distribution of air pressure and winds. Figure 19.3A, an idealized diagram of Earth's general circulation, helps visualize the relationship. Heated air in the pressure belt known as the *equatorial low* rises to great heights (usually between 15 and 20 kilometers) and then spreads out. As the upper-level flow reaches 20° to 30° latitude, north or south, it sinks toward the surface. Air that rises through the atmosphere expands and cools, a process that leads to the development of clouds and precipitation. For this reason, the areas under the influence of the equatorial low are among the rainiest on Earth. Just the opposite is true for the regions in the vicinity of 30° north and south latitude, where high pressure predominates. Here, in the zones known as the *subtropical highs,* air is subsiding. When air sinks, it is compressed and warmed. Such conditions are just opposite of

FIGURE 19.1 A scene in Organ Pipe Cactus National Monument near the Arizona-Mexico border. The appearance of desert landscapes varies a great deal from place to place. (Photo by Marek Zuk/Alamy)

what is needed to produce clouds and precipitation. Consequently, these regions are known for their clear skies, sunshine, and ongoing drought (Figure 19.3B).

Middle-Latitude Deserts

Unlike their low-latitude counterparts, middle-latitude deserts and steppes are not controlled by the subsiding air masses associated with high pressure. Instead, these dry lands exist principally because they are sheltered in the deep interiors of large landmasses. They are far removed from the ocean, which is the ultimate source of moisture for cloud formation and precipitation. One well-known example is the Gobi Desert of central Asia, shown on the map north of India.

The presence of high mountains across the paths of prevailing winds further separates these areas from water-bearing, maritime air masses; plus, the mountains force the air to lose much of its water. The mechanism is simple: As pre-

vailing winds meet mountain barriers, the air is forced to ascend. When air rises, it expands and cools, a process that can produce clouds and precipitation. The windward sides of mountains, therefore, often have high precipitation. By contrast, the leeward sides of mountains are usually much drier (Figure 19.4). This

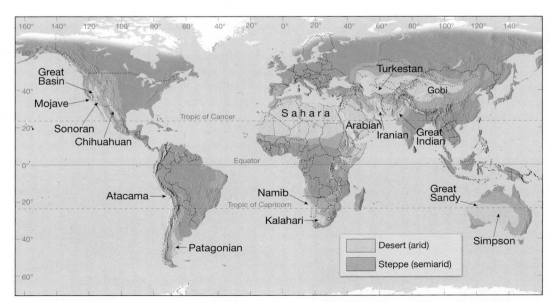

FIGURE 19.2 Arid and semiarid climates cover about 30 percent of Earth's land surface. No other climate group covers so large an area.

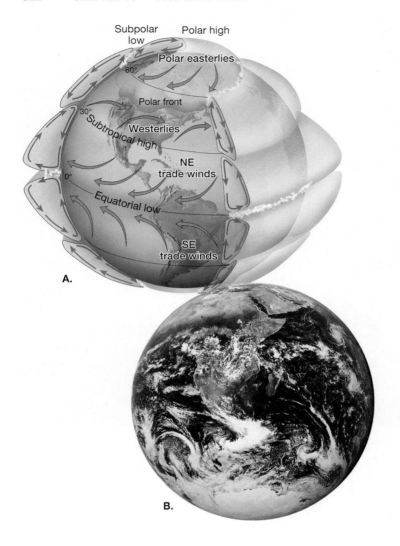

A.

B.

the image shows a full earth from space

FIGURE 19.3　**A.** Idealized diagram of Earth's general circulation. The deserts and steppes that are centered in the latitude belt between 20° and 30° north and south coincide with the subtropical high-pressure belts. Here dry, subsiding air inhibits cloud formation and precipitation. By contrast, the pressure belt known as the equatorial low is associated with areas that are among the rainiest on Earth. **B.** In this view of Earth from space, North Africa's Sahara Desert, the adjacent Arabian Desert, and the Kalahari and Namib deserts in southern Africa are clearly visible as tan-colored, cloud-free zones. The band of clouds that extends across central Africa and the adjacent oceans coincides with the equatorial low-pressure belt. (Photo courtesy of NASA)

FIGURE 19.4　Many deserts in the middle latitudes are rainshadow deserts. As moving air meets a mountain barrier, it is forced to rise. Clouds and precipitation on the windward side often result. Air descending the leeward side is much drier. The mountains effectively cut the leeward side off from the sources of moisture, producing a rainshadow desert. The Great Basin desert is a rainshadow desert that covers nearly all of Nevada and portions of adjacent states. (Photo on left by Dean Pennala/Shutterstock. Photo on right by Dennis Tasa.)

UNDERSTANDING EARTH

BOX 19.1

What Is Meant by "Dry"?

Albuquerque, New Mexico, in the southwestern United States, receives an average of 20.7 centimeters (8.07 inches) of rainfall annually. As you might expect, because Albuquerque's precipitation total is modest, the station is classified as a desert when the commonly used Köppen climate classification is applied. The Russian city of Verkhoyansk is a remote station located near the Arctic Circle in Siberia. The yearly precipitation total there averages 15.5 centimeters (6.05 inches), about 5 centimeters less than Albuquerque's. Although Verkhoyansk receives less precipitation than Albuquerque, its classification is that of a humid climate. How can this occur?

We all recognize that deserts are dry places, but just what is meant by the term *dry*? That is, how much rain defines the boundary between humid and dry regions? Sometimes it is arbitrarily defined by a single rainfall figure, for example, 25 centimeters (10 inches) per year of precipitation. However, the concept of dryness is a relative one that refers to any situation in which a water deficiency exists. Hence, climatologists define *dry climate* as one in which yearly precipitation is not as great as the potential loss of water by evaporation. Dryness, then, not only is related to annual rainfall totals but is also a function of evaporation, which in turn is closely dependent upon temperature.

As temperatures climb, potential evaporation also increases. Fifteen to 25 centimeters of precipitation may be sufficient to support coniferous forests in northern Scandinavia or Siberia, where evaporation into the cool, humid air is slight and a surplus of water remains in the soil. However, the same amount of rain falling on New Mexico or Iran supports only a sparse vegetative cover because evaporation into the hot, dry air is great. So, clearly, no specific amount of precipitation can serve as a universal boundary for dry climates.

To establish the boundary between dry and humid climates, the widely used Köppen classification system uses formulas that involve three variables: average annual precipitation, average annual temperature, and seasonal distribution of precipitation. The use of average annual temperature reflects its importance as an index of evaporation. The amount of rainfall defining the humid–dry boundary will be larger where mean annual temperatures are high, and smaller where temperatures are low. The use of seasonal precipitation distribution as a variable is also related to this idea. If rain is concentrated in the warmest months, loss to evaporation is greater than if the precipitation is concentrated in the cooler months.

Table 19.A summarizes the precipitation amounts that divide dry and humid climates. Notice that a station with an annual mean of 20 °C (68 °F) and a summer rainfall maximum of 68 centimeters (26.5 inches) is classified as dry. If the rain falls primarily in winter, however, the station must receive only 40 centimeters (15.6 inches) or more to be considered humid. If the precipitation is more evenly distributed, the figure defining the humid–dry boundary is between the other two.

TABLE 19.A   Average Annual Precipitation Defining the Boundary between Dry and Humid Climates

Average Annual Temp. (C°)	Winter Rainfall Maximum (centimeters)	Even Distribution (centimeters)	Summer Maximum (centimeters)
5	10	24	38
10	20	34	48
15	30	44	58
20	40	54	68
25	50	64	78
30	60	74	88

situation exists because air reaching the leeward side has lost much of its moisture, and if the air descends, it is compressed and warmed, making cloud formation even less likely. The dry region that results is often referred to as a **rainshadow desert.** Because many middle-latitude deserts occupy sites on the leeward sides of mountains, they can also be classified as rainshadow deserts. In North America, the foremost mountain barriers to moisture from the Pacific are the Coast Ranges, Sierra Nevada, and Cascades. (Figure 19.4). In Asia, the great Himalayan chain prevents the summertime monsoon flow of moist Indian Ocean air from reaching the interior (Box 19.2).

Because the Southern Hemisphere lacks extensive land areas in the middle latitudes, only a small areas of desert and steppe occurs in this latitude range, existing primarily near the southern tip of South America in the rainshadow of the towering Andes.

The middle-latitude deserts provide an example of how tectonic processes affect climate. Rainshadow deserts exist by virtue of the mountains produced when plates collide. Without such mountain-building episodes, wetter climates would prevail where many dry regions exist today.

GEOLOGIC PROCESSES IN ARID CLIMATES

GEODe EARTH

DESERTS AND WINDS
▶ Common Misconceptions about Deserts

The angular hills, the sheer canyon walls, and the desert surface of pebbles or sand contrast sharply with the rounded hills and curving slopes of more humid places. To a visitor from a humid region, a desert landscape may seem to have been shaped by forces different from those operating in well-watered areas. However, although the contrasts may be striking, they do not reflect different processes. They merely disclose the differing effects of the same processes that operate under contrasting climatic conditions.

PEOPLE AND THE ENVIRONMENT

The Disappearing Aral Sea—A Large Lake Becomes a Barren Wasteland

The Aral Sea lies on the border between Uzbekistan and Kazakhstan in central Asia (Figure 19.A). The setting is the Turkestan desert, a middle-latitude desert in the rain-shadow of Afghanistan's high mountains. In this region of interior drainage, two large rivers, the Amu Darya and the Syr Darya, carry water from the mountains of northern Afghanistan across the desert to the Aral Sea. Water leaves the sea by evaporation. Thus, the size of the water body depends upon the balance between river inflow and evaporation.

In 1960 the Aral Sea was one of the world's largest inland water bodies, with an area of about 67,000 square kilometers (26,000 square miles). Only the Caspian Sea, Lake Superior, and Lake Victoria were larger. By the year 2008, the area of the Aral Sea was about 10 percent of its 1960 size. Its volume was reduced by nearly 90 percent, from 708 cubic kilometers to 75 cubic kilometers. The shrinking Aral Sea is depicted in Figure 19.B. All that remains are three shallow remnants.

What caused the Aral Sea to dry up? The answer is that the flow of water from the mountains that supplied the sea was significantly reduced and then all but eliminated. The waters of the Amu Darya and Syr Darya were diverted to supply a major expansion of irrigated agriculture in this dry realm.

The intensive irrigation greatly increased agricultural productivity, but not without significant costs. The deltas of the two major rivers have lost their wetlands, and wildlife has disappeared. The once thriving fishing industry is dead, and the 24 species of fish that once lived in the Aral Sea are no longer there. The shoreline is now tens of kilometers from the towns that were once fishing centers.

The shrinking sea has exposed millions of acres of former seabed to sun and wind. The surface is encrusted with salt and with

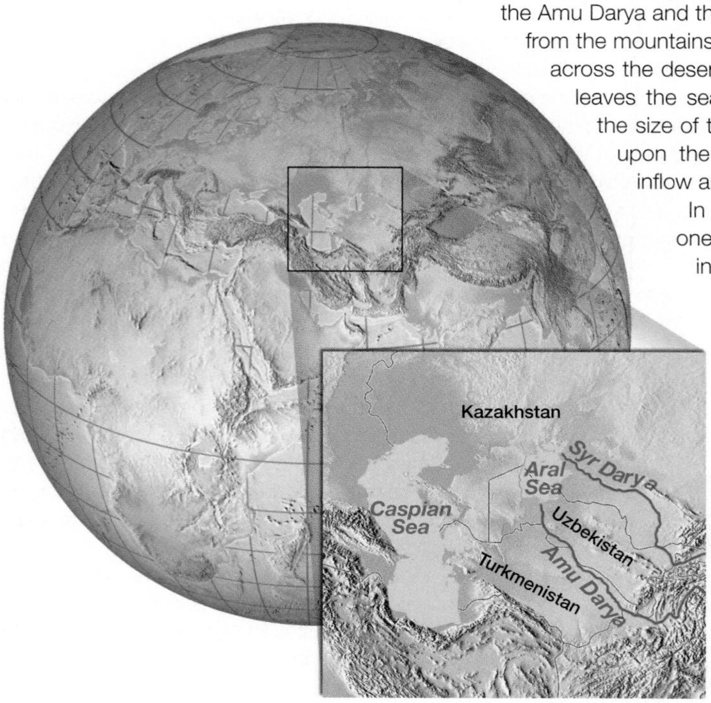

FIGURE 19.A The Aral Sea lies east of the Caspian Sea in the Turkestan Desert. Two rivers, the Amu Darya and Syr Darya, bring water from the mountains to the south.

Students Sometimes Ask . . .

Are all deserts hot?

No, but many deserts do experience some very high temperatures. For instance, the highest authentically recorded temperature in the United States—as well as the entire Western Hemisphere—is 57 °C (134 °F), measured at Death Valley, California, on July 10, 1913. The world-record high temperature of nearly 59 °C (137 °F) was recorded in Azizia, Libya, in North Africa's Sahara Desert on September 13, 1922.

Despite these remarkably high figures, cold temperatures are also experienced in desert regions. For example, the average daily minimum temperature in January in Phoenix, Arizona, is 1.7 °C (35 °F), just barely above freezing. At Ulan Bator in Mongolia's Gobi Desert the average *high* temperature on January days is only −19 °C (−2 °F). Dry climates are found from the tropics poleward to the high middle latitudes. Although tropical deserts lack a cold season, deserts in the middle latitudes do experience seasonal temperature changes, which cause some to get quite cold.

Weathering

In humid regions, relatively fine-textured soils support an almost continuous cover of vegetation that mantles the surface and the slopes and rock edges are rounded, reflecting the strong influence of chemical weathering in a humid climate. By contrast, much of the weathered debris in deserts consists of unaltered rock and mineral fragments—the result of mechanical weathering processes. In dry lands, rock weathering of any type is greatly reduced because of the lack of moisture and the scarcity of organic acids from decaying plants. However, chemical weathering is not completely lacking in deserts. Over long spans of time, clays and thin soils do form, and many iron-bearing silicate minerals oxidize, producing the rust-colored stain that tints some desert landscapes.

The Role of Water

Permanent streams are normal in humid regions, but practically all desert streambeds are dry most of the time (Figure 19.5A). Deserts have **ephemeral streams,** (*ephemero* = short-lived) which

agrochemicals brought by the rivers. Strong winds routinely pick up and deposit thousands of tons of material every year. This process has not only contributed to a significant reduction in air quality for people living in the region but has also appreciably affected crop yields due to the deposition of salt-rich sediments on arable land.

The shrinking Aral Sea has had a noticeable impact on the region's climate. Without the moderating effect of a large water body, there are greater extremes of temperature, a shorter growing season, and reduced local precipitation. These changes have caused many farms to switch from growing cotton to growing rice, which demands even more diverted water.

Could this crisis be reversed if enough fresh water were to once again flow into the Aral Sea? Prospects appear grim. Experts estimate that restoring the Aral

FIGURE 19.B   The shrinking Aral Sea.

Sea to about twice its present size would require stopping all irrigation from the two major rivers for decades. This could not be done without ruining the economies of the countries that rely on that water. One effort has improved the situation in the northern portion of the Aral. In November 2005, a large earthen dam was completed that blocked the southward outflow of water. Prior to this structure, the modest amount of water contributed by the Syr Darya was lost as it flowed southward and evaporated. The dam allowed water from the Syr Darya to recharge and partially restore this portion of the water body. The North Aral Sea is now larger and less salty than before.

The decline of the Aral Sea is a major environmental disaster that, sadly, is of human making.

A.

FIGURE 19.5   A. Most of the time, desert stream channels are dry. B. An ephemeral stream shortly after a heavy shower. Although such floods are short-lived, large amounts of erosion occur. (Photos by E. J. Tarbuck)

B.

means they carry water only in response to specific episodes of rainfall. A typical ephemeral stream might flow only a few days or perhaps just a few hours during the year. In some years the channel might carry no water at all.

This fact is obvious even to the casual traveler who notices numerous bridges with no streams beneath them or numerous dips in the road where dry channels cross. However, when the rare heavy showers do come, so much rain falls in such a short time that all of it cannot soak in (Figure 19.6). Because desert vegetative cover is sparse, runoff is largely unhindered and consequently rapid, often creating flash floods along valley floors (Figure 19.5B). These floods are quite unlike floods in humid regions. A flood on a river

FIGURE 19.6   Desert thunderstorm over Tucson, Arizona. There are often many weeks, months, or occasionally even years separating periods of rain in the desert. When rains do occur, they are often heavy and of relatively short duration. Because the rainfall intensity is high, all of the water cannot soak in and rapid runoff results. (Photo by Andrew Brown/Ecoscience/Corbis)

almost 3000 kilometers of the Sahara without a single tributary. By contrast, in humid regions the discharge of a river grows as it flows downstream because tributaries and groundwater contribute additional water along the way.

It should be emphasized that *running water, although infrequent, nevertheless does most of the erosional work in deserts* (Figure 19.7). This is contrary to the common belief that wind is the most important erosional agent sculpturing desert landscapes. Although wind erosion is indeed more significant in dry areas than elsewhere, most desert landforms are carved by running water. As you will see shortly, the main role of wind is in the transportation and deposition of sediment, which creates and shapes the ridges and mounds we call dunes.

like the Mississippi may take several days to reach its crest and then subside. But desert floods arrive suddenly and subside quickly. Because much surface material in a desert is not anchored by vegetation, the amount of erosional work that occurs during a single short-lived rain event is impressive.

In the dry, western United States, different names are used for ephemeral streams, including *wash* and *arroyo*. In other parts of the world, a dry desert stream may be a *wadi* (Arabia and North Africa), a *donga* (South America), or a *nullah* (India).

Humid regions are notable for their integrated drainage systems. But in arid regions, streams usually lack an extensive system of tributaries. In fact, a basic characteristic of desert streams is that they are small and die out before reaching the sea. Because the water table is usually far below the surface, few desert streams can draw upon it as streams do in humid regions (see Figure 17.7, p. 468). Without a steady supply of water, the combination of evaporation and infiltration soon depletes the stream.

The few permanent streams that do cross arid regions, such as the Colorado and Nile rivers, originate *outside* the desert, often in well-watered mountains. Here the water supply must be great to compensate for the losses occurring as the stream crosses the desert. For example, after the Nile leaves its headwaters in the lakes and mountains of central Africa, it traverses

BASIN AND RANGE: THE EVOLUTION OF A DESERT LANDSCAPE

DESERTS AND WINDS
GEODe EARTH   ▶ Reviewing Landforms and Landscapes

Because arid regions typically lack permanent streams, they are characterized as having **interior drainage.** This means that they have a discontinuous pattern of intermittent streams that do not flow out of the desert to the ocean. In the United States, the dry Basin and Range region provides an excellent example. The region includes southern Oregon, all of Nevada, western Utah, southeastern California, southern Arizona, and southern New Mexico. The name Basin and Range is an apt description for this almost 800,000-square-kilometer region because it is characterized by more than 200 relatively small mountain ranges that rise 900 to 1500 meters (3000 to 5000 feet) above the basins that separate them.

In this region, as in others like it around the world, erosion mostly occurs without reference to the ocean (ultimate base level) because the interior drainage never reaches the sea. Even where permanent streams flow to the ocean, few tributaries exist, and thus only a narrow strip of land adjacent to the stream has sea level as its ultimate level of land reduction.

Because the coarsest material is dropped first, the head of the fan is steepest, having a slope of perhaps 10 to 15 degrees. Moving down the fan, the size of the sediment and the steepness of the slope decrease and merge imperceptibly with the basin floor. An examination of the fan's surface would likely reveal a braided channel pattern because of the water shifting its course as successive channels became choked with sediment. Over the years a fan enlarges, eventually coalescing with fans from adjacent canyons to produce an apron of sediment called a **bajada** along the mountain front.

On the rare occasions of abundant rainfall, streams may flow across the bajada to the center of the basin, converting the basin floor into a shallow **playa lake.** Playa lakes are temporary features that last only a few days or at best a few weeks before evaporation and infiltration remove the water. The dry, flat lake bed that remains is called a **playa.** Playas are typically composed of fine silts and clays and are occasionally encrusted with salts precipitated during evaporation (see Figure 7.18, p. 212). These precipitated salts may be unusual. A case in point is the sodium borate (better known as borax) mined from ancient playa lake deposits in Death Valley, California.

With the ongoing erosion of the mountain mass and the accompanying sedimentation, the local relief continues to diminish. Eventually, nearly the entire mountain mass is gone. Thus, by the late stages of erosion, the mountain areas are reduced to a few large bedrock knobs projecting above the surrounding sediment-filled basin. These isolated erosional remnants on a late-stage desert landscape are called **inselbergs,** a German word meaning "island mountains" (Box 19.3).

Each of the stages of landscape evolution in an arid climate depicted in Figure 19.8 can be observed in the Basin and Range region. Recently uplifted mountains in an early stage of erosion are found in southern Oregon and northern Nevada. Death Valley, California, and southern Nevada fit into the more advanced middle stage, whereas the late stage, with its inselbergs, can be seen in southern Arizona.

Figure 19.9 is a satellite image of a portion of Death Valley. Many of the features that were just described are visible. This February 2005 image shows the area shortly after a rare heavy rain. As has occurred many times over thousands of years, a

FIGURE 19.7 Desert rains are infrequent, but when they occur, erosion can be significant. The dry channels of ephemeral streams are conspicuous on this aerial view of a parched desert surface in northern Arizona. (Photo by Michael Collier)

The block models in Figure 19.8 depict how the landscape has evolved in the Basin and Range region. During and following the uplift of the mountains, running water begins carving the elevated mass and depositing large quantities of debris in the basin. During this early stage, relief is greatest, because as erosion lowers the mountains and sediment fills the basins, elevation differences gradually diminish.

When the occasional torrents of water produced by sporadic rains move down the mountain canyons, they are heavily loaded with sediment. Emerging from the confines of the canyon, the runoff spreads over the gentler slopes at the base of the mountains and quickly loses velocity. Consequently, most of its load is dumped within a short distance. The result is a cone of debris at the mouth of a canyon known as an **alluvial fan.**

Students Sometimes Ask . . .

Where is the driest desert on Earth?

The Atacama Desert of Chile has the distinction of being the world's driest desert. This relatively narrow belt of arid land extends for about 1200 kilometers (750 miles) along South America's Pacific Coast (see Figure 19.2). It is said that some portions of the Atacama have not received rain for more than 400 years! One must view such pronouncements skeptically. Nevertheless, for places where records have been kept, Arica, Chile, in the northern part of the Atacama, has experienced a span of 14 years without measurable rainfall.

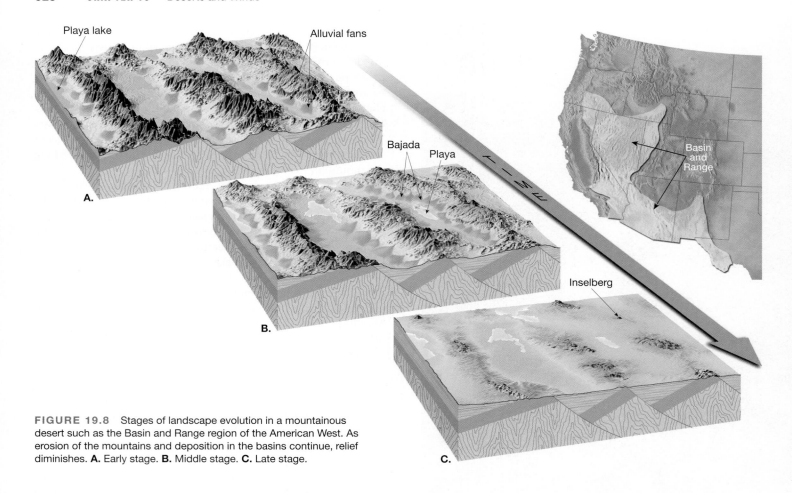

FIGURE 19.8 Stages of landscape evolution in a mountainous desert such as the Basin and Range region of the American West. As erosion of the mountains and deposition in the basins continue, relief diminishes. **A.** Early stage. **B.** Middle stage. **C.** Late stage.

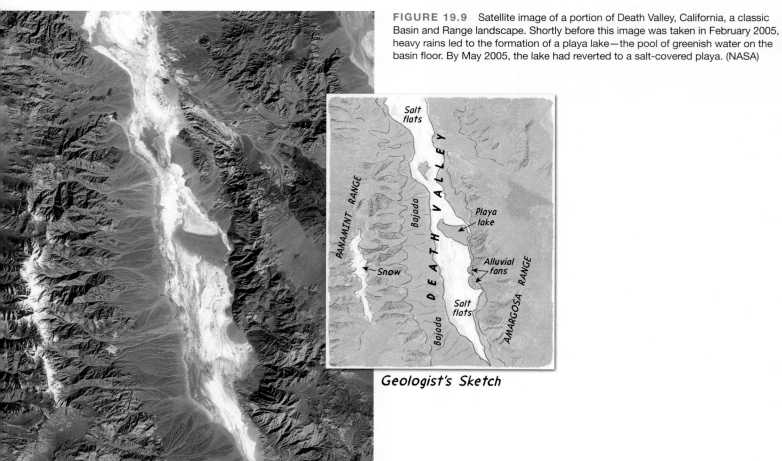

FIGURE 19.9 Satellite image of a portion of Death Valley, California, a classic Basin and Range landscape. Shortly before this image was taken in February 2005, heavy rains led to the formation of a playa lake—the pool of greenish water on the basin floor. By May 2005, the lake had reverted to a salt-covered playa. (NASA)

Geologist's Sketch

UNDERSTANDING EARTH

BOX 19.3

Australia's Mount Uluru

When travelers contemplating a trip to Australia consult brochures and other tourist literature, they are bound to see a photograph or read a description of Mount Uluru (formerly Ayers Rock). As Figure 19.C illustrates, this well-known attraction is a massive feature that rises steeply from the surrounding plain. Located in Uluru—Kata Tjuta National Park, southwest of Alice Springs in the dry center of the continent, the roughly circular monolith is more than 350 meters (1200 feet) high, and its base is more than 9.5 kilometers (6 miles) in circumference. Its summit is flattened, its sides furrowed. The rock type is sandstone, and the hues of red and orange change with the light of day. In addition to being a striking geological attraction, Mount Uluru is of interest because it is a sacred place for the aboriginal tribes of the region.

Mount Uluru is a spectacular example of a feature known as an inselberg. *Inselberg* is a German word meaning "island mountain" and seems appropriate because these masses clearly resemble rocky islands standing above the surface of a broad sea. Similar features are scattered throughout many other arid and semiarid regions of the world. Mount Uluru is a special type of inselberg that consists of a very resistant rock mass exhibiting a rounded or domed form. Such masses are termed *bornhardts* for the 19th-century German explorer Wilhelm Bornhardt, who described similar features in parts of Africa.

Bornhardts form in regions where massive or resistant rock such as granite or sandstone is surrounded by rock that is more susceptible to weathering. The greater susceptibility of the adjacent rock is often the result of its being more highly jointed. Joints allow water, and therefore weathering processes, to penetrate to greater depths. When the adjacent, deeply weathered rock is stripped away by erosion, the far less weathered rock mass remains standing high. After a bornhardt forms, it tends to shed water. By contrast, the land surrounding a bornhardt helps to perpetuate its existence by reinforcing the processes that created it. In fact, masses such as Mount Uluru can remain a part of the landscape for tens of millions of years.

Bornhardts are more common in the lower latitudes because the weathering that is responsible for their formation proceeds more rapidly in warmer climates. In regions that are now arid or semiarid, bornhardts may reflect times when the climate was wetter than it is today.

FIGURE 19.C Mount Uluru (formerly Ayers Rock) rises conspicuously above the dry plains of central Australia. It is a type of inselberg known as a *bornhardt*. As erosion gradually lowers the surface, the less weathered massive rock remains standing high above the more jointed and more easily weathered rock that surrounds it. (Photo by Art Wolfe, Inc.)

wide, shallow playa lake formed in the lowest spot. By May 2005, only three months after the storm, the valley floor had returned to being a dry, salt-encrusted playa.

TRANSPORTATION OF SEDIMENT BY WIND

Moving air, like moving water, is turbulent and able to pick up loose debris and transport it to other locations. Just as in a stream, the velocity of wind increases with height above the surface. Also like a stream, wind transports fine particles in suspension while heavier ones are carried as bed load. However, the transport of sediment by wind differs from that of running water in two significant ways. First, wind's lower density compared to water renders it less capable of picking up and transporting coarse materials. Second, because wind is not confined to channels, it can spread sediment over large areas, as well as high into the atmosphere.

Bed Load

The **bed load** carried by wind consists of sand grains. Observations in the field and experiments using wind tunnels indicate that windblown sand moves by skipping and bouncing along the surface—a process termed **saltation.** The term is not a reference to salt, but instead derives from the Latin word meaning "to jump."

The movement of sand grains begins when wind reaches a velocity sufficient to overcome the inertia of the resting particles. At first the sand rolls along the surface. When a moving sand grain strikes another grain, one or both of them may jump into the air. Once in the air, the grains are carried forward by the wind until gravity pulls them back toward the surface. When the sand hits the surface, it either bounces back into the air or dislodges other grains, which then jump upward. In this manner, a chain reaction is established, filling the air near the ground with saltating sand grains in a short period of time (Figure 19.10).

Bouncing sand grains never travel far from the surface. Even when winds are very strong, the height of the saltating sand seldom exceeds a meter and usually is no greater than a half meter. Some sand grains are too large to be thrown into the air by impact from other particles. When this is the case, the energy provided by the impact of the smaller saltating grains drives the larger grains forward. Estimates indicate that between 20 and 25 percent of the sand transported in a sandstorm is moved in this way.

Suspended Load

Unlike sand, finer particles of dust can be swept high into the atmosphere by the wind. Because dust is often composed of rather flat particles that have large surface areas compared to their weight, it is relatively easy for turbulent air to counterbalance the pull of gravity and keep these fine particles airborne for hours or even days. Although both silt and clay can be carried in suspension, silt commonly makes up the bulk of the **suspended load** because the reduced level of chemical weathering in deserts provides only small amounts of clay.

Fine particles are easily carried by the wind, but they are not so easily picked up to begin with. The reason is that the wind velocity is practically zero within a very thin layer close to the ground. Thus, the wind cannot lift the sediment by itself. Instead, the dust must be ejected or spattered into the moving air by bouncing sand grains or other disturbances. This idea is illustrated nicely by a dry, unpaved country road on a windy day. Left undisturbed, little dust is raised by the wind. However, as a

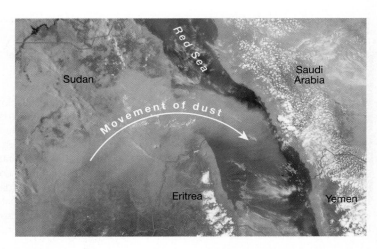

FIGURE 19.11   This satellite image shows thick plumes of dust from the Sahara Desert blowing across the Red Sea on June 30, 2009. Such dust storms are common in arid North Africa. In fact, this region is the largest dust source in the world. Satellites are an excellent tool for studying the transport of dust on a global scale. They show that dust storms can cover huge areas and that dust can be transported great distances. (Image courtesy of NASA)

car or truck moves over the road, the layer of silt is kicked up, creating a thick cloud of dust.

Although the suspended load is usually deposited relatively near its source, high winds are capable of carrying large quantities of dust great distances (Figure 19.11). In the 1930s, silt that was picked up in Kansas was transported to New England and beyond into the North Atlantic. Similarly, dust blown from the Sahara has been traced as far as the West Indies.

WIND EROSION

DESERTS AND WINDS
▶ Common Misconceptions about Deserts

Compared to running water and glaciers, wind is a relatively insignificant erosional agent. Recall that even in deserts, most erosion is performed by intermittent running water, not by the wind. Wind erosion is more effective in arid lands than in humid areas because in humid places moisture binds particles together and vegetation anchors the soil. For wind to be an effective erosional force, dryness and scanty vegetation are important prerequisites. When such circumstances exist, wind may pick up, transport, and deposit great quantities of fine sediment. During the 1930s, parts of the Great Plains experienced vast dust storms. The plowing under of the natural vegetative cover for farming, followed by severe drought, exposed the land to wind erosion and led to the area's being labeled the Dust Bowl.

Deflation and Blowouts

One way that wind erodes is by **deflation** (*de* = out, *flat* = blow), the lifting and removal of loose material. Deflation sometimes is difficult to notice because the entire surface is being lowered at the same time, but it can be significant. In portions of the 1930s

FIGURE 19.10   A cloud of saltating sand grains moving up the gentle slope of a dune. (Photo by Stephen Trimble)

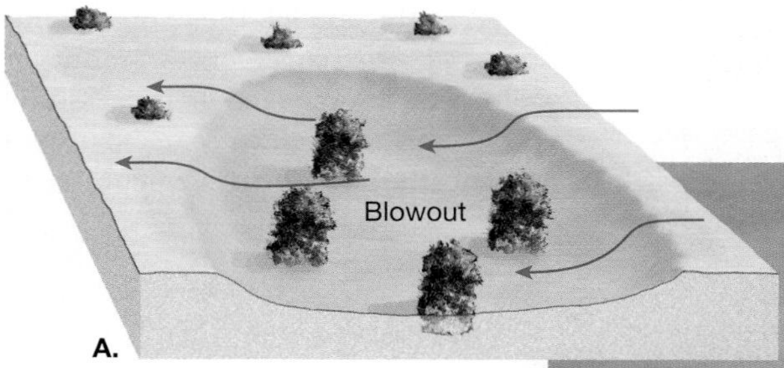

FIGURE 19.12 **A.** Blowouts are depressions created by deflation. Land that is dry and largely unprotected by anchoring vegetation is particularly susceptible. **B.** In this example, deflation has removed about 4 feet of soil—the distance from the man's outstretched arm to his feet. (Photo courtesy of U.S.D.A./Natural Resources Conservation Service)

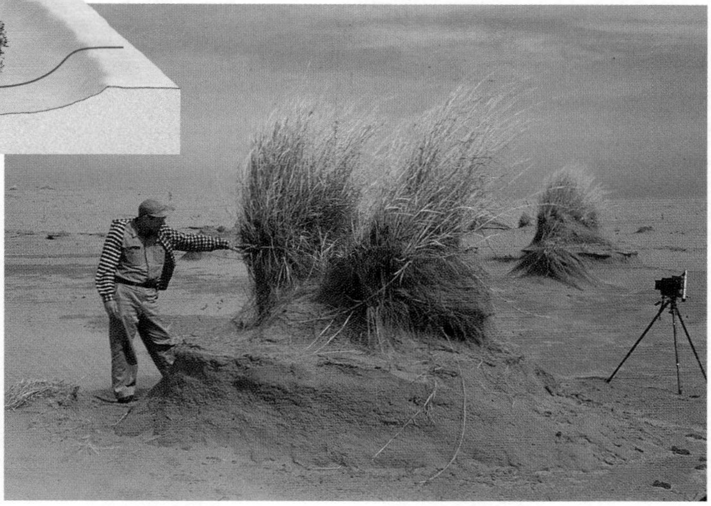

A.

Blowout

B.

(Figure 19.13). Beneath is a layer containing a significant proportion of silt and sand. When desert pavement is present, it is an important control on wind erosion because pavement stones are too large for deflation to remove. When this armor is disturbed, wind can easily erode the exposed fine silt.

For many years, the most common explanation for the formation of desert pavement was that it develops when wind removes sand and silt within poorly sorted surface deposits. As Figure 19.14A illustrates, the concentration of larger particles at the surface gradually increases as the finer particles are blown away. Eventually the surface is completely covered with pebbles and cobbles too large to be moved by the wind.

Studies have shown that the process depicted in Figure 19.14A is not an adequate explanation for all environments in which desert pavement exists. For example, in many places, desert pavement is underlain by a relatively thick layer of silt that contains few if any pebbles and cobbles. In such a setting, deflation of fine sediment could not leave behind a layer of coarse particles. Studies also showed that in some areas the pebbles and cobbles composing desert pavement have all been exposed at the surface for about the same length of time. This would not be the case for the process shown in Figure 19.14A.

Dust Bowl, vast areas of land were lowered by as much as a meter in only a few years.

The most noticeable results of deflation in some places are shallow depressions appropriately called **blowouts** (Figure 19.12). In the Great Plains region, from Texas north to Montana, thousands of blowouts are visible on the landscape. They range from small dimples less than a meter deep and 3 meters wide to depressions that approach 50 meters in depth and several kilometers across. The factor that controls the depths of these basins (that is, acts as base level) is the local water table. When blowouts are lowered to the water table, damp ground and vegetation prevent further deflation.

Desert Pavement

In portions of many deserts, the surface consists of a closely packed layer of coarse particles. This veneer of pebbles and cobbles, called **desert pavement,** is only one or two stones thick

FIGURE 19.13 Desert pavement consists of a closely packed veneer of pebbles and cobbles that is only one or two stones thick. Beneath the pavement is material containing a significant proportion of finer particles. If left undisturbed, desert pavement will protect the surface from deflation. (Photo by Bobbé Christopherson)

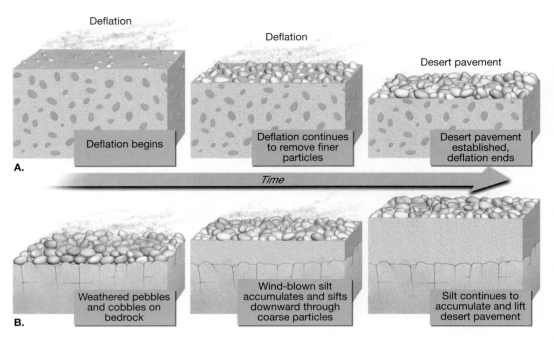

Deflation

Deflation

Desert pavement

| Deflation begins | Deflation continues to remove finer particles | Desert pavement established, deflation ends |

A.

Time

| Weathered pebbles and cobbles on bedrock | Wind-blown silt accumulates and sifts downward through coarse particles | Silt continues to accumulate and lift desert pavement |

B.

FIGURE 19.14   A. This model portrays an area with poorly sorted surface deposits. Coarse particles gradually become concentrated into a tightly packed layer as deflation lowers the surface by removing sand and silt. Here desert pavement is the result of wind erosion. **B.** This model shows the formation of desert pavement on a surface initially covered with coarse pebbles and cobbles. Wind-blown dust accumulates at the surface and gradually sifts downward through spaces between coarse particles. Infiltrating rainwater aids the process. This depositional process raises the surface and produces a layer of coarse pebbles and cobbles underlain by a substantial layer of fine sediment.

Here, the coarse particles that make up the pavement reach the surface over an extended time span as deflation gradually removes the fine material.

As a result, an alternate explanation for desert pavement was formulated (Figure 19.14B). This hypothesis suggests that pavement develops on a surface that initially consists of coarse particles. Over time, protruding cobbles trap fine, windblown grains that settle and sift downward through the spaces between the larger surface stones. The process is aided by infiltrating rainwa-

ter. In this model, the cobbles composing the pavement were never buried. Moreover, it successfully explains the lack of coarse particles beneath the desert pavement.

Ventifacts and Yardangs

Like glaciers and streams, wind also erodes by **abrasion** (*ab* = away *radere* = to scrape). In dry regions as well as along some beaches, windblown sand cuts and polishes exposed rock surfaces. Abrasion sometimes creates interestingly shaped stones called **ventifacts** (Figure 19.15A). The side of the stone exposed to the prevailing wind is abraded, leaving it polished, pitted, and with sharp edges. If the wind is not consistently from one direction, or if the pebble becomes reoriented, it may have several faceted surfaces.

Unfortunately, abrasion is often given credit for accomplishments beyond its capabilities. Such features as balanced rocks that stand high atop narrow pedestals, and intricate detailing on tall pinnacles, are not the results of abrasion. Sand seldom travels more than a meter above the surface, so the wind's sandblasting effect is obviously limited in vertical extent.

In addition to ventifacts, wind erosion is responsible for creating much larger features, called yardangs (from the Turkistani word *yar*, meaning "steep bank"). A **yardang** is a streamlined, wind-sculpted landform that is oriented parallel to the prevailing wind (Figure 19.15B). Individual yardangs are generally small features

A.

B.

FIGURE 19.15   A. Ventifacts are rocks that are polished and shaped by sandblasting (Photo by Stephen Trimble). **B.** Yardangs are usually small, wind-sculpted landforms that are aligned parallel with the wind. (Photo by Peter M. Wilson/CORBIS)

that stand less than 5 meters (16 feet) high and no more than about 10 meters (32 feet) long. Because the sand-blasting effect of wind is greatest near the ground, these abraded bedrock remnants are usually narrower at their base. Sometimes yardangs are large features. Peru's Ica Valley contains yardangs that approach 100 meters (330 feet) in height and several kilometers in length. Some in the desert of Iran reach 150 meters (nearly 500 feet) in height.

WIND DEPOSITS

DESERTS AND WINDS
▶ Reviewing Landforms and Landscapes

Although wind is relatively unimportant in producing *erosional* landforms, significant *depositional* landforms are created by the wind in some regions. Accumulations of windblown sediment are particularly conspicuous in the world's dry lands and along many sandy coasts. Wind deposits are of two distinctive types: (1) mounds and ridges of sand from the wind's bed load, which we call dunes, and (2) extensive blankets of silt, called loess, that once were carried in suspension.

Sand Deposits

As is the case with running water, wind drops its load of sediment when velocity falls and the energy available for transport diminishes. Thus, sand begins to accumulate wherever an obstruction across the path of the wind slows its movement. Unlike many deposits of silt, which form blanketlike layers over large areas, winds commonly deposit sand in mounds or ridges called **dunes** (Figure 19.16).

As moving air encounters an object, such as a clump of vegetation or a rock, the wind sweeps around and over it, leaving a shadow of slower-moving air behind the obstacle, as well as a smaller zone of quieter air just in front of the obstacle. Some of the saltating sand grains moving with the wind come to rest in these wind shadows. As the accumulation of sand continues, it becomes a more imposing barrier to the wind and thus a more efficient trap for even more sand. If there is a sufficient supply of sand and the wind blows steadily for a long enough time, the mound of sand grows into a dune.

Many dunes have an asymmetrical profile, with the leeward (sheltered) slope being steep and the windward slope more gently inclined. The dunes in Figure 19.16 are a good example. Sand moves up the gentler slope on the windward side by saltation. Just beyond the crest of the dune, where the wind velocity is reduced, the sand accumulates. As more sand collects, the slope steepens and eventually some of it slides under the pull of gravity. In this way, the leeward slope of the dune, called the **slipface,** maintains an angle of about 34 degrees, the angle of repose for loose dry sand. (Recall from Chapter 15 that the angle of repose is the steepest angle at which loose material remains stable.) Continued sand accumulation, coupled with periodic slides down the slipface, results in the slow migration of the dune in the direction of air movement.

As sand is deposited on the slipface, layers form that are inclined in the direction the wind is blowing. These sloping lay-

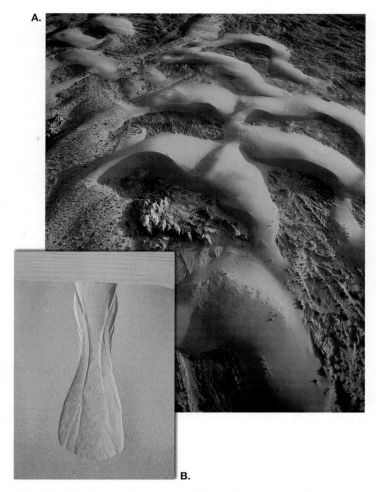

FIGURE 19.16 Part **A** shows the White Point dunes near Preston Mesa, Arizona. Strong winds move sand up the more gentle windward slopes. As sand accumulates near the dune crest, the slope becomes steeper. Eventually, some of the sand slides down the *slip face* as occurred in part **B**. (Photos by Michael Collier)

ers are called **cross beds** (Figure 19.17). When the dunes are eventually buried under other layers of sediment and become part of the sedimentary rock record, their asymmetrical shape is destroyed, but the cross beds remain as testimony to their origin. Nowhere is cross-bedding more prominent than in the sandstone walls of Zion Canyon in southern Utah (Figure 19.17).

For some areas, moving sand is troublesome. In Figure 19.18, dunes are advancing across irrigated fields in Egypt. In portions of the Middle East, valuable oil rigs must be protected from encroaching dunes. In some cases, fences are built sufficiently upwind of the dunes to stop their migration. As sand continues to collect, however, the fences must be built higher. In Kuwait protective fences extend for almost 10 kilometers around one important oil field. Migrating dunes can also pose a problem to the construction and maintenance of highways and railroads that cross sandy desert regions. For example, to keep a portion of Highway 95 near Winnemucca, Nevada, open to traffic, sand must be taken away about three times a year. Each time, between 1500 and 4000 cubic meters of sand are removed. Attempts at stabilizing the dunes by planting different varieties of grasses have been unsuccessful because the meager rainfall cannot support the plants.

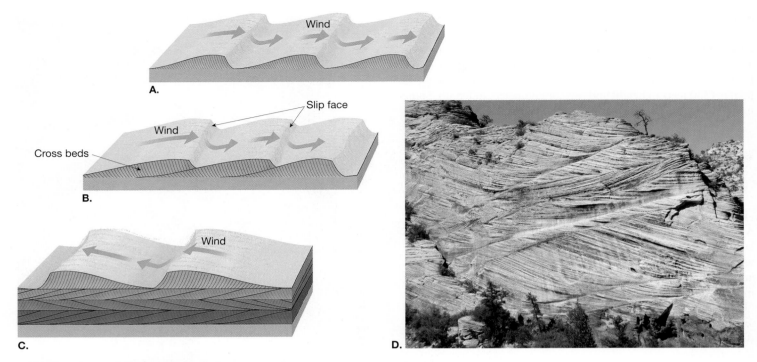

FIGURE 19.17   As parts **A** and **B** illustrate, dunes commonly have an asymmetrical shape. The steeper leeward side is called the *slipface.* Sand grains deposited on the slipface at the angle of repose create the cross-bedding of the dunes. **C.** Over time, a complex pattern develops. Also notice that when dunes are buried and become part of the sedimentary record, the cross-bedded structure is preserved. **D.** Cross beds are an obvious characteristic of the Navajo Sandstone in Zion National Park, Utah. (Photo by Dennis Tasa)

Types of Sand Dunes

Dunes are not just random heaps of windblown sediment. Rather, they are accumulations that usually assume patterns that are surprisingly consistent. Addressing this point, a leading early investigator of dunes, the British engineer R. A. Bagnold, observed: "Instead of finding chaos and disorder, the observer never fails to be amazed at a simplicity of form, an exactitude of repetition, and a geometric order. . . ." A broad assortment of dune forms exists, generally simplified to a few major types for discussion.

Of course, gradations exist among different forms as well as irregularly shaped dunes that do not fit easily into any category. Several factors influence the form and size that dunes ultimately assume. These include wind direction and velocity, availability of sand, and the amount of vegetation present. Six basic dune types are shown in Figure 19.19, with arrows indicating wind directions.

Barchan Dunes   Solitary sand dunes shaped like crescents and with their tips pointing downwind are called **barchan dunes** (Figure 19.19A). These dunes form where supplies of sand are limited and the surface is relatively flat, hard, and lacking vegetation. They migrate slowly with the wind at a rate of up to 15 meters (50 feet) per year. Their size is usually modest, with the largest barchans reaching heights of about 30 meters (100 feet) while the maximum spread between their horns approaches 300 meters (1000 feet). When the wind direction is nearly constant, the crescent form of these dunes is nearly symmetrical. However, when the wind direction is not perfectly fixed, one tip becomes larger than the other.

FIGURE 19.18   These desert dunes (called *barchans*) in Egypt are advancing from right to left across irrigated fields. (Photo by Georg Gerster/Photo Researchers, Inc.)

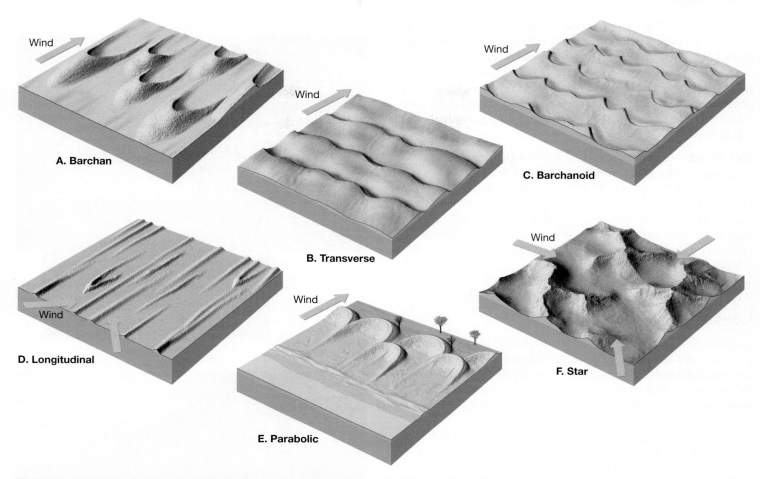

FIGURE 19.19 Sand dune types. **A.** Barchan dunes. **B.** Transverse dunes. **C.** Barchanoid dunes. **D.** Longitudinal dunes. **E.** Parabolic dunes. **F.** Star dunes.

Transverse Dunes   In regions where the prevailing winds are steady, sand is plentiful, and vegetation is sparse or absent, the dunes form a series of long ridges that are separated by troughs and oriented at right angles to the prevailing wind. Because of this orientation, they are termed **transverse dunes** (Figure 19.19B). Typically, many coastal dunes are of this type. In addition, transverse dunes are common in many arid regions where the extensive surface of wavy sand is sometimes called a *sand sea*. In some parts of the Sahara and Arabian deserts, transverse dunes reach heights of 200 meters, are 1 to 3 kilometers across, and can extend for distances of 100 kilometers or more.

There is a relatively common dune form that is intermediate between isolated barchans and extensive waves of transverse dunes. Such dunes, called **barchanoid dunes,** form scalloped rows of sand oriented at right angles to the wind (Figure 19.19C). The rows resemble a series of barchans that have been positioned side by side.

Longitudinal Dunes   **Longitudinal dunes** are long ridges of sand that form more or less parallel to the prevailing wind and where sand supplies are moderate (Figure 19.19D). Apparently the prevailing wind direction must vary somewhat but still remain in the same quadrant of the compass. Although the smaller types are only 3 or 4 meters high and several dozens of meters long, in some large deserts longitudinal dunes can reach great size. For example, in portions of North Africa, Arabia, and central Australia, these dunes may approach a height of 100 meters (300 feet) and extend for distances of more than 100 kilometers (62 miles).

Parabolic Dunes   Unlike the other dunes that have been described thus far, **parabolic dunes** form where vegetation partially

Students Sometimes Ask . . .

Aren't deserts mostly covered with sand dunes?

A common misconception about deserts is that they consist of mile after mile of drifting sand dunes. It is true that sand accumulations do exist in some areas and may be striking features. But, perhaps surprisingly, sand accumulations worldwide represent only a small percentage of the total desert area. For example, in the Sahara—the world's largest desert—accumulations of sand cover only *one-tenth* of its area. The sandiest of all deserts is the Arabian, one-third of which consists of sand.

covers the sand. The shape of these dunes resembles the shape of barchans except that their tips point into the wind rather than downwind (Figure 19.19E). Parabolic dunes often form along coasts where there are strong onshore winds and abundant sand. If the sand's sparse vegetative cover is disturbed at some spot, deflation creates a blowout. Sand is then transported out of the depression and deposited as a curved rim, which grows higher as deflation enlarges the blowout.

Star Dunes   Confined largely to parts of the Sahara and Arabian deserts, **star dunes** are isolated hills of sand that exhibit a complex form (Figure 19.19F). Their name is derived from the fact that the bases of these dunes resemble multipointed stars. Usually three or four sharp-crested ridges diverge from a central high point that in some cases may approach a height of 90 meters (Figure 19.20). As their form suggests, star dunes develop where wind directions are variable.

Loess (Silt) Deposits

In some parts of the world the surface topography is mantled with deposits of windblown silt, called **loess**. Over periods of perhaps thousands of years, dust storms deposited this material. When loess is breached by streams or road cuts, it tends to maintain vertical cliffs and lacks any visible layers, as you can see in Figure 19.21.

The distribution of loess worldwide indicates that there are two primary sources for this sediment: deserts and glacial outwash deposits. The thickest and most extensive deposits of loess on Earth occur in western and northern China. They were blown there from the extensive desert basins of central Asia. Accumulations of 30 meters (100 feet) are common, and thicknesses of more than 100 meters have been measured. It is this fine, buff-colored sediment that gives the Yellow River (Huang Ho) its name.

A.

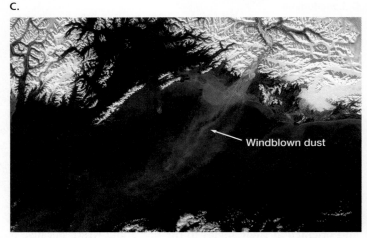

B.

C.

Windblown dust

FIGURE 19.21   A. This vertical loess bluff near the Mississippi River in southern Illinois is about 3 meters high. (Photo by James E. Patterson) **B.** In parts of China loess has sufficient structural strength to permit the excavation of dwellings. (Photo by Christopher Liu/ChinaStock Photo Library) **C.** This satellite image from March 13, 2003, shows streamers of windblown dust moving southward into the Gulf of Alaska. It illustrates a process similar to the one that created many loess deposits in the American Midwest during the Ice Age. Fine silt is produced by the grinding action of glaciers, then transported beyond the margin of the ice by running water and deposited. Later, the fine silt is picked up by strong winds and deposited as loess. (NASA)

In the United States, deposits of loess are significant in many areas, including South Dakota, Nebraska, Iowa, Missouri, and Illinois, as well as portions of the Columbia Plateau in the Pacific Northwest. The correlation between the distribution of loess and important farming regions in the Midwest and eastern Washington State is not just a coincidence because soils derived from this wind-deposited sediment are among the most fertile in the world.

Unlike the deposits in China, which originated in deserts, the loess in the United States (and Europe) is an indirect prod-

FIGURE 19.20   Star dune in the Namib Desert in southwestern Africa. (Photo by Comstock)

uct of glaciation. Its source is deposits of stratified drift. During the retreat of the ice sheets, many river valleys were choked with sediment deposited by meltwater. Strong westerly winds sweeping across the barren floodplains picked up the finer sediment and dropped it as a blanket on the eastern sides of the valleys. Such an origin is confirmed by the fact that loess deposits are thickest and coarsest on the lee side of such major glacial drainage outlets as the Mississippi and Illinois rivers and rapidly thin with increasing distance from the valleys. Furthermore, the angular, mechanically weathered particles composing the loess are essentially the same as the rock flour produced by the grinding action of glaciers.

Students Sometimes Ask . . .

Where are the largest sand dunes found, and how big are they?

The highest dunes in the world are located along the southwest coast of Africa in the Namib Desert. In places, these huge dunes reach heights of 300 to 350 meters (1000 to 1167 feet). The dunes at Great Sand Dunes National Park in southern Colorado are the highest in North America, rising over 210 meters (700 feet) above the surrounding terrain.

CHAPTER 19 DESERTS AND WINDS IN REVIEW

- The *concept of dryness is relative*; it refers to any situation in which a water deficiency exists. Dry regions encompass about 30 percent of Earth's land surface. Two climatic types are commonly recognized: *desert,* which is arid, and *steppe* (a marginal and more humid variant of desert), which is semiarid. *Low-latitude deserts* coincide with the zones of subtropical highs in lower latitudes. On the other hand, *middle-latitude deserts* exist principally because of their positions in the deep interiors of large landmasses far removed from the ocean. Many are in the rainshadow of mountains.

- The same geologic processes that operate in humid regions also operate in deserts, but under contrasting climatic conditions. In dry lands *rock weathering of any type is greatly reduced* because of the lack of moisture and the scarcity of organic acids from decaying plants. Much of the weathered debris in deserts is the result of *mechanical weathering.* Practically all desert streams are dry most of the time and are said to be *ephemeral.* Stream courses in deserts are seldom well integrated and lack an extensive system of tributaries. Nevertheless, *running water is responsible for most of the erosional work in a desert.* Although wind erosion is more significant in dry areas than elsewhere, the main role of wind in a desert is in the transportation and deposition of sediment.

- Because arid regions typically lack permanent streams, they are characterized as having *interior drainage.* Many of the landscapes of the Basin and Range region of the western and southwestern United States are the result of streams eroding uplifted mountain blocks and depositing the sediment in interior basins. *Alluvial fans, playas,* and *playa lakes* are features often associated with these landscapes. In the late stages of erosion, the mountain areas are reduced to a few large bedrock knobs, called *inselbergs,* projecting above sediment-filled basins.

- The transport of sediment by wind differs from that by running water in two ways. First, wind has a low density compared to water; thus, it is not capable of picking up and transporting coarse materials. Second, because wind is not confined to channels, it can spread sediment over large areas. The *bed load* of wind consists of sand grains skipping and bouncing along the surface in a process termed *saltation.* Fine dust particles are capable of being carried by the wind great distances as *suspended load.*

- Compared to running water and glaciers, wind is a less significant erosional agent. *Deflation,* the lifting and removal of loose material, often produces shallow depressions called *blowouts.* Wind also erodes by *abrasion,* often creating interestingly shaped stones called *ventifacts. Yardangs* are narrow, streamlined, wind-sculpted landforms.

- *Desert pavement* is a thin layer of coarse pebbles and cobbles that covers some desert surfaces. Once established, it protects the surface from further deflation. Depending upon circumstances, it may develop as a result of deflation or deposition of fine particles.

- Wind deposits are of two distinct types: (1) *mounds and ridges of sand,* called *dunes,* which are formed from sediment that is carried as part of the wind's bed load, and (2) extensive *blankets of silt,* called *loess,* that once were carried by wind in *suspension.* The profile of a dune shows an asymmetrical shape with the leeward (sheltered) slope being steep and the windward slope more gently inclined. The *types of sand dunes* include (1) *barchan dunes;* (2) *transverse dunes;* (3) *barchanoid dunes;* (4) *longitudinal dunes;* (5) *parabolic dunes;* and (6) *star dunes.* The thickest and most extensive deposits of loess occur in western and northern China. Unlike the deposits in China, which originated in deserts, the loess in the United States and Europe is an indirect product of glaciation.

KEY TERMS

abrasion (p. 532)
alluvial fan (p. 527)
bajada (p. 527)
barchan dune (p. 534)
barchanoid dune (p. 535)
bed load (p. 529)
blowout (p. 531)
cross beds (p. 533)

deflation (p. 530)
desert (p. 520)
desert pavement (p. 531)
dune (p. 533)
ephemeral stream (p. 524)
inselberg (p. 527)
interior drainage (p. 526)
loess (p. 536)

longitudinal dune (p. 535)
parabolic dune (p. 535)
playa (p. 527)
playa lake (p. 527)
rainshadow desert (p. 523)
saltation (p. 529)
slipface (p. 533)
star dune (p. 536)

steppe (p. 520)
suspended load (p. 530)
transverse dune (p. 535)
ventifact (p. 532)
yardang (p. 532)

QUESTIONS FOR REVIEW

1. How extensive are the desert and steppe regions of Earth?

2. What is the primary cause of subtropical deserts? Of middle-latitude deserts?

3. In which hemisphere (Northern or Southern) are middle-latitude deserts most common?

4. Why is the amount of precipitation that is used to determine whether a place has a dry climate or a humid climate a variable figure? (see Box 19.1, p. 523)

5. *Deserts are hot, lifeless, sand-covered landscapes shaped largely by the force of wind.* The preceding statement summarizes the image of arid regions that many people hold, especially those living in more humid places. Is it an accurate view?

6. Why is rock weathering reduced in deserts?

7. As a permanent stream such as the Nile River crosses a desert, does discharge increase or decrease? How does this compare to a river in a humid region?

8. What is the most important erosional agent in deserts?

9. Why is sea level (ultimate base level) not a significant factor influencing erosion in desert regions?

10. Why is the Aral Sea shrinking? (see Box 19.2, p. 524)

11. Describe the features and characteristics associated with each of the stages in the evolution of a mountainous desert. Where in the United States can these stages be observed?

12. Describe the way in which wind transports sand. During very strong winds, how high above the surface can sand be carried?

13. Why is wind erosion relatively more important in arid regions than in humid areas?

14. What factor limits the depths of blowouts?

15. Briefly describe two hypotheses used to explain the formation of desert pavement.

16. Describe the process by which the dunes in Figure 19.16 (p. 533) migrate. In what direction are these dunes moving—toward the top of the photo or toward the bottom?

17. List three factors that influence the form and size of a sand dune.

18. Six major dune types are recognized. Indicate which type of dune is associated with each of the following statements.

    a. dunes whose tips point into the wind

    b. long sand ridges oriented at right angles to the wind

    c. dunes that often form along coasts where strong winds create a blowout

    d. solitary dunes whose tips point downwind

    e. long sand ridges that are oriented more or less parallel to the prevailing wind

    f. an isolated dune consisting of three or four sharp-crested ridges diverging from a central high point

    g. scalloped rows of sand oriented at right angles to the wind

19. Although sand dunes are the best-known wind deposits, accumulations of loess are very significant in some parts of the world. What is loess? Where are such deposits found? What are the origins of this sediment?

COMPANION WEBSITE

The *Earth 10e* Web site uses the resources and flexibility of the Internet to aid in your study of the topics in this chapter. Written and developed by the authors and other geology instructors, this site will help improve your understanding of geology. Visit www.mygeoscienceplace.com in order to:

- **Review** key chapter concepts.

- **Read** with links to the eBook and to chapter-specific web resources.

- **Visualize** and comprehend challenging topics using learning activities in *GEODe Earth*.

- **Test** yourself with online quizzes.

GEODe EARTH

GEODe Earth is a valuable and easy-to-use learning aid that can be accessed from your book's Companion Web site **(www .mygeoscienceplace.com)**. It is a dynamic instructional tool that promotes understanding and reinforces important concepts by using tutorials, animations, and exercises that actively engage the student.

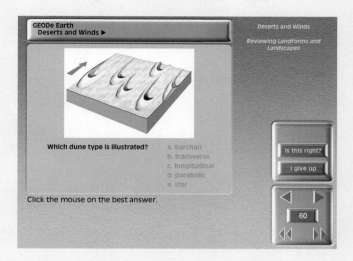

20

SHORELINES

Hatteras Island, North Carolina is a barrier island. When storm waves strike this beach, these homes are obviously vulnerable.

(PHOTO BY MICHAEL COLLIER)

541

The restless waters of the ocean are constantly in motion. Winds generate surface currents, the gravity of the Moon and Sun produces tides, and density differences create deep-ocean circulation. Further, waves carry the energy from storms to distant shores, where their impact erodes the land.

Shorelines are dynamic environments. Their topography, geologic makeup, and climate vary greatly from place to place. Continental and oceanic processes converge along coasts to create landscapes that frequently undergo rapid change. When it comes to the deposition of sediment, they are transition zones between marine and continental environments.

THE SHORELINE: A DYNAMIC INTERFACE

Nowhere is the restless nature of the ocean's water more noticeable than along the shore—the dynamic interface among air, land, and sea. An *interface* is a common boundary where different parts of a system interact. This is certainly an appropriate designation for the coastal zone. Here we can see the rhythmic rise and fall of tides and observe waves constantly rolling in and breaking. Sometimes the waves are low and gentle. At other times they pound the shore with awesome fury (Figure 20.1).

Although it may not be obvious, the shoreline is constantly being modified by waves. For example, along Cape Cod, Massachusetts, wave activity is eroding cliffs of poorly consolidated glacial sediment so aggressively that the cliffs are retreating inland up to 1 meter per year (Figure 20.2A). By contrast, at Point Reyes, California, the far more durable bedrock cliffs are less susceptible to wave attack and therefore are retreating much more slowly (Figure 20.2B). Along both coasts, wave activity is moving sediment along the shore and building narrow sandbars that protrude into and across some bays.

The nature of present-day shorelines is not just the result of the relentless attack of the land by the sea. Indeed, the shore has a complex character that results from multiple geologic processes. For example, practically all coastal areas were affected by the worldwide rise in sea level that accompanied the melting of glaciers at the close of the Pleistocene epoch (see Figure 18.29, p. 508). As the sea encroached landward, the shoreline retreated, becoming superimposed upon existing landscapes that had resulted from such diverse processes as stream erosion, glaciation, volcanic activity, and the forces of mountain building.

Today the coastal zone is experiencing intensive human activity. Unfortunately, people often treat the shoreline as if it were a stable platform on which structures can safely be built. This attitude inevitably leads to conflicts between people and nature. As you will see, many coastal landforms, especially beaches and barrier islands, are relatively fragile, short-lived features that are inappropriate sites for development.

THE COASTAL ZONE

In general conversation a number of terms are used when referring to the boundary between land and sea. In the preceding section, the terms *shore, shoreline, coastal zone,* and *coast* were all used. Moreover, when many think of the land—sea interface, the word *beach* comes to mind. Let's take a moment to clarify these terms and introduce some other terminology used by those who study the land—sea boundary zone. You will find it helpful to refer to Figure 20.3, which is an idealized profile of the coastal zone.

FIGURE 20.1 Wind is responsible for creating the ocean waves that modify shorelines. These waves are crashing along the coast of the Hawaiian island of Oahu. (Photo by Douglas Peebles/Corbis)

Basic Features

The **shoreline** is the line that marks the contact between land and sea. Each day, as tides rise and fall, the position of the shoreline migrates. Over longer time spans, the average position of the shoreline gradually shifts as sea level rises or falls.

The **shore** is the area that extends between the lowest tide level and the highest elevation on land that is affected by storm waves. By contrast, the **coast** extends inland from the shore as far as ocean-related features can be found. The **coastline** marks the coast's seaward edge, whereas the inland boundary is not always obvious or easy to determine.

As Figure 20.3 illustrates, the shore is divided into the *foreshore* and the *backshore*. The **foreshore** is the area exposed when the tide is out (low tide) and submerged when the tide is in (high tide). The **backshore** is landward of the high-tide shoreline. It is usually dry, being affected by waves only during storms. Two other zones are commonly identified. The **nearshore zone** lies between the low-tide shoreline and the line where waves break at low tide. Seaward of the nearshore zone is the **offshore zone.**

Beaches

For many, a beach is the sandy area where people lie in the sun and walk along the water's edge. Technically, a **beach** is an accumulation of sediment found along the landward margin of the ocean or a lake. Along straight coasts, beaches may extend for tens or hundreds of kilometers. Where coasts are irregular, beach formation may be confined to the relatively quiet waters of bays.

Beaches consist of one or more **berms,** which are relatively flat platforms often composed of sand that are adjacent to coastal dunes or cliffs and marked by a change in slope at the seaward edge. Another part of the beach is the **beach face,** which is the wet, sloping surface that extends from the berm to the shoreline. Where beaches are sandy, sunbathers usually prefer the berm, whereas joggers prefer the wet, hard-packed sand of the beach face.

Beaches are composed of whatever material is locally abundant. The sediment for some beaches is derived from the erosion of adjacent cliffs or nearby coastal mountains. Other beaches are built from sediment delivered to the coast by rivers.

A.

B.

FIGURE 20.2 **A.** This satellite image includes the familiar outline of Cape Cod. Boston is to the upper left. The two large islands off the south shore of Cape Cod are Martha's Vineyard (left) and Nantucket (right). Although the work of waves constantly modifies this coastal landscape, shoreline processes are not primarily responsible for creating it. Rather, the present size and shape of Cape Cod result from the positioning of moraines and other glacial materials deposited during the Pleistocene epoch. (Satellite image courtesy of Earth Satellite Corporation/Science Photo Library/Photo Researchers, Inc.) **B.** High-altitude image of the Point Reyes area north of San Francisco, California. The 5.5-kilometer-long south-facing cliffs at Point Reyes (bottom of photo) are exposed to the full force of the waves from the Pacific Ocean. Nevertheless, this promontory retreats slowly because the bedrock from which it formed is very resistant. (Image courtesy of USDA-ASCS)

Although the mineral makeup of many beaches is dominated by durable quartz grains, other minerals may be dominant. For example, in areas such as southern Florida, where there are no mountains or other sources of rock-forming minerals nearby, most beaches are composed of shell fragments and the remains of organisms that live in coastal waters. Some beaches on volcanic islands in the open ocean are composed of weathered grains of the basaltic lava that comprise the islands, or of coarse debris eroded from coral reefs that develop around islands in low latitudes (Figure 20.4).

Regardless of the composition, the material that comprises the beach does not stay in one place. Instead, crashing waves are constantly moving it. Thus, beaches can be thought of as material in transit along the shore.

WAVES

SHORELINES
▶ Waves and Beaches

Ocean waves are energy traveling along the interface between ocean and atmosphere, often transferring energy from a storm far out at sea over distances of several thousand kilometers. That's why even on calm days the ocean still has waves that travel across its surface. When observing waves, always remember that you are watching *energy* travel through a medium (water). If you make waves by tossing a pebble into a pond, or by splashing in a pool, or by blowing across the surface of a cup of coffee, you are imparting *energy* to the liquid, and the waves you see are just the visible evidence of the energy passing through.

Wind-generated waves provide most of the energy that shapes and modifies shorelines. Where the land and sea meet, waves that may have traveled unimpeded for hundreds or thousands of kilometers suddenly encounter a barrier that will not allow them to advance farther and must absorb their energy. Stated another way, the shore is the location where a practically irresistible force confronts an almost immovable object. The conflict that results is never-ending and sometimes dramatic.

Wave Characteristics

Most ocean waves derive their energy and motion from the wind. When a breeze is less than 3 kilometers (2 miles) per hour, only small wavelets appear. At greater wind speeds, more stable waves gradually form and advance with the wind.

Characteristics of ocean waves are illustrated in Figure 20.5, which shows a simple, nonbreaking waveform. The tops of the waves are the *crests,* which are separated by *troughs.* Halfway between the crests and troughs is the *still water level,* which is the level the water would occupy if there were no waves. The vertical distance between trough and crest is called the **wave height,** and the horizontal distance between successive crests (or troughs) is the **wavelength.** The time it takes one full wave—one wavelength—to pass a fixed position is the **wave period.**

The height, length, and period that are eventually achieved by a wave depend on three factors: (1) the wind speed, (2) the

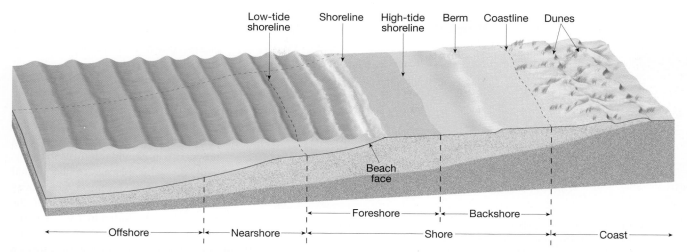

FIGURE 20.3 The coastal zone consists of several parts. The beach is an accumulation of sediment on the landward margin of the ocean or a lake. It can be thought of as material in transit along the shore.

FIGURE 20.4 The black sand at this beach is derived from dark volcanic rock. (Photo by E. J. Tarbuck)

length of time the wind has blown, and (3) the **fetch,** or distance that the wind has traveled across open water. As the quantity of energy transferred from the wind to the water increases, the height and steepness of the waves increase as well. Eventually a critical point is reached where waves grow so tall that they topple over, forming ocean breakers called *whitecaps.*

For a particular wind speed, there is a maximum fetch and duration of wind beyond which waves will no longer increase in size. When the maximum fetch and duration are reached for a given wind velocity, the waves are said to be "fully developed." The reason that waves can grow no further is that they are losing as much energy through the breaking of whitecaps as they are receiving from the wind.

When wind stops or changes direction, or if waves leave the stormy area where they were created, they continue on without relation to local winds. The waves also undergo a gradual change to *swells,* which are lower and longer and may carry a storm's energy to distant shores. Because many independent wave systems exist at the same time, the sea surface acquires a complex, irregular pattern. Hence, the sea waves we watch from the shore are often a mixture of swells from faraway storms and waves created by local winds.

Circular Orbital Motion

Waves can travel great distances across ocean basins. In one study, waves generated near Antarctica were tracked as they traveled through the Pacific Ocean basin. After more than 10,000 kilometers (over 6000 miles), the waves finally expended their energy a week later along the shoreline of the Aleutian Islands of Alaska. The water itself doesn't travel the entire distance, but the waveform does. As the wave travels, the water passes the energy along by moving in a circle. This movement is called *circular orbital motion.*

Observation of an object floating in waves reveals that it moves not only up and down but also slightly forward and backward with each successive wave. Figure 20.6 shows that a floating object

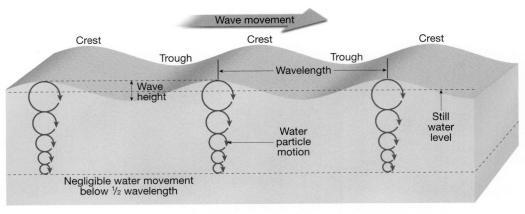

FIGURE 20.5   Diagrammatic view of an idealized nonbreaking ocean wave showing the basic parts of a wave as well as the movement of water particles at depth. Negligible water movement occurs below a depth equal to one half the wavelength (lower dashed line).

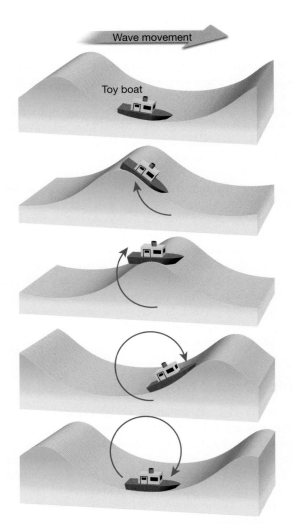

FIGURE 20.6   The movements of the toy boat show that the wave form advances, but the water does not advance appreciably from the original position. In this sequence, the wave moves from left to right as the boat (and the water in which it is floating) rotates in an imaginary circle.

moves up and backward as the crest approaches, up and forward as the crest passes, down and forward after the crest, down and backward as the trough approaches, and rises and moves backward again as the next crest advances. When the movement of the toy boat shown in Figure 20.6 is traced as a wave passes, it can be seen that the boat moves in a circle and it returns to essentially the same place. Circular orbital motion allows a waveform (the wave's shape) to move forward *through the water* while the individual water particles that transmit the wave move in a circle. Wind moving across a field of wheat causes a similar phenomenon: The wheat itself doesn't travel across the field, but the waves do.

The energy contributed by the wind to the water is transmitted not only along the surface of the sea but also downward. However, beneath the surface the circular motion rapidly diminishes until, at a depth equal to one half the wavelength measured from still water level, the movement of water particles becomes negligible. This depth is known as the *wave base*. The dramatic decrease of wave energy with depth is shown by the rapidly diminishing diameters of water-particle orbits in Figure 20.5.

Waves in the Surf Zone

As long as a wave is in deep water, it is unaffected by water depth (Figure 20.7, *left*). However, when a wave approaches the shore, the water becomes shallower and influences wave behavior. The wave begins to "feel bottom" at a water depth equal to its wave base. Such depths interfere with water movement at the base of the wave and slow its advance (Figure 20.7, *center*).

As a wave advances toward the shore, the slightly faster waves farther out to sea catch up, decreasing the wavelength. As the speed and length of the wave diminish, the wave steadily grows higher. Finally, a critical point is reached when the wave

Students Sometimes Ask . . .

What are tidal waves?

Tidal waves, more accurately known as *tsunami* (*tsu* = harbor, *nami* = wave), have nothing to do with the tides! They are long-wavelength, fast-moving, often large, and sometimes destructive waves that originate from sudden changes in the topography of the sea-floor. They are caused by underwater fault slippage, underwater avalanches, or underwater volcanic eruptions. Since the mechanisms that trigger tsunami are frequently seismic events, tsunami are appropriately termed *seismic sea waves*. For more information about characteristics of tsunami and their destructive effects, see Chapter 11, "Earthquakes."

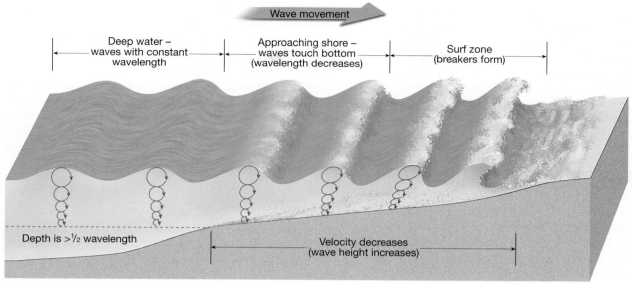

Wave movement

| Deep water – waves with constant wavelength | Approaching shore – waves touch bottom (wavelength decreases) | Surf zone (breakers form) |

Depth is >½ wavelength

Velocity decreases (wave height increases)

FIGURE 20.7 Changes that occur when a wave moves onto shore. The waves touch bottom as they encounter water depths less than half a wavelength. The wave speed decreases, and the waves stack up against the shore, causing the wavelength to decrease. This results in an increase in wave height to the point where the waves pitch forward and break in the surf zone.

is too steep to support itself and the wave front collapses, or *breaks* (Figure 20.7, *right*), causing water to advance up the shore.

The turbulent water created by breaking waves is called **surf.** On the landward margin of the surf zone the turbulent sheet of water from collapsing breakers, called *swash,* moves up the slope of the beach. When the energy of the swash has been expended, the water flows back down the beach toward the surf zone as *backwash.*

WAVE EROSION

SHORELINES
▶ Wave Erosion

During calm weather, wave action is minimal. However, just as streams do most of their work during floods, so too do waves accomplish most of their work during storms. The impact of high, storm-induced waves against the shore can be awesome in its violence (Figure 20.8). Each breaking wave may hurl thousands of tons of water against the land, sometimes causing the ground to literally tremble. The pressures exerted by Atlantic waves in wintertime, for example, average nearly 10,000 kilograms per square meter

(more than 2000 pounds per square foot). The force during storms is even greater.

It is no wonder that cracks and crevices are quickly opened in cliffs, seawalls, breakwaters, and anything else that is subjected to these enormous shocks. Water is forced into every opening, causing air in the cracks to become highly compressed by the thrust of crashing waves. When the wave subsides, the air expands rapidly, dislodging rock fragments and enlarging and extending fractures.

FIGURE 20.8 When waves break against the shore, the force of the water can be powerful and the erosional work that is accomplished can be great. These storm waves are breaking along the coast of Wales. (The Photolibrary Wales/Alamy)

In addition to the erosion caused by wave impact and pressure, **abrasion**—the sawing and grinding action of the water armed with rock fragments—is also important. In fact, abrasion is probably more intense in the surf zone than in any other environment. Smooth, rounded stones and pebbles along the shore are obvious reminders of the relentless grinding action of rock against rock in the surf zone (Figure 20.9A). Further, such fragments are used as "tools" by the waves as they cut horizontally into the land (Figure 20.9B).

SAND MOVEMENT ON THE BEACH

Beaches are sometimes called "rivers of sand." The reason is that the energy from breaking waves often causes large quantities of sand to move along the beach face and in the surf zone roughly parallel to the shoreline. Wave energy also causes sand to move perpendicular to (toward and away from) the shoreline.

Movement Perpendicular to the Shoreline

If you stand ankle deep in water at the beach, you will see that swash and backwash move sand toward and away from the shoreline. Whether there is a net loss or addition of sand depends on the level of wave activity. When wave activity is relatively light (less energetic waves), much of the swash soaks into the beach, which reduces the backwash. Consequently, the swash dominates and causes a net movement of sand up the beach face toward the berm.

When high-energy waves prevail, the beach is saturated from previous waves, so much less of the swash soaks in. As a result, the berm erodes because backwash is strong and causes a net movement of sand down the beach face.

Along many beaches, light wave activity is the rule during the summer. Therefore, a wide sand berm gradually develops. During winter, when storms are frequent and more powerful, strong wave activity erodes and narrows the berm. A wide berm that may have taken months to build can be dramatically narrowed in just a few hours by the high-energy waves created by a strong winter storm.

Wave Refraction

The bending of waves, called **wave refraction,** plays an important part in shoreline processes (Figure 20.10). It affects the distribution of energy along the shore and thus strongly influences where and to what degree erosion, sediment transport, and deposition will take place.

Waves seldom approach the shore straight on. Rather, most waves move toward the shore at an angle. However, when they reach the shallow water of a smoothly sloping bottom, they are bent and tend to become parallel to the shore. Such bending occurs because the part of the wave nearest the shore reaches shallow water and slows first, whereas the end that is still in deep water continues forward at its full speed. The net result is a wave front that may approach nearly parallel to the shore regardless of the original direction of the wave.

FIGURE 20.9   A. Abrasion can be intense in the surf zone. Smooth, rounded rocks along the shore are an obvious reminder of this fact. (Photo by Michael Collier) **B.** Sandstone cliff undercut by wave erosion at Gabriola Island, British Columbia, Canada. (Photo by Fletcher and Baylis/Photo Researchers, Inc.)

A.

B.

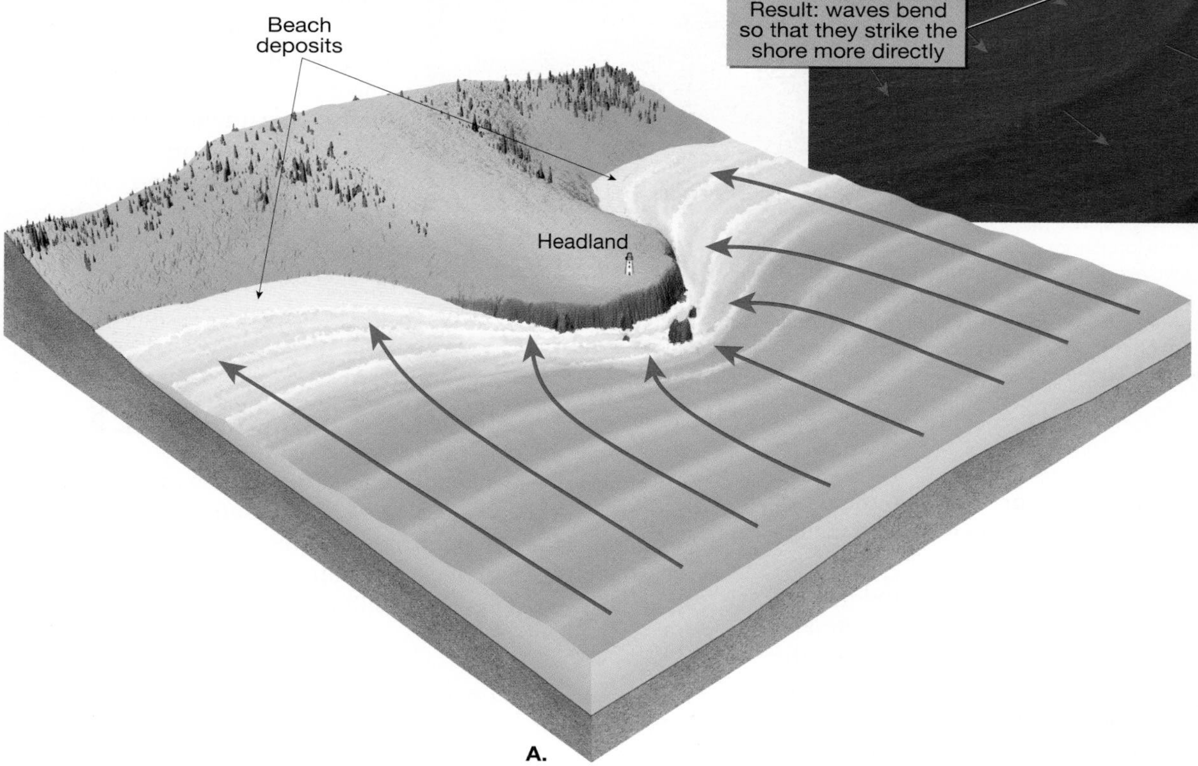

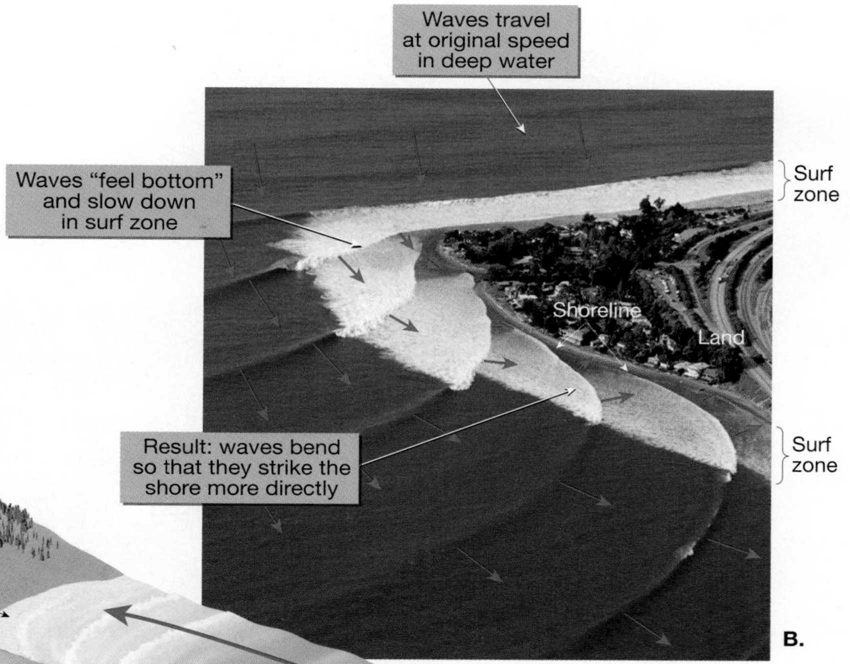

FIGURE 20.10   As waves first touch bottom in the shallows along an irregular coast, they are slowed, causing them to bend (refract) and align nearly parallel to the shoreline. **A.** In this diagram, waves approach nearly straight on. Refraction causes wave energy to be concentrated at headlands (resulting in erosion) and dispersed in bays (resulting in deposition). **B.** Wave refraction at Rincon Point, California. (Photo by Woody Woodworth/Creation Captured)

Waves travel at original speed in deep water

Waves "feel bottom" and slow down in surf zone

Surf zone

Shoreline

Land

Result: waves bend so that they strike the shore more directly

Surf zone

B.

Beach deposits

Headland

A.

Because of refraction, wave impact is concentrated against the sides and ends of headlands that project into the water, whereas wave attack is weakened in bays. This differential wave attack along irregular coastlines is illustrated in Figure 20.10. As the waves reach the shallow water in front of the headland sooner than they do in adjacent bays, they are bent more nearly parallel to the protruding land and strike it from all three sides. By contrast, refraction in the bays causes waves to diverge and expend less energy. In these zones of weakened wave activity, sediments can accumulate and form sandy beaches. Over a long period, erosion of the headlands and deposition in the bays will straighten an irregular shoreline.

Students Sometimes Ask . . .

During heavy wave activity, where does the sand from the berm go?

The orbital motion of waves is too shallow to move sand very far offshore. Consequently, the sand accumulates just beyond where the surf zone ends and forms one or more offshore sandbars called longshore bars.

Longshore Transport

Although waves are refracted, most still reach the shore at some angle, however slight. Consequently, the uprush of water from each breaking wave (the swash) is at an oblique angle to the shoreline. However, the backwash is straight down the slope of the beach. The effect of this pattern of water movement is to transport sediment in a zigzag pattern along the beach face (Figure 20.11). This movement is called **beach drift,** and it can

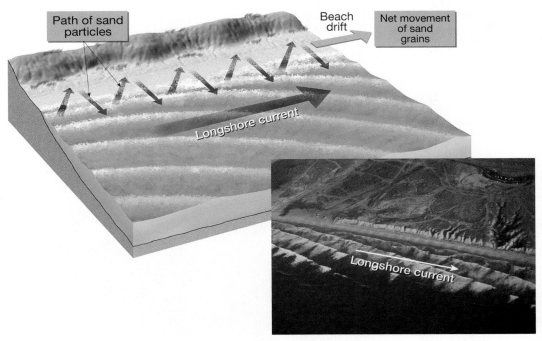

FIGURE 20.11 Beach drift and longshore currents are created by obliquely breaking waves. Beach drift occurs as incoming waves carry sand obliquely up the beach, while the water from spent waves carries it directly down the slope of the beach. Similar movements occur offshore in the surf zone to create the longshore current. These processes transport large quantities of material along the beach and in the surf zone. In the photo, waves approaching the beach at a slight angle near Oceanside, California, produce a longshore current moving from left to right. (Photo by John S. Shelton)

transport sand and pebbles hundreds or even thousands of meters each day. However, a more typical rate is 5 to 10 meters per day.

Waves that approach the shore at an angle also produce currents within the surf zone that flow parallel to the shore and move substantially more sediment than beach drift. Because the water here is turbulent, these **longshore currents** easily move the fine suspended sand and roll larger sand and gravel along the bottom. When the sediment transported by longshore currents is added to the quantity moved by beach drift, the total amount can be very large. At Sandy Hook, New Jersey, for example, the quantity of sand transported along the shore over a 48-year period averaged almost 750,000 tons annually. For a 10-year period in Oxnard, California, more than 1.5 million tons of sediment moved along the shore each year.

Both rivers and coastal zones move water and sediment from one area (*upstream*) to another (*downstream*). As a result, the beach has often been characterized as a "river of sand." Beach drift and longshore currents, however, move in a zigzag pattern, whereas rivers flow mostly in a turbulent, swirling fashion. Additionally, the direction of flow of longshore currents along a shoreline can change, whereas rivers flow in the same direction (downhill). Longshore currents change direction because the direction that waves approach the beach changes seasonally. Nevertheless, longshore currents generally flow southward along both the Atlantic and Pacific shores of the United States.

Rip Currents

Rip currents are concentrated movements of water that flow in the *opposite* direction of breaking waves.* Most of the backwash from spent waves finds its way back to the open ocean as an unconfined flow across the ocean bottom called *sheet flow*. However, sometimes a portion of the returning water moves seaward in the form of surface rip currents. These currents do not travel far beyond the surf zone before breaking up and can be recognized by the way they interfere with incoming waves or by the sediment that is often suspended within the rip current (Figure 20.12). They can be a hazard to swimmers, who, if caught in them, can be carried out away from shore. The best strategy for exiting a rip current is to swim *parallel* to the shore for a few tens of meters.

SHORELINE FEATURES

A fascinating assortment of shoreline features can be observed along the world's coastal regions. These shoreline features vary depending on the type of rocks exposed along the shore, the intensity of wave activity, the nature of coastal currents, and whether the coast is stable, sinking, or rising. Features that owe

---

*Sometimes rip currents are incorrectly called *rip tides,* although they are unrelated to tidal phenomena.

FIGURE 20.12 A rip current (red arrow) extends outward from shore and interferes with incoming waves. As the warning sign indicates, rip currents can be dangerous. (Photo by A. P. Trujillo/APT Photos)

Professional Profile ---Rob Thieler

Marine Geologist

Growing up on the New Jersey shore, Rob Thieler learned that beaches are literally as changeable as the weather. As a teenage lifeguard, he watched entire beaches run short on sand, leaving waves to break perilously close to buildings and streets.

Today, Thieler's coastal observations have expanded considerably. As a research geologist with the U.S. Geological Survey, he studies how sandy coastlines shift and change over time.

"We don't have good models to predict where the shoreline's going to be in a hundred years. A beach can lose 30 meters of sand in a single storm, and we certainly can't predict when and where storms will hit for the next century. We have a really incomplete understanding of how sand moves around on beaches and the inner continental shelf," Thieler states. The problems aren't simply academic; the answers have major implications for the estimated 160 million Americans who live near the coast.

In past centuries, Thieler claims, residents threatened by disappearing beaches merely moved their houses further up the dunes. But for modern homeowners, that's no longer an option—higher ground generally means impinging upon someone else's property.

Most years, Thieler spends about a month at sea gathering information about coastal geology. The ship tows sidescan sonar, seismic reflection profilers, and bathymetric gear over the shallow waters of the continental shelf in a back-and-forth pattern that resembles the path of a lawnmower. The echoes reveal the texture, topography, and structure of bottom sediments in exquisite detail.

> **"I love going to sea. There's nothing like exploring a part of the ocean no one has seen before."**

"I love going to sea. There's nothing like exploring a part of the ocean no one has seen before," Thieler says. Much of the technology Thieler uses to study ocean topography was only invented in the last few decades. As a result, the majority of the seafloor remains uncharted territory.

Thieler combines his marine data with land observations such as the historic locations of shorelines and the elevation of bluffs and beaches. His results can help explain why waves break in certain locations, and how currents shape nearshore marine topography.

Thieler has also put this data to use analyzing how vulnerable the U.S. coastlines are to sea-level rise. Already some planners are using his findings to adapt to rising seas. "Before, we've often chosen to hold the line to protect property. As we improve our understanding of how beaches and coasts evolve in a time of rising sea levels and changing climate, the choice is no longer clear," he says.

---Kathleen Wong

Dr. Rob Thieler, marine geologist standing along the Massachusetts shore near his U.S. Geological Survey office in Woods Hole. Dr. Thieler is in one of his favorite places—the coast. Dr. Thieler studies coastal processes as part of a team of scientists assessing the vulnerability of the United States coastline to rising sea level. (Courtesy of Rob Thieler and the U.S. Geological Survey)

FIGURE 20.13 Wave-cut platform and marine terrace. A wave-cut platform is exposed at low tide along the California coast at Bolinas Point near San Francisco. A wave-cut platform has been uplifted to create the marine terrace. (Photo by John S. Shelton)

their origin primarily to the work of erosion are called *erosional features*, whereas accumulations of sediment produce *depositional features*.

Erosional Features

Many coastal landforms owe their origin to erosional processes. Such erosional features are common along the rugged and irregular New England coast and along the steep shorelines of the West Coast of the United States.

Wave-cut Cliffs, Wave-cut Platforms, and Marine Terraces   **Wave-cut cliffs,** as the name implies, originate by the cutting action of the surf against the base of coastal land. As erosion progresses, rocks overhanging the notch at the base of the cliff crumble into the surf and the cliff retreats. A relatively flat, benchlike surface, called a **wave-cut platform,** is left behind by the receding cliff (Figure 20.13 *left*). The platform broadens as wave attack continues. Some debris produced by the breaking waves remains along the water's edge as sediment on the beach, while the remainder is transported farther seaward. If a wave-cut platform is uplifted above sea level by tectonic forces, it becomes a **marine terrace** (Figure 20.13, *right*). Marine terraces are easily recognized by their gentle seaward-sloping shape and are often desirable sites for coastal roads, buildings, or agriculture.

Sea Arches and Sea Stacks   Headlands that extend into the sea are vigorously attacked by waves because of refraction. The surf erodes the rock selectively, wearing away the softer or more highly fractured rock at the fastest rate. At first, sea caves may

form. When two caves on opposite sides of a headland unite, a **sea arch** results (Figure 20.14). Eventually the arch falls in, leaving an isolated remnant, or **sea stack,** on the wave-cut platform (Figure 20.14). In time, it too will be consumed by the action of the waves.

Depositional Features

Sediment eroded from the beach is transported along the shore and deposited in areas where wave energy is low. Such processes produce a variety of depositional features.

Spits, Bars, and Tombolos   Where beach drift and longshore currents are active, several features related to the movement of sediment along the shore may develop. A **spit** (*spit* = spine) is an elongated ridge of sand that projects from the land into the mouth of an adjacent bay. Often the end in the water hooks landward in response to the dominant direction of the longshore current (Figure 20.15). The term **baymouth bar** is applied to a sandbar that completely crosses a bay, sealing it off from the open ocean (Figure 20.15). Such a feature tends to form across bays where currents are weak, allowing a spit to extend to the other side. A **tombolo** (*tombolo* = mound), a ridge of sand that connects an island to the mainland or to another island, forms in much the same manner as a spit.

Barrier Islands   The Atlantic and Gulf Coastal Plains are relatively flat and slope gently seaward. The shore zone is characterized by **barrier islands.** These low ridges of land parallel the coast at distances from 3 to 30 kilometers offshore. From Cape Cod,

FIGURE 20.14   Sea arch and sea stack at the tip of Mexico's Baja Peninsula. (Photo by Mark A. Johnson/The Stock Market)

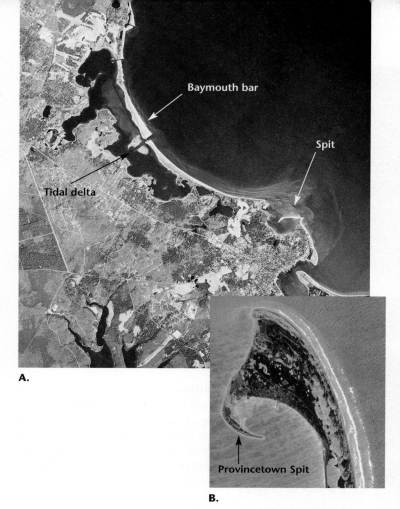

A.

B.

FIGURE 20.15   A. High-altitude image of a well-developed spit and baymouth bar along the coast of Martha's Vineyard, Massachusetts. (Image courtesy of USDA-ASCS) **B.** This photograph, taken from the International Space Station, shows Provincetown Spit located at the tip of Cape Cod. Can you pick this feature out in the satellite image in Figure 20.2A? (NASA image)

Massachusetts, to Padre Island, Texas, nearly 300 barrier islands rim the coast (Figure 20.16).

Most barrier islands are from 1 to 5 kilometers wide and between 15 and 30 kilometers long. The tallest features are sand dunes, which usually reach heights of 5 to 10 meters; in a few areas, unvegetated dunes are more than 30 meters high. The lagoons separating these narrow islands from the shore represent zones of relatively quiet water that allow small craft traveling between New York and northern Florida to avoid the rough waters of the North Atlantic.

Barrier islands probably form in several ways. Some originate as spits that were subsequently severed from the mainland by wave erosion or by the general rise in sea level following the last episode of glaciation. Others are created when turbulent waters in the line of breakers heap up sand scoured from the bottom. Because these sand barriers rise above normal sea level, the piling of sand likely is the result of the work of storm waves at high tide. Finally, some barrier islands may be former sand-dune ridges that originated along the shore during the last glacial period, when sea level was lower. When the ice sheets melted, sea level rose and flooded the area behind the beach-dune complex.

The Evolving Shore

A shoreline continually undergoes modification regardless of its initial configuration. At first most coastlines are irregular, although the degree of and reason for the irregularity may vary considerably from place to place. Along a coastline that is characterized by varied geology, the pounding surf may at first increase its irregularity because the waves will erode the weaker rocks more easily than the stronger ones. However, if a shoreline remains stable, marine erosion and deposition will eventually produce a straighter, more regular coast. Figure 20.17 illustrates the evolution of an initially irregular coast. As waves erode the headlands, creating cliffs and a wave-cut platform, sediment is carried along the shore. Some material is deposited in the bays, while other debris is formed into spits and baymouth bars. At the same time, rivers fill the bays with sediment. Ultimately a generally straight, smooth coast results.

STABILIZING THE SHORE

SHORELINES
▶ Waves and Beaches

Today the coastal zone teems with human activity. Unfortunately, people often treat the shoreline as if it were a stable platform on which structures can be built safely. This approach jeopardizes both people and the shoreline because many coastal landforms are relatively fragile, short-lived features that are easily damaged by development. And as anyone who has endured a tropical storm knows, the shoreline is not always a safe place to live. We will examine this latter idea more closely in the section on "Hurricanes—The Ultimate Coastal Hazard."

Compared with natural hazards such as earthquakes, volcanic eruptions, and landslides, shoreline erosion is often perceived to be a more continuous and predictable process that appears to cause relatively modest damage to limited areas. In reality, the shoreline is a dynamic place that can change rapidly in response to natural forces. Exceptional storms are capable of eroding beaches and cliffs at rates that are far in excess of the long-term average. Such bursts of accelerated erosion not only have a significant impact on the natural evolution of a coast but can also have a profound impact on people who reside in the coastal zone (Figure 20.18). Erosion along our coasts causes significant property damage. Huge sums are spent annually, not only to repair damage but also to prevent or control erosion. Already a problem at many sites, shoreline erosion is certain to become an increasingly serious problem as extensive coastal development continues.

Although the same processes cause change along every coast, not all coasts respond in the same way. Interactions among different processes and the relative importance of each process depend on local factors. The factors include (1) the proximity of a coast to sediment-laden rivers, (2) the degree of tectonic activity, (3) the topography and composition of the land, (4) prevailing winds and weather patterns, and (5) the configuration of the coastline and nearshore areas.

FIGURE 20.16   Nearly 300 barrier islands rim the Gulf and Atlantic coasts. The islands along the coast of North Carolina are excellent examples. (Photo by Michael Collier)

natural migration of sand is an ongoing struggle in many coastal areas. Such interference can result in unwanted changes that are difficult and expensive to correct.

Hard Stabilization

Structures built to protect a coast from erosion or to prevent the movement of sand along a beach are known as **hard stabilization.** Hard stabilization can take many forms and often results in predictable yet unwanted outcomes. Hard stabilization includes jetties, groins, breakwaters, and seawalls.

Jetties   From relatively early in America's history a principal goal in coastal areas was the development and maintenance of harbors. In many cases, this involved the construction of jetty systems. **Jetties** are usually built in pairs and extend into the ocean at the entrances to rivers and harbors. With the flow of water confined to a narrow zone, the ebb and flow caused by the rise and fall of the tides keep the sand in motion and prevent deposition in the channel. However, as illustrated in Figure 20.19, the jetty may act as a dam against which the longshore current and beach drift deposit sand. At the same time, wave activity removes sand on the other side. Because the other side is not receiving any new sand, there is soon no beach at all.

Groins   To maintain or widen beaches that are losing sand, groins are sometimes constructed. A **groin** (*groin* = ground) is a barrier built at a right angle to the beach to trap sand that is moving parallel to the shore. Groins are usually constructed of large rocks but may also be composed of wood. These structures often do their job so effectively that the longshore current beyond the groin becomes sand starved. As a result, the current erodes sand from the beach on the downstream side of the groin.

To offset this effect, property owners downstream from the structure may erect a groin on their property. In this manner, the

During the past 100 years, growing affluence and increasing demands for recreation have brought unprecedented development to many coastal areas. As both the number and the value of buildings have increased, so too have efforts to protect property from storm waves by stabilizing the shore. Also, controlling the

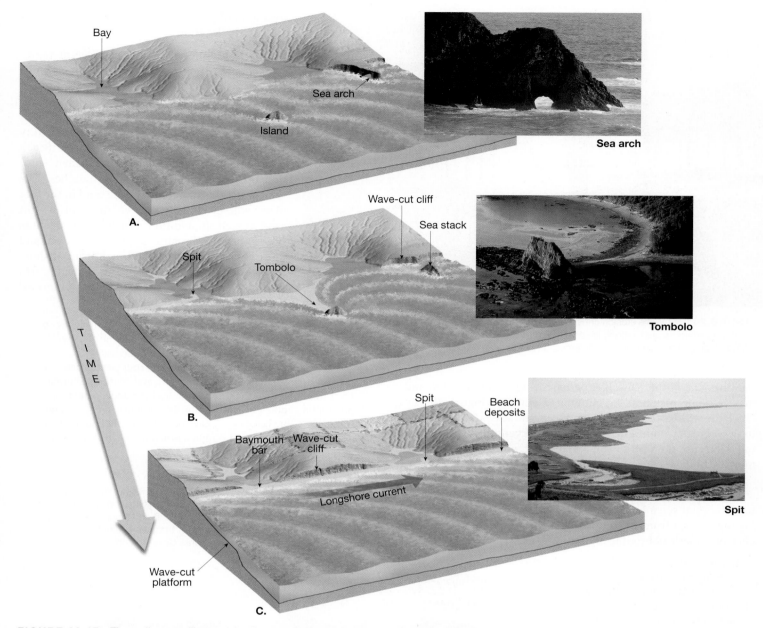

FIGURE 20.17 These diagrams illustrate the changes that can take place through time along an initially irregular coastline that remains relatively stable. The coastline shown in part **A** gradually evolves to **B,** and then **C.** The diagrams also serve to illustrate many of the features described in the section on shoreline features. (Photos A and C by E. J. Tarbuck; Photo B by Michael Collier)

number of groins multiplies, resulting in a *groin field* (Figure 20.20). An example of such proliferation is the shoreline of New Jersey, where hundreds of these structures have been built. Because it has been shown that groins often do not provide a satisfactory solution, they are no longer the preferred method of keeping beach erosion in check.

Breakwaters and Seawalls Hard stabilization can also be built parallel to the shoreline. One such structure is a **breakwater,** the purpose of which is to protect boats from the force of large breaking waves by creating a quiet water zone near the shoreline. However, when this is done, the reduced wave activity along the shore behind the structure may allow sand to accumu-

late. If this happens, the marina will eventually fill with sand while the downstream beach erodes and retreats. At Santa Monica, California, where the building of a breakwater created such a problem, the city uses a dredge to remove sand from the protected quiet water zone and deposit it downstream where longshore currents and beach drift continue to move the sand down the coast (Figure 20.21).

Another type of hard stabilization built parallel to the shoreline is a **seawall,** which is designed to armor the coast and defend property from the force of breaking waves. Waves expend much of their energy as they move across an open beach. Seawalls cut this process short by reflecting the force of unspent waves seaward. As a consequence, the beach to the

FIGURE 20.18 Wave erosion caused by strong storms forced abandonment of this highly developed shoreline area in Long Island, New York. Coastal areas are dynamic places that can change rapidly in response to natural forces. (Photo by Mark Wexler/Woodfin Camp & Associates)

tives to hard stabilization include beach nourishment and relocation.

Beach Nourishment **Beach nourishment** represents an approach to stabilizing shoreline sands without hard stabilization. As the term implies, this practice involves the addition of large quantities of sand to the beach system (Figure 20.23). By building the beaches seaward, beach quality and storm protection are both improved. Beach nourishment, however, is not a permanent solution to the problem of shrinking beaches. The same processes that removed the sand in the first place will eventually remove the replacement sand as well. In addition, beach nourishment is very expensive because huge volumes of sand must be transported to the beach from offshore areas, nearby rivers, or other source areas. Orrin Pilkey, a respected coastal scientist, described the situation this way:

> Nourishment has been carried out on beaches on both sides of the continent, but by far the greatest effort in dollars and sand volume has been expended on the East Coast barrier islands between the southern shore of Long Island, N.Y., and southern Florida. Over this shoreline reach, communities have added a total of 500 million cubic yards of sand on 195 beaches in 680 separate instances. Some beaches . . . have been renourished more than 20 times each since 1965. Virginia Beach has been renourished

seaward side of the seawall experiences significant erosion and may in some instances be eliminated entirely (Figure 20.22). Once the width of the beach is reduced, the seawall is subjected to even greater pounding by the waves. Eventually this battering will cause the wall to fail, and a larger, more expensive wall must be built to take its place.

The wisdom of building temporary protective structures along shorelines is increasingly questioned. The opinions of many coastal scientists and engineers are expressed in the following excerpt from a position paper that grew out of a conference on America's Eroding Shoreline:

> It is now clear that halting the receding shoreline with protective structures benefits only a few and seriously degrades or destroys the natural beach and the value it holds for the majority. Protective structures divert the ocean's energy temporarily from private properties, but usually refocus that energy on the adjacent natural beaches. Many interrupt the natural sand flow in coastal currents, robbing many beaches of vital sand replacement.*

Alternatives to Hard Stabilization

Armoring the coast with hard stabilization has several potential drawbacks, including the cost of the structure and the loss of sand on the beach. Alterna-

FIGURE 20.19 Jetties are built at the entrances to rivers and harbors and are intended to prevent deposition in the navigation channel. Jetties interrupt the movement of sand by beach drift and longshore currents. Beach erosion often occurs downcurrent from the site of the structure. The photo is an aerial view of jetties at Santa Cruz, California. (Photo by U.S. Army Corps of Engineers)

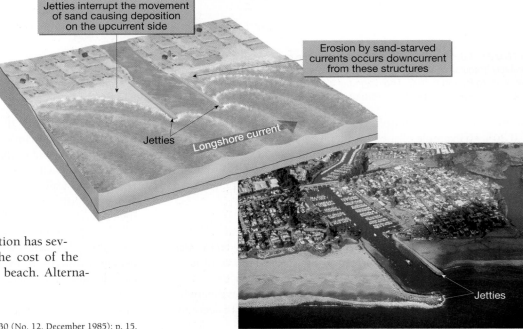

Jetties interrupt the movement of sand causing deposition on the upcurrent side

Erosion by sand-starved currents occurs downcurrent from these structures

Jetties

Longshore current

Jetties

*"Strategy for Beach Preservation Proposed," *Geotimes* 30 (No. 12, December 1985): p. 15.

FIGURE 20.20 A series of groins along the shoreline near Chichester, Sussex, England. (Photo by Sandy Stockwell/London Aerial Photo Library/Corbis)

more than 50 times. Nourished beaches typically cost between $2 million and $10 million per mile.*

In some instances, beach nourishment can lead to unwanted environmental effects. For example, beach replenishment at Waikiki Beach, Hawaii, involved replacing coarse calcareous sand with softer, muddier calcareous sand. Destruction of the soft beach sand by breaking waves increased the water's turbidity and killed offshore coral reefs. At Miami Beach increased turbidity also damaged local coral communities.

Beach nourishment appears to be an economically viable long-range solution to the beach preservation problem only in areas where there exists dense development, large supplies of sand, relatively low wave energy, and reconcilable environmental issues. Unfortunately, few areas possess all these attributes.

Relocation Instead of building structures such as groins and seawalls to hold the beach in place, or adding sand to replenish eroding beaches, another option is available. Many coastal scientists and planners are calling for a policy shift from defending and rebuilding beaches and coastal property in high hazard areas to *relocating* storm-damaged buildings in those places and letting nature reclaim the beach (see Box 20.1). This approach is similar to that adopted by the federal government for river floodplains following the devastating 1993 Mississippi River floods in which vulnerable structures are abandoned and relocated on higher, safer ground.

*"Beaches Awash with Politics," in *Geotimes,* July 2005, pp. 38–39.

Such proposals, of course, are controversial. People with significant nearshore investments shudder at the thought of not rebuilding and defending coastal developments from the erosional wrath of the sea. Others, however, argue that with sea level rising, the impact of coastal storms will only get worse in the decades to come. This group advocates that oft-damaged structures be abandoned or relocated to improve personal safety and to reduce costs. Such ideas will no doubt be the focus of much study and debate as states and communities evaluate and revise coastal land-use policies.

EROSION PROBLEMS ALONG U.S. COASTS

The shoreline along the Pacific Coast of the United States is strikingly different from that characterizing the Atlantic and Gulf coast regions. Some of the differences are related to plate tectonics. The West Coast represents the leading edge of the North American plate, and because of this it experiences active uplift and deformation. By contrast, the East Coast is a tectonically quiet region that is far from any active plate boundary. Because of this basic geological difference, the nature of shoreline erosion problems along America's opposite coasts is different.

Atlantic and Gulf Coasts

Much of the coastal development along the Atlantic and Gulf coasts has occurred on barrier islands. Typically, barrier islands, also termed *barrier beaches* or *coastal barriers,* consist of a wide beach that is backed by dunes and separated from the mainland by marshy lagoons. The broad expanses of sand and exposure to the ocean have made barrier islands exceedingly attractive sites for development. Unfortunately, development has taken place more rapidly than our understanding of barrier island dynamics.

Because barrier islands face the open ocean, they receive the full force of major storms that strike the coast. When a storm

FIGURE 20.21 Aerial view of a breakwater at Santa Monica, California. The breakwater appears as a faint line in the water behind which many boats are anchored. The construction of the breakwater disrupted longshore transport and caused the seaward growth of the beach. (Photo by John S. Shelton)

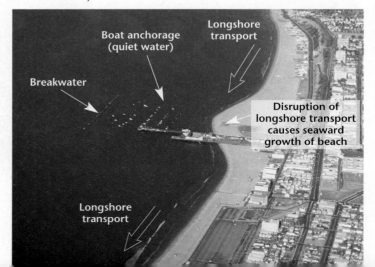

Seawall

FIGURE 20.22   Seabright in northern New Jersey once had a broad, sandy beach. A seawall 5 to 6 meters (16 to 20 feet) high and 8 kilometers (5 miles) long was built to protect the town and the railroad that brought tourists to the beach. As you can see, after the wall was built, the beach narrowed dramatically. (Photo by Rafael Macia/Photo Researchers, Inc.)

Just as a flexible reed may survive a wind that destroys an oak tree, so the barriers survive hurricanes and nor' easters not through unyielding strength but by giving before the storm.

This picture changes when a barrier is developed for homes or as a resort. Storm waves that previously rushed harmlessly through gaps between the dunes now encounter buildings and roadways. Moreover, since the dynamic nature of the barriers is readily perceived only during storms, homeowners tend to attribute damage to a particular storm, rather than to the basic mobility of coastal barriers. With their homes or investments at stake, local residents are more likely to seek to hold the sand in place and the waves at bay than to admit that development was improperly placed to begin with.*

occurs, the barriers absorb the energy of the waves primarily through the movement of sand. This process and the dilemma that results have been described as follows:

> Waves may move sand from the beach to offshore areas or, conversely, into the dunes; they may erode the dunes, depositing sand onto the beach or carrying it out to sea; or they may carry sand from the beach and the dunes into the marshes behind the barrier, a process known as overwash. The common factor is movement.

Pacific Coast

In contrast to the broad, gently sloping coastal plains of the Atlantic and Gulf Coasts, much of the Pacific Coast is characterized by relatively narrow beaches that are backed by steep cliffs and mountain ranges. Recall that America's western margin is a more rugged and tectonically active region than the eastern margin. Because uplift continues, a rise in sea level in the West is not so readily apparent. Nevertheless, like the shoreline erosion problems facing the East's barrier islands, West Coast difficulties also stem largely from the alteration of a natural system by people.

A major problem facing the Pacific shoreline, and especially portions of southern California, is a significant narrowing of many beaches. The bulk of the sand on many of these beaches is supplied by rivers that transport it from the mountains to the coast. Over the years this natural flow of material to the coast has been interrupted by dams built for irrigation and flood control. The reservoirs effectively trap the sand that would otherwise nourish the beach environment. When the beaches were wider, they served to protect the cliffs behind them from the force of storm waves. Now, however, the waves move across the narrowed beaches without losing much energy and cause more rapid erosion of the sea cliffs.

A.

B.

FIGURE 20.23   Miami Beach. **A.** Before beach nourishment and **B.** After beach nourishment. (Courtesy of the U.S. Army Corps of Engineers, Vicksburg District)

*Frank Lowenstein, "Beaches or Bedrooms— The Choice as Sea Level Rises," *Oceanus* 28 (No. 3, Fall 1985): p. 22.

PEOPLE AND THE ENVIRONMENT

The Move of the Century—Relocating the Cape Hatteras Lighthouse*

BOX 20.1

In spite of efforts to protect structures that are too close to the shore, they can still be in danger of being destroyed by receding shorelines and the destructive power of waves. Such was the case for one of the nation's most prominent landmarks, the candy-striped lighthouse at Cape Hatteras, North Carolina, which is 21 stories tall—the nation's tallest lighthouse.

The lighthouse was built in 1870 on the Cape Hatteras barrier island 457 meters (1500 feet) from the shoreline to guide mariners through the dangerous offshore shoals known as the "Graveyard of the Atlantic." As the barrier island began migrating toward the mainland, its beach narrowed. When the waves began to lap just 36 meters (120 feet) from its brick and granite base, there was concern that even a moderate-strength hurricane could trigger beach erosion sufficient to topple the lighthouse.

In 1970 the U.S. Navy built three groins in front of the lighthouse in an effort to protect the beach from further erosion. The groins initially slowed erosion but disrupted sand flow in the surf zone, which caused the flattening of nearby dunes and the formation of a bay south of the lighthouse. Attempts to increase the width of the beach in front of the lighthouse included beach nourishment and artificial offshore beds of seaweed, both of which failed to widen the beach substantially. In the 1980s the Army Corps of Engineers proposed building a massive stone seawall around the lighthouse but decided the eroding coast would eventually move out from under the structure, leaving it stranded at sea on its own island. In 1988 the National Academy of Sciences determined that the shoreline in front of the lighthouse would retreat so far as to destroy the lighthouse and recommended relocation of the tower as had been done with smaller lighthouses. In 1999 the National Park Service, which owns the lighthouse, finally authorized moving the structure to a safer location (Figure 20.A).

Moving the lighthouse, which weighs 4395 metric tons (4830 short tons), was accomplished by severing it from its foundation and carefully hoisting it onto a platform of steel beams fitted with roller dollies. Once on the platform, it was slowly rolled along a specially designed steel track using a series of hydraulic jacks. A strip of vegetation was cleared to make a runway along which the lighthouse traveled 1.5 meters (5 feet) at a time, with the track picked up from behind and reconstructed in front of the tower as it moved. In less than a month, the lighthouse was gingerly transported 884 meters (2900 feet) from its original location, making it one of the largest structures ever successfully moved.

After its $12 million move, the lighthouse now resides in a scrub oak and pine woodland (Figure 20.B). Although it now stands farther inland, the light's slightly higher elevation makes it visible just as far out to sea, where it continues to warn mariners of the hazardous shoals.

*This box was prepared by Professor Alan P. Trujillo, Palomar College.

FIGURE 20.A When North Carolina's historic Cape Hatteras Lighthouse was threatened by coastal erosion, it was moved in the summer of 1999. Before it was moved, the lighthouse was only about 36 meters (120 feet) from the water. (Photo by Don Smetzer/ PhotoEdit Inc.)

Former location of lighthouse

FIGURE 20.B Cape Hatteras Lighthouse after being moved 884 meters (2900 feet) from the shoreline. At its new location it should be safe for 50 years or more. (Photo courtesy of Drew Wilson, Virginian-Pilot, copyright 1999)

A.

B.

FIGURE 20.24 Satellite images of Hurricane Katrina in late August 2005 shortly before it devastated the Gulf Coast. **A.** Color-enhanced infrared image from the *GOES-East* satellite. The most intense activity is associated with red and orange. **B.** A more traditional satellite image from NASA's *Terra* satellite. (NASA)

sible consequences of global warming in Chapter 21 on "Global Climate Change."

HURRICANES—THE ULTIMATE COASTAL HAZARD

The whirling tropical cyclones that occasionally have wind speeds exceeding 300 kilometers (185 miles) per hour are known in the United States as *hurricanes*—the greatest storms on Earth (Figure 20.24). In the western Pacific they are called *typhoons*, and in the Indian Ocean they are simply called *cyclones*. No matter which name is used, these storms are among the most destructive of natural disasters. When a hurricane reaches land, it is capable of annihilating coastal areas and killing tens of thousands of people.

Fortunately, when Hurricane Ike made landfall along the Texas coast near Galveston on September 13, 2008, loss of life was relatively small. Because forecasts were accurate and warnings were timely, tens of thousands were safely evacuated from the most vulnerable areas before the storm arrived. Nevertheless, damages to the coastal zone and inland areas were substantial and costly (Figure 20.25).

On the positive side, hurricanes provide essential rainfall over many areas they cross. Consequently, a resort owner along the Florida coast may dread the coming of hurricane season, whereas a farmer in Japan may welcome its arrival.

The vast majority of hurricane-related deaths and damage are caused by relatively infrequent yet powerful storms. Table 20.1 lists the deadliest hurricanes to strike the United States between 1900 and 2009. The storm that pounded an unsuspecting Galveston, Texas, in 1900 was not just the deadliest U.S. hurricane ever, but the deadliest natural disaster *of any kind* to affect the United States. Of course, the deadliest and most costly storm in recent memory occurred in August 2005, when Hurricane Katrina devastated the Gulf Coast of Louisiana, Mississippi, and Alabama. Although hundreds of thousands fled before the storm made landfall, thousands of others were caught by the storm. In addition to the human suffering and tragic loss of life that were left in the wake of Hurricane Katrina, the financial losses caused by the storm are practically incalculable. Up until August 2005, the $25 billion in damages associated with Hurricane Andrew in 1992 represented the costliest natural disaster in U.S. history. This figure was exceeded many times over when Katrina's economic impact was calculated. Although imprecise, some suggest that the final accounting could exceed $100 billion.

Although the retreat of the cliffs provides material to replace some of the sand impounded behind dams, it also endangers homes and roads built on the bluffs. In addition, development atop the cliffs aggravates the problem. Urbanization increases runoff, which, if not carefully controlled, can result in serious bluff erosion. Watering lawns and gardens adds significant quantities of water to the slope. This water percolates downward toward the base of the cliff, where it may emerge in small seeps. This action reduces the slope's stability and facilitates mass wasting.

Shoreline erosion along the Pacific Coast varies considerably from one year to the next, largely because of the sporadic occurrence of storms. As a consequence, when the infrequent but serious episodes of erosion occur, the damage is often blamed on the unusual storms and not on coastal development or the sediment-trapping dams that may be great distances away. If, as predicted, sea level rises at an increasing rate in the years to come, increased shoreline erosion and sea-cliff retreat should be expected along many parts of the Pacific Coast. Coastal vulnerability to sea-level rise is examined in some detail as part of a discussion of the pos-

FIGURE 20.25 Crystal Beach, Texas, on September 16, 2008, three days after Hurricane Ike came ashore. At landfall the storm had sustained winds of 165 kilometers (105 miles) per hour. The extraordinary storm surge caused much of the damage pictured here. (Photo by Earl Nottingham/Associated Press)

TABLE 20.1	The 10 Deadliest Hurricanes to Strike the U.S. Mainland 1900–2009			
Rank	Hurricane	Year	Category	Deaths
1.	Texas (Galveston)	1900	4	8000*
2.	SE Florida (Lake Okeechobee)	1928	4	2500–3000
3.	Katrina	2005	4	1833
4.	Audrey	1957	4	At least 416
5.	Florida Keys	1935	5	408
6.	Florida (Miami)/Mississippi/Alabama/Florida (Pensacola)	1926	4	372
7.	Louisiana (Grande Isle)	1909	4	350
8.	Florida Keys/South Texas	1919	4	287
9. (tie)	Louisiana (New Orleans)	1915	4	275
9. (tie)	Texas (Galveston)	1915	4	275

*This number may actually have been as high as 10,000-12,000.

Source: National Weather Service/National Hurricane Center NOAA Technical Memorandum NWS TPC-5.

Our coasts are vulnerable. People are flocking to live near the ocean. The proportion of the U.S. population residing within 75 kilometers (45 miles) of a coast in 2010 is well in excess of 50 percent. The concentration of such large numbers of people near the shoreline means that hurricanes place millions at risk. Moreover, the potential costs of property damage are incredible.

Profile of a Hurricane

Most hurricanes form between the latitudes of 5 degrees and 20 degrees over all the tropical oceans except those of the South Atlantic and eastern South Pacific (Figure 20.26). The North Pacific has the greatest number of storms, averaging 20 per year. Fortunately for those living in the coastal regions of the southern and eastern United States, fewer than five hurricanes, on the average, develop annually in the warm sector of the North Atlantic.

Although many tropical disturbances develop each year, only a few reach hurricane status. By international agreement, a hurricane has wind speeds in excess of 119 kilometers (74 miles) per hour and a rotary circulation. Mature hurricanes average 600 kilometers (375 miles) in diameter and often extend 12,000 meters (40,000 feet) above the ocean surface. From the outer edge to the center, the barometric pressure has, on occasion, dropped 60 millibars, from 1,010 millibars to 950 millibars. The lowest pressures ever recorded in the Western Hemisphere are associated with these storms.

A steep pressure gradient generates the rapid, inward-spiraling winds of a hurricane (Figure 20.27). As the air rushes toward the center of the storm, its velocity increases. This occurs for the same reason that skaters with their arms extended spin faster as they pull their arms in close to their bodies.

As the inward rush of warm, moist surface air approaches the core of the storm, it turns upward and ascends in a ring of cumulonimbus towers (Figure 20.27). This doughnut-shaped wall of intense convective activity surrounding the center of the storm is called the **eye wall.** It is here that the greatest wind speeds and heaviest rainfall occur. Surrounding the eye wall are curved bands of clouds that trail away in a spiral fashion. Near the top of the hurricane, the airflow is outward, carrying the rising air away from the storm center, thereby providing room for more inward flow at the surface.

At the very center of the storm is the **eye** of the hurricane (Figure 20.27). This well-known feature is a zone about 20 kilometers (12.5 miles) in diameter where precipitation ceases and winds subside. It offers a brief but deceptive break from the extreme weather in the enormous curving wall clouds that surround it. The air within the eye gradually descends and heats by compression, making it the warmest part of the storm. Although many people believe that the eye is characterized by clear blue skies, such is usually not the case because the subsidence in the eye is seldom strong enough to produce cloudless conditions. Although the sky appears much brighter in this region, scattered clouds at various levels are common.

Hurricane Destruction

The amount of damage caused by a hurricane depends on several factors, including the size and population density of the area affected and the shape of the ocean bottom near the shore. The most significant factor, of course, is the strength of the storm itself. By studying past storms, a scale has been established to rank the relative intensities of hurricanes. As Table 20.2 indicates, a *category 5* storm is the worst possible, whereas a *category 1* hurricane is least severe.

During the hurricane season it is common to hear scientists and reporters alike use the numbers from the *Saffir-Simpson Hurricane Scale.* When Hurricane Katrina made landfall, sustained winds were 225 kilometers (140 miles) per hour, making it a strong category 4 storm. Storms that fall into category 5 are rare. Hurricane Camille, a 1969 storm that caused catastrophic damage along the coast of Mississippi, is one well-known example.

Damage caused by hurricanes can be divided into three categories: (1) storm surge, (2) wind damage, and (3) inland flooding.

Storm Surge   Without question, the most devastating damage in the coastal zone is caused by the storm surge. It not only accounts for a large share of coastal property losses but is also responsible for a high percentage of all hurricane-caused deaths. A **storm surge** is a dome of water 65 to 80 kilometers (40 to 50 miles) wide that sweeps across the coast near the point

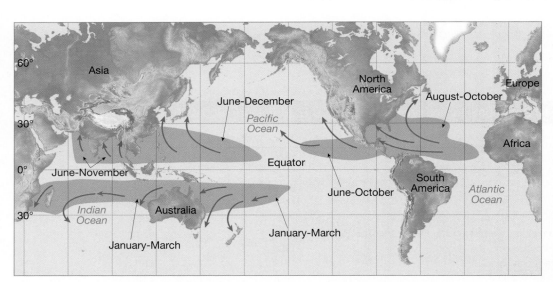

FIGURE 20.26   This world map shows the regions where most hurricanes form as well as their principal months of occurrence and the most common tracks they follow. Hurricanes do not develop within about 5° of the equator because the Coriolis effect (a force related to Earth's rotation that gives storms their "spin") is too weak. Because warm surface ocean temperatures are necessary for hurricane formation, they seldom form poleward of 20° latitude nor over the cool waters of the South Atlantic and the eastern South Pacific.

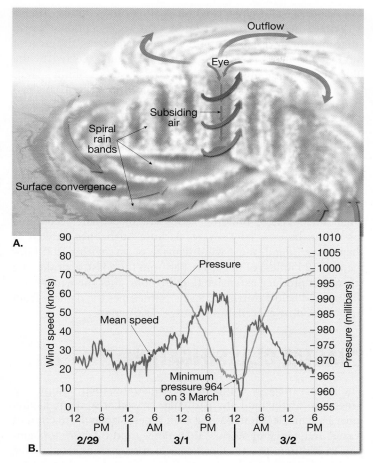

A.

B.

FIGURE 20.27   A. Cross section of a hurricane. Note that the vertical dimension is greatly exaggerated. The eye, the zone of relative calm at the center of the storm, is a distinctive hurricane feature. Sinking air in the eye warms by compression. Surrounding the eye is the eye wall, the zone where winds and rain are most intense. Tropical moisture spiraling inward creates rain bands that pinwheel around the storm center. Outflow of air at the top of the hurricane is important because it prevents the convergent flow at lower levels from "filling in" the storm. (After NOAA) **B.** Measurements of surface pressure and wind speed during the passage of Cyclone Monty at Mardie Station in Western Australia between February 29 and March 2, 2004. (Hurricanes are called "cyclones" in this part of the world.) The strongest winds are associated with the eye wall, and the weakest winds and lowest pressure are found in the eye. (Data from World Meteorological Organization)

where the eye makes landfall. If all wave activity were smoothed out, the storm surge would be the height of the water above normal tide level. In addition, tremendous wave activity is superimposed on the surge. We can easily imagine the damage that this surge of water could inflict on low-lying coastal areas (see

Figure 20.25). The worst surges occur in places like the Gulf of Mexico, where the continental shelf is very shallow and gently sloping. In addition, local features such as bays and rivers can cause the surge height to double and increase in speed.

As a hurricane advances toward the coast in the Northern Hemisphere, storm surge is always most intense on the right side of the eye where winds are blowing *toward* the shore. In addition, on this side of the storm, the forward movement of the hurricane also contributes to the storm surge. In Figure 20.28, assume a hurricane with peak winds of 175 kilometers (109 miles) per hour is moving toward the shore at 50 kilometers (31 miles) per hour. In this case, the net wind speed on the right side of the advancing storm is 225 kilometers (140 miles) per hour. On the left side, the hurricane's winds are blowing opposite the direction of storm movement, so the net winds are *away* from the coast at 125 kilometers (78 miles) per hour. Along the shore facing the left side of the oncoming hurricane, the water level may actually decrease as the storm makes landfall.

Wind Damage   Destruction caused by wind is perhaps the most obvious of the classes of hurricane damage. Debris such as signs, roofing materials, and small items left outside become dangerous flying missiles in hurricanes. For some structures, the force of the wind is sufficient to cause total ruin. Mobile homes are particularly vulnerable. High-rise buildings are also susceptible to hurricane-force winds. Upper floors are most vulnerable because wind speeds usually increase with height. Recent research suggests that people should stay below the 10th floor but remain above any floors at risk for flooding. In regions with good building codes, wind damage is usually not as catastrophic as storm-surge damage. However, hurricane-force winds affect a much larger area than storm surge and can cause huge economic losses. For example, in 1992 it was

TABLE 20.2	Saffir-Simpson Hurricane Scale					
Scale Number (Category)	Central Pressure (Millibars)	Wind Speed (KPH)	Wind Speed (MPH)	Storm Surge (Meters)	Storm Surge (Feet)	Damage
1	≥980	119–153	74–95	1.2–1.5	4–5	*Minimal*
2	965–979	154–177	96–110	1.6–2.4	6–8	*Moderate*
3	945–964	178–209	111–130	2.5–3.6	9–12	*Extensive*
4	920–944	210–250	131–155	3.7–5.4	13–18	*Extreme*
5	<920	>250	>155	>5.4	>18	*Catastrophic*

FIGURE 20.28 Winds associated with a Northern Hemisphere hurricane advancing toward the coast. This hypothetical storm, with peak winds of 175 kilometers (109 miles) per hour, is moving toward the coast at 50 kilometers (31 miles) per hour. On the right side of the advancing storm, the 175-kilometer-per-hour winds are in the same direction as the movement of the storm (50 kilometers per hour). Therefore, the *net* wind speed on the right side of the storm is 225 kilometers (140 miles) per hour. On the left side, the hurricane's winds are blowing opposite the direction of storm movement, so the *net* winds of 125 kilometers (78 miles) per hour are away from the coast. Storm surge will be greatest along that part of the coast hit by the right side of the advancing hurricane.

largely the winds associated with Hurricane Andrew that produced more than $25 billion of damage in southern Florida and Louisiana.

Hurricanes sometimes produce tornadoes that contribute to the storm's destructive power. Studies have shown that more than half of the hurricanes that make landfall produce at least one tornado. In 2004 the number of tornadoes associated with tropical storms and hurricanes was extraordinary. Tropical Storm Bonnie and five landfalling hurricanes—Charley, Frances, Gaston, Ivan, and Jeanne—produced nearly 300 tornadoes that affected the southeast and mid-Atlantic states.

Heavy Rains and Inland Flooding The torrential rains that accompany most hurricanes represent a third significant threat: flooding. Whereas the effects of storm surge and strong winds are concentrated in coastal areas, heavy rains may affect places hundreds of kilometers from the coast for up to several days after the storm has lost its hurricane-force winds.

In September 1999, Hurricane Floyd brought flooding rains, high winds, and rough seas to a large portion of the Atlantic seaboard. More than 2.5 million people evacuated their homes from Florida north to the Carolinas and beyond. It was the largest peacetime evacuation in U.S. history up to that time. Torrential rains falling on already saturated ground created devastating inland flooding. Altogether Floyd dumped more than 48 centimeters (19 inches) of rain on Wilmington, North Carolina, 33.98 cm (13.38 inches) in a single 24-hour span.

Another well-known example is Hurricane Camille (1969). Although this storm is best known for its exceptional storm surge and the devastation it brought to coastal areas, the greatest number of deaths associated with this storm occurred in the Blue Ridge Mountains of Virginia two days after Camille's landfall. Many places received more than 25 centimeters (10 inches) of rain.

Summary To summarize, extensive damage and loss of life in the coastal zone can result from storm surge, strong winds, and torrential rains. When loss of life occurs, it is commonly caused by the storm surge, which can devastate entire barrier islands or zones within a few blocks of the coast. Although wind damage is usually not as catastrophic as the storm surge, it affects a much larger area. Where building codes are inadequate, economic losses can be especially severe. Because hurricanes weaken as they move inland, most wind damage occurs within 200 kilometers (125 miles) of the coast. Far from the coast, a weakening storm can produce extensive flooding long after the winds have diminished below hurricane levels. Sometimes the damage from inland flooding exceeds storm-surge destruction.

COASTAL CLASSIFICATION

The great variety of shorelines demonstrates their complexity. Indeed, to understand any particular coastal area, many factors must be considered, including rock types, size and direction of waves, frequency of storms, tidal range, and offshore topography. In addition, practically all coastal areas were affected by the worldwide rise in sea level that accompanied the melting of Ice Age glaciers at the close of the Pleistocene epoch. Finally, tectonic events that elevate or drop the land or change the volume of ocean basins must be taken into account. The large number of factors that influence coastal areas make shoreline classification difficult.

Students Sometimes Ask . . .

Why are hurricanes given names, and who picks the names?

Actually, the names are given once the storms reach tropical-storm status (winds between 61 and 119 kilometers per hour). Tropical storms are named to provide ease of communication between forecasters and the general public regarding forecasts, watches, and warnings. Tropical storms and hurricanes can last a week or longer, and two or more storms can be occurring in the same region at the same time. Thus, names can reduce the confusion about what storm is being described.

The World Meteorological Organization (affiliated with the United Nations) creates the lists of names. The names for Atlantic storms are used again at the end of a six-year cycle unless a hurricane was particularly noteworthy. Such names are retired to prevent confusion when storms are discussed in future years.

Many geologists classify coasts based on the changes that have occurred with respect to sea level. This commonly used classification divides coasts into two general categories: emergent and submergent. **Emergent coasts** develop either because an area experiences uplift or as a result of a drop in sea level. Conversely, **submergent coasts** are created when sea level rises or the land adjacent to the sea subsides.

Emergent Coasts

In some areas, the coast is clearly emergent because rising land or a falling water level exposes wave-cut cliffs and platforms above sea level. Excellent examples include portions of coastal California, where uplift has occurred in the recent geological past (see Figure 20.13, p. 552). The marine terrace shown in Figure 20.13 also illustrates this situation. In the case of the Palos Verdes Hills, south of Los Angeles, seven different terrace levels exist, indicating seven episodes of uplift. The ever persistent sea is now cutting a new platform at the base of the cliff. If uplift follows, it too will become an elevated marine terrace.

Other examples of emergent coasts include regions that were once buried beneath great ice sheets. When glaciers were present, their weight depressed the crust, and when the ice melted, the crust began gradually to spring back. Consequently, prehistoric shoreline features may now be found high above sea level. The Hudson Bay region of Canada is such an area, portions of which are still rising at a rate of more than a centimeter annually.

Submergent Coasts

In contrast to the preceding examples, other coastal areas show definite signs of submergence. Shorelines that have been submerged in the relatively recent past are often highly irregular because the sea typically floods the lower reaches of river valleys flowing into the ocean. The ridges separating the valleys, however, remain above sea level and project into the sea as headlands. These drowned river mouths, which are called **estuaries** (*aestus* = tide) characterize many coasts today. Along the Atlantic coastline, the Chesapeake and Delaware bays are examples of large estuaries created by submergence (Figure 20.29).

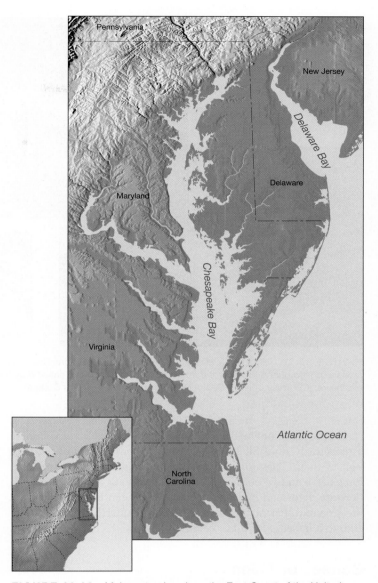

FIGURE 20.29 Major estuaries along the East Coast of the United States. The lower portions of many river valleys were submerged by the rise in sea level that followed the end of the Ice Age, creating large estuaries such as Chesapeake and Delaware bays.

The picturesque coast of Maine, particularly in the vicinity of Acadia National Park, is another excellent example of an area that was flooded by the post-glacial rise in sea level and transformed into a highly irregular coastline.

Keep in mind that most coasts have complicated geologic histories. With respect to sea level, many have, at various times, emerged and then submerged. Each time they may retain some of the features created during the previous situation.

TIDES

Tides are daily changes in the elevation of the ocean surface. Their rhythmic rise and fall along coastlines have been known since antiquity. Other than waves, they are the easiest ocean movements to observe (Figure 20.30).

Students Sometimes Ask . . .

Where are the world's largest tides?

The world's largest *tidal range* (the difference between successive high and low tides) is found in the northern end of Nova Scotia's 258-kilometer-(160-mile-) long Bay of Fundy. During maximum spring tide conditions, the tidal range at the mouth of the bay (where it opens to the ocean) is only about 2 meters (6.6 feet). However, the tidal range progressively increases from the mouth of the bay northward because the natural geometry of the bay concentrates tidal energy. In the northern end of Minas Basin, the maximum spring tidal range is about 17 meters (56 feet). This extreme tidal range leaves boats high and dry during low tide (see Figure 20.30).

FIGURE 20.30 High tide and low tide on Nova Scotia's Minas Basin in the Bay of Fundy. The areas exposed during low tide and flooded during high tide are called *tidal flats.* Tidal flats here are extensive. (Courtesy of Nova Scotia Department of Tourism and Culture)

Although known for centuries, tides were not explained satisfactorily until Sir Isaac Newton applied the law of gravitation to them. Newton showed that there is a mutual attractive force between two bodies and that since oceans are free to move, they are deformed by this force. Hence, ocean tides result from the gravitational attraction exerted upon Earth by the Moon and, to a lesser extent, by the Sun.

Causes of Tides

It is easy to see how the Moon's gravitational force can cause the water to bulge on the side of Earth nearest the Moon. In addition, however, an equally large tidal bulge is produced on the side of Earth directly opposite the Moon (Figure 20.31).

Both tidal bulges are caused, as Newton discovered, by the pull of gravity. Gravity is inversely proportional to the square of the distance between two objects, meaning simply that it quickly weakens with distance. In this case, the two objects are the Moon and Earth. Because the force of gravity decreases with distance, the Moon's gravitational pull on Earth is slightly greater on the near side of Earth than on the far side. The result of this differential pulling is to stretch (elongate) the "solid" Earth very slightly. In contrast, the world ocean, which is mobile, is deformed quite dramatically by this effect, producing the two opposing tidal bulges.

Because the position of the Moon changes only moderately in a single day, the tidal bulges remain in place while Earth rotates "through" them. For this reason, if you stand on the seashore for 24 hours, Earth will rotate you through alternating areas of deeper and shallower water. As you are carried into each

tidal bulge, the tide rises, and as you are carried into the intervening troughs between the tidal bulges, the tide falls. Therefore, most places on Earth experience two high tides and two low tides each day.

Further, the tidal bulges migrate as the Moon revolves around Earth about every 29 days. As a result, the tides, like the time of moonrise, shift about 50 minutes later each day. After 29 days the cycle is complete and a new one begins.

In many locations, there may be an inequality between the high tides during a given day. Depending on the position of the Moon, the tidal bulges may be inclined to the equator as in Figure 20.31. This figure illustrates that one high tide experienced by an observer in the Northern Hemisphere is considerably higher than the high tide half a day later. In contrast, a Southern Hemisphere observer would experience the opposite effect.

Monthly Tidal Cycle

The primary body that influences the tides is the Moon, which makes one complete revolution around Earth every 29 and a half days. The Sun, however, also influences the tides. It is far larger than the Moon, but because it is much farther away, its effect is considerably

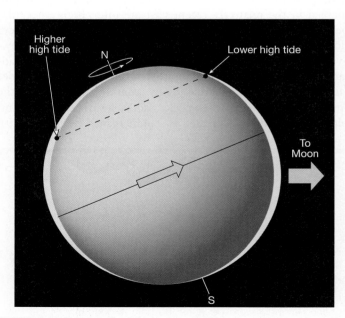

FIGURE 20.31 Idealized tidal bulges on Earth caused by the Moon. If Earth were covered to a uniform depth with water, there would be two tidal bulges: one on the side of Earth facing the Moon (right), and the other on the opposite side of Earth (left). Depending on the Moon's position, tidal bulges may be inclined relative to Earth's equator. In this situation, Earth's rotation causes an observer to experience two unequal high tides during a day.

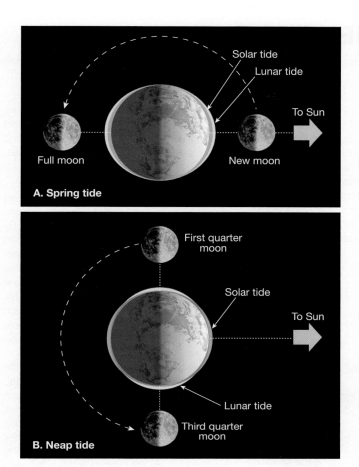

FIGURE 20.32 Earth-Moon-Sun positions and the tides. **A.** When the Moon is in the full or new position, the tidal bulges created by the Sun and Moon are aligned, there is a large tidal range on Earth, and *spring tides* are experienced. **B.** When the Moon is in the first- or third-quarter position, the tidal bulges produced by the Moon are at right angles to the bulges created by the Sun. Tidal ranges are smaller, and *neap tides* are experienced.

less. In fact, the Sun's tide-generating effect is only about 46 percent that of the Moon's.

Near the times of new and full moons, the Sun and Moon are aligned and their forces are added together (Figure 20.32A). Accordingly, the combined gravity of these two tide-producing bodies causes larger tidal bulges (higher high tides) and deeper tidal troughs (lower low tides), producing a large tidal range. These are called the **spring tides** (*springen* = to rise up), which have no connection with the spring season but occur twice a month during the time when the Earth—Moon—Sun system is aligned. Conversely, at about the time of the first and third quarters of the Moon, the gravitational forces of the Moon and Sun act on Earth at right angles, and each partially offsets the influence of the other (Figure 20.32B). As a result, the daily tidal range is less. These are called **neap tides** (*nep* = scarcely or barely touching), which also occur twice each month. Each month, then, there are two spring tides and two neap tides, each about one week apart.

Tidal Patterns

So far, the basic causes and types of tides have been explained. Keep in mind, however, that these theoretical considerations cannot be used to predict either the height or the time of actual tides at a particular place. This is because many factors—including the shape of the coastline, the configuration of ocean basins, and water depth—greatly influence the tides. Consequently, tides at various locations respond differently to the tide-producing forces. This being the case, the nature of the tide at any coastal location can be determined most accurately by actual observation. The predictions in tidal tables and tidal data on nautical charts are based on such observations.

Three main tidal patterns exist worldwide. A **diurnal tidal pattern** (*diurnal* = daily) is characterized by a single high tide and a single low tide each tidal day (Figure 20.33). Tides of this type

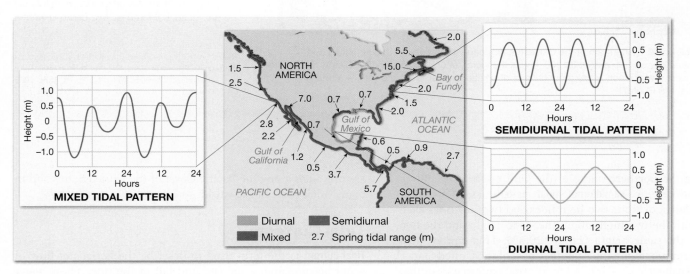

FIGURE 20.33 Tidal patterns and their occurrence along portions of the coastlines of North and South America. A diurnal tidal pattern (lower right) shows one high and one low tide each tidal day. A semidiurnal pattern (upper right) shows two highs and two lows of approximately equal heights during each tidal day. A mixed tidal pattern (left) shows two highs and two lows of unequal heights during each tidal day.

PEOPLE AND THE ENVIRONMENT

Hurricane Forecasting

A location only a few hundred kilometers from a hurricane—just a day's striking distance away—may experience clear skies and virtually no wind. Before the age of weather satellites, such a situation made it very difficult to warn people of impending storms. The deadly hurricane that devastated Galveston, Texas, in 1900 is a tragic example (Figure 20.C). It was a complete surprise to the unsuspecting residents of this vulnerable barrier island.

Today we have the benefit of numerous observational tools. Using input from satellites, aircraft reconnaissance, coastal radar, and remote data buoys in conjunction with sophisticated computer models, meteorologists monitor and forecast the movements and intensity of hurricanes. The goal is to issue timely watches and warnings. In the United States, early warning systems have greatly reduced the number of hurricane deaths. At the same time, however,

an astronomical rise has occurred in the amount of property damage. The primary reason for this latter trend is the rapid population growth and accompanying development in coastal areas.

Hurricane Watches and Warnings

A *hurricane watch* is an announcement aimed at specific coastal areas that a hurricane poses a possible threat, generally within 36 hours. By contrast, a *hurricane warning* is issued when sustained winds of 119 kilometers (74 miles) per hour or higher are expected within a specified coastal area in 24 hours or less. A hurricane warning can remain in effect when dangerously high water or a combination of dangerously high water and exceptionally high waves continue, even though winds may be less than hurricane force.

Two factors are especially important in the watch-and-warning decision process. First, adequate lead time must be provided to protect life, and to a lesser degree, property. Second, forecasters must attempt to keep overwarning at a minimum. This, however, can be a difficult task. Clearly, the decision to issue a warning represents a delicate balance between the need to protect the public on the one hand and the desire to minimize the degree of overwarning on the other.

Hurricane Forecasting*

Hurricane forecasts are a basic part of any warning program. There are several aspects that can be part of such a forecast. We certainly want to

FIGURE 20.C   Aftermath of the Galveston hurricane that struck an unsuspecting and unprepared city on September 8, 1900. It was the worst natural disaster in U.S. history. Entire blocks were swept clean, while mountains of debris accumulated around the few remaining buildings. (AP Photo)

---

*Based on "Hurricane Forecasting in the United States: An Information Statement of the AMS," in *Bulletin of the American Meteorological Society*, Vol. 88, No. 6, June 2007, pp. 950–953.

---

occur along the northern shore of the Gulf of Mexico, among other locations. A **semidiurnal tidal pattern** (*semi* = twice, *diurnal* = daily) exhibits two high tides and two low tides each tidal day, with the two highs about the same height and the two lows about the same height (Figure 20.33). This type of tidal pattern is common along the Atlantic Coast of the United States. A **mixed tidal pattern** is similar to a semidiurnal pattern except that it is characterized by a large inequality in high water heights, low water heights, or both (Figure 20.33). In this case, there are usually two high and two low tides each day, with high tides of different heights and low tides of different heights. Such tides

are prevalent along the Pacific Coast of the United States and in many other parts of the world.

Tidal Currents

Tidal current is the term used to describe the *horizontal* flow of water accompanying the rise and fall of the tide. These water movements induced by tidal forces can be important in some coastal areas. Tidal currents flow in one direction during a portion of the tidal cycle and reverse their flow during the remainder. Tidal currents that advance into the coastal zone as the tide

BOX 20.2

know where the storm is headed. The predicted path of a storm is called the *track forecast*. Of course, there is also interest in knowing the intensity (strength of the winds), probable rainfall amounts, and the likely size of the storm surge.

The track forecast is probably the most basic information because accurate prediction of other storm characteristics is of little value if there is significant uncertainty as to where the storm is going. Accurate track forecasts are important because they can lead to timely evacuations from the surge zone where the greatest number of deaths usually occur. Fortunately, track forecasts have been steadily improving. During the span 2001–2005, forecast errors were roughly half of what they were in 1990. During the very active 2004 and 2005 Atlantic hurricane seasons, 12–72 hour track forecast accuracy was at or near record levels. Consequently, the length of official track forecasts issued by the National Hurricane Center was extended from three days to five days (Figure 20.D). Current five-day track forecasts are now as accurate as three-day forecasts were 15 years ago. Much of the progress is due to improved computer models and to a dramatic increase in the quantity of satellite data from over the oceans.

Despite improvements in accuracy, forecast uncertainty still requires that hurricane warnings be issued for relatively large coastal areas. During the span 2000–2005, the average length of coastline under a hurricane warning in the United States was 510 kilometers (316 miles). This represents a significant improvement over the preceding decade when the average was 730 kilometers (452 miles). Nevertheless, only about one-quarter of an average warning area experiences hurricane conditions.

In contrast to the improvements in track forecasts, errors in forecasts of hurricane intensity (wind speeds) have not changed significantly in 30 years. Accurate predictions of rainfall as hurricanes make landfall also remain elusive. However, accurate predictions of impending storm surge are possible when good information regarding the storm's track and surface wind structure are known and reliable data regarding coastal and offshore (underwater) topography are available.

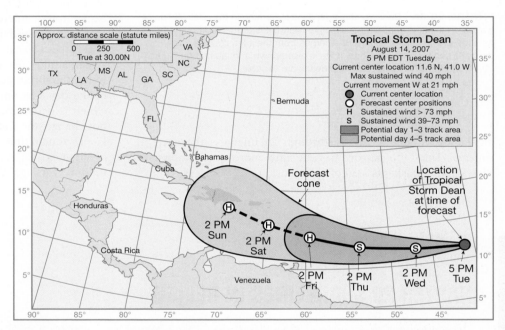

FIGURE 20.D Five-day track forecast for Tropical Storm Dean issued at 5 PM EDT, Tuesday, August 14, 2007. When a hurricane track forecast is issued by the National Hurricane Center, it is termed a *forecast cone*. The cone represents the probable track of the center of the storm and is formed by enclosing the area swept out by a set of circles along the forecast track (at 12, 24, 36 hours, etc.). The size of each circle gets larger with time. Based on statistics from 2003–2007 the entire track of an Atlantic tropical cyclone can be expected to remain entirely within the cone roughly 60 to 70 percent of the time. (National Weather Service/National Hurricane Center)

rises are called **flood currents.** As the tide falls, seaward-moving water generates **ebb currents.** Periods of little or no current, called *slack water,* separate flood and ebb. The areas affected by these alternating tidal currents are **tidal flats** (see Figure 20.30). Depending on the nature of the coastal zone, tidal flats vary from narrow strips seaward of the beach to extensive zones that may extend for several kilometers.

Although tidal currents are not important in the open sea, they can be rapid in bays, river estuaries, straits, and other narrow places. Off the coast of Brittany in France, for example, tidal currents that accompany a high tide of 12 meters (40 feet) may

attain a speed of 20 kilometers (12 miles) per hour. While tidal currents are not generally major agents of erosion and sediment transport, notable exceptions occur where tides move through narrow inlets. Here they constantly scour the small entrances to many good harbors that would otherwise be blocked.

Sometimes deposits called **tidal deltas** are created by tidal currents (Figure 20.34). They may develop either as *flood deltas* landward of an inlet or as *ebb deltas* on the seaward side of an inlet. Because wave activity and longshore currents are reduced on the sheltered landward side, flood deltas are more common and more prominent (see Figure 20.15A). They form after the

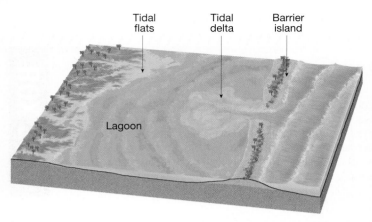

Tidal flats   Tidal delta   Barrier island

Lagoon

FIGURE 20.34   Because this tidal delta is forming in the relatively quiet waters on the landward side of a barrier island, it is termed a flood delta. As a rapidly moving tidal current emerges from the inlet, it slows and deposits sediment. Such a tidal delta is also shown and labeled on Figure 20.15A, p. 553. The shapes of tidal deltas are variable.

tidal current moves rapidly through an inlet. As the current emerges from the narrow passage into more open waters, it slows and deposits its load of sediment.

Tides and Earth's Rotation

The tides, by friction against the floor of the ocean basins, act as weak brakes that are steadily slowing Earth's rotation. The rate of slowing, however, is not great. Astronomers, who have precisely measured the length of the day over the past 300 years, have found that the day is increasing by 0.002 second per century. Although this may seem inconsequential, over millions of years this small effect will become very large.

If Earth's rotation is slowing, the length of each day must have been shorter and the number of days per year must have been greater in the geologic past. One method used to investigate this phenomenon involves the microscopic examination of shells of certain invertebrates. Clams and corals, as well as other organisms, grow a microscopically thin layer of new shell material each day. By studying the daily growth rings of some well-preserved fossil specimens, we can determine the number of days in a year. Studies using this ingenious technique indicate that early in the Cambrian period, about 540 million years ago, the length of the day was only 21 hours. Because the length of the year, which is determined by Earth's revolution about the Sun, does not change, the Cambrian year contained 424 21-hour days. By late Devonian time, about 365 million years ago, a year consisted of about 410 days, and as the Permian period opened, about 290 million years ago, there were 390 days in a year.

CHAPTER 20 SHORELINES IN REVIEW

- The *shore* is the area extending between the lowest tide level and the highest elevation on land that is affected by storm waves. The *coast* extends inland from the shore as far as ocean-related features can be found. The shore is divided into the *foreshore* and *backshore*. Seaward of the foreshore are the *nearshore* and *offshore* zones.

- A *beach* is an accumulation of sediment found along the landward margin of the ocean or a lake. Among its parts are one or more *berms* and the *beach face*. Beaches are composed of whatever material is locally abundant and should be thought of as material in transit along the shore.

- *Waves are moving energy* and *most ocean waves are initiated by the wind.* The three factors that influence the *height, length,* and *period* of a wave are (1) *wind speed,* (2) *length of time the wind has blown,* and (3) *fetch,* the distance the wind has traveled across open water. Once waves leave a storm area, they are termed *swells,* which are symmetrical, longer-wavelength waves.

- As waves travel, *water particles transmit energy by circular orbital motion,* which extends to a depth equal to one half the wavelength. When a wave travels into shallow water, it experiences physical changes that can cause the wave to collapse, or *break,* and form *surf.*

- Wave erosion is caused by *wave impact pressure* and *abrasion* (the sawing and grinding action of water armed with rock fragments). The bending of waves is called *wave refraction.* Owing to refraction, wave impact is concentrated against the sides and ends of headlands.

- Most waves reach the shore at an angle. The uprush (swash) and backwash of water from each breaking wave moves the sediment in a zigzag pattern along the beach. This movement, called *beach drift,* can transport sand hundreds or even thousands of meters each day. Oblique waves also produce *longshore currents* within the surf zone that flow parallel to the shore.

- Features produced by *shoreline erosion* include *wave-cut cliffs* (which originate from the cutting action of the surf against the base of coastal land), *wave-cut platforms* (relatively flat, benchlike surfaces left behind by receding cliffs), *sea arches* (formed when a headland is eroded and two caves from opposite sides unite), and *sea stacks* (formed when the roof of a sea arch collapses).

- Some of the depositional features that form when sediment is moved by beach drift and longshore currents are *spits* (elongated ridges of sand that project from the land into the mouth of an adjacent bay), *baymouth bars* (sandbars that completely cross a bay), and *tombolos* (ridges of sand that connect an island to the mainland or to another island). Along the Atlantic and Gulf Coastal Plains, the shore zone is characterized by *barrier islands,* low ridges of sand that parallel the coast at distances from 3 to 30 kilometers offshore.

- Local factors that influence shoreline erosion are (1) the proximity of a coast to sediment-laden rivers, (2) the degree of tectonic activity, (3) the topography and composition of the land, (4) prevailing winds and weather patterns, and (5) the configuration of the coastline and nearshore areas.

- *Hard stabilization* involves building hard, massive structures in an attempt to protect a coast from erosion or prevent the movement of sand along the beach. Hard stabilization includes *groins* (short walls constructed at a right angle to the shore to trap moving sand), *breakwaters* (structures built parallel to the shore to protect it from the force of large breaking waves), and *seawalls* (armoring the coast to prevent waves from reaching the area behind the wall). *Alternatives to hard stabilization* include *beach nourishment,* which involves the addition of sand to replenish eroding beaches, and *relocation* of damaged or threatened buildings.

- Because of basic geological differences, the *nature of shoreline erosion problems along America's Pacific and Atlantic coasts is very different.* Much of the development along the Atlantic and Gulf coasts has occurred on barrier islands, which receive the full force of major storms. Much of the Pacific Coast is characterized by narrow beaches backed by steep cliffs and mountain ranges. A major problem facing the Pacific shoreline is a narrowing of beaches caused by the natural flow of materials to the coast being interrupted by dams built for irrigation and flood control.

- Although damages caused by a hurricane depend on several factors, including the size and population density of the area affected and the nearshore bottom configuration, the most significant factor is the strength of the storm itself. The *Saffir-Simpson* scale ranks the relative intensities of hurricanes. A category 5 storm is most severe and a category 1 storm is least severe. Damage caused by hurricanes can be divided into three categories: (1) *storm surge,* which is most intense on the right side of the eye where winds are blowing toward the shore, occurs when a dome of water sweeps across the coast near the point where the eye makes landfall; (2) *wind damage;* and (3) *inland flooding,* which is caused by torrential rains that accompany most hurricanes.

- One commonly used classification of coasts is based upon changes that have occurred with respect to sea level. *Emergent coasts,* often with wave-cut cliffs and wave-cut platforms above sea level, develop either because an area experiences uplift or as a result of a drop in sea level. Conversely, *submergent coasts,* with their drowned river mouths, called *estuaries,* are created when sea level rises or the land adjacent to the sea subsides.

- *Tides,* the daily rise and fall in the elevation of the ocean surface at a specific location, are caused by the *gravitational attraction* of the Moon and, to a lesser extent, the Sun. The Moon and the Sun each produce a pair of *tidal bulges* on Earth. These tidal bulges remain in fixed positions relative to the generating bodies as Earth rotates through them, resulting in alternating high and low tides. *Spring tides* occur near the times of new and full moons when the Sun and Moon are aligned and their bulges are added together to produce especially high and low tides (a *large daily tidal range*). Conversely, *neap tides* occur at about the times of the first and third quarters of the Moon when the bulges of the Moon and Sun are at right angles, producing a *smaller daily tidal range.*

- *Three main tidal patterns exist worldwide. A diurnal tidal pattern* exhibits one high and low tide daily; a *semidiurnal tidal pattern* exhibits two high and low tides daily of about the same height; and a *mixed tidal pattern* usually has two high and low tides daily of different heights.

- *Tidal currents* are horizontal movements of water that accompany the rise and fall of tides. *Tidal flats* are the areas that are affected by the advancing and retreating tidal currents. When tidal currents slow after emerging from narrow inlets, they deposit sediment that may eventually create *tidal deltas.*

KEY TERMS

abrasion (p. 548)
backshore (p. 543)
barrier island (p. 552)
baymouth bar (p. 552)
beach (p. 543)
beach drift (p. 549)
beach face (p. 543)
beach nourishment (p. 556)
berm (p. 543)
breakwater (p. 555)

coast (p. 543)
coastline (p. 543)
diurnal tidal pattern (p. 567)
ebb current (p. 569)
emergent coast (p. 565)
estuary (p. 565)
eye (p. 562)
eye wall (p. 562)
fetch (p. 545)
flood current (p. 569)

foreshore (p. 543)
groin (p. 554)
hard stabilization (p. 554)
jetty (p. 554)
longshore current (p. 550)
marine terrace (p. 552)
mixed tidal pattern (p. 568)
neap tide (p. 567)
nearshore zone (p. 543)
offshore zone (p. 543)

rip current (p. 550)
sea arch (p. 552)
sea stack (p. 552)
seawall (p. 555)
semidiurnal tidal pattern (p. 568)
shore (p. 543)
shoreline (p. 543)
spit (p. 552)
spring tide (p. 567)

storm surge (p. 562)
submergent coast (p. 565)
surf (p. 547)
tidal current (p. 568)

tidal delta (p. 569)
tidal flats (p. 569)
tide (p. 565)
tombolo (p. 552)

wave-cut cliff (p. 552)
wave-cut platform (p. 552)
wave height (p. 544)

wavelength (p. 544)
wave period (p. 544)
wave refraction (p. 548)

QUESTIONS FOR REVIEW

1. Distinguish among shore, shoreline, coast, and coastline.

2. What is a beach? Briefly distinguish between beach face and berm. What are the sources of beach sediment?

3. List three factors that determine the height, length, and period of a wave.

4. Describe the motion of a floating object as a wave passes (see Figure 20.6, page 546).

5. Describe the physical changes that occur to a wave's speed, wavelength, and height as it moves into shallow water and breaks.

6. Describe two ways in which waves cause erosion.

7. What is wave refraction? What is the effect of this process along irregular coastlines (see Figure 20.10, page 549)?

8. Why are beaches often called "rivers of sand"?

9. Describe the formation of the following features: wave-cut cliff, wave-cut platform, marine terrace, sea stack, spit, bay-mouth bar, and tombolo.

10. List three ways that barrier islands may form.

11. In what direction is beach drift and longshore currents moving sand in Figure 20.20, p. 557? Is it moving toward the top left or toward the bottom right of the photo?

12. List some examples of hard stabilization and describe what each is intended to do. What effect does each one have on the distribution of sand on the beach?

13. List two alternatives to hard stabilization, indicating potential problems with each one.

14. Relate the damming of rivers to the shrinking of beaches at many locations along the West Coast of the United States. Why do narrower beaches lead to accelerated sea-cliff retreat?

15. Hurricane damage can be divided into three broad categories. Name them. Which category is responsible for the greatest number of hurricane-related deaths?

16. What observable features would lead you to classify a coastal area as emergent?

17. Are estuaries associated with submergent or emergent coasts? Explain.

18. Discuss the origin of ocean tides. Explain why the Sun's influence on Earth's tides is only about half that of the Moon's, even though the Sun is so much more massive than the Moon.

19. Explain why an observer can experience two unequal high tides during one day (see Figure 20.31, page 566).

20. How do diurnal, semidiurnal, and mixed tidal patterns differ?

21. Distinguish between flood current and ebb current.

22. How have tides affected Earth's rotation? How did geologists substantiate this idea?

COMPANION WEBSITE

The *Earth 10e* website uses the resources and flexibility of the Internet to aid in your study of the topics in this chapter. Written and developed by the authors and other geology instructors, this site will help improve your understanding of geology. Visit www.mygeoscienceplace.com in order to:

• **Review** key chapter concepts

• **Read** with links to the eBook and to chapter-specific web resources

• **Visualize** and comprehend challenging topics using learning activities in *GEODe Earth*

• **Test** yourself with online quizzes

GEODe EARTH

GEODe Earth is a valuable and easy to use learning aid that can be accessed from your book's Companion Website (www .mygeoscienceplace.com). It is a dynamic instructional tool that promotes understanding and reinforces important concepts by using tutorials, animations, and exercises that actively engage the student.

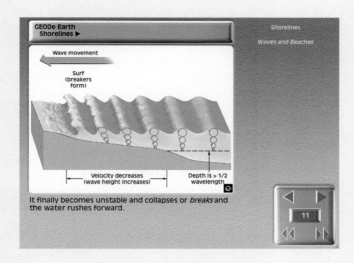

GLOBAL CLIMATE CHANGE

Sea ice in the Arctic Ocean. Sea ice is frozen seawater. The area covered by Arctic sea ice at the end of the summer melt period has been declining. This is considered a sign of global warming.

(PHOTO BY SCOTT DARSNEY/ALASKASTOCK.COM)

Climate is the long-term aggregate of weather. It is more than just an expression of average atmospheric conditions. In order to accurately portray the character of a place or an area, variations and extremes must also be included. Anyone who has the opportunity to travel around the world will find such an incredible variety of climates that it is hard to believe they could all occur on the same planet. Climate strongly influences plant and animal life, the soil, and many of Earth's surface processes. Climate influences people as well.

CLIMATE AND GEOLOGY ARE LINKED

Climate has a profound impact on many geologic processes. When climate changes, these processes respond. A glance back at the rock cycle in Chapter 1 (p. 33) reminds us about many of the connections. Of course rock weathering has an obvious climate connection, as do processes that operate in arid, tropical, and glacial landscapes. Events such as debris flows and river flooding are often triggered by atmospheric happenings such as periods of extraordinary rainfall. Clearly the atmosphere is a basic link in the hydrologic cycle. Other climate–geology connections involve the impact of internal processes on the atmosphere. For example, the particles and gases emitted by volcanoes can change the composition of the atmosphere, and mountain building can have a significant impact on regional temperature, precipitation, and wind patterns.

Climate not only varies from place to place, but it is naturally variable over time. Over the great expanse of Earth history, and long before humans were roaming the planet, there were many shifts—from warm to cold and from wet to dry, and back again. The geologic record is a storehouse of data that confirms this fact.

The study of sediments, sedimentary rocks, and fossils clearly demonstrates that, through the ages, practically every place on our planet has experienced wide swings in climate, such as from ice ages to conditions associated with subtropical coal swamps or desert dunes. Chapter 22, "Earth's Evolution through Geologic Time," reinforces this fact. Time scales for climate change vary from decades to millions of years.

What factors have been responsible for climate variations during Earth history? One of the topics treated in this chapter involves natural causes of climate change.

Today, global climate change is more than just a topic that is of "academic interest" to a group of scientists who are curious about Earth history. Rather, the subject is making headlines. Many members of the general public are not only curious but concerned about the possibilities. Why is climate change newsworthy? The reason is that research focused on human activities and their impact on the environment has demonstrated that people are inadvertently changing the climate. Unlike changes in the past, which were natural variations, modern climate change is dominated by human influences that are sufficiently large that they exceed the bounds of natural variability. Moreover, these changes are likely to continue for many centuries. The effects of this venture into the unknown with climate could be very disruptive, not only to humans but to many other life forms as well.

How are the details of past climates determined? How is this information useful to us now? In what ways are humans changing global climate? What are the possible consequences?

THE CLIMATE SYSTEM

Throughout this book you have been reminded that Earth is a multidimensional system that consists of many interacting parts. A change in any one part can produce changes in any or all of the other parts—often in ways that are neither obvious nor immediately apparent. This fact is certainly true when it comes to the study of climate and climate change.

To understand and appreciate climate, it is important to realize that climate involves more than just the atmosphere (Figure 21.1):

> The atmosphere is the central component of the complex, connected, and interactive global environmental system upon which all life depends. Climate may be broadly defined as the long-term behavior of this environmental system. To understand fully and to predict changes in the atmospheric component of the climate system, one must understand the sun, oceans, ice sheets, solid earth, and all forms of life.*

*The American Meteorological Society and the University Corporation for Atmospheric Research, "Weather and the Nation's Well-Being," *Bulletin of the American Meteorological Society,* 73, No. 12 (December 2001), p. 2038.

Indeed, we must recognize that there is a **climate system** that includes the atmosphere, hydrosphere, geosphere, biosphere, and cryosphere. (The *cryosphere* refers to the ice and snow that exist at Earth's surface.) The climate system *involves the exchanges of energy and moisture that occur among the five spheres.* These exchanges link the atmosphere to the other spheres so that the whole functions as an extremely complex, interactive unit. Changes to the climate system do not occur in isolation. Rather, when one part of it changes, the other components also react. The major components of the climate system are shown in Figure 21.2.

HOW IS CLIMATE CHANGE DETECTED?

High-technology and precision instrumentation are now available to study the composition and dynamics of the atmosphere. But such tools are recent inventions and, therefore, have been providing data for only a short time span. To understand fully the behavior of the atmosphere and to anticipate future climate change, we must somehow discover how climate has changed over broad expanses of time.

FIGURE 21.1 Reflection Lake in Washington's Mount Rainier National Park. Glaciers are still sculpting this large volcanic mountain. The five major parts of the climate system are represented in this image. (Photo by Art Wolfe)

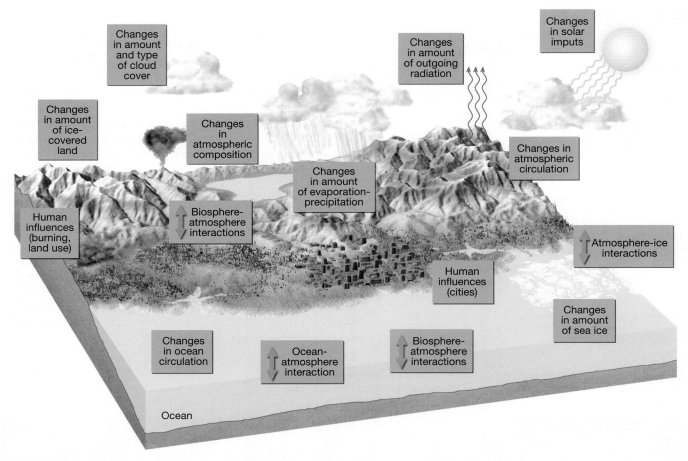

FIGURE 21.2 Schematic view showing several components of Earth's climate system. Many interactions occur among the various components on a wide range of space and time scales, making the system extremely complex.

Instrumental records go back only a couple of centuries at best, and the further back we go, the less complete and more unreliable the data become. To overcome this lack of direct measurements, scientists must decipher and reconstruct past climates by using indirect evidence. Such **proxy data** comes from natural recorders of climate variability, such as seafloor sediments, glacial ice, fossil pollen, and tree-growth rings, as well as from historical documents. Scientists who analyze proxy data and reconstruct past climates are engaged in the study of **paleoclimatology.** The main goal of such work is to understand the climate of the past in order to assess the current and potential future climate in the context of natural climate variability. In the following discussion, we will briefly examine some of the important sources of proxy data.

Seafloor Sediment—A Storehouse of Climate Data

We know that the parts of the Earth system are linked so that a change in one part can produce changes in any or all of the other parts. In this example you will see how changes in atmospheric and oceanic temperatures are reflected in the nature of life in the sea.

Most seafloor sediments contain the remains of organisms that once lived near the sea surface (the ocean–atmosphere interface). When such near-surface organisms die, their shells slowly settle to the floor of the ocean, where they become part of the sedimentary record (Figure 21.3). These seafloor sediments are useful recorders of worldwide climate change because the numbers and types of organisms living near the sea surface change with the climate:

> We would expect that in any area of the ocean/atmosphere interface the average annual temperature of the surface water of the ocean would approximate that of the contiguous atmosphere.

Students Sometimes Ask . . .

What's the difference between weather and climate?

The term weather refers to the state of the atmosphere at a given time and place. Changes in the weather are frequent and sometimes seemingly erratic. Climate is a description of aggregate weather conditions based on observations over many decades. Climate is often defined simply as "average weather," but this definition is inadequate because variations and extremes are also part of a climate description.

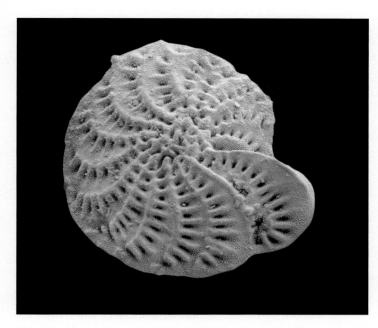

FIGURE 21.3 Image from a scanning electron microscope of the shell of a foraminifera. These tiny, single-celled organisms are sensitive to even small fluctuations in temperature. Seafloor sediments containing fossils such as this are useful recorders of climate change. (Photo by Andrew Syred/Photo Researchers, Inc.)

The temperature equilibrium established between surface seawater and the air above it should mean that . . . changes in climate should be reflected in changes in organisms living near the surface of the deep sea. . . . When we recall that the seafloor sediments in vast areas of the ocean consist mainly of shells of pelagic foraminifers, and that these animals are sensitive to variations in water temperature, the connection between such sediments and climate change becomes obvious.*

Thus, in seeking to understand climate change as well as other environmental transformations, scientists are tapping the huge reservoir of data in seafloor sediments. The sediment cores gathered by drilling ships and other research vessels have provided invaluable data that have significantly expanded our knowledge and understanding of past climates (Figure 21.4).

One notable example of the importance of seafloor sediments to our understanding of climate change relates to unraveling the fluctuating atmospheric conditions of the Ice Age. The records of temperature changes contained in cores of sediment from the ocean floor have proven critical to our present understanding of this recent span of Earth history.**

Oxygen Isotope Analysis

Oxygen isotope analysis is based on precise measurement of the ratio between two isotopes of oxygen: ^{16}O, which is the most common, and the heavier ^{18}O. A molecule of H_2O can form from either ^{16}O or ^{18}O. But the lighter isotope, ^{16}O, evaporates

more readily from the oceans. Because of this, precipitation (and hence the glacial ice that it may form) is enriched in ^{16}O. This leaves a greater concentration of the heavier isotope, ^{18}O, in the ocean water. Thus, during periods when glaciers are extensive, more of the lighter ^{16}O is tied up in ice, so the concentration of ^{18}O in seawater increases. Conversely, during warmer interglacial periods when the amount of glacial ice decreases dramatically, more ^{16}O is returned to the sea, so the proportion of ^{18}O relative to ^{16}O in ocean water also drops. Now, if we had some ancient recording of the changes of the $^{18}O/^{16}O$ ratio, we could determine when there were glacial periods and, therefore, when the climate grew cooler.

Fortunately, we do have such a recording. As certain microorganisms secrete their shells of calcium carbonate ($CaCO_3$), the prevailing $^{18}O/^{16}O$ ratio is reflected in the composition of these hard parts. When the organisms die, their hard parts settle to the ocean floor, becoming part of the sediment layers there. Consequently, periods of glacial activity can be determined from variations in the oxygen isotope ratio found in shells of certain microorganisms buried in deep-sea sediments.

The $^{18}O/^{16}O$ ratio also varies with temperature. More ^{18}O is evaporated from the oceans when temperatures are high, and less is evaporated when temperatures are low. Therefore, the heavy isotope is more abundant in the precipitation of warm eras

FIGURE 21.4 The *Chikyu* (meaning "Earth" in Japanese), the world's most advanced scientific drilling vessel. It can drill as deep as 7,000 meters (nearly 23,000 feet) below the seabed in water as deep as 2,500 meters (8,200 feet). It is part of the Integrated Ocean Drilling Program (IODP). (AP Photo/Itsuo Inouye)

---

*Richard F. Flint, *Glacial and Quaternary Geology* (New York: John Wiley & Sons, 1971), p. 718.

**For more information on this topic, see "Causes of Glaciation" in Chapter 18, page 512.

and less abundant during colder periods. Using this principle, scientists studying the layers of ice and snow in glaciers have been able to produce a record of past temperature changes.

Climate Change Recorded in Glacial Ice

Ice cores are an indispensable source of data for reconstructing past climates (see Figure 18.37, p. 515). Research based on vertical cores taken from the Greenland and Antarctic ice sheets has changed our basic understanding of how the climate system works.

Scientists collect samples with a drilling rig, like a small version of an oil drill. A hollow shaft follows the drill head into the ice, and an ice core is extracted. In this way, cores that sometimes exceed 2000 meters (6500 feet) in length, and may represent more than 200,000 years of climate history, are acquired for study (Figure 21.5A).

The ice provides a detailed record of changing air temperatures and snowfall. Air bubbles trapped in the ice record variations in atmospheric composition. Changes in carbon dioxide and methane are linked to fluctuating temperatures. The cores also include atmospheric fallout such as wind-blown dust, volcanic ash, pollen, and modern-day pollution.

Past temperatures are determined by *oxygen isotope analysis*. Using this technique, scientists are able to produce a record of past temperature changes. A portion of such a record is shown in Figure 21.5B.

Tree Rings—Archives of Environmental History

If you look at the end of a log, you will see that it is composed of a series of concentric rings. Each of these *tree rings* becomes larger in diameter outward from the center (Figure 21.6). Every year, in temperate regions, trees add a layer of new wood under the bark. Characteristics of each tree ring, such as size and density, reflect the environmental conditions (especially climate) that prevailed during the year when the ring formed. Favorable growth conditions produce a wide ring; unfavorable ones produce a narrow ring. Trees growing at the same time in the same region show similar tree-ring patterns.

Because a single growth ring is usually added each year, the age of the tree when it was cut can be determined by counting the rings. If the year of cutting is known, the age of the tree and the year in which each ring formed can be determined by counting back from the outside ring. Scientists are not limited to working with trees that have been cut down. Small, nondestructive core samples can be taken from living trees.

To make the most effective use of tree rings, extended patterns known as *ring chronologies* are established. They are produced by comparing the patterns of rings among trees in an area. If the same pattern can be identified in two samples, one of which has been dated, the second sample can be dated from the first by matching the ring patterns common to both. This technique, called *cross dating*, is illustrated in Figure 21.7. Tree-ring chronologies extending back for thousands of years have been established for some regions. To date a timber sample of unknown age, its ring pattern is matched against the reference chronology.

Tree-ring chronologies are unique archives of environmental history and have important applications in such disciplines as climate, geology, ecology, and archaeology. For example, tree rings are used to reconstruct climate variations within a region for spans of thousands of years prior to human historical records. Knowledge of such long-term variations is of great value in making judgments regarding the recent record of climate change.

A.

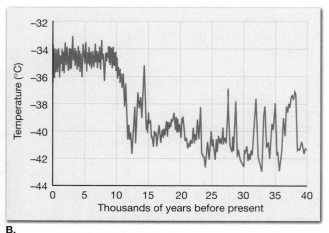

B.

FIGURE 21.5   A. The National Ice Core Laboratory is a physical plant for storing and studying cores of ice taken from glaciers around the world. These cores represent a long-term record of material deposited from the atmosphere. The lab provides scientists the capability to conduct examinations of ice cores, and it preserves the integrity of these samples in a repository for the study of global climate change and past environmental conditions. (Photo by USGS/National Ice Core Laboratory)
B. This graph, showing temperature variations over the past 40,000 years, is derived from oxygen isotope analysis of ice cores recovered from the Greenland ice sheet. (After U.S. Geological Survey)

FIGURE 21.6 **A.** Ancient bristlecone pines in California's White Mountains. The study of tree-growth rings is one way that scientists reconstruct past climates. Some of these trees are more than 4000 years old. (Photo by Dennis Flaherty/Photo Researchers, Inc.) **B.** Each year a growing tree produces a layer of new cells beneath the bark. If the tree is felled and the trunk examined (or if a core is taken, to avoid cutting the tree), each year's growth can be seen as a ring. Because the amount of growth (thickness of a ring) depends upon precipitation and temperature, tree rings are useful recorders of past climates. (Photo by Stephen J. Krasemann/Photo Researchers, Inc.)

FIGURE 21.7 Cross dating is a basic principle in dendrochronology. Here it was used to date an archaeological site by correlating tree-ring patterns for wood from trees of three different ages. First, a tree-ring chronology for the area is established using cores extracted from living trees. This chronology is extended further back in time by matching overlapping patterns from older, dead trees. Finally, cores taken from beams inside the ruin are dated using the chronology established from the other two sites.

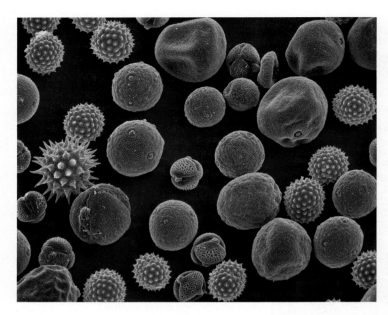

FIGURE 21.8   This false color image from an electron microscope shows an assortment of pollen grains. Note how the size, shape, and surface characteristics differ from one species to another. Analysis of the types and abundance of pollen in lake sediments and peat deposits provides information on how climate has changed over time. (Photo by David Scharf/Science Photo Library)

Other Types of Proxy Data

Other sources of proxy data that are used to gain insight into past climates include fossil pollen, corals, and historical documents.

Fossil Pollen   Climate is a major factor influencing the distribution of vegetation, so knowing the nature of the plant community occupying an area is a reflection of the climate. Pollen and spores are parts of the life cycles of many plants, and because they have resistant walls, they are often the most abundant, easily identifiable, and best-preserved plant remains in sediments (Figure 21.8). By analyzing pollen from accurately dated sediments, it is possible to obtain high-resolution records of vegetational changes in an area. From such information, past climates can be reconstructed.

Corals   Coral reefs consist of colonies of corals that live in warm shallow waters and form atop the hard material left behind by past corals. Corals build their hard skeletons from calcium carbonate ($CaCO_3$) extracted from seawater. The carbonate contains isotopes of oxygen that can be used to determine the temperature of the water in which the coral grew.

The portion of the skeleton that forms in winter has a density that is different from that formed in summer because of variations in growth rates related to temperature and other environmental factors. Thus, corals exhibit seasonal growth bands very much like those observed in trees. The accuracy and reliability of the climate data extracted from corals has been established by comparing recent instrumental records to coral records for the same period. Oxygen isotope analysis of coral growth rings can also serve as a proxy measurement for precipitation, particularly in areas where large variations in annual rainfall occur.

Think of coral as a paleothermometer that enables us to answer important questions about climate variability in the world's oceans. The graph in Figure 21.9 is a 350-year sea-surface temperature record based on oxygen isotope analysis of a core from the Galapagos Islands.

Historical Data   Historical documents sometimes contain helpful information. Although it might seem that such records should readily lend themselves to climate analysis, such is not the case. Most manuscripts were written for purposes other than climate description. Furthermore, writers understandably neglected periods of relatively stable atmospheric conditions and mention only droughts, severe storms, memorable blizzards, and other extremes. Nevertheless, records of crops, floods, and the migration of people have furnished useful evidence of the possible influences of changing climate (Figure 21.10).

SOME ATMOSPHERIC BASICS

In order to better understand climate change, it is helpful to possess some knowledge about the composition and structure of the atmosphere and the process by which it is heated—the *greenhouse effect*.

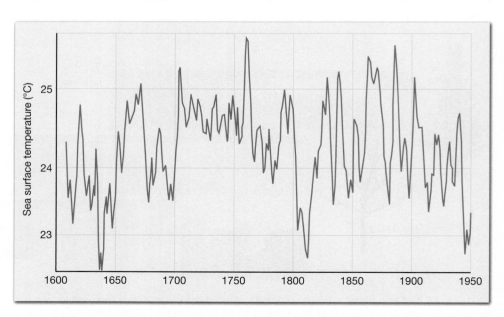

FIGURE 21.9   This graph shows a 350-year record of sea-surface temperatures obtained by oxygen isotope analysis of coral from the Galapagos Islands. (After National Climatic Data Center/NOAA)

FIGURE 21.10 Historical records can sometimes be helpful in the analysis of past climates. The date for the beginning of the grape harvest in the fall is an integrated measure of temperature and precipitation during the growing season. These dates have been recorded for centuries in Europe and provide a useful record of year-to-year climate variations. (Photo by Owen Franken/CORBIS)

Composition of the Atmosphere

Air is *not* a unique element or compound. Rather, air is a *mixture* of many discrete gases, each with its own physical properties, in which varying quantities of tiny solid and liquid particles are suspended.

Clean Dry Air   As you can see in Figure 21.11, clean, dry air is composed almost entirely of two gases—78 percent nitrogen and 21 percent oxygen. Although these gases are the most plentiful components of air and are of great significance to life on Earth, they are of little or no importance in affecting weather phenomena. The remaining 1 percent of dry air is mostly the inert gas argon (0.93 percent) plus tiny quantities of a number of other gases. Carbon dioxide, although present in only minute amounts (0.0387 percent, or 387 parts per million), is nevertheless an important constituent of air, because it has the ability to absorb heat energy radiated by Earth and thus influences the heating of the atmosphere.

Air includes many gases and particles that vary significantly from time to time and from place to place. Important examples include water vapor, ozone, and tiny solid and liquid particles.

Water Vapor   The amount of water vapor in the air varies considerably, from practically none at all up to about 4 percent by volume. Why is such a small fraction of the atmosphere so significant? Certainly the fact that water vapor is the source of all clouds and precipitation would be enough to explain its importance. However, water vapor has other roles. Like carbon dioxide, it has the ability to absorb heat energy given off by Earth as well as some solar energy. It is, therefore, important when we examine the heating of the atmosphere.

Ozone   Another important component of the atmosphere is *ozone*. It is a form of oxygen that combines three oxygen atoms into each molecule (O_3). Ozone is not the same as the oxygen we breathe, which has two atoms per molecule (O_2). There is very little of this gas in the atmosphere, and its distribution is not uniform. It is concentrated well above Earth's surface in a layer called the *stratosphere,* at an altitude of between 10 and 50 kilometers (6 and 31 miles). The presence of the ozone layer in our atmosphere is crucial to those of us who dwell on Earth. The reason is that ozone absorbs the potentially harmful ultraviolet (UV) radiation from the Sun. If ozone did not filter a great deal of the ultraviolet radiation, and if the Sun's UV rays reached the surface of Earth undiminished, our planet would be uninhabitable for most life as we know it.

Aerosols   The movements of the atmosphere are sufficient to keep a large quantity of solid and liquid particles suspended within it. Although visible dust sometimes clouds the sky, these relatively large particles are too heavy to stay in the air for very long. Still, many particles are microscopic and remain suspended for considerable periods of time. They may originate

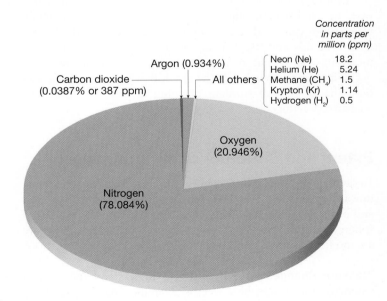

FIGURE 21.11 Proportional volume of gases composing dry air. Nitrogen and oxygen obviously dominate.

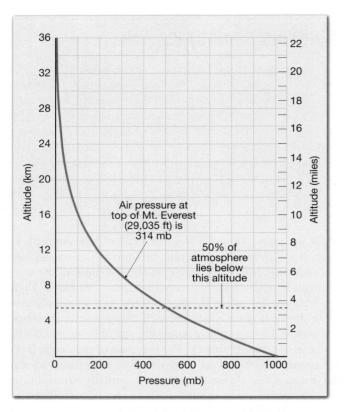

FIGURE 21.12 Atmospheric pressure changes with altitude. Pressure decreases rapidly near Earth's surface and more gradually at greater heights.

from many sources, both natural and human made, and include sea salts from breaking waves, fine soil blown into the air, smoke and soot from fires, pollen and microorganisms lifted by the wind, ash and dust from volcanic eruptions, and more. Collectively, these tiny solid and liquid particles are called **aerosols.**

From a meteorological standpoint, these tiny, often invisible particles can be significant. First, many act as surfaces on which water vapor can condense, an important function in the formation of clouds and fog. Second, aerosols can absorb or reflect incoming solar radiation. Thus, when an air pollution episode is occurring, or when ash fills the sky following a volcanic eruption, the amount of sunlight reaching Earth's surface can be measurably reduced.

Extent and Structure of the Atmosphere

To say that the atmosphere begins at Earth's surface and extends upward is obvious. However, where does the atmosphere end and outer space begin? There is no sharp boundary; the atmosphere rapidly thins as you travel away from Earth, until there are too few gas molecules to detect.

Pressure Changes with Height   To understand the vertical extent of the atmosphere, let us examine changes in atmospheric pressure with height. Atmospheric pressure is simply the weight of the air above. At sea level, the average pressure is slightly more than 1,000 millibars. This corresponds to a weight

of slightly more than 1 kilogram per square centimeter (14.7 pounds per square inch). Obviously, the pressure at higher altitudes is less (Figure 21.12).

One half of the atmosphere lies below an altitude of 5.6 kilometers (3.5 miles). At about 16 kilometers (10 miles), 90 percent of the atmosphere has been traversed, and above 100 kilometers (62 miles), only 0.00003 percent of all the gases making up the atmosphere remains. Even so, traces of our atmosphere extend far beyond this altitude, gradually merging with the emptiness of space.

Temperature Changes   In addition to vertical changes in air pressure, there are also changes in air temperature as we ascend through the atmosphere. Earth's atmosphere is divided vertically into four layers on the basis of temperature (Figure 21.13).

The bottom layer in which we live is characterized by a decrease in temperature with an increase in altitude and is called the *troposphere*. The term literally means the region where air "turns over," a reference to the appreciable vertical mixing of air in this lowermost zone. The troposphere is the chief focus of meteorologists, because it is in this layer that essentially all important weather phenomena occur.

The temperature decrease in the troposphere is called the *environmental lapse rate*. Its average value is 6.5 °C per kilometer (3.5 °F per 1,000 feet), a figure known as the *normal lapse rate*. It

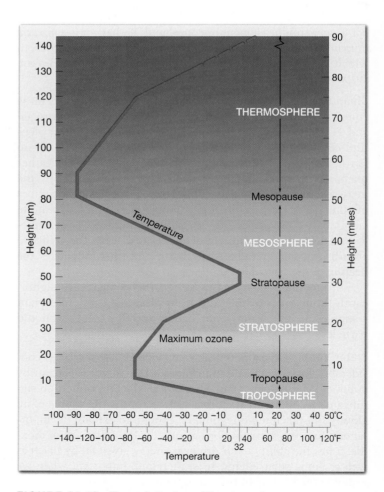

FIGURE 21.13 Thermal structure of the atmosphere.

should be emphasized, however, that the environmental lapse rate is not a constant, but rather can be highly variable, and must be regularly measured. To determine the actual environmental lapse rate as well as to gather information about vertical changes in pressure, wind, and humidity, radiosondes are used. The *radiosonde* is an instrument package that is attached to a balloon and transmits data by radio as it ascends through the atmosphere.

The thickness of the troposphere is not the same everywhere; it varies with latitude and the season. On the average, the temperature drop continues to a height of about 12 kilometers (7.4 miles). The outer boundary of the troposphere is the *tropopause.*

Beyond the tropopause is the *stratosphere.* In the stratosphere, the temperarature remains constant to a height of about 20 kilometers (12 miles) and then begins a gradual increase that continues until the *stratopause,* at a height of nearly 50 kilometers (30 miles) above Earth's surface. Below the tropopause, atmospheric properties like temperature and humidity are readily transferred by large-scale turbulence and mixing. Above the tropopause, in the stratosphere, they are not. Temperatures increase in the stratosphere because it is in this layer that the atmosphere's ozone is concentrated. Recall that ozone absorbs ultraviolet radiation from the Sun. As a consequence, the stratosphere is heated.

In the third layer, the *mesosphere,* temperatures again decrease with height until, at the *mesopause,* more than 80 kilometers (50 miles) above the surface, the temperature approaches $-90\,°C$ ($-130\,°F$). The coldest temperatures anywhere in the atmosphere occur at the mesopause.

The fourth layer extends outward from the mesopause and has no well-defined upper limit. It is the *thermosphere,* a layer that contains only a tiny fraction of the atmosphere's mass. In the extremely rarefied air of this outermost layer, temperatures again increase, owing to the absorption of very short-wave, high-energy solar radiation by atoms of oxygen and nitrogen.

Temperatures rise to extremely high values of more than $1000\,°C$ in the thermosphere. But such temperatures are not comparable to those experienced near Earth's surface. Temperature is defined in terms of the average speed at which molecules move. Because the gases of the thermosphere are moving at very high speeds, the temperature is very high. But the gases are so sparse that, collectively, they process only an insignificant quantity of heat.

Energy from the Sun

Nearly all of the energy that drives Earth's variable weather and climate comes from the Sun.

From our everyday experience, we know that the Sun emits light and heat as well as the ultraviolet rays that cause suntan. Although these forms of energy comprise a major portion of the total energy that radiates from the Sun, they are only part of a large array of energy called *radiation* or *electromagnetic radiation.* This array, or spectrum, of electromagnetic energy is shown in Figure 21.14. All radiation—whether X-rays, microwaves, or radio waves—transmits energy through the vacuum of space at 300,000 kilometers (186,000 miles) per second and only slightly slower through our atmosphere. When an object absorbs any form of radiant energy, the result is an increase in molecular motion, which causes a corresponding increase in temperature.

To better understand how the atmosphere is heated, it is useful to have a general understanding of the basic laws governing radiation.

1. *All objects, at whatever temperature, emit radiant energy.* Hence, not only hot objects like the Sun but also Earth, including its polar ice caps, continually emit energy.
2. *Hotter objects radiate more total energy per unit area than do colder objects.*
3. *The hotter the radiating body, the shorter the wavelength of maximum radiation.* The Sun, with a surface temperature of about 5700 °C, radiates maximum energy at 0.5 micrometer, which is in the visible range. The maximum radiation for Earth occurs at a wavelength of 10 micrometers, well within the infrared (heat) range. Because the maximum Earth radiation is roughly 20 times longer than the maximum solar radiation, Earth radiation is often called *longwave radiation,* and solar radiation is called *shortwave radiation.*
4. *Objects that are good absorbers of radiation are good emitters as well.* Earth's surface and the Sun approach being perfect radiators because they absorb and radiate with nearly 100 percent efficiency for their respective temperatures. On the other hand, *gases are selective absorbers and emitters of radiation.* For some wavelengths the atmosphere is nearly transparent (little radiation absorbed). For others, however, it is nearly opaque (a good absorber). Experience tells us that the atmosphere is transparent to visible light; hence, these wavelengths readily reach Earth's surface. This is not the case for the longer wavelength radiation emitted by Earth.

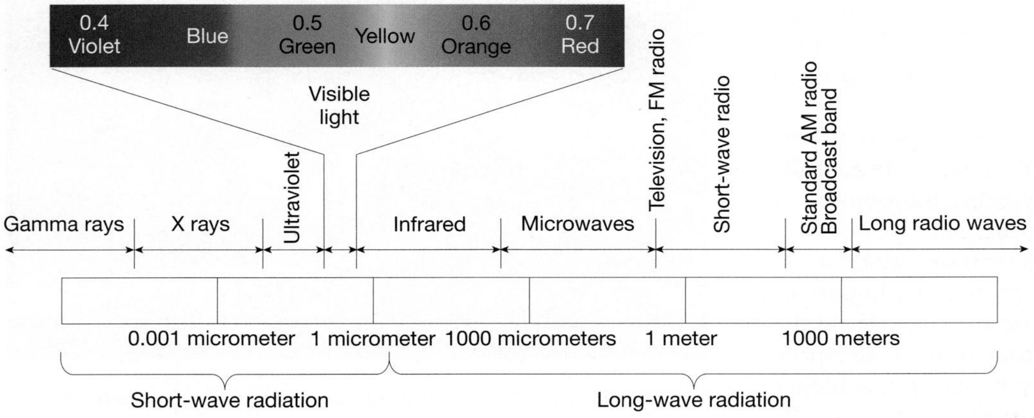

FIGURE 21.14 The electromagnetic spectrum, illustrating the wavelengths and names of various types of radiation.

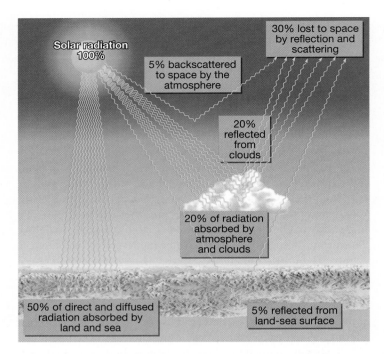

FIGURE 21.15   Average distribution of incoming solar radiation by percentage. More solar energy is absorbed by Earth's surface than by the atmosphere. Consequently, the air is not heated directly by the Sun, but is heated indirectly from Earth's surface. These percentages can vary. For example, if cloud cover increases or the surface is brighter, the percentage of light reflected increases. This situation leaves less solar energy to take the other two paths.

Heating the Atmosphere: The Greenhouse Effect

If Earth had no atmosphere, it would experience an average surface temperature far below freezing. But the atmosphere warms the planet and makes Earth livable. The extremely important role the atmosphere plays in heating Earth's surface has been named the **greenhouse effect.**

As discussed earlier, cloudless air is largely transparent to incoming shortwave solar radiation and, hence, transmits it to Earth's surface. By contrast, a significant fraction of the long-wave radiation emitted by Earth's land–sea surface is absorbed by water vapor, carbon dioxide, and other trace gases in the atmosphere. This energy heats the air and increases the rate at which it radiates energy, both out to space and back toward Earth's surface. The energy that is emitted back to the surface causes it to heat up more, which then results in greater emissions from the surface. This complicated game of "pass the hot potato" keeps Earth's average temperature much warmer than it would otherwise be (Figure 21.17). Without these absorptive gases in our atmosphere, Earth would not provide a suitable habitat for humans and other life forms.

This natural phenomenon was named the greenhouse effect because it was once thought that greenhouses were heated in a similar manner. The glass in a greenhouse allows shortwave solar radiation to enter and be absorbed by the objects inside. These objects, in turn, radiate energy but at longer wavelengths, to which glass is nearly opaque. The heat, therefore, is "trapped" in the greenhouse. It has been shown, however, that air inside

The Fate of Incoming Solar Energy

Figure 21.15 shows the fate of incoming solar radiation averaged for the entire globe. Notice that the atmosphere is quite transparent to incoming solar radiation. On average, about 50 percent of the solar energy reaching the top of the atmosphere passes through the atmosphere and is absorbed at Earth's surface. Another 20 percent is absorbed directly by clouds and certain atmospheric gases (including oxygen and ozone) before reaching the surface. The remaining 30 percent is reflected back to space by the atmosphere, clouds, and reflective surfaces such as snow and ice. The fraction of the total radiation that is reflected by a surface is called its *albedo*. Thus, the albedo for Earth as a whole (the *planetary* albedo) is 30 percent.

What determines whether solar radiation will be transmitted to the surface, scattered, or reflected outward? It depends greatly on the wavelength of the energy being transmitted, as well as on the nature of the intervening material.

The numbers shown in Figure 21.15 represent global averages. The actual percentages can vary greatly. An important reason for much of this variation has to do with changes in the percentage of light reflected and scattered back to space (Figure 21.16). For example, if the sky is overcast, a higher percentage of light is reflected back to space than when the sky is clear.

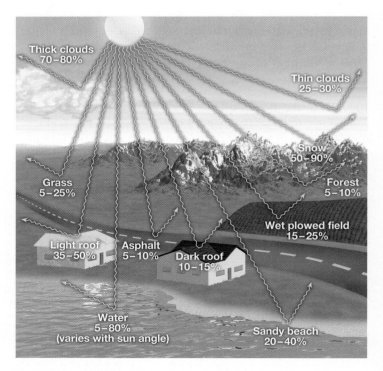

FIGURE 21.16   Albedo (reflectivity) of various surfaces. In general, light-colored surfaces tend to be more reflective than dark-colored surfaces and thus have higher albedos.

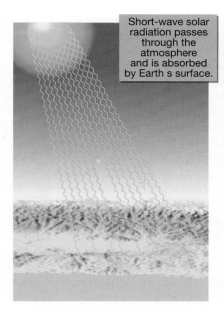
Short-wave solar radiation passes through the atmosphere and is absorbed by Earth's surface.

Earth's surface emits longwave radiation which is absorbed by greenhouse gases.

H_2O

CO_2

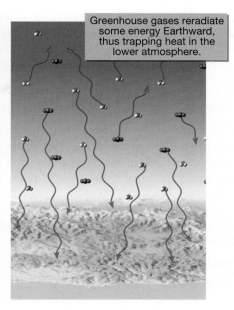
Greenhouse gases reradiate some energy Earthward, thus trapping heat in the lower atmosphere.

FIGURE 21.17   The heating of the atmosphere. Most of the short-wavelength radiation from the Sun passes through the atmosphere and is absorbed by Earth's land–sea surface. This energy is then emitted from the surface as longer-wavelength radiation, much of which is absorbed by certain gases in the atmosphere. Some of the energy absorbed by the atmosphere will be reradiated Earthward. This so-called greenhouse effect is responsible for keeping Earth's surface much warmer than it would be otherwise.

greenhouses attains higher temperatures than outside air mainly because greenhouses restrict the exchange of air between the inside and outside. Nevertheless, the term "greenhouse effect" remains.

NATURAL CAUSES OF CLIMATE CHANGE

A great variety of hypotheses have been proposed to explain climate change. Several have gained wide support, only to lose it and then sometimes to regain it again. Some explanations are controversial. This is to be expected, because planetary atmospheric processes are so large-scale and complex that they cannot be reproduced physically in laboratory experiments. Rather, climate and its changes must be simulated mathematically (modeled) using powerful computers.

In Chapter 18, the section on "Causes of Glaciation" (pp. 512–515) described two "natural" mechanisms of climate change. Recall that the movement of lithospheric plates gradu-

ally moves Earth's continents closer to or farther from the equator. Although these shifts in latitude are slow, they can have a dramatic impact on climate over spans of millions of years. Moving landmasses can also lead to significant shifts in ocean circulation, which influences heat transport around the globe.*

A second natural mechanism of climate change discussed in Chapter 18 involved variations in Earth's orbit. Changes in the shape of the orbit (*eccentricity*), variations in the angle that Earth's axis makes with the plane of its orbit (*obliquity*), and the wobbling of the axis (precession) cause fluctuations in the seasonal and latitudinal distribution of solar radiation. These variations, in turn, contributed to the alternating glacial-interglacial episodes of the Ice Age.

In this section, we describe two additional hypotheses that have earned serious consideration from the scientific community. One involves the role of volcanic activity. How do the gases and particles emitted by volcanoes impact climate? A second natural cause of climate change discussed in this section involves solar variability. Does the Sun vary in its radiation output? Do sunspots affect the output?

After this look at natural factors, we will examine human-made climate changes, including the effects of rising levels of carbon dioxide and other trace gases.

As you read this section, you will find that more than one hypothesis may explain the same climate change. In fact, several mechanisms may interact to shift climate. Also, no single hypothesis can explain climate change on all time scales. A proposal that explains variations over millions of years generally

Students Sometimes Ask . . .

If Earth's atmosphere had no greenhouse gases, what would surface-air temperatures be like?

Cold! Earth's average surface temperature would be a chilly −18 °C (−0.4 °F) instead of the relatively comfortable 14.5 °C (58 °F) that it is today.

---

*For more on this, see the section titled "Supercontinents and Climate" in Chapter 22, p. 620.

cannot explain fluctuations over hundreds of years. If our atmosphere and its changes ever become fully understood, we will probably see that climate change is caused by many of the mechanisms discussed here, plus new ones yet to be proposed.

Volcanic Activity and Climate Change

The idea that explosive volcanic eruptions might alter Earth's climate was first proposed many years ago. It is still regarded as a plausible explanation for some aspects of climatic variability. Explosive eruptions emit huge quantities of gases and fine-grained debris into the atmosphere (Figure 21.18). The greatest eruptions are sufficiently powerful to inject material high into the atmosphere, where it spreads around the globe and remains for many months or even years.

The Basic Premise The basic premise is that this suspended volcanic material will filter out a portion of the incoming solar

FIGURE 21.18 Mount Etna, a volcano on the island of Sicily, erupting in late October 2002. Mount Etna is Europe's largest and most active volcano. **A.** This photo of Mount Etna looking southeast was taken by a member of the International Space Station. It shows a plume of volcanic ash streaming southeastward from the volcano. **B.** This image from the Atmospheric Infrared Sounder on NASA's *Aqua* satellite shows the sulfur dioxide (SO_2) plume in shades of purple and black. Climate may be affected when large quantities of SO_2 are injected into the atmosphere. (Images courtesy of NASA)

A.

B.

radiation, which in turn will lower temperatures in the troposphere. More than 200 years ago Benjamin Franklin used this idea to argue that material from the eruption of a large Icelandic volcano could have reflected sunlight back to space and therefore might have been responsible for the unusually cold winter of 1783–1784.

Perhaps the most notable cool period linked to a volcanic event is the "year without a summer" that followed the 1815 eruption of Mount Tambora in Indonesia. The eruption of Tambora is the largest of modern times. During April 7–12, 1815, this nearly 4000-meter-high (13,000-foot) volcano violently expelled an estimated 100 cubic kilometers (24 cubic miles) of volcanic debris. The impact of the volcanic aerosols on climate is believed to have been widespread in the Northern Hemisphere. From May through September 1816 an unprecedented series of cold spells affected the northeastern United States and adjacent portions of Canada. There was heavy snow in June and frost in July and August. Abnormal cold was also experienced in much of Western Europe. Similar, although apparently less dramatic, effects were associated with other great explosive volcanoes, including Indonesia's Krakatoa in 1883.

Three major volcanic events have provided considerable data and insight regarding the impact of volcanoes on global temperatures. The eruptions of Washington state's Mount St. Helens in 1980, the Mexican volcano El Chichón in 1982, and the Philippines' Mount Pinatubo in 1991 have given scientists an opportunity to study the atmospheric effects of volcanic eruptions with the aid of more sophisticated technology than had been available in the past. Satellite images and remote-sensing instruments allowed scientists to monitor closely the effects of the clouds of gases and ash that these volcanoes emitted.

Mount St. Helens When Mount St. Helens erupted, there was immediate speculation about the possible effects on our climate. Could such an eruption cause our climate to change? There is no doubt that the large quantity of volcanic ash emitted by the explosive eruption had significant local and regional effects for a short period. Still, studies indicated that any longer-term lowering of hemispheric temperatures was negligible. The cooling was so slight, probably less than 0.1 °C (0.2 °F), that it could not be distinguished from other natural temperature fluctuations.

El Chichón Two years of monitoring and studies following the 1982 El Chichón eruption indicated that its cooling effect on global mean temperature was greater than that of Mount St. Helens, on the order of 0.3 to 0.5 °C (0.5 to 0.9 °F). The eruption of El Chichón was *less explosive* than the Mount St. Helens blast, so why did it have a greater impact on global temperatures? The reason is that the material emitted by Mount St. Helens was largely fine ash that settled out in a relatively short time. El Chichón, on the other hand, emitted far greater quantities of sulfur dioxide gas (an estimated 40 times more) than Mount St. Helens. This gas combines with water vapor in the stratosphere to produce a dense cloud of tiny sulfuric acid particles (Figure 21.19A). The particles, called *aerosols,* take several years to settle out completely. They lower the troposphere's mean temperature because they reflect solar radiation back into space (Figure 21.19B).

We now understand that volcanic clouds that remain in the stratosphere for a year or more are composed largely of sulfuric-acid droplets and not of dust, as was once thought. Thus, the volume of fine debris emitted during an explosive event is not an accurate criterion for predicting the global atmospheric effects of an eruption.

Mount Pinatubo   The Philippines' volcano, Mount Pinatubo, erupted explosively in June 1991, injecting 25–30 million tons of sulfur dioxide into the stratosphere. The event provided scientists with an opportunity to study the climatic impact of a major explosive volcanic eruption using NASA's spaceborne Earth Radiation Budget Experiment. During the next year the haze of tiny aerosols increased reflectivity and lowered global temperatures by 0.5 °C (0.9 °F).

Students Sometimes Ask . . .

Could a meteorite colliding with Earth cause the climate to change?

Yes, it is possible. For example, the most strongly supported hypothesis for the extinction of dinosaurs (about 65 million years ago) is related to such an event. When a large (about 10 kilometers in diameter) meteorite struck Earth, huge quantities of debris were blasted high into the atmosphere. For months the encircling dust cloud greatly restricted the amount of light reaching Earth's surface. Without sufficient sunlight for photosynthesis, delicate food chains collapsed. When the sunlight returned, more than half of the species on Earth, including the dinosaurs and many marine organisms, had become extinct. There is more about this in Chapter 22.

The impact on global temperature of eruptions like El Chichón and Mount Pinatubo is relatively minor, but many scientists agree that the cooling produced could alter the general pattern of atmospheric circulation for a limited period. Such a change, in turn, could influence the weather in some regions. Predicting, or even identifying, specific regional effects still presents a considerable challenge to atmospheric scientists.

The preceding examples illustrate that the impact on climate of a single volcanic eruption, no matter how great, is relatively small and short-lived. The graph in Figure 21.19B reinforces that point. Therefore, if volcanism is to have a pronounced impact over an extended period, many great eruptions, closely spaced in time, need to occur. If this happens, the stratosphere would be loaded with enough gases and volcanic dust to seriously diminish the amount of solar radiation reaching the surface. Because no such period of explosive volcanism is known to have occurred in historic times, it is most often mentioned as a possible contributor to prehistoric climatic shifts. Another way in which volcanism may influence climate is described in Box 21.1.

Solar Variability and Climate

Among the most persistent hypotheses of climate change have been those based on the idea that the Sun is a variable star and that its output of energy varies through time. The effect of such changes would seem direct and easily understood: Increases

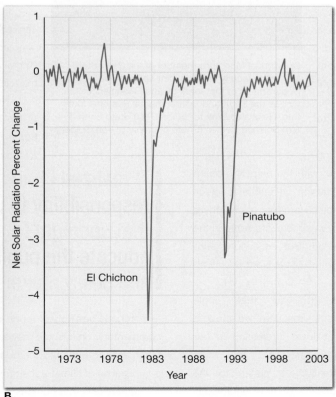

FIGURE 21.19   A. This satellite image shows a plume of white haze from Anatahan Volcano, blanketing a portion of the Philippine Sea following a large eruption in April 2005. The haze is *not* volcanic ash. Rather, it consists of tiny droplets of sulfuric acid formed when sulfur dioxide from the volcano combined with water in the atmosphere. The haze is bright and reflects sunlight back to space. (NASA image) **B.** Net solar radiation at Hawaii's Mauna Loa Observatory relative to 1970 (zero on the graph). The eruptions of El Chichón and Mt. Pinatubo clearly caused a temporary drop in solar radiation reaching the surface. (After Earth System Research Laboratory/NOAA)

Climate Change Scientist

In the last few decades of the twentieth century, evidence of global warming was growing too obvious to ignore. Vast ice shelves in the Antarctic were breaking into floating isles, glaciers worldwide were in full retreat, and each turn of the calendar seemed to bring a new record for the hottest year yet measured. But whether this warming was due to humans or nature, no one was sure. Researchers needed to study a longer history of Earth's climate than just the twentieth century.

Michael Mann, of Pennsylvania State University, knew where to find the answers. Ice cores, sediments, coral skeletons, and tree rings all contain natural records of past temperature fluctuations. However, Mann, a meteorology professor, points out that "it's a fuzzy record, not as precise as thermometer readings." For example, while individual annual layers of ice in ice cores provide a direct reading of ancient atmospheric conditions, deciphering which layers belonged to which years can be difficult. And snow might have fallen only in winter in some areas, versus year round in others, making it difficult to compare cores from different areas.

These variables can make interpreting such proxy data a complex affair. Even so, with the help of modern statistical methods and climate models, Mann was able to piece together a reconstruction of the average temperature of the Northern Hemisphere going back 1,000 years. Mann was most interested in the surface temperature patterns he uncovered. But colleagues suggested he also plot out the average temperature of the Northern Hemisphere.

Dr. Michael Mann is a member of the Penn State University faculty, holding joint appointments in the Departments of Meteorology and Geosciences, as well as the Earth and Environmental Systems Institute. He is also Director of the Penn State Earth System Science Center. (Photo courtesy of Jon Golden)

"It was that single result—the item we thought was least interesting in our work—that got the most attention. It spoke to the question of how anomalous recent warming was," Mann says. The graph revealed an alarming trend. Temperatures stayed relatively warm during medieval times, cooled after that, but zoomed sharply higher starting around 1900—the same period when humans began burning large quantities of fossil fuels.

Known as the "hockey stick graph" for its distinctive shape, Mann's 1998 paper caused a sensation. Such vivid evidence of human contributions to global warming helped catapult both Mann and the issue of climate change onto the public stage. Meanwhile, some commentators encouraged their peers to "break the hockey stick" and discredit the idea of a warming Earth. The attacks convinced Mann to spread the word about climate change research. "I realized I had a responsibility to use that spotlight to educate the public and policy makers, to try to do some good with that opportunity," he says.

In 2001, Mann's research became a centerpiece of the Third Report of the Intergovernmental Panel on Climate Change (IPCC). These reports, written by scientists, are summaries of the most current and important climate research. Mann served as a lead author of the Third Report—and received a singular honor for his labors. For their work to raise awareness of the threat of human-caused climate change, the IPCC and former Vice President Al Gore were awarded the 2007 Nobel Peace Prize.

Mann continues to spread the word about global warming today. In addition to his scientific research, he has helped found the website RealClimate.org, featuring commentary on climate news by climate researchers themselves. Mann also co-authored the book *Dire Predictions,* which explains the complexities of climate-change science to the public.

Mann says that humans can still steer a course that could evade the worst consequences of a warming globe. "I like to think that together, those of us who are engaging the public on this issue are making some positive steps forward," he says. His main message is that "we hold the future in our own hands."

> "**I realized I had a responsibility to use that spotlight to educate the public and policy makers . . .**"

> "**It was that single result—the item we thought was least interesting in our work—that got the most attention.**"

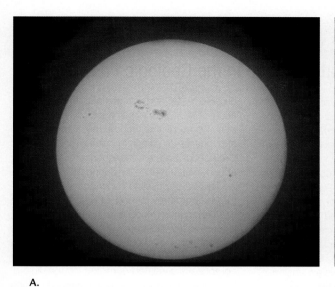

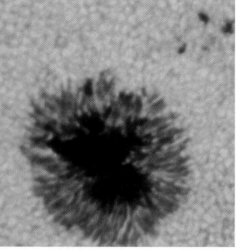

A. **B.**

FIGURE 21.20 **A.** Large sunspot group on the solar disk. (Celestron 8 photo courtesy of Celestron International) **B.** Sunspots having visible umbra (dark central area) and penumbra (lighter area surrounding umbra). (Courtesy of National Optical Astronomy Observatories)

in solar output would cause the atmosphere to warm, and reductions would result in cooling. This notion is appealing because it can be used to explain climate change of any length or intensity. However, no major *long-term* variations in the total intensity of solar radiation have yet been measured outside the atmosphere. Such measurements were not even possible until satellite technology became available. Now that it is possible, we will need many years of records before we begin to sense how variable (or invariable) energy from the Sun really is.

Several proposals for climate change, based on a variable Sun, relate to sunspot cycles. The most conspicuous and best-known features on the surface of the Sun are the dark blemishes called **sunspots** (Figure 21.20). Sunspots are huge magnetic storms that extend from the Sun's surface deep into the interior. Moreover, these spots are associated with the Sun's ejection of huge masses of particles that, on reaching Earth's upper atmosphere, interact with gases to produce auroral displays such as the Aurora Borealis, or Northern Lights, in the Northern Hemisphere.

Along with other solar activity, the numbers of sunspots seem to increase and decrease in a regular way, creating a cycle of about 11 years. The graph in Figure 21.21 shows the annual number of sunspots, beginning in the early 1700s. However, this pattern does not always occur. There have been periods when the Sun was essentially free of sunspots. In addition to the well-known 11-year cycle, there is also a 22-year cycle. This longer cycle is based on the fact that the magnetic polarities of sunspot clusters reverse every successive 11 years.

Interest in possible Sun-climate effects has been sustained by an almost continuous effort to find correlations on time scales ranging from days to tens of thousands of

years. Two widely debated examples are briefly described here.

Sunspots and Temperature

Studies indicate prolonged periods when sunspots have been absent or nearly so. Moreover, these events correspond closely with cold periods in Europe and North America. Conversely, periods characterized by plentiful sunspots have correlated well with warmer times in these regions.

Referring to these matches, some scientists have suggested that such correlations make it appear that changes on the Sun are an important cause of climate change. But other scientists seriously question this notion. Their hesitation stems, in part, from subsequent investigations using different climate records from around the world that failed to find a significant correlation between solar activity and climate. Even more troubling is that no testable physical mechanism exists to explain the purported effect.

Sunspots and Drought A second possible Sun—climate connection, on a time scale different from the preceding example, relates to variations in precipitation rather than temperature. An extensive study of tree rings revealed a recurrent period of about 22 years in the pattern of droughts in the western United States. This periodicity coincides with the 22-year magnetic cycle of the Sun mentioned earlier.

Commenting on this possible connection, a panel of the National Research Council pointed out:

> No convincing mechanism that might connect so subtle a feature of the sun to drought patterns in limited regions has yet

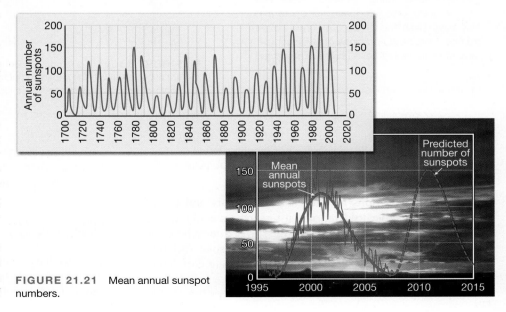

FIGURE 21.21 Mean annual sunspot numbers.

EARTH AS A SYSTEM

BOX 21.1

A Possible Link between Volcanism and Climate Change in the Geologic Past

The Cretaceous Period is the last period of the Mesozoic Era, the era of *middle life* that is often called the "age of dinosaurs." It began about 145.5 million years ago and ended about 65.5 million years ago with the extinction of the dinosaurs (and many other life forms as well).*

The Cretaceous climate was among the warmest in Earth's long history. Dinosaurs, which are associated with mild temperatures, ranged north of the Arctic Circle. Tropical forests existed in Greenland and Antarctica, and coral reefs grew as much as 15 degrees latitude closer to the poles than at present. Deposits of peat that would eventually form widespread coal beds accumulated at high latitudes. Sea level was as much as 200 meters (650 feet) higher than today, indicating that there were no polar ice sheets.

What was the cause of the unusually warm climates of the Cretaceous Period? Among the significant factors that may have contributed was an enhancement of the greenhouse effect due to an increase in the amount of carbon dioxide in the atmosphere.

Where did the additional CO_2 come from that contributed to the Cretaceous warming? Many geologists suggest that the probable source was volcanic activity. Carbon dioxide is one of the gases emitted during volcanism, and there is now considerable geologic evidence that the Middle Cretaceous was a time when there was an unusually high rate of volcanic activity. Several huge oceanic lava plateaus were produced on the floor of the western Pacific during this span. These vast features were associated with hot spots that may have been the product of large mantle plumes. Massive outpourings of lava over millions of years would have been accompanied

by the release of huge quantities of CO_2, which in turn would have enhanced the atmospheric greenhouse effect. *Thus, the warmth that characterized the Cretaceous may have had its origins deep in Earth's mantle.*

There were other probable consequences of this extraordinarily warm period that are linked to volcanic activity. For example, the high global temperatures and enriched atmospheric CO_2 in the Cretaceous led to increases in the quantity and types of phytoplankton (tiny, mostly microscopic plants such as algae) and other life forms in the ocean. This expansion in marine life is reflected in the widespread chalk deposits associated with the Cretaceous Period (Figure 21.A). Chalk consists of the calcite-rich hard parts of microscopic marine organisms. Oil and gas originate from the alteration of biological remains (chiefly phytoplankton). Some of the world's most important oil and gas fields occur in marine sediments of Cretaceous age, a consequence of the greater abundance of marine life during this warm time.

This list of possible consequences linked to the extraordinary period of volcanism during the Cretaceous Period is far from complete, yet it serves to illustrate the interrelationships among parts of the Earth system. Materials and processes that at first might seem to be completely unrelated turn out to be linked. Here you have seen how processes originating deep in Earth's interior are connected directly or indirectly to the atmosphere, the oceans, and the biosphere.

FIGURE 21.A   The famous chalk deposits, known as the White Cliffs of Dover, are associated with the expansion of marine life that occurred during the exceptional warmth of the Cretaceous Period. (Photo by Jon Arnold/Getty Images)

*For more on the end of the Cretaceous, see Chapter 22.

---

appeared. Moreover, the cyclic pattern of droughts found in tree rings is itself a subtle feature that shifts from place to place within the broad region of the study.*

Possible connections between solar variability and climate would be much easier to determine if researchers could identify physical links between the Sun and the lower atmosphere. But despite much research, no connection between solar variations and weather has yet been well established. Apparent correlations

have almost always faltered when put to critical statistical examination or when tested with different data sets. As a result, the subject has been characterized by ongoing controversy and debate.

HUMAN IMPACT ON GLOBAL CLIMATE

So far we have examined four potential causes of climate change that are natural. In this section we discuss how humans contribute to global climate change (Box 21.2). One impact largely

Solar Variability, Weather, and Climate (Washington, D.C.: National Academy Press, 1982), p. 7.

results from the addition of carbon dioxide and other greenhouse gases to the atmosphere. A second impact is related to the addition of human-generated aerosols to the atmosphere.

Human influence on regional and global climate did not just begin with the onset of the modern industrial period. There is good evidence that people have been modifying the environment over extensive areas for thousands of years. The use of fire and the overgrazing of marginal lands by domesticated animals have both reduced the abundance and distribution of vegetation. By altering ground cover, humans have modified such important climatological factors as surface albedo, evaporation rates, and surface winds. Commenting on this aspect of human-induced climate modification, the late astronomer Carl Sagan noted: "In contrast to the prevailing view that only modern humans are able to alter climate, we believe it is more likely that the human species has made a substantial and continuing impact on climate since the invention of fire."**

Subsequently, these ideas were reinforced and expanded upon by a study that used data collected from Antarctic ice cores. This research suggested that humans may have started to have a significant impact on atmospheric composition and global temperatures thousands of years ago.

> Humans started slowly ratcheting up the thermostat as early as 8000 years ago, when they began clearing forests for agriculture, and 5000 years ago with the arrival of wet-rice cultivation. The greenhouse gases carbon dioxide and methane given off by these changes would have warmed the world. . . .†

CARBON DIOXIDE, TRACE GASES, AND CLIMATE CHANGE

Earlier you learned that carbon dioxide (CO_2) represents only about 0.038 percent of the gases that make up clean, dry air. Nevertheless, it is a very significant component meteorologically. Carbon dioxide is influential because it is transparent to incoming short-wavelength solar radiation, but it is not transparent to some of the longer-wavelength outgoing Earth radiation. A portion of the energy leaving the ground is absorbed by atmospheric CO_2. This energy is subsequently reemitted, part of it back toward the surface, thereby keeping the air near the ground warmer than it would be without CO_2.

Thus, along with water vapor, carbon dioxide is largely responsible for the *greenhouse effect* of the atmosphere. Carbon dioxide is an important heat absorber, and it follows logically that any change in the air's CO_2 content could alter temperatures in the lower atmosphere.

CO_2 Levels Are Rising

Earth's tremendous industrialization of the past two centuries has been fueled—and still is fueled—by burning fossil fuels: coal, natural gas, and petroleum (see Figure 23.4, p. 647). Combustion of these fuels has added great quantities of carbon dioxide to the atmosphere.

The use of coal and other fuels is the most prominent means by which humans add CO_2 to the atmosphere, but it is not the only way. The clearing of forests also contributes substantially because CO_2 is released as vegetation is burned or decays. Deforestation is particularly pronounced in the tropics, where vast tracts are cleared for ranching and agriculture or subjected to inefficient commercial logging operations (Figure 21.22). According to U.N. estimates, nearly 10.2 million hectares (25.1 million acres) of tropical forest were permanently destroyed each year during the decade of the 1990s. Between the years 2000 and 2005 the average figure increased to 10.4 million hectares (25.7 million acres) per year.

Some of the excess CO_2 is taken up by plants or is dissolved in the ocean. It is estimated that 45 to 50 percent remains in the atmosphere. Figure 21.23 is a graphic record of changes in atmospheric CO_2 extending back 400,000 years. Over this long span, natural fluctuations varied from about 180 to 300 ppm. As a result of human activities, the present CO_2 level is about 30 percent higher than its highest level over at least the last 650,000 years. The rapid increase in CO_2 concentrations since the onset of industrialization is obvious. The annual rate at which atmospheric CO_2 concentrations are growing has been increasing over the last several decades (Figure 21.24).

The Atmosphere's Response

Given the increase in the atmosphere's carbon dioxide content, have global temperatures actually increased? The answer is yes. According to a 2007 report by the Intergovernmental Panel on Climate Change (IPCC), "Warming of the climate system is unequivocal, as is now evident from observations of increases in global average air and ocean temperatures, widespread melting of snow and ice, and rising global sea level."†† Most of the observed increase in global average temperatures since the mid-twentieth century is *very likely* due to the observed increase in

---

*P. Foukal, et al. "Variations in solar luminosity and their effect on the Earth's climate," *Nature*, Vol. 443, 14 September, 2006, pp. 161–66.

**Carl Sagan et al., "Anthropogenic Albedo Changes and the Earth's Climate," *Science*, 206, No. 4425 (1980), p. 367.

†An Early Start for Greenhouse Warming?", *Science*, Vol. 303, 16 January, 2004. This article is a report on a paper given by paleoclimatologist William Ruddiman at a meeting of the American Geophysical Union in December 2003.

††IPCC, Summary for Policy Makers. In *Climate Change 2007: The Physical Science Basis*. p.4. Cambridge University Press, Cambridge, United Kingdom and New York, NY.

A.   **B.**

FIGURE 21.22 Clearing the tropical rain forest is a serious environmental issue. In addition to the loss of biodiversity, tropical deforestation is a significant source of carbon dioxide. **A.** This August 2007 satellite image shows deforestation in the Amazon rain forest in western Brazil. Intact forest is dark green, whereas cleared areas are tan (bare ground) or light green (crops and pasture). (NASA) **B.** Fires are frequently used to clear the land. This scene is also in Brazil's Amazon Basin. (Photo by Pete Oxford/Nature Picture Library)

human-generated greenhouse gas concentrations. (As used by the IPCC, *very likely* indicates a probability of 90–99 percent.) Global warming since the mid-1970s is now about 0.6 °C (1 °F) and total warming in the past century is about 0.8 °C (1.4 °F). The upward trend in surface temperatures is shown in Figure 21.25A. The world map in Figure 21.25B compares surface temperatures for 2008 to the base period (1951–1980). You can see that the greatest warming has been in the Arctic and neighboring high latitude regions. Here are some related facts:

- When we consider the time span for which there are instrumental records (since 1850), thirteen of the last fourteen years (1995–2008) rank among the fourteen warmest (Figure 21.26).
- Global mean temperature is now higher than at any time in at least the past 500 to 1000 years.
- The average temperature of the global ocean has increased to depths of at least 3000 meters (10,000 feet).

Are these temperature trends caused by human activities, or would they have occurred anyway? The scientific consensus of the IPCC is that human activities were *very likely* responsible for most of the temperature increase since 1950.

What about the future? Projections for the years ahead depend in part on the quantities of greenhouse gases that are emitted. Figure 21.27 shows the best estimates of global warming for several different scenarios. The 2007 IPCC report also states that if there is a doubling of the preindustrial level of carbon dioxide (280 ppm) to 560 ppm, the "likely temperature increase will be in the range of 2 to 4.5 °C (3.5 to 8.1 °F). The increase is "very unlikely" (1 to 10 percent probability) to be less than 1.5 °C (2.7 °F) and values higher than 4.5 °C (8.1 °F) cannot be excluded.

The Role of Trace Gases

Carbon dioxide is not the only gas contributing to a global increase in temperature. In recent years atmospheric scientists have come to realize that the industrial and agricultural activities of people are causing

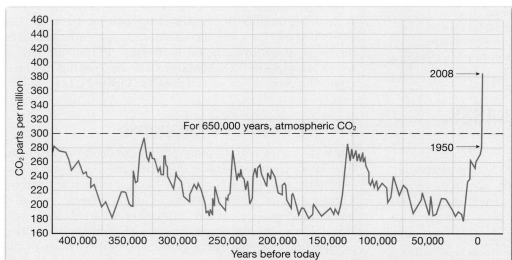

FIGURE 21.23 Carbon dioxide concentrations over the past 400,000 years. Most of the data comes from the analysis of air bubbles trapped in ice cores. The record since 1958 comes from direct measurements of atmospheric CO_2 taken at Mauna Loa Observatory, Hawaii. The rapid increase in CO_2 concentrations since the onset of the industrial revolution is obvious.

Annual CO₂ Contribution of an Average American*

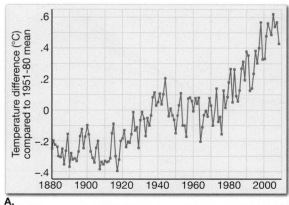

17,000 pounds of CO₂ by using 1,100 kilowatt-hours of electricity per month

8,800 pounds of CO₂ by using 6,300 cubic feet of natural gas per month

1,000 pounds of CO₂ by creating 4.5 pounds of trash per day

8,900 pounds of CO₂ by driving 160 miles per week

1,000 pounds of CO₂ by flying 1,900 miles per year

FIGURE 21.24 Americans are responsible for about 25 percent of the world's greenhouse-gas emissions. That amounts to 24,300 kilograms (54,000 pounds) of carbon dioxide each year for an average American, which is about five times the emissions of the average global citizen. This diagram represents some of the ways that people in the United States contribute. (Data from various U.S. government agencies; Photo by Jerry Schad/Photo Researchers, Inc.)

a buildup of several trace gases that also play a significant role. The substances are called *trace gases* because their concentrations are so much smaller than that of carbon dioxide. The trace gases that are most important are methane (CH_4), nitrous oxide (N_2O), and chlorofluorocarbons (CFCs). These gases absorb wavelengths of outgoing radiation from Earth that would otherwise escape into space. Although individually their impact is modest, taken together the effects of these trace gases play a significant role in warming the troposphere.

Methane is present in much smaller amounts than CO_2, but its significance is greater than its relatively small concentration would indicate (Figure 21.28). The reason is that methane is about 20 times more effective than CO_2 at absorbing infrared radiation emitted by Earth.

Methane is produced by *anaerobic* bacteria in wet places where oxygen is scarce (anaerobic means "without air," specifically oxygen). Such places include swamps, bogs, wetlands, and the guts of termites and grazing animals like cattle and sheep. Methane is also generated in flooded paddy fields ("artificial swamps") used for growing rice. Mining of coal and drilling for oil and natural gas are other sources because methane is a product of the formation of these fuels (Figure 21.29).

The concentration of methane in the atmosphere has risen rapidly since 1800, an increase that has been in step with the growth in human population. This relationship reflects the close link between methane formation and agriculture. As population has risen, so have the number of cattle and rice paddies.

Nitrous oxide, sometimes called "laughing gas," is also building in the atmosphere, although not as rapidly as methane (see Figure 21.28). The increase results primarily from agricultural activity. When farmers use nitrogen fertilizers to boost crop yields, some of the nitrogen enters the air as nitrous oxide. This gas is also produced by high-temperature combustion of fossil fuels. Although the annual release into the atmosphere is small, the lifetime of a nitrous oxide molecule is about 150 years! If the use of nitrogen fertilizers and fossil fuels grows at

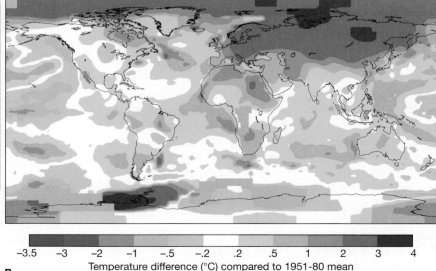

FIGURE 21.25 **A.** The graph depicts global temperature change in °C since the year 1880. **B.** The world map shows how temperatures in 2008 deviated from the mean for the 1951–80 base period. The high latitudes in the Northern Hemisphere clearly stand out. (After NASA/Goddard Institute for Space Studies)

Temperature difference (°C) compared to 1951-80 mean

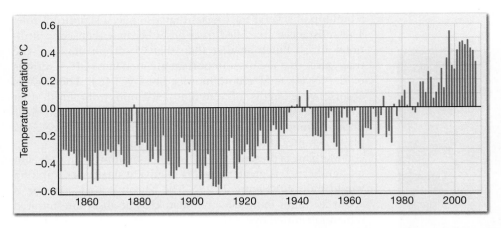

FIGURE 21.26 Annual average global temperature variations for the period 1850 to 2008. The basis for comparison is the average for the 1961–90 period (the 0.0 line on the graph). Each narrow bar on the graph represents the departure of the global mean temperature from the 1961–90 average for a particular year. For example, the global mean temperature for 1862 was more than 0.5 °C below the 1961–90 average, whereas the global mean for 1998 was more than 0.5 °C above.

projected rates, nitrous oxide may make a contribution to greenhouse warming that approaches half that of methane.

Unlike methane and nitrous oxide, chlorofluorocarbons (CFCs) are not naturally present in the atmosphere. CFCs are manufactured chemicals with many uses that have gained notoriety because they are responsible for ozone depletion in the stratosphere. The role of CFCs in global warming is less well known. CFCs are very effective greenhouse gases. They were not developed until the 1920s and were not used in great quantities until the 1950s, but they already contribute to the greenhouse effect at a level equal to methane. Although corrective action has been taken, CFC levels will *not* drop rapidly. CFCs remain in the atmosphere for decades, so even if all CFC emissions were to stop immediately, the atmosphere would not be free of them for many years.

Carbon dioxide is clearly the most important single cause for the projected global greenhouse warming. However, it is not

Students Sometimes Ask . . .

What is the Intergovernmental Panel on Climate Change?

Recognizing the problem of potential global climate change, the World Meteorological Organization and the United Nation's Environment Program established the *Intergovernmental Panel on Climate Change (IPCC,* for short) in 1988. The IPCC assesses the scientific, technical, and socioeconomic information that is relevant to an understanding of human-induced climate change. This authoritative group provides advice to the world community through periodic reports that assess the state of knowledge of causes of climate change. More than 1250 authors and 2500 scientific reviewers from more than 130 countries contributed to the IPCC's most recent report, *Climate Change 2007: The Fourth Assessment Report.*

the only contributor. When the effects of all human-generated greenhouse gases other than CO_2 are added together and projected into the future, their collective impact significantly increases the impact caused by CO_2 alone.

Sophisticated computer models show that the warming of the lower atmosphere caused by CO_2 and trace gases will not be the same everywhere. Rather, the temperature response in polar regions could be two to three times greater than the global average. One reason is that the polar troposphere is very stable, which suppresses vertical mixing and thus limits the amount of surface heat that is transferred upward. In addition, the expected reduction in sea ice will also contribute to the greater temperature increase. This topic will be explored more fully in the next section.

CLIMATE-FEEDBACK MECHANISMS

Climate is a very complex interactive physical system. Thus, when any component of the climate system is altered, scientists must consider many possible outcomes. These possible outcomes are

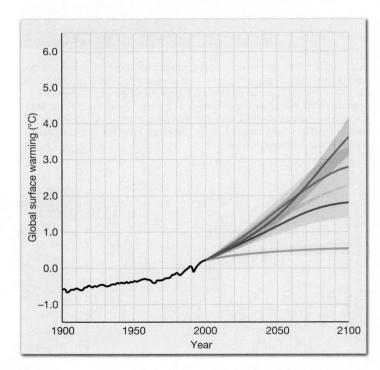

FIGURE 21.27 The left half of the graph (black line) shows global temperature changes for the twentieth century. The right half shows projected global warming in different emissions scenarios. The shaded zone adjacent to each colored line shows the uncertainty range for each scenario. The basis for comparison (0.0 on the vertical axis) is the global average for the period 1980–1999. The orange line represents the scenario in which carbon dioxide concentrations were held constant at the values for the year 2000. (After IPCC, 2007)

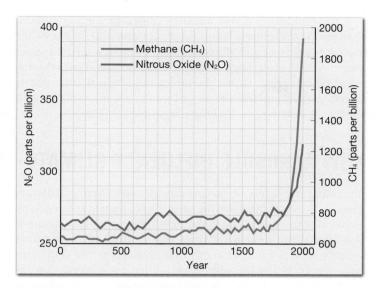

FIGURE 21.28 Increases in the concentrations of two trace gases, methane (CH_4) and nitrous oxide (N_2O), that contribute to global warming. The sharp rise during the industrial era is obvious. (After U.S. Global Change Research Program)

called **climate-feedback mechanisms.** They complicate climate-modeling efforts and add greater uncertainty to climate predictions.

What climate-feedback mechanisms are related to carbon dioxide and other greenhouse gases? One important mechanism is that warmer surface temperatures increase evaporation rates. This in turn increases the atmosphere's water vapor content. Remember that water vapor is an even more powerful absorber of radiation emitted by Earth than is carbon dioxide. Therefore, with more water vapor in the air, the temperature increase caused by carbon dioxide and the trace gases is reinforced.

Recall that the temperature increase at high latitudes may be two to three times greater than the global average. This assumption is based in part on the likelihood that the area covered by sea ice will decrease as surface temperatures rise. Because ice reflects a much larger percentage of incoming solar radiation than does open water, the melting of the sea ice replaces a highly reflecting surface with a relatively dark surface (Figure 21.30). The result is a substantial increase in the solar energy absorbed at the surface. This in turn feeds back to the atmosphere and magnifies the initial temperature increase created by higher levels of greenhouse gases.

A.

B.

FIGURE 21.29 **A.** Coal mining and drilling for oil and natural gas are sources of methane. This large flame is a flare of methane being burned at an oil well. **B.** Methane is also produced by anaerobic bacteria in wet places, where oxygen is scarce (anaerobic means "without air," specifically oxygen). Such places include swamps, bogs, wetlands, and the guts of termites and grazing animals, like cattle and sheep. Methane is also generated in flooded paddy fields ("artificial swamps") used for growing rice. (Photo on left by Kim Steele/Getty Images; photo on right by Moodboard/CORBIS)

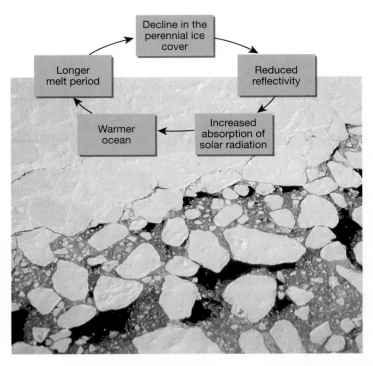

FIGURE 21.30 This satellite image shows the springtime breakup of sea ice near Antarctica. The inset shows a likely feedback loop. A reduction in sea ice acts as a positive-feedback mechanism because surface reflectivity would decrease and the amount of energy absorbed at the surface would increase. (Reproduced with permission from "Science" (Cover, January 25, 2002). Copyright American Association for the Advancement of Science. Photo: D. N. Thomas.)

So far, the climate-feedback mechanisms discussed have magnified the temperature rise caused by the buildup of carbon dioxide. Because these effects reinforce the initial change, they are called **positive-feedback mechanisms.** However, other effects must be classified as **negative-feedback mechanisms** because they produce results that are just the opposite of the initial change and tend to offset it.

One probable result of a global temperature rise would be an accompanying increase in cloud cover due to the higher moisture content of the atmosphere. Most clouds are good reflectors of solar radiation. At the same time, however, they are also good absorbers and emitters of radiation emitted by Earth. Consequently, clouds produce two opposite effects. They are a negative-feedback mechanism because they increase the reflection of solar radiation and thus diminish the amount of solar energy available to heat the atmosphere. On the other hand, clouds act as a positive-feedback mechanism by absorbing and emitting radiation that would otherwise be lost from the troposphere.

Which effect, if either, is stronger? Atmospheric modeling shows that the negative effect of a higher reflectivity is dominant. Therefore, the net result of an increase in cloudiness should be a decrease in air temperature. The magnitude of this negative feedback, however, is not believed to be as great as the positive feedback caused by added moisture and decreased sea ice. Thus, although increases in cloud cover may partly offset a global temperature increase, climate models show that the ultimate effect of the projected increase in CO_2 and trace gases will still be a temperature increase.

The problem of global warming caused by human-induced changes in atmospheric composition continues to be one of the most studied aspects of climate change. Although no models yet incorporate the full range of potential factors and feedbacks, the scientific consensus is that the increasing levels of atmospheric carbon dioxide and trace gases will lead to a warmer planet with a different distribution of climate regimes.

HOW AEROSOLS INFLUENCE CLIMATE

Increasing the levels of carbon dioxide and other greenhouse gases in the atmosphere is the most direct human influence on global climate. But it is not the only impact. Global climate is also affected by human activities that contribute to the atmosphere's aerosol content. Recall that *aerosols* are the tiny, often microscopic, liquid and solid particles that are suspended in the air. Unlike cloud droplets, aerosols are present even in relatively dry air. Atmospheric aerosols are composed of many different materials, including soil, smoke, sea salt, and sulfuric acid. Natural sources are numerous and include such phenomena as dust storms and volcanoes.

Presently the human contribution of aerosols to the atmosphere *equals* the quantity emitted by natural sources. Most human-generated aerosols come from the sulfur dioxide emitted during the combustion of fossil fuels and as a consequence of burning vegetation to clear agricultural land. Chemical reactions in the atmosphere convert the sulfur dioxide into sulfate aerosols, the same material that produces acid precipitation.

How do aerosols affect climate? Aerosols act directly by reflecting sunlight back to space and indirectly by making clouds "brighter" reflectors. The second effect relates to the fact that many aerosols (such as those composed of salt or sulfuric acid) attract water and thus are especially effective as cloud condensation nuclei. The large quantity of aerosols produced by human activities (especially industrial emissions) trigger an increase in the number of cloud droplets that form within a cloud. A greater number of small droplets increases the cloud's brightness—that is, more sunlight is reflected back to space.

By reducing the amount of solar energy available to the climate system, aerosols have a net cooling effect. Studies indicate that the cooling effect of human-generated aerosols offsets a portion of the global warming caused by the growing quantities of greenhouse gases in the atmosphere. Unfortunately, the magnitude and extent of the cooling effect of aerosols is highly uncertain. This uncertainty is a significant hurdle in advancing our understanding of how humans alter Earth's climate.

It is important to point out some significant differences between global warming by greenhouse gases and aerosol cooling. After being emitted, greenhouse gases, such as carbon dioxide, remain in the atmosphere for many decades. By contrast, aerosols released into the troposphere remain there for only a few days or, at most, a few weeks before they are "washed out" by precipitation. Because of their short lifetime in the troposphere, aerosols are distributed unevenly over the globe. As expected, human-generated aerosols are concentrated near the areas that produce them, namely industrialized regions that

UNDERSTANDING EARTH

BOX 21.2

Computer Models of Climate: Important Yet Imperfect Tools

Earth's climate system is amazingly complex. Comprehensive state-of-the-science climate simulation models are among the basic tools used to develop possible climate-change scenarios. They are based on fundamental laws of physics and chemistry and incorporate human and biological interactions. The models simulate many variables, including temperature, rainfall, snow cover, soil moisture, winds, clouds, sea ice, and ocean circulation over the entire globe through the seasons and over spans of decades.

In many other fields of study, hypotheses can be tested by direct experimentation in the laboratory or by observations and measurements in the field. However, this is often not possible in the study of climate. Rather, scientists must construct computer models of how our planet's climate system works. If we understand the climate system correctly and construct the model appropriately, then the behavior of the model climate system should mimic the behavior of Earth's climate system (Figure 21.B).

What factors influence the accuracy of climate models? Clearly, mathematical models are *simplified* versions of the real Earth and cannot capture its full complexity, especially at smaller geographic scales. Moreover, when computer models are used to simulate future

climate change, many assumptions have to be made that significantly influence the outcome. They must consider a wide range of possibilities for future changes in population, economic growth, consumption of fossil fuels, technological development, improvements in energy efficiency, and more.

Despite many obstacles, our ability to use supercomputers to simulate climate continues to improve. Although today's models are far from infallible, they are powerful tools for understanding what Earth's future climate might be like.

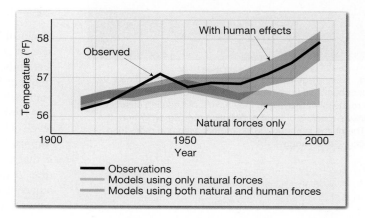

FIGURE 21.B   The blue band shows how global average temperatures would have changed due to natural forces only, as simulated by climate models. The red band shows model projections of the effects of human and natural forces combined. The black line shows actual observed global average temperatures. As the blue band indicates, without human influences, temperature over the past century would actually have first warmed and then cooled slightly over recent decades. Bands of color are used to express the range of uncertainty. (After U.S. Global Change Research Program)

---

burn fossil fuels and land areas where vegetation is burned (Figure 21.31).

Because their lifetime in the atmosphere is short, the effect of aerosols on today's climate is determined by the amount emitted during the preceding couple of weeks. By contrast, the carbon dioxide and trace gases released into the atmosphere remain for much longer spans and thus influence climate for many decades.

SOME POSSIBLE CONSEQUENCES OF GLOBAL WARMING

What consequences can be expected if the carbon dioxide content of the atmosphere reaches a level that is twice what it was early in the 20th century? Because the climate system is so complex, predicting the distribution of particular regional changes is speculative. It is not yet possible to pinpoint specifics, such as

where or when it will become drier or wetter. Nevertheless, plausible scenarios can be given for larger scales of space and time.

As noted, the magnitude of the temperature increase will not be the same everywhere. The temperature rise will probably be smallest in the tropics and increase toward the poles. As for precipitation, the models indicate that some regions will experience significantly more precipitation and runoff. However, others will experience a decrease in runoff due to reduced precipitation or greater evaporation caused by higher temperatures.

Table 21.1 summarizes some of the more likely effects and their possible consequences. The table also provides the IPCC's estimate of the probability of each effect. Levels of confidence for these projections vary from "*likely*" (67 to 90 percent probability) to "*very likely*" (90 to 99 percent probability) to *virtually certain* (greater than 99 percent probability). Box 21.3 looks at key findings regarding the possible consequences of climate change for the United States in the 21st century.

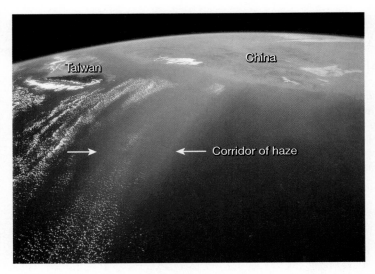

FIGURE 21.31 Human-generated aerosols are concentrated near the areas that produce them. Because aerosols reduce the amount of solar energy available to the climate system, they have a net cooling effect. This satellite image shows a dense blanket of pollution moving away from the coast of China. The plume is about 200 kilometers wide and more than 600 kilometers long. (NASA Image)

Students Sometimes Ask . . .
What are scenarios, and why are they used?

A scenario is an example of what might happen under a particular set of assumptions. Scenarios are a way of examining questions about an uncertain future. For example, future trends in fossil-fuel use and other human activities are uncertain. Therefore, scientists have developed a set of scenarios for how climate may change based on a wide range of possibilities for these variables.

Sea-Level Rise

A significant impact of a human-induced global warming is a rise in sea level (Figure 21.32). As this occurs, coastal cities, wetlands, and low-lying islands could be threatened with more frequent flooding, increased shoreline erosion, and saltwater encroachment into coastal rivers and aquifers.

How is a warmer atmosphere related to a global rise in sea level? The most obvious connection, the melting of glaciers, is important, but not the only factor. An equally important factor is that a warmer atmosphere causes an increase in ocean volume due to thermal expansion. Higher air temperatures warm the adjacent upper layers of the ocean, which in turn causes the water to expand and sea level to rise.

TABLE 21.1   Projected Changes and Effects of Global Warming in the 21st Century

Projected changes and estimated probability[*]	Examples of projected impacts
Higher maximum temperatures; more hot days and heat waves over nearly all land areas (*virtually certain*).	Increased incidence of death and serious illness in older age groups and urban poor. Increased heat stress in livestock and wildlife. Shift in tourist destinations. Increased risk of damage to a number of crops. Increased electric cooling demand and reduced energy supply reliability.
Higher minimum temperatures; fewer cold days, frost days, and cold waves over nearly all land areas (*virtually certain*)	Decreased cold-related human morbidity and mortality. Decreased risk of damage to a number of crops, and Increased risk to others. Extended range and activity of some pest and disease vectors. Reduced heating energy demand.
Frequency of heavy precipitation events increases over most areas (*very likely*)	Increased flood, landslide, avalanche, and debris flow damage. Increased soil erosion. Increased flood runoff could increase recharge of some floodplain aquifers. Increased pressure on government and private flood insurance systems and disaster relief.
Area affected by drought increases (*likely*)	Decreased crop yields. Increased damage to building foundations caused by ground shrinkage. Decreased water-resource quantity and quality. Increased risk of forest fire.
Intense tropical cyclone activity Increases (*likely*)	Increased risks to human life, risk of Infectious-disease epidemics, and many other risks. Increased coastal erosion and damage to coastal buildings and infrastructure. Increased damage to coastal ecosystems, such as coral reefs and mangroves.

[*]*Virtually certain* indicates a probability greater than 99 percent, *very likely* indicates a probability of 90–99 percent, and *likely* indicates a probability of 67–90 percent.
Source: IPCC, 2001, 2007

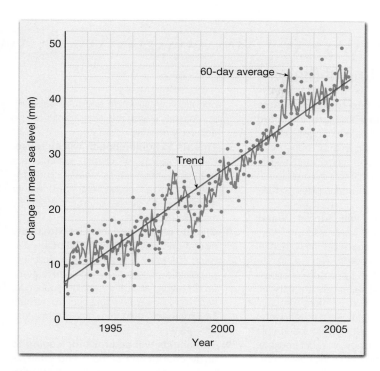

FIGURE 21.32 Using data from satellites and floats, it was determined that sea level rose an average of 3 millimeters (0.1 inch) per year between 1993 and 2005. Researchers attributed about half the rise to melting glacial ice and the other half to thermal expansion. Rising sea level can adversely affect some of the most densely populated areas on Earth. (NSAS/Jet Propulsion Laboratory)

Research indicates that sea level has risen between 10 and 23 centimeters (4 and 8 inches) over the past century and that trend will continue at an accelerated rate. Some models indicate that the rise may approach or even exceed 50 centimeters (20 inches) by the end of the twenty-first century. Such a change may seem modest, but scientists realize that any rise in sea level along a *gently* sloping shoreline, such as the Atlantic and Gulf coasts of the United States, will lead to significant erosion and severe, permanent inland flooding (Figure 21.33). If this happens, many beaches and wetlands will be eliminated and coastal civilization would be severely disrupted.

Because rising sea level is a gradual phenomenon, it may be overlooked by coastal residents as an important contributor to shoreline erosion problems. Rather, the blame may be assigned to other forces, especially storm activity. Although a given storm may be the immediate cause, the magnitude of its destruction may result from the relatively small sea level rise that allowed the storm's power to cross a much greater land area (Figure 21.33).

As mentioned, a warmer climate will cause glaciers to melt. In fact, a portion of the 10- to 25-centimeter (4- to 8-inch) rise in sea level over the past century is attributed to the melting of mountain glaciers. This contribution is projected to continue through the twenty-first century. Of course, if the Greenland and Antarctic ice sheets were to experience a significant increase in melting, it would lead to a much greater rise in sea level and a major encroachment by the sea in coastal zones. Is this possible? In their 2007 report, the IPCC indicates that such changes

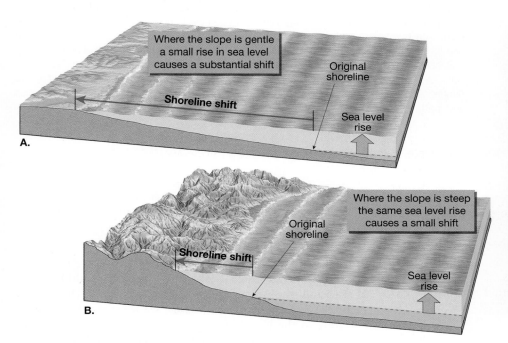

FIGURE 21.33 The slope of a shoreline is critical to determining the degree to which sea-level changes will affect it. **A.** When the slope is gentle, small changes in the sea level cause a substantial shift. **B.** The same sea-level rise along a steep coast results in only a small shoreline shift. **C.** As sea level gradually rises, the shoreline retreats, and structures that were once thought to be safe from wave attack are exposed to the force of the sea. (Photo by Kenneth Hasson)

EARTH AS A SYSTEM

BOX 21.3

Global Climate Change Impacts on the United States

The U.S. Global Change Research Program (USGCRP) coordinates and integrates federal research on changes in the global environment and their implications for society. Thirteen departments and agencies participate. In addition, the USGCRP collaborates with several other national and international science programs.

In June 2009, the USGCRP published an authoritative scientific report, *Global Change Impacts in the United States* (www. globalchange.gov). The report is written in plain language and summarizes the science of climate change and its impact on the United States now and in the future. It integrates research from the USGCRP with related research from around the world. What follows is a brief list of the report's *key findings*.

1. **Global warming is unequivocal and primarily human-induced.** Global temperature has increased over the past 50 years. This observed increase is due primarily to human-induced emissions of heat-trapping gases.

2. **Climate changes are underway in the United States and are projected to grow.** Climate-related changes are already observed in the United States and its coastal waters. These include increases in heavy downpours, rising temperature and sea level, rapidly retreating glaciers, thawing permafrost, lengthening growing seasons, lengthening ice-free seasons in the ocean and on lakes and rivers, earlier snowmelt, and alterations in river flows. These changes are projected to grow.

3. **Widespread climate-related impacts are occurring now and are expected to increase.** Climate changes are already affecting water, energy, transportation, agriculture, ecosystems, and health. These impacts are different from region to region and will grow under projected climate change.

4. **Climate change will stress water resources.** Water is an issue in every region, but the nature of the potential impacts varies. Drought, related to reduced precipitation, increased evaporation, and increased water loss from plants, is an important issue in many regions, especially in the West. Floods and water quality problems are likely to be amplified by climate change in most regions. Declines in mountain snowpack are important in the West and Alaska where snowpack provides vital natural water storage.

5. **Crop and livestock production will be increasingly challenged.** Agriculture is considered one of the sectors most adaptable to changes in climate. However, increased heat, pests, water stress, diseases, and weather extremes will pose adaptation challenges for crop and livestock production.

6. **Coastal areas are at increasing risk from sea-level rise and storm surge.** Sea-level rise and storm surge place many U.S. coastal areas at increasing risk of erosion and flooding, especially along the Atlantic and Gulf Coasts, Pacific Islands, and parts of Alaska. Energy and trans-

portation infrastructure and other property in coastal areas are very likely to be adversely affected.

7. **Threats to human health will increase.** Health impacts of climate change are related to heat stress, waterborne diseases, poor air quality, extreme weather events, and diseases transmitted by insects and rodents.

8. **Climate change will interact with many social and environmental issues.** Climate change will combine with pollution, population growth, overuse of resources, urbanization, and other social, economic, and environmental stresses to create larger impacts than from any of these factors alone.

9. **Thresholds will be crossed, leading to large changes in climate and ecosystems.** There are a variety of thresholds in the climate system and ecosystems. These thresholds determine, for example, the presence of sea ice and permafrost, and the survival of species, from fish to insect pests, with implications for society. With further climate change, the crossing of additional thresholds is expected.

10. **Future climate change and its impacts depend on choices made today.** The amount and rate of future climate change depend primarily on current and future human-caused emissions of heat-trapping gases and airborne particles. Responses involve reducing emissions to limit future warming, and adapting to the changes that are unavoidable.

---

may occur over time scales of thousands of years. However, the report goes on to say that a more rapid rise on century time scales *cannot* be excluded.*

The Changing Arctic

A recent study of climate change in the Arctic began with the following statement:

> For nearly 30 years, Arctic sea ice extent and thickness have been falling dramatically. Permafrost temperatures are rising and coverage is decreasing. Mountain glaciers and the Greenland ice sheet are shrinking. Evidence suggests we are witness-

ing the early stage of an anthropogenically induced global warming superimposed on natural cycles, reinforced by reductions in Arctic ice.**

Arctic Sea Ice Climate models are in general agreement that one of the strongest signals of global warming should be a loss of sea ice in the Arctic. This is indeed occurring. The map in Figure 21.34A compares the average sea ice extent for September 2007 to the long-term average for the period 1979–2000. The minimum extent of sea ice in September 2005 is also shown. September represents the end of the melt period when the area covered by sea ice is at a minimum. In September 2007

*IPCC, *Summary for Policymakers of the Synthesis Report of the IPCC Fourth Assessment Report*, p. 13.

**J. T. Overpeck, et al. "Arctic System on Trajectory to New, Seasonally Ice-Free States," *EOS, Transactions, American Geophysical Union*, Vol. 86, No. 34, 23 August 2005, p. 309.

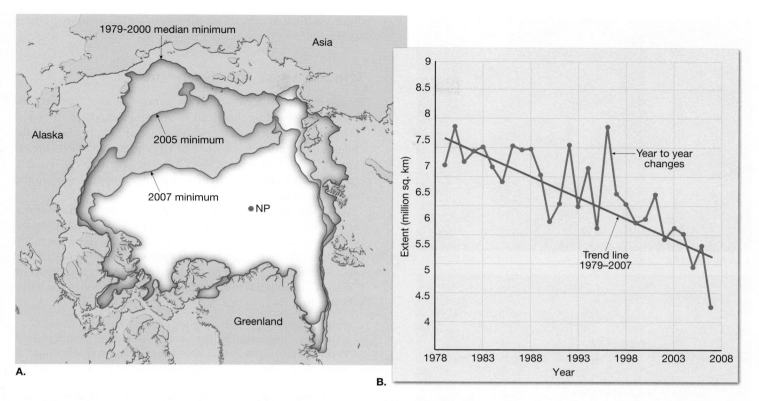

FIGURE 21.34   **A.** A comparison showing the extent of sea ice at the end of the summer melting periods in 2005 and 2007 compared to the average for the 1979–2000 period. The extent of sea ice in September 2007 was 39 percent below the long-term average for 1979–2000. (After NASA) **B.** The graph depicts the decline in Arctic sea ice from 1979 to 2007. The September rate of sea ice decline since 1979 is 10 percent per decade. (National Snow and Ice Data Center)

sea ice was 39 percent below the long-term average. The rate of sea ice decline between 1979 and 2007 was about 10 percent per decade or 72,000 square kilometers (28,000 square miles) per year. The trend is also clear when you examine the graph in Figure 21.34B. Is it possible that this trend may be part of a natural cycle? Yes, but it is more likely that the sea-ice decline represents a combination of natural variability and human-induced global warming, with the latter becoming increasingly evident in coming decades. As was noted in the section on "Climate Feedback Mechanisms," a reduction in sea ice represents a positive feedback mechanism that reinforces global warming.

Permafrost   During the past decade, evidence has mounted to indicate that the extent of permafrost in the Northern Hemisphere has decreased, as would be expected under long-term warming conditions. Figure 21.35 presents one example that such a decline is occurring.

In the Arctic, short summers thaw only the top layer of frozen ground. The permafrost beneath this *active layer* is like the cement bottom of a swimming pool. In summer, water cannot percolate downward, so it saturates the soil above the permafrost and collects on the surface in thousands of lakes. However, as Arctic temperatures climb, the bottom of the "pool" seems to be "cracking." Satellite imagery shows that over a 20-year span, a significant number of lakes have shrunk or disappeared altogether. As the permafrost thaws, lake water drains deeper into the ground.

Thawing permafrost represents a potentially significant positive feedback mechanism that may reinforce global warming. When vegetation dies in the Arctic, cold temperatures inhibit its total decomposition. As a consequence, over thousands of years a great deal of organic matter has become stored in the permafrost. When the permafrost thaws, organic matter that may have been frozen for millennia comes out of "cold storage" and decomposes. The result is the release of carbon dioxide and methane—greenhouse gases that contribute to global warming.

Increasing Ocean Acidity

The human-induced increase in the amount of carbon dioxide in the atmosphere has some serious implications for ocean chemistry and for marine life. Recent studies show that about one-third of the human-generated carbon dioxide currently ends up in the oceans. As a result, the ocean's pH becomes lowered, making seawater more acidic. The pH scale is shown and briefly described in Figure 6.C, p. 183

When atmospheric CO_2 dissolves in seawater (H_2O), it forms carbonic acid (H_2CO_3). This lowers the ocean's pH and changes the balance of certain chemicals found naturally in seawater. In fact, the oceans have already absorbed enough carbon dioxide for surface waters to have experienced a pH decrease of 0.1 pH units since preindustrial times, with an additional pH

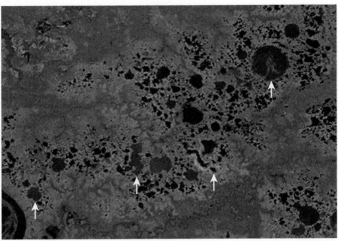

A. June 27, 1973

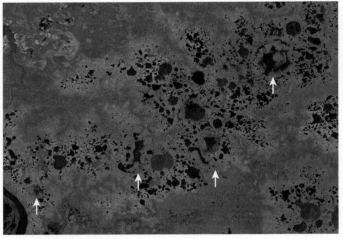

B. July 2, 2002

FIGURE 21.35 This image pair shows lakes dotting the tundra in northern Siberia in 1973 and 2002. The tundra vegetation is colored a faded red, whereas lakes appear blue or blue-green. Many lakes have clearly disappeared or shrunk considerably between 1973 and 2002. Compare the areas highlighted by the white arrowheads in each image. After studying satellite imagery of about 10,000 large lakes in a 500,000-square-kilometer area in northern Siberia, scientists documented an 11 percent decline in the number of lakes, with at least 125 disappearing completely.

decrease likely in the future. Moreover, if the current trend in carbon dioxide emissions continues, by the year 2100 the ocean will experience a pH decrease of at least 0.2 pH units, which represents a change in ocean chemistry that has not occurred for millions of years. This shift toward acidity and the changes in ocean chemistry that result make it more difficult for certain marine creatures to build hard parts out of calcium carbonate. The decline in pH thus threatens a variety of calcite-secreting organisms as diverse as microbes and corals, which concerns marine scientists because of the potential consequences for

other sea life that depend on the health and availability of these organisms.

The Potential for "Surprises"

In summary, you have seen that climate in the 21st century, unlike the preceding thousand years, is not expected to be stable. Rather, a constant state of change is very likely. Many of the changes will probably be gradual environmental shifts, imperceptible from year to year. Nevertheless, the effects, accumulated over decades, will have powerful economic, social, and political consequences.

Despite our best efforts to understand future climate shifts, there is also the potential for "surprises." This simply means that, due to the complexity of Earth's climate system, we might experience relatively sudden, unexpected changes or see some aspects of climate shift in an unexpected manner. The report on *Climate Change Impacts on the United States* describes the situation like this:

> Surprises challenge humans' ability to adapt, because of how quickly and unexpectedly they occur. For example, what if the Pacific Ocean warms in such a way that El Niño events become much more extreme? This could reduce the frequency, but perhaps not the strength, of hurricanes along the East Coast, while on the West Coast, more severe winter storms, extreme precipitation events, and damaging winds could become common. What if large quantities of methane, a potent greenhouse gas currently frozen in icy Arctic tundra and sediments, began to be released to the atmosphere by warming, potentially creating an amplifying "feedback loop" that would cause even more warming? We simply do not know how far the climate system or other systems it affects can be pushed before they respond in unexpected ways.
>
> There are many examples of potential surprises, each of which would have large consequences. Most of these potential outcomes are rarely reported, in this study or elsewhere. Even if the chance of any particular surprise happening is small, the chance that at least one such surprise will occur is much greater. In other words, while we can't know which of these events will occur, it is likely that one or more will eventually occur.*

The impact on climate of an increase in atmospheric carbon dioxide and trace gases is obscured by some uncertainties. Yet, climate scientists continue to improve our understanding of the climate system and the potential impacts and effects of global climate change. Policy makers are confronted with responding to the risks posed by emissions of greenhouse gases knowing that our understanding is imperfect. However, they are also faced with the fact that climate-induced environmental changes cannot be reversed quickly, if at all, owing to the lengthy time scales associated with the climate system.

---

*National Assessment Synthesis Team, *Climate Change Impacts on the United States: The Potential Consequences of Climate Variability and Change.* Washington, D.C.: U.S. Global Research Program, 2000, p. 19.

CHAPTER 21 GLOBAL CLIMATE CHANGE IN REVIEW

- The *climate system* includes the atmosphere, hydrosphere, geosphere, biosphere, and cryosphere (the ice and snow that exists at Earth's surface). The system involves the exchanges of energy and moisture that occur among the five spheres.

- Techniques for analyzing Earth's climate history on a scale of hundreds to thousands of years include evidence from *seafloor sediments* and *oxygen isotope analysis.* Seafloor sediments are useful recorders of worldwide climate change because the numbers and types of organic remains included in the sediment are indicative of past sea-surface temperatures. Using oxygen isotope analysis, scientists can use the $^{18}O/^{16}O$ ratio found in the shells of microorganisms in sediment and layers of ice and snow to detect past temperatures. Other sources of data used for the study of past climates (called *proxy data*) include the growth rings of trees, pollen contained in sediments, coral reefs, and information contained in historical documents.

- Air is a mixture of many discrete gases, and its composition varies from time to time and place to place. After water vapor, dust, and other variable components are removed, two gases, *nitrogen* and *oxygen,* make up 99 percent of the volume of the remaining clean, dry air. *Carbon dioxide,* although present in only minute amounts (0.0387 percent or 387 parts per million), is an efficient absorber of energy emitted by Earth and thus influences the heating of the atmosphere.

- Two important variable components of air are water vapor and aerosols. Like carbon dioxide, water vapor can absorb heat given off by Earth. *Aerosols* (tiny solid and liquid particles) are important because these often invisible particles act as surfaces on which water vapor can condense and are also good absorbers and reflectors (depending on the particles) of incoming solar radiation.

- *Electromagnetic radiation* is energy emitted in the form of rays, or waves, called electromagnetic waves. All radiation is capable of transmitting energy through the vacuum of space. One of the most important differences among electromagnetic waves is their *wavelengths,* which range from very long *radio waves* to very short *gamma rays. Visible light* is the only portion of the electromagnetic spectrum we can see. Some of the basic laws that govern radiation as it heats the atmosphere are (1) all objects emit radiant energy, (2) hotter objects radiate more total energy than do colder objects, (3) the hotter the radiating body, the shorter the wavelengths of maximum radiation, and (4) objects that are good absorbers of radiation are good emitters as well. Gases are selective absorbers, meaning that they absorb and emit certain wavelengths but not others.

- Because the atmosphere gradually thins with increasing altitude, it has no sharp upper boundary but simply blends into outer space. Based on temperature, the atmosphere is divided vertically into four layers. The *troposphere* is the lowermost layer. In the troposphere, temperature usually decreases with increasing altitude. This *environmental lapse rate* is variable, but averages about 6.5 °C per kilometer (3.5 °F per 1,000 feet). Essentially all important weather phenomena occur in the troposphere. Above the troposphere is the *stratosphere,* which exhibits warming because of absorption of ultraviolet radiation by ozone. In the mesosphere, temperatures again decrease. Upward from the mesosphere, is the *thermosphere,* a layer with only a tiny fraction of the atmosphere's mass, and no well-defined upper limit.

- Approximately 50 percent of the solar energy that strikes the top of the atmosphere reaches Earth's surface. About 30 percent is reflected back to space. The remaining 20 percent of the energy is absorbed by clouds and the atmosphere's gases. The wavelengths of the energy being transmitted, as well as the size and nature of the absorbing or reflecting substance, determine whether solar radiation will be scattered and reflected back to space, or absorbed.

- Radiant energy that is absorbed heats Earth and eventually is reradiated skyward. Because Earth has a much lower surface temperature than the Sun, its radiation is in the form of long-wave infrared radiation. Because the atmospheric gases, primarily water vapor and carbon dioxide, are more efficient absorbers of terrestrial (longwave) radiation, the atmosphere is heated from the ground up. The transmission of shortwave solar radiation by the atmosphere, coupled with the selective absorption of Earth radiation by atmospheric gases, results in the warming of the atmosphere and is referred to as the *greenhouse effect.*

- Several explanations have been formulated to explain climate change. Current hypotheses for the "natural" mechanisms (causes unrelated to human activities) of climate change include (1) plate tectonics, rearranging Earth's continents closer or farther from the equator, (2) variations in Earth's orbit, involving changes in the shape of the orbit (*eccentricity*), angle that Earth's axis makes with the plane of its orbit (*obliquity*), and/or the wobbling of the axis (*precession*), (3) volcanic activity, reducing the solar radiation that reaches the surface, and (4) changes in the Sun's output associated with *sunspots.*

- Humans have been modifying the environment for thousands of years. By altering ground cover with the use of fire and the overgrazing of land, people have modified such important climatological factors as surface reflectivity (*albedo*), evaporation rates, and surface winds.

- By adding carbon dioxide and other trace gases (methane, nitrous oxide, and chlorofluorocarbons) to the atmosphere, modern humans are contributing to global climate change in a significant way.

When any component of the climate system is altered, scientists must consider the many possible outcomes, called *climate-feedback mechanisms*. Changes that reinforce the initial change are called *positive-feedback mechanisms*. For example, warmer surface temperatures cause an increase in evaporation, which further increases temperature as the additional water vapor absorbs more radiation emitted by Earth. On the other hand, *negative-feedback mechanisms* produce results that are the opposite of the initial change and tend to offset it. An example would be the negative effect that increased cloud cover has on the amount of solar energy available to heat the atmosphere.

- Global climate is also affected by human activities that contribute to the atmosphere's *aerosol* (tiny, often microscopic, liquid and solid particles that are suspended in air) content. By reflecting sunlight back to space, aerosols have a net

cooling effect. The effect of aerosols on today's climate is determined by the amount emitted during the preceding couple of weeks, while carbon dioxide remains for much longer spans and influences climate for many decades.

- Because the climate system is so complex, predicting specific regional changes that may occur as the result of increased levels of carbon dioxide in the atmosphere is speculative. However, some possible consequences of greenhouse warming include (1) altering the distribution of the world's water resources, (2) a probable rise in sea level, (3) a greater intensity of tropical cyclones, and (4) changes in the extent of Arctic sea ice and permafrost.

- Due to the complexity of the climate system, not all future shifts can be foreseen. Thus, "surprises" (relatively sudden unexpected changes in climate) are possible.

KEY TERMS

aerosols (p. 584)
climate-feedback mechanism (p. 597)
climate system (p. 577)
greenhouse effect (p. 586)

negative-feedback mechanism (p. 598)
oxygen isotope analysis (p. 579)

paleoclimatology (p. 578)
positive-feedback mechanism (p. 598)

proxy data (p. 578)
sunspots (p. 591)

QUESTIONS FOR REVIEW

1. List the five parts of the climate system.

2. What are *proxy data*? List several examples. Why are such data necessary in the study of climate change?

3. Why are seafloor sediments useful in the study of past climates?

4. Briefly describe why tree rings are helpful in studying the geologic past.

5. What are the major components of clean, dry air? List two significant variable components of the atmosphere.

6. The atmosphere is divided vertically into four layers on the basis of temperature. List these layers in order from bottom to top.

7. Compared to Earth, does the Sun emit most of its energy as longer or shorter wavelengths of electromagnetic radiation?

8. What are the three basic paths taken by incoming solar radiation? On average, what percentage takes each path? What might cause their percentages to vary?

9. Explain why the atmosphere is heated chiefly by radiation from Earth's surface rather than by direct solar radiation.

10. Which gases are the primary heat absorbers in the lower atmosphere?

11. Describe or outline the greenhouse effect.

12. The volcanic eruptions of El Chichón in Mexico and Mount Pinatubo in the Philippines had measurable short-term effects on global temperatures. Describe and briefly explain these effects.

13. List two examples of possible climate change linked to solar variability. Are these Sun-climate connections widely accepted?

14. Why has the carbon dioxide level of the atmosphere been rising for more then 150 years?

15. How are temperatures in the lower atmosphere likely to change as carbon dioxide levels continue to increase?

16. Aside from carbon dioxide, what other trace gases are contributing to a future global temperature change?

17. What are climate-feedback mechanisms? Give some examples.

18. What are the main sources of human-generated aerosols? What effect do these aerosols have on temperatures in the troposphere? How long do aerosols remain in the lower atmosphere before they are removed?

19. List four potential consequences of global warming.

COMPANION WEBSITE

The *Earth 10e* Web site uses the resources and flexibility of the Internet to aid in your study of the topics in this chapter. Written and developed by the authors and other geology instructors, this site will help improve your understanding of geology. Visit **www.mygeoscienceplace.com** in order to:

- **Review** key chapter concepts.

- **Read** with links to the eBook and to chapter-specific web resources.
- **Test** yourself with online quizzes.

EARTH'S EVOLUTION THROUGH GEOLOGIC TIME

Maroon Bells in Autumn, Colorado Rockies.

(PHOTO BY TIM FITZHARRIS/ MINDEN PICTURES)

Earth's history is long and complex. Time and again, the splitting and colliding of continents formed new ocean basins and created great mountain ranges. Meanwhile, the assemblage of organisms inhabiting our planet experienced dramatic changes.

Many of the changes on planet Earth occur at a "snail's pace," geologically speaking—generally too slow for people to perceive. Thus, human understanding of evolution is fairly recent. Evolutionary processes are not exclusively applicable to life-forms, as all of Earth's "spheres" have evolved together: the atmosphere, hydrosphere, geosphere, and biosphere (Figure 22.1). Changes have occurred in the air we breathe, in the composition of the world's oceans, the ponderous movements of crustal plates that give rise to mountains, and the evolution of a vast array of life-forms. As each of Earth's spheres evolve, they dramatically influence the others.

IS EARTH UNIQUE?

There is only one place in the universe, as far as we know, that can support life—a modest-sized planet called Earth that orbits an average-sized star, the Sun. Life on Earth is ubiquitous; it is found in boiling mudpots and hot springs, in the deep abyss of the ocean, and even under the Antarctic ice sheet. Living space on our planet however, is significantly limited when we consider the needs of individual organisms, particularly humans. The global ocean covers 71 percent of Earth's surface, but only a few hundred meters below the water's surface pressures are so intense that human lungs cannot function. In addition, many continental areas are too steep, too high, or too cold for us to inhabit (Figure 22.2). Nevertheless, based on what we know about other bodies in the solar system—and the hundreds of planets recently discovered orbiting around other stars—Earth is still, by far, the most accommodating.

What fortuitous events produced a planet so hospitable to life? Earth was not always as we find it today. During its formative years, our planet became hot enough to support a magma ocean. It also survived a several-hundred-million-year period of extreme bombardment, to which the heavily cratered surfaces of Mars, the Moon, and asteroids testify. The oxygen-rich atmosphere that makes higher life-forms possible developed relatively recently. Serendipitously, Earth seems to be the right planet, in the right location, at the right time.

The Right Planet

What are some of the characteristics that make Earth unique among the planets? Consider the following:

1. If Earth were considerably larger (more massive) its force of gravity would be proportionately greater. Like the giant planets, Earth would have retained a thick, hostile atmosphere consisting of ammonia and methane, and possibly hydrogen and helium.

2. If Earth were much smaller, oxygen, water vapor, and other volatiles would escape into space and be lost forever. Thus, like the Moon and Mercury, both of which lack an atmosphere, Earth would be void of life.

3. If Earth did not have a rigid lithosphere overlaying a weak asthenosphere, plate tectonics would not operate. The continental crust (Earth's "highlands") would not have formed without the recycling of plates. Consequently, the entire planet would likely be covered by an ocean a few kilometers deep. As the author Bill Bryson so aptly stated, "There might be life in that lonesome ocean, but there certainly wouldn't be baseball."

4. Most surprisingly, perhaps, is the fact that if our planet did not have a molten metallic core, most of the life-forms on Earth would not exist. Fundamentally, without the flow of iron in the

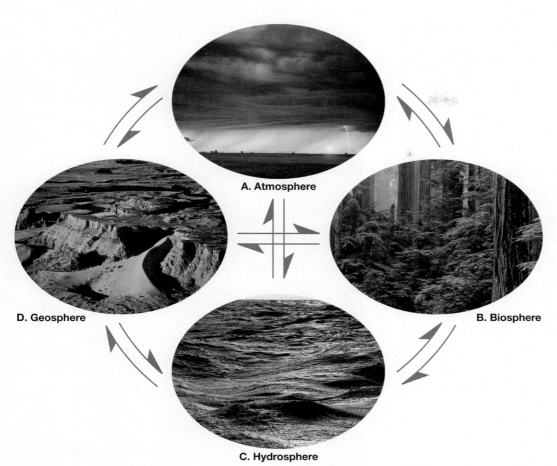

FIGURE 22.1 Earth's spheres have evolved together through the long expanse of geologic time. (Photos by **A.** Momatiuk/Animals Animals–Earth Scenes, **B.** Dean Pennala/ Shutterstock, and **C.** and **D.** Michael Collier)

A. Atmosphere

B. Biosphere

C. Hydrosphere

D. Geosphere

FIGURE 22.2 Climbers near the top of Mount Everest. At this altitude the level of oxygen is only one-third the amount available at sea level. (Photo courtesy of Woodfin Camp and Associates)

core, Earth could not support a magnetic field. It is the magnetic field which prevents lethal cosmic rays (the solar wind) from showering Earth's surface.

The Right Location

One of the primary factors that determine whether a planet is suitable for higher life-forms is its location in the solar system. The following scenarios substantiate Earth's right position:

1. If Earth were about 10 percent closer to the Sun, like Venus, our atmosphere would consist mainly of the greenhouse gas, carbon dioxide. As a result, Earth's surface temperature would be too hot to support higher life-forms.
2. If Earth were about 10 percent farther from the Sun, the problem would be reversed—it would be too cold. The oceans would freeze over and Earth's active water cycle would not exist. Without liquid water all life would perish.
3. Earth is near a star of modest size. Stars like the Sun have a life span of roughly 10 billion years. During most of this time, radiant energy is emitted at a fairly constant level. Giant stars, on the other hand, consume their nuclear fuel at very high rates and "burn out" in a few hundred million years. Therefore, Earth's proximity to a modest sized star allowed enough time for the evolution of humans, who first appeared on this planet only a few million years ago.

The Right Time

The last, but certainly not the least, fortuitous factor is timing. The first organisms to inhabit Earth were extremely primitive and came into existence roughly 3.8 billion years ago. From this point in Earth's history innumerable changes occurred—life-forms came and went along with changes in the physical environment of our planet. Two of many timely, Earth-altering events include:

1. The development of our modern atmosphere. Earth's primitive atmosphere is thought to have been composed mostly of water vapor and carbon dioxide, with small amounts of other gases, but no free oxygen. Fortunately, microorganisms evolved that released oxygen into the atmosphere by the process of *photosynthesis*. About 2.2 billion years ago an atmosphere with free oxygen came into existence. The result was the evolution of the forbearers of the vast array of organisms that occupy Earth today.
2. About 65 million years ago our planet was struck by an asteroid 10 kilometers in diameter. This impact caused a mass extinction during which nearly three-quarters of all plant and animal species were obliterated—including dinosaurs (Figure 22.3). Although this may not seem fortuitous, the extinction of dinosaurs opened new habitats for small mammals that survived the impact. These habitats, along with evolutionary forces, led to the development of many large mammals that occupy our modern world. Without this event, mammals may not have evolved beyond the small rodent-like creatures that live in burrows.

As various observers have noted, Earth developed under "just right" conditions to support higher life-forms. Astronomers refer to this as the *Goldilocks scenario*. Like the classic

FIGURE 22.3   Paleontologist excavating the back leg of a huge (20 tons and 70 feet long) dinosaur, *Jobaria tiguidensis.* The scene is the Sahara Desert in the Republic of Niger. (Photo © Didier Duthell/Sygma/Corbis)

Goldilocks and the Three Bears fable, Venus is too hot (Papa Bear's porridge), Mars is too cold (Mama Bear's porridge), but Earth is just right (Baby Bear's porridge). Do these "just right" conditions exist purely by chance as some researchers suggest, or might Earth's hospitable environment have developed for the evolution and survival of higher life-forms as others have argued?

The remainder of this chapter will focus on the origin and evolution of planet Earth—the one place in the Universe we know fosters life. As you learned in Chapter 9, researchers utilize many tools to interpret the clues about Earth's past. Using these tools, and clues contained in the rock record, scientists continue to unravel many complex events of the geologic past. The goal of this chapter is to provide a brief overview of the history of our planet and its life-forms—a journey that takes us back about 4.6 billion years to the formation of Earth and its atmosphere. Later, we will consider how our physical world assumed its present state and how Earth's inhabitants changed through time. We suggest that you reacquaint yourself with the *geologic time scale* presented in Figure 22.4 and refer to it throughout the chapter.

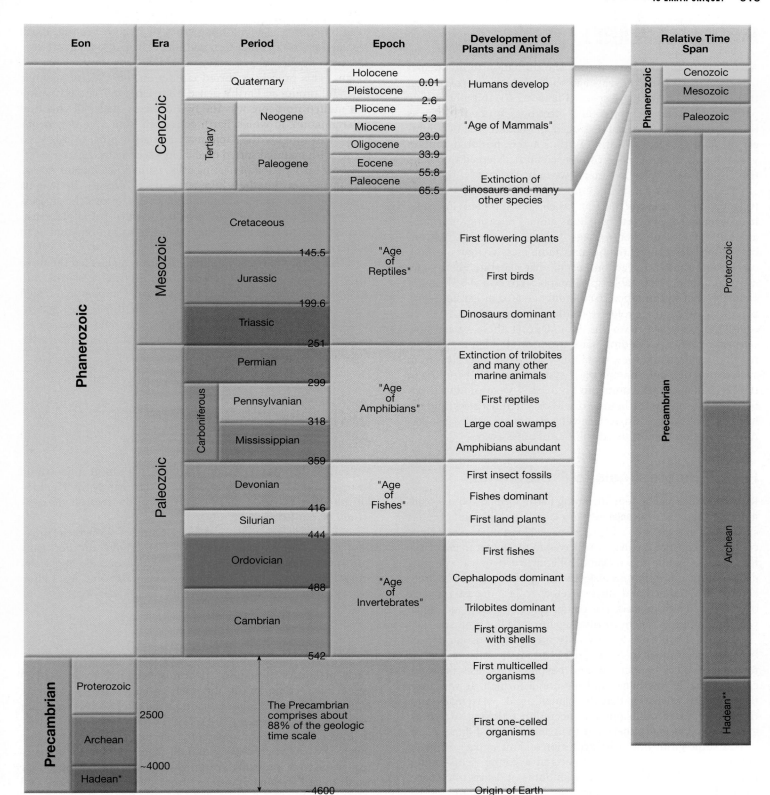

* Hadean is the informal name for the span that begins at Earth s formation and ends with Earth's earliest-known rocks.

FIGURE 22.4   The geologic time scale. Numbers represent time in millions of years before the present. These dates were added long after the time scale had been established using relative dating techniques. The Precambrian accounts for about 88 percent of geologic time.

BIRTH OF A PLANET

According to the Big Bang theory, the formation of our planet began about 13.7 billion years ago with a cataclysmic explosion that created all matter and space (Figure 22.5). Initially, atomic particles (protons, neutrons, and electrons) formed. Later, as this debris cooled, atoms of hydrogen and helium, the two lightest elements, began to form. Within a few hundred million years, clouds of these gases condensed and coalesced into stars that compose the galactic systems we now observe.

As these gases contracted to become the first stars, heating triggered the process of *nuclear fusion*. Within stars' interiors, hydrogen atoms convert to helium atoms, releasing enormous amounts of radiant energy (heat, light, cosmic rays). Astronomers have determined that in stars more massive than our Sun, other thermonuclear reactions occur that generate all the elements on the periodic table up to number 26, iron. The heaviest elements (beyond number 26) are only created at extreme temperatures during the explosive death of a star perhaps 10 to 20 times more massive than the Sun. During these cataclysmic **supernova** events, exploding stars produce all of the elements heavier than iron and spew them into interstellar space. It is from such debris that our Sun and solar system formed. According to the Big Bang scenario, atoms in your body were produced billions of years ago in the hot interior of now defunct stars, and the formation of the gold in your jewelry was triggered by a supernova explosion that occurred trillions of miles away.

From Planetesimals to Protoplanets

Recall that the solar system, including Earth, formed about 4.6 billion years ago from the **solar nebula**, a large rotating cloud of interstellar dust and gas (see Chapter 1, page 21). As the solar nebula contracted, most of the matter collected in the center to create the hot *protosun*, while the remainder became a flattened spinning disk. Within this spinning disk, matter gradually coalesced into clumps that collided and stuck together to become asteroid-size objects called **planetesimals.** The composition of each planetesimal was determined largely by its distance from the hot protosun.

Near the orbit of Mercury only metallic grains condensed from the solar nebula. Near Earth's orbit, metallic as well as rocky substances condensed, and beyond Mars, ices of water, carbon dioxide, methane, and ammonia formed. It was from these clumps of matter that the planetesimals were made and, through repeated collisions and accretion (sticking together), grew into eight **protoplanets** and their moons (see Figure 22.5).

At some point in Earth's evolution a giant impact occurred between a Mars-sized planetesimal and a young, semi-molten Earth. This collision ejected huge amounts of debris into space, some of which coalesced to form the Moon (see Figure 22.5 J, K, L).

Earth's Early Evolution

As material continued to accumulate, the high-velocity impact of interplanetary debris (planetesimals) and the decay of radioactive elements caused the temperature of our planet to steadily increase. During this period of intense heating, Earth became hot enough that iron and nickel began to melt. Melting produced liquid blobs of heavy metal that sank under their own weight. This process occurred rapidly on the scale of geologic time and produced Earth's dense iron-rich core. The formation of a molten iron core was the first of many stages of chemical differentiation, in which Earth converted from a homogeneous body, with roughly the same matter at all depths, to a layered planet with material sorted by density (see Figure 22.5).

This early period of heating also resulted in a magma ocean, perhaps several hundred kilometers deep. Within the magma ocean, buoyant masses of molten rock rose toward the surface, and eventually solidified to produce a thin, primitive crust. Earth's first crust was likely basaltic in composition, similar to modern oceanic crust.

This period of chemical differentiation established the three major divisions of Earth's interior—the iron-rich *core,* the thin *primitive crust,* and Earth's thickest layer, the *mantle,* located between the core and the crust. In addition, the lightest materials, including water vapor, carbon dioxide, and other gases, escaped to form a primitive atmosphere and, shortly thereafter, the oceans (Figure 22.6).

ORIGIN OF THE ATMOSPHERE AND OCEANS

We can be thankful for our atmosphere; without it there would be no greenhouse effect and Earth would be nearly 60 °F colder. Earth's water bodies would be frozen and the hydrologic cycle would be nonexistent.

The air we breathe is a stable mixture of 78 percent nitrogen, 21 percent oxygen, about 1 percent argon (an inert gas), and small amounts of gases such as carbon dioxide and water vapor. However, our planet's original atmosphere 4.6 billion years ago was substantially different.

Earth's Primitive Atmosphere

Early in Earth's formation, its atmosphere likely consisted of gases most common in the early solar system—hydrogen, helium, methane, ammonia, carbon dioxide, and water vapor (see Chapter 24, pages 675–677). The lightest of these gases, hydrogen and helium, apparently escaped into space because Earth's gravity was too weak to hold them. Most of the remaining gases were probably scattered into space by strong *solar winds* (a vast stream of particles) from a young, active Sun. (All stars, including the Sun, apparently experience a highly active stage early in their evolution known as the *T-Tauri phase,* during which their solar winds are very intense.)

Earth's first enduring atmosphere was generated by a process called **outgassing,** through which gases trapped in the planet's interior are released. Outgassing from hundreds of active volcanoes still remains an important planetary function worldwide (Figure 22.7). However, early in Earth's history,

FIGURE 22.5   Major events that led to the formation of early Earth.

FIGURE 22.6 Artistic depiction of Earth over 4 billion years ago. This was a time of intense volcanic activity that produced Earth's primitive atmosphere and oceans, while early life-forms produced mound-like structures called stromatolites.

when massive heating and fluid-like motion occurred in the mantle, the gas output must have been immense. Based on our understanding of modern volcanic eruptions, Earth's primitive atmosphere probably consisted of mostly water vapor, carbon dioxide, and sulfur dioxide with minor amounts of other gases, and minimal nitrogen. Most importantly, free oxygen was not present.

Oxygen in the Atmosphere

As Earth cooled, water vapor condensed to form clouds, and torrential rains began to fill low-lying areas which became the oceans. In those oceans, nearly 3.5 billion years ago, photosynthesizing bacteria began to release oxygen into the water. During *photosynthesis,* the Sun's energy is used by organisms to produce organic material (energetic molecules of sugar containing hydrogen and carbon) from carbon dioxide (CO_2) and water (H_2O). The first bacteria probably used hydrogen sulfide (H_2S) as the source of hydrogen rather than water. One of the earliest bacteria, *cyanobacteria* (once called blue-green algae), began to produce oxygen as a byproduct of photosynthesis.

Initially, the newly released oxygen was readily consumed by chemical reactions with other atoms and molecules (particularly iron) in the ocean. It seems that large quantities of iron were released into the early ocean through submarine volcanism and associated hydrothermal vents. Iron has tremendous affinity for oxygen. When these two elements join, they become iron oxide (rust). As it accumulated on the seafloor, these early iron oxide deposits created alternating layers of iron-rich rocks and chert, called **banded iron formations** (Figure 22.8). Most banded iron deposits accumulated in the Precambrian between 3.5 and 2 billion years ago, and represent the world's most important reservoir of iron ore.

As the number of oxygen-generating organisms increased, oxygen began to build in the atmosphere. Chemical analysis of rocks suggest that a significant amount of oxygen appeared in the atmosphere as early as 2.2 billion years ago, and increased steadily until it reached stable levels about 1.5 billion years ago. Obviously, the availability of free oxygen had a positive impact on the development of life.

FIGURE 22.7 Earth's first enduring atmosphere was formed by a process called outgassing, which continues today from hundreds of active volcanoes worldwide. (Photo by Game McGimsey/CORBIS)

FIGURE 22.8 These layered, iron-rich rocks, called banded iron formations, were deposited during the Precambrian. Much of the oxygen generated as a by-product of photosynthesis was readily consumed by chemical reactions with iron to produce these rocks. (Courtesy Spencer R. Titley)

Another significant benefit of the "oxygen explosion" is that oxygen (O_2) molecules readily absorb ultraviolet radiation, and rearrange themselves to form *ozone* (O_3). Today ozone is concentrated above the surface in a layer called the *stratosphere* where it absorbs much of the ultraviolet radiation that strikes the upper atmosphere. For the first time, Earth's surface was protected from this type of solar radiation, which is particularly harmful to DNA. Marine organisms had always been shielded from ultraviolet radiation by the oceans, but the development of the atmosphere's protective ozone layer made the continents more hospitable.

Evolution of the Oceans

About 4 billion years ago, as much as 90 percent of the current volume of seawater was contained in the ocean basins. Because the primitive atmosphere was rich in carbon dioxide, sulfur dioxide, and hydrogen sulfide, the earliest rainwater was highly acidic—to an even greater degree than the acid rain that damaged lakes and streams in eastern North America during the latter part of the 20th century. Consequently, Earth's rocky surface weathered at an accelerated rate. The products

released by chemical weathering included atoms and molecules of various substances, including sodium, calcium, potassium, and silica, that were carried into the newly formed oceans. Some of these dissolved substances precipitated to become chemical sediment that mantled the ocean floor. Other substances formed soluble salts which increased the salinity of seawater. Research suggests that the salinity of the oceans increased rapidly at first, but has remained constant over the last two billion years.

Earth's oceans also serve as a depository for tremendous volumes of carbon dioxide, a major constituent in the primitive atmosphere. This is significant because carbon dioxide is a greenhouse gas that strongly influences the heating of the atmosphere. Venus, once thought to be very similar to Earth, has an atmosphere composed of 97 percent carbon dioxide that produced a "runaway" greenhouse effect. As a result, its surface temperature is 475 °C (900 °F).

Carbon dioxide is readily soluble in seawater where it often joins other atoms or molecules to produce various chemical precipitates. The most common compound generated by this process is calcium carbonate ($CaCO_3$), which makes up limestone, the most abundant chemical sedimentary rock. About 542 million years ago, marine organisms began to extract calcium carbonate from seawater to make their shells and other hard parts. Included were trillions of tiny marine organisms such as foraminifera, whose shells were deposited on the seafloor at the end of their life-cycle. Today, some of these deposits can be observed in the chalk beds exposed along the White Cliffs of Dover, England shown in Figure 22.9. By "locking up" carbon dioxide, these limestone deposits store this greenhouse gas so it cannot easily re-enter the atmosphere.*

---

*For more on this, see Box 7.1, "The Carbon Cycle and Sedimentary Rocks," on p. 214.

FIGURE 22.9 Prominent chalk deposit, the White Chalk Cliffs, Sussex, England. Similar deposits are also found in northern France. (Photo by Art Wolf)

PRECAMBRIAN HISTORY: THE FORMATION OF EARTH'S CONTINENTS

Earth's first 4 billion years are encompassed in the time span called the *Precambrian*. Representing nearly 90 percent of Earth's history, the Precambrian is divided into the *Archean eon* ("ancient age") and the *Proterozoic eon* ("early life"). Our knowledge of this ancient time is limited because much of the early rock record has been obscured by the very Earth processes you have been studying, especially plate tectonics, erosion, and deposition. Most Precambrian rocks lack fossils, which hinders correlation of rock units. In addition, rocks this old are metamorphosed and deformed, extensively eroded, and frequently concealed by younger strata. Indeed, Precambrian history is written in scattered, speculative episodes, like a long book with many missing chapters.

Earth's First Continents

More than 95 percent of Earth's human population lives on the continents—not included are people living on volcanic islands such as the Hawaiian Islands and Iceland. These islanders inhabit pieces of oceanic crust that were thick enough to rise above sea level.

What differentiates continental crust from oceanic crust? Recall that oceanic crust is a relatively dense (3.0 g/cm^3), homogeneous layer of basaltic rocks derived from partial melting of the rocky, upper mantle. In addition, oceanic crust is thin, averaging only 7 kilometers in thickness. Continental crust, on the other hand, is composed of a variety of rock types, has an average thickness of nearly 40 kilometers, and contains a large percentage of low-density (2.7 g/cm^3), silica-rich rocks such as granite.

The significance of these differences cannot be overstated in a review of Earth's geologic evolution. Oceanic crust, because it is relatively thin and dense, is found several kilometers below sea level—unless, of course it has been pushed onto a landmass by tectonic forces. Continental crust, because of its great thickness and lower density, extends well above sea level. Also, recall that oceanic crust of normal thickness readily subducts, whereas thick, buoyant blocks of continental crust resist being recycled into the mantle.

Making Continental Crust Earth's first crust was probably basalt, similar to that generated at modern oceanic ridges, but because physical evidence no longer exists, we are not certain.

The hot, turbulent mantle that most likely existed during the Archean eon recycled most of this material back into the mantle. In fact, it may have been continuously recycled, much like the "crust" that forms on a lava lake is repeatedly replaced with fresh lava from below (Figure 22.10).

The oldest preserved continental rocks occur as small, highly deformed terranes, which are incorporated within somewhat younger blocks of continental crust (Figure 22.11). The oldest of these is the 4 billion-year-old Acasta gneiss located in the Slave Province of Canada's Northwest Territories.

The formation of continental crust is simply a continuation of the gravitational segregation of Earth materials that began during the final accretionary stage of our planet. After the metallic core and rocky mantle evolved, low density, silica-rich minerals were gradually extracted from the mantle to form continental crust. This is an on-going, multi-stage process during which partial melting of ultramafic mantle rocks (peridotite) generates basaltic rocks. Next, the melting of basaltic rocks produces magmas that crystallize to form felsic, quartz-bearing rocks (see Chapter 4). However, little is known about the details of the mechanisms that generated these silica-rich rocks during the Archean.

Most geologists agree that some type of plate-like motion operated early in Earth's history. In addition, hot spot volcanism likely played a role as well. However, because the mantle was hotter in the Archean than it is today, both of these phenomena would have progressed at higher rates than their modern counterparts. Hot spot volcanism is thought to have created immense shield volcanoes as well as oceanic plateaus. Simultaneously, subduction of oceanic crust generated volcanic island arcs. Collectively, these relatively small crustal fragments represent the first phase in creating stable, continent-size landmasses.

FIGURE 22.10   Rift pattern on lava lake. The crust covering this lava lake is continually being replaced with fresh lava from below, much like Earth's crust was recycled early in its history. (Photo by Juerg Alean/www.stromboli.net)

nental crust was generated, a substantial amount of crustal material was destroyed as well. Crust can be lost by either weathering and erosion, or direct reincorporation into the mantle through subduction. Evidence suggests that during much of the Archean, thin slabs of continental crust were eliminated, mainly by subduction into the mantle. However, by about 3 billion years ago, cratons grew sufficiently large and thick to resist subduction. After that time, weathering and erosion became the primary processes of crustal destruction. By the close of the Precambrian, an estimated 85 percent of the modern continental crust had formed.

FIGURE 22.11 These rocks at Isua, Greenland, some of the world's oldest, have been dated at 3.8 billion years. (Photo courtesy of James L. Amos/CORBIS)

From Continental Crust to Continents According to one model, the growth of large continental masses was accomplished through collision and accretion of various types of terranes as illustrated in Figure 22.12. This type of collision tectonics deformed and metamorphosed sediments caught between converging crustal fragments, thereby shortening and thickening the developing crust. Within the deepest regions of these collision zones, partial melting of the thickened crust generated silica-rich magmas that ascended and intruded the rocks above. The result was the formation of large crustal provinces that, in turn, accreted with others to form even larger crustal blocks called **cratons.** (The portion of a modern craton that is exposed at the surface is referred to as a *shield.*) The assembly of a large craton involves the accretion of several crustal blocks that cause major mountain building episodes similar to India's collision with Asia. Figure 22.13 shows the extent of crustal material that was produced during the Archean and Proterozoic eons. This was accomplished by the collision and accretion of many thin, highly mobile terranes into nearly recognizable continental masses.

Although the Precambrian was a time when much of Earth's conti-

The Making of North America

North America provides an excellent example of the development of continental crust and its piecemeal assembly into a continent. Notice in Figure 22.14 that very little continental crust older than 3.5 billion years remains. In the late Archean, between 3–2.5 billion years ago, there was a period of major continental growth. During this span, the accretion of numerous island arcs and other fragments generated several large crustal provinces. North America contains some of these crustal

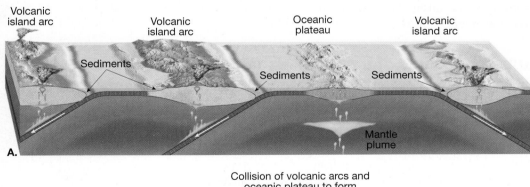

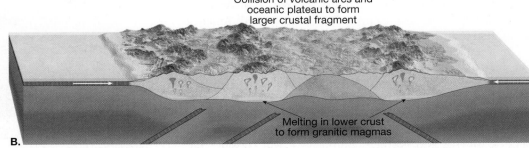

FIGURE 22.12 According to one model, the growth of large continental masses was accomplished through the collision and accretion of various types of terranes.

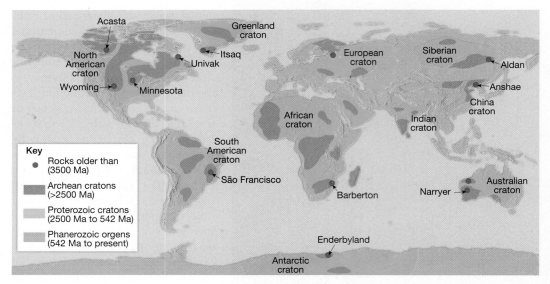

FIGURE 22.13   Illustration showing the extent of crustal material remaining from the Archean and Proterozoic eons.

fragments are gradually reassembled into a new supercontinent with a different configuration is called the **supercontinent cycle.** The assembly and dispersal of supercontinents had a profound impact on the evolution of Earth's continents. In addition, this phenomenon greatly influenced global climates and contributed to periodic episodes of rising and falling sea level.

Supercontinents and Climate

As continents move, the patterns of ocean currents and global winds change, which influences the global distribution of temperature and precipitation. One example of how a supercontinent's dispersal influenced climate is the formation of the Antarctic ice sheet. Although eastern Antarctica remained over the South Pole for more than 100 million years, it was not glaciated until about 25 million years ago. Prior to this period of glaciation, South America was connected to the Antarctic Peninsula. This arrangement of landmasses helped maintain a circulation pattern in which

units, including the Superior and Hearne/Rae cratons shown in Figure 22.14. It remains unknown where these ancient continental blocks formed.

About 1.9 billion years ago these crustal provinces collided to produce the Trans-Hudson mountain belt (Figure 22.14). (This mountain-building episode was not restricted to North America, because ancient deformed strata of similar age are also found on other continents.) This event built the North American craton, around which several large and numerous small crustal fragments were later added. These late arrivals include Blue Ridge and Piedmont provinces of the Appalachians. Additionally, several terranes were added to the western margin of North America during the Mesozoic and Cenozoic eras to generate the mountainous North American Cordillera.

Supercontinents of the Precambrian

Supercontinents are large landmasses that contain all, or nearly all, the existing continents. Pangaea was the most recent, but certainly not the only, supercontinent to exist in the geologic past. The earliest well-documented supercontinent, *Rodinia,* formed during the Proterozoic eon about 1.1 billion years ago. Although its reconstruction is still being researched, it is clear that Rodinia's configuration was quite different from Pangaea (Figure 22.15). One obvious distinction is that North America was located near the center of this ancient landmass.

Between 800 and 600 million years ago, Rodinia gradually split apart. By the end of the Precambrian, many of the fragments reassembled producing a large landmass in the Southern Hemisphere called *Gondwana,* comprised mainly of present-day South America, Africa, India, Australia, and Antarctica (Figure 22.16). Other continental fragments also developed—North America, Siberia, and northern Europe. We will consider the fate of these Precambrian landmasses later in the chapter.

Supercontinent Cycle   The idea that rifting and dispersal of one supercontinent is followed by a long period during which the

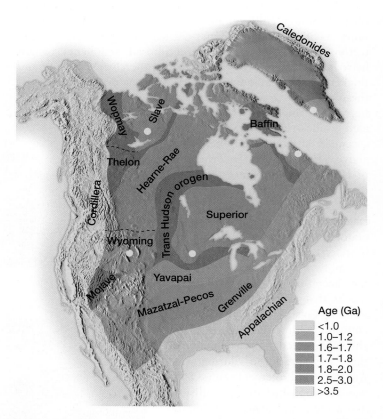

FIGURE 22.14   Map showing the major geological provinces of North America and their ages in billions of years (Ga). It appears that North America was assembled from crustal blocks that were joined by processes very similar to modern plate tectonics. These ancient collisions produced mountainous belts that include remnant volcanic island arcs trapped by the colliding continental fragments.

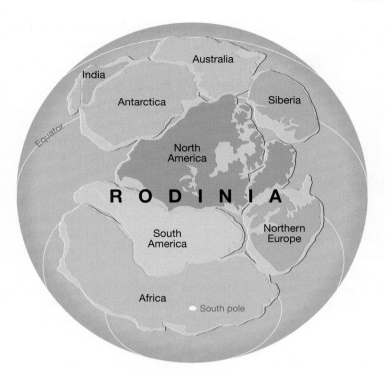

FIGURE 22.15 Simplified drawing showing one of several possible configurations of the supercontinent Rodinia. For clarity, the continents are drawn with somewhat modern shapes, which was not the case 1 billion years ago. (After P. Hoffman, J. Rogers, and others)

warm ocean currents reached the coast of Antarctica as shown in Figure 22.17A. This is similar to how the modern Gulf Stream keeps Iceland mostly ice free, despite its name. However, as South America separated from Antarctica, it moved northward, permitting ocean circulation to flow from west to east around the entire continent of Antarctica (Figure 22.17B). This cold current, called the West Wind Drift, effectively isolated the entire Antarctic coast from the warm, poleward-directed currents in the southern oceans. As a result, most of the Antarctic landmass became covered with glacial ice.

Local and regional climates have also been impacted by large mountain systems created by the collision of large cratons. Because of their high elevations, mountains exhibit markedly lower average temperatures than surrounding lowlands. In addition, when air rises over these lofty structures, lifting "squeezes" moisture from the air, leaving the region downwind relatively dry. A modern analogy is the wet, heavily forested western slopes of the Sierra Nevada compared to the dry climate of the Great Basin desert that lies directly downwind (see Figure 19.4, p. 522)

Supercontinents and Sea Level Changes Significant and numerous sea level changes have been documented in geologic history, many of which appear related to the assembly and dispersal of supercontinents. If sea level rises, shallow seas advance onto the continents. Evidence for periods when the seas advanced onto the continents include thick sequences of ancient marine sedimentary rocks that blanket large areas of modern landmasses—including much of the eastern two-thirds of the United States.

The supercontinent cycle and sea level changes are directly related to rates of *seafloor spreading*. When the rate of spreading is rapid, as it is along the East Pacific Rise today, the production

of warm oceanic crust is also high. Because warm oceanic crust is less dense (takes up more space) than cold crust, fast spreading ridges occupy more volume in the ocean basins than slow spreading centers. (Think of getting into a tub filled with water.) As a result, when the rates of seafloor spreading increase, sea level rises. This, in turn, causes shallow seas to advance onto the low-lying portions of the continents.

GEOLOGIC HISTORY OF THE PHANEROZOIC: THE FORMATION OF EARTH'S MODERN CONTINENTS

The time span since the close of the Precambrian, called the *Phanerozoic eon,* encompasses 542 million years and is divided into three eras: *Paleozoic, Mesozoic,* and *Cenozoic.* The beginning

A. Continent of Gondwana

B. Continents not a part of Gondwana

FIGURE 22.16 Reconstruction of Earth as it may have appeared in late Precambrian time. The southern continents were joined into a single landmass called Gondwana. Other landmasses that were not part of Gondwana include North America, northwestern Europe, and northern Asia. (After P. Hoffman, J. Rogers, and others)

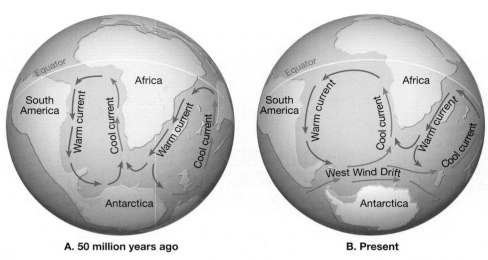

A. 50 million years ago

B. Present

FIGURE 22.17 Comparison of the oceanic circulation pattern 50 million years ago with that of the present. When South America separated from Antarctica, the West Wind Drift developed, which effectively isolated the entire Antarctic coast from the warm, poleward-directed currents in the southern oceans. This led to the eventual covering of much of Antarctica with glacial ice.

record provided invaluable information for deciphering ancient environments.

Paleozoic History

As the Paleozoic era opened, North America hosted no living things, neither plant nor animal. There were no Appalachian or Rocky Mountains; the continent was largely a barren lowland. Several times during the early Paleozoic, shallow seas moved inland, then receded from the continental interior and left behind thick deposits of limestones, shales, and clean sandstones which mark the shorelines of these previous mid-continent shallow seas.

Formation of Pangaea One of the major events of the Paleozoic was the formation of the supercontinent of Pangaea, that began with a series of collisions that gradually joined North America, Europe, Siberia, and other smaller crustal fragments (Figure 22.19). These events eventually generated a large northern continent called *Laurasia*. This tropical landmass supported warm wet conditions that led to the formation of vast swamps that eventually converted to coal.

of the Phanerozoic is marked by the appearance of the first life-forms with hard parts such as shells, scales, bones, or teeth—all of which greatly enhance the possibility of an organism being preserved in the fossil record (Figure 22.18).* Thus, the study of Phanerozoic crustal history was aided by the availability of fossils, which improved our ability to date and correlate geologic events (Box 22.1). Moreover, because every organism is associated with its own particular niche, the greatly improved fossil

Simultaneously, the vast southern continent of *Gondwana* encompassed five continents—South America, Africa, Australia, Antarctica, India, and perhaps portions of China. Evidence of an extensive continental glaciation places this landmass near the South Pole. By the end of the Paleozoic, Gondwana had migrated northward and collided with Laurasia, culminating in the formation of the supercontinent *Pangaea*.

*For more about this see the discussion on "Conditions Favoring Preservation" in Chapter 9, p. 266)

FIGURE 22.18 Natural cast of a trilobite. Trilobites dominated the early Paleozoic ocean, scavenging food from the bottom. (Photo by Ed Reschke/Peter Arnold, Inc.)

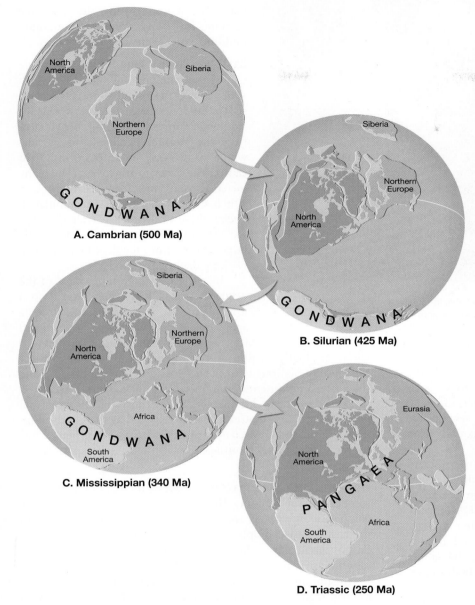

FIGURE 22.19 During the late Paleozoic, plate movements were joining the major landmasses to produce the supercontinent of Pangaea. (After P. Hoffman, J. Rogers, and others)

The accretion of Pangaea spans more than 300 million years and resulted in the formation of several mountain belts. The collision of northern Europe (mainly Norway) with Greenland produced the Caledonian Mountains, whereas the joining of northern Asia (Siberia) and Europe created the Ural Mountains. Northern China is also thought to have accreted to Asia by the end of the Paleozoic, whereas southern China may not have become part of Asia until after Pangaea had begun to rift. (Recall that India did not begin to accrete to Asia until about 50 million years ago.)

Pangaea reached its maximum size about 250 million years ago as Africa collided with North America (Figure 22.19D). This event marked the final episode of growth in the long history of the Appalachian Mountains (see Chapter 14).

Mesozoic History

Spanning about 186 million years, the Mesozoic era is divided into three periods: the *Triassic, Jurassic,* and *Cretaceous.* Major geologic events of the Mesozoic include the breakup of Pangaea and the evolution of our modern ocean basins.

The Mesozoic era began with much of the world's continents above sea level. The exposed Triassic strata are primarily red sandstones and mudstones that lack marine fossils, features that indicate a terrestrial environment. (The red color in sandstone comes from the oxidation of iron.)

As the Jurrasic period opened, the sea invaded western North America. Adjacent to this shallow sea, extensive continental sediments were deposited on what is now the Colorado Plateau. The most prominent is the Navajo Sandstone, a cross-bedded, quartz-rich layer, that in some places approaches 300 meters (1000 feet) thick. These remnants of massive dunes indicate that an enormous desert occupied much of the American Southwest during early Jurassic times (Figure 22.20). Another well-known Jurassic deposit is the Morrison Formation—the world's richest storehouse of dinosaur fossils. Included are the fossilized bones of massive dinosaurs such as Apatosaurus (formerly Brontosaurus), Brachiosaurus, and Stegosaurus.

As the Jurassic period gave way to the Cretaceous, shallow seas again encroached upon much of western North America, as well as the Atlantic and Gulf coastal regions. This led to the formation of "coal swamps" similar to those of the Paleozoic era. Today, the Cretaceous coal deposits in the western United States and Canada are economically important. For example, on the Crow Native American reservation in Montana, there are nearly 20 billion tons of high-quality, Cretaceous age coal.

Another major event of the Mesozoic era was the breakup of Pangaea (see Box 2.3, p. 63). About 165 million years ago a rift developed between what is now North America and western Africa, which marked the birth of the Atlantic Ocean. As Pangaea gradually broke apart, the westward-moving North American plate began to override the Pacific basin. This tectonic event triggered a continuous wave of deformation that moved inland along the entire western margin of North America. By Jurassic times, subduction of the Farallon plate had begun to produce the chaotic mixture of rocks that exist today in the Coast Ranges of California. Further inland, igneous activity was widespread, and for over 100 million years volcanism was rampant as huge masses of magma rose within a few miles of Earth's surface. The remnants of this activity include the granitic plutons of the Sierra Nevada as well as the Idaho batholith, and British Columbia's Coast Range batholith.

Tectonic activity initiated in the Jurassic continued throughout the Cretaceous. Compressional forces moved huge

FIGURE 22.20 These massive, cross-bedded sandstone cliffs in Zion National Park are the remnants of ancient sand dunes. (Photo by Ruth Tomlinson/Robert Harding)

rock units in a shingle-like fashion toward the east. Across much of North America's western margin, older rocks were thrust eastward over younger strata, for distances exceeding 150 kilometers (90 miles). Ultimately, this activity was responsible for creating the vast Northern Rockies that extend from Wyoming to Alaska.

Toward the end of the Mesozoic, the southern portions of the Rocky Mountains developed. This mountain-building event, called the *Laramide Orogeny,* occurred when large blocks of deeply buried Precambrian rocks were lifted nearly vertically along steeply dipping faults, upwarping the overlying younger sedimentary strata. The mountain ranges produced by the Laramide Orogeny include Colorado's Front Range, the Sangre de Cristo of New Mexico and Colorado, and the Bighorns of Wyoming (see Box 14.2, p. 398).

Cenozoic History

The Cenozoic era, or "era of recent life," encompasses the last 65.5 million years of Earth history. It was during this span that the physical landscapes and life-forms of our modern world came into existence. The Cenozoic era represents a considerably smaller fraction of geologic time than either the Paleozoic or the Mesozoic. Nevertheless, much more is known about this time span because the rock formations are more widespread and less disturbed than those of any preceding era.

Most of North America was above sea level during the Cenozoic era. However, the eastern and western margins of the continent experienced markedly contrasting events because of their different plate boundary relationships. The Atlantic and Gulf coastal regions, far removed from an active plate boundary, were tectonically stable. By contrast, western North America was the leading edge of the North American plate. As a result, plate interactions during the Cenozoic account for many events of mountain building, volcanism, and earthquakes.

Eastern North America The stable continental margin of eastern North America was the site of abundant marine sedimentation. The most extensive deposition surrounded the Gulf of Mexico, from the Yucatan Peninsula to Florida, where a massive buildup of sediment caused the crust to downwarp. In many instances, faulting created structures in which oil and natural gas accumulated. Today, these and other petroleum traps are the Gulf Coast's most economically important resource, evident by numerous offshore drilling platforms.

By early Cenozoic time, most of the original Appalachians had eroded to a low plain. Later, isostatic adjustments again raised the region and rejuvenated its rivers. Streams eroded with renewed vigor, gradually sculpting the surface into its present-day topography. Sediments from this erosion were deposited along the eastern continental margin, where they accumulated to a thickness of many kilometers. Today, portions of the strata deposited during the Cenozoic are exposed as the gently sloping Atlantic and Gulf coastal plains, where a large percentage of the eastern and southeastern United States population resides.

Western North America In the West, the Laramide Orogeny responsible for building the southern Rocky Mountains was coming to an end. As erosional forces lowered the mountains, the basins between uplifted ranges began to fill with sediment. East of the Rockies, a large wedge of sediment from the eroding mountains created the Great Plains.

Beginning in the Miocene epoch about 20 million years ago, a broad region from northern Nevada into Mexico experienced crustal extension that created more than 150 fault-block mountain ranges. Today, they rise abruptly above the adjacent basins, creating the Basin and Range Province (see Chapter 14).

As the Basin and Range Province was developing, the entire western interior of the continent gradually uplifted. This event re-elevated the Rockies and rejuvenated many of the West's major rivers. As the rivers became incised, many spectacular gorges were created, including the Grand Canyon of the Colorado River, the Grand Canyon of the Snake River, and the Black Canyon of the Gunnison River.

Volcanic activity was also common in the West during much of the Cenozoic. Beginning in the Miocene epoch, great volumes of fluid basaltic lava flowed from fissures in portions of present-day Washington, Oregon, and Idaho. These eruptions built the extensive (1.3 million square miles) Columbia Plateau. Immediately west of the Columbia Plateau, volcanic activity was different in character. Here, more viscous magmas with higher silica contents erupted explosively, creating the Cascades, a chain of stratovolcanoes extending from northern California into Canada, some of which are still active (Figure 22.21).

A final episode of deformation occurred in the late Cenozoic, creating the Coast Ranges that stretch along the Pacific Coast. Meanwhile, the Sierra Nevada were faulted and uplifted along their eastern flank, forming the impressive mountains we see today.

As the Cenozoic was drawing to a close, the effects of mountain building, volcanic activity, isostatic adjustments, and extensive erosion and sedimentation created the physical landscape we know today. All that remained of Cenozoic time was the final 2.6 million year episode called the Quaternary period. During this most recent, and on-going, phase of Earth's history, in which humans evolved, the action of glacial ice, wind, and running water added the finishing touches to our planet's long, complex geologic history.

EARTH'S FIRST LIFE

The oldest fossils provide evidence that life on Earth was established at least 3.5 billion years ago. Microscopic fossils similar to modern cyanobacteria (formerly known as blue-green algae) have been found in silica-rich chert deposits worldwide. Notable examples include southern Africa, where rocks date to more than 3.1 billion years, and the Lake Superior region of western Ontario and northern Minnesota where the Gunflint Chert contains some fossils that are older than 2 billion years. Chemical traces of organic matter in rocks of greater age have led paleontologists to strongly suggest that life may have existed 3.8 billion years ago.

How did life begin? This question sparks considerable debate, and hypotheses abound. Requirements for life, assuming the presence of a hospitable environment, include the chemical raw materials that form the essential molecules of DNA, RNA, and proteins. These substances require organic compounds called *amino acids*. The first amino acids may have been synthesized from methane and ammonia, both of which were plentiful in Earth's primitive atmosphere. Some scientists suspect these gases could have been easily reorganized into useful organic molecules by ultraviolet light, while others consider lightning the impetus, as the well-known experiments conducted by Stanley Miller and Harold Urey attempted to demonstrate.

Other researchers suggest that amino acids arrived "ready-made," delivered by asteroids or comets that collided with a young Earth. A group of meteorites (debris from asteroids and comets that strike Earth) called *carbonaceous chrondrites* known to contain amino acid-like organic compounds, led some to hypothesize that early life may have had an extraterrestrial beginning.

Yet another hypothesis proposes that the organic material needed for life came from the methane and hydrogen sulfide that spews from deep-sea hydrothermal vents (black smokers). Is it also possible that life originated within a hot spring similar to those in Yellowstone National Park (Figure 22.22)? Some origin-of-life researchers consider this scenario highly improbable as the scalding temperatures would have destroyed any early types of self-replicating molecules. Their supposition is that life would have first appeared along sheltered stretches of ancient beaches, where waves and tides brought together various organic materials formed in the Precambrian oceans.

Regardless of where or how life originated, it is clear that the journey from "then" to "now" involved change (Figure 22.23). The first known organisms were single-cell bacteria called **prokaryotes,** which means their genetic material (DNA) is *not separated* from the rest of the cell by a nucleus. Because oxygen was absent from Earth's early atmosphere and oceans, the first organisms employed anaerobic (without oxygen) metabolism to extract energy from "food." Their food source was likely organic molecules in their surroundings, but the supply was very limited. Later, bacteria evolved that used solar energy to synthesize organic compounds (sugars). This event was an important turning point in evolution—for the first time organisms had the capability of producing food for themselves as well as for other life-forms.

Recall that photosynthesis by ancient cyanobacteria, a type of prokaryote, contributed to the gradual rise in the level of oxygen, first in the ocean and later in the atmosphere. Thus, these early organisms dramatically transformed our planet. Fossil evidence for the existence of these microscopic bacteria includes distinctively layered mounds of calcium carbonate,

FIGURE 22.21   Mount Hood, Oregon. This volcano is one of several large composite cones that comprise the Cascade Range. (Photo by John M. Roberts/CORBIS/Stock Market)

UNDERSTANDING EARTH

BOX 22.1

The Burgess Shale

The possession of hard parts greatly enhances the likelihood of organisms being preserved in the fossil record. Nevertheless, there have been rare occasions in geologic history when large numbers of soft-bodied organisms have been preserved. The Burgess Shale is one well-known example. Located in the Canadian Rockies near the town of Field in southeastern British Columbia, the site was discovered in 1909 by Charles D. Walcott of the Smithsonian Institution.

The Burgess Shale is a site of exceptional fossil preservation with unmatched animal diversity (Figure 22.A). The animals of the Burgess Shale lived shortly after the *Cambrian explosion,* a time when there had been a huge expansion of marine biodiversity. Its exquisitely preserved fossils represent our most complete and authoritative snapshot of Cambrian life, far better than deposits containing only fossils of organisms with hard parts. To date, more than 100,000 unique fossils have been found.

The animals preserved in the Burgess Shale inhabited a warm, shallow sea adjacent to a large reef that was part of the continental margin of North America. During the Cambrian, the North American continent was in the tropics astride the equator. Life was restricted to the ocean, and the land was barren and uninhabited.

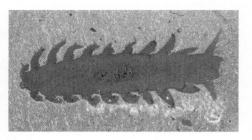

FIGURE 22.A Two examples of Burgess Shale fossils. *Thumatilon walcotti* (left) was a relatively large (up to 20 centimeters, or 8 inches, long) leaflike animal. (Photo by Simon Conway Morris, University of Cambridge) *Aysheaia pedunculata* (right) was an ancient relative of modern velvet worms and may have clung to soft sponges with tiny hooks on its feet. (Photo with permission of the Royal Ontario Museum © ROM. Photo Credit: D.H. Collins)

What were the circumstances that led to the preservation of the many life-forms found in the Burgess Shale? The animals lived in muddy marine sediments that had accumulated on the elevated portion of a reef. Periodically, the muds became unstable and moved rapidly down the steep slope of the reef as turbidity currents. These flows transported the animals in turbulent clouds of sediment to the base of the reef where they were buried. In this environment which lacked oxygen, the buried carcasses were protected from scavengers and decomposing bacteria. This process occurred repeatedly over several million years, building a thick sequence of fossil-rich sedimentary layers. Beginning about 175 million years ago, mountain-building forces elevated these strata from the seafloor and moved them many kilometers eastward along huge faults to their present location in the Canadian Rockies.

The Burgess Shale is one of the most important fossil discoveries of the 20th century. Its layers preserve an intriguing glimpse of early animal life more than a half billion years old.

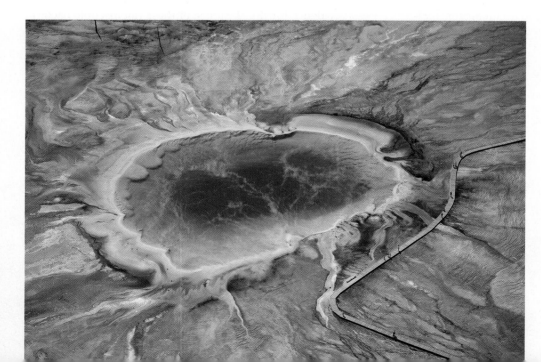

FIGURE 22.22 The Grand Prismatic Pool, Yellowstone National Park, Wyoming. This hot-water pool gets its blue color from several species of heat-tolerant cynobacteria. (Photo by George Steinmetz/CORBIS)

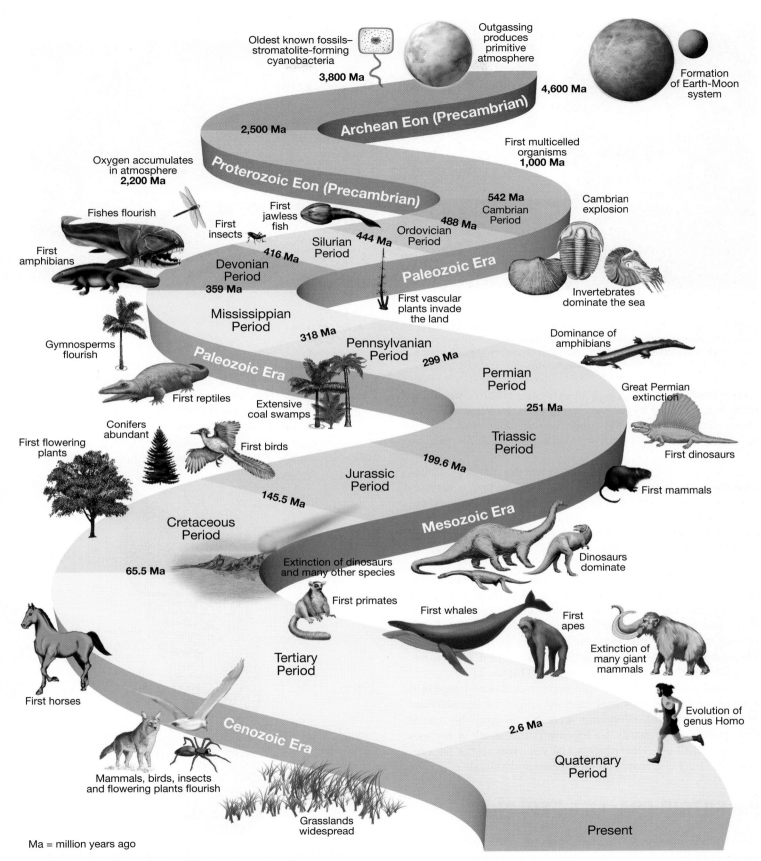

FIGURE 22.23 The evolution of life through geologic time.

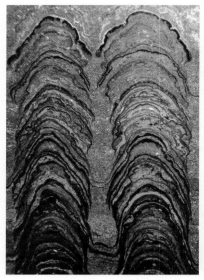

A.

B.

FIGURE 22.24   Stromatolites are among the most common Precambrian fossils. **A.** Precambrian fossil stromatolites composed of calcium carbonate deposited by algae. (Photo by Sinclair Stammers/Photo Researchers, Inc.) **B.** Modern stromatolites growing in shallow seas, western Australia. (Photo by Bill Bachman/Photo Researchers, Inc.)

called **stromatolites** (Figure 22.24A). Stromatolites are limestone mats built up by lime-accreting bacteria. What is known about these ancient fossils comes mainly from modern stromatolites like those found in Shark Bay, Australia (Figure 22.24B).

The oldest fossils of more advanced organisms, called **eukaryotes,** are about 2.1 billion years old. The first eukaryotes were microscopic, water-dwelling organisms. Unlike prokaryotes, the cellular structures of eukaryotes contain nuclei. This distinctive structure is what all multicellular organisms that now inhabit our planet—trees, birds, fishes, reptiles, and humans—have in common.

During much of the Precambrian, life consisted exclusively of single-celled organisms. It wasn't until perhaps 1.5 billion years ago that multicelled eukaryotes evolved. Green algae, one of the first multicelled organisms, contained chloroplasts (used in photosynthesis) and were the ancestors of modern plants. The first primitive marine animals did not appear until somewhat later, perhaps 600 million years ago (Figure 22.25).

Fossil evidence suggests that organic evolution progressed at an excruciatingly slow pace until the end of the Precambrian. At this time, Earth's continents were barren, and the oceans were populated primarily by organisms too small to be seen with the naked eye. Nevertheless, the stage was set for the evolution of larger and more complex plants and animals.

PALEOZOIC ERA: LIFE EXPLODES

The Cambrian period marks the beginning of the Paleozoic era, a time span that saw the emergence of a spectacular variety of new life-forms. All major invertebrate (animals lacking backbones) groups made their appearance, including jellyfish, sponges, worms, mollusks (clams, snails), and arthropods

(insects, crabs). This huge expansion in biodiversity is often referred to as the *Cambrian explosion* (Box 22.1, page 626).

But did the Cambrian explosion really happen? Recent research suggests that these life-forms may have gradually diversified late in the Precambrian, but were not preserved in the

FIGURE 22.25   Ediacaran fossil. The Ediacarans are a group of sea-dwelling animals that may have come into existence about 600 million years ago. These soft-bodied organisms were up to one meter in length and are the oldest animal fossils so far discovered. (Photo courtesy of the South Australian Museum)

FIGURE 22.26 During the Ordovician period (488–444 million years ago), the shallow waters of an inland sea over central North America contained an abundance of marine invertebrates. Shown in this reconstruction are straight-shelled cephalopods, trilobites, brachiopods, snails, and corals. (© The Field Museum, Neg. # GEO80820c, Chicago)

fossil record. Considering the Cambrian period marked the first time organisms developed hard parts, is it possible that the Cambrian event was simply an explosion of animal forms that grew in size and became "hard" enough to be fossilized?

Paleontologists may never answer that question definitively. They know, however, that hard parts clearly served many useful purposes and aided lifestyle adaptations. Sponges, for example, developed a network of fine, interwoven silica spicules that allowed them to grow larger and more erect, and thus capable of extending above the seafloor in search of food. Clams and snails secreted external shells of calcium carbonate that provided protection and allowed bodies to function in a more controlled environment. The successful trilobites developed a flexible exoskeleton of a protein called chitin (similar to a human fingernail), which permitted them to be mobile and search for food by burrowing through soft sediment (see Figure 22.18, page 622).

Early Paleozoic Life-Forms

The Cambrian period was the golden age of *trilobites*. More than 600 genera of these mud-burrowing scavengers flourished worldwide. The Ordovician marked the appearance of abundant cephalopods—mobile, highly developed mollusks that became the major predators of their time (Figure 22.26). Descendants of these cephalopods include the squid, octopus, and chambered nautilus that inhabit our modern oceans. Cephalopods were the first truly large organisms on Earth, including one species that reached a length of nearly 10 meters (30 feet).

The early diversification of animals was driven, in part, by the emergence of predatory lifestyles. The larger mobile cephalopods preyed on trilobites that were typically smaller than a child's hand. The evolution of efficient movement was often associated with development of greater sensory capabilities and more complex nervous systems. These early animals developed sensory devices for detecting light, odor, and touch.

Approximately 400 million years ago, green algae that had adapted to survive at the water's edge gave rise to the first multicellular land plants. The primary difficulty of sustaining plant life on land was obtaining water and staying upright despite gravity and winds. These earliest land plants were leafless, vertical spikes about the size of a human index finger (Figure 22.27). The fossil record indicates that by the end of the Devonian, 40 million years later, there were forests with trees tens of meters tall—clear evidence that evolutionary processes were in full swing.

In the ocean, fishes perfected an internal skeleton as a new form of support, and were the first creatures to have jaws. Armor-plated fishes that evolved during the Ordovician continued to adapt. Their armor plates thinned to lightweight scales that increased their speed and mobility. Other fishes evolved during the Devonian, including primitive sharks with cartilage skeletons and bony fishes—the groups in which many modern fishes are classified. Fishes, the first large vertebrates, proved to be faster swimmers than invertebrates and possessed more acute senses and larger brains. They became the dominant predators of the sea, which is why the Devonian period is often referred to as the "Age of the Fishes."

629

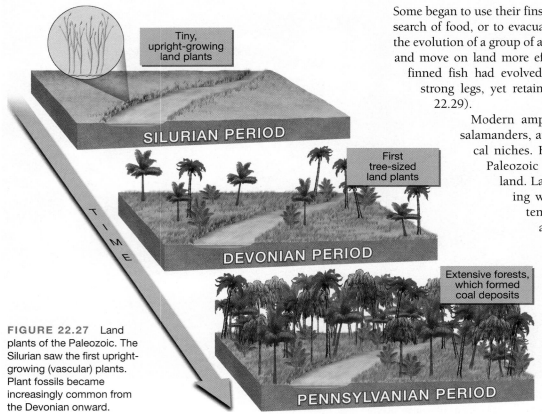

FIGURE 22.27 Land plants of the Paleozoic. The Silurian saw the first upright-growing (vascular) plants. Plant fossils became increasingly common from the Devonian onward.

Labels in figure: Tiny, upright-growing land plants — SILURIAN PERIOD — TIME — First tree-sized land plants — DEVONIAN PERIOD — Extensive forests, which formed coal deposits — PENNSYLVANIAN PERIOD

Vertebrates Move to Land

During the Devonian, a group of fishes called the *lobe-finned fish* began to adapt to terrestrial environments (Figure 22.28). Lobe-finned fishes had sacks that could be filled with air to supplement their "breathing" through gills. The first lobe-finned fish probably occupied freshwater tidal flats or small ponds near the ocean.

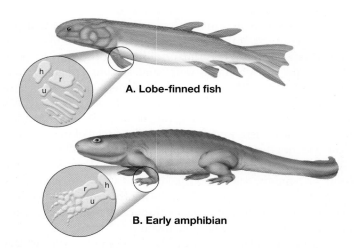

A. Lobe-finned fish

B. Early amphibian

FIGURE 22.28 Comparison of the anatomical features of the lobe-finned fish and early amphibians. **A.** The fins on the lobe-finned fish contained the same basic elements (*h*, humerus, or upper arm; *r*, radius; and *u*, ulna, or lower arm) as those of the amphibians. **B.** This amphibian is shown with the standard five toes, but early amphibians had as many as eight toes. Eventually the amphibians evolved to have a standard toe count of five.

Some began to use their fins to move from one pond to another in search of food, or to evacuate a deteriorating pond. This favored the evolution of a group of animals able to stay out of water longer and move on land more efficiently. By the late Devonian, lobe-finned fish had evolved into air-breathing amphibians with strong legs, yet retained a fish-like head and tail (Figure 22.29).

Modern amphibians, such as frogs, toads, and salamanders, are small and occupy limited biological niches. However, conditions during the late Paleozoic were ideal for these newcomers to land. Large tropical swamps that were teeming with large insects and millipedes extended across North America, Europe, and Siberia (Figure 22.30). With virtually no predatory risks, amphibians diversified rapidly. Some even took on lifestyles and forms similar to modern reptiles such as crocodiles.

Despite their success, amphibians were not fully adapted to life out of water. In fact, amphibian means "double life," because these animals need both the water from where they came and the land to which they moved. Amphibians are born in water, as exemplified by tadpoles, complete with gills and tails. These features disappear during the maturation process, resulting in air-breathing adults with legs.

The Great Permian Extinction

By the close of the Permian period, a mass extinction destroyed 70 percent of all land-dwelling vertebrate species, and perhaps 90 percent of all marine organisms. The late Permian extinction was the most significant of at least five mass extinctions that occurred over the past 500 million years. Each extinction wreaked havoc with the existing biosphere, wiping out large numbers of species. In each case, however, survivors entered new biological communities that were ultimately more diverse. Therefore, mass extinctions actually invigorated life on Earth, as the few hardy survivors eventually filled more environmental niches than those left behind by the victims.

Several mechanisms have been proposed to explain these ancient mass extinctions. Initially, paleontologists believed they were gradual events caused by a combination of climate change and biological forces, such as predation and competition. In the 1980s, a research team proposed the mass extinction of 65 million years ago was caused by the explosive impact of a large asteroid. This event, which caused the swift extinction of dinosaurs, is described in Box 22.2.

Was the Permian extinction caused by a giant impact? For many years, researchers thought this was a possible explanation. However, scant evidence has been found to substantiate this hypothesis.

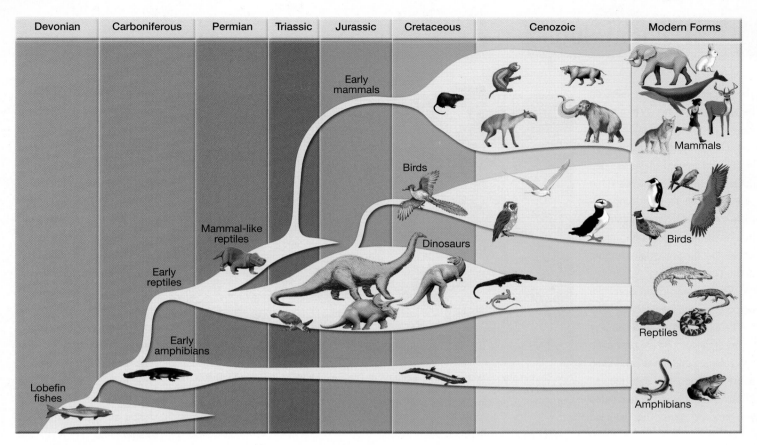

FIGURE 22.29 Relationships of various vertebrates and their evolution from a fish-like ancestor.

FIGURE 22.30 Restoration of a Pennsylvanian-age coal swamp (318 million to 299 million years ago). Shown are scale trees (left), seed ferns (lower left), and scouring rushes (right). Also note the large dragonfly. (© The Field Museum, Neg. # GEO85637c, Chicago. Photographer John Weinstein)

Paleontologist

The process of *discovery* relies on people with relentlessly inquisitive minds. Neil Shubin is such a person. One focus of Shubin's work is directed toward trying to "discover one of the key stages in the shift from fish to land-living animals." Further, he understands that such findings "offer clues about the fundamental structure of our bodies."

Recognizing that over "99 percent of all species that ever lived are now extinct, and that only a very small fraction are preserved as fossils, and that an even smaller fraction are ever found," Shubin has pursued a calculated yet daunting path. The path led

Neil Shubin stands with a cast of *Taktaalik* at The Field Museum in Chicago, Illinois. Neil and Ted Daeschler co-conceived this project in 1999 and have been working the Arctic with Farish A. Jenkins, Jr. since then. (Photo by John Weinstein, The Field Museum)

> ## "One focus of Shubin's work is directed toward trying to discover one of the key stages in the shift from fish to land-living animals."

Shubin and a research team to Ellesmer Island, in the Canadian Arctic, where their efforts resulted in a significant discovery, but only after several expensive and exhausting attempts. Shubin and his team knew their efforts had to focus on: rocks of a specific age (375 million years old); rocks that would likely contain fossils; and rocks that were exposed at the surface. Challenges faced by the team included extreme arctic weather, wildlife threats, space and time restrictions, and the knowledge that they needed to pinpoint a very small site within a 1500 kilometer-wide expanse of Canadian tundra. Over the span of six years and four expeditions, Shubin and his colleagues were deeply engaged in a search for a fossil that would bridge a major evolutionary gap. Their hard work paid off.

The fossil they discovered, named *Tiktaalik* (meaning "large freshwater fish"), exhibits characteristics of both a water-dwelling fish and a land-living animal. Its scales and fins led some to identify it as a fish, whereas its flat head with eyes on the top led others to see it as akin to a crocodile or lizard. *Tiktaalik's* head moved freely from its shoulder, in contrast to fish in which bones attach the skull to the shoulder. (A simple nod or twist of the head demonstrates the concept of the head moving

> ## "99 percent of all species that ever lived are now extinct and . . . only a very small fraction are preserved as fossils"

independently from the rest of the body.) Shubin's research also demonstrated *Tiktaalik's* links to amphibians, reptiles, birds, and mammals (including humans). Skeletal features such as wrists, ribs, and ears can be traced to the evolutionary transition observed in *Tiktaalik*.

As professionals in many fields will attest, expertise is often associated with the ability to "wear many hats." This is certainly true for Neil Shubin. At the University of Chicago he holds an endowed professorship and is Associate Dean in the Department of Organismal Biology and Anatomy. He is also Provost at Chicago's Field Museum. An unexpected assignment "directing the human anatomy course" at the UC medical school, coupled with his research on *Tiktaalik* led Shubin to "explore a profound connection" between fossils, genes, embryos, and human anatomy. Wearing the hats of research paleontologist, professor, and author, Shubin blended these seemingly unrelated topics into a monograph, *Your Inner Fish*. A national bestseller, this book provides readers with a remarkable and captivating explanation of the human body as it relates to fossils, genes, and embryos. A good measure of humor enhances the book's appeal to a diverse audience.

In the afterword to *Your Inner Fish*, Shubin explains that the study of *Tiktaalik* is ongoing and that fieldwork continues in an effort to obtain additional fossils. As is true in many scientific endeavors, Shubin readily acknowledges that questions are posed more frequently than they are answered. Advancements in science rely on relentlessly inquisitive minds like those of Neil Shubin and his research team.

Additional information regarding Neil Shubin and *Tiktaalik* can be found at http://www.neilshubin.com and http://tiktaalik.uchicago.edu/

---Joanne Bannon

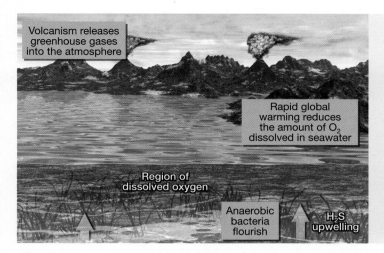
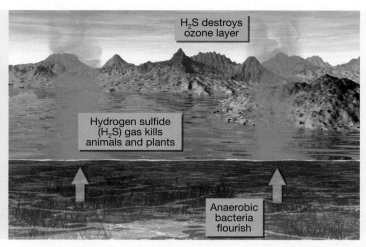

FIGURE 22.31 Model for the "Great Permian Extinction." Extensive volcanism released greenhouse gases which resulted in extreme global warming. This condition reduced the amount of oxygen dissolved by seawater. This, in turn, favored "oxygen-hating" anaerobic bacteria, which generated toxic hydrogen sulfide as a waste gas. Eventually, the concentration of hydrogen sulfide reached a critical threshold and great bubbles of this toxin exploded into the atmosphere, wreaking havoc on organisms on land, but oxygen-breathing marine life was hit the hardest.

Another possible mechanism to explain the Permian extinction was voluminous eruptions of basaltic lavas which began about 250 million years ago and covered thousands of square kilometers of the continents. (This period of volcanism produced the Siberian Traps located in northern Russia.) The release of carbon dioxide would certainly have enhanced greenhouse warming, and the emissions of sulfur dioxide probably resulted in copious amounts of acid rain.

A group of researchers have taken the global warming hypothesis still further. Although they agree that rapid greenhouse warming occurred, it alone would not have destroyed most plants because they are heat tolerant and consume CO_2 in photosynthesis. They contend, instead, that the trouble begins in the ocean rather than on land.

Most organisms, including humans, use oxygen to metabolize food. However, some forms of bacteria employ *anaerobic* (without oxygen) metabolism. Under normal conditions, oxygen from the atmosphere is readily dissolved in seawater, and is then evenly distributed to all depths by deep-water currents. This "oxygen-rich" water relegates "oxygen-hating" anaerobic bacteria to anoxic (oxygen free) environments found in deep-water sediments.

An intense period of greenhouse warming caused by an extreme episode of volcanism would have heated the ocean surface, thereby significantly reducing the amount of oxygen absorbed by seawater (Figure 22.31). This condition favors deep-sea anaerobic bacteria, which generate toxic hydrogen sulfide as a waste gas. As these organisms proliferated, the amount of hydrogen sulfide dissolved in seawater would have steadily increased. Eventually, the concentration of hydrogen sulfide would reach a critical threshold and release large toxic bubbles into the atmosphere (Figure 22.31). On land, hydrogen sulfide is lethal to both plants and animals, but it is most destructive to oxygen-breathing marine life.

How plausible is this scenario? Remember that these ideas represent a *hypothesis*, a tentative explanation regarding a particular set of observations. Additional research about this

and other hypotheses that relate to the Permian extinction continues.

MESOZOIC ERA: AGE OF THE DINOSAURS

As the Mesozoic era dawned, its life-forms were the survivors of the great Permian extinction. These organisms diversified in many ways to fill the biological voids created at the close of the Paleozoic. On land, conditions favored those that could adapt to drier climates. Among plants, gymnosperms were one such group. Unlike the first plants to invade the land, seed-bearing gymnosperms did not depend on freestanding water for fertilization. Consequently, these plants were not restricted to a life near water's edge.

The gymnosperms quickly became the dominant trees of the Mesozoic. They included: cycads that resembled a large pineapple plant; ginkgoes that had fan-shaped leaves, much like their modern relatives; and the largest plants, the conifers, whose modern descendants include the pines, firs, and junipers. The best-known fossil occurrence of these ancient trees is in northern Arizona's Petrified Forest National Park. Here, huge petrified logs lie exposed at the surface, having been weathered from rocks of the Triassic Chinle Formation (Figure 22.32).

Among the animals, reptiles readily adapted to the drier Mesozoic environment, thereby relegating amphibians to the wetlands where most remain today. Reptiles were the first true terrestrial animals with improved lungs for active lifestyles and "waterproof" skin that helped prevent the loss of body fluids. Most importantly, reptiles developed shell-covered eggs laid on land. The elimination of a water-dwelling stage (like the tadpole stage in frogs) was an important evolutionary advancement.

Of interest is the fact that the watery fluid within the reptilian egg closely resembles seawater in chemical composition. Because the reptile embryo develops in this watery

FIGURE 22.32   Petrified log of Triassic age in Arizona's Petrified Forest National Park. (Photo by Michael Collier)

environment, the shelled egg has been characterized as a "private aquarium" in which the embryos of these land vertebrates spend their water-dwelling stage of life. With this "sturdy egg," the remaining ties to the oceans were broken, and reptiles moved inland.

The first reptiles were small, but larger forms evolved rapidly, particularly the dinosaurs. One of the largest was *Apatosaurus*, which weighed more than 30 tons and measured over 25 meters (80 feet) from head to tail. Some of the largest dinosaurs were carnivorous (*Tyrannosaurus*), whereas others were herbivorous (like ponderous *Apatosaurus*).

Reptiles made perhaps the most spectacular adaptive radiations in all of Earth history. One group, the pterosaurs, became air-borne. These "dragons of the sky" possessed huge membranous wings that allowed them rudimentary flight. How the largest pterosaurs (some had wingspans of 26 feet and weighed 200 pounds) took flight is still unknown. Another group, exemplified by the fossil *Archaeopteryx*, led to more successful flyers—birds (Figure 22.33). Other reptiles returned to the sea, including fish-eating plesiosaurs and ichthyosaurs (Figure 22.34). These reptiles became proficient swimmers, but retained their reptilian teeth and breathed by means of lungs.

For nearly 160 million years, dinosaurs reigned supreme. However, by the close of the Mesozoic, like many reptiles, they became extinct. Select reptile groups survived to recent times, including turtles, snakes, crocodiles, and lizards. The huge, land-dwelling dinosaurs, the marine plesiosaurs, and the fly-

ing pterosaurs are known only through the fossil record. What caused this great extinction? (See Box 22.2.)

CENOZOIC ERA: AGE OF MAMMALS

During the Cenozoic, mammals replaced reptiles as the dominant land animals. At nearly the same time, angiosperms (flowering plants with covered seeds) replaced gymnosperms as the dominant plants. The Cenozoic is often called the "Age of Mammals," but can also be considered the "Age of Flowering Plants" because, in the plant world, angiosperms enjoy a similar status as mammals have in the animal world.

The development of flowering plants strongly influenced the evolution of both birds and mammals that feed on seeds and fruits. During the middle of the Cenozoic, another type of angiosperm, grasses, developed rapidly and spread over the plains (Figure 22.35). This fostered the emergence of herbivorous (plant-eating) mammals which, in turn, provided the evolutionary foundation for large, predatory mammals.

During the Cenozoic, the ocean was teeming with modern fish such as tuna, swordfish, and barracuda. In addition, some mammals, including seals, whales, and walruses returned to the sea.

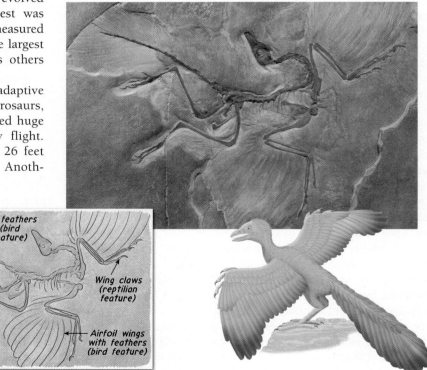

Geologist's Sketch

FIGURE 22.33   Paleontologists think that flying reptiles similar to *Archaeopteryx* were the ancestors of modern birds. Fossil evidence indicates that *Archaeopteryx* was capable of powerful flight but retained many characteristics of nonflight reptiles. Lower right is an artist's reconstruction of *Archaeopteryx*, perhaps the first birds. (Photo by Michael Collier)

FIGURE 22.34 Marine reptiles such as this *Ichthyosaur* were the most spectacular of sea animals. (Photo by Chip Clark)

From Reptiles to Mammals

The earliest mammals coexisted with dinosaurs for nearly 100 million years but were small rodentlike creatures that gathered food at night when dinosaurs were less active. Then, about 65 million years ago, fate intervened when a large asteroid collided with Earth and dealt a crashing blow to the reign of the dinosaurs. This transition, during which one dominant group is replaced by another, is clearly visible in the fossil record.

Mammals are distinct from reptiles in that they give birth to live young which suckle on milk and are warm-blooded. This latter adaptation allowed mammals to lead more active lives and to occupy more diverse habitats than reptiles because they could survive in cold regions. (Most modern reptiles are dormant during cold weather.) Other mammalian adaptations included the development of insulating body hair and more efficient organs such as hearts and lungs.

With the demise of the large Mesozoic reptiles, Cenozoic mammals diversified rapidly. The many forms that exist today evolved from small primitive mammals that were characterized by short legs; flat, five-toed feet, and small brains. Their development and specialization took four principal directions: (1) increase in size, (2) increase in brain capacity, (3) specialization of teeth to better accommodate their diet, and (4) specialization of limbs to be better equipped for a particular lifestyle or environment.

Marsupial and Placental Mammals Two groups of mammals, the marsupials and the placentals, evolved and diversified during the Cenozoic. The groups differ principally in their modes of reproduction. Young marsupials are born live at a very early stage of development. At birth, the tiny and immature young enter the mother's pouch to suckle and

FIGURE 22.35 Angiosperms, commonly known as flowering plants, are seed-plants that have reproductive structures called flowers and fruits. **A.** The most diverse and widespread of modern plants, many angiosperms display easily recognizable flowers. **B.** Some angiosperms, including grasses, have very tiny flowers. The expansion of the grasslands during the Cenozoic era greatly increased the diversity of grazing mammals and the predators that feed on them. (Photo by **A.** Mike Potts/Nature Picture Library and **B.** Torleif/CORBIS)

A.

B.

UNDERSTANDING EARTH

BOX 22.2

Demise of the Dinosaurs

The boundaries between divisions on the geologic time scale represent times of significant geological and/or biological change. Of special interest is the boundary between the Mesozoic era ("middle life") and the Cenozoic era ("recent life"), about 65.5 million years ago. During this transition about three-quarters of all plant and animal species died out in a *mass extinction.* This

A.

B.

FIGURE 22.B Dinosaurs dominated the Mesozoic landscape until their extinction at the close of the Cretaceous period. **A.** Dinosaurs such as Allosaurus were fearsome predators. (Image by Joe Tuccia-rone/Photo Researchers, Inc.) **B.** These dinosaur footprints near Cameron, Arizona were originally made in mud that eventually became sedimentary rock. (Photo by Tom Bean/Corbis)

complete their development. Today, marsupials are found primarily in Australia, where they underwent a separate evolutionary expansion largely isolated from placental mammals. Modern marsupials include kangaroos, opossums, and koalas (Figure 22.36).

Placental mammals, conversely, develop within the mother's body for a much longer period, so birth occurs when the young are comparatively mature. Most modern mammals, including humans, are placental.

In South America, primitive marsupials and placentals coexisted in isolation for about 40 million years after the breakup of Pangaea. Evolution and specialization of both groups continued undisturbed until about 3 million years ago when the Panamanian land-bridge connected the two American continents. This event permitted the exchange of fauna between the two continents—monkeys, armadillos, sloths, and opossums arrived in North America, while various types of horses, bears, rhinos, camels, and

boundary marks the end of the era in which dinosaurs and other reptiles dominated the landscape and the beginning of the era when mammals assumed that role (Figure 22.B). Because the last period of the Mesozoic is the Cretaceous (abbreviated K to avoid confusion with other "C" periods), and the first period of the Cenozoic is the Tertiary (abbreviated T), the time of this mass extinction is called the *Cretaceous–Tertiary* or *KT boundary.**

The extinction of the dinosaurs is generally attributed to their collective inability to adapt to radical change in environmental conditions. What event could have triggered the rapid extinction of one of the most successful groups of land animals?

The most strongly supported hypothesis proposes that about 65 million years ago, our planet was struck by a large carbonaceous meteorite, a relic from the formation of the solar system. The errant mass of rock was approximately 10 kilometers in diameter and traveled at about 90,000 kilometers per hour at the time of impact. It collided with the southern portion of North America in a shallow tropical sea— now Mexico's Yucatan Peninsula (Figure 22.C). The energy released by the impact is estimated to have been equivalent to 100 million megatons (*mega* = million) of explosives.

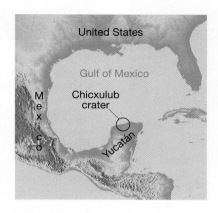

FIGURE 22.C   Chicxulub crater is a giant impact crater that formed about 65 million years ago and has since been filled with sediments. About 180 kilometers (110 miles) in diameter, Chicxulub crater is regarded by some researchers to be the impact site that resulted in the demise of the dinosaurs.

Following the impact, suspended dust greatly reduced sunlight reaching Earth's surface, which resulted in global cooling ("impact winter"), and inhibited photosynthesis, disrupting food production. Long after the dust settled, carbon dioxide, water vapor, and sulfur oxides, added to the atmosphere by the blast, remained. Sulfate aerosols, because of their high reflectivity perpetuated the cooler surface temperatures for a few more years. Eventually, sulfate aerosols dissipated from the atmosphere as acid precipitation. With the aerosols gone, but carbon dioxide still present in large quantities, an enhanced greenhouse effect would have led to a long-term rise in average global temperatures. The likely result was that some of the plant and animal life that survived the initial impact eventually fell victim to stresses associated with global cooling, followed by acid precipitation and global warming.

The extinction of the dinosaurs opened habitats for the small mammals that survived. These new habitats, along with evolutionary forces, led to the development of the large mammals that occupy our modern world.

What evidence points to such a catastrophic collision 65 million years ago? First, a thin layer of sediment nearly 1 centimeter thick has been discovered at the KT boundary, worldwide. This sediment contains a high level of the element *iridium,* rare in Earth's crust but found in high proportions in stony meteorites. Could this layer be the scattered remains of the meteorite responsible for the environmental changes that led to the demise of many reptile groups?

Despite substantial evidence and significant scientific support, some researchers disagree with the impact hypothesis. They suggest instead that huge volcanic eruptions may have led to a breakdown in the food chain. To support this hypothesis, they cite enormous outpourings of lavas in the Deccan Plateau of northern India about 65 million years ago.

Regardless of what caused the KT extinction, its outcomes provide valuable lessons in understanding the role that catastrophic events play in shaping our planet's physical landscape and life-forms.

---

wolves migrated southward. Many animals that had been unique to South America disappeared completely after this event, including hoofed mammals, rhino-sized rodents, and a number of carnivorous marsupials. Because this period of extinction coincided with the formation of the Panamanian land-bridge, it was thought that advanced carnivores from North America were responsible. However, recent research suggests that other factors, including climatic changes, may have played a significant role.

Large Mammals and Extinction

During the rapid mammal diversification of the Cenozoic era, some groups became very large. For example, by the Oligocene epoch, a hornless rhinoceros evolved that stood nearly 5 meters (16 feet) high. It is the largest land mammal known to have existed. As time passed, many other mammals evolved to larger sizes—more, in fact, than now exist. Many of these large forms

FIGURE 22.36 After the breakup of Pangaea, the Australian marsupials evolved differently than their relatives in the Americas. (Photo by Martin Harvey/Peter Arnold Inc.)

were common as recently as 11,000 years ago. However, a wave of late Pleistocene extinctions rapidly eliminated these animals from the landscape.

North America experienced the extinction of mastodons and mammoths, both huge relatives of the modern elephant

FIGURE 22.37 Mammoths, related to modern elephants, were among the large mammals that became extinct at the close of the Ice Age. (Image courtesy of Jonathan Blair/Woodfin Camp)

(see Figure 9.12B, p. 265). In addition, saber-toothed cats, giant beavers, large ground sloths, horses, giant bison, and others died out (Figure 22.37). In Europe, late Pleistocene extinctions included woolly rhinos, large cave bears, and Irish elk. Scientists remain puzzled about the reasons for this recent wave of extinctions that targeted large animals. Having survived several major glacial advances and interglacial periods, it is difficult to ascribe extinctions of these animals to climate change. Some scientists hypothesize that early humans hastened the decline of these mammals by selectively hunting large forms (Figure 22.38).

FIGURE 22.38 Cave painting of animals that early humans encountered about 17,000 years ago. (Photo courtesy of Sisse Brimberg/ National Geographic Society)

CHAPTER 22 EARTH'S EVOLUTION THROUGH GEOLOGIC TIME IN REVIEW

- The history of Earth began about 13.7 billion years ago when the first elements were created during the *Big Bang*. It was from this material, plus other elements ejected into interstellar space by now defunct stars, that Earth, along with the rest of the solar system, formed. As material collected, high velocity impacts of chunks of matter called *planetesimals* and the decay of radioactive elements caused the temperature of our planet to steadily increase. Iron and nickel melted and sank to form the metallic core, while rocky material rose to form the mantle and Earth's initial crust.

- Earth's primitive atmosphere, which consisted mostly of water vapor and carbon dioxide, formed by a process called *outgassing*, which resembles the steam eruptions of modern volcanoes. About 3.5 billion years ago, photosynthesizing bacteria began to release oxygen, first into the oceans and then into the atmosphere. This began the evolution of our modern atmosphere. The oceans formed early in Earth's history as water vapor condensed to form clouds, and torrential rains filled low-lying areas. The salinity in seawater came from volcanic outgassing and from elements weathered and eroded from Earth's primitive crust.

- The *Precambrian*, which is divided into the *Archean* and *Proterozoic eons*, spans nearly 90 percent of Earth's history, beginning with the formation of Earth about 4.6 billion years ago and ending approximately 542 million years ago. During this time, much of Earth's stable continental crust was created through a multi-stage process. First, partial melting of the mantle generated magma that rose to form volcanic island arcs and oceanic plateaus. These thin crustal fragments collided and accreted to form larger crustal provinces, which in turn assembled into larger blocks called *cratons*. Cratons, which form the core of modern continents, were created mainly during the Precambrian.

- Supercontinents are large landmasses that consist of all, or nearly all, existing continents. *Pangaea* was the most recent supercontinent, but other massive continents including an even larger one, *Rodinia*, preceded it. The splitting and reassembling of supercontinents have generated most of Earth's major mountain belts. In addition, the movements of these crustal blocks have profoundly affected Earth's climate, and caused sea level to rise and fall.

- The time span following the close of the Precambrian, called the *Phanerozoic eon*, encompasses 542 million years and is divided into three eras: *Paleozoic, Mesozoic,* and *Cenozoic.* The Paleozoic era was dominated by continental collisions as the supercontinent of Pangaea assembled—forming the Caledonian, Appalachian, and Ural Mountains. Early in the Mesozoic, much of the land was above sea level. However, by the middle Mesozoic, seas invaded western North America. As Pangaea began to break up, the westward-moving North American plate began to override the Pacific plate, causing crustal deformation along the entire western margin of North America. Owing to their different relations with

plate boundaries, the eastern and western margins of the continent experienced contrasting events. The stable eastern margin was the site of abundant sedimentation as isostatic adjustment raised the modern Appalachians, causing streams to erode with renewed vigor and deposit their sediment along the continental margin. In the West, the *Laramide Orogeny* (responsible for building the Rocky Mountains) was coming to an end, the Basin and Range Province was forming, and volcanic activity was extensive.

- The first known organisms were single-celled bacteria, *prokaryotes,* which lack a nucleus. One group of these organisms, called cyanobacteria, used solar energy to synthesize organic compounds (sugars). For the first time, organisms had the ability to produce their own food. Fossil evidence for the existence of these bacteria includes layered mounds of calcium carbonate called *stromatolites.*

- The beginning of the Paleozoic is marked by the *appearance of the first life-forms with hard parts* such as shells. Therefore, abundant fossils occur, and a far more detailed record of Paleozoic events can be constructed. Life in the early Paleozoic was restricted to the seas and consisted of several invertebrate groups, including trilobites, cephalopods, sponges, and corals. During the Paleozoic, organisms diversified dramatically. Insects and plants moved onto land, and lobe-finned fishes that adapted to land became the first amphibians. By the Pennsylvanian period, large tropical swamps, which became the major coal deposits of today, extended across North America, Europe, and Siberia. At the close of the Paleozoic, a mass extinction destroyed 70 percent of all vertebrate species on land and 90 percent of all marine organisms.

- The Mesozoic era, literally the era of middle life, is often called the *"Age of Reptiles."* Organisms that survived the extinction at the end of the Paleozoic began to diversify in spectacular ways. *Gymnosperms* (cycads, conifers, and ginkgoes) became the dominant trees of the Mesozoic because they could adapt to the drier climates. Reptiles became the dominant land animals. The most awesome of the Mesozoic reptiles were the *dinosaurs.* At the close of the Mesozoic, many large reptiles, including the dinosaurs, became extinct.

- The Cenozoic is often called the *"Age of Mammals"* because these animals replaced the reptiles as the dominant vertebrate life-forms on land. Two groups of mammals, the marsupials and the placentals, evolved and expanded during this era. One tendency was for some mammal groups to become very large. However, a wave of late *Pleistocene* extinctions rapidly eliminated these animals from the landscape. Some scientists suggest that early humans hastened their decline by selectively hunting the larger animals. The Cenozoic could also be called the *"Age of Flowering Plants."* As a source of food, flowering plants (angiosperms) strongly influenced the evolution of both birds and herbivorous (plant-eating) mammals throughout the Cenozoic era.

KEY TERMS

banded iron formations (p. 616)
cratons (p. 619)
eukaryotes (p. 628)
outgassing (p. 614)
planetesimals (p. 614)
prokaryotes (p. 625)
protoplanets (p. 614)
solar nebula (p. 614)
stromatolites (p. 628)
supercontinent (p. 620)
supercontinent cycle (p. 620)
supernova (p. 614)

QUESTIONS FOR REVIEW

1. Why is Earth's molten, metallic core important to humans living today?

2. What two elements made up most of the very early universe?

3. What is the cataclysmic event called in which an exploding star produces all of the elements heavier than iron?

4. Briefly describe the formation of the planets from the solar nebula.

5. What is meant by outgassing and what modern phenomenon serves that role today?

6. Outgassing produced Earth's early atmosphere, which was rich in what two gases?

7. Why is the evolution of a type of bacteria that employed photosynthesis to produce food important to most modern organisms?

8. What was the source of water for the first oceans?

9. How does the ocean remove carbon dioxide from the atmosphere? What role do tiny marine organisms, such as foraminifera, play?

10. Explain why Precambrian history is more difficult to decipher than more recent geological history.

11. Briefly describe how cratons come into being.

12. What is the supercontinent cycle?

13. How can the movement of continents trigger climate change?

14. Match the following words and phrases to the most appropriate time span. Select from the following: *Precambrian, Paleozoic, Mesozoic, Cenozoic.*

    a. Pangaea came into existence.

    b. First trace fossils.

    c. The era that encompasses the least amount of time.

    d. Earth's major cratons formed.

    e. "Age of Dinosaurs."

    f. Formation of the Rocky Mountains.

    g. Formation of the Appalachian Mountains.

    h. Coal swamps extended across North America, Europe, and Siberia.

    i. Gulf Coast oil deposits formed.

    j. Formation of most of the world's major iron-ore deposits.

    k. Massive sand dunes covered a large portion of the Colorado Plateau region.

    l. The "Age of the Fishes" occurred during this span.

    m. Pangaea began to break apart and disperse.

    n. "Age of Mammals."

    o. Animals with hard parts first appeared in abundance.

    p. Gymnosperms were the dominant trees.

    q. Stromatolites were abundant.

    r. Fault-block mountains formed in the Basin and Range region.

15. Contrast the eastern and western margins of North America during the Cenozoic era in terms of their relationships to plate boundaries.

16. What did plants have to overcome to move onto land?

17. What group of animals is thought to have left the ocean to become the first amphibians?

18. Why are amphibians not considered "true" land animals?

19. What major development allowed reptiles to move inland?

20. What event is thought to have ended the reign of the dinosaurs?

COMPANION WEBSITE

The *Earth 10e* Web site uses the resources and flexibility of the Internet to aid in your study of the topics in this chapter. Written and developed by the authors and other geology instructors, this site will help improve your understanding of geology. Visit **www.mygeoscienceplace.com** in order to:

- **Review** key chapter concepts.

- **Read** with links to the eBook and to chapter-specific web resources.
- **Test** yourself with online quizzes.

23

ENERGY AND MINERAL RESOURCES

Potash mine evaporation ponds west of Moab, Utah. Read the "Students Sometimes Ask" on p. 667 to learn more about this place.

(PHOTO BY MICHAEL COLLIER)

Materials that we extract from Earth are the basis of modern civilization (Figure 23.1). Mineral and energy resources from the crust are the raw materials from which the products used by society are made. Like most people who live in highly industrialized nations, you may not realize the quantity of resources needed to maintain your present standard of living. Figure 23.2 shows the annual per capita consumption of several important metallic and nonmetallic mineral resources for the United States. This is each person's prorated share of the materials required by industry to provide the vast array of homes, cars, electronics, cosmetics, packaging, and so on that modern society demands. Figures for other highly industrialized countries, such as Canada, Australia, and several nations in Western Europe are comparable.

The number of different mineral resources required by modern industries is large. Although some countries, including the United States, have substantial deposits of many important minerals, no nation is entirely self-sufficient. This reflects the fact that important deposits are limited in number and localized in occurrence. All countries must rely on international trade to fulfill at least some of their needs.

RENEWABLE AND NONRENEWABLE RESOURCES

Resources are commonly divided into two broad categories—renewable and nonrenewable. **Renewable resources** can be replenished over relatively short time spans such as months, years, or decades. Common examples are plants and animals for food, natural fibers for clothing, and trees for lumber and paper. Energy from flowing water, wind, and the Sun are also considered renewable.

By contrast, **nonrenewable resources** continue to be formed in Earth, but the processes that create them are so slow that significant deposits take millions of years to accumulate. For human purposes, Earth contains fixed quantities of these substances. When the present supplies are mined or pumped from the ground, there will be no more. Examples are fuels (coal, oil, natural gas) and many important metals (iron, copper, uranium, gold). Some of these nonrenewable resources, such as aluminum, can be used over and over again; others, such as oil, cannot be recycled.

Occasionally some resources can be placed in either category, depending on how they are used. Groundwater is one such example. Where it is pumped from the ground at a rate that can be replenished, groundwater can be classified as a renewable resource. However, in places where groundwater is withdrawn faster than it is replenished, the water table drops steadily. In this case, the groundwater is being "mined" just like other non-renewable resources.*

Figure 23.3 shows that the population of our planet is growing rapidly. Although the number did not reach 1 billion until the beginning of the 19th century, just 130 years later the population doubled to 2 billion. Between 1930 and 1975 the figure doubled again to 4 billion, and by 2015 more than 7 billion people will inhabit the planet. Clearly, as population grows, the demand for resources expands as well. However, the rate of mineral-and-energy resource usage has climbed faster than population growth. This results from an increasing standard of living. In the United States only 6 percent of the world's population uses approximately 30 percent of the world's annual production of mineral and energy resources!

How long can our remaining resources sustain the rising standard of living in today's industrialized countries and still provide for the growing needs of developing regions? How much environmental deterioration are we willing to accept in pursuit of resources? Can alternatives be found? If we are to cope with an increasing per capita demand and a growing world population, we must understand our resources and their limits.

---

*The problem of declining water-table levels is discussed in Chapter 17.

FIGURE 23.1 As this scene of Houston, Texas, reminds us, mineral and energy resources are the basis of modern civilization. (Photo by H. R. Bramaz/Peter Arnold, Inc.)

ENERGY RESOURCES

Coal, petroleum, and natural gas are the primary fuels of our modern industrial economy (Figure 23.4). About 84 percent of the energy consumed in the United States today comes from these basic fossil fuels. Although major shortages of oil and gas will not occur for many years, proven reserves are declining. Despite new exploration, even in very remote regions and severe environments, new sources are not keeping pace with consumption.

Unless large, new petroleum reserves are discovered (which is possible, but not likely), a greater share of our future needs will have to come from coal and from alternative energy sources such as nuclear, geothermal, solar, wind, tidal, and hydroelectric power (Box 23.1). Two fossil-fuel alternatives—oil sands and oil shale—are sometimes mentioned as promising sources of liquid fuels. In the following sections, we will briefly examine the fuels that have traditionally supplied our energy needs, as well as sources that will provide an increasing share of our future requirements.

Coal

Along with oil and natural gas, coal is commonly called a **fossil fuel.** Such a designation is appropriate because each time we burn coal we are using energy from the Sun that was stored by plants many millions of years ago. We are indeed burning a "fossil."

Coal has been an important fuel for centuries. In the 19th and early 20th centuries, cheap and plentiful coal powered the Industrial Revolution. By 1900 coal was providing 90 percent of the energy used in the United States. Although still important, coal currently accounts for about 23 percent of the nation's energy needs (see Figure 23.4).

Until the 1950s, coal was an important domestic heating fuel as well as a power source for industry. However, its direct use in the home has been largely replaced by oil, natural gas, and electricity. These fuels are preferred because they are more readily available (delivered via pipes, tanks, or wiring) and cleaner to use.

Nevertheless, coal remains the major fuel used in power plants to generate electricity, and it is therefore indirectly an important source of energy for our homes. Coal supplies more than half the electricity consumed in the United States. As oil reserves gradually diminish in the years to come, the use of coal may increase. Expanded coal production is possible because the world has enormous reserves and the technology to mine coal efficiently. In the United States, coal fields are widespread and contain supplies that should last for hundreds of years (Figure 23.5).

Although coal is plentiful, its recovery and use present a number of problems. Surface mining can turn the countryside into a scarred wasteland if careful (and costly) reclamation is not carried out to restore the land. (Today all U.S. surface mines must reclaim the land.) Although underground mining does not

Nonmetallic Resources

5713 kg (12695 lbs)
Stone

4025 kg (8945 lbs)
Sand and gravel

360 kg (790 lbs)
Cement

137 kg (304 lbs)
Clays

178 kg (395 lbs)
Salt

162 kg (361 lbs)
Phosphate rock

302 kg (672 lbs)
Other nonmetals

Metallic Resources

35 kg (77 lbs)
Aluminum

6 kg (14 lbs)
Lead

6 kg (13 lbs)
Manganese

249 kg (553 lbs)
Iron

11 kg (25 lbs)
Copper

5 kg (11 lbs)
Zinc

9 kg (20 lbs)
Other metals

Energy Resources

3500 kg (7700 lbs)
Petroleum

3700 kg (8140 lbs)
Coal

3850 kg (8470 lbs)
Natural gas

FIGURE 23.2 The annual per capita consumption of nonmetallic and metallic mineral resources for the United States is about 11,000 kilograms (12 tons)! About 97 percent of the materials used are nonmetallic. The per capita use of oil, coal, and natural gas exceeds 11,000 kilograms. (After U.S. Geological Survey)

scar the landscape to the same degree, it has been costly in terms of human life and health.

Moreover, underground mining long ago ceased to be a pick-and-shovel operation and is today a highly mechanized and computerized process (Figure 23.6). Strong federal safety regulations have made U.S. mining quite safe. However, the hazards of collapsing roofs, gas explosions, and working with heavy equipment remain.

Air pollution is a major problem associated with the burning of coal. Much coal contains significant quantities of sulfur. Despite efforts to remove sulfur before the coal is burned, some remains; when the coal is burned, the sulfur is converted into noxious sulfur oxide gases. Through a series of complex chemical reactions in the atmosphere, the sulfur oxides are converted to sulfuric acid, which then falls to Earth's surface as rain or snow. This acid precipitation can have adverse ecological effects over widespread areas (see Box 6.2, pp. 182–83).

As is the case when other fossil fuels are burned, the combustion of coal produces carbon dioxide. This major "greenhouse gas" plays a significant role in the heating of our atmosphere. Chapter 21, "Global Climate Change," examines this issue in some detail.

OIL AND NATURAL GAS

Petroleum and natural gas are found in similar environments and frequently occur together. Both consist of various hydrocarbon compounds (compounds consisting of hydrogen and carbon mixed together). They may also contain small quantities of other elements, such as sulfur, nitrogen, and oxygen. Like coal, petroleum and natural gas are biological products derived from the remains of organisms. However, the environments in which they form are very different, as are the organisms. Coal is formed mostly from plant material that accumulated in a swampy environment above sea level. Oil and gas are derived from the remains of both plants and animals having a marine origin.

Petroleum Formation

Petroleum formation is complex and not completely understood. Nonetheless, we know that it begins with the accumulation of sediment in ocean areas that are rich in plant and animal remains. These accumulations must occur where biological activity is high, such as in nearshore areas. However, most marine environments are oxygen-rich, which leads to the decay of organic remains before they can be buried by other sediments. Therefore, accumulations of oil and gas are not as widespread as are the marine environments that support abundant biological activity. This limiting factor notwithstanding, large quantities of organic matter are buried and protected from oxidation in many offshore sedimentary basins. With increasing burial over millions of years, chemical reactions gradually transform some of the original organic matter into the liquid and gaseous hydrocarbons we call petroleum and natural gas.

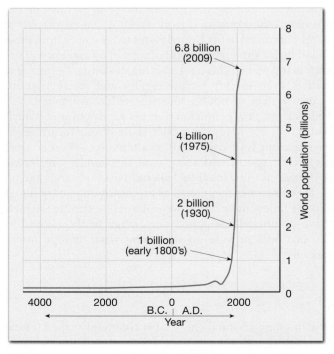

FIGURE 23.3 Growth of world population. It took until 1800 for the number to reach 1 billion. By the year 2015, more than 7 billion people may inhabit the planet. The demand for basic resources is growing faster than the rate of population increase. (Data from the Population Reference Bureau)

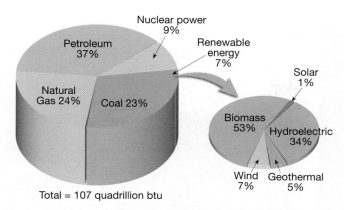

FIGURE 23.4 U.S. energy consumption, 2008. The total was nearly 107 quadrillion Btu. A quadrillion, by the way, is 10 raised to the 12th power, or a million million—a quadrillion Btu is a convenient unit for referring to U.S. energy use as a whole. (Source: U.S. Department of Energy, Energy Information Administration)

Unlike the organic matter from which they formed, the newly created petroleum and natural gas are mobile. These fluids are gradually squeezed from the compacting, mud-rich layers where they originate into adjacent permeable beds such as sandstone, where openings between sediment grains are larger. Because this occurs under water, the rock layers containing the oil and gas are saturated with water. But oil and gas are less dense than water, so they migrate upward through the water-filled pore spaces of the enclosing rocks. Unless something halts this upward migration, the fluids will eventually reach the surface, at which point the volatile components will evaporate.

Oil Traps

Sometimes the upward migration is halted. A geologic environment that allows for economically significant amounts of oil and gas to accumulate underground is termed an **oil trap.** Several geologic structures may act as oil traps, but all have two basic conditions in common: a porous, permeable **reservoir rock** that will yield petroleum and natural gas in sufficient quantities to make drilling worthwhile; and a **cap rock,** such as shale, that is virtu-

ally impermeable to oil and gas. The cap rock halts the upwardly mobile oil and gas and keeps the oil and gas from escaping at the surface (Figure 23.7).

Figure 23.8 illustrates some common oil and natural-gas traps. One of the simplest traps is an *anticline,* an uparched series of sedimentary strata (Figure 23.8A). As the strata are bent, the rising oil and gas collect at the apex (top) of the fold. Because of its lower density, the natural gas collects above the oil. Both rest upon the denser water that saturates the reservoir rock. One of the world's largest oil fields, El Nala in Saudi Arabia, is the result of an anticlinal trap, as is the famous Teapot Dome in Wyoming.

Fault traps form when strata are displaced in such a manner as to bring a dipping reservoir rock into position opposite an impermeable bed, as shown in Figure 23.8B. In this case the upward migration of the oil and gas is halted where it encounters the fault.

In the Gulf coastal plain region of the United States, important accumulations of oil occur in association with *salt domes.* Such areas have thick accumulations of sedimentary strata, including layers of rock salt. Salt occurring at great depths has been forced to rise in columns by the pressure of overlying beds. These rising salt columns gradually deform the overlying strata. Because oil and gas migrate to the highest level possible, they accumulate in the upturned sandstone beds adjacent to the salt column (Figure 23.8C).

Yet another important geologic circumstance that may lead to significant accumulations of oil and gas is termed a *stratigraphic trap.* These oil-bearing structures result primarily from the original pattern of sedimentation rather than structural deformation. The stratigraphic trap illustrated in Figure 23.8D exists because a sloping bed of sandstone thins to the point of disappearance.

When the lid created by the cap rock is punctured by drilling, the oil and natural gas, which are under pressure,

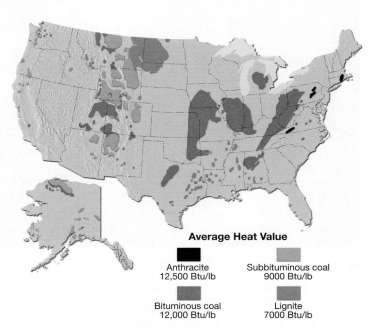

FIGURE 23.5 Coal fields of the United States. (Data from the U.S. Geological Survey)

A.

B.

FIGURE 23.6 **A.** Modern underground coal mining is highly mechanized and relatively safe. (Photo by Melvin Grubb/Grubb Photo Services, Inc.) **B.** Strip mining of coal at Black Mesa, Arizona. Surface mining is common when coal seams are near the surface. (Photo by Richard W. Brooks/Photo Researchers, Inc.)

A.

B.

FIGURE 23.7 Oil accumulates in oil traps that consist of porous, permeable *reservoir rock* overlain by an impermeable *cap rock.* **A.** Modern offshore oil production platform in the North Sea. (Photo by Peter Bowater/Photo Researchers, Inc.) **B.** The first successful oil well was completed by Edwin Drake (right) on August 27, 1859, near Titusville, Pennsylvania. The oil-bearing reservoir rock was encountered at a depth of 21 meters (69 feet). (Photo by CORBIS/Bettman)

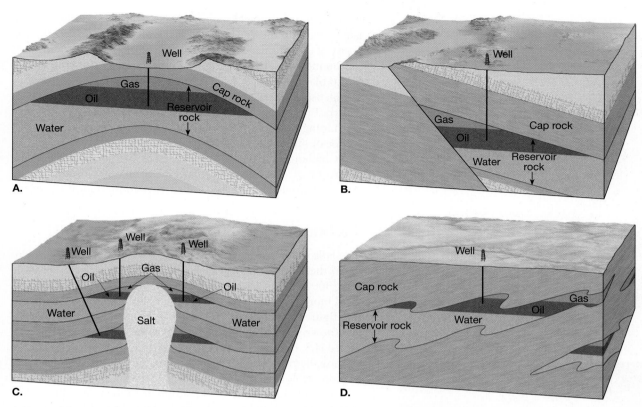

FIGURE 23.8 Common oil traps. **A.** Anticline. **B.** Fault trap. **C.** Salt dome. **D.** Stratigraphic (pinch-out) trap. Also see Figure 12.4, p. 336.

migrate from the pore spaces of the reservoir rock to the drill hole. On rare occasions, when fluid pressure is great, it may force oil up the drill hole to the surface, causing a "gusher" or oil fountain at the surface. Usually, however, a pump is required to lift the oil out.

A drill hole is not the only means by which oil and gas can escape from a trap. Traps can be broken by natural forces. For example, Earth movements may create fractures that allow the hydrocarbon fluids to escape. Surface erosion may breach a trap with similar results. The older the rock strata, the greater the chance that deformation or erosion has affected a trap. Indeed, not all ages of rock yield oil and gas in the same proportions. The greatest production comes from the youngest rocks, those of Cenozoic age. Older Mesozoic rocks produce considerably less, followed by even smaller yields from the still older Paleozoic strata. There is virtually no oil produced from the most ancient rocks, those of Precambrian age.

OIL SANDS AND OIL SHALE— PETROLEUM FOR THE FUTURE?

In years to come, world oil supplies will dwindle. When this occurs, a lower grade of hydrocarbon may have to be substituted. Are the fuels derived from oil sands and oil shales good candidates?

Oil Sands

Oil sands are usually mixtures of clay and sand combined with water and varying amounts of a black, highly viscous tar known as *bitumen*. The use of the term *sand* can be misleading, because not all deposits are associated with sands and sandstones. Some occur in other materials, including shales and limestones. The oil in these deposits is very similar to heavy crude oils pumped from wells. The major difference between conventional oil reservoirs and oil sand deposits is in the viscosity (resistance to flow) of the oil they contain. In oil sands, the oil is much more viscous and cannot simply be pumped out.

Substantial oil sand deposits occur in several locations around the world. The two largest are the Athabasca Oil Sands in the Canadian province of Alberta and the Orinoco River deposit in Venezuela (Figure 23.9).*

Currently, oil sands are mined at the surface in a manner similar to the strip mining of coal. The excavated material is then heated with pressurized steam until the bitumen softens and rises. Once collected, the oily material is treated to remove impurities and then hydrogen is added. This last step upgrades the material to a synthetic crude, which can then be refined. Extracting and refining oil sands requires a great deal of energy—nearly

---

*An interesting account describing Canada's oil sands industry is "Unconventional Crude," by Elizabeth Kolbert in *The New Yorker*, November 12, 2007, pp. 46–51.

FIGURE 23.9 **A.** In North America, the largest oil sand deposits occur in the Canadian province of Alberta. Known as the Athabasca Oil Sands, these deposits cover an area of more than 42,000 square kilometers (16,300 square miles). Alberta's major oil sands deposits contain more than 1.7 trillion barrels of bitumen. However, much of the bitumen cannot be recovered at reasonable cost. With current technology, it is estimated that only about 300 billion barrels are recoverable. **B.** Mining of oil sands in Alberta. Notice the "tar" on the hands of the worker. (Photo by Burkard/Bilderberg/Peter Arnold, Inc.)

half as much as the end product yields! Nevertheless, oil sands from Alberta's vast deposits are the source of about 45 percent of Canada's oil production.

Obtaining oil from oil sand has significant environmental drawbacks. Substantial land disturbance is associated with mining huge quantities of rock and sediment. Moreover, large quantities of water are required for processing, and when processing is completed, contaminated water and sediment accumulate in toxic disposal ponds.

About 80 percent of the oil sands in Alberta are buried too deeply for surface mining. Oil from these deep deposits must be

recovered by *in situ* techniques. Using drilling technology, steam is injected into the deposit to heat the oil sand, reducing the viscosity of the bitumen. The hot, mobile bitumen migrates towards producing wells, which pump it to the surface, while the sand is left in place ("*in situ*" is Latin for "in place"). *In situ* technology is expensive and requires certain conditions like a nearby water source. Production using *in situ* techniques already rivals open pit mining and in the future will replace mining as the main source of bitumen production from the oil sands.

Challenges facing *in situ* processes include increasing the efficiency of oil recovery, management of water used to make steam, and reducing the costs of energy required for the process.

Oil Shale

Oil shale contains enormous amounts of untapped oil. Worldwide, the U.S. Geological Survey estimates that there are more than 3000 billion barrels of oil contained in shales that would yield more than 38 liters (10 gallons) of oil per ton of shale. But this figure is misleading because fewer than 200 billion barrels are known to be recoverable with present technology. Still, estimated U.S. resources are about 14 times as great as those of conventionally recoverable oil, and estimates will probably increase as more geological information is gathered.

Roughly half of the world supply is in the Green River Formation in Colorado, Utah, and Wyoming (Figure 23.10). Within this region the oil shales are part of sedimentary layers that accumulated at the bottoms of two vast, shallow lakes during the Eocene epoch (57 million to 36 million years ago).

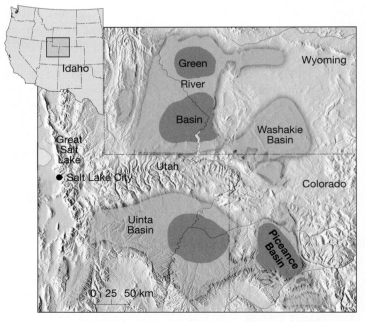

FIGURE 23.10 Distribution of oil shale in the Green River Formation of Colorado, Utah, and Wyoming. The areas shaded with the darker color represent the richest deposits. Government and industry have invested large sums to make these oil shales an economic resource, but costs have always been higher than the price of oil. However, as prices of competing fuels rise, these vast deposits may someday become economically more attractive. (After D. C. Duncan and V. E. Swanson, U.S. Geological Survey Circular 523, 1965)

UNDERSTANDING **EARTH**

BOX 23.1

Gas Hydrates— A Fuel from Ocean- Floor Sediments

Gas hydrates are unusually compact chemical structures made of water and natural gas. The most common type of natural gas is methane, which produces *methane hydrate*. Gas hydrates occur beneath permafrost areas on land and under the ocean floor at depths below 525 meters (1720 feet).

Most oceanic gas hydrates are created when bacteria break down organic matter trapped in seafloor sediments, producing methane gas

FIGURE 23.A This sample retrieved from the ocean floor shows layers of white, ice-like gas hydrate mixed with mud. (Photo courtesy of GEOMAR Research Center, Kiel, Germany)

with minor amounts of ethane and propane. These gases combine with water in deep-ocean sediments (where pressures are high and temperatures are low) in such a way that the gas is trapped inside a latticelike cage of water molecules.

Vessels that have drilled into gas hydrates have retrieved cores of mud mixed with chunks or layers of gas hydrates (Figure 23.A) that fizzle and evaporate quickly when they are exposed to the relatively warm, low-pressure conditions at the ocean surface. Gas hydrates resemble chunks of ice but ignite when lit by a flame because methane and other flammable gases are released as gas hydrates vaporize (Figure 23.B).

Some estimates indicate that as much as 20 quadrillion cubic meters (700 quadrillion cubic feet) of methane are locked up in sediments containing gas hydrates. This is equivalent to about *twice* as much carbon as Earth's coal, oil, and conventional gas reserves combined, so gas hydrates have great potential. One major drawback in exploiting reserves of gas hydrate is that they rapidly decompose at surface temper-

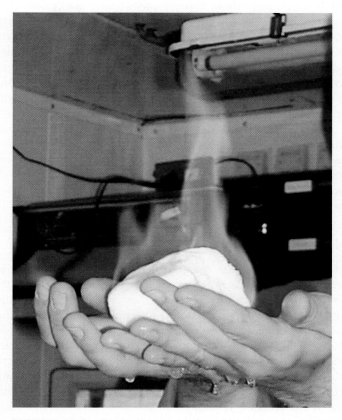

FIGURE 23.B Gas hydrates evaporate when exposed to surface conditions and release natural gas, which can be ignited. (Photo courtesy of GEOMAR Research Center, Kiel, Germany)

atures and pressures. In the future, however, these vast seafloor reserves of energy may help power modern society.

Oil shale has been suggested as a partial solution to dwindling fuel supplies. However, the heat energy in oil shale is about one-eighth that in crude oil, owing to the large proportion of mineral matter in the shales.

Students Sometimes Ask . . .

Are "oil sands" and "tar sands" the same thing?

Yes, the term "tar sand" is often used to describe bitumen deposits, but it is inaccurate because tar is a man-made substance produced by the distillation of organic material. Bitumen looks like tar, but it is a naturally-occurring substance. "Oil sands" is a more appropriate term for bitumen deposits.

This mineral matter adds cost to mining, processing, and waste disposal. Producing oil from oil shale has the same problems as producing oil from oil sands. Surface mining causes widespread land disturbance and presents significant waste-disposal problems. Moreover, processing requires large amounts of water, something that is in short supply in the semiarid region occupied by the Green River Formation.

At present, oil is plentiful and relatively inexpensive on world markets. Therefore, with current technologies, most oil shales are not worth mining. Oil shale research-and-development efforts have been nearly abandoned by industry. Nevertheless, the U.S. Geological Survey suggests that the large amount of oil that potentially can be extracted from oil shales in the United States probably ensures its eventual inclusion in the national energy mix.

ALTERNATE ENERGY SOURCES

An examination of Figure 23.4 clearly shows that we live in the era of fossil fuels. About 84 percent of United State's energy needs comes from these nonrenewable resources. Present estimates indicate that the amount of recoverable fossil fuels may equal 10 trillion barrels of oil, which is enough to last 170 years at the present rate of consumption. Of course, as world population soars, the rate of consumption will climb. Thus, reserves will eventually be in short supply. In the meantime, the environmental impact of burning huge quantities of fossil fuels will undoubtedly have an adverse effect.

How can a growing demand for energy be met without radically affecting the planet we inhabit? Although no clear answer has yet been formulated, the need to rely more heavily on alternate energy sources must be considered. In this section, we will examine several possible sources, including nuclear, solar, wind, hydroelectric, geothermal, and tidal energy.

Nuclear Energy

About 9 percent of the energy demand of the United States is being met by nuclear power plants. The fuel for these facilities comes from radioactive materials that release energy by the process of **nuclear fission.** Fission is accomplished by bombarding the nuclei of heavy atoms, commonly uranium-235, with neutrons. This causes the uranium nuclei to split into smaller nuclei and to emit neutrons and heat energy. The ejected neutrons, in turn, bombard the nuclei of adjacent uranium atoms, producing a *chain reaction*. If the supply of fissionable material is sufficient and if the reaction is allowed to proceed in an uncontrolled manner, an enormous amount of energy would be released in the form of an atomic explosion.

In a nuclear power plant, the fission reaction is controlled by moving neutron-absorbing rods into or out of the nuclear reactor. The result is a controlled nuclear chain reaction that releases great amounts of heat. The energy produced is transported from the reactor and used to drive steam turbines that turn electrical generators, which is similar to what occurs in most conventional power plants.

Uranium   Uranium-235 is the only naturally occurring isotope that is readily fissionable and is therefore the primary fuel used in nuclear power plants.* Although large quantities of uranium ore have been discovered, most contain less than 0.05 percent uranium. Of this small amount, 99.3 percent is the nonfissionable isotope uranium-238 and just 0.7 percent consists of the fissionable isotope uranium-235. Because most nuclear reactors operate with fuels that are at least 3 percent uranium-235, the two isotopes must be separated in order to concentrate the fissionable uranium-235. The process of separating the uranium isotopes is difficult and substantially increases the cost of nuclear power.

Although uranium is a rare element in Earth's crust, it does occur in enriched deposits. Some of the most important occur-

FIGURE 23.11   Diablo Canyon nuclear power plant near San Luis Obispo, California. Reactors are in the dome-shaped buildings, and cooling water is being released to the ocean. The setting of this facility was controversial because of its close proximity to faults capable of producing potentially damaging earthquakes. (Photo by Comstock)

rences are associated with what are believed to be ancient placer deposits in stream beds.** For example, in Witwatersrand, South Africa, grains of uranium ore (as well as rich gold deposits) were concentrated by virtue of their high density in rocks made largely of quartz pebbles. In the United States, the richest uranium deposits are found in Jurassic and Triassic sandstones in the Colorado Plateau and in younger rocks in Wyoming. Most of these deposits have formed through the precipitation of uranium compounds from groundwater. Here, precipitation of uranium occurs as a result of a chemical reaction with organic matter, as evidenced by the concentration of uranium in fossil logs and organic-rich black shales.

Obstacles to Development   At one time nuclear power was heralded as the clean, cheap source of energy to replace fossil fuels. However, several obstacles have emerged to hinder the development of nuclear power as a major energy source. Not the least of these is the skyrocketing costs of building nuclear facilities that contain numerous safety features. More important, perhaps, is the concern over the possibility of a serious accident at one of the nearly 200 nuclear plants in existence worldwide (Figure 23.11). The 1979 accident at Three Mile Island near Harrisburg, Pennsylvania, helped bring this point home. Here, a malfunction led the plant operators to believe there was too much water in the primary system instead of too little. This confusion allowed the reactor core to lie uncovered for several hours. Although there was little danger to the public, substantial damage to the reactor resulted.

Unfortunately, the 1986 accident at Chernobyl in the former Soviet Union was far more serious. In this incident, the reactor ran out of control and two small explosions lifted the roof off the structure, allowing pieces of uranium to be thrown

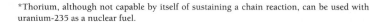

*Thorium, although not capable by itself of sustaining a chain reaction, can be used with uranium-235 as a nuclear fuel.

**Placer deposits are discussed later in the chapter.

over the immediate area. During the 10 days that it took to quench the fire that ensued, high levels of radioactive material were carried by the atmosphere and detected as far away as Norway. In addition to the 18 people who died within six weeks of the accident, many thousands more face an increased risk of death from cancers associated with the fallout.

It should be emphasized that the concentrations of fissionable uranium-235 and the design of reactors are such that nuclear power plants cannot explode like an atomic bomb. The dangers arise from the possible escape of radioactive debris during a meltdown of the core or other malfunction. In addition, hazards such as the disposal of nuclear waste and the relationship that exists between nuclear energy programs and the proliferation of nuclear weapons must be considered as we evaluate the pros and cons of employing nuclear power.

Among the "pros" for nuclear energy is the fact that nuclear power plants do not emit carbon dioxide—the greenhouse gas that contributes significantly to global warming (see Chapter 21). By contrast, the generation of electricity from fossil fuels produces large quantities of carbon dioxide. Thus, substituting nuclear power for power generated by fossil fuels represents one option for reducing carbon emissions.

Solar Energy

The term *solar energy* generally refers to the direct use of the Sun's rays to supply energy for the needs of people. The simplest and perhaps most widely used *passive solar collectors* are south-facing windows. As sunlight passes through the glass, its energy is absorbed by objects in the room. These objects in turn radiate heat that warms the air. In the United States we often use south-facing windows, along with better insulated and more airtight construction, to reduce heating costs substantially.

More elaborate systems used for home heating involve an *active solar collector.* These roof-mounted devices are usually large, blackened boxes that are covered with a transparent material. The heat they collect can be transferred to where it is needed by circulating air or fluids through piping. Solar collectors are also used successfully to heat water for domestic and commercial needs. For example, solar collectors provide hot water for more than 80 percent of Israel's homes.

Research is currently underway to improve the technologies for concentrating sunlight. One method uses parabolic troughs as solar energy collectors. Each collector resembles a large tube that has been cut in half. Their highly polished surfaces reflect sunlight onto a collection pipe (Figure 23.12A). A fluid (usually oil) runs through the pipe and absorbs the concentrated sunlight. The fluid is heated to over 400 °F and is typically used to make steam that drives a turbine to produce electricity. Parabolic troughs usually have a north-south alignment and gradually rotate to track the Sun.

Another type of collector uses photovoltaic (solar) cells that convert the Sun's energy directly into electricity. Photovoltaic cells are usually connected together to create solar panels in which sunlight knocks electrons into higher energy states, to produce electricity. For many years solar cells were used mainly to power calculators and novelty devices. Today, however, large photovoltaic power stations are connected to electric grids to supplement other power generating facilities. The leading countries in photovoltaic capacity are Germany, Japan, Spain, and the United States, and large power plants have recently been completed in

A.

B.

FIGURE 23.12 **A.** Parabolic troughs, like the one shown here, are very efficient solar collectors. The troughs focus sunlight onto collection pipes filled with a fluid. This heat is typically used to make steam that drives turbines used to generate electricity. (Photo by Schott AG/Commercial Handout) **B.** This solar facility near Los Angeles consists of a 20,000-dish array using Stirling dish technology. (Photo by Randy J. Montoya/Sandia National Laboratories)

Australia and Portugal. The high cost of solar cells makes generating solar electricity more expensive than electricity created by other sources. As the cost of fossil fuels continues to increase, advances in photovoltaic technology should narrow the price differential.

Southern California Edison is nearing completion of a massive 4,500-acre solar-generating facility located 70 miles northeast of Los Angeles. The technology being employed, called the *Stirling dish,* converts thermal energy to electricity by using a mirror array to focus the Sun's rays on the receiver end of a Stirling engine (Figure 23.12B). The internal side of a receiver then heats hydrogen gas, causing it to expand. The pressure created by the expanding gas drives a piston, which turns a small electric generator. When completed in 2011 this facility will consist of a 20,000-dish array that will produce 500 megawatts of electricity, enough to serve 278,000 homes.

Wind Energy

Air has mass, and when it moves (that is, when the wind blows), it contains the energy of that motion—kinetic energy. A portion of that energy can be converted into other forms—mechanical force or electricity—that we can use to perform work (Figure 23.13).

Mechanical energy from wind is commonly used for pumping water in rural or remote places. The "farm windmill," still a familiar sight in many rural areas, is an example. Mechanical energy converted from wind can also be used for other purposes, such as sawing logs, grinding grain, and propelling sailboats. By contrast, wind-powered electric turbines generate electricity for homes, businesses, and for sale to utilities.

Today, modern wind turbines are being installed at breakneck speed. In fact, worldwide, in 2008 the installed wind power capacity increased by more than 47 percent. The year before, capacity had increased 32 percent. Much of this growth was in the United States, China, India, Germany, and Spain. In 2008 the United States passed Germany to become the world's leading producer of wind energy, generating 25,170 megawatts.* Within the next decade, China is expected to produce the most wind-generated electricity.

*One megawatt is enough electricity to supply 250–300 average American households.

A. **B.**

FIGURE 23.13 **A.** Farm windmills are still familiar sights in some areas. Mechanical energy from wind is commonly used to pump water. (Photo by Darren Bennett/Animals Animals–Earth Scenes) **B.** These wind turbines are operating near Palm Springs, California. Although California is the state where significant wind-power development got its start, by 2009 it had been surpassed by Texas and Iowa. (Photo by John Mead/ Science Photo Library/Photo Researchers, Inc.)

TABLE 23.1 The Leading States for Wind Energy Potential

Rank	State	Potential*
1	North Dakota	1,210
2	Texas	1,190
3	Kansas	1,070
4	South Dakota	1,030
5	Montana	1,020
6	Nebraska	868
7	Wyoming	747
8	Oklahoma	725
9	Minnesota	657
10	Iowa	551
11	Colorado	481
12	New Mexico	435
13	Idaho	73
14	Michigan	65
15	New York	62
16	Illinois	61
17	California	59
18	Wisconsin	58
19	Maine	56
20	Missouri	52

*The total amount of electricity that could potentially be generated each year, measured in *billions* of kilowatt hours. A typical American home would use several hundred kilowatt hours per month.
Source: U.S. Department of Energy

Wind speed is a crucial element in determining whether a place is a suitable site for installing a wind-energy facility. Generally a minimum average wind speed of 21 kilometers (13 miles) per hour is necessary for a large-scale wind-power plant to be profitable. A small difference in wind speed results in a large difference in energy production, and therefore a large difference in the cost of the electricity generated. For example, a turbine operating on a site with an average wind speed of 12-mph would generate about 33 percent more electricity than one operating at 11-mph. Also, there is little energy to be harvested at low wind speeds—6-mph winds contain less than one-eighth the energy of 12-mph winds.

The United States has tremendous wind energy resources. Currently 34 states have commercial facilities that produce electricity from wind power. In early 2009 the leading states in installed capacity were Texas, number one, followed by Iowa, California, and Minnesota. Despite the fact that California gave birth to the modern U.S. wind industry, 16 states have greater wind potential. As shown in Table 23.1, the top five states for wind energy potential include North Dakota, Texas, Kansas, South Dakota, and Montana. Although only a small fraction of

U.S. electrical generation currently comes from wind energy, it has been estimated that wind energy potential equals more than twice the total electricity currently consumed (Figure 23.14).

Recent technological improvements have made the cost of wind-generated electricity more competitive. As a result, the growth of installed capacity has grown dramatically. Worldwide, the total amount of installed wind power grew from 7636 megawatts in 1997 to 120,791 megawatts in 2008. This is enough electricity to supply more than 25 million average American households.

The U.S. Department of Energy has announced a goal of obtaining 5 percent of U.S. electricity from wind by the year 2020—a goal that seems consistent with the current growth rate of wind energy nationwide. Thus wind-generated electricity appears to be shifting from being an "alternative" to a "mainstream" energy source.

Hydroelectric Power

Falling water has been an energy source for centuries. Through much of human history, the mechanical energy produced by waterwheels was used to power mills and other machinery. Today, the power generated by falling water is used to drive turbines that produce electricity, hence the term **hydroelectric power.** In the United States, hydroelectric power plants contribute about 5.5 percent of the country's demand. Most of this energy is produced at large dams, which allow for a controlled flow of water (Figure 23.15). The water impounded in a reservoir is a form of stored energy that can be released at any time to produce electricity.

Although water power is considered a renewable resource, the dams built to provide hydroelectricity have finite lifetimes.

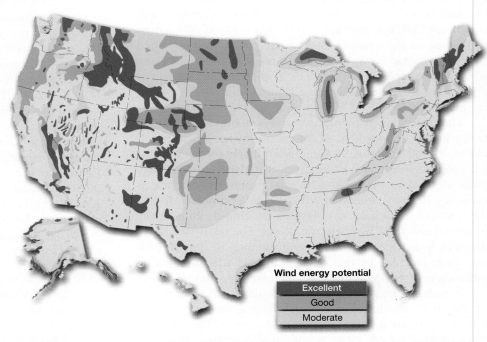

Wind energy potential

| Excellent |
| Good |
| Moderate |

FIGURE 23.14 Wind energy potential for the United States. Large wind energy systems require average wind speeds of about 6 meters per second (13 miles per hour). In the key, "moderate" refers to regions that experience wind speeds of 6.4–6.9 meters per second (m/s), "good" means 7–7.4 m/s, and "excellent" means 7.5 m/s and higher. (After U.S. Department of Energy)

FIGURE 23.15 Lake Powell is the reservoir that was created when Glen Canyon Dam was built across the Colorado River. As water in the reservoir is released, it drives turbines and produces electricity. (Photo by Michael Collier)

reservoir is available to drive turbines and produce electricity to supplement the power supply.

Geothermal Energy

Geothermal energy is harnessed by tapping natural underground reservoirs of steam and hot water. These occur where subsurface temperatures are high, owing to relatively recent volcanic activity. Geothermal energy is put to use in two ways: The steam and hot water are used for heating and to generate electricity.

Iceland is a large volcanic island with current volcanic activity (Figure 23.16). In Iceland's capital, Reykjavik, steam and hot water are pumped into buildings throughout the city for space heating. They also warm greenhouses, where fruits and vegetables are grown all year. In the United States, localities in several western states use hot water from geothermal sources for space heating.

As for generating electricity geothermally, the Italians were first to do so in 1904, so the idea is not new. In 2008, more than 250 geothermal power plants in 24 countries were producing more than 10,000 megawatts (million watts). These plants provide power to more than 60 million people. The leading producers of geothermal power are listed in Table 23.2.

The first commercial geothermal power plant in the United States was built in 1960 at The Geysers, north of San Francisco (Figure 23.17). The Geysers remains the world's largest geothermal power plant, generating nearly 1000 megawatts of U.S. geothermal power. In addition to The Geysers, geothermal development is occurring elsewhere in the western United States, including Nevada, Utah, and the Imperial Valley in southern California. The U.S. geothermal generating capacity of about 3000 megawatts is enough to supply more than 3 million homes. This is an amount of electricity comparable to burning about 60 million barrels of oil each year.

What geologic factors favor a geothermal reservoir of commercial value?

1. *A potent source of heat,* such as a large magma chamber deep enough to ensure adequate pressure and slow cooling, yet not so deep that the natural water circulation is inhibited. Such magma chambers are most likely in regions of recent volcanic activity.
2. *Large and porous reservoirs with channels connected to the heat source,* near which water can circulate and then be stored in the reservoir.

All rivers carry suspended sediment that is deposited behind the dam as soon as it is built. Eventually the sediment will completely fill the reservoir. This takes 50 to 300 years, depending on the quantity of suspended material transported by the river. An example is Egypt's huge Aswan High Dam, which was completed in the 1960s. It is estimated that half of the reservoir will be filled with sediment from the Nile River by the year 2025.

The availability of appropriate sites is an important limiting factor in the development of large-scale hydroelectric power plants. A good site provides a significant height for the water to fall and a high rate of flow. Hydroelectric dams exist in many parts of the United States, with the greatest concentrations occurring in the Southeast and the Pacific Northwest. Most of the best U.S. sites have already been developed, limiting the future expansion of hydroelectric power. The total power produced by hydroelectric sources might still increase, but the relative share provided by this source will likely decline because other alternate energy sources will increase at a faster rate.

In recent years a different type of hydroelectric power production has come into use. Called a *pumped water-storage system,* it is actually a type of energy management. During times when demand for electricity is low, unneeded power produced by nonhydroelectric sources is used to pump water from a lower reservoir to a storage area at a higher elevation. Then, when demand for electricity is great, the water stored in the higher

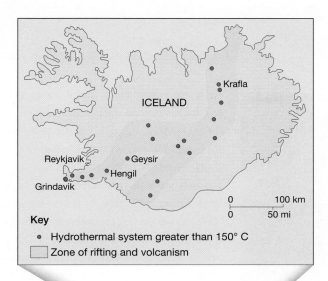

Key

● Hydrothermal system greater than 150° C

▢ Zone of rifting and volcanism

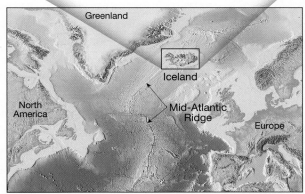

FIGURE 23.16 Iceland straddles the Mid-Atlantic Ridge. This divergent plate boundary is the site of numerous active volcanoes and geothermal systems. Because the entire country consists of geologically young volcanic rocks, warm water can be encountered in holes drilled almost anywhere. More than 45 percent of Iceland's energy comes from geothermal sources. The photo shows a power station in southwestern Iceland. The steam is used to generate electricity. Hot (83 °C) water from the plant is sent via an insulated pipeline to Reykjavik for space heating. (Photo by Simon Fraser/Science Photo Library/Rhoto Researchers, Inc.)

TABLE 23.2 Worldwide Geothermal Power Production, 2007

Producing Country	Megawatts
United States	2687
Philippines	1970
Indonesia	992
Mexico	953
Italy	810
Japan	535
New Zealand	472
Iceland	421
El Salvador	204
Costa Rica	163
All others	525
Total	9732

Source: Geothermal Resources Council.

3. *A cap of low-permeability rocks* that inhibits the flow of water and heat to the surface. A deep, well-insulated reservoir contains much more stored energy than a similar but uninsulated reservoir.

As with other alternative methods of power production, geothermal sources are not expected to provide a high percentage of the world's growing energy needs. Nevertheless, in regions where its potential can be developed, its use will continue to grow.

FIGURE 23.17 The Geysers, near the city of Santa Rosa in northern California, is the world's largest electricity-generating geothermal development. Most of the steam wells are about 3000 meters deep. (AP Photo/Calpine)

Climate and Energy Scientist

Sally Benson has spent her career researching solutions to the most pressing environmental problems of our time. In the mid-1970s, as a young scientist with Lawrence Berkeley National Laboratory, she tackled the first oil shortage by investigating ways to harness the power of geothermal energy.

Ten years later, she was elbow-deep in the mud of Kesterson Reservoir in California's Central Valley. Irrigation runoff had caused selenium from local soil to accumulate in the water, causing local birds to hatch chicks with horrifying deformities. Benson's studies made her realize microbes could be environmentally friendly tools to clean up other sites with toxic metal contamination.

"The experience was so positive because I was doing this cutting edge science at the same time regulators had to make a decision about how to clean up the site. I got an idea of the impact that science can make, and how research could get results," Benson says.

By the mid-1990's, Benson was directing all Earth science research at the laboratory. "I talked with many people, read a lot, and came to the conclusion that climate change was the most significant issue facing the world." Benson organized research programs that developed regional models of climate change to help residents plan for droughts and temperature shifts, and studied the carbon cycle of the oceans, among other projects.

In 2007, Benson was appointed director of the Global Climate and Energy Project at Stanford University, which seeks to develop energy sources that release fewer greenhouse gases. "Energy efficiency in lighting, heating, and cooling systems, and autos, makes all

Professor Sally Benson is Executive Director of Stanford University's Global Climate and Energy Project, a $225 million-dollar effort to develop energy resources that are not harmful to the environment, especially energy sources that release fewer greenhouse gases. Dr. Benson has a multidisciplinary background with degrees in geology, materials science, and minerals engineering, and with applied research in hydrology and reservoir engineering. (Courtesy Dr. Sally Benson)

the sense in the world. But at the end of the day, we need to do a lot more than that. If our current understanding is correct, we need to cut overall emissions by 80 percent of today's levels," Benson states.

Benson sees many promising ways to achieve that goal. One is renewable energy. "Today's biofuels don't provide much advantage in terms of carbon dioxide emissions. But alternatives such as cellulosic ethanol (producing ethanol with plant fibers), where there are low emissions in the process of growing and making them, are incredibly important," Benson opines.

Another is to continue using some fossil fuels to run power plants and ships, but to capture the greenhouse gases they emit. "More than half of the electricity worldwide is produced by burning coal, a plentiful resource in many places. We need to find a way to make it carbon neutral," Benson says. One way to remove such emissions from the atmosphere altogether would be to inject them in aquifers deep within Earth. Benson herself has studied how to use technology developed by the oil and gas industry to select sites where the rock offers a good seal, inject the gas, and monitor the area for leaks.

"I get to do important work solving critical problems, get to be outdoors, do experiments at scale, and have fun while I do it. Interacting with the huge number of people impacted by these issues makes the Earth sciences very rewarding."

> **"I get to do important work . . . and have fun while I do it.**

---Kathleen Wong

Tidal Power

Several methods of generating electrical energy from the oceans have been proposed, but the ocean's energy potential remains largely untapped. The development of tidal power is the principal example of energy production from the ocean.

Tides have been used as a source of power for centuries. Beginning in the 12th century, waterwheels driven by the tides were used to power gristmills and sawmills. During the 17th and 18th centuries, much of Boston's flour was produced at a tidal mill. Today far greater energy demands must be satisfied,

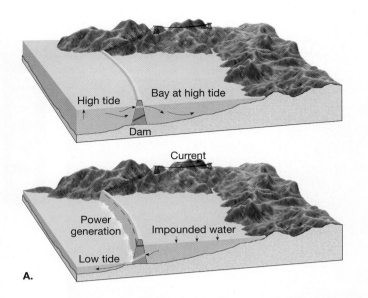

B.

FIGURE 23.18 **A.** Simplified diagram showing the principle of the tidal dam. Electricity is generated only when a sufficient water-height difference exists between the bay and the ocean. **B.** Tidal power plant at Annapolis Royal, Nova Scotia, on the Bay of Fundy. (Photo by James P. Blair/ National Geographic Image Collection)

and more sophisticated ways of using the force created by the perpetual rise and fall of the ocean must be employed.

Tidal power is harnessed by constructing a dam across the mouth of a bay or an estuary in a coastal area having a large tidal range (Figure 23.18A). The narrow opening between the bay and the open ocean magnifies the variations in water level that occur as the tides rise and fall. The strong in-and-out flow that results at such a site is then used to drive turbines and electrical generators.

Tidal energy utilization is exemplified by the tidal power plant at the mouth of the Rance River in France. By far the largest yet constructed, this plant went into operation in 1966 and produces enough power to satisfy the needs of Brittany and also contribute to the demands of other regions. Much smaller experimental facilities have been built near Murmansk in Russia and near Taliang in China, and on the Annapolis River estuary,

an arm of the Bay of Fundy in the Canadian province of Nova Scotia (Figure 23.18B).

Along most of the world's coasts it is not possible to harness tidal energy. If the tidal range is less than 8 meters (25 feet), or if narrow, enclosed bays are absent, tidal power development is uneconomical. For this reason the tides will never provide a very high portion of our ever increasing electrical energy requirements. Nevertheless, the development of tidal power may be worth pursuing at feasible sites because electricity produced by the tides consumes no exhaustible fuels and creates no noxious wastes.

MINERAL RESOURCES

Earth's crust is the source of a wide variety of useful and essential substances. In fact, practically every manufactured product contains substances derived from minerals. Table 23.3 lists some important examples.

Mineral resources are the endowment of useful minerals ultimately available commercially. Resources include already identified deposits from which minerals can be extracted profitably, called **reserves**, as well as known deposits that are not yet economically or technologically recoverable. Deposits inferred to exist but not yet discovered are also considered as mineral resources. In addition, the term **ore** is used to denote those useful metallic minerals that can be mined at a profit. In common usage the term *ore* is also applied to some nonmetallic minerals

TABLE 23.3	Occurrence of Metallic Minerals	
Metal	**Principal Ores**	**Geological Occurrences**
Aluminum	Bauxite	Residual product of weathering
Chromium	Chromite	Magmatic segregation
Copper	Chalcopyrite Bornite Chalcocite	Hydrothermal deposits; contact metamorphism; enrichment by weathering processes
Gold	Native gold	Hydrothermal deposits; placers
Iron	Hematite Magnetite Limonite	Banded sedimentary formations; magmatic segregation
Lead	Galena	Hydrothermal deposits
Magnesium	Magnesite Dolomite	Hydrothermal deposits
Manganese	Pyrolusite	Residual product of weathering
Mercury	Cinnabar	Hydrothermal deposits
Molybdenum	Molybdenite	Hydrothermal deposits
Nickel	Pentlandite	Magmatic segregation
Platinum	Native platinum	Magmatic segregation; placers
Silver	Native silver Argentite	Hydrothermal deposits; enrichment by weathering processes
Tin	Cassiterite	Hydrothermal deposits; placers
Titanium	Ilmenite Rutile	Magmatic segregation; placers
Tungsten	Wolframite Scheelite	Pegmatites; contact metamorphic deposits; placers
Uranium	Uraninite (pitchblende)	Pegmatites; sedimentary deposits
Zinc	Sphalerite	Hydrothermal deposits

PEOPLE AND THE ENVIRONMENT

Utah's Bingham Canyon Mine

In the photo, a mountain once stood where there is now a huge pit (Figure 23.C). This is the world's largest open-pit mine, Bingham Canyon copper mine, about 40 kilometers (25 miles) southwest of Salt Lake City, Utah. The rim is nearly 4 kilometers (2.5 miles) across and covers almost 8 square kilometers (3 square miles). Its depth is 900 meters (3000 feet). If a steel tower were erected at the bottom, it would have to be five times taller than the Eiffel Tower to reach the top of the pit!

It began in the late 1800s as an underground mine for veins of silver and lead. Later copper was discovered. Similar deposits occur at several sites in the American Southwest and in a belt from southern Alaska to northern Chile.

As in other places in this belt, the ore at Bingham Canyon is disseminated throughout *porphyritic* igneous rocks; hence, the name *porphyry copper deposits.* The deposit formed after magma was intruded to shallow depths. Following this, shattering created extensive fractures that were penetrated by hydrothermal solutions from which the ore minerals precipitated.

Although the percentage of copper in the rock is small, the total volume of copper is huge. Ever since open-pit operations started in 1906, some 5 billion tons of material have been removed, yielding more than 12 million tons of

FIGURE 23.C Aerial view of Bingham Canyon copper mine near Salt Lake City, Utah. This huge open-pit mine is about 4 kilometers across and 900 meters deep. Although the amount of copper in the rock is less than 1 percent, the huge volumes of material removed and processed each day (about 200,000 tons) yield significant quantities of metal. (Photo by Michael Collier)

copper. Significant amounts of gold, silver, and molybdenum have also been recovered.

The ore body is far from exhausted today. Over the next 25 years, plans call for an additional 3 billion tons of material to be removed and processed. This largest of artificial excavations has generated most of Utah's mineral production for more than 80 years and has been called the "richest hole on Earth."

Like many older mines, the Bingham pit was unregulated during most of its history. Development occurred prior to the present-day awareness of the environmental impacts of mining and prior to effective environmental legislation. In recent years, problems of groundwater and surface water contamination, air pollution, solid and hazardous wastes, and land reclamation have received needed attention at Bingham Canyon.

---

such as fluorite and sulfur. However, materials used for such purposes as building stone, road aggregate, abrasives, ceramics, and fertilizers are not usually called ores; rather, they are classified as industrial rocks and minerals.

Recall that more than 98 percent of the crust is composed of only eight elements. Except for oxygen and silicon, all other elements make up a relatively small fraction of common crustal rocks (see Figure 3.27, p. 92). Indeed, the natural concentrations of many elements are exceedingly small. A deposit containing the average percentage of a valuable element is worthless if the cost of extracting it exceeds the value of the material recovered. To be considered valuable, an element must be concentrated above the level of its average crustal abundance. Generally, the lower the crustal abundance, the greater the concentration.

For example, copper makes up about 0.0135 percent of the crust. However, for a material to be considered a copper ore, it must contain a concentration that is about 50 times this amount. Aluminum, in contrast, represents 8.13 percent of the crust and must be concentrated to only about four times its average crustal percentage before it can be extracted profitably.

It is important to realize that a deposit may become profitable to extract or lose its profitability because of economic changes. If demand for a metal increases and prices rise, the status of a previously unprofitable deposit changes, and it becomes an ore. The status of unprofitable deposits may also change if a technological advance allows the useful element to be extracted at a lower cost than before. This occurred at the copper-mining operation located at Bingham Canyon, Utah, one of the largest open-pit mines on Earth (Box 23.2). Mining was halted here in 1985 because out-moded equipment had driven the cost of extracting the copper beyond the selling price. The owners responded by replacing an antiquated 1000-car railroad with conveyor belts and pipelines for transporting the ore and waste. These devices achieved a cost reduction of nearly 30 percent and returned this mining operation to profitability.

Over the years, geologists have been keenly interested in learning how natural processes produce localized concentrations of essential metallic minerals. One well-established fact is that occurrences of valuable mineral resources are closely related to the rock cycle. That is, the mechanisms that generate igneous, sedimentary, and metamorphic rocks, including the processes of weathering and erosion, play a major role in producing concentrated accumulations of useful elements. Moreover, with the development of the plate tectonics theory, geologists added yet another tool for understanding the processes by which one rock is transformed into another.

MINERAL RESOURCES AND IGNEOUS PROCESSES

Some of the most important accumulations of metals, such as gold, silver, copper, mercury, lead, platinum, and nickel, are produced by igneous processes (see Table 23.3). These mineral resources, like most others, result from processes that concentrate desirable materials to the extent that extraction is economically feasible.

Magmatic Segregation

The igneous processes that generate some of these metal deposits are quite straightforward. For example, as a large magma body cools, the heavy minerals that crystallize early tend to settle to the lower portion of the magma chamber. This type of magmatic segregation is particularly active in large basaltic magmas where chromite (ore of chromium), magnetite, and platinum are occasionally generated. Layers of chromite, interbedded with other heavy minerals, are mined from such deposits in the Stillwater Complex of Montana. Another example is the Bushveld Complex in South Africa, which contains over 70 percent of the world's known reserves of platinum.

Magmatic segregation is also important in the late stages of the magmatic process. This is particularly true of granitic magmas in which the residual melt can become enriched in rare elements and heavy metals. Further, because water and other volatile substances do not crystallize along with the bulk of the magma body, these fluids make up a high percentage of the melt during the final phase of solidification. Crystallization in a fluid-rich environment, where ion migration is enhanced, results in the formation of crystals several centimeters, or even a few meters, in length. The resulting rocks, called **pegmatites,** are composed of these unusually large crystals (Figure 4.9, p. 114).

Most pegmatites are granitic in composition and consist of unusually large crystals of quartz, feldspar, and muscovite. Feldspar is used in the production of ceramics, and muscovite is used for electrical insulation and glitter. Further, pegmatites often contain some of the least abundant elements. Thus, in addition to the common silicates, some pegmatites include semiprecious gems such as beryl, topaz, and tourmaline. Moreover, minerals containing the elements lithium, cesium, uranium, and the rare earths* are occasionally found (Figure 23.19). Most pegmatites are located within large igneous masses or as dikes or veins that cut into the host rock surrounding the magma chamber (Figure 23.20).

FIGURE 23.19 This pegmatite in the Black Hills of South Dakota was mined for its large crystals of spodumene, an important source of lithium. Arrows are pointing to impressions left by crystals. Note person in upper center of photo for scale. (Photo by James G. Kirchner)

Not all late-stage magmas produce pegmatites, nor do all have a granitic composition. Rather, some magmas become enriched in iron or occasionally copper. For example, at Kiruna, Sweden, magma composed of over 60 percent magnetite solidified to produce one of the largest iron deposits in the world.

Diamonds

Another economically important mineral with an igneous origin is diamond. Although best known as gems, diamonds are used extensively as abrasives. Diamonds originate at depths of nearly

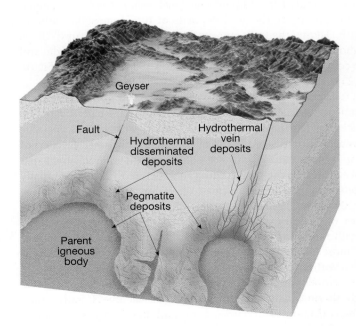

FIGURE 23.20 Illustration of the relationship between a parent igneous body and the associated pegmatite and hydrothermal deposits.

---

*The rare earths are a group of 15 elements (atomic numbers 57 through 71) that possess similar properties. They are useful catalysts in petroleum refining and are used to improve color retention in television picture tubes.

FIGURE 23.21 Native copper from northern Michigan's Keweenaw Peninsula is an excellent example of a hydrothermal deposit. At one time this area was an important source of copper, but it is now largely depleted. (Photo by E. J. Tarbuck)

200 kilometers, where the confining pressure is great enough to generate this high-pressure form of carbon. Once crystallized, the diamonds are carried upward through pipe-shaped conduits that increase in diameter toward the surface. In diamond-bearing pipes, nearly the entire pipe contains diamond crystals that are disseminated throughout an ultramafic rock called *kimberlite*. The most productive kimberlite pipes are those found in South Africa. The only equivalent source of diamonds in the United States is located near Murfreesboro, Arkansas, but this deposit is exhausted and today serves merely as a tourist attraction.

Hydrothermal Solutions

Among the best-known and most important ore deposits are those generated from **hydrothermal (hot-water) solutions.** Included in this group are the gold deposits of the Homestake Mine in South Dakota; the lead, zinc, and silver ores near Coeur d'Alene, Idaho; the silver deposits of the Comstock Lode in Nevada; and the copper ores of the Keweenaw Peninsula in Michigan (Figure 23.21).

The majority of hydrothermal deposits originate from hot, metal-rich fluids that are remnants of late-stage magmatic processes. During solidification, liquids plus various metallic ions accumulate near the top of the magma chamber. Because of their mobility, these ion-rich solutions can migrate great distances through the surrounding rock before they are eventually deposited, usually as sulfides of various metals (see Figure 23.20). Some of this fluid moves along openings such as fractures or bedding planes, where it cools and precipitates the metallic ions to produce **vein deposits** (Figure 23.22). Many of the most productive deposits of gold, silver, and mercury occur as hydrothermal vein deposits.

Another important type of accumulation generated by hydrothermal activity is called a **disseminated deposit.** Rather than being concentrated in narrow veins and dikes, these ores are distributed as minute masses throughout the entire rock mass. Much of the world's copper is extracted from disseminated deposits,

including those at Chuquicamata, Chile, and the huge Bingham Canyon copper mine in Utah (see Box 23.2). Because these accumulations contain only 0.4 to 0.8 percent copper, between 125 and 250 kilograms of ore must be mined for every kilogram of metal recovered. The environmental impact of these large excavations, including the problem of waste disposal, is significant.

Some hydrothermal deposits have been generated by the circulation of ordinary groundwater in regions where magma was emplaced near the surface. The Yellowstone National Park area is a modern example of such a situation (Figure 23.23). When groundwater invades a zone of recent igneous activity, its temperature rises, greatly enhancing its ability to dissolve minerals. These migrating hot waters remove metallic ions from intrusive igneous rocks and carry them upward, where they may be deposited as an ore body. Depending on the conditions, the resulting accumulations may occur as vein deposits, as disseminated deposits, or, where hydrothermal solutions reach the surface in the form of hot springs or geysers, as surface deposits.

With the development of the plate tectonics theory it became clear that some hydrothermal deposits originated along ancient oceanic ridges. A well-known example is found on the island of Cyprus, where copper has been mined for more than 4000 years. Apparently, these deposits represent ores that formed on the seafloor at an ancient oceanic spreading center.

Since the mid-1970s, active hot springs and metal-rich sulfide deposits have been detected at several sites, including study areas along the East Pacific Rise and the Juan de Fuca Ridge. The deposits are forming where heated seawater, rich in dissolved metals and sulfur, gushes from the seafloor as particle-filled clouds called *black smokers*. As shown in Figure 23.24, seawater infiltrates the hot oceanic crust along the flanks of the ridge. As the water moves through the newly formed material, it is heated and chemically interacts with the basalt, extracting and transporting sulfur, iron, copper, and other metals. Near the ridge axis, the hot, metal-rich fluid rises along faults. Upon reaching the seafloor, the spewing liquid mixes with the cold seawater, and the sulfides precipitate to form massive sulfide deposits.

FIGURE 23.22 Gneiss laced with quartz veins at Diablo Lake Overlook, North Cascades National Park, Washington. (Photo by James E. Patterson)

FIGURE 23.23 Some hydrothermal deposits are formed by the circulation of heated groundwater in regions where magma is near the surface, as in the Yellowstone National Park area. (Photo by Rob Tilley/ www.DanitaDelimont)

MINERAL RESOURCES AND METAMORPHIC PROCESSES

The role of metamorphism in producing mineral deposits is frequently tied to igneous processes. For example, many of the most important metamorphic ore deposits are produced by contact metamorphism. Here the host rock is recrystallized and chemically altered by heat, pressure, and hydrothermal solutions emanating from an intruding igneous body. The extent to which the host rock is altered depends on the nature of the intruding igneous mass as well as the nature of the host rock.

Some resistant materials, such as quartz sandstone, may show very little alteration, whereas others, including limestone, might exhibit the effects of metamorphism for several kilometers from the igneous pluton. As hot, ion-rich fluids move through limestone, chemical reactions take place, which produce useful minerals such as garnet and corundum. Further, these reactions release carbon dioxide, which greatly facilitates the outward migration of metallic ions. Thus, extensive aureoles of metal-rich deposits commonly surround igneous plutons that have invaded limestone strata.

The most common metallic minerals associated with contact metamorphism are sphalerite (zinc), galena (lead), chalcopyrite (copper), magnetite (iron), and bornite (copper). The hydrothermal ore deposits may be disseminated throughout the altered zone or exist as concentrated masses that are located either next to the intrusive body or along the margins of the metamorphic zone.

Regional metamorphism can also generate useful mineral deposits. Recall that at convergent plate boundaries the oceanic crust, along with sediments that have accumulated at the continental margins, are carried to great depths. In these high-temperature, high-pressure environments the mineralogy and texture of the subducted materials are altered, producing deposits of nonmetallic minerals such as talc and graphite.

WEATHERING AND ORE DEPOSITS

Weathering creates many important mineral deposits by concentrating minor amounts of metals that are scattered through unweathered rock into economically valuable concentrations. Such a transformation is often termed **secondary enrichment** and takes place in one of two ways. In one situation chemical weathering coupled with downward-percolating water removes undesirable materials from decomposing rock, leaving the desirable elements enriched in the upper zones of the soil. The second way is basically the reverse of the first. That is, the desirable elements that are found in low concentrations near the surface are removed and carried to lower zones, where they are redeposited and become more concentrated.

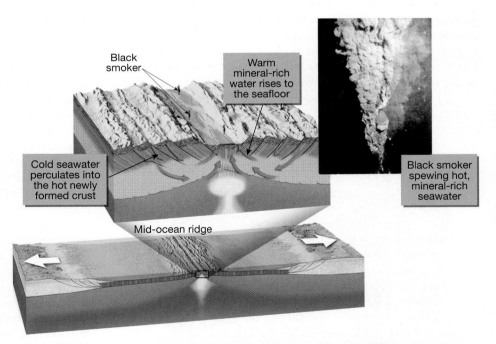

FIGURE 23.24 Massive sulfide deposits can result from the circulation of seawater through the oceanic crust along active spreading centers. As seawater infiltrates the hot basaltic crust, it leaches sulfur, iron, copper, and other metals. The hot, enriched fluid returns to the seafloor near the ridge axis along faults and fractures. Some metal sulfides may be precipitated in these channels as the rising fluid begins to cool. When the hot liquid emerges from the seafloor and mixes with cold seawater, the sulfides precipitate to form massive deposits. Photo shows a close-up view of a black smoker spewing hot, mineral-rich seawater along the East Pacific Rise. (Photo © Robert Ballard, Woods Hole Oceanographic Institution)

Bauxite

The formation of *bauxite,* the principal ore of aluminum, is one important example of an ore created as a result of enrichment by weathering processes (Figure 23.25). Although aluminum is the third most abundant element in Earth's crust, economically valuable concentrations of this important metal are not common, because most aluminum is tied up in silicate minerals from which it is extremely difficult to extract.

Bauxite forms in rainy tropical climates in association with laterites. (In fact, bauxite is sometimes referred to as aluminum laterite.) When aluminum-rich source rocks are subjected to the intense and prolonged chemical weathering of the tropics, most of the common elements, including calcium, sodium, and silicon, are removed by leaching. Because aluminum is extremely insoluble, it becomes concentrated in the soil as bauxite, a hydrated aluminum oxide. Thus, the formation of bauxite depends both on climatic conditions in which chemical weathering and leaching are pronounced and on the presence of an aluminum-rich source rock. Important deposits of nickel and cobalt are also found in laterite soils that develop from igneous rocks rich in ferromagnesian silicate minerals.

Other Deposits

Many copper and silver deposits result when weathering processes concentrate metals that are deposited through a low-grade primary ore. Usually such enrichment occurs in deposits containing pyrite (FeS_2), the most common and widespread sulfide mineral. Pyrite is important because when it chemically weathers, sulfuric acid forms, which enables percolating waters to dissolve the ore metals. Once dissolved, the metals gradually migrate downward through the primary ore body until they are precipitated. Deposition takes place because of changes that occur in the chemistry of the solution when it reaches the

FIGURE 23.25 Bauxite is the ore of aluminum and forms as a result of weathering processes under tropical conditions. Its color varies from red or brown to nearly white. (Photo by E. J. Tarbuck)

groundwater zone (the zone beneath the surface where all pore spaces are filled with water). In this manner the small percentage of dispersed metal can be removed from a large volume of rock and redeposited as a higher-grade ore in a smaller volume of rock.

This enrichment process is responsible for the economic success of many copper deposits, including one located in Miami, Arizona. Here the ore was upgraded from less than 1 percent copper in the primary deposit to as much as 5 percent copper in some localized zones of enrichment. When pyrite weathers (oxidizes) near the surface, residues of iron oxide remain. The presence of these rusty-colored caps at the surface indicates the possibility of an enriched ore below, and this represents a visible sign for prospectors.

PLACER DEPOSITS

Sorting typically results in like-sized grains being deposited together. However, sorting according to the specific gravity of particles also occurs. This latter type of sorting is responsible for the creation of **placers,** which are deposits formed when heavy minerals are mechanically concentrated by currents. Placers associated with streams are among the most common and best known, but the sorting action of waves can also create placers along the shore. Placers deposits usually involve minerals that are not only heavy but also durable (to withstand physical destruction during transportation) and chemically resistant (to endure weathering processes). Placers form because heavy minerals settle quickly from a current, whereas less dense particles remain suspended and are carried onward. Common sites of accumulation include point bars on the insides of meanders as well as cracks, depressions, and other irregularities on stream beds.

Many economically important placer deposits exist, with accumulations of gold the best known. Indeed, it was the placer deposits discovered in 1848 that led to the famous California gold rush. Years later similar deposits created a gold rush to Alaska as well (Figure 23.26). Panning for gold by washing sand and gravel from a flat pan to concentrate the fine "dust" at the bottom was a common method used by early prospectors to recover the precious metal, and it is a process similar to that which created the placer in the first place.

In addition to gold, other heavy and durable minerals form placers. These include platinum, diamonds, and tin. The Ural Mountains contain placers rich in platinum, and placers are important sources of diamonds in southern Africa. Significant portions of the world's supply of cassiterite, the principal ore of tin, have come from placer deposits in Malaysia and Indonesia. Cassiterite is often widely disseminated through granitic igneous rocks. In this state the mineral is not sufficiently concentrated to be extracted profitably. However, as the enclosing rock dissolves and disintegrates, the heavy and durable cassiterite grains are set free. Eventually the freed particles are washed to a stream, where they are deposited in placers that are significantly more concentrated than the original deposit. Similar circumstances and events are common to many minerals mined from placers.

In some cases, if the source rock for a placer deposit can be located, it too may become an important ore body. By following

A.

B.

FIGURE 23.26   **A.** It was placer deposits that led to the 1848 California gold rush. Here, a prospector in 1850 swirls his gold pan, separating sand and mud from flecks of gold. (Photo courtesy of Seaver Center for Western History Research, Los Angeles County Museum of Natural History) **B.** This historic image from 1915 shows a large dredge mining gold from a placer deposit on the Klondike River. (Photo Library of Congress)

placer deposits upstream, one can sometimes locate the original deposit. This is how the gold-bearing veins of the Mother Lode in California's Sierra Nevada batholith were found, as well as the famous Kimberly diamond mines of South Africa. The placers were discovered first, their source at a later time.

NONMETALLIC MINERAL RESOURCES

Earth materials that are not used as fuels or processed for the metals they contain are referred to as **nonmetallic mineral resources.** Realize that use of the word "mineral" is very broad in this economic context and is quite different from the geologist's strict definition of mineral found in Chapter 3. Nonmetallic mineral resources are extracted and processed either for the nonmetallic elements they contain or for the physical and chemical properties they possess.

People often do not realize the importance of nonmetallic minerals because they see only the products that resulted from their use and not the minerals themselves. That is, many nonmetallics are used up in the process of creating other products. Examples include the fluorite and limestone that are part of the steelmaking process, the abrasives required to make a piece of machinery, and the fertilizers needed to grow a food crop (see Table 23.4).

The quantities of nonmetallic minerals used each year are enormous. A glance at Figure 23.2 reminds us that the per capita consumption of nonfuel resources in the United States totals more than 11 metric tons, of which about 94 percent are nonmetallics. Nonmetallic mineral resources are commonly divided into two broad groups—*building materials* and *industrial minerals*. Because some substances have many different uses, they are found in both categories. Limestone, perhaps the most versatile and widely used rock of all, is the best example. As a building material, it is used not only as crushed rock and building stone but also in the making of cement. Moreover, as an industrial mineral, limestone is an ingredient in the manufacture of steel and is used in agriculture to neutralize soils.

Building Materials

Natural aggregate consists of crushed stone, sand, and gravel. From the standpoint of quantity and value, aggregate is a very important building material. The United States produces nearly 2 billion tons of aggregate per year, which represents about one half of the entire nonenergy mining volume in the country. It is

TABLE 23.4	Occurrences and Uses of Nonmetallic Minerals	
Mineral	Uses	Geological Occurrences
Apatite	Phosphorus fertilizers	Sedimentary deposits
Asbestos	Incombustible fibers	Metamorphic alteration (chrysotile)
Calcite	Aggregate; steelmaking; soil conditioning; chemicals; cement; building stone	Sedimentary deposits
Clay minerals	Ceramics; china	Residual product of weathering (kaolinite)
Corundum	Gemstones; abrasives	Metamorphic deposits
Diamond	Gemstones; abrasives	Kimberlite pipes; placers
Fluorite	Steelmaking; aluminum refining; glass; chemicals	Hydrothermal deposits
Garnet	Abrasives; gemstones	Metamorphic deposits
Graphite	Pencil lead; lubricant; refractories	Metamorphic deposits
Gypsum	Plaster of Paris	Evaporite deposits
Halite	Table salt; chemicals; ice control	Evaporite deposits; salt domes
Muscovite	Insulator in electrical applications	Pegmatites
Quartz	Primary ingredient in glass	Igneous intrusions; sedimentary deposits
Sulfur	Chemicals; fertilizer manufacture	Sedimentary deposits; hydrothermal deposits
Sylvite	Potassium fertilizers	Evaporite deposits
Talc	Powder used in paints, cosmetics, etc.	Metamorphic deposits

produced commercially in every state and is used in nearly all building construction and in most public works projects (Figure 23.27).

Besides aggregate, other important building materials include gypsum for plaster and wallboard, clay for tile and bricks, and cement, which is made from limestone and shale. Cement and aggregate go into the making of concrete, a material that is essential to practically all construction. Aggregate gives concrete its strength and volume, and cement binds the mixture into a rock-hard substance. Building just 2 kilometers of four-lane highway requires more than 85 metric tons of aggregate. On a smaller scale, 90 tons of aggregate are needed just to build an average six-room house.

Because most building materials are widely distributed and present in almost unlimited quantities, they have little intrinsic value. Their economic worth comes only after the materials are removed from the ground and processed. Since their per-ton value compared with metals and industrial minerals is low, mining and quarrying operations are usually undertaken to satisfy local needs. Except for special types of cut stone used for building and monuments, transportation costs greatly limit the distance most building materials can be moved.

Industrial Minerals

Many nonmetallic resources are classified as industrial minerals. In some instances these materials are important because they are sources of specific chemical elements or compounds. Such minerals are used in the manufacture of chemicals and the production of fertilizers. In other cases their importance is related to the physical properties they exhibit. Examples include minerals such as corundum and garnet, which are used as abrasives. Although supplies are generally plentiful, most industrial minerals are not nearly as abundant as building materials. Moreover, deposits are far more restricted in distribution and extent. As a result, many of these nonmetallic resources must be transported considerable distances, which of course adds to their cost. Unlike most building materials, which need a minimum of process-

FIGURE 23.27 Crushed stone and sand and gravel are primarily used for aggregate in the construction industry, especially in cement concrete for residential and commercial buildings, bridges, and airports, and as cement concrete or bituminous concrete (asphalt) for highway construction. A large percentage is used without a binder as road base, for road surfacing, and as railroad ballast. (Photo by Robert Ginn/PhotoEdit)

FIGURE 23.28 Large open-pit phosphate mine in Florida. The phosphorus-bearing mineral apatite is a calcium phosphate associated with bones and teeth. Fish and other marine organisms extract phosphate from sea-water to form apatite. These sedimentary deposits are associated with the floor of a shallow sea. (Photo by C. Davidson/Comstock)

pal use is in the manufacture of phosphate fertilizer, sulfuric acid has a large number of other applications as well. Sources include deposits of native sulfur associated with salt domes and volcanic areas, as well as common iron sulfides such as pyrite. In recent years an increasingly important source has been the sulfur removed from coal, oil, and natural gas in order to make these fuels less polluting.

Salt Common salt, known by the mineral name *halite,* is another important and versatile resource. It is among the more prominent nonmetallic minerals used as a raw material in the chemical industry. In addition, large quantities are used to "soften" water and to keep streets and highways free of ice. Of course, most people are aware that it is also a basic nutrient and a part of many food products.

Salt is a common evaporite, and thick deposits are exploited using conventional underground mining techniques. Subsurface deposits are also tapped, using brine wells in which a pipe is introduced into a salt deposit and water is pumped down the pipe. The salt dissolved by the water is brought to the surface through a second pipe. In addition, seawater continues to serve as a source of salt as it has for centuries. The salt is harvested after the Sun evaporates the water.

ing before they are ready to use, many industrial minerals require considerable processing to extract the desired substance at the proper degree of purity for its ultimate use.

Fertilizers The growth in world population toward 7 billion requires that the production of basic food crops continues to expand. Therefore, fertilizers—primarily nitrate, phosphate, and potassium compounds—are extremely important to agriculture. The synthetic nitrate industry, which derives nitrogen from the atmosphere, is the source of practically all the world's nitrogen fertilizers. The primary source of phosphorus and potassium, however, remains Earth's crust. The mineral apatite is the primary source of phosphate. In the United States most production comes from marine sedimentary deposits in Florida and North Carolina (Figure 23.28). Although potassium is an abundant element in many minerals, the primary commercial sources are evaporite deposits containing the mineral sylvite. In the United States, deposits near Carlsbad, New Mexico, have been especially important. The chapter-opening photo shows a mining operation in southern Utah that produces potassium (commonly called potash).

Sulfur Because it has many uses, sulfur is an important nonmetallic resource. In fact, the quantity of sulfur used is considered one index of a country's level of industrialization. More than 80 percent is used to produce sulfuric acid. Although its princi-

Students Sometimes Ask . . .

Can you tell us a little more about the brightly-colored ponds in the chapter-opening photo?

Sure. Potash (potassium chloride) at this site is obtained using a system that combines solution mining and solar evaporation. Water from the nearby Colorado River is injected into potash deposits that are about 3000 feet below the surface. The mineral-rich water (brine) is brought to the surface and piped to 400 acres of shallow ponds. The water evaporates leaving behind deposits of potash and common salt (sodium chloride). A blue dye is added to assist with the evaporation process. The crystals are harvested and sent to a mill where the potash is separated from the salt by a flotation method. The solar evaporation process works well in the dry and sunny climate of southern Utah.

CHAPTER 23 ENERGY AND MINERAL RESOURCES IN REVIEW

- *Renewable resources* can be replenished over relatively short time spans. Examples include natural fibers for clothing and trees for lumber. *Nonrenewable resources* form so slowly that, from a human standpoint, Earth contains fixed supplies. Examples include fuels such as oil and coal, and metals such as copper and gold. A rapidly growing world population and

the desire for an improved living standard are causing nonrenewable resources to become depleted at an increasing rate.

- *Coal, petroleum,* and *natural gas,* the *fossil fuels* of our modern economy, are all associated with sedimentary rocks. Coal originates from large quantities of plant remains that accumulate in an oxygen-deficient environment, such as a swamp. More than 70 percent of present-day coal usage is for the generation of electricity. Air pollution produced by the sulfur oxide gases that form from burning most types of coal is a significant environmental problem.

- Oil and natural gas, which commonly occur together in the pore spaces of some sedimentary rocks, consist of various *hydrocarbon compounds* (compounds made of hydrogen and carbon mixed together). Petroleum formation is associated with the accumulation of sediment in ocean areas that are rich in plant and animal remains that become buried and isolated in an oxygen-deficient environment. As the mobile petroleum and natural gas form, they migrate and accumulate in adjacent permeable beds such as sandstone. If the upward migration is halted by an impermeable rock layer, referred to as a *cap rock*, a geologic environment that allows for economically significant amounts of oil and gas to accumulate underground, called an *oil trap,* develops. The two basic conditions common to all oil traps are (1) a porous, permeable *reservoir rock* that will yield petroleum and/or natural gas in sufficient quantities, and (2) a cap rock.

- When conventional petroleum resources are no longer adequate, fuels derived from *oil sands* and *oil shale* may become substitutes. Presently, oil sands from the province of Alberta are a significant contributor to Canada's oil production. Oil from oil shale is presently uneconomical to produce. Oil production from both oil sands and oil shale has significant environmental drawbacks.

- About 84 percent of our energy is derived from fossil fuels. In the United States the most important alternative energy sources are *nuclear energy* and *hydroelectric power.* Other alternative energy sources are locally important but collectively provide about 1 percent of the U.S. energy demand. These include *solar power, geothermal energy, wind energy,* and *tidal power.*

- *Mineral resources* are the endowment of useful minerals ultimately available commercially. Resources include already identified deposits from which minerals can be extracted profitably, called *reserves,* as well as known deposits that are not yet economically or technologically recoverable. Deposits inferred to exist but not yet discovered are also considered mineral resources. The term *ore* is used to denote those useful metallic minerals that can be mined for a profit,

as well as some nonmetallic minerals, such as fluorite and sulfur, that contain useful substances.

- Some of the most important accumulations of metals, such as gold, silver, lead, and copper, are produced by igneous processes. The best-known and most important ore deposits are generated from *hydrothermal* (hot-water) *solutions.* Hydrothermal deposits originate from hot, metal-rich fluids that are remnants of late-stage magmatic processes. These ion-rich solutions move along fractures or bedding planes, cool, and precipitate the metallic ions to produce *vein deposits.* In a *disseminated deposit* (e.g., much of the world's copper deposits) the ores from hydrothermal solutions are distributed as minute masses throughout the entire rock mass.

- Many of the most important metamorphic ore deposits are produced by contact metamorphism. Extensive aureoles of metal-rich deposits commonly surround igneous bodies where ions have invaded limestone strata. The most common metallic minerals associated with contact metamorphism are sphalerite (zinc), galena (lead), chalcopyrite (copper), magnetite (iron), and bornite (copper). Also of economic importance are the metamorphic rocks themselves. In many regions, slate, marble, and quartzite are quarried for a variety of construction purposes.

- Weathering creates ore deposits by concentrating minor amounts of metals into economically valuable deposits. The process, often called *secondary enrichment,* is accomplished by either (1) removing undesirable materials and leaving the desired elements enriched in the upper zones of the soil, or (2) removing and carrying the desirable elements to lower zones where they are redeposited and become more concentrated. *Bauxite,* the principal ore of aluminum, is one important ore created as a result of enrichment by weathering processes. In addition, many copper and silver deposits result when weathering processes concentrate metals that were formerly dispersed through low-grade primary ore.

- Earth materials that are not used as fuels or processed for the metals they contain are referred to as *nonmetallic resources.* Many are sediments or sedimentary rocks. The two broad groups of nonmetallic resources are *building materials* and *industrial minerals.* Limestone, perhaps the most versatile and widely used rock of all, is found in both groups.

KEY TERMS

cap rock (p. 647)
disseminated deposit (p. 662)
fossil fuel (p. 645)
geothermal energy (p. 656)
hydroelectric power (p. 655)
hydrothermal solution (p. 662)

mineral resource (p. 659)
nonmetallic mineral resource (p. 665)
nonrenewable resource (p. 644)
nuclear fission (p. 652)

oil trap (p. 647)
ore (p. 659)
pegmatite (p. 661)
placer (p. 664)
renewable resource (p. 644)
reserve (p. 659)

reservoir rock (p. 647)
secondary enrichment (p. 663)
vein deposit (p. 662)

QUESTIONS FOR REVIEW

1. Contrast renewable and nonrenewable resources. Give one or more examples of each.

2. What is the estimated world population for the year 2015? How does this compare to the figures for 1930 and 1975? Is demand for resources growing as rapidly as world population?

3. More than 70 percent of present-day coal usage is for what purpose?

4. What is an oil trap? List two conditions common to all oil traps.

5. List two drawbacks associated with the processing of oil sands recovered by surface mining.

6. The United States has huge oil shale deposits but does not produce oil shale commercially. Explain.

7. What is the main fuel for nuclear fission reactors?

8. List two obstacles that have hindered the development of nuclear power as a major energy source. What environmental advantage does nuclear power have compared to fossil fuels?

9. Briefly describe two methods by which solar energy might be used to produce electricity.

10. Explain why dams built to provide hydroelectricity do not last indefinitely.

11. What advantages does tidal power production offer? Is it likely that tides will ever provide a significant proportion of the world's electrical energy requirements?

12. Contrast *resource* and *reserve*.

13. What might cause a mineral deposit that had not been considered an ore to be reclassified as an ore?

14. List two general types of hydrothermal deposits.

15. Metamorphic ore deposits are often related to igneous processes. Provide an example.

16. Name the primary ore of aluminum and describe its formation.

17. A rusty-colored zone of iron oxide at the surface may indicate the presence of a copper deposit at depth. Briefly explain.

18. Briefly describe the way in which minerals accumulate in placers. List four minerals that are mined from such deposits.

19. Which is greater, the per capita consumption of metallic resources or that of nonmetallic mineral resources?

20. Nonmetallic resources are commonly divided into two broad groups. List the two groups and some examples of materials that belong to each. Which group is most widely distributed?

COMPANION WEBSITE

The *Earth 10e* Web site uses the resources and flexibility of the Internet to aid in your study of the topics in this chapter. Written and developed by the authors and other geology instructors, this site will help improve your understanding of geology. Visit www.mygeoscienceplace.com in order to:

- **Review** key chapter concepts.

- **Read** with links to the eBook and to chapter-specific web resources.

- **Test** yourself with online quizzes.

PLANETARY GEOLOGY*

Artists' concept of Cassini Orbiter after crossing Saturn's rings.

(COURTESY OF NASA)

*This chapter was revised with the assistance of Teresa Tarbuck and Mark Watry.

Planetary geology is the study of the formation and evolution of the bodies in our solar system—including the eight planets and a myriad of smaller objects: moons, asteroids, comets, and meteoroids. Studying these objects provides valuable insights into the dynamic processes that operate on Earth. Understanding how other atmospheres evolve helps scientists build better models for predicting climate change; while studying tectonic processes on other planets helps us appreciate how these complex interactions alter Earth. In addition, seeing how erosional forces work on other bodies allows us to observe the many ways landscapes are created. Finally, the uniqueness of Earth, a body that harbors life, is revealed through investigations of other planetary bodies.

Many questions remain unanswered. Is it possible for humans to live on another planet? Can we extract resources from asteroids or the Moon and deliver them to Earth? Is Earth on a collision course with other planetary bodies? In addition to answering questions such as these, interplanetary space exploration seeks to expand our understanding of the processes and events that created and modified the solar system.

OUR SOLAR SYSTEM: AN OVERVIEW

The Sun is at the center of a revolving system, trillions of miles wide, consisting of eight planets, their satellites, and numerous smaller asteroids, comets, and meteoroids (Figure 24.1). An estimated 99.85 percent of the mass of our solar system is contained within the Sun. Collectively, the planets account for most of the remaining 0.15 percent. Starting from the Sun, the planets are Mercury, Venus, Earth, Mars, Jupiter, Saturn, Uranus, and Neptune (Figure 24.1). Pluto was recently reclassified as a member of a new class of solar system bodies called *dwarf planets*.

Tethered to the Sun by gravity, all of the planets travel in the same direction on slightly elliptical orbits (Table 24.1). Gravity causes objects nearest the Sun to travel fastest. Therefore, Mercury has the highest orbital velocity, 48 kilometers per second, and the shortest period of revolution around the Sun, 88 Earth-days. By contrast, the distant dwarf planet Pluto has an orbital speed of just 5 kilometers per second and requires 248 Earth-years to complete one revolution. Most large bodies orbit the Sun approximately in the same plane. The planets' inclination with respect to the Earth-Sun orbital plane, known as the *ecliptic*, is shown in Table 24.1.

Nebular Theory: Formation of the Solar System

The **Nebular theory,** which explains the formation of the solar system, states that the Sun and planets formed from a rotating cloud of interstellar gases (mainly hydrogen and helium) and dust called the **solar nebula.** As the solar nebula contracted due to gravity, most of the material collected in the center to form the hot *protosun.* The remaining materials formed a thick, flattened, rotating disk, within which matter gradually cooled and condensed into grains and clumps of icy, rocky material. Repeated collisions resulted in most of the material eventually collecting into asteroid-sized objects called **planetesimals.**

The composition of planetesimals was largely determined by their proximity to the protosun. As you might expect, temperatures were highest in the inner solar system and decreased toward the outer edge of the disk. Therefore, between the present orbits of Mercury and Mars, the planetesimals were composed of materials with high melting temperatures—metals and rocky substances. Then, through repeated collisions and accretion (sticking together) these asteroid-sized rocky bodies combined to form the four **protoplanets** that eventually became Mercury, Venus, Earth, and Mars.

The planetesimals that formed beyond the orbit of Mars, where temperatures are low, contained high percentages of ices—water, carbon dioxide, ammonia, and

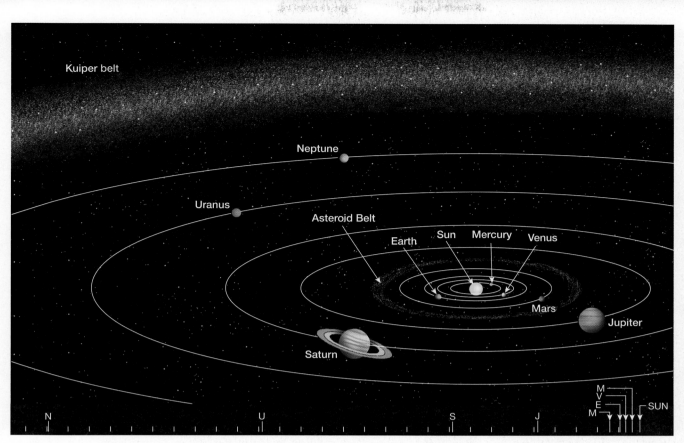

FIGURE 24.1 Orbits of the planets. Positions of the planets are shown to scale along bottom of diagram.

TABLE 24.1	Planetary Data							
		Mean Distance from Sun					Orbital Velocity	
Planet	Symbol	AU*	Millions of Miles	Millions of Kilometers	Period of Revolution	Inclination of Orbit	mI/s	km/s
Mercury	☿	0.39	36	58	88^d	7°00′	29.5	47.5
Venus	♀	0.72	67	108	225^d	3°24′	21.8	35.0
Earth	⊕	1.00	93	150	365.25^d	0°00′	18.5	29.8
Mars	♂	1.52	142	228	687^d	1°51′	14.9	24.1
Jupiter	♃	5.20	483	778	12^{yr}	1°18′	8.1	13.1
Saturn	♄	9.54	886	1427	30^{yr}	2°29′	6.0	9.6
Uranus	♅	19.18	1783	2870	84^{yr}	0°46′	4.2	6.8
Neptune	♆	30.06	2794	4497	165^{yr}	1°46′	3.3	5.3

Planet	Period of Rotation	Diameter		Relative Mass (Earth = 1)	Average Density (g/cm^3)	Polar Flattening (%)	Eccentricity†	Number of Known Satellites††
		Miles	Kilometers					
Mercury	59^d	3015	4878	0.06	5.4	0.0	0.206	0
Venus	243^d	7526	12,104	0.82	5.2	0.0	0.007	0
Earth	23^{h}56^{m}04^s	7920	12,756	1.00	5.5	0.3	0.017	1
Mars	24^{h}37^{m}23^s	4216	6794	0.11	3.9	0.5	0.093	2
Jupiter	9^{h}56^m	88,700	143,884	317.87	1.3	6.7	0.048	63
Saturn	10^{h}30^m	75,000	120,536	95.14	0.7	10.4	0.056	61
Uranus	17^{h}14^m	29,000	51,118	14.56	1.2	2.3	0.047	27
Neptune	16^{h}07^m	28,900	50,530	17.21	1.7	1.8	0.009	13

*AU = astronomical unit, Earth's mean distance from the Sun.

†Eccentricity is a measure of the amount an orbit deviates from a circular shape. The larger the number, the less circular the orbit.

††Includes all satellites discovered as of August 2009.

Mercury

Venus

Earth

Mars

Sun

Jupiter

Saturn

Uranus

Neptune

FIGURE 24.2   The planets drawn to scale.

methane—as well as small amounts of rocky and metallic debris. It was mainly from these planetesimals that the four outer planets eventually formed. The accumulation of ices accounts, in part, for the large sizes and low densities of the outer planets. The two most massive planets, Jupiter and Saturn, had surface gravities sufficient to attract and retain large quantities of hydrogen and helium, the lightest elements.

It took roughly a billion years after the protoplanets formed for the planets to gravitationally accumulate most of the interplanetary debris. This was a period of intense bombardment as the planets cleared their orbits of much of the leftover material. The "scars" of this period are still evident on the Moon's surface. Small bodies were flung into planet-crossing orbits, or into interstellar space. The small fraction of interplanetary matter that escaped this violent period became asteroids, comets, and meteoroids. By comparison, the present-day solar system is a much quieter place, although many of these processes continue today at a reduced pace.

The Planets: Internal Structures, Atmospheres, and Weather

The planets fall into two groups based on location, size, and density; the **terrestrial** (Earth-like) **planets** (Mercury, Venus, Earth, and Mars), and the **Jovian** (Jupiter-like) **planets** (Jupiter, Saturn, Uranus, and Neptune). Because of their relative locations, the four terrestrial planets are also known as "*inner planets*" and the four Jovian planets are known as "*outer planets.*" A correlation exists between planetary locations and sizes—the inner planets are substantially smaller than the outer planets, also known as *gas giants*. For example, the diameter of Neptune (the smallest Jovian planet) is three times larger than the diameter of Earth or Venus. Further, Neptune's mass is 17 times greater than that of Earth or Venus (Figure 24.2).

Other properties that differ include densities, chemical compositions, orbital periods, and numbers of satellites. Variations in the chemical composition of planets are largely responsible for their density differences. Specifically, the average density of the terrestrial planets is about five times the density of water, whereas the average density of the Jovian planets is only 1.5 times that of water. Saturn has a density only 0.7 times that of water, which means that it would float if placed in a large enough tank of water. The outer planets are also characterized by long orbital periods and numerous satellites.

Internal Structures   Recall that shortly after Earth formed, the segregation of material resulted in the formation of three major layers defined by their chemical composition—the crust, mantle, and core. This type of chemical separation occurred in the other planets as well. However, because the terrestrial planets are compositionally different than the Jovian planets, the nature of these layers differs between these two groups (Figure 24.3).

The terrestrial planets are dense, having relatively large cores of iron, iron compounds, and nickel. From their centers outward, the amount of metallic iron decreases while the amount of rocky silicate minerals increase. The outer cores of Earth and Mercury are liquid, whereas the cores of Venus and Mars are thought to be partially molten. This difference is attributable to Venus and Mars having lower internal temperatures than those of Earth and Mercury. Silicate minerals and other lighter compounds make up the mantles of the terrestrial planets. Finally, the silicate crusts of terrestrial planets are relatively thin compared to their mantles.

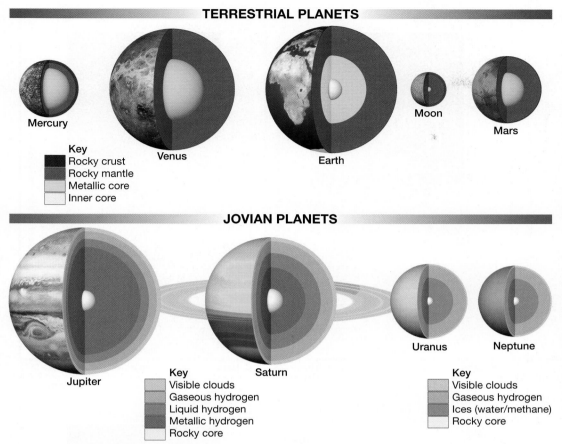

TERRESTRIAL PLANETS

Key
Rocky crust
Rocky mantle
Metallic core
Inner core

Mercury

Venus

Earth

Moon

Mars

JOVIAN PLANETS

Jupiter

Saturn

Uranus

Neptune

Key
Visible clouds
Gaseous hydrogen
Liquid hydrogen
Metallic hydrogen
Rocky core

Key
Visible clouds
Gaseous hydrogen
Ices (water/methane)
Rocky core

FIGURE 24.3 Comparing the internal structures of the planets.

The two largest Jovian planets, Jupiter and Saturn, have small metallic inner cores consisting of iron compounds at extremely high temperatures and pressures. The outer cores of these two giants are thought to be liquid metallic hydrogen, whereas the mantles are comprised of liquid hydrogen and helium. The outermost layers are gases and ices of hydrogen, helium, water, ammonia, and methane—which account for the low densities of these planets. Uranus and Neptune also have small metallic cores but their mantles are likely hot dense water and ammonia. Above their mantles, the amount of hydrogen and helium increases, but exists in much lower concentrations than those of Jupiter and Saturn.

All planets, except Venus and Mars, have significant magnetic fields generated by flow in their liquid outer cores, or liquid mantles. Venus has a weak field due to the interaction between the solar wind and its uppermost atmosphere (ionosphere), while Mars' weak magnetic field is thought to be a remnant from when its interior was hotter. Magnetic fields play an important role in determining the nature of a planet's atmosphere. In addition, a planets' magnetic field can protect its surface from bombardment by charged particles of the solar wind—a necessary condition for the survival of life-forms.

The Atmospheres of the Planets The compositions of planetary atmospheres are shown in Figure 24.4. The Jovian planets have very thick atmospheres composed mainly of hydrogen and helium,

with lesser amounts of water, methane, ammonia, and other hydrocarbons. The Jovian atmospheres are so thick that there is not a clear boundary between "atmosphere" and "planet." By contrast, the terrestrial planets, including Earth, have relatively meager atmospheres composed of carbon dioxide, nitrogen, and oxygen.

Two factors explain these significant differences—solar heating (temperature) and gravity (Figure 24.5). These variables determine what planetary gases, if any, were captured by planets during the formation of the solar system and which were ultimately retained.

During planetary formation, the inner regions of the developing solar system were too hot for ices and gases to condense. By contrast, the Jovian planets formed where temperatures were low and solar heating of planetesimals was minimal. This allowed water vapor, ammonia, and methane to condense into ices. Hence, the gas giants contain large amounts of these volatiles. As the planets grew, the largest Jovian planets, Jupiter and Saturn, also attracted large quantities of the lightest gases, hydrogen and helium.

How did Earth acquire water and other volatile gases? It seems that early in the history of the solar system, gravitational tugs by the developing protoplanets sent planetesimals into very eccentric orbits. As a result, Earth was bombarded with icy objects that originated beyond the orbit of Mars. This was a fortuitous event for organisms that currently inhabit our planet.

Mercury, our Moon, and numerous other small bodies lack significant atmospheres even though they certainly would have been bombarded by icy bodies early in their development. Airless bodies develop where solar heating exceeds a certain level, which depends on the strength of the body's gravity (see Figure 24.5). Simply stated, *less massive planets* have a better chance of losing their atmosphere because gas molecules need less speed to escape their weak gravities. Comparatively warm bodies with small surface gravity, such as our Moon, are unable to hold even heavy gases such as carbon dioxide and nitrogen. Mercury holds trace amounts of gas.

The slightly larger terrestrial planets, Earth, Venus, and Mars, retain some heavy gases including water vapor, nitrogen, and carbon dioxide. However, their atmospheres are miniscule compared to their total mass. Early in their development, the terrestrial planets probably had much thicker atmospheres. Over time, however, these primitive atmospheres gradually

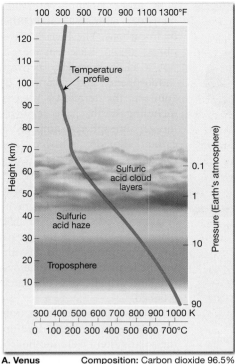

A. Venus Composition: Carbon dioxide 96.5%
Nitrogen 3.5%

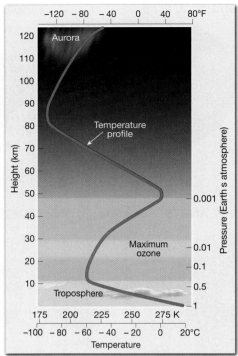

B. Earth Composition: Nitrogen 78%
Oxygen 21%

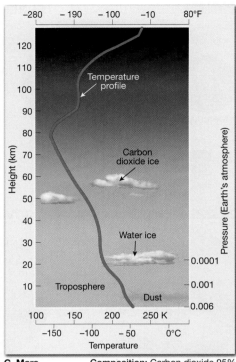

C. Mars Composition: Carbon dioxide 95%
Nitrogen 2.7%
Argon 1.6%

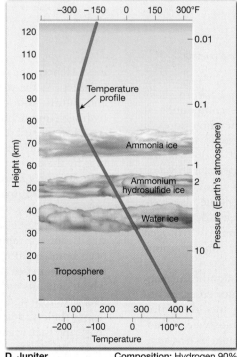

D. Jupiter Composition: Hydrogen 90%
Nitrogen 10%

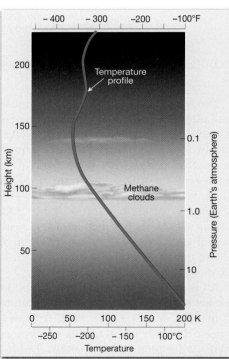

E. Uranus Composition: Hydrogen 83%
Helium 15%

FIGURE 24.4 Comparing the atmospheres of the planets. Jupiter's atmosphere is very similar to Saturn's, whereas the structure of Uranus' atmosphere is thought to be similar to Neptune's.

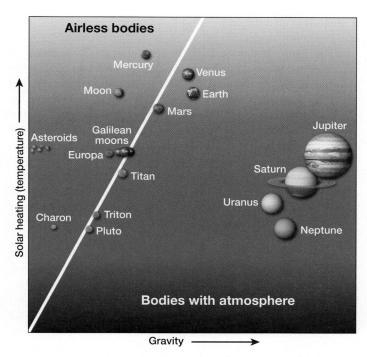

FIGURE 24.5 The factors that explain why some bodies have thick atmospheres, whereas others are airless include: solar heating (temperature) and gravity. Airless worlds are relatively warm and have weak gravity. Bodies with significant atmospheres have weak heating and strong gravity.

changed as certain gases trickled away into space. For example, Earth's atmosphere continues to leak hydrogen and helium (the two lightest gases) into space. This phenomenon occurs near the top of Earth's atmosphere where air is so tenuous that nothing stops the fastest moving ions from flying off into space. The speed required to escape a planet's gravity is called **escape velocity.** Because hydrogen is the lightest gas, it most easily reaches the space needed to overcome Earth's gravity.

Sometime in the distant future, the loss of hydrogen (one of the components of water) will eventually "dry out" Earth's oceans, ending its hydrologic cycle. Life, however, may still hang on in Earth's polar regions.

Because Mars and Venus lack significant magnetic fields, their upper atmospheres are exposed to the brute force of the solar wind, which consists of fast moving charged particles. Without a magnetic field to shield their atmospheres, the solar wind picks up ionic gases and carries them out to space. Mars' atmosphere is enriched in heavy isotopes of both nitrogen and carbon (carbon dioxide), suggesting that it has lost as much as 90 percent of its primitive atmosphere.

Because they have strong gravitational and magnetic fields, the massive Jovian planets have a better chance of retaining their atmospheres. Furthermore, because of their great distances from the Sun, the temperatures that occur in their upper atmospheres are incredibly cold. For example, at its cloud tops, Neptune's atmosphere has a temperature of about $-218\,°C$ ($-360\,°F$)—one of the coldest places in the solar system. Because the molecular motion of a gas is temperature dependent, even hydrogen and helium move too slowly to escape the gravitational pull of these large planets. This partly explains why the outer planets have been able to retain their thick atmospheres.

Weather The planets, with the exception of Mercury, have clouds, atmospheric circulation, and storms. The wide range of cloud compositions is shown in Figure 24.4. The clouds on Venus are composed of sulfuric acid (H_2SO_4) and sulfur dioxide (SO_2). Earth has clouds of water (H_2O), while icy clouds of ammonia (NH_3), methane (CH_4), and water (H_2O) occur in the atmospheres of the Jovian planets.

Wind speeds vary considerably among the planets and are strongly affected by the energy received from the Sun, or in the case of the Jovian planets, from their interiors. Planetary surfaces are unevenly heated (more energy is received at the equator), which leads to dynamic weather systems and fast moving air currents called *jet streams.*

On the Jovian planets, wind speeds of more than 360 kilometers (220 miles) per hour are common. Lightning and whistlers (a phenomenon associated with lightning) have been detected on all the planets except Mercury and Mars, which have very thin atmospheres.

Earth has two types of well-organized storms—mid-latitude cyclones and hurricanes. Mars has dust storms that occasionally cover almost the entire planet, and the Jovian planets have huge rotating storm systems. Jupiter's Great Red Spot is infamous as an on-going counterclockwise-rotating storm trapped between two jet stream-like bands flowing in opposite directions (Figure 24.6). This enormous hurricane-like storm (twice Earth's size) completes a rotation about once every 12 days and has been going on for at least 300 years.

Saturn hosts a rapidly rotating storm called the Great White Spot. Neptune's Great Dark Spot was discovered in 1989

FIGURE 24.6 Jupiter with the Great Red Spot visible in its southern hemisphere. Earth for scale. (Courtesy of NASA)

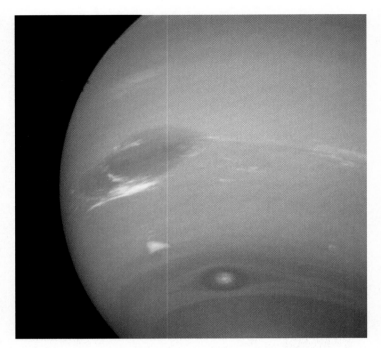

FIGURE 24.7   This image of Neptune was constructed from two images. At the top is the Great Dark Spot, accompanied by bright white clouds. To the south is a bright white area thought to be high cloud tops. Still farther south is a second dark spot with a bright core. (Courtesy of NASA/JPL)

(Figure 24.7). Subsequent observations of the Great Dark Spot by the Hubble Space Telescope in 1994 revealed that this storm disappeared, only to be replaced by another dark spot in the planet's northern hemisphere. Unlike Jupiter's Great Red Spot, this structure appears to be a hole in Neptune's methane cloud deck—similar to the ozone hole on Earth. Winds around the spot were measured at 2400 kilometers (1500 miles) per hour, the fastest in the solar system.

PLANETARY IMPACTS

Planetary impacts have occurred throughout the history of the solar system. On bodies that have little or no atmosphere, such as the Moon and Mercury, even the smallest pieces of interplanetary debris (meteorites) can produce microscopic cavities on individual mineral grains. By contrast, large **impact craters** are the result of collisions with massive bodies, such as asteroids and comets.

Planetary impacts were considerably more common in the early history of the solar system than they are today, with the heaviest bombardment occurring 3.8 to 4.1 billion years ago. Following that period, the rate of cratering diminished dramatically and now remains essentially constant. Because weathering and erosion are almost non-existent on the Moon and Mercury, evidence of their cratered past is clearly evident.

On larger bodies, thick atmospheres may cause the impacting objects to break up and/or decelerate. For example, Earth's atmosphere causes meteoroids with masses of less than 10 kilograms (22 pounds) to lose up to 90 percent of their speed as they penetrate the atmosphere. Therefore, impacts of low-mass bodies produce only small craters on Earth. Earth's atmosphere

is much less effective in slowing large bodies—fortunately, they make very rare appearances.

The formation of a large impact crater is illustrated in Figure 24.8. The meteoroid's high-speed impact compresses the material it strikes, causing an almost instantaneous rebound which ejects material from the surface. Craters excavated by objects that are several kilometers across often exhibit a central peak, such as the one in the large crater in Figure 24.9. Much of the material expelled, called *ejecta*, lands in or near the crater, where it accumulates to form a rim. Large meteoroids may generate sufficient heat to melt some of the impacted rock. Samples of glass beads produced in this manner, as well as rocks consisting of broken fragments welded by the heat of impacts have been collected from the Moon, allowing planetary geologists to learn about such events.

Earth's surface has only a few dozen easily recognizable impact craters. The vast majority have been erased by erosional processes, or tectonic forces that altered crustal rocks. Marble-sized meteors are generally large enough to reach Earth's surface. However, meteors often break apart, or partially vaporize in the atmosphere, with only a few tiny pieces eventually reaching the surface. Consequently, dust sized particles make up the majority of meteorites that impact Earth's surface.

EARTH'S MOON: A CHIP OFF THE OLD BLOCK

The Earth-Moon system is unique because the Moon is the largest satellite relative to its planet. Mars is the only other terrestrial planet with moons, but its tiny satellites are likely captured asteroids. Most of the 150 or so satellites of the Jovian planets are composed of low-density rock-ice mixtures, none of which resemble the Moon. As we will see later, our unique planet-satellite system is closely related to the mechanism that created it.

The diameter of the Moon is 3,475 kilometers (2,160 miles), about one-fourth of Earth's 12,756 kilometers (7,926 miles). The Moon's surface temperature averages about 107 °C (225 °F) for daylight hours and −153 °C (−243 °F) for night. Because its period of rotation on its axis equals its period of revolution around Earth, the same lunar hemisphere always faces Earth. All of the landings of manned *Apollo* missions were confined to the side of the Moon facing Earth.

The Moon's density is 3.3 times that of water, comparable to that of *mantle* rocks on Earth, but considerably less than Earth's average density (5.5 times that of water). The Moon's relatively small iron core is thought to account for much of this difference.

The Moon's low mass relative to Earth results in a lunar gravitational attraction that is one-sixth that of Earth. A person weighing 150 pounds on Earth weighs only 25 pounds on the Moon, although their mass remains the same. This difference allows an astronaut to carry a heavy life-support system with relative ease. If not burdened with such a load, an astronaut could jump six times higher on the Moon than on Earth. The Moon's small mass (and low gravity) is the primary reason it was not able to retain an atmosphere.

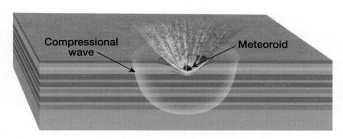

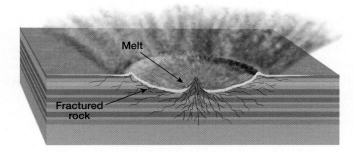

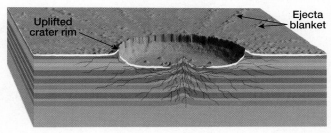

FIGURE 24.8 Formation of an impact crater. The energy of the rapidly moving meteoroid is transformed into heat and compressional waves. The rebound of the compressed rock causes debris to be ejected from the crater. Heat melts some material, producing glass beads. Small secondary craters are formed by the material "splashed" from the impact crater. (After E. M. Shoemaker)

How Did the Moon Form?

Until recently, the origin of the Moon—our nearest planetary neighbor—was a topic of considerable debate among scientists. Current models show that Earth is too small to have formed

with a moon, particularly one so large. Furthermore, a captured moon would likely have an eccentric orbit similar to the captured moons that orbit the Jovian planets.

The current consensus is that the Moon formed as the result of a collision between a Mars-sized body and a youthful, semi-molten Earth about 4.5 billion years ago. (Collisions of this type were probably frequent at that time.) During this explosive event, some of the ejected debris was thrown into Earth's orbit and gradually coalesced to form the Moon. Computer simulations show that most of the ejected material would have come from the rocky mantle of the impactor while its core became part of the growing Earth. This *impact model* is consistent with the Moon's low density and internal structure which consists of a large mantle and a small iron-rich core.

The Lunar Surface   When Galileo first pointed his telescope toward the Moon, he observed two different types of terrain— dark lowlands and brighter, highly cratered highlands (Figure 24.10). Because the dark regions appeared smooth, resembling seas on Earth, they were called **maria** (*mar* = sea, singular *mare*). The *Apollo 11* mission showed conclusively that the maria are exceedingly smooth plains composed of basaltic lavas. These vast plains are strongly concentrated on the side of the Moon facing Earth and cover about 16 percent of the lunar surface. The lack of large volcanic cones on these surfaces is evidence of high eruption rates of very fluid basaltic lavas similar to the Columbia Plateau flood basalts on Earth.

By contrast, the Moon's light-colored areas resemble Earth's continents, so the first observers dubbed them **terrae** (Latin for "land"). These areas are now generally referred to as the **lunar highlands,** because they are elevated several kilometers above the maria. Rocks retrieved from the highlands are mainly breccias, pulverized by massive bombardment early in the Moon's history. The arrangement of terrae and maria result in the legendary "face" of the "man in the moon."

Some of the most obvious lunar features are impact craters. The larger craters shown in Figure 24.10 are about 250 kilometers (150 miles) in diameter, roughly the width of Indiana. A meteoroid 3 meters (10 feet) in diameter can blast out a crater 50 times larger, or about 150 meters (500 feet) in diameter. Larger craters, such as Kepler and Copernicus (32 and 93 kilometers in diameter, respectively) were created from bombardment by bodies 1 kilometer or more in diameter (see Figure 24.10). These two craters are thought to be relatively young because of the bright *rays* (lightly-colored ejected material) that radiate from them for hundreds of kilometers.

History of the Lunar Surface   The evidence used to unravel the history of the lunar surface comes primarily from radiometric dating of rocks returned from *Apollo* missions and studies of crater densities—counting the number of craters per unit area. The greater the crater density, the older the feature. Such evidence suggests that, after the Moon coalesced it passed through the following four phases: (1) formation of the original crust and lunar highlands; (2) excavation of the large impact basins; (3) filling of mare basins, and; (4) formation of rayed craters.

During the late stages of its accretion, the Moon's outer shell was most likely completely melted—literally a magma ocean.

FIGURE 24.9 The 20-kilometer-wide lunar crater Euler in the southwestern part of Mare Imbrium. Clearly visible are the bright rays, central peak, secondary craters, and the large accumulation of ejecta near the crater rim. (Courtesy of NASA)

Then, about 4.4 billion years ago, the magma ocean began to cool and underwent magmatic differentiation (see Chapter 4). Most of the dense minerals, olivine and pyroxene, sank, while less dense silicate minerals floated to form the Moon's primitive crust. The highlands are made of these igneous rocks that rose buoyantly like "scum" from the crystallizing magma. The most common highland rock-type is *anorthosite,* which is composed mainly of calcium-rich plagioglase feldspar.

Once formed, the lunar crust was continually impacted as the Moon swept up debris from the solar nebula. During this time, several large impact basins were created. Then, about 3.8 billion years ago, the Moon, as well as the rest of the solar system, experienced a sudden drop in the rate of meteoritic bombardment.

The Moon's next major event was the filling of the large impact basins created at least 300 million years earlier (Figure 24.11). Radiometric dating of the maria basalts puts their age between 3.0 billion and 3.5 billion years, considerably younger than the initial lunar crust.

The Mare basalts are thought to have originated at depths between 200 and 400 kilometers. They were likely generated by a slow rise in temperature attributed to the decay of radioactive elements. Partial melting probably occurred in several isolated pockets as indicated by the diverse chemical makeup of the rocks retrieved during the *Apollo* missions. Recent evidence suggests that some mare-forming eruptions may have occurred as recently as one billion years ago.

Other lunar surface features related to this period of volcanism include small shield volcanoes (8–12 kilometers in diameter), evidence of pyroclastic eruptions, rilles formed by localized lava channels, and grabens.

The last prominent features to form were rayed craters, as exemplified by the 90 kilometer-wide Copernicus crater shown in Figure 24.10. Material ejected from these younger depressions blankets the maria surfaces and many older, rayless craters. The relatively young Copernicus crater is thought to be about one billion years old. Had it formed on Earth, weathering and erosion would have long since obliterated it.

Today's Lunar Surface: Weathering and Erosion The Moon's small mass and low gravity account for its lack of an atmosphere or flowing water. Therefore, the processes of weathering and erosion that continually modify Earth's surface are absent on the Moon. In addition, tectonic forces are no longer active on the Moon, so quakes and volcanic eruptions have ceased. Because the Moon is unprotected by an atmosphere, erosion is dominated by the impact of tiny particles from space (*micrometeorites*) that continually bombard its surface and gradually smooth the landscape. This activity has crushed and repeatedly mixed the upper portions of the lunar crust.

Both the maria and terrae are mantled with a layer of gray, unconsolidated debris derived from a few billion years of meteoric bombardment (Figure 24.12). This soil-like layer, properly called **lunar regolith** (*rhegos* = blanket, *lithos* = stone) is composed of igneous rocks, breccia, glass beads, and fine *lunar dust.* The lunar regolith is anywhere from 2 to 20 meters thick depending on the age of the surface.

FIGURE 24.10 Telescopic view of the lunar surface from Earth. The major features are the dark maria and the light, highly cratered highlands. (UCO/Lick Observatory Image)

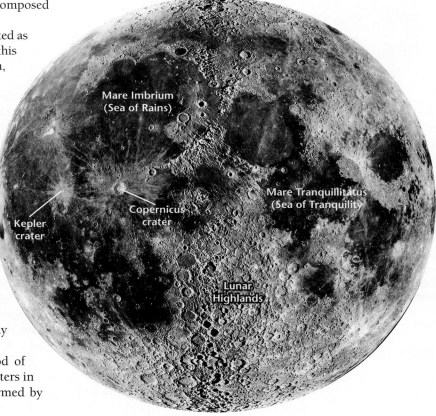

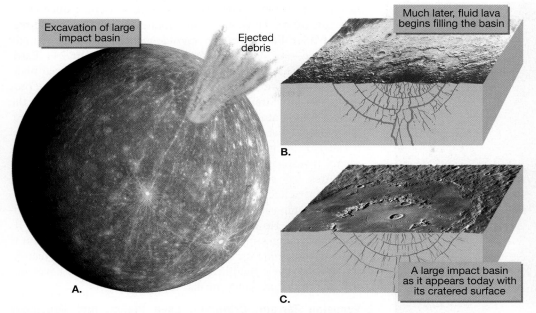

Excavation of large impact basin

Ejected debris

Much later, fluid lava begins filling the basin

B.

A large impact basin as it appears today with its cratered surface

A.

C.

FIGURE 24.11 Formation and filling of large impact basins. **A.** Impact of an asteroid-size mass produced a huge crater hundreds of kilometers in diameter and disturbed the lunar crust far beyond the crater. **B.** Filling of the impact area with fluid basalts, perhaps derived from partial melting deep within the lunar mantle. **C.** Today, such basins make up the lunar maria and a few similar large structures on Mercury.

PLANETS AND MOONS

Mercury: The Innermost Planet

Mercury, the innermost and smallest planet, revolves around the Sun quickly (88 days) but rotates slowly on its axis. Mercury's day-night cycle, which lasts 176 Earth-days, is very long compared to Earth's 24-hour cycle. One "night" on Mercury is roughly equivalent to three months on Earth, and is followed by the same duration of daylight. Mercury has the greatest temperature extremes of any planet. Nighttime temperatures drop as low as $-173\,°C$ ($-280\,°F$) while noontime temperatures exceed $427\,°C$ ($800\,°F$), hot enough to melt tin and lead. These extreme temperatures make life "as we know it" impossible on Mercury.

Mercury absorbs most of the sunlight that strikes it, reflecting only 6 percent into space, a characteristic of terrestrial bodies that have little or no atmosphere. The minuscule amount of gas present on Mercury may have originated from several sources: ionized gas emitted from the Sun; ices that vaporized during a recent comet impact; and/or outgassing of the planet's interior.

Although Mercury is small and scientists expected the planet's interior to have already cooled, the *Messenger* spacecraft detected a magnetic field. This finding suggests that Mercury has a large core that remains hot and fluid enough to generate a magnetic field.

Mercury resembles Earth's Moon in that it has very low reflectivity, no sustained atmosphere, numerous volcanic features, and a heavily cratered terrain (Figure 24.13). The youngest and largest (1300 kilometers in diameter) known impact crater on

FIGURE 24.12 Astronaut Harrison Schmitt sampling the lunar surface. Notice the footprint (inset) in the lunar "soil." (Courtesy of NASA)

FIGURE 24.13   Mercury. This view of Mercury looks similar to Earth's Moon. (Courtesy of NASA)

Mercury is Caloris Basin. Images and other data gathered by *Mariner 10* show evidence of volcanism in and around Caloris Basin and a few other smaller basins. Also like our Moon, Mercury has smooth plains that cover nearly 40 percent of the area imaged by *Mariner 10*. Most of these smooth areas are associated with large impact basins, including Caloris Basin, where lava partially filled the basins and the surrounding lowlands. Consequently, these smooth plains appear to be similar in origin to lunar maria. Hopefully, data gathered by *Messenger* during its orbit around Mercury in 2011 will shed additional light on the relationship between cratering and volcanism.

Unique to Mercury are hundreds of *lobate scarps* (Figure 24.14). Viewed from space, they appear as scalloped-edged cliffs. These cliffs, which cut across numerous craters, are thousands of kilometers in length and some are elevated as much as 3 kilometers above the surrounding landscape. They may be the result of crustal shortening as the interior of the planet cooled, shrinking the planet. As the planet contracted, compressional forces displaced large slabs of crustal rocks over one another along large thrust faults (Figure 24.14).

Venus: The Veiled Planet

Venus, second only to the Moon in brilliance in the night sky, is named for the Roman "Goddess of Love and

Beauty." It orbits the Sun in a nearly perfect circle once every 225 Earth-days. However, Venus rotates in the opposite direction of the other planets (*retrograde motion*) at an agonizingly slow pace—one Venus day is equivalent to about 244 Earth days. Venus has the densest atmosphere of the terrestrial planets, consisting mostly of carbon dioxide (97 percent)—the prototype for an extreme *greenhouse effect*. As a consequence, the surface temperature of Venus averages about 450 °C (900 °F). Temperature variations at the surface are generally minimal because of the intense mixing within the planet's dense atmosphere. Investigations of the extreme and uniform surface temperatures led scientists to more fully understand how the greenhouse effect operates on Earth.

The composition of the Venusian interior is probably similar to Earth's. However, Venus' weak magnetic field means its internal dynamics must be very different. Mantle convection is thought to operate on Venus, but the processes of plate tectonics, which recycle rigid lithosphere, do not appear to have contributed to the present Venusian topography.

Venusian Terrain: Cratering, Lava Plains, and Volcanoes

The surface of Venus is completely hidden from view by a thick cloud layer composed mainly of tiny sulfuric acid droplets. In the 1970s, despite extreme temperatures and pressures, four Russian spacecraft landed successfully and obtained surface images. (As expected, however, all of the probes were crushed by the planet's immense atmospheric pressure within an hour of landing.) Using radar imaging, the unmanned spacecraft *Magellan* mapped Venus' surface in amazing detail (Figure 24.15).

Approximately 1000 impact craters have been identified on Venus, far fewer than Mercury and Mars, but more than Earth. Researchers, who expected that Venus would show evidence of extensive cratering from the heavy bombardment period, found

FIGURE 24.14   Lobate scarps (cliffs) on Mercury. These scarps, which are often more than 1.6 kilometers (1 mile) high, formed when Mercury's crust was contracting as the planet cooled. Image from the *Messenger* orbiter. (Courtesy of NASA)

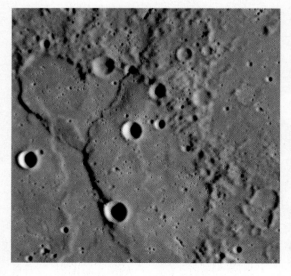

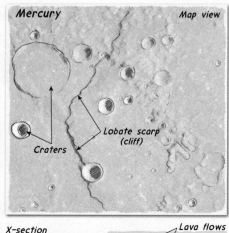

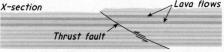

Geologist's Sketch

FIGURE 24.15 This global view of the surface of Venus is computer-generated from years of investigations culminating with the *Magellan* mission. The twisting bright features that cross the globe are highly fractured ridges and canyons of the eastern Aphrodite highland. (Courtesy of NASA/JPL)

instead that a period of extensive volcanism was responsible for resurfacing Venus. Its thick atmosphere also limits the number of impacts by breaking up large incoming meteoroids, and incinerating most of the small debris.

About 80 percent of the Venusian surface consists of low lying plains covered by lava flows, some of which traveled along lava channels that extend for hundreds of kilometers (Figure 24.16). Venus' Baltis Vallis, the longest known lava channel in the solar system, meanders 6,800 kilometers (4,255 miles) across the planet. More than 100 large volcanoes have been identified on Venus. However, high surface temperatures and pressures inhibit explosive volcanism. In addition, Venus' extreme conditions result in volcanoes that tend to be shorter and wider than those on Earth or Mars (Figure 24.17). Maat Mons, the largest volcano on Venus is about 8.5 kilometers high and 400 kilometers wide. By comparison, Mauna Loa, the largest volcano on Earth is about 9 kilometers high and only 120 kilometers wide.

Venus also has major highlands that consist of plateaus, ridges, and topographic rises that stand above the plains. The rises are thought to have formed where hot mantle plumes encountered the base of the planet's crust, causing uplift. Much like mantle plumes on Earth, abundant volcanism is associated with mantle upwelling on Venus (Figure 24.18). Recent data

collected by the European Space Agency's *Venus Express* suggest that Venus' highlands contain silica-rich granitic rock. As such, these elevated landmasses resemble Earth's continents, albeit on a much smaller scale.

Venus versus Earth Similar to Earth in size, density, mass, and location in the solar system, Venus was once referred to as "Earth's twin." Like Earth, mantle upwelling and downwelling occur on Venus and drive tectonic activity at the surface. These processes have created the dozen or so major Venusian highlands that are extensively deformed by both compressional and tensional forces. Furthermore, flow in the mantle produced pervasive linear fractures and *wrinkle ridges*, best described as simple compressional folds.

Unlike Earth, plate tectonics does not operate on Venus. The current view is that Venus' outer shell is too strong to break into plates. This can be explained by the gradual loss of water which also contributed to the dominance of CO_2 in the Venusian atmosphere. Remember that water lowers the melting temperature of rock. Therefore, the rocks that compose Venus' dry lithosphere are farther from their melting points and, consequently, stiffer and stronger than similar "wet" rocks of Earth's lithosphere.

The reason for Venus' loss of water is not fully understood. One hypothesis suggests that its weak magnetic field exposed its atmosphere to erosion by the solar wind. Hydrogen, an ingredient of water, has long since been stripped away by solar radiation.

Some researchers believe that Venus was once a more pleasant place—a second Earth-like blue marble covered with water. If this was the case, what led to it becoming the hellish, uninhabitable planet we know today? Scientists anticipate addressing this question with future space explorations designed to shed light on interactions among a planet's atmosphere, surface, and interior.

Mars: The Red Planet

Mars, approximately one-half the diameter of Earth, revolves around the Sun in 687 Earth-days. Mean surface temperatures range from lows of −140 °C (−220 °F) in the winter to highs of 20 °C (68 °F) in the summer. Although seasonal temperature variations are similar to Earth's, daily temperature variations are greater due to its very thin atmosphere (only 1 percent as dense as Earth's). The tenuous Martian atmosphere consists primarily of carbon dioxide (95 percent), with small amounts of nitrogen, oxygen, and water vapor.

Mars' Topography Mars, like the Moon, is pitted with impact craters. The smaller craters are usually filled with wind-blown dust—confirming that Mars is a dry, desert world. The reddish color of the Martian landscape is iron oxide (rust). Large impact

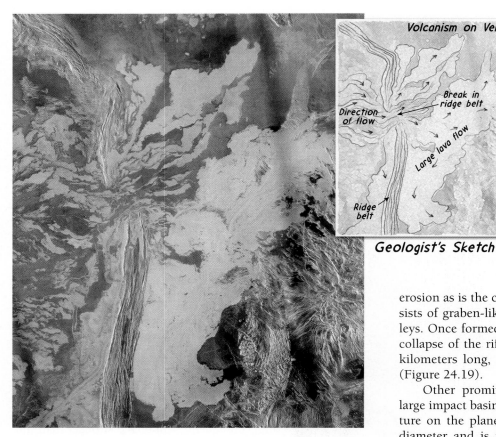

FIGURE 24.16 Extensive lava flows on Venus. This *Magellan* radar image shows a system of lava flows that orignated from a volcano named Ammavaru, that lies approximately 300 kilometers (186 miles) west of the scene. The lava, which appears bright in this radar image, has rough surfaces, whereas the darker flows are smooth. Upon breaking through the ridge belt (left of center), the lava collected in a 100,000 square kilometer pool. (Courtesy of NASA)

craters provide information about the nature of the Martian surface. For example, if the surface is composed of dry dust and rocky debris, ejecta similar to that surrounding lunar craters is to be expected. But the ejecta surrounding some Martian craters has a different appearance—that of a muddy slurry that was splashed from the crater. Planetary geologists believe that a layer of permafrost (frozen, icy soil) lies below portions of the Martian surface and that impacts heated and melted the ice to produce the fluid-like appearance of these ejecta.

About two-thirds of the surface of Mars consists of heavily cratered highlands, concentrated mostly in its southern hemisphere (Figure 24.19). The period of extreme cratering occurred early in the planet's history and ended about 3.8 billion years ago, as it did in the rest of the solar system. Thus, Martian highlands are similar in age to the lunar highlands.

Based on relatively low crater counts, the northern plains, which account for the remaining one-third of the planet, are younger than the highlands (Figure 24.19). The plains' relatively flat topography is consistent with vast outpourings of fluid basaltic lavas. Visible on these plains are volcanic cones, some with summit pits and lava flows with wrinkled edges.

Located along the Martian equator is an enormous elevated region, about the size of North America, called the *Tharsis bulge* (Figure 24.19). This feature, about 10 kilometers high, appears to have been uplifted and capped with a massive accumulation of volcanic rock that includes the planet's five largest volcanoes. A much smaller volcanic center (bulge) also exists.

The tectonic forces that created the Tharsis region also produced fractures that radiate from its center, like spokes on a bicycle wheel. Along the eastern flanks of the bulge, a series of vast canyons called *Valles Marineris* (Mariner Valleys) developed. Valles Marineris is so vast it can be seen in the image of Mars in Figure 24.19. This canyon network was largely created by downfaulting, not by stream erosion as is the case for Arizona's Grand Canyon. Thus, it consists of graben-like valleys similar to the East African Rift Valleys. Once formed, Valles Marineris grew by water erosion and collapse of the rift walls. The main canyon is more than 5000 kilometers long, 7 kilometers deep, and 100 kilometers wide (Figure 24.19).

Other prominent features on the Martian landscape are large impact basins. Hellas, the largest identifiable impact structure on the planet, is about 2300 kilometers (1400 miles) in diameter and is the planet's lowest elevation (Figure 24.19). Debris ejected from this basin contributed to the elevation of the adjacent highlands. Other buried crater basins that are even larger than Hellas probably exist.

FIGURE 24.17 Venus' Sapas Mons (center) is a broad volcano 400 kilometers (250 miles) wide. The bright areas in the foreground are lava flows. Another large volcano, Maat Mons, is in the background. This computer generated view is constructed from data acquired by the *Magellan* spacecraft. (Courtesy of NASA/JPL)

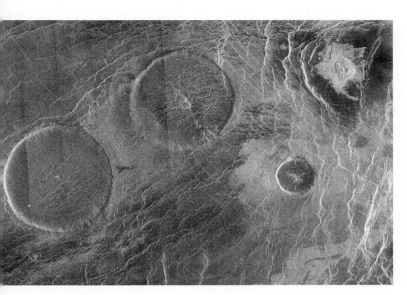

FIGURE 24.18 The prominent circular features are volcanic domes, 65 kilometers (39 miles) in diameter, with broad, flat tops less than one kilometer (0.6 mile) in height. Sometimes referred to as pancake domes, they represent a unique category of volcanic extrusions on Venus that formed from viscous (sticky) lava. The cracks and pits commonly found in these features result from the cooling and withdrawl of lava. (Courtesy of NASA/JPL)

Volcanism was prevalent on Mars during most of its history. The scarcity of impact craters on some volcanic surfaces suggests that the planet is still active. Mars has several of the largest known volcanoes in the solar system, including the largest, Olympus Mons, which is about the size of Arizona and stands nearly three times higher than Mount Everest. This gigantic volcano was last active about 100 million years ago and resembles Earth's Hawaiian shield volcanoes (Figure 24.20).

How did the volcanoes on Mars grow so much larger than similar structures on Earth? The largest volcanoes on the terrestrial planets tend to form where plumes of hot rock rise from deep within their interiors. On Earth, moving plates keep the crust in constant motion. Consequently, mantle plumes tend to produce a chain of volcanic structures, like the Hawaiian Islands. By contrast, on Mars, plate tectonics is absent so successive eruptions build up in the same location. As a result, enormous volcanoes such as Olympus Mons form rather than a string of smaller ones.

Currently, the dominant force shaping the Martian surface is wind erosion. Extensive dust storms, with winds up to 270 kilometers (170 miles) per hour, can persist for weeks. Dust devils have also been photographed. Most of the Martian landscape resem-

bles Earth's rocky deserts, with abundant dunes and low areas partially filled with dust.

Yes, Water Ice on Mars! Liquid water does not appear to exist anywhere on the Martian surface. However, poleward of about 30 degrees latitude ice can be found within a meter of the surface. In the polar regions it forms small permanent ice caps along with carbon dioxide ice. In addition, considerable evidence indicates that in the first billion years of the planet's history, liquid water flowed on the surface creating stream valleys and related features.

One location where running water was involved in carving valleys can be seen in the Mars *Reconnaissance Orbiter* image in Figure 24.21. Researchers have proposed that melting of subsurface ice caused spring-like seeps to emerge along the valley wall, slowly creating the gullies—a process that may still be active today.

Other channels have stream-like banks and contain numerous teardrop-shaped islands (Figure 24.22). These valleys appear to have been cut by catastrophic floods with discharge rates that were more than 1,000 times greater than those of the Mississippi River. Most of these large flood channels emerge from areas of chaotic topography that appear to have formed when the surface collapsed. The most likely source of water for these flood-created valleys was the melting of subsurface ice. If the melt water was trapped beneath a thick layer of permafrost, pressure could mount until a catastrophic release occurred. As the water escaped, the overlying surface would collapse, creating the chaotic terrain.

Not all Martian valleys appear to be the result of water released in this manner. Some exhibit branching, tree-like patterns that resemble dendritic drainage networks on Earth. In addition, the *Opportunity* rover investigated structures similar to features created by water on Earth. These included layered sedimentary rocks, playas (salt flats), and lake beds. Minerals that

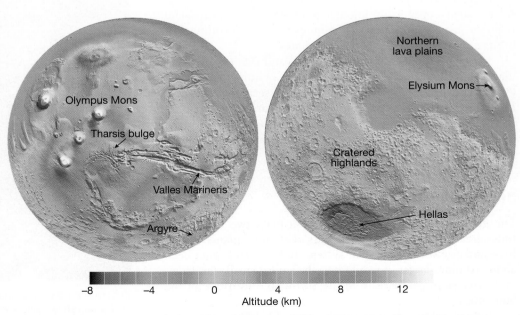

FIGURE 24.19 Two computer-generated globes of Mars with some major surface features labeled. Color represents height above (or below) the mean planetary radius—white is about 12 kilometers above average and dark blue 8 kilometers below average. (Courtesy of NASA/JPL)

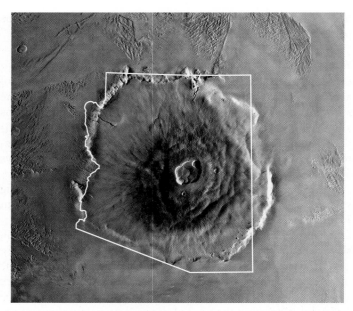

FIGURE 24.20 Image of Olympus Mons, an inactive shield volcano on Mars that covers an area about the size of the state of Arizona. (Courtesy of the U.S. Geological Survey)

form only in the presence of water such as hydrated sulfates were also detected. Small spheres of hematite, dubbed "blueberries," were found that probably precipitated from water to form lake sediments. Nevertheless, with the exception of the polar regions, water does not appear to have significantly altered the topography of Mars for more than a billion years.

The Martian Satellites Phobos and Deimos, the two tiny satellites of Mars—24 and 15 kilometers in diameter, respectively—were not discovered until the late 1800s because their sizes made them nearly impossible to view telescopically. Phobos is nearer to its parent than any other natural satellite in the solar system—only 5,500 kilometers (3,400 miles) away—and requires just 7 hours and 39 minutes for one revolution. It orbits *around* Mars faster than Mars rotates!

Mariner 9 revealed that both Phobos and Deimos are irregularly shaped and have surfaces that display numerous impact craters (Figure 24.23). It is likely that these moons are captured celestial bodies. Gravitational forces are lowering Phobos's orbit and, as a result, it will either impact Mars, or break up into a planetary ring that will eventually rain on the planet's surface.

Jupiter: Lord of the Heavens

The giant among planets, Jupiter has a mass two and a half times greater than the combined mass of all other planets, satellites, and asteroids in the solar system. In fact, had Jupiter been about 10 times larger, it would have become a small star. However, it pales in comparison to the Sun with only 1/800 of the Sun's mass.

Jupiter orbits the Sun once every 12 Earth-years, and rotates more rapidly than any other planet, completing one rota-

tion in slightly less than 10 hours. When viewed telescopically, the effect of this fast spin is noticeable. The bulge of the equatorial region and the contraction of the polar dimension are evident (see the Polar Flattening column in Table 24.1).

Jupiter's appearance is mainly attributable to the colors of light reflected from its three main cloud layers (see Figure 24.4). The warmest, and lowest, layer is composed mainly of water ice and appears blue-gray—generally not seen in visible-light images. A little higher, where temperatures are cooler, is a layer of brown to orange-brown clouds of ammonium hydrosulfide droplets. These colors are thought to be by-products of chemical reactions occurring in Jupiter's atmosphere. Near the top of its atmosphere lie white wispy clouds of ammonia ice.

Because of its immense gravity, Jupiter is shrinking a few centimeters each year. This contraction generates most of the heat that drives Jupiter's atmospheric circulation. Thus, unlike winds on Earth, which are driven by solar energy, the heat ema-

FIGURE 24.21 This image was obtained by the *Mars Reconnaissance Orbiter* and shows gullies emanating from rocky cliffs. The meandering and braided patterns are typical of water-carved channels. (Courtesy of NASA/JPL)

and cooling, whereas the dark belts represent cool material that is sinking and warming. This convective circulation, along with Jupiter's rapid rotation, generates the high-speed, east-west flow observed between the belts and zones.

The largest storm on the planet is the Great Red Spot. This enormous, anticylonic storm that is twice the size of Earth, has been known for 300 years. In addition to the Great Red Spot, there are various white and brown oval-shaped storms (see Figure 24.24). The white ovals are the cold cloud tops of huge storms many times larger than hurricanes on Earth. The brown storm clouds reside at lower levels in the atmosphere. Lightning in various white oval storms has been photographed by the *Cassini* spacecraft, but the strikes appear to be less frequent than on Earth.

Jupiter's magnetic field, the strongest in the solar system, is probably generated by a rapidly rotating, liquid metallic hydrogen layer surrounding its core. Bright auroras, associated with the magnetic field, have been photographed over Jupiter's poles (Figure 24.25). Unlike Earth's auroras that occur only in conjunction with heightened solar activity, Jupiter's auroras are continous. The charged particles needed to fuel the auroras come mainly from volcanic activity on Jupiter's moon, Io. The gases trapped by Jupiter's magnetic field are ionized by ultraviolet radiation from the Sun. The interaction of these energetic ions with other atmosphereic molecules is what generates these colorful displays.

Jupiter's Moons: Jupiter's satellite system, consisting of 63 moons discovered thus far, resembles a miniature solar system. Galileo discovered the four largest satellites, referred to as

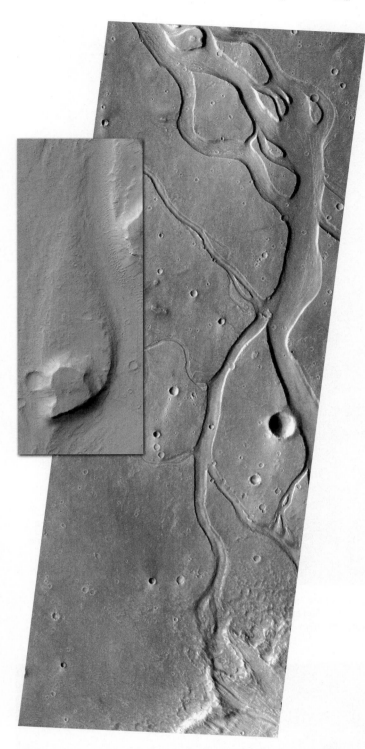

FIGURE 24.22 Stream-like channels are strong evidence that Mars once had flowing water. Inset shows close-up of streamlined island where running water encountered resistant material along its channel. (Courtesy of NASA/JPL)

nating from Jupiter's interior produces the huge convection currents observed in its atmosphere.

Jupiter's convective flow produces alternating dark-colored *belts* and light-colored *zones* as shown in Figure 24.24. The light clouds (*zones*) are regions where warm material is ascending

FIGURE 24.23 Phobos, one of Mars's two small moons. Its small size and irregular shape mean Phobos is likely an asteroid captured by Mars. The most prominent feature on Phobos is the large crater in the lower right. (Courtesy of NASA)

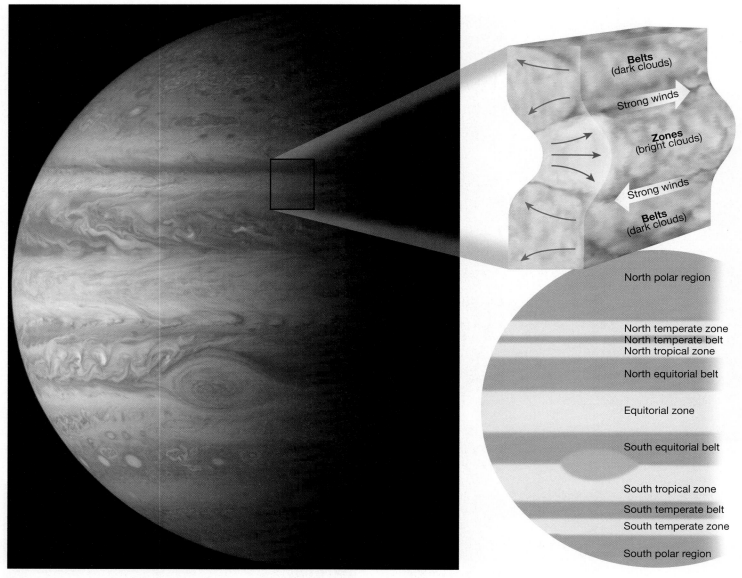

FIGURE 24.24   The structure of Jupiter's atmosphere. The areas of light clouds (*zones*) are regions where gases are ascending and cooling. Sinking dominates the flow in the darker cloud layers (*belts*). This convective circulation, along with the rapid rotation of the planet, generates the high-speed winds observed between the belts and zones.

Galilean satellites, in 1610 (Figure 24.26). The two largest, Ganymede and Callisto, are roughly the size of Mercury, whereas the two smaller ones, Europa and Io, are about the size of Earth's Moon. The eight largest moons appear to have formed around Jupiter as the solar system condensed.

Jupiter also has many very small satellites (about 20 kilometers in diameter) that revolve in the opposite direction (*retrograde motion*) of the largest moons, and have eccentric orbits steeply inclined to the Jovian equator. These satellites appear to be asteroids or comets that passed near enough to be gravitationally captured by Jupiter, or are remnants of the collisions of larger bodies.

The Galilean moons can be observed with binoculars or a small telescope and are interesting in their own right. Images from *Voyagers 1* and *2* revealed, to the surprise of most geoscien-

FIGURE 24.25   View of Jupiter's aurora taken by the Hubble Space Telescope. This phenomenon is produced by high energy electrons racing along Jupiter's magnetic field. The electrons excite atmospheric gases and make them glow. (Courtesy of NASA/John Clark)

FIGURE 24.26 Jupiter's four largest moons (from left to right) are called the Galilean moons because they were discovered by Galileo. **A.** The innermost moon, Io, is one of only three volcanically active bodies known to exist in the solar system. **B.** Europa, smallest of the Galilean moons, has an icy surface that is criss-crossed by many linear features. **C.** Ganymede, the largest Jovian satellite, exhibits cratered areas, smooth regions, and areas covered by numerous parallel grooves. **D.** Callisto, the outermost of the Galilean satellites, is densely cratered, much like Earth's moon. (Courtesy of NASA/NGS Image Collection)

A. Io   B. Europa   C. Ganymede   D. Callisto

tists, that each of the four Galilean satellites is a unique world (Figure 24.26). The *Galileo* mission also unexpectedly revealed that the composition of each satellite is strikingly different, implying a different evolution for each. For example, Ganymede has a dynamic core that generates a strong magnetic field not observed in other satellites.

The innermost of the Galilean moons, Io, is perhaps the most volcanically active body in our solar system. In all, more than 80 active, sulfurous volcanic centers have been discovered. Umbrella-shaped plumes have been observed rising from Io's surface to heights approaching 200 kilometers (Figure 24.27A). The heat source for volcanic activity is tidal energy generated by a relentless "tug of war" between Jupiter and the other Galilean satellites—with Io as the rope. The gravitational field of Jupiter and the other nearby satellites pull and push on Io's tidal bulge as its slightly eccentric orbit takes it alternately closer to, then farther from Jupiter. This gravitational flexing of Io is transformed into heat (similar to the back-and-forth bending of a piece of sheet metal) and results in Io's spectacular sulfurous volcanic eruptions. Moreover, lava, thought to be mainly composed of silicate minerals, regularly erupts on its surface (Figure 24.27B).

Jupiter's Rings One of the surprising aspects of the *Voyager 1* mission was the discovery of Jupiter's ring system. More recently, the ring system was thoroughly investigated by the *Galileo* mission. By analyzing how these rings scatter light, researchers determined that the rings are composed of fine, dark particles, similar in size to smoke particles. Furthermore, the faint nature of the rings indicates that these minute particles are

widely dispersed. The main ring is composed of particles believed to be fragments blasted from the surfaces of Metis and Adrastea, two small moons of Jupiter. Impacts on Jupiter's moons Amalthea and Thebe are believed to be the source of the debris from which the outer Gossamer ring formed.

Students Sometimes Ask . . .

Besides Earth, do any other bodies in the solar system have liquid water?

The planets closer to the Sun than Earth are considered too warm to contain liquid water, and those farther from the Sun are generally too cold (although some features on Mars indicate that it probably had abundant liquid water at some point in its history). The best prospects of finding liquid water within our solar system lie beneath the icy surfaces of some of Jupiter's moons. For instance, an ocean of liquid water is possibly hidden under Europa's outer covering of ice. Detailed images from *Galileo* have revealed that Europa's icy surface is quite young and exhibits cracks apparently filled with dark fluid from below. This suggests that under its icy shell, Europa must have a warm, mobile interior—perhaps an ocean. Because liquid water is a necessity for life as we know it, there has been considerable interest in sending an orbiter to Europa—and eventually a lander capable of launching a robotic submarine—to determine if it harbors life.

UNDERSTANDING EARTH

Pathfinder—The First Geologist on Mars

On July 4, 1997, the *Mars Pathfinder* bounced onto a rock-littered valley of Mars (Ares Vallis) and deployed its wheeled companion, *Sojourner.* Over the next three months the lander sent three gigabits of data back to Earth, including 16,000 images and 20 chemical analyses. The landing site, a vast rolling landscape apparently carved by ancient floods, was selected in anticipation that a variety of rock types would be available for *Sojourner* to examine.

Sojourner carried an alpha photon X-ray spectrometer (APXS) used to determine the chemical composition of Martian rocks and "soil" (regolith) at the landing site (Figure 24.A). Researchers, using close-up images of the rocks, determined that the Martian rocks retrieved were igneous. One hard, white, flat object named Scooby Doo, that was originally thought to be a sedimentary rock, is probably well-cemented soil.

Currently, three orbiting spacecraft and two operational rovers, *Spirit* and *Opportunity,* continue to gather data including daily weather reports, and occasionally record geologic events. Figure 24.B, an image captured by the *Mars Reconnaissance Orbiter,* shows dust clouds rising immediately following a Martian avalanche near its North Pole.

FIGURE 24.A   *Pathfinder*'s rover *Sojourner* (left) obtaining data on the chemical composition of a Martian rock known as Yogi. (Courtesy of NASA)

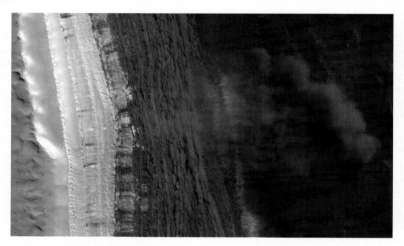

FIGURE 24.B   Martian avalanche speeds down a steep slope about 700 meters (2300 feet) high. The reddish layers are known to be rich in water ice. The avalanche is thought to consist of more ice than rock. (Courtesy of NASA)

Saturn: The Elegant Planet

Requiring more than 29 Earth-years to make one revolution, Saturn is almost twice as far from the Sun as Jupiter, yet their atmospheres, compositions, and internal structures are remarkably similar. The most striking feature of Saturn is its system of rings, first observed by Galileo in 1610 (Figure 24.28). With his primitive telescope, the rings appeared as two small bodies adjacent to the planet. Their ring nature was determined 50 years later by the Dutch astronomer Christian Huygens.

Saturn's atmosphere, like Jupiter's, is dynamic (Figure 24.28). Although the bands of clouds are fainter and wider near the equator, rotating "storms" similar to Jupiter's Great Red Spot occur in Saturn's atmosphere as does intense lightning. Although the atmosphere is nearly 75 percent hydrogen and 25 percent helium, the clouds are composed of ammonia, ammonia

hydrosulfide, and water, each segregated by temperature. Like Jupiter, the atmosphere's dynamics are driven by the heat released by gravitational compression.

Saturn's Moons   The Saturnian satellite system consists of 61 known moons of which 53 have been named. The moons vary significantly in size, shape, surface age, and origin. Twenty-three of the moons are "original" satellites that formed in tandem with their parent planet. At least two (Dione and Tethys) show evidence of tectonic activity, where internal forces have ripped apart their icy surfaces. Others, like Hyperion, are so porous that impacts punch into their surfaces, and Rhea may have its own rings (Figure 24.29). Many of Saturn's smallest moons have irregular shapes and are only a few tens of kilometers in diameter.

Saturn's largest moon, Titan, is larger than Mercury and is the second-largest satellite in the solar system. Titan, and Nep-

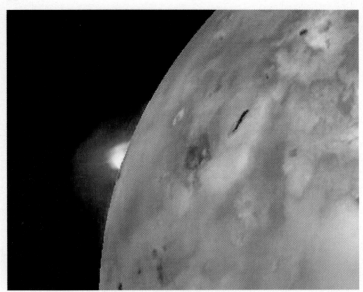

FIGURE 24.27 A volcanic eruption on Jupiter's Moon Io. **A.** This plume of volcanic gases and debris is rising more than 100 kilometers (60 miles) above Io's surface. **B.** The bright red area on the left side of the image is newly erupted lava. (Courtesy of NASA)

tune's Triton, are the only satellites in the solar system known to have substantial atmospheres. Titan was visited and photographed by the *Huygens* probe in 2005. The atmospheric pressure at Titan's surface is about 1.5 times that at Earth's surface, and the atmospheric composition is about 98 percent nitrogen and 2 percent methane with trace organic compounds. Titan has Earth-like geological landforms and geological processes, such as dune formation and stream-like erosion caused by methane "rain." In addition, the northern latitudes appear to have lakes of liquid methane.

Enceladus is another unique satellite of Saturn—one of the few where active eruptions have been observed (Figure 24.30). The outgassing, comprised mostly of water, is thought to be the main source replenishing the material in Saturn's E ring. The

geyser-like activity occurs in an area called "tiger stripes" that consists of four large fractures with ridges on either side.

Saturn's Ring System In the early 1980s, the nuclear-powered *Voyagers 1* and 2 explored Saturn within 100,000 miles of its surface. More information was collected about Saturn in that short time than had been acquired since Galileo first viewed this "elegant planet" in the early 1600s. More recently, observations from ground-based telescopes, the Hubble Space Telescope, and the *Cassini-Huygens* spacecraft have added to our knowledge of Saturn's ring system. In 1995 and 1996, when the positions of Earth and Saturn allowed the rings to be viewed edge-on, Saturn's faintest rings and satellites became visible. (The rings were visible edge-on again in 2009.)

Saturn's ring system is more like a large rotating disk of varying density and brightness than a series of independent ringlets. Each ring is composed of individual particles—mainly water ice with lesser amounts of rocky debris—that circle the planet while regularly impacting one another. There are only a few gaps—most of the areas that look like empty space either contain fine dust particles, or coated ice particles that are inefficient reflectors of light.

Most of Saturn's rings fall into one of two categories based on density. Saturn's main (bright) rings, designated A and B, are tightly packed and contain particles that range in size from a few centimeters (pebble-size) to tens of meters (house-size), with most of the particles being roughly the size of a large snowball (see Figure 24.28). In the dense rings, particles collide frequently as they orbit the planet. Although Saturn's main rings (A and B) are 40,000 kilometers wide, they are very thin, only 10 to 30 meters from top to bottom.

At the other extreme are the faint rings (C, D, F, G, and E). Saturn's outermost ring, not visible in Figure 24.28, is composed of widely dispersed, smoke-size particles. The faint rings are thought to be thicker (100 to 1000 meters) than Saturn's bright rings, but conclusive evidence is not yet available.

Studies have shown that the gravitational tugs of nearby moons tend to shepherd the ring particles by gravitationally altering their orbits (Figure 24.31). For example, the F ring, which is very narrow, appears to be the work of satellites located on either side that confine the ring by pulling back particles that try to escape. Whereas the Cassini Division, a clearly visible gap in Figure 24.28, arises from the gravitational pull of Jupiter's moon, Mimas.

Some of the ring particles are believed to be debris ejected from the moons embedded in them. It is also possible that material is continually recycled between the rings and the ring moons. The ring moons gradually sweep up particles, which are subsequently ejected by collisions with large chunks of ring material, or perhaps by energetic collisions with other moons. It seems then, that planetary rings are not the timeless features that we once thought—rather, they are continually recycled.

The origin of planetary ring systems is still being debated. Perhaps the rings formed simultaneously and from the same material as the planets and moons—condensing from a flattened cloud of dust and gases that encircled the parent planet. Or, perhaps the rings formed later, when a moon or large asteroid was

FIGURE 24.28 Image taken by the Earth-orbiting Hubble Space Telescope shows Saturn's dynamic ring system. The two bright rings, called A ring (outer) and B ring (inner) are separated by the Cassini division. A second small gap (Encke gap) is also visible as a thin line in the outer portion of the A ring. (Courtesy of NASA)

gravitationally pulled apart after straying too close to a planet. Yet another hypothesis suggests that a foreign body collided catastrophically with one of the planet's moons; the fragments of which would tend to jostle one another and form a flat, thin ring. Researchers expect more light to be shed on the origin of planetary rings as the *Cassini* spacecraft continues its tour of Saturn.

Uranus and Neptune: Twins

While Earth and Venus have many similar traits, Uranus and Neptune are perhaps more deserving of being called "twins." They are nearly equal in diameter (both about four times the size of Earth), and are both bluish in appearance—a result of methane in their atmospheres. Their days are nearly the same length and their cores are made of rocky silicates and iron—similar to the other gas giants. Their mantles, comprised mainly of water, ammonia, and methane, are thought to be very different from Jupiter and Saturn (see Figure 24.3). One of the most pronounced differences between Uranus and Neptune is the time they take to complete one revolution around the Sun—84 and 165 Earth-years, respectively.

Uranus: The Sideways Planet Unique to Uranus is its axis of rotation, which lies nearly parallel to the ecliptic ("lying on its side"). Its rotational motion, therefore, resembles a rolling ball instead of a spinning toy top (Figure 24.32). This unusual characteristic of Uranus is likely due to a huge impact that essentially knocked the planet sideways from its original orbit early in its evolution.

FIGURE 24.29 Saturn's impact-pummeled satellite, Hyperion, imaged by the *Cassini Orbiter.* Planetary geologists think Hyperion's surface is so weak and porous that impacts punch into its surface. (Courtesy of NASA/JPL)

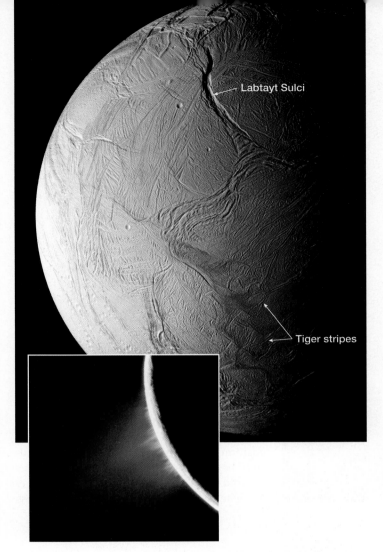

recently geologically active—most likely driven by gravitational heating, as occurs on Io.

Uranus' Rings   A surprise discovery in 1977 showed that Uranus has a ring system. The find occurred as Uranus passed in front of a distant star and blocked its view, a process called *occultation* (*occult* = hidden). Observers saw the star "wink" briefly five times (meaning five rings) before the primary occultation and again five times afterward (see Figure 24.32). More recent ground- and space-based observations indicate that Uranus has at least ten sharp-edged, distinct rings orbiting its equatorial region. Interspersed among these distinct structures are broad sheets of dust.

Neptune: The Windy Planet   Because of its great distance from Earth, astronomers knew very little about Neptune until 1989. Twelve years and nearly 3 billion miles of *Voyager 2* travel provided investigators an amazing opportunity to view the outermost planet in the solar system.

Neptune has a dynamic atmosphere, much like that of the other Jovian planets (Figure 24.34). Record wind speeds that exceed 2400 kilometers (1500 miles) per hour encircle the planet, making Neptune one of the windiest places in the solar system. Recall that Neptune exhibits large dark spots thought to be rotating storms similar to Jupiter's Great Red Spot. However, Neptune's storms appear to have comparatively short life spans—usually only a few years. Another feature that Neptune has in common with the other Jovian planets are layers of white, cirrus-like clouds (probably frozen methane) about 50 kilometers above the main cloud deck.

Neptune's Moons   Neptune has 13 known satellites, the largest of which is the moon Triton—the remaining 12 are small, irregularly shaped bodies. Triton is the only large moon in the

FIGURE 24.30   NASA's *Cassini Orbiter* captured this mosiac of Saturn's tectonically active, icy satellite Enceladus. The northern hemisphere contains a one-kilometer deep chasm, while linear features, called tiger stripes, are visible in the lower right. Inset image show jets spurting ice particles, water vapor, and organic compounds from the area of the tiger stripes. (Courtesy of NASA/JPL)

Uranus, once thought to be weatherless, shows evidence of huge storm systems the size of the United States. Recent photographs from the Hubble Space Telescope also reveal banded clouds composed mainly of ammonia and methane ice—similar to the cloud systems of the other gas giants.

Uranus' Moons   Spectacular views from *Voyager 2* showed that Uranus' five largest moons have varied terrains. Some have long, deep canyons and linear scars, whereas others possess large, smooth areas on otherwise crater-riddled surfaces (Figure 24.33). Studies conducted at California's Jet Propulsion Laboratory suggest that Miranda, the innermost of the five largest moons, was

FIGURE 24.31   Two of Saturn's ring moons. **A.** Pan is a small moon about 30 kilometers in diameter that orbits in the Encke gap, located in the A ring. It is responsible for keeping the Encke gap open. **B.** Prometheus, a potato-shaped moon, acts as a ring shepard. Its gravity helps confine the moonlets in Saturn's thin F ring. (Courtesy of NASA/JPL)

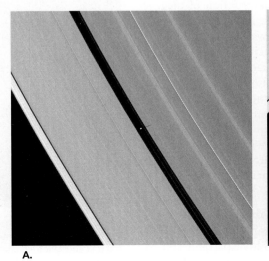

A.

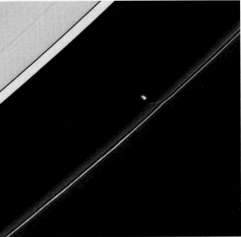

B.

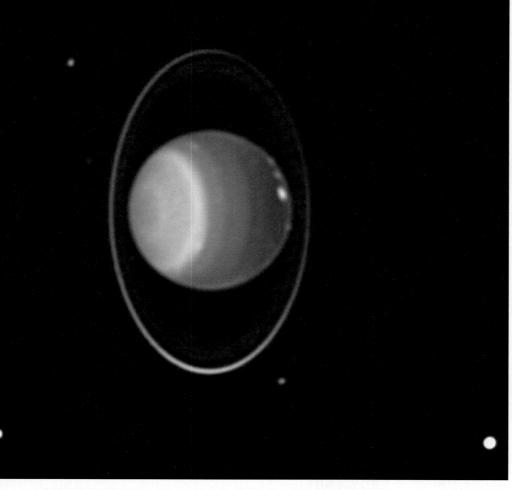

Jupiter's in that they appear faint, which suggests they are composed mostly of dust-size particles. Neptune's rings also display red colors that indicate the dust is composed of organic compounds.

MINOR MEMBERS OF THE SOLAR SYSTEM

There are countless chunks of debris in the vast spaces separating the eight planets and in the outer reaches of the solar system. In 2006, the International Astronomical Union organized solar system objects not classified as planets or moons into two broad categories: (1) dwarf planets and (2) small solar system bodies that include the *asteroids, comets,* and *meteoroids.* The newest grouping, dwarf planets, includes Ceres, the largest known object in the asteroid belt, and Pluto, a former planet.

Asteroids and meteoroids are composed of rocky and/or metallic material with compositions somewhat like the terrestrial planets. They are distinguished according to size: asteroids are larger than 100 meters in diameter, whereas meteoroids have diameters less

FIGURE 24.32 Uranus surrounded by its major rings and 10 of its 17 known moons. Also visible are cloud patterns and several oval storm systems. This false-color image was generated from data obtained by Hubble's Near Infrared Camera. (Image by Hubble Space Telescope courtesy of NASA)

solar system that exhibits retrograde motion, indicating that it most likely formed independently, and was later gravitationally captured by Neptune (Figure 24.35).

Triton and a few other icy moons erupt "fluid" ices—an amazing manifestation of volcanism. **Cryovolcanism** (from the Greek *Kryos,* meaning "frost") describes the eruption of magmas derived from the partial melting of ice instead of silicate rocks. Triton's icy magma is a mixture of water-ice, methane, and probably ammonia. When partially melted, this mixture behaves as molten rock does on Earth. In fact, upon reaching the surface these magmas can generate quiet outpourings of ice lavas, or occasionally, produce explosive eruptions. An explosive eruptive column can generate the ice equivalent of volcanic ash. In 1989, *Voyager 2* detected active plumes on Triton that rose 8 kilometers above the surface and were blown downwind for more than 100 kilometers. In other environments, ice lavas develop that can flow great distances from their source—similar to the fluid basaltic flows on Hawaii.

Neptune's Rings Neptune has five named rings, two of which are broad, and three that are narrow, perhaps no more than 100 kilometers wide. The outermost ring appears to be partially confined by the satellite Galatea. Neptune's rings are most similar to

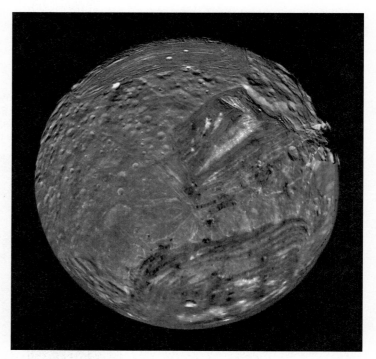

FIGURE 24.33 Image of Uranus' moon, Miranda, obtained by *Voyager 2.* Miranda's surface consists of an old, heavily cratered terrain and a contrasting young terrain characterized by chasms, cliffs, and ridges. (Courtesy of NASA/USGS)

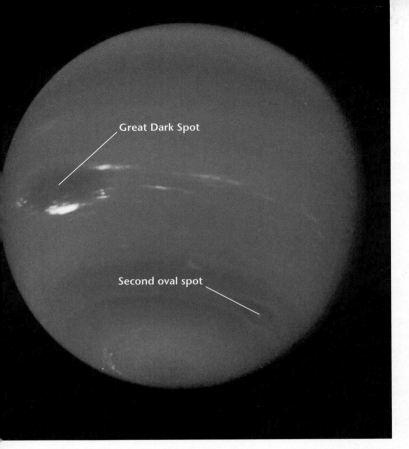

FIGURE 24.34   This image of Neptune shows the Great Dark Spot (left center). Also visible are bright cirrus-like clouds that travel at high speed around the planet. A second oval spot is at south latitude on the east side of the planet. (Courtesy of the Jet Propulsion Laboratory)

than 100 meters. Comets, on the other hand, are loose collections of ices, dust, and small rocky particles that originate in the outer reaches of the solar system.

Asteroids: Leftover Planetesimals

Asteroids are small bodies (planetesimals) remaining from the formation of the solar system, making them about 4.6 billion years old. Most asteroids orbit the Sun between Mars and Jupiter in the region known as the **asteroid belt** (Figure 24.36). There are only five that are more than 400 kilometers in diameter, but the solar system hosts an estimated 1 to 2 million asteroids larger than 1 kilometer, and many millions that are smaller. Some travel along eccentric orbits that take them very near the Sun, and others regularly pass close to Earth and the Moon (Earth crossing asteroids). Many of the recent large-impact craters on the Moon and Earth were probably the result of collisions with asteroids. About 2000 Earth crossing asteroids are known, one-third of which are more than one kilometer in diameter. Inevitably, Earth-asteroid collisions will occur again (Box 24.2).

Because most asteroids have irregular shapes, planetary geologists initially speculated that they might be fragments of a broken planet that once orbited between Mars and Jupiter (Figure 24.37). However, the combined mass of all asteroids is now estimated to be only 1/1000 of the modest-sized Earth. Today, most researchers agree that asteroids are leftover debris from the solar nebula. Asteroids have lower densities than scien-

tists originally thought, suggesting they are porous bodies, like "piles of rubble," loosely bound together.

In February 2001 an American spacecraft became the first visitor to an asteroid. Although it was not designed for landing, *NEAR—Shoemaker* landed successfully on Eros and collected information that has planetary geologists both intrigued and perplexed. Images obtained as the spacecraft drifted toward the surface of Eros revealed a barren, rocky surface composed of particles ranging in size from fine dust to boulders up to 10 meters (30 feet) across (Figure 24.37). Researchers unexpectedly discovered that fine debris tends to concentrate in the low areas where it forms flat deposits resembling ponds. Surrounding the low areas, the landscape is marked by an abundance of large boulders.

One of several hypotheses to explain the boulder-strewn topography is seismic shaking, which would cause the boulders to move upward as the finer materials sink. This is analogous to what happens when a jar of sand and various sized pebbles is shaken—the larger pebbles rise to the top while the smaller sand grains settle to the bottom.

Indirect evidence from meteorites suggests that some asteroids might have been heated by a large impact event. A few large asteroids may have completely melted, causing them to differentiate into a dense iron core, and a rocky mantle. In November 2005, the Japanese probe, *Hayabusa*, landed on a small near-Earth asteroid named 25143 Itokawa and is scheduled to return samples to Earth by June 2010.

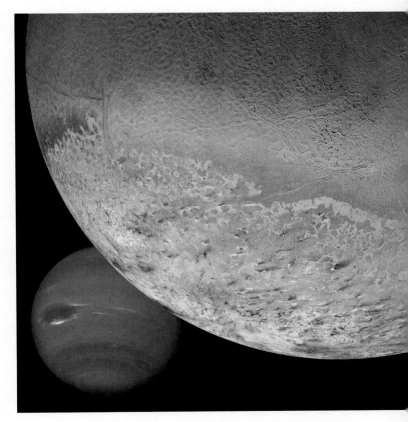

FIGURE 24.35   This montage shows Triton, Neptune's largest moon, with Neptune in the background. The bottom of the image shows Triton's wind and sublimation-eroded south polar cap. Sublimation is the process whereby a solid (ice) changes directly to a gas. (Courtesy of NASA/JPL)

EARTH AS A SYSTEM

BOX 24.2

Is Earth on a Collision Course?

The solar system is cluttered with asteroids, active comets, and extinct comets. These fragments travel at great speeds and can strike Earth with an explosive force many times greater than a powerful nuclear weapon.

In recent decades, it has become increasingly clear that comets and asteroids collide with Earth far more frequently than previous estimates, evidenced by the many large impact structures that have been identified (Figure 24.C). (Many impact craters were once mistaken for volcanic structures.) Most impact structures are so old and highly eroded that they were not discovered until satellite images became available (Figure 24.D). One notable exception is a very fresh-looking crater near Winslow, Arizona, known as Meteor Crater (see Figure 24.42, page 700). This crater was produced by a relatively small body, about the size of an Olympic swimming pool (50 meters in diameter).

About 65 million years ago, a large asteroid about 10 kilometers (6 miles) in diameter collided with Earth off the Yucatan peninsula in Mexico. This impact is thought to have caused

the demise of the dinosaurs, as well as the extinction of nearly 50 percent of all plant and animal species (see Chapter 22).

More recently, a spectacular explosion has been attributed to the collision of an asteroid or comet with our planet. In 1908, in a remote region of Siberia, a "fireball" that appeared more brilliant than the Sun exploded violently. The shock waves rattled windows and triggered reverberations heard up to 1,000 kilometers away. Called the *Tunguska event,* it scorched, delimbed, and flattened trees up to 30 kilometers from the point of impact. Surprisingly, expeditions to the area found no evidence of an impact crater or any metallic fragments. Evidently, the explosion, which equaled at least a 10-megaton nuclear bomb, occurred several kilometers above the surface. Why it exploded prior to impact is uncertain.

The dangers of living with these comparatively small, but deadly objects from space came to public attention again in 1989 when an asteroid nearly 1 kilometer across shot past Earth in a "near miss." Traveling at 70,000 kilometers (44,000 miles) per hour, it could have produced a crater 10 kilometers (6 miles) wide,

FIGURE 24.D   Manicouagan, Quebec, is a 200-million-year-old eroded impact structure. The lake outlines the crater remnant, which is 70 kilometers (42 miles) across. Fractures related to this event extend outward for an additional 30 kilometers. (Courtesy of U.S. Geological Survey)

and perhaps 2 kilometers (1.2 miles) deep. As an observer noted, "Sooner or later it will be back." Statistics show that collisions with bodies larger than 1 kilometer should be expected every few hundred thousand years. Collisions with bodies larger than 6 kilometers, resulting in mass extinctions, are anticipated every 100 million years.

NASA scientists continually track *near-Earth objects.* When asteroids or comets pass closely to any large body in the solar system, their orbits may be altered by the gravitational interaction, which may send them toward Earth. As of May 2009, more than 6000 near-Earth objects have been discovered, of which slightly more than 1000 have been classified as potentially hazardous asteroids.

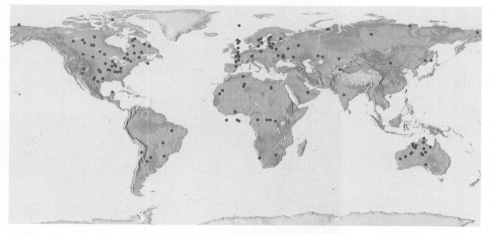

FIGURE 24.C   World map of major impact structures. Additional structures are being identified every year. (Data from Griffith Observatory)

Comets: Dirty Snowballs

Comets, like asteroids, are leftover material from the formation of the solar system. They are loose collections of rocky material, dust, water ice, and frozen gases (ammonia, methane, and carbon dioxide), thus the nickname "dirty snowballs." Recent space missions to comets have shown their surfaces to be dry and dusty, which indicates their ices are hidden beneath a rocky layer.

Most comets reside in the outer reaches of the solar system and take hundreds of thousands of years to complete a single orbit around the Sun. However, a smaller number of *short-period comets* (those having orbital periods of less than 200 years), such as the famous Halley's Comet, make regular encounters with the inner solar system (Figure 24.38). The shortest period comet (Encke's Comet) orbits around the Sun once every three years.

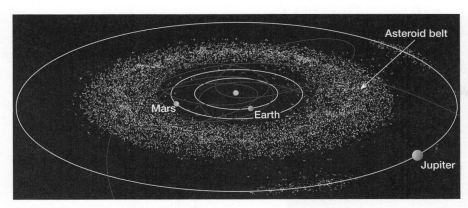

FIGURE 24.36 The orbits of most asteroids lie between Mars and Jupiter. Also shown are the orbits of a few known near-Earth asteroids.

All of the phenomena associated with comets come from a small central body called the **nucleus.** These structures are typically 1 to 10 kilometers in diameter, but nuclei 40 kilometers across have been observed. When comets reach the inner solar system, solar energy begins to vaporize their ices. The escaping gases carry dust from the comet's surface, producing a highly reflective halo called the **coma** (Figure 24.39). Within the coma, the small glowing nucleus with a diameter of only a few kilometers can sometimes be detected.

As comets approach the Sun, most develop tails that can extend for millions of kilometers. The tail of a comet points away from the Sun in a slightly curved manner (see Figure 24.38) which led early astronomers to believe that the Sun has a repulsive force that pushes away particles of the coma to form the tail. Scientists have identified two solar forces known to contribute to tail formation. One is *radiation pressure* caused by radiant energy (light) emitted by the Sun, and the second is the *solar wind,* a stream of charged particles ejected from the Sun. Sometimes a single tail composed of both dust and ionized gases is produced, but two tails are often observed (Figure 24.40). The heavier dust particles produce a slightly curved tail that follows the comets orbit; whereas the extremely light ionized gases are "pushed" directly away from the Sun, forming the second tail.

As a comet's orbit carries it away from the Sun, the gases forming the coma recondense, the tail disappears, and the comet returns to cold storage. Material that was blown from the coma to form the tail is lost forever. When all the gases are expelled, the inactive comet, which closely resembles an asteroid, continues its orbit without a coma or tail. It is believed that few comets remain active for more than a few hundred close orbits of the Sun.

The very first samples from a comet's coma (Comet Wild 2) were returned to Earth in January of 2006 by NASA's *Stardust* spacecraft (Figure 24.41). Images from *Stardust* show that the comet's surface was riddled with flat-bottomed depressions and appeared dry, although at least 10 gas jets were active. Laboratory studies revealed that the coma contained a wide range of organic compounds and substantial amounts of silicate crystals.

Most comets originate in one of two regions—the *Kuiper belt* or the *Oort cloud.* Named in honor of astronomer Gerald Kuiper, who predicted its existence, the **Kuiper belt** hosts comets

that orbit in the outer solar system, beyond Neptune (see Figure 24.1). This disc-shaped structure is thought to contain about a billion objects over one kilometer in size. However, most comets are too small and too distant to be observed from Earth, even using the Hubble Space Telescope. Like the asteroids in the inner solar system, most Kuiper belt comets move in slightly elliptical orbits that lie roughly in the same plane as the planets. A chance collision between two Kuiper belt comets, or the gravitational influence of one of the Jovian planets, occasionally alters their orbits sufficiently to send them into our view.

Halley's Comet originated in the Kuiper belt. Its orbital period averages 76 years, and every one of its 29 appearances since 240 BC has been recorded by Chinese astronomers—testimony to their dedication as astronomical observers and the endurance of Chinese culture. When seen in 1910, Halley's Comet had a tail nearly 1.6 million kilometers (1 million miles) long and was visible during the daytime.

The 1986 showing of Halley's Comet was unspectacular and disappointed many Northern Hemisphere residents who anticipated a tremendous show. Yet it was during this most recent visit that the European probe *Giotto* approached to within 600 kilometers of the comet's nucleus and obtained the first close-up images of a comet's structure. We now know that the nucleus of Halley's Comet is potato-shaped, roughly 15 by 8 kilometers in size. Its surface is irregular and full of craterlike pits. When observed by *Giotto,* only 10 percent of the surface area contained active jets, all of which were seen when sunlight struck

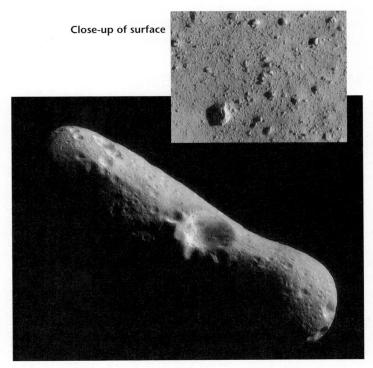

Close-up of surface

FIGURE 24.37 Image of asteroid Eros obtained by the *NEAR-Shoemaker* probe. Inset shows a close-up of Eros's barren rocky surface. (Courtesy of NASA)

FIGURE 24.38   Orientation of a comet's tail as it orbits the Sun.

the surface. The remaining surface area of the comet is covered with a dark layer thought to be organic material. This dark layer allows absorption of solar energy that drives its outgassing.

Named for Dutch astronomer Jan Oort, the **Oort cloud** consists of comets that are distributed in all directions from the Sun, forming a spherical shell around the solar system. Most Oort cloud comets orbit the Sun at distances greater than 10,000 times the Earth-Sun distance. The gravitational effect of a distant passing star may send an occasional Oort cloud comet into a highly eccentric orbit that carries it toward the Sun. However, only a tiny fraction of Oort cloud comets have orbits that bring them into the inner solar system.

Meteoroids: Visitors to Earth

Nearly everyone has seen a **meteor,** commonly (but inaccurately) called "shooting stars." These streaks of light can be observed in as little as the blink-of-an-eye, or can last as "long" as a few seconds. They occur when a small solid particle, a **meteoroid,** enters Earth's atmosphere from interplanetary space. Heat, created by friction between the meteoroid and the air, produces the light we see trailing across the sky. Most meteoroids originate from one of the following three sources: (1) interplanetary debris missed by the gravitational sweep of the planets during formation of the solar system; (2) material that is continually being ejected from the asteroid belt, or; (3) the rocky and/or metallic remains of comets that once passed through Earth's orbit. A few meteoroids are probably fragments of the Moon, Mars, or possibly Mercury, ejected by a violent asteroid impact. Before *Apollo* astronauts brought Moon rocks back to Earth, meteorites were the only extraterrestrial materials that could be studied in the laboratory.

Meteoroids less than about a meter in diameter generally vaporize before reaching Earth's surface. Some, called *micrometeorites,* are so tiny and their rate of fall so slow that they

drift to Earth continually as space dust. Researchers estimate that thousands of meteoroids enter Earth's atmosphere every day. After sunset on a clear, dark night, many are bright enough to be seen with the naked eye anywhere on Earth.

Occasionally, meteor sightings increase dramatically to 60 or more per hour. These displays, called **meteor showers,** result when Earth encounters a swarm of meteoroids traveling in the same direction at nearly the same speed as Earth. The close association of these swarms to the orbits of some short-term comets strongly suggests that they represent material lost by these comets (Table 24.2). Some swarms, not associated with the orbits of known comets, are probably the scattered remains of the nucleus of a long-defunct comet. The notable *Perseid meteor shower* that occurs each year around August 12 is likely material ejected from of the Comet *Swift-Tuttle* on previous approaches to the Sun.

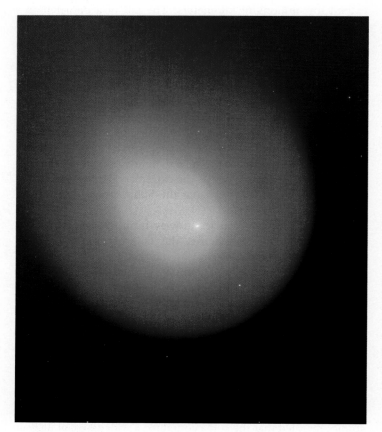

FIGURE 24.39   Coma of comet Holmes as it orbits the Sun. The nucleus of the comet is within the bright spot in the center. Comet Holmes, which orbits the Sun every six years, was uncharacteristically active during its most recent entry into the inner solar system. (Image by Spitzer Space Telescope, courtesy of NASA)

Ion tail

Dust tail

FIGURE 24.40 Comet Hale-Bopp. The two tails seen in the photograph are between 10 million and 15 million miles long. (Peoria Astronomical Society photograph by Eric Clifton and Craig Neaveill)

Most meteoroids large enough to survive probably originate among the asteroids, where chance collisions or gravitational interactions with Jupiter modify their orbits and project them toward Earth. Earth's gravity does the rest.

A few very large meteoroids have blasted craters on Earth's surface that strongly resemble those on our Moon. At least 40 terrestrial craters exhibit features that could only be produced by an explosive impact of a large asteroid, or perhaps even a comet nucleus. More than 250 others may be of impact origin. Most notable is Arizona's Meteor Crater, a huge cavity more than one kilometer wide, 170 meters (560 feet) deep, with an upturned rim that rises above the surrounding countryside (Figure 24.42). More than 30 tons of iron fragments have been found in the immediate area, but attempts to locate the main body have been unsuccessful. Based on the amount of erosion observed on the crater rim, the impact likely occurred within the last 50,000 years.

The remains of meteoroids, when found on Earth, are referred to as **meteorites** (Figure 24.43). Classified by their composition, meteorites are either: (1) *irons*, mostly iron with 5–20 percent nickel; (2) *stony* (also called *chrondites*), silicate minerals with inclusions of other minerals, or; (3) *stony–irons*, mixtures of the two. Although stony meteorites are the most common, irons are found in large numbers because metallic meteorites withstand impacts better, weather more slowly, and are easily distinguished from terrestrial rocks. Iron meteorites are probably fragments of once molten cores of large asteroids or small planets.

One type of stony meteorite, called a *carbonaceous chondrite,* contains organic compounds and occasionally simple amino acids which are the basic building blocks of life. This discovery confirms similar findings in observational astronomy, which indicate that numerous organic compounds exist in interstellar space.

Meteorites discovered accidentally are called *finds,* whereas those found because the event was observed are called *falls.* Although meteorites have been found in many places, one of the richest hunting grounds is the Antarctic ice sheet. Because there are few sources of rocks on this vast ice sheet, it is likely that almost any rock found on its surface is a meteorite.

Data from meteorites have been used to ascertain the internal structure of Earth and the age of the solar system. If meteorites represent the composition of the terrestrial planets, as some planetary geologists suggest, our planet must contain a much larger percentage of iron than is indicated by surface rocks. This is one reason that geologists think Earth's core is mostly iron and nickel. In addition, radiometric dating of meteorites indicates the age of our solar system is about 4.6 billion years. This "old age" has been confirmed by data obtained from lunar samples.

Dwarf Planets

Since its discovery in 1930, Pluto has been a mystery to astronomers who were searching for another planet in order to explain irregularities in Neptune's orbit. At the time of its discovery, Pluto was thought to be the size of Earth—too small to significantly alter Neptune's orbit. Later, estimates of Pluto's diameter, adjusted because of improved satellite images, indicated that it was less than half Earth's diameter. Then, in 1978, astronomers realized that Pluto appeared much larger than it really is because of the brightness of its newly discovered satellite, Charon (Figure 24.44). Most recently, calculations based on images obtained by the Hubble Space Telescope show that Pluto's diameter is 2300 kilometers (1430 miles), about one-fifth the diameter of Earth and less than half that of Mercury (long considered the solar system's "runt"). In fact, seven moons in the solar system, including Earth's, are larger than Pluto.

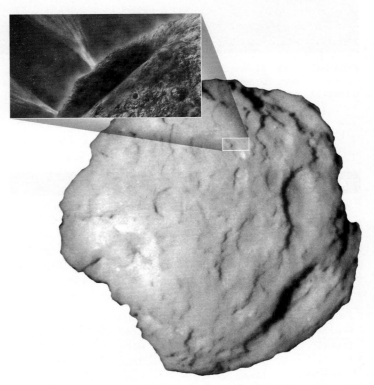

FIGURE 24.41 Comet Wild 2, as seen by NASA's *Stardust* spacecraft, during its flyby of the comet. Inset, artist's concept depicting jets of gas and dust erupting from comet Wild 2. (Courtesy of NASA)

FIGURE 24.42   Meteor Crater, near Winslow, Arizona. This cavity is about 1.2 kilometers (0.75 mile) across and 170 meters (560 feet) deep. The solar system is cluttered with meteoroids and other objects that can strike Earth with explosive force. (Photo by Michael Collier)

TABLE 24.2	Major Meteor Showers	
Shower	Approximate Dates	Associated Comet
Quadrantids	January 4–6	
Lyrids	April 20–23	Comet 1861 I
Eta Aquarids	May 3–5	Halley's comet
Delta Aquarids	July 30	
Perseids	August 12	Comet 1862 III
Draconids	October 7–10	Comet Giacobini-Zinner
Orionids	October 20	Halley's comet
Taurids	November 3–13	Comet Encke
Andromedids	November 14	Comet Biela
Leonids	November 18	Comet 1866 I
Geminids	December 4–16	

Even more attention was given to Pluto's status as a planet when astronomers discovered another large icy body in orbit beyond Neptune. Soon, over a thousand of these *Kuiper belt objects* were discovered forming a band of objects—a second asteroid belt, but located at the outskirts of the solar system. The Kuiper belt objects are rich in ices and have physical properties similar to those of comets. Many other planetary objects, some perhaps larger than Pluto, are thought to exist in this belt of icy worlds beyond Neptune's orbit.

The International Astronomical Union, the group responsible for naming and classifying celestial objects, voted in 2006 to create a new class of solar system objects called **dwarf planets.** These are celestial bodies that orbit the Sun, are essentially round due to their own gravity, but are not large enough to sweep their orbits clear of other debris. By this definition, Pluto is recognized as a dwarf planet and the prototype of this new category of planetary objects. Other dwarf planets include Eros, a Kuiper belt object, and Ceres, the largest known asteroid.

Pluto's reclassification was not the first such "demotion." In the mid-1800s, astronomy textbooks listed as many as 11 plan-

FIGURE 24.43   Iron meteorite found near Meteor Crater, Arizona. (Courtesy of Meteor Crater Enterprises, Inc.)

FIGURE 24.44   Pluto with its three known moons. Image by the Hubble Space Telescope. (Courtesy of NASA)

ets in our solar system, including the asteroids Vesta, Juno, Ceres, and Pallas. Astronomers continued to discover dozens of other "planets," a clear signal that these small bodies represent a class of objects separate from the planets.

Researchers now recognize that Pluto was unique among the classical planets—completely different from the four rocky, innermost planets, as well as the four gaseous giants. The new classification will give a home to the hundreds of additional dwarf planets astronomers assume exist in the solar system. *New Horizons,* the first spacecraft designed to explore the outer solar system, was launched in January 2006. Scheduled to fly by Pluto in July 2015, and later explore the Kuiper belt, *New Horizons* carries tremendous potential for aiding researchers in further understanding the solar system.

CHAPTER 24 PLANETARY GEOLOGY IN REVIEW

- The planets are arranged into two groups: the *terrestrial* (Earth-like) *planets* (Mercury, Venus, Earth, and Mars) and the *Jovian* (Jupiter-like) *planets* (Jupiter, Saturn, Uranus, and Neptune). When compared to the Jovian planets, the terrestrial planets are smaller, more dense, contain proportionally more rocky material, have slower rates of rotation, and meager atmospheres.

- The lunar surface exhibits several types of features. *Impact craters* were produced by the collision of rapidly moving interplanetary debris (*meteoroids*). Bright, densely cratered *highlands* make up most of the lunar surface. The dark, fairly smooth lowlands are called *maria.* Maria basins are enormous impact craters that were later flooded with layer upon layer of very fluid basaltic lava. All lunar terrains are mantled with a soil-like layer of gray, unconsolidated debris, called *lunar regolith,* which has been derived from a few billion years of meteoric bombardment. One hypothesis for the Moon's origin suggests that a Mars-sized object collided with Earth to produce the Moon. Scientists conclude that the *lunar surface evolved in four stages:* (1) *formation of the original crust (highlands),* (2) *excavations of the large impact basins,* (3) *filling of maria basins,* and (4) *formation of youthful rayed craters.*

- *Mercury* is a small, dense planet that has virtually no atmosphere and exhibits the greatest temperature extremes of any planet. *Venus,* the brightest planet in the sky, has a thick, heavy atmosphere composed of 97 percent carbon dioxide, a surface of relatively subdued plains and inactive volcanic features, a surface atmospheric pressure 90 times that of Earth's, and surface temperatures of 475 °C (900 °F). *Mars,* the Red Planet, has a carbon dioxide atmosphere only 1 percent as dense as Earth's, extensive dust storms, numerous inactive volcanoes, many large canyons, and several valleys exhibiting drainage patterns similar to stream valleys on Earth. *Jupiter,* the largest planet, rotates rapidly, has a banded appearance caused by huge convection currents driven by the planet's interior heat, a *Great Red Spot* that varies in size, a thin ring system, and at least 63 moons (one of the moons, *Io,* is perhaps the most volcanically active body in the solar system). *Saturn* is best known for its system of rings. It also has a dynamic atmosphere with strong winds up to 930 miles per hour and storms similar to Jupiter's Great Red Spot. *Uranus* and *Neptune* are often called "the twins" because of their similar structure and composition. A unique feature of Uranus is the fact that it rotates on its side. Neptune has white, cirrus-like clouds above its main cloud deck and large dark spots, assumed to be large rotating storms similar to Jupiter's Great Red Spot.

- The minor members of the solar system include *asteroids, comets, meteoroids,* and *dwarf planets.* Most asteroids lie between the orbits of Mars and Jupiter. Asteroids are leftover rocky and metallic debris from the solar nebula that never accreted into a planet. Comets are made of ices (water, ammonia, methane, carbon dioxide, and carbon monoxide) with small pieces of rocky and metallic material. Comets are thought to reside in the outer solar system in one of two locations: the *Kuiper belt* or the *Oort cloud.* Meteoroids, small solid particles that travel through interplanetary space, become *meteors* when they enter Earth's atmosphere and vaporize with a flash of light. *Meteor showers* occur when Earth encounters a swarm of meteoroids, probably material lost by a comet. *Meteorites* are the remains of meteoroids found on Earth. Recently, Pluto was placed into a new class of solar system objects called *dwarf planets.*

KEY TERMS

asteroids (p. 695)
asteroid belt (p. 695)
coma (p. 697)
comet (p. 696)
cryovolcanism (p. 694)
dwarf planet (p. 701)
escape velocity (p. 677)

impact craters (p. 678)
Jovian planet (p. 674)
Kuiper belt (p. 697)
lunar highlands (p. 679)
lunar regolith (p. 680)
maria (p. 679)
meteor (p. 698)

meteorite (p. 699)
meteoroid (p. 698)
meteor shower (p. 699)
Nebular theory (p. 672)
nucleus (p. 697)
Oort cloud (p. 698)
planetesimals (p. 672)

protoplanets (p. 672)
solar nebula (p. 672)
terrae (p. 679)
terrestrial planet (p. 674)

QUESTIONS FOR REVIEW

1. By what criteria are planets considered either Jovian or terrestrial?

2. What accounts for the large density differences between the terrestrial and Jovian planets?

3. Explain why the terrestrial planets have meager atmospheres, as compared to the Jovian planets.

4. How is crater density used in the relative dating of surface features on the Moon?

5. Briefly outline the history of the Moon.

6. How are maria on the Moon similar to the Columbia Plateau in the Pacific Northwest?

7. Venus was once referred to as "Earth's twin." How are these two planets similar? How do they differ?

8. What surface features do Mars and Earth have in common?

9. Why are the largest volcanoes on Earth so much smaller than the largest ones on Mars?

10. Why might astrobiologists be intrigued by evidence that groundwater has seeped onto the surface of Mars?

11. Explain why the two moons of Mars are likely captured asteroids?

12. What is the nature of Jupiter's Great Red Spot?

13. Why are the Galilean satellites of Jupiter so named?

14. What is distinctive about Jupiter's satellite, Io?

15. Why are many of Jupiter's small satellites thought to have been captured?

16. How are Jupiter and Saturn similar?

17. What two roles do ring moons play in the nature of planetary ring systems?

18. How are Saturn's satellite Titan and Neptune's satellite Triton similar?

19. Name three bodies in the solar system that exhibit active volcanism.

20. Where are most asteroids found?

21. What do you think would happen if Earth passed through the tail of a comet?

22. Where are most comets thought to reside? What eventually becomes of comets that orbit close to the Sun?

23. Compare meteoroid, meteor, and meteorite.

24. What are the three main sources of meteoroids?

25. Why are impact craters more common on the Moon than on Earth, even though the Moon is a much smaller target and has a weaker gravitational field?

26. It has been estimated that Halley's Comet has a mass of 100 billion tons. Further, this comet is estimated to lose 100 million tons of material during the few months that its orbit brings it close to the Sun. With an orbital period of 76 years, what is the maximum remaining life span of Halley's Comet?

COMPANION WEBSITE

The *Earth 10e* Web site uses the resources and flexibility of the Internet to aid in your study of the topics in this chapter. Written and developed by the authors and other geology instructors, this site will help improve your understanding of geology. Visit www.mygeoscienceplace.com in order to:

- **Review** key chapter concepts.

- **Read** with links to the eBook and to chapter-specific web resources.

- **Test** yourself with online quizzes.

METRIC AND ENGLISH UNITS COMPARED

Units

1 kilometer (km) = 1000 meters (m)
1 meter (m) = 100 centimeters (cm)
1 centimeter (cm) = 0.39 inch (in.)
1 mile (mi) = 5280 feet (ft)
1 foot (ft) = 12 inches (in.)
1 inch (in.) = 2.54 centimeters (cm)
1 square mile (mi^2) = 640 acres (a)
1 kilogram (kg) = 1000 grams (g)
1 pound (lb) = 16 ounces (oz)
1 fathom = 6 feet (ft)

Conversions

When you want
to convert: multiply by: to find:

Length

inches	2.54	centimeters
centimeters	0.39	inches
feet	0.30	meters
meters	3.28	feet
yards	0.91	meters
meters	1.09	yards
miles	1.61	kilometers
kilometers	0.62	miles

Area

square inches	6.45	square centimeters
square centimeters	0.15	square inches
square feet	0.09	square meters
square meters	10.76	square feet
square miles	2.59	square kilometers
square kilometers	0.39	square miles

Volume

cubic inches	16.38	cubic centimeters
cubic centimeters	0.06	cubic inches
cubic feet	0.028	cubic meters
cubic meters	35.3	cubic feet
cubic miles	4.17	cubic kilometers
cubic kilometers	0.24	cubic miles
liters	1.06	quarts
liters	0.26	gallons
gallons	3.78	liters

Masses and Weights

ounces	28.35	grams
grams	0.035	ounces
pounds	0.45	kilograms
kilograms	2.205	pounds

Temperature

When you want to convert degrees Fahrenheit (°F) to degrees Celsius (°C), subtract 32 degrees and divide by 1.8.

When you want to convert degrees Celsius (°C) to degrees Fahrenheit (°F), multiply by 1.8 and add 32 degrees.

When you want to convert degrees Celsius (°C) to Kelvins (K), delete the degree symbol and add 273. When you want to convert Kelvins (K) to degrees Celsius (°C), add the degree symbol and subtract 273.

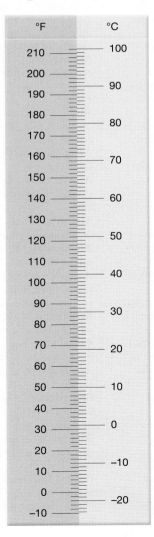

FIGURE A.1 A comparison of Fahrenheit and Celsius temperature scales.

Glossary

Aa flow A type of lava flow that has a jagged, blocky surface.

Ablation A general term for the loss of ice and snow from a glacier.

Abrasion The grinding and scraping of a rock surface by the friction and impact of rock particles carried by water, wind, and ice.

Abyssal plain Very level area of the deep-ocean floor, usually lying at the foot of the continental rise.

Accretionary wedge A large wedge-shaped mass of sediment that accumulates in subduction zones. Here sediment is scraped from the subducting oceanic plate and accreted to the overriding crustal block.

Active continental margin Usually narrow and consisting of highly deformed sediments. Such margins occur where oceanic lithosphere is being subducted beneath the margin of a continent.

Active layer The zone above the permafrost that thaws in summer and refreezes in winter.

Aerosols Tiny solid and liquid particles suspended in the atmosphere.

Aftershock A smaller earthquake that follows the main earthquake.

Alluvial channel A stream channel in which the bed and banks are composed largely of unconsolidated sediment (alluvium) that was previously deposited in the valley.

Alluvial fan A fan-shaped deposit of sediment formed when a stream's slope is abruptly reduced.

Alluvium Unconsolidated sediment deposited by a stream.

Alpine glacier A glacier confined to a mountain valley, which in most instances had previously been a stream valley.

Andesitic composition See *Intermediate composition*.

Angle of repose The steepest angle at which loose material remains stationary without sliding downslope.

Angular unconformity An unconformity in which the older strata dip at an angle different from that of the younger beds.

Antecedent stream A stream that continued to downcut and maintain its original course as an area along its course was uplifted by faulting or folding.

Anthracite A hard, metamorphic form of coal that burns cleanly and hot.

Anticline A fold in sedimentary strata that resembles an arch.

Aphanitic texture A texture of igneous rocks in which the crystals are too small for individual minerals to be distinguished without the aid of a microscope.

Aquifer Rock or sediment through which groundwater moves easily.

Aquitard An impermeable bed that hinders or prevents groundwater movement.

Archean eon The first eon of Precambrian time. The eon preceding the Proterozoic. It extends between 4.5 and 2.5 billion years ago.

Arête A narrow, knifelike ridge separating two adjacent glaciated valleys.

Arkose A feldspar-rich sandstone.

Artesian well A well in which the water rises above the level where it was initially encountered.

Assimilation In igneous activity, the process of incorporating country rock into a magma body.

Asteroid One of thousands of small planet-like bodies, ranging in size from a few hundred kilometers to less than one kilometer across. Most asteroids' orbits lie between those of Mars and Jupiter.

Asthenosphere A subdivision of the mantle situated below the lithosphere. This zone of weak material exists below a depth of about 100 kilometers and in some regions extends as deep as 700 kilometers. The rock within this zone is easily deformed.

Atmosphere The gaseous portion of a planet, the planet's envelope of air. One of the traditional subdivisions of Earth's physical environment.

Atoll A coral island consisting of a nearly continuous ring of coral reef surrounding a central lagoon.

Atom The smallest particle that exists as an element.

Atomic mass unit A mass unit equal to exactly one-twelfth the mass of a carbon-12 atom.

Atomic number The number of protons in the nucleus of an atom.

Atomic weight The average of the atomic masses of isotopes for a given element.

Aureole A zone or halo of contact metamorphism found in the country rock surrounding an igneous intrusion.

Backarc basin A basin that forms on the side of a volcanic arc away from the trench.

Backshore The inner portion of the shore, lying landward of the high-tide shoreline. It is usually dry, being affected by waves only during storms.

Backswamp A poorly drained area on a floodplain resulting when natural levees are present.

Bajada An apron of sediment along a mountain front created by the coalescence of alluvial fans.

Banded iron formations A finely layered iron and silica-rich (chert) layer deposited mainly during the Precambrian.

Bar Common term for sand and gravel deposits in a stream channel.

Barchan dune A solitary sand dune shaped like a crescent with its tips pointing downwind.

Barchanoid dune Dunes forming scalloped rows of sand oriented at right angles to the wind. This form is intermediate between isolated barchans and extensive waves of transverse dunes.

Barrier island A low, elongate ridge of sand that parallels the coast.

Basal slip A mechanism of glacial movement in which the ice mass slides over the surface below.

Basalt A fine-grained igneous rock of mafic composition.

Basaltic composition A compositional group of igneous rocks indicating that the rock contains substantial dark silicate minerals and calcium-rich plagioclase feldspar.

Base level The level below which a stream cannot erode.

Basin A circular downfolded structure.

Batholith A large mass of igneous rock that formed when magma was emplaced at depth, crystallized, and subsequently exposed by erosion.

Bathymetry The measurement of ocean depths and the charting of the topography of the ocean floor.

Baymouth bar A sandbar that completely crosses a bay, sealing it off from the main body of water.

Beach An accumulation of sediment found along the landward margin of the ocean or a lake.

Beach drift The transport of sediment in a zigzag pattern along a beach, caused by the uprush of water from obliquely breaking waves.

Beach face The wet, sloping surface that extends from the berm to the shoreline.

Beach nourishment Process in which large quantities of sand are added to the beach system to offset losses caused by wave erosion. Building beaches seaward improves beach quality and storm protection.

Bed See *Strata*.

Bedding plane A nearly flat surface separating two beds of sedimentary rock. Each bedding plane marks the end of one deposit and the beginning of another having different characteristics.

Bed load Sediment moved along the bottom of a stream by moving water, or particles moved along the ground surface by wind.

Bedrock channel A channel in which a stream is cutting into solid rock. Such channels typically form in the headwaters or river systems where gradients are high.

Benioff zone See *Wadati—Benioff zone.*

Berm The dry, gently sloping zone on the backshore of a beach at the foot of the coastal cliffs or dunes.

Biochemical A type of chemical sediment that forms when material dissolved in water is precipitated by water-dwelling organisms. Shells are common examples.

Biogenous sediment Seafloor sediments consisting of material of marine-organic origin.

Biosphere The totality of life forms on Earth.

Bituminous coal The most common form of coal, often called soft, black coal.

Black smoker A hydrothermal vent on the ocean floor that emits a black cloud of hot, metal-rich water.

Block lava Lava having a surface of angular blocks associated with material having andesitic and rhyolitic compositions.

Blowout A depression excavated by wind in easily eroded materials.

Body wave A seismic wave that travels through Earth's interior.

Bottomset bed A layer of fine sediment deposited beyond the advancing edge of a delta and then buried by continued delta growth.

Bowen's reaction series A concept proposed by N. L. Bowen that illustrates the relationships between magma and the minerals crystallizing from it during the formation of igneous rocks.

Braided stream A stream consisting of numerous intertwining channels.

Breakwater A structure protecting a nearshore area from breaking waves.

Breccia A sedimentary rock composed of angular fragments that were lithified.

Burial metamorphism Low-grade metamorphism that occurs in the lowest layers of very thick accumulations of sedimentary strata.

Caldera A large depression typically caused by collapse or ejection of the summit area of a volcano.

Caliche A hard layer, rich in calcium carbonate, that forms beneath the *B* horizon in soils of arid regions.

Calving Wastage of a glacier that occurs when large pieces of ice break into the water.

Capacity The total amount of sediment a stream is able to transport.

Capillary fringe A relatively narrow zone at the base of the zone of aeration. Here water rises from the water table in tiny, threadlike openings between grains of soil or sediment.

Cap rock A necessary part of an oil trap. The cap rock is impermeable and hence keeps upwardly mobile oil and gas from escaping at the surface.

Cassini gap A wide gap in the ring system of Saturn between the *A* ring and the *B* ring.

Catastrophism The concept that Earth was shaped by catastrophic events of a short-term nature.

Cavern A naturally formed underground chamber or series of chambers most commonly produced by solution activity in limestone.

Cementation One way in which sedimentary rocks are lithified. As material precipitates from water that percolates through the sediment, open spaces are filled and particles are joined into a solid mass.

Cenozoic era A time span on the geologic time scale beginning about 65.5 million years ago, following the Mesozoic era.

Chemical bond A strong attractive force that exists between atoms in a substance. It involves the transfer or sharing of electrons that allows each atom to attain a full valence shell.

Chemical sedimentary rock Sedimentary rock consisting of material that was precipitated from water by either inorganic or organic means.

Chemical weathering The processes by which the internal structure of a mineral is altered by the removal and/or addition of elements.

Cinder cone A rather small volcano built primarily of ejected lava fragments that consist mostly of pea- to walnut-size lapilli.

Cirque An amphitheater-shaped basin at the head of a glaciated valley produced by frost wedging and plucking.

Clastic texture A sedimentary rock texture consisting of broken fragments of preexisting rock.

Cleavage The tendency of a mineral to break along planes of weak bonding.

Climate feedback mechanism Because the atmosphere is a complex interactive physical system, several different possible outcomes may result when one of the system's elements is altered. These various possibilities are called *climate-feedback mechanisms.*

Climate system The exchanges of energy and moisture occurring among the atmosphere, hydrosphere, lithosphere, biosphere, and cryosphere.

Closed system A system that is self-contained with regard to matter—that is, no matter enters or leaves.

Coarse-grained See *Phaneritic texture.*

Coast A strip of land that extends inland from the coastline as far as ocean-related features can be found.

Coastline The coast's seaward edge. The landward limit of the effect of the highest storm waves on the shore.

Col A pass between mountain valleys where the headwalls of two cirques intersect.

Color A phenomenon of light by which otherwise identical objects may be differentiated.

Column A feature found in caves that is formed when a stalactite and stalagmite join.

Columnar joints A pattern of cracks that forms during cooling of molten rock to generate columns.

Coma The fuzzy, gaseous component of a comet's head.

Comet A small body that generally revolves about the Sun in an elongated orbit.

Compaction A type of lithification in which the weight of overlying material compresses more deeply buried sediment. It is most important in the fine-grained sedimentary rocks such as shale.

Competence A measure of the largest particle a stream can transport; a factor dependent on velocity.

Composite cone A volcano composed of both lava flows and pyroclastic material.

Compound A substance formed by the chemical combination of two or more elements in definite proportions and usually having properties different from those of its constituent elements.

Compressional mountains Mountains in which great horizontal forces have shortened and thickened the crust. Most major mountain belts are of this type.

Compressional stress Differential stress that shortens a rock body.

Concordant A term used to describe intrusive igneous masses that form parallel to the bedding of the surrounding rock.

Conduction The transfer of heat through matter by molecular activity.

Conduit A pipelike opening through which magma moves toward Earth's surface. It terminates at a surface opening called a vent.

Cone of depression A cone-shaped depression in the water table immediately surrounding a well.

Confined aquifer An aquifer that has impermeable layers (aquitards) both above and below.

Confining pressure Stress that is applied uniformly in all directions.

Conformable layers Rock layers that were deposited without interruption.

Conglomerate A sedimentary rock composed of rounded, gravel-size particles.

Contact metamorphism Changes in rock caused by the heat from a nearby magma body.

Continental drift A hypothesis, credited largely to Alfred Wegener, that suggested all present continents once existed as a single supercontinent. Further, beginning about 200 million years ago, the supercontinent began breaking into smaller continents, which then "drifted" to their present positions.

Continental margin That portion of the seafloor adjacent to the continents. It may include the continental shelf, continental slope, and continental rise.

Continental rift A linear zone along which continental lithosphere stretches and pulls apart. Its creation may mark the beginning of a new ocean basin.

Continental rise The gently sloping surface at the base of the continental slope.

Continental shelf The gently sloping submerged portion of the continental margin, extending from the shoreline to the continental slope.

Continental slope The steep gradient that leads to the deep-ocean floor and marks the seaward edge of the continental shelf.

Continental volcanic arc   Mountains formed in part by igneous activity associated with the subduction of oceanic lithosphere beneath a continent. Examples include the Andes and the Cascades.

Convection   The transfer of heat by the mass movement or circulation of a substance.

Convergent plate boundary   A boundary in which two plates move together, resulting in oceanic lithosphere being thrust beneath an overriding plate, eventually to be reabsorbed into the mantle. It can also involve the collision of two continental plates to create a mountain system.

Coral reef   Structure formed in a warm, shallow, sunlit ocean environment that consists primarily of the calcite-rich remains of corals as well as the limy secretions of algae and the hard parts of many other small organisms.

Core   The innermost layer of Earth based on composition. It is thought to be largely an iron-nickel alloy with minor amounts of oxygen, silicon, and sulfur.

Correlation   Establishing the equivalence of rocks of similar age in different areas.

Corrosion   The process by which soluble rock is gradually dissolved by flowing water.

Covalent bond   A chemical bond produced by the sharing of electrons.

Crater   The depression at the summit of a volcano, or that which is produced by a meteorite impact.

Craton   That part of the continental crust that has attained stability; that is, it has not been affected by significant tectonic activity during the Phanerozoic eon. It consists of the shield and stable platform.

Creep   The slow downhill movement of soil and regolith.

Crevasse   A deep crack in the brittle surface of a glacier.

Cross-bedding   Structure in which relatively thin layers are inclined at an angle to the main bedding. Formed by currents of wind or water.

Cross-cutting   A principle of relative dating. A rock or fault is younger than any rock (or fault) through which it cuts.

Crust   The very thin outermost layer of Earth.

Crystal   Any natural solid with an ordered, repetitive atomic structure.

Crystalline   See *Crystal*.

Crystal shape   See *Habit*.

Crystal settling   During the crystallization of magma, the earlier-formed minerals are denser than the liquid portion and settle to the bottom of the magma chamber.

Crystalline texture   See *Nonclastic texture*.

Crystallization   The formation and growth of a crystalline solid from a liquid or gas.

Curie point   The temperature above which a material loses its magnetization.

Cut bank   The area of active erosion on the outside of a meander.

Cutoff   A short channel segment created when a river erodes through the narrow neck of land between meanders.

Darcy's law   An equation stating that groundwater discharge depends on the hydraulic gradient, hydraulic conductivity, and cross-sectional area of an aquifer.

Dark silicate   Silicate minerals containing ions of iron and/or magnesium in their structure. They are dark in color and have a higher specific gravity than nonferromagnesian silicates.

Daughter product   An isotope resulting from radioactive decay.

Debris flow   A flow of soil and regolith containing a large amount of water. Most common in semiarid mountainous regions and on the slopes of some volcanoes.

Debris slide   See *Rockslide*.

Decompression melting   Melting that occurs as rock ascends due to a drop in confining pressure.

Deep-ocean basin   The portion of seafloor that lies between the continental margin and the oceanic ridge system. This region comprises almost 30 percent of Earth's surface.

Deep-ocean trench   A narrow, elongated depression of the seafloor.

Deep-sea fan   A cone-shaped deposit at the base of the continental slope. The sediment is transported to the fan by turbidity currents that follow submarine canyons.

Deflation   The lifting and removal of loose material by wind.

Deformation   General term for the processes of folding, faulting, shearing, compression, or extension of rocks as the result of various natural forces.

Delta   An accumulation of sediment formed where a stream enters a lake or an ocean.

Dendritic pattern   A stream system that resembles the pattern of a branching tree.

Density   A property of matter defined as mass per unit volume.

Desalination   The removal of salts and other chemicals from seawater.

Desert   One of the two types of dry climate; the driest of the dry climates.

Desert pavement   A layer of coarse pebbles and gravel created when wind removes the finer material.

Detachment fault   A nearly horizontal fault that may extend for hundreds of kilometers below the surface. Such faults represent a boundary between rocks that exhibit ductile deformation and rocks that exhibit brittle deformation.

Detrital sedimentary rocks   Rocks that form from the accumulation of materials that originate and are transported as solid particles derived from both mechanical and chemical weathering.

Diagenesis   A collective term for all the chemical, physical, and biological changes that take place after sediments are deposited and during and after lithification.

Differential stress   Forces that are unequal in different directions.

Differential weathering   The variation in the rate and degree of weathering caused by such factors as mineral makeup, degree of jointing, and climate.

Dike   A tabular-shaped intrusive igneous feature that cuts through the surrounding rock.

Dip   The angle at which a rock layer or fault is inclined from the horizontal. The direction of dip is at a right angle to the strike.

Dip-slip fault   A fault in which the movement is parallel to the dip of the fault.

Discharge   The quantity of water in a stream that passes a given point in a period of time.

Disconformity   A type of unconformity in which the beds above and below are parallel.

Discontinuity   A sudden change of depth in one or more of the physical properties of the material making up Earth's interior. The boundary between two dissimilar materials in Earth's interior as determined by the behavior of seismic waves.

Discordant   A term used to describe plutons that cut across existing rock structures, such as bedding planes.

Disseminated deposit   Any economic mineral deposit in which the desired mineral occurs as scattered particles in the rock but in sufficient quantity to make the deposit an ore.

Dissolution   A common form of chemical weathering, it is the process of dissolving into a homogeneous solution, as when an acidic solution dissolves limestone.

Dissolved load   That portion of a stream's load carried in solution.

Distributary   A section of a stream that leaves the main flow.

Divergent plate boundary   A boundary in which two plates move apart, resulting in upwelling of material from the mantle to create new seafloor.

Divide   An imaginary line that separates the drainage of two streams, often found along a ridge.

D″ layer   A region in roughly the lowermost 200 kilometers of the mantle where P-waves experience a sharp decrease in velocity.

Dome   A roughly circular upfolded structure.

Drainage basin   The land area that contributes water to a stream.

Drawdown   The difference in height between the bottom of a cone of depression and the original height of the water table.

Drift   See *Glacial drift*.

Drumlin   A streamlined symmetrical hill composed of glacial till. The steep side of the hill faces the direction from which the ice advanced.

Dry climate   A climate in which yearly precipitation is less than the potential loss of water by evaporation.

Ductile deformation   A type of solid-state flow that produces a change in the size and shape of a rock body without fracturing. Occurs at depths where temperatures and confining pressures are high.

Dune   A hill or ridge of wind-deposited sand.

Earthflow   The downslope movement of water-saturated, clay-rich sediment. Most characteristic of humid regions.

Earthquake   Vibration of Earth produced by the rapid release of energy.

Ebb current   The movement of tidal current away from the shore.

Echo sounder   An instrument used to determine the depth of water by measuring the time interval between emission of a sound signal and the return of its echo from the bottom.

Elastic deformation   Rock deformation in which the rock will return to nearly its original size and shape when the stress is removed.

Elastic rebound   The sudden release of stored strain in rocks that results in movement along a fault.

Electron   A negatively charged subatomic particle that has a negligible mass and is found outside an atom's nucleus.

Element   A substance that cannot be decomposed into simpler substances by ordinary chemical or physical means.

Eluviation   The washing out of fine soil components from the *A* horizon by downward-percolating water.

Emergent coast   A coast where land formerly below sea level has been exposed by crustal uplift or a drop in sea level or both.

End moraine   A ridge of till marking a former position of the front of a glacier.

Energy levels or shells   Spherically shaped, negatively charged zones that surround the nucleus of an atom.

Environment of deposition   A geographic setting where sediment accumulates. Each site is characterized by a particular combination of geologic processes and environmental conditions.

Eon   The largest time unit on the geologic time scale, next in order of magnitude above era.

Ephemeral stream   A stream that is usually dry because it carries water only in response to specific episodes of rainfall. Most desert streams are of this type.

Epicenter   The location on Earth's surface that lies directly above the focus of an earthquake.

Epoch   A unit of the geologic time scale that is a subdivision of a period.

Era   A major division on the geologic time scale; eras are divided into shorter units called periods.

Erosion   The incorporation and transportation of material by a mobile agent, such as water, wind, or ice.

Eruption column   Buoyant plumes of hot, ash-laden gases that can extend thousands of meters into the atmosphere.

Escape velocity   The initial velocity an object needs to escape from the surface of a celestial body.

Esker   Sinuous ridge composed largely of sand and gravel deposited by a stream flowing in a tunnel beneath a glacier near its terminus.

Estuary   A funnel-shaped inlet of the sea that formed when a rise in sea level or subsidence of land caused the mouth of a river to be flooded.

Eukaryotes   An organism whose genetic material is enclosed in a nucleus; plants, animals, and fungi are eukaryotes.

Evaporite   A sedimentary rock formed of material deposited from solution by evaporation of the water.

Evapotranspiration   The combined effect of evaporation and transpiration.

Exfoliation dome   Large, dome-shaped structure, usually composed of granite, formed by sheeting.

Exotic stream   A permanent stream that traverses a desert and has its source in well-watered areas outside the desert.

External process   Process such as weathering, mass wasting, or erosion that is powered by the Sun and contributes to the transformation of solid rock into sediment.

Extrusive   Igneous activity that occurs at Earth's surface.

Facies   A portion of a rock unit that possesses a distinctive set of characteristics that distinguishes it from other parts of the same unit.

Fall   A type of movement common to mass-wasting processes that refers to the freefalling of detached individual pieces of any size.

Fault   A break in a rock mass along which movement has occurred.

Fault-block mountain   A mountain formed by the displacement of rock along a fault.

Fault creep   Gradual displacement along a fault. Such activity occurs relatively smoothly and with little noticeable seismic activity.

Fault scarp   A cliff created by movement along a fault. It represents the exposed surface of the fault prior to modification by weathering and erosion.

Felsic composition   See *Granitic composition*.

Ferromagnesian silicate   See *Dark silicate*.

Fetch   The distance that the wind has traveled across the open water.

Fine-grained   See *Aphanitic texture*.

Fiord   A steep-sided inlet of the sea formed when a glacial trough was partially submerged.

Firn   Granular, recrystallized snow. A transitional stage between snow and glacial ice.

Fissility   The property of splitting easily into thin layers along closely spaced, parallel surfaces, such as bedding planes in shale.

Fission (nuclear)   The splitting of a heavy nucleus into two or more lighter nuclei, caused by the collision with a neutron. During this process a large amount of energy is released.

Fissure   A crack in rock along which there is a distinct separation.

Fissure eruption   An eruption in which lava is extruded from narrow fractures or cracks in the crust.

Flood   The overflow of a stream channel that occurs when discharge exceeds the channel's capacity. The most common and destructive geologic hazard.

Flood basalts   Flows of basaltic lava that issue from numerous cracks or fissures and commonly cover extensive areas to thicknesses of hundreds of meters.

Floodplain   The flat, low-lying portion of a stream valley subject to periodic inundation.

Flood current   The tidal current associated with the increase in the height of the tide.

Flow   A type of movement common to mass-wasting processes in which water-saturated material moves downslope as a viscous fluid.

Flowing artesian well   An artesian well in which water flows freely at Earth's surface because the pressure surface is above ground level.

Fluorescence   The absorption of ultraviolet light, which is reemitted as visible light.

Focus (earthquake)   The zone within Earth where rock displacement produces an earthquake.

Fold   A bent layer or series of layers that were originally horizontal and subsequently deformed.

Fold-and-thrust belts   Regions within compressional mountain systems where large areas have been shortened and thickened by the processes of folding and thrust faulting, as exemplified by the Valley and Ridge Province of the Appalachains.

Foliated texture   A texture of metamorphic rocks that gives the rock a layered appearance.

Foliation   A term for a linear arrangement of textural features often exhibited by metamorphic rocks.

Footwall block   The rock surface below a fault.

Forearc basin   The region located between a volcanic arc and an accretionary wedge where shallow-water marine sediments typically accumulate.

Foreset bed   An inclined bed deposited along the front of a delta.

Foreshocks   Small earthquakes that often precede a major earthquake.

Foreshore   That portion of the shore lying between the normal high and low water marks; the intertidal zone.

Fossil   The remains or traces of organisms preserved from the geologic past.

Fossil fuel   General term for any hydrocarbon that may be used as a fuel, including coal, oil, natural gas, bitumen from tar sands, and shale oil.

Fossil magnetism   See *Paleomagnetism*.

Fossil succession   Fossil organisms succeed one another in a definite and determinable order, and any time period can be recognized by its fossil content.

Fracture   Any break or rupture in rock along which no appreciable movement has taken place.

Fracture zone   Linear zone of irregular topography on the deep-ocean floor that follows transform faults and their inactive extensions.

Fragmental texture   See *Pyroclastic texture*.

Frost wedging   The mechanical breakup of rock caused by the expansion of freezing water in cracks and crevices.

Fumarole   A vent in a volcanic area from which fumes or gases escape.

Gaining stream Streams that gain water from the inflow of groundwater through the streambed.

Geologic time scale The division of Earth history into blocks of time—eons, eras, periods, and epochs. The time scale was created using relative dating principles.

Geology The science that examines Earth, its form and composition, and the changes that it has undergone and is undergoing.

Geosphere The solid Earth; one of Earth's four basic spheres.

Geothermal energy Natural steam used for power generation.

Geothermal gradient The gradual increase in temperature with depth in the crust. The average is 30 °C per kilometer in the upper crust.

Geyser A fountain of hot water ejected periodically from the ground.

Glacial budget The balance, or lack of balance, between ice formation at the upper end of a glacier, and ice loss in the zone of wastage.

Glacial drift An all-embracing term for sediments of glacial origin, no matter how, where, or in what shape they were deposited.

Glacial erratic An ice-transported boulder that was not derived from the bedrock near its present site.

Glacial striations Scratches and grooves on bedrock caused by glacial abrasion.

Glacial trough A mountain valley that has been widened, deepened, and straightened by a glacier.

Glacier A thick mass of ice originating on land from the compaction and recrystallization of snow that shows evidence of past or present flow.

Glass (volcanic) Natural glass produced when molten lava cools too rapidly to permit recrystallization. Volcanic glass is a solid composed of unordered atoms.

Glassy A term used to describe the texture of certain igneous rocks, such as obsidian, that contain no crystals.

Gneissic texture A texture of metamorphic rocks in which dark and light silicate minerals are separated, giving the rock a banded appearance.

Gondwanaland The southern portion of Pangaea consisting of South America, Africa, Australia, India, and Antarctica.

Graben A valley formed by the downward displacement of a fault-bounded block.

Graded bed A sediment layer characterized by a decrease in sediment size from bottom to top.

Graded stream A stream that has the correct channel characteristics to maintain exactly the velocity required to transport the material supplied to it.

Gradient The slope of a stream, generally expressed as the vertical drop over a fixed distance.

Granitic composition A compositional group of igneous rocks indicating the rock is composed almost entirely of light-colored silicates.

Gravitational collapse The gradual subsidence of mountains caused by lateral spreading of weak material located deep within these structures.

Greenhouse effect The transmission of shortwave solar radiation by the atmosphere coupled with the selective absorption of longer-wavelength terrestrial radiation, especially by water vapor and carbon dioxide, resulting in warming of the atmosphere.

Groin A short wall built at a right angle to the seashore to trap moving sand.

Groundmass The matrix of smaller crystals within an igneous rock that has porphyritic texture.

Ground moraine An undulating layer of till deposited as an ice front retreats.

Groundwater Water in the zone of saturation.

Guyot A submerged, flat-topped seamount.

Habit Refers to the common or characteristic shape of a crystal, or aggregate of crystals.

Half graben A tilted fault block in which the higher side is associated with mountainous topography and the lower side is a basin that fills with sediment.

Half-life The time required for one-half of the atoms of a radioactive substance to decay.

Hanging valley A tributary valley that enters a glacial trough at a considerable height above the floor of the trough.

Hanging wall block The rock surface immediately above a fault.

Hardness A mineral's resistance to scratching and abrasion.

Head (stream) The beginning or source area for a stream. Also called the headwaters.

Headward erosion The extension upslope of the head of a valley due to erosion.

Historical geology A major division of geology that deals with the origin of Earth and its development through time. Usually involves the study of fossils and their sequence in rock beds.

Hogback A narrow, sharp-crested ridge formed by the upturned edge of a steeply dipping bed of resistant rock.

Horizon A layer in a soil profile.

Horn A pyramid-like peak formed by glacial action in three or more cirques surrounding a mountain summit.

Horst An elongate, uplifted block of crust bounded by faults.

Hot spot A concentration of heat in the mantle, capable of producing magma that, in turn, extrudes onto Earth's surface. The intraplate volcanism that produced the Hawaiian Islands is one example.

Hot spot tracks Chain of volcanic structures produced as a lithospheric plate moves over a mantle plume.

Hot spring A spring in which the water is 6–9 °C (10–15 °F) warmer than the mean annual air temperature of its locality.

Humus Organic matter in soil produced by the decomposition of plants and animals.

Hydraulic conductivity A factor relating to groundwater flow; it is a coefficient that takes into account the permeability of the aquifer and the viscosity of the fluid.

Hydraulic gradient The slope of the water table. It is determined by finding the height difference between two points on the water table and dividing by the horizontal distance between the two points.

Hydroelectric power Electricity generated by falling water that is used to drive turbines.

Hydrogenous sediment Seafloor sediment consisting of minerals that crystallize from seawater. An important example is manganese nodules.

Hydrologic cycle The unending circulation of Earth's water supply. The cycle is powered by energy from the Sun and is characterized by continuous exchanges of water among the oceans, the atmosphere, and the continents.

Hydrolysis A chemical weathering process in which minerals are altered by chemically reacting with water and acids.

Hydrosphere The water portion of our planet; one of the traditional subdivisions of Earth's physical environment.

Hydrothermal metamorphism Chemical alterations that occur as hot, ion-rich water circulates through fractures in rock.

Hydrothermal solution The hot, watery solution that escapes from a mass of magma during the latter stages of crystallization. Such solutions may alter the surrounding country rock and are frequently the source of significant ore deposits.

Hypocenter See *Focus* (earthquake).

Hypothesis A tentative explanation that is then tested to determine if it is valid.

Ice cap A mass of glacial ice covering a high upland or plateau and spreading out radially.

Ice-contact deposit An accumulation of stratified drift deposited in contact with a supporting mass of ice.

Ice sheet A very large, thick mass of glacial ice flowing outward in all directions from one or more accumulation centers.

Ice shelf Forming where glacial ice flows into bays, it is a large, relatively flat mass of floating ice that extends seaward from the coast but remains attached to the land along one or more sides.

Igneous rock Rock formed from the crystallization of magma.

Immature soil A soil lacking horizons.

Impact metamorphism Metamorphism that occurs when meteorites strike Earth's surface.

Incised meander Meandering channel that flows in a step, narrow valley. These features form either when an area is uplifted or when base level drops.

Inclusion A piece of one rock unit contained within another. Inclusions are used in relative dating. The rock mass adjacent to the one containing the inclusion must have been there first in order to provide the fragment.

Index fossil A fossil that is associated with a particular span of geologic time.

Index mineral A mineral that is a good indicator of the metamorphic environment in which it formed. Used to distinguish different zones of regional metamorphism.

Inertia Objects at rest tend to remain at rest, and objects in motion tend to stay in motion unless either is acted upon by an outside force.

Infiltration The movement of surface water into rock or soil through cracks and pore spaces.

Infiltration capacity The maximum rate at which soil can absorb water.

Inner core The solid innermost layer of Earth, about 1216 kilometers (754 miles) in radius.

Inner planets The innermost planets of our solar system, which include Mercury, Venus, Earth, and Mars. Also known as the terrestrial planets because of their Earth-like internal structure and composition.

Inselberg An isolated mountain remnant characteristic of the late stage of erosion in a mountainous arid region.

Intensity (earthquake) A measure of the degree of earthquake shaking at a given locale based on the amount of damage.

Interface A common boundary where different parts of a system interact.

Interior drainage A discontinuous pattern of intermittent streams that do not flow to the ocean.

Intermediate composition A compositional group of igneous rocks, indicating that the rock contains at least 25 percent dark silicate minerals. The other dominant mineral is plagioclase feldspar.

Internal process A process such as mountain building or volcanism that derives its energy from Earth's interior and elevates Earth's surface.

Intraplate volcanism Igneous activity that occurs within a tectonic plate away from plate boundaries.

Intrusion See *Pluton.*

Intrusive rock Igneous rock that formed below Earth's surface.

Ion An atom or molecule that possesses an electrical charge.

Ionic bond A chemical bond between two oppositely charged ions formed by the transfer of valence electrons from one atom to the other.

Iron meteorite One of the three main categories of meteorites. This group is composed largely of iron with varying amounts of nickel (5–20 percent). Most meteorite finds are irons.

Island arc See *Volcanic island arc.*

Isostasy The concept that Earth's crust is "floating" in gravitational balance upon the material of the mantle.

Isotatic adjustment Compensation of the lithosphere when weight is added or removed. When weight is added, the lithosphere will respond by subsiding, and when weight is removed, there will be uplift.

Isotopes Varieties of the same element that have different mass numbers; their nuclei contain the same number of protons but different numbers of neutrons.

Jetties A pair of structures extending into the ocean at the entrance to a harbor or river that are built for the purpose of protecting against storm waves and sediment deposition.

Joint A fracture in rock along which there has been no movement.

Jovian planet One of the Jupiter-like planets, Jupiter, Saturn, Uranus, and Neptune. These planets have relatively low densities.

Kame A steep-sided hill composed of sand and gravel, originating when sediment collected in openings in stagnant glacial ice.

Kame terrace A narrow, terracelike mass of stratified drift deposited between a glacier and an adjacent valley wall.

Karst A type of topography formed on soluble rock (especially limestone) primarily by dissolution. It is characterized by sinkholes, caves, and underground drainage.

Kettle holes Depressions created when blocks of ice become lodged in glacial deposits and subsequently melt.

Klippe A remnant or outlier of a thrust sheet that was isolated by erosion.

Kuiper belt A region outside the orbit of Neptune where most short-period comets are thought to originate.

Laccolith A massive igneous body intruded between preexisting strata.

Lag time The amount of time between a rainstorm and the occurrence of flooding.

Lahar Debris flows on the slopes of volcanoes that result when unstable layers of ash and debris become saturated and flow downslope, usually following stream channels.

Laminar flow The movement of water particles in straight-line paths that are parallel to the channel. The water particles move downstream without mixing.

Lateral moraine A ridge of till along the sides of a valley glacier composed primarily of debris that fell to the glacier from the valley walls.

Laterite A red, highly leached soil type found in the tropics that is rich in oxides of iron and aluminum.

Laurasia The northern portion of Pangaea, consisting of North America and Eurasia.

Lava Magma that reaches Earth's surface.

Lava dome A bulbous mass associated with an old-age volcano, produced when thick lava is slowly squeezed from the vent. Lava domes may act as plugs to deflect subsequent gaseous eruptions.

Lava tube Tunnel in hardened lava that acts as a horizontal conduit for lava flowing from a volcanic vent. Lava tubes allow fluid lavas to advance great distances.

Law A formal statement of the regular manner in which a natural phenomenon occurs under given conditions, (e.g., the "law of superposition").

Law of Constancy of Interfacial Angles A law stating that the angle between equivalent faces of the same mineral are always the same.

Law of Superposition In any undeformed sequence of sedimentary rocks, each bed is older than the one above and younger than the one below.

Leaching The depletion of soluble materials from the upper soil by downward-percolating water.

Light silicate Silicate minerals that lack iron and/or magnesium. They are generally lighter in color and have lower specific gravities than dark silicates.

Liquefaction The transformation of a stable soil into a fluid that is often unable to support buildings or other structures.

Lithification The process, generally cementation and/or compaction, of converting sediments to solid rock.

Lithosphere The rigid outer layer of Earth, including the crust and upper mantle.

Lithospheric plate A coherent unit of Earth's rigid outer layer that includes the crust and upper unit.

Local base level See *Temporary (local) base level.*

Loess Deposits of windblown silt, lacking visible layers, generally buff-colored, and capable of maintaining a nearly vertical cliff.

Longitudinal dunes Long ridges of sand oriented parallel to the prevailing wind; these dunes form where sand supplies are limited.

Longitudinal profile A cross section of a stream channel along its descending course from the head to the mouth.

Longshore current A nearshore current that flows parallel to the shore.

Long (L) waves These earthquake-generated waves travel along the outer layer of Earth and are responsible for most of the surface damage. L waves have longer periods than other seismic waves.

Losing stream Stream that loses water to the groundwater system by outflow through the streambed.

Lower mantle See *Mesosphere.*

Low-velocity zone A subdivision of the mantle located between 100 and 250 kilometers and discernible by a marked decrease in the velocity of seismic waves. This zone does not encircle Earth.

Lunar breccia A lunar rock formed when angular fragments and dust are welded together by the heat generated by the impact of a meteoroid.

Lunar regolith A thin, gray layer on the surface of the Moon, consisting of loosely compacted, fragmented material believed to have been formed by repeated meteoritic impacts.

Luster The appearance or quality of light reflected from the surface of a mineral.

Mafic composition See *Basaltic composition.*

Magma A body of molten rock found at depth, including any dissolved gases and crystals.

Magma mixing The process of altering the composition of a magma through the mixing of material from another magma body.

Magmatic differentiation The process of generating more than one rock type from a single magma.

Magnetic reversal A change in Earth's magnetic field from normal to reverse or vice versa.

Magnetometer A sensitive instrument used to measure the intensity of Earth's magnetic field at various points.

Magnitude (earthquake) An estimate of the total amount of energy released during an earth-quake, based on seismic records.

Manganese nodules A type of hydrogenous sediment scattered on the ocean floor, consisting mainly of manganese and iron and usually containing small amounts of copper, nickel, and cobalt.

Mantle One of Earth's compositional layers. The solid rocky shell that extends from the base of the crust to a depth of 2900 kilometers.

Mantle plume A mass of hotter-than-normal mantle material that ascends toward the surface, where it may lead to igneous activity. These plumes of solid yet mobile material may originate as deep as the core—mantle boundary.

Maria The smooth areas on our Moon's surface that were incorrectly thought to be seas.

Massive An igneous pluton that is not tabular in shape.

Mass number The sum of the number of neutrons and protons in the nucleus of an atom.

Mass wasting The downslope movement of rock, regolith, and soil under the direct influence of gravity.

Meander A looplike bend in the course of a stream.

Mechanical weathering The physical disintegration of rock, resulting in smaller fragments.

Medial moraine A ridge of till formed when lateral moraines from two coalescing alpine glaciers join.

Melt The liquid portion of magma excluding the solid crystals.

Mercalli intensity scale See *Modified Mercalli intensity scale.*

Mesosphere The part of the mantle that extends from the core–mantle boundary to a depth of 660 kilometers. Also known as the lower mantle.

Mesozoic era A time span on the geologic time scale between the Paleozoic and Cenozoic eras—from about 248 to 65.5 million years ago.

Metallic bond A chemical bond present in all metals that may be characterized as an extreme type of electron sharing in which the electrons move freely from atom to atom.

Metamorphic facies A group of associated minerals that are used to establish the pressures and temperatures at which rocks undergo metamorphism.

Metamorphic rock Rock formed by the alteration of preexisting rock deep within Earth (but still in the solid state) by heat, pressure, and/or chemically active fluids.

Metamorphism The changes in mineral composition and texture of a rock subjected to high temperatures and pressures within Earth.

Meteor The luminous phenomenon observed when a meteoroid enters Earth's atmosphere and burns up; popularly called a "shooting start."

Meteorite Any portion of a meteoroid that survives its traverse through Earth's atmosphere and strikes the surface.

Meteoroid Any small, solid particle that has an orbit in the solar system.

Meteor shower Numerous meteoroids traveling in the same direction and at nearly the same speed. They are thought to be material lost by comets.

Micrometeorite A very small meteorite that does not create sufficient friction to burn up in the atmosphere but slowly drifts down to Earth.

Microcontinents Relatively small fragments of continental crust that may lie above sea level, such as the island of Madagascar, or be submerged, as exemplified by the Campbell Plateau located near New Zealand.

Mid-ocean ridge A continuous mountainous ridge on the floor of all the major ocean basins and varying in width from 500 to 5000 kilometers (300 to 3000 miles). The rifts at the crests of these ridges represent divergent plate boundaries.

Migmatite A rock exhibiting both igneous and metamorphic rock characteristics. Such rocks may form when light-colored silicate minerals melt and then crystallize, while the dark silicate minerals remain solid.

Mineral A naturally occurring, inorganic crystalline material with a unique chemical structure.

Mineralogy The study of minerals.

Mineral resource All discovered and undiscovered deposits of a useful mineral that can be extracted now or at some time in the future.

Modified Mercalli intensity scale A 12-point scale developed to evaluate earthquake intensity based on the amount of damage to various structures.

Mohorovičić discontinuity (Moho) The boundary separating the crust and the mantle, discernible by an increase in seismic velocity.

Mohs scale A series of 10 minerals used as a standard in determining hardness.

Moment magnitude A more precise measure of earthquake magnitude than the Richter scale that is derived from the amount of displacement that occurs along a fault zone.

Monocline A one-limbed flexure in strata. The strata are usually flat-lying or very gently dipping on both sides of the monocline.

Mouth The point downstream where a river empties into another stream or water body.

Mud crack A feature in some sedimentary rocks that forms when wet mud dries out, shrinks, and cracks.

Mudflow See *Debris flow.*

Natural levee An elevated landform composed of alluvium that parallels some streams and acts to confine their waters, except during floodstage.

Neap tide The lowest tidal range, occurring near the times of the first and third quarters of the Moon.

Nearshore The zone of a beach that extends from the low-tide shoreline seaward to where waves break at low tide.

Nebular theory A model for the origin of the solar system that supposes a rotating nebula of dust and gases that contracted to form the Sun and planets.

Negative feedback mechanism As used in climatic change, any effect that is opposite of the initial change and tends to offset it.

Neutron A subatomic particle found in the nucleus of an atom. The neutron is electrically neutral, with a mass approximately equal to that of a proton.

Nonclastic texture A term for the texture of sedimentary rocks in which the minerals form a pattern of interlocking crystals.

Nonconformity An unconformity in which older metamorphic or intrusive igneous rocks are overlain by younger sedimentary strata.

Nonferromagnesian silicate See *Light silicate.*

Nonflowing artesian well An artesian well in which water does not rise to the surface because the pressure surface is below ground level.

Nonfoliated Metamorphic rocks that do not exhibit foliation.

Nonmetallic mineral resource A mineral resource that is not a fuel or processed for the metals it contains.

Nonrenewable resource A resource that forms or accumulates over such long time spans that it must be considered as fixed in total quantity.

Normal fault A fault in which the rock above the fault plane has moved down relative to the rock below.

Normal polarity A magnetic field the same as that which presently exists.

Nuclear fission The splitting of atomic nuclei into smaller nuclei, causing neutrons to be emitted and heat energy to be released.

Nucleus The small, heavy core of an atom that contains all of its positive charge and most of its mass.

Nuée ardente Incandescent volcanic debris buoyed up by hot gases that moves downslope in an avalanche fashion.

Numerical date The number of years that have passed since an event occurred.

Occultation The disappearance of light resulting when one object passes behind an apparently larger one. For example, the passage of Uranus in front of a distant star.

Oceanic ridge   See *Mid-ocean ridge*.

Oceanic plateau   An extensive region on the ocean floor composed of thick accumulations of pillow basalts and other mafic rocks that, in some cases, exceed 30 kilometers in thickness.

Octet rule   Atoms combine in order that each may have the electron arrangement of a noble gas; that is, the outer energy level contains eight neutrons.

Offshore   The relatively flat submerged zone that extends from the breaker line to the edge of the continental shelf.

Oil trap   A geologic structure that allows for significant amounts of oil and gas to accumulate.

Oort cloud   A spherical shell composed of comets that orbit the Sun at distances generally greater than 10,000 times the Earth–Sun distance.

Open system   A system in which both matter and energy flow into and out of the system. Most natural systems are of this type.

Ophiolite complex   The sequence of rocks that make up the oceanic crust. The three-layer sequence includes an upper layer of pillow basalts, a middle zone of sheeted dikes, and a lower layer of gabbro.

Ore   Usually a useful metallic mineral that can be mined at a profit. The term is also applied to certain nonmetallic minerals such as fluorite and sulfur.

Organic sedimentary rock   Sedimentary rock composed of organic carbon from the remains of plants that died and accumulated on the floor of a swamp. Coal is the primary example.

Original horizontality   Layers of sediment that are generally deposited in a horizontal or nearly horizontal position.

Orogenesis   The processes that collectively result in the formation of mountains.

Outer core   A layer beneath the mantle about 2270 kilometers (1410 miles) thick, which has the properties of a liquid.

Outer planets   The outermost planets of our solar system, which include Jupiter, Saturn, Uranus, Neptune, and Pluto. Except for Pluto these bodies are known as the Jovian planets.

Outgassing   The escape of dissolved gases from molten rocks.

Outlet glacier   A tongue of ice normally flowing rapidly outward from an ice cap or ice sheet, usually through mountainous terrain to the sea.

Outwash plain   A relatively flat, gently sloping plain consisting of materials deposited by melt-water streams in front of the margin of an ice sheet.

Oxbow lake   A curved lake produced when a stream cuts off a meander.

Oxidation   The removal of one or more electrons from an atom or ion. So named because elements commonly combine with oxygen.

Oxygen isotope analysis   A method of deciphering past temperatures based on the precise measurement of the ratio between two isotopes of oxygen, ^{16}O and ^{18}O. Analysis is commonly made of seafloor sediments and cores from ice sheets.

Pahoehoe flow   A lava flow with a smooth-to-ropy surface.

Paleoclimatology   The study of ancient climates; the study of climate and climate change prior to the period of instrumental records using proxy data.

Paleomagnetism   The natural remnant magnetism in rock bodies. The permanent magnetization acquired by rock that can be used to determine the location of the magnetic poles and the latitude of the rock at the time it became magnetized.

Paleontology   The systematic study of fossils and the history of life on Earth.

Paleozoic era   A time span on the geologic time scale between the Precambrian and Mesozoic eras—from about 542 million to 251 million years ago.

Pangaea   The proposed supercontinent that 200 million years ago began to break apart and form the present landmasses.

Parabolic dune   A sand dune similar in shape to a barchan dune except that its tips point into the wind. These dunes often form along coasts that have strong onshore winds, abundant sand, and vegetation that partly covers the sand.

Parasitic cone   A volcanic cone that forms on the flank of a larger volcano.

Parent material   The material upon which a soil develops.

Parent rock   The rock from which a metamorphic rock formed.

Partial melting   The process by which most igneous rocks melt. Since individual minerals have different melting points, most igneous rocks melt over a temperature range of a few hundred degrees. If the liquid is squeezed out after some melting has occurred, a melt with a higher silica content results.

Passive continental margin   A margin that consists of a continental shelf, continental slope, and continental rise. They are not associated with plate boundaries and therefore experience little volcanism and few earthquakes.

Pater noster lakes   A chain of small lakes in a glacial trough that occupies basins created by glacial erosion.

Pegmatite   A very coarse-grained igneous rock (typically granite) commonly found as a dike associated with a large mass of plutonic rock that has smaller crystals. Crystallization in a water-rich environment is believed to be responsible for the very large crystals.

Pegmatitic texture   A texture of igneous rocks in which the interlocking crystals are all larger than one centimeter in diameter.

Perched water table   A localized zone of saturation above the main water table, created by an impermeable layer (aquiclude).

Peridotite   An igneous rock of ultramafic composition thought to be abundant in the upper mantle.

Period   A basic unit of the geologic time scale that is a subdivision of an era. Periods may be divided into smaller units called epochs.

Periodic table   An arrangement of the elements in which atomic number increases from the left to right and elements with similar properties appear in columns called families or groups.

Permafrost   Any permanently frozen subsoil. Usually found in the subarctic and arctic regions.

Permeability   A measure of a material's ability to transmit water.

Phaneritic texture   An igneous rock texture in which the crystals are roughly equal in size and large enough so the individual minerals can be identified without the aid of a microscope.

Phanerozoic eon   That part of geologic time represented by rocks containing abundant fossil evidence. The eon extending from the end of the Proterozoic eon (540 million years ago) to the present.

Phenocryst   Conspicuously large crystal embedded in a matrix of finer-grained crystals.

Physical geology   A major division of geology that examines the materials of Earth and seeks to understand the processes and forces acting beneath and upon Earth's surface.

Piedmont glacier   A glacier that forms when one or more alpine glaciers emerge from the confining walls of mountain valleys and spread out to create a broad sheet in the lowlands at the base of the mountains.

Pillow basalts   Basaltic lava that solidifies in an underwater environment and develops a structure that resembles a pile of pillows.

Pipe   A vertical conduit through which magmatic materials have passed.

Placer   Deposit formed when heavy minerals are mechanically concentrated by currents, most commonly streams and waves. Placers are sources of gold, tin, platinum, diamonds, and other valuable minerals.

Planetesimal   A solid celestial body that accumulated during the first stages of planetary formation. Planetesimals aggregated into increasingly larger bodies, ultimately forming the planets.

Plastic flow   A type of glacial movement that occurs within the glacier, below a depth of approximately 50 meters, in which the ice is not fractured.

Plate   See *Lithospheric plate*.

Plate resistance   A force that counteracts plate motion as a subducting plate scrapes against an overriding plate.

Plate tectonics   The theory that proposes that Earth's outer shell consists of individual plates that interact in various ways and thereby produce earthquakes, volcanoes, mountains, and the crust itself.

Playa   The flat central area of an undrained desert basin.

Playa lake A temporary lake in a playa.

Pleistocene epoch An epoch of the Quaternary period beginning about 1.8 million years ago and ending about 10,000 years ago. Best known as a time of extensive continental glaciation.

Plucking The process by which pieces of bedrock are lifted out of place by a glacier.

Pluton A structure that results from the emplacement and crystallization of magma beneath the surface of Earth.

Plutonic rock Igneous rocks that form at depth. After Pluto, the god of the lower world in classical mythology.

Pluvial lake A lake formed during a period of increased rainfall. For example, this occurred in many nonglaciated areas during periods of ice advance elsewhere.

Point bar A crescent-shaped accumulation of sand and gravel deposited on the inside of a meander.

Polymorphs Two or more minerals having the same chemical composition but different crystalline structures. Exemplified by the diamond and graphite forms of carbon.

Porosity The volume of open spaces in rock or soil.

Porphyritic texture An igneous rock texture characterized by two distinctively different crystal sizes. The larger crystals are called phenocrysts, whereas the matrix of smaller crystals is termed the groundmass.

Porphyroblastic texture A texture of metamorphic rocks in which particularly large grains (porphyroblasts) are surrounded by a fine-grained matrix of other minerals.

Porphyry An igneous rock with a porphyritic texture.

Positive feedback mechanism As used in climatic change, any effect that acts to reinforce the initial change.

Pothole A depression formed in a stream channel by the abrasive action of the water's sediment load.

Precambrian All geologic time prior to the Phanerozoic eon. A term encompassing both the Archean and Proterozoic eons.

Primary (P) wave A type of seismic wave that involves alternating compression and expansion of the material through which it passes.

Principal shell The shell or energy level an electron occupies.

Principle of faunal succession Fossil organisms succeed one another in a definite and determinable order, and any time period can be recognized by its fossil content.

Principle of original horizontality Layers of sediment are generally deposited in a horizontal or nearly horizontal position.

Prokaryotes Refers to the cells or organisms such as bacteria whose genetic material is not enclosed in a nucleus.

Proterozoic eon The eon following the Archean and preceding the Phanerozoic. It extends between 2500 and 542 million years ago.

Proton A positively charged subatomic particle found in the nucleus of an atom.

Protoplanets A developing planetary body that grows by the accumulation of planetesimals.

Proxy data Data gathered from natural recorders of climate variability such as tree rings, ice cores, and ocean-floor sediments.

Pumice A light-colored, glassy vesicular rock commonly having a granitic composition.

P wave The fastest earthquake wave, which travels by compression and expansion of the medium.

Pyroclastic texture An igneous rock texture resulting from the consolidation of individual rock fragments that are ejected during a violent volcanic eruption.

Pyroclastic flow A highly heated mixture, largely of ash and pumice fragments, traveling down the flanks of a volcano or along the surface of the ground.

Pyroclastic material The volcanic rock ejected during an eruption. Pyroclastics include ash, bombs, and blocks.

Quarrying The removal of loosened blocks from the bed of a channel during times of high flow rates.

Radial pattern A system of streams running in all directions away from a central elevated structure, such as a volcano.

Radioactive decay The spontaneous decay of certain unstable atomic nuclei.

Radioactivity See *Radioactive decay*.

Radiocarbon (carbon-14) dating The radioactive isotope of carbon is produced continuously in the atmosphere and used in dating events from the very recent geologic past (the last few tens of thousands of years).

Radiometric dating The procedure of calculating the absolute ages of rocks and minerals that contain certain radioactive isotopes.

Rainshadow desert A dry area on the lee side of a mountain range. Many middle-latitude deserts are of this type.

Rapids A part of a stream channel in which the water suddenly begins flowing more swiftly and turbulently because of an abrupt steepening of the gradient.

Rays Bright streaks that appear to radiate from certain craters on the lunar surface. The rays consist of fine debris ejected from the primary crater.

Recessional moraine An end moraine formed as the ice front stagnated during glacial retreat.

Rectangular pattern A drainage pattern characterized by numerous right angle bends that develops on jointed or fractured bedrock.

Recurrence interval The average time interval between occurrences of hydrological events such as floods of a given or greater magnitude.

Refraction See *Wave refraction*.

Regional metamorphism Metamorphism associated with large-scale mountain building.

Regolith The layer of rock and mineral fragments that nearly everywhere covers Earth's land surface.

Rejuvenation A change in relation to base level, often caused by regional uplift, which causes the forces of erosion to intensify.

Relative dating Rocks and structures are placed in their proper sequence or order. Only the chronological order of events is determined.

Renewable resource A resource that is virtually inexhaustible or that can be replenished over relatively short time spans.

Reserve Already identified deposits from which minerals can be extracted profitably.

Reservoir rock The porous, permeable portion of an oil trap that yields oil and gas.

Residual soil Soil developed directly from the weathering of the bedrock below.

Return period See *Recurrence interval*.

Reverse fault A fault in which the material above the fault plane moves up in relation to the material below.

Reverse polarity A magnetic field opposite to that which presently exists.

Richter scale A scale of earthquake magnitude based on the amplitude of the largest seismic wave.

Ridge push A mechanism that may contribute to plate motion. It involves the oceanic lithosphere sliding down the oceanic ridge under the pull of gravity.

Rift valley A long, narrow trough bounded by normal faults. It represents a region where divergence is taking place.

Rills Tiny channels that develop as unconfined flow begins producing threads of current.

Rip current A strong, narrow surface or near-surface current of short duration and high speed that moves seaward through the breaker zone at nearly a right angle to the shore.

Ripple marks Small waves of sand that develop on the surface of a sediment layer by the action of moving water or air.

River A general term for a stream that carries a substantial amount of water and has numerous tributaries.

Roche moutonnée An asymmetrical knob of bedrock formed when glacial abrasion smoothes the gentle slope facing the advancing ice sheet and plucking steepens the opposite side as the ice overrides the knob.

Rock A consolidated mixture of minerals.

Rock avalanche The very rapid downslope movement of rock and debris. These rapid movements may be aided by a layer of air trapped beneath the debris, and they have been known to reach speeds in excess of 200 kilometers per hour.

Rock cleavage The tendency of rocks to split along parallel, closely spaced surfaces. These surfaces are often highly inclined to the bedding planes in the rock.

Rock cycle A model that illustrates the origin of the three basic rock types and the interrelatedness of Earth materials and processes.

Rock flour Ground-up rock produced by the grinding effect of a glacier.

Rockslide   The rapid slide of a mass of rock downslope along planes of weakness.

Rock structure   All features created by the processes of deformation from minor fractures in bedrock to a major mountain chain.

Runoff   Water that flows over the land rather than infiltrating into the ground.

Salinity   The proportion of dissolved salts to pure water, usually expressed in parts per thousand (0/000).

Saltation   Transportation of sediment through a series of leaps or bounces.

Salt flat   A white crust on the ground produced when water evaporates and leaves its dissolved materials behind.

Schistosity   A type of foliation characteristic of coarser-grained metamorphic rocks. Such rocks have a parallel arrangement of platy minerals such as the micas.

Scoria   Vesicular ejecta that is the product of basaltic magma.

Scoria cone   See *Cinder cone.*

Sea arch   An arch formed by wave erosion when caves on opposite sides of a headland unite.

Seafloor spreading   The hypothesis, first proposed in the 1960s by Harry Hess, which suggested that new oceanic crust is produced at the crests of mid-ocean ridges, which are the sites of divergence.

Seamount   An isolated volcanic peak that rises at least 1000 meters (3300 feet) above the deep-ocean floor.

Sea stack   An isolated mass of rock standing just offshore, produced by wave erosion of a headland.

Seawall   A barrier constructed to prevent waves from reaching the area behind the wall. Its purpose is to defend property from the force of breaking waves.

Secondary enrichment   The concentration of minor amounts of metals that are scattered through unweathered rock into economically valuable concentrations by weathering processes.

Secondary (S) wave   A seismic wave that involves oscillation perpendicular to the direction of propagation.

Sediment   Unconsolidated particles created by the weathering and erosion of rock by chemical precipitation from solution in water, or from the secretions of organisms, and transported by water, wind, or glaciers.

Sedimentary environment   See *Environment of deposition.*

Sedimentary rock   Rock formed from the weathered products of preexisting rocks that have been transported, deposited, and lithified.

Seismic gap   A segment of an active fault zone that has not experienced a major earthquake over a span when most other segments have. Such segments are probable sites for future major earthquakes.

Seismic reflection profile   A method of viewing the rock structure beneath a blanket of sediment by using strong, low-frequency sound waves that penetrate the sediments and reflect off the contacts between rock layers and fault zones.

Seismic sea wave   A rapidly moving ocean wave, generated by earthquake activity, which is capable of inflicting heavy damage in coastal regions.

Seismogram   The record made by a seismograph.

Seismograph   An instrument that records earthquake waves.

Seismology   The study of earthquakes and seismic waves.

Settling velocity   The speed at which a particle falls through a still fluid. The size, shape, and specific gravity of particles influence settling velocity.

Shadow zone   The zone between 105 and 140 degrees' distance from an earthquake epicenter, which direct waves do not penetrate because of refraction by Earth's core.

Shear   Stress that causes two adjacent parts of a body to slide past one another.

Sheeted dike complex   A large group of nearly parallel dikes.

Sheet flow   Runoff moving in unconfined thin sheets.

Sheeting   A mechanical weathering process characterized by the splitting off of slablike sheets of rock.

Shelf break   The point at which a rapid steepening of the gradient occurs, marking the outer edge of the continental shelf and the beginning of the continental slope.

Shield   A large, relatively flat expanse of ancient igneous and metamorphic rocks within the craton.

Shield volcano   A broad, gently sloping volcano built from fluid basaltic lavas.

Shore   Seaward of the coast, this zone extends from the highest level of wave action during storms to the lowest tide level.

Shoreline   The line that marks the contact between land and sea. It migrates up and down as the tide rises and falls.

Silicate mineral   Any one of numerous minerals that have the silicon-oxygen tetrahedron as their basic structure.

Silicon-oxygen tetrahedron   A structure composed of four oxygen atoms surrounding a silicon atom that constitutes the basic building block of silicate minerals.

Sill   A tabular igneous body that was intruded parallel to the layering of preexisting rock.

Sinkhole   A depression produced in a region where soluble rock has been removed by groundwater.

Slab-pull   A mechanism that contributes to plate motion in which cool, dense oceanic crust sinks into the mantle and "pulls" the trailing lithosphere along.

Slab suction   One of the driving forces of plate motion, it arises from the drag of the subducting plate on the adjacent mantle. It is an induced mantle circulation that pulls both the subducting and overriding plates toward the trench.

Slaty cleavage   The type of foliation characteristic of slates in which there is a parallel arrangement of fine-grained metamorphic minerals.

Slide   A movement common to mass-wasting processes in which the material moving downslope remains fairly coherent and moves along a well-defined surface.

Slip face   The steep, leeward surface of a sand dune that maintains a slope of about 34 degrees.

Slump   The downward slipping of a mass of rock or unconsolidated material moving as a unit along a curved surface.

Snowfield   An area where snow persists throughout the year.

Snowline   The lower limit of perennial snow.

Soil   A combination of mineral and organic matter, water, and air; that portion of the regolith that supports plant growth.

Soil horizon   A layer of soil that has identifiable characteristics produced by chemical weathering and other soil-forming processes.

Soil profile   A vertical section through a soil, showing its succession of horizons and the underlying parent material.

Soil taxonomy   A soil classification system consisting of six hierarchical categories based on observable soil characteristics. The system recognizes 12 soil orders.

Solar nebula   The cloud of interstellar gas and/or dust from which the bodies of our solar system formed.

Solifluction   Slow, downslope flow of water-saturated materials common to permafrost areas.

Solum   The *O, A,* and *B* horizons in a soil profile. Living roots and other plant and animal life are largely confined to this zone.

Sonar   Instrument that uses acoustic signals (sound energy) to measure water depths. Sonar is an acronym for *so*und *na*vigation and *r*anging.

Sorting   The degree of similarity in particle size in sediment or sedimentary rock.

Specific gravity   The ratio of a substance's weight to the weight of an equal volume of water.

Speleothem   A collective term for the dripstone features found in caverns.

Spheroidal weathering   Any weathering process that tends to produce a spherical shape from an initially blocky shape.

Spit   An elongate ridge of sand that projects from the land into the mouth of an adjacent bay.

Spreading center   See *Divergent plate boundary.*

Spring   A flow of groundwater that emerges naturally at the ground surface.

Spring tide   The highest tidal range. Occurs near the times of the new and full moons.

Stable platform That part of the carton that is mantled by relatively undeformed sedimentary rocks and underlain by a basement complex of igneous and metamorphic rocks.

Stalactite The icicle-like structure that hangs from the ceiling of a cavern.

Stalagmite The columnlike form that grows upward from the floor of a cavern.

Star dune An isolated hill of sand that exhibits a complex form and develops where wind directions are variable.

Steno's Law See *Law of Constancy of Interfacial Angles.*

Steppe One of the two types of dry climate. A marginal and more humid variant of the desert that separates it from bordering humid climates.

Stock A pluton similar to but smaller than a batholith.

Stony-iron meteorite One of the three main categories of meteorites. This group, as the name implies, is a mixture of iron and silicate minerals.

Stony meteorite One of the three main categories of meteorites. Such meteorites are composed largely of silicate minerals with inclusions of other minerals.

Storm surge The abnormal rise of the sea along a shore as a result of strong winds.

Strain An irreversible change in the shape and size of a rock body caused by stress.

Strata Parallel layers of sedimentary rock.

Stratified drift Sediments deposited by glacial meltwater.

Stratovolcano See *Composite cone.*

Streak The color of a mineral in powdered form.

Stream A general term to denote the flow of water within any natural channel. Thus, a small creek and a large river are both streams.

Stream Water flowing in a channel regardless of size.

Stream piracy The diversion of the drainage of one stream, resulting from the headward erosion of another stream.

Stream valley The channel, valley floor, and sloping valley walls of a stream.

Stress The force per unit area acting on any surface within a solid.

Striations (glacial) Scratches or grooves in a bedrock surface caused by the grinding action of a glacier and its load of sediment.

Strike The compass direction of the line of intersection created by a dipping bed or fault and a horizontal surface. Strike is always perpendicular to the direction of dip.

Strike-slip fault A fault along which the movement is horizontal.

Stromatolites Distinctively layered mounds of calcium carbonate, which are fossil evidence for the existence of ancient microscopic bacteria.

Subduction The process by which oceanic lithosphere plunges into the mantle along a convergent zone.

Subduction zone A long, narrow zone where one lithospheric plate descends beneath another.

Subduction zone metamorphism High pressure, low temperature metamorphism that occurs where sediments are carried to great depths by a subducting plate.

Submarine canyon A seaward extension of a valley that was cut on the continental shelf during a time when sea level was lower, or a canyon carved into the outer continental shelf, slope, and rise by turbidity currents.

Submergent coast A coast whose form is largely the result of the partial drowning of a former land surface due to a rise of sea level or subsidence of the crust, or both.

Subsoil A term applied to the *B* horizon of a soil profile.

Sunspot A dark area on the Sun associated with powerful magnetic storms that extend from the Sun's surface deep into the interior.

Supercontinent A large landmass that contains all, or nearly all, of the existing continents.

Supercontinent cycle The idea that the rifting and dispersal of one supercontinent is followed by a long period during which the fragments gradually reassemble into a new supercontinent.

Supernova An exploding star that increases its brightness many thousands of times.

Superposed stream A stream that cuts through a ridge lying across its path. The stream established its course on uniform layers at a higher level without regard to underlying structures and subsequently downcut.

Superposition, law of In any undeformed sequence of sedimentary rocks, each bed is older than the one above and younger than the one below.

Surf A collective term for breakers; also the wave activity in the area between the shoreline and the outer limit of breakers.

Surface waves Seismic waves that travel along the outer layer of Earth.

Surge A period of rapid glacial advance. Surges are typically sporadic and short-lived.

Suspended load The fine sediment carried within the body of flowing water or air.

Suture The zone along which two crustal fragments are jointed together. For example, following a continental collision the two continental blocks are sutured together.

S wave An earthquake wave, slower than a P wave, that travels only in solids.

Swells Wind-generated waves that have moved into an area of weaker winds or calm.

Syncline A linear downfold in sedimentary strata; the opposite of anticline.

System A group of interacting or interdependent parts that form a complex whole.

Tablemount See *Guyot.*

Tabular Describing a feature such as an igneous pluton having two dimensions that are much longer than the third.

Talus An accumulation of rock debris at the base of a cliff.

Tarn A small lake in a cirque.

Tectonic plate See *Lithospheric plate.*

Tectonic structure A basic geologic feature, such as a fold, fault, or rock foliation, that results from forces associated with the interaction of tectonic plates.

Tectonics The study of the large-scale processes that collectively deform Earth's crust.

Temporary (local) base level The level of a lake, resistant rock layer, or any other base level that stands above sea level.

Tenacity Describes a mineral's toughness or its resistance to breaking or deforming.

Tensional stress The type of stress that tends to pull a body apart.

Terminal moraine The end moraine marking the farthest advance of a glacier.

Terrace A flat, benchlike structure produced by a stream, which was left elevated as the stream cut downward.

Terrane A crustal block bounded by faults, whose geologic history is distinct from the histories of adjoining crustal blocks.

Terrestrial planet One of the Earthlike planets: Mercury, Venus, Earth, and Mars. These planets have similar densities.

Terrigenous sediment Seafloor sediments derived from terrestrial weathering and erosion.

Texture The size, shape, and distribution of the particles that collectively constitute a rock.

Theory A well-tested and widely accepted view that explains certain observable facts.

Thermal metamorphism See *Contact metamorphism.*

Thrust fault A low-angle reverse fault.

Tidal current The alternating horizontal movement of water associated with the rise and fall of the tide.

Tidal delta A deltalike feature created when a rapidly moving tidal current emerges from a narrow inlet and slows, depositing its load of sediment.

Tidal flat A marshy or muddy area that is alternately covered and uncovered by the rise and fall of the tide.

Tide Periodic change in the elevation of the ocean surface.

Till Unsorted sediment deposited directly by a glacier.

Tillite A rock formed when glacial till is lithified.

Tombolo A ridge of sand that connects an island to the mainland or to another island.

Topset bed An essentially horizontal sedimentary layer deposited on top of a delta during flood-stage.

Transform fault A major strike-slip fault that cuts through the lithosphere and accommodates motion between two plates.

Transform fault boundary A boundary in which two plates slide past one another without creating or destroying lithosphere.

Transpiration   The release of water vapor to the atmosphere by plants.

Transported soil   Soils that form on unconsolidated deposits.

Transverse dunes   A series of long ridges oriented at right angles to the prevailing wind; these dunes form where vegetation is sparse and sand is very plentiful.

Travertine   A form of limestone ($CaCO_3$) that is deposited by hot springs or as a cave deposit.

Trellis drainage pattern   A system of streams in which nearly parallel tributaries occupy valleys cut in folded strata.

Trench   See *Deep-ocean trench.*

Truncated spurs   Triangular-shaped cliffs produced when spurs of land that extend into a valley are removed by the great erosional force of a valley glacier.

Tsunami   The Japanese word for a seismic sea wave.

Turbidite   Turbidity current deposit characterized by graded bedding.

Turbidity current   A downslope movement of dense, sediment-laden water created when sand and mud on the continental shelf and slope are dislodged and thrown into suspension.

Turbulent flow   The movement of water in an erratic fashion often characterized by swirling, whirlpool-like eddies. Most streamflow is of this type.

Ultimate base level   Sea level; the lowest level to which stream erosion could lower the land.

Ultramafic composition   A compositional group of igneous rocks containing mostly olivine and pyroxene.

Unconformity   A surface that represents a break in the rock record, caused by erosion and nondeposition.

Uniformitarianism   The concept that the processes that have shaped Earth in the geologic past are essentially the same as those operating today.

Unsaturated zone   The area above the water table where openings in soil, sediment, and rock are not saturated but filled mainly with air.

Valence electron   The electrons involved in the bonding process; the electrons occupying the highest principal energy level of an atom.

Valley glacier   See *Alpine glacier.*

Valley train   A relatively narrow body of stratified drift deposited on a valley floor by meltwater streams that issue from the terminus of an alpine glacier.

Vein deposit   A mineral filling a fracture or fault in a host rock. Such deposits have a sheetlike, or tabular, form.

Vent   The surface opening of a conduit or pipe.

Ventifact   A cobble or pebble polished and shaped by the sandblasting effect of wind.

Vesicles   Spherical or elongated openings on the outer portion of a lava flow that were created by escaping gases.

Vesicular texture   A term applied to aphanitic igneous rocks that contain many small cavities called vesicles.

Viscosity   A measure of a fluid's resistance to flow.

Volatiles   Gaseous components of magma dissolved in the melt. Volatiles will readily vaporize (form a gas) at surface pressures.

Volcanic   Pertaining to the activities, structures, or rock types of a volcano.

Volcanic bomb   A streamlined pyroclastic fragment ejected from a volcano while still semimolten.

Volcanic island arc   A chain of volcanic islands generally located a few hundred kilometers from a trench where there is active subduction of one oceanic plate beneath another.

Volcanic neck   An isolated, steep-sided, erosional remnant consisting of lava that once occupied the vent of a volcano.

Volcano   A mountain formed from lava and/or pyroclastics.

Wadati–Benioff zone   The narrow zone of inclined seismic activity that extends from a trench downward into the asthenosphere.

Waterfall   A precipitous drop in a stream channel that causes water to fall to a lower level.

Water gap   A pass through a ridge or mountain in which a stream flows.

Water table   The upper level of the saturated zone of groundwater.

Wave-cut cliff   A seaward-facing cliff along a steep shoreline formed by wave erosion at its base and mass wasting.

Wave-cut platform   A bench or shelf along a shore at sea level, cut by wave erosion.

Wave height   The vertical distance between the trough and crest of a wave.

Wavelength   The horizontal distance separating successive crests or troughs.

Wave of oscillation   A water wave in which the wave form advances as the water particles move in circular orbits.

Wave of translation   The turbulent advance of water created by breaking waves.

Wave period   The time interval between the passage of successive crests at a stationary point.

Wave refraction   A change in direction of waves as they enter shallow water. The portion of the wave in shallow water is slowed, which causes the waves to bend and align with the underwater contours.

Weathering   The disintegration and decomposition of rock at or near the surface of Earth.

Welded tuff   A pyroclastic deposit composed of particles fused together by the combination of heat still contained in the deposit after it has come to rest and the weight of overlying material.

Well   An opening bored into the zone of saturation.

Wetted perimeter   The total distance in a linear cross-section of a stream that is in contact with water.

Wilson Cycle   See *Supercontinent cycle.*

Wind gap   An abandoned water gap. These gorges typically result from stream piracy.

Xenolith   An inclusion of unmelted country rock in an igneous pluton.

Xerophyte   A plant highly tolerant of drought.

Yardang   A streamlined, wind-sculpted ridge having the appearance of an inverted ship's hull that is oriented parallel to the prevailing wind.

Yazoo tributary   A tributary that flows parallel to the main stream because a natural levee is present.

Zone of accumulation   The part of a glacier characterized by snow accumulation and ice formation. The outer limit of this zone is the snowline.

Zone of fracture   The upper portion of a glacier consisting of brittle ice.

Zone of saturation   The zone where all open spaces in sediment and rock are completely filled with water.

Zone of soil moisture   A zone in which water is held as a film on the surface of soil particles and may be used by plants or withdrawn by evaporation. The uppermost subdivision of the unsaturated zone.

Zone of wastage   The part of a glacier beyond the snowline where annually there is a net loss of ice.

Index

A

Aa flows, 142
Ablation, 496
Abrasion, 437, 498, 532, 548
Abyssal plains, 29, 362–63
Accretionary wedge, 361, 388
Acid mine drainage, 181, 182
Acid precipitation, 182–83
Active continental margin, 361–62
Active solar collector, 653
Adams, Frank D., 5
Adobe layers, 189
Aerosols, 583–84
African superplume, 347
Aftershock, 306
Agassiz, Louis, 512
Agate, 210, 211
Aggregate, 75
Agricultural drought, 470, 471
Ahlgren, Andrew, 12
Air pollution, 646
Airy, George, 401
Aleutian arc, 384, 385
Alluvial channels, 440–41
Alluvial fan, 220, 419, 449, 527
Alluvium, 439
Alpha particle, 268
Alpha photon X-ray spectrometer (APXS), 690
Alpine glaciers, 490
Alpine-Himalayan belt, 316
Alternate energy sources, 652–59
    geothermal, 656–57
    hydroelectric, 655–56
    nuclear, 652–53
    solar, 653–54
    tidal power, 658–59
    wind, 654–55
American Philosophical Society, 45
Amino acids, 625
Amphibians, 613, 627, 630, 631
Amphiboles, 98
Amphibolite, 240, 248
Andalusite, 250
Andean-type plate margins, 386
    mountain building along, 386–89
Anders, Bill, 14
Andesite, 119
Andesitic composition, 111, 112, 119, 128
Andesitic magmas, formation of, 126–27
Angiosperms, 634, 635
Angle of repose, 409
Angularity, 205

Angular unconformity, 259–60, 261
Anions, 83
Antarctic Mapping Mission, 11
Antecedent stream, 450
Anthracite, 212–13, 239
Anticline, 286, 287
Apatite, 666, 667
Apatosaurus, 634
Aphanitic texture, 113
Apollo, 698
Appalachian Mountains, 392–94
Aquifer, 468
Aquitard, 468
Aral Sea, 524–25
Archaeology, 266
Archaeopteryx, 634
Archean eon, 271–72
Arches National Park, 297
Arctic sea ice, 602–3
Arête, 502–3
Arid climates, geologic processes in, 523–26
    water, role of, 524–26
    weathering, 524
Aristotle, 5, 6
Arkose, 206
Artesian, defined, 476
Artesian wells, 476–77
Artificial levees, 454
Asbestos, 85
Assimilation, 123–24
Asteroid belt, 695
Asteroids, 695–96
Asthenosphere, 24, 51, 339
Aswan High Dam, 656
Athabasca Oil Sands, 650
Atlantic Ocean, 372–73
Atmosphere, 15–16
Atmospheric basics, 582–87
    composition, 582–84
    extent and structure of, 584–85
    greenhouse effect, 586–87
    solar energy, incoming, 585–86
Atolls, 364
Atomic number, 77, 268
Atomic structure, 268
Atoms, 76–80
    defined, 76
Augite, 96, 98
Aureole, 241
Average annual precipitation, 523
Axial plane, 286

B

Backarc basin, 385
Backarc region, 385
Backshore, 543

Back swamp, 449
Bacon, Francis, 54
Bagnold, R. A., 534
Bajada, 527
Banded iron formations, 616
Bar, 220, 445
Barchan dunes, 534, 535
Barchanoid dunes, 535
Barrier islands, 220, 552–53
Barrier reef, 364
Basal slip, 494
Basalt, 23, 29, 119
Basaltic composition, 110–11
Basaltic magma, formation of, 124–26
Base level, 441–42
    graded streams and, 441–43
Basin, 288, 289
Basin and Range Province, 396–97
Batholiths, 131–32, 387
Bathymetry, 356
Bauxite, 664
Baymouth bar, 552
Beach drift, 549–50
Beaches, 220, 543–44
    sand movement on, 548–50
Beach face, 543
Beach nourishment, 556–57
Bedding planes, 221
Bed load, 437, 438, 529–30
Bedrock channels, 440
Beds, 30, 221
Benson, Sally, 658
Berm, 543
Beta particle, 268
Big Bang Theory, 20–21
Big Sur, 15
Bingham Canyon Mine, Utah, 660
Biochemical, 207
Biological activity, 178–79
Biomass, 645
Biosphere, 16
Biotite, 96, 98
Bird-foot delta, 447, 448
Birth and Development of the Geological Sciences, The (Adams), 5
Blackhawk landslide, 416
Black smokers, 369, 370
Block lava, 143
Blowout, 531
Blue planet, 15
Blueschist facies, 248, 249
Body wave, 310
Bonneville salt flats, 212
Bornhardt, Wilhelm, 529
Bottomset beds, 446, 447
Bowen, N. L., 122–23, 127

Bowen's reaction series, 122–23, 127, 185
Braided channels, 440–41
Breaks, 547
Breakwater, 555
Breccia, 207
Brittle deformation, 283
    structures formed by, 290–97
Bronowski, Jacob, 12
Building materials, 665–66
Bullard, Edward, 41
Bureau of Labor Statistics, 6
Burgess Shale, 626
Burial metamorphism, 244

C

Calderas, 156–57
    crater lake-type, 156
    defined, 156
    Hawaiian-type, 156
    Yellowstone-type, 157, 158
Callisto, 688, 689
Calving, 496
Cambrian explosion, 626, 627, 628
Camille, Hurricane, 564
Capacity, 438
Cape Hatteras Lighthouse, relocating, 559
Capillary fringe, 465
Capital Reef National Park, 201
Cap rock, 647
Carat, 80
Carbon-14, 270
Carbonaceous chrondrites, 625, 637, 699
Carbonate reefs, 208–9
Carbon cycle, 19
Carbon dioxide, rising levels of, 593
Carbonic acid, 482
Carbonization, 265
Cascadia subduction zone, 387
Cassini Orbiter, 693
Catastrophism, 5
Cations, 83
Cavern, 482–83
Cementation, 30, 213
Cenozoic era, 271, 634–38
Cephalopods, 629
Chalk, 209
Chamberlain, R. T., 45
Channeled Scablands, 511
Channelization, 454, 456
Charakusa Valley, 393
Charon, 677, 700, 701
Chemical bond, 78, 80
Chemical compounds, 77

Chemically active fluids, 230, 234
Chemical sedimentary rocks, 30
 chert, 210, 211
 defined, 202
 dolostone, 210
 evaporites, 210–12
 identification of, 216
 limestone, 208–10
Chemical weathering, 175–76
Chert, 210, 211
Chief Mountain, 294, 295
Chikyu, 64, 355, 579
Chrons, 49–50
Cinder cones, 149–51
Circular orbital motion, 545–46
Circum-Pacific belt, 316
Cirque, 500
Clastic texture, 214–15, 216
Clasts, 202
Clay minerals, 97–98
Clean dry air, 583
Cleavage, 89, 90
Cleveland Volcano, 58
Climate change, causes of, 587–96
 aerosol, 598–99
 atmospheric, 582–87, 593–94
 carbon dioxide levels, 593
 geological, 576
 global warming, 599–604
 human impact, 592–93
 natural, 587–92
 solar variability, 589, 591–92
 trace gases, 594–96
 volcanic activity, 588–89
Climate change, detecting, 577–82
 coral reefs and, 582
 fossil pollen and, 582
 glacial ice recording, 580
 historical data and, 582
 oxygen isotope analysis and, 579–80
 seafloor sediment and, 578–79
 tree rings and, 580–82
Climate feedback mechanisms, 596–98
Climate system, 577
Closed systems, 17–18
Coal, 212, 213, 645–46
Coarse-grained texture, 113
Coast, 543
Coastal classification, 564–65
Coastal zone, 542–44
 beaches, 543–44
 features of, 543
Coastline, 543
Coast Ranges, 389–90
Col, 502
Collapse pits, 156
Collier, Michael, 3, 286, 287
Collisional mountain belts, 390–94
 Appalachian Mountains, 392–94

continental collisions, 391
 Himalayas, 391–92
 terranes and mountain building, 390–91
Color, 87
Colorado River, 439, 446
Columbia Plateau, 159
Columnar joints, 130–31, 296, 297
Coma, 697
Comets, 696–98
Compaction, 30, 213
Competence, 438–39
Composite cones, 152–55
Composition, 206
Compressional mountains, 382
Compressional stress, 281
Conchoidal fractures, 89, 90
Concordant (instrusion), 128
Conduction, 344
Conduit, 146
Cone of depression, 475
Confined aquifer, 476
Confining pressure, 233, 281, 284
Conformable rock layers, 258
Conglomerate, 206–7
Conservative margins. *See* Transform fault boundary
Constancy of Interfacial Angles, Law of (Steno's Law), 83
Constructive margins. *See* Divergent boundaries
Contact metamorphism, 31, 232, 241–42
Continental collisions, 390–94
Continental-continental convergence, 59–60
Continental crust, 337
Continental drift, 14, 40–44
 ancient climates as evidence of, 44
 fossils as evidence of, 41–43
 hypothesis rejection, 45
 paleomagnetism and, 45–48
 rock types as evidence of, 43–44
 scientific method and, 45
 shorelines as evidence of, 40–41
Continental environments, 220
Continental margins, 28, 358–62
 active, 361–62
 passive, 358, 360–61
Continental rifts, 56, 371–75
 mechanisms for, 373–75
Continental rise, 28, 361
Continental shelf, 28, 358, 360
Continental slope, 28, 360
Continental volcanic arc, 57–58, 162, 384
Continents, 25–28
Convection, 67–68, 343–44

Convergence and subducting plates, 384–85
Convergent boundaries, 56–60
 continental-continental, 59–60
 defined, 55
 oceanic continental, 57–58
 oceanic–oceanic, 58
Convergent plate boundaries volcanoes and, 162, 164
Coquina, 209
Coral atolls, 364
Coral reefs, 217, 364, 582
Core, 23, 24, 67, 341–42
Core–mantle boundary, 340, 341
Correlation, 261–62
Corrosion, 437
Covalent bond, 79, 80
Crater, 146
Crater Lake, 152
Cratons, 27, 619
Creep, 422
Cretaceous climate, 592
Cretaceous–Tertiary or KT boundary, 637
Crevasse, 495
Cross-bedding, 222, 533
Cross-cutting relationships, principle of, 258
Cross-over, 339
Crust, 23–24, 337
Cryovolcanism, 694
Crystal, 81–83
 mineral forming and, 81–82
 structures, 82–83
Crystalline, 81
Crystalline texture, 215
Crystallization, 32, 81–82, 109
Crystal settling, 123
Crystal shape, 88
Crystal structures, 82–83
Curie point, 46
Cut bank, 440
Cutoff, 440
Cyanobacteria, 616, 625

D

Daeschler, Ted, 632
Dam-failure floods, 453
Darcy, Henri, 469
Darcy's law, 469
Dark silicates, 98–99, 110
Darwin, Charles, 54, 364
Dead Sea, 81
Death Valley, California, 528
DeBari, Susan, 359
Debris flow, 419–20
Debris slide, 417–19
Decompression melting, 121–22, 366–67
Dee double-prime layer, 24, 340
Deep marine, 220
Deep-ocean basins, 29, 362–63
 abyssal plains, 362–63

deep-ocean trenches, 362
 guyots, 363
 oceanic plateaus, 363
 seamounts, 363
Deep-ocean trenches, 29, 56–57, 362, 384–85
Deep Sea Drilling Project, 62
Deep-sea fan, 361
Deflation, 530–31
Deformation, 280
 brittle, 283
 ductile, 283
 elastic, 283
 of rock, 282–85
 rock strength and, 284–85
Delta, 220, 446–47
Denali National Park, 383, 499
Dendritic drainage pattern, 449–50
Density, 89–90
Depositional landforms, 445–49
 alluvial fans, 449
 deltas, 446–47
 Mississippi delta, 447–48
 natural levees, 448, 449
Deposition of sediment, 440
Desert pavement, 531–32
Deserts
 basin and range, 526–29
 sediment transport by wind, 529–30
 wind deposits, 533–37
 wind erosion, 530–33
Destructive margins. *See* Convergent boundaries
Detachment fault, 294
Detrital sedimentary rocks, 30
 breccia, 207
 conglomerate, 206–7
 defined, 202
 identification of, 216
 particle size classification for, 203
 sandstone, 204–6
 shale, 203–4
Detrital sediments, 30
Devil's Tower National Monument, 297
Devonian period, 629, 630
Diagenesis, 213
Diamonds, 661–62
Diatreme, 160
Dietz, Robert, 54
Differential stress, 233, 281, 282
Differential weathering, 186
Dikes, 128–31
Dike swarms, 129
Diorite, 119
Dip, 298–99
Dipolar field, 348
Dip-slip fault, 292–94
Discharge, 435
Disconformity, 260

Discontinuous reaction series, 127
Discordant (instrusion), 128
Disseminated deposit, 662
Dissolution, 179–80
Dissolved load, 437
Distributary, 446–47
Diurnal tidal pattern, 567–68
Divergent boundaries, 55–56
    continental rifting and, 56
    defined, 54
    oceanic ridges and, 55–56
    volcanoes and, 162, 165
Divide, 432–33
D″ layer, 24, 340
Dolomitization, 210
Dolostone, 210
Dome, 288–89
Donnellan, Andrea, 309
Dot diagrams, 79
Double refraction, 91
Drainage basin, 432–33
Drainage patterns, 449–52
    headward erosion and, 450
    stream piracy and, 452
    water gap formation, 450
Drawdown, 474–75
Dripstone, 207
Droughts, 470, 471
Drumlin, 505–6
Ductile deformation, 283
    structures formed by, 285–90
Dunes, 205, 533–36
Dust Bowl, 194, 195, 530, 531
Du Toit, Alexander, 45
Dwarf planets, 700–701

E

Earth. See also Life on Earth
    face of, 25–29
    internal structure of, 23–25
    layered structure formation
        of, 22, 23–24
    origin of, 20–22
    spheres, 14–17
    studying, from space, 11
    system, 17–20
Earth, evolution through geologic
        time, 610–38
    birth of planet, 614–17
    Cenozoic era, 634–38
    life, 625–28
    Mesozoic era, 633–34
    Paleozoic era, 628–33
    Phanerozoic eon, 621–25
    Precambrian history, 618–21
    unique characteristics, 610–13
Earthflow, 421–22
Earthquakes
    aftershocks and foreshocks,
        306–7
    belts and plate boundaries,
        316–17
    causes of, discovering, 305–6
    defined, 304–5

East of Rockies, 329
    faults and faulting, 307–8
    notable, 324
    plate tectonics and, 327–28
    predicting, 322–25
    sources of, locating, 312–13
Earthquakes, destruction caused
        by, 317–22
    fire and, 320
    landslides/ground subsidence
        and, 320
    seismic vibrations and, 317–19
    tsunamis and, 320–22
Earthquakes, measuring, 313–16
    intensity scales and, 313–14
    magnitude scales and, 314–16
Earth's crust, 337–39
Earth's interior, seismic waves
        and, 334–37
Earth's layers, 337–42
    core, 341–42
    crust, 337–39
    mantle, 339–41
Earth's magnetic field, 45–46,
        347–50
    geodynamo and, 348
    magnetic reversals and, 349–50
    measurement of, 349
Earth's moon, 678–80
    formation of, 679–80
Earth's temperature, 342–45
    heat flow mechanisms, 343–44
    profile, 344–45
    thermal stages, 343
Earth's three dimensional
        structures, 345–50
    gravity and, 346–47
    magnetic field and, 347–50
    seismic tomography and, 347
Earth system, 17–20
    cycles in, 19
    defined, 17–18
    energy sources for, 19
    feedback mechanisms and, 18
    humans and, 20
    linked parts of, 20
    science, 17–18
East African Rift, 52, 56, 371
Ebb current, 569
Echo sounder, 356
Eisinger, Chris, 163
Eismitte, 45
Elastic deformation, 283
Elastic rebound, 306, 307
El Chichón, 588–89
Electrical charge, 77
Electromagnetic radiation, 605
Electrons, 76–77, 268
Element, 77
    periodic table of, 78
Emergent coasts, 565
End moraine, 504–5, 506
Energy resources, 645–46. See also
        Fossil fuel

Environmental Protection Agency
        (EPA), 85
Environment of deposition. See
        Sedimentary environment
Eon, 271, 272
Ephemeral stream, 524–26
Epicenter, 304, 305, 312, 313
Epoch, 271, 272
Era, 271, 272
Erosion, 174, 436–37
    rates of, 194–95
    sedimentation/chemical
        pollution and, 195
    of shorelines, 552
Eruption columns, 141
Eruptions, volcanic, 140–42
    anatomy of, 141
    materials extruded during,
        142–46
    volatiles role in, 140–41
Escape velocity, 677
Esker, 506, 507
Eskola, Pennti, 248
Estuaries, 565
Etna, Mount, 4
Eukaryotes, 628
Europa, 688, 689
Evaporite deposit, 211
Evaporites, 210–12
Evapotranspiration, 430–31
Everest, George, 401
Exfoliation domes, 178
External processes, 174
Extinction
    of large mammals, 637–38
    late Pleistocene, 638
    mass, 589, 612, 630
    Permian, 630–31
Extrusive igneous rocks, 29, 110
Eye of hurricane, 562
Eye wall, 562

F

Facies, 220–21
Fall (motion type), 413
Falls (meteors), 699
Farallon plate, 347, 376–77
Fault, 290, 292, 304, 307
Fault-block mountains, 293,
        394–98
    basin and range province,
        396–97
Fault breccia, 244
Fault creep, 307
Fault gouge, 244
Fault scarp, 292, 293
Fault slip, 308
Feedback mechanisms, 18
Feldspar group, 95–97
Felsic compositions, 110–11
Ferromagnesian, 98–99, 110
Fertilizers, 667
Fetch, 545
Filchner Ice Shelf, 492

Finds (meteorites), 699
Fine-grained texture, 113
Finger Lakes (New York), 509,
        510
Fiord, 502
Firn, 494
Fissility, 204
Fissure eruptions, 159
Fissures, 157, 159
Fixed carbon, 212
Flash floods, 453, 455
Flint, 210, 211
Flood basalts, 159
Flood control, 454, 456
Flood-control dams, 454
Flood current, 569
Floodplain, 220, 444
Floods, 452–56
    control of, 454, 456
    defined, 452
    types of, 452–53
Flow, 414
Flowing artesian well, 477
Flow velocity, 434
Floyd, Hurricane, 564
Focus, 304, 305
Fold, 285–90
Fold-and-thrust belts, 391
Foliated texture, 31
Foliation, 235
Footwall block, 292
Forearc basin, 388–89
Foreshock, 307
Foreshore, 543
Fossil fuel, 645–51
    coal, 645–46
    oil and natural gas, 646–49
    oil sands and shales, 649–51
Fossil magnetism, 46
Fossil pollen, 582
Fossils, 9, 224, 262
    correlation and, 267
    Glossopteris, 42–43
    Mesosaurus, 42
    preservation conditions, 266
    types of, 265–66
Fossil succession, principle
        of, 9, 267
Fracture, 89, 90
Fracture zone, 60
Fragmental texture, 114
Frost heave, 176
Frost wedging, 176
Fujiyama, 152
Fumaroles, 147

G

Gabbro, 119
Gaining stream, 467
Garnet, 99, 100
Gases. See Volatiles
Gas giants, 674
Gas hydrates, 651
Gemstones, 102

Geode, 82
Geodynamo, 348
Geological Society of America, 6
Geologic time scale, 7–10, 271–73, 613
   difficulties in dating, 273–74
   fossils as evidence and, 9, 262–67
   magnitude of, 7–8
   Precambrian time and, 271–73
   radioactivity and, 267–71
   relative dating and, 8–9, 256–61
   rock layer correlation and, 261–62
   structure of, 271
Geology
   career opportunities in, 6
   defined, 2
   environmental issues and, 2–4
   historical notes about, 5–7
   science of, 2
   studies of, 3
Geoscientists, 6
Geosphere, 16–17
Geotherm. See Geothermal gradient
Geothermal energy, 656–57
Geothermal gradient, 121, 232, 344
Geysers, 473–74
Giants Causeway National Park, 131
Gilbert, G. K., 132
Giotto, 698
Glacial budget, 496–98
   defined, 496
   Greenland, 496, 498
Glacial deposits, 503–6
   drumlins, 505–6
   moraine, 503–5
   till, 503
Glacial drift, 503
Glacial erosion, 498–99
Glacial erosion landforms, 499–503
   arêtes, 502–3
   glaciated valleys, 499–502
   horns, 502–3
   roche moutonnée, 503
   troughs, 499, 501
Glacial erratics, 503
Glacial striations, 498
Glacier, Switzerland, 13
Glacier National Park, 295
Glaciers
   causes of, 512–15
   defined, 490
   formation of, 494
   ice age and, 512
   movement of, 13, 494–96
Glaciers, effects of, 507–12
   crustal subsidence and rebound, 508

pluvial lakes, 510, 511
proglacial lake creation, 509–10
river and valley changes, 508–9
sea-level changes, 508
Glass, 113
Glass, making from minerals, 76
Glassy texture, 113
Glen Canyon Dam, 656
Global Positioning System (GPS), 67
Global warming, consequences of, 599–604
   arctic, changing, 602–3
   ocean acidity, increasing, 603–4
   sea-level rise, 600–602
Glomar Challenger, 62
Glossopteris, 42–43
Gneiss, 31, 240, 662
Gneissic texture, 236
Gobi Desert, 521
Goldilocks scenario, 612
Gondwana, 620, 621, 622
Gore, Al, 590
Graben, 293
Graded beds, 222
Graded stream, 442
Gradient, 435
Grand Canyon, 8, 32, 204, 206, 207, 256–59, 260, 263, 284, 298, 408, 439
Grand Prismatic Pool, 626
Granite, 23, 29, 116–17
Granitic composition, 110–11
Granodiorite, 23
Gravitational collapse, 399
Gravity, 346–47
   layered planets and, 334
   mass wasting and, 405–7
   specific, 89–90
   tides and, 566
Gravity Recovery and Climate Experiment (GRACE), 498
Graywacke, 206
Great Barrier Reef, 208
Great Dark Spot, 678
Great Salt Lake, 81
Great Sand Dunes National Park, 205
Great Valley, 389–90
Greenhouse effect, 586–87
Greenhouse gases, 18
Greenland Ice Sheet, 45
Green River Formation, 650
Groin, 554–55
Groundmass, 113
Ground moraine, 504, 506
Groundwater
   artesian wells, 476–77
   basic resource, as, 464–65
   defined, 465
   distribution of, 465–66
   geologic work of, 482–85

geysers, 473–74
hot springs, 472–73
importance of, 462–63
springs, 472
water table and, 466–67
wells, 474–75
Groundwater contamination, 480–82
Groundwater movement, 468–71
   Darcy's law and, 469
   droughts and, 470, 471
   scales of, 470, 471
Groundwater storage, 467–68
   permeability and, 468
   porosity and, 467–68
Groundwater withdrawal problems, 477–82
   contamination and, 480–82
   nonrenewable resource, treating as, 477–78
   saltwater contamination and, 478, 480
Guadalupe Mountains National Park, 208
Gutenberg, Beno, 340
Guyots, 149, 363

H
Habit, 88
Half-graben, 293–94
Half-life, 269
Halley's Comet, 697–98
Hanging valley, 500
Hanging wall block, 292
Hardness, 88–89
Hardpans, 189
Hard stabilization, 554–56
Hawaiian Island–Emperor Seamount chain, 64
Hayabusa, 696
Head, 435
Headward erosion, 450, 452
Headwaters, 435
Henry Mountains, 132
Hess, Harry, 48–49, 50, 54, 366
Himalayas, 391–92
Hinge line, 286
Historical geology, 2
HMS Challenger, 356, 357
Hogback, 290
Holmes, Arthur, 45, 54
Horizons, 189
Horn, 502–3
Hornblende, 96, 98
Hornfels, 242
Horst, 293
Hot spot, 64, 66, 166
Hot spot track, 64
Hot springs, 472–73
Hubbard Glacier, 497
Hubble Space Telescope, 678, 691, 701
Humus, 187

Hurricanes, 560–64
   deadliest, 561
   destruction caused by, 562–64
   forecasting, 568–69
   naming, 564
   profile of, 562
Hurricane warning, 568
Hurricane watch, 568
Hutton, James, 6–7, 256, 259–60
Hydraulic conductivity, 469
Hydraulic gradient, 469
Hydroelectric power, 655–56
Hydrological drought, 471
Hydrologic cycle, 19, 430–32
Hydrolysis, 181
Hydrosphere, 15
Hydrosphere fresh water, 463
Hydrothermal metamorphism, 31, 242, 369
Hydrothermal solution, 242, 662
Hypocenter, 304
Hypothesis, 12

I
Ice Age, 512
Iceberg, 496–97
Ice cap, 493
Ice-contact deposits, 507
Ice-jam floods, 453
Ice sheets, 490–92
Ice shelf, 492
Igneous compositions, 110–12
   granitic vs. basaltic, 110–11
   other groups of, 111
   silica content as indicator of, 111–12
Igneous processes, 110
Igneous rocks, 29, 32, 108, 109
Igneous rocks, naming, 115–20
   classification of, 115–16
   felsic (granitic), 116–19
   intermediate (andesitic), 119
   mafic (basaltic), 119
   pyroclastic, 119–20
Igneous texture, 112–15
   defined, 112
   types of, 113–15
Ike, Hurricane, 561
Impact craters, 678, 679
Impact metamorphism, 245, 247
Incised meander, 445, 446
Inclusions, 258
Index fossil, 267
Index mineral, 246
Inertia, 310
Infiltration, 430
Inner core, 24, 341–42
Inner planets, 674
Inorganic limestone, 209–10
Inselberg, 527
Integrated Ocean Drilling Program (ICDP), 64, 579
Intensity, 313
Intensity scales, 313–14

Interface, 19, 186, 542
Intergovernmental Panel on Climate Change (IPCC), 590, 593, 596
Interior drainage, 526
Intermediate composition, 111
Internal processes, 174
International Committee on Stratigraphy (ICS), 273
International Mineralogical Association, 92
International Union of Geological Sciences, 273
Intraplate volcanism, 163–67
Intrusions, 128
Intrusive igneous activity, 128–32
    massive bodies, 131–32
    nature of bodies, 128
    tabular bodies, 128–31
Intrusive igneous rocks, 29, 110
Ionic bond, 79–80
Ionic compounds, 77
Ions, 79
Island arcs, 58, 162, 384
Isostasy, 399
Isostatic adjustment, 399
Isotopes, 81
Izu Bonin trench, 359

J

Jasper, 210, 211
Jetty, 554
JOIDES Resolution, 62, 64
Joint, 290, 296–97
Jovian planet, 674, 675
Juan de Fuca ridge, 61
    Jupiter, 686–89
    moons of, 687–89
    rings of, 689

K

Kame, 507
Kame terrace, 507
Kaolinite, 97
Karat, 80
Karst topography, 483–85
Katrina, Hurricane, 560
Kettle, 507
Kilauea, Hawaii, 148–49
Kilauea Volcano, 109
Klippe, 294
Knik River, 442
Kuiper belt, 697

L

La Brea tar pits, 265
Laccoliths, 132
La Conchita massive debris flow, 410–11
Lagoon Nebula, 22
Lagoons, 220
Lahar, 154, 419–20

Lake Missoula, 511
Lambert Glacier, 11
Laminae, 204
Laminar flow, 434
Larsen B Ice Shelf, 492
Lateral moraine, 504
Laterite, 193
Lava, 108
Lava dome, 159–60
Lava flows, 142–43
    aa, 142–43
    pahoehoe, 143
Lava Lake, 232
Lava plateaus, 29
Lava tubes, 143, 144
Law of superposition, 8–9
Layering at 660 kilometers, 69
Leaching, 188, 190
Leibniz, Gottfried W., 54
Lewis Overthrust Fault, 295
Libby, Willard F., 270
Life on Earth, 625–38
    beginning, 625–28
    Cenozoic era, 634–38
    Mesozoic era, 633–34
    Paleozoic era, 628–33
Light silicates, 95–98, 110
Lignite, 212
Limbs, 285
Limestone, 30, 75, 208–10
    carbonate reefs, 208–9
    coquina and chalk, 209
    inorganic, 209–10
Liquefaction, 319, 412
Liquid layer, 24
Lithification, 32, 213
Lithified, 30
Lithosphere, 24, 51, 339
Lithospheric mantle, 339
Lithospheric plate, 51–54
Lobate scarps, 682
Local base level, 442
Loess deposits, 536–37
Longitudinal dunes, 535
Longitudinal profile, 435
Longshore current, 550
Longshore transport, 549–50
Long Valley Caldera, 158
Losing stream, 467
Lower mantle, 24, 340
Low-latitude deserts, 520–21
Lunar highlands, 679
Lunar regolith, 680
Luster, 87
Lyell, Charles, 256

M

Mafic, 111
Magma, 29, 108–10
    composition of, 140
    crystallization of, 108–10
    evolvement of, 122–24
    nature of, 108

origin of, 120–22
    partial melting and, 124–28
Magma mixing, 124
Magmatic differentiation, 123
Magmatic segregation, 661
Magnetic reversal, 49–50
Magnetic time scale, 49–50
Magnetite, 45–46
Magnetometer, 50
Magnitude, 313
Magnitude scales, 314–16
Malaspina Glacier, 493
Mammoth Hot Springs, 474, 475
Mann, Michael, 590
Mantle, 24, 339
Mantle plumes, 64–66, 166, 373–74
    hot spots and, 64–65
    plate motion and, 65–66
Marble, 31–32, 240
Maria, 679
Marine depositional environments, 220
Mariner 10, 682
Marine terrace, 552
Mars, 683–86
Mars Pathfinder, 690
Martha's Vineyard, 544, 553
Massive (instrusion), 128
Mass number, 81
Mass wasting, 174
    debris flow and, 419–20
    earthflow, 421–22
    landform development and, 406–7
    permafrost landscape and, 423–24
    rockslide and, 417–19
    slope change and, 407
    slow movements, 422–23
    slump and, 416–17
    submarine landslides and, 424–25
Mass wasting, classifications of, 413–16
    material type, 413
    motion type, 413–14
    movement rate, 414–16
Mass wasting, controls and triggers of, 407–13
    earthquakes, 411–12
    oversteepened slopes, 409
    water's role as, 408–9
Matthews, D. H., 49–50, 54
Mauna Loa, 148
Mazama, Mount (Crater Lake), 125
McKinley, Mount, 383. *See also* Denali National Park
McKinley, William, 383
Meandering channels, 440
Meanders, 440
Mechanical weathering, 175
Medial moraine, 504

Megathrust earthquakes, 316
Melt, 108
Menard, H. W., 54
Mercalli, Giuseppe, 313
Mercury, 681–82
Mesosaurus, 42
Mesozoic era, 271, 633–34
Messenger, 682
Metallic bonds, 79, 80
Metallic substances, 77
Metamorphic agents, 230–35
    chemically active fluids, 234
    heat, 231–32
    parent rocks and, 234–35
    pressure, 233–34
Metamorphic environments, 241–45
    burial, 244
    contact or thermal, 241–42
    fault zones as, 244–45
    hydrothermal, 242–43
    impact areas as, 245, 247
    interpreting, 248–50
    regional, 244
    subduction zone, 244
Metamorphic facies, 248
Metamorphic rocks, 31, 32–34, 238–41
    foliated, 238–40
    nonfoliated, 239, 240–41
Metamorphic textures, 235–38
    foliation, 235–37
    nonfoliated, 237–38
Metamorphic zones, 245–48
    index minerals and, 246, 248
    textural variations and, 245–46
Metasomatism, 234
Meteor, 698
Meteor Crater, 247, 696, 700, 701
Meteorites, 22, 245, 699
Meteoroids, 698–700
Meteorological drought, 470, 471
Meteor shower, 699, 700
Methane hydrate, 651
Michigan Basin, 291
Microcontinent, 390
Microplates, 54
Middle-latitude deserts, 521–23
Mid-ocean ridge, 29, 365
Migmatite, 248
Milankovitch, Milutin, 513–14
Milky Way Galaxy, 17
Mineral classes, 92
Mineralogy, 74
Mineral phase change, 334
Mineral physics, 338
Mineral resources, 659–64. *See also* Nonmetallic mineral resources
    igneous processes and, 661–62
    metamorphic processes and, 663
    weathering and, 663

Minerals, 74–76
  characteristics of, 74–75
  classes of, 92
  compositional variations in, 83–86
  defined, 29, 74
  dietary supplement comparison to, 76
  glass making from, 76
  naming and classifying, 91–93
  nonsilicate, 99–103
  silicate, 93–98
  structural variations in, 86–87
Minerals, physical properties of, 87–91
  crystal shape or habit, 88
  density and specific gravity, 89–90
  optical, 87–88
  strength, 88–89
Mineral species, 92
Mississippi delta, 447–48
  coastal wetlands on, 451
Mixed tidal pattern, 568
Modified Mercalli Intensity Scale, 313–14
Moho, 337, 339
Mohorovčič, Andrija, 337, 339
Mohs scale, 88–89
Molecules, 77
Moment magnitude, 315–16
Monocline, 290
Monthly tidal cycle, 566–67
Monument Valley, 175
Mooney, Walter, 322
Moraine, 503–5
  end and ground, 504–5
  lateral and medial, 503–4
Morenci, Arizona, 5
Morley, L. W., 54
Mountain belts, 26–28
Mountain building, 382–84
  subduction and, 386–90
Mount Rushmore National Memorial, 110
Mouth, 435
Mud cracks, 221, 224
Mudflow, 419
Muscovite, 96, 97
Mylonites, 245

N

Nantucket, 544
National Hurricane Center, 569
National Ice Core Laboratory, 580
Natural gas, 646–49
  oil traps and, 647–48
  petroleum formation and, 647
Natural levee, 449
Navajo Sandstone, 205, 297
Neap tide, 567
Nearshore zone, 543
Nebular theory, 21, 672–74

Negative-feedback mechanism, 18, 598
Negative gravity anomaly, 347
Neptune, 692–94
Neutrons, 76–77, 268
Nevado del Ruiz, 155
New Horizons, 701
Newton, Isaac, 54
Nonclastic texture, 215, 216
Nonconformity, 260–61
Nonferromagnesian. *See* Light silicates
Nonflowing artesian well, 476–77
Nonfoliated texture, 31, 237
Nonmetallic mineral resources, 665–67
  building materials, 665–66
  industrial materials, 666–67
  occurrences and uses, 666
Nonrenewable resources, 644
Nonsilicate minerals, 93, 99–103
  common classes of, 100
Normal faults, 292–93, 307, 308
Normal polarity, 49
North American Cordillera, 391, 392
Nuclear energy, 652–53
  uranium and, 652
Nuclear fission, 652
Nucleus, 76, 268, 697
Nuée ardentes, 153–54
  St. Pierre destroyed by, 154
Numerical date, 256

O

Oblate ellipsoid, 346
Obsidian, 75, 76, 113, 118, 119
Occultation, 693
Occupational Outlook Handbook, 6
Ocean floor
  mapping, 356–58
  provinces of, 358
  viewing, from space, 358
Oceanic continental convergence, 57–58
Oceanic crust, 337, 367–71
  forming of, 368–69
  seawater interactions with, 369, 371
Oceanic lithosphere, 375–77
  plate subduction and, 375–76
Oceanic-oceanic convergence, 58
Oceanic plateau, 362, 363
Oceanic ridges, 29, 55–56
  anatomy of, 365–66
  seafloor spreading and, 366–67
Oceanic ridge system, 48
Offshore zone, 543
Oil, 646–49
  oil traps and, 647–49
  petroleum formation and, 647
Oil sands, 649–50
Oil shale, 650–51

Oil trap, 647–49
Old Faithful, 473
Oldham, Richard Dixon, 340
Old Man of the Mountain, 178
Olivine, 96, 98, 334, 339, 369
Oolitic limestone, 210
Oort, Jan, 398
Oort cloud, 22, 697, 698
Opaque, 87
Open systems, 18
Ophiolite complex, 368
Ore, 659–60
Organic sedimentary rocks
  coal, 212
  defined, 202
  identification of, 216
Organ Pipe Cactus National Monument, 521
Original horizontality, principle of, 257–58
Orogenesis, 382
Osorno Volcano, 57
Outcrops, 298
Outer core, 24, 341
Outer planets, 674
Outgassing, 614, 616
Outlet glacier, 493
Outwash plain, 506–7
Oversteepened slopes, 409
Over the Mountains (Collier), 3
Oxbow lake, 440, 442
Oxidation, 181
Oxygen isotope analysis, 579–80
Ozone, 583

P

Pahoehoe flow, 143
Paleoclimatology, 578
Paleomagnetism, 45–48
  apparent polar wandering and, 46, 48
  defined, 46
  Earth's magnetic field and, 45–46
  plate motion and, 66–67
Paleontology, 262, 265
Paleoseismology, 325
Paleozoic era, 271, 628–33
  life forms, 629–30
  lobe-finned fish and, 630
  Permian extinction, 630–31, 633
Pangaea, 40, 41, 63
Parabolic dunes, 535–36
Parasitic cone, 146
Parent material, 187
Parent rock, 230, 234–35
Parícutin, 150–51
Partial melting, 57, 124. *See also* Decompression melting
Passive continental margin, 358, 360–61
Pasteur, Louis, 12

Pater noster lakes, 500
"Peach Garden Land of Immortals" (painting), 485
Peat, 212
Pegmatites, 114–15, 661
Pegmatitic texture, 114–15
Pele's hair, 114
Pele's tears, 114
Perched water table, 472
Peridotite, 24, 339, 340
Period, 271, 272
Periodic table of elements, 77, 78
Permafrost, 423–24, 603
Permeability, 468
Perrault, Pierre, 472
Perseid meteor shower, 699
Petroleum formation, 647
Phaneritic texture, 113
Phanerozoic eon, 271, 621–25
Phase change, 87
Phenocrysts, 113
Phyllite, 238
Physical geology, 2
Piccard, Jacques, 362
Piedmont glacier, 493
Pillow basalts, 369
Pillow lavas, 143
Pinatubo, Mount, 589
Pipe, 146, 160
Placer deposits, 664–65
Placers, 664
Planetary geology, 681–94
  atmospheres of, 675–77
  internal structures of, 674–75
  Jupiter, 686–89
  Mars, 683–86
  Mercury, 681–82
  orbits of, 673
  planetary data, 673
  Saturn, 690–92
  Uranus and Neptune, 692–94
  Venus, 682–83
  weather of, 677–78
Planetary impacts, 678
Planetesimals, 22, 25, 614, 672
Plastic flow, 494
Plate-mantle convection, 67–69
  models of, 68–69
Plate motion
  drivers of, 67–69
  mantle plumes and, 65–66
  measurement of, 65–67
  paleomagnetism and, 66–67
Plates, 51, 52–53
Plate tectonics, 51–54, 513
  changes to, 62
  defined, 51
  glaciation and, 513
  importance of, 69
  mantle plumes and, 64–66
  map of, 52–53
  measurement of plate motion, 65–66

Plate tectonics (*cont.*)
ocean drilling and, 62, 64
plate boundaries and, 53–54
testing, model, 62–65
theory of, 14, 40
Playa, 527
Playa lake, 527
Pleistocene epoch, 512
Plucking, 498
Plunge, 286
Plutonic rocks, 110
Plutons, 128
Pluvial lake, 510, 512
Point bar, 440
Polarizing microscope, 118
Polar wandering, 46–47, 48
Polymerization, 93
Polymorphs, 86, 92
Pore space, 204
Porosity, 467–68
Porphyritic texture, 113
Porphyroblastic texture, 237
Porphyry, 113
Positive-feedback mechanism, 18, 598
Positive gravity anomaly, 346–47
Potassium-argon, 269–70
Pothole, 437
Powell, John Wesley, 256, 257, 441–42
Pratt, J. H., 401
Precambrian history, 272–73, 618–21
Primary (P) waves, 310, 311
Principal shells, 77
Proglacial lakes, 509
Prokaryotes, 625
Proterozoic eon, 272
Protons, 76–77, 268
Protoplanets, 614, 672
Protosun, 21
Proxy data, 578
Pulido, Dionisio, 150
Pumice, 75, 118–19, 145, 146
Pyroclastic flows, 153
Pyroclastic materials, 144–45
Pyroclastic texture, 114
Pyroxenes, 98, 369

Q
Quarrying, 437
Quartz, 75, 76, 96, 97
Quartzite, 240–41
Quartz sandstone, 206
Qui Ying, 485

R
Radial drainage pattern, 450
Radioactive decay, 81
Radioactivity, 268
Radiocarbon dating, 270
Radiogenic heat, 343

Radiometric dating, 268, 269–71
error sources in, 270
importance of, 270–71
isotopes used in, 270
potassium-argon and, 269–70
Rainier, Mount, 155
Rainshadow desert, 523
Rasely, Bob, 421
Rate of movement, 414–16
Recessional moraine, 504
Reconnaissance Orbiter, 685
Recrystallization, 213
Rectangular drainage pattern, 450
Red Deer River, 446
Redoubt, Mount, 420
Redoubt Volcano, 388
Red Sea, 371
Reefs, 208–9, 217
Refracted waves, 339
Regional floods, 452–53
Regional metamorphism, 31, 244
Regolith, 187
Reid, H. F., 305–6
Relative dating, 8, 257, 264
Renewable resources, 644
Reserve, 659
Reservoir rock, 647
Residual soils, 187
Reverse faults, 294, 307, 308
Reverse polarity, 49
Rhône Glacier, 13
Rhyolite, 117–18
Richter, Charles, 314–15
Richter scale, 314–15
Ridge push, 68
Rift valley, 55, 365–66
Rills, 432
Ring of Fire, 160–62, 362
Rip currents, 550
Ripple marks, 223–24
River discharge, 436
Rivers, 432
River systems, 433–34
Roche moutonnée, 503
Rock avalanche, 415–16
Rock cleavage, 236
Rock cycle, 19, 32–34
sedimentary rock formation and, 202
Rock deformation, 282–85
Rock flour, 498
Rocks, 29–32. *See also specific types*
defined, 75
evidence of continental drift in, 43–44
igneous, 29
metamorphic, 31–32
sedimentary, 29–31
Rockslide, 417–19
Rock strength, 284–85
confining pressure and, 284
rock type and, 284

temperature and, 284
time and, 284–85
Rock structure, 280, 285
mapping, 297–98
naming, 298
Rocky Mountains, 398
Roof pendant, 242
Ross Ice Shelf, 492
Runcorn, S. K., 46
Runoff, 430

S
Saffir-Simpson Hurricane Scale, 562, 563
Sagan, Carl, 593
Saint Elias National Park, 505
St. Helens, Mount, 20, 126, 138–39, 152, 154–55, 160, 419, 420, 588
Salt, 667
Saltation, 438, 529–30
Salt crystal growth, 176
Salt flats, 212
Salt River Canyon, 130
Saltwater contamination and groundwater, 478, 480
San Andreas Fault, 326–27
Sand deposits, 533–34
Sand dunes, 534–36
Sand movement on beach, 548–50
longshore transport, 549–50
perpendicular to shoreline, 548
rip currents and, 550
wave refraction and, 548–49
Sandstone, 30, 204–6
San Joaquin Valley, 479
San Rafael Swell, 291
Santorini island, 155
Sarychev Peak, 169
Saturn, 690–92
Schist, 31, 238, 240
Schistosity, 237
Scientific inquiry, nature of, 9–11
plate tectonics and, 14
Scientific law, 14
Scientific method, 12–14
Scoria, 145, 146
Scoria cones, 149–51
Scott, W. B., 45
Sea arch, 552
Seafloor mapping, 356–58
Seafloor sediment, 578–79
Seafloor spreading, 48–51, 366–67
Harry Hess and, 48–49
magnetic reversal and, 49–51
oceanic ridges and, 55–56
Seamounts, 29, 363
Sea stack, 552
Seawall, 555–56
Secondary enrichment, 663
Secondary (S) waves, 310, 311
Sediment, 29, 32, 193

Sedimentary environment, 216–21
defined, 216
facies as description of, 220–21
types of, 217–20
Sedimentary rocks, 29, 32
carbon cycle and, 214–15
chemical, 207–12
classification of, 214–16
common characteristics of, 221
defined, 201–2
detrital, 202–7
diagenesis and, 213–14
identification of, 216
importance of, 200
lithification and, 213–14
organic, 212–16
origins of, 200–202
rock cycle of, 202
Sedimentary structures, 221–24
Seiches, 319
Seismic gaps, 324
Seismic reflection profile, 357
Seismic sea wave, 320
Seismic tomography, 346, 347
Seismic vibrations, 317–19
Seismic waves, 24, 304, 319
Seismogram, 310, 312
Seismograph, 310, 311
Seismology
defined, 309
seismographs and, 310–12
Semidiurnal tidal pattern, 568
Settling velocity, 438
Shale, 30, 203–4
Shallow marine, 220
Shear, 281
Sheep Mountain, 3, 289
Sheeted dike complex, 368, 369
Sheet erosion, 192
Sheet flow, 432
Sheeting, 176–78
Shields, 27
Shield volcanoes, 147–49
defined, 147–48
Kilauea, Hawaii, 148–49
Mauna Loa, 148
volcanic islands and, 149
Shock metamorphism, 245, 247
Shore, 543
Shoreline erosion, 557–60
Atlantic and Gulf coasts, 557–58
Pacific coast, 558–60
Shoreline features, 550, 552–53
depositional, 552–53
erosional, 552
evolving, 553
Shorelines, stabilizing, 553–57
alternative, 556–57
hard, 554–56
Shubin, Neil, 632
Sierra Nevada, 389–90

Silicate, 93–98
    minerals, common, 95–98
    silicon–oxygen tetrahedron, 93–94
    structures, joining, 94–95
Silicon–oxygen tetrahedron, 93–94
Sills, 128–31
Sinkhole, 462, 483
Sinks, 483
Slab pull, 68
Slate, 31, 238
Slaty cleavage, 236
Slickensides, 292
Slide, 414
Slipface, 533
Slope orientation, 189
Slump, 416–17
Smith, William, 267
Snake River, 447
Snowline, 494
Socioeconomic drought, 471
Soil, 187
    classifying, 191
    climate and, 188
    defined, 17, 186–87
    erosion, 191–95
    formation, control of, 187–89
    as interface, 186–87
    orders, 191–92
    plants/animals and, 188–89
    profile, 189–90
    time and, 188
    topography and, 189
Soil Taxonomy, 191
Sojourner, 690
Solar energy, 653–54
Solar nebula, 21, 614, 672
Solar system, 672–78
    minor members of, 694–701
    Nebular theory, 672–74
    overview, 672
    planets and, 674–78
Solifluction, 422–23
Solum, 190
Sonar, 356
Sorting, 205–6, 439
Specific gravity, 89–90
Specific retention, 468
Specific yield, 468
Speleothem, 482
Sphericity, 205
Spheroidal weathering, 184–85
Spinel, 334
Spits, 220, 552
Spreading center, 55
Springs, 472
Spring tide, 567
Stable platforms, 28
Stair Hole, 281
Stalactite, 482
Stalagmite, 483
Star dunes, 535, 536

Stardust, 697
Steno, Nicolas, 83, 257
Steno's Law, 83
Steppe, 520
Stocks, 131–32
Storm surge, 562–63
Strain, 281–82
Strata, 221
Stratified drift, 506–7
Stratovolcanoes, 151–52
Streak, 87–88
Streak plate, 87–88
Stream, 432
Stream channels, 439–41
    alluvial, 440–41
    bedrock, 440
Stream erosion, 436–37
Streamflow, 434–36
    changes downstream and, 435–36
    deposition of sediment by, 440
    discharge and, 435, 436
    flow velocity and, 434
    gradient and, 435
    sediment transport by, 437–39
Stream piracy, 452
Stream valleys, 443–45
    deepening, 443–44
    defined, 443
    incised meanders and terraces, 445
    widening, 444
Stress, 280–81, 282
Strike, 298–99
Strike-slip faults, 294–96, 307, 308
Stromatolites, 625, 628
Structural geology, 280
    mapping, 297–99
Structures, ductile deformation, 285–90
Subduction and mountain building, 386–90
    along Andean-type margins, 386–89
    volcanic island arcs, 386
Subduction zone, 56
    dynamics, 385
    features, 384–85
Subduction zone metamorphism, 244
Submergent coasts, 565
Subsystems, 17
Sulfur, 667
Sunset Crater National Monument, 30
Sunspots, 591–92
    drought and, 591–92
    temperature and, 591
Supercontinent, 40, 620–21
Supercontinent cycle, 620
Supernova, 21, 614
Superplume, 400

Superposed stream, 450
Superposition, law of, 8–9, 257, 258
Surf, 547
Surface subsidence and groundwater, 478, 479
Surface wave, 310
Surge, 495–96
Surtsey, island of, 149
Suspended load, 437, 438, 530
Suture, 391
Syncline, 286–88
System, 17

T
Tablemount, 363
Tabular (instrusion), 128
Taku Glacier, 491
Tarn, 500
Taylor, F. B., 54
Tectonic plate, 51–54
Tectonic structures, 280, 285
Tektites, 247
Temporary base level, 442
Tenacity, 88
Tensional stress, 281
Terminal moraine, 504
Terrace, 445, 447
Terrae, 679
Terrane, 390
Terrestrial planet, 674, 675
Tetiaroa Atoll, 364
Texture, 112, 235
Theory, 12
Theory of the Earth (Hutton), 6–7
Thermal expansion, 178
Thermal metamorphism, 31, 241–42
Thieler, Rob, 551
Thin section, 118
Thoreau, Henry David, 507
Thousand Springs, 472
Thrust fault, 294
Tidal delta, 569–70
Tidal flats, 220, 569, 570
Tidal power, 658–59
Tides, 565–70
    causes of, 566
    defined, 565–66
    Earth's rotation and, 570
    monthly tidal cycle, 566–67
    tidal currents, 568–69
    tidal patterns, 567–68
Tiktaalik, 632
Till, 503
Tillite, 513
Tombolo, 552
Topsoil, 189
Trace elements, 86
Track forecast, 569
Transactions of the Royal Society of Edinburgh, 7

Trans-Alaskan oil pipeline, 325
Transform fault boundary, 55, 60–61
Transform faults, 296, 307
Transitional environments, 220
Transition zone, 24, 339–40
Translucent, 87
Transparent, 87
Transpiration, 430
Transported soils, 187
Transverse dunes, 535
Travertine, 210
Tree rings, 580–82
Trellis drainage pattern, 450
Triangulation, 312, 313
Trigger, 407
Tropical Rainfall Measuring Mission (TRMM), 11
Tropical rain forests, 193
Truncated spur, 499
Tsunami, 149, 316, 320–22
    Alaska 1964, 321
    defined, 320
    Indonesia 2004, 322
    warning system, 322
Tube worms, 370
Tuff, 119–20
Tunguska event, 696
Turbulent flow, 434

U
Ultimate base level, 442
Ultramafic rocks, 111, 112
Uluru, Mount, 529
Unconformity, 258–61
Uniformitarianism, 6
Unit cells, 83
U.S. Department of Labor, 6
U.S. Geological Survey, 256, 326
U.S. Global Change Research Program (USGCRP), 602
Unloading, 176
Unsaturated zone, 465–66
Upper mantle, 24, 339
Upper Mississippi drainage basin, 508–9
Uranus, 692–94
Ussher, James, 5

V
Vaiont Dam disaster, 415
Valence electrons, 77
Valley glaciers, 490
Valley trains, 506–7
Variegated Glacier, 496
Vegetation removal and mass wasting, 409–11
Vein deposit, 662
Venetz, Ignaz, 512
Vent, 146
Ventifacts, 532–33
Venus, 682–83
Venus Express, 683

Vertical crust movement, 399–401
  isostasy and, 399
  mantle convection and, 399–400
Vesicles, 113
Vesicular texture, 113
Vesuvius, 152–53
Vine, Fred, 49–50, 54
Viscosity, 114, 138, 344
  factors affecting, 138, 140
Vishnu Schist, 32
Vog, 147
Volatiles, 108, 140, 143–44
Volcanic breccia, 120
Volcanic eruptions, 138–42
  materials extruded during, 142–46
  monitoring, 168–69
  plate tectonics and, 160–67
Volcanic island arcs, 29, 58, 162, 384, 386
Volcanic rocks, 110
Volcanic structures, 146–60
  living with, 167–69
Volcano
  activity of, climate change and, 588–89
  air pollution caused by, 147

anatomy of, 146–47
  hazards of, 167–68
  neck of, 160
Voyager 2, 693

W
Wadati–Benioff zone, 328
Walden Pond, 507
Wallace, Alfred, 54
Walsh, Don, 362
Water, running, 432–34
  drainage basins, 432–33
  river systems, 433–34
  sediment transport by, 437–39
  stream erosion and, 436–37
Water gap, 450
Water table, 466–67
  defined, 465
  groundwater interaction and, 467
  map of, 467
  variations in, 466–67
Water vapor, 583
Watkins, N. D., 54
Wave-cut cliff, 552
Wave-cut platform, 552
Wave fronts, 335

Wavelength, 544
Wave period, 544
Wave refraction, 548–49
Waves, 544–47
  characteristics of, 544–45
  circular orbital motion and, 545–46
  erosion and, 547–48
  height of, 544
  surf zone, in, 546–47
Weathering, 32
  defined, 174
  products of, 184
  rates of, 185–86
  spheroidal, 184–85
  types of, 174–85
Wegener, Alfred, 40–45, 47, 54, 63
Welded tuff, 114, 120, 145
Wells, 474–75
Wetted perimeter, 435
White Chalk Cliffs, 209, 617
Whole-mantle convection, 69
Wilson, J. Tuzo, 60
Wind deposits, 533–37
  loess deposits and, 536–37
  sand deposits and, 533–34
  sand dunes and, 534–36

Wind energy, 654–55
  leading states for, 655
Wind erosion, 530–33
  deflation and blowouts, 530–31
  desert pavement, 531–32
  ventifacts and yardangs, 532–33
World population, 5

X
Xenoliths, 132

Y
Yardangs, 532–33
Yazoo tributary, 449
Yellowstone National Park, 663
Yellowstone River, 444
Yosemite National Park, 30

Z
Zion National Park, 31, 205
Zone of accumulation, 496, 497
Zone of fracture, 495
Zone of saturation, 465
Zone of soil moisture, 465
Zone of wastage, 496, 497